Complete Casting Handbook

Complete Casting Handbook
Metal Casting Processes, Metallurgy, Techniques and Design

Second Edition

John Campbell

Emeritus Professor of Casting Technology,
Department of Metallurgy and Materials,
University of Birmingham, UK

AMSTERDAM • BOSTON • HEIDELBERG • LONDON
NEW YORK • OXFORD • PARIS • SAN DIEGO
SAN FRANCISCO • SINGAPORE • SYDNEY • TOKYO

Butterworth-Heinemann is an imprint of Elsevier

Acquiring Editor: Christina Gifford
Editorial Project Manager: Jeff Freeland
Project Manager: Priya Kumaraguruparan
Cover Designer: Greg Harris

Butterworth-Heinemann is an imprint of Elsevier
The Boulevard, Langford Lane, Kidlington, Oxford OX5 1GB, UK
225 Wyman Street, Waltham, MA 02451, USA

ISBN: 978-0-444-63509-9

British Library Cataloguing in Publication Data
A catalogue record for this book is available from the British Library

Library of Congress Cataloging-in-Publication Data
A catalog record for this book is available from the Library of Congress

For Information on all Butterworth-Heinemann publications
visit our website at http://store.elsevier.com/

To Robert Puhakka
For his dedication to the 10 rules and the living proof that they work.

Contents

CASTING MANUFACTURE

SECTION 1

SECTION 2 FILLING SYSTEM DESIGN

Preface

In this first update of the *Handbook*, the major revisions are probably those relating to running system design in which the vestiges of filling defects have finally been eliminated from castings.

Thus, the powerful benefits of contact pouring (in which the universal conical trumpet decorating all traditional filling systems is now eliminated) is finally shown to have been hugely underestimated by a number of foundries. Contact pouring has probably been the most important (and the most simple and zero-cost) initiative to revolutionise quality in castings. In addition, the adoption of various forms of tangential filter designs to gates has finally eliminated the problem of the entrainment of priming bubbles. These residual bubbles have long impaired the benefits of previous filling systems.

Gravity pouring has now advanced to the point at which I find myself having to admit that it starts to threaten my cherished and favoured casting production system: countergravity.

This is seen to be especially true for those low-pressure systems which use a refractory lining for the pressurised furnace. I only recently discovered the hugely damaging emission of bubbles from these linings during depressurisation of the furnace. This problem has clearly been a major source of impaired castings in the low-pressure casting industry and has hampered this industry since its beginnings.

The use of my pneumatic pump is described for the first time. It would lower costs and solve most of the problems of this industry. Thus, I continue to stand by countergravity as the optimum casting system where it can be used. My hope is that it will be teamed up with a good melting and metal handling system. Only careful foundry design will minimise bifilm populations in metals. Only when castings can be produced substantially free from bifilms will we enjoy the full benefits of castings, and metals in general, resistant to hot tearing, cracking, blisters, corrosion pitting and attack of grain boundaries, plus the benefits of extraordinary mechanical properties, potentially eliminating future failure by fracture or fatigue.

These are heady predictions. However, early results in foundries are already indicating that beautiful defect-free castings with revolutionary metallurgical benefits appear to be routinely attainable. Despite challenges from the undoubtedly unique benefits of such new processes as additive manufacture, my hope for the future for castings is based on the adoption of simple principles which could not only secure the future of our casting industry, but improve the welfare and environment of all of us whose lives depend on it.

JC
Ledbury, Herefordshire, England
02 April 2015

Introduction

CASTINGS HANDBOOK, 2ND EDITION, 2015

When *Castings Handbook* first appeared in 2011, I had not expected to revise it so soon. However, the latest findings require publicising as quickly as possible—there is still a long way for the industry to go! The message of the book, summarised in the section "Bifilm-free Properties", is that a quality improvement of astonishing scale is possible now. When I first started to experiment with novel filling systems for castings, there were naturally many disappointments. However, those days are long gone. The concepts of entrainment and bifilm creation laid out in the book are now proven. Some foundries are already being designed to take advantage of a unique and easily affordable quality revolution and scrap reduction. More need to follow. The risks are minimal and the rewards are great.

With regard to improved casting techniques which can reduce or even eliminate the usual vast populations of bifilms in our metals, I have always been aware of the potential benefit of contact pouring, but had completely underestimated its effects. It achieves miraculous improvements to castings by eliminating the 50% air mixing step. Contact pouring is strongly recommended in this volume as a major but low-cost step forward.

The ultimate step forward is countergravity casting which should always be targeted if possible. However, although I discuss the traditional pumping techniques, I present here for the first time my new pneumatic pump. It is another low-cost, reliable technique which enjoys uniquely low turbulence and might allow a rapid takeup of this unique technology.

Turning to the seriousness of the current position in casting, and in metallurgical engineering as a whole, the fact that most current metals can fail by cracking should alert us to the glaring inconsistency in our metallurgical thinking because many of our metals and alloys are ductile, so failure by cracking should be impossible. In a ductile metal an attempt to propagate a crack should merely result in the crack tip blunting, preventing propagation.

In the absence of any other viable alternative mechanism, that metals do crack is a strong indication that cracks pre-exist in metals in the form of bifilms formed in the liquid state during pouring of the metal to make a casting. The poor practice is almost universally associated with pouring methods which ensure that the molten metal is mixed and emulsified with at least 50 volume % of air during its journey into the mould. I defy anyone to make a respectable casting from such a disgracefully inappropriate and inept technique. This book presents the case that this need not be so; metals need not contain bifilms, and thus need not contain those Griffith cracks which can initiate failure by cracking. To achieve this, we simply have to improve our casting technology.

It remains the case that bifilms are still lamentably researched, so that this book has to resort to sifting through inconclusive and fragmentary evidence from researchers who were not looking for bifilms. Unfortunately, researchers up until now have not been aware of their presence, and certainly did not suspect their overwhelming influence on their results. Although a welcome start is being made by a few workers, I remain impatient for more definitive research to be carried out.

In the meantime, while researchers slowly get around to proving the background theory, founders need not wait for answers. Practical low bifilm casting techniques have already been developed and are described here. They promise the quality improvement and cost reduction that the casting industry so badly needs.

Acknowledgements

It is a pleasure to acknowledge the significant help and encouragement I have received from many good friends. John Grassi has been my close friend and associate at Alotech, the company promoting the new, exciting ablation castings process. Ken Harris has been an inexhaustible source of knowledge on silicate binders, aggregates and recycling. His assistance is clear in Chapter 15. Clearly, the casting industry needs more chemists like him. Bob Puhakka has been the first regular user of my casting recommendations for the production of large aluminium and steel castings, which has provided me with inspirational confirmation of the soundness of the technology described in this book. In addition, the practical feedback and warm friendship over several years from those at the UK steel foundry Furniss and White is a pleasure to record. Murat Tiryakioglu has been a loyal supporter and critic, and provides the elegantly written publications that have provided welcome scientific underpinning. He has provided generous and invaluable help with the important section *The Statistics of Failure*. Naturally, many other acknowledgements are deserved among friends and students whose research has been a privilege to supervise. I do not take these for granted. Even if not listed here, they are not forgotten.

The American Foundry Society is thanked for the use of a number of illustrations from *Transactions*.

CASTING METALLURGY

THE MELT

Some liquid metals may be really pure liquid. Such metals may include pure liquid gold, possibly some carbon-manganese steels while in the melting furnace at a late stage of melting. These, however, are rare.

Many liquid metals are actually so full of sundry solid phases floating about that they begin to more closely resemble slurries than liquids. This slurry type nature can be seen quite often as some metals are poured; the melt overflows the lip of the melting furnace as though it were a cement mixture. In the absence of information to the contrary, this awful condition of a liquid metal should be assumed to be correct. Thus many of our models of liquid metals that are formulated to explain the occurrence of defects neglect to address this fact. As techniques have improved over recent years, there has been growing evidence for the real internal structure of liquid metals, revealing melts to be crammed with defects. Some of this evidence is described in this chapter. Much evidence applies to aluminium and its alloys where the greatest effort has been focussed, but evidence for some steels and all Ni-base alloys is already impressive and is growing steadily.

It is sobering to realise that many of the strength-related properties of metals can *only* be explained by assuming that the original melt was full of defects. Classical physical metallurgy and solidification science that has considered metals as merely pure metals currently cannot explain aspects of the important properties of cast materials such as the effect of dendrite arm spacing; it cannot explain the existence of pores and their area density; it cannot explain the reason for the cracking of strong, crack-resistant precipitates formed from the melt. These key aspects of cast metals will be seen to arise naturally from the assumption of a population of a particular type of defect: the bifilm.

Any attempt to quantify the number and size distribution of these defects is a non-trivial task. McClain and co-workers (2001) and Godlewski and Zindel (2001) have drawn attention to the unreliability of results taken from polished sections of castings. A technique for liquid aluminium involves the collection of inclusions by forcing up to 2 kg of melt through a fine filter, as in the porous disc filtration analysis and Prefil tests. The method overcomes some of the sampling problem by concentrating the inclusions by a factor of about 10,000 times (Enright and Hughes, 1996 and Simard et al., 2001). The layer of inclusions remaining on the filter can be studied on a polished section. The total quantity of inclusions is assessed as the area of the layer as seen under the microscope, divided by the quantity of melt that has passed through the filter. The unit is therefore the curious quantity mm^2kg^{-1}. (It is to be hoped that at some future date this unhelpful unit will, by universal agreement, be converted into some more meaningful quantity such as volume of inclusions per volume of melt. In the meantime, the standard provision of the diameter of the filter in reported results would at least allow a reader the option to do this.)

To gain some idea of the huge range of possible inclusion contents, an impressively dirty melt might reach $10 \ mm^2kg^{-1}$, whereas an alloy destined for a commercial extrusion might be in the 0.1–1 range, foil stock might reach 0.001, and computer discs 0.0001 mm^2kg^{-1}. For a filter of 30 mm diameter, these figures approximately encompass the range of volume fraction 10^{-3} (0.1%) down to 10^{-7} (0.1 part per million by volume).

Other techniques for the monitoring of inclusions in Al alloy melts include Liquid Metal Cleanness Analyser (LiMCA) (Syvertsen and Engh 2001), in which the melt is drawn through a narrow tube. The voltage drop applied along the length of the tube is measured. The entry of an inclusion of different electrical conductivity into the tube causes the voltage differential to rise by an amount that is assumed to be proportional to the size of the inclusion. The technique is generally thought to be limited to inclusions approximately in the 10–100 μm range or so.

Although widely used for the casting of wrought alloys, the author regrets that the LiMCA technique has to be viewed with great reservation. Inclusions in light alloys are often oxide bifilms up to a 10 mm diameter, as will become clear.

Complete Casting Handbook. http://dx.doi.org/10.1016/B978-0-444-63509-9.00001-7

Such inclusions do find their way into the LiMCA tube, where they tend to hang, caught up at the mouth of the tube, and rotate into spirals like a flag tied to the mast by only one corner. These are torn free from time to time and sediment in the bottom of the sampling crucible of the LiMCA probe, where they have the appearance of a heap of spiral Italian noodles (Asbjornsonn, 2001). It unfortunate that most workers using LiMCA have been unaware of these serious problems. Because of the air enfolded into the bifilm, the defects in the LiMCA probe have often been thought to be bubbles, which, probably, they sometimes partly are and sometimes completely are. One can see the confusion.

Ultrasonic reflections have been used from time to time to investigate the quality of melt. The early work by Mountford and Calvert (1959) is noteworthy, and has been followed up by considerable development efforts in Al alloys (Mansfield, 1984), Ni alloys and steels (Mountford et al., 1992). Ultrasound is efficiently reflected from oxide bifilms (almost certainly because the films are double, and the elastic wave cannot cross the intermediate layer of air, and therefore is efficiently reflected). However, the reflections may not give an accurate idea of the size of the defects because their irregular, crumpled form and their tumbling action in the melt. The tiny mirror-like facets of a large, scrambled defect reflects back to the source only when they happen to rotate to face the beam. The result is a general scintillation effect, apparently from many minute and separate particles. It is not easy to discern whether the images correspond to many small or a few large defects.

Neither LiMCA nor the various ultrasonic probes can distinguish any information on the types of inclusions that they detect. In contrast, the inclusions collected by pressurised (forced) filtration can be studied in some detail, although even here the areas of film defects are often difficult to discern. In addition to films, many different inclusions can be found as listed in Table 1.1.

Nearly all of these foreign materials will be deleterious to products intended for such products as foil or computer discs. However, for shaped castings, those inclusions such as carbides and borides may not be harmful at all. This is because having been precipitated from the melt, so they are usually therefore in excellent atomic contact with the matrix. These well-bonded non-metallic phases are thereby unable to act as initiators of other defects such as pores and cracks. Conversely, they may act as grain refiners. Furthermore, their continued good bonding with the solid matrix is expected to confer on them a minor or negligible influence on mechanical properties. (However, we should not forget that it is possible that they may have some influence on other physical or chemical properties such as machinability or corrosion.)

Generally, therefore, this book concentrates on those inclusions that have a major influence on mechanical properties, and that can be the initiators of other serious problems such as pores and cracks. Thus the attention will centre on *entrained surface films* which exhibit unbonded interfaces in the melt and lead to a spectrum of problems. Usually, these inclusions will be oxides. However, carbon films are also common, and occasionally nitrides, sulphides and other compounds. That these films are necessarily entrained into the melt as double films is a key feature of their structure. This aspect of their make-up will be a central feature of the book.

Table 1.1 Types of Inclusions in Al Alloys

Inclusion Type		Possible Origin
Carbides	Al_4C_3	Pot cells from Al smelters
Boro-carbides	Al_4B_4C	Boron treatment
Titanium boride	TiB_2	Grain refinement
Graphite	C	Fluxing tubes, rotor wear, entrained film
Chlorides	NaCl, KCl, $MgCl_2$, etc.	Chlorine or fluxing treatment
Alpha alumina	α-Al_2O_3	Entrainment after high-temperature melting
Gamma alumina	γ-Al_2O_3	Entrainment during pouring
Magnesium oxide	MgO	Higher Mg containing alloys
Spinel	$MgOAl_2O_3$	Medium Mg containing alloys

The pressurised filtration tests can find many of these entrained solids, and the analysis of the inclusions present on the filter can help to identify the source of many inclusions in a melting and casting operation. However, the only inclusions that remain undetectable but are enormously important are the newly entrained films that occur on a clean melt as a result of surface turbulence. These films are commonly entrained during the pouring of castings. They are typically only 20 nm thick, and so remain invisible under an optical microscope, especially if draped around a piece of refractory filter that when sectioned will appear many thousands of times thicker. The only detection technique for such inclusions is the lowly reduced pressure test. This test opens the films (because they are always double and contain air, as will be explained in detail in Chapter 3) so that they can be seen. Metallographic sections (or radiographs) of the cast test pieces clearly reveal the size, shape and numbers of such important inclusions, as has been shown by Fox and Campbell (2000). The test will be discussed in detail again and again.

1.1 REACTIONS OF THE MELT WITH ITS ENVIRONMENT

A liquid metal is a highly reactive chemical. It will react both with the gases above it and, if there is any kind of slag or flux floating on top of the melt, it will probably react with that, too. Many melts also react with their containers such as crucibles and furnace linings.

The driving force for these processes is the striving of the melt to come into equilibrium with its surroundings. Its success in achieving equilibrium is, of course, limited by the rate at which reactions can happen and by the length of time available.

Thus reactions in the crucible or furnace during the melting of the metal are clearly seen to be serious because there is usually plenty of time for extensive changes. Hydrogen being picked up from damp refractories is common. Similar troubles are often found with metals that are melted in furnaces heated by the burning of hydrocarbon fuels such as gas or oil.

We can denote the chemical composition of hydrocarbons as C_xH_y and thus represent the straight chain compounds such as methane CH_4, ethane C_2H_6 and so on, or aromatic ring compounds such as benzene C_6H_6 etc. (Other more complicated molecules may contain other constituents such as oxygen, nitrogen and sulphur, not counting impurities which may be present in fuel oils such as arsenic and vanadium.)

For our purposes, we will write the burning of fuel taking methane as an example:

$$CH_4 + 2O_2 = CO_2 + 2H_2O \tag{1.1}$$

Clearly, the products of combustion of hydrocarbons contain water, so the hot waste gases from such furnaces are effectively wet.

Even electrically heated furnaces are not necessarily free from the problem of wet environment: an electric resistance furnace that has been allowed to stand cold over a weekend will have had the chance to absorb considerable quantities of moisture in its lining materials. Most furnace refractories are hygroscopic and will absorb water up to 5 or 10% of their weight. This water is released into the body of the furnace over the next few days of operation. It has to be assumed that the usual clay/graphite crucible materials commonly used for melting nonferrous alloys are quite permeable to water vapour and/or hydrogen because they are designed to be approximately 40% porous. Additionally, hydrogen permeates freely through most materials, including steel, at normal metallurgical operating temperatures of around 700°C and higher.

The moisture from linings or atmosphere can react in turn with the metal M:

$$M + H_2O = MO + H_2 \tag{1.2}$$

Thus a little metal is sacrificed to form its oxide, and the hydrogen is released to equilibrate itself between the gas and metal phases. Whether it will, on average, enter the metal or the gas above the metal will depend on the relative partial pressure of hydrogen already present in both of these phases. The molecular hydrogen has to split into atomic hydrogen (sometimes called 'nascent' hydrogen) before it can be taken into solution, as is described by the simple relation:

$$H_2 = 2H \tag{1.3}$$

The equation predicting the partial pressure of hydrogen in equilibrium with a given concentration of hydrogen in solution in the melt is:

$$[H]^2 = kP_{H_2} \tag{1.4}$$

where the constant k has been the subject of many experimental determinations for a variety of gas-metal systems (Brandes, 1992; Ransley and Neufeld, 1948). It is found to be affected by alloy additions (Sigworth and Engh, 1982) and temperature. The relation is a statement of the famous Sievert law, which describes the squared relation between a diatomic gas concentration and its pressure; for instance, if the gas concentration in solution is doubled, its equilibrium pressure is increased by 4 times. A further point to note is that when the partial pressure of hydrogen $P = 1$ atm, it is immediately clear that k is numerically equal to the solubility of hydrogen in the metal at that temperature. Figure 1.1(a) shows how the solubility of hydrogen in aluminium increases with temperature.

Figure 1.1(b) shows that although many metals dissolve more hydrogen than aluminium, it is aluminium that suffers most from hydrogen porosity because of the huge difference in solubility between the liquid and the solid. The solid can hold only about 1/20th of the gas in the liquid (corresponding to a partition coefficient of 0.05) so that there is a major driving force for the rejection of nearly all the hydrogen on solidification, having the potential to create significant porosity. This contrasts with magnesium and many other metals, where the partition coefficients are closer to 1.0, so that their higher hydrogen content is, in general, not such a problem to drive the nucleation of pores (even though it may contribute significantly to the growth of pores because of its high rate of diffusion, draining the hydrogen content from significantly greater volumes of the alloy). These factors are discussed at length in Section 7.2 relating to the growth of gas porosity.

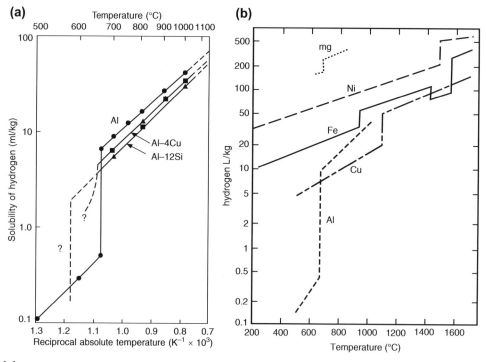

FIGURE 1.1

(a) Hydrogen solubility in aluminium and two of its alloys, showing the abrupt fall in solubility on solidification; (b) hydrogen solubility for a number of metals.

Moving on to the concepts of equilibrium, it is vital to understand fully the concept of an equilibrium gas pressure associated with the gas in solution in a liquid. We shall digress to present a few examples to illustrate the concept.

Consider a liquid containing a certain amount of hydrogen atoms in solution. If we place this liquid in an evacuated enclosure, then the liquid will find itself out of equilibrium with respect to the environment above the liquid; it is supersaturated with respect to its environment. It will then gradually lose its hydrogen atoms from solution, and these will combine on its surface to form hydrogen molecules which will escape into the enclosure as hydrogen gas. The gas pressure in the enclosure will therefore gradually build up until the rate of loss of hydrogen from the surface of the liquid becomes equal to the rate of gain, reconverting the hydrogen molecule to individual atoms on the surface and re-entering solution in the liquid. The liquid can then be said to have come into equilibrium with its environment. The effective hydrogen pressure in the liquid has become equal to the hydrogen pressure in the region over the melt.

Similarly, if a liquid containing little or no gas (and therefore having a low equilibrium gas pressure) were placed in an environment of high gas pressure, then the net transfer would, of course, be from gas phase to liquid phase until the equilibrium partial pressures in both phases were equal. Figure 1.2 illustrates the case of three different initial concentrations of hydrogen in a copper alloy melt, showing how initially high concentrations fall, and initially low concentrations rise, all finally reaching the same concentration which is in equilibrium with the environment.

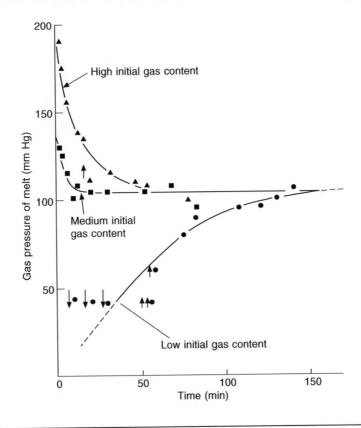

FIGURE 1.2

Hydrogen content of liquid aluminium bronze held in a gas-fired furnace, showing how the melt equilibrates with its surroundings.

Data from Ostrom et al. (1975).

This equilibration with the external surroundings is relatively straightforward to understand. What is perhaps less easy to appreciate is that the equilibrium gas pressure in the liquid is also effectively in operation *inside* the liquid.

This concept can be grasped by considering bubbles of gas which have been introduced into the liquid by stirring or turbulence, or which are adhering to fragments of surface films or other inclusions that are floating about. Atoms of gas in solution migrate across the free surface of the bubbles and into their interior to establish an equilibrium pressure inside.

On a microscopic scale, a similar behaviour will be expected between the individual atoms of the liquid. As they jostle randomly with their thermal motion, small gaps open momentarily between the atoms. These embryonic bubbles will also therefore come into equilibrium with the surrounding liquid.

It is clear, therefore, that the equilibrium gas pressure of a melt applies both to the external and internal environments of the melt.

We have so far not touched on those processes that control the rate at which reactions can occur. The kinetics of the process can sometimes over-ride the thermodynamics and can exert control over the reaction.

Consider, for instance, the powerful reaction between the oxygen in dry air and liquid aluminium: no disastrous burning takes place; the reaction is held in check by the surface oxide film which forms, slowing the rate at which further oxidation can occur. This is a beneficial interaction with the environment. Other beneficial passivating (i.e. inhibiting) reactions are seen in the melting of magnesium under a dilute SF_6 (sulphur hexafluoride) gas, as described, for instance, by Fruehling and Hanawalt (1969).

A further example is the beneficial effect of water vapour in strengthening the oxide skin on the zinc alloy during hot-dip galvanising so as to produce a smooth layer of solidified alloy free from 'spangle'. Without the water vapour, the usual protective atmosphere formed from a clean hydrogen-nitrogen mix provides a thin, delicate oxide, with the result that the growth of surface crystals disrupts the smoothness of the surface film on the liquid zinc to reveal the sharply delineated crystal patterns (Hart et al., 1984).

Water vapour is also known to stabilise the protective gamma alumina film on aluminium (Cochran et al., 1976 and Impey et al., 1993), reducing the rate of oxidation in moist atmospheres. Theile first saw this effect in 1962. His results are replotted in Figure 1.3. Although his curve for oxidation in moist air is seen to be generally lower than the curves for air and oxygen (which are closely similar), the most important feature is the very low *initial* rate, the rate at very short

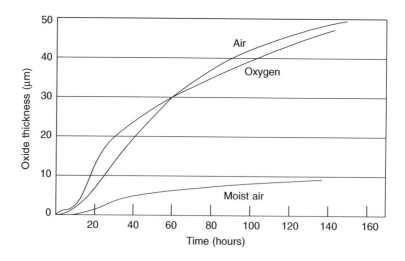

FIGURE 1.3

Growth of oxide on 99.9 Al at 800°C in a flow of oxygen, dry air and moist air.

Data from Theile (1962).

times. Entrainment events usually create new surface that is folded in within milliseconds. Obtaining oxidation data for such short times is a problem.

The kinetics of surface reactions can also be strongly influenced on the atomic scale by surface-active solutes that segregate preferentially to the surface. Only a monolayer of atoms of sulphur will slow the rate of transfer of nitrogen across the surface of liquid iron. Interested readers are referred to the nicely executed work by Hua and Parlee (1982).

1.2 **TRANSPORT OF GASES IN MELTS**

Gases in solution in liquids travel most quickly when the liquid is moving because, of course, they are simply carried by the liquid.

However, in many situations of interest, the liquid is stationary, or nearly so. This is the case in the boundary layer at the surface of the liquid. The presence of a solid film on the surface will hold the surface stationary, and because of the effect of viscosity, this stationary zone will extend for some distance into the bulk liquid, although, of course, the thickness of the boundary layer will be reduced if the bulk of the liquid is violently stirred. However, within the stagnant liquid of the boundary layer the movement of solutes can occur only by the slow process of diffusion, i.e. the migration of populations of atoms by the process of each atom carrying out one random atomic jump at a time.

Another region where diffusion is important is in the partially solidified zone of a solidifying casting, where the bulk flow of the liquid is normally a slow drift.

In the solid state, of course, diffusion is the *only* mechanism by which solutes can spread.

The average distance d to which an element can diffuse (whether in a liquid, solid or gas) in time t is given by the simple order of magnitude relation

$$d = (Dt)^{1/2} \tag{1.5}$$

where D is the coefficient of diffusion, measured in m/s^2. This simple formula, together with the values of D taken from Figure 1.4, is often extremely useful to estimate, if only relatively roughly, the distances involved in reactions. We shall return to the use of this equation many times throughout this book.

There are two broad classes of processes of diffusion processes, each having quite a different value of D: one is *interstitial diffusion* and the other is *substitutional diffusion*. *Interstitial diffusion* is the squeezing of small atoms through the *interstices* between the larger matrix atoms. This is a relatively easy process and thus interstitial diffusion is relatively rapid, characterised by a high value of D. *Substitutional diffusion* is the exchange, or *substitution*, of the solute atom for a similar-sized matrix atom. This process is more difficult (i.e. has a higher activation energy) because the solute atom has to wait for a gap of sufficient size to be created by a sufficiently large atomic fluctuation before it can jostle its way among the crowd of similar-sized individuals to reach the newly created space. Thus for substitutional diffusion D is relatively small.

Figure 1.4 shows the rates of diffusion of various alloying elements in the pure metals, aluminium, copper and iron. Clearly, hydrogen is an element that can diffuse interstitially because of its small size. In iron, the elements C, N and O all behave interstitially, although significantly more slowly than hydrogen.

The common alloying elements in aluminium, Mg, Zn and Cu clearly all behave as substitutional solutes. Other substitutional elements form well-defined groups in melts of copper and iron (Figure 1.4(b) and (c)).

However, there are a few elements that appear to act in an intermediate fashion. Oxygen in copper occupies an intermediate position. The elements sulphur and phosphorous in iron occupy an interesting intermediate position; a curious behaviour that does not appear to have been widely noticed.

Figure 1.4(c) also illustrates the other important feature of diffusion in the various forms of iron: the rate of diffusion in the open body centred cubic lattice (alpha and delta phases) is faster than in the more closely packed face centred cubic (gamma phase) lattice. Furthermore, in the liquid phase diffusion is fastest of all, and differences between the rates of diffusion of elements that behave widely differently in the solid become less marked.

These relative rates of diffusion using Eqn (1.5) and the data from Figure 1.4 will be referred to often in relation to many different phenomena throughout this book.

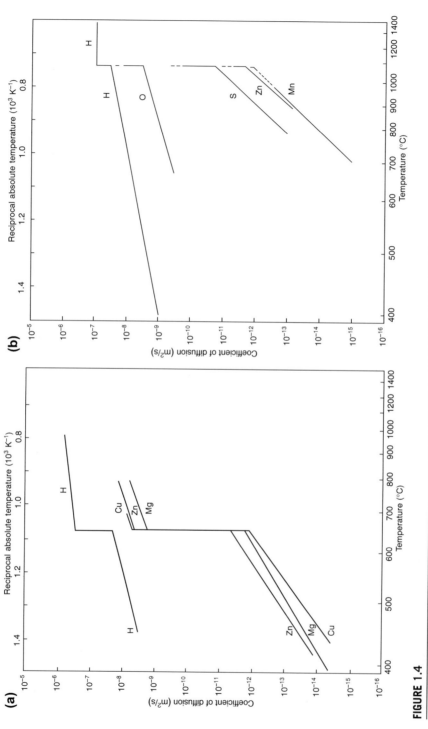

Data sources listed in 'Castings' 2nd Edition 2003.

FIGURE 1.4

Diffusion coefficients for elements in solution in (a) Al, (b) Cu and (c) Fe.

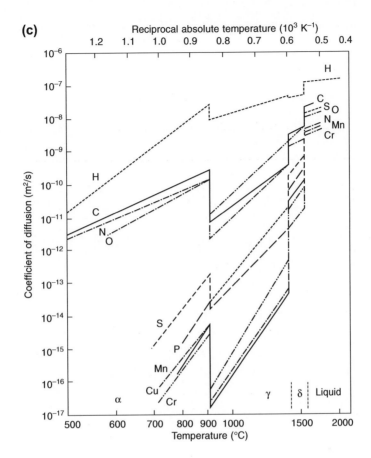

FIGURE 1.4 Cont'd

1.3 SURFACE FILM FORMATION

When the hot metal interacts with its environment, many of the reactions result in products that dissolve rapidly in the metal and diffuse away into its interior. Some of these processes have already been described. In this section, we shall focus our attention on the products of reactions that remain on the surface. Such products are usually films.

Whether there is any tendency for a film to form or not depends on its stability, which can be quantified by its free energy of formation. A diagram for oxides showing this energy as a function of temperature was famously promoted by Ellingham and is shown in Figure 1.5. The extremely stable oxides are at the base, and those easily reduced back to their component metals are high on the graph. This concept of stability is based on an estimate of thermodynamic equilibrium.

In reality of course, the kinetics of the formation of oxides (and other compounds) depends on the rate at which components can arrive, and the rate at which they can be processed. The processing rate depends in turn on the structure of the crystal lattice as it develops.

Oxide films usually start as simple amorphous (i.e. non-crystalline) layers, such as Al_2O_3 on Al, or MgO on Mg and Al-Mg alloys (Cochran et al., 1977). Their amorphous structure probably derives necessarily from the amorphous melt on which they nucleate and grow. However, they quickly convert to crystalline products as they thicken, and later often develop into a bewildering complexity of different phases and structures. Many examples can be seen in the studies reviewed by Drouzy and Mascre (1969) and in the various conferences devoted to oxidation (for instance, Microscopy of Oxidation, 1993). Some films remain thin, some grow thick. Some are strong, some are weak. Some grow slowly, others quickly. Some are smoothly uniform in composition, whereas others are heterogeneous and complex in their structure, being lumpy mixtures of different phases.

The nature of the film on a liquid metal in a continuing equilibrium relationship with its environment needs to be appreciated. In such a situation the melt will always be covered with the film. For instance, if the film is skimmed off it will immediately re-form. A standard foundry complaint about the surface film on certain casting alloys is that 'you can't get rid of it!'

Furthermore, it is worth bearing in mind that the two most common film-forming reactions, the formation of oxide films from the decomposition of moisture and the formation of graphitic films from the decomposition of hydrocarbons, both result in the increase of hydrogen in the metal. The comparative rates of diffusion of hydrogen and other elements in solution in various metals are shown in Figure 1.4. These reactions will be dealt with in detail later.

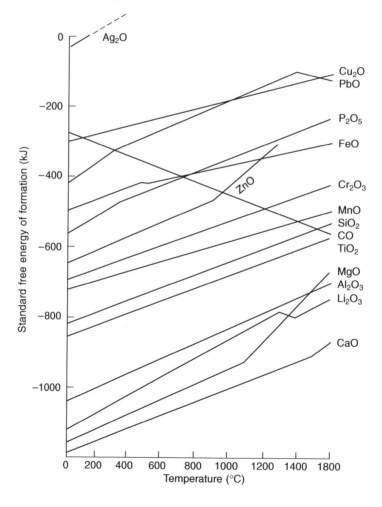

FIGURE 1.5

The Ellingham diagram, illustrating the free energy of formation of oxides as a function of temperature.

The noble metals such as gold, platinum and iridium are, for all practical purposes, totally film-free. These are, of course, all metals that are high on the Ellingham diagram, reflecting the relative instability of their oxides and thus the ease with which they are reduced back to the metal.

Iron is an interesting case, occupying an intermediate position in the Ellingham diagram. Liquid irons and steels therefore have a complicated behaviour, having a film which may be liquid or solid, depending on the composition of the alloy and its temperature. Its behaviour is considered in detail in Section 6.5 devoted to cast irons and steels.

Liquid silver and copper both dissolve oxygen. In terms of the Ellingham diagram (Figure 1.5), it is seen that silver oxide, Ag_2O, is just stable at room temperature, causing silver to tarnish (together with some help from the presence of sulphur in the atmosphere to form sulphides), as every jeweller will know! However, the free energy of formation of the oxide is positive at higher temperatures, therefore appearing above zero on the figure. This means that the oxide is unstable at higher temperatures. It would therefore not be expected to exist at higher temperatures except in cases of transient non-equilibrium.

The light alloys aluminium and magnesium have casting alloys characterised by the stability of the products of their surface reactions. Although one reaction product is hydrogen, which diffuses away into the interior, the noticeable remaining reaction product is a surface oxide film. The oxides of the light alloys are so stable that once formed, in normal circumstances, they cannot be decomposed back to the metal and oxygen. The oxides become permanent features for good or ill, depending on where they come to final rest on or in the cast product. This is, of course, our central theme once again.

A wide range of other important alloys exist whose main constituents would not cause any problem in themselves, but which form troublesome films in practice because their composition includes just enough of the above highly reactive metals. These are discussed later in the metallurgical section, Chapter 6.

Al-Mg alloy family, where the magnesium level can be up to 10 weight percent, are widely known as being especially difficult to cast. Along with aluminium bronze, those aluminium alloys containing 5–10% Mg share the dubious reputation of being the world's most uncastable casting alloys! This notoriety is, as we shall see, ill-deserved. If well cast, these alloys have enviable ductility and toughness and take a bright anodised finish much favoured by the food industry and those markets in which decorative finish is all-important.

Aluminium bronze itself can contain up to 10% Al, and the casting temperature is of course much higher than that of aluminium alloys. The high aluminium level and high temperature combine to produce a thick and tenacious film of Al_2O_3 that makes aluminium bronze one of the most difficult of all foundry alloys. Some other high-strength brasses and bronzes that contain aluminium are similarly difficult.

Ductile irons (otherwise known as spheroidal graphite or nodular irons) are markedly more difficult to cast free from oxides and other defects when compared with grey (otherwise known as flake graphite) cast iron. This is the result of the minute concentration of magnesium that is added to spherodise the graphite, resulting in a solid magnesium silicate surface film that is easily entrained during pouring to create bifilms and dross.

In the course of this work, we shall see how in a few cases the chemistry of the surface film can be altered to convert the film from a solid to a liquid, thus reducing the dangers that follow from an entrainment event. This is a really valuable reaction if it can be achieved. More usually, however, the film can neither be liquefied nor eliminated. It simply has to be lived with. A surface entrainment event therefore usually ensures the creation of a permanent defect.

Entrained films form the major defect in cast materials. Our ultimate objective to avoid films in cast products cannot be achieved by eliminating the formation of films. The only practical solution to the elimination of entrainment defects is the elimination of entrainment. The simple implementation of an improved filling system design can completely solve the problem. This apparently obvious solution is so self-evident that has succeeded to escape the attention of most of the casting community for the last several thousand years.

The techniques to avoid entrainment during the production of cast material represent an engineering challenge that occupies much of the second volume of this book.

1.4 **VAPORISATION**

When melting and casting metals, temperatures are often sufficiently high that some alloys' components will be evaporating all the time. The evaporation of elements from melts can be severe, and has consequences of which it is useful to be aware. Although examples will occur repeated throughout the book, several instances are gathered here to illustrate how common the effect is.

Figure 1.6 illustrates how volatile sodium is, so that additions to Al-Si alloys to modify the eutectic are only short-lived, because of the sodium evaporating off from the melt within 15 or 20 min.

Zinc similarly evaporates from Cu-Zn alloys, and can oxidise, creating the familiar zinc flare. The wind of vapour blowing away from the melt seems responsible for the lack of gas porosity problems in the zinc-containing brasses because gases such as hydrogen in the environment are continuously flushed away. This interesting and useful phenomenon is dealt with in more detail in Section 6.4.

When melting and holding magnesium alloys in a low-pressure casting machine, it is essential to suppress the evolution of vapour by the presence of some air, or other actively protective gas or flux above the melt. The experimental use of an inert gas, argon, above the melt to avoid oxidation led to the atmosphere becoming high in magnesium vapour such that when the unfortunate operator opened the door to charge more ingots, admitting air, he was killed in the explosion. When vaporisation is properly suppressed, the use of magnesium alloys in low-pressure casting machines is perfectly safe.

In the production of ductile iron, magnesium is actually above its boiling point when added to cast iron. The addition reaction is therefore so energetic that the boiling action requires special techniques often involving special reactor vessels to prevent the melt erupting out of the container.

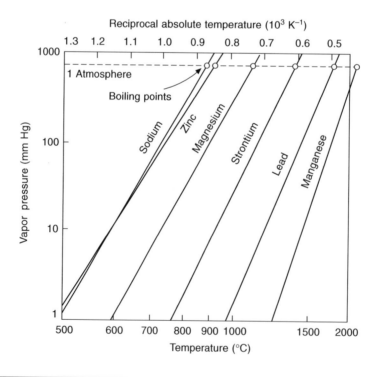

FIGURE 1.6

Increase in the pressure of vapour of some of the more volatile elements as temperature increases.

Data from Brandes (1983).

Alloys of copper and iron that contain lead are a particular problem because of the toxicity of lead. Thus leaded copper alloys and leaded free-cutting steel are now being phase out. The casting of these alloys into sand moulds led to increasing amounts of lead that had condensed in the sand, causing the contamination of nearly everything in the foundry.

Manganese vapour from manganese steels similarly condenses in the moulding sand and is thought to be responsible for the enhanced wetting and penetration of moulds by manganese steels.

Evaporation is a particular problem for alloys melted in vacuum. The charge make-up has to allow for the losses by evaporation. Furthermore, the condensation of the metal vapours on the cold walls of the vacuum chamber usually ignite when the chamber is opened to the air; the fine black metallic dust then burns, with a flame that licks its way around the chamber walls. If not burned on each occasion, the dust can accumulate to a thickness that might become dangerous. Even when the dust is oxidised, it is black and dirty and can be a health hazard. Thus vacuum melting and casting requires special personal care. Melting and casting in an inert gas such as argon greatly reduces the rate of evaporation and consequently reduces these problems.

ENTRAINMENT

If perfectly clean water is poured, or is subject to a breaking wave, the newly created liquid surfaces fall back together again, and so impinge and mutually assimilate. The body of the liquid re-forms seamlessly. We do not normally even think to question such an apparently self-evident process.

However, the same is not true for many common liquids, the surface of which is not a liquid, but a solid, often invisible film of extreme thinness. Aqueous liquids often exhibit films of proteins or other large molecular compounds.

Liquid metals are a special case. The surface of most liquid metals comprises an oxide film. If the surface happens to fold, by the action of a breaking wave, or by droplets forming and falling back into the melt, the surface oxide becomes entrained in the bulk liquid (Figure 2.1).

The *entrainment process* can be a folding action, as seen in Figure 2.1. Alternatively, also shown in the figure, parts of the flow can impinge, as droplets falling back into the liquid. In both cases, the film necessarily comes together dry side to dry side. The submerged surface films are therefore *necessarily always double*.

Also, of course, because of the negligible bonding across the dry opposed interfaces, the defect now *necessarily resembles and acts as a crack*. Turbulent pouring of liquid metals can therefore quickly fill the liquid with cracks. The cracks have a relatively long life, and in many alloys can survive long enough to be frozen into the casting. We shall see how they have a key role in the creation of other defects during the process of freezing and how they degrade the properties of the final casting.

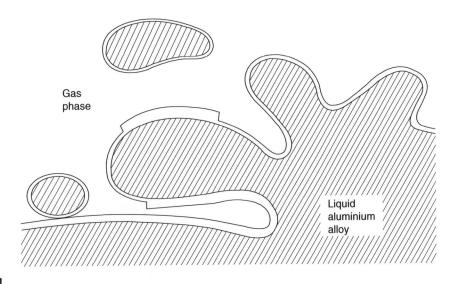

Gas phase

Liquid aluminium alloy

FIGURE 2.1

Sketch of a surface entrainment event.

Complete Casting Handbook. http://dx.doi.org/10.1016/B978-0-444-63509-9.00002-9

Entrainment does not necessarily occur only by the dramatic action of a breaking wave, as seen in Figure 2.1. It can occur simply by the contraction of a 'free liquid' surface. In the case of a liquid surface that contracts in area, its area of oxide, being a solid, is not able to contract. Thus the excess area is forced to fold in a concertina-like fashion. Considerations of buoyancy (in all but the most rigid and thick films) confirm that the fold will be inward, and therefore entrained (Figure 2.2). Such loss of surface is common during rather gentle undulations of the surface, the slopping and surging that can occur during the filling of moulds. Such gentle folding might be available to unfold again during a subsequent expansion, so that the entrained surface might almost immediately detrain once again. This potential for reversible entrainment may not be important compared to the probability that much enfolded material will remain enfolded and entrained. Masses of entrained oxides will entangle and adhere to cores and moulds, but more severe bulk turbulence may tear it away and transport it elsewhere.

With regard to all film-forming alloys, accidental entrainment of the surface during pouring is, unfortunately, only to be expected. This phenomenon of the degradation of liquid metals by pouring is perfectly natural and fundamental to the quality and reliability issues for cast metals. Because these defects are inherited by wrought metals, nearly all of our

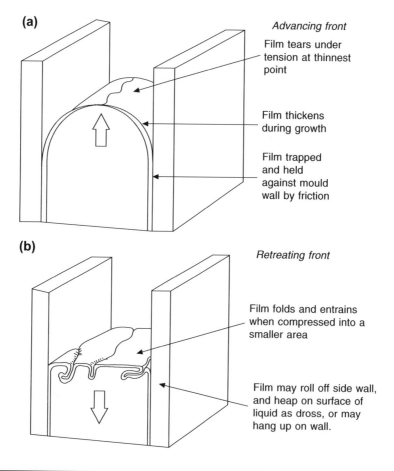

(a)

Advancing front

Film tears under tension at thinnest point

Film thickens during growth

Film trapped and held against mould wall by friction

(b)

Retreating front

Film folds and entrains when compressed into a smaller area

Film may roll off side wall, and heap on surface of liquid as dross, or may hang up on wall.

FIGURE 2.2

Modes of filling (a) a liquid metal advancing by the splitting of its surface oxide (this may occur via a transverse unzipping wave); (b) the retreat of a surface illustrating the consequent entrainment of the surface oxide. (Compare the flow behaviour of cast iron: Figure 6.34).

engineering metals are degraded, too. It is amazing that such a simple mechanism could have arrived at the twenty-first century having escaped notice of thousands of workers, researchers and teachers for the past 6000 years.

In any case, it is now clear that the entrained film has the potential to become one of the most severely damaging defects in cast products (and, as we shall see, in wrought products too). It is essential, therefore, to understand film formation and the way in which films can become incorporated into a casting so as to damage its properties. These are vitally important issues.

It is worth repeating that a surface film is not harmful while it continues to stay on the surface. In fact, in the case of the oxide on liquid aluminium in air, it is doing a valuable service in protecting the melt from catastrophic oxidation. This is clear when compared with liquid magnesium in air. Because magnesium oxide is not as protective, the liquid magnesium can burn, generating its characteristic brilliant flame until the whole melt is converted to oxide. In the meantime, so much heat is evolved that the liquid melts its way through the bottom of the crucible, through the base of the furnace, and will continue down through a concrete floor, taking oxygen from the concrete to sustain the oxidation process until all the metal is consumed. This is the incendiary bomb effect. Oxidation reactions can be impressively energetic!

A solid film grows from the surface of the liquid, atom by atom, as each metal atom combines with newly arriving atoms or molecules of the surrounding gas. Thus for an alumina film on the surface of liquid aluminium, the underside of the film is in perfect atomic contact with the melt, and can be considered to be 'well wetted' by the liquid. (Care is needed with the concept of *wetting* as used in this instance. Here it refers merely to the perfection of the atomic contact which is evidently automatic when the film is grown in this way. The concept contrasts with the use of the term *wetting* for the case in which a sessile drop is placed on an alumina substrate. Perfect atomic contact is now unlikely to exist where the liquid covers the substrate, so that at its edges the liquid will form a large contact angle with the substrate, indicating, in effect, that it does not wish to be in contact with such a surface. Technically, the creation of the liquid/solid interface raises the total energy of the system. The *wetting* in this case is said to be poor.)

The problem with the surface film only occurs when it becomes *entrained* and thus submerged in the bulk liquid.

When considering submerged oxide films, it is important to emphasise that the side of the film which was originally in contact with the melt will continue to be well wetted, i.e. it will enjoy its perfect atomic contact with the liquid. As such, it will adhere well and be an unfavourable nucleation site for volume defects such as cracks, gas bubbles or shrinkage cavities. When the metal solidifies, the metal-oxide bond will be expected to continue to be strong, as in the perfect example of the oxide on the surface of all solid aluminium products, especially noticeable in the case of anodised aluminium.

The upper surface of the solid oxide as grown on the liquid is of course dry. On a microscale, it is known to have some degree of roughness. In fact the upper surfaces of oxide films can be extremely rough. Some, like MgO, being microscopically akin to a concertina, others like a rucked carpet or ploughed field, or others, like the spinel Al_2MgO_4, are an irregular jumble of crystals.

The other key feature of surface films is the great speed at which they can grow. Thus in the fraction of a second (probably between 10 and 100 ms) that it takes to cause a splash or to enfold the surface, the expanding surface, newly creating additional area of liquid, will react with its environment to cover itself in new film. The reaction is so fast to be effectively instantaneous for the formation of oxides.

Other types of surface films on liquid metals are of interest to casters. Liquid oxides such as silicates are sometimes beneficial because they can detrain by balling-up under the action of surface tension and then easily float out, leaving no harmful residue in the casting. Solid graphitic films seem to be common when liquid metals are cast in hydrocarbon-rich environments. In addition, there is some evidence that other films such as sulphides and oxychlorides are important in some conditions. Fredriksson (1996) describes TiN films on alloys of Fe containing Ti, Cr and C when melted in a nitrogen atmosphere. Oxide films are common, but nitride films are to be expected in circumstances were oxygen has been consumed in a submerged crack. Raiszadeh and Griffiths (2008) have done excellent work to illustrate this in aluminium alloys.

In passing, in the usual case of an alloy with a solid oxide film, it is of interest to examine whether the presence of oxide in a melt necessarily implies that the oxide is double. For instance, why cannot a single piece of oxide be simply taken and immersed in a melt to give a single (i.e., non-double) interface with the melt? The reason is that as the piece of oxide is pushed through the surface of the liquid, the surface film on the liquid is automatically pulled down either side of the introduced oxide, coating both sides with a double film, as illustrated schematically in Figure 2.3(a). Thus the entrainment mechanism necessarily results in a submerged film that is at least double. If the surface film is solid, it therefore always has the nature of a crack. Figures 2.3(b) and 2.3(c) illustrate the problem of introducing solid particles to a melt when attempting to manufacture a metal/matrix composite (MMC). Each particle, or cluster of particles, only succeeds in penetrating the surface if it takes with it a 'paper bag' of surrounding oxide. The dry side of the oxide faces the introduced particle, enclosing a remnant of air. Thus the introduced particle is not actually in contact with the liquid, but remains effectively surrounded by a layer of air enclosed within an oxide envelope. Bonding between the particle and the melt is therefore difficult, or even impossible, greatly limiting the mechanical properties of such MMCs.

Finally, it is worth warning about widespread inaccurate and vague concepts that are heard from time to time, and where clear thinking would be a distinct advantage. Two of these are discussed in the following section.

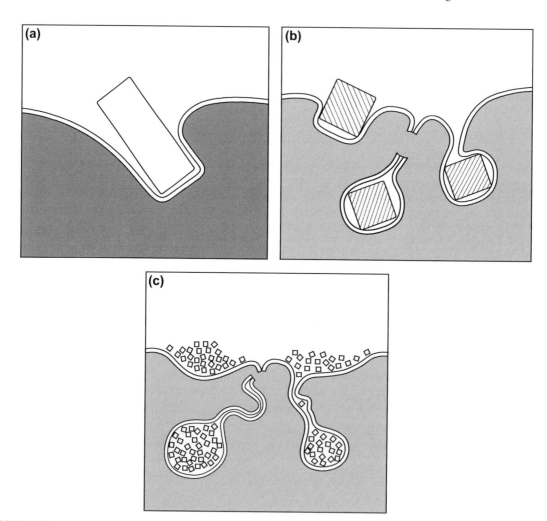

FIGURE 2.3

Entrainment of solids into liquid metals (a) the introduction of melt charge materials; (b) optimum production of MMCs; (c) usual production of MMCs.

For instance one often hears about 'the breaking of the surface tension'. What can this mean? Surface tension is a physical force in the surface of the liquid that arises as a result of the atoms of the liquid pulling their neighbours in all directions. Atoms deep in the liquid experience forces in all directions results, of course, in zero net force. However, for atoms at the surface, there are no neighbours above the surface, so that these atoms experience a net inward force from atoms below in the bulk. This net inward force is the force we know as surface tension. It is always present. It cannot make any sense to consider it being 'broken'.

Another closely related misconception describes 'the breaking of the surface oxide' implying that this is some kind of problem. However, the surface oxide, if a solid film, is always being broken during normal filling. This must occur as the liquid surface expands to form waves and droplets. However, the film is being continuously reformed as fresh liquid surface is created. As the melt fills a mould, rising up between its walls, an observer looking down at the metal will see its surface oxide tear apart and slide sideways across the meniscus toward the walls of the casting, eventually becoming the skin of the casting (Figure 2.2(a)). However, of course, the surface oxide is immediately and continuously reforming, as though capable of infinite expansion. This is a natural and protective mode of advancement of the liquid metal front. It is to be encouraged by good design of filling systems.

As a fine point of logic, it is to be noted that the tearing and sliding across the open surface are driven by the friction of the casting skin, pressed by the liquid against the microscopically rough mould wall. Because this part of the film is trapped and cannot move, and because the melt is forced to rise, the film on the top surface is forced to yield by tearing. This mode of advance is the secret of success of many beneficial products that enhance the surface finish of castings. For instance, coal dust replacements in moulding sands encourage the graphitic film on the surface of liquid cast irons, and the graphitic film (not the relatively weak surface tension) mechanically bridges surface imperfections with a smooth solid film, improving surface finish, as will be detailed later.

As we have explained previously, the mechanism of entrainment is the folding over of the surface to create a sub-merged, doubled-over oxide defect. This is the central problem. The folding action can be macroscopically dramatic, as in the pouring of liquid metals, or the overturning of a wave or the reentering of a droplet. Alternatively, it may be hardly noticeable, like the contraction of a gently undulating liquid surface.

The concept of the entrainment of the surface to form double films (bifilms) are so vital, and so central to the whole problem of the manufacture of castings, that they have been brought to the front of this publication. I found myself unable to write this text without introducing this issue first.

2.1 **ENTRAINMENT DEFECTS**
2.1.1 **BIFILMS**

The entrainment mechanism is a folding-in of the surface or impingement action between two liquid surfaces. Figure 2.4 illustrates how entrainment can result in a variety of submerged defects. In the case of the case of the folding-in of a solid film from the surface of the liquid the defect will be called a *bifilm*. This convenient short-hand denotes the *double* nature of the film defect, as in the word *bicycle*. The name is also reminiscent of the type of marine shellfish; the *bivalve*, whose two leaves of its shell are hinged, allowing it to open and close. (It was never my intention that the word should be hyphenated as 'bi-film' which appears to have been adopted by some. In addition, the pronunciation is suggested to be similar to 'bicycle' and 'bivalve', so that 'bifilm' would be pronounced '*bi*film' as in 'bye-film' and not with a short 'i', that might suggest the word was 'biff-ilm'.)

If the entrained surface is a solid film the resulting defect is a crack (Figure 2.4(a)) that may be only a few nanometres thick, and so be invisible to most inspection techniques. Schaffer (2004) finds the thickness of the oxide on an atomised powder is in the range 5–15 nm. The rapid cooling of the power would correspond well with the short time for oxides to grow when new liquid surface is exposed during turbulence, so this figure is probably a good measure of initially entrained oxides.

Figure 2.4(a) illustrates the important feature that entrained films usually constitute a main bifilm on which transverse bifilms are situated at intervals. This structure is a consequence of the two impinging surfaces being of different areas, the side having the larger area contracts and so forms the transverse folds. Naturally, the transverse bifilms are mainly on the

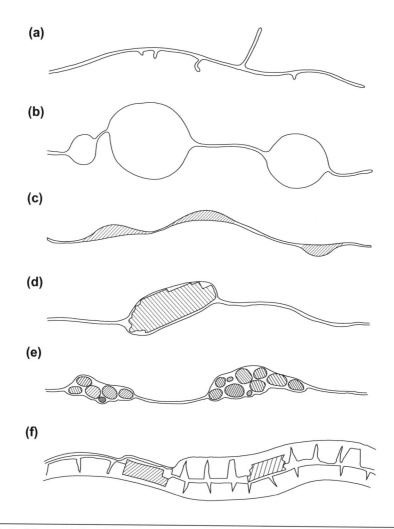

FIGURE 2.4

Entrainment defects: (a) a new bifilm; (b) bubbles entrained as an integral part of the bifilm; (c) liquid flux trapped in a bifilm; (d) surface debris entrained with the bifilm; (e) sand inclusions entrained in the bifilm; (f) an entrained surface film containing integral debris.

one side, the side enclosed by the film of larger area. This expected natural behaviour has interesting consequences for the structure of faults in some important second phases as we shall see.

Other consequences of the entrainment of the surface oxide film are seen in Figure 2.4 to be (b) the entrainment of bubbles as integral features of the bifilm being simply local regions where the two films have not come together because of the presence of entrapped gas; or (c) the entrainment of a surface liquid flux, or (d) solid debris, or (e) moulding sand, or finally (f) an entrained old oxide, possibly from the surface of the ladle or furnace, in which sundry debris has accumulated by falling on to the surface from time to time and has been incorporated into the oxide structure.

It is also important to keep in mind that a wide spectrum of sizes and thicknesses of oxide film exist. With the benefit of the hindsight provided by much intervening research, a speculative guess dating from the first edition of Castings

Table 2.1 Forms of Oxide in Liquid Aluminium Alloys

Growth Time	Thickness	Type	Description	Possible Source
0.01 s	1 nm—1 μm	New	Confetti-like fragments	Pour and mould fill
10 s—1 min	10 μm	Old 1	Flexible, extensive films	Transfer ladles
10 min—1 h	100 μm	Old 2	Thicker films, less flexible	Melting furnace
10 h—10 days	1000 μm	Old 3	Rigid lumps and plates	Holding furnace

(1991) is seen to be a reasonable description of the real situation (Table 2.1). The only modification to the general picture is the realisation that the very new films can be even thinner than was first thought. Thicknesses of only 20 nm seem common for films in many aluminium alloys. The names 'new' and 'old' (probably better stated 'young' and 'old') are a rough and ready attempt to distinguish the types of film, and have proved to be a useful way to categorise the two main types of film.

The limited thickness of a newly created surface film is a consequence of its rapid formation, the film having little time to grow and thicken. Research quoted by Birch (2000) on Zn-Al pressure die casting defects using a cavity fill time from 20 to 100 ms can be back-extrapolated to indicate that in the case of this alloy the surface films formed in times of about 10 ms. Most recent work on entrained young oxides find a limiting thinness in the region of 20 nm. Thus these films are only about 10 molecules thick.

To emphasise the important characteristic crack-like feature of the folded-in defect, the reader will notice that it will be often referred to as a 'bifilm crack' or 'oxide crack'. A typical entrained film is seen in Figure 2.5(a), showing its convoluted nature. This irregular form, repeatedly folding back on itself, distinguishes it from a crack resulting from stress in a solid; its morphology distinguishes it as a defect that could only have formed in a turbulent liquid. Returning from a first session on the scanning electron microscope (SEM) to examine bifilms to ascertain whether they were double or not, Nick Green was amazed to find that the gap between the two films looked like a bottomless canyon (Figure 2.5(b)). This layer of air (or other mould gas) is nearly always present, trapped partly by the irregular folds and partly by the microscopic roughness of the film as it folds over.

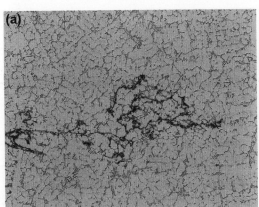 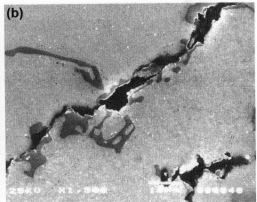

FIGURE 2.5

(a) Convoluted bifilm in Al-7Si-0.4Mg alloy; (b) close-up reported by Green (Green and Campbell, 1994) to appear 'canyon-like' in the SEM.

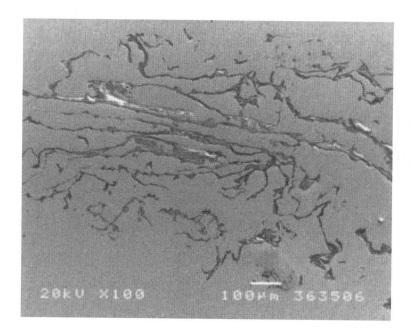

FIGURE 2.6

Polished surface of Al-7Si-0.4Mg alloy breaking into a bifilm, showing upper part of the double film removed, revealing the inside of the lower film (Divandari, 2000).

Figure 2.6 is an unusual polished section photographed by SEM in the author's laboratory by Divandari (2000). It shows the double nature of the bifilm, because by chance, the section happened to be at precisely the level to take away part of the top film, revealing underneath the smooth, glossy surface of a second, clearly unbonded film. Such confirmations of the double nature of the bifilm, and its unbonded central interface, are relatively common, but in the past have generally been overlooked.

As we have mentioned, the surface can be entrained simply by a contraction to create a problem of too much area of film for the area of surface available. However, if more severe disturbance of the surface is experienced, as typically occurs during the pouring of liquid metals, pockets of air can be accidentally trapped by chance creases and folds at random locations in the double film because the surface turbulence event is usually chaotic; waves in a storm rarely resemble sine waves, and a gentle stream contrasts with the 'white water' of a turbulent torrent. The turbulent torrent, its 'whiteness' arising from its content of millions of bubbles, is unfortunately an accurate description of many designs of filling systems for liquid metals.

The resultant random scattering of porosity in castings seems nearly always to originate from these pockets of entrained air. This appears to be the most common source of porosity in castings (so-called 'shrinkage', and so-called 'gas' precipitating from solution are only additive effects that may or may not contribute additional growth). The creation of this source of porosity has now been regularly observed in the study of mould filling using X-ray radiography. It explains how this rather random distribution of porosity typical in many castings has confounded the efforts of computers programmed to simulate only solidification. Such porosity is commonly misidentified as 'shrinkage porosity'. In reality, it is nearly always tangled masses of oxide bifilms. For such defects, often observed on radiographs, we should *not* say 'it is shrinkage' but we should all practise the phrase 'it *appears* to be shrinkage', preferably followed by the phrase 'but most likely is *not* shrinkage'. Gradually, we shall appreciate that the porosity observed on radiographs is *rarely* shrinkage because in general foundry personnel know how to feed castings, so that in general, genuine shrinkage problems are not expected.

Once entrained, the film may sink or float, depending on its relative density. For films on dense alloys such as copper-based and ferrous materials, the entrained bifilms float. In very light materials such as magnesium alloys and lithium alloys, the films generally sink. For aluminium oxide in liquid aluminium, the situation is rather balanced, with the oxide being denser than the liquid, but its entrained air, entrapped between the two halves of the film, often brings its density close to neutral buoyancy. The behaviour of oxides in aluminium is therefore more complicated and worth considering in detail.

Initially of course, enclosed air will aid buoyancy, assisting the films to float to the top surface of the melt. However, as will be discussed later, the enclosed air will be slowly consumed by the continuing slow oxidation (followed by the slow nitridation as discovered by Raiszadeh and Griffiths in 2008) of the surfaces of the central interfaces exposed to the entrapped air. Thus the air is gradually consumed and the buoyancy of the films will slowly be lost providing the hydrogen content of the melt is low. These authors found that when the hydrogen content of the melt was 8 mlkg^{-1} or more the diffusion of hydrogen into the bifilm more than compensated for the loss of air, so that the bifilm would expand.

This complicated behaviour of the bifilm explains a commonly experienced sampling problem because the consequential distribution of defects in suspension at different depths in aluminium furnaces makes it problematic to obtain good quality metal out of a furnace. Initially, a significant density of defects will collect just under the top surface of the liquid, but as their buoyancy is lost, they will gradually tumble out of this population to make their way to the bottom of the melt. Naturally, this makes sampling of the better quality material at intermediate depth rather difficult.

In fact, the centre of the melt would be expected to have a transient population of oxides that, for a time, were just neutrally buoyant. Thus these films would leave their position at the top and would circulate for a time in the convection currents, finally taking up residence on the bottom. Furthermore, any disturbance of the top would be expected to augment the central population, producing a shower, perhaps a storm, of defects that had become too heavy, easily dislodged from the support of their neighbours, and which would then tumble out of the high level clouds. Thus in many furnaces, although the mid-depth of the melt would probably be the best material, it would not be expected to be completely free from defects.

2.1.2 BUBBLES

Small bubbles of air entrapped between films (Figure 2.4(b)) are often the source of **microporosity** observed in castings. Round micropores would be expected to decorate a bifilm, the bifilm itself often being not visible on a polished microsection. Samuel (1993) reports reduced pressure test samples of aluminium alloy in which bubbles in the middle of the reduced pressure test casting are clearly seen to be prevented from floating up by the presence of oxide films.

Large bubbles are another matter, as illustrated in Figure 2.7. The entrainment of larger bubbles is envisaged as possible only if fairly severe surface turbulence occurs. The powerful buoyancy of those larger pockets of entrained air, generally larger than 5 mm in diameter, will give them a life of their own. They may be sufficiently energetic to drive their way through the morass of other films as schematically shown in Figure 2.7. They may even be sufficiently buoyant to force a path through partially solidified regions of the casting, powering their way through the dendrite mesh, bending and breaking dendrites. Large bubbles have sufficient buoyancy to continuously break the oxide skin on their crowns, powering an ascent, overcoming the drag of the *bubble trail* which it leaves in its wake. Bubble trails are especially damaging consequence of the entrainment process and are dealt with later. Large bubbles that are entrained during the pouring of the casting are rarely retained in the casting. This is because they arrive quickly at the top surface of the casting before any freezing has had time to occur. Because their buoyancy is sufficient to split the oxide at its crown, it is similarly sufficient to burst the oxide skin of the casting that constitutes the last barrier between them and the atmosphere, and so escape. The detrainment of the bubble itself is welcome, but leaves the legacy of the bubble trail as a damaging feature, often leading to leak defects.

Bubble trails are a kind of elongated bifilm. However, when scrambled and ravelled together they are practically indistinguishable from ordinary bifilms, as will be described later (Figure 2.34).

Masses of bubbles can be introduced to the casting by a poor filling system design, so that bubbles arriving later are trapped in the tangled mesh of trails left by earlier bubbles. Thus a mess of oxide trails and bubbles is the result. I have

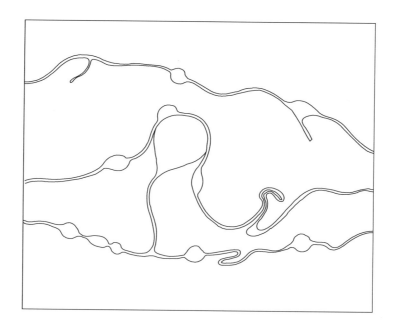

FIGURE 2.7

Schematic illustration of bifilms with their trapped microbubbles and active buoyant macrobubbles.

called this mixture of bubbles and oxide bubble trails 'bubble damage'. In my experience, bubble damage is the most common defect in castings, accounting for perhaps 80% of all casting defects. It is no wonder that the current computer simulations cannot predict the problems in many castings. (In fact, it seems that relatively few of our important defects are attributable to the commonly blamed 'gas' or 'shrinkage'.)

Pockets of air, as bubbles, are commonly an integral feature of the bifilm as we have seen. However, because the bifilm is itself an entrainment feature, there is a possibility that the bifilm or its bubble trail can form a leak path connecting to the outside world, allowing the bubble to deflate if the pressure in the surrounding melt rises. Such collapsed bubbles are particularly noticeable in some particulate metal matrix composites as shown in the work of Emamy and the author (1997), and illustrated in Figures 2.8 and 2.9. The collapsed bubble then becomes an integral part of the original bifilm, but is characterised by (1) a thicker oxide film because of its longer exposure to a plentiful supply of air and (2) a characteristically convoluted shape within the ghost outline of the original bubble.

Larger entrained bubbles are always somewhat crumpled, like a prune. The reason is almost certainly the result of the deformation of the bubble during the period of intense turbulence while the mould is filling. A spherical bubble would have a minimum surface area. However, when deformed, its area necessarily increases, increasing the area of oxide film on its surface. On attempting to regain its original spherical shape, the additional area of film is now too large for the bubble, so that the skin becomes wrinkled. As the forces of turbulence hammer the bubble, each deformation of its shape would be expected to add additional area, adding to the deformation of the bubble. A further factor, perhaps less important, may be the reduction in volume of the bubble as the system cools, and as air is consumed by ongoing oxidation and nitridation. In this case, the analogy with a large shiny round plum shrinking to become a smaller wrinkled prune, may be relatively accurate.

The growth of the area of oxide as the surface deforms seems a general feature of entrainment. It is a one-way, irreversible process. Once formed, the area of the oxide cannot be reduced, so that as the liquid surface attempts to contract; for instance, as the disturbance passes, the oxide has no choice but to wrinkle. The consequent crinkling and folding of the surface is a necessary characteristic of entrained films formed on a turbulent liquid surface, and is the

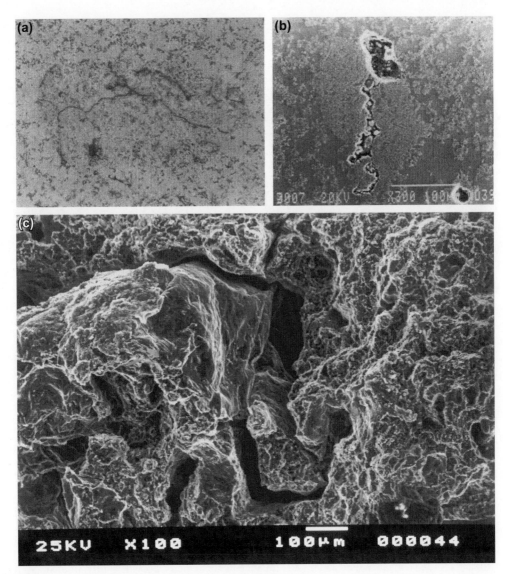

FIGURE 2.8

Collapsed bubbles in Al-TiB$_2$ MMC (a) and (b) show ghost outlines of collapsed bubbles; (c) the resulting bifilm intersecting a fracture surface (Emamy and Campbell, 1997).

common feature that assists to identify films on fracture surfaces. Figure 2.10(a) is a good example of a thin, probably young, film on an Al-7Si-0.4Mg alloy. Figure 2.10(b) is typical of the much thicker films on an Al-5 Mg alloy, displaying the concertina-like corrugations of a film that has grown too large for the area available. A typical thick and granular older film on an Al-11 Si alloy is seen in Figure 2.10(c). The extreme thinness of some films can be seen on a fracture

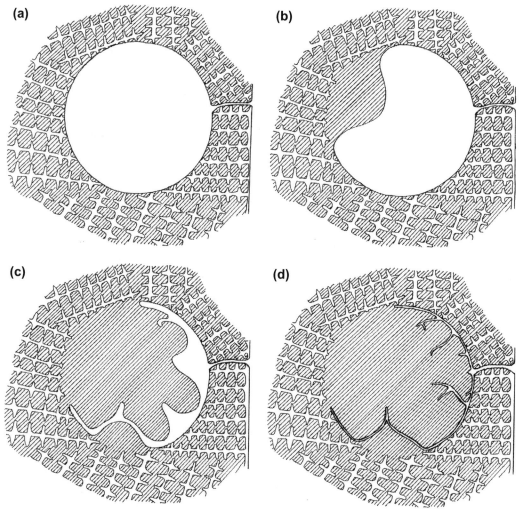

FIGURE 2.9

Schematic illustration of stages of (a) the collapse of a bubble, losing its entrained air via a leak path to the surface, showing (b) and (c) the penetration of residual liquid from the dendrite mesh, and finally (d) a residual bifilm surrounded by penetrated liquid free from dendrites, creating a ghostly outline of the original bubble.

surface of an Al-7Si-0.4Mg alloy (Figure 2.11) that reveals a film folded multiple times but in its thinnest part measures just 20 nm thick.

The irregular shape of bubbles has led to them often being confused with shrinkage pores. Furthermore, perfectly round gas bubbles have been observed by video X-ray radiography of solidifying castings to form initiation sites for shrinkage porosity; bubbles appear to expand by a 'furry' growth of interdendritic porosity as residual liquid is drawn away from their surface in a poorly fed region of a casting. Such growth processes create complicated shapes which obscure the key role of the entrained bubble as the originating source of the problem.

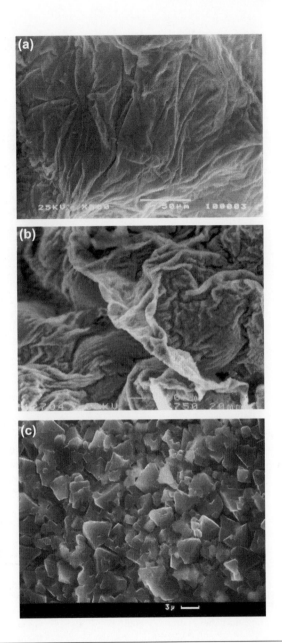

FIGURE 2.10

Fracture surfaces showing (a) a fairly thin film on an Al-7Si-0.4Mg alloy; (b) a thick but flexible film on Al-5 Mg alloy and (c) a thick granular spinel film on Al-7Si-0.6Mg alloy.

(b) Courtesy Divandari, 2000; (c) Courtesy Philip Bell, 1981.

FIGURE 2.11

Multiply folded film on the fracture surface of an Al-7Si-0.4Mg alloy.

Courtesy N R Green.

2.1.3 EXTRINSIC INCLUSIONS

In addition to bifilms (cracks) and bubbles (porosity), there are a number of other, related defects that can be similarly entrained (i.e. introduced into the matrix from the outside). These externally introduced features are variously but aptly called *entrained inclusions, extrinsic inclusions or external inclusions* (they are also occasionally known by the equally accurate but awkward term *exogenous inclusions* which we shall avoid). Such features are, of course, completely different in character to second phases that precipitate from the melt as part of the cooling and solidification processes. The differences warrant careful examination and are discussed later.

Flux and slag inclusions

Fluxes containing chlorides or fluorides are relatively commonly found on machined surfaces of light alloy cast components. Such fluxes are deliquescent, so that when opened to the air in this way they absorb moisture, leading to localised pockets of corrosion and the unsightly exudation of corrosion products on machined surfaces. During routine examination of fracture surfaces, the elements chlorine and fluorine are quite often found as chlorides or fluorides on aluminium and magnesium alloys. The most common flux inclusions to be expected are NaCl and KCl.

However, chlorine and fluorine, and their common compounds, the chlorides and fluorides, are insoluble in aluminium, presenting the problem of how such elements came to be located in the matrix. This can probably only be explained by assuming that such materials were originally on the surface of the melt, but have been mechanically entrained, wrapped in the surface oxide film (Figure 2.4(c)). Thus flux inclusions on a fracture surface indicate the presence of a bifilm, probably of considerable area, although the presence of its single remaining half on one of the fracture surfaces will probably be not easily seen on the fracture.

Also, because flux inclusions are commonly liquid at the temperatures of molten Al and Mg alloys, and perfectly immiscible, why do they appear in the matrix? Why do they not spherodise and rapidly float out, becoming lost from the surface of the melt? It seems that rapid detrainment does not occur. The explanation lies once again in the presence of the bifilm: the flux would originally have sat on the oxide surface of the melt. Surface turbulence folded in the surface, forming a flux sandwich in an entrained oxide bifilm. This package is slow to float or settle because of its extended area. The consequential long residence times allow transport of these contaminants over long distances in melt transfer and launder systems. Sokolowski and colleagues (2000) find that chloride flux inclusions travel distances of tens of metres in launders, whereas if they were simply spherical droplets, they would be expected to completely separate from the melt within a metre or so of travel along a launder. They found that the electromagnetic pumps at the end of the launder system were eventually blocked by a mix of oxides and fluxes depositing in the working interiors of the pumps. It is likely that the inclusions are forced out of suspension by the combined centrifugal and electromagnetic body forces in the pump. It is hardly conceivable that fluxes themselves, as relatively low-viscosity liquids, could form a blockage. However, the accretion of a mixture of solid oxide films bonded with a sticky liquid flux would be expected to be highly effective in choking the system.

When bifilms are created by folding into the melt, the presence of a surface liquid flux would be expected to have a powerful effect by effectively causing the two halves of the film to adhere together by viscous adhesion. This may be one of the key mechanisms explaining why fluxes are so effective in reducing the porosity in aluminium alloy castings. The bifilms may be glued shut. At room temperature, the bonding by the solidified flux may aid strength and ductility to some extent. On the other hand, the observed benefits to strength and ductility in flux-treated alloys may be the result of the reduction in films by agglomeration (because of their sticky nature in the presence of a liquid flux) and flotation. These factors will require much research to disentangle.

Whether fluxes are completely successful to prevent oxidation of the surface of light alloys does not seem to be clear. It may be that a solid oxide film always underlies the simple chloride fluxes and possibly some of the fluoride fluxes. Even if an underlying oxide film were unstable, and gradually dissolving in the flux, the early entrainment of the melt surface prior to the complete solution of the oxide would explain the oxide presence despite any thermodynamic instability.

Even now, some foundries successfully melt light alloys without fluxes. Whether such practice is really more beneficial deserves to be thoroughly investigated. What is certain is that the environment would benefit from the reduced dumping of flux residues from so-called cleansing treatments.

There are circumstances when the flux inclusions may not be associated with a solid oxide film simply because the oxide is rapidly soluble in the flux. Such fluxes include cryolite $AlF_3.3NaF$ as used to dissolve alumina during the electrolytic production of aluminium, and the family of other fluorides K_2TiF_6, KBF_4, K_3TaF_7 and K_2ZrF_6. Thus the surface layer may be a uniform liquid phase in these special cases, provided the temperature is sufficiently high for the flux itself to be above its melting point.

For irons and steels when a liquid slag layer is present it is normally expected to be liquid throughout. Where a completely liquid slag surface is entrained, Figure 2.12 shows the expected detrainment of the slag and the accidentally entrained gases and entrained liquid metal. Such spherodisation of fluid phases is expected to occur in seconds as a result of the high surface energy of liquid metals and their rather low viscosity.

A classical and spectacular breakup of a film of liquid is seen in the case of the granulation of liquid ferro-alloys (Figure 2.13). A ladle of alloy is poured onto a ceramic plate sited centrally above a bath of water. On impact with the plate, the jet of metal spreads into a film that thins with distance from the centre. The breakup of the metal film is seen to occur by the nucleation of holes in the film, followed by the thinning of ligaments between holes, and finally by the breakup of the ligaments into droplets. This is the classical mode of break-up of a thinning liquid film.

A similar pattern of fragmentation of thin films may not only occur for liquid films. It seems possible that alumina is first entrained into liquid steel as solid films (the melting point of alumina being higher than 2000°C). However, once entrained, the films would be expected to spherodise because the solid films are extremely thin, and the temperature will be sufficiently high to encourage rapid surface diffusion, leading to breakup of the films into roughly spherical particles. The spherodisation will be driven by the high interfacial energy that is typical of high-temperature metallic systems. The clusters and stringers of isolated alumina inclusions in rolled steels might have formed by this process, the rolling action being merely to further separate and align them along the rolling direction (Figure 2.14(a)). This suggestion is supported by observations of Way (2001) that show the blockage of nozzles of steels in continuous casting plants to be the result of arrays of separate granular alumina inclusions (Figure 2.14(b)). It seems certain that such inclusions were originally

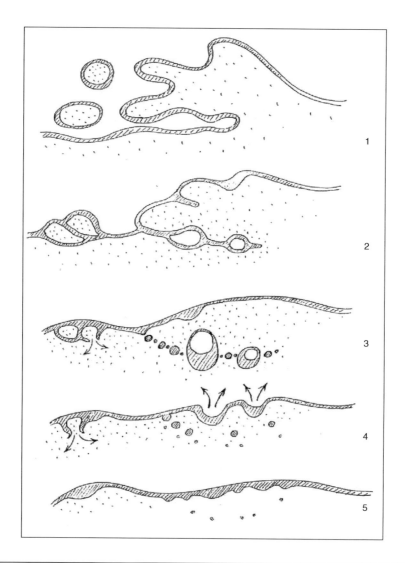

FIGURE 2.12

Entrainment of a liquid surface film, showing the subsequent detrainment of metal, gases and most of the entrained droplets of liquid. No bifilm is formed. The cast metal remains essentially free from any serious permanent entrained damage.

alumina films (separate inclusions seem hardly conceivable) formed by the annular backflow of oxidising air inside the lower regions of the nozzle where the flow detaches from the nozzle wall. Properly tapered nozzles (instead of parallel bore nozzles) avoid this problem. Why all nozzles are not tapered because they would then be easier and less costly to mould, and would save the steel industry huge costs, is a mystery.

In contrast, a film of liquid flux on aluminium alloys is not, in general, expected to have this simple completely liquid structure. It seems certain that in most instances a solid alumina film will exist under the liquid flux. Thus this composite

FIGURE 2.13

Granulation of liquid ferromanganese poured on to a flat ceramic target, showing the characteristic mode of fragmentation of a film into ligaments, then into droplets, and finally being cooled in a bath of water.

Courtesy Uddeholm Technology, Sweden 2001.

film will behave as a sticky solid layer, and will be capable of permanently entraining bubbles and liquid flux. Spherodisation of the liquid flux will be impossible.

Old oxides

Figures 2.3 and 2.4(d) represent one of many different kinds of extrinsic or external inclusion, such as a piece of refractory from the wall of a furnace or ladle or a piece of old surface oxide from the same source. External inclusions can only be entrained together with a wrapping of young surface film. Attempts to introduce inoculants of various kinds into aluminium alloys have been made from time to time. All of these attempts are seen to be undermined by the problem of the surrounding young oxide with its associated layer of air, preventing the introduced particle from ever making contact with the melt as in Figure 2.3. This behaviour necessarily occurs despite attempts by such authors as Mohanty and Gruzleski (1995) to add TiB_2 particles to liquid aluminium beneath the surface of the melt. They found that such particles did not nucleate grains, but were pushed by dendrites to the grain boundaries. This is completely unexpected behaviour for a TiB_2 particle in good atomic contact with the melt. A particle of TiB_2 in contact with the matrix is expected to be incorporated into the growing solid, and actually nucleate new solid grains. However, being pushed ahead of the solidification front is the behaviour expected for any particle microscopically separated from its surrounding melt by a layer of air held in place by a film of oxide, and thus effectively completely non-wetted (i.e. not in contact with the melt, and so actually remaining 'dry').

Similarly, the vortex technique for the addition of particles to make a particulate metal matrix composite (MMC) is similarly troubled. It is to be expected that the method would drag the oxide film down into the vortex, together with the

(a)

(b)

FIGURE 2.14

(a) Alumina stringers in deep-etched rolled steel; (b) alumina particles from a blocked tundish nozzle from a steel continuous casting plant (Way, 2001).

heaps of particles supported on the film, so that the clumping of particles, now nicely and collectively packaged, wrapped in film, is probably unavoidable. The 'clumping' of particles observed in such MMCs survives even long periods of vigorous stirring (Figure 2.3(c)). Such features are seen in the work by Nai and Gupta (2002) in which the central particles in clumps of stirred-in SiC particles are observed to have fallen out during the preparation of the polished sample, leaving small cave-like structures lined with SiC particles that are still held in place, probably bonded or mechanically keyed to the oxide film. In addition, whether in clumps or not, during the preparation of the mixture the particles are almost certainly held in suspension by the network of bifilms that will have been introduced, in the same way that networks of films have been observed to hold gas bubbles in suspension (Samuel 1993). It seems also likely that the dense oxide network generated in the massive turbulence of the vortex will reduce the fluidity of the material, and reduce the achievable mechanical properties, as discovered by Emamy et al. (2009).

In an additional example, the manufacture of aluminium alloy MMCs based on aluminium alloy containing SiC particles introduced under high vacuum were investigated by Emamy and Campbell (1997). These authors found that under the hydrostatic tension produced by poor feeding conditions the SiC/matrix interface frequently decohered. This contrasted with the behaviour of an Al-TiB$_2$ MMC in which the TiB$_2$ particles were created by a flux reaction and added to the matrix alloy without ever having been in contact with the air. These particles exhibited much better adhesion at the particle/matrix interface when subjected to tension; relatively few pores were initiated when subjected to the hydrostatic tensile stress generated by the poor feeding.

Figure 2.4(d) illustrates the entrainment of a piece of *old oxide*, probably from an earlier part of the melting or holding process. Such oxide, if very old, will have grown thick and no longer justifies description as a film. It can be crusty or even plate-like. It is simply just another external inclusion. A typical example is seen in Figure 2.10(c). If entrained during pouring, its passage through the liquid surface will ensure that it becomes wrapped in a new thin film, its entrained air film separating the old inclusion from contact with the melt. This wrapping will allow the matrix to 'decohere' easily from the inclusion, leading to the initiation of porosity or cracks. However, if the old oxide was entrained some time previously during the melting or holding process, its entrained wrapping will itself become old, having lost the majority of its air layer by continued reaction. In this case, it will have become fairly 'inert' in the sense that it cannot easily initiate porosity; all its surfaces will have reacted with the melt and thus have come into good atomic contact; its air layer will have become largely lost and welded closed.

Sand inclusions

Inclusions of moulding sand are perhaps the most common extrinsic inclusion (Figures 2.3(c) and 2.4(e)), but whose mechanism of entrainment is probably rather complicated. It is not easy to envisage how minute sand grains could penetrate the surface of a liquid metal against the repulsive action of surface tension and the presence of an oxide film acting as a mechanical barrier. The penetration of the liquid surface would require the grain to be fired at the surface at high velocity, like a bullet. However, of course, such a dramatic mechanism is unlikely to occur in reality. Although the following description appears complicated at first sight, a sand entrainment mechanism can occur easily, involving little energy, as described next.

In a well-designed filling system for a sand casting, the liquid metal entirely fills the system, and its hydrostatic pressure acts against the walls of the channel to gently support the mould, holding sand grains in place. Thus the mould surface will become hot. If bonded with a resin binder, the binder will often first soften, then harden and strengthen as volatiles are lost. The sand grains in contact with the metal will finally have their binder degraded (pyrolysed) to the point at which only carbon remains, now hardened and rigid, like coke, forming strong mechanical links between the grains. The carbon layer remaining on the grains has a high refractoriness. It continues to protect the grains because it in turn is protected from oxidation because, at this late stage, most of the local oxygen has been consumed to form carbon monoxide. The carbon forms a non-wetted interface with the metal, thus enhancing protection from penetration and erosion.

The situation is different if the filling system has been poorly designed, allowing a mix of air to accompany the melt. This problem commonly accompanies the use of filling systems that have over-sized cross-section areas, and thus remain unfilled with metal. In such systems, the melt can ricochet backwards and forwards in the channel. The mechanical impact, akin to the cavitation effect on a ship's propeller, is a factor that assists erosion. However, other factors are also important. The contact of the melt with the wall of the mould heats the sand surface. As the melt bounces back from the wall, air is drawn through the mould surface, and on the return bounce, the air flow is reversed. In this way air is pumped backwards and forwards though the heated sand, and so burns away the binder. The burning action is intensified akin to the air pumped by bellows in the blacksmith's forge. When the carbon is finally burned off the surface of the grains, the sand is no longer bonded to its fellow grains in the mould. Furthermore, the oxide on the surface of the melt can now react with the freshly exposed silica surface of the grains, and thus adhere to them. If the melt now ricochets off the sand surface once again, the oxidised liquid surface peeling away from the mould is now covered with adhering grains of sand. As the surface folds over, the grains are thereby enfolded into an envelope of oxide film, as wrapped in a paper bag (Figure 2.15). Under the microscope, such oxide films can practically always be seen enfolding clusters of sand inclusions. Similarly, sand inclusions are often found on the inner surfaces of bubbles.

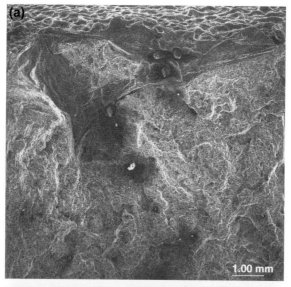

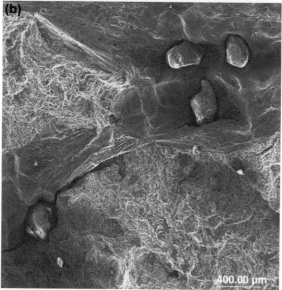

FIGURE 2.15

Fracture surface of Al-7Si-0.4Mg alloy showing (a) the sand-cast surface (top) with a surface-connected oxide film; (b) close-up showing the sand inclusions entrained in their 'paper bag' of oxide film.

Courtesy Fox, (2002).

Sand inclusions are therefore a sure sign that the filling system design is poor, involving significant surface turbulence. Conversely, sand inclusions can nearly always be eliminated by attention to the filling system, not the strength nor the variety of the sand binder. A rash of sand inclusions is a common ailment in foundries. The best solution lies neither with the sand plant operator nor the binder supplier, but in the hands of the designer of the filling system of the casting.

Finally, Figure 2.4(f) is intended to give a glimpse of the complexity that is likely to be present in many bifilms. Part of the film is new and thin, forming an asymmetrical thick/thin double film. (Reactions of the melt with the bifilm, as discussed at many points later in this work, often lead to asymmetrical precipitation of reactants if one side of the bifilm is more favourable than the other.) It would be expected that other parts of the defect would be fairly symmetrical thin double films. Elsewhere both halves may be old, thick and heavily cracked. In addition, the old film contains debris that has become incorporated, having fallen onto the surface of the melt at various times in its progress through the melting and distribution system.

Entrained material is probably always rather messy.

2.2 ENTRAINMENT PROCESSES

The entrainment of surface films, leading to the creation of a double film in the liquid, can occur in very different ways, and lead to quite different types of bifilm crack:

1. by surface turbulence, in which the bulk liquid and the liquid meniscus are in general travelling in at least approximately the same direction; the motion of the liquid front ensure that the breaking and re-forming of the surface film gives limited time for film growth, so these entrained films are thin, small, irregular and numerous, and occur widely during all types of pouring;
2. by submergence of charge material used to make the melt; these films are often thick, with large area, especially if returns from a sand casting operation are employed; they are probably most easily removed by treatment with small bubbles as in rotary degassing;
3. by laminar flow, in which the bulk liquid is in general flowing parallel to the meniscus and the meniscus is stationary. The stationary meniscus then grows a thick surface film even though the melt may be travelling at high speed. The thick asymmetric films form large horizontal cracks or vertical tubes, and are therefore also common in top-gated castings;
4. by the passage of bubbles or air or mould gases through the melt, creating the long bubble trail as a collapsed tube. Because most running systems entrain up to 50% air, there is no shortage of these damaging leak paths between ingates and the top of the casting.

2.2.1 SURFACE TURBULENCE

As liquid metal rises up a vertical plate mould cavity, with the meniscus constrained between the two walls of the mould, an observer looking down from above (suitably equipped with protective shield, of course!) will see that the liquid front advances upwards by splitting the surface film along its centre. This occurs because the earlier area of the film is trapped between the wall of the mould and the liquid surface, effectively held in place by friction, and therefore cannot rise with the rising liquid. Thus film on the rising surface is torn apart, moving to each side, to become the skin of the casting, which is pinned in place against the mould walls (Figure 2.2(a)). As the surface tearing happens, of course, the surface film is continuously renewed so that the process is effectively continuous.

It seems that this vertical advance actually occurs by the action of one or more horizontal transverse travelling waves (I call them 'unzipping waves') that wander backwards and forwards along a meniscus rising between walls. This interesting phenomenon will be discussed in detail later.

The important feature of this form of advance of the liquid front is that the surface film never becomes entrained in the bulk of the liquid. Thus the casting remains potentially perfect. If this is not already perfectly understood at this point, all readers are encouraged to learn this fact by heart before proceeding to the remainder of the book.

(a)   **(b)**

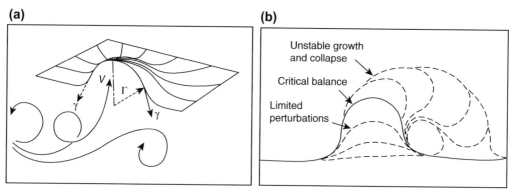

FIGURE 2.16

(a) The balance of forces in the liquid surface; (b) perturbations of the surface by inertial forces in the melt remain reversible up to the critical hemispherical limit, after which the surface becomes unstable and irreversibly deforms, introducing the danger of the entrainment of the surface.

A problem arises if the velocity of flow is sufficiently high at some locality for the melt to rise enough above the general level of the liquid surface so that it subsequently falls back under gravity, thus entrapping a portion of its own surface. This is the action of a breaking wave. It is represented in Figure 2.16. This process of entrainment of the liquid surface, and in particular, its surface film, has been modelled by computer simulation by a number of workers. The various approaches have been reviewed by the author (Campbell, 2006).

To gain some insight into the physics involved, we can carry out the following analysis. The pressure tending to perturb the surface is the inertial pressure $\rho V^2/2$ in which ρ is the density of the melt and V is the local flow velocity. The action of the surface tension γ tends to counter any perturbation by tensioning the surface, pulling the surface flat. The maximum pressure that surface tension can hold back is achieved when the surface is deformed into a hemisphere of radius r, giving a pressure $2\gamma/r$ resisting the inertial pressure (Figure 2.16). Thus in the limit that the velocity is just sufficiently high to push the surface beyond the hemisphere, and so just exceeding the maximum restraint that the surface tension can provide, we have $\rho V^2/2 = 2\gamma/r$, so that the critical velocity V_{crit} at which the surface will become unstable and start to suffer *surface turbulence* (my name for the phenomenon of breaking waves and entrainment of the surface into the bulk) is

$$V_{crit} = 2(\gamma/r\rho)^{1/2} \tag{2.1}$$

Subsequent experiments have shown that in spite of its rough derivation, this is a surprisingly accurate result. However, the approach is unsatisfactory in the sense that the radius r has to be assumed, but there is little wrong with the physics, so the derivation is reproduced here because of its directness and simplicity.

Equation (2.1) is not an especially convenient equation because the critical radius of curvature r of the front at which instability occurs is not known independently. However, we can guess that it will not be far from a natural radius of the front as defined, for instance, by a sessile drop. (The word 'sessile' means 'sitting', and is used in contrast to the 'glissile' or 'gliding' drop.) The sessile drop sitting on a non-wetted surface exhibits a specific height h defined by the balance between the forces of gravity, tending to flatten it, and surface tension, tending to pull the drop together to make it as spherical as possible. We can derive an approximate expression for h considering a drop on a non-wetted substrate as illustrated in Figure 2.17. The pressure to expand the drop is the average pressure $\rho g h/2$ acting over the central area hL, giving the net force $\rho g h^2 L/2$. For a drop sitting quietly, with its forces in equilibrium, this force is equal to the net force because of surface tension acting over the length L on the top and bottom surfaces of the drop, $2\gamma L$. Hence

$$\rho g h^2 L/2 = 2\gamma L$$

$$\text{giving} \quad h = 2(\gamma/\rho g)^{1/2} \tag{2.2}$$

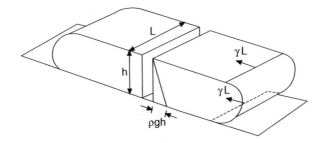

FIGURE 2.17

Balance of forces in a sessile drop sitting on a non-wetted substrate.

Assuming that $h = 2r$ approximately, and eliminating r from Eqn (2.1), we have

$$V_{crit} = 2(\gamma g/\rho)^{1/4} \qquad (2.3)$$

This is the more elegant way to derive the critical velocity involving no assumptions and no adjustable parameters. It is analogous to the concept illustrated in Figure 2.18 in which liquid metal rises through a vertical ingate to enter a mould.

Clearly, for case (a) in Figure 2.18(a), when the velocity at the entrance to the gate is zero, the liquid is perfectly safe. There is no danger of enfolding the liquid surface. However, of course, the zero velocity filling condition is unfortunately not particularly useful for the manufacture of castings.

In contrast, case (c) can be envisaged in which the velocity is so high that the liquid behaves as a jet. The liquid is then in great danger of enfolding its surface during its subsequent fall. Regrettably, this condition is common in castings.

The interesting intermediate case (b) illustrates the situation in which the melt travels at sufficient velocity to rise to a level that is just supported by its surface tension. This is the case of a sessile drop. Any higher, and the melt would subsequently have to fall back, and would be in danger of enfolding its surface. Thus the critical velocity required is that which is only just sufficient to project the liquid to the height h of the sessile drop. This height is simply that derived from conservation of energy, equating kinetic energy gain to potential energy loss, $mV^2/2 = mgh$, we obtain $h = V^2/2g$. Substituting this value in the value for the height of the sessile drop shown previously in Eqn (2.2) gives the same result for the critical velocity in Eqn (2.3).

From Eqn (2.2), we can find the height from which the metal can fall, accelerated by gravity, before it reaches the critical velocity given by Eqn (2.3). When falling from greater heights, the surface of the liquid will be in danger of being entrained in the bulk liquid. The critical fall heights for entrainment of the surface are listed in Table 2.2. (Physical properties of the liquids have been taken at their melting points from Brandes and Brook, 1992 because the values change negligibly with temperature. The values will be expected to be influenced somewhat by alloying of course.)

At first sight, it seems surprising that nearly all liquids, including water, have a critical velocity in the region of 0.3 to 0.5 ms^{-2}. This is a consequence of the fact that the velocity is a function of the *ratio* of surface tension and density, and, in general, both these physical constants change in the same direction from one liquid to another. Furthermore, the effect of the one-quarter power in Eqn (2.3) further suppresses differences. Most of the important engineering metals have critical velocities close to 0.5 ms^{-1}, so that I now generally assume this value in the calculation of filling systems for castings of all metals and alloys.

The value of 0.5 ms^{-1} for liquid aluminium has been confirmed experimentally by Runyoro et al. (1992). He observed by video the emergence of liquid aluminium at increasing speeds from an ingate in a sand mould. An outline of some of melt profiles is given in Figure 2.18(b). At just over the critical velocity, at 0.55 ms^{-1}, the metal falls back only slightly, forming its first fold, taking on the appearance of a traditional English bread called a cottage loaf. At progressively higher velocities, the potential for damage during the subsequent fall of the jets is evident.

(a)

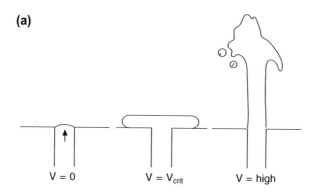

(b)

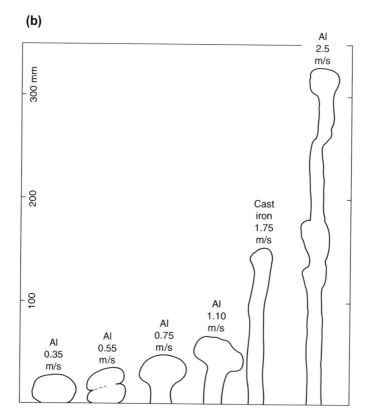

FIGURE 2.18

(a) Concept of liquid emerging into a mould at zero, critical and high velocities; (b) experimentally observed profiles of liquid Al and cast iron emerging from an ingate at various speeds (Runyoro et al., 1992).

Table 2.2 The Critical Heights and Velocities of Some Liquids

Liquid	Density (kgm^{-3})	Surface Tension (Nm^{-1})	Critical Height h (mm)	Critical Velocity (ms^{-1})
Ti	4110	1.65	12.8	0.50
Al	2385	0.914	12.5	0.50
Mg	1590	0.559	16.0	0.42
Fe	7015	1.872	10.4	0.45
Ni	7905	1.778	9.6	0.43
Cu	8000	1.285	8.1	0.40
Zn	6575	0.782	7.0	0.37
Pt	19,000	1.8	6.2	0.35
Au	17,360	1.14	5.2	0.32
Pb	10,678	0.468	4.2	0.29
Hg	13,691	0.498	3.9	0.27
Water	1000	0.072	5.4	0.33

Densities and surface tensions adopted from Brandes and Brook (1992).

In further work shown in Figure 2.19, Runyoro (1992) observed a dramatic rise in surface cracks identified by dye penetrant testing, and the precipitous fall in bend strength of castings filled above 0.5 ms^{-2}. I was so shaken by the alarming nature and clarity of this result that I completely changed the direction of research work at the University of Birmingham for the next decade, focussing on the role of surface turbulence generated by filling system design on the mechanical performance of castings.

For instance, the critical velocity for Cu-10 Al alloy at 0.4 ms^{-1} was subsequently confirmed by Helvaee and Campbell (1996), and a similar value for ductile cast irons shown in Figure 2.20 was reported by the author (2000). The critical velocity for aluminium has been explored by computer simulation (Lai et al., 2002) in which area of the melt surface was computed as a function of increasing ingate velocity. Assuming the same values for physical constants as used here, a value of precisely 0.5 ms^{-1} was found (Figure 2.21), providing reassuring confirmation of the concept of critical velocity by a totally independent technique.

Since those early days, further confirmations of the critical entry velocity into moulds has been found for magnesium (Bahreinian 2005) to be in the range 0.25–0.50 ms^{-1}. Mirak et al. (2011) located a value for the Mg-9Al-1Zn alloy 0.30 ± 0.05 ms^{-1}. Furthermore, in much subsequent industrial work in foundries, in which filling systems have been redesigned to reduce ingate velocities to near 0.5 ms^{-1}, the improvements in quality of castings has been demonstrated in metals and alloys of all types, from light alloys to high temperature alloys, steels and titanium-based alloys.

Weber number

The concept of the critical velocity for the entrainment of the surface by what the author has called *surface turbulence* is enshrined in the *Weber number, We*. This elegant dimensionless quantity is defined as the ratio of the inertial pressure in the melt, assessed as $\rho V^2/2$, with the pressure resulting from surface tension $\gamma(1/r_1 + 1/r_2)$ in which r_1 and r_2 are the radii of curvature of the surface in two perpendicular directions. If only curved in one plane, the pressure becomes γ/r, or for a hemisphere where the orthogonal radii are equal, $2\gamma/r$. Thus

$$We = \rho L V^2 / \gamma \tag{2.4}$$

Only the dimensioned quantities are included in the ratio, the scalar numbers (the factors of 2) are neglected, as is usual for a dimensionless number used over many orders of magnitude. The radius term becomes the so-called 'characteristic

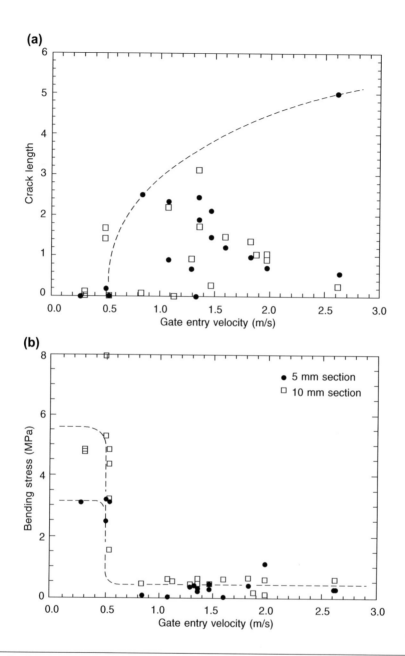

FIGURE 2.19

(a) Crack lengths by dye penetrant testing in Al alloys cast at different ingate velocities; (b) the bend strength of the same plate castings (Runyoro et al., 1992).

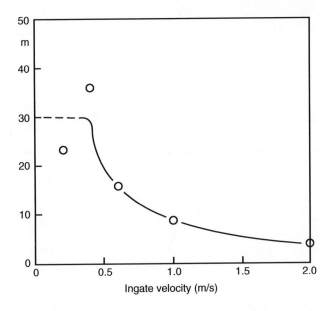

FIGURE 2.20

Reliability of ductile iron cast at different ingate velocities in terms of the two-parameter Weibull modulus m (Campbell, 2000a).

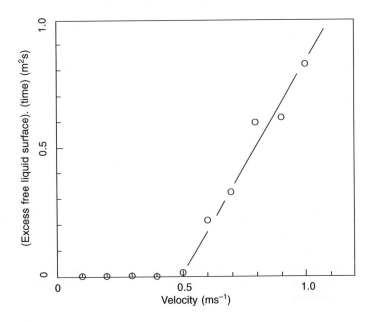

FIGURE 2.21

Computer simulation of the filling of a mould with Al, showing the excess free area of the melt as a function of ingate velocity (Lai et al., 2002).

length' parameter, L. For flows in channels, L is typically chosen to be the hydraulic radius, defined as the ratio of the occupied area of the channel to its wetted perimeter. For a filled cylindrical channel, this is half the radius of the channel (diameter/4). For a filled thin slot, it is approximately half of the slot width. The hydraulic radius is a concept closely allied to that of 'geometrical modulus' as commonly used by ourselves as casting engineers.

When $We = 1$, the inertial and surface forces are roughly balanced. From the definition of We in Eqn (2.4), and assuming for liquid aluminium a height of 12.5 mm for a sessile drop, giving a radius $L = 12.5/2$ mm, we find $V = 0.25$ ms^{-1}, a value only a factor of 2 different from the critical velocity found earlier. When it is recalled that the real value of dimensionless numbers lies in their use to define numbers to the nearest order of magnitude (i.e. the nearest factor of 10), this is impressively close agreement. The agreement is only to be expected, of course because the condition that $We < 1$ defines those conditions where the liquid surface is tranquil, and $We > 1$ defines the conditions in which surface turbulence is expected.

The We number has been little used in casting research. This is a major omission on the part of researchers. It is instructive to look at the values that We can adopt in some casting processes.

Work on aluminium bronze casting has shown empirically that these alloys retain their quality when the metal enters the mould cavity through vertical gates at speeds of up to 60 mms^{-1}. However, this extremely cautious velocity is almost certainly in error as a result of confusion introduced from other faulty parts of the filling system (notably problems with the pouring basin, and the well at the sprue/runner junction). Helvaee (1996) confirmed the theoretical predicted critical velocity for aluminium bronze of 0.4 ms^{-1}. However, if we accept the low velocity 0.060 ms^{-1} into the mould cavity, we can assume that the radius of the free liquid surface is close to 5 mm (corresponding to half of the height of the sessile drop given in Table 2.2), the density approximately 8000 kgm^{-3}, and surface tension is approximately 2.3 Nm^{-1}, then We in this case is seen to be approximately 0.1 representing a condition dominated by surface tension, and thus ultrasafe from any danger of surface turbulence.

In a mould of section thickness of 5 mm fitted with a glass window, Nieswaag observed that a rising front of liquid aluminium just starts to break up into jets above a speed of approximately 0.37 ms^{-1}. This is reasonably close to our theoretical prediction 0.5 ms^{-1} (Table 2.2). Assuming that the radius of the liquid front is 2.5 mm, and taking 2400 kgm^{-3} for the density of liquid aluminium, the corresponding critical value of We at which the breakup of the surface first occurs is therefore seen to be approximately 0.9. The users of vertically injected squeeze casting machines now commonly use 0.4 ms^{-1} as a safe filling speed (for instance, Suzuki 1989), corresponding to We close to 2.0.

UK work in the 1950s (IBF Subcommittee, 1960) showed that cast iron issues from a horn gate as a fountain when the sprue is only about 125 mm high and the gate about 12 mm in diameter (deduced from careful inspection of the photograph of the work). We can calculate, therefore, that We is approximately 80 for this well-advanced condition of surface turbulence.

Going on to consider the case of a pressure die casting of an aluminium alloy in a die of section thickness 5 mm and for which the velocity V might be 40 ms^{-1}, We becomes approximately 100,000. It is known that the filling conditions in a pressure die casting are, in fact, characterised by extreme surface fragmentation, and possibly represent an upper limit which might be expected whilst making a casting. Owen (1966) noted from examination of the surfaces of zinc alloy pressure die castings that the metal advanced into the mould in a series of filaments or jets. These advanced splashes were subsequently overtaken by the bulk of the metal, although they were never fully assimilated. These observations are typical of much work carried out on the filling of the die during the injection of pressure die castings.

In summary, it seems that We numbers in the range of 0–2.0 define the range of flow conditions that are free from surface turbulence. Although 1.0 is theoretically perfectly safe, it seems that a value nearer 2.0 appears to be transitional, indicating that a safe working critical velocity might be as high as 1.0 ms^{-1}. By the time the We reaches 100, surface turbulence is certainly a problem, with jumping and splashing to a height of 100 mm or so. However, at values of 100,000 the concept of surface turbulence has probably given way to modes of flow characterised by jets and sprays, and eventually, atomisation.

Froude number

The Froude number (Fr) is another useful dimensionless quantity. It assesses the propensity of conditions in the liquid for making gravity waves. Thus it compares the inertial pressures $\rho V^2/2$ that are perturbing the surface with the restraining action of gravity, ρgh, in which g is the acceleration resulting from gravity and h is the height of the wave. Thus the Froude number is defined as

$$Fr = V^2/gh \qquad (2.5)$$

It is important to emphasise the regimes of application of the We and Fr numbers. The We deals with small-scale surface effects such as ripples, droplets and jets. This is in contrast to the Fr number, which deals with larger scale gravity waves, including such effects as slopping and surging in more open vessels, and in the 'see-saw' oscillations of flow observed in U-tube conditions of some filling systems. We shall see later how Fr is useful to assess conditions for the formation of the hydraulic jump, a troublesome feature of the flow in large runners.

In the meantime, however, Fr is useful to assess conditions where the undulation of the surface may cause its surface area to contract, forcing an entrainment condition. This concept has been so far overlooked as a major potential role in entrainment. The continued expansion of the advancing front is a necessary and sufficient condition for the avoidance of entrainment. Conversely, the contraction of the front in the sense of a loss of area of the 'free surface' of the liquid is a necessary and sufficient condition for the entrainment of bifilms.

Interestingly, the We condition, indicating the onset of surface turbulence in the form of a potentially breaking wave, is a sufficient but not necessary condition for entrainment of bifilms. However, it is most probably a necessary condition for the entrainment of bubbles and perhaps some other surface materials. It is also likely that the surface turbulence condition aids the permanent entrainment of material entrained by surface contraction that might otherwise detrain. Thus the surface turbulence condition in the form of the critical velocity criterion remains a major influence on casting quality.

Reynolds number

Many texts deal with turbulence, especially in running and gating systems, citing the Reynolds number (Re):

$$Re = \rho VL/\mu \qquad (2.6)$$

in which V is the velocity of the liquid, ρ is the density of the liquid, L is a characteristic linear dimension of the geometry of the flow path and μ is the viscosity. The definition of Re follows from a comparison of inertial pressures $C_d\rho V^2/2$ (in which C_d is the drag coefficient) and viscous pressures $\mu V/L$. The reader can confirm that the ratio of these two forces does give the dimensionless number Re (neglecting, of course, the presence of numerical factors $2C_d$). The characteristic length L is, as before, usually defined as the hydraulic radius.

Clearly, the inertial forces are a measure of those effects that would cause the liquid to flow in directions dictated by its momentum. These effects are resisted by the effect of viscous drag from the walls of the channel containing the flow. At values of Re below about 2000, viscous forces prevail, causing the flow to be smooth and laminar, i.e. approximately parallel to the walls. At values of Re higher than this, flow tends to become turbulent; the walls are too far away to provide effective constraint, so that momentum overcomes viscous restraint, causing the flow to degenerate into a chaos of unpredictable swirling patterns.

It is vital to understand, however, that all of this flow behaviour takes place in the bulk of the liquid, underneath the surface. During such turbulence in the bulk liquid, therefore, the surface of the melt may remain relatively tranquil. Turbulence as predicted and measured by Re is therefore strictly *bulk turbulence*. It does not apply to the problem of assessing whether the surface film will be incorporated into the melt. This can only be assessed by the concept of *surface turbulence* associated with We and Fr.

Flow in running systems is typically turbulent as defined by Re. Even for very narrow running systems designed to keep surface turbulence to a minimum, bulk turbulence would still be expected. In this case for a small aluminium alloy casting in which $V = 2$ ms^{-1}, in a runner only 4 mm high, hence $L = 2$ mm, Re is still close to 7000. For a fairly small steel casting weighing a few kilograms in which the melt is flowing at 2 ms^{-1} in a runner 25 mm high, giving approximately $L = 10$ mm, and assuming $\mu = 5.5 \times 10^{-3}$ Nsm^{-2} and $\rho = 7000$ kgm^{-3}, Re is approximately 20,000. For a large casting a metre or more high and weighing several tonnes, it is quickly shown that Re is in the region of 2,000,000. Thus bulk turbulence exists in running systems of all types, and can be severe.

Only in the very narrow pores of ceramic foam filters is flow expected to be laminar as indicated in Figure 2.22. For velocities of 2 ms^{-1} for liquid aluminium in a filter of pore diameter 1 mm $Re = 2000$. For liquid steel, which is

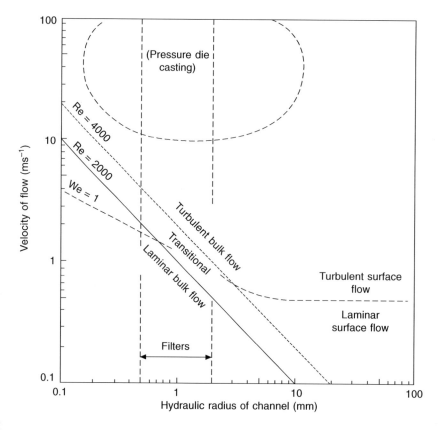

FIGURE 2.22

Map of flow regimes in castings.

somewhat more viscous, $Re = 500$. These values show that the flow is not far from an unstable turbulent condition even in the narrowest of channels found in most running systems. Only moderate increases in speed of flow would promote turbulence even in these very small channels.

The map of flow regimes presented in Figure 2.22 is a guide to our thinking and our research about the very different flow behaviour that liquids can adopt when subjected to different conditions of speed and geometry.

2.2.2 OXIDE SKINS FROM MELT CHARGE MATERIALS

Before melting, the charge materials need to be selected and weighed. The oxide originally on the charge material (we shall call such oxides 'skins') becomes necessarily submerged, to become part of the melt when the underlying solid melts. For dense metals such as irons and steels the oxide skins submerged in this way often separate quickly to form a dross or slag that can be removed from the top of the melt.

For Al alloys for which the submerged oxide necessarily entrains with it a layer of air as a bifilm, the oxide skins are nearly neutrally buoyant so the melt clears only with great reluctance. Problems of the introduction of skins from the surfaces of charge materials are most commonly seen in the melting of aluminium and its alloys. Whether or not the skins on the charge materials find their way into the liquid metal depends on the type of furnace used for melting.

Furnaces in which the solid charge materials are added directly into a melting furnace or into a liquid pool produce a poor quality of melt. Problem melters include crucible furnaces among others. For such melting practice, all the oxide skins on the charge necessarily finish floating about in the melt.

In the case of charge materials such as ingots that have been chill cast into metal moulds, the surface oxide skin introduced in this way is relatively thin. However, charges that are made from sand castings that are to be recycled (usually called 'foundry returns') represent the worst case. The oxide skins on sand castings have grown thick during the extended cooling period of the casting in the aggressively moist and oxidising environment of the sand mould. The author has seen complete skins of cylinder head castings dragged out of the liquid metal, complete with clearly recognisable combustion chambers and ports. The melt can become so bad as to resemble a slurry of old sacks. Unfortunately, this is not unusual.

In a less severe case in which normal melting was carried out repeatedly on 99.5% pure aluminium, Panchanathan (1965) found that progressively poorer mechanical properties were obtained. By the time the melt had been recycled eight times, the high elongation of relatively pure Al had fallen from approximately 30% to 20%. This is easily understood if the oxide skin content of the metal is progressively increased by repeated casting.

Dry hearth furnaces can, in principle, give a significantly improved quality. Such units are often based on a tower or shaft design to provide a classical counterflow heat exchange process; the charge is preheated during its course down the tower, whereas furnace waste gases flow upwards. The base of such a tower is usually a gentle slope of refractory on which the charge in the tower is supported. The slope is known as the dry hearth. Such a design of melting unit is not only thermally efficient, but has the advantage that the oxide skins on the surface of the metallic charge materials are mainly left in place on the dry hearth. The liquid metal flows, practically at its melting point, out of the skin of the melting materials, down the hearth, travelling through an extended oxide tube, and into the bulk of the melt. Here it needs to be brought up to a useful casting temperature. The oxide tube forms as the liquid flows, and finally collapses, more or less empty, when the flow is finished. The oxide content of melts from this variety of furnace is low. (The melting losses and gas content might be expected to be a little higher in this design of melter, but because of the protective nature of the oxide that forms in fuel waste gas atmosphere with high water content, the reaction between the melt and the waste gases is surprisingly low).

The additional advantage of a dry hearth furnace for aluminium alloys is that foundry returns that contain iron or steel cast-in inserts (such as the iron liners of cylinder blocks or valve seats in cylinder heads) can be recycled. The insets remain on the hearth and can, from time to time, be raked clear, together with all the dross of oxide skins from the charge materials, via a side door into a skip. (A dross consists of oxides with entrapped liquid metal. Thus most dross contains between 50% and 80% metal, making the recovery of aluminium from dross economically valuable.)

The benefits of melting in a dry hearth furnace are, of course, eliminated at a stroke by the misguided enthusiasm of the operator, who, thinking he is keeping the furnace clean and tidy, and that the heap of remaining oxide debris sitting on the hearth will all make good castings, shoves the heap into the melt. Unfortunately, it is probably slightly less effort to push the dross downhill, rather than rake it out of the furnace through the dross door. The message is clear, but requires restating frequently. Good technology alone will not produce good castings. Good training and vigilant management remain essential.

2.2.3 POURING

During the pouring of some alloys, the surface film on the liquid grows so quickly that it forms a tube around the falling stream. The author calls this an oxide flow tube.

A patent dating from 1928 by Beck describes how liquid magnesium can be transferred from a ladle into a mould by arranging for the pouring lip of the ladle to be as close as possible to the pouring cup of the mould, and to be in a relatively fixed position so that the semi-rigid oxide pipe which forms automatically around the jet is maintained unbroken, and thus protects the metal from contact with the air (Figure 2.23(a)).

A similar phenomenon is seen in the pouring of aluminium alloys and other metals such as aluminium bronze, ductile iron and stainless steels.

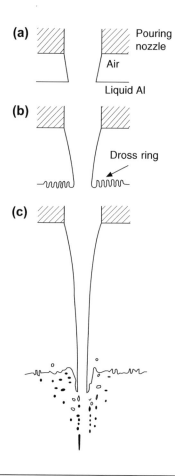

FIGURE 2.23

Effect of increasing height of a falling stream of liquid illustrating (a) the oxide flow tube remaining intact; (b) the oxide flow tube successively detached and accumulating to form a dross ring; and (c) the oxide film and air being entrained in the bulk liquid.

However, if the length of the falling stream is increased, then the shear force of the falling liquid against the inner surface of the oxide tube increases. This drag may become so great that after a second or so the oxide tears, allowing the tube to detach from the lip of the ladle. The tube then accompanies the metal into the mould, only to be immediately replaced by a second tube, and so on. A typical 10 kg aluminium alloy casting poured in about 10 s can be observed to carry an area of between 0.1 and 2.0 m^2 of oxide into the melt in this way. This is an impressive area of oxide to be dispersed in a casting of average dimensions only 100 × 200 × 500 mm, especially when it is clear that this is only one source. The oxide in the original metal, together with the oxides entrained by the surface turbulence of the pour, will be expected to augment the total significantly.

The critical fall height

The critical fall height is that height from which the melt falls to gain just sufficient energy to enfold its surface, creating an entrainment defect. This height happens to be identical to the height of the sessile drop of the melt.

Why is the critical fall height the same as the height of a sessile drop? The critical fall height can be seen to be a kind of re-statement of the critical velocity condition. It is because the critical velocity V is that required to propel the metal vertically upwards to the height at which it is still just supported by surface tension (Figure 2.18). This is the same velocity V that the melt would have acquired by falling from that height. (A freely travelling particle of melt starting from a vertical ingate would execute a parabola, with its upward starting and downward finishing velocities identical.)

When melts are transferred by pouring from heights less than the critical heights predicted in Table 2.2 (the heights of the sessile drop), there is no danger of the formation of entrainment defects. Surface tension is dominant in such circumstances, and can prevent the folding inward of the surface and thus prevent entrainment defects (Figure 2.23(a)). It is unfortunate that the critical fall height is such a minute distance. Most falls that an engineer might wish to design into a melt handling system, or running system, are nearly always greater, if not vastly greater. However, the critical fall height is one of those extremely inconvenient facts that we casting engineers have to learn to live with.

However, even above this theoretical height, in practice the melt may not be damaged by the pouring action. The mechanical support of the liquid by the surface film in the form of its surrounding oxide tube can still provide freedom from entrainment, although the extent of this additional beneficial regime is perhaps not great. For instance, if the surface tension is effectively increased by a factor of 2 or 3 by the presence of the film, the critical height may increase by a factor $3^{1/4} = 1.3$. Thus perhaps 30% or so may be achievable, taking the maximum fall from about 12 to 16 mm for aluminium. This seems negligible for most practical purposes.

At slightly higher speed of the falling stream, the tubes of oxide concertina together to form a dross ring (Figure 2.23(b)). Although this represents an important loss of metal on transferring liquid aluminium and other dross-forming alloys, it is not clear whether defects are also dragged beneath the surface and thereby entrained.

At higher speeds still, the dross is definitely carried under the surface of the liquid, together with entrained air, as shown in Figure 2.23(c). Turner (1965) has reported that, above a pouring height of 90 mm, air begins to be taken into an Al alloy melt with the stream, to reappear as bubbles on the surface. Experiments by Din and the author (2003) on Al-7Si-0.4Mg alloy have demonstrated that in practice the damage to the tensile properties of castings caused by falls up to 100 mm appears controlled and reproducible provided the melt has first gone through a ceramic foam filter. (However, above 200 mm, random damage was certain.)

These heights are well above the critical fall heights predicted previously, and almost certainly are a consequence of the some stabilisation of the surface of the falling jet by the presence of a film. The mechanical rigidity of the tubular film holds the jet in place, and effectively delays the onset of entrainment by the plunging action greatly in excess of the predicted 30%. It may be that the surface film stabilises the surface, maintaining the smoothness of the jet, so avoiding entrainment of the surface oxide as it plunges into the melt as described below. Clearly, more work is required to clarify the allowable fall heights of different alloys.

In a study of water models, Goklu and Lange (1986) found that the quality of the pouring nozzle affects the surface smoothness of the plunging jet, which in turn influences the amount of air entrainment. They found that the disturbance to the surface of the falling jet is mainly controlled by the turbulence ahead of and inside the nozzle that forms the jet.

During the pouring of a casting from the lip of a ladle via an offset stepped basin kept properly full of metal, then the previously discussed benefit will apply: the oxide will probably not be entrained in the pouring basin if the pouring head is sufficiently low, as is achievable during lip pouring. However, in practice it seems that for fall distances of more than perhaps 50 or 100 mm freedom from damage cannot be relied upon. The pouring basin will always represent a threat to a filling system from this source.

Clearly, the benefits of defect-free pouring are easily lost if the pouring speed into the entry point of the filling system is too high. This is often observed when pouring castings from unnecessary height. A succession of increasingly bad pouring conditions can be listed as current examples: In aluminium foundries, pouring is often by robot. In iron foundries, it is commonly via automatic pouring systems from fixed launders sited over the line of moulds. In steel foundries, it is common to pour from bottom poured ladles that contain over a metre depth of steel above the exit nozzle (the situation for steel from bottom-teemed ladles is further complicated by the depth of metal in the ladle decreasing progressively). In a practical instance of a jet plunging at 10 ms^{-1} into steel held in a 4 m diameter ladle, Guthrie (1989) found that the Weber Number was 2.7×10^6, whereas the Fr number was only 2.5. Thus despite very little slopping and surging, the surface forces were being overwhelmed by inertial forces by nearly 2 million times greater, causing the creation of a very dirty re-oxidised steel.

In all types of foundries, the surface oxide is automatically entrained and carried into the casting if a simple conical pouring bush is used to funnel the liquid stream into the sprue. In this case, of course, practically all of the oxide formed on the stream will enter the casting. It has to be admitted that pouring basins of sophisticated design are not without their problems and their limitations, but the current widespread use of conical pouring basins undermines all other attempts to improve casting quality. Conical basins have to be eliminated if casting quality can have any chance to improve.

In summary, it is clear that the lower the pour heights the less damage is suffered by the melt. In addition, of course, less metal is oxidised, thus directly saving the costs of unnecessary melt losses. Ultimately, however, it is, of course, best to avoid pouring altogether. In this way, losses are reduced to a minimum and the melt is maintained free from damage.

The adoption of countergravity designs for foundries, in which metal is never poured at any point of the process, is no longer a pipe dream, but a practical and economical technique for casting production that is already in use in a number of foundries worldwide. It is an ultimate solution that cannot be recommended too strongly.

2.2.4 THE OXIDE LAP DEFECT I: SURFACE FLOODING

The steady, progressive rise of the liquid metal in a mould may be interrupted for a number of reasons. There could be (1) an inadvertent break during pouring, or (2) an overflow of the melt (called elsewhere in this work a 'waterfall effect') into a deep cavity at some other location in the mould or (3) the arrival of the front at a very much enlarged area, thus slowing the rate of rise nearly to a stop.

If the melt stops its advance, the thickness of the oxide on the melt surface is no longer controlled by the constant splitting and re-growing action. It now simply thickens. If the delay to its advance is prolonged, the surface oxide may become a rigid crust.

When filling restarts (for instance, when pouring resumes or the overflow cavity is filled), the fresh melt may be unable to break through the thickened surface film. When it eventually builds up enough pressure to force its way through at a weak point, the new melt will flood over the old, thick film, sealing it in place. Because the newly arriving melt will roll over the surface, laying down its own new, thin film, a double film defect will be created, creating a bifilm. The bifilm will be highly asymmetrical, consisting of a lower thick film and an upper thin film.

Asymmetric films are interesting, in that precipitates sometimes prefer one film as a substrate for formation and growth, but not the other. Examples are briefly described later in the sections relating to confluence welds and oxide flow tubes. The asymmetric bifilm is a common defect in castings.

A key aspect of stopping the front is that the double film defect that is thereby created is a single, huge planar defect, extending completely through the product. Also, its orientation is perfectly horizontal. (Notice it is quite different from the creation of double film defects by surface turbulence. In this chaotic process, the defects are random in shape, size, orientation and location in the casting.) Flooding over the surface in this way is relatively common during the filling of castings, especially during the slow filling of alloys which form strong films.

From an exterior view of the casting, the newly arriving melt flooding over the thick stationary oxide will only have the pressure of its own sessile drop height as it attempts to run into the tapering gap left between the old meniscus and the mould wall. Thus this gap is imperfectly filled, leaving a horizontal lap defect clearly visible around the perimeter of the casting. This surface lap defect corresponding to the arrest of the progress of the melt is the only external sign of a major defect that extends probably completely through the casting.

Notice that in this way (assuming oxidising conditions) we have created an *oxide lap*. If the arrest of the advance of the melt had been further delayed or if the solidification of the melt had been accelerated (as near a metal chill, or in a metal mould), the meniscus could have lost so much heat that it had become partially or completely solid. In this case, the lap would take on the form of a *cold lap* (the name 'cold shut' is recommended to be consigned to the scrap heap of unhelpful non-descriptive jargon. I hate jargon.).

The distinction between oxide laps and cold laps is sometimes useful because although both may be eliminated by avoiding any arrest of progress of the liquid front, only the cold lap may be cured by increasing the casting temperature, whereas the oxide lap may become worse.

For mould surfaces which are horizontal, the advance of the front takes an unstable dendritic form, with narrow streams, like dendrites, progressing freely ahead of the rest of the melt. This is because, while the molten metal advances quickly in the mould, the surface film is being repeatedly burst and moved aside. The faster the metal advances in one location, the thinner and weaker the film, so that the rate of advance of the front becomes less impeded. If another part of the front slows, then the film has additional time to strengthen, further retarding the local rate of advance. Thus in film-forming conditions fast-moving parts of the advancing front advance faster, and slow-moving parts advance slower, causing the advance of the liquid front to become unstable (Campbell, 1988). This is a classic type of instability condition that gives rise to finger-like dendritic extensions on an advancing front.

Figure 2.24 shows the filling pattern of a thin walled box casting such as an automotive sump or oil pan. The vertical walls fill nicely because of the organising effect of gravity. However, problems arise on the horizontal top section. If the streams continue to flow, to fill eventually the whole of the horizontal section, the confluence welds (see Section 2.2.5) abutting the oxides on the sides of the streams will constitute cracks through the complete thickness of the casting. When highly strained, such castings are known to fail along the lines of the confluence welds outlining the paths of the original filling streams. In effect, each of the flowing streams generates a kind of oxide flow tube as is discussed in the next section.

For the case of vertical filling, when the advance of the front has slowed to near zero, or has actually momentarily stopped, then the strength of the film and its attachment to the mould will prevent further advance at that location. If the filling pressure continues to build up, the metal will burst through at a weak point, flooding over the stationary front. In a particular locality of the casting, therefore, the advance of the metal will be a succession of arrests and floods, each new flood burying a double oxide film (Figure 2.25).

This very deleterious mode of filling can be avoided by increasing the rate of filling of the mould.

The problem can, in some circumstances, also be tackled by reducing the film-forming conditions. This is perhaps not viable for the very stable oxides such as alumina and titania when casting in air. It is the reason for vacuum casting those alloys which are troubled by films; although the oxides cannot be prevented from forming by casting in vacuum, their rate of formation is reduced. The thin double film is expected, in principle, to constitute a crack as serious as that of thick double film (because the entrained layer of air is expected to have the same negligible strength). However, there are

Oxide flow tube defects
from horizontal filling

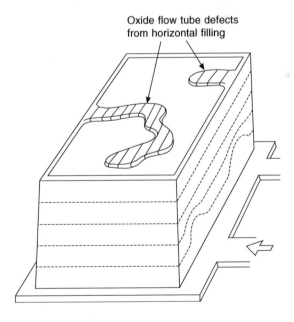

FIGURE 2.24

Filling of a thin-walled oil pan casting, showing the gravity controlled rise in the walls, but the unstable meandering flow across horizontal areas leading to the danger of confluence weld type through wall cracks.

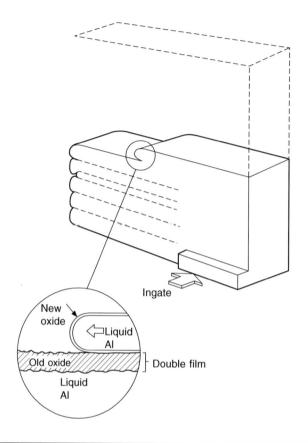

New
oxide

Ingate

Liquid
Al

Old oxide

Double film

Liquid
Al

FIGURE 2.25

Unstable filling of a mould showing the formation of laps containing an asymmetric bifilm crack.

additional reasons why the thin film may be less damaging. A film that is mechanically less strong is more easily torn and more easily ravelled into a more compact form. Internal turbulence in the melt will tend to favour the settling of the defect into stagnant corners of the mould. Here it will be quickly frozen in to the casting before it has chance to unfurl significantly.

Films on cast iron for instance are controllable to some extent by casting temperature and by additions to the sand binder to control the environment in the mould (Sections 4.5 and 6.5). Films on some steels are controllable by minor changes to the deoxidation practice (Section 6.6).

2.2.5 OXIDE LAP DEFECT II: THE CONFLUENCE WELD

Even in those castings where the metal is melted and handled perfectly, so that no surface film is created and submerged, the geometry of the casting may mean that the metal stream has to separate and subsequently join together again at some distant location. This separation and rejoining necessarily involves the collision of the films on the advancing fronts of both streams, with the consequent danger of the streams having difficulty in rejoining successfully. I called this junction a confluence weld (Campbell, 1988). Most complex castings necessarily contain dozens of confluence welds, and it is important to recognise that the confluence weld does not necessarily, or even commonly, lead to any significant defect. However, occasionally it does lead to a severe defect. The conditions in which these issues arise are discussed below.

The author recalls that in the early days of the Cosworth Process, a small aluminium alloy pipe casting was made for very high-pressure service conditions (Figure 2.26). At that time, it was assumed that the mould should be filled as slowly as possible, gating the metal in at the bottom of the pipe, and arriving at the top of the pipe just as the melt was about to freeze to encourage directional feeding. When the pipe was finally cast, it looked perfect. It passed radiographic and dye penetrant tests. However, it failed catastrophically under a simulated service test by splitting longitudinally, exactly along its top, where the metal streams were assumed to join. The problem defeated our team of casting experts, but was solved instantly by our foundry manager, George Wright, the company's very own dyed-in-the-wool traditional foundryman. He simply turned up the filling rate (neglecting the niceties of setting up favourable temperature gradients to assist feeding). The problem never occurred again. Readers will note a moral (or two) in this story.

Figure 2.26 shows various situations where confluence problems occur in castings. Such locations have been shown to be predictable in interesting detail by computer simulation (Barkhudarov and Hirt, 1999). The weld ending in a point illustrated in Figure 2.27 is commonly seen in thin-walled aluminium alloy sand castings; the point often has the appearance of a dark, upstanding pip. The dark colour is usually the result of the presence of sand grains, impregnated with metal. The metal penetration of the mould occurs at this point as a result of the conservation of momentum of the flow, concentrated and impacted at this point. (The effect is analogous to the implosion of bubbles on the propeller of a ship: the bubble collapses as a jet, concentrating the momentum of the in-falling liquid. The repeated impacts of the jet cause fatigue cracks to initiate in the metal surface, finally causing failure in the form of cavitation damage.). The problem has been called in US foundries 'the black plague defect', which seems somewhat over-melodramatic, but does emphasise the founder's exasperation of having to deal with the unsightly marks by additional surface dressing.

Returning to the issue of the confluence weld, as with many phenomena relating to the mechanical effects of double oxide films, the understanding is rather straightforward. The concept is illustrated in Figure 2.28.

In the case of two liquid fronts that progress toward each other by the splitting and re-forming of their surface films, the situation just after the instant of contact is fascinating and key to the understanding of this problem. At this moment, the splitting will occur at the point of contact because the film is necessarily thinnest at this point but no oxygen can access the microscopic area where the metal is in contact. As the streams continue to engage, the oxide on the surfaces of the two menisci continue to slide back from the point of contact, but because of the exclusion of oxygen from the contact region, no new film can form here. Remnants of the double film occupy a quarter to a third of the outer part of the casting section, existing as a possible crack extending inward from each surface. This is most unlikely to result in a defect because such films will be thin because of their short growth period. Having little rigidity, being more akin to tissue paper of gossamer lightness, it will be folded against the oxide skin of the casting by the random gusts and gales of internal turbulence. There it will attach, adhering as a result of little-understood atomic forces; probably van der Waal forces. Any such forces, if they exist, are likely to be only weak but the vanishingly thin and weak films will not need strong forces to ensure their permanent capture. Thus, finally, the weld is seen to be perfect. This situation is expected to be common in castings.

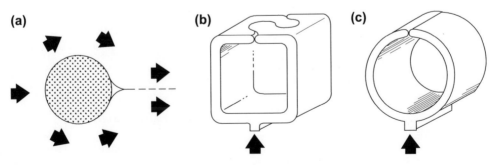

FIGURE 2.26

Confluence weld geometries (a) at the side of a round core; (b) randomly irregular joins on the top of a bottom-gated box; (c) a straight join on the top of a bottom-gated round pipe (Campbell, 1988).

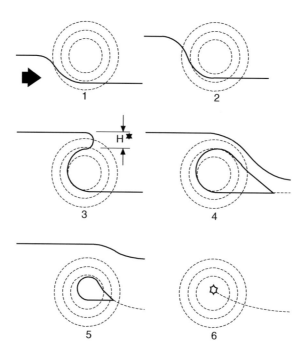

FIGURE 2.27

Local thin area denoted by concentric contours in an already thin wall, leading to the creation of a filling instability, and a confluence weld ending in a point discontinuity (Campbell, 1988).

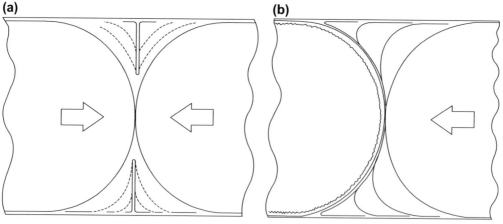

FIGURE 2.28

Mechanism of the confluence weld (a) in the case of the meeting of two moving fronts which fuse to create a perfect weld (the residual entrained bifilms flatten against the walls of the casting); (b) a through-thickness crack because one front has stopped.

The case contrasts with the approach of two liquid fronts, in which one front comes to a stop, while the other continues to advance toward it. In this case the stationary front builds up the thickness of its oxide layer to become strong and rigid. When the 'live' front meets it, the thin film on the newly arriving front is now pinned against the rigid, thick film and is held in place by friction. Thus the continuously advancing stream expands around the rigidised meniscus, forcing its oxide film to split and expand, as the moving front wraps itself around the stationary front, causing a layer of thin new film to be laid down on the old thick substrate. Clearly, a double film defect constituting a crack has been created completely across the wall of the casting. Again, the double film is asymmetrical.

Note once again that for the conditions in which one of the fronts is stationary, the final defect is a lap defect in which the crack is usually in a vertical plane (although, of course, other geometries can be envisaged). This contrasts with the surface flooding defect, lap defect type I, in which the orientation of the crack is substantially horizontal.

The meandering cracks formed in the conditions of filling shown in Figure 2.24 are long confluence weld defects. Some other examples of different kinds of confluence welds are given in the following section.

A location in an Al alloy casting where a confluence weld was known to occur was found to result in a crack. When observed under the scanning electron microscope, the original thick oxide could be seen trapped against the tops of dendrites that had originally flattened themselves against the double film. The poor feeding in that locality had drained residual liquid away from the defect, sucking large areas of the film deeply into the dendrite mesh. One of the remaining islands of film pinned in its original place by the dendrites is shown in Figure 2.29. The draped appearance suggesting the dragging action of the surrounding film as it was pulled and torn away while the surrounding film disappeared into the depths.

Garcia and colleagues (2007) studied the problem of severe leakage defects in an Al alloy cylinder head casting, discovering the problem to be a confluence weld formed as a result of the backpressure of gas in the mould causing the advancing front to stop momentarily. They noted alpha-Fe intermetallics had nucleated on the stopped front, but the 'live' front was clear of precipitates. (Their successful identification of a confluence weld defect contrasts with the study by

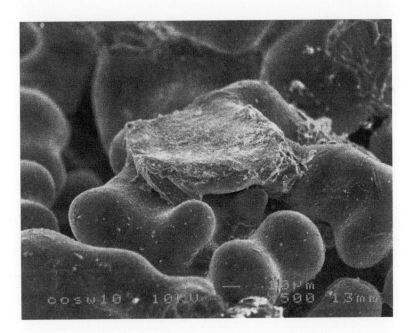

FIGURE 2.29

SEM fractograph of a confluence weld at a stopped front. An island of thick oxide remains, the rest having withdrawn into the depths of the dendrite mesh, retreating with residual liquid because of local feeding problems in the casting.

Lett (2009) in which neither advancing liquid front was designed to stop. The result was no clear identification of the formation of confluence welds.)

Pellerier and Carpentier (1988) are among the few who have studied a confluence weld defect in a cast iron, almost certainly formed from a carbon film on the advancing meniscus (although it seems likely that the carbon film had itself formed on a precursor oxide film as illustrated in Figure 6.34). Their thin-walled ductile iron casting was poured in a mould containing cores bonded with a urethane resin. They found a thin film (but did not appear to have noticed whether the film might have been double) of graphite and oxides through the casting at a point where two streams met. The bulk metal matrix structure was ferritic (indicating an initial low carbon content in solution), but close to the film was pearlitic, indicating that some carbon from the film was going into solution. The authors did not go on to explore conditions under which the defect could be avoided.

In summary, if the two fronts can be kept 'alive', the confluence is expected to be a perfect weld. If one fronts stops, the result is a crack. Thus the problem can be eliminated by keeping the liquid fronts moving. This is simply ensured by casting at a rate which is sufficiently high. Care is needed of course to avoid casting at too high a rate at which surface turbulence may become an issue. Providentially, there is usually a comfortably wide operational window in which the fill rate can meet the requirements to avoid all defects.

2.2.6 THE OXIDE FLOW TUBE

The oxide tube around a flowing jet, if encapsulated in the casting, can form a cylindrical bifilm as the melt rises around it to entrain it. It becomes therefore a major geometrical crack. The cylindrical crack is an almost unbelievably unexpected form. Any metallurgical engineer familiar with normal fairly planar stress cracks would be hard pressed to believe it, and even more hard pressed to explain it.

The stream might be a falling jet, commonly generated in a waterfall condition in the mould, as in Figure 2.30. The stream does not need to fall vertically. Streams can be seen that have slid down gradients in such processes as tilt casting when carried out under poor control. Part of the associated flow tube is often visible along the surface of the casting.

Alternatively, a wandering horizontal stream can define the flow tube, as is commonly seen in the spread of liquid across a horizontal surface. Figure 2.24 shows how, in a thin horizontal section, the banks of the flowing stream remain stationary whilst the melt continues to flow. When the flow finally fills the section, coming to rest against the now-rigid banks of the stream, the banks will constitute long meandering bifilms as cracks, following the original line of the flow.

Oxide flow tube defect from a fall

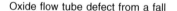
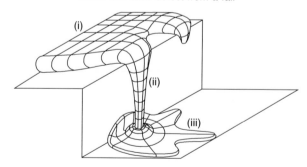

FIGURE 2.30

Waterfall effect in a casting causing (i) a stationary top surface forming a horizontal lap and crack defect; (ii) a falling jet creating a cylindrical oxide flow tube leading to a cylindrical crack; and (iii) random surface turbulence damage in the lower levels of the casting.

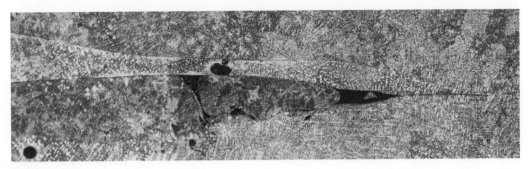

FIGURE 2.31

Oxide flow tubes separating regions of the solidifying structure of a high pressure Al alloy die casting.

Courtesy Ghomashchi and Chadwick (1986).

The jets of flow in high pressure die castings can be seen to be leave permanent legacies as oxide tubes as seen in section in Figure 2.31. Many authors have noted such defects: Lavelle (1962) found such a density of oxide laps in a high pressure die casting that the casting fractured with a fibrous appearance. More recently, Ahmed and Kato (2008) using a scanning acoustic microscope noted the damage to fatigue properties by 'cold flakes' that could detach to reveal a crack.

All these examples illustrate how unconstrained (i.e. free from contact with guiding walls) *gravity filling* or *horizontal filling* both risk the formation of serious defects. (The unconstrained filling of moulds without risk can only be achieved by *countergravity* filling.)

Both falling and horizontal streams exhibit surfaces that are effectively stationary, and thus grow a thick oxide. When the rising melt finally entrains such features, the new thin oxide that arrives, rolling up against the old thick oxide, creates a characteristic asymmetrical double film. On such a double film in a vacuum cast Ni-based superalloy, the author has seen sulphide precipitates formed only on one side of the defect, indicating that only one side of the double film was favourable to nucleation and growth (microphotographs were not released for security reasons). Too little work was carried out to know whether the thick or thin side of the bifilm was the active substrate in this case.

In all cases it will be noticed that such interruptions to flow, where, for any reason, the surface of the liquid locally stops its advance, a large, asymmetric double film defect is created. These defects are always large, and always have a recognisable, predictable geometrical form (i.e. they are cylinders, planes, meandering streams etc.). They are quite different to the random size and shape of double films formed in the chaos of surface turbulence.

2.2.7 MICROJETTING

Where the advance of the liquid front occurs in very narrow sections, the smooth form of the meniscus can disappear, becoming chaotic on a microscale. Jets of liquid issue from the front, only to be caught up within a fraction of a second by the general advance of the front, and so become incorporated back into the bulk liquid. The jets, of course, become oxidised, so that the advancing liquid will naturally be expected to become contaminated with a random assortment of tangled double films.

Such behaviour was observed during the casting of Al-7Si-0.4Mg alloy in plaster moulds (Evans and Campbell, 1997). In this experiment, the wall thickness of the castings was progressively reduced to increase the effect of surface tension to constrain the flow, reducing surface turbulence, and thus increasing reliability. As predicted, properties were clearly improved as the section was reduced from 6 to 3 mm. However, as the section was reduced further to increase the benefit, instead of the reliability increasing further as expected, it fell dramatically.

An explanation of this surprising behaviour came from direct video observation of the meniscus of the casting as it travelled toward the camera in these smaller sections. The smooth profile of the meniscus was observed to be punctured by cracks, through which tiny jets of metal spurted ahead, only to be quickly engulfed by the following flow. The image could be likened to advancing spaghetti.

It seems likely that the effect is the result of the strength of the oxide film on the advancing front in thin-section castings. In thin sections, the limited area of the front limits the number of defects present in the film. (The effect seems analogous to the behaviour of metal whiskers, whose remarkable strength derives from the fact that they are too small to contain any significant defects.) Following this logic, a small area of film may contain no significant defect and so may resist failure. Pressure therefore builds up behind the film, until finally it ruptures, the split releasing a jet of liquid. (To explain further, the effect is not observed in thick sections because the greater area of film assures the presence of plenty of defects, so the film splits easily under only gentle pressure so that the advance of the melt is therefore smooth.)

Similar microjets have been observed to occur during the filling of Al alloy castings via 2 mm thick ingates. Single or multiple narrow jets, a millimetre or so in diameter, have been seen to shoot 100 mm or more across the mould cavity from the narrow slot ingates before the arrival of the main flow of metal (Cunliffe, 1994).

Microjetting has also been observed in detail under the scanning electron microscope (Storaska and Howe, 2004). When small droplets of Al-Si alloy have been melted in the electron beam, the expansion of the liquid causes the oxide skin of the droplet to be subject to a massive stress in the region of 15 GPa. Some droplets are sufficiently small to have a good chance of not containing any defect in their skin, with the result that the skin slowly creeps to gradually reduce the internal pressure. Others, however, are seen to fracture suddenly, shooting out the compressed liquid as a jet.

The microjetting mechanism of advance of liquid metals in castings has so far only been observed in aluminium alloys, and the precise conditions for its occurrence are not yet known. It does not seem to occur in all narrow channels. The gaseous environment surrounding the flow may be critical to the behaviour of the oxide film and its failure mechanism. Also, the effect may only be observed in conditions where not only the thickness but the width of the channel is also limited, thus discouraging the advance of the front by the steady motion of transverse waves (the unzipping mode of advance to be discussed later).

Where it does occur, however, the mechanical properties of the casting are seriously impaired. The reliability falls by up to a factor of 3 because casting sections reduce from 3 to 1 mm. Unless microjetting can be understood and controlled, the effect might impose an ultimate limit on the reliability of very thin section castings. This would be a bitter disappointment and hard to accept. To avoid the risk of this outcome more research is needed.

2.2.8 BUBBLE TRAILS

Entrainment defects can be caused by the folding action of the (oxidised) liquid surface. Sometimes only oxides are entrained, as doubled-over film defects that we call bifilms. Sometimes the bifilms themselves contain small pockets of accidentally enfolded air, so that the bifilm is decorated by arrays of trapped bubbles. Much, if not all, of the microporosity observed in castings either is or has originated from a bifilm.

Sometimes, however, the folded-in packet of air is so large that its buoyancy confers on it a life of its own. This oxide-wrapped bubble is a massive entrainment defect that can become important enough to power its way through the liquid and sometimes through the dendrites. In this way, it develops its own distinctive damage pattern in the casting. Most bad filling system designs are capable of generating such bubbles, and not just one; sometimes hundreds of large bubbles are generated. Unfortunately, this is all too common.

The passage of a single bubble through an oxidisable melt is likely to result in the creation of a bubble trail as an oxide crack, a kind of long bifilm, in the liquid. A bubble trail is the name I coined to describe the defect that that was predicted to remain in a film-forming alloy after the passage of a bubble through the melt (Campbell, 1991). Therefore, even though the bubble may be lost by bursting at the liquid surface, the trail remains as permanent damage in the casting.

The bubble trail occurs because the trail necessarily starts at the point where the bubble was first entrained in the liquid. The enclosing shroud of oxide film covering the crown of the bubble attempts to hinder its motion. This restraint is so large that bubbles of diameter less than about 5 mm are held back, unable to float freely to the surface. They remain tethered like balloons on a string. However, if the bubble is larger, its upward buoyancy force will split this restraining cover. Immediately, of course, the oxide re-forms on the crown, and splits and re-forms repeatedly. The expanding region of film on the crown effectively slides around the surface of the bubble, continuing to expand until the equator of the bubble is reached. At this point, the area of the film is a maximum. Although the film was able to expand by splitting and re-forming, it is, of course, not able to contract its area since the film is a solid. Further progress of the bubble causes the film to continue sliding around the bubble, gathering together in a mass of longitudinal pleats under the bubble, to form a trail that leads back to the distant point at which the bubble was first entrained as a packet of gas. Any spiralling motion of the bubble will twist and additionally tighten the rope-like tether that continues to lengthen as the bubble rises.

The structure of the trail is a kind of collapsed tube. In section it is star-like but with a central portion that has resisted complete collapse because of the small residual rigidity of the oxide film (Figure 2.32). This is expected to form an excellent leak path if it joins opposing surfaces of the casting, or if cut into by machining. The leak path is non-trivial because it can stretch from the bottom to the top of the casting. In addition, of course, the coming together of the opposite skins of the bubble during the formation of the trail ensure that the films make contact dry side to dry side, and so constitute our familiar classical bifilm crack, but cutting through the casting as a sword through butter.

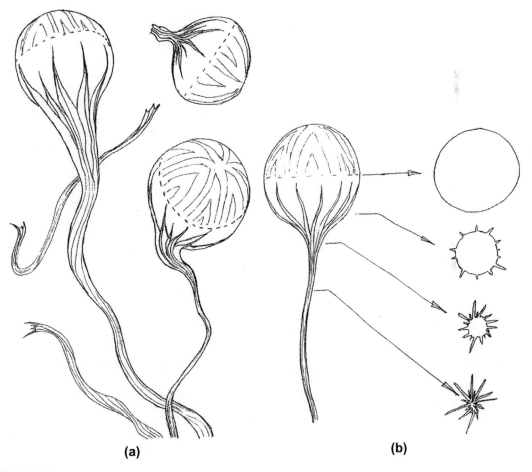

(a)   (b)

FIGURE 2.32

(a) Schematic illustration of rising bubbles and associated trails; (b) cross-sections illustrating the progressive collapse of the bubble trail.

Because air, water vapour and other core gases are normally all highly oxidising to the liquid metal, a bubble of any of these gases will react aggressively, oxidising the metal as it progresses, and leaving in its wake the collapsed tube of oxide like an old, long sack.

Occasionally the bubble trail may break. At the top of Figure 2.32 is a bubble that has been torn free from its trail. Such bubbles, with the stump of their trail showing the bubble to be tumbling irregularly as it rises, have been directly observed by X-ray video.

In cast irons containing resin-bonded cores, core gases initially contain significant quantities of water vapour so that bubble trails originally form from oxides. These oxides are mainly silicates and have rather low melting points, thus eventually succumbing, ravelling and compacting into formless heaps of slag in the tornado of escaping gases that blasts its way through the wall of the casting to form a permanent open link: a leakage defect. However, after the core has become sufficiently hot to drive off all its moisture, the supply of oxidising gases is exhausted and the evolving gases become carbon-containing volatiles. At that stage of gas evolution, graphite bubble trails start to form (Campbell, 2008). Examples of such complicated bubble damage in cast irons are given in Section 10.5.

Bubble damage

Poor designs of filling systems can result in the entrainment of much air into the liquid stream during its travel through the filling basin, during its fall down the sprue and during its journey along the runner. In this way, dozens or even hundreds of bubbles can be introduced into the mould cavity.

When so many bubbles are involved the later bubbles have problems to rise through the maze of bubble trails that remain after the passage of the first bubbles. Thus the escape of late-arriving bubbles is hampered by the tangled accumulation of earlier trails. If the density of trails is sufficiently great, fragments of bubbles remain entrapped as permanent features of the casting. This messy mixture of bifilm trails and bubbles is collectively called '*bubble damage*'. In the experience of the author, bubble damage is probably the most common defect in castings, but up to now has been almost universally unrecognised.

Bubble damage is nearly always mistaken for shrinkage porosity as a result of its irregular form, usually with characteristic cusp-like morphology. When seen on polished sections, the cusp forms that characterise bubble damage are often confused with cusps that are associated with the shapes of interdendritic shrinkage porosity. However, they can nearly always be distinguished with complete certainty by their difference in size. Careful inspection of the dendrite arm spacing will usually reveal the cusps that would have formed around dendrites as the residual interdendritic liquid is sucked into the dendrite mesh. These interdendritic cusps are usually up to 10 times smaller than the cusps that are caused by the folds of oxide in bubble trails (Figure 2.33). Clearly, the two are totally unrelated and quite distinct.

Bubble damage is commonly observed just inside and above the first (or sometimes the last) ingate from the runner (Figure 2.33). The large bubbles have sufficient buoyancy to escape up the first ingate, but smaller bubbles can be carried the length of the runner, to appear through the farthest ingate. Alternatively, they can even be carried back once again to an earlier ingate if there is a back wave. This non-uniform distribution associated sometimes with first and sometimes with some other ingate position is a common feature of bubble damage from running systems. However, sometimes strong flows of metal inside the mould cavity can cause the bubble path to deviate a long way from a direct vertical path to the surface. Highly indirect paths are commonly observed in video radiography studies. Nevertheless, the common feature of bubble damage is its non-uniform distribution; it tends to be clustered along favoured flow paths.

If the bubbles completely escape, the remaining trails can float around, finally settling sometimes at great distances from their source. Irregular masses of oxides in odd corners of castings have been positively identified as groups of tangled bubble trails (Figure 2.34). The work by Divandari confirms that bubble trails, in filtered melts known to be free from oxide tangles, can detach and float freely, finally appearing at distant locations. The bubbles have moved on and escaped from the casting, but their trails have remained in suspension. They have broken free from their moorings (the point at which the bubble was first entrained) and have travelled, tumbling and ravelling as they go, carried by the sweeping and circulating flow of the liquid during the filling process. Texan founders will recognise an analogy with tumbleweed.

Another common feature of bubble damage is the entrapment of small bubbles just under the cope skin of the casting (Figure 2.33). They are prevented from escaping only by the thickness of the oxide skin on the casting and their own

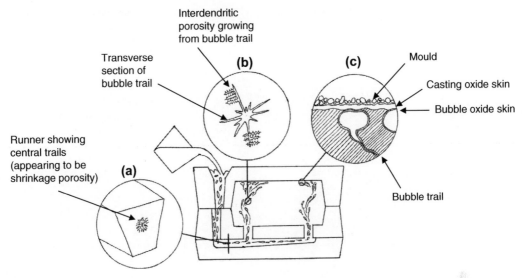

FIGURE 2.33

Pattern of bubble damage in a casting; (a) dendrites push trails into the centres of runners (misleadingly appearing as centre-line shrinkage porosity); (b) trails invisible in radiography are usually visible on transverse sections; (c) entrained air bubbles smaller than about 5 mm diameter appear as immediately sub-surface porosity, having insufficient buoyancy to break the double oxide barrier to escape to atmosphere.

oxide skin (a double oxide film feature as usual!). Both of these films must be broken before the bubble can escape to the outside world. (This is achieved by larger bubbles because of their greater buoyancy forces, but not by smaller bubbles. The dividing line between large and small bubbles seems to be in the region of 5 mm diameter for both light and dense alloys.) Such small bubbles are commonly observed in video radiographic studies to float up and sit under the top surface, resting only a double thickness of oxide depth under the skin of the casting. This sub-surface porosity is, of course, easily broken into when shot blasting, or when taking the first machining cut. For carbon steels, they are revealed after the intense oxidation suffered by the casting during heat treatment.

Close optical examination of the interiors of bubbles and bifilms in an aluminium alloy casting often reveals some shiny dendrite tips characteristic of shrinkage porosity. This adds to confusion of identification because the action of a fall in pressure resulting from solidification shrinkage will often expand an existing bubble, impaling it on surrounding dendrites, to give it a dendritic appearance. Alternatively, shrinkage may act to expand an existing bifilm, unfurling and opening it, and finally sucking its oxide films deep into the dendrite mesh. Whether the origin of the cavity was a bubble or a bifilm subsequent examination will usually reveal only fragments of the originating oxides among the dendrites (Figure 2.29).

This process has been observed in video radiographic studies of castings. An unfed casting has been seen to draw in air bubbles at a hot spot on its surface. The bubbles floated up in steady succession every few seconds, initially bursting and disappearing at the top of the casting. However, later bubbles became trapped during their rise by dendrites growing in from the walls. As solidification progressed, shrinkage caused the originally spherical air bubbles to gradually convert to shrinkage cavities. The perfectly round and sharp radiographic images were seen to become 'furry' and indistinct as the liquid meniscus was sucked into the surrounding mesh of dendrites. Finally, the defect resembled extensive shrinkage sponge; its origin as a gas bubble no longer discernible.

Other real-time radiography has shown bubbles entrained in the runner and swept through the gate and into the casting. The upward progress of one bubble in the region of 5 mm diameter appeared to be arrested, the bubble

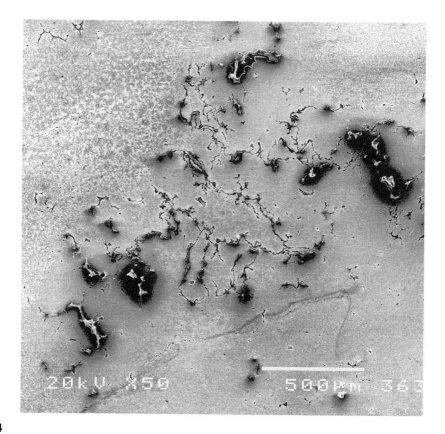

FIGURE 2.34

SEM metallographic section showing apparently normal entrained bifilms which are in fact a cluster of bubble trails (Divandari, 2000).

circulating in the centre of the casting, behaving like a balloon on a string. The string, of course, was the bubble trail acting as a tether. Other bubbles of various sizes up to about 5 mm diameter in the same casting were observed to float to the top of the casting, coming to rest under the oxide skin of the cope surface. These bubbles had clearly broken free from their tethers, probably as a result of the extreme turbulence during the early part of the filling process. The central bubble was marginally just too small to tear free from its trail. In addition, it may have lost some buoyancy as a result of loss of oxygen during its rise, or perhaps more likely, it ascended as far as it did because of assistance from the force of the flow of the melt. When the force of the flow abated higher in the mould cavity, its buoyancy alone was insufficient to split its crown of oxide skin, so that its upward progress was halted.

A scenario such as that shown in Figure 2.35 is common. The bubbles are entrained early in the filling system, arriving as a welter of defects, boiling up with the liquid as it fills the mould cavity. Many of the early bubbles will have sufficient buoyancy to escape. When the casting is finally solidified the appearance on a radiograph is expected to be something like that shown in Figure 2.35(b). Porosity and cracks may be detected in regions above the ingate. The large number of remaining trails is likely to be invisible. The reader will recall, this mix of residual bubbles, oxides and cracks is usefully termed *bubble damage*.

When many bubbles have passed into the mould, a cross-section of the runner or ingate will reveal some central porosity. Close examination will confirm that this porosity is not shrinkage porosity, but a mass of double oxide films, the

(a)

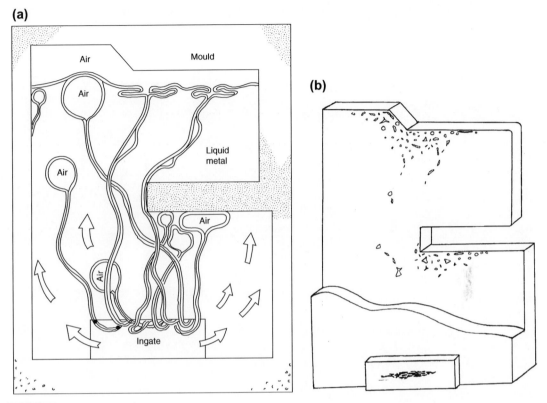

FIGURE 2.35

Schematic illustration of (a) bubbles entering a casting, and (b) the resulting permanent damage of residual bubbles and cracks. The residual bifilms remain invisible. The bubble trails in the runner, which were originally at the top of the section are pushed and concentrated by the freezing front into the centre of the section.

bubble trails. The trails are pushed ahead of the dendrites growing out from the walls and so are concentrated in the centre of the ingate section. In Al alloys, they appear as a series of dark, nonreflective oxidised surfaces interleaved like the flaky, crumpled pages of an old newspaper, sepia-coloured as a result of stains from mould gases.

Divandari (1999) was the first to study systematically the formation of bubble trails in aluminium castings by X-ray video technique. He introduced air bubbles artificially into a casting and was subsequently able to pinpoint the location of the trails and fracture the casting to reveal the defect. Figure 2.36(a) shows the inside of a trail in Al-7Si-0.4Mg alloy. The longitudinally folded film is clear, as is the presence of shrinkage cavities that have expanded away from the defect because the casting was not provided with a feeder. The small amount of shrinkage has sucked back the residual liquid among the dendrites seen in Figure 2.36(b), and in places stretching the film over the dendrites as seen in Figure 2.36(c). The thicker, older part of the film is seen pulled in tight creases, and where torn is seen to be replaced by newer, thinner film.

Figure 2.37(a) shows a bronze casting that has suffered a highly turbulent filling process. The bubbles entrained by this turbulence can be seen to be grouped along a horizontal line (a site of a poor joint between cope and drag, allowing a cooling fin to create a line of dendrites inside the casting that had caught many of the bubbles) and toward the top of the casting in the 'skeins of geese' mode that characterises these defects in bronzes. The skeins are probably the

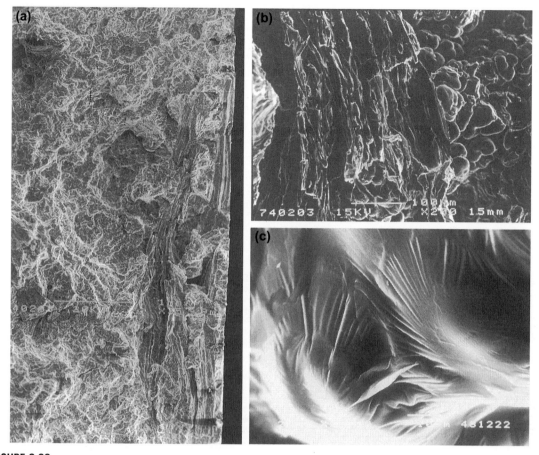

FIGURE 2.36

SEM fractographs of (a) bubble trail in Al-7Si-0.4Mg alloy; (b) close-up showing area of shrinkage porosity grown from the trail; (c) the oxide film of the trail draped over dendrites, on the point of being sucked back into the dendrite mesh because of a slight shrinkage condition in the casting (Divandari, 2000).

directionality to the visible defects provided by the arrays of invisible oxide bifilms, streaming like tattered banners in the wind, and funnelling bubbles along their surfaces to create the 'skeins' on machined surfaces. Figure 2.37(b) is a close up of the top of the same casting after defects have been revealed by machining.

A radiograph of a nearly pure copper casting that has also suffered bubble entrainment in its filling system is shown in Figure 2.38. This casting was thin-walled and extensive, so that it continued to receive bubbles late into pouring, with the result that they have become trapped in the solidifying casting. The author has seen similar extensive arrays of bubbles in aluminium alloy oil pan castings.

During the radiography of some stainless steels, I have witnessed an area initially identified as shrinkage porosity, but under the microscope was seen to be a mixture of bubbles and cracks. Such classic bubble damage is a remarkable combination that without the concept of the bubble trail would be extremely difficult to explain; tensile stress cannot have been present in the liquid phase because the bubbles would simply have expanded to relieve the stress, thus how can stress cracks have arisen? The answer, of course, is that the cracks are not due to stress but are bifilms in the form of

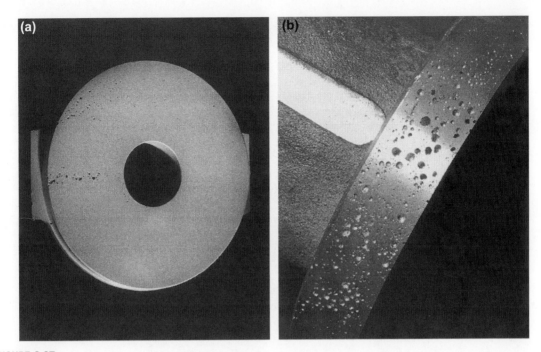

FIGURE 2.37

Bronze casting after machining revealing entrained air bubbles on (a) the front face and (b) the top edge of the flange.

bubble trails, and have no problem to sit side by side with bubbles. In these strong steels, the high cooling strain leads to high stresses that open up the bifilms to reveal their true nature as cracks.

In grey iron cylinder blocks, the bubbles and their trails are sometimes formed from an oxide (a glassy silicate) and sometimes from a lustrous carbon film, depending on the composition of the gases in the bubble (Campbell, 1970). The carbon film appears to be somewhat more rigid than most oxide films, and so resists to some extent the complete collapse of the trail, and retains a more open centre. In effect, the bubbles punch holes through the cope surfaces of the casting, so that their trails form highly efficient leak paths. The subject of core blows in cast iron is dealt with in more detail under Rule 5 for good castings in Section 10.5.

Bubble trails are known in low-pressure casting if the riser tube is not pressure-tight, allowing bubbles of the pressurising gas to leak through, rising up the riser tube and so directly entering the casting. The author has observed such a trail in a radiograph of an aluminium alloy casting. In the quiescent conditions after filling, the resulting trail was smooth and straight, rising through the complete height of the casting. When concentrating on the examination of the radiograph for minute traces of porosity or inclusions, such extensive geometric features are easy to overlook, appearing to be the shadows of integral structural parts of the casting!

In general, the bubble trail is a collapsed, or nearly closed, tube. However, completely open bubble trails have been observed by Divandari in pressure die castings (Figure 2.39). In this process, the very high injection velocities, of the order of 10–100 times the critical velocity for entrainment, naturally entrains considerable quantities of air and mould gases. The very high pressure (up to 100 MPa, or 15,000 psi) applied during casting are mainly used to compress these unwanted gases to persuade them to take up the minimum volume in the casting. If, however, the die is opened before the casting is fully solidified, as is usual to maximise productivity, the entrained bubbles may experience a reduction in their surrounding mechanical support. The bubbles and their trails containing the highly pressurised gas can be suddenly

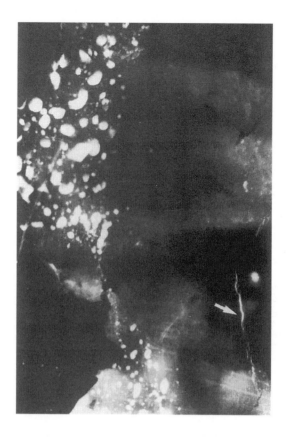

FIGURE 2.38

Radiograph of a thin-wall copper statue showing extensive bubble damage.

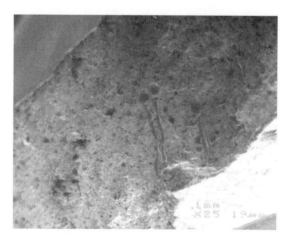

FIGURE 2.39

SEM fractograph of a zinc alloy pressure die casting revealing a bubble with an open trail (Divandari, 2000).

released from their constraining pressure, and so expand, opening like powerful springs, while the casting is still in a plastic state. Such open bubble trails in pressure die cast components are expected to be serious sources of leakage, particularly when broken into by machining operations. In more extreme cases, it is common for a high pressure die casting to explode if the die is opened too soon. These problems are of course greatly reduced (although perhaps never quite eliminated) by sacrificing some productivity, allowing the castings to solidify more completely before opening the die.

Finally, it is worth considering what length a bubble trail might reach. If we assume a bubble of 10 mm diameter rising through liquid aluminium, and if aluminium oxide grows to a thickness of 20 nm, it is not difficult to estimate that the 20% oxygen in the air is used up after creating a trail of about 0.5 m. This is quite sufficient to cause a major problem in most castings. However, this estimate is uncertain because Raiszadeh and Griffiths (2008) show that after the oxygen is consumed in a bubble, the nitrogen then starts to react to form aluminium nitride. Thus the maximum trail length might be nearer 2.5 m.

For castings poured in vacuum (actually corresponding only to dilute air of course!), vacuum bubbles are still to be expected to form, although the situation is a little more complicated. The entrained atmosphere in this case will be at a pressure somewhere between 10^{-3} and 10^{-6} atm ($1-10^{-3}$ torr). (The local vacuum inside the mould at the instant of pouring is likely to be much higher than the pressure indicated on the vacuum gauge of the furnace as a result of mould outgassing.) Thus, considering the case of the vacuum casting of Ni-base superalloys, at a nominal depth of 100 mm in the liquid of density close to 8000 kgm^{-3}, a bubble of 20 mm diameter will collapse down to somewhere in the region of 5–0.5 mm diameter before its internal pressure rises to equal that of the surrounding melt. (We are neglecting the somewhat balancing effects of (1) a small reduction in size because of surface tension and (2) a small increase in size from the expansion of the mould gases because moulds for investment casting of Ni-based alloys are already at 1000°C or more.) The bubbles will be smaller at greater depths and larger toward the top of the casting of course. Their considerably reduced concentration of oxygen will reduce the potential lengths of trails, or tend to generate more nitride film based trails. (It is known that Ni-based superalloys suffer from time to time from nitride problems.) It is not easy to predict what form a bubble trail may take in these circumstances, if the bubble is able to rise at all. There seems no shortage of research to do yet.

2.3 FURLING AND UNFURLING

Throughout its life, the bifilm undergoes a series of geometrical rearrangements. An understanding of these different forms is essential to the understanding of the properties of castings. The stages in the lifecycle of the bifilm are as follows.

1. Rapid entrainment by surface turbulence;
2. immediate and rapid furling by bulk turbulence to a compact, ravelled shape; and
3. slow unfurling in the stillness of the liquid after the filling of the casting is complete, slowly re-expanding to its original size.

The verb '*to furl*' is used in the same sense that a sailing boat will *furl* its sails, gathering them in, or it may *unfurl* them, letting them out to adopt their full area and fill with wind.

We shall show later that if the surface of the liquid suffers turbulence (i.e. the *We* significantly in excess of unity), the liquid is likely to enfold our familiar bifilm defects. Their size will probably depend on the strength (i.e. the tear resistance) of the film and the power of the turbulent flow. For some high-duty stainless steels and Ni-base alloys, the size might be 100 mm across. For Al-Si alloys containing low levels of Mg the bifilms seem to be more usually in the range 0.1–1 mm and occasionally up to a 10 or 15 mm diameter.

Once submerged, the bifilms will be subjected to the conditions of bulk turbulence beneath the surface of the liquid. This is because *Re* is nearly always over the critical value of 2000 in liquid metal filling systems as is clear from Figure 2.22. (Exceptions may be counter gravity and tilt pouring systems.) The launching of our delicate, gossamer-thin double film into a maelstrom of vortices in this heavy liquid will ensure that it is pummelled and ravelled into a compact, convoluted form almost immediately. In this form, it will be effectively reduced in shape and size from a planar crack of diameter perhaps 1–10 mm to a small ragged ball of perhaps a tenth of this size, i.e. a diameter in the 0.1–1 mm range.

These are the sizes of small shadows of crumpled dross-like defects seen by Fox (2000) in the X-ray radiograph, Figure 2.40(a), and commonly seen tangled arrays on polished metallographic sections of aluminium alloy castings (Figure 2.5(a)).

It is worth emphasising that although the bifilm now has a highly contorted form; crumpled, convoluted and ravelled in an untidy random manner it has not lost its crack-like character; its form contrasts sharply with the types of cracks familiar to the metallurgist and engineer. Clearly this bears no resemblance to a crack formed by the application of stress. It is a crack morphology unique to liquid metals and castings.

It seems also worth underlining that the double film is unlikely to become separated once again into single films. This is because where the two halves are in contact, interfacial forces will be important, and inter-film friction will ensure that both films of the bifilm will move as one. In addition, the continued oxidation, thickening the films, will reduce the volume of air in the inter-film gap, reducing the pressure between the films. The exterior atmospheric and metallostatic pressures will thereby pressurise the two halves together, ensuring that the films continue their association as a pseudo-single entity, despite the absence, or near-absence, of bonding forces between the halves.

In this compact form, bifilms are able to pass through filters of normal pore size in the region of a 1 mm diameter. Some will be arrested as a result of untidy trails of film that may become caught by the filter because the compact forms will be winding and unwinding continuously in a random manner in the turbulent flow. Inside the filter the constraint of the narrow channels probably promotes laminar flow (Figure 2.22). It is possible that many compact bifilms will be unravelled and flattened against the internal surfaces of the filter if they become hooked up somewhere in a laminar flow stream. This seems likely to be a potentially important filtration mechanism whereby compacted bifilms become caught at one point and straightened by the flowing stream.

At practically all other locations, during the whole of the voyage into the mould, the bulk turbulence will ensure that the bifilms are continuously tumbled, entering the mould in their compact form.

Thus the casting initially finds itself with a distribution of compact, convoluted bifilms. Their tensile strength will, on average, be about half that of a pore of the same average diameter as the bifilm sphere because of the effect of the mechanical interlinking of the crumpled crack as seen in Figure 2.41. If the convoluted crack is subject to a tensile stress it cannot come apart easily. There is much interlocking of sound metal, so that much plastic deformation and shearing will be involved in the failure.

Because of the combination of a certain amount of strength together with their compact size, the convoluted bifilms are rendered as harmless as possible at this stage. As a result the casting will enjoy maximum strength, even though this is perhaps significantly less than a 'perfect' casting—if such a phenomenon were achievable.

However, once the filling of the casting is complete, the bifilm now finds itself in a new, quiescent environment. If there is any movement of liquid because of solute or thermally driven convection, such movements are relatively gentle, occurring at low values of Re.

Conditions are now right for the bifilm to unfurl, opening, unfolding progressively, fold after fold, and effectively growing in size by about a factor of 10 on average, to its original size in the range of 1–10 mm or so. By this action, the defect now re-establishes itself as a planar crack, and so can impair the strength and ductility of the metal to the maximum extent. This astonishing ability of the bifilm to open like a flower explains many of the problems of castings, as will become clear throughout this book.

What causes the bifilm to open again?

Interestingly, there are a number of potential driving forces. Although they will be considered in detail in later sections of the book, they are listed briefly here. The evidence mainly comes from recent studies of aluminium alloys where the effects have been most thoroughly studied to date. They include the following.

1. The precipitation of hydrogen in the gas film between the oxide interfaces, thus inflating the defect. The inflation is likely to take place in two main stages. The first we may call microinflation, with short lengths inflating individually. These short lengths are defined and separated from adjacent lengths by the crease lines of adjacent folds. When sufficiently inflated, these short lengths will start to exert an unfolding pressure on adjacent lengths. As gas continues to inflate the bifilm, the bifilm will unfold progressively from one fold to the next fold, and so on. Thus the unfurling acting will probably be in the form of irregular, leisurely jerks, until the whole bifilm is unfolded

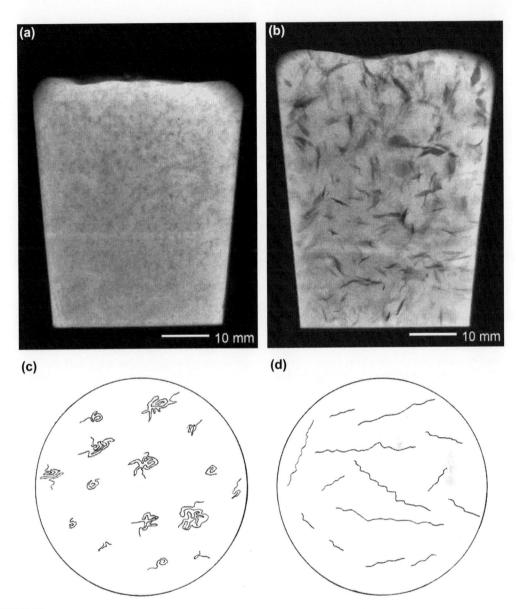

FIGURE 2.40

Radiographs of reduced pressure test samples of the same as-melted Al-7Si-0.4Mg alloy solidified under pressure of (a) 1 atm and (b) 0.01 atm (Fox and Campbell, 2000). (c) Bifilm cracks furled so as to be fairly harmless; (d) the same bifilms unfurled to show their damaging potential resembling engineering cracks as in (b).

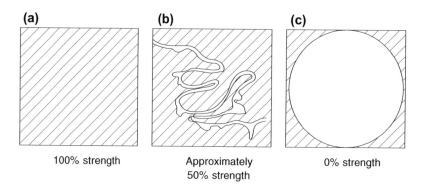

(a)   **(b)**   **(c)**

100% strength   Approximately   0% strength
                       50% strength

FIGURE 2.41

Relative strengths of (a) solid matrix; (b) convoluted bifilms, and (c) pores.

flat. At this stage, of course, if enough hydrogen continues to precipitate, the planar bifilm crack will eventually inflate into a spherical pore. The stages of growth are illustrated in Figure 2.42(a) for a singly folded bifilm. Figure 2.42(b) illustrates the stages for a more convoluted bifilm.

If much gas precipitates, either because there is much gas in solution or because there is plenty of time for diffusion, thereby aiding the collection of gas from a wide region, the gas pore may eventually outgrow its original bifilm to become a large spherical pore if growing freely in the liquid as shown in Figure 2.42(c), or if growing later during freezing, appearing now as an interdendritic pore as in Figure 2.42(d). Note that in this situation, the two very different morphologies (c) and (d) are both hydrogen gas pores. On sectioning the pore, it is probable that the originating fragment of bifilm may never be found.

Interestingly, at the stage at which the pore starts to outgrow the boundaries of its originating bifilm, it will extend its own oxide film using its own residual oxygen gas (probably derived from the original bifilm). The new area of the defect will then consist of new, very thin film. It is possible to imagine that if all the oxygen were consumed in this way that eventually the pore would grow by consumption of the residual nitrogen in the air as discovered by Raiszadeh and Griffiths (2008). If this were also entirely consumed, growth might occur by adding clean, virgin metal surface. The many round pores in cast metals indicate that the linear dimensions of many bifilms are probably smaller than the diameters of such pores. It is a reminder of the widespread of bifilm sizes to be expected.

2. Shrinkage acts to unfurl bifilms because it can reduce the pressure in a local region of the casting. Thus in this region it may suck the two halves of the bifilm apart. If this happened at an early stage of freezing, the defect would grow no differently to a hydrogen pore, the inflation process proceeding from one fold to the next. Exactly this effect is seen in the transition from Figure 2.40(a) and (b) in which small furled bifilms are seen to unfurl, becoming significant cracks up to 15 mm long. If the shrinkage effect was severe, the final stage of unfurling would be followed by the bifilm eventually growing to be more or less spherical, as seen in Figure 2.42(a) and (c). If the opening took place at a late stage of freezing the defect would attempt to open when surrounded by dendrites as shown in Figure 2.42(d).

Both of these modes of opening could be driven by either gas or shrinkage or both driving forces working in cooperation. Thus the rounded and interdendritic forms are not reliable indicators of the driving force for the growth of pores. In reality, the round form is a reliable indicator that the pore grew before the arrival of dendrites, whereas the dendritic pore is a reliable indicator that this pore was a late arrival. No more can be concluded.

In the case of the dendritic morphology, withdrawal of the residual liquid (being either pushed by gas from inside the pore or pulled by reduced pressure outside the pore) would stretch the film over surrounding dendrites (Figure 2.36(c)), and pull the bifilm halves into the interdendritic regions. The originating films might eventually be detected as fragments draped over some dendrites (Figure 2.29), or may be sucked completely out of sight deep into the dendrite mesh (Figure 2.42(d)).

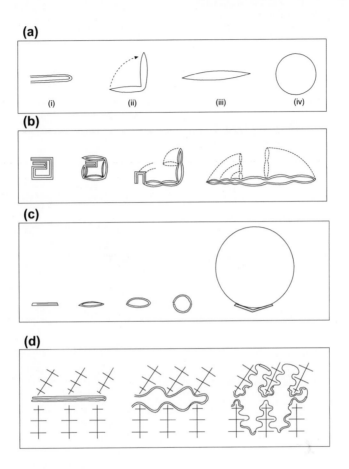

FIGURE 2.42

(a–d) Stages of unfurling and inflation of bifilms.

3. Intermetallics in Al-Si alloys, such as Si- and Fe-rich phases, nucleate and grow on the outer, wetted surfaces of the bifilm. The βFe particles, Al_5FeSi, are particularly good examples. Initially, when the βFe crystals are no more than a few nanometres thick, the crystals can follow the curvature and creases of the crumpled double film. However, as the crystals grow in thickness they quickly develop rigidity (because the rigidity of a beam in bending mode is proportional to its thickness to the third power). Thus as the precipitates thicken and rigidise, they force the straightening of the bifilm. The result is the familiar βFe plate seen on polished sections, straight as a needle. (Occasional curved βFe plates are probably the result of restraint at the two ends of the plate, so that the plate has been unable to straighten fully, remaining stressed like an archer's bow.) The βFe plates sometimes exhibit a crack along their centreline (if βFe has precipitated on both sides of the bifilm) or between the βFe and the matrix, showing apparent decohesion from the matrix (if the βFe has precipitated only on one side of the bifilm). It seems likely that all βFe particles are cracked or decohered in this way because of the presence of the originating bifilm. The straightening of bifilms by intermetallics in Al alloys is dealt with in detail in Section 6.3.6.

The cracks may not always be seen on a polished section, especially if the casting is well fed. Keeping up the pressure on the bifilms during freezing will ensure that the two halves of the bifilm remain closely pressed together and thus becoming an effectively invisible crack. Moderately poor feeding, or some hydrogen in solution, will part the films, allowing them to be seen on a polished metallurgical specimen. Research on this topic requires, paradoxically, to be carried out on poorly fed sample castings, in contrast to the foundry person who requires his castings to at least appear as sound as possible.

4. Oxides are pushed ahead of growing dendrites essentially because they are unable to grow through the layer of air between the films. The result is the bifilms are automatically unravelled and flattened, effectively teased out and organised into planar areas among dendrite arrays. The teasing action is all the more effective because the bifilms are usually attached to the side walls of the casting, adjacent to the roots of the dendrites. Thus as the dendrites grow, pushing ahead the heap of films, those that are attached near their roots are stretched into flatness between the rows of dendrites (Figure 2.43). In addition, oxides are pushed into interdendritic and grain boundary regions. This is an

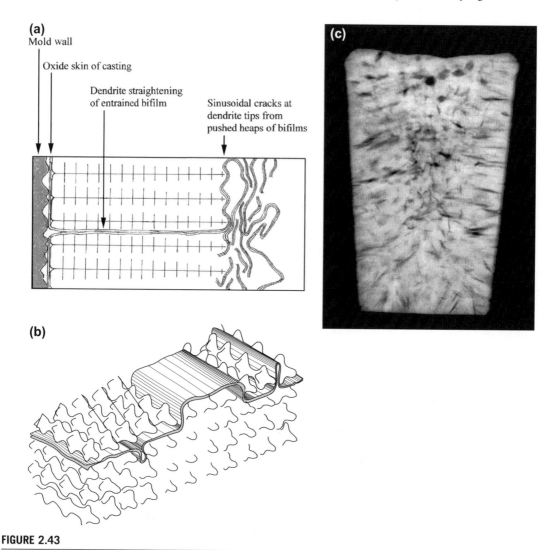

FIGURE 2.43

Dendrite flattening of bifilms (a) the pushing action, (b) the flattened bifilm, (c) radiograph of RPT showing dendrite-flattened cracks grown inward from the surface of the cast sample.

important mechanism because it can occur in all cast alloys that have solid surface films. The pushing sideways or ahead results in large flat facets on fracture surfaces, or, occasionally, regularly undulating hilly regions of a fracture surface marking the tips of dendrites separated from the matrix by the heap of bifilm cracks (Figure 2.43(b)).

A room temperature fracture surface is seen in Figure 2.44 for an Al-4.5 Cu alloy (Mi, 2000) in which the fracture surface is completely covered in thin alumina films, as 'half' of the original double film, or bifilm. The films have been pushed from 'east' and 'west' by the advance of dendrites. The heaps of excess film pushed ahead of the dendrites are clearly seen forming the central 'north-south' ridge. Images (a), (b), (c) and (d) of this figure show fracture surfaces of test bars taken from the same casting; test bar failing by fracture (a) was observed by video radiography to suffer a major turbulent breaking wave during mould filling, whereas (c) was observed to fill smoothly. The surface of (c) exhibited only rather small bifilms from the population that existed in original melt and

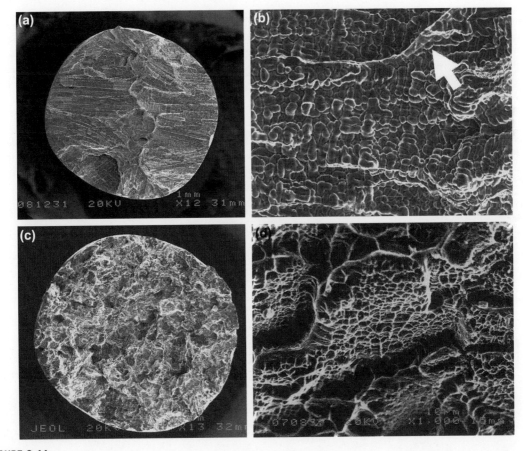

FIGURE 2.44

(a) Tensile fracture of Al-4.5 Cu alloy that had suffered an entrainment effect observed by video X-ray radiography, showing large new bifilms straightened by large grains grown in from both sides, and heaps of spare film pushed into central areas. (b) The close-up shows a thin doubly folded area (arrowed). (c) Elsewhere in the same casting shows evidence of a population of small bifilm defects interspersed between ductile matrix, showing (d) ductile dimples on fracture.

Courtesy Mi (2000).

caused the scattered microporosity between regions of ductile dimple fracture surface. The ductile dimples indicated that there had been real metal-to-metal contact between the two halves of the fracture surface, in complete contrast to surface (a) where not a single ductile dimple could be found, indicating no metal-to-metal contact over its whole area. In fact, its surface had been entirely covered with bifilms and was therefore effectively completely pre-cracked. This explained its elongation of only 0.3%, compared to the 3% of specimen (c).

Close examination of (a) was required to reveal tiny clues that the dendrite surface was actually covered with an oxide film derived from the turbulent liquid surface, and which had constituted half of the original bifilm; the arrow in Figure 2.44(b) indicates a small area where the film had been folded over twice, confirming that the oxide film was not merely an oxide grown on the dendrites in the solid state (for instance, after fracture). Other samples from this research exhibited rather thicker oxide bifilms that were much easier to discern (Figure 2.45). In fact, it seems in this image that a second oxide lies beneath the top fractured oxide, thus possibly originally constituting a *double bifilm* (i.e. originally four layers of oxide).

In Figure 2.46, an AlN or Al_2O_3 film is probably the cause of the planar boundaries seen in the fracture surface of a vacuum cast and hot isostatic pressing (hipping) Ni-base alloy IN939 (Cox, 2000). In the case of the vacuum-cast alloy, these facets were only observed in casting which had suffered surface turbulence during filling; no similar planar features were observed in quiescently poured controls. The planar features were observed to be associated with poor tensile ductility. Further examples of planar defects created by dendrite straightening are shown for cast irons (Section 6.5) and steels (Section 6.6).

In all cases in which large bifilms are found to have experienced dendrite straightening, the grain size always appears to be large. It seems reasonable to suppose that in a solidifying casting in which large bifilms are present, the natural action of convection would be suppressed. Thus there would not be the normal eddies of high- and low-temperature metal destabilising the growth of dendrites causing their arms to melt off and thus promoting grain multiplication. These barriers to flow would be present across the complete casting section, in contrast to other

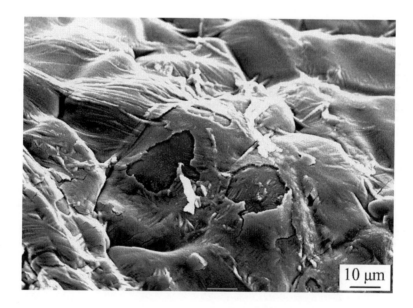

FIGURE 2.45

An area of the fracture seen in Figure 2.44(a) in which the film is thicker, revealing a second bifilm underneath. The parallel markings are deformation twins always seen in stretched alumina film.

Courtesy Mi (2000).

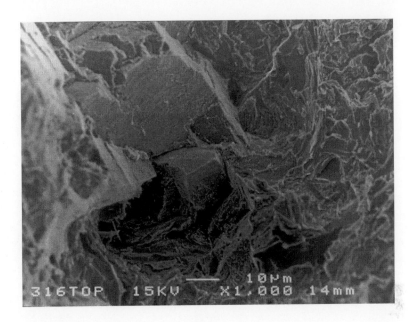

FIGURE 2.46

Tensile fracture surface of a turbulently cast, vacuum-melted and cast hipped Ni-base alloy, showing apparently brittle grain boundaries. Such features were not observed in quiescently cast material.

regions of the casting which contained only confetti-like bifilms in suspension. These would convect freely, so that remelting and fragmentation of dendrites would naturally generate a refined grain size.

Clearly, the driving forces for the unfurling of bifilms are all those factors that are already known to the casting metallurgist as precisely those factors that impair ductility, including hydrogen in solution, shrinkage, iron contamination in Al alloys, and large grain size. Thus the only common factor implicated in the loss of ductility is the presence of bifilms.

However, in practice, we need to bear in mind that it will take time for the unfurling processes to occur. In a casting that is frozen quickly, the bifilms are frozen in to the casting in their compact form, with the result that most casting alloys that are chilled rapidly are strong and ductile, as all foundry engineers know.

Conversely, castings that have suffered lengthy solidification times exhibit shrinkage problems, higher levels of gas porosity or large grain size. In addition, all exhibit reduced properties, particularly reduced ductility. The action of slow freezing impairing ductility is, once again, well known to the casting metallurgist.

In summary, it is clear that all the factors that refine dendrite arm spacing (DAS) improve the mechanical properties of cast aluminium alloys (Figure 2.47). This is almost certainly not primarily the result of action of the DAS alone, as has been commonly supposed. The DAS is mainly a measure of the local freezing time, which is the time available for the inflation and unfurling of defects. If the growth of area of the defect was not bad enough, its inflation, causing its surfaces to separate, transforming it from a crack that can enjoy some tensile strength into a pore which has none, makes the defect even more damaging. Thus its strength falls further as is illustrated in Figure 2.47. It is both (1) the growth of area and (2) the growth of volume of the defects that combine to inflict so much damage to the mechanical properties as solidification time increases. Increases in DAS are not the cause of the loss of ductility; the DAS is merely an indicator of the time involved. It is the built-in, independent clock.

Returning to the pivotal role of the entrainment defects, it seems clear from practical work in foundries that the highest velocities and maximum turbulence is in the sprue, runner and gates, where Re of 10^4–10^5 are easily attained,

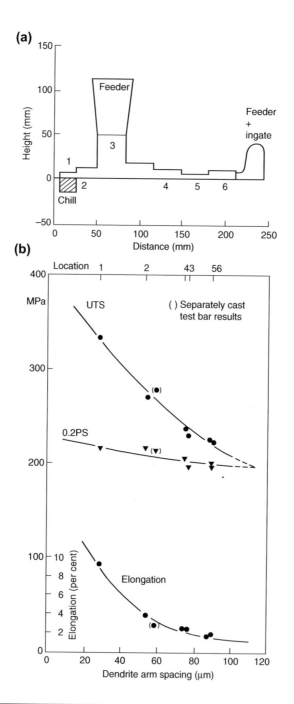

FIGURE 2.47

(a) Al-7Si-0.4Mg alloy casting and (b) its mechanical properties.

Data from Miguelucci (1985).

indicating severe bulk turbulence. Bifilms entrained in these parts of the running system will therefore be highly compacted, and, at least initially, have less damaging consequences for mechanical properties.

The less severe surface turbulence and bulk turbulence in the mould cavity are, paradoxically, a concern. *Re* in a nicely filled mould cavity falls to values in the region of 10^3, so that little energy is available to make the films more compact. Thus any bifilm entrained in the mould cavity itself will probably remain substantially open, maximising its area. This in itself will cause maximum damaging effect, but will be additionally enhanced by the hydrogen diffusion to the enlarged area, inflating the defect, and so further reducing any limited load carrying capacity it may once have enjoyed. It follows that the control of surface turbulence in the mould cavity itself is the most important factor to yield reliable properties.

It is almost certain that in some circumstance the bifilms will be unable to unfurl despite the action of some or all of these driving forces. This is because they will sometimes be glued shut. Such glues will include liquid fluxes. If present on the surface of the melt, they will find themselves folded into the bifilm, and so form a sticky centre to the sandwich, holding the bifilm closed, but the excess flux weeping or squeezed to its outer edges will aid the pinning of folds. The extreme thinness of bifilms will mean that even weak glues such as molten fluxes will be effective welds. Such action may explain part of the beneficial effect of fluxes. The melt may not be significantly cleaned by the flux, but its inclusions merely rendered less damaging by being unable to unfurl.

A similar action may explain the effect of low melting point additions such as bismuth to aluminium alloys. Although Papworth and Fox (1998) attribute the benefits of the Bi addition to the disruption of the integrity of alumina films, it seems more likely that a low melting point Bi-rich liquid will be inert toward the oxide, and that its action is passive, being merely an adhesive.

Although bifilms remain a feature of our processing techniques for aluminium alloys, fluxes or low melting point metals might be a useful ploy, even if it is effectively 'papering over the cracks' of our current metallurgical inadequacies.

For the future, the very best cast material will be made free from bifilms in foundries specifically designed to deliver perfect quality metal. Such concepts are already on the drawing board and may be realised soon. When this utopia is achieved, we shall be able to make castings with any DAS, solidified fast or slow, all with perfect properties. There is already evidence that this prediction is accurate, as is explained in Section 9.4, Figure 9.26. Its routine achievement is a day to which we can look forward.

2.4 DEACTIVATION OF ENTRAINED FILMS

Once submerged, the inside of the bifilm will act as a thin reservoir of air. Air will exist in pockets, in bubbles trapped between the films, and in folds of the films, especially for films that have grown somewhat thicker, because their greater rigidity will bridge larger cavities among the folds. In addition, most oxides are microscopically rough, so that the opposed roughness of the surfaces themselves will constitute part of the reservoir, and assist to convey the gas to all parts of the inter-layer. A particularly rough oxide that takes a corrugated form is MgO as seen on Al-5 Mg alloy (Figure 2.10(b)).

Nevertheless, of course, the bifilm is not likely to know that it is submerged. The oxides on its interior surfaces will therefore be expected to continue their growth regardless, under the impression that they are still in the fresh air. Thus, the layer of air in the bifilm will be gradually consumed as the oxygen is used up. As soon as all the oxygen has reacted, the nitrogen will then be converted to nitrides as demonstrated by Raiszadeh and Griffiths (2008).

In aluminium alloys, although the nitride of aluminium is not as stable as the oxide, and will therefore not form in competition with the oxide, it is nevertheless very stable. It is therefore likely to grow, if only slowly on Al alloys. In contrast, on molten ferrous alloys aluminium nitride forms rapidly at the much higher temperatures. Thus in a bifilm the layer of air is expected to disappear in time, but depending on the alloy and its impurities, at different rates. In steels, this would be rapid, but in aluminium alloys it would be expected to be slow. In either case the 1% remnant of argon in the air would continue to remain as a gas because of its nearly perfect insolubility in molten metals (Boom et al., 2000).

During the final stages of the disappearance of the gas layer the opposed oxides would be expected to grow together to some extent. El-Sayed et al. (2014) find the loss of air fastest with the growth of MgO films in Al-Mg alloy, taking only 6 min. The time was 9 min in pure Al with its Al_2O_3 film, but 25 min for the formation of the spinel oxide in Al-7Si-0.3Mg alloy. The rate of loss of gas can, of course, be slowed, stopped or reversed depending on the amount of hydrogen present. Any resultant bonding may not be particularly strong, but may confer some improvement in strength compared with the layer of air. Additional limited strength would be expected from the mechanical inter-locking of the opposed crystalline or otherwise rough surfaces. Raiszadeh et al. (2012, 2013) have reported increases in bonding of Al alloy bifilms with time.

The presence of hydrogen in solution in Al alloys can cause the rate of deactivation of the bifilm by bonding to slow, or even go into reverse by inflating the double film, moving the two halves out of contact. This is found for Al alloys with higher magnesium contents (Shafaei and Raiszadeh, 2014). Toda (2014) found that when Al alloys were fractured in vacuum, hydrogen was released causing him to conclude that Al alloys contained a high density of 'micropores' filled with hydrogen. We can conclude that the two gases, hydrogen and argon, will both resist the deactivation of the bifilm by bonding, thus preserving the crack-like nature of the defect despite, on occasions, significant plastic working.

If the gas layer is eventually consumed, and if some diffusion bonding has occurred, then the most deleterious effects of the bifilm will have been removed. In effect, the potential of the defect for causing leaks, or nucleating bubbles, cavities or cracks, will have been greatly reduced. This has been confirmed by Nyahumwa, Green and the author (1998, 2000) who found that porosity in the fatigue fracture surfaces of Al-7Si-0.4Mg alloy appeared to be linked often with new oxides, but not old oxides. Raiszadeh and Griffiths (2007) observed bridging between the leaves of the bifilm after solution treatment.

Further evidence for the growing together of bifilms with time is provided by the work of Huang et al. (2000) who examined fragments of bifilms detached by ultrasonic vibration from polished metallographic sections. These revealed that the bifilms of alumina, in the process of transforming to spinel, appeared to be extensively fused together. Figure 2.48 is a reinterpretation of their observations. Their work is separately interesting because it indicates that bifilms of different numbers and sizes are associated with different alloys.

The automatic deactivation of bifilms with time may be relatively fast for new, thin films. This would explain the action of good running systems, where, after the fall of the liquid in the down-runner, the metal is allowed to reorganise for perhaps a second or so before entering the gates into the mould cavity. In general, such castings do appear to be good, although this speculation has not been followed by any formal study to explore whether the effect is real or not.

However, the deactivation of bifilms seems certain to be encouraged in cases where the metal is subjected to pressure because mechanical properties of the resulting castings are so much improved, becoming significantly more uniform and repeatable, and pores and cracks of all types appear to be greatly reduced if not eliminated. Even if the air film in bifilms is not completely eliminated, it seems inevitable that the oxygen and nitrogen will react at a higher rate and so be more completely reacted, and any remaining gases be compressed. In addition, if the pressure is sufficient, liquid metal may be forced through the permeable oxide so as to fuse with, and thus weld to, metal exuding through faults in the far side of the film. The crack-like discontinuity would then have effectively been 'stitched' together, or possibly 'tack welded'. Pressurising treatments that would promote such assisted deactivation include:

1. The solidification of the casting under an applied pressure. This is well seen in the case of squeeze casting, where pressures of 50–150 MPa (500–1500 atm) are used. However, significant benefits are still reported for sand and investment castings when solidified under pressures obtainable from a normal compressed-air supply of only 0.1–0.7 MPa (1–7 atm). Berry and Taylor (1999) review these attempts.

   It is interesting to reflect that the reduced pressure test, commonly used as a porosity test in aluminium alloy casting, further confirms the importance of this effect but uses exactly the same principle in the opposite direction: the pressure on the solidifying casting is lowered to maximise porosity so that it can be seen more easily (Dasgupta et al., 1998, Fox and Campbell, 2000).

2. Hot isostatic pressing (hipping) is a solid state deactivation process, which, by analogy with the liquid state, appears to offer clues concerning the mechanism of deactivation processes in the liquid. The hipping of (solid) castings is carried out at temperatures close to their melting point to soften the solid as far as possible. Pressures of up to 200 MPa (2000 atm) are applied in an attempt to compress flat all the internal volume defects and weld together the walls of the defects by diffusion bonding. It is clear that in the case of hipping aluminium alloys, the aluminium

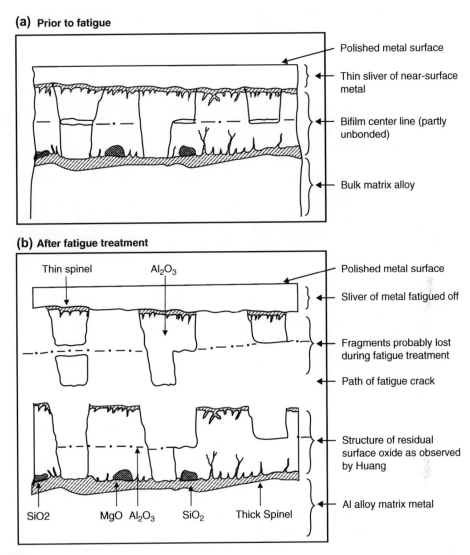

(a) Prior to fatigue

- Polished metal surface
- Thin sliver of near-surface metal
- Bifilm center line (partly unbonded)
- Bulk matrix alloy

(b) After fatigue treatment

Thin spinel   Al$_2$O$_3$

- Polished metal surface
- Sliver of metal fatigued off
- Fragments probably lost during fatigue treatment
- Path of fatigue crack
- Structure of residual surface oxide as observed by Huang
- Al alloy matrix metal

SiO2   MgO Al$_2$O$_3$ SiO$_2$ Thick Spinel

FIGURE 2.48

Interpretation of observations by Huang et al. (2000) showing (a) a section of double film near the surface of the cast Al-7Si-0.4Mg alloy with (b) fragments separated by fatigue when applying ultrasonics.

oxide that encases the gas film will not weld to itself because the melting point of the oxide is higher than 2000°C and a temperature close to this will be required to cause any significant diffusion bonding.

The fact that hipping is successful at much lower temperatures, for instance approximately 530°C in the case of Al-7Si-0.4Mg alloy, indicates that some additional processes are at work. For instance, some diffusion bonding of oxides may occur in the presence of reactions in the oxide. In the case of Al alloys that have less than 0.10 wt% Mg, the oxide will be a variety of alumina, and so very stable and unlikely to bond. For higher Mg alloys, above about 1.0 wt% Mg (particularly the Al-5 Mg and Al-10 Mg alloys), the oxide involved in this case is MgO, and so, like alumina, is stable and unlikely to take part in any bonding action. For alloys with an intermediate Mg content in the range approximately 0.1–1.0 wt% Mg, after time at temperature, the initially formed alumina film absorbs Mg from the matrix, converting the oxide to a spinel structure. During the atomic rearrangements required for the reconstruction of this quite different lattice, some diffusion bonding will be favoured.

In an elegant experiment, Aryafar et al. (2010) showed that no bonding across the Al_2O_3 of bifilms occurred until all of the oxygen and nitrogen was consumed. After this, some weak patchy bonding resulted from the conversion of Al_2O_3 to spinel, $MgAl_2O_4$, in a short time. After longer times, stronger bonding resulted from the conversion of spinel to MgO. Bakhtiarani and Raiszadeh (2010) studied Al-4.5 Mg melts, finding that after entrainment the bifilms continued to grow for several hours, the oxide spinel gradually converting to MgO. The atomic rearrangement resulted in some bonding across impinging regions of the bifilm, a process resembling the action of diffusion bonding, but occurring, of course, at temperatures hundreds of degrees lower than that at which true diffusion bonding of oxides could occur.

3. The closure and elimination of bifilms might be expected to occur as a result of severe plastic working, by forging, rolling or extrusion. However, quite the converse appears to be true. For the 8000–70,000 pores/mm^3 in wrought Al alloys, both Talbot (1975) and Toda (2009) find that pores do not close during subsequent extensive plastic deformation. Also for the hot rolling of steel, Joo (2014) finds that some pores appear to close but bonding is rare or not at all.

4. Finally, if the entrained solid film is partially liquid, some kind of bonding is likely to occur much more readily, as has been noted in Section 2.1. This may occur in light alloy systems where fluxes have been used in cleaning processes, or contamination may have occurred from traces of chloride or fluoride fluxes from the charge materials or from the crucible. Such fluxes will be expected to cause the surfaces of the oxide to adhere by the mechanism of viscous adhesion when liquid, and as a solid binder when cold. The beneficial action of fluxes in the treatment of liquid Al alloys may therefore be not the elimination of bifilms but their partial deactivation by assisting bonding. That flux inclusions can appear commonly on machined surfaces of Al alloy castings is an indication that the rapid detrainment of fluxes by flotation as would be expected for a liquid inclusion has been inhibited by the flux being wrapped in a much larger bifilm.

Similarly, as we have noted previously, the observations by Papworth (1998) might indicate that Bi acts as a kind of solder in the bifilms of Al alloys. The metal exists as a liquid to 271°C, so that when subjected to the pressure of squeeze casting, it would be forced to percolate and in-fill bifilms. The consequential resistance of bifilms to deformation and separation would be expected to be improved if only by their generally increased mechanical rigidity (i.e. not by any chemical bonding effect). If proven to be true, this effect could be useful.

In the case of higher temperature liquid alloys, particularly steels, some oxides act similarly, forming low melting point eutectic mixtures. These valuable systems will be discussed in detail in Chapter 6.

2.5 SOLUBLE, TRANSIENT FILMS

Although many films such as alumina on aluminium are extremely stable and completely insoluble in the liquid metal, there are some alloy/film combinations in which the film is soluble. This is of course especially true for the higher melting point metals in which the higher temperatures and the reduced stability of the films encourage this behaviour.

Transient films are to be expected in many cases in which the arrival of film-forming elements (such as oxygen or carbon) at the surface of the liquid exceeds the rate at which the elements can diffuse away into the bulk metal. If such a film is then entrained, during its brief life the double film may create a crack, a hot tear, or may initiate shrinkage porosity. However, after the initiation of these consequential 'secondary' defects, the originating bifilm then quietly goes into solution, never to be seen again. Thus the defect in the form of a crack, hot tear or pore continues to exist but exhibiting no clue about its origins.

It is to be expected that many oxide and nitride films might be soluble in some irons and steels.

Such behaviour is commonly seen in grey cast irons. If a lustrous carbon film forms on the iron and is entrained in the melt, after some time it will dissolve. The rate of dissolution will be rather slow because the iron is already nearly saturated with carbon. The longer the time available the greater is the chance that the film will go into solution. Thus there are various conditions that can be envisaged:

1. In a thin section, the lustrous carbon will be frozen in place, constituting a carbon-based bifilm defect to reduce strength or create a leak path.
2. In an intermediate section, the entrained bifilm may last just long enough to initiate some other longer lived and more serious defect, such as a hot tear or shrinkage porosity, prior to its disappearance.
3. In heavy section, grey iron castings entrained lustrous carbon films are usually never seen, having plenty of time to go into solution.

Titanium alloys, including titanium aluminide alloys, have a high solubility for oxygen, so that oxide films, if they form, are almost certainly soluble. These issues are discussed in the section on Ti alloys (Section 6.8).

2.6 DETRAINMENT

In a liquid metal subjected to surface turbulence, many of the defects *entrained* by the folding and impingement actions can find themselves eliminated (*detrained*) from the melt at a later stage. Detrainment events take a variety of forms.

If the oxide surface film is particularly strong, it is possible that even if entrained by a folding action of the surface, the folded film may not be free to be carried off by the flow. It is likely to be attached to part of the surface that remains firmly attached to some piece of hardware such as the sides of a launder, or the wall of a sprue. Thus the entrained film might be detrained by being pulled clear from the main flow, effectively trapped against the side walls. This detrainment process is so quick that the film hardly has time to consider itself entrained. A strong film, strongly attached to the wall of the mould may simply remain hanging in place, possibly flapping in the flow in the melt delivery system, but fortunately never entering the casting, and so remaining harmless.

Beryllium added at levels of only 0.005% to reduce oxidation losses on Al-Mg alloys probably acts in this way. On attempting to eliminate Be for environmental and health reasons, difficulties have been found in the successful production of wrought alloys by continuous (direct chill) casting. The beneficial action of Be that was originally unsuspected, but subsequently highlighted by this problem, is thought to be the result of the strengthening of the film by the low levels of Be, thus encouraging the hanging up of entrained films in the delivery system to the mould rather than their release into the flowing stream.

The detrainment of large bubbles is relatively easy because they have sufficient buoyancy to break their own oxide film, and the casting surface oxide skin (once again, the two films constituting a double oxide barrier of course; the double film phenomenon appearing to be a fundamental and recurring film condition in casting technology) between them and the outside world at the top of the casting. They can thus detrain (Figure 2.33). If successful, this detrainment is not without trace, however, because of the presence of the bubble trail that remains to impair the casting.

Small bubbles of up to about 5 mm diameter have more difficulty to detrain. They are commonly trapped immediately under the top skin of the casting, having insufficient buoyancy to break through the double film barrier. Thus they are unable to allow their contents to escape to atmosphere. They are the bane of the life of the machinist because they lodge just under the oxide skin at the top of a casting, and become visible only after the first machining cut. In iron and steel castings requiring heat treatment, this thin surface oxidises, scaling off to reveal the sub-surface porosity. Similarly, shot blasting will also often reveal such defects as a result of their nearness to the outside surface.

The most complete and satisfactory detrainment is achieved if the liquid metal has a liquid oxide film. This is because, once entrained, the liquid oxide spherodises to form droplets, and floats out with maximum speed (Figure 2.12). On arrival at the liquid oxide surface, the liquid oxide droplets are simply reassimilated in the surface liquid layer, and disappear. This is the mechanism by which steels finally deoxidised with Ca + Al achieve such high levels of cleanness in comparison with steels deoxidised with the usual Si, Mn and Al. The oxides CaO and Al_2O_3 form a low melting point eutectic which, when entrained, simply balls up and floats out. Those very small fragments of the entrained oxides that do not have sufficient Stokes velocity to reach the surface remain entrapped in the steel as minute, nicely spherical droplets. Steel metallurgists call this 'inclusion shape control', usually overlooking the main benefit of the detrainment of the larger inclusions.

Liquid fluxes in Al alloys probably behave somewhat differently. The liquid flux, if composed of a chloride mixture, does not dissolve the extremely stable aluminium oxide. Thus the liquid flux will remain on the surface of the melt, sitting on top of the solid oxide film. When entrained during stirring or pouring, the solid film wraps up some of the flux, preventing it from spherodising. In addition, the much larger area of the film greatly reducing its Stokes velocity. Thus alumina films containing a central flux layer can remain in suspension for long periods. This is known from the experience in the Nemak, Windsor foundry. Mould filling was automated by electromagnetic pumps, sited at the ends of long launder systems that delivered liquid metal to the casting stations. The pumps would commonly sit in the melt without problem for several months. However, when chlorine gas was introduced in the rotary degassing units the pumps blocked within 24 h from a mixture of chloride flux and oxides. The chloride flux alone could never have made the long journey (several tens of metres) from the degassers because detrainment would have been expected for spherical droplets within seconds or minutes, clearing the melt within a metre or so of progress along the launder system. However, if entrained in an alumina film the flux could remain in suspension for hours. It would finally be detrained by the high body forces experienced by the liquid metal in the pump, and thus building up on the inner walls of the pump to block it with maximum efficiency.

The benefit of using fluxes in liquid aluminium alloys is therefore curious. How do the mechanical properties improve if the aluminium oxide is not eliminated or dissolved? It seems likely that the flux, originally on the top surface of the oxide film on the liquid, becomes entrapped inside the oxide bifilm. Thus the bifilm is effectively 'glued' together, unable to open to create cracks or pores, effectively improving the properties of the metal. However, the exudation of fluxes from machined surfaces, causing unsightly corrosion pustules as the salts absorb moisture from the air, is an indication that they probably are contained within bifilms because without the bifilm liquid globules of salt would have detrained within seconds or minutes and thus not become a permanent feature of the casting.

It may be that the beneficial effect of boron added to steels might be due to this mechanism. The boron will form low melting point liquid borates on the steel surface, but will probably not take into solution stable films such as Al_2O_3 or Cr_2O_3, and will therefore simply float on the surface of these solid stable films. On pouring, the stable oxide films are entrained, sandwiching the liquid borate which will freeze later, once again acting as a 'glue'. The absence of the residual air layer between the films will greatly increase thermal conductivity across the bifilm, possibly explaining the great effect which boron additions make to the hardenability of steels.

Detrainment of bifilms in suspension in the melt can be encouraged by a number of means:

1. Filtration. The wide use of filters indicates there are benefits, even though the effectiveness of filters is extremely variable in different situations. The problems are discussed in more detail in Section 12.8.
2. Regassing: Chegini and Raiszadeh (2014) find evidence that bifilms in Al alloys are inflated and floated out by the introduction of excess hydrogen.
3. Rotary degassing treatment for Al alloys. As for filters, rotary degassing is a mixed blessing. Although a key benefit has been the stabilisation of melt quality in foundries, its action probably far from being fully optimised at the time of writing (see Section 14.4).
4. Sedimentation. The simple holding of melts and doing nothing but wait is an effective means to detrain many different kinds of melts in many different kinds of casting operation. Sedimentation can be encouraged by precipitating heavy second phases on bifilms. This technique is found to be particularly useful as the key benefit of grain refinement of Al alloys.

This important natural benefit is enjoyed by steelmakers in the steel-making furnace, but who go to lengths to ruin their metal by pouring it through a fall of several metres into a ladle. Subsequently, thanks to the large density difference between the steel and its oxides, the quality is rescued by detrainment of many of the entrained inclusions by flotation. This detrainment can be so rapid that much entrained material has floated out to the surface of the ladle, adding to the surface slag, during the several minutes that are required for the ladle to be lifted out of the casting pit and transported to the casting station. This rapid detrainment cannot occur for Al-deoxidised steels because the stable bifilms of alumina remain in suspension for much longer times, and create problems in the cast material.

In Al alloys, the significant benefits of sedimentation after pouring from a furnace into a transfer ladle are not achieved because of the long periods of near-zero Stokes velocity of suspended alumina bifilms. However, if grain refining additions are made, the relatively heavy TiB particles precipitate on the bifilms, dropping them out like stones. If the sediment of oxides and Ti-rich compounds is not disturbed at the bottom of the melt, the extremely clean metal above this can be decanted to create castings of high mechanical properties. Note that much if not nearly all of the improvement is due not to the grain refinement but to the absence of bifilms. This important effect is described in detail in Section 6.3.

2.7 EVIDENCE FOR BIFILMS

The evidence for bifilms actually constitutes the entire theme of this book. Nevertheless, it seems worthwhile to devote some time to highlight some of the more direct evidence.

It is important to bear in mind that the double oxide film defects are everywhere in metals. We are not describing occasional single 'dross' or 'slag' defects or other occasional accidental extrinsic types of inclusions. The bifilm defect occurs naturally, and in copious quantities, every time a metal is poured. Many metals are crammed with bifilm defects. That they are usually so thin has allowed them to evade detection for so long. Until recently, the ubiquitous presence of these very thin double films has not been widely accepted because until recently no single metal quality test has been able to resolve such thin but extensive defects.

Even so, over recent years there have been many significant confirmations.

The presence of bifilms is easily seen on fracture surfaces, particularly if a good SEM is available. If a bifilm is present, its component halves appear on each half of the fracture with nice butterfly wing symmetry (Figure 2.49). Griffiths and Lai (2007) show such symmetry on all of their fractured tensile bars of Mg alloy. Each oxide film is characterised by absence of metallic features and appearance of folds and creases, even though the folds and creases may be so thin as to be difficult to see except under the best operating conditions of an SEM.

Another clear example of bifilms is seen in the use of the reduced pressure test (RPT) for aluminium alloys. The technique is also known, with slight variations in operating procedure, as the Straube Pfeiffer test (Germany), Foseco Porotec test (UK) and IDECO test (Germany). At low gas contents, many operators have been puzzled by the appearance of hairline cracks, often extending over the whole section of the RPT casting. They had problems understanding

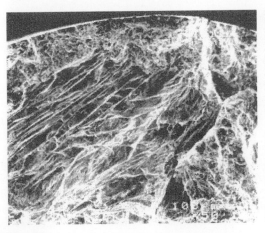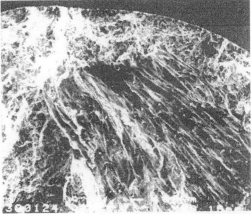

FIGURE 2.49

The secondary electron image of the near-mirror-images of the spinel bifilm halves on each half of the tensile failure surface of an Al-11.5 Si alloy.

Courtesy Cao (2010).

the cracks because the test is commonly only assumed to be an assessment of hydrogen porosity. However, once the phenomenon of bifilms is accepted, the cracks become immediately understandable. An example of cracks is seen in Figure 2.40. However, as gas content rises, the crack-like defects expand to become lens-shaped, and finally, if expansion continues, become completely spherical, fulfilling at last their expected appearance as hydrogen pores. Such an effect of thin cracks transforming into spherical bubbles has been widely observed many foundry people many times (e.g. Rooy, 1992), and can only be explained reasonably by the presence of bifilms. Figure 2.50 illustrates examples of cracks and pores from different melts.

In a variant of the reduced pressure test to determine the quantity known as the density index, two small samples of a melt are solidified in thin-walled steel crucibles—one in air and one under a partial vacuum, respectively. A comparison of the densities of the two samples gives the so-called density index. However, in this simple form, the density index is not particularly reproducible. The comparison is complicated as a result of the development of shrinkage porosity in the sample solidified in air. A better comparison is found by taking the lower half of the air-solidified sample and discarding the top half containing the shrinkage porosity. The fairly sound base is then compared with the sample frozen under vacuum. This gives a more accurate assessment of the porosity because of the combined effect of gas and bifilms. Without bifilms the hydrogen cannot precipitate, leading to a sound test casting, so that filtered metal always appears less 'gassy', giving the curious (and of course misleading) impression that the hydrogen can be 'filtered' out of liquid aluminium.

A novel development of the reduced pressure test has been made that allows direct observation of bifilms (Fox and Campbell, 2000). The rationale behind the use of this test is as follows. The bifilms are normally difficult or impossible to see by X-ray radiography when solidified at less than 1 atm pressure. If, however, the melt is subjected to a reduced pressure of only 0.1 atm, the entrained layer of air should expand by 10 times. At less than 0.01 atm, the layer should expand 100 times etc. In this way, it should be possible to see the entrained bifilms by radiography. A result is shown in Figure 2.40.

In this work, a novel reduced pressure test machine was constructed so that tests could also be carried out using chemically bonded sand moulds to make test castings as small slabs with overall dimensions approximately 50 mm high, 40 mm wide and 15 mm thick. The parallel faces of the slabs allowed X-ray examination without further preparation. The slower solidification provided by the sand mould was advantageous to allow a good unfurling response throughout the sample. (The placing of the usual thin steel cup on an insulating ceramic paper base acts almost as well.)

Figure 2.40 shows radiographs of plate castings from a series of tests that were carried out on metal from a large gas-fired melting furnace in a commercial foundry. Figure 2.40(a) shows a sample that was solidified in air indicating evidence of fine-scale, low-density regions appearing as dark, faint compact images of the order of 1 mm in diameter. As the test pressures were lowered, the compact regions unfolded and grew into progressively longer and thicker streaks, finally reaching 10–15 mm in length at 0.01 atm, Figure 2.40(b).

The 'streak-like' appearance of the porosity is due to an edge-on view of an essentially planar defect (although residual creases of the original folds are still clear in some bifilm images). That these defects are shown in such high contrast at the lowest test pressure suggests that they almost completely penetrate the full 15 mm thickness of the casting, and may only be limited in size by the 15 mm thickness of the test mould. The more extensive areas of lower density porosity are a result of defects lying at different angles to the major plane of the casting.

At 0.01 atm, the thickness of bifilms as measured on the radiographs for those defects lying in the line of sight of the radiation was in the range 0.1–0.5 mm. This indicates that the original thickness of bifilms at 1 atm was approximately 1–5 µm.

These samples containing large bifilms are shown here for clarity. They contrast with more usual samples in which the bifilms appear to be often less than 1 mm in size, and when solidified at 1 atm pressure are barely visible on radiographs. The work by Fox and the author (2000) on increasing the hydrogen content of such melts in the RPT at a constant reduced pressure typically reveals the inflation of clouds of bifilms, first becoming unfurled and slightly expanded by the internal pressure of hydrogen gas, and finally resulting in the complete inflation of the defects into expanded spheres at high hydrogen levels. Figure 2.50 shows a variety of RPT results for different melts, showing a range of different bifilm populations.

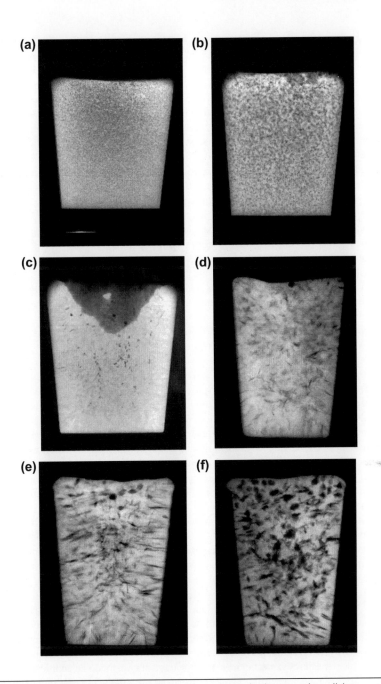

FIGURE 2.50

The examples illustrate (a) and (b) high defect density, with moderate hydrogen spherodising pores; (c) and (d) reduced hydrogen so defects, particularly aligned bifilms from the sides, are nearly invisible; (e) and (f) higher hydrogen to expand and clarify edge and central bifilms.

Courtesy S Fox.

A much earlier but critical observation of the action of bifilms was so many years ahead of its time that it remained unappreciated until recently. In 1959, at Rolls-Royce, UK, Mountford and Calvert observed the echoes of ultrasonic waves that they directed into liquid aluminium alloys held in a crucible. What appeared to be an entrapped layer of air was observed as a mirror-like reflection of ultrasound from floating debris. (Reflections from other fully wetted solid phases would not have been so clear; only a discontinuity like a crack containing a layer of air which elastic shear waves in the melt could not traverse could result in completely reflected waves, yielding such strong echoes.) Some larger particles could be seen to rotate, flashing reflections like a beacon when turning to face the receiver after each revolution.

Immediately after stirring, the melts became opaque with a fog of particles. However, after a period of 10–20 min the melt was seen to clear, with the debris forming a deep layer at the base of the crucible. If the melt was stirred again the phenomenon could be repeated. Stirred melts were found to give castings containing oxide debris together with associated porosity. It is clear that the macroscopic pores observed on their polished sections appear to have grown from traces of micropores observable along the length of the immersed films. Melts that were allowed to settle and then carefully decanted from their sediment gave castings clear of porosity.

Other interesting features that were observed included the precipitation of higher melting-point heavy phases, such as those containing iron and titanium, onto the floating oxides as the temperature was lowered. This caused the oxides to drop rapidly to the bottom of the crucible. Such precipitates were not easy to get back into suspension again. However, they could be poured during the making of a casting if a determined effort was made to disturb the accumulated sludge from the bottom of the container. The resulting defects had a characteristic appearance of large, coarse crystals of the heavy intermetallic phases, together with entrained oxide films and associated porosity. These observations have been confirmed more recently by Cao and Campbell (2000) on other Al alloys.

Over subsequent years, there has been significant development of the ultrasonic technique for the study of bifilms in Al alloys and steels (Mountford and Sommerville, 1997 and 1993, respectively).

It is clear, therefore, from all that has been presented so far that a melt cannot be considered to be merely a liquid metal. In fact, the casting engineer must think of it as a slurry of various kinds of debris, mostly bifilms of various kinds, all containing entrained layers of air or other gases.

In a definitive piece of research into the fatigue of filtered and unfiltered Al-7Si-0.4Mg alloy by Nyahumwa, Green and Campbell (1998, 2000), test bars were cast by a bottom filling technique and were sectioned and examined by optical metallography. The filtered bars were relatively sound. However, for unfiltered castings, extensively tangled networks of oxide films were observed to be randomly distributed in almost all polished sections. Figure 2.5 shows an example of such a network of oxide films in which micropores (assumed to be residual air from the chaotic entrainment process) were frequently observed to be present. In the oxide film networks, it was observed that oxide film defects constitute cracks showing no bond developed across the oxide–oxide interface. In the higher magnification view of Figure 2.5, the width between the two dry surfaces of folded oxide film is seen to vary between 1 and 10 μm, in confirmation of the low-pressure test results described previously. However, widths of cracks in places approached 1 mm where the bifilm had opened sufficiently to be considered as a pore.

A polished section of a cast aluminium alloy breaking into a tangled bifilm is presented in Figure 2.6. The top half of the bifilm happens to meet the polished surface and has been polished away, revealing the glossy inside surface of the underlying remaining half of the bifilm, demonstrating the oxide film to be a *double* film, with little or no bonding between the two films.

The detachment of the top halves of bifilms to reveal the underlying half is a technique used to find bifilms by Huang et al. (2000). They subjected polished surfaces of aluminium alloy castings to ultrasonic vibration in a water bath. Parts of bifilms that were attached only weakly were fatigued off, revealing strips or clouds of glinting marks and patches when the polished surface was observed by reflected light. They found that increasing the Si content of the alloys reduced the lengths of the strips and the size of the clouds, but increased the number of marks. The addition of 0.5 and 1.0 Mg reduced the both the number and size of marks. Their fascinating polished sections of the portions of the bifilms that had detached revealed fragmentary remains of the double films of alumina apparently bonded together in extensive patches, appearing to be in the state of partially transforming to spinel (Figure 2.48).

The SEM has been a powerful tool that has revealed much detail of bifilms in recent years. One such example by Green (1995) is seen in Figure 2.11, revealing a film folded many times on the fracture surface of a Al-7Si-0.4Mg alloy casting. Its composition was confirmed by microanalysis to be alumina. The thickness of the thinnest part appeared to be close to 20 nm. It was so thin that despite its multiple folds the microstructure of the alloy remained clearly visible through the film.

There are varieties of bifilms in some castings that are clear for all to see. These occur in lost foam castings and are appropriately known as fold defects. Some of these are clearly pushed by dendrites into interdendritic spaces of the as-cast structure (Tschapp at el., 2000). The advance of the liquid metal into the foam is usually sufficiently slow that the films grow thick and the defects huge, and easily visible to the unaided eye. Other clear examples, but on a finer scale, are seen in high-pressure die castings. Ghomashchi (1995) has recorded that the solidified structure is quite different on either side of such features. For instance, the jets of metal that have formed the casting are each surrounded by oxides (their 'oxide flow tubes' as discussed in Section 2.2.6) seen in Figure 2.31. Between the various flowing jets, each bounded by its film, the boundaries naturally and necessarily come together as double films, or bifilms. They form effective barriers between different regions of the casting, permitting different regions to solidify at different rates, each region therefore having its own appropriate dendrite arm spacing.

As an 'opposite' or 'inverted' defect to a flow tube is the bubble trail. It constitutes a long bifilm of rather special form. The passage of air bubbles though aluminium alloy melts has been observed by video radiography (Divandari, 1998). The bubble trail caused by the flow path of the bubbles has been initially invisible on the video radiographic images. However, the prior solidification of the outer edges of the observed plate casting imposed a tensile stress on the interior of the plate that increased with time. After a few minutes, the level of stress in the central region of the plate had risen sufficiently to open the bifilm; it flashed into view in a fraction of a second, expanding as a long crack, following the path taken by the bubbles, carving a path through what had appeared previously to be featureless solidifying metal.

Evidence for bifilms in zinc-based alloys, copper-based alloys, grey iron, ductile iron, steels and high temperature nickel-based alloys will be presented in later portions of the book.

Finally, it is salutary to consider the potential for metals if bifilms were not present. Gold is one such example. It can be beaten or pounded into foil (known as gold leaf) thinner than that of any other metal. This is because it is uniquely free of oxides, its malleability is not impaired. Thus thicknesses of 0.1 μm are possible, which are so thin that it is possible to see through the metal, even though the golden lustre of the metal is not lost. This contrasts with metals such as aluminium where foils are generally in the range 200 μm down to 4 μm, but below about 25 μm the foil starts to show pinholes as a result of the presence of inclusions, particularly oxide bifilms. In highly specialised applications in ribbon microphones, Al foil might be as thin as about 1 μm, which is still 10 times thicker than the thinnest gold. If other face-centred cubic metals (aluminium, silver, copper etc.) could be made free from entrained oxides, foils of thickness similar to that of gold would, I assume, be easily produced. From Figure 2.51, lead could be predicted to become the world's most malleable metal if it were not for its content of bifilms.

The evidence for bifilms and the benefits of bifilm-free metals have been with us all for many years.

Ductile																						Brittle
Liquid	Pb	Au	Nb	Pt	Pd	Hf	Ag	Al	Cu	Zr	Ti	Ni	Co	Fe	Mg	Mo	Nd	W	Re	Ir	Cr	Be
μ/B																						
0	0.12	0.15	0.22	0.22	0.23	0.27	0.29	0.35	0.35	0.39	0.42	0.43	0.45	0.48	0.49	0.48	0.50	0.52	0.54	0.56	0.72	1.42
v																						
0.50	0.44	0.42	0.40	0.39	0.39	0.37	0.37	0.34	0.34	0.33	0.32	0.31	0.30	0.29	0.29	0.29	0.28	0.28	0.26	0.26	0.21	0.02

FIGURE 2.51

The engineering metals listed according to the ratio elastic/bulk modulus and Poisson ratio showing the range of ductile to brittle behaviour. From this figure, lead (Pb) might be expected to be the most malleable element, but is denied this place as a result of its bifilm content. Gold (Au) is the world's most malleable metal at this time.

2.8 THE IMPORTANCE OF BIFILMS

Although the whole of this work is given over to the concept of bifilms, so that, naturally, much experimental evidence is presented as a matter of course, this short section lists the compelling logic of the concept and the inescapable and important consequences.

Because the folded oxides and other films constitute cracks in the liquid, and are known to be of all sizes and shapes, they can become by far the largest defects in the final casting and can be dauntingly numerous. They can easily be envisaged as reaching from wall to wall of a casting, causing a leakage defect in a component required to be leak-tight, or causing a major structural weakness in a product requiring strength or fatigue resistance.

In addition to constituting defects in their own right, if they are given the right conditions during the cooling of the casting, the loosely encapsulated gas film can act as an excellent initiation site for the subsequent growth of gas bubbles, shrinkage cavities, hot tears, cold cracks etc. The nucleation and growth of such consequential damage will be considered in later sections. The important message to take on board at this stage is that without the presence of bifilms, such defects would probably be impossible. Liquid metals free from bifilms could make castings free from defects. This is a message never to be forgotten.

Entrainment creates bifilms that may never come together properly and so constitute air bubbles immediately; alternatively, they may be opened (to become thin cracks, or opened so far as to become bubbles) by a number of mechanisms:

1. Precipitation of gas from solution creating gas porosity;
2. Hydrostatic strain, creating shrinkage porosity;
3. Uniaxial (tensile) strain, creating hot tears or cold cracks;
4. In-service strain, causing failure by fracture in service.

Thus bifilms can be seen to simplify and rationalise the main features of the problems of castings. For those who wish to see the logic laid out formally this is done in Figure 2.51 for metals (1) without films, such as liquid gold, (2) with films that are liquid, (3) with films that are partially solid and (4) with films that are fully solid.

Note that the defects on the right of Figure 2.51 cannot, in general, be generated without starting from the bifilm defect on the left. The necessity for the bifilm initiator follows from the near impossibility of generating volume defects by other mechanisms in liquid metals, as will be discussed in later sections of this book. The classical approach using nucleation theory predicts that nucleation of any type (homogeneous or heterogeneous) is almost certainly impossible. Only surface-initiated defects (i.e. porosity or cracks growing in to the casting from locations on the casting surface) appear to be possible without the action of bifilms. In contrast to the difficulty of homogeneous or heterogeneous nucleation of defects, the initiation of defects by the simple mechanical action of the opening of bifilms requires nearly zero driving force; it is so easy that in all practical situations it is the only initiating mechanism to be expected.

We are therefore forced to the fascinating and enormously significant conclusion that in the absence of bifilms castings cannot generate defects that reduce strength or ductility. In other words, defects that lead to mechanical impairment are not produced by solidification, but only by casting. To restate once again for emphasis, in general, there is no such thing as a solidification defect. There are only casting defects.

As noted previously, this hugely important fact has to be tempered only very slightly because porosity, and possibly cracks, can also be generated easily by surface initiation if a moderate pressurisation of the interior of the casting is not provided by adequate feeders. However, of course, adequate feeding of the casting is widely understood and applied in foundries, and we can therefore assume its application here.

The wide application of good feeding, especially since the advent of good computer simulations, means that shrinkage porosity is *not* to be expected in castings. In fact, nearly all porosity I see in castings or on radiographs is usually *never* shrinkage, despite its appearance. It is usually a mess of bifilms and their associated tangled layers of air resulting from a poor filling technique. It is extremely important therefore for the reader never to utter the phrase 'It is shrinkage'. The reader is urged to learn by heart and repeat often the phrase 'It *appears* to be shrinkage' but then add the rider 'but probably is not!'

The author has the pleasant memories of the early days (c. 1980) of the development of the Cosworth Process, when the melt in the holding furnace had the benefit of days to settle, becoming clear from bifilms because production at that time did not occupy more than a few production shifts per week. The melt was therefore unusually free from bifilms, and the castings were found to be completely free from porosity. As the production rate increased during the early years the settling time was progressively reduced to only a few hours, causing a disappointing reappearance of microporosity, and a corresponding reduction of mechanical properties. This link between melt cleanness and freedom from porosity is well known. One of the first demonstrations of this fact was the simple and classic experiment by Brondyke and Hess (1964) that showed that filtered metal exhibited reduced porosity.

An important point to note is that a subsequently generated defect which may be large in extent, may be simply initiated by and grow from a small bifilm. On the other hand, the bifilm itself may be large, so that any consequential defect such as a pore or a hot tear actually is the bifilm, but simply opened up. In the latter case, no growth of area of the subsequent defect is involved, only separation of the two halves of the bifilm. Both situations seem possible in castings.

Standing back for a moment to view the larger scene of the commercial supply of castings, it is particularly sobering that there is a proliferation of standards and procedures throughout the world to control the observable defects such as gas porosity and shrinkage porosity in castings. Although once widely known as quality control the practice is now more accurately named quality assurance. However, as we have seen, the observable porosity and shrinkage defects are often negligible compared to the likely presence of bifilms, that are difficult, if not impossible, to detect with any degree of reliability. They are likely to be more numerous, more extensive in size, and have more serious consequences.

The significance of bifilms is clear, and worth repeating. They are often not detectable by normal non-destructive testing techniques, but can be more important than observable defects. They are often so numerous and/or so large that they can control the properties of castings, sometimes outweighing the effects of alloying and heat treatment.

The conclusion is inescapable: it is more important to specify and control the casting process to *avoid* the formation of bifilms than to employ apparently rigorous quality assurance procedures, searching retrospectively (and possibly without success) for any defects they may or may not have caused.

Castings free from bifilms probably do not exist at this time. However, in principle, technologies exist that would enable the production of products with significantly reduced bifilm contents, possibly actually bifilm-free. Such products would be expected to be significantly stronger, totally reliable, and ultimately lower in cost than those made by our current production techniques. The first steps are being demonstrated for a number of alloy systems as I write.

Furthermore, there are good reasons for believing that Griffith's cracks, universally blamed for the start of failures of all our engineering metals, may cease to exist if bifilms can be eliminated because it seems atomic-sized voids and cracks are not easily formed in metals because of the extremely high interatomic forces. There are good reasons for believing that every Griffith's crack originates from a bifilm, as I have argued in detail elsewhere (Campbell, 2011). No other volume defect seems available or possible. This conclusion would revolutionise the science of metallurgy and the attainments of engineering. It is an awesome and fascinating thought, and even now possibly within the grasp of currently available technology, to produce metals which are resistant to failure by cracking and fatigue, and resistant to failure by concentrated forms of corrosion such as pitting corrosion. Our new generation of casting engineers will probably have the pleasure and satisfaction to achieve this long-sought goal.

2.9 THE FOUR COMMON POPULATIONS OF BIFILMS

Summarising the efforts to obtain a clean melt, it is important to remember that there are commonly four sources of bifilms nearly always present and in general require to be dealt with separately. The following considerations are targeted mainly at the melting of aluminium alloys, but will be seen to apply analogously to other metal alloys.

1. The most serious crop of bifilms are what I call the primary oxide skins. They arise during melting because the charge materials are gradually submerged beneath the rising liquid metal. The liquid rolls up the sides of the chunks of charge, rolling up against the thick oxide already in place on the surface of the charge to form huge bifilms of the exact size and shape of the individual pieces of the charge. These massive bifilms can be the size of newspapers, floating about in the melt like plastic bags, with compositions of their oxides denoting their age. The dry hearth

furnace is invaluable in this respect since the 'primary oxide skins' are automatically separated from the melt by remaining on the dry hearth of the furnace, from where they can be scraped off at intervals. When melting in crucible furnaces these massive defects must be removed by other techniques such as, perhaps, rotary degassing; it is expected that the removal of primary oxide skins is probably the most valuable action of the rotary degassing treatment for liquid Al alloys. If not removed before attempting to pour a casting, and if a filter is in place in the filling system, it can be imagined that these huge and strong skins will block filters with great efficiency.

2. An unpredictable population of oxide bifilms in metals is that which is naturally inherited from the previous melting and casting operations which the metal has experienced. This population is often dense, but the oxides are often unpredictable size, in the range of centimetres down to micrometres, and probably characterised by a chemistry typical of 'older' oxides such as spinels.

3. The third source of oxide bifilms are those manufactured by turbulent processes during metal transfers such as the filling of the ladle and the filling of the mould. Those made in the channels of the filling system are likely to be limited in size as a result of the powerful churning action of the high velocity metal in these regions, probably shredding and tearing oxides many of which will be partially attached to the walls of the channels. These bifilms will be thin and 'young'.

4. Those bifilms made in the mould cavity itself, where the melt is starting to reduce in velocity, will not suffer such forces, and so will retain their large size, measured in centimetres, but not large in thickness as a result of their rapid submergence in the turbulence and splashing. These bifilms will also retain their pure chemistries associated with their 'youth'.

It is common to overlook the presence of the primary skins. Such oversights were unknowingly routine in the Al casting industry for many years, explaining the horrendous failures which were occasionally reported. The finer population of inherited oxide bifilms requires to be dealt with by sedimentation (possibly aided by heavy element addition) or, so far as possible, by filtration. The use of filters can only address bifilms, usually in their compact form, of size larger than the filter pore size, and even this can only take place at low velocities (filters in filling systems for castings are of practically no use for filtration because of the overwhelmingly high metal velocity). Also, of course, the bifilm is likely to unfurl from its compact form after passing through the filter, so as to maximise its area as a crack in the casting.

Having cleaned the melt from populations one and two so far as possible, only countergravity casting is usually completely reliable to avoid reintroducing bifilms as in two and four. However, gravity pouring can be improved significantly, especially if air is excluded by contact pouring, and if the geometry of the filling system corresponds to a good naturally pressurised design, delivering metal into the mould at or near 0.5 m/s.

FLOW

3

As the liquid metal flows into an empty mould cavity, its surface can exert considerable influence over the filling behaviour.

As a pure metal surface (for instance, if we were casting pure gold), the surface tension would be important when attempting to penetrate into small cavities or thin walls. This is the classical phenomenon of capillary repulsion (the opposite effect of capillary attraction that draws water up a glass tube occurs because water wets glass; but moulds are designed to be non-wetting by liquid metals).

If the metal reacts with its environment to create some kind of surface film, the action of the surface becomes even more dominant. Penetration of narrow sections and small holes is now even more problematic. Furthermore, the whole mode of advance of the liquid into the mould now has to change because of the mechanical strength of the surface film which is to some extent tending to hold back the advance.

The *unzipping wave* is the fascinating behaviour of the *vertical* advance of a liquid front only by *sideways* travel of *transverse waves*. Fortunately, it appears this type of advance is usually harmless to the casting, even though each horizontal traverse of the wave leaves a horizontal witness on the surface of the casting. It seems likely that all vertical advances of film-forming liquids occur by unzipping waves, although their action is not always obvious.

The elusive phenomenon of the *rolling wave* is examined and found only to occur when the underlying matrix gains strength by freezing. Fortunately therefore is it not a common mode of advance of the liquid front. If rolling can be induced to occur, the entrainment of the oxide creates a serious bifilm, otherwise known as a lap defect, and possibly in this case, more specifically, a cold lap defect.

Finally, as sections become very thin, any advance of the front becomes increasingly difficult. The small area of the surface oxide ensures, statistically, that few defects are contained, so the surface oxide starts to develop the high strength of a defect-free material. When the pressure becomes sufficiently high to burst the film, the miniscule jet that issues through the fracture in the film can appear as single thin, high speed stream. Alternatively, successive fractures under pressure can accumulate to become a forest of jets, each separated from the others by its own cylindrical oxide film. This is the curious and little-researched phenomenon of *microjetting* described in Section 2.2.7.

3.1 EFFECT OF SURFACE FILMS ON FILLING
3.1.1 EFFECTIVE SURFACE TENSION

When the surface of the liquid is covered with a film, especially a strong solid film, what has happened to the concept of surface tension?

It is true that when the surface is at rest the whole surface is covered by the film, and any tension applied to the surface will be borne by the surface film (not the surface of the liquid. Actually, there will be a small contribution towards the bearing of the tension in the surface by the effect of the interfacial tension between the liquid and the film, but this can probably be neglected for most practical purposes). This is a common situation for the melt when it is arrested by capillary repulsion at the entrance to a narrow section. Once stopped, the surface film will thicken, growing to be a mechanical barrier holding back the liquid.

The case of a stationary meniscus is commonly observed when multiple ingates are provided from a runner into a variety of sections, as in some designs of fluidity test. The melt fills the runner and is arrested at the entrances to the narrower sections, the main liquid supply diverting to fill the thicker sections that do not present any significant capillary repulsion. During this period, the melt is held stationary at the entrances to the thinner sections initially by the repulsive effect of surface tension, but with increasing time the meniscus develops a thick strong oxide layer that acts as a barrier to further progress. A short time later, when the heavier sections are finally filled, the melt in the runner pressurises. By this time the thin sections require an additional tension in their surfaces to overcome the tensile strength of the thickened oxide film before the metal can burst through. For this reason fluidity tests with multiple sections from a single runner are always found to give an effective surface tension typical of a stationary surface, being two or three times greater than the surface tension of the liquid. Results of such tests are described in Section 3.2.4.

Turning now to the dynamic situation where the front of the melt is moving, new surface is continuously being created as the old surface is pinned against the mould wall by friction, becoming the outer skin of the casting (as in an unzipping type of propagation as described below). The film on the advancing surface continuously splits, and is continuously replaced. Thus any tension in the surface of the melt will now be supported by a strong chain (the surface film) but with weak links (the fresh liquid metal) in series. The maximum stress in the surface is therefore limited to the strength of the weak link, the surface tension of the liquid, in this instance. The strong solid film merely rides as pieces of loose floating debris on the surface, as rather flat icebergs on a polar sea. Thus normal surface tension would be expected to apply in the case of a dynamically expanding surface, as applies for instance to the front of an advancing liquid.

During the turbulent filling of a casting, the dynamic surface tension is the one that is applicable because new casting surface is being created with great rapidity. It is clear that the critical velocities for liquid metals calculated using the dynamic surface tension actually agree accurately with experimental determinations, lending confidence to the use of surface tension of the liquid for expanding liquid surfaces.

3.1.2 THE ROLLING WAVE

Lap type defects are rather commonly observed on castings that have been filled slowly (Figure 2.25). It was expected therefore that a lap-type defect would be caused by the melt rolling over the horizontal, oxidised liquid surface, creating an extensive horizontal double film defect (Figure 3.1(b)). Interestingly, an experiment set up to investigate the effect (Evans et al., 1997) proved the expectation wrong.

As a background to the thinking behind this search, notice the difference between the target of the work and various similar defects. The authors were not looking for (1) a cold lap (the old name 'cold shut' is an unhelpful piece of jargon, and is not recommended) because no freezing had necessarily occurred. They were not searching for (2) a randomly incorporated film as generated by surface turbulence nor (3) a rolling backwards wave seen in runners, where the tumbling of the melt over a fast under-jet causes much turbulent entrainment of air and oxides.

This was a careful study of several aluminium alloys over a wide range of filling speeds. It seems conclusive that a rolling surface wave to cause an oxide lap does not exist in most situations of interest to the casting engineer. Although Loper and Newby (1994) do appear to claim that they observe a rolling wave in their experiments on steel the description of their work is not clear on this point. It does seem that they observed unzipping waves (see the following section). A repeat of this work would be useful.

The absence of the rolling wave at the melt surface of aluminium alloys is strong evidence that the kind of laps shown in Figures 2.25 and 3.1(b) must be cold laps. Rolling waves that form cold laps in aluminium alloy castings can probably only form when the metal surface has developed sufficient strength by solidification to support the weight of the wave. Whether this is a general rule for all cast metals is not yet clear. It does seem to be true for steels, and possibly aluminium alloys, continuously cast into direct chill moulds as described in the following section.

3.1.3 THE UNZIPPING WAVE

Continuing our review of the experiment by Evans (1997) to investigate surface waves using the mould design shown in Figure 3.1(a), the mould initially filled rather quickly but became progressively slower as the meniscus approached the top of the mould as is normal. However, at a level partway up the mould an unexpected discontinuous filling behaviour was recorded. Surprisingly, the upper surface of the liquid metal was observed to be horizontally level and stationary, and

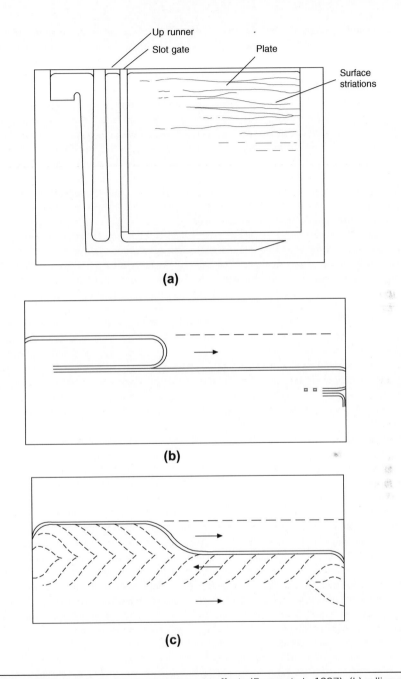

FIGURE 3.1

(a) Side-gated plate casting used to explore transverse wave effects (Evans et al., 1997); (b) rolling wave that may only occur on a partially frozen surface; and (c) an unzipping wave that may leave no internal defect, only a faint superficial witness on the casting surface.

its upward advance occurred not upward, but sideways in discrete steps. A transverse wave travelled swiftly along the length of the meniscus (Figure 3.1(c)), raising the level by the height of the wave, until reaching the most distant point. The speed of propagation of the wave was of the order of 100 mms^{-1}. On arrival at the end of the plate, the wave was observed to be reflected back. Waves coming and going appeared to cross without difficulty, simply adding their height as they passed.

What was unexpected was the character of the waves. Instead of breaking through and rolling over the top of surface, the wave broke through the oxide from underneath, and propagated by splitting the surface oxide as though opening a zip (Figure 3.1(c)). The splitting action occurred only locally at the point where the wave increased its height. Elsewhere, both ahead and behind the wave front, the melt surface was pinned in place by its oxide film, unable to advance vertically.

The propagation of these meniscus-unzipping waves was observed to be the origin of faint lines on the surface of the casting that indicated the level of the meniscus from time to time during the filling process. They probably occurred by the transverse wave causing the thickened oxide on the meniscus to be split, and subsequently displaced to lie flat against the surface of the casting. The overlapping and tangling of these striations appeared to be the result of the interference between waves and out-of-phase reflections of earlier waves.

The surface markings are, in general, quite clear to the unaided eye, but are too faint to be captured on a photograph. A general impression is given by the sketch in Figure 3.1. The first appearance of the striations seems to occur when the velocity of rise the advancing meniscus in liquid 99.8% purity Al falls to 60 ± 20 mm/s or less. At that speed, the advance of the liquid changes from being smooth and steady to the unstable discontinuous mode. Most of the surface of the melt is pinned in place by its surface oxide, and its vertical progress occurs only by the passage of successive horizontal transverse waves. At the wave front the surface oxide on the meniscus is split, being opened out and laid against the surface of the casting, where it is faintly visible as a transverse striation.

Since this early work, the author has seen the unzipping wave travelling in a constant direction around the circumference of a cylindrical feeder, the wave spiralling its way to the top. Even more recently, the surfaces of continuously cast cylindrical ingots of 300 mm diameter have been observed to be covered with spirals. Some of these are grouped, showing that there were several waves travelling helically around the circumference, leaving the trace of a 'multistart thread'. Clearly, different alloys produced different numbers of waves, indicating a different strength of the oxide film. The cylindrical geometry represents an ideal way of studying the character of the wave in different alloys. Such studies have yet to be carried out.

In the meantime, it is to be noted that there are a great many experimental and theoretical studies of the meniscus marks on steels. Particularly fascinating are the historical observations by Thornton (1956). He records the high luminosity surface oxide promoting a jerky motion to the meniscus, and the radiant heat of the melt causing the boiling of volatiles from the mould dressing, creating a wind that seemed to blow the oxide away from the mould interface. The oxide and the interfacial boiling are also noted by Loper and Newby (1994).

Much more work has been carried out recently on the surface ripples on continuously cast steels. However, in this case the continuously cast strand is withdrawn stepwise by pinch rolls, and the ripples are strongly correlated with the frequency and length of the withdrawal stroke. Here the surface striations are not merely superficial. They often take the form of cracks, and have to be removed by scalping the ingot before any working operations can be carried out. It seems that in at least some cases, as a result of the presence of the direct chill mould, the meniscus freezes, promoting a rolling wave, and a rolled-in double oxide crack. Thus the defect is a special kind of cold lap.

An example is presented from some microalloyed steels that are continuously cast. Cracking during the straightening of the cast strand has been observed by Mintz and co-workers (1991). The straightening process occurs in the temperature range 1000 down to 700°C, which coincides with a ductility minimum in laboratory hot tensile tests. The crack appears to initiate from the oscillation marks on the surface of the casting, and extends along the grain boundaries of the prior austenite to a distance of at least 5–8 mm beneath the surface. The entrainment event in this case is the rolling over of the (lightly oxidised) meniscus onto the heavily oxidised and probably partly solidified meniscus in contact with the mould. Evidence that entrainment occurs by a rolling wave is provided by the inclusion of traces of mould powder in the crack. The problem is most noticeable with microalloyed grades containing niobium. The fracture surfaces of laboratory samples of this material are found to be facetted by grain boundaries, and often contain mixtures of AlN, NbN and sulphides, highly suggestive of second phases formed on an oxide bifilm.

Other typical early researches are those by Tomono (1983) and Saucedo (1983). The problems of the solidifying meniscus are considered by Takeuchi and Brimacombe (1984). Later work is typified by that of Thomas (1995) who has considered the complexities introduced by the addition of flux powder (which, when molten, acts as a lubricant) to the mould, and the effects of the thermal and mechanical distortion of the solidified shell.

All this makes for considerable complication. However, the role of the non-metallic surface film on the metal being entrained to form a crack seems in general to have been overlooked as a potential key defect forming mechanism. In addition, the liquid steel melt in the mould cannot be seen under its cover of molten flux, so that any wave travelling around the inside of the mould, if present, is perfectly concealed. In addition, it must be remembered that continuously cast steels are usually actively withdrawn from the mould on a regular cycle, compared with continuously cast Al alloys which are normally withdrawn from the mould smoothly and continuously. Thus the spiral traces of the travelling waves, so clearly seen on Al ingots are not to be expected on steels, where the travelling waves, if present, will be forced to travel intermittently, along horizontal paths, leaving horizontal traces on the cast steel strand.

Finally, as a general feature of all vertical advance of melts in the apparent absence of any obvious unzipping phenomena, it is possible that all vertical advance of the meniscus may be the action of mini, ultrashort unzipping events, in which the oxide is split in some regions, while repairing in other, closely adjacent regions. Thus on a microscale the vertical advance may be 'patchy', or rather, may oscillate between closely adjacent regions. Such a concept avoids the obvious conceptual difficulties associated with the assumption that the mechanism involved in a rising meniscus is the simultaneous splitting and repairing of the surface film. Here is another question for which research would be illuminating.

3.2 MAXIMUM FLUIDITY (THE SCIENCE OF UNRESTRICTED FLOW)

Getting the liquid metal out of the crucible or melting furnace and into the mould is a critical step when making a casting: it is likely that most casting scrap arises during the few seconds of pouring of the casting.

The series of funnels, pipes and channels to guide the metal from the ladle into the mould constitutes our liquid metal plumbing, and is known as *the running system*. Its design is crucial; so crucial, that this important topic requires treatment at length, describing the practical aspects of making castings. Thus Chapters 10–13 are required reading for all casting engineers. As a result of learning these chapters by heart, the reader will know how to introduce the melt into the running system, so that the running system is now nicely primed, having excluded all the air, allowing the melt to arrive at the gate, ready to enter the mould cavity at a safe speed. Up to this point, everything has performed well.

The question now is, 'Will the metal fill the mould?'

There is always the concern that the filling system might freeze prematurely, starving the casting of metal. Alternatively, the melt may not reach all parts of the mould before it loses its heat and stops.

For these reasons, as has been mentioned previously, immediately after the pouring of a new casting, everyone assembles around the mould to see the mould opened for the first time. The casting engineer that designed the filling system, and the pourer, are both present, about to have their expertise subjected to the ultimate acid test, while colleagues, some sceptical, some hopeful, all look on. There is question asked every time, reflecting the general hush of concern amid the foundry din, and asked for the benefit of defusing any high expectations and preparing for the worst: 'Is it all there?'

This is the aspect of flow dealt with in this section. The ability of the metal to flow is influenced by many factors, including films, both those on the surface and those entrained, and by the rate of heat flow and the metallurgy of solidification. In different ways all these factors limit the distance to which the metal can progress without freezing. We shall examine them in turn.

The ability of the metal to run some distance to fill a mould is known as its *fluidity*. We shall see how pure metals and eutectics have high fluidity because of their smooth freezing fronts, allowing the melt to slip past an advancing freezing front with minimum hindrance, and so maximising fluidity.

Conversely, long freezing range alloys have only a fraction, perhaps a half or a quarter of the fluidity of their pure metal components because of the friction to flow provided by the solidification front, now no longer smooth, but dendritic, and extending up to twice the distance into the flow.

Oxides in suspension in the melt can partially or completely block the entrance to thin sections, thus directly reducing flow in the thin section. However, in general, the role of oxides is more complicated. For instance in Al-Si alloys oxides will act as substrates for the growth of Si particles and beta-iron particles. These particles grow as rigid plate-like solids which, naturally, greatly interfere with the flow, reducing fluidity. When Sr (or Na) is added to inhibit precipitation of these pre-eutectic phases, the solidification now occurs at a lower temperature with a smooth eutectic front, greatly increasing fluidity. However, the oxide bifilms, now redundant as favoured substrates, float aimlessly in the melt possibly raising the viscosity a little, but the effect is probably negligible because the bifilms are so flimsy and ravelled by the turbulence of the flow, so posing little restriction to the liquid. However, at a later stage of freezing, the bifilms can now impair interdendritic flow by bridging between dendrite meshes to form impenetrable barriers (analogous to the blocking of flow into thin sections), while inflating to initiate porosity. Thus interdendritic porosity becomes a danger. This complicated scenario is described in more detail in Section 6.3.5. It seems likely that the role of oxides has contributed (probably together with variations in test pieces, and the effect of poor filling system designs) to the wide scatter in experimental results that is evident historically.

The provision of unstable foundations for the growth of dendrites allows them to break off, and tumble along with the flow, thereby reducing the overall impediment to flow, as when acetylene smoke is applied to the mould (to be described later in Section 3.2.6) or possibly when vibration is applied to the mould.

Clearly, anything that impedes flow will reduce fluidity (it seems necessary to state the obvious from time to time). Thus one can see that the concept of *good fluidity* is seen to be simply the attainment of *unimpeded flow*. The more impediments we can remove, and the smoother the flow channel that can be provided, the greater the fluidity distance.

Ultimately, therefore, careful application of casting science should allow us now to know that not only will the casting be all there, but it will be all right.

FLUIDITY DEFINITION

The ability of the molten metal to continue to flow while it continues to lose temperature and even while it is starting to solidify is a valuable feature of the casting process. There has been much research into this property, with the result that a quantitative concept has evolved, which has been called '*fluidity*'. In terms of casting alloys, the fluidity is defined as the *maximum* distance to which the metal will flow in a standard mould. Thus fluidity is simply a length denoted here as L_f, measured, for instance, in millimetres. (The use of the foundry term *fluidity* should not be confused with its use in physics, where fluidity is defined as the reciprocal of viscosity.)

A second valuable quantitative concept that has often been overlooked was introduced by Feliu (1962). It is the parameter L_c that Feliu called the critical length. Here, however, we shall use the name *continuous-fluidity length*, to emphasise its relation to L_f, to which we could similarly give the full descriptive name of *maximum-fluidity length* (although, in common with general usage, we shall keep the name 'fluidity' for short!). In this section, we shall confine our attention to L_f, dealing with the equally important L_c in a following section.

Although fluidity (actually maximum fluidity, remember!) has therefore been measured as the (maximum) length to which the metal will flow in a long horizontal channel, this type of mould is inconveniently long for regular use in the foundry. If the channel is wound into a spiral, then the mould becomes compact and convenient, and less sensitive to levelling errors. Small pips on the pattern at regular spacings of approximately 50 mm along the centreline of the channel assist in the measurement of length. Figure 3.2 shows a typical fluidity spiral with a basin provided with a channel to a cup-shaped overflow basin to ensure a constant head height during the pouring process. Also shown is a horizontal channel that has been used for laboratory investigations into the fluidity of metals in narrow glass tubes. In this case, the transparent mould allows the progress of the flow to be observed from start to finish.

There has been much work carried out over decades on the fluidity of a variety of casting alloys in various types of fluidity test moulds, mostly of the spiral type. The wide variety of mould types and the sensitivity of fluidity to numerous variables have prevented fluidity from achieving the status of an internationally accepted material parameter. This is not too much of a disadvantage because most fluidity work is at least internally consistent. Fundamental insights have been gained mainly from the great bulk of work carried out at Massachusetts Institute of Technology under the direction of Merton C. Flemings. He has given useful reviews of his work several times, notably in 1964, 1974 and 1987.

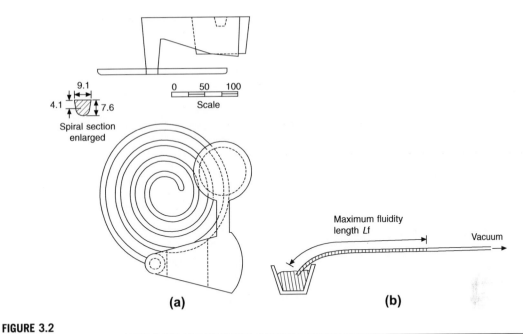

FIGURE 3.2

Typical fluidity tests for (a) foundry and (b) laboratory use.

The work of the many different experimenters in this field is difficult to review in any comprehensive way because almost each new worker has in turn introduced some new variant of the spiral test. A noteworthy effort to reduce problems so far as possible is seen in the test by Sabatino and others (2005). Other workers have gone further, introducing completely new tests usually with the emphasis on the casting of thin sections.

Here we attempt to review the data once again, with the aim on this occasion of emphasising the unity of the subject because of the basic, common underlying concepts. The various fluidity tests of the spiral, tube, or strip types are shown not to be in opposition. On the contrary, if proper allowance is made for the surface tension and modulus effects, these very different tests are shown to give exactly equivalent information.

It will become clear that the fluidity of metals is mainly controlled by the effects of fluid mechanics, solidification, and surface tension. This is a reassuring demonstration of good science.

3.2.1 MODE OF SOLIDIFICATION

Flemings (1974) demonstrated that the fluidity of pure metals and eutectics, which freeze at a single temperature, is different to that of alloys that freeze over a range of temperatures. These two different solidification types we shall call skin freezing and pasty freezing for short.

For a skin-freezing material the mode of solidification of the stream in a fluidity test appears to be, as one might expect, by planar front solidification from the walls of the mould towards the centre (Figure 3.3(a)). As the liquid metal travels down the channel, gradually cooling as it progresses, at the point at which the metal has finally lost its initial superheat (the excess temperature above its melting point) freezing first starts. The solidified region actually migrates downstream somewhat as the leading edge of the material is re-melted by the incoming hot metal, and re-freezing occurs further downstream. It is clear that the stream can continue to flow until the moment at which the freezing fronts meet, closing the flow channel. Note that this choking of flow happens far back from the flow front. In addition, solidification

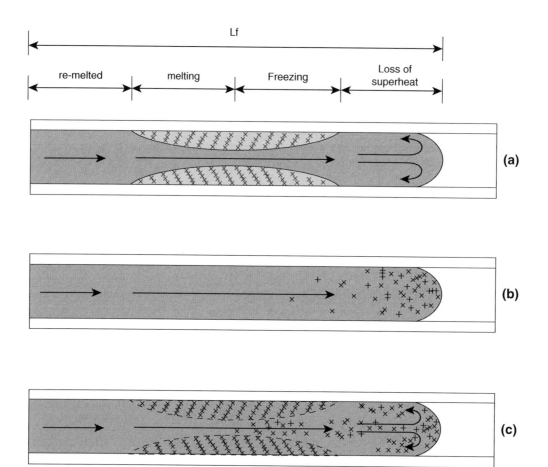

FIGURE 3.3

(a) Flow arrest in pure metals and coupled eutectics by complete solidification *(after Matsuda and Ohmi, 1981)*; (b) early model for long freezing range alloys; (c) new model proposed here.

needs to be 100% complete at this location for flow to stop. Assuming the liquid has an approximately constant velocity V, then if its freezing time in that section is t_f, we have the flow distance

$$L_f = V \cdot t_f \qquad (3.1)$$

This is an approximate equation first proposed by Flemings and co-workers (1963). It might be expected to over-estimate L_f because the velocity does reduce somewhat as the channel becomes constricted. Nevertheless Eqn (3.1) is surprisingly accurate, probably as a result of the acceleration of freezing as the channel closes, as described in Section 5.1 (Figure 5.10). We shall assume the applicability of Eqn (3.1) here.

The pattern of solidification of short-freezing-range material can be confirmed from the evidence of the microstructure of cross-sections of the castings. Columnar crystals are seen to grow from the sides of the mould, angled inwards against the flow, and meeting in the middle of the section. Confirmation of the presence of liquid cut off by the meeting of the solidification fronts is clear from the shrinkage cavity at the tip of the casting; a pipe (or collapsed exterior) forms in the unpressurised region near the liquid tip. An internal pipe (or a general collapse of the casting) usually grows back nearly to the point at which the flow was blocked. Thus the mechanism of arrest of flow for skin freezing materials is known with some certainty and in some detail.

The same situation does not apply to long freezing range alloys.

The pattern of solidification of longer-freezing-range materials is not so easily interpreted by metallography of the cast test piece. It turns out to be quite different to that of the skin-freezing material described previously. Clearly, the dendrites growing from the mould wall at an early stage of freezing are somehow fragmented by the flow so that the stream develops as a slurry of tumbling dendritic fragments. These equiaxed dendrites are carried in the central flow to congregate at the liquid front (Figure 3.3(b)). It has been assumed that when the amount of solid in suspension exceeds a critical percentage, the dendrites start to interlock, making the mixture unflowable (Flemings et al., 1961). This theory gives some cause for concern because the critical concentration to stop flow appears to depend on the applied head of metal and varies over the wide range 20–50%. The arrest of flow occurs therefore when there is only approximately 20–50% solidification, thus reducing fluidity by this corresponding amount. Thus there is wide potential inaccuracy in the predictions that this theory is unable to explain. Furthermore, the concept of a clutter of equiaxed grains at the tip of the flow overlooks the fact that such grains usually originate from the fragmentation of dendrites (although it is just possible that the effect might occur if effective grain refinement were carried out). Because dendrites usually grow into the flow from the mould surface, the presence of stationary features in the flow would be expected to be much more of a hindrance to flow than grains freely tumbling along.

A possibly more accurate arrest mechanism is therefore proposed in Figure 3.3(c). Here the freezing front proceeds to advance towards the centre of flow, outlined by the broken line. However, of course, the freezing region is now not solid, but is a permeable array of dendrites. This permeable array is a fixed hazard to traffic. If flow were steady, the dendrites would angle into the flow as they grew, as elegantly researched by Turchin (2007a), but would continue to grow stably and steadily. However, flow is rarely stable. Hot and cold vortices will swirl through the mesh, together perhaps with the devastating mechanical disruption caused by occasional bubbles (to be expected in most fluidity tests). The dendrite arms will therefore melt off during pulses of heat and shear forces, releasing fragments into the stream. (As an aside, the breakup of dendrites at only low velocities in the region of 0.2 m/s seen by Turchin (2007b) is clearly the result of the extreme turbulence developed by his displacement of the experimentally observed volume into a recess, instead of being aligned with the streamlines of the flow.)

In a more conventional fluidity test channel, the fragments would be carried forward by the flow as illustrated in Figure 3.3(c), concentrating near the front of the stream where they are particularly visible. Their concentration in this region has been previously assumed to arrest flow. Much more probably the flow is slowed by the viscous drag in the fixed mesh of dendrites far back from the flow front, although the equiaxed grains accumulating near the flow front may make some minor contribution. Some discriminating research would clarify this, probably revealing that the answer is different for different alloys.

Conditions in the permeable dendrite mesh are likely to be complicated, if not chaotic. It is possible that the restraining mesh might partially collapse, consolidating the dendrites and impeding flow even further. Free-floating crystals might jam in the mesh. Oxide films would wrap around dendrites or might bridge and thereby block whole sections of the mesh, thus causing much more restriction than their volume fraction might suggest. Thus for pasty freezing systems, the prior turbulence in the filling system, creating areas of oxide bifilms that would help to block the mesh, would be expected to have a disproportionately high effect. Such complications might help to explain the significant differences between different fluidity research results. Overall, the development of the mesh and its disintegration into fragments may not be a particularly reproducible phenomenon.

Whatever the detailed mechanism for arresting flow, it follows that for long freezing range alloys flow stops when much less total solid has formed. Equation (3.1) therefore becomes

$$L_f = 0.2V \cdot t_f \quad to \quad 0.5V \cdot t_f \tag{3.2}$$

Equation (3.2) reflects a feature that appears to be common to all alloy systems, and yet does not seem to have been widely recognised; long freezing range alloys have only one fifth to one half of the fluidity enjoyed by skin freezing materials. These are big effects. The precise values should be predictable from the geometry of the dendrite mesh, but this also has yet to be researched. As an instance, Figure 3.4 illustrates the profound effect of a small amount of Sn on pure Al, causing the excellent fluidity of pure Al to fall by a factor between 3 and 4.

A similar effect is seen in Figure 3.5 illustrating the lead/tin system at 50°C above its liquidus. The fluidity of the intermediate alloys is dramatically different from the straight line interpolation of the fluidity of the two pure elements. The curve joining the lower data points is based on the fluidity of the long freezing range alloys being approximately one fifth of the short freezing range alloys when lead-rich, from the available data points. There are no equivalent data points on the tin-rich side of the diagram, so here we have guessed that the longer freezing range alloys might have only approximately half fluidity, because the freezing range is narrower, and the one data point is consistent with this. More experimental points would have removed this uncertainty.

The fluidity of the eutectic in Figure 3.5 appears to be more than 50% higher than the straightforward method of mixtures of its components (i.e. the straight line joining the fluidities of the two pure metals). This may be understandable if the pure metals may be exhibiting some dendritic growth, probably because of the presence of some impurity. This would suppress the fluidity of the pure constituents, but not necessarily the eutectic (especially if the main 'impurities' in each of the pure constituents consisted of the other 'pure constituent', i.e. if Pb were contaminated with Sn, and Sn was contaminated with Pb).

The Al-Zn system is more in line with expectations: the eutectic fluidity is very close to the straight line interpolation of the pure constituents (Figure 3.6), but the constant temperature result gives the interesting appearance of enhanced fluidity as predicted later in this section. Exasperatingly, the widely used commercial pressure die casting alloy Zn-4%Al alloy resides almost in the centre of the fluidity minimum. What is worse, Friebel and Roe (1963) find that the fluidity falls further as a result of the level of 3%Cu in the alloy, normally present to add strength. It seems that the efforts of

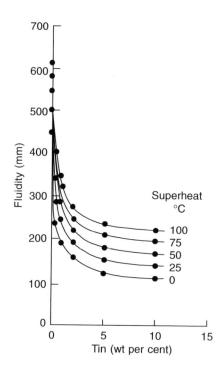

FIGURE 3.4

Variation of fluidity with composition of Al-Sn alloys.

Data from Feliu et al. (1960).

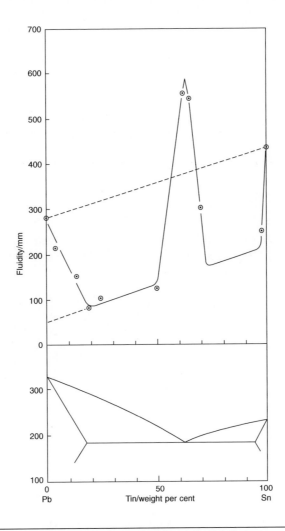

FIGURE 3.5

Fluidity of Pb-Sn alloys at 50°C superheat determined in the glass tube fluidity test.

Data from Ragone et al. (1956).

metallurgists to optimise alloys have paid minimal attention to the potentially huge benefits to castability from more fluid compositions, allowing thinner and lighter castings to be made, currently the one of the biggest disadvantages currently faced by the zinc pressure die casting industry.

Impressively early work by Portevin and Bastien (1934) suggests the shape of precipitating solid crystals affect the flow of the remaining liquid; smooth crystals of solidifying intermetallic compounds creating less friction than dendritic crystals. Their famous result for the Sb-Cd system poured at a constant superheat of 100°C is shown in Figure 3.7 appears to illustrate this convincingly. The greater fluidity of intermetallic compounds and eutectics with respect to their pure constituent elements seems to be true for this system in agreement with the Pb-Sn result but not with the Al-Zn fluidity. Perhaps more careful work with purer, bifilm-free constituents might help to resolve the question whether eutectics do have some special advantage over their constituent elements.

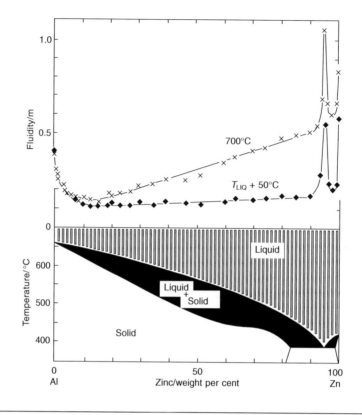

FIGURE 3.6

Al-Zn alloys poured into a straight cast iron channel showing fluidity at constant superheat, and the enhanced fluidity of the eutectic when cast at constant temperature (Lang, 1972).

Incidentally, the details of this classic work are rarely questioned, but deserve close attention. For instance, the peritectic reaction shown in Figure 3.7 at approximately 63%Cd is found not to exist in later versions of this binary phase diagram (Brandes and Brook, 1992). Thus the small step in fluidity, carefully depicted at this point, seems likely to be attributable to experimental error. Furthermore, at the Sb-rich end of the figure, the fall in fluidity will probably be steeper than that shown, so that the plateau minimum will be reached sooner than the limit of solid solution as assumed by the authors. This will be as a result of non-equilibrium freezing. More importantly, given the existence of some modest scatter in the results, it seems likely that the high fluidity peak is actually not centred on the compound SbCd but is generated by the adjacent eutectic. However, these are minor quibbles of an otherwise monumental and enduring piece of work, years ahead of its time.

It needs to be emphasised that the fluidity of a complete alloy system at constant temperature will always be expected to favour the eutectic fluidity; their effectively higher superheat enhances the advantage to eutectics, although reduces the advantage to intermetallic compounds. This effect is seen later in Figures 3.8 and 3.9.

We can usefully generalise these important effects of composition in the following way. Figure 3.8 shows a schematic illustration of a simple binary eutectic system. The fluidity at zero superheat is linear with composition for the skin freezing metals and alloys (the pure metal A, the AB eutectic and the pure metal B), whereas it is assumed to be about half of this for the long freezing alloys. The resulting relationship of fluidity with composition is therefore seen to be the cusped line in figure (i). This would be the expected form of the fluidity/composition relation for a constant superheat.

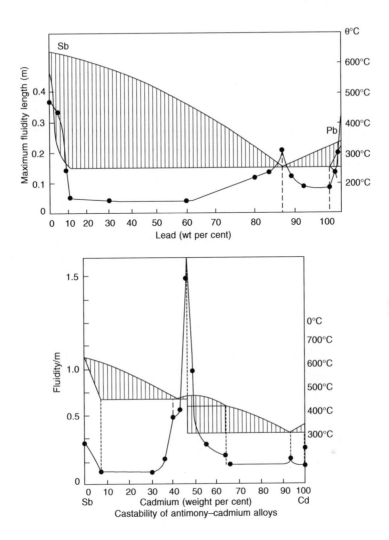

FIGURE 3.7

Fluidity of Sb-Pb and Sb-Cd alloys showing the high fluidity of the eutectics determined by casting at a constant superheat into a cast iron fluidity spiral (Portevin and Castien, 1934).

If, however, the alloys are cast at a constant temperature, T_c, there is an additional contribution to fluidity from the superheat that now varies with composition as indicated in Figure 3.8 (ii). To find the total fluidity as a function of composition, curves (i) and (ii) are added (The addition in these simplified illustrative examples neglects any differences of scale between the two contributions). The effect seen in (iii) is to greatly enhance the fluidity at the eutectic composition.

The same exercise has been carried out for a high melting point intermetallic compound, AB, in Figure 3.9. The disappointing total fluidity predicted for the intermetallic (summing (i) and (ii)) explains some of the problems that are found in practice with attempts to make shaped castings in such alloys. It explains why intermetallic compounds are rarely used as natural casting alloys, compared to eutectics and near-eutectics, which are common.

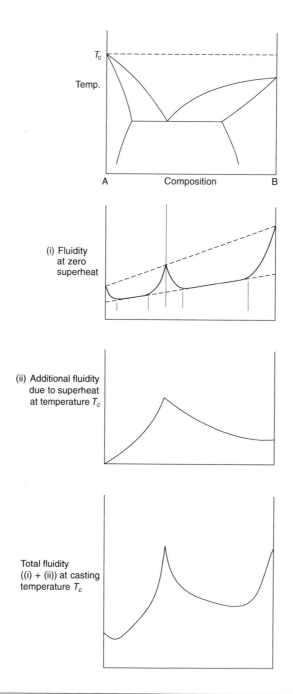

FIGURE 3.8

Schematic illustration of the different behaviours of eutectics when tested at constant superheat or constant temperature.

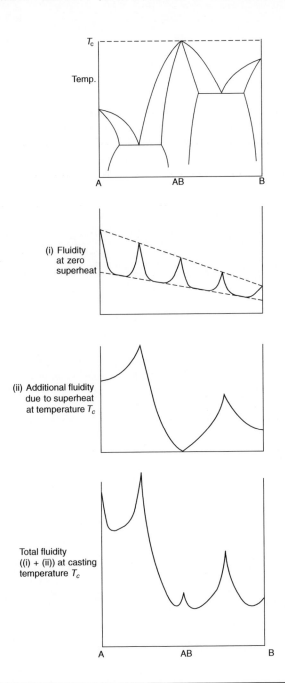

FIGURE 3.9

Schematic illustration of the different behaviours of high melting point intermetallics when tested at constant superheat or constant temperature.

The expected poor fluidity of intermetallics underlines the suggestion by Portevin and Bastien that the high fluidity of intermetallics that they observe, although at constant superheat as seen in Figure 3.7, is nevertheless in fact strongly influenced by some other factor, such as the shape of the solidifying crystals. Alternatively, as suggested previously, they have mistakenly attributed the high fluidity to the intermetallic, whereas perhaps it was really the result of the adjacent eutectic.

Another interesting and important lesson is to be gained from the predictions illustrated in Figures 3.8 and 3.9. The peaks in fluidity are impressively narrow. Thus for certain eutectic alloys the fluidity is awesomely sensitive to small changes to composition, particularly when it is realised that Figures 3.8 and 3.9 are simple relationships. Many alloy systems are far more complicated, crowding the cusps of fluidity closer and so making their slopes steeper. Dramatic changes of fluidity performance are to be expected therefore with only minute changes in composition. Perhaps this is the reason why fluidity research earns its reputation for being not reproducible. Perhaps it is also the reason why some castings can sometimes be filled, and at other times not; some batches of alloy cast well, but some nominally identical alloys cast poorly. Clearly, to maximise reproducibility we must improve the targeting of peak performance. The unattractive alternative is to forsake the peaks and devise more easily reproducible alloys in the troughs of mediocre fluidity. A good compromise, where possible, would be to target the broader peaks, where the penalties of missing the peak are less severe.

The approach illustrated in Figures 3.8 and 3.9 can be used on more complex ternary alloy systems. In more complex multi-component alloy systems the understanding that pure metals and eutectics exhibit a fluidity at least twice that of their long freezing range intermediate alloys is a key allowing even sparse and apparently scattered data on alloy systems to be rationalised. This was the background of Figure 3.10 (a, b and c) that showed results for the Al-Cu-Si system that could not be drawn by its investigators, but was only unravelled in a subsequent independent effort (Campbell, 1991). However, I learned only later, while my 1991 paper was in the process of publication, that Portevin and Bastien had already accomplished a similar evaluation of the Sn-Pb-Bi ternary system as early as 1934. In this prophetic work, they had constructed a three-dimensional wax model of their fluidity response as a function of composition, showing ridges, peaks and valleys.

In practice, it is important to realise that the improved fluidity of short- versus long-freezing-range materials forms the basis of much foundry technology.

The enormous use of cast iron compared to cast steel is in part a reflection of the fact that cast iron is a eutectic or near-eutectic alloy and so has excellent fluidity. Steel, in general, has a longer freezing range and relatively poor fluidity. Additionally, of course, the higher casting temperatures of steel greatly increase the practical problems of melting and casting, and cause the liquid metal to lose its heat at a faster rate than iron, further reducing its fluidity by lowering its effective fluid life t_f. For instance, Figure 3.11(a) shows that a 1%C steel has zero fluidity at approximately 1450°C whereas a cast iron will run like water over a distance of 1.3 m at this temperature. Alternatively, the figure illustrates that if a fluidity of 500 mm is required to fill a mould, a grey iron would achieve this at a temperature lower than 1300°C whereas a 0.5%C steel would require 1600°C.

The work by Porter and Rosenthal (1952) nicely illustrates the peak of fluidity at the eutectic composition, approximately 4.5 wt% carbon equivalent (Figure 3.12). As in many fluidity studies, this work is beautifully self-consistent, probably as a result of carefully carried out experiments with their own particular variety of fluidity test mould and own particular set of compositions of their iron alloys. The results therefore do not fit particularly well with the earlier results presented in Figure 3.11, but this is probably as good agreement as one can expect. The essential features, including the high peak at the eutectic, are supported by all studies.

Fluidity can be affected by changes in composition of the alloy in other ways. For instance, the effect of phosphorus additions to grey iron are well known: the wonderful artistic castings, of statues, fountains, railing and gates produced in the nineteenth and early twentieth century Europe were made in high-phosphorus iron because of its excellent fluidity. The effect is quantified in Figure 3.13(a). The powerful effect of phosphorus on cast iron is solely the result of its action to reduce its freezing point. This is proven by Evans (1951), who found that when plotted as a function of superheat (the casting temperature minus the liquidus), the phosphorus addition hardly affects the fluidity (Figure 3.13(b)).

Flemings has taken up this point, suggesting that the good fluidity of cast irons compared with steels is only a function of the higher superheat which can be used for cast iron. However, this is only part of the truth. Figure 3.11 shows

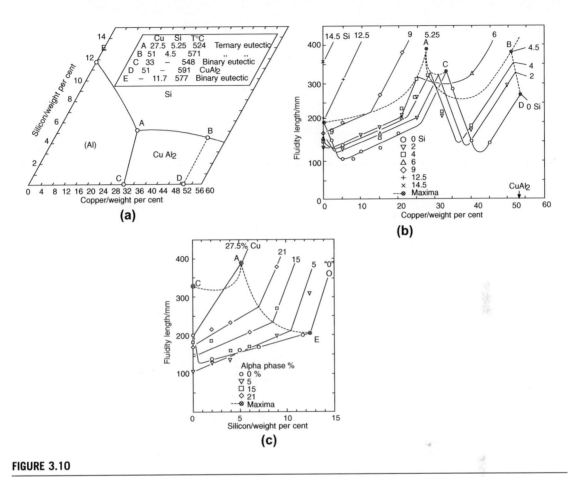

FIGURE 3.10

Phase diagram and fluidity of the Al-Si-Cu system.

Data from Garbellini et al. (1990); interpretation by JC (1991).

data re-plotted from early work by Evans (1951) and Porter and Rosenthal (1952) on grey iron and Andrew (1936) on steels. The reasonably linear plots of fluidity against casting temperature (Figure 3.11(a)) are important and interesting in themselves. However, they can be redrawn as in Figure 3.11(b). Here the horizontal flat portions of the curves show that the long-freezing-range alloys have constant fluidity at a given superheat, in agreement with Flemings and confirming the similar effect of phosphorus that we have already witnessed; any increase in fluidity is only the result of increases in superheat as the composition changes. However, as either the pure metal, Fe, or the eutectic at 4.3% carbon, is approached there is clear evidence of enhanced flowability, showing that there is an additional effect at work here, almost certainly relating to the mode of solidification as we have discussed previously. This effect was found as long ago as 1932 by Berger. Other binary alloy systems exhibit a similar special enhancement of fluidity at the eutectic as illustrated in Figures 3.5–3.7. The effect is also seen in a ternary system in Figure 3.10.

Interestingly, the effect of freezing range on fluidity is not confined to metals. Bastien et al. (1962) have shown that the effect is also clearly present in molten nitrate mixtures and in mixtures of organic compounds. It is to be expected that the effect will be significant in other solidifying systems such as water-based solutions, and molten ceramics, etc.

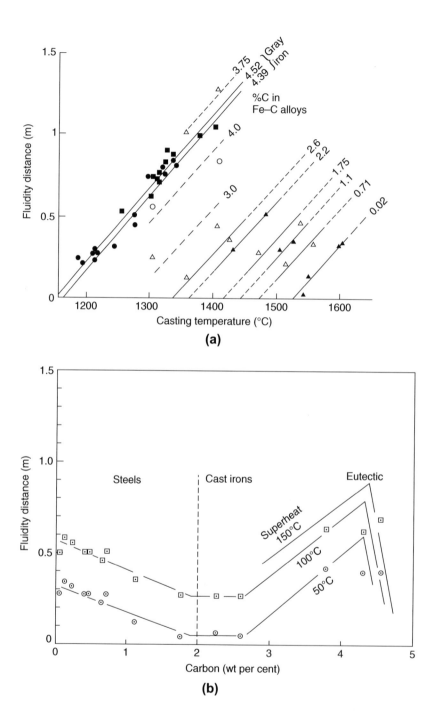

FIGURE 3.11

(a) Effect of temperature and carbon content on the fluidity of Fe-C alloys; (b) the same data plotted as a function of composition.

Data from Berger (1932), Andrew et al. (1936), Evans (1951), Porter and Rosenthal (1952).

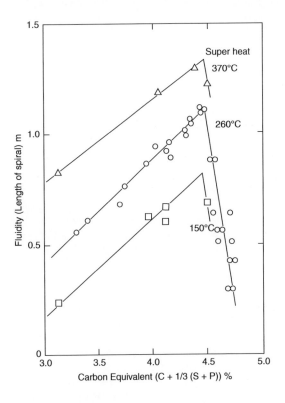

FIGURE 3.12

The excellent fluidity results for grey iron by Porter and Rosenthal (1952).

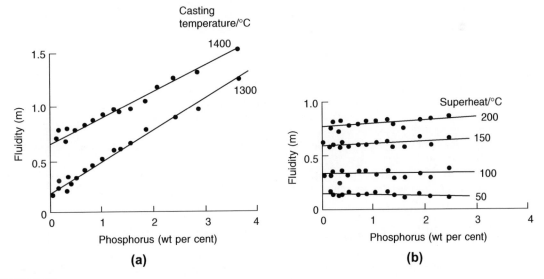

FIGURE 3.13

Effect of phosphorus on the fluidity of grey iron plotted as a function of (a) phosphorus addition; and (b) superheat above the liquidus (Evans, 1951).

3.2.2 EFFECT OF VELOCITY

The velocity is explicit in the Eqns (3.1) and (3.2) for fluidity. It is all the more surprising therefore that it has not been the subject of greater attention. However, most workers in the field have used fluidity tests with relatively small head heights, usually not exceeding 100 mm, so that velocities during test are usually fixed, and rather modest, under 2 ms^{-1}.

It might seem reasonable to expect that fluidity would increase linearly with increase in velocity, particularly in view of the forecasts of Eqns (3.1) and (3.2). However, with regret, this appears to be a mistake. It is this erroneous belief that causes many casting engineers to assume that to fill thin-section casting the melt has to be thrown into the mould at maximum speed. These attempts to improve filling usually fail to produce good castings because of the complicating effects of bulk and surface turbulence.

For instance, as head height h of a casting is increased in a filling system free from friction effects, we might expect that because $V = (2gh)^{1/2}$, fluidity would increase proportionally to $h^{0.5}$. However, the resistance to flow from turbulence in the bulk of the liquid rise according to $(1/2)\rho V^2$. Thus losses rise rapidly with increased velocity V, causing it to be increasingly difficult to create high velocities, particularly in narrow channels. In fact Tiryakioglu and co-authors (1993) find experimentally that fluidity of 355 aluminium alloy rises at a much lower rate which seems proportional to only $h^{0.1}$. This is almost certainly the result of the rapidly increasing drag effects on the motion of the liquid as velocity increases. Thus attempts to increase fluidity by increasing pressurisation and/or velocity are subject to the law of rapidly diminishing returns.

Furthermore, at high speeds, it becomes increasingly difficult to keep surface turbulence under control. Thus the jetting and splashing of the flow entrains air, and creates oxide laps. The oxide laps occur because of the disintegration of the flow front. Splashes or surges of melt ahead of the general liquid front have to await the arrival of the main flow in order to be re-assimilated into the casting. During this time their surface oxidises, so that re-assimilation is difficult, or impossible, resulting in an oxide lap. If the arrival of the main flow takes longer, the awaiting material might freeze. Thus the splash and its droplets might become a cold lap. On seeing these problems, the reaction of some casting engineers is to increase the filling speed, with predictably lamentable results.

In contrast, at more modest speeds, if the melt is introduced so that the liquid front remains smooth and intact, the air is not mixed in, but is pushed ahead, and the front has no isolated fragments, the very integrity of the front acts to keep itself warm. Both oxide laps and cold laps are avoided because the front probably advances by harmless unzipping perturbations. The casting is as sound as the filling system can make it.

In confirmation of these problems, Zadeh and the author (2001) have demonstrated how the fall of melts down sprues 500 mm high have resulted in good fluidity lengths, with apparently good castings judging by their external appearance. We thought we had success until we viewed the test pieces by X-ray radiography. To our dismay, the fluidity strip castings were found to be full of entrained bubbles and films. They would have been totally unsaleable as commercial products. It was necessary to introduce ceramic filters with an integral bubble trap to regain control over the integrity of the flow to give good quality castings. Having done this, the fluidity distance was then no better than if the melt had been poured from 100 mm. I have never forgotten the lesson of this experience.

Thus, in general, up to small head heights of the order of 100 mm fluidity will be expected to rise linearly with increase in speed of flow. However, further increases in speed with current filling system designs appear to be counter-productive if castings relatively free from defects are required.

In summary of the effects of velocity, therefore, it is essential to fill the mould without surface turbulence for the majority of alloy systems. The importance of this fact cannot be overstated. There is no substitute for a good filling system that will give a controlled advance of the liquid front. Essentially, this means it is counter-productive to attempt to increase the velocity of filling for the production of sound castings.

3.2.3 EFFECT OF VISCOSITY (INCLUDING EFFECT OF ENTRAINED BIFILMS)

Loper (1992) made the point that the variations of viscosity of liquid metals with temperature and with composition are relatively minor, and not capable of explaining the variations of fluidity seen across alloy systems. Furthermore, the viscosity of the melt does not even appear in the fundamental equations for fluidity, Eqns (3.1) and (3.2). Thus the theoretical prediction is that fluidity should not depend on viscosity. This surprising prediction has been shown to be remarkably accurate.

It is instructive to compare the fluidity observed for semi-solid (strictly 'partly solid') and metal-matrix composite (MMC) cast materials. For concentrations Φ of solids up to about 20% (0.2 fraction), the formula by Einstein dating from 1906 and in its modified form from 1911, for the viscosity η of a mixture of liquid plus solid, is sufficiently accurate for our purpose

$$\eta = (1 + 0.5\Phi)/(1 - \Phi)^2$$

A mix with 20% solid would be expected to have a viscosity about double that of the pure liquid. This is the concentration level of many MMCs. However, their fluidity as measured by the spiral or other fluidity test is hardly impaired. In fact, quite extensive and relatively thin-walled castings of most MMCs can be poured by gravity without too much difficulty. For instance, Kolsgaard and Brusethaug (1994) are among many who have found that the fluidity of well made MMCs of Al-7Si-0.3Mg alloy containing up to 20 volume percent SiC particles remained unchanged. ('Well made' in this instance refers to the particulate component of the MMC being mixed in to the liquid metal under extremely high vacuum, so as reduce the entrainment of surface oxides.)

In contrast to the nearly zero effect of added particles or changes in alloy composition that might affect viscosity, there is a profound effect quickly generated by the presence of films in suspension in the liquid. This appears to be associated with the large effect on flow resulting from the large area of films, despite the negligible volume fraction of material that the films represent. Thus when melts become highly viscous, taking on the behaviour and appearance of porridge (a Scottish dish of oats boiled in water and strongly flavoured with salt) that cannot be the effect merely of the addition of solid particles because this can only explain a rise in viscosity by a factor of 2. If water is taken as unit viscosity, most metals have a coefficient of viscosity in the region of 2–5. All metals and alloys therefore are expected to flow like water.

When the melt becomes as viscous as a thick syrup, the viscosity is now in the region of 1000–10,000 times that of water. Many MMCs take on the appearance of such flow behaviour. This viscous behaviour cannot be explained merely by the presence of solid particles in suspension. Such high viscosities are only explicable if copious quantities of films are present in suspension in the liquid. With such extremely high viscosity, fluidity as measured by a spiral flow test is now impaired a little, but not as seriously as might be expected. Timelli and Bonollo (2007) studied pressure die casting alloys, confirming the high percentage of foundry returns in a charge did lower fluidity; the effect was due to oxide films as identified by metallographic studies.

Fluidity to fill castings does start to be reduced because the entrained films can bridge dendrite meshes, and can bridge the entrance to narrow sections of castings, forming impenetrable barriers. Emamy (2009a,b) found that fluidity of his MMCs produced by vortex stirring was limited by oxides introduced by the stirring process. The oxides, as bifilms of course, also reduced the mechanical properties of the cast mixtures.

It was found by Groteke (1985) that a filtered aluminium alloy was 20% more fluid than a 'dirty' melt. This finding was even more impressively confirmed by Adufeye (1997) who added a ceramic foam filter to the filling system of his fluidity mould. He expected to find a reduction in fluidity because of the added resistance of the filter. To his surprise, he observed on average approximately 100% increase of fluidity.

It seems, therefore, in line with common sense, that the presence of films in the liquid causes filling problems for the casting. In particular, the entry of metal from thick sections into thin sections would be expected to require good-quality liquid metal, because of the possible accumulation and blocking action of the solids, particularly films, at the entry into the thin section. Thin sections attached to thick, and filled from the thick section are in danger of breaking off rather easily as a result of layers of films across the junction. Metal flash on castings will often break off cleanly if there is a high concentration of bifilms in a melt. Flash adhering to a casting but which is resistant to be broken off because of its ductility might therefore be a useful quick test of the cleanness and resulting properties of the alloy. As an aside, an abrupt change of section, of the order of 10 mm down to 1–2 mm, is such an efficient film concentrating location that it is a good way to locate and research the structure of films on fracture surfaces by simply breaking off the thin section. Bifilms are easily revealed and opened by this technique.

Dirty alloys suffer from a wide dispersion of sizes of inclusions. It is almost certainly the large films that are effective in bridging, and therefore blocking, the entrances to narrow sections. It is common to see that dirty melts produce castings free from flash, despite a relatively poor mould assembly resulting from ill-fitting cores. Conversely, the casting of clean metal produces a casting bristling with flash, and which bends rather than breaks. It is therefore a pain to get off. Good metal quality is going to require good tooling to obtain nicely fitting core assemblies if costly dressing of flash is to be avoided.

After grain refinement additions to an aluminium melt the casting often exhibits significantly more flash. This is the result of the melt cleaning action. The heavy titanium-rich intermetallics precipitate on the oxide bifilms suspended in the melt and cause them to precipitate out, rapidly forming a layer of sludge on the bottom of the furnace or ladle. Thus the most important aspect of the grain refinement of Al alloys is the cleaning of the melt. This effect almost certainly exceeds the effect of grain refinement itself. There is no doubt, therefore, that the filling of extremely thin sections and fine detail is strongly dependent on the cleanness of the melt.

In support of this conclusion, it is known from experience, if not from controlled experiment, that melts of the common aluminium alloy Al-7Si-0.4Mg that have been subjected to treatments that are expected to increase the oxide content have suffered reduced fluidity as a result. The treatments include:

1. Repeated re-melting of the same material;
2. High content of foundry returns (especially sand cast materials) in the charge;
3. The pouring of excess metal from casting ladles back into the bulk metal in the casting furnace, particularly if the fall height is great;
4. Recycling in pumped systems where the returning melt has been allowed to fall back into the bulk melt (recirculating pumped systems that operate entirely submerged probably do not suffer to nearly the same extent, providing air is not entrained from the surface);
5. Degassing with nitrogen from a submerged lance for an extended period of time (for instance a 1000 kg furnace subjected to degassing for several days).

These treatments, if carried out to excess, can cause the melt to become so full of dispersed films that the liquid assumes the consistency of concrete slurry. As a consequence the fluidity is reduced. More serious still, the mechanical properties, particularly the ductility, of the resulting castings is nothing short of catastrophic.

Treatments of the melts to reduce their oxide content by flushing with argon, or treatments with powdered fluxes introduced in a carrier gas via an immersed lance, are reported to return the situation to normality. Much work is required to be carried out in this important area to achieve a proper understanding of these problems.

Finally, it seems likely that computer models, renowned for having to increase their viscosity values by a factor of up to 1000 to achieve a more realistic flow simulation. This apparently shocking fudge seems likely to be the result of assuming a value for pure metals, but have to adjust the viscosity to that of a real metal, usually full of oxide bifilms. When we founders learn to cast purer melts the computer models will have to revert to the viscosity values for pure metals.

3.2.4 EFFECT OF SOLIDIFICATION TIME t_f

From Eqn (3.1), it is clear that the greater the solidification time, t_f, the further the metal will run before freezing. This is a far more satisfactory way to improve the filling of castings than increasing the velocity because control over surface turbulence is not necessarily lost.

For conditions in which the mould material controls the rate of heat loss from the casting, such as in sand casting, quoting Flemings (1963), the freezing time is approximately

$$t_f = \frac{\pi}{16K_m\rho_m C_m}\left\{\frac{a\rho_s H}{(T_M - T_o)}\right\}^2 \tag{3.3}$$

Where subscripts s and m refer to solid metal and mould, respectively; K is the thermal conductivity, ρ is density, C is specific heat, H is latent heat of solidification, T_M is casting temperature, T_o mould temperature and $a/2$ is the geometric modulus (the volume V to surface area A ratio) of the channel.

Showman (2006) shows how small localised patches of lightweight aggregate of low heat diffusivity $K_m\rho_mC_m$ in a sand mould can significantly extend fluidity, as expected from the previous equation. These 'patches' are blown as cores, and incorporated into the mould by inserting into them into the mould core boxes and blowing them integrally with the mould cores.

For conditions of freezing in chill moulds where the metal/mould interface resistance to heat flow h is dominant, we may quote the alternative analytical equation for the freezing time per Flemings

$$t_f = \rho_S HV/\{h(T_M - T_o)A\} \tag{3.4}$$

The main factors that increase t_f are clear in these equations. The factors are dealt with individually in the following subsections.

3.2.4.1 Modulus

Greaves in 1936 made the far-sighted suggestion that fluidity length was a function of modulus m.

The effect of the thickness, or shape of the channel, on the time of freezing of the metal in the channel, is nicely accounted for by Chvorinov's rule (implicit in Eqns (3.3) and (3.4)). We can define the modulus m as the ratio 'volume/cooling area' of the casting, which is in turn of course equivalent to the ratio 'area/cooling perimeter' of the cross-section of a long channel. Remember that modulus has dimensions of length, and is most conveniently quoted, therefore, in millimetres. Restating Eqn (3.3) gives

$$t_f = k_m m^2 \tag{3.5}$$

The relation applies where the heat flow from the casting is regulated by the thermal conductivity of the mould (as is normal for a chunky casting in a reasonably insulating sand mould). The value of the constant k_m can be determined by experiment in some convenient units, such as $s \cdot mm^{-3}$ To gain some idea of the range of k_m in normal castings: for steel in greensand moulds, it is approximately 0.40; for Al-7Si alloys in dry sands, it is 5.8, and for Al-3Cu-5Si it is 11.0 $s \cdot mm^{-3}$ It is worth noting that Eqn (3.5) applies nicely to chunky castings in sand moulds, and other reasonably insulating moulds such as investment moulds.

For the case where interfacial heat transfer dominates heat flow, as in chill moulds, Eqn (3.4) becomes

$$t_f = k_i m/h \tag{3.6}$$

where k_i is a constant, and h is the heat transfer coefficient: the rate of flow of heat across the air gap for a given temperature difference across the gap, and for a given area of interface. We shall see later how h starts to dominate in thin section castings, so that Eqn (3.6) is often found to apply not only to chill castings but also to thin section sand castings.

In his experiments on sand moulds, Feliu (1964) appeared to find a relation between fluidity and (modulus)$^{3/2}$ for thin section sand castings in the range 0.7–2 mm modulus (section thickness 1.4–4 mm). However, he did not allow for the effect of surface tension in reducing the effective head of metal. When this is allowed for, as will be illustrated in the exercise in the following section, it indicates that a linear relation between fluidity and section thickness as suggested by Eqn (3.6) actually applies reasonably well. Barlow (1970) confirms that over a range of Pyrex tubes from 1 to 4 mm diameter the fluidity of both Al-10Si and Cu-10Al is closely linear with channel section.

Restating Eqn (3.1) in terms of the relations (3.5) and (3.6), we have for sand moulds

$$L_f/m = kVm \tag{3.7}$$

and for metal moulds (and thin section sand moulds)

$$L_f/m = k'V/h \tag{3.8}$$

We now have two simple but powerful formulae to assist in the prediction of whether a mould will fill. Interestingly, it seems that Eqn (3.8) has rather general validity. Clearly, interfacially controlled heat flow is rather perhaps more common than is generally thought, especially in thin sections, where freezing times are short.

Applying Eqn (3.8) to thin walled sand castings shows that the ratio of fluidity to modulus, L_f/m, is a constant for a given mould material. For instance in Al alloys cast into thin section sand moulds, shop floor experience indicates the *fluidity to section thickness* ratio is around 100 (i.e. a section thickness of 4 mm should allow the metal to flow 400 mm before arrest by freezing). This is a really useful relation. For instance, if the ratio of thickness to distance to be run is only 50 the casting is no problem to make. When the ratio is 100, the casting is not to be underestimated; it is near the limits. When the ratio is 150 the casting cannot be made by flowing metal through the mould cavity from a single gate entry point; a completely different filling system is required. This is therefore a useful guide as a quick check which can be made at the drawing stage as to whether a design of casting might be castable or not. We can restate the critical *fluidity to section thickness ratio* = 100 in terms of the more universal *fluidity to modulus ratio* = 200, allowing us to transfer this useful ratio to any shape of interest to a fair approximation.

The value of the concept of modulus, m, is that any shape of channel can be understood and compared with any other shape. For instance straight tubes can be compared with trapezoidal spirals, or with thin flat strips. All that matters for the control of solidification time is the modulus (we are neglecting for the moment the refinements that allow for the slight inaccuracies of the simple modulus approach).

3.2.4.2 Heat-transfer coefficient

A reduction in the rate of heat transfer will benefit fluidity (as will become clearer when we look into the relation between the rate of heat transfer and the rate of solidification in Chapter 4). For this reason, insulating ceramic coatings are applied to all permanent moulds (gravity and low-pressure dies). The work of Rivas and Biloni (1980) illustrates the benefit to fluidity from the application of a white oxide–based coating to a steel die. This is confirmed by Hiratsuka et al. (1994) from work in which metal is drawn into tubes of quartz, copper and stainless steel.

For sand moulds, acetylene black applied from a sooty flame (Flemings et al., 1959) gives very substantial increases in fluidity, by a factor of 2 or 3. This dramatic improvement in fluidity is used in some precision sand foundries, allowing thin-walled castings to be filled that could not otherwise be cast. Similar results are reported for permanent mould foundries casting Al alloys and casting cast irons. Although attempts have been made to develop other fluidity-enhancing coatings, none has even approached a similar effectiveness. However, the unfortunate greasy black pollution caused by the soot in the foundry is unwelcome, causing the whole foundry environment to become blackened by soot. However, this severe disadvantage continues to be tolerated whilst alternatives are unavailable because many castings cannot be filled without this practice.

The action of the carbon deposit was for many years thought to be an insulating effect; the carbon acting to reduce the heat transfer coefficient. This is now known not to be true as explained later in Section 3.3.5.

3.2.4.3 Superheat $(T_M\text{-}T_0)$

Superheat is defined here as the excess of casting temperature over liquidus temperature. Raised superheat benefits fluidity in a direct way, as predicted by Eqns (3.3) and (3.4), but the linear relationships clear in Figures 3.15–3.17 and many other examples in this book confirm the interesting fact that only Eqn (3.4) applies accurately for all types of mould, indicating that in all the rather slim fluidity test castings the heat transfer at the interface is the key parameter controlling heat flow during solidification.

The effect of superheat is therefore valuable in itself. In addition, however, it can help us understand the special fluid properties of eutectics, and the poor flow capabilities of most high melting point intermetallics as explained earlier in this section. As a reminder, in Figure 3.6 obtained by Lang (1972) for the Al-Zn system, at constant superheat the fluidity of the eutectic is almost exactly that expected from the rule of mixtures, interpolating between the fluidities of the pure elements Al and Zn. (A rule of mixtures would have predicted the fluidity of the pure elements and the eutectic to lie on a straight line.) When determined at a constant temperature of 700°C, however, the eutectic now has the advantage of a large effective superheat, and the fluidity of the eutectic is correspondingly enhanced, becoming significantly higher than that of either Al or Zn as explained in the derivation of Figure 3.8.

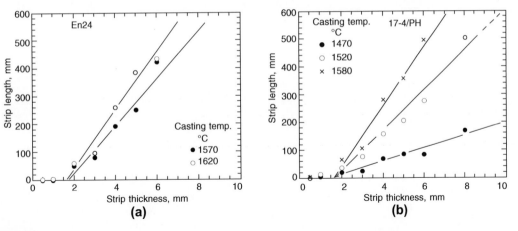

FIGURE 3.14

(a) Fluidity data for a low alloy steel; and (b) a stainless steel poured in a straight channel, furan bonded sand mould (Boutorabi et al., 1990).

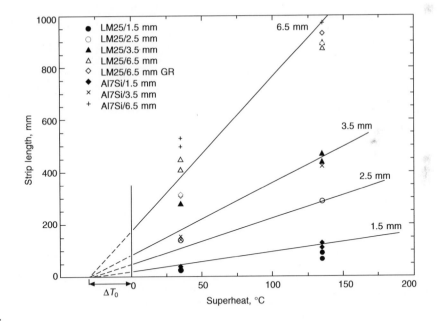

FIGURE 3.15

Fluidity of a variety of Al-7Si alloys, one grain refined (GR), showing linear increase with section thickness and casting temperature (Boutorabi et al., 1990).

3.2.4.4 Latent heat H

The latent heat given up on solidification will take time to diffuse away, thereby delaying solidification, and extending fluidity. The time taken to lose the superheat explains how many instances have been recorded of useful amounts of fluidity even at the freezing temperature for the metal. Much has been made of this point (see for instance Arnold et al.

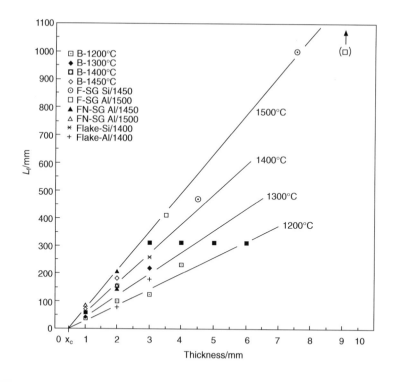

FIGURE 3.16

Fluidity of a variety of grey and ductile irons showing linear increase with section thickness and casting temperature (Boutorabi et al., 1990).

1963) in explaining the claims for the especially good fluidity of the hypereutectic Al-Si alloys because pure silicon has a latent heat of solidification 4.65 times greater than that of aluminium (data from Brandes, 1992).

Equation (3.4) is probably the most appropriate equation to describe fluidity, but of course is not likely to be particularly accurate. Nevertheless, when used in a comparative way, it is likely to give somewhat better results. Thus to find the improvement, for instance, in fluidity when changing from pure Al to pure Si, we can see that the comparative freezing times are simply given by the ratio

$$\frac{t_{Si}}{t_{Al}} = \left\{\frac{T_{Al} - T_0}{T_{Si} - T_0}\right\}\left\{\frac{\rho_{Si}}{\rho_{Al}}\right\}\left\{\frac{H_{Si}}{H_{Al}}\right\} \tag{3.9a}$$

$$= 0.460 \times 0.867 \times 4.65 \tag{3.9b}$$

$$= 1.85 \tag{3.9c}$$

Thus from Eqn (3.9b), pure Si would have been expected to have 4.65 times the fluidity of pure aluminium as a result of its higher latent heat. However silicon's rate of heat loss, seen to be about two times faster as a result of its higher freezing temperature, reduces this significantly. The low density of silicon also slightly reduces the effect further. The final result, Eqn (3.9c), is that pure silicon would be expected to deliver effectively 1.85 times the fluidity of pure aluminium. (Rather paradoxically, this conclusion will only be accurate at low concentrations of Si in Al because the other terms in Eqn (3.3) then are not significantly changed and thus properly cancel to give Eqn (3.9) accurately. Also, in the first edition of this book, Eqn (3.3) was selected to generate these comparative figures, although subsequently it has

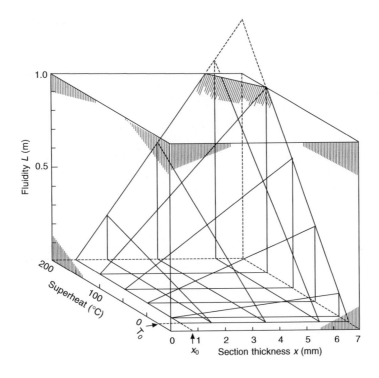

FIGURE 3.17

Fluidity results for Al-7Si alloys from Figure 3.1 re-presented in a three-dimensional format.

become clear that the use of Eqn (3.4) is probably more accurate, significantly reducing the estimates of the effect of Si at 22 as given in the earlier text, down to 4.65 here.)

We can now construct a relation to examine the effect of the latent heat of Si on the fluidity of Al-Si alloys in sand moulds as in Figure 3.18 using the excellent experimental data from Lang (1972) in which he studies the effect of the addition of sodium (Na) to Al-Si alloys. The equilibrium eutectic composition of Al-Si alloys is generally accepted to be somewhere near 11.5%Si but Lang's fluidity data indicate peaks well above this value, somewhere between 14.5 and 20%Si.

Although an upward trend of fluidity assignable to the latent heat effect is predicted by previous authors to be a major effect responsible for the good fluidity of the high Si alloys, it is clearly seen that the effect (shown as broken lines for maximum fluidity and 50% fluidity) is practically negligible compared to the massive peaks assignable to the eutectic.

The unmodified non-equilibrium eutectic shows a fall in fluidity to approximately half of the fluidity of the pure aluminium. This will correspond to fluidity of a mix of dendrites and Si particles (formed on the oxide bifilms in suspension) exerting a drag on the flowing alloy. The rather broad fluidity peak somewhere between 17%Si and 18%Si corresponds to the minimum drag offered by the rather messy non-coupled eutectic consisting of Si plates protruding from a smooth aluminium matrix.

The addition of 0.05 ± 0.02%Na prevents Si formation on bifilms, stabilising the modified coupled eutectic with its extremely smooth advancing front, and therefore creates a clear, sharp peak corresponding to a eutectic at approximately 14.5%Si. The existence of a maximum in fluidity at about 14.5%Si has been confirmed by a number of researchers, for instance Pan and Hu (1996) and Adefuye (1997). In addition, the presence of a peak can be interpolated in the work of Parland (1987) who missed the peak, but whose data points on either side corroborate its existence. Reassuringly, the tips of the peaks fall closely on the interpolated line of fluidity of the pure elements, Al and Si, assuming the ratio 4.65 for their fluidities due merely to the effect of latent heat.

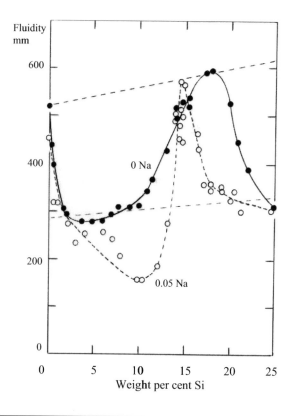

FIGURE 3.18

The fluidity of Al-Si alloys showing that the increase in fluidity resulting from the high latent heat of Si is negligible compared with the height of the eutectic peak.

Re-plotted from data by Lang 1972.

These various values for the eutectic composition of Al-Si are undoubtedly the result of confusion between the eutectics of (1) coarse Si particles formed on bifilms, possibly formed *above* the equilibrium eutectic temperature (as shown in Figure 6.24(c)) as a result of the presence of the favoured oxide substrate; and (2) fine, modified, coupled growth eutectic which should probably be considered the 'true' eutectic composition at 14.5%Si, which might form *at* the equilibrium eutectic temperature (paradoxically therefore not undercooled at all!). Further research to clarify the behaviour of this important eutectic system would be valuable.

After the addition of Na, the dip in the fluidity data to well below predicted values for alloy of intermediate levels of Si, if true, is extraordinary. It is tempting to interpret this as being the effect of the release of the bifilms from their role as substrates for Si particles. Their presence, now floating freely in the melt, and becoming tangled with the mesh of aluminium dendrites, would provide a maximum resistance to flow.

The apparent fall in fluidity at high Si levels beyond the peaks in Figure 3.18 seems doubtful. Some authors find this (e.g. Pan and Hu) but others find the good fluidity to be maintained; for instance Adefuye, and the result on the Al-Cu-Si system in Figure 3.10, both show that the maximum is not a cuspoid peak but merely a modest change of slope. Such discrepancies are probably to be expected when working at such high superheats in this difficult, high Si range of the Al-Si system. In particular, the rapid formation of oxides would be troubling problem.

The use of the extraordinary Al-17Si alloy, developed around 1970 under the leadership of John Jorstad in the United States, with phosphorus additions to aid the nucleation of primary silicon particles is well known for its use on automotive engine blocks, even though less use has been made of the alloy for blocks than it appears to deserve. It has, however, found many other uses for which it was never designed, and for which it is not perhaps really optimum, for instance for gearbox housings and other castings as a probable result of its availability (Jorstad, 1996). The more rational use of this and other alloys would be welcome.

It is a further curious illogicality that most of the common Al-Si casting alloys are in the range of 4–10%Si, and are therefore concentrated in the region of mediocre fluidity properties between the high fluidity of pure Al and the high fluidity of the non-equilibrium eutectic. Contrary to the claims of practically every writer on this subject, it is quite clear that the silicon *reduces* the fluidity at these intermediate compositions. (Almost certainly a confusion has arisen with the well known, but less easily defined concept of *castability*. The improved castability of Al-Si alloys is probably a result of the benefits of their greater quantity of eutectic, and possibly a more benign oxide film, compared to many other Al alloy systems. Again, more research to clarify the contributions of these factors would be really helpful.)

The real benefit of silicon in relation to fluidity is only seen nearer to 14.5%Si and above, corresponding, probably, to the composition of the non-equilibrium eutectic. It is possible to modify these so-called hypereutectic alloys with sodium or strontium to stabilise the eutectic form of the silicon (avoiding the usual addition of phosphorus, and so avoiding the usual chunky primary silicon particles that give difficulties for machining). However, these compositions have not yet been promoted by suppliers nor have been taken up by users (Smith, 1981).

In an interesting parallel with Zn-rich alloys from the Zn-Al-Cu system, as mentioned previously, Friebel and Roe (1963) point out from their fluidity results that the commonly used pressure die casting alloys at 4%Al hardly benefit from the huge fluidity peak at 5%Al, whereas the famous ZA series alloys with 8, 12 and 27%Al all have completely missed all trace of benefit, and consequently have only mediocre flow properties. The ZA series have clearly sacrificed fluidity for strength, but the pressure die casting alloys, particularly the Al-Si alloys, have no such excuse; for their market, targeted at the manufacture of castings requiring intricate detail, they simply have a really disappointing non-optimum composition.

From time to time, I cannot help reflecting on the irony that the great interest in fluidity exhibited by foundry people and researchers is curiously at odds with customers, who persist in ignoring the huge benefits in fluidity on offer at no cost, but perversely select and continue to use alloys exhibiting the most mediocre flow properties.

Even in the twenty-first century, it seems that casting compositions with promise of good fluidity, mechanical properties, and machinability, remain to be properly developed. They promise exciting potential. The challenge will be to sell these substantial benefits to the conservative customer.

3.2.4.5 Mould temperature

For most casting processes, mould temperature is fixed at or close to room temperature. For such processes, there is little or nothing that can be done to mould temperature to affect fluidity. Although many sand moulds for irons and steels are surface dried prior to casting, the relatively small rise in temperature is unlikely to benefit fluidity (even though it is vital for reduction of volatile gases of course).

The intermediate temperatures of gravity and high pressure dies, usually at around 300–400°C, do contribute modestly to the increase of fluidity in these casting processes. Die temperatures are sometimes raised a little to gain a little extra filling capability. However, the natural working temperature of a die is only changed by significant determined effort, and is therefore unpopular and rarely resorted to.

However, for investment casting the ceramic shell allows a complete range of temperatures to be chosen without difficulty. From Eqn (3.3) it is predicted that the freezing time will be proportional to the difference between the freezing point of the melt and the temperature of the mould. The few tests of this prediction are reasonably well confirmed, for instance by Campbell and Olliff in 1971 (Figures 3.19–3.21).

One important prediction is that when the mould temperature is raised to the melting point of the alloy, the fluidity becomes infinite; i.e. the melt will run for ever! Actually of course, this self-evident conclusion needs to be tempered by

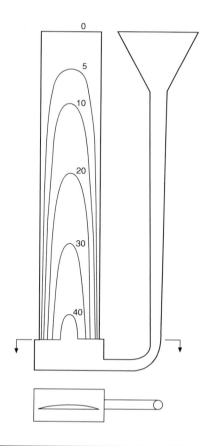

FIGURE 3.19

Aerofoil fluidity test for investment casting. The outlines of the cast shape are computed for increasing values of $\gamma/\rho gh$, units in millimetres (Campbell and Olliff, 1971).

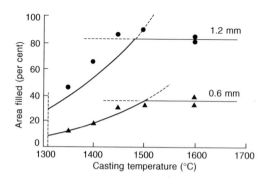

FIGURE 3.20

Results from the aerofoil test for a Ni-base superalloy. Lines denote theoretical predictions; points are experimental data.

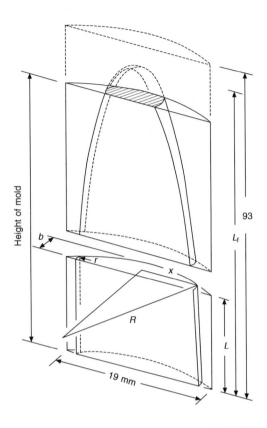

FIGURE 3.21

Geometry of a typical aerofoil cast test piece.

the realisation that the melt will run until stopped by some other force, such as gravity, surface tension, or the mould wall! All this corresponds to common sense. Even so, the raising of the mould temperature to the point of eliminating fluidity limitations is an important feature widely used in the casting of thin-walled aluminium alloy investment castings. With ceramic moulds it is easy to raise their temperature in excess of the freezing point of aluminium alloys at approximately 600°C. Single crystal turbine blades in nickel-based alloys are also cast into moulds heated 1450°C or more, again well above the freezing point of the alloy.

Any problems of fluidity are thereby avoided. Having this one concern removed, the founder is then left with only the dozens of additional important factors that require to be controlled for the success of the casting. Solving one problem completely is a valued step forward, but still leaves plenty of challenges for the casting engineer!

3.2.5 EFFECT OF SURFACE TENSION

If metals wetted the moulds into which they were cast, then the metal would be drawn into the mould by the familiar action of capillary attraction, as water wets and thus climbs up a narrow bore glass tube.

In general, however, metals do not wet moulds. In fact, mould coatings and release agents are designed to resist wetting by metals. Thus the curvature of the meniscus at the liquid metal front leads to capillary repulsion; the metal experiences a back pressure resisting entry into the mould. The back pressure due to surface tension, P_{ST}, can be quantified by the simple relation, where r and R are the two orthogonal radii which characterise the local shape of the surface, and γ is the surface tension

$$P_{ST} = 2\gamma\{(1/r) + (1/R)\} \tag{3.10}$$

When the two radii are equal, $R = r$, as when the metal is in a cylindrical tube, then the liquid meniscus takes on the shape of a sphere of radius r, and Eqn (3.10) takes on the familiar form

$$P_{ST} = 2\gamma/r \tag{3.11}$$

Alternatively, if the melt is filling a thin, wide strip, so that R is large compared with r, then 1/R becomes negligible and back pressure becomes dominated by only one radius of curvature, r, because the liquid meniscus now approximates to the shape of a cylinder

$$P_{ST} = \gamma/r \tag{3.12}$$

At the point at which the back pressure due to capillary repulsion equals or exceeds the hydrostatic pressure, ρgh, to fill the section, the liquid will not enter the section. This condition in the thin, wide strip is

$$\rho gh = \gamma/r \tag{3.13}$$

This simple pressure balance across a cylindrical meniscus is useful to correct the head height, to find the net available head pressure for filling a thin-walled casting. In the case of the filling of a circular section tube (with a spherical meniscus) do not forget the factor 2 for both the contributions to the total curvature as in Eqn (3.11). In the case of an irregular section, an estimate may need to be made of both radii, as in Eqn (3.10).

The effect of capillary repulsion, repelling metal from entering thin sections is clearly seen by the positive intercept in Figure 3.14 for a medium alloy steel and a stainless steel, in Figure 3.16 for cast iron and in Figure 3.22 for a zinc alloy.

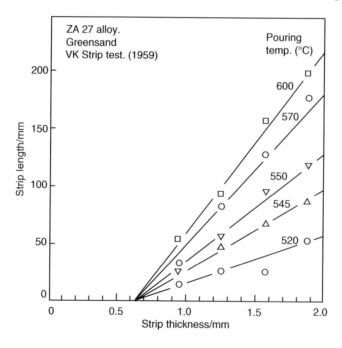

FIGURE 3.22

Fluidity of Zn-27Al alloy cast in greensand using the VK fluidity strip test.

Data from Sahoo and Whiting (1984).

Thus the effect appears to be quite general, as would be expected. The effective surface tension can be worked out in all these cases from an equation such as 3.13. In each case it is found to be around twice the value to be expected for the pure metal in a vacuum. Again, this high effective value is to be expected as explained in Section 3.1.1.

In larger round or square sections, where the radii R and r both become large, in the range of 10–20 mm, the effects of surface tension become sufficiently small to be neglected for most purposes. Large sections are therefore filled easily.

Flowability and fillability

In the filling of many castings, the sections to be filled are not uniform; the standard complaint in the foundry is 'the sections are thick and thin'. This does sometimes give its problems. This is especially true where the sections become so thin in places that they become difficult to fill because of the resistance presented by surface tension. Aerofoils on propellers and turbine blades are typical examples.

To investigate the filling of aerofoil sections that are typical of many investment casting problem shapes, an aerofoil test mould was devised as shown in Figure 3.19. (This test mould also included some tensile test pieces whose combined volume interfered to some extent with the filling of the aerofoil itself; in later work the tensile test pieces were removed, giving considerably improved reproducibility of the fluidity test.)

Typical results for a vacuum-cast nickel-based superalloy are given in Figure 3.20 (Campbell and Olliff, 1971). Clearly, the 1.2 mm section fills more fully than the 0.6 mm section. However, it is also clear that at low casting temperature the filling of both sections is limited by the ability of the metal to flow prior to freezing. At these low casting temperatures the fluidity improves as temperature increases, as expected.

However, above a metal casting temperature of approximately 1500°C further increases of temperature do not further improve the filling. As the metal attempts to enter the diminishing sections of the mould, the geometry of the liquid front is closely defined as a simple cylindrical surface. Thus it is not difficult to calculate the thickness of the mould at any point. Half of this thickness is taken as the radius of curvature of the liquid metal meniscus (Figure 3.21). It is possible to predict, therefore, that the degree of filling is dictated by the local balance at every point around the perimeter of the meniscus between the filling pressure due to the metal head, and the effective back pressure due to the local curvature of the metal surface. In fact, if momentarily overfilled because of the momentum of the metal as it flowed into the mould, the repulsion effect of surface tension would cause the metal to 'bounce back', oscillating either side of its equilibrium filling position, finally settling at its balanced, equilibrium state of fullness.

The authors of this work emphasise the twin aspects of filling such thin sections; *flowability* limited by heat transfer and *fillability* limited by surface tension.

At low mould and/or metal temperatures, the first type of filling, flowability, turns out to be simply classical fluidity as we have discussed previously. Metallographic examination of the structures of aerofoils cast at lower temperatures shows columnar grains grown at an angle into the direction of flow, typical of solidification occurring while the metal was flowing. The flow length is controlled by solidification, and thus observed to be a function of superheat and other thermal factors, as we have seen.

The second type of filling, fillability, occurs at higher mould and/or metal temperatures where the heat content of the system is sufficiently high that solidification is delayed until after filling has come to a stop. Studies of the microstructure of the castings confirm that the grains are large and randomly oriented, as would be expected if the metal were stationary during freezing. Filling is then controlled by a mechanical balance of forces. The mode of solidification and further increases of temperature of the metal and the mould play no part in this phase of filling.

In a fluidity test of simpler geometry consisting of straight strips of various thickness, the linear plots of fluidity L_f versus thickness x and superheat ΔT_S are illustrated in Figures 3.15 and 3.16 for Al 7Si alloy and cast iron in sand moulds. It is easy to combine these plots giving the resultant 3-dimensional pyramid plot, as is shown for the Al alloy data in Figure 3.17. In terms of the pressure head h, and the intercepts ΔT_o and x_o defined on fluidity plots 3.15 and 3.16, the equation describing the slightly skewed surface of the pyramid is

$$L_f = C(\Delta T_s + \Delta T_o)(x - (2\gamma/\rho gh)) \tag{3.14}$$

where C is a constant with dimensions of reciprocal temperature. For the Al alloy, C is found from Figure 3.15 to have a value of about 1.3 ± 0.1 K^{-1} and $\Delta T_o = 30 \pm 5$ K. The surface tension of pure liquid Al is close to 1.0 Nm^{-1} but is seen in this work to be effectively $\gamma = 2$ Nm^{-1} because of the contribution of the oxide film to the apparent value of the surface tension. Other values are $\rho = 2500$ kgm^{-3}, $g = 10$ ms^{-2} and $h = 0.10$ m. We can then write an explicit equation for fluidity (mm) in terms of superheat (degrees Celsius) and section thickness (mm)

$$L_f = 1.3(\Delta T_s + 30)(x - 1.6)$$

For a superheat $\Delta T_s = 100°C$ and section thickness $x = 2$ mm, we can achieve a flow distance $L_f = 68$ mm for Al-7Si in a sand mould. If the head h were increased, fluidity would be higher, as indicated by Eqn (3.14) (but noting the limitations discussed in Section 3.2.2).

As we have seen, in these thin-section moulds both heat transfer and surface tension contribute to limit the filling of the mould, their relative effects differ in different circumstances. This action of both effects causes the tests to be complicated, but, as we have seen, not impossible to disentangle. Further practical examples of the simultaneous action of heat transfer and surface tension will be considered in Section 3.2.7.

3.2.6 EFFECT OF AN UNSTABLE SUBSTRATE

In 1980, Southin and Romeyn investigated the amazing benefit to fluidity that followed from the blackening of moulds by the soot from a lazy acetylene flame, and the similar benefit from painting the mould with a hexachloroethane-containing coating. They found a complicated behaviour. For aluminium alloys up to 1%Cu the use of carbon black or hexa-chloroethane mould coatings *reduced* the fluidity because the planar (or cellular) growth front appeared to be stable, so that the strong, compact freezing front easily stopped the flow at an early stage as a result of an *enhanced* thermal transfer of the carbon coating. However, at higher copper contents the fluidity was increased because the dendritic front was fragmented by the flow, and the dendritic fragments were carried along in the flowing stream, giving a fine grain size and good fluidity.

The authors suggested that these results were only explained if (1) the rate of heat transfer was increased and (2) the coating mechanically destabilised the growing dendrites. This action was thought to occur by the unstable support of the growing dendrites, because of the mechanical collapse of the coating as the heavy melt ran over the fragile carbon layer, probably aiding the oxidation of this fluffy deposit so as to further weaken support. The collapse of the substrate on which the dendrites were growing would destabilise the growing dendrites, causing them to deform plastically, recrystallise, and release dendrite fragments as the melt wetted and penetrated the newly formed grain boundaries. The mechanism is that proposed by Vogel, Doherty and Cantor (1977) who considered the plastic deformation at the dendrite root would lead to recrystallisation, with the result that the newly formed high angle grain boundary that formed across the root was favourably wetted and therefore penetrated by the melt, allowing the dendrite arm to detach. Thus instead of the dendrites growing into the stream to obstruct its flow, they were released to tumble along with the stream. The increased rate of heat loss, although usually reducing fluidity, was in this case more than countered by the effect of the undermining of the dendrite barrier.

Cupini and Prates (1977) similarly investigated the effect of hexachloroethane mould coatings. The release of chlorine from the breakdown of this chemical was found to refine the grain size of 99.44Al from the range of between 2 and 20 mm down to 0.2 mm. The effect seemed to be the result of the microscopic disturbance of the surface of the casting, where the dendritic grains were starting to grow, in general supporting the work by Southin and Romeyn, although they did not go on to study the fluidity of the alloy.

The work by Southin and Romeyn seems to me to be critical to the proper understanding of the fluidity of long freezing range alloys. In Section 3.2.1, the standard assumption was the long freezing alloys arrest flow when the grains in the partly molten slurry start to impinge to form a solid network. We are now in a position to reassess this theory.

Southin and Romeyns' work shows that the real blockage to flow is not the cluster of equiaxed grains near the flow front (Figure 3.3(b)), but the forest of fixed dendrites (Figure 3.3(c)), probably growing roughly in the same geometrical envelop as the solidification front of the short freezing range alloys (Figure 3.3(a)).

It seems that the real scenario is that arms are melted off the dendrites and so tumble along with the stream, but while tumbling freely along do not constitute any serious threat to fluidity. Although such a slurry may have a high effective viscosity, as we have seen, viscosity has very little effect on fluidity. These tumbling grains are swept more or less effortlessly along the central channel and conveyed to the front of the flow where they have been thought to block further flow when their density rose sufficiently high. Thus the equiaxed grains immediately behind the flow tip, studied by experimenters for years on the assumption they constituted the blockage, appear to be largely irrelevant as obstacles to flow. I summarise what seems to be a more realistic model of flow blockage in Figure 3.3(c).

The carbon blacking technique is widely used in precision sand foundries casting thin-walled parts in aluminium and magnesium alloys. However, the author has also seen its use for the casting of cast iron in grey cast iron moulds, coated with a zircon wash, followed by a topping of carbon black from an acetylene flame. Incidentally, the technique of blacking moulds and cores in this way requires good local extraction of smoke. Otherwise, after a year or so, the whole foundry simply goes a deeper shade of black.

Evidence put forward from time to time that grain refinement increases fluidity (e.g. Alsem, 1992) can perhaps be understood. Two mechanisms are possible, and both may be contributory. (1) Although the addition of grain refiners might increase the solids content of the flowing melt we now know that this has a very limited effect on fluidity. At the same time, it will reduce or eliminate solidification in the form of arrays of dendrites growing from the walls which would have exerted significant drag on the flow. (2) The cleaning effect of grain refinement, causing the Ti and B intermetallics to precipitate on bifilms in the furnace or ladle, causing the bifilms to drop out of suspension prior to casting is probably an important factor (Tiryakioglu and Campbell, 2007). The clean melt, unhindered by its normal population of bifilms can now flow more freely.

3.2.7 COMPARISON OF FLUIDITY TESTS

Kondic (1959) proposed the various thin-section cast strip tests (called here the Voya Kondic (VK) strip test) as an alternative fluidity test because it seemed to him that the spiral test was subject to unacceptable scatter (Betts and Kondic, 1961).

There has been some justification for these concerns because the effect of pouring head, and sprue shape, among other parameters, were never well optimised. This situation continues to this day, although workers in Norway (Sabatino 2005, 2006) have made some progress to improve the specification of the test.

For a proper interpretation of all types of strip or spiral test results, they need to be corrected for the back-pressure from surface tension at the liquid front. As we have seen, this effectively reduces the available head pressure applied from the height of the sprue. The resulting cast length will correspond to that flow distance controlled by heat transfer, appropriate to that effective head and that section thickness. These results are worked through as an example that follows.

Figure 3.22 shows the results by Sahoo and Whiting (1984) on a Zn-27Al alloy cast into strips, 17 mm wide, and of thickness 0.96, 1.27, 1.58 and 1.88 mm.

The results for the ZA27 alloy indicate the minimum strip thickness that can be entered by the liquid metal using the pressure head available in this test is 0.64 ± 0.04 mm. Using Eqn (3.13), assuming that the metal head is close to 0.1 m, $R = 17/2$ mm and $r = 0.64/2$ mm, and liquid density close to 5720 kgm^{-3}, we obtain the effective surface tension $\gamma = 1.90 \text{ Nm}^{-1}$. (If the $R = 17/2$ curvature is neglected, the surface tension then works out to be 1.98 Nm^{-1} and therefore is negligibly different for our purpose.) This is an interesting value, over double that found for the surface tension of pure Zn or pure Al. It almost certainly reflects the presence of a strong oxide film.

It suggests that the liquid front was, briefly, held up by surface tension at the entry to the thin sections, so that an oxide film was grown that assisted to hold back the liquid even more. The delay is typical of castings where the melt is given a choice of routes, but all initially resisting entry, so that the sprue and runner have to fill completely before pressure is raised sufficiently to break through the surface oxide. If the melt had arrived without any choice, and without any delay to

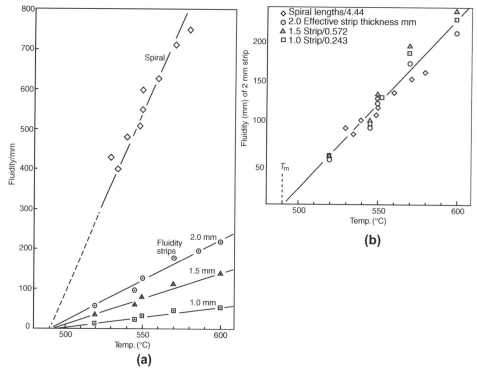

FIGURE 3.23

(a) Results from 3.22 replotted to show the effect of superheat explicitly, as though from strips of thickness 1.0, 1.5 and 2.0 mm, together with results from the spiral test. (b) Data from the spiral and strip tests shown in (a) reduced by the factors shown in the figure to simulate results as though all the tests had been carried out in a similar size mould of section 2 × 17 mm. All the results are seen to agree, confirming the validity of the comparison.

pressurisation, the melt would probably have entered with a resistance due only to surface tension. In such a condition, γ would be expected to have been close to $1.0\ Nm^{-1}$.

It suggests that, to be safe, values of at least double the surface tension be adopted when allowing for the possible loss of metal head in filling thin section castings.

The ability to extrapolate back to a thickness that will not fill is a valuable feature of the VK fluidity strip test. It allows the estimation of an effective surface tension. This cannot be derived from tests, such as the spiral test, that only uses one flow channel. The knowledge of the effective surface tension is essential to allow the comparison of the various fluidity tests that is suggested next.

The data from Figure 3.22 is cross-plotted in Figure 3.23(a) at notional strip thickness of 1.0, 1.5 and 3.0 mm. (These rounded values are chosen simply for convenience of reporting. The accurate values were employed for calculation purposes.) The individual lengths in each section have been plotted separately, not added together to give a total as originally suggested by Kondic. (Totalling the individual lengths seems to be a valid procedure, but does not seem to be helpful, and simply adds to the problem of disentangling the results.) Interestingly all the results extrapolate back to a common value for zero fluidity at the melting point for the alloy, 490°C. This is a surprising finding for this alloy. Most alloys extrapolate to a finite fluidity at zero superheat because the metal still takes time to give up its latent heat, allowing the metal time to flow. The apparent zero fluidity at the melting point in this alloy requires further investigation.

Also shown in Figure 3.23(a) are fluidity spiral results. An interesting point is, that despite his earlier concerns, I am sure VK would have been reassured that the percentage scatter in the data was not significantly different to the percentage scatter in the strip test results.

The further obvious result from Figure 3.23(a) shows how the fluidity length measurements of the spiral are considerably higher than those of the strip tests. In a qualitative way this is only to be expected because of the great difference in the cross-sections of the fluidity channels. We can go further, though, and demonstrate the quantitative equivalence of these results.

In Figure 3.23(b), the spiral and strip results are all reduced to the value that would have been obtained if the spiral and the strip tests all had sections of 2 × 17 mm.

This is achieved by reducing the spiral results by a factor 4.44 to allow for the effect of surface tension and modulus, making the results equivalent to those in the 2 mm thick cast strip. The 2 mm section experimental results remain unchanged of course. The 1.5 and 1.0 mm results are increased by factors 1.75 and 4.12, respectively. These adjustment factors are derived next.

Taking Eqn 3.1 (Eqn 3.2 can be used in its place, because we are about to take ratios), together with Eqns (3.6) and (3.7), and remembering that the velocity is given approximately by $(2gH)^{1/2}$ then we have for sand moulds

$$L_f = km^n(2gH)^{1/2}$$
$$= km^n(2g(H - (\gamma/r\rho g)))^{1/2} \tag{3.15}$$

where n is 1 for interface controlled heat flow, such as in metal dies and thin sand moulds, and n is 2 for mould control of heat flow, such as in thick sand moulds.

Returning now the comparison of fluidity tests, then by taking a ratio of Eqn (3.15) for two tests numbered 1 and 2, we obtain

$$\frac{L_1}{L_2} = \left(\frac{m_1}{m_2}\right)^n \left\{\frac{H_1 - (\gamma/r_1\rho g)}{H_2 - (\gamma/r_2\rho g)}\right\}^{1/2} \tag{3.16}$$

For the work carried out by Sahoo and Whiting on both the spiral and strip tests, the ratio given in Eqn (3.16) applies as accurately as possible because the liquid metal and the moulds were the same in each case. Assuming the moduli were 1.74 and 0.985 mm respectively, and the radii were 4 and 1 mm, respectively, $\gamma = 1.9$ Nm^{-1}, and $\rho = 5714$ kgm^{-3}, and the height of the sprue in each case approximately 0.1 m, it follows

$$\frac{L_{f1}}{L_{f2}} = \left\{\frac{1.74}{0.895}\right\}^2 \left\{\frac{0.1 - 0.00847}{0.1 - 0.0339}\right\}^{1/2}$$
$$= 3.77 \times 1.18$$
$$= 4.44$$

The calculation is interesting because it makes clear that the largest contribution towards increased fluidity in these thin section castings derives from their modulus (i.e. their increased solidification time). The effect of the surface tension is less important in the case of the comparison of the spiral with the 2 mm section. If the spiral of modulus 1.74 mm had been compared with a thin section fluidity test piece of only 1 mm thick, then

$$L_{f1}/L_{f2} = 13.6 \times 1.68 = 9.25$$

Thus although the surface tension factor has risen in importance from 1.18 to 1.68, the effect of freezing time is still completely dominant, rising from 3.77 to 13.6.

The dominant effect of modulus over surface tension appears to be a general phenomenon in sand moulds as a result of the (usually) small effect of surface tension compared to the head height.

The accuracy with which the spiral data is seen to fit the fluidity strip test results for the Zn-27Al alloy when all are adjusted to the common section thickness of 2 × 17 mm (Figure 3.23(b)) indicates that, despite the arguments that have raged over the years, both tests are in fact measuring the same physical phenomenon, that we happen to call fluidity, and both are in agreement.

3.2.8 EFFECT OF VIBRATION

The use of vibration to aid the filling of moulds has a reasonably long history, but seems to have been only half heartedly taken up at any one time, and appears to be little used at the time of writing. This technique should probably be reassessed, taking note of what appear to be significant advantages, and taking note of occasionally reported problems. The short history presented below suggests interesting potential.

Rostoker and Berger (1953) were among the first users in the Western world. They noted that vibration was useful in getting liquid Al alloy to enter small holes, the effect increasing with increasing acceleration up to around $4g$ (where g is the gravitational acceleration). Intensities higher than this caused metal to jump out of the mould and caused hot tearing in the casting. In retrospect, the hot tearing could almost certainly have been eliminated by a good casting technique.

Walther and colleagues (1954) confirm the hot tearing experiences and note outgassing and surface imperfections in permanent mould castings at what appear to be nodes in the 60 Hz vibration pattern. It is noteworthy that the hot tearing and outgassing reports are limited to early work, probably using poor quality metal and poor degassing techniques.

Levinson and co-workers (1955) using Al-7Si-0.4Mg alloy investment moulds in steel flasks, bolted to a table vibrating at 60 Hz, found that as vibration intensity increased from zero to $4g$, the smallest hole filled by the liquid metal reduced from close to 1 mm down to 0.25 mm. They attribute this demonstration of improved filling to the fracturing of the oxide skin on the meniscus of the liquid metal.

Next came an interesting study carried out with pressure die casting by Weber and Rearwin in 1961 in which they bolted a sonic vibrator to the die. An Al alloy was found to have a finer, denser structure and improved properties. In addition they report that the vibration allowed the gate to be opened up and the injection velocity lowered, allowing them to reduce a 30–40% scrap rate on an Al alloy casting to zero. A zinc alloy also benefited with improved properties, and exhibited a dramatic improvement in its surface condition, reducing electrolytic plating rejects by 20%. These authors are also enthusiastic about the use of ultrasonic vibrations to the gate area of dies and to cores, allowing them to withdraw a zero-draft core from a completely solidified casting. They also make reference to core breakages when ultrasonics are applied, but such fatigue fractures should be resisted by cleaner, modern steels.

Work at the Massachusetts Institute of Technology by Flemings and co-workers spread over many years (1962 and 1987) using 60 Hz on Al alloys in greensand moulds has confirmed the action of vibration in helping to define detail of moulds, even though the accelerations were modest, being limited to 0.37–1.3g (corresponding respectively to amplitudes 25–90 μm). These authors cite the work of Angelov (1969) who claims to have increased the fluidity of Al-12Si alloy as much as a factor of three by applying vibration 100 Hz at the base of the sprue. At this frequency the increase in fluidity was roughly proportional to the amplitude of the vibration.

It seems that vibration could be useful and deserves to be used more.

3.3 EXTENDED FLUIDITY

We have examined metallurgical techniques in which fluidity can be extended, such as cleaning the melt from suspended oxides; reducing the rate of heat transfer by mould material or mould coat; and providing a coating to confer an unstable substrate for freezing. All these approaches involve flow of liquid in channels that are sufficiently narrow that the flow is constrained to be essentially unidirectional.

The casting technique for extending fluidity presented in this section involves multi-directional flow in which the cool metal at the flow tip is diverted away from the general direction of advance, allowing fresh, hot metal to take its place at the flow tip so that the total flow distance can be extended.

Such an effect is common in castings, and happens naturally if the flow channel is sufficiently wide so that the melt arriving at the front is diverted sideways to freeze, leaving a central flow channel open (Figures 5.24(b) and 3.24(a)). This flow channel phenomenon is easily identified in the structure of the casting as a lengthy region of coarse grains, and

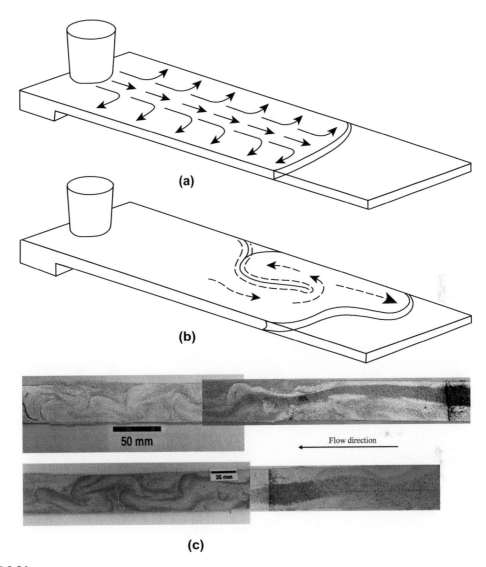

(a)

(b)

50 mm

Flow direction

25 mm

(c)

FIGURE 3.24

Flow patterns of Al-Si alloy on (a) a 4 mm thick plate cast at 700°C; (b) a 3 mm thick fluidity strip cast at 750°C; (c) photographs of the strip cast in (b) showing flow channel behaviour and lap defects formed by the irregularities of the slowing flow.

Courtesy A Habibollahzadeh (2003).

sometimes some porosity, in contrast with the sides of the casting which are fine grained and dense. Tol and co-workers (1997) notice this effect in their computational fluid dynamic studies of flow in a narrow strip casting; as the number of cells across the thickness was increased the predicted fluidity distance increased because of the formation of a flow channel along the centre of the strip (Figure 3.24(a)).

A less desirable multi-directional effect can occur in some fluidity test sections if they are sufficiently wide to allow an oxide flow tube to progress up one side of the fluidity channel, breaking through to the other side and flowing to back-fill (Figure 3.24(b)). This diverted flow takes away some of the cold metal from the tip, allowing the front to move ahead once again, but such extension to the fluidity cannot be recommended because a bifilm crack may be formed, extending through the cast section, effectively separating the flow channels. In the absence of close walls to constrain the flow front in this fluidity channel, which at 40 mm is unusually wide, the irregular bursting of the oxide on the front dictates a randomness that gravity is powerless to control in this horizontal channel. The result can at times become quite chaotic as seen in Figure 3.24(c).

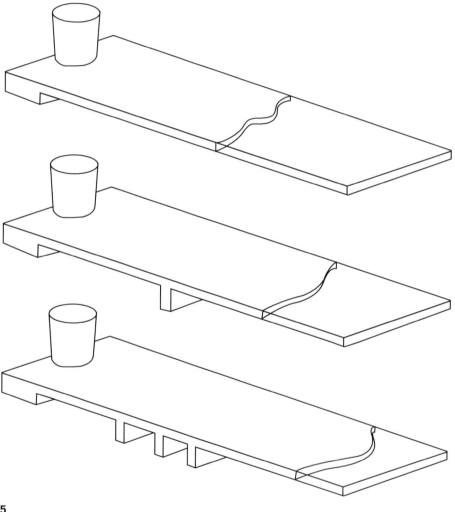

FIGURE 3.25

Flow distances of grey iron in a long plate enhanced by the provision of flow-offs to take away the cool front metal, allowing it to be replaced by hotter metal (Hiratsuka et al., 1998).

Hiratsuka and colleagues (1998) describe an artificial technique for extending fluidity, by providing overflows or 'flow-offs' at intervals along the length of a long casting. Their purpose was to extend the grey zone in a long grey cast iron casting, 3 mm thick × 60 mm wide, pushing the white, chilled iron beyond the length required for good machinability. Figure 3.25 shows the way in which one overflow of 20 mm length increased the flow distance by 30 mm, showing a positive gain in efficiency of the addition. However, the three overflows totalling 60 mm of overflow increased the flow length of the casting by almost exactly 60 mm.

One of the reasons that Hiratsuka's extension technique was successful was the extreme narrowness of the sections, so that problems of circulating or reverse flow that might threaten to bring the cold liquid back into the casting, were avoided.

The positioning of the overflows in Hiratsuka's work is curious, and would not be expected to result in an optimum effect. The authors give no reason for this. It seems there is plenty of scope for research to find rules for optimum positioning of overflows to obtain maximum benefit.

3.4 CONTINUOUS FLUIDITY

In a series of papers published in the early 1960s, Feliu (13) introduced a concept of the volume of flow through a section before flow was arrested. He carried out this investigation on, among other methods, a spiral test pattern, moulded in green sand. He made a number of moulds, cutting a hole through the drag by hand to shorten the spiral length. The metal that poured through the escape hole was collected in a crucible placed underneath, and weighed together with the length of the cast spiral. As the flow distance was progressively reduced by cutting progressively nearer to the start on a succession of moulds, he discovered that at a critical flow distance the metal would continue flowing indefinitely (Figure 3.26). Clearly, any metal that had originally solidified in the flow channel was subsequently re-melted by the continued passage of hot metal.

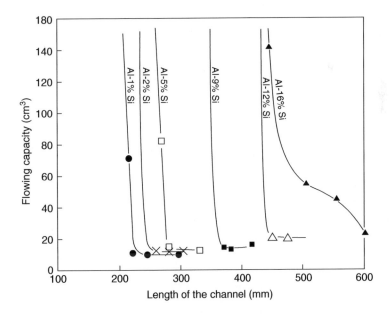

FIGURE 3.26

Flow capacity of the channel as a function of the length of the channel (Feliu, 1962).

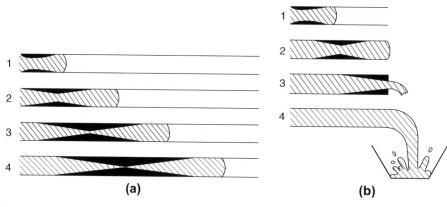

FIGURE 3.27

Concepts of (a) maximum fluidity length showing the stages of freezing leading to the arrest of the flow in a long mould; and (b) the continuous flow that can occur if the length of the mould does not exceed a critical length, defined as the continuous fluidity length.

The conditions for re-melting in the channel so as to allow continuous flow are illustrated in Figure 3.27. The concept is essential to the understanding of running systems, whose narrow sections would otherwise prematurely block with solidified metal. It is also clearly important in those cases where a casting is filled by running through a thin section into more distant heavy sections.

Because of its importance, I have coined the name 'continuous fluidity length' for this measurement of a flow distance for which flow can continue to take place indefinitely. It contrasts with the normal fluidity concept, which to be strict, should perhaps by more accurately named as 'maximum fluidity length'.

The results by Feliu shown in Figure 3.28(a) seem typical. The *maximum* fluidity length has a finite value at zero superheat. This is because the liquid metal has latent heat, at least part of which has to be lost into the mould before the metal ceases to flow. *Continuous fluidity*, on the other hand, has zero value until the superheat rises to some critical level. (Note that in Figures 3.28(a–c), the liquidus temperature T_m has been reduced from that of the pure metal by 5–10°C to allow for the presence of impurities.)

Figures 3.28(a–c) display three zones: (1) is a zone in which the flow distance is sufficiently short, and/or the temperature sufficiently high, that flow continues indefinitely; (2) is a region between the maximum and the continuous fluidity thresholds where flow will not flow indefinitely, but will continue for increasingly long periods as distance decreases, or temperature rises; and (3) is a zone in which the flow distance cannot be achieved, bounded on its lower edge by the maximum fluidity threshold.

Examining the implications of these three zones in turn; zone (1) is the regime in which most running systems operate; zone (2) is the regime in many castings, particularly if they have thin walls; zone (3) is the regime of bitter experience of costly redesigns, often after all the budget has been expended on the patternwork, and it is finally acknowledged that the casting cannot be made. Fluidity really can therefore be important to the casting designer and to the founder. It is a pleasure and relief to acknowledge that experience (3) can now be avoided by good computer simulation.

As a detail that may have some importance, whereas the curves derived by Feliu exhibit abrupt vertical turns when the continuous fluidity starts (Figure 3.26), this behaviour is not shown for rather pure metals (and possibly eutectics) as

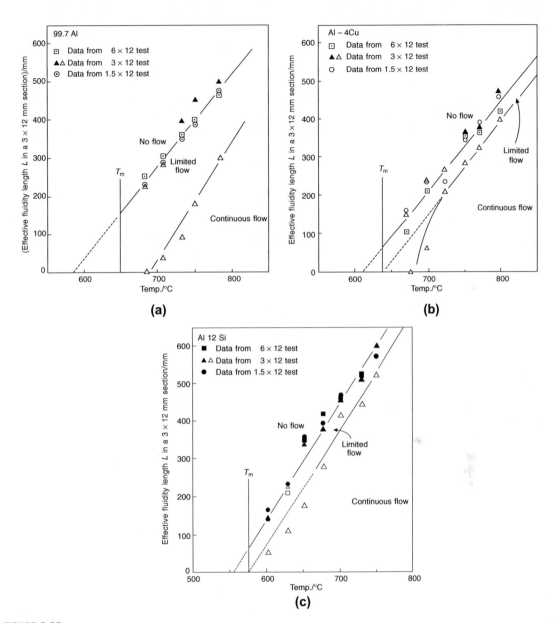

FIGURE 3.28

(a) Maximum and continuous fluidity data by Feliu for 99.7Al cast into greensand moulds of section 6 × 12, 3 × 12 and 1.5 × 12, all reduced for presentation in this figure as though cast only in a section 3 × 12 mm. (b) Data for Al-4Cu alloy by Feliu recalculated as though only from section 3 × 12 mm. (c) Data for Al-12Si alloy by Feliu recalculated as though only from section 3 × 12 mm.

seen in Feliu's original publication (1962) for both Al-Si and Al-Cu alloys. This is probably the result of the smoother growth front, consisting of planar or cellular growth being more resistant to re-melting, whereas heat transfer to an array of dendrites is probably fast, promoting rapid re-melting.

The author is aware of little other experimental work relating to continuous fluidity. An example worth quoting because of its rarity is that of Loper and LeMahieu on white irons in green sand dating from 1971. (Even so, the interested reader should take care to note that freezing time is not measured directly in this work.)

There is a nice computer simulation study carried out at Aachen University (Wu and Sahm 1997) that confirms the principles outlined here. More work is required in this important but neglected field.

MOULDS AND CORES

4

As in so much technology, the production of moulds and cores is a complex subject that we can only touch upon here. Although this text is about the metallurgy of the casting, the mould can be profoundly influential. The mould and the casting co-exist for sufficiently long, and at a sufficiently high temperature, that the two cannot fail to have an important mutual impact. Essentially, this section is about the *science* of mould and casting interactions.

In the later part of the book, *Casting Manufacture*, we concentrate on the *technology* of moulds and castings, outlining the main benefits and problems of the various types of moulds and cores to allow a user to make a more informed choice of manufacturing route.

4.1 MOULDS: INERT OR REACTIVE

Very few moulds and cores are really inert towards the material being cast into them. However, some are very nearly so, especially at lower temperatures. This short section lists those types of moulds and cores that in general do not react with their cast metals.

The usefulness of a relatively inert mould is emphasised by the work of Stolarczyk (1960), who suffered 4.5% porosity in gunmetal test bars cast in greensand moulds compared with 0.5% porosity for identical castings into steel-lined moulds. This simple experiment confirms one of the most important advantages of metal moulds: they are impressively inert towards their liquid metal and cast product. Thus all of the gravity and low pressure dies (permanent moulds) are reassuringly inert.

Unfortunately, the inertness of high pressure die casting dies and some squeeze casting dies is compromised by die coolants and lubricants which definitely impair the casting to varying degrees.

Greensand moulds that have been dried in an oven (i.e. dry sand moulds) have been found to be largely inert, as shown by Locke and Ashbrook (1950).

This behaviour contrasts with that of the original greensand in its various forms. The water and hydrocarbon contents of greensand mixes lead to masses of outgassing and very rapid attack, leading to oxidation or deposit of carbon on the liquid metal front, plus the generation of copious amounts of lower hydrocarbons (such as methane) and even more hydrogen. At first sight, it is perhaps amazing that any useful casting could be produced by such a reactive system. However, greensand remains, justifiably, the most important volume producer of good castings worldwide. Its many benefits, particularly its unbeatable moulding speed, are discussed in more detail in Chapter 15.

Whereas the various resin binders used for sand moulds do not outgas as impressively as greensand, the quantities and nature of the various volatiles that are attempting to escape during the dramatic early seconds and minutes of attack by the melt are seen in Figures 4.1 and 4.2. Clearly, there is plenty of chemistry involved, and the high temperatures ensure that this is energetic.

In recent years, several synthetic aggregates have become available in tonnage quantities. The hardness, wear and fracture resistance of these new moulding materials has meant that their high cost can sometimes be justified because of the very high efficiency with which they can be recycled because of low losses. The materials are mainly based on highly stable, high melting point oxides such as alumina and silica. A mixture of these two oxides to give a crystal composition such as mullite has the great benefit of avoiding the alpha/beta quartz transition, and so has the potential to improve the accuracy of castings. The synthetic grains so far available are produced in two ways, both of which result in beautifully spherical grains which have excellent flowability. However, they can have quite different results for some castings.

Complete Casting Handbook. http://dx.doi.org/10.1016/B978-0-444-63509-9.00004-2

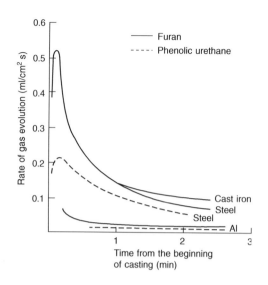

FIGURE 4.1

Measured gas evolution rates from castings of aluminium, iron and steel in chemically bonded sand moulds (Bates and Monroe, 1981).

1. Aggregates can be produced in the solid state, aggregating powders with an organic binder, by rolling around on oscillating trays. The weakly bonded spherical grains are then sintered, eliminating the temporary binder, and forming strong grains but containing several percent of porosity, some closed and some continuously linked through the grain. My experience with this variety of aggregate is that the mould binder tends to enter the pores of the grains, unfortunately boiling and exploding when contacted with liquid aluminium alloy, thus punching vapour pores into the surface of the casting. It is not known whether such 'microblows' occur with other metals and alloys. Such microblows will, of course, occur in addition to the chemical reactions normally to be expected between the binder and the liquid metal.
2. Aggregates can be produced by a melting route, one of which employs a jet of liquid oxide which is atomised by a blast of air. Such grains tend to be perfectly sound and can still be very respectably spherical. Moulds made from such aggregates would be expected to react only by reason of the presence of the aggregate binder.

4.2 TRANSFORMATION ZONES

As the hot metal is poured into a greensand mould, the blast of heat from the melt at temperature greatly exceeding the boiling point of water, very rapidly heats the surface of the mould, boiling off the water (and other volatiles). As the heat continues to advance into the mould, the moisture continues to migrate away, only to condense again in the deeper, cooler parts of the mould. As the heat continues to diffuse in, the water evaporates again and migrates further. This is of course a continuous process. Dry and wet zones travel through the mould like weather systems in the atmosphere. The evaporation of water in greensand moulds has been the subject of much research.

Looking at these in detail, four zones can be distinguished, as shown in Figure 4.3.

1. The dry zone is where the temperature is high and all moisture has been evaporated from the binder. It is noteworthy that this very high temperature region will continue to retain a relatively stagnant atmosphere composed of nearly 100% water vapour as a hot dry gas. However, of course, some of this superheated and very dry steam will be reacting at the casting surface to produce oxide and free hydrogen.

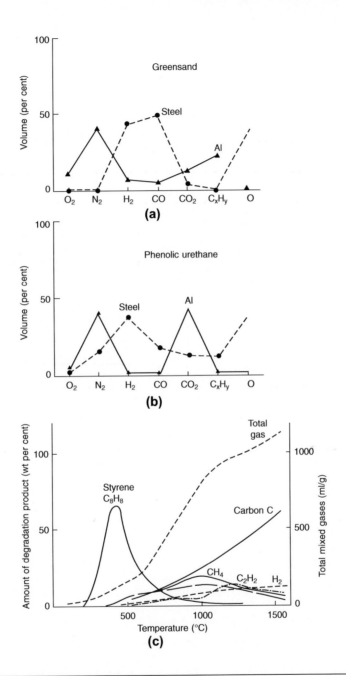

FIGURE 4.2

Composition of mould gases (a) from greensand (Chechulin, 1965); (b) phenolic urethane (Bates and Monroe, 1981) and (c) thermally decomposing expanded polystyrene (Goria et al., 1986).

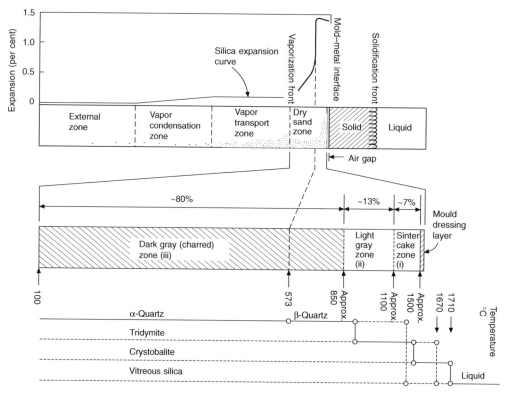

FIGURE 4.3

Structure of the heated surface of a greensand mould against a steel casting and the forms of silica (after Sosman, 1927), with solid lines denoting stable states and broken lines denoting unstable states.

2. The vapour transport zone, essentially at a uniform temperature of 100°C, and at a roughly constant content of water, in which steam is migrating away from the casting.
3. The condensation zone, where the steam re-condenses. This zone was for many years the subject of some controversy as to whether it was a narrow zone or whether it was better defined as a front. The definitive theoretical model by Kubo and Pehlke (1986) has provided an answer where direct measurement has proved difficult; it is in fact a zone, confirming the early measurements by Berry, Kondic and Martin (1959). This zone gets particularly wet. The raised water content usually greatly reduces the strength of greensand moulds, so that mechanical failure is most common in this zone.
4. The external zone where the temperature and water content of the mould remain as yet unchanged.

It is worth taking some space to describe the structure of the dry sand zone. When casting light alloys and other low-temperature materials, the dry sand layer has little discernible structure.

However, when casting steel it becomes differentiated into various layers that have been detailed from time to time (e.g. Polodurov, 1965; Owusu and Draper, 1978). These are, counting the mould coating as number zero:

1. Dressing layer of usually no more than 0.5 mm thickness, and having a dark metallic lustre as a result of its high content of metal oxides.
2. Sinter cake zone, characterised by a dark brown or black colour. It is mechanically strong, being bonded with up to 20% fayalite, the reaction product of iron oxide and silica sand. The remaining silica exists as shattered quartz

grains partially transformed to tridymite and cristobalite, which is visible as glittering crystals (explaining the origin of the name cristobalite). This layer is largely absent when casting grey iron at ordinary casting temperatures.

3. Light-grey zone, with few cracked quartz grains and little cristobalite. What iron oxides are present are not alloyed with the silica grains. This zone is only weakly bonded, and disintegrates on touch.
4. Charred zone, of dark-grey colour, of intermediate strength, containing unchanged quartz grains but significant levels of iron oxide. Polodurov speculates that this must have been blown into position by mould gases.

The pattern of these zones is further complicated by convection effects inside the mould or core for a relatively long period after casting, carrying carbonaceous vapours back into the heated zones, chemically 'cracking' the compounds to release hydrogen and depositing carbon. The inner layers therefore become black, with the sand grains seen under the microscope to be coated with a kind of fibrous, furry layer of graphite. Churches and Rundman (1995) studying a phenolic urethane mould found this reaction to occur between 15 min and 3 h, and at temperatures down to about 540°C for their grey iron castings. Highly heated cores had lost up to 50% of their carbon after finally cooling to room temperature.

The changes in form of the silica sand during heating are complicated. An attempt to illustrate these relations graphically is included in Figure 4.3. This complexity, and particularly the expansion accompanying the phase change from alpha to beta quartz, has prompted a number of foundries to abandon silica sand in favour of more predictable moulding aggregates. This advantageous move is expected to become more widespread in future especially as interesting new synthetic aggregates become available.

4.3 EVAPORATION AND CONDENSATION ZONES

As the heat diffuses from the solidifying casting into the mould (Figure 4.4), the transformation zones migrate deeper into the mould. We can follow the progress of the advance of the zones by considering the distance d that a particular isotherm reaches as a function of time t. The solution to this simple one-dimensional heat-flow problem is

$$d = (Dt)^{1/2} \tag{4.1}$$

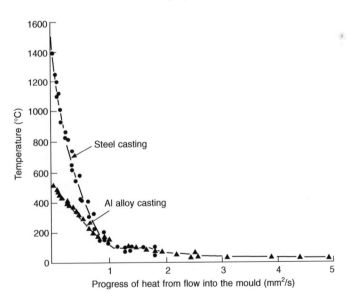

FIGURE 4.4

Temperature distribution in a greensand mould on casting an aluminium alloy (Ruddle and Mincher, 1949, 1950) and a steel (Chvorinov, 1940).

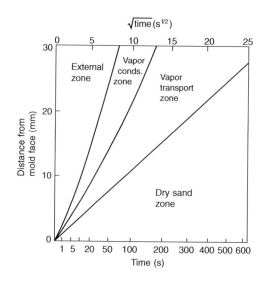

FIGURE 4.5

Position of the vapour zones after the casting aluminium in a greensand mould.

Data from Kubo and Pehlke (1986).

where D is the coefficient of diffusion. This solution is of course equivalent to the solute diffusion given earlier (Eqn (1.5)).

In the case of the evaporation front, the isotherm of interest is that at 100°C. We can see from Figures 4.4 and 4.5 that the value of D is close to 1 mm^{2}s^{-1}. This means that the evaporation front at 1 s has travelled 1 mm, at 100 s has travelled 10 mm, and requires 10,000 s (nearly 3 h!) to travel 100 mm. It is clear that the same is true for aluminium, as well as steel. (This is because we are considering a phenomenon that relies only on the rate of heat flow in the mould—the metal and its temperature are not involved.)

For the condensation zone the corresponding value of D is approximately 3 mm^{2}s^{-1}, so that the position of the front at 1, 100, and 10,000 s is 1.7, 17, and 170 mm, respectively.

These figures are substantiated to within 10 or 20% by the theoretical model by Tsai et al. (1988). This work adds interesting details such as that the rate of advance of the evaporation front depends on the amount of water present in the mould, higher water contents making slower progress. This is to be expected, because more heat will be required to move the front, and this extra heat will require extra time to arrive. The extra ability of the mould to absorb heat is also reflected in the faster cooling rates of castings made in moulds with high water content. Measurements of the thermal conductivity of various moulding sands by Yan et al. (1989) have confirmed that the apparent thermal conductivity of the moisture-condensation zone is about three or four times as great as that of the dry sand zone.

An earlier computer model by Cappy et al. (1974) also indicates interesting data that would be difficult to measure experimentally. They found that the velocity of the vapour was in the range of 10–100 mms^{-1} over the conditions they investigated. Their result for the composition and movement of the zones is given in Figure 4.6. Kubo and Pehlke calculate flow rates of 20 mms^{-1}. These authors go on to show that moisture vapourises not only at the evaporation front, but also in the transportation and condensation zones. Even in the condensation zone a proportion of the water vapourises again at temperatures below 100°C (Figure 4.5).

The pressure of water vapour at the evaporation front will only be slightly above atmospheric pressure in a normal greensand mould. However, because the pressure must be the same everywhere in the region between the mould-metal interface and the evaporation front, it follows that the dry sand zone must contain practically 100% water vapour. This is

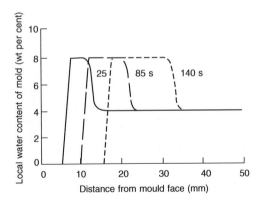

FIGURE 4.6

Water content of the vapour transport zone with time and position (smoothed computed results of Capp et al., 1974).

surprising at first sight. However, a moment's reflection will show that there is no paradox here. The water vapour is very dry and hot, reaching close to the temperature of the mould-metal interface. At such high temperatures, water vapour is highly oxidising. There is no need to invoke theories of additional mechanisms to get oxygen to this point to oxidise the metal—there is already an abundance of highly oxidising water vapour present (the breakdown of the water vapour also providing a high-hydrogen environment, of course, to enter the metal, and to increase the rate of heat transfer in the dry sand zone).

Kubo and Pehlke (1986) confirm that gas in the dry sand and transportation zones consists of nearly 100% water vapour. In the condensation zone, the percentage of air increases, until it reaches nearly 100% air in the external zone (water vapour would be expected to be present at its equilibrium vapour pressure, 32 mmHg or 42 mb at 30°C).

It is found that similar evaporation and condensation zones are present for other volatiles in the greensand mould mixture. Marek and Keskar (1968) have measured the movement of the vapour transport zone for benzene and xylene. The evaporation and condensation fronts of these more volatile materials travel somewhat faster than those of water. When such additional volatiles are present they will, of course, contribute to the 1 atm of gas pressure in the dry zone, helping to dilute the oxidising effect of water vapour and helping to explain part of the beneficial effect of such additives. In the following section, we will see that many organics decompose at these high temperatures, providing a deposit of carbon, which further assists, in the case of such metals as cast iron, in preventing oxidation and providing a non-wetting metal surface of carbon which contains the liquid iron and helps to prevent its contact with the sand mould.

It is to be expected that vapour transport zones will also be present to various degrees in chemically bonded sands. The zones will be expected to have traces of water mixed with other volatiles such as organic solvents. Little work appears to have been carried out for such binder systems, so it is not easy to conclude how important the effects are, if any. In general, however, the volatiles in such dry sand systems usually total less than 10% of the total volatiles in greensand, so that the associated condensation zones will be expected to be less than one-tenth of those occurring in greensand. It may be, therefore, that they will be unimportant. However, at the time of writing we cannot be sure; it would be nice to know.

What is certain is that silicate binders appear to have a high chilling effect in ferrous castings because of their water content. The loss of the water will create evaporation and condensation zones that will carry heat away from the casting.

All of these considerations on the rate of advance of the moisture assume no other flows of gases through the mould. This is probably fairly accurate in the case of the drag mould, where the flow of the liquid metal over the surface of the mould effectively seals the surface against any further ingress of gases. A certain amount of convection is expected in the mould, but this will probably not affect the conditions in the drag significantly.

In the vertical walls of the mould, however, convection may be significant. Close to the hot metal, hot gases are likely to diffuse upwards and out of the top of the mould, their place being taken by cold air being drawn in from the surroundings at the base of the mould, or the outer regions of the cope.

General conditions in the cope, however, are likely to be more complicated. It was Hofmann in 1962 that first emphasised the different conditions experienced during the heating up and outgassing of the cope. He pointed out that the radiated heat from the rising melt would cause the cope surface of the mould to start to dry out before the moment of contact with the melt. During this pre-contact period two different situations can arise:

1. If the mould is open, as the cope surface heats up, the water vapour can easily escape through the mould cavity and out via the opening (Figure 4.7). The rush of water vapour through an open feeder can easily be demonstrated by holding a piece of cold metal above the opening. It quickly becomes covered with condensate. The water vapour starts its life at a temperature of only 100°C. It is therefore a relatively cool gas, and is thus most effective in cooling the surface of the mould as it travels out through the surface of the cope on its escape route.

2. If the mould is closed, the situation is quite different. The air being displaced and expanded by the melt will force its way through the mould, carrying away the vapour from the interface (Figure 4.7). The rate of flow of air is typically in the $10-100 \ ls^{-1}m^{-2}$ range (the reader is encouraged to confirm this for typical castings and casting rates). This is in the same range of flow rate as the transport of vapour given in computer models. Thus if the casting rate is relatively low, then the vapour transport zone is likely to be relatively unaffected, although perhaps a little accelerated in its progress. When the casting rate is relatively high, then the vapour transport zone will be effectively blown away, diluted with the gale of air so that no condensation can occur.

Because the water vapour is driven away from the surface and into the interior of the mould, its beneficial cooling effect at the surface is not felt, with the result that the surface reaches much higher temperatures as a result of the direct

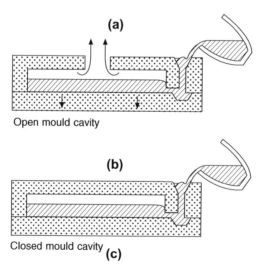

(a)

Open mould cavity

(b)

Closed mould cavity **(c)**

FIGURE 4.7

Three conditions of vapour transport in moulds: (a) unrestricted free evaporation from the cope; (b) evaporation from the cope constrained to occur via the cope mould; (c) evaporation from the drag confined by the cover of metal, and possibly confined by the substrate on which the mould sits, and possibly at its sides by a moulding box, leaving no option but for bubbling through the metal.

Partly from Hofman (1962).

radiation of heat from the melt, as is seen in Figure 4.8. The prospect of the failure of the cope surface by expansion and spalling of the sand is therefore much enhanced.

However, the rate of heating of the surface by radiation from the melt may be reduced by a white mould coat, such as a zircon- or alumina-based mould wash, now widely applied for large castings of iron and steel. Boenish (1967) confirmed that light-coloured mould coats resist scabbing for up to 400% longer than the jet-black graphitic surfaces commonly used for cast iron.

One final aspect of vapour transport in the mould is worth noting. There has been much discussion over the years about the contribution of the thermal transpiration effect to the flow of gases in moulds.

Although it appears to have been widely disputed, the effect is certainly real. It follows from the kinetic theory of gases and essentially is the effect of heated gases diffusing away from the source of heat, allowing cooler gases to diffuse up the temperature gradient. In this way it has been argued that oxygen from the air can arrive continuously at the casting to oxidise the surface to a greater degree than would normally have been expected.

Williams (1970) described an experiment that demonstrated this effect. He took a sample of clay approximately 50 mm long in a standard 25 mm diameter sand sampling tube. When one end was heated to 1000°C and the other was at room temperature, he measured a pressure difference of 10 mmHg if one end was closed, or a flow rate of 20 ml per min if both ends were open. If these results are typical of those that we might expect in a sand mould, then we can make a comparison as follows. The rate of thermal transpiration is easily shown to convert to $0.53 \ \mathrm{ls^{-1}m^2}$ for the conditions of temperature gradient and thickness of sample used in the experiment. From the model of Cappy et al. (1974), we obtain an estimate of the rate of transport of vapour of $100 \ \mathrm{ls^{-1}m^2}$ at approximately this same temperature gradient through a similar thickness of mould. Thus thermal transpiration is seen to be less than 1% of the rate of vapour transport.

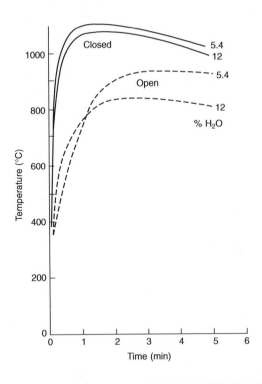

FIGURE 4.8

Temperature in the cope surface seen to be significantly lowered by open moulds and high moisture levels.

Data from Hofman (1962).

Additional flows such as the rate of volume displacement during casting and the rate of thermal convection in the mould will further help to swamp thermal transpiration.

Thermal transpiration does seem to be a small contributor to gas flow in moulds. It is possible that it may be more important in other circumstances. More work is required to reinstate it to its proper place, or lay it to rest as an interesting but unimportant detail.

4.4 MOULD ATMOSPHERE
4.4.1 COMPOSITION

On the arrival of liquid iron or steel in the mould, a rich soup of gases boils from the surface of the mould and the cores. The air originally present in the cavity dilutes the first gases given off, but this is quickly expelled through vents or feeders or may diffuse out through the cope. After the air is forced out, the composition of the mould gas is relatively constant. The speed at which the original gas in the mould is expelled by the arrival of the metal is the reason why attempts to fill the mould with inert gases to avoid oxidising reactions are practically always a waste of effort. The casting of Al alloys is rather less dramatic, and the original air is expelled only gradually as seen from experiments described in the following section.

In the case of steel being cast into greensand moulds, the mould gas mixture has been found to contain up to 50% hydrogen (Figure 4.2). The content of hydrogen depends almost exactly on the percentage water in the sand binder, with dry sand moulds having practically no hydrogen as found by Locke and Ashbrook (1950, 1972). Other changes brought about by increased moisture in the sand were a decrease in oxygen, an increase in the CO/CO_2 ratio and the appearance of a few% of paraffins (it is not clear whether these originated from the use of lubricant sprayed on to the pattern). The presence of cereals in the binder was found to provide some oxygen, even though the concentration of oxygen in the atmosphere fell because of dilution with other gases (Locke and Ashbrook, 1950). Chechulin (1965) describes the results for greensand when aluminium alloys, cast irons and steels are cast into them. His results are given in Figure 4.2(a). Irons and steels produce rather similar mould atmospheres, so only his results for steel are presented.

The high oxygen and nitrogen content of the atmosphere in the case of moulds filled with aluminium simply reflects the high component of residual air (originally, of course, at approximately 20% oxygen and 80% nitrogen). The low temperature of the incoming metal is insufficient to generate enough gas and expand it to drive out the original atmosphere. This effective replacement of the atmosphere is only achieved in the case of iron and steel castings.

The atmosphere generated when ferrous alloys are cast into chemically bonded sand moulds is, perhaps rather surprisingly, not so different from that generated in the case of greensand (Figure 4.2(b)). The mixture consists mainly of hydrogen and carbon monoxide.

The kinetics of gas evolution have been studied by a number of researchers. Jones and Grim (1959) found some clays evolved significant water at just over 100°C, particularly Western bentonite, although Southern bentonite results went wildly off scale. A second peak of evolution at around 550°C was relatively unimportant. Scott and Bates (1975) found that hydrogen evolution peaked within 4–5 min for most chemical binders. However, for the sodium silicate binder, a rapid burst of hydrogen was observed which peaked in less than 1 min.

Scarber et al. (2006) measured 200 ml gas from about 100 g (they do not give the exact weight of their samples) of a core bonded with some resin binders which rose to more than 500 ml if the core was coated with a water-based wash even though the core had been thoroughly dried. Even this paled into insignificance when compared with the outgassing of greensand which they found to be nearly 10 times that from resin-bonded sands. In addition, they found multiple peaks associated with solvent evaporation, and, probably, water evaporation. The process seemed to be complicated by the re-condensation of these volatiles in the cooler centres of cores, and their final re-evaporation (a re-enactment of the condensation and evaporation zones previously discussed in Section 4.3). Yamamoto et al. (1980) described the variation in output they observed as (1) peak I attributed to the expansion of air plus the evaporation of free moisture and other volatiles in the core when the core was first covered with liquid metal; (2) peak II after about 30 s identified as the release of combined water (water of crystallisation) in the binder and/or aggregate; and finally (3) a broad smooth peak III, constituting a nearly constant pressure output period for about 50–80 s, attributed to the general breakdown of the organics in the binder. They confirmed that peak II was much higher for sodium silicate/CO_2 cores.

Lost-foam casting, where the mould cavity is filled with polystyrene foam (the 'full mould' process), is a special case. Here it is the foam that is the source of gases as it is vapourised by the molten metal. At aluminium casting temperatures the polymerised styrene merely breaks down into styrene, but little else happens, as is seen in Figure 4.2(c). It seems that the liquid styrene soaks into the ceramic surface coating on the foam, so that the coating will temporarily become completely impermeable. This unhelpful behaviour probably accounts for many of the problems suffered by aluminium alloy castings made by the lost foam process.

At the casting temperatures appropriate for cast iron, more complete breakdown occurs, with the generation of hydrogen and considerable quantities of free carbon. Rao and Lee (1984) show how even methane is largely decomposed at these temperatures, forming less than 1% of a $C–H_2–O_2$ mixture at 1 atm. The carbon deposits on the advancing metal front as a pyrolytic form of carbon widely known as 'lustrous carbon'. Once formed, the layer is rather stable at iron-casting temperatures, and can therefore lead to serious defects if entrained in the metal. The problem has impeded the successful introduction of lost foam technology. For steel casting the temperature is sufficiently high to cause the carbon to be taken into solution. Steel castings of low or intermediate carbon content are therefore contaminated by pockets of high-carbon alloy. This problem has prevented lost-foam technology in the form of the full mould process being used for low- and medium-carbon steel castings.

Lost-foam iron castings are not the only type of ferrous castings to suffer from lustrous-carbon defects. The defect is also experienced in cast iron made in phenolic urethane-bonded moulds, and at times can be a serious headache. The absence of carbon is therefore a regrettable omission from the work reported in Figure 4.2(a). At the time Chechulin carried out this study, the problem would not have been known. Later, Gloria (1986) does report carbon as a product of decomposition of polystyrene foam.

It seems reasonable to expect that carbon may also be produced from the pyrolysis of other binder systems. More work is required to check this important point.

4.4.2 MOULD GAS EXPLOSIONS

The various reactions of the molten metal with the volatile constituents of the mould, particularly the water in many moulding materials, would lead to explosive reactions if it were not for the fact that the reactions are dampened by the presence of masses of sand in mould materials or cores. Thus although the reactions in the mould are fierce, and not to be underestimated, in general they are not of explosive violence because the 90% or more of the materials involved are inert (simply sand and nitrogen) and have considerable thermal inertia. Taleyarkhan (1998) draws attention to the important role of non-condensable gases (e.g. nitrogen) in the suppression of explosive reactions. Outgassing reactions are therefore rather steady and sustained.

When carrying out a series of experiments in about 1992, liquid aluminium was poured while the interior of a polyurethane-bonded mould was recorded by video though a glass window, Helvaee was surprised to note that in every case, on arrival of the metal in the mould, a flicker of flame licked across the mould cavity, sometimes more energetically than others. On no occasion was there an explosion. Much earlier, in 1948, Johnson and Baker recorded similar flashes of light when recording the filling of open greensand moulds by molten steel at 1650°C. It seems such phenomena might be common or even normal.

It is noteworthy that the gases from the outgassing of an aggregate mould may contain a number of potentially flammable or explosive gases. These include a number of vapours such as hydrocarbons such as methane, other organics such as alcohols and a number of reaction products such as hydrogen and carbon monoxide.

Because of the presence of these gases, explosions sometimes occur and sometimes not. The reasons have never been properly investigated. This is an unsatisfactory situation because the explosion of a mould during casting can be an unpleasant event. The author has witnessed explosions a number of times in furan-bonded boxless moulds when casting an aluminium alloy casting weighing more than 50 kg: there was a muffled explosion, and large parts of the sand mould together with liquid metal flew apart in all directions. After several repeat performances, the operators developed ways of pouring this component at the end of long-handled ladles to keep as far away as possible, never knowing whether the mould would explode. The cause always remained a mystery. Everyone was relieved when the job came to an end.

Explosions in and around moulds containing iron or steel castings are relatively common. One of the most common is from under the mould, between the mould and its base plate, after the casting has solidified, so that there is less danger either to personnel or casting. The sound of muffled explosions from the mix of CO and air under moulds is common in many greensand foundries.

With subsequent experience, and in the absence of any other suggestions, the following is suggested as a possible cause of the problem in the case of the light alloy casting.

Explosions can, of course, only happen when the flammable components of the gas mix with an oxidising component such as oxygen from the air. The mixing has to be efficient, which suggests that turbulence is important. Also, the mix often has to be within close compositional limits, otherwise either no reaction occurs, or only slow burning takes place. The limits for the carbon monoxide, oxygen and inert (carbon dioxide and nitrogen) gas mixtures are shown in Figure 4.9.

In the first edition of this book the mixing with air, which is essential for explosions, was thought to occur only under certain conditions including feeder heads open to the air and the use of two sprues which would be not easily synchronised. Subsequent experience has modified these conclusions, even though definitive reasons are still not known. It is now clear that at least 50% of the mixture arriving at the base of the sprue is air, scrambled in with the hot metal, as a result of the pumping action of the conical trumpet entrance to the sprue, and the fact that most sprues are oversized and therefore allow the ingress and entrainment of air. Thus bubbles of air, together with hot splashes of metal, mixing turbulently with the mould gases are a recipe for explosive conditions.

We can surmise that contact pouring, together with a slim 'naturally pressurised' filling system design would eliminate explosions as a result of the melt displacing the air in the filling system, then filling the mould uphill, with minimal mixing of the hot mould gases in contact with the melt; melt and gases will both tend to rise steadily through the mould as progressive layers. In a system filling quiescently from the bottom upwards, the outgassing of the mould and cores will provide a spreading blanket of gas over the liquid. There will be almost no air in this cover, so no burning or explosion can occur. The air will be displaced ahead and will diffuse out of the upper parts of the mould. Where the flammable gas blanket meets the air, it is expected to be cool, well away from the liquid metal. Thus any slight mixture that will occur at the interface between these layers of gases is not likely to ignite to cause an explosion.

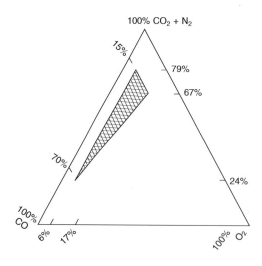

FIGURE 4.9

Shaded region defines the explosive regime for mixtures of CO, O_2, and a mixture of $CO_2 + N_2$ (Ellison and Wechselblatt, 1966).

Occasionally, Al-Si alloys which have been usually docile during pouring have been known to explode after the addition of Sr as a modifier. This is thought to be the enhanced reactivity of the Sr, converting moisture to hydrogen gas. The elimination of the trumpet intake and the provision of a properly calculated filling system should both help, as would the reduction or elimination of the Sr addition.

4.5 MOULD SURFACE REACTIONS

4.5.1 PYROLYSIS

When the metal has filled a sand mould, the mould becomes hot. A common misconception is to assume that the sand binder then burns; however, this is not true. It simply becomes hot. There is insufficient oxygen to allow any significant burning. What little oxygen is available is consumed in a minor, transient oxidation which quickly comes to a stop. What happens then to the binder is not burning, but pyrolysis.

Pyrolysis is the decomposition of compounds, usually organic compounds, simply by the action of heat. Oxygen is absent, so that no burning (i.e. high temperature oxidation) takes place. Pyrolysis of various kinds of organic binder components to produce carbon is one of the more important reactions that take place in the mould surface. Carbon is poorly wetted by many liquid metals, so the formation of carbon on the grains of sand, as a pyrolysed residue of the sand binder, produces a non-wetted mould surface, which can lead to an improved surface finish to the casting (although, as will become clear, the effect on the surface film on the liquid metal is probably more important).

This non-wetting feature of residual carbon on sand grains is at first sight curious, because carbon is soluble in many metals, and so should react, and should therefore wet. Cast iron would be a prime candidate for this behaviour. Why does this not happen?

In the case of ductile iron, sand cores do not need a core coat because the solid magnesium oxide rich film (probably a magnesium silicate) on the surface is a mechanical barrier that prevents penetration of the metal into the sand.

In the case of grey cast iron in a greensand mould the atmosphere may be oxidising, causing the melt surface to grow a film of liquid silicate. This is highly wetting to sand grains, so that the application of a core coating such as a refractory wash may be necessary to prevent penetration of the metal into the core.

However, in the case of grey iron cast in a mould rich in hydrocarbons (i.e. greensand with sufficient additions of coal dust or certain resin-bonded sands), metal penetration is prevented when the hydrocarbons in the atmosphere of the mould pyrolyse on contact with the surface of the hot liquid metal to deposit a thin film of solid carbon on the liquid. Thus the reason for the robust non-wetting behaviour is that a solid pyrolysed carbon film on the liquid contacts a solid, pyrolysed carbon layer on the sand grains of the mould surface. This carbon-against-carbon behaviour explains the excellent surface finish that grey irons can achieve if bottom-gated and filled in a laminar fashion (Figure 6.34).

For the casting of iron, powdered coal additions or coal substitutes are usually added to greensands to improve surface finish in this way, providing a carbon layer to both the sand grains and the liquid surface. The reactions in the pyrolysis of coal were originally described by Kolorz and Lohborg (1963):

1. The volatiles are driven out of the coal to form a reducing atmosphere.
2. Gaseous hydrocarbons break down on the surface of the liquid metal. (Kolorz and Lohborg originally thought the hydrocarbons broke down on the sand grains to form a thin skin of graphite, but this is now known to be not true.)
3. The coal swells, and on account of its large expansion, is driven into the pores of the sand. This plastic phase of the coal addition appears to plasticise the binder temporarily and thereby eases the problems associated with the expansion of the silica, allowing its expansion to be accommodated without the fracture of the surface. As the temperature increases further and the final volatiles are lost, the mass becomes rigid, converting to a semi-coke. The liquid metal is prevented from contacting and penetrating the sand by this in-filling of carbon that acts as a non-wetted mechanical barrier.

Kolorz and Lohborg recommend synthetically formulated coal dusts with a high tendency to form anthracitic carbon, of good coking capacity and with good softening properties. They recommend that the volatile content be near 30% and sulphur less than 0.8% if no sulphur contamination of the surface is allowable. If some slight sulphurisation is permissible, then 1.0–1.2% sulphur could be allowed. Peterson and Blanke (1980) emphasise the bituminous nature and low ash content of desirable coals.

In the case of phenolic urethane and similar organic chemical binders based on resin systems, the thermal breakdown of the binder assists the formation of a good surface finish to cast irons and other metals largely in the manner described previously. The binder usually goes through its plastic stage prior to rigidising into a coke-like layer. The much smaller volume fraction of binder, however, does not provide for the swelling of the organic phase to seal the pores between grains. So in principal, the sand remains somewhat vulnerable to penetration by the liquid metal. The second aspect of non-wettability is discussed next.

The founder should be aware of binder pyrolysis reactions that can have adverse effects on the surface quality of the casting. These are more likely to occur with binders that contain nitrogen (phenol urethanes and some furans) which can be converted to cyanide and amine gases. Different reactions occur in the case of furans, where problems may arise from the use of the sulphonic acid or phosphoric acid catalysts which become cumulative, being least when new sand is used and progressively more concentrated as the content of mechanically reclaimed sand in the mix increases. Sulphonic acids react with iron and iron oxide residues in the sand, with feldspars, with limestone often present in poor quality sand, and with certain types of coatings, to form the corresponding sulphonates. These will in turn be reduced to sulphides during casting and these may then cause sulphide damage to cast parts.

Phosphoric acid damage is caused by a different mechanism because phosphates are stable under casting conditions. It seems that the damage caused is primarily from the reaction of its vapour with iron[II] oxide (or chromite) dust, leading to the formation of iron[II] phosphate which coats the sand grains. Both the phosphoric acid vapour and the iron[II] phosphate can interact with components in the (ferrous) metal being cast.

4.5.2 LUSTROUS CARBON FILM

The carbonaceous gases evolved from the binder complete their breakdown at the white hot surface of the advancing front of liquid metal, giving up carbon and hydrogen to the advancing liquid front. For steel, the carbon dissolves quickly and usually causes relatively little problem. For cast iron, the carbon dissolves hardly at all, because the temperature is lower, and the metal is already nearly saturated with carbon; thus the carbon cannot dissolve away into the liquid iron as quickly as it arrives, so that it accumulates on the surface as a film. The time for dissolution seems to be about the same as the time for mould filling and solidification; thus the film has a life sufficiently long to affect the flow of the liquid for good or ill, depending on the circumstances.

If the metal is rising nicely in the mould, the surface film migrates across the liquid surface as it splits and reforms and then thickens with time on its approach to the side walls. The rising metal causes the liquid iron to rolls out on the surface of the mould, rolling out and laying down its surface film as the liquid advances. This laying down of the surface film can be likened to the laying down of the tracks of a track-laying vehicle. The film confers a non-wetting behaviour on the liquid itself because the liquid is effectively sealed in a non-wetting, dry solid skin. The skin forms a mechanical barrier between the liquid and the mould. This barrier is laid down as the liquid progresses because of friction between the liquid and the mould, the friction effectively stretching the film and tearing it near the centre of the meniscus where it immediately re-forms to continue the process (Figures 2.2 and 6.34). The strength and rigidity of the carbon film helps the liquid surface to bridge unsupported regions between sand grains or other imperfections in the mould surface. By this mechanism the surface of the casting is smoothed to a significantly finer finish that could be achieved by surface tension alone attempting to bridge between the sand grains.

The benefits of laying out a solid film as the liquid iron rolls out on the mould surface is only achieved if the liquid is encouraged to fill progressively, as a rolling action up against the mould surface. The mechanism for the improvement of surface finish can only operate effectively if the progress of the meniscus is steady and controlled, i.e. in the absence of surface turbulence. The casting surface can then take on an attractive, smooth glossy shine. It makes sense to fill the casting nicely using a properly designed bottom-gated technique. However, if the melt is top-poured, splashing and hammering the mould, no benefit is obtained; in fact, the liquid penetration can be severe giving a casting surface so spiky, sharp and rough to take the skin off your hands.

In addition to surface finish problems, a further problem arises if surface turbulence causes the carbon film to become entrained in the liquid. The film will necessarily become entrained as a double film, a carbon bifilm (because single films cannot be entrained) constituting a serious crack-like defect. Fortunately, in heavy-section castings, the entrained double film defect has time to dissolve, and never seems to be a problem. Light-section castings should only be attempted using binder systems that produce little lustrous carbon because defects will not have time to dissolve and will therefore be effectively permanent. Alternatively, a preferable strategy would of course be the use of a running system that could guarantee the absence of surface turbulence.

The lustrous carbon film, in contrast to all other surface films on other metals that I know, appears to detach easily from the iron during cooling. This curious behaviour means that after breaking the casting out of the mould, the film is sometimes seen attached to the surface of the mould, covering the surface as a smooth, glossy sheet, leading many observers to the mistaken conclusion that the film had initially formed on the mould surface (Naro, 2004). This is clearly not possible, of course, because pyrolysis reactions can only occur in high temperature conditions, such as on the melt surface, in contrast to the sand grains which are relatively cold.

4.5.3 SAND REACTIONS

Other reactions in the mould surface occur with the sand grains themselves. The most common of sand reactions is the reaction between silica (SiO_2) and iron oxide (wustite, FeO) to produce various iron silicates, of which the most commonly quoted (but not necessarily the most common nor the most important) is fayalite (Fe_2SiO_4). This happens frequently at the high temperatures required for the casting of irons and steels. It causes the grains to fuse and collapse as they melt into each other, because the melting point of some of the silicates is below the casting temperature. (The much quoted melting point of fayalite at only 1205°C may be an error because fayalite may have a significantly higher melting temperature. A more likely candidate to melt is grunerite, $FeSiO_3$). The reacted grains adhere to the surface of the casting because of the presence of the low-melting-point liquid 'glue'. This is known as burn-on.

The common method of dealing with this problem is to prevent the iron oxidising to form FeO in the first place. This is usually achieved by adding reducing agents to the mould material, such as powdered coal to greensand or aluminium powder additions to mould washes and the like. The problem is also reduced in other sands that contain less silica, such as chromite sand. However, the small amounts of silica present in chromite can still give trouble in steel castings, where the extreme temperature causes the residual silica to fuse with the clay. At these temperatures, the chromite may break down, releasing FeO or even droplets of Fe on the surface of the chromite sand grains. Metal penetration usually follows as the grains melt into each other, and the mould surface generally collapses. The molten, fused mass is sometimes known as 'chromite glaze'. It is a kind of burn-on, and is difficult to remove from steel castings (Petro and Flinn, 1978). Again, carbon compounds added to the moulding material are useful in countering this problem (Dietert et al., 1970).

Peterson and Blanke (1980) draw attention to the potential role of hydrogen released during the pyrolysis of coal dust in greensand, proposing that hydrogen plays an important part in avoiding the formation of oxides and silicates, and which may explain its contribution to good surface finish and the avoidance of burn-on. If hydrogen is actually valuable in this way, it clearly cannot arise from the main carbon constituent of coal, but must be generated from its minor hydrocarbon content, strongly suggesting that coal might be eliminated in favour of a more targeted binder addition.

4.5.4 MOULD CONTAMINATION

There are a few metallic impurities that find their way into moulding sands as a result of interaction between the cast metal and the mould. We are not thinking for the moment of the odd spanner or tonnes of iron filings from the steady wearing away of sand plant. Such ferrous contamination is retrieved in most sand plants by a powerful magnet located at some convenient point in the recirculating sand system. (The foundry maintenance crew always have interesting stories to tell of items found from time to time attached to the magnet.) Nor are we thinking of the pieces of tramp metal such as flash and other foundry returns. Our concern is with the microscopic traces of metallic impurities that lead to a number of problems, particularly because of the need to protect the environment from contamination.

Foundries that cast brasses find that the grains of their moulding sand become coated with a zinc-rich layer, with lead-rich nodules on the surface of the zinc (Mondloch et al., 1987). The metals are almost certainly lost from the casting by evaporation from the surface after casting. The vapour condenses among the cool sand grains in the mould as either particles of metallic alloy, or reacts with the clay present, particularly if this is bentonite, to produce Pb-Al silicates. If there is no clay present, as in chemical binder systems such as furan resins, then no reaction is observed so that metallic lead remains (Ostrom et al., 1982). Thus ways of reducing this problem are: (1) the complete move, where possible in simple castings, to metal moulds; (2) the complete move, where possible, from lead-containing alloys; or (3) the use of chemical binders, together with the total recycling of sand in-house. This policy will contain the problem, but the sand will have a fair degree of toxicity. If the metallic lead can be separated from the sand in the sand recycling plant, the proceeds might provide a modest economic return, and the sand toxicity could be limited.

There has been a suggestion that iron can evaporate from the surface of a ferrous casting in the form of iron carbonyl $Fe(CO)_5$. This suggestion appears to have been eliminated on thermodynamic grounds; Svoboda and Geiger (1969) show that the compound is not stable at normal pressures at the temperature of liquid iron. Similar arguments eliminated the carbonyls of nickel, chromium and molybdenum. These authors carry out a useful survey of the existing knowledge of the vapour pressures of the metal hydroxides and various sub-oxides but find conclusions difficult because the data is sketchy and contradictory. Nevertheless, they do produce evidence that indicates vapour transport of iron and manganese occurs by the formation of the sub-oxides $(FeO)_2$ and $(MnO)_2$. The gradual transfer of the metal by a vapour phase, and its possible reduction back to the metal on arrival on the sand grains coated in carbon might explain some of the features of metal penetration of the mould, which is often observed to be delayed, and then occur suddenly. More work is required to test such a mechanism.

The evaporation of manganese from the surface of castings of manganese steel is an important factor in the production of castings. The surface depletion of manganese seriously reduces the surface properties of the steel. In a study of this problem, Holtzer (1990) found that the surface concentration of manganese in the casting was depleted to an impressive non-trivial depth of 8 mm and the concentration of manganese silicates in the surface of the moulding sand was increased.

The process of Mn evaporation can be observed to occur from a drop of liquid steel placed on a water-cooled copper substrate. A 'halo' is seen to develop around the drop, indicating mainly Mn condensation, although Cr and Fe can also contribute (Nolli and Cramb, 2008).

Figure 6.26 confirms that the vapour pressure of manganese is significant at the casting temperature of steel. However, the depth of the depleted surface layer is nearly an order of magnitude larger than can be explained by diffusion alone. It seems necessary to assume, therefore, that the transfer occurs mainly while the steel is liquid, and that some mixing of the steel is occurring in the vicinity of the cooling surface.

It is interesting that a layer of zircon wash on the surface of the mould reduces the manganese loss by about half. This seems likely to be the result of the thin zircon layer heating up rapidly, thereby reducing the condensation of the vapour. In addition, it will form a barrier to the progress of the manganese vapour, keeping the concentration of vapour near the equilibrium value close to the casting surface. Both mechanisms will help to reduce the rate of loss. If, however, the protective wash is applied after the moulding sand has already become significantly contaminated with Fe and Mn oxides in the recycling process, the underlying sand may partially melt and collapse (Kruse, 2006). This instability of the underlying sand will cause mechanical penetration of the zircon wash, and extensive permeation of the metal into the underlying partially melted sand. The carryover of such contamination should be eliminated by careful control of the recycling process or revised selection of moulding aggregate (see the section on mould aggregates).

Gravity die casters that use sand cores (semi-permanent moulds) will be all too aware of the serious contamination of their moulds from the condensation of volatiles from the breakdown of resins in the cores. The buildup of these products can be so severe as to cause the breakage of cores, and the blocking of vents. Both lead to the scrapping of castings. The blocking of vents by tar-like deposits in permanent moulds is the factor that controls the length of a production run prior to the mould being taken out of service for cleaning. On carousels of dies in Al alloy cylinder head production, a die may need to be taken out of service every 10th or 15th casting. The absence of such problems in sand moulds is a natural advantage, aiding the already high productivity of sand moulding that is usually overlooked.

4.5.5 MOULD PENETRATION

Svoboda (1994) reviews the wide field of mould penetration by melts, and concludes that (1) mechanical balance be-tween the driving pressure and capillary effects lead to liquid state penetration in 75% of cases, (2) chemical reactions cause 20% of cases of penetration, and (3) vapour-state reactions may control 5% of penetration problems. We shall examine his categories in detail next. In addition, the time dependence of penetration is an interesting behaviour that requires explanation.

4.5.5.1 Liquid penetration: effect of surface tension and pressure

Any liquid will immediately start to impregnate a porous solid if the solid is wetted by the liquid. The effect is the result of capillary attraction. For this reason, moulding materials are selected for their non-wetting behaviour, so that pene-tration by liquid metals is resisted; the resistance is known as capillary repulsion.

It was the French workers Portevin and Bastien (1936) that first proposed that capillary forces should be significant for the penetration of liquid metals into sand. However, this prediction was neglected until the arrival of Hoar and Atterton (1950) who first set out the basic physics, creating a quantitative model in which the surface tension γ of the melt would hold back the penetration of a liquid subjected to gradually increasing pressure against the surface of a porous, non-wetted aggregate. The liquid interface bridging the inter-particle spaces would gradually swell out, its bulging action progressively confined to a steadily smaller radius until it became hemispherical (radius r in Figure 4.11). Up to this critical value, the melt could be held back, but beyond this condition, further advance of the meniscus would cause the radius of curvature to increase as indicated by the formula that follows, lowering the resistance. The balance condition at the critical pressure for penetration is that quoted in Eqn (3.11), $P = 2\gamma/r$.

If we substitute $\rho g H$ for pressure from depth H in a metal of density ρ, we have the critical depth of metal for penetration

$$H = 2\gamma/r\rho g$$

Thus, beyond this critical point, penetration would instantly occur as a runaway effect; Figure 4.10 shows the meniscus expanding rapidly away after the narrowest point between sand grains is passed. Penetration might subse-quently only be stopped by the advancing front losing its heat and freezing. Such elementary physics is helpful to un-derstand why finer grain size of the aggregate or the provision of any extremely fine-grained surface coating are both helpful to reduce penetration for any fixed value of H by this simple mechanical model.

Resistance to penetration until some critical pressure is achieved has been demonstrated many times. For instance, Draper and Gaindhar (1975) find a metal depth of approximately 300 mm is required for steel to penetrate consistently into greensand moulds in their experiment. This critical depth was influenced to some degree by mould compaction, oxidising conditions and the temperature of mould hot-spots. If we assume a density of 7000 kg/m^3, surface tension approximately 2 N/m and acceleration from gravity of about 10 m/s^2, we find the radius r is approximately 0.02 mm, indicating the interparticle diameter to be roughly 0.04 mm (40 μm) in their moulds which seems a reasonable value.

All of the discussion so far has assumed perfect non-wetting between the metal and the mould (i.e. a contact angle of 180°). Hayes et al. (1998) go to some lengths to emphasise the effect of conditions intermediate between wetting and non-wetting, taking account of the contact angle between the melt and its substrate. They find a good correlation between the contact angle and the penetration of liquid steel into silica sand moulds. Petterson (1951) suggests that the penetration of steel into sand moulds would be aided by a gradual fall in the contact angle as is commonly seen between liquids and substrates. There may be some truth in this in some situations, particularly for some alloys where vapour transport or chemical reactions are occurring. However, Hayes and colleagues measure the change in contact angle in their work and find the change insignificant. In their careful study the penetration appeared to be a simple mechanical process involving varying degrees of capillary repulsion.

Levelink and Berg have investigated and described conditions in which they suggested dynamic conditions were important (Figure 4.11). They claimed that iron castings in greensand moulds were subject to a problem that they suggested was a water explosion. This led to a severe but highly localised form of mould penetration by the metal. However, careful evaluation of their work indicates that it seems most likely that they were observing a simple

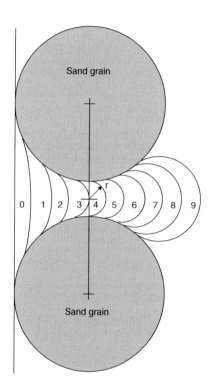

FIGURE 4.10

Metal penetration between two cylinders to simulate two sand grains. Surface tension resists the applied pressure up to the minimum radius *r*. Beyond this curvature penetration is a run-away instability.

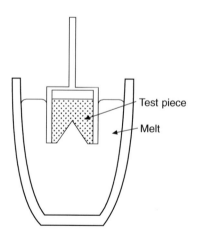

FIGURE 4.11

Water hammer (momentum effect) test piece (Levelink and Berg, 1971).

conservation of momentum effect. As the liquid metal fills their conical mould, it accelerates as the area of the cone decreases. As the melt nears the tip it reaches its maximum velocity, slamming the melt into the ever-decreasing space. The result is the generation of a high shock pressure, forcing metal into the sand. The effect is similar to a cavitation damage event associated with the collapse of a bubble against a ship's propeller. Although they cite the presence of bubbles as evidence of some kind of explosion, the oxides and bubbles present in many of their tests seem to be the result of entrainment in their rather poor filling system, rather than associated with any kind of explosion.

The impregnation of the mould with metal in last regions to fill is commonly observed in all metals in sand moulds. A pressure pulse generated by the filling of a boss in the cope will often also cause some penetration in the opposite drag surface of a thin-wall casting. The point discontinuity shown in Figure 2.27 will be a likely site for metal penetration into the mould. If the casting is thin-walled, the penetration on the front face will also be mirrored on its back face. Such surface defects in thin-walled aluminium alloy castings in sand moulds are highly unpopular, because the silvery surface of an aluminium alloy casting is spoiled by these dark spots of adhering black sand (sometimes called 'the black plague'!) and thus will require the extra expense of being removed by hand, or blasting with shot or grit.

Levelink and Berg (1968) report that the problem is increased in greensand by the use of high-pressure moulding. This may be the result of the general rigidity of the mould accentuating the concentration of momentum (weak moulds will yield more generally, and thus dissipate the pressure over a wider area). They list a number of ways in which this problem can be reduced:

1. Reduce mould moisture.
2. Reduce coal and organics.
3. Improve permeability or local venting; gentle filling of mould to reduce final filling shock.
4. Retard moisture evaporation at critical locations by local surface drying or the application of local oil spraying.

The reduction in the mechanical forces involved by reduced pouring rates or by local venting is understandable as reducing the final impact forces. Similarly, the use of a local application of oil will reduce permeability, causing the air to be compressed, acting as a cushion to decelerate the flow more gradually.

The other techniques in their list seem less clear in their effects, and raise the concern that they may possibly be counter-productive! It seems there is plenty of scope for additional studies to clarify these problems.

Work over a number of years at the University of Alabama, Tuscaloosa (Stefanescu et al., 1996), has clarified many of the special issues relating to the penetration of sand moulds by cast iron. Essentially, this work concludes that hot spots in the casting, corresponding to regions of isolated residual liquid, are localised regions in which high pressures can be generated in the residual liquid inside the casting by the expansion of graphite, forcing the liquid out via microscopic patches of unfrozen surface and causing local penetration of the mould. The pressure can be relieved by careful provision of 'feed paths' to allow the expanding liquid to be returned to the feeder. The so-called feed paths are, of course, allowing residual liquid to escape, working in reverse of normal feeding.

Naturally, any excess pressure inside the casting will assist in the process of mould penetration. Thus large steel castings are especially susceptible to mould penetration because of the high metallostatic pressure. This factor is in addition to the other potential high temperature reactions listed previously. This is reason for the widespread adoption in steel foundries of the complete coating of moulds with a ceramic wash as an artificial barrier to penetration.

4.5.5.2 Natural barriers to penetration

We have seen previously that in some conditions cast iron can develop a strong carbon film (the lustrous carbon film) on its advancing meniscus which is pulled down into the gap between the mould and the metal. Here it effectively separates the liquid from the aggregate, providing a mechanical barrier to prevent penetration, and retain a smooth shiny surface on the casting. Also, as we shall see, oxidising conditions can eliminate the carbon film replacing it with a liquid silicate. Penetration in oxidising conditions can then be practically guaranteed.

In copper-based foundries, the alloy aluminium bronze is renowned for its ability to resist penetration of the mould. The reason is also certainly the mechanical barrier presented by the strong alumina film; the high content of Al, in the 5–10% range, and the high temperature combine to create one of the most tenacious films in the casting world. In contrast, those bronzes not protected by a strong film in the liquid state, tin-, lead- and phosphor-bronze, all suffer penetration problems.

Similarly, Gonya and Ekey (1951) compared the behaviour of the common copper-base alloy 85-5-5 to 5-5 and the Al-5Si alloy, investigating a number of moulding variables. However, their most significant result related to pressure. They gradually increased pressure, finding penetration in the copper alloy at a critical pressure 17 kPa (2.5 psi), whereas the Al alloy continued to resist penetration up to the maximum they were able to provide in their experiment, about 30 kPa (4.5 psi). These pressures correspond to head of liquid metal of approximately 230 and 1240 mm, respectively. Clearly the low density of Al combined with the presence of the alumina film on the surface of the liquid alloy is a major benefit in avoiding mould penetration.

4.5.5.3 Temperature and time dependence of penetration

Clearly, the metal cannot penetrate the mould if the metal has solidified. Taking this impressively undeniable basic logic to heart, Brookes et al. (2007) draw attention to the importance of the mould temperature. By both computer model and experiment, they were able to demonstrate that as the mould surface temperature increased on contact with the melt and later decreased during cooling, penetration of liquid steel did not occur until the mould temperature exceeded a critical value. The penetration subsequently continued, causing the penetration effect to worsen while the temperature remained above the critical temperature. Penetration finally stopped when the temperature fell once again below the critical value. It is interesting that this relatively simple model appears to provide an explanation for the time dependence of penetration as has been known for many years (Jones, 1948; Shirey and Williams, 1968).

It should be noted, however, that the reduction of contact angle, i.e. progressively increasing effectiveness of wetting, also has been observed for many metal/mould combinations to be a function of time (for instance Wu et al., 1997; Shen et al., 2009). Also, of course, mould atmosphere greatly affects wetting ability; this too is a function of time. Thus a reducing or neutral atmosphere is useful to reduce penetration in low carbon steels in greensand moulds, whereas an oxidising atmosphere encouraged penetration as found by Draper and Gaindhar in 1975.

4.5.5.4 Chemical interactions

There seems little doubt that the contact angle between the mould particles and the melt can change with both time and temperature. Hayes et al. (1998) noted a reduction of the contact angle of liquid steel on silica sand from 110° to 93° over a 30 minute period. However, this lengthy period and relatively small change are not likely to greatly affect most steel castings because the majority will have frozen in this time.

To effect a major change in behaviour would require a major change in contact angle over a relatively short period such as a few minutes. The penetration of grey irons into silica sand moulds in oxidising condition is exactly such a candidate as has already been noted previously.

4.5.5.5 Effects of vapour transport

For a general overview of the knowledge of vapour transport, the reader is referred to the excellent review by Svoboda and Geiger (1969). It is salutary to reflect that it seems relatively little additional knowledge on this important subject has been gained since this early date.

On a microscale, the effects of vapour transport are likely to be complicated. Ahn and Berghezan (1991) studied the infiltration of liquid Sn, Pb and Cu inside metal capillaries using a scanning electron microscope. They found evidence of the deposition of metal vapours over a region up to 0.1 mm ahead of the advancing front. Clearly, the presence of this freshly deposited metal influenced the effect of the wetting of the liquid that followed closely behind.

Shen et al. (2009) drew similar conclusions from experiments with sessile drops in vacuum. They found that the contact angle of a zirconium/copper-based alloy on an alumina substrate fell from about 90° to 0° in about 10 min. They found Zr adsorption at the liquid/solid interface, followed by a Zr-Cu precursor film that accounted for the excellent wetting.

On a macroscale, vapour transport from metals into moulds is a common feature in foundries and will be referred to repeatedly elsewhere in this book. To give just two contrasting instances here:

a. Foundries casting magnesium into plaster moulds filled with vacuum assistance (Sin et al., 2006) find that Mg diffuses into the plaster, reacting with the SiO_2, taking the oxygen to form MgO, and reducing the Si to Mg_2Si according to the reaction

$$4Mg + SiO_2 = 2MgO + Mg_2Si$$

b. Bronze foundries casting classical tin-bronzes and lead-bronzes find that both alloys suffer mould penetration. However, only lead has any significant vapour pressure at the casting temperature, its pressure being between 100 and 1000 times greater than tin (see Figure 6.26). We can conclude therefore that in this particular case, vapour pressure is seen to be of minor, if any, importance; a high vapour pressure does not necessarily lead to enhanced penetration. The expected absence of any significant mould reactions or any strong surface films leaves only the relatively weak action of surface tension to inhibit penetration. Thus mould penetration for these alloys seems possible only by mechanical means: the provision of a fine mould aggregate, or a ceramic surface coat.

4.6 METAL SURFACE REACTIONS

Easily the most important metal/mould reaction is the reaction of the metal with water vapour to produce a surface oxide and hydrogen, as discussed in Chapter 1.

However, the importance of the release of hydrogen and other gases at the surface of the metal, leading to the growth of porosity in the casting, is to be dealt with in Chapter 6. Here we shall devote ourselves to the many remaining reactions. Some are reviewed by Bates and Scott (1977). These and others are listed briefly next.

4.6.1 OXIDATION

Oxidation of the casting skin is common for low carbon–equivalent cast irons and for most low-carbon steels. It is likely that the majority of the oxidation is the result of reaction with water vapour from the mould, and not from air which is expelled at an early stage of mould filling as shown earlier. Carbon additions to the mould help to reduce the problem.

The catastrophic oxidation of magnesium during casting, leading to the casting (and mould) being consumed by fire, is prevented by the addition of so-called inhibitors to the mould. These include sulphur, boric acid, and other compounds such as ammonium borofluoride. More recently, much use has been made of the oxidation-inhibiting gas, sulphur hexafluoride (SF_6) which is used diluted to about 1–2% in air or other gas such as CO_2 to prevent the burning of magnesium during melting and casting. However, since its identification as a powerful ozone-depleting agent, SF_6 is being discontinued for good environmental reasons. A return is being made to dilute mixtures of SO_2 in CO_2 and other more environmentally friendly atmospheres are now under development.

In any case, the burning of Mg alloys during pouring of the mould is almost certainly the result of surface turbulence. For instance, if liquid Mg can be introduced into a mould quiescently, so that it rises steadily with a substantially flat liquid surface in the mould, the huge heat of oxidation released on the surface is rapidly conveyed away into the bulk of the melt so that the surface temperature never rises to a dangerous level. Such a quiescent fill is therefore safe. Conversely, if the melt is jumping and splashing, the heat of oxidation will diffuse into the thin liquid splash from both sides of the splash surface, and quickly heats the small amount of metal in the splash. The ignition temperature of the melt is quickly exceeded, with disastrous results. There is a saying, which regrettably probably reflects some truth, that the few Mg foundries in existence nowadays are the result of most of them having burned down. True or not, the turbulent handling of Mg alloys is clearly dangerous but mostly avoidable. The use of an offset step basin and stopper, and a naturally pressurised filling system would make a huge difference to Mg foundries.

Titanium and its alloys are also highly reactive. Despite being cast under vacuum into moulds of highly stable ceramics such as zircon, alumina or yttria, the metal reacts to reduce the oxides of the mould, contaminating the surface of the casting with oxygen, thereby stabilising the alpha phase of the alloy. The 'alpha case' usually has to be removed by chemical machining.

4.6.2 CARBON (PICKUP AND LOSS)

Mention has already been made of the problem of casting titanium alloy castings in carbon-based moulds. The carburisation of the surface, again results in the stabilisation of the alpha phase, and requires to be subsequently removed.

The difficulty is found with stainless steel of carbon content less than 0.3% cast in Croning resin-bonded shell moulds (McGrath, 1973). The relatively high resin binder content of these moulds, generally in the region of 2.5–3.0 wt%, causes steels to suffer a carburised layer between 1 and 2 mm deep. Tordoff (1996) also found significant carbon pickup in stainless steels from phenolic urethane binders which rarely exceed 1.0–1.2 wt% based on sand. The carburisation, of course, appears more severe the lower the carbon content of the steel. Naturally, the problem is worse on drag than on cope faces as a result of the casting sitting hard down on the drag face of the mould, and contracting away from the upper surfaces. Tordoff also found that iron oxide in the moulding sand reduced the effect, whereas the use of a furan resin eliminated pickup altogether. The author could not find a record of the effect of silicate-based binders, although their usually low content of organics would be expected to give minimal problems.

Carbon pickup is the principal reason why low carbon steel castings are not produced by the lost-foam process. The atmosphere of styrene vapour, which is created in the mould as the polystyrene decomposes, causes the steel to absorb carbon (and presumably hydrogen). The carbon-rich regions of the casting are easily seen on an etched cross-section as swathes of pearlite in an otherwise ferritic matrix.

In controlled tests of the rate of carburisation of low carbon steel in hydrocarbon/nitrogen mixtures at 925°C (Kaspersma, 1982), methane was the slowest and acetylene the fastest of the carburising agents tested, and hydrogen was found to enhance the rate, possibly by reducing adsorbed oxygen on the surface of the steel. At high ratios of H_2/CH_4 at this temperature, hydrogen decarburises steel by forming methane (CH_4). This may be the important reaction in the casting of steel in greensand sand moulds containing only low carbonaceous additions.

Decarburisation can also be a problem. For instance, surface decarburisation of steels is often noted with acid-catalysed furan resin binders (Naro and Wallace, 1992).

In the investment casting of steel in air, the decarburisation of the surface layer is particularly affected because atmospheric oxygen persists in the mould as a consequence of the inert character of the mould and its permeability to the surrounding environment. Doremus and Loper (1970) have measured the thickness of the decarburised layer on a low carbon steel investment casting and find that it increases mainly with mould temperature and casting modulus. The placing of the mould immediately after casting into a bin filled with charcoal helps to recarburise the surface. However, Doremus and Loper point out the evident danger that if the timing and extent of recarburisation is not correct, the decarburised layer will still exist below a recarburised layer!

In iron castings, the decarburisation of the surface gives a layer free from graphite. This adversely affects machinability, giving pronounced tool wear, especially in large castings such as the bases of machine tools. The decarburisation seems to be mainly the result of oxidation of the carbon by water vapour because dry moulds reduce the problem. An addition of 5–6% coal dust to the mould further reduces it. Rickards (1975) found that the reaction seems to start at about the freezing point of the eutectic, 1150°C, and proceeds little further after the casting has cooled to 1050°C (Figure 4.12). Stefanescu et al. (2009) find that the depth of the casting skin in ductile iron is controlled by diffusion to only about 0.5 mm if the liquid in the casting is relatively quiescent, whereas the strong convection currents during solidification of larger castings causes mixing, increasing the skin depth to nearly 3 mm.

4.6.3 NITROGEN

Nitrogen pickup in grey cast irons appears to be directly related to the nitrogen content of mould binders (Graham et al., 1987). Ammonia is released during the pyrolysis of urea and amines contained in hot box and Croning shell systems when they become recycled into a greensand system. Ammonia appears to be reversibly absorbed by the bentonite clays and is released on heating. The pyrolysis of ammonia releases nascent nitrogen and nascent hydrogen by the simple decomposition

$$NH_3 = N + 3H$$

Graham and coworkers confirmed that subsurface porosity and fissures in irons do not correlate well with the total nitrogen content of the sand, but were closely related to the total ammonia content. Lee (1987) confirms the usefulness of an ammoniacal nitrogen test which in his work pointed to wood flour as a major contributor of ammonia in his greensand system.

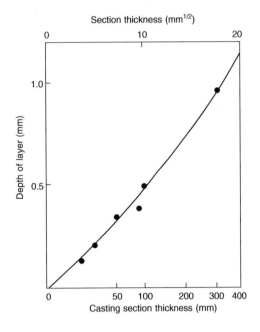

FIGURE 4.12

Depth of decarburisation in grey iron plates cast in greensand.

Data from Rickards (1975).

The link of ammonia and the so-called nitrogen fissures in iron castings suggests the formation of nitride bifilms which might be opened, becoming visible, by inflation with the copious amounts of hydrogen released by the decomposition of ammonia.

At the other end of the casting value system, vacuum-cast Ni-based superalloys suffer severely from nitrogen pickup from the moment the door of the vacuum furnace is opened, allowing air to rush in and react while the casting is still hot. Casting returns in such foundries are known to contaminate the new melts, although the contamination would be expected not to be nitrogen in solution, but nitride bifilms in suspension. These conjectures require more research to clarify the situation.

4.6.4 SULPHUR

The use of moulds bonded with furane resin catalysed with sulphuric and/or sulphonic acid causes problems for ferrous castings because of the pickup of sulphur in the surface of the casting (Naro and Wallace, 1992). This is especially serious for ductile iron castings, because the graphite reverts from spheroidal back to flake form in the high-sulphur surface layer. This has a serious impact on the fatigue resistance of the casting.

4.6.5 PHOSPHORUS

The use of moulds bonded with furane resin catalysed with phosphoric acid leads to the contamination of the surfaces of ferrous castings with phosphorus (Naro and Wallace, 1992). In grey iron, the presence of the hard phosphide phase in the surface causes machining difficulties because of rapid tool wear.

4.6.6 SURFACE ALLOYING

There has been some Russian (Fomin, 1965) and Japanese work (Uto, 1967) on the alloying of the surface of steel castings by the provision of materials such as ferrochromium or ferromanganese in the facing of the mould. Because the alloyed layers that have been produced have been up to 3 or 4 mm deep, it is clear once again that not only is diffusion involved but also some additional transport of added elements must be taking place by mixing in the liquid state. Omel'chenko further describes a technique to use higher-melting-point alloying additions such as titanium, molybdenum and tungsten by the use of exothermic mixes. Predictably enough, however, there appear to be difficulties with the poor surface finish and the presence of slag inclusions. Until this difficult problem is solved, the technique does not have much chance of attracting any widespread interest.

4.6.7 GRAIN REFINEMENT

The use of cobalt aluminate ($CoAl_2O_4$) in the primary mould coat for the grain refinement of nickel and cobalt alloy investment castings is now widespread as described in early reports by Rappoport (1964) and Watmough (1980). The mechanism of refinement is not yet understood. It seems unlikely that the aluminate as an oxide phase can wet and nucleate metallic grains. That the surface finish of grain refined castings is somewhat rougher than that of similar castings without the grain refiner indicates that some wetting action has occurred. This suggests that the particles of $CoAl_2O_4$ decompose to some metallic form, possibly CoAl. This phase has a melting point of 1628°C. It would therefore retain its solid state at the casting temperatures of Ni-based alloys. In addition it has an identical face-centred–cubic crystal structure to nickel. On being wetted by the liquid alloy, it would be expected to constitute an excellent substrate for the initiation of grains.

Watmough also investigates a number of other additives to coatings for Ni-base high temperature alloy castings including cobalt oxide, CoO. Because oxides are almost certainly not effective nuclei for solid formation and because CoO is not especially stable (Llewelyn and Ball, 1962), it is likely that the compound decomposes at casting temperature forming metallic cobalt which would be expected to be an effective nucleus for the alloy. Although its melting point is not significantly higher than the alloy, that the Co particles sit at the mould wall will ensure that they remain cool and therefore will resist melting and continue to act as nucleating particles.

It is to be expected that there can be no refinement if all the CoAl particles are either melted or dissolved, or if the newly nucleated grains are themselves re-melted. Unfortunately, such re-melting problems are common in investment castings as a result of (1) casting at too high a temperature and (2) convection problems that have so far been overlooked. It is regrettable that the beautifully fine grain structure produced by these techniques is often lost by subsequent convection in the casting, conveying so much heat from hotter regions that parts of the surface in the path of the flow re-melt, destroying the nuclei and the early grains, and replacing these with massive grains of uncontrolled size. In addition, the small depth of grain refinement (Watmough reports only 1.25 mm) is another probable consequence of uncontrolled convection. It is expected that control and suppression of convection in investment castings would greatly improve cast structures in polycrystalline high temperature alloy castings. These issues are dealt with in *Casting Manufacture* in Casting Rule 7 'Avoiding Convection'.

Cobalt addition to a mould coat is also reported to grain-refine malleable cast iron (Bryant, 1971), presumably by a similar mechanism to that enjoyed by the vacuum-cast Ni-base alloys.

The use of zinc in a mould coat by Bryant and Moore to achieve a similar aim in iron castings must involve a quite different mechanism, because the temperature of liquid iron (1200–1450°C) greatly exceeds not only the melting point, but even the boiling point (907°C) of zinc! It may be that the action of the zinc boiling at the surface of the solidifying casting disrupts the formation of dendrites, detaching them from the surface so that they become freely floating nuclei within the melt. Thus the grain refining mechanism in this case is grain multiplication rather than nucleation.

The zinc-containing coating and others listed next all appear to induce grain multiplication as a result of the rafts of dendrites attempting to grow on an unstable, moving and collapsing substrate. The effect seems analogous to that described in Section 3.2.6 for the enhancement of the fluidity of Al alloy castings by coatings of acetylene black or hexachlorethane on moulds.

Nazar et al. (1979) report the use of hexachlorethane-containing coatings for the grain refinement of Al alloys.

4.6.8 MISCELLANEOUS

Boron has been picked up in the surfaces of stainless steel castings from furane bonded moulds that contain boric acid as an accelerator (McGrath, 1973).

Tellurium is sometimes deliberately added as a mould wash to selected areas of a grey iron casting. This element is a strong carbide former, and will locally convert the structure of the casting from grey to a fully carbidic white iron. Chen et al. (1989) describe how TeCo surface alloying is useful to produce wear-resistant castings.

In other work, the carbide-promoting action of Te is said to reduce local internal shrinkage problems, although its role in this respect seems difficult to understand. It has been suggested that a solid skin is formed rapidly, equivalent to a thermal chill (Vandenbos, 1985). The effect needs to be used with caution: tellurium and its fumes are toxic, and the chilled region causes difficulties in those parts requiring to be machined.

The effect of tellurium converting grey to white irons is used to good purpose in the small cups used for the thermal analysis of cast irons. Tellurium is added as a wash on the inside of the cup. During the pouring of the iron, it seems to be well distributed into the bulk of the sample, not just the surface, so that the whole test piece is converted from grey to white iron. This simplifies the interpretation of the cooling curve, allowing the composition of the iron to be deduced.

4.7 MOULD COATINGS
4.7.1 AGGREGATE MOULDS

Although we have dealt at length with reactions that can occur between the metal and the mould, the purpose of a mould coating is to prevent such happenings by keeping the two apart. It has to be admitted that for many formulations, these attempts are of only limited success. Some useful reviews of coatings from which the author has drawn are given by Vingas (1986), Wile et al. (1988) and Beeley (2001). Vingas in particular describes techniques for measuring the thickness of coatings in the liquid and solid states. All describe the various ways in which coatings can be applied by dipping, swabbing, brushing and flow-over. These issues will not be dealt with here.

With regard to the definition of a coating, it is a creamy mixture made up from of a number of constituents:

1. *Refractory,* usually oxides of many kinds ground to a particle size of 50 μm or less, but for cast irons carbonaceous material is common in the form of graphite or ground coke etc.
2. *Carrier* nowadays becoming more commonly water (alcohols and chlorinated hydrocarbons have environmental problems). This has to be removed after the coating has been applied, usually requiring the application of heat.
3. *Suspension agent* required to maintain a uniform dispersion of the refractory. Commonly used materials include sodium bentonite clay or cellulose compounds. Beeley (2001) describes a problem with the clay addition, as a result of the water soaking into the mould surface, but being hindered during the subsequent drying because of the presence of the impervious clay coat. The development of a coating without the clay addition doubled the rate of drying.
4. *Binder* which is often the same as that used for bonding moulds. Thus, organic resins, colloidal silica and sodium silicate-based binders are common. Interesting incompatibilities are sometime experienced, such as the fact that magnesite coatings for steel castings may be difficult to use with sand bonded with furan resins because of the acid content of the resin, whereas magnesite coatings perform excellently when used in conjunction with silicate-bonded moulds and cores.
5. *Sundry chemicals* including surfactants to improve wettability, antifoaming agents and bactericides.

The advantages of coatings include the following

1. Reduction of cleaning costs because of improved surface finish (finer surface, reduced or eliminated veining, reduced penetration and/or burn-on and reduced reactions between metal and mould, for instance between (a) manganese steel and silica sand, and (b) binder gases such as sulphur compounds from furan binders)

2. Improved shake-out because of improved sand peel.
3. Reduced machining time and tool wear.

However, an important aspect to bear in mind, as with the provision of feeders, the best action (if possible) is to avoid coatings. There are many reasons for avoiding coatings. Disadvantages include the following

1. Cost of materials, especially those based on expensive minerals such as zircon.
2. Potential loss of accuracy because of difficulty of controlling coat thickness and coating penetration.
3. Possibility of cosmetic defects from runs and drops.
4. Floor space for coating station.
5. Energy cost to dry and possible capital cost and floor space for drying ovens and extraction ducting and fans.
6. Floor space to dry if dried naturally (this might exceed moulding space because drying is slow).
7. Drying time can severely reduce productivity.
8. Cores appear to be never fully dried, despite all attempts, after the application of a coating, so that there is enhanced danger of blow defects if the core cannot be vented to the atmosphere.
9. Environmental problems: at this time many coatings are still alcohol (ethanol) based, and so burned off rather than dried. This reduces floor space and energy, but does add to the loading of volatile organic compounds in the environment. This approach is likely to be banned under future legislation, forcing the use of water-based coatings.

The water-based coatings pose a significantly increased drying problem. Puhakka (2009) describes the novel use of a remote infra-red camera to monitor the drying process; while water is still evaporating the surface will remain cool. Only when fully dry will the surface temperature of the mould rise to room temperature.

Coating costs for the foundry are usually justified on (1) and (2) of the previous advantages and should be based on the dry refractory deposited. Naturally, costs are directly related to the surface area to volume ratio of the castings, and so will vary from foundry to foundry. Alternatively, surface area per ton of castings is a useful measure.

Coatings are not required in general for the lower melting point metals such as the zinc-based alloys and the light metals, Al- and Mg-based alloys, but are widely used for cast iron, copper-based alloys and steels.

The use of coatings on cores to prevent core blows is usually a mistake. Although the coating does reduce the permeability of the surface, assisting to keep in the expanding gases, the additional volatiles from the coat, which appear to be never completely removed by drying, are usually present in excess to overwhelm the coating barrier.

Another interesting mistake is the use of coatings to prevent sand erosion. The presence of sand defects in a casting is never, in my experience, the fault of the sand. It immediately signals that the filling system is faulty. A turbulent filling system that entrains air will splatter and hammer against the sand surface, oxidising away any binder, and so erode away cores and moulds. Although, of course, a coating will reduce the problem, it is often less costly and more effective to upgrade the filling system to a naturally pressurised design. In this way, the moulds and cores are at all times gently pressurised by the melt, causing the sand grains of the mould to be held safely in place. An additional bonus is a better casting, enjoying more reliable properties and freedom from other defects such as porosity and cracks.

4.7.2 PERMANENT MOULDS AND METAL CHILLS

Permanent moulds or chills in grey iron or steel are practically inert, so any mould coat is not required to prevent chemical reactions between the two. A coat will help to

1. protect the mould from thermal shock;
2. avoid the premature chilling of the metal that might result in cold lap defects and
3. confer some 'surface permeability' on the impermeable surface to allow the melt to flow better over the surface, and allowing the escape of any volatiles or condensates (particularly from the surface of chills).

It is often seen that two different mould coatings are used on a permanent mould. A thin smooth coating is applied to the mould cavity which forms the casting, whereas a thick, rough, insulating coating is applied to the running system in an effort to avoid temperature losses during mould filling. This is a mistake. The metal is in the running system for perhaps only 1 to 2 s (because its velocity is in the range of metres per second, and the distance involved is usually only a metre or less) that any loss of temperature is negligible. Furthermore, the roughness of this coat endangers the flow because of the consequential turbulence. It is far better, much less trouble, less time-consuming and less costly simply to coat the whole die with the same thin, smooth coat. The castings will probably be improved.

4.7.3 DRY COATINGS

Dry coatings are of course an attractive concept because no drying is involved and there is no danger of introducing additional volatiles into moulds or cores.

Greensand moulds have benefited somewhat from the use of dusting powders of various kinds and more recently by the electrostatic precipitation of dry zircon flour (precoated with an extremely thin layer of thermosetting resin, activated by the heat of the metal during casting). This development appears to have been resistant to extension to dry sand processes because of the generally poorer electrical conductivity of dry sand moulds particularly when bonded with a phenol-urethane resin. It would be interesting to know whether more success might be expected from moulds made with materials having higher electrical conductivity such as sand bonded with acid-catalysed resins, silicates or alkaline phenolics, or with chromite sand.

A successful technique for the application of dry coatings to dry sand moulds has therefore proved elusive. Simply dusting on dry powder, or attempting to apply it in a fluidised bed, suffers from most of the powder being blown off as clouds of dust when the mould is cleaned by blowing out with compressed air prior to closure.

A Japanese patent (Kokai, 1985) describes how fine refractory powder in a fluid bed is drawn into the surface pores of an aggregate mould or core by applying a vacuum via some far location on the mould or core. However, of course, despite probably being useful for simple geometries, this technique fails for complex shapes. Another Japanese patent describes how fine powder in a fluid bed is forced into the pores of a mould by a mechanical brushing action. A Canadian automotive foundry has developed a fluid bed containing a mixture of the fine coating powder together with relatively heavy zircon grains. The jostling of the zircon grains effectively hammers the powder into the mould pores.

These dry coating techniques deserve wider use. When correctly applied, the coating simply fills the interstices between the surface grains of the mould, not adding any thickness of deposit. Thus the technique has the advantage of preserving the accuracy of the mould while improving surface finish with no drying time penalty.

SOLIDIFICATION STRUCTURE

5

In this section, we consider how the metal changes state from the liquid to the solid and develops its structure.

It is a widely accepted piece of dogma, often quoted, that the properties of the casting are controlled by its structure. This seems to me to be largely untrue. For instance, in meeting mechanical property specifications of a casting its solidification structure, its grain size or dendrite arm spacing (DAS) bears only a superficial relation to the properties. The feature really in control is the bifilm population. That the grain size and DAS appear to be important seems mainly in relation to the way in which they influence the unfurling of bifilms and the bifilm distribution. More of this later.

In a later chapter, we consider the problems of the volume deficit on solidification, and the so-called shrinkage problems that lead to a set of void phenomena, sometimes appearing as porosity, even though, as we shall see, such problems are actually relatively rare in castings (mainly because foundry people understand how to avoid shrinkage porosity). Most of the problems attributed to shrinkage are either not shrinkage problems at all, or are only secondarily shrinkage problems. Most problems that *appear* to be 'shrinkage' are actually oxides, actually bifilms together with their entrained layers of air and bubbles.

When we understand (1) how shrinkage voids can form, we are then in a better position to understand the development of (2) gas pores. Both of these sets of defects generally grow from bifilms.

These issues highlight the problem faced by the author. The problem is how to organise the descriptions of the complex but inter-related phenomena that occur during the solidification of a casting into a logical presentation necessary for a book. This book could be organised in many different ways. For instance, naturally, the gas and shrinkage contributions to the overall pore structure are complimentary and additive, and both rely on the presence and character of the pre-existing population of bifilms.

The reader is requested to be vigilant to see this integration. I am conscious that, although spelling out the detail in a didactic dissection of phenomena, emphasising the separate physical mechanisms, the holistic vision for the reader is easily lost. Take care not to lose the overall coherence of our fascinating field.

5.1 HEAT TRANSFER
5.1.1 RESISTANCES TO HEAT TRANSFER

The hot liquid metal takes time to lose its heat and solidify. The rate at which the heat escapes is controlled by a number of resistances described by Flemings (1974). We shall follow his clear treatment in this section.

The five main resistances to heat flow from the interior of the casting (starting at zero for convenience of numbering later) are:

0. The liquid.
1. The solidified metal.
2. The metal-mould interface.
3. The mould.
4. The surroundings of the mould.

All these resistances add, as though in series, as shown schematically in Figure 5.1.

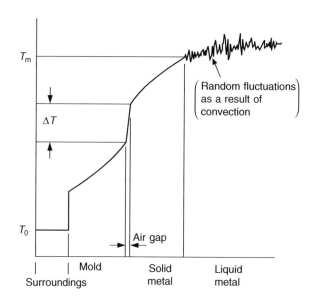

FIGURE 5.1

Temperature profile across a casting freezing in a mould, showing the effect of the addition of thermal resistances that control the rate of loss of heat.

As it happens, in nearly all cases of interest, resistance (0) is negligible as a result of stirring by forced convection during filling and thermal convection during cooling. The turbulent flow and mixing quickly transport heat and so smooth out temperature gradients. This happens quickly because the viscosity of the melt is usually low, so that the flow of the liquid is fast, and the heat is transported out of the centre of large ingots and castings in a time that is short compared to that required by the remaining resistances, whose rates are controlled by much slower diffusion processes.

In many instances, resistance (4) is also negligible in practice. For instance, for normal sand moulds the environment of the mould does not affect solidification because the mould becomes hardly warm on its outer surface by the time the casting has solidified inside.

However, there are, of course, a number of exceptions to this general rule, all of which relate to various kinds of thin-walled moulds that, because of the thinness of the mould shell, are somewhat sensitive to their environment. Iron castings made in Croning shell moulds (the Croning shell process is one in which the sand grains are coated with a thermosetting resin, which is cured against a hot pattern to produce a thin, biscuit-like mould) solidify faster when the shell is thicker, or when the shell is thin but backed up by steel shot. Conversely, the freezing of steel castings in investment shell moulds can be delayed by a thick backing of thermal insulation around the shell, all preheated to high temperature. Alternatively, without a backing, cooling is relatively fast, radiating heat away freely to cooler surroundings. Naturally, aluminium alloys in iron or steel dies can be cooled even faster when the dies are water cooled.

Nevertheless, despite such useful ploys for coaxing greater productivity, it remains essential to understand that in general the major fundamental resistances to heat flow from castings are items (1), (2) and (3). For convenience we shall call these resistances 1, 2 and 3.

The effects of all three simultaneously can nowadays be simulated with varying degrees of success by computer. However, the problem is both physically and mathematically complex, especially for castings of complex geometry.

There is therefore still much understanding and useful guidance to be obtained by a less ambitious approach, whereby we look at the effect of each resistance in isolation, considering only one dimension (i.e. uni-directional heat flow). In this way, we can define some valuable analytical solutions that are surprisingly good approximations to casting problems. We shall continue to follow the approach by Flemings.

Resistance 1: The casting

The rate of heat flow through the casting, helping to control the freezing time of the casting, applies in the case of such examples as Pb-Sb alloy cast into steel dies for the production of battery grids and terminals; or the casting of steel into metal moulds; or the casting of hot wax into metal dies as in the injection of wax patterns for investment casting. It would be of wide application in the plastics industry.

However, it has to be admitted that this type of freezing regime does not apply for many metal castings of high thermal conductivity such as the light alloys or Cu-based alloys.

For the unidirectional flow of heat from a metal poured exactly at its melting point T_m against a mould wall initially at temperature T_0, the transient heat flow problem is described by the partial differential equation, where α_s is the thermal diffusivity of the solid

$$\frac{\partial T}{\partial t} = \alpha_s \frac{\partial^2 T}{\partial x^2} \tag{5.1}$$

The boundary conditions are $x = 0$, $T = T_0$; at $x = S$, $T = T_m$, and at the solidification front the rate of heat evolution must balance the rate of conduction down the temperature gradient, i.e.

$$H\rho_s \left(\frac{\partial S}{\partial t}\right) = K_s \left(\frac{\partial T}{\partial x}\right)_{x=s} \tag{5.2}$$

where K_s is the thermal conductivity of the solid, H is the latent heat of solidification, and for which the solution is:

$$S = 2\gamma\sqrt{\alpha_s t} \tag{5.3}$$

The reader is referred to Flemings for the rather cumbersome relation for γ. The important result to note is the parabolic time law for the thickening of the solidified shell. This agrees well with experimental observations. For instance, the thickness S of steel solidifying against a cast iron ingot mould is found to be:

$$S = at^{1/2} - b \tag{5.4}$$

where the constants a and b are of the order of 3 and 25, respectively, when the units are millimetres and seconds. The result is seen in Figure 5.2.

The apparent delay in the beginning of solidification shown by the appearance of the constant b is a consequence of the following: (1) The turbulence of the liquid during and after pouring, resulting in the loss of superheat from the melt, and so slowing the start of freezing. (2) The finite interface resistance further slows the initial rate of heat loss. Initially, the solidification rate will be linear, as described in the next section (and hence giving the initial curve in Figure 5.2 because of this plot using the square root of time). Later, the resistance of the solidifying metal becomes dominant, giving the parabolic relation (shown, of course, as a straight line in Figure 5.2 because of using the square root plot of time).

Resistance 2: The metal-mould interface

In many important casting processes, heat flow is controlled to a significant extent by the resistance at the metal-mould interface. This occurs when both the metal and the mould have reasonably good rates of heat conductance, leaving the boundary between the two the dominant resistance. The interface becomes overriding in this way when an insulating mould coat is applied, or when the casting cools and shrinks away from the mould (and the mould heats up, expanding away from the metal), leaving an air gap separating the two. These circumstances are common in the die casting of light alloys.

For unidirectional heat flow, the rate of heat released during solidification of a solid of density ρ_s and latent heat of solidification H is simply:

$$q = -\rho_s H A \frac{\partial S}{\partial t} \tag{5.5}$$

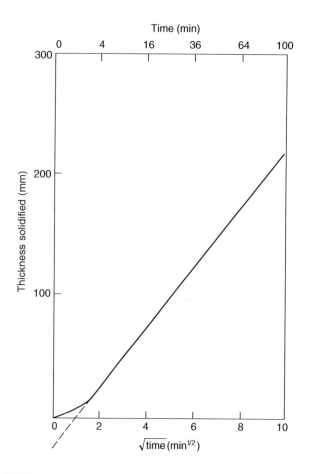

FIGURE 5.2

Unidirectional solidification of pure iron against a cast iron mould coated with a protective wash.

From Flemings (1974).

This released heat has to be transferred to the mould. The heat transfer coefficient h across the metal/mould interface is simply defined as the rate of transfer of energy q (usually measured in watts) across unit area (usually a square metre) of the interface, per unit temperature difference across the interface. This definition can be written:

$$q = -hA(T_m - T_0) \tag{5.6}$$

assuming the mould is sufficiently large and conductive not to allow its temperature to increase significantly above T_0, effectively giving a constant temperature difference $(T_m - T_0)$ across the interface. Hence equating (5.5) and (5.6) and integrating from $S = 0$ at $t = 0$ gives:

$$S = \frac{h(T_m - T_0)}{\rho_s H} \cdot t \tag{5.7}$$

It is immediately apparent that because shape is assumed not to alter the heat transfer across the interface, Eqn (5.7) may be generalised for simple shaped castings to calculate the solidification time t_f in terms of the volume V to cooling surface area A ratio (the geometrical modulus) of the casting:

$$t_f = \frac{\rho_s H}{h(T_m - T_0)} \cdot \frac{V}{A} \tag{5.8}$$

All of these calculations assume that h is a constant. As we shall see later, this is perhaps a tolerable approximation in the case of gravity die (permanent mould) casting of aluminium alloys where an insulating die coat has been applied. In most other situations h is highly variable, and is particularly dependent on the geometry of the casting.

The air gap

As the casting cools and the mould heats up, the two remain in good thermal contact while the casting is still liquid. When the casting starts to solidify, it rapidly gains strength and can contract away from the mould. In turn, as the mould surface increases in temperature it will expand. Assuming for a moment that this expansion is *homogeneous*, we can estimate the size of the gap d as a function of the diameter D of the casting:

$$d/D = \alpha_c\{T_f - T\} + \alpha_m\{T_{mi} - T_0\}$$

where α is the coefficient of thermal expansion, and subscripts c and m refer to the casting and mould respectively. The temperatures T are T_f the freezing point, T_{mi} the mould interface, and T_0 the original mould temperature.

The benefit of the gap equation is that it shows how straightforward the process of gap formation is. It is simply a thermal contraction-expansion problem, directly related to interfacial temperature. It indicates that for a steel casting ($\alpha_c = 14 \times 10^{-6}$ K^{-1}) of 1 m diameter that is allowed to cool to room temperature the gap would be expected to be of the order of 10 mm at each of the opposite sides simply from the contraction of the steel, neglecting any expansion of the mould. This is a substantial gap by any standards!

Despite the usefulness of the elementary formula in giving some order-of-magnitude guidance on the dimensions of the gap, there are a number of interesting reasons why this simple approach requires further sophistication.

In a thin-walled aluminium alloy casting of section only 2 mm the room temperature gap would be only 10 μm. This is only one-twentieth of the size of an average sand grain of 200 μm diameter. Thus the imagination has some problem in visualising such a small gap threading its way amid the jumble of boulders masquerading as sand grains. It really is not clear whether it makes sense to talk about a gap in this situation.

Woodbury and co-workers (2000) lend support to this view for thin wall castings. In horizontally sand cast aluminium alloy plates of 300 mm square and up to 25 mm thickness, they measured the rate of transfer of heat across the metal-mould interface. They confirmed that there appeared to be no evidence for an air gap. Our equation would have predicted a gap of 25 μm. This small distance could easily be closed by the slight inflation of the casting because of two factors. (1) The internal metallostatic pressure provided by the filling system (no feeders were used). (2) The precipitation of a small amount of gas; for instance, it can be quickly shown that 1% porosity would increase the thickness of the plate by at least 70 μm. Thus the plate would swell by creep with minimal difficulty under the combined internal pressure from head height and the growth of gas pores. The 25 μm movement from thermal contraction would be so comfortably overwhelmed that a gap would probably never have chance to form.

Our simple air gap formula assumes that the mould expands *homogeneously*. This may be a reasonable assumption for the surface of a greensand mould, which will expand into its surrounding cool bulk material with little resistance. A more rigid chemically bonded sand would be subject to rather more restraint, thus preventing the surface from expanding so freely. The surface of a metal die will, of course, be most constrained of all by the surrounding metal at lower temperature, but the higher conductivity of the mould will raise the temperature of the whole die more uniformly, giving a better approximation once again to homogeneous expansion.

Also, the sign of the mould movement for the second half of the equation is only positive if the mould wall is allowed to move outwards because of small mould restraint (i.e. a weak moulding material) or because the interface is concave. A rigid mould and/or a convex interface will tend to cause inward expansion, reducing the gap, as shown in Figure 5.3. It might be expected that a flat interface will often be unstable, buckling either way. However, Ling, Mampaey and co-workers (2000) found that both theory and experiment agreed that the walls of their cube-like mould poured with white cast iron distorted always outwards in the case of greensand moulds, but always inwards in the case of the more rigid chemically bonded moulds.

There are further powerful geometrical effects to upset our simple linear temperature relation. Figure 5.4 shows the effect of linear contraction during the cooling of a shaped casting. Clearly, anything in the way of the contraction of the straight lengths of the casting will cause the obstruction to be forced hard against the mould. This happens in the corners

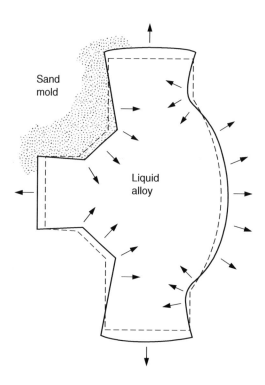

FIGURE 5.3

Movement of mould walls, illustrating the principle of inward expansion in convex regions and outward expansion in concave regions.

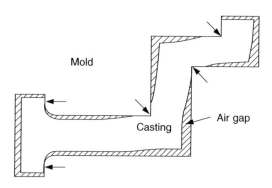

FIGURE 5.4

Variable air gap in a shaped casting: arrows denote the probable sites of zero gap.

at the ends of the straight sections. Gaps cannot form here. Similarly, gaps will not occur around cores that are surrounded with metal, and onto which the metal contracts during cooling. Conversely, large gaps open up elsewhere. The situation in shaped castings is complicated and is only just being tackled with some degree of success by computer models. Even so, Kron (2004) is sceptical of the computer models to date. Few attempt to allow for the mechanical factors influencing the air gap. Even where such allowance is attempted, there is no agreement on a suitable constitutive equation for mechanical deformation of solids at temperatures near their melting points. The swelling of the casting in the mould from the hydrostatic pressure in the liquid, and the friction of the rather soft casting against the mould are additional complicating factors. Large square section ingots of several tons in weight solidifying in ingot moulds are known to contract first from the mould corners, but under the high internal metallostatic pressure the casting balloons out to form a good contact over the central areas of the flat sides for much longer, causing these areas of the mould to suffer cracking by thermal fatigue.

Richmond and Tien (1971) and Tien and Richmond (1982) demonstrate via a theoretical model how the formation of the gap is influenced by the internal hydrostatic pressure in the casting and by the internal stresses that occur within the solidifying solid shell. Richmond (1990) goes on to develop his model further, showing that the development of the air gap is not uniform, but is patchy. He found that air gaps were found to initiate adjacent to regions of the solidified shell that were thin, because, as a result of stresses within the solidifying shell, the casting-mould interface pressure first dropped to zero at these points. Conversely, the casting-mould interface pressure was found to be raised under thicker regions of the solid shell, thereby enhancing the initial non-uniformity in the thickness of the solidifying shell. Growth becomes unstable, automatically moving away from uniform thickening. This rather counter-intuitive result may help to explain the large growth perturbations that are seen from time to time in the growth fronts of solidifying metals. Richmond reviews a considerable amount of experimental evidence to support this model. All the experimental data seem to relate to solidification in metal moulds. It is possible that the effect is less severe in sand moulds.

Lukens et al. (1991) confirmed that increased feeder height increased the heat transfer at the base of the casting. Furthermore, for a horizontal cylinder 91 mm diameter in a chromite greensand mould, as the cylinder contracted away from the mould, gravity caused the cylinder to sit at the bottom of the mould, therefore contacting the mould purely along the line of contact along its base. Thus for this rather larger casting than the 25 mm thick plate cast by Woodbury, it seems an air gap definitely occurs in a sand mould. This result confirms Lukens earlier result (1990) in which most heat appears to be extracted through the drag, and increased head pressure enhances metal/mould contact.

Attempts to measure the gap formation directly (Isaac et al., 1985; Majumdar and Raychaudhuri, 1981) are difficult to carry out accurately. Results averaged for aluminium cast into cast iron dies of various thickness reveal the early formation of the gap at the corners of the die where cooling is fastest, and the subsequent spread of the gap to the centre of the die face. A surprising result is the reduction of the gap if thick mould coats are applied. (The results in Figure 5.5 are plotted as straight lines. The apparent kinks in the early opening of the gap reported by these authors may be artefacts of their experimental method.)

It is not easy to see how the gap can be affected by the thickness of the coating. The effect may be the result of the creep of the solid shell under the internal hydrostatic pressure of the feeder. This is more likely to be favoured by thicker mould coats as a result of the increased time available and the increased temperature of the solidified skin of the casting. If this is true, then the effect is important because the hydrostatic head in these experiments was modest, only about 200 mm. Thus, for aluminium alloys that solidify with higher heads and times as long or longer than a minute or so, this mechanism for gap reduction will predominate. It seems possible, therefore, that in gravity die casting of aluminium the die coating will have the major influence on heat transfer, giving a large and stable resistance across the interface. The air gap will be a small and variable contributor. For computational purposes, therefore, it is attractive to consider the great simplification of neglecting the air gap in the special case of gravity die casting of aluminium.

It is probably helpful to draw attention to the fact that the name 'air gap' is perhaps a misnomer. The gap will contain almost everything except air. As we have seen previously, mould gases are often high in hydrogen, containing typically 50%. At room temperature, the thermal conductivity of hydrogen is approximately 5.9 times higher than that of air, and at 500°C the ratio rises to 7.7. Thus, the conductivity of a gap at the casting-mould interface containing a 50:50 mixture of air and hydrogen at 500°C can be estimated to be approximately a factor of four higher than that of air. In the past, therefore, most investigators in this field have probably chosen the wrong value for the conductivity of the gap, and by a substantial margin!

This effect has been used by Doutre (1998) who injected helium gas, with a conductivity nearly identical to hydrogen (approximately seven times the conductivity of air), into the air gap of a large Al alloy intake manifold cast in a permanent mould. The gas was introduced 30 s after pouring at a rate in the region of 10 mLs^{-1}. In a full-scale works trial, the production rate of the casting was increased by 25%. An even larger potential productivity increase, 45%, was found

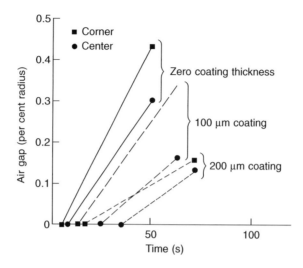

FIGURE 5.5

Results averaged from various dies (Isaac et al., 1985), illustrating the start of the air gap at the corners, and its spread to the centre of the mould face. Increased thickness of mould coating is seen to delay solidification and to reduce the growth of the gap.

by Grandfield's team (2007) when casting horizontal ingots in an open iron ingot mould. Gebelin and Griffiths (2007) found a reduced effect when attempting to cast resin-bonded sand moulds in an evacuated chamber back-filled with He. The explanation of the poor result in this latter case is almost certainly the result of the outgassing of sand moulds under vacuum. This outward wind of volatiles would prevent the ingress of He.

In passing, it seems worth commenting that He is expensive, and world supplies are limited. Naturally pure hydrogen introduced into the gap would perform similarly, but involve an unacceptable danger in the work place. Almost the same effect would be expected if steam, or better still, a water mist, were introduced into the air gap of a permanent mould. It would react with the metal to form hydrogen in situ, precisely where it is needed. Alternatively, many sand moulds, such as greensand or sodium silicate bonded sands, have sufficient water that the hydrogen atmosphere is provided automatically, and free of charge.

The heat transfer coefficient

The authors Ho and Pehlke (1984) from the University of Michigan have reviewed and researched this area thoroughly. We shall rely mainly on their analytical approach for an understanding of the heat transfer problem.

When the metal first enters the mould the macroscopic contact is good because of the conformance of the molten metal to the detailed shape of the mould. Gaps exist on a microscale between high spots as shown in Figure 5.6. At the high spots themselves, the high initial heat flux causes nucleation of the solid metal by local severe undercooling (Prates and Biloni, 1972). The nucleated solid then spreads to cover most of the surface of the casting because the thin layer of liquid adjacent to the cool mould surface will be expected to be undercooled. Conformance and overall contact between the surfaces is expected to remain good during all of this early period, even though both the casting and mould will now be starting to move rapidly because of distortion.

After the creation of a solidified layer with sufficient strength, further movements of both the casting and the mould are likely to cause the good fit to be broken, so that contact is maintained across only a few widely spaced random high spots (Figure 5.6(b)). The total transfer of heat across the interface h_t may now be written as the sum of three components:

$$h_t = h_s + h_c + h_r$$

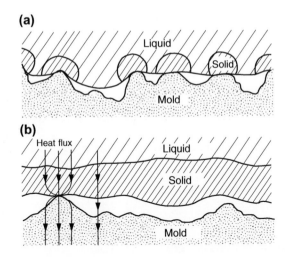

FIGURE 5.6

Metal-mould interface at an early stage when solid is nucleating at points of good thermal contact. Overall macroscopic contact is good at this stage (a). Later (b) the casting gains strength, and casting and mould both deform, reducing contact to isolated points at greater separations on non-conforming rigid surfaces.

where h_s is the conduction through the *solid* contacts, h_c is the *conduction* through the gas phase and h_r is that transferred by *radiation*. Ho and Pehlke produce analytical equations for each of these contributors to the total heat flux. We can summarise their findings as follows:

1. While the casting surface can conform, the contribution of solid–solid conduction is the most important. In fact, if the area of contact is enhanced by the application of pressure, then values of h_t up to 60,000 $Wm^{-2}K^{-1}$ are found for aluminium in squeeze casting. Such high values are quickly lost as the solid thickens and conformance is reduced, the values falling to more normal levels of 100–1000 $Wm^{-2}K^{-1}$ (Figure 5.7).

2. When the interface gap starts to open, the conduction via any remaining solid contacts becomes negligible. The point at which this happens is clear in Figure 5.7(b). (The actual surface temperature of the casting and the mould in this figure are reproduced from the results calculated by Ho and Pehlke.) The rapid fall of the casting surface

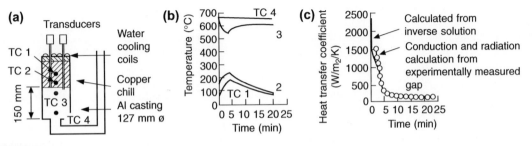

FIGURE 5.7

The experiment by Ho and Pehlke (1984) showing (a) their bottom gated mould; (b) the temperature history recorded across a casting-chill interface; and (c) the inferred heat transfer coefficient.

temperature is suddenly halted, and reheating of the surface starts to occur. An interesting mirror image behaviour can be noted in the surface temperature of the mould that, now out of contact with the casting, starts to cool. The estimates of heat transfer are seen to simultaneously reduce from over 1000 to around 100 $Wm^{-2}K^{-1}$ (Figure 5.7(c)).

3. After solid conduction diminishes, the important mechanism for heat transfer becomes the conduction of heat through the gas phase. This is calculated from:

$$h_c = k/d$$

where k is the thermal conductivity of the gas and d is the thickness of the gap. An additional correction is noted by Ho and Pehlke for the case where the gap is smaller than the mean free path of the gas molecules, which effectively reduces the conductivity. Thus heat transfer now becomes a strong function of gap thickness. As we have noted previously, it will also be a strong function of the composition of the gas. Even a small component of hydrogen will greatly increase the conductivity. Note also that it is assumed, almost certainly accurately, that the gas is stationary, providing heat flow by conduction only, not by convection.

For the case of light alloys, Ho and Pehlke find that the contribution to heat transfer from radiation is of the order of 1% of that from conduction by gas. Thus radiation can be safely neglected at these temperatures.

For higher-temperature metals, results by Jacobi (1976) from experiments on the casting of steels in different gases and in vacuum indicate that radiation becomes important to heat transfer at these higher temperatures.

Turning now to experimental work on the effect of die coatings on permanent moulds on heat transfer coefficients, a comprehensive review has been made by Nyamekye et al. (1994). Chiesa (1990) found that the conductance of a black coat was roughly twice that of a white coat of moderate thickness in the region of 120 μm. Also, the insulating effect of a white coat increased only marginally with thickness. Their findings that coats with high surface roughness were more effective insulators have been confirmed by the calculations of Hallam and Griffiths (2000) for the case of Al alloy castings. They demonstrate excellent predictions based on the assumption that the resistance of the die coating is mainly from the gas voids between the casting and the coating surface. Thus the structure of the coating surface was a highly influential factor in determining the heat transfer across the casting/mould interface.

The effect of gravity on the contact between the casting and mould has already been discussed previously. Woodbury (1998) finds that for a lightweight horizontal plate casting of only 6 mm thickness the heat transfer settles to a constant level of 70 $Wm^{-2}K^{-1}$. However, for a 25 mm thick plate the heat transfer from the top is approximately unchanged at 70, but the value from the underside of this heavier plate is now approximately doubled at 140 $Wm^{-2}K^{-1}$.

For sand moulds the use of pressure to enhance metal/mould contact is of course limited by penetration of the metal into the sand. The use of pressure has no such limitation in the case of metal moulds. Tadayon (1992) reports the freezing time of a squeeze casting to be 84 s at zero applied pressure, but reducing to 56 s at 5 MPa. A further increase of pressure to 10 MPa reduced the time only minimally further to 54 s.

Finally, it seems necessary to draw attention to the comprehensive review by Woolley and Woodbury (2007). These authors critically assess the vast literature on the determination of heat transfer coefficients, concluding that they are sceptical of the accuracy of all of the published data, and cite a number of key reasons for unreliability. In a later paper (2009), for instance, these authors find that the use of thermocouples to measure heat flow in these experiments alone introduces an error of 65%. It seems, we still have a long way to go to achieve reliable heat transfer coefficients.

Resistance 3: The mould

The rate of freezing of castings made in silica sand moulds is generally controlled by the rate at which heat can be absorbed by the mould. In fact, compared with many other casting processes, the sand mould acts as an excellent insulator, keeping the casting hot. However, of course, ceramic investment and plaster moulds are even more insulating, avoiding premature cooling of the metal, and aiding fluidity to give the excellent ability to fill thin sections for which these casting processes are renowned. It is a pity that the extremely slow cooling generally contributes to rather poorer mechanical properties, but this is to some extent a self-inflicted problem. If the metal were free from bifilms it is predicted that mechanical properties would be not noticeably affected by slower rates of freezing (see Chapter 9.4).

Considering the simplest case of unidirectional conditions once again, and metal poured at its melting point T_m against an infinite mould originally at temperature T_0, but whose surface is suddenly heated to temperature T_m at $t = 0$, and that has thermal diffusivity α_m, we now have:

$$\frac{\partial T}{\partial t} = \alpha_m \frac{\partial^2 T}{\partial x^2} \tag{5.9}$$

Following Flemings, the final solution is:

$$S = \frac{2}{\sqrt{\pi}} \underbrace{\left(\frac{T_m - T_0}{\rho_s H} \right)}_{metal} \underbrace{\sqrt{K_m \rho_m C_m}}_{mould} \sqrt{t} \tag{5.10}$$

This relation is most accurate for the highly conducting non-ferrous metals, aluminium, magnesium and copper. It is less good for iron and steel, particularly those ferrous alloys that solidify to the austenitic (face centred cubic) structure that has especially poor conductivity. The relation quantifies a number of interesting outcomes as discussed next.

Note that at a high temperature heat is lost more quickly, so that a casting in steel should solidify faster than a similar casting in grey iron. This perhaps surprising conclusion is confirmed experimentally, as seen in Figure 5.8.

Low heat of fusion of the metal, H, similarly favours rapid freezing because less heat has to be removed. Magnesium castings therefore freeze faster than similar castings in aluminium despite their similar freezing points (Table 5.1).

The product $K_m \rho_m C_m$ is a useful parameter to assess the rate at which various moulding materials can absorb heat. The reader needs to beware that some authorities have called this parameter the heat diffusivity, and this definition was followed in CASTINGS (Campbell, 1991). However, originally the definition of heat diffusivity b was $(K_m \rho_m C_m)^{1/2}$ as described for instance by Ruddle (1950). In subsequent years, the square root seems to have been overlooked in error. Ruddle's definition is therefore accepted and followed here. However, of course, both b and b^2 are useful quantitative measures. What we call them is merely a matter of definition. (I am grateful to John Berry of Mississippi State University for pointing out this fact. As a further aside from Professor Berry, the units of b are even more curious than the units of toughness; see Table 5.2.)

For simple shapes, if we assume that we may replace S with V_s/A where V_s is the volume solidified at a time t, and A is the area of the metal-mould interface (i.e. the cooling area of the casting), then when $t = t_f$ where t_f is the total freezing time of a casting of volume V we have:

$$\frac{V}{A} = \frac{2}{\sqrt{\pi}} \left(\frac{T_m - T_0}{\rho_s H} \right) \sqrt{K_m \rho_m C_m} \sqrt{t_f} \tag{5.11}$$

and so:

$$t_f = B(V/A)^2 \tag{5.12}$$

where B is a constant for given metal and mould conditions. The ratio (V/A) is the useful parameter generally known as the *modulus m;* thus, Eqn (5.12) indicates that the parameter m^2 is the important factor that controls the solidification time of the casting. Approximate values of m for simple shaped castings as illustrated in Table 5.3 are usefully memorised.

Equation (5.12) is the famous Chvorinov rule. Convincing demonstrations of its accuracy have been made many times. Chvorinov himself showed in his paper published in 1940 that it applied to steel castings from 12 to 65,000 kg weight made in greensand moulds. This superb result is presented in Figure 5.9. Experimental results for other alloys are illustrated in Figure 5.8.

Chvorinov's rule is one of the most useful guides to the student. It provides a powerful general method of tackling the feeding of castings to ensure their soundness.

However, the previous derivation of Chvorinov's rule is open to criticism in that it uses one-dimensional theory but goes on to apply it to three-dimensional castings. In fact, it is quickly appreciated that the flow of heat into a concave

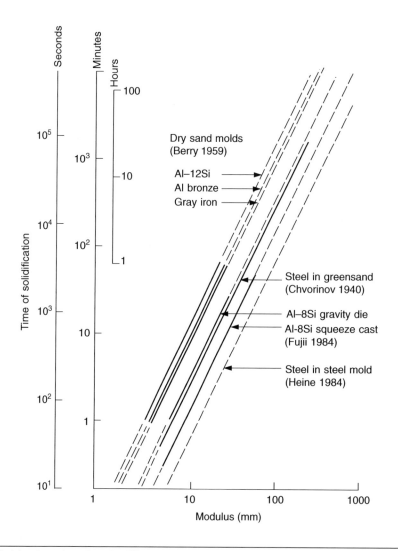

FIGURE 5.8

Freezing times of plate shaped castings in different alloys and moulds.

mould wall will be divergent, and so will be capable of carrying away heat more rapidly than in a one-dimensional case. We can describe this exactly (without the assumption of one-dimensional heat flow), following Flemings once again:

$$\frac{\partial T}{\partial t} = \alpha_m \left(\frac{\partial^2 T}{\partial r^2} + \frac{n \partial T}{r \partial r} \right) \tag{5.13}$$

where $n = 0$ for a plane, 1 for a cylinder, and 2 for a sphere. The casting radius is r. The solution to this equation is:

$$\frac{V}{A} = \left(\frac{T_m - T_0}{\rho_s H} \right) \left(\frac{2}{\sqrt{\pi}} \sqrt{K_m \rho_m C_m} \sqrt{t_f} + \frac{n K_m t_f}{2r} \right) \tag{5.14}$$

Table 5.1 Mould and Metal Constants

Material	Melting Point (m.p.) (°C)	Latent Heat Melting (J/g)	Liquid–Solid Contraction (%)		Specific Heat (J/kg K)			Density (kg/m³)			Thermal Conductivity (J/m K s)		
					Solid		Liquid	Solid		Liquid	Solid		Liquid
			a	b	20°C	m.p.	m.p.	20°C	m.p.	m.p.	20°C	m.p.	m.p.
Pb	327	23	3.22	3.20	130	(138)	152	11,680	11,020	10,678	39.4	(29.4)	15.4
Zn	420	111	4.08	4.08	394	(443)	481	7140	(6843)	6575	119	95	9.5
Mg	650	362	4.2	4.21	1038	(1300)	1360	1740	(1657)	1590	155	(90)?	78
Al	660	388	7.14	6.92	917	(1200)	1080	2700	(2550)	2385	238	–	94
Cu	1084	205	5.30	4.78	386	(480)	495	8960	8382	8000	397	(235)	166
Fe	1536	272	3.16	3.56	456	(1130)	795	7870	7265	7015	73	(14)?	–
Graphite	–	–	–	–	1515	–	–	2200	–	–	147	–	–
Silica sand	–	–	–	–	1130	–	–	1500	–	–	0.0061	–	–
Mullite	–	–	–	–	750	–	–	1600	–	–	0.0038	–	–
Plaster	–	–	–	–	840	–	–	1100	–	–	0.0035	–	–

References: Brandes (1991), Flemings (1974)
[a]*Wray (1976)*
[b]*From densities presented here.*

Table 5.2 Thermal Properties of Mould and Chill Materials at Approximately 20°C

Material	Heat Diffusivity $(K\rho C)^{1/2}$ $(\mathrm{Jm^{-2}K^{-1}s^{-1/2}})$	Thermal Diffusivity $K/\rho C$ $(\mathrm{m^2s^{-1}})$	Heat Capacity per Unit Volume ρC $(\mathrm{JK^{-1}m^{-3}})$
Silica sand	3.21×10^3	3.60×10^{-9}	1.70×10^6
Investment	2.12×10^3	3.17×10^{-9}	1.20×10^6
Plaster	1.8×10^3	3.79×10^{-9}	0.92×10^6
Magnesium	16.7×10^3	85.8×10^{-6}	1.81×10^6
Aluminium	24.3×10^3	96.1×10^{-6}	2.48×10^6
Copper	37.0×10^3	114.8×10^{-6}	3.60×10^6
Iron (pure Fe)	16.2×10^3	20.3×10^{-6}	3.94×10^6
Graphite	22.1×10^3	44.1×10^{-6}	3.33×10^6

Table 5.3 Moduli of Some Common Shapes

Shape			Modulus 100% Cooled Area		Modulus Base Uncooled	
Sphere			$D/6$	0.167D	—	—
Cube			$D/6$	0.167D	$D/5$	0.200D
Cylinder		H/D				
		1.0	$D/6$	0.167D	$D/5$	0.200D
		1.5	$3D/16$	0.188D	$3D/14$	0.214D
		2.0	$D/5$	0.200D	$2D/9$	0.222D
Infinite cylinder		∞	$D/4$	0.250D	—	—
Infinite plate			$D/2$	0.500D	—	—

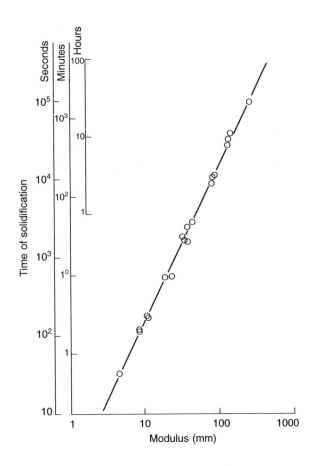

FIGURE 5.9

Freezing time of steel castings in greensand moulds as a function of modulus (Chvorinov, 1940).

The effect of the divergency of heat flow predicts that for a given value of the ratio V/A (i.e. a given modulus m) a sphere will freeze quickest, the cylinder next, and the plate last. Katerina Trbizan (2001) provides a useful study, confirming these relative freezing rates for these three shapes. For aluminium in sand moulds, Eqn (5.14) indicates these differences to be close to 20%. This is part of the reason for the safety factor 1.2 recommended when applying Chvorinov's feeding rule, because the feeding rules tacitly assume that all shapes with the same modulus freeze at the same time.

The simple Chvorinov link between modulus and freezing time is capable of great sophistication. One of the great exponents of this approach has been Wlodawer (1966), who produced a famous volume devoted to the study of the problem for steel castings. This has been a source book for the steel castings industry ever since.

The subject has been advanced further by the work of Tiryakioglu in 1997 (interestingly using his deceased father's excellent doctoral research at the University of Birmingham, UK, in 1964) that showed secondary, but important, effects of shape, volume and superheat on the freezing time of the casting.

A final aspect relating to the divergency of heat flow is important. For a planar freezing front, the rate of increase of thickness of the solidified metal is parabolic, gradually slowing with thickness, as described by Equations such as 5.3 and 5.4 relating to 1-dimensional heat flow. However, for more compact shapes such as cylinders, spheres, cubes, etc., the heat flow from the casting is 3-dimensional. Thus initially for such shapes, when the solidified layer is relatively thin, the solid thickens parabolically. However, at a much later stage of freezing, when little liquid remains in the centre of

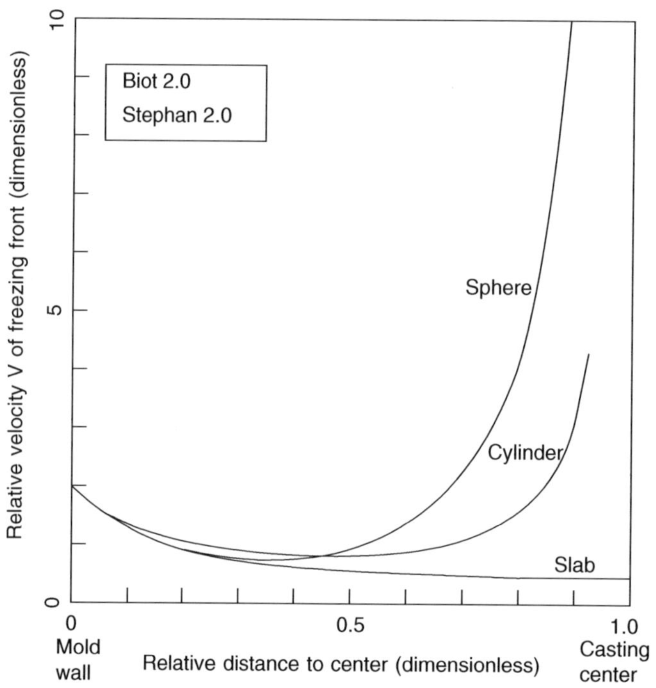

FIGURE 5.10

Acceleration of the freezing front in compact castings as a result of three-dimensional extraction of heat (sphere and cylinder curves calculated from Santos and Garcia, 1998).

the casting, the extraction of heat in all three directions greatly accelerates the rate of freezing. Santos and Garcia (1998) show that the effect, accurately predicted theoretically by Adams in 1956, is general. Whereas in a slab casting the velocity of the front slows progressively with distance according to the well-known parabolic law, for cylinders and spheres the growth rate is similar until the front has progressed to about 40% of the radius. From then onwards the front accelerates rapidly (Figure 5.10).

This increase of the rate of freezing in the interior of many castings explains the otherwise baffling observation of 'inverse chill' as seen in cast irons. Normal intuition would lead the caster to expect fast cooling near the surface of the casting, and this is true to a modest degree in all castings. From this point onwards the front slows progressively in uniform plate-like sections. But in bars and cylinders, as the residual liquid shrinks in size towards the centre of the casting, the front speeds up dramatically, causing grey iron to change to carbidic white iron. The accelerated rate has been demonstrated experimentally by Santos and Garcia on a Zn-4Al alloy by measurement of the increasing fineness of dendrite arm spacing towards the centre of a cylindrical casting.

It is interesting that the effect of accelerated freezing appears never to have been seen in Al alloys. This seems to be the result of the high thermal conductivity of these alloys, causing dendritic freezing over the whole cross-section of the casting, and thus smoothing and obscuring the acceleration of solidification towards the centre of castings.

5.1.2 INCREASED HEAT TRANSFER

In practice, the casting engineer can manipulate the rate of heat extraction from a casting using a number of tricks. These include the placement of blocks of metal in the mould, adjacent to the casting, to act as local chills. Similarly, fins or pins attached to the casting increase the local surface area from which heat can be conducted deep into the mould where it can be dissipated. These techniques will be described next.

The action of chills, fins and pins can provide localised cooling of the casting to assist directional solidification of the casting towards the feeder, thus assisting in the achievement of soundness. This is one of the important actions of these chilling devices. It is, however, not the only action, as discussed next.

The use of any of these chilling techniques acts to increase the ductility and strength of that locality of the casting. It seems most probable that this occurs because the faster solidification freezes in the bifilms in their compact form before they have chance to unfurl. (Recall that the bifilms are compacted by the extreme bulk turbulence during pouring and during their travel through the running system. However, they subsequently unravel if they have sufficient time or sufficient driving force, gradually opening up in the mould cavity when conditions in the melt become quiet once again, so as to gradually reduce properties.) However, there is a small contribution towards strength and ductility from the refinement of structure. Al-Si alloys and Mg-based alloys particularly benefit. The remaining Al alloys and other fcc structured metals such as Cu-based alloys and austenitic steels would not be expected to benefit usefully from finer grain or finer DAS.

The interesting corollary of this fact is that if chills are seen to increase ductility and strength of these alloys, it confirms that the cast material is defective, containing a high percentage of bifilms. Another interesting corollary is that if the alloy can be cast without bifilms, its properties will already be high, so that chilling should not increase its properties further. This rather surprising prediction is fascinating, and, if true, indicates the huge potential for the increase of the properties of cast alloys. All castings without bifilms are therefore predicted to have extraordinary ductility and strength. It also explains our lamentable current condition in which most of us constantly struggle in our foundries to achieve minimum mechanical properties for castings even when the casting is plastered in chills. The common experience when attempting to achieve minimum properties is that some days we win, other days we continue to struggle. The message is clear, we need to focus on technologies for the production of castings with reduced bifilm content, preferably zero bifilm content. The rewards are huge.

Another action of chills is to straighten bifilms. This unfortunate action occurs because the advancing dendrites cannot grow through the air layer between the double films, and so push the bifilms ahead. Those that are somehow attached to the wall will be partially pushed, straightened and unravelled by the gentle advance of grains. This effect is reported in Section 2.3 on furling and unfurling of bifilms. Thus although a large percentage of bifilms will be pushed ahead of the chilled region, concentrating (and probably reducing the properties) in the region immediately ahead, some bifilms will remain aligned in the dendrite growth direction, and so be largely perpendicular to the mould wall. The mechanism is presented in Figure 2.43(a) and (b), and an example is seen in the radiograph in Figure 2.43(c).

The overall effects on mechanical properties of the pushing action are therefore not easily predicted. The reduction in density of defects close to the chill will raise properties, but the presence of occasional bifilms aligned at right angles to the surface of the casting would be expected to be severely detrimental. These complicated effects require to be researched. However, we can speculate that they seem likely to be the cause of troublesome edge cracking in the rolling of cast materials of many types, leading to the expense of machining off the surface of many alloys before rolling can be attempted. The superb formability of electroslag remelted (ESR) compared to vacuum arc remelted (VAR) alloys is almost certainly explained in this way. The ESR process produces an extremely clean material because oxide films are dissolved during remelting under the layer of liquid slag, and will not re-form in the solidifying ingot. In contrast, the relatively poor vacuum of the VAR process ensures that the lapping of the melt over the liquid meniscus at the mould wall will create excellent double oxide films. If considerable depths of the VAR ingot surface are not first removed (a process sometimes called 'peeling') oxide lap defects will open as surface cracks when subjected to forging or rolling.

Although, as outlined previously, the chilling action of chills and fins is perhaps more complicated than we first thought, the chilling action itself on the rate of solidification is well documented and understood. It is this thermal aspect of their behaviour that is the subject of the remainder of this section.

External chills

In a sand mould, the placing of a block of metal on the pattern, and subsequently packing the sand around it to make the rest of the mould in the normal way, is a widely used method of applying localised cooling to that part of the casting.

A similar procedure can be adopted in gravity and low-pressure die-casting by removing the die coat locally to enhance the local rate of cooling. In addition, in dies of all types, this effect can be enhanced by the insertion of metallic inserts into the die to provide local cooling, especially if the die insert is highly conductive (such as made from copper alloy) and/or artificially cooled, for instance by air, oil or water.

Such chills placed as part of the mould, and that act against the outside surface of the casting are strictly known as external chills, to distinguish them from internal chills that are cast in, and become integral with, the casting.

In general terms, the ability to chill is a combination of ability to absorb heat and to conduct it away. It is quantitatively assessed by

$$\text{heat diffusivity} = (K\rho C)^{1/2}$$

where K the thermal conductivity, ρ the density, and C is the specific heat of the mould. It has complex units $\text{Jm}^{-2}\text{K}^{-1}\text{s}^{1/2}$. Take care not to confuse with

$$\text{thermal diffusivity} = K/\rho C$$

normally quoted in units of m^2s^{-1}.

From the room temperature data in Table 5.1 (unfortunately, high temperature values are less easily obtained) we can obtain some comparative data on the chilling power of various mould and chill materials, shown in Table 5.2. It is clear that the various refractory mould materials—sand, investment and plaster—do not act effectively as chilling materials. The various chill materials are all in a league of their own, having chilling powers orders of magnitude higher than the refractory mould materials. They improve marginally, within a mere factor of 5, in the order steel, graphite and copper.

The heat diffusivity value indicates the action of the material to absorb heat when it is infinitely thick, being unconstrained in the amount of heat it can conduct away and store in itself i.e. as would be reasonably well approximated by constructing a thick-walled mould from such material.

This behaviour contrasts with that of a relatively small lump of cast iron or graphite used as an external chill in a sand mould. A small chill does not develop its full potential for chilling as promised by the heat diffusivity because it has limited capacity for heat. Thus although the initial rate of freezing of a metal may be in the order given by the previous list, for a chill of limited thickness its cooling effect quickly becomes limited because it becomes saturated with heat; after a time it can absorb no more. The amount of heat that it can absorb is defined as its heat capacity. We can formulate the useful concept of heat capacity per unit volume ρC in terms of its density ρ and its specific heat C, so that the heat capacity of a chill of volume V is simply

$$\text{volumetric heat capacity} = V\rho C$$

In the SI system, its units are JK^{-1}. Figure 5.11 illustrates the fact that if chills are limited by their relatively low heat capacity, there is little difference between copper, graphite and iron. However, for larger chills that are able to conduct heat away without saturating, copper is by far the best material. The next best, graphite, is only half as good, and iron is only a quarter as effective. For situation intermediate between small and large chills conditions partway between the extremes can be read off the nomogram.

Aluminium chills are interesting, in that if rather small, they are relatively poor compared with steel, graphite and copper, whereas larger blocks capable of carrying heat away without saturation are better than steel or graphite, but still only half as good as copper.

The results by Rao and Panchanathan (1973) on the casting of 50 mm thick plates in Al-5Si-3Cu reveals that the casting is insensitive to whether it is cooled by copper, graphite or steel chills, provided that the volumetric heat capacity of the chill is taken into account. We can conclude that their chills were rather limited in size, and so limited by their heat capacity.

These authors show that for a steel chill 25 mm thick its heat capacity is 900 JK^{-1}. A chill with identical capacity in copper they claimed to be 32 mm thick, and in graphite 36 mm. These values originally led the author to conclude that copper may therefore not always the best chill material (CASTINGS 1991). However, using somewhat more accurate data (Figure 5.11) copper is found, after all, to be the most effective whether limited by heat diffusivity or heat capacity.

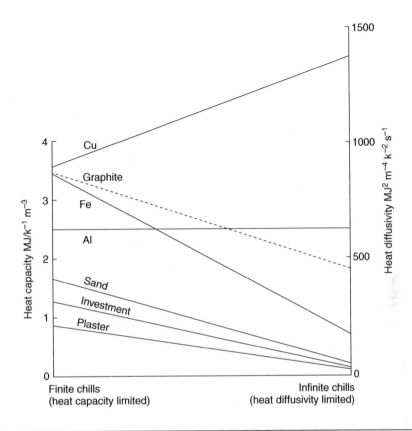

FIGURE 5.11

Relative diffusivities (ability to diffuse heat away if a large chill) and heat capacities (ability to absorb heat if relatively small) of chill materials.

Lerner and Kouznetsov (2004) confirm for copper and iron chills on a 20 mm thick Al alloy casting the rather surprising fact that the volume of external chills is far more important than their surface area of contact with the casting.

Figure 5.12 illustrates that the chills are effective over a considerable distance, the largest chills greatly influencing the solidification time of the casting even up to 200 mm (four times the section thickness of the casting) distant. This large distance is perhaps typical of such a thick-section casting in an alloy of high thermal conductivity, providing excellent heat transfer along the casting. A steel casting would respond less at this distance. This work by Rao and Panchanathan (1973) reveals the widespread sloppiness of much present practice on the chilling of castings. General experience of the chills generally used in foundrywork nowadays shows that chill size and weight are rarely specified, and that chills are in general too small to be fully effective in any particular job. It clearly matters what size of chill is added.

Computational studies by Lewis and colleagues (2002) have shown that the number, size and location of chills can be optimised by computer. These studies are among the welcome first steps towards the intelligent use of computers in casting technology.

Finally, in detail, the action of the chill is not easy to understand nor to predict. The surface of the casting against the chill will often contract, distorting away and thus opening up an air gap. The chilled casting surface may then reheat to such an extent that the surface of the casting remelts. The exudation of eutectic is often seen between the casting and the

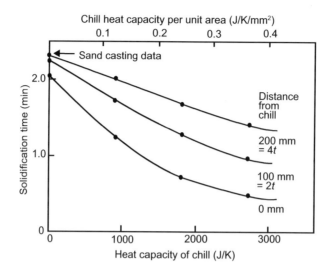

FIGURE 5.12

Freezing times of a plate 225 × 150 × 50 mm in Al-5Si-3Cu alloy at various distances from the chill is seen to shrink steadily as the chill is approached and as the chill size is increased (Rao and Panchanathan, 1973).

chill (Figure 5.13). The new contact between the eutectic and chill probably then starts a new burst of heat transfer and thus a new rapid phase of solidification of the casting. Thus the history of cooling in the neighbourhood of a chill may be a succession of stop/start, or slow/fast events.

Internal chills

The placing of chills inside the mould cavity with the intention of casting them in place is an effective way of localised cooling. Such chills are usually carefully chosen to be of the same chemical composition as the melt so that they will be essentially invisible when finally frozen into the casting. The technique is an excellent solution to the challenge of making sound a heavy boss or thick section in the centre of a complex casting that cannot be accessed by conventional feeding, especially if the centre of the boss, together with the internal chill, is to be subsequently machined out.

Internal chills take a great variety of shapes as illustrated in Figure 5.14.

When estimating what size of chill might be required (i.e. its weight), the simple method of mixtures approach (Campbell and Caton, 1977) indicates that to cool superheated pure liquid iron to its freezing point, and freezing a proportion of it, will require various levels of addition of cold, solid low carbon steel depending on the extent that the addition itself actually melts. These estimations take no account of other heat losses from the casting. Thus for normal castings the predictions are likely to be incorrect by up to a factor of 2. This is broadly confirmed by Miles (1956), who top-poured steel into dry sand moulds 75 mm square and 300 mm tall. In the centre of the moulds was positioned a variety of steel bars ranging from 12.5 mm round to 25 mm square, covering a range of chilling from 2% to 11% solid addition. His findings reveal that the 2% solid addition nearly melted, compared with the predicted value for complete melting of 3.5% solid. The 11% solid addition caused extensive (possibly total) freezing of the casting judging by the appearance of the radial grain structure in his macrosections. He found 5% addition to be near optimum; it had a reasonable chilling effectiveness and caused relatively few defects.

These additions are plotted in graphical form in Figure 5.15. They illustrate that additions up to about 3.5% completely melt and disappear (although clearly cool the melt by this process). Further additions linearly increase the amount of liquid that is solidified, finally becoming completely frozen at about 10% addition.

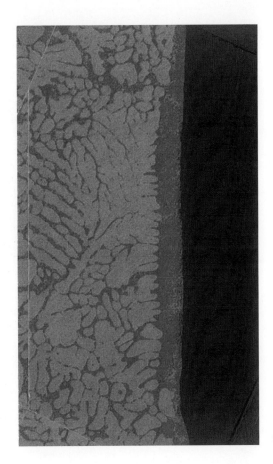

FIGURE 5.13

Al-Si eutectic liquid segregation by exudation at a chilled interface of an Al-7Si alloy.

In the case of additions higher than 10%, where the heat input is not sufficient to melt the chill, the fusing of the chill surface into the casting has to be the result of a kind of diffusion bonding process. This would emphasise the need for cleanness of the surface, requiring the minimum presence of oxide films or other debris against the chill during the filling of the mould. If Miles had used a better bottom gated filling technique he may have reduced the observed filling defects further, and found that higher percentages were practical.

The work by Miles does illustrate the problems generally experienced with internal chills. If the chills remain for any length of time in the mould, particularly after it is closed, and more particularly if closed overnight, then condensation is likely to occur on the chill, and blow defects will be caused in the casting. Blows are also common from rust spots or other impurities on the chill such as oil or grease. The matching of the chemical composition of the chill and the casting is also important; mild steel chills will, for instance, usually be unacceptable in an alloy steel casting.

Internal chills in aluminium alloy castings have not generally been used, almost certainly as a consequence of the difficulty introduced by the presence of the oxide film on the chill. This appears to be confirmed by the work of Biswas et al. (1985), who found that at 3.5% by volume of chill and at superheats of only 35°C the chill was only partially melted and retained part of its original shape. It seems that over this area it was poorly bonded. At superheats of more than 75°C, or at only 1.5% by volume, the chill was more extensively melted, and was useful in reducing internal porosity and in

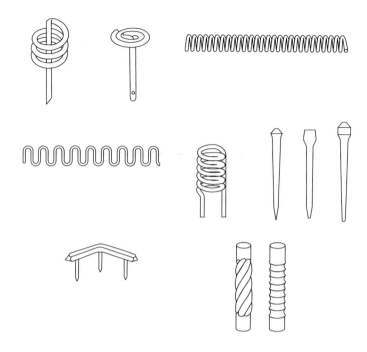

FIGURE 5.14

A variety of internal chills.

From Heine and Rosenthal (1955).

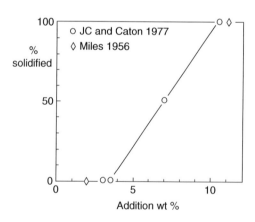

FIGURE 5.15

The addition of increasing percentage of cold mild steel particles to liquid mild steel cools the melt by melting of the additions up to about 4% addition, then starts to promote freezing, finally causing complete freezing of the melt at about 10% addition.

raising mechanical properties. The lingering presence of the oxide film from the chill (now having floated off to lurk elsewhere in the casting) remains a concern however.

The development of a good bond between the internal chill and the casting is a familiar problem with the use of chaplets – the metal devices used to support cores against sagging because of weight, or floating because of buoyancy. A one-page review of chaplets is given by Bex (1991). To facilitate the bond for a steel chaplet in an iron or steel casting the chaplet is often plated with tin. The tin serves to prevent the formation of rust, and its low melting point (232°C) and solubility in iron assists the bonding process.

There has been much work carried out on the casting of steel inserts into cast irons. Xu (2007) studied thin steel chill plates in a heavy grey iron block casting, measuring the diffusion of alloying elements between them. He notes that carbon content is smoothed across the interface, but the silicon concentration forms a sharp step at the interface. This is behaviour to be expected in terms of the rates of diffusion given in Figure 1.4(c).

Noguchi (1993) used a thermally sprayed Ni-based self-fluxing alloy on mild steel inserts in both flake and ductile iron. Although he reports that the bonding of inserts is sensitive to the volume ratio of the poured metal and insert (as we have seen in Figure 5.15) and the pouring temperature, the thermal spray greatly expands the regime of successful bonding. In later work, Noguchi (2005) cast steel sleeves and end rings into cast irons, avoiding the use of sprayed interlayers. In this work he raised the temperature of the insert higher than could be achieved by simple immersion. He arranged for an extended period of flow of hot metal around the insert, in some cases running the excess metal into flow-offs. He found that if he could raise the interface temperature to 1150°C, the melting point of the iron, he could achieve what appeared to be a perfect bond over large areas of the insert.

The bond between steel and titanium inserts in Al alloy castings has been investigated in Japan (Noguchi et al., 2001) who found only a 10 µm silver coating was effective to achieve a good bond, although even this took up to 5 min to develop at the Al-Ag eutectic temperature 566°C. Attempts to achieve a bond with gold plating and Al-Si sprayed alloy were largely unsuccessful.

Biswas and co-workers (1994) have researched Al alloy chills in Al alloys as a function of relative volume and casting superheat. However, these authors overlook the problem of the oxide films, not appreciating that it represents an ever-present danger. It will persist as a double film (having acquired its second layer during the immersion of the chill by the liquid) and so pose the risk of leakage or crack formation. Such risks are only acceptable for low duty products.

Brown and Rastall (1986) take advantage of the oxide on the surface of heavy aluminium inserts in aluminium castings, using this to avoid any bonding between the casting and the insert. They use a cast aluminium alloy core inside an aluminium alloy casting to form re-entrant details that could not easily be provided in a pressure die cast product. Also, of course, because the freezing time is shortened, productivity is enhanced. The internal core is subsequently removed by disassembly or part machining, or by mechanical deformation, peeling apart the oxide bifilm separating the core and the casting.

Fins

If we add an appendage such as a fin to a casting, we inevitably form a T-junction between the two. Depending on the thickness of the appendage, the junction may be either a hot spot or a cold spot. Therefore, before we look specifically at fins on castings, it is worth spending some time to consider the concepts involved in junctions of all types.

Kotschi and Loper (1974) were among the first to evaluate junctions. Their results are summarised in Figures 5.16 and 5.17 and further interpreted in Figure 5.18 to show the complete range of junctions and their effect on the residual liquid in the main cast plate. Considering the range in Figure 5.18, starting at the thinnest appendage:

1. When the wall forming the upright of the T is thin, it acts as a cooling fin, chilling the junction and the adjacent wall (the top cross of the T) of casting. We shall return to a more detailed consideration of fins shortly.
2. When the upright of the T-section has increased to a thickness of half the casting section thickness, then the junction is close to thermal balance, the cooling effect of the fin balancing the hot-spot effect of the concentration of metal in the junction.
3. By the time that the upright of the T has become equal to the casting section, the junction is a hot spot. This is common in castings. Foundry engineers are generally aware the 1:1 T-junction is a problem. It is curious therefore

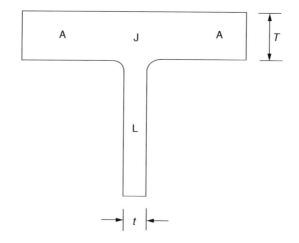

FIGURE 5.16

Geometry of a T-shaped junction.

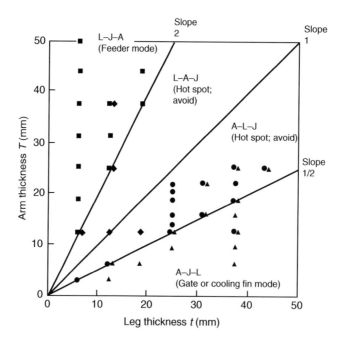

FIGURE 5.17

Solidification sequence for T-shaped castings (A = arm, J = junction, L = leg).

Experimental data from Hodjat and Mobley (1984).

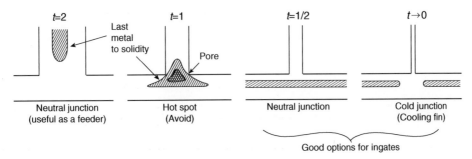

FIGURE 5.18

Array of T-junctions showing the thermal effects at the junction with different relative thickness of casting.

that castings with uniform wall thickness are said to be preferred, and that designers are encouraged to design them. Such products necessarily contain 1:1 junctions that will be hot spots. (Techniques for dealing with these are dealt with later when considering the feeding of castings; Section 6.)

4. When finally the section thickness of the upright of the T is twice the casting section, then the junction is balanced once again, with the casting now acting as the mild chill to counter the effect of the hot spot at the junction. We have considered these junctions merely in the form of the intersections of plates. However, we can extend the concept to more general shapes, introducing the use of the geometric modulus $m = $ (volume)/(cooling area). It subsequently follows that an additional requirement when a feeder forms a T-junction on a casting is that the feeder must have a modulus two times the modulus of the casting. The hot spot is then moved out of the junction and into the feeder, with the result that the casting is sound. This is the basis behind Feeding Rule 4 discussed later (Chapter 10.6.3).

Pellini (1953) was one of the first experimenters to show that the siting of a thin 'parasitic' plate on the end of a larger plate could improve the temperature gradient in the larger plate. However, the parasitic plate that he used was rather thick, and his experiments were carried out only on steel, whose conductivity is poor, reducing useful benefits.

Figure 5.19 shows the results from Kim et al. (1985) of pour-out tests carried out on 99.9% pure aluminium cast into sand moulds. The faster advance of the freezing front adjacent to the junction with the fin is clearly shown. (As an aside, this simple result is a good test of some computer simulation packages. The simulation of a brick-shaped casting with a cast-on fin should show the cooling effect by the fin. Some relatively poor computer algorithms do not take into account the conduction of heat in the casting, thus predicting, erroneously, the appearance of the junction as a hot spot, clearly revealed by the contours near the junction curving the wrong way.)

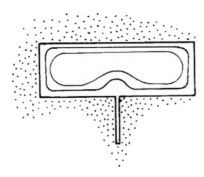

FIGURE 5.19

T-junction casting in 99.9Al by Kim et al. (1985) showing successive positions of the solidification front.

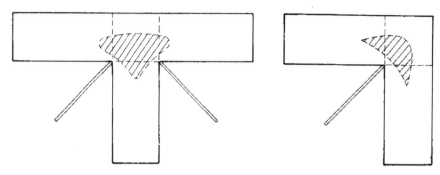

FIGURE 5.20

T- and L-junctions in pure aluminium cast in oil-bonded greensand. The shape of porosity in these junctions is shown, and the region of the junction used to calculate the percentage porosity is shown by the broken lines. The position of fins added to eliminate the porosity is shown. Results are presented in Figure 5.21.

Creese and Sarfaraz (1987) demonstrate the use of a fin to chill a hot spot in pure Al castings that was difficult to access in other ways. They cast on fins to T- and L-junctions as shown in Figure 5.20. The reduction in porosity achieved by this technique is shown in Figure 5.21. For these casting sections of 50 mm there was no apparent difference between fins of 2.5 and 3.3 mm thickness so these results are treated together in this figure. These fins at 5 and 6% of the casting section happen to be close to optimum as is confirmed later below. The reason that they conduct away perhaps less effectively than might be expected is because of their unfavourable location at 45° between two hot components of the junction.

Returning to the case where the upright of the T is sufficiently thin to act as a cooling fin, one further case that is not presented in Figure 5.18 is the case where the fin is so thin that it does not exist. This, you will say, is a trivial case. But think what it tells us. It proves that the fin can be too thin to be effective, because it will have insufficient area to carry away enough heat. Thus we arrive at the important conclusion that there is an optimum thickness of fin for a given casting section.

Similarly, an identical argument can be made about the fin length. A fin of zero length will have zero effect. As length increases, effectiveness will increase, but beyond a certain length, additional length will be of reducing value. Thus the length of fins will also have an optimum.

These questions have been addressed in a preliminary study by Wright and the author (1997) on a horizontal plate with a symmetrical fin (Figure 5.22). Symmetry was chosen so that thermocouple measurements could be taken along the centre line (otherwise the precise thermal centre was not known so that the true extension in freezing time may not have been measured accurately). In addition the horizontal orientation of the plate was selected to suppress any complicating effects of convection so far as possible. The thickness of the fin was t.H and the length L.H where t and L are the dimensionless numbers to quantify the fin in terms of H, the thickness of the plate. From this study it was discovered that there was an optimum thickness of a fin, and this was less than one tenth of H. Figure 5.22(a) interpolates an optimum in the region of 5% of the casting section thickness. The optimum length was 2H, and longer lengths were not significantly more effective (Figure 5.22(b)). For the optimum conditions the freezing time of the casting was increased by approximately 10 times. Thus the effect is useful. However, the effect is also rather localised, so that it needs to be used with caution. Eventually, non-symmetrical results for a chill on one side of the plate would be welcome.

Even so, the practical benefits to the use of a fin as opposed to a chill are interesting, and possibly even compelling. They are.

1. The fin is always provided on the casting because it is an integral part of the tooling. Thus, unlike a chill, the placing of it cannot be forgotten.

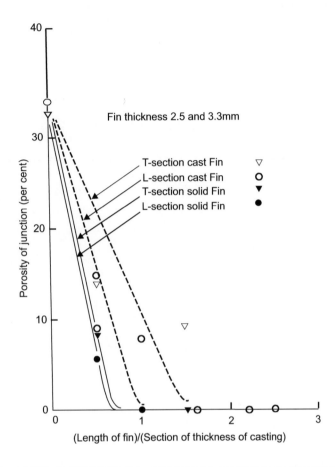

FIGURE 5.21

Results from Creese and Sarfaraz (1987, 1988) showing the reduction in porosity as a result of increasing length of fins.

2. It is always exactly in the correct place. It cannot be wrongly sited before the making of the mould. (The incorrect positioning of a chill is easily appreciated, because although the location of the chill is normally carefully painted on the pattern, the application of the first coat of mould release agent usually does an effective job in eliminating all traces of this. Furthermore of course, the chill can easily move during the filling of the moulding box with sand.)

3. It cannot be displaced or lifted during the making of the mould. If a chill lifts slightly during the filling of the tooling with sand, the resulting sand penetration under the edges of the chill, and the casting of additional metal into the roughly shaped gap, makes an unsightly local mess of the casting surface. Displacement or complete falling out of the chill from the mould is a common danger, sometimes requiring studs to support the chill if awkwardly angled or on a vertical face. Displacement commonly results in sand inclusion defects around the chill or can add to defects elsewhere. All this is costly to dress off.

4. An increase in productivity has been reported as a result of not having to find, place and carefully tuck in a block chill into a sand mould (Dimmick, 2001).

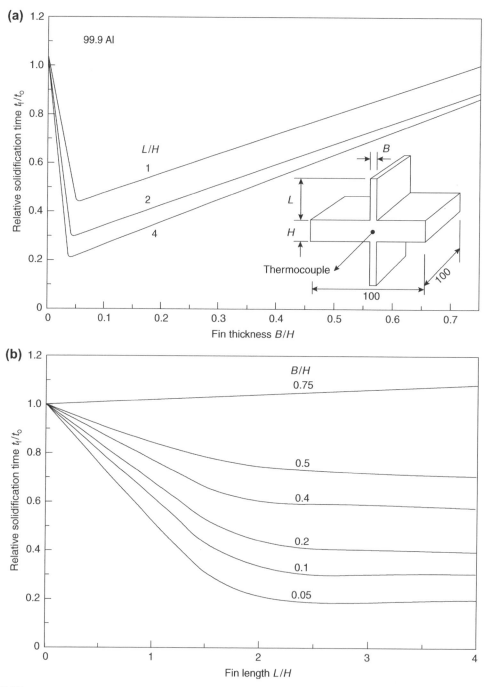

FIGURE 5.22

The effect of a symmetrical fin on the freezing time at the centre of a cast plate of 99.9Al as a function (a) the thickness, and (b) the length of the fin.

Averaged results of simulation and experiment from Wright and Campbell (1997).

5. The fin is easily cut off. In contrast, the witness from a chill also usually requires substantial dressing, especially if the chill was equipped with v-grooves, or if it became misplaced during moulding, as mentioned previously.

6. The fin does not cause scrap castings because of condensation of moisture and other volatiles, with consequential blow defects, as is a real danger from chills.

7. The fin does not require to be retrieved from the sand system, cleaned by shot blasting, stored in special bins, on special shelves, re-located, counted, losses made up by re-ordering new chills, casting new chills (particularly if the chill is shaped) and finally ensuring that the correct number in good condition, re-coated, and dried, is delivered to the moulder on the required date.

8. The fin does not wear out. Old chills become rounded to the point that they are effectively worn out. In addition, in iron and steel foundries, grey iron chills are said to 'lose their nature' after some use. This seems to be the result of the oxidation of the graphite flakes in the iron, thus impairing the thermal conductivity of the chill. This is understandable in terms of Figure 5.11 because graphite has significantly better conductivity than pure iron or steel which would materially affect sufficiently large chills.

9. Sometimes it is possible to solve a localised feeding problem (the typical example is the isolated boss in the centre of the plate) by chilling with a fin instead of providing a local supply of feed metal. In this case the fin is enormously cheaper than the feeder, and sometimes the feeder would be located where it cannot be subsequently removed, and its continued presence would be objectionable to the customer. The customer might accept a fin remaining in an obscure part of the casting, or the fin might be more easily removed.

This lengthy list represents considerable costs attached to the use of chills that are not easily accounted for, so that the real cost of chills is often underestimated.

Even so, the chill may be the correct choice for technical reasons. Fins perform poorly for metals of low thermal conductivity such as zinc, Al-bronze, iron and steel. The computer simulation result in Figure 5.23 illustrates for the rather low thermal conductivity material, Al-bronze, that there are extensive conditions in which the chill is far more effective. Lerner and Kouznetsov (2004) compare the action of chills and fins, concluding that the two are sometimes interchangeable but the action of the fin is necessarily rather localised, whereas the chill give a better distribution of cooling effect.

The kind of result shown in Figure 5.23 would be valuable if available for a variety of casting alloys varying from high to low thermal conductivity, so that an informed choice could be made whether a chill or fin were best in any particular case. Such data have yet to be worked out and published.

Fins are most easily provided on a joint line of the mould or around core prints. Sometimes, however, there is no alternative but to mould them at right angles to the joint. From a practical point of view, these upstanding fins on patternwork are of course vulnerable to damage. Dimmick (2001) records that fins made from flexible and tough vinyl plastic solved the damage problem in their foundry. They would carry out an initial trial with fins glued onto the pattern. If successful, the fins would then be permanently inserted into the pattern. In addition, only a few standard fins were found to be satisfactory for a wide range of patterns; a fairly wide deviation from the optimum ratios did not seem to be a problem in practice.

Safaraz and Creese (1989) investigated an interesting variant of the cast-on fin. They applied loose metal fins to the pattern, and rammed them up in the sand as though applying a normal external chill, in the manner shown in Figure 5.20. The results of these 'solid' or 'cold' fins (so-called to distinguish them from the empty cavity that would, after filling with liquid metal, effectively constituting a 'cast' or 'hot' fin) are also presented in Figure 5.21. It is seen that the cold fins are more effective than the cast fins in reducing the porosity in the junction castings. This is the consequence of the heat capacity of the fin being used in addition to its conducting role. This effect clearly over-rides any disadvantage of heat transfer resistance across the casting/chill interface.

The cold fin is, of course, really a chill of rather slim shape. It raises the interesting question, that as the geometry of the fin and the chill is varied, which can be the most effective. This question has been tackled in the author's laboratory (Wen and colleagues, 1997) by computer simulation. The results are summarised in Figure 5.23. Clearly, if the cast fin is sufficiently thin, it is more effective than a thin chill. However, for normal chills that occupy a large area of the casting (effectively approaching an 'infinite' chill as shown in the figure), as opposed to a slim contact line, the chill is massively more effective in speeding the freezing of the casting.

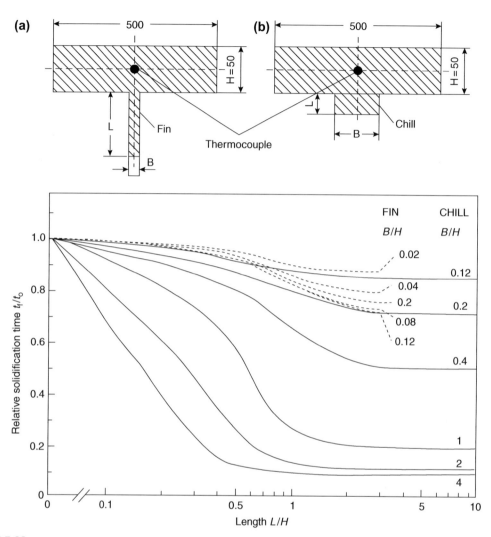

FIGURE 5.23

Comparison of the action of (a) cooling fins, and (b) chills on the rate of freezing of the aluminium bronze alloy AB1 (Wen, Jolly and Campbell 1997).

Other interesting lessons to be learned from Figure 5.23 are that a chill has to be at least equal to the section thickness of the casting to be really effective. A chill of thickness up to twice the casting section is progressively more valuable. However, beyond twice the thickness, increasingly thick chills show progressively reducing benefit.

It is to be expected that in alloys of higher thermal conductivity than aluminium bronze, a figure such as Figure 5.23 would show a greater regime of importance for fins compared to chills. The exploration of these effects for a variety of materials would be instructive and remains as a task for the future.

The business of getting the heat away from the casting as quickly as possible is taken to a logical extreme by Czech workers (Kunes et al., 1990) who show that a heat pipe can be extremely effective for a steel casting. Canadian workers (Zhang et al., 2003) explore the benefits of heat pipes for aluminium alloys. The conditions for successful application of the principle are not easy, however, so I find myself reluctant at this stage to recommend the heat pipe as a general purpose technique in competition to fins or chills. In special circumstances, however, it could be ideal.

Pins

Pins are analogous to fins and can be surprisingly effective. They take the form of simple pencil-like projections from a casting. Their narrow form allows them to be accommodated in spaces where chills or fins have insufficient room. Furthermore, compared to fins, they are fundamentally more effective as a result of the powerful local effect of their doubled rate of local heat extraction; they lose heat radially (in x and y dimensions, if the fin axis is the z dimension) as opposed to the fin which can lose its heat only unidirectionally (only in the single x dimension).

The geometry of pins, their diameter and length in terms of the thickness of section to be cooled, has never been investigated. This is a loss to the foundry engineer because pins have all the advantages listed for fins, plus the additional special features that can make them uniquely useful. Cooling pins deserve wide application in castings, and deserve the immediate attention of researchers to research and publish optimised use for different cast alloys. Even steel castings which do not respond to fins might benefit from pins. This would be a welcome benefit.

5.1.3 CONVECTION

Convection is the bulk movement of the liquid under the driving force of density differences in the liquid. In Section 5.3.4 we shall consider the problems raised by convection driven by solutes; heavy solutes cause the liquid to sink, and the lighter solutes cause flotation. In this section, we shall confine our discussion simply to the effects of temperature: hot liquid will expand, becoming less dense, and will rise; cool liquid will contract, becoming denser, and so will sink.

The existence of convection has been cited as important because it affects the columnar to equiaxed transition (Smith et al., 1990). There may be some truth in this. However, in most castings, grain structure is of no importance. Hardly any customer specifies the grain structure of their castings, the usual only critical features being soundness and leak tightness. Only in very few castings is grain size specified or is in any way noticeably significant in terms of affecting properties. Turbine blades are the exceptional example.

It is of course understood that soundness is of vital importance to nearly every casting, but it is not well known that in certain circumstances, convection can give severe unsoundness problems. This is especially true in counter-gravity systems, and sometimes in investment castings. This important phenomenon is dealt with at length in Chapter 10, Rule 7 for the manufacture of good castings. It is recommended reading.

5.1.4 REMELTING

When considering the solidification of castings it is easy to think simply of the freezing front as advancing. However, there are many times when the front goes into reverse! Melting is common during the filling and solidification of castings and needs to be considered at many stages.

On a microscale, melting is known to occur at different points on the dendrite arms. In a temperature gradient along the main growth direction of the dendrite the secondary arms can migrate down the temperature gradient by the remelting of the hot side of the arms and the freezing of the cold side. Allen and Hunt (1979) show how the arms can move several arm spacings. Similar microscopic remelting occurs as the small arms shrink and the larger arms grow to reduce the overall energy of the system. This is the mechanism of dendrite arm coarsening, leading to the dendrite arm spacing (DAS) being a useful indicator of the solidification time of a casting.

Slightly more serious thermal perturbations can cause the secondary dendrite arms to become detached when their roots are remelted (Jackson et al., 1966). The separated secondaries are then free to float away into the melt to become nuclei for the growth of equiaxed grains. If, however, there is too much heat available, then the growth front stays in reverse, with the result that the nuclei vanish, having completely remelted!

On a larger scale in the casting, the remelting of large sections of the solidification front can occur. This can happen as heat flows are changed as a result of changes in heat transfer at the interface, as the casting flexes and moves in the mould, changing its contact points and pressures at different locations and at different times. It is likely that this can happen as parts of the mould, such as an undersized chill, become saturated with heat, while cooling continues elsewhere. Thus the early rapid solidification in that locality is temporarily reversed.

Local remelting of the solid is seen to occur as a result of the influx of fresh quantities of heat from forced convection during filling. The so-called 'flow lines' seen on the radiographs of magnesium alloy castings are clearly a result of the local washing away of the solidification front, as a curving river erodes its outer bank. These features appear quite distinct from other linear defects such as hot tears or cracks as a result of their smooth outlines and gradual shading of radiographic density (Skelly and Sunnucks 1954; Lagowski 1967). Interestingly, no damaging loss of properties has ever been reported from such linear features of castings, nor would any loss of properties be expected.

The existence of continuous fluidity is a widely seen effect resulting directly from the remelting of the metal which has solidified in the filling system, keeping the metal flowing despite an unfavourable modulus. Without the benefits of this phenomenon it would be difficult to make castings at all!

Other convective flows produced by solute density gradients in the freezing zone take time to get established. Thus channels are formed by the remelting action of low-melting-point liquid flowing at a late stage of the freezing process. The A and V segregated channels in steel ingots, and freckle defects in nickel- and cobalt-based alloys, are good examples of this kind of defect.

5.1.5 FLOW CHANNEL STRUCTURE

Consider the direct-gated vertical plate shown in Figure 5.24(a). If this casting is filled quickly, the twin rotating vortices ensure that the plate fills with liquid at a uniform temperature. This may now be a problem to feed, and might exhibit the usual problems of distributed microporosity or surface sinks etc.

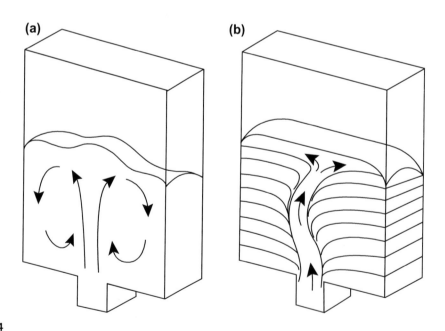

FIGURE 5.24

(a) Fast filling showing mixing by vortices, giving a uniform temperature in the filled plate. (b) Slow filling, showing layered freezing during filling and the development of a central flow channel.

This scenario contrasts with that of the slow filling of the casting, perhaps being filled slowly to reduce the potential for surface turbulence. In this case slow filling means *filling in a time commensurate with the solidification time of the section being filled*. If the filling rate were reduced to the point that the metal just reached the top of the mould by the time the metal had just cooled to its freezing point, then it might be expected that the top of the casting would be at its coldest, and freezing would then progress steadily down the plate, from the top to the gate. Nothing could be further from the truth.

In reality, the slow filling of the plate causes metal to flow sideways from the gate into the sides of the plate (Figure 5.24(b)), cooling as it goes, and freezing near the walls. Thus while more distant parts of the casting are freezing, layers of fresh hot metal continue to arrive through the gate. The successive positions of the freezing front are shown in Figure 5.24(b). The final effect is a flow path kept open by the hot metal through a casting that by now has mainly solidified. The well-fed panels either side of the flow channel are usually extremely sound and fine-grained, and contrast with the flow channel. The final freezing of the flow channel is slow because of the preheated mould around the path, and so its structure is coarse and porous. The porosity will be encouraged by the enhanced gas precipitation under the conditions of slow cooling, and shrinkage may contribute if local feeding is poor because either the flow path is long or it happens to be distant from a source of feed metal. Rabinovich (1969) describes these patterns of flow in thin vertical plates, calling them *jet streams*. *Flow channel* is suggested as a good name, if somewhat less dramatic.

Figure 5.25 shows an extreme example. This figure is a radiograph of an Al-7Si-0.4Mg alloy plate cast in an acid catalysed furan bonded silica sand mould. The flow channel is outlined by minute bubbles (most probably of hydrogen) that appear to have formed just under the surface of the casting as a result of reaction with the mould, but have floated

(a)

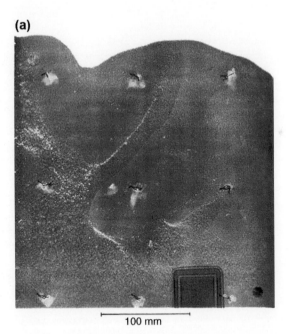

100 mm

FIGURE 5.25

(a) Radiograph of an Al-7Si-0.4Mg alloy vertical plate filled via a side riser and slot gate shown in Figure 3.1(a), showing unexpected flow channel filling behaviour when cast particularly cool. Remains of thermocouples can be seen (Runyoro, 1992). (b) Computer simulation of two flow channels developing in the walls of a grey iron casting despite the provision of five ingates in an effort to spread heat as evenly as possible.

Courtesy Puhakka (2009).

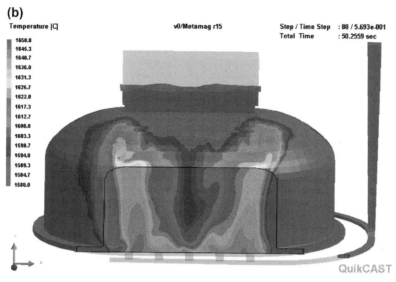

FIGURE 5.25 Cont'd

upwards to rest against and decorate the upper surface of the channel. They have probably grown sufficiently large to become significantly buoyant because of the extra time and temperature for reaction in the flow channel region. Those parts of the casting that have solidified quickly do not show bubbles on the radiograph. The flow channel in this case arises because the casting was made at a particularly low pouring temperature, so that the metal was flowing into the mould as a partly solidified slurry. The flow therefore adopted the form of a magma vent in the earth's crust, the pasty material resembling viscous lava, forming a volcano-like structure at the top surface.

Although Figure 5.25 illustrates a clear example because of the metal/mould reaction, in general the defect is not easily recognisable. It can occasionally be seen as a region of coarse grain and fine porosity in radiographs of large plate-like parts of castings. The structure contrasts with the extensive areas of clear, defect-free regions of the plate on either side. It is possible that many so-called shrinkage problems (for which more or less fruitless attempts are made to provide a solution by extra feeders or other means) are actually residual flow channels that might be cured by changing ingate position or size, or raising fill rate. No research appears to have been carried out to guide us out of this difficulty.

The flow channel structure is a standard feature of castings that are filled slowly from their base. This serious limitation to structure control seems to have been largely overlooked, and is one of the phenomena that limit the choice of a very long filling time.

Nevertheless, in general, the problem is reduced by filling faster, if that is possible without introducing other problems. More precisely, the velocity into the mould is initially controlled below approximately 0.5 m/s to avoid jetting through the ingate. After the base of the mould is covered to a sufficient depth the velocity can be ramped up. This is easily accomplished with a counter-gravity system, or even gravity system using a surge control design of filling system. A sufficiently high velocity will drive large circulating eddies (Figure 5.24) and finer scale circulating eddies (bulk turbulence) beneath the relatively tranquil surface. In this way the temperature of a large area, if not the whole casting, will be relatively uniform, completely free of a flow channel.

However, even fast filling does not cure the other major problem of bottom gating, which is the adverse temperature gradient, with the coldest metal being at the top and the hottest at the bottom of the casting, particularly concentrated in the ingate. Where feeders are placed at the top of the casting this thermal regime is clearly unfavourable for efficient feeding. However, of course, feeding can be made to be effective by oversized, if inefficient, feeders.

The problems of directly gating into the casting arise because the hot metal has to travel through the casting to reach all parts. One solution is *not* to gate into the casting but create a separate flow path called an *up-runner, outside* the casting. The melt is transferred by sideways flow off the up-runner via a slot gate into the mould cavity. The technique is described in Section 12.4.16. Another technique to counter adverse temperature gradient as a result of filling is the inversion of the casting after filling. The 'roll over' technique, such as used in the Cosworth casting process, is highly effective and especially recommended for volume production.

It is well to remember that on occasions remelting to create flow channels can occur without the transfer of heat. The channels seen commonly in Ni-base superalloys, known as 'freckles' because of their appearance when etched revealing the random orientations of fragments of grains, and the 'A' and 'V' segregates in heavy steel castings and ingots, have been formed by residual liquid, strongly segregated in C, S and P, that has a lower melting point than the matrix (Section 5.3.4). The matrix therefore melts when in contact with this liquid as a result of the diffusion of alloying elements into the matrix, lowering its melting point. Melting therefore occurs in this case by the transfer of solutes rather than the transfer of heat.

5.2 DEVELOPMENT OF MATRIX STRUCTURE
5.2.1 GENERAL

The liquid phase can be regarded as a randomly close-packed heap of atoms, in ceaseless random thermal motion, with atoms vibrating, shuffling, and jostling a meandering route, shoulder to shoulder, among and between their neighbours. The haphazard motion and random overall direction of travel of individual atoms has been termed 'the drunkard's walk'.

In contrast, the solid phase is an orderly array, or lattice, of atoms arranged in more or less close-packed rows and layers called a lattice. Atoms arranged in lattices constitute the solid bodies we call crystals. Iron in its alpha phase takes on the body-centred-cubic lattice known as ferrite (Figure 5.26(a)). This is a rather less close packed lattice than the face-centred-cubic lattice known as the gamma-phase, or as austenite. Figure 5.26(b) shows only a single '*unit cell*' of the lattice. The concept of the lattice is that it repeats such *unit cells*, replicating the symmetry into space millions of times in all directions. Macroscopic lattices can sometimes be seen in castings as crystals having sizes from 1 µm to 100 mm, representing arrays 10^3–10^8 atoms across.

The transition from liquid to solid, the process of solidification, is not always easy, however. For instance, in the case of glass, the liquid continues to cool, gradually losing the thermal motion of its atoms, to the point at which it becomes incapable of undergoing sufficient atomic rearrangements for it to convert to a lattice. It has therefore become a supercooled liquid, capable of remaining in this state for ever.

Metals, too, are sometimes seen to experience this reluctance to convert to a solid, despite on occasions being cooled hundreds of degrees Celsius below their equilibrium freezing temperature. This is easily demonstrated for clean metals in a clean container, for instance liquid iron in an alumina crucible.

(a) 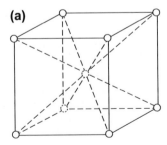 **(b)**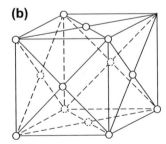

FIGURE 5.26

Body-centred cubic form of α-iron (a) exists up to 910°C. Above this temperature, iron changes to (b) the face-centred cubic form. At 1390°C the structure changes to β-iron, which is bcc once again (a).

If and when the conversion from liquid to solid occurs, it is by a process first of nucleation, and then of growth.

Nucleation is the process of the aggregation of clusters of atoms that represent the first appearance of the new solid phase. *Growth* is, self-evidently, getting bigger. However this process is subject to factors that encourage or discourage it as we shall discuss.

In fact, the complexities of the real world dictate not only that the main solid phase appears during solidification, but also that alloys and impurities concentrate in ways to trigger the nucleation and growth of other phases. These include solid and liquid phases that we call second phases or inclusions, and gas or vapour phases which we call gas pores or shrinkage pores. It is convenient to treat the solid and liquid phases together as condensed (i.e. practically incompressible) matter that we shall consider in this Section 5. The gas and vapour phases, constituting the non-condensed (and very compressible) matter such as gas and shrinkage porosity respectively, will be treated separately in Section 7.

For those readers who are enthusiastic about nucleation theory, there are many good formal accounts, some highly mathematical. A readable introduction relating to the solidification of metals is presented in Flemings (1974). We shall consider only a few basic aspects here, enough to enable us to understand how the structures of castings originate.

5.2.2 NUCLEATION OF THE SOLID

At first, as the temperature of a liquid is reduced below its freezing point, nothing happens. This is because in clean liquids the conversion to the solid phase involves a nucleation problem. We can gain an insight into the nature of the problem as follows.

As the temperature continues to fall, the thermal agitation of the atoms of the liquid reduces, allowing small random aggregations of atoms into crystalline regions. For a small cubic cluster of size d the net energy to form this new phase is reduced in proportion to its volume d^3 because of the lower free energy per unit volume ΔG_v of the solid. At the same time however the creation of new surface area $6d^2$ involves extra energy because of the interfacial energy γ per unit area of surface. The net energy to form our little cube of solid is therefore:

$$\Delta G = 6d^2\gamma - d^3\Delta G_v \qquad (5.15)$$

Figure 5.27 shows plots of the factors in this equation, showing that the net energy to grow the embryo increases at first, reaching a maximum. Embryos that do not reach the maximum require more energy to grow, they are unstable, so normally they will shrink and redissolve in the liquid.

Only when the temperature is sufficiently low to allow a chance chain of random additions to grow an embryo to the critical size will further growth be encouraged by a reduction in energy; thus growth will enter a 'runaway' condition. The temperature at which this event can occur is called the *homogeneous nucleation temperature*. For metals like iron and nickel, it is hundreds of degrees Celsius below the equilibrium freezing point as has been demonstrated many years ago by Walker (1961) and by many more recent researchers such as Valdez and colleagues (2007).

Such low temperatures may in fact be attained when making castings, because the liquid metal sits inside its own surface oxide film, never making contact with its container. At microscopic points of contact of the metal with the container, for moulds with high conductivity, the cooling through the surface film may be so intense that homogeneous nucleation in the liquid may occur. Nucleation is not likely to be a heterogeneous event in this situation because the liquid contained in its oxide skin is not in actual atomic contact with the surface of the mould, and its oxide is not a favourable nucleating substrate. (We shall see later that although oxides, such as some crucible materials, are *not* good heterogeneous nuclei to initiate nucleation of the solid, they are *excellent* nuclei for many second phases, and are efficient, indirectly as bifilms, for the initiation of porosity and cracks. These two entirely different behaviours of oxides are a major factor in the development of cast structure which will be returned to repeatedly.)

Dispinar and Campbell (2007) have reported evidence of massive undercoolings in comparatively large aluminium alloy castings. The observations that drew our attention were studies of solidified filters from the filling systems of castings, in which adjacent regions of melt, separated by an apparent grain boundary, had vastly different DAS on either side of the boundary (Figure 5.28). This seemingly indicated the impossible situation of two different rates of freezing in adjacent regions of metal. The fine microstructure was finally interpreted as undercooled regions isolated from the bulk

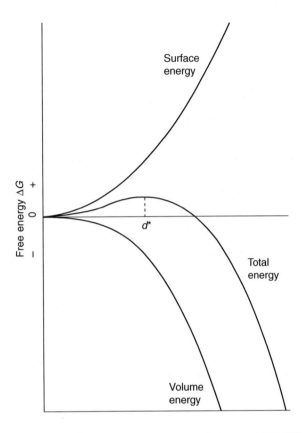

FIGURE 5.27

Surface and volume energies of an embryo of solid growing in a liquid give the total energy as shown. Below the critical size d^* any embryos will tend to shrink and disappear. Above d^* increasing size reduces the total energy, so growth will be increasingly favoured, becoming a runaway process.

melt by oxide bifilms. Thus some of these separated regions had a high probability of containing no suitable inclusion on which to nucleate. The surrounding matrix, having solidified some time previously and so much cooler, acted as an efficient chill when freezing finally occurred. The enclosed droplet therefore exhibits its quenched-in, rapidly frozen fine dendritic structure. This phenomenon, first seen in regions of melt trapped among oxide film inclusions, has now been identified in many castings poured turbulently (including, for instance, in Ni-base superalloys cast in vacuum), where droplets from splashes cannot be re-assimilated by the melt because of their surface (double) oxide, and consequently have their own independent freezing behaviour.

It is more common for the liquid to contain other solid particles in suspension on which new embryo crystal can form. In this case the interfacial energy component of Eqn (5.15) can be reduced or even eliminated. Thus the presence of foreign nuclei in a melt can give a range of heterogeneous nucleation temperatures; the more effective nuclei requiring less undercooling. It is even conceivable that some solid might exist on an extremely favourable substrate at temperature above its freezing point. This seems to be the case, for instance, with silicon particles nucleating on AlP particles on oxide bifilms in Al-Si alloys. Graphite flakes appear to form similarly above the general Fe-graphite eutectic in cast irons. These examples will be discussed further in Chapter 6.

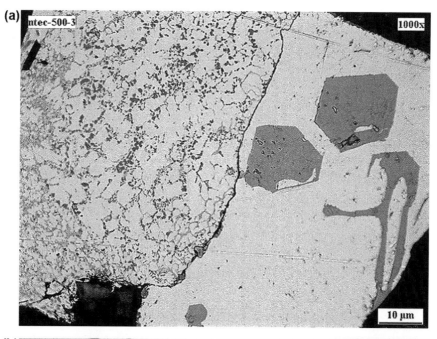

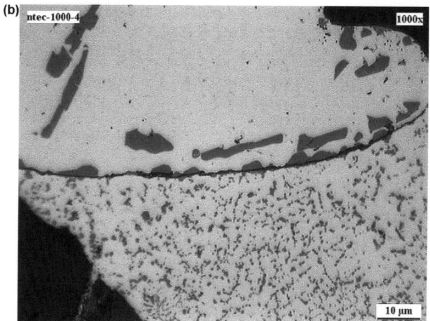

FIGURE 5.28

The separation and isolation of regions of Al-7Si alloy by oxide bifilms, resulting in (a) widely different solidification conditions, and (b) formation of Si particles on the bifilms (Dispinar, 2007).

Nevertheless, it is important to keep in mind that not all foreign particles in liquids are favourable nuclei for the formation of the solid phase. In fact it is likely that the liquid is indifferent to the presence of much of this debris. Only rarely will particles be present that reduce the interfacial energy term in Eqn (5.15). Thus, as far as most metals are concerned, oxides are not good nuclei. It is worth noting that it makes no difference whether the lattice structure and spacing of the oxide and metal are closely matched. The oxides are not wetted. This indicates that their electronic contribution to the interfacial energy with the metal is not favourable for nucleation. The covalently bonded oxides sit reluctantly against the metallic bonds of the metals, taking no part in the liquid to solid transformation of the metal.

Grain refinement

Materials with more metallic properties are good nuclei for the initiation of the solid phase. For the nucleation of steels, these include some borides, nitrides and carbides. For Al alloys, an intermetallic compound, $TiAl_3$, is the key inoculant, together with TiB_2. These details are discussed in the section on casting alloys in Chapter 6.

It seems, however, that the action of an effective grain refining addition to a melt is not only that of nucleating new grains. An important secondary role is that of inhibiting the rate of growth of grains, thereby allowing more grains the opportunity to nucleate.

Even so, as a cautionary note, the addition of TiB_2 to Fe-3Si alloys (Campbell and Bannister, 1975) exhibited a profound grain refinement action, almost certainly enhanced by the thick layer of borides that surrounded each grain. The mechanical properties were expected to be seriously impaired by this brittle grain boundary phase, illustrating the probability that not all grain refinement of cast alloys is beneficial.

For more complex systems, where many solutes are present, the rate of growth of grains was assumed by Greer and colleagues (2001) to be controlled by the rate at which solute can diffuse through the segregated region ahead to the advancing front. They carried out a detailed exercise for aluminium alloys, using a *growth restriction parameter*. This concept was refined by Australian workers (StJohn et al., 2007) as reported later in Section 6.3.3.

Although the grain refinement of Al alloys now appears to be well understood, the same cannot be said for the hexagonal close packed (hcp) alloys of Zn and Mg. Liu et al. (2013) admit that current theories cannot explain the grain refinement of Zn, and Birol (2012) suggests the same conclusion for Mg, even though StJohn and colleagues (2013) suggest that Fe and Mn impair the grain refinement of Mg, explaining contradictory previous outcomes. It does not seem clear whether we really understand the grain refinement of the common hcp metals. The presence of bifilms has so far been overlooked in these researches even though it is not evident that bifilms could play any part in the nucleation of the primary solid phase. The extent of our ignorance leaves wide scope for further research.

5.2.3 GROWTH OF THE SOLID

Once nucleated, the initial grow of the solid will release latent heat, the ΔG_v term in Eqn (5.15), causing the temperature of the melt to rise rapidly to near its equilibrium freezing point. This is called 'recalescence' meaning 'reheating'. The primary solid will spread relatively quickly through the undercooled liquid but will be slowed by recalescence. After this point, further growth will be controlled by the much slower loss of heat from the casting.

Commonly in castings it is only the metal in contact with the face of the mould which experiences any cooling below its equilibrium freezing temperature. Thus when nucleation occurs in this thin layer, a solid skin is quickly grown which envelops the casting. The interesting question now is 'How does it then continue its progress into the melt'?

Progress will only occur if heat is extracted through the solidified layer, cooling the freezing front below the equilibrium freezing point. The actual amount of undercooling experienced at the freezing front is usually only a few degrees Celsius. If the rate of heat extraction is increased, the temperature of the solidification front will fall further, and the velocity of advance, V_s, of the solid will increase correspondingly.

For pure metals (assuming relative freedom from over-enthusiastic grain refining additions), as the driving force for solidification increases, so the front is seen to go through a series of transitions. Initially, it is planar; at higher rates of advance it develops deep intrusions, spaced rather regularly over the front. These are parts of the front that have been left far behind. At higher velocities still, this type of growth transforms into cigar-like projections called cells, which finally

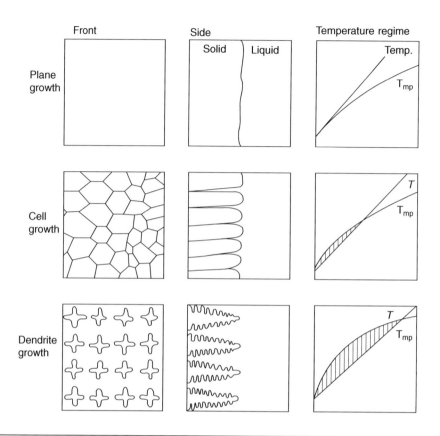

FIGURE 5.29

Transition of growth morphology from planar, to cellular, to dendritic, as compositionally induced undercooling increases (equivalent to G/V being reduced).

develop complex geometry involving side branches (Figure 5.29). These tree-like forms have given them the name dendrites (after the Greek word for tree, *Dendros*).

For the more important case of alloys, however, the three growth forms (planar, cellular and dendritic) are similarly present (Figure 5.29). However, the driving force for promoting the instability of planar growth, encouraging cellular or dendritic growth, is a kind of effective undercooling that arises because of the segregation of alloying elements ahead of the front. The presence of this extra concentration of alloying elements reduces the melting point of the liquid. If this reduction is sufficient to reduce the melting point to below the actual local temperature, then the liquid is said to be locally *constitutionally undercooled* (that is, effectively undercooled because of a change in the constitution of the liquid).

Figure 5.30 shows how detailed consideration of the phase diagram can explain the relatively complicated effects of segregation during freezing. It is worth examining the logic carefully.

The original melt of composition C_o starts to freeze at the liquidus temperature T_L. The first solid to appear has composition kC_0 where k is known as the partition coefficient. This coefficient k usually has a value less than 1 (although the reader needs to be aware of the existence of the less common but important cases where k is greater than 1). For instance, for $k = 0.1$ the first solid has only 10% of the concentration of alloy compared with the original melt; the first solid to appear is therefore usually rather pure.

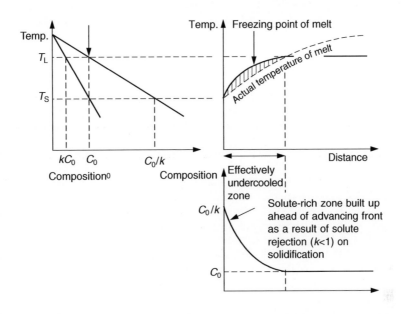

FIGURE 5.30

Link between the constitutional phase diagram for a binary alloy, and constitutional undercooling on freezing.

In general, k defines how the solute alloy partitions between the solid and liquid phases. Thus:

$$k = C_S/C_L \qquad (5.16)$$

For those equilibrium phase diagrams for which the solidus and liquidus lines are straight, k is accurately constant for all compositions. However, even where they are curved, the relative matching of the curvatures often means that k is still reasonably constant over wide ranges of composition. When k is close to 1, the close spacing of the liquidus and solidus lines indicates little tendency towards segregation. When k is small, then the wide horizontal separation of the liquidus and solidus lines warns of a strongly partitioning alloying element.

On forming the solid that contains only kC_0 amount of alloy, the alloy remaining in the liquid has to be rejected ahead of the advancing front. Thus although the liquid was initially of uniform composition Co, after an advance of about a millimetre or so the composition of the liquid ahead of the front builds up to a peak value of C_0/k. The build-up effect is like that of snow ahead of a snow plough. This is the steady-state condition shown in Figure 5.30.

In common with all other diffusion-controlled spreading problems, we can estimate the spread of the solute layer ahead of the front by the order-of-magnitude relation for the thickness d of the layer. If the front moves forward by d in time t, this is equivalent to a rate V_s. We then have:

$$d = (Dt)^{1/2} \quad \text{where} \quad V_s = d/t$$

so:

$$d = D/V_s \qquad (5.17)$$

where D is the coefficient of diffusion of the solute in the liquid. It follows that constitutional undercooling will occur when the temperature gradient, G, in the liquid at the front is:

$$G \leq -\frac{T_L - T_S}{D/V_S} \qquad (5.18)$$

or

$$G/V_S \leq -(T_L - T_S)/D \qquad (5.19)$$

from Figure 5.30, assuming linear gradients. Again, from elementary geometry which the reader can quickly confirm, assuming straight lines on the equilibrium diagram, we may eliminate T_L and T_S and substitute C_0, k and m, where m is the slope of the liquidus line, to obtain the equivalent statement:

$$\frac{G}{V_S} = \frac{-mC_0(1-k)}{kD} \qquad (5.20)$$

which is the classical solution derived from more rigorous diffusion theory by Chalmers in 1953, nicely summarised by Flemings (1974). This famous result marked the breakthrough in the history of the understanding of solidification by the application of physics. It marked the historical revolution from *qualitative description* to *quantitative prediction*. Computers have encouraged an acceleration of this radical transformation.

Figure 5.29 illustrates how the progressive increase in constitutional undercooling causes progressive instability in the advancing front, so that the initial planar form changes first to form cells, and with further instability ahead of the front will be finally provoked to advance as dendrites.

Notice that the growth of dendrites is in response to an *instability condition* in the environment ahead of the growing solid, not the result of some influence of the underlying crystal lattice (although, of course, the crystal structure will subsequently influence the details of the shape of the dendrite). In the same way, stalactites will grow as dendrites from the roof of a cave as a result of the destabilising effect of gravity on the distribution of moisture on the roof. Icicles are a similar example; their forms being, course, independent of the crystallographic structure of ice. Droplets running down window panes are a similar unstable-advance phenomenon that can owe nothing to crystallography. There are numerous other natural examples of dendritic advance of fronts that are not associated with any long-range crystalline internal structure. It is interesting to look out for such examples. Remember also the converse situation that the planar growth condition also effectively suppresses any influence that the crystal lattice might have. It is clear therefore that the constitutional undercooling, assessed by the ratio G/R, is the factor that measures the degree of stability of the growth conditions, and so controls the type of growth front (*not*, primarily, the crystal structure).

Figure 5.31 shows a transition from planar, through cellular, to dendritic solidification in a low-alloy steel that had been directionally solidified in a vertical direction. The speeding up of the solidification front has caused increasing instability. Figures 5.32 and 5.33 show different types of dendritic growth. Both types are widely seen in metallic alloy systems. In fact, dendritic solidification is the usual form of solidification in castings.

A columnar dendrite nucleated on the mould wall of a casting will grow both forwards and sideways, its secondary arms generating more primaries, until an extensive 'raft' has formed (Figure 5.34). All these arms will be parallel, reflecting the internal alignment of their atomic planes. Thus on solidification the arms will 'knit' together with almost atomic perfection, forming a single-crystal lattice known as a grain. A grain may consist of thousands of dendrites in a raft. Alternatively a grain may consist merely of a single primary arm, or, in the extreme, merely an isolated secondary arm.

The boundaries formed between rafts of different orientation, originating from different nucleation events, are known as grain boundaries. Usually these are high-angle grain boundaries, so-called to distinguish them from the low-angle boundaries within grains. Low angle boundaries result from small imperfections in the way the separate arms of the raft may grow, or suffer slight mechanical damage, so that their lattices join slightly imperfectly, at small but finite angles.

Given a fairly pure melt, and extremely quiescent conditions, it is not difficult to grow an extensive dendrite raft sufficient to fill a mould having dimensions of 100 mm or so, producing a single crystal. Nordland (1967) describes an unusual and fascinating experiment in which he solidifies bismuth at high undercooling and high rates, but preserves the fragile dendrite in one piece. He achieves this by adding weights to the furnace that contained his

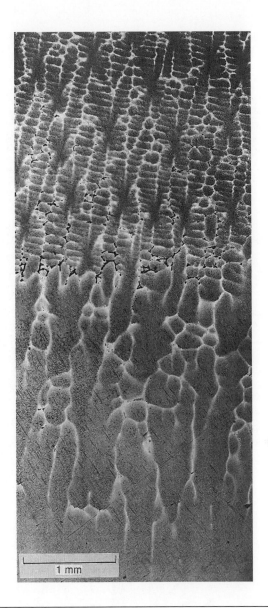

FIGURE 5.31

Structure of a low-alloy steel subjected to accelerating freezing from bottom to top, changing from planar, through cellular, to dendritic growth.

sample of solidifying metal, and suspends the whole assembly in **mid-air**, using long lengths of polypropylene tubing from the walls and ceiling of the room. In this way he was able to absorb and dampen any outside vibrations.

In a review of the effects of vibration on solidifying metals, the author (1981) confirms that Nordland's results fall into a regime of frequency and amplitude where the vibrational energy is too low for damage to occur to the dendrites (Figure 5.35).

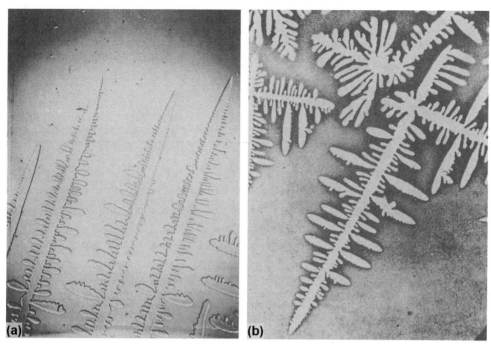

FIGURE 5.32

Transparent organic alloy showing dendritic solidification. Columnar growth (a) and equiaxed growth (b) with a modification to the alloy by the addition of a strongly partitioning solute, with k « 1 which can be seen to be segregated ahead of the growing front.

Courtesy J. D. Hunt; see Jackson et al. (1966).

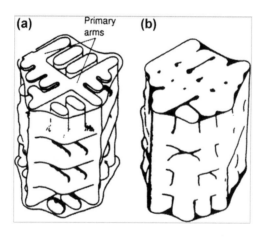

FIGURE 5.33

Rather irregular dendrites common in aluminium alloys at (a) 50% and (b) 90% solidified. The secondary arms spread laterally, joining to form continuous plates.

After Singh et al. (1970).

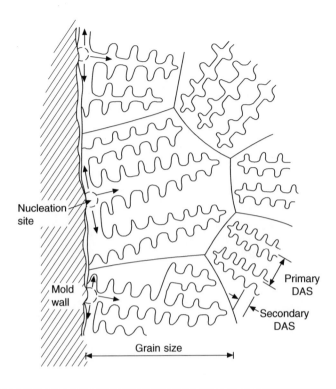

FIGURE 5.34

Schematic illustration of the formation of a raft of dendrites to make grains. The dendrite stems within any one raft or grain are all crystallographically related to a common nucleus.

Dendrite arm spacing

In the metallurgy of wrought materials, it is the *grain size* of the alloy that is usually the important structural feature. Most metallurgical textbooks therefore emphasise the importance of grain size.

In castings, however, grain size is sometimes important (as will be discussed later), but more often it is the *secondary dendrite arm spacing* (often shortened to *dendrite arm spacing or DAS*) that appears to be the most important structural length parameter.

The mechanical properties of most cast alloys are usually seen to be strongly dependent on secondary arm spacing. As DAS increases, so ultimate strength, ductility and elongation fall. Also, because homogenisation heat treatments are dependent on the time required to diffuse a solute over a given average distance d, if the coefficient of diffusion in the solid is D, then from the order-of-magnitude relation, Eqn 1.5, we predict, quantitatively, the finer DAS to give us specifically shorter homogenisation times, or better homogenisation in similar times; the cast material is more responsive to heat treatment, giving better properties or faster treatments.

It is now known that the secondary DAS is controlled by a coarsening process, in which the dendrite arms first grow at very small spacing near the tip of the dendrite. As time goes on, the dendrite attempts to reduce its surface energy by reducing its surface area. Thus small arms preferentially go into solution whilst larger arms grow at their expense, increasing the average spacing between arms. The rate of this process appears to be limited by the rate of diffusion of solute in the liquid as the solute transfers between dissolving and growing arms. From a relation such as Eqn 1.5, and assuming the alloy solidifies in a time t_f, we would expect that DAS would be proportional to $t_f^{1/2}$, because t_f is the time available for coarsening. In practice, it has been observed that DAS is actually proportional to t^n where n usually lies between 0.3 and 0.4 (Young and Kirkwood 1975). Figure 5.36 shows the magnificent research result, illustrating the

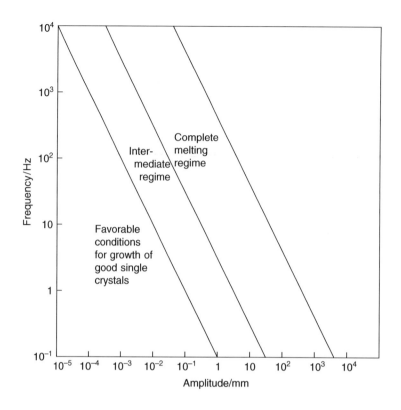

FIGURE 5.35

Grain refinement threshold as a function of amplitude and frequency of vibration (Campbell, 1981).

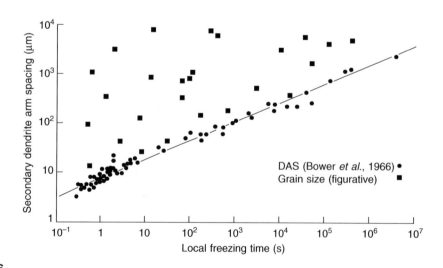

FIGURE 5.36

Relation between dendrite arm spacing (DAS), grain size, and local solidification time for Al-4.5Cu alloy.

relation between DAS and t_f continuing to hold for Al-4.5Cu alloy over eight orders of magnitude (Bower et al., 1966). Interestingly, however, a plot of grain size on the same figure shows that grain size is completely scattered above the DAS line. Clearly, grain size is completely independent of solidification time. In addition, of course, a grain cannot be smaller than a single dendrite arm, but can grow to unlimited size in some situations.

The primary DAS is of course significantly coarser than the secondary DAS, and appears controlled by rather different parameters. Young and Kirkwood (1975) proposed it is proportional to the parameter $(GV)^{-1/2}$ where G is the temperature gradient and V is the rate of advance of the solidification front. A little later, Hunt and Thomas (1977) present more accurate analytical solutions as a function of G, V and k, which were fitted to the results generated by a sophisticated computer program. Readers are recommended to the original papers for details of this piece of exemplary work.

In summary, primary DAS appears linked to solidification parameters and secondary DAS is controlled by solidification time. Grain size, on the other hand, is controlled by a number of quite separate processes, some of which are discussed further in the following section.

5.2.4 DISINTEGRATION OF THE SOLID (GRAIN MULTIPLICATION)

As the growers of single-crystal turbine blades know only too well, a single knock or other slight disturbance during freezing can damage the growing dendrite, breaking off secondary arms that then constitute nuclei for new separate grains. The growing crystal is especially vulnerable when strongly partitioning alloys are present that favour the growth of dendrites with secondary arms with weak roots. Thus some single crystals are fragile, and much more difficult to grow than others.

Figure 5.35 indicates that for any disturbance that can be characterised by a vibration of frequency f and/or amplitude a, a critical threshold exists at which the grains are fragmented. In a review of the mechanism of fragmentation by vibration (Campbell, 1981) it was not clear whether the dendrite roots melted, or whether they were mechanically sheared, because these two processes could not be distinguished by the experimental results. Whatever mechanism was operating, the experimental results on a wide variety of metals that solidify in a dendritic mode from Al alloys to steels, could be summarised to a close approximation by

$$f \cdot a = 0.10 \text{ mms}^{-1}$$

This relation describes the product of frequency (Hz) and amplitude (mm) that represents a critical velocity threshold for grain fragmentation. It seems to be valid over the complete range of experimental conditions ever tested, from subsonic to ultrasonic frequencies, and from amplitudes of micrometres to centimetres. Thus at ultrasonic frequency 25 kHz the amplitude for refinement is only 4 nm, whereas at the frequency of mains electricity 50 Hz the amplitude needs to be 2 μm.

In single-crystal growth, it seems that the damage to a dendrite arm may not be confined to breaking off the arm. Simply bending an arm will cause that part of the crystal to be misaligned with respect to its neighbours. Its subsequent growth might be in a direction favourable for its continued growth, causing it to grow to the size of a significant defect. Vogel and colleagues (1977) propose that given a sufficiently large angle of bend, the plastically deformed material will recrystallise rapidly. The newly formed high angle grain boundary, having a high energy, will be preferentially wetted by the melt, so that liquid will therefore propagate along the boundary and detach the arm. The arm now becomes a free-floating grain.

In the pouring of conventional castings, Ohno (1987) has drawn attention to the way in which the grains of some metals grow from the nucleation site on the mould wall. The grains grow from narrow stems that are vulnerable to plastic deformation and detachment. Thus as metal washes over the mould surface, thousands of crystals are swept into the melt. The nucleation sites with their remnant of dendrite root continue to be attached to the mould wall, quickly re-growing to 'seed' strings of replacement crystals, one after another. There is an element of runaway catastrophe in this process; as one dendrite is felled, it will lean on its neighbours and encourage their fall.

The fragments of crystals that are detached in this way may dissolve once again as they are carried off into the interior of the melt if the casting temperature is too high. The interior of the casting may therefore become free of so-called equiaxed grains. If so, the structure of the casting will consist only of columnar grains that grow inwards from the mould wall.

However, if the casting temperature is not too high, then the detached crystals will survive, forming the seeds of grains that subsequently grow freely in the melt. The lack of directionality and the *equal* length of the *axes* of these crystals has given them the name '*equiaxed*' grains. At very low casting temperatures, perhaps together with sufficient bulk turbulence, the whole of the casting may solidify with an equiaxed structure.

In mixed situations where modest quantities of equiaxed grains exist, they may be caught up as isolated grains in the growing forest of columnar dendrites. The directional heat flow that they will then experience will grow them unidi-rectionally, converting them to columnar grains. However, a sufficient deluge of equiaxed grains will swamp the progress of the columnar zone, converting the structure to equiaxed.

The *columnar-to-equiaxed transition* has been the subject of much solidification research. In summary. it seems that the transition is controlled by the numbers of equiaxed grains that are available. The recent model by Bisuola and Martorano (2008) indicates that the advance of the columnar zone is blocked when only 20% of the liquid ahead has transformed to equiaxed grains. The modelling work by Spittle and Brown (1989) illuminates the concepts admirably (Figure 5.37).

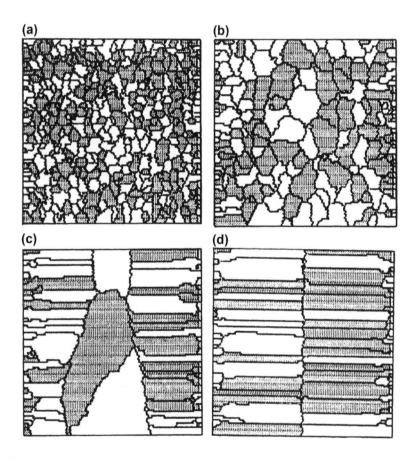

FIGURE 5.37

Computer simulated macrostructure of growth inwards from the sides of an ingot for progressively increasing casting temperature (a) to (d).

Reprinted with permission from J. Materials Science, Chapman & Hall, London.

In large steel ingots the columnar grains can reach lengths of 200 mm or more. These long cantilevered projections from the mould wall are under considerable stress as a result of their weight, and the additional weight of equiaxed grains, en route to the bottom of the ingot, that happen to settle, clustering on their tips. Under this weight, the grains will therefore bend by creep, possibly recrystallising at the same time, allowing the grains that grow beyond a certain length to sag downwards at various angles. This mechanism seems consistent with the structure of the columnar zone in large castings, and explains the so-called *branched columnar zone*. The straight portions of the columnar crystals near the base have probably resisted bending by the support provided by secondary arms, linked to form transverse web-like walls, providing the excellent rigidity of a box-girder supporting structure.

The bending of dendrites under their own weight when growing horizontally is not confined to steel ingots. Newell and colleagues (2009) found that dendrites of Ni-base superalloys growing across the horizontal platforms of turbine blade single crystal castings sagged under their own weight, but remaining straight as if creating a 'hinge' at their base, causing misalignments of up to $10°$.

Growing dendrites can be damaged or fragmented in other ways to create the seeds of new grains.

Mould coatings that contain materials that release gas on solidification, and so disturb the growing crystals, are found to be effective grain refiners (Cupini et al., 1980). Although these authors do not find any apparent increase of gas take-up by the casting, it seems prudent to view the 10% increase in strength as hardly justifying such a risk until the use of volatile mould dressings is assessed more rigorously.

The application of vibration to solidifying alloys is also successful in refining grain size. The author admits that it was hard work reviewing the vast amount of work in this area (Campbell, 1981). It seems that all kinds of vibration, whether subsonic, sonic, or ultrasonic, are effective in refining the grain size of most dendritically freezing materials providing the energy input is sufficient (Figure 5.35). The product of frequency and amplitude has to exceed 0.01 mms^{-1} for 10% refinement, 0.02 mms^{-1} for 50% refinement and 0.1 mms^{-1} for 90% refinement (Campbell, 1980). It is possible that at the free liquid surface of the metal the energy required to fragment dendrites is much less than this, as Ohno (1987) points out. This is sometimes known as *shower nucleation*, as proposed by the Australian researcher, Southin (1967), although it is almost certainly not a nucleation process at all. Most probably it is a dendrite fragmentation and multiplication process, resulting from the damage to dendrites growing across the cool surface liquid. These are possibly actually attached to the floating oxide film, or growing from the side walls, but are disturbed by the washing effect of the surface waves.

If we return to Figure 5.36 grain sizes are dotted randomly all over the upper half of the diagram, above the DAS size line. Occasionally some grains will be as small as one dendrite arm, and so will lie on the DAS line. No grain size can be lower than the line. This is because, if we could imagine a population of grains smaller than the DAS, and which would therefore find itself below the line, then in the time available for freezing, the population would have coarsened, reducing its surface energy to grow its average grain size up to the predicted size corresponding to that available time. Thus although grains cannot be smaller than the dendrite arm size, the grain size is otherwise independent of solidification time. Clearly, totally independent factors control the grain size.

It is clear, then, that the size of grains in castings results not only from nucleation events, such as homogeneous events on the side walls, or from chance foreign nuclei, or intentionally added grain refiners. Grain nuclei are also subject to further chance events such as re-solution. Further complications, mostly in larger-grained materials, result from chance events of damage or fragmentation from a variety of causes.

A further effect should be mentioned. The grains formed during solidification may not continue to exist down to room temperature. Many steels, for instance, as discussed in Section 6.6, undergo phase changes during cooling. Even in those materials that are single phase from the freezing point down to room temperature can experience grain boundary migration, grain growth or even wholesale recrystallisation. Figure 5.38 shows an example of grain boundary migration in an aluminium alloy.

It bears emphasising once again that dendrite arm spacing is controlled principally by freezing time, whereas grain size is influenced by many independent factors.

Before leaving the subject of the as-cast structure, it is worth giving a warning of a few confusions concerning nomenclature in the technical literature.

First, there is a widespread confusion between the concept of a grain and the concept of a dendrite. It is necessary to be on guard against this.

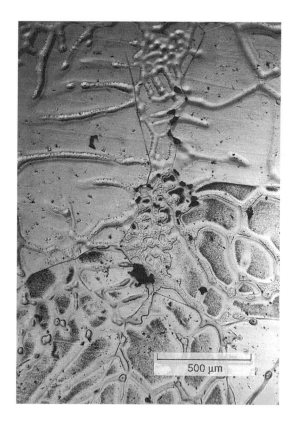

FIGURE 5.38

Micrograph of Al-0.2Cu alloy showing porosity and interdendritic segregation. Some grain boundary migration during cooling is clear. (Electropolished in perchloric and acetic acid solution and etched in ferric chloride. Dark areas are etch pitted.)

Second, the word 'cell' has a number of distinct technical meanings that need to be noted:

1. A cell can be a general growth form of the solidification front, as used in this book.
2. Cell is the term used to denote graphite 'rosettes' in grey cast irons. Strictly, these are graphite grains; crystals of graphite which have grown from a single nucleation event. They grow within, and appear crystallographically unrelated to, the austenite grains that form the large dendritic rafts of the grey iron structure.
3. The term 'cell count' or 'cell size' is sometimes used as a measure of the fineness of the microstructure, particularly in aluminium alloys. In Al alloys the distinction between primary arm, secondary arm, and grain is genuinely difficult to make in randomly oriented grains, where primary and secondary arms are not clearly differentiated (see Figure 5.33). To avoid the problem of having to make any distinction, a count is made of the number of rounded, bright features (that could be primary or secondary arms or grains) in a measured length. This is arbitrarily called 'the cell count', giving a measure of something called a 'cell size'. In these difficult circumstances it is perhaps the only practical quantity that can be measured, whatever it really is!

5.3 SEGREGATION

Segregation may be defined as any departure from uniform distribution of the chemical elements in the alloy. Because of the way in which the solutes in alloys partition between the solid and the liquid during freezing, it follows that all alloy castings are segregated to some extent.

Some variation in composition occurs on a microscopic scale between dendrite arms, known as microsegregation. It can usually be significantly reduced by a homogenising heat treatment lasting only minutes or hours because the distance over which diffusion has to take place to redistribute the alloying elements is sufficiently small, usually in the range 10–100 μm. Equation 1.5 can assist to estimate times required for homogenisation for different spacings.

Macrosegregation cannot be removed. It occurs over distances ranging from 1 cm to 1 m, and so cannot be removed by diffusion without geological time scales being available! In general, therefore, whatever macrosegregation occurs has to be lived with.

In this section, we shall consider explicitly only the case for which the distribution coefficient k (the ratio of the solute content of the solid compared to the solute content of the liquid in equilibrium) is less than 1. This means that solute is rejected on solidification and therefore builds up ahead of the advancing front. The analogy, used repeatedly before, is the build-up of snow ahead of the advancing snow plough.

(It is worth keeping in mind that all the discussion can, in fact, apply in reverse, where k is greater than 1. In this case, extra solute is taken into solution in the advancing front, and a depleted layer exists in the liquid ahead. The analogy now is that of a domestic vacuum cleaner advancing on a dusty floor, and sucking up dust that lies ahead.)

For a rigorous treatment of the theory of segregation the interested reader should consult the standard text by Flemings (1974), which summarises the pioneering work in this field by the team of researchers at the Massachusetts Institute of Technology. We shall keep our treatment here to a minimum, just enough to gain some insight into the important effects in castings.

5.3.1 PLANAR FRONT SEGREGATION

There are two main types of normal segregation that occur when the solid is freezing on a planar front: one that results from the freezing of quiescent liquid and the other of stirred liquid. Both are important in solidification, and give rise to quite different patterns of segregation.

Figure 5.39 shows the way in which the solute builds up ahead of the front if the liquid is still (quiescent liquid, case 1). The initial build-up to the steady-state situation is called the initial transient. This is shown rather spread out for clarity. Flemings (1974) shows that for small k the initial transient length is approximately $D/V_s k$ where D is the co-efficient of diffusion of the solute in the liquid and V_s is the velocity of the solidification front. In most cases, the transient length is only of the order of 0.1–1 mm.

After the initial build-up of solute ahead of the front, the subsequent freezing to solid of composition C_0 takes place in a steady, continuous fashion until the final transient is reached, at which the liquid and solid phases both increase in segregate. The length of the final transient is even smaller than that of the initial transient because it results simply from the impingement of the solute boundary layer on the end wall of the container. Thus its length is of the same order as the thickness of the solute boundary layer D/R. For many solutes, this is therefore between 5 and 50 times thinner than the initial transient.

For the case in which the liquid is stirred, moving past the front at such a rate to sweep away any build-up of solute, Figure 5.39 (stirred liquid case 2) shows that the solid continues to freeze at its original low composition kC_0. The slow rise in concentration of solute in the solid is, of course, only the result of the bulk liquid becoming progressively more concentrated.

The important example of the effect of normal segregation, building up as an initial transient, is that of subsurface porosity in castings. The phenomenon of porosity being concentrated in a layer approximately 1 mm beneath the surface of the casting is a clear example of the build-up of solutes. In this case the solute is a dissolved gas, which only after 1 mm

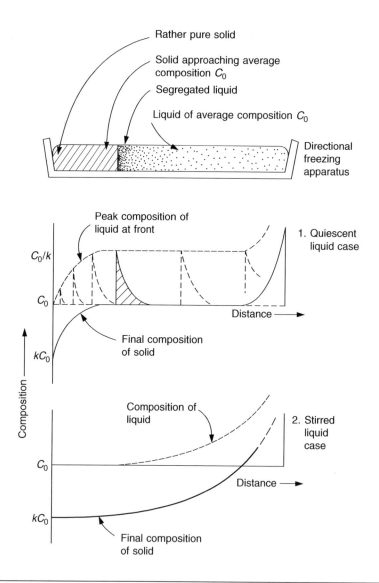

FIGURE 5.39

Directional solidification on a planar front giving rise to two different patterns of segregation depending on whether solute is allowed to build up at the advancing front or is swept away by stirring.

or so advance of solidification its concentration finally exceeds the threshold to nucleate pores. The nucleation and growth of gas pores is discussed in Section 7.

Moving on now to consider an example of segregation where the liquid is rapidly stirred, the classic case was that of the rimming steel ingot seen in Figure 5.40. Although rimming steels are now a phenomenon of the past, their behaviour is instructive. During the early stages of freezing, the high temperature gradient against the mould walls favoured a planar front. The rejection of carbon and oxygen resulted in bubbles of carbon monoxide. These detached from the planar front and rose to the surface, driving a fast upward flow of metal, effectively scouring the interface clean of any solute that

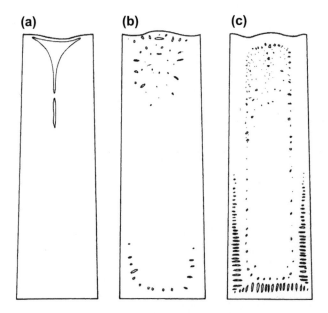

FIGURE 5.40

Ingot structures: (a) a killed steel; (b) a balanced steel; and (c) a rimming steel.

attempted to build up. Thus the solid continued to freeze with its original low impurity content, forming the pure iron 'rim', giving the steel its name. At lower levels in the ingot there was a lower density of bubbles to scour the front, so some bubbles succeeded in remaining attached, explaining the array of wormhole-like cavities in the lower part of the ingot. During this period, the incandescent spray from the tops of the ingots as the bubbles emerged and burst at the surface was one of the great spectacles of the old steelworks, almost ranking in impressiveness with the blowing of the Bessemer converters. A good spray was said to indicate a good rimming action. I have fond memories of these times from my early days casting ingots in the steel industry. Continuous casting of steels is, I regret, comparatively boring (although it is necessary to admit its impressive productivity and quality of product).

As the rim thickened, the temperature gradient fell so that the front started to become dendritic, retaining both bubbles and solute. Thus the composition then adjusted sharply to the average value characteristic of the remaining liquid, which was then concentrated in carbon, sulphur and phosphorus.

Rimming steel was widely used for rolling into strips, and for such purposes as deep drawing, where the softness and ductility of the rim assisted the production of products with high surface finish.

The oxygen levels in rimming steels were in excess of 0.02%, and were strongly dependent on the carbon and manganese contents. Typically these were 0.05–0.20%C and 0.1–0.6%Mn, giving a useful range of hardness, ductility and strength.

With the development of continuously cast steel, the casting of steel into ingots has now become part of steelmaking history. The challenge to the deep-drawing qualities was to regain so far as possible the benefits of the rimmed steel from a continuous casting process, where the rimming action could no longer be used. This has been achieved by attention to the surface quality and inclusion content of the new steels.

As an interesting diversion it is worth including at this point the two other major classes of steel that were produced as ingots, because these still have lessons for us as producers of shaped castings. The two other types of ingots were produced as (a) killed steels and (b) balanced steels (Figure 5.40).

The balanced, or semi-killed steel, was one that, after partially deoxidising, contained 0.01–0.02% oxygen. This was just enough to cause some evolution of carbon monoxide towards the end of freezing, to counter the effect of solidification shrinkage, thus yielding a substantially level top. (The deep shrinkage pipe typical of fully killed steels required to be cropped off and remelted). The whole ingot could be utilised. The great advantage of this quality of steel was the high yield on rolling, because the dispersed cavities in the ingot tended to weld up. For this reason bulk constructional steel could be produced economically. However, it was a difficult balancing act to maintain such precise control of the chemistry of the metal. It was only because balanced steels were so economical that such feats were routinely attempted. Some steelmakers were declared foolhardy for attempting such tasks!

In contrast, killed steels were easy to manufacture. They included the high-carbon steels and most alloy steels. They contain low levels of free oxygen, normally less than 0.003% because of late additions of deoxidisers such as Si, Mn and Al. The Al addition was often made together with other components such as Ca, and sometimes Ti etc. Consequently, there was no evolution of carbon monoxide on freezing, and a considerable shrinkage cavity was formed as seen in Figure 5.40. If allowed to form in this way, the cavity opened up on rolling as a fish-tail, and had to be cropped and discarded. Alternatively the top of the ingot was maintained hot during solidification by special hot-topping techniques. Either way, the shrinkage problem involved expense above that required for balanced or rimming steels. Fully killed steels were generally therefore reserved for higher priced, low and medium alloy applications.

5.3.2 MICROSEGREGATION

As the dendrite grows into the melt, and as secondary arms spread from the main dendrite stem, solute is rejected. The solute is effectively pushed aside and concentrates in the tiny regions enclosed by the secondary dendrite arms. Because these liquid regions are smaller than the diffusion distance, we may consider them more or less uniform in composition. The situation, therefore, is closely modelled by Figure 5.39, case 2. Remember, the uniformity of the liquid phase in this case results from diffusion within its small size, rather than any bulk motion of the liquid.

The interior of the dendrite therefore has an initial composition close to kC_0, whereas, towards the end of freezing, the centre of the residual interdendritic liquid has a composition corresponding to the peak of the final transient. This gradation of composition from the inside to the outside of the dendrite earned its common description as 'coring' because, on etching a polished section of such dendrites, the progressive change in composition is revealed, appearing as onion-like layers around a central core. The concentration of chromium and nickel in the interdendritic regions of the low-alloy steel shown in Figure 5.31 has caused these regions to be relatively 'stainless', resisting the etch treatment, and so causing them to be revealed in the micrograph.

Some diffusion of solute in the dendrite will tend to smooth the initial as-cast coring. This is often called back-diffusion. Additional smoothing of the original segregation can occur as a consequence of other processes such as the remelting of secondary arms as the spacing of the arms coarsens.

The partial homogenisation resulting from back-diffusion and other factors means that, for rapidly diffusing elements such as carbon in steel, homogenisation is rather effective. For plain carbon steels, therefore, the final composition in the dendrite and in the interdendritic liquid is not far from that predicted from the equilibrium phase diagram. The maximum freezing range from the phase diagram is clearly at about 2.0% carbon and would be expected to apply.

Even so, in steels where the carbon is in association with more slowly diffusing carbide-forming elements, the carbon is not free to homogenise: the resulting residual liquid concentrates in carbon to the point at which the eutectic is formed at carbon contents well below those expected from the phase diagram. In steels that contained between 1.3 and 2.0% manganese the author found that the eutectic phase first appeared between 0.8 and 1.3% carbon (Campbell, 1969) as shown in Figure 5.41. Similarly, in 1.5Cr-1C steels, Flemings et al. (1970) found that the eutectic phase first appeared at about 1.4% carbon (Figure 5.42). This point was also associated with a peak in the segregation ratio, S, the ratio of the maximum to the minimum composition; this is found between the interdendritic liquid and the centre of the dendrite arm. (Note: The interpretation of these diagrams as two separate curves intersecting in a cusp rather than a smoothly curved maximum is based on the fact that the two parts of the curve are expected to follow different laws. The first part represents the solidification of a solid solution; the second part represents the solidification of a solid solution plus some eutectic. As we have seen before, this is much more common in freezing problems than appears to have been generally recognised.)

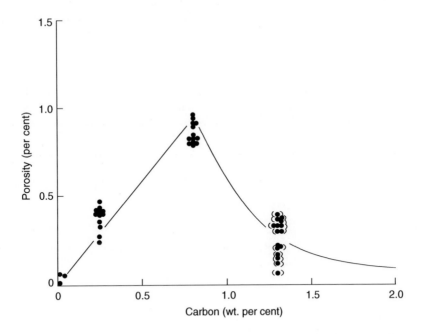

FIGURE 5.41

Porosity in Fe-C-Mn alloys, showing the reduction associated with the presence of non-equilibrium eutectic liquid (data points in brackets) (Campbell, 1969).

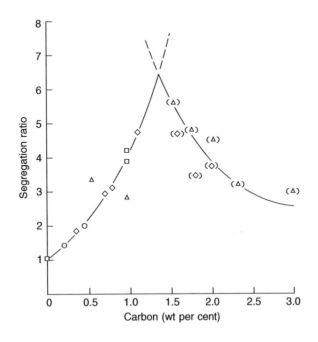

FIGURE 5.42

Severity of microsegregation in C-Cr–bearing steels, illustrating the separate regime for structures containing eutectic (points in brackets).

Data from Flemings et al. (1970).

The segregation ratio S is a useful parameter when assessing the effects of treatments to reduce microsegregation. Thus the progress of homogenising heat treatments can be followed quantitatively.

It is important to realise that S is only marginally affected by changes to the rate of solidification in terms of the rates that can be applied in conventional castings. This is because although the dendrite arm spacing will be reduced at higher freezing rates, the rate of back diffusion is similarly reduced. Both are fundamentally controlled by diffusion, so that the effects largely cancel. (During any subsequent homogenising heat treatment, however, the shorter diffusion distances of the material frozen at a rapid rate will be a useful benefit in reducing the time for treatment.)

Where microsegregation results in the appearance of a new liquid interdendritic phase, there are a number of consequences that may be important:

1. The appearance of a eutectic phase reduces the problem for fluid flow through the dendrite mesh because the most difficult part of the mesh, the dendrite root region, is eliminated by the planar front of the advancing eutectic. Shrinkage porosity is thereby reduced after the arrival of the eutectic, as seen in Figure 5.41. This important effect on porosity is discussed in greater detail in Chapter 7.
2. The alloy may now be susceptible to hot tearing, especially if there is only a very few percent of the liquid phase. This effect is discussed further in Chapter 8.
3. A low-melting-point phase may limit the temperature at which the material can be heat treated.
4. A low-melting-point phase may limit the temperature at which an alloy can be worked because it may be weakened, disintegrating during working because of the presence of liquid in its structure.

5.3.3 DENDRITIC SEGREGATION

Figure 5.43 shows how microsegregation, the sideways displacement of solute as the dendrite advances, can lead to a form of macrosegregation. As freezing occurs in the dendrites, the general flow of liquid that is necessary to feed solidification shrinkage in the depths of the pasty zone carries the progressively concentrating segregate towards the roots of the dendrites.

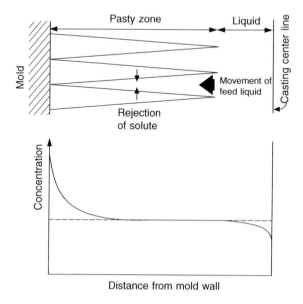

FIGURE 5.43

Normal dendritic segregation (usually misleadingly called inverse segregation) arising as a result of the combined actions of solute rejection and shrinkage during solidification in a temperature gradient.

In the case of a freely floating dendrite in the centre of the ingot that may eventually form an equiaxed grain, there will be some flow of concentrated liquid towards the centre of the dendrite if in fact any solidification is occurring at all. This may be happening if the liquid is somewhat undercooled. However, the effect will be small, and will be separate for each equiaxed grain. Thus the build-up of long-range segregation in this situation will be negligible.

For the case of dendritic growth against the wall of the mould, however, the temperature gradient will ensure that all the flow is in the direction towards the wall, concentrating the segregation here. Thus the presence of a temperature gradient is necessary for a significant build-up of segregation.

It will by now be clear that this type of segregation is in fact the usual type of segregation to be expected in dendritic solidification. The phenomenon has in the past suffered the injustice of being misleadingly named '*inverse segregation*' on account of it appearing anomalous in comparison to planar front segregation and the normal pattern of positive segregation seen in the centres of large ingots. In this book we shall refer to it simply as '*dendritic segregation*'. It is perfectly normal and to be expected in the normal conditions of dendritic freezing.

Dendritic segregation is observable but is not normally severe in sand castings because the relatively low temperature gradients allow freezing to occur rather evenly over the cross-section of the casting; little directional freezing exists to concentrate segregates in the direction of heat flow.

In castings that have been made in metal moulds, however, the effect is clear and makes the chill casting of specimens for chemical analysis a seriously questionable procedure. Chemists should beware! The effect of positively segregating solutes such as carbon, sulphur and phosphorus in steel is clearly seen in Figure 5.44 as the high concentration around the edges and the base of the ingot; all those surfaces in contact with the mould.

In some alloys with very long freezing ranges, such as tin bronze (liquidus temperature below 1000°C and solidus close to 800°C), the contraction of the casting in the solid state and/or the pressure in the liquid from the metallostatic head, plus perhaps the evolution of dissolved gases in the interior of the casting causes eutectic liquid to be squeezed out on to the surface of the casting. This exudation is known as *tin sweat*. It was described by Biringuccio in the year 1540 as a feature of the manufacture of bronze cannon. Similar effects can be seen in many other materials; for instance when making sand castings in the commonly-used Al-7Si-0.3Mg alloy, eutectic (Al-11Si) is often seen to exude against the surface of external chills (Figure 5.13). The surface exudation of eutectic liquid gives problems during the manufacture of Ni-base superalloy single crystal turbine blades as is discussed in the section on Ni-based alloys.

5.3.4 GRAVITY SEGREGATION

In the early years of attempting to understand solidification, the presence of a large concentration of positive segregation in the head of a steel ingot was assumed to be merely the result of normal segregation. It was simply assumed to be the same mechanism as illustrated in Figure 5.39, case 1, where the solute is concentrated ahead of a planar front.

This assumption overlooked two key factors: (1) the amount of solute that can be segregated in this way is negligible compared to the huge quantities of segregate found in the head of a conventional steel ingot and (2) this type of positive segregation applies only to planar front freezing. In fact, having now realised this, if we look at the segregation that should apply in the case of dendritic freezing then an opposite pattern (previously called inverse segregation) applies such as that shown in Figure 5.43! Clearly, there was a serious mismatch between theory and fact. That this situation had been overlooked for so long illustrates how easy it is for us to be unaware of the most glaring anomalies. It is a lesson for us all in the benefits of humility!

This problem was brilliantly solved by McDonald and Hunt (1969). In work with a transparent model, they observed that the segregated liquid in the dendrite mesh moved under the influence of gravity. It had a density that was in general different from that of the bulk liquid. Thus the lighter liquid floated, and the heavier sank.

In the case of steel, they surmised that as the residual liquid travels towards the roots of the dendrites to feed the solidification contraction, the density will tend to rise as a result of falling temperature. Simultaneously, of course, its density will tend to decrease as a result of becoming concentrated in light elements such as carbon, sulphur and phosphorus. The compositional effects outweigh the temperature effects in this case, so that the residual liquid will tend to rise. Because of its low melting point, the liquid will tend to dissolve dendrites in its path as solutes from the stream diffuse into and reduce the melting point of the dendrites. Thus as the stream progresses it reinforces its channel, as a flooding river carves obstructions from its path. This slicing action causes the side of the channel that contains the flow to be straighter, and its

(a)

(b)

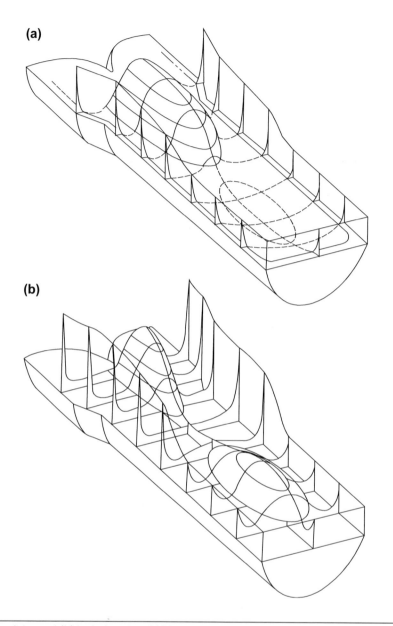

FIGURE 5.44

Segregation of (a) solutes and (b) inclusions in a 3000 kg sand cast ingot.

Information mainly from Nakagawa and Momose (1967).

opposite side to be somewhat ragged. It was noted by Northcott (1941) when studying steel ingots that the edge nearest to the wall (i.e. the upper edge) was straighter. This confirms the upward flow of liquid in these segregates.

The 'A' segregates in a steel ingot are formed in this way (Figure 5.45). They constitute an array of channels at roughly mid-radius positions and are the rivers that empty segregated, low density liquid into the sea of segregated liquid floating at the top of the ingot.

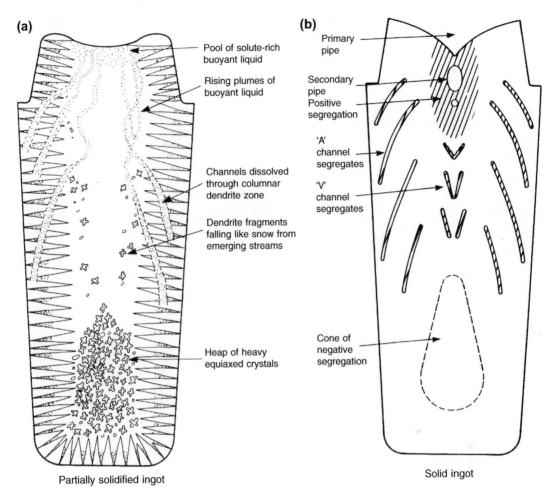

(a)

- Pool of solute-rich buoyant liquid
- Rising plumes of buoyant liquid
- Channels dissolved through columnar dendrite zone
- Dendrite fragments falling like snow from emerging streams
- Heap of heavy equiaxed crystals

Partially solidified ingot

(b)

- Primary pipe
- Secondary pipe
- Positive segregation
- 'A' channel segregates
- 'V' channel segregates
- Cone of negative segregation

Solid ingot

FIGURE 5.45

Development of segregation in a killed steel ingot (a) during solidification and (b) in the final ingot.

At the same time these channels are responsible for emptying the debris from partially melted dendrites into the bulk liquid in the centre of the ingot. These fragments tumble along the channels, finally emerging in the sea of segregation in the head of the ingot, from where they subsequently fall down the centre of the ingot at a rate somewhere between that of a stone and a snowflake. They are likely to grow as they fall if they travel through the undercooled liquid just ahead of the growing columnar front, possibly by rolling or tumbling down this front. The heap of such fragments at the base of the ingot has a characteristic cone shape. In some ingots, as a result of their width, there are heaps on either side, forming a double cone. Because such cones are composed of dendritic fragments their average composition is that of rather pure iron, having less solute than the average for the ingot. The region is therefore said to exhibit *negative segregation*. It is clearly seen in Figure 5.44(a). The equiaxed cone at the base of ingots is a variety of gravity segregation arising as a result of the sedimentation of the solid, in contrast with most other forms of gravity segregation that arise because of the gravitational response of the liquid.

A further contributing factor to the purity of the equiaxed cone region probably arises as a result of the divergence of the flow of residual liquid through this zone at a late stage in solidification, as suggested by Flemings (1974).

The 'V' segregates are found in the centre of the ingot. They are characterised by a sharply delineated edge on the opposite side to that shown by the A segregates. This clue confirms the pioneering theoretical work by Flemings and co-workers that indicated that these channels were formed by liquid flowing downwards. It seems that they form at a late stage in the freezing of the ingot, much later than the formation of the 'A' segregates, when the segregated pool of liquid floating at the top of the ingot is drawn downwards to feed the solidification shrinkage in the centre and lower parts of the ingot.

On sectioning the ingot transversely, and etching to reveal the pattern of segregation, the A and V segregates appear as a fairly even distribution of clearly defined spots, having a diameter in the range 2–10 mm. Probably depending on the size and shape of the ingot, they may be concentrated at mid-radial to central positions in zones, or evenly spread. The central region of positive segregation is seen as a diffuse area of several hundred millimetres in diameter. In both areas, the density of inclusions is high. These channel segregates, seen as spots on the cross-section, survive extensive processing of the ingot, and may be still be seen even after the ingot has been rolled and finally drawn down to wire!

It is interesting to note that in alloys such as tool steels that contain high percentages of tungsten and molybdenum, the segregated liquid is higher in density than the bulk liquid, and so sinks, creating channel segregates that flow in the opposite direction to those in conventional carbon steels. The heavy concentrated liquid then collects at the base of the ingot, giving a reversed pattern to that shown in Figure 5.44.

In nickel-based and high-alloy steel castings the presence of partially melted and collapsed crystals in the channels has the effect of a localised grain refinement, so that on etching longitudinal sections of the castings, the channels seem to sparkle with numerous grains at different angles. In this industry, channel segregates are therefore widely known as 'freckle defects'. The production rate of nickel-based ingots weighing many tonnes produced by secondary remelting processes, such as electroslag and vacuum arc processes, is limited by the unwanted appearance of freckles. It is their locally enhanced concentration of inclusions such as sulphides, oxides and carbides etc. that make freckles particularly undesirable.

Channel segregates are also observed in Al-Cu alloys. In fact workers at Sheffield University (Bridge et al., 1982) have carried out real-time radiography on solidifying Al-21Cu alloy. They were able to see that channels always started to form from defects in the columnar dendrite mesh. These defects were regions of liquid partially entrapped by either the sideways growth of a dendrite arm, or the agglomeration of equiaxed crystals at the tips of the columnar grains. Channels developed both downstream and upstream of these starting points.

The author has even observed channel defects on radiographs of castings in Al-7Si-0.3Mg alloy. Despite the small density differences in this system, the conditions for the formation of these defects seem to be met in sand castings of an approximately 50 mm cross-section.

Although few ingots are cast in modern steelworks, large steel castings continue to be made in steel foundries. Such castings are characterised by the presence of channel segregates, in turn causing extensive and troublesome macrosegregation.

From the point of view of understanding channel formation, Valdes (2010) proposes a condition to predict freckle formation based on the assumptions (1) the Rayleigh number; (2) Darcy flow through the pasty zone; and (3) the condition by Flemings and Nereo (1967) that the liquid flow from colder to hotter regions is faster than the rate of crystal growth. This comprehensive approach to this complex problem appears to yield impressive results. Over succeeding years, there have been a number of excellent studies of the formation of channel defects following Valdes, leading to what appears to be a mature understanding of these defects. Torabi Rad, Kotas and Beckermann (2013) use similar fundamental theory to achieve good predictions calibrated against 27 different steel compositions.

In summary, in terms of parameters which may be within the foundry engineer's control, channel segregates may be controlled by:

1. Decreasing the time available for their formation by increasing the rate of solidification. This action also reduces the spacing of the primary dendrites creating conditions of higher drag forces on the movement of liquid, thus suppressing the rate of formation and growth of channels.
2. Adjusting the chemical composition of the alloy to give a solute-rich liquid that has a more nearly neutral buoyancy at the temperature within the freezing zone.

In practice, both these approaches have been used successfully.

CASTING ALLOYS

The metallurgy of the casting alloys is presented here, interpreted according to bifilm theory. There will be those educated as physicists and metallurgists, like myself, who may be uncomfortable with this new approach. It is with regret that I view the new interpretation as revolutionary. I request the readers' patience to study and understand the approach.

It has to be recognised that at the time of writing the bifilm interpretation is more speculative than I would wish. However, it is a theory that fits the facts with uncanny accuracy. It has to be kept in mind that theories are useful only if they can explain and if they can predict. As a well-known instance, the flat Earth theory was useful, being accurate to predict both distances and directions while humans could only walk limited distances on the Earth's surface. It is a theory that served mankind for thousands of years until the arrival of global travel. The bifilm approach is similarly presented as a useful theory until the time that it is proven inadequate. But at this time, it seems more than adequate to provide many useful insights into the behaviour of liquid metals and castings.

The reader is invited to form his or her own opinion as the metals are described in turn throughout the chapter. For traditional reviews of the metals listed in this section, but not including bifilm theory, nor the benefits of improved methoding techniques such as naturally pressurised filling systems, the reader is recommended the interesting and detailed accounts in the ASM Handbook 2008, Volume 15 'Casting'.

The very low melting point metals such as Pb, Sn and Bi etc. are not dealt with here. We shall concentrate on the structural engineering metals. Even so, in passing, we should note that the casting of lead and its alloys for battery grids is currently a major application that should not be overlooked. The new area of lead-free solders is also of interest, noting that a Sn-based solder containing Ag and Cu exhibits planar voids at the base of ductile dimples on a fracture surface, suggestive that the bifilms in this alloy had been straightened by their formation on the planar intermetallic Ag_3Sn at the base of each dimple. It seems probable that for all metals, not only the high melting point engineering metals, the theory of the control of microstructure and properties of castings by bifilms remains applicable.

6.1 ZINC ALLOYS

For a general introduction to zinc casting alloys, particularly covering the practical details of routine melting and casting, the reader is strongly recommended to the readable account by Arthur C. Street in his famous work *The Diecasting Book* (1986) because approximately 80% of zinc alloys are cast by high pressure die casting (HPDC).

Relatively little is continuously cast, or gravity cast into permanent or sand moulds. The alloy Kirksite is used for prototypes for cast tooling for sheet metal working and plastic injection moulding. The higher strength zinc–aluminium (ZA) alloys are similarly sometimes used for low-volume products cast in permanent or sand moulds. Graphite moulds are sometimes machined from solid blocks for ease and relatively low cost for low volume moulds.

Zinc alloys were first cast by pressure die casting in about 1914. However, these early days were plagued by quality problems which undermined public confidence. The problems were eventually claimed to be brought under some kind of control from research carried out by Brauer and Peirce and published in 1923. As a result, modern zinc-based casting alloys are made from 99.99% pure Zn in an attempt to guard against the problems of contamination from the heavy elements including Pb, Bi, Tl, Cd, In and Sn that were found migrate to grain boundaries and seemed to be associated with the loss of mechanical properties, particularly under conditions of heat or moisture. We shall return to this issue later in the chapter.

Complete Casting Handbook. http://dx.doi.org/10.1016/B978-0-444-63509-9.00006-6

A feature of most Zn-based alloys is that practically all contain Al. The two major hot chamber casting alloys were developed in the 1920s as the ZAMAK 3 and ZAMAK 5 alloys (Z = zinc; A = aluminium; MA = magnesium; K = kopper, the German for copper). These are denoted alloys A and B, respectively, in UK specifications. They all contain 4% Al. Alloy A is most common, but alloy B contains a little Cu and is significantly harder. The nominal compositions of the casting alloys are given in Table 6.1. In the US alloy number system, nos. 2, 3, 5 and 7 coincide approximately at nos. 3 and 5 with ZAMAK 3 and 5 and alloys A and B.

The addition of Al to the hot chamber casting alloys appears to have originally been designed to reduce the attack of the mild steel crucible and swan neck components of the pressure die casting machine because of the development of the thin, protective oxide formed by the Al. Also, the 4%Al addition produced an alloy close to the Zn/Al eutectic at 5%Al, and thus making a useful reduction in the melting temperature from 419°C to 382°C, and at the same time giving a valuable increase in strength.

The International Zinc Association have recently publicised a recent development in an effort to achieve even thinner-walled castings, to aid its competition with the light pressure die cast alloys based on Al and Mg. Whereas the traditional 4%Al alloys are claimed to be limited to 0.75 mm thickness sections, the new composition claims 0.3 mm section thickness potential (Goodwin, 2009). Clearly, some benefit will have arisen from the new composition at 4.5%Al being nearer to the eutectic at 5%Al as is clear from Figure 6.1. Furthermore, the limit imposed on the copper content at only 0.07% seems also to be intended to limit the development of a freezing range which would have inhibited fluidity as explained in Chapter 3. This alloy has yet to prove itself commercially at this time, but it has a good chance, because the logic of its design seems sound.

The ZA hyper-eutectic series of alloys containing 8%, 12% and 27% Al were a significant advance achieved by research carried out in the late 1960s. Their high strength, toughness and bearing properties increasing with increasing Al content, make these alloys among the strongest and toughest of commonly available low-cost casting alloys. ZA27 can achieve 440 MPa in the as-cast condition, which is higher than practically all Al and Mg alloys even after expensive heat treatments.

The microsegregation and macrosegregation accompanying the solidification of the higher ZA alloys is impressive and can be troublesome. ZA27 starts as a homogenous liquid of 27%Al, but the first metal to solidify

Table 6.1 Zn Casting Alloys

Alloy	Nominal Composition (wt%)			Comments
Alloy 2	4Al	0.03Mg	3Cu	'Kirksite' if gravity cast
Zamak 3 (alloy A)	4Al	0.03Mg		Common hot chamber alloy
Zamak 5 (alloy B)	4Al	0.05Mg	1Cu	Common hot chamber alloy
Alloy 7	4Al	0.01Mg	0.01Ni	
Internat. Zinc Assoc	4.5Al	0.01Mg	0.07Cu	Eutectic fluid alloy
ZA8	8Al	0.02Mg	1Cu	Hot chamber or gravity cast
ZA12	11Al	0.02Mg	1Cu	Cold chamber or gravity in permanent mould or sand
ZA27	27Al	0.015Mg	2Cu	
AlCuZinc 5	4Al	<0.05Mg	5Cu	General Motors creep-resistant alloys
AlCuZinc 10	4Al	<0.05Mg	10Cu	
Superplastic zinc	22Al	0.01Mg	0.5Cu	(Wrought alloy)
ZM 11	22Al	1.5Si 0.5Cu 0.3Mn		Mitsubishi, high strength, high damping, lightweight alloys
ZM 3	40Al	3.0Si 1.0Cu 0.3Mn		
Super cosmal	60Al	6.0Si 1.0Cu 0.3Mn		

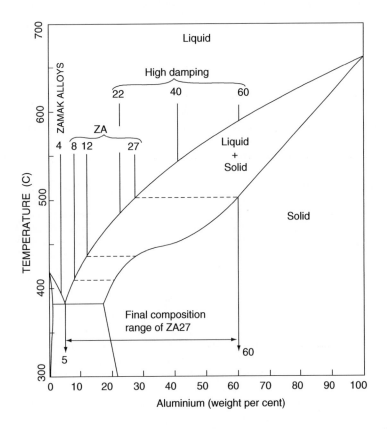

FIGURE 6.1

The Zn-Al phase diagram with some common zinc-based casting alloys.

contains 60%Al, whereas the final liquid to solidify contains 5%Al (Figure 6.1). These wide differences lead to a highly cored dendritic structure that a homogenising heat treatment is sometimes used to encourage some limited redistribution of Al.

Although ZA8 can be cast by the hot chamber pressure die casting process, the higher temperatures required for ZA12 and ZA27 would result in unacceptable rates of attack of the casting machine. The liquidus temperature for ZA27 is 500°C.

The superplastic zinc should not really be included in the casting alloys because it needs plastic working and heat treatment to reduce its grain size before it can develop its full superplastic properties. It is included here simply to illustrate its interesting similarity with Zn casting alloys. The spectacular 1000% elongation that this alloy can develop is limited by microscopic cavitation, almost certainly the result of oxide bifilms entrained during melting and casting. Better melting and casting might deliver even better properties. Its composition is nevertheless close to other casting alloys, indicating that perhaps heat treatment might deliver some part of the superplastic properties, giving a huge benefit to toughness. This does not seem to be generally recognised.

One of the standard problems with Zn-based alloys is their poor creep resistance; the relaxation of steel bolts in holes threaded in Zn-based castings, especially at slightly raised temperatures in the region of 100–150°C has been a central issue for automotive application in the engine compartment of vehicles. Research by General Motors (Rashid and Hanna, 1989) has resulted in the two high copper alloys in Table 6.1 that exhibit improved creep performance.

A related problem is the continued ageing of zinc alloys at room temperature over weeks or months. This change of structure is accompanied by changes in properties, and by minute changes in the size of the casting. The size issues are discussed in Chapter 18.

In a separate quest, this time to optimise noise reduction in vehicles by improving the damping properties of zinc alloys, Mitsubishi has proposed three new alloys at very high levels of Al, 22%, 40% and 60% (Mae and Sakonooka, 1987). These are, of course, really Al-based alloys, their volume additions corresponding approximately to 40%, 60% and 80%, respectively, explaining their improved strength and increased lightness.

Unfortunately, however, from the point of view of the casting quality, the formation of the alumina film on the surface of the liquid alloys gives the standard problems, involving the entrainment of air bubbles and films during the extreme surface turbulence of filling. Although the melting and casting temperatures of zinc pressure die casting alloys are low, therefore restricting the rate of thickening of films, making them even less visible than in Al alloys, this effectively merely sweeps the problem out of sight. The films have to be double, necessarily have an unbonded inner interface, and therefore still perform efficiently as cracks, even though less detectable than those in Al alloys.

The high level of Al in the three Mitsubishi alloys probably means the alloys contain high levels of aluminium oxide in the form of bifilms, so it may be the bifilms that are the reason for the high damping properties; during the shearing across the bifilm as the matrix is strained (see Section 9) the friction involved will dissipate energy in the form of heat. The elastic stiffness (Young's Modulus) of the alloy is similarly significantly reduced by bifilms (as demonstrated by Hall and Shippen (1994) for a bronze alloy). It seems reasonable to conclude therefore that although the damping properties may be found to disappear if the alloys are cleaned from oxides, the alloys will benefit in other ways: their tensile and corrosion properties are likely to improve. They might then enjoy a wider commercial acceptance.

Bifilms seem almost certainly present in Figure 2.31, as witnessed by the sharp changes in microstructure across the boundaries formed by oxide flow tubes around the jets of metal that filled the die. In many situations, the continued flow over previous jets that have solidified can re-melt the solid, allowing it to lose its grip on its oxide tube which is stripped away in the flow. The prior solidified jet then continues to re-melt back somewhat, blurring the original sharp divisions between jets. The longitudinal layering of structures in HPDC products is common, being a characteristic feature of HPDC zinc pressure die cast alloys (Goodwin, 2008).

Romankiewicz (1976) found that zinc alloys can contain up to 1% or 2% of oxides. These reduced mechanical properties. Furthermore, in damp conditions, they became a source of selective corrosion. He offered no explanation of this behaviour. Clearly, in terms of bifilms, the answer is that the films, when entrained by turbulence, are actually double and act as cracks and will usually be pushed by dendrite growth into the interdendritic or intergranular regions. In damp conditions, corrodants such as rainwater can penetrate into the matrix along the unbonded interiors of the bifilms. The corrosion will be enhanced by the precipitation of other elements on the back of the films, explaining the effects of heavy elements such as Pb in promoting the often catastrophic degradation of properties over time. The degradation is often called intercrystalline or intergranular corrosion. It is easily demonstrated after only 10 days of testing in steam (Colwell, 1973) or perhaps as long as a year in tropical locations.

The deterioration of properties with time eventually cause the casting to fail is an experience most of us have with such items as zip fasteners on travel goods, handles on brief cases and window fasteners in the home. Thus although the research of Brauer and Pierce made huge improvements to Zn alloys, the problems are clearly not totally solved.

Other Canadian workers (Dionne et al., 1984) found that toughness of ZA27 was limited by brittle $CuZn_4$ following high temperature homogenisation treatment. This is typical behaviour of a compound precipitating on bifilms (otherwise the failure by cracking is difficult to explain).

Blisters commonly form on the surface of pressure die cast alloys immediately after opening the die. Kaiser and Groenveld (1975) described how when the die is opened, the wall of the hot casting may be insufficiently strong to withstand the pressure of gases entrapped in the casting during die filling. Although the authors report that the defect can be controlled by reducing metal and die temperatures to increase the strength of the casting wall, they do not recommend this practice. Temperatures may rise in service to allow the alloy to creep, slowly forming a blister.

Blisters are known to form on Zn alloys at ambient temperature in the open air after a period varying from 1 to 6 months. The blisters almost certainly grow by the pressure of gases trapped in defects just under the surface of the casting. The defects are most likely bifilms and/or porosity (there really is practically no difference between the two!). The origin of the high pressure of gas inside the defect is less certain. Possibilities include:

1. high pressures of air or other gases entrained during the turbulence of die filling.
2. high levels of hydrogen from plating processes are common in many plated zinc castings
3. hydrogen generated from a surface corrosion processes can diffuse into the metal. The presence of moisture creates a gently corroding environment, releasing hydrogen that will diffuse into the metal, precipitating in the centres of bifilms and/or bubbles.

The blister may grow by the gradual precipitation and increasing pressure in the defect, leading to the gradual creep of the alloy. A flow of hydrogen into the defect seems necessary, because after the initial expansion of the blister, the pressure will be expected to fall rapidly with increasing volume, bringing the rate of growth to a stop. In practice, the blisters appear to continue to grow over a period of months or years.

That Birch (2000) found that the properties fall as dendrite arm spacing increases is further evidence that Zn die casting alloys, particularly alloy 3, contain bifilms that have chance to straighten, reducing properties, as the time for solidification lengthens. This effect is discussed in more detail in Section 9.4.3.

A quite different problem arises in heavy section castings. As the Al content of the ZA series of alloys increases some casting problems emerge which are unique, and which is found in all these alloys, but particularly ZA27. During solidification, the Al-rich dendrites form first and float to the top of the casting. Thus the Zn-rich liquid, being much denser than the Al, is displaced to the bottom of the casting. Because of its much lower melting point, the Zn-rich liquid freezes much later and therefore forms a shrinkage cavity on the base of the casting (completely opposite to a normal casting where the shrinkage would be concentrated in a top feeder). It is known as *underside shrinkage*. Naturally, this problem is more severe in heavier sections and where freezing times are long because of the use of a sand mould as opposed to a metal die.

Canadian workers have explored a variety of alloy additions that appear capable of eliminating underside shrinkage, of which the most practical and cost-effective is Sr (Sahoo et al., 1984, 1985). This worked well—provided an adequate feeder was sited on the top of the casting. These authors were not able to explain why Sr was effective, but surmised that a skin of metal formed on the base of the casting which was sufficiently strong to withstand collapse (even though an internal cavity was now likely to occur in place of the external sink). The pressurisation of the casting interior by a generous feeder will assist to prevent collapse during the early stages of solidification. Furthermore, the pressurisation of the casting skin against the drag surface of the mould will increase the rate of heat transfer, helping to thicken and strengthen the skin. The detailed action of Sr remains a mystery, even though it is tempting to draw an analogy with the modification of Al-Si alloys in which Sr acts to suppress nucleation of the solid ahead of the freezing front, thereby straightening the front, and so forming a stronger solidified skin.

The hot chamber pressure die cast Zn-based alloys yield relatively small components of high accuracy, good finish requiring no machining and thin walls. It also has unequalled productivity and fulfils the requirements of a wide market. Larger castings at somewhat slower rates are produced by the cold chamber pressure die technique. Both types of HPDC parts have practically unique features that make them popular with designers, such as integrally cast rivets or cast threads to facilitate joining, or integrally cast steel inserts to eliminate assembly.

However, even at this date, it is clear that there remains great scope for the improvement of the properties and behaviour of Zn alloys. In my opinion, this centres on the problem to reduce oxide bifilms by improved melt handling and casting. Here once again there is plenty of work to do.

6.2 MAGNESIUM ALLOYS

Magnesium is the lightest structural metal with a density of only 1738 kgm^{-3} at room temperature. Its light weight drives exploration of the use of its alloys for automotive and aerospace in competition with aluminium-based alloys. The competition is finely balanced, with Mg having the highest specific stiffness, and relatively high specific strength, with good damping characteristics (i.e. quiet castings), but remains limited in some applications because its relatively low strength means that full advantage cannot usually be taken of its attractive low density. Numerous other factors weigh against Mg compared with Al.

Historically, the relative volatility of the cost of Mg has been a serious disadvantage in comparison with Al for volume applications. Even so, the aircraft industry in particular finds Mg alloys essential to its designs of numerous castings, and the electronics and communications industries (e.g. mobile telephones) opt for Mg for weight saving of chassis and frames.

Another major factor hampering the wider use of Mg at this time is the difficulty of recycling Mg alloys. Rapid oxidation and the numerous alloys, all in relatively small quantities, make recycling problematic. This contrasts with Al alloys for which an efficient worldwide recycling industry is already in place, making a big contribution towards the reduction in costs of Al castings using these so-called 'secondary alloys', meaning recycled alloys.

The relatively poor corrosion resistance of some Mg alloys means that expensive surface protection is required for some components. However, the more recent introduction of a purer variety of AZ91 alloy (Mg-9Al-1Zn) has made a major contribution towards the growth potential of a volume Mg market. It seems likely that if Mg could be cast without bifilm defects corrosion resistance would be enhanced further at practically zero cost. This prediction remains to be put into practice.

In the meantime, for readers who want a wealth of practical advice on the melting and casting of Mg alloys, two works stand out: *Principles of Magnesium Technology* by Emley (1966) and *Product Design and Development for Magnesium Die Castings* by the Dow Chemical Company (1985). For a more detailed and more up-to-date account of the metallurgy of Mg alloys, the excellent text by Polmear (2006) is recommended. Even so, it is necessary to bear in mind that these renowned texts neither include the concept of bifilms nor such concepts as naturally pressurised filling systems for castings. Thus the reader needs caution.

In fact, when starting to consider the metallurgy of the light casting alloys, Al- and Mg-based, the role of oxide films appears so fundamental that it seems sensible to make a start with first understanding the oxides to be expected on such melts.

6.2.1 FILMS ON LIQUID Mg ALLOYS + PROTECTIVE ATMOSPHERES

In their comprehensive review of the oxidation of liquid metals made in 1969, Drouzy and Mascre mentioned some of the benefits of oxidation, but caution 'oxidation is something of a hindrance in the handling of liquid metals, particularly in casting'. This masterpiece of understatement 'something of a hindrance' seemed worth quoting. Such disarming and admirable restraint summarises the awesome, towering importance of perhaps the central problem in metallurgy.

The oxidation of magnesium metal in air is not easily studied, because the metal often ignites before it reaches its melting point. This emphasises that for foundries, the metal has to be handled at all times in an atmosphere and in a manner that helps to suppress ignition.

The oxide, MgO, has a density greater than that of solid Mg, with the result that the prediction based on the Pilling-Bedworth ratio is less than 1, with the consequence that the oxide does not cover the metal from which it forms, and therefore does not protect it. Clearly, these considerations only strictly apply to solid Mg and its alloys, but require to be kept in mind during the heating of the Mg charge in the furnace before melting.

Once molten, it has been traditional to protect the metal by fluxes based on chlorides and fluorides. However, this practice is less popular today as a result of the widespread problems of the entrainment of flux inclusions in the castings, and as a result of the contamination of the environment with fluorides and other spent fluxes.

There has therefore been a major move to so-called 'dry melting', using protection from gases, plus possibly a little sprinkled sulphur. One of the gases used to suppress the burning of Mg has been sulphur hexafluoride (SF_6). At a dilution of only 0.1–1.0% in air or other gas such as CO_2, the mixture was extremely effective. Xiong and Liu (2007) found that the film of MgO on the melt surface was supplemented by islands of MgF_2 under the oxide. The islands gradually spread to about 50% of the total area. Emami and co-workers (2014) confirmed the formation of a stable MgF_2 film in a mixture of air and 0.5% SF_6, but find that this takes up to 7 min to achieve good protection.

Although SF_6 is now rightly outlawed because of its huge effect on the Earth's ozone layer, other chlorine- and/or fluorine-containing gases are being actively evaluated that are less harmful to the environment, and their mode of action may be similar (International Magnesium Association, 2006). Polmear (2006) describes the use of the refrigerant gas HFC 134a, one of the several forms of tetrafluoroethane. Another promising new protective atmosphere contains boron trifluoride; the toxicity problems of this gas are addressed by using the gas in extreme dilutions generated in line from a solid fluoroborate (Revankar, 2000). Other developments are under way and appear to be having some success to refine further the effectiveness of the traditional dilute SO_2 mixtures in CO_2. Perhaps the old remedies will be proven best after all.

The action of CO_2 alone is perhaps more active than might be expected because it is reduced by Mg to form MgO in two stages: (1) CO_2 is reduced to CO and (2) CO is reduced to carbon (Wightman and Fray, 1983). It is possible that the deposition of carbon may help to suppress the rate of oxidation although the formation of magnesium carbide would also be expected to complicate the situation further. Cochran and colleagues (1977) confirmed that CO_2 strongly inhibited the onset of breakaway oxidation although the mechanism appeared mysterious; they looked for but failed to find any carbon-containing phase in the surface of the metal after solidification. For low-Mg alloys, they found even a small amount of CO_2 was valuable to reduce oxidation, although the amount required was sensitive to the Mg content and the temperature of the melt.

An important detail is that magnesium alloys are known to give off magnesium vapour at normal casting temperatures, the oxide film growing by oxidation of the vapour, effectively growing 'from the top downwards'. This mechanism seems to apply not only for magnesium-based alloys (Sakamoto, 1999) but also for Al alloys containing as little as 0.4 wt% Mg (Mizuno, 1996).

The microscopic structure of MgO films on AZ91 alloy has been studied by Mirak and colleagues (2007) in the limited oxidation conditions offered by the interior of bubbles trapped in the melt. The films are thick and are characterised by much fine-scale wrinkling, clearly capable of entrapping much air. Their general appearance is indistinguishable from that shown for the Al-5Mg alloy in Figure 2.10(b).

In some admirably comprehensive experiments to compare protective atmospheres for melting Mg alloys, Frueling and Hanawalt (1969) noted that a CO_2 atmosphere was better than N_2 or Ar in preventing the 'smoking' of Mg. The smoke is almost certainly the condensation of Mg vapour to Mg metal and its oxidation to MgO (or possibly the direct oxidation of the vapour to oxide) in the atmosphere over the melt. In fact, the attempted protection of molten Mg by Ar is potentially hazardous. I recall a tragic fatal accident in which an experimental low-pressure furnace containing Mg alloy was provided with an atmosphere of Ar. Unknown to anyone, including the operator, the furnace had filled with Mg vapour. On opening the door to top up the furnace with additional Mg alloy ingots, air entered and mixed with the vapour, triggering a massive explosion. Such accidents do not happen when the furnace atmosphere comprises some oxidising fraction, such as air, CO_2, SO_2 and/or gases such as freon containing chlorine and/or fluorine. The clear lesson is (no apologies for spelling this out) the treatment of Mg requires knowledge, expertise, possibly and preferably humility, but certainly caution.

A rather different approach to the protection of Mg melts from oxygen in the air has been proposed by Rossmann (1982) who used Ar in a way that appears to be free from the dangers mentioned previously. He pours liquid argon directly on to the surface of the melt contained in a crucible in the open air. One litre of liquid argon will generate 836 L of argon gas at 1 atm and 15°C, so that very small volumes of liquid argon are required to entirely eliminate oxygen from the surface of the melt. The heaviness of argon is a help to keep it on the surface of the melt, but may not be totally proof against air. At −233°C, the density of Ar gas is 1.66, at 20°C it is 1.28 and at 700°C it is 1.12 kgm^{-3}. Thus it becomes a little less dense than air at 20°C, approximately 1.20 kgm^{-3}. Even so, the laminar spreading of the liquid Ar over the melt, and its action as a wind to keep air away as it boils, may continue to be effective in keeping air away from the melt surface.

Other approaches to developing resistance to burning include the addition of Ca to the Mg alloy at levels in the region of 1% (Sakamoto, 1996). The oxide film on the melt consisted of two layers: a mix of MgO and CaO on the lower layer and a CaO-rich layer on the top. The development of the modified oxide increased the ignition temperature of the alloy by approximately 250 K. It seems curious that two metals that both fail the Pilling-Bedworth criterion when mixed, appear to satisfy it nicely when present on liquid Mg. It is not clear, and perhaps is not likely, that the benefits of such alloying additions can be extended to the more useful engineering alloys such as AZ91. Although such benefits might be welcome, there are, as we have seen, other approaches to the control of Mg ignition that are already available.

For sand moulds, the addition of inhibitors to prevent the ignition of Mg is universal. In particular, for silica-based greensands containing perhaps 4% clay and 3–4% water, additions of 2% boric acid plus 2% sulphur are typical. For a self-setting resin-bonded silica sand, the level of inhibition is similar. However, of course, heavier castings might require more and lighter casting less (Mandal, 2000). Furthermore, the skill of the designer of the filling system is also likely to influence the required amount of inhibition, as described later.

One final aspect of the ignition of Mg alloys is critically important. It is predictable that if a mould was filled with a liquid Mg alloy, but the filling was carried out without surface turbulence, so that the melt surface was at all times undisturbed, the alloy would probably not ignite. This proposal is based on the fact that the huge heat of formation of the oxide is easily conducted away into the bulk of the melt, limiting the rise of temperature of the metal surface to some trivial few degrees. However, if the melt surface is subjected to jumping and splashing, as occurs often when poured under gravity, each splash and droplet has a mass of only a few grams, and is only a few millimetres in thickness or diameter. Thus a sliver of liquid metal when travelling through the air would experience oxidation from both sides, the huge heat of oxidation now saturating the small mass and limited diffusion distance, with the result that the sliver, jet or droplet now reaches a huge temperature, well above the ignition temperature. Thus surface turbulence initiates burning.

If the melt starts to burn, its reactivity will cause it to take oxygen from the silica sand of the mould, and thus will consume its way through mould, possibly continuing through the concrete floor, taking oxygen from the concrete, until the Mg is entirely oxidised. Only then will the burning stop. The next clear lesson for the handling of Mg alloys is that it is absolutely necessary to ensure that burning of the metal does not start.

We shall turn now to the behaviour of Al-Mg alloys in the range of approximately 2–20% Mg. These Al-Mg alloys are so dominated by the oxidation of Mg that they act practically as though they were pure Mg. We shall therefore include them in this section on Mg.

Provided burning can be successfully suppressed, as would be normal for the competent handling of Al-Mg alloys, the surface oxide initially develops as a thin layer of amorphous MgO (Figure 6.2(a)). After an incubation time that seems to depend on chance events to nucleate a change, the oxide crystallises to MgO, and some spinel, the name for the mixed MgO and Al_2O_3 oxide, magnesium aluminate, which can be written $MgAl_2O_4$ (Cochran, 1977). This stage of oxidation is fast (Figure 6.3), particularly if the atmosphere is moist air, creating so much oxide that the film has to corrugate as in a concertina (Figure 6.2(b)). The development of the oxidation process proposed in Figure 6.2 is based on an interpretation of the observations by Rault (1996) and Haginoya (1976). These authors found that the hydrogen released into the melt by the surface oxidation causes some of the sub-surface bifilms to inflate and float, coming into contact with the underside of the surface film. Here they can break open the surface leading to irregular masses of spinel formation. However, Haginoya noticed that those films that were more deeply immersed in the liquid, and for some reason unable to float, perhaps being attached to a submerged surface of the casting, remained as films. This is perhaps to be expected because they will probably require at least a part of the bifilm to be within the diffusion distance to the surface to be successful to gain any hydrogen.

6.2.2 STRENGTHENING Mg ALLOYS

Magnesium metal, like most pure metals, is naturally rather weak, and requires strengthening for most engineering applications. The Mg-based alloys can be complicated (Unsworth, 1988), and only a brief summary can be included here. In general, in common with the other family of low melting point alloys based on zinc, the magnesium based alloys are not only mainly wanting in strength, but tend to creep rapidly as temperatures are raised.

Alloying with Al and Zn has produced a series of useful alloys, known as the AZ series including AZ31, AZ63 and AZ91. The latter alloy (its letters standing for 9 wt% aluminium and 1 wt% zinc) is widely popular. Many of these alloys contain up to 0.4% Mn, as also is the case for the Al and Mn containing alloys AM50 (Mg-5Al-0.4Mn) and AM60 (Mg-6Mg-0.4Mn). Other Al- and/or Zn-containing alloys include Mg-Al-Si, Mg-Al-Rare Earth and Mg-Zn-Cu. The introduction of high purity variants of these alloys, with lower levels of Fe, Cu and Ni has significantly improved corrosion resistance. The common sand-casting alloy AZ91C has now been largely replaced by AZ91E, its high-purity equivalent which has about 100 times better corrosion resistance. The alloy ZC63 (Mg-6Zn-3Cu-0.5Mn) was developed as an easy-to-process material for engine castings such as cylinder blocks and oil pans.

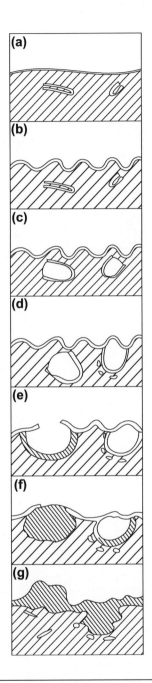

FIGURE 6.2

The stages in oxidation of Al-5Mg alloy. (a) MgO film on surface and folded into the melt; (b) growth of surface film leads to corrugations; (c) in moist air, OH^- ions diffuse through film, giving H^+ ions in the melt, swelling bifilms with H_2; (d) bubbles rise, disrupt surface film, starting break-away oxidation; (e) spinel $MgAl_2O_4$ (dark shaded) forms; and (f, g) complete conversion to thick, irregular spinel film.

Experimental data from Rault et al. (1996).

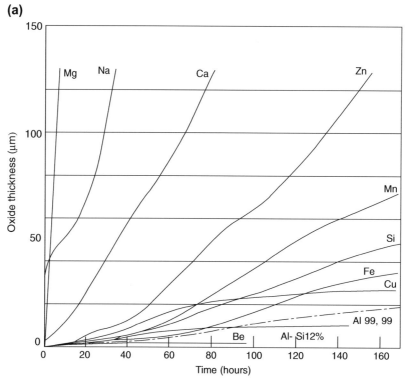

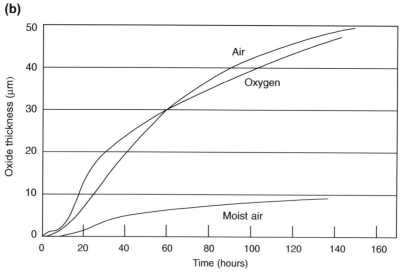

FIGURE 6.3

(a) Growth of oxide on Al and its alloys each containing 1 atomic percent alloying element at 800 °C. (b) Growth of oxide on 99.9Al at 800 °C in a flow of oxygen, and dry and moist air.

Data from Theile (1962).

So far as possible, it is important to grain refine the structure of Mg alloy castings. A typical grain refining action might reduce the grain size from 250 μm to 50 μm. The reason for the importance of grain refinement is that the yield strength σ_y in Mg alloys is strongly related to the grain size by the Hall–Petch relationship:

$$\sigma_y = \sigma_o + K_y d^{-1/2} \tag{6.1}$$

The relationship is strong for Mg alloys because of their hexagonal close packed crystal structure. The only plane allowing relatively easy slip is the basal 0001 plane. All other directions in the crystal represent significant difficulty to initiate slip. Thus, because it will be unlikely for two adjacent grains to have their basal planes aligned, propagating slip from one grain to the next will normally be difficult. Thus the K_y factor in the Hall–Petch equation for Mg alloys is unusually large. (This contrasts with αAl alloys, in which the effect of grain size is relatively weak because K_y is small. This follows because propagating slip from one grain to another is easy in the aluminium face centred cubic [fcc] lattice as a result of its high symmetry; an fcc crystal has many slip possibilities in all directions.)

Unfortunately, the grain refinement of Mg alloys is an unsatisfactory list of recipes of traditional practices, with no convincing science to provide explanations. For instance, the mysterious nature of the grain refinement of Mg-Al alloys by superheating to around 850°C is widely practised, but not understood. The role of carbon dissolved from the steel crucibles has been suspected. A rather wider range of alloys can be grain refined by carbon additions, most popularly by hexachloroethane plunged into the melt to effect some simultaneous de-gassing action. Other black art to grain refine these alloys is listed by Emley (1966). Zhang et al. (2010) described a new grain refining technique for AZ91 consisting of bubbling a mixture of Ar + CO_2 through a melt contained in a rotary de-gassing unit. The treatment takes 30 min, after which the grain size reduction seems better than competitive treatments. Although the authors suggest aluminium carbide, Al_4C_3, may be the effective nucleating agent, they neglect that the melt is almost certainly full of oxide films. Despite this, they report increased tensile strengths increased by up to 20% and elongation by up to 100%.

The Mg-Zr alloys

A completely different class of Mg alloys was opened up by the discovery of the Mg-Zr alloys by Sauerwald in 1947. This has been an outstanding development for Mg, giving alloys with enhanced properties at both room and elevated temperatures. The alloys derive their properties from the remarkable grain refining action of Zr.

Zirconium is only slightly soluble in Mg having a limit of only 0.60%Zr, and it is necessary to saturate the melt with Zr because only the Zr in solution at the time of casting is effective in refining the grain size. On solidification, undergoing a peritectic transformation close to 650°C (the available phase diagrams are not clear in this region), it appears that Zr-rich particles precipitate in the melt. The Zr-rich particles form an extremely effective nucleus for Mg as a result of their similar hexagonal lattice structure, and their nearly identical lattice spacings. The particles are often seen at the centres of Mg grains, confirming their role as nuclei for grain refinement. (The action of Zr is only useful in those alloys that do not contain Al because Al and Zr form a stable compound that effectively eliminates all free Zr from solution. Similar reactions occur with Mn, Si, Fe, Ni, Co, Sn and Sb.)

The main effect of Zr on the strength of Mg alloys (in those alloys free from the previous list of elements) is probably not the result of the usual mechanisms of precipitation hardening or solute hardening etc., but its impressive action on the refinement of grains according to Eqn (6.1). Grain size is typically reduced from 2 to 0.05 mm. This dramatic grain refining action appears to be the result not only of the presence of effective nuclei, but of a significant contribution to the subsequent restriction of the growth of the grains by other solutes, as made clear by StJohn (2005).

Historically, the addition of Zr to a Mg melt has proved to be difficult because of the low solubility of Zr and its loss by reaction with air and reaction with Fe from the steel crucible, particularly as the temperature increases from 730° to 780°C (Cao, Qian et al., 2004). Traditionally, one of the most successful techniques for the addition of Zr has been through the use of a master alloy containing Mg and 30%Zr, plus residual heavy flux, 'weighted down' often with a dense chloride such as $BaCl_2$. The alloy and its content of entrained chlorides and fluorides has to be 'puddled' (i.e. mechanically pummelled or stirred, and then left to settle again. It sits in excess at the base of the crucible to act as a reservoir of Zr, maintaining saturation as nearly as possible, replacing that being continuously lost by oxidation and by reaction with the iron crucible. The melt is poured leaving 10–20% of the melt with its residual flux undisturbed at the bottom of the crucible. (This material can be recycled in other Mg-Zr alloys.)

During studies of the separation of flux from Mg alloys, Reding (1968) confirmed that the flux settles to the bottom of the melt, but remains as separate droplets which refuse to coalesce into a single drop. This seems consistent with the drop being enclosed within a tenacious oxide skin. The SEM EDX studies revealing high Fe and Zr content at the drop/matrix interface corroborating the presence of an oxide skin on which Fe-rich and possibly Zr-rich intermetallics precipitate as on a favoured substrate (see the analogous behaviour in Al alloys, Section 6.3). The report by Reding makes sobering reading and is recommended to readers as a wakeup call in the problems of making Mg-Zr alloys. It is a concern that in the Mg-Zn-Zr alloy that he studies the flux droplets appear not to settle but remain dispersed uniformly through the melt. This suggests the presence of a network of large films, probably of MgO, that are resisting the settling of the flux. Whether this behaviour is exclusive to this alloy seems unlikely, and may reflect poor handling or melting practice. We have much to learn about Mg alloys.

In view of these uncertainties it is perhaps surprising that the flux technique for the addition of Zr has been more or less successful for the production of Mg-Zr alloys for many years (Emley, 1966). A more recent development by (Qian, 2003) showing Zr dissolution working well at only 730°C, stirring for only 2 min, and using melts protected from contamination from their iron crucibles by a simple wash of boron nitride promises future hope of a more controlled manufacture of these useful alloys.

6.2.3 MICROSTRUCTURE

The microstructural development of the very many Mg-based alloys is a complex and vast subject that is beyond our scope. However, some general points can be made about the simplest alloys, particularly the Mg-Al alloys.

The solid grains forming in a liquid Mg alloy melt have a flat (i.e. two-dimensional), hexagonal form, closely resembling the forms of snowflakes. These contrast with the dendritic forms of the cubic structured metals, whose dendrites are three-dimensional stars, with three sets of axes at right angles, their positive and negative directions creating six primary arms.

Slow cooling of the Mg-Al alloys causes a coarse precipitate of $Mg_{17}Al_{12}$ to precipitate at grain and dendrite boundaries (Figure 6.4). This curiously complex phase is a common feature of many Mg alloys. At even lower temperatures, the particles become surrounded by a fine lamellar eutectoid precipitate of Mg and $Mg_{17}Al_{12}$.

Whereas with Al alloys that have enjoyed massive attention from researchers, so that the central role of oxide bifilms in the control of microstructure and properties has now been established beyond doubt, Mg alloys have hardly started on

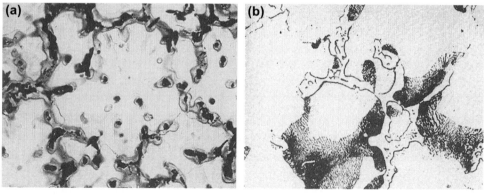

FIGURE 6.4

(a) Microstructure of chill cast Mg-8Al-0.5Zn-0.25Mn. The common intermetallic compound $Mg_{17}Al_{12}$ (dark) and interdendritic coring (grey) is shown by the electrolytic polish. (b) Slowly cooled alloy showing eutectoid precipitate from decomposition of Al-rich solid solution (nital etch) (Emley, 1966).

this path. Nevertheless, it is interesting to take an overview of an authoritative text such as Emley's *Principles of Magnesium Technology* (1966) to see that most of the defects appear to be films. At that time, of course, it was not known that all such films were in fact bifilms. It is tantalising therefore to consider that Mg alloys might similarly be controlled by the presence of bifilms because the bifilms are to be expected in Mg alloys. A few early hints that this may be so are listed later.

At this time, there is little evidence that $Mg_{17}Al_{12}$ nucleates on the outer surfaces of bifilms, even though closely analogous behaviour in Al-based and other alloy systems would lead us to expect this. Also, the presence of central linear features, including voids, and the well known brittleness of this compound are not easily explained in the absence of bifilms.

Srinivasan and colleagues (2006) investigated the structure and properties of AZ91 alloy to which had been added 0.5Si in an effort to improve high temperature strength and creep resistance by the precipitation of Mg_2Si particles. They found that the Mg_2Si formed as coarse 'Chinese script' and actually slightly reduced properties, but an 0.1Sr addition made a radical change to the microstructure, changing the Mg_2Si from Chinese script to a refined form, evenly distributed along the grain boundaries.

It is tempting to draw a comparison with Al alloys, where it is fairly certain that Sr deactivates oxide bifilms, preventing the nucleation and growth of other phases on the bifilms, and thereby forcing the formation of these phases at lower temperatures in eutectic growth modes. In this case, it seems probable therefore that Mg_2Si is precipitating on bifilms and taking a Chinese script form. After Sr addition, Mg_2Si no longer precipitates on bifilms and thus seeks the next most energetically favourable mode of precipitation, which appears to be a eutectic form associated with the grain boundaries. Other work on the addition of Sr at levels of only 0.01–0.02 wt% to AZ91 was reported by Aliravci (1992) reporting a number of changes such as finer grain size, reduced porosity and improved mechanical properties. These issues are unexplained and clearly require a significant further research effort.

Over the past few years, work has started to reveal the true role of bifilms in Mg metallurgy. As is now well known in Al metallurgy, in Mg metallurgy bifilms are also in control. Griffiths and Lai (2007) find large butterfly mirror images of bifilms on every tensile fracture surface and facets aligned by dendrites, suggesting bifilms cover practically the whole of their fracture surfaces. Using Weibull statistics to quantify the scatter of strength properties, they find the modulus of strength of top-filled castings only four compared with a modulus 12 for bottom-gated castings, confirming a three-fold reduction in scatter of properties for carefully poured castings. Other recent work by the author on large Mg alloy castings for the frames of jet turbines for aero engines has clearly demonstrated the benefit of naturally pressurised filling systems which reduce the entrainment of bifilms and bubbles.

As in all cast metals at this time, the attainment of zero bifilms for Mg alloys remains an as-yet unrealised, but probably realisable, target. The first steps are being taken to reduce bifilm damage. It seems likely therefore that when founders start to cast Mg alloys entirely without bifilms, we should expect surprises.

6.2.4 INCLUSIONS

Flux inclusions can be troublesome in Mg alloys, reacting with moisture in the air and growing to create unsightly pustules of corrosion products on both the cast and machined surfaces. It seems likely that during surface turbulence, with a layer of flux on the surface, sitting on top of the MgO film, the flux inclusion always takes the form of a MgO bifilm which contain entrapped flux; it is probably a 'flux sandwich'.

Sand inclusions in Mg alloy sand castings are expected to be common if the filling system is sufficiently turbulent. However, because of an energetic mould–metal reaction Mg alloys can contain a unique defect: a reacted sand inclusion. Lagowski (1979) carried out a detailed study of these defects and their effects. It seems that the sand grains are detached from the mould wall and swept into the interior of the melt. The majority sink and sit on the drag surface of the casting. Here the 0.1 to 0.2 mm diameter grains react with the magnesium, swelling and dissolving until finally only a nebulous halo of the order of 1 mm diameter remains. On a radiograph the residue of a multitude of such inclusions can look like a snow storm. However, despite the alarming appearance, the mechanical properties appear to be hardly affected, provided the radiographic rating does not deteriorate beyond 4 on the ASTM radiographic standards. (The ASTM standard radiographs represent increasing severity of defects ranging from the least severe, rated 1, equivalent to a single inclusion of approximately 1 mm diameter in the representative 50 mm × 50 mm area, to the most severe, rated 8, depicting a blizzard.) Entrained sand inclusions are, of course, a clear signal that the filling system is turbulent; thus the problem is easily avoided by substituting a non-turbulent mode of filling.

6.3 ALUMINIUM

A special feature of the Al alloy market is that the metal is divided into primary and secondary markets in which the primary alloys are those made from metal freshly won from the ore, whereas secondary alloys are based on recycled material, recycled alloy economises on the huge energy required to smelt Al from its ore because melting it for the second time requires only about 5% of that energy. This economy is of course offset by the greater oxide and impurity contents, particularly iron. The Al alloy casting alloys for most applications are based on secondary metal. Only relatively few applications for aerospace and safety-critical items might specify primary alloys. Otherwise, most primary metal goes for wrought applications such as sheet and foil.

The Al alloys have enjoyed a huge expansion over the past few decades as a result of their accessible casting temperatures and light weight. Although the alloys have been presumed easily processable, probably as a result of their low melting temperature compared with cast iron, the industry has paid a high price for this presumption. Processing the alloys in the way that cast iron has been traditionally handled has resulted in major problems for Al castings and for the fledgling Al casting industry; the surface oxide has become entrained, leading to numerous and serious bifilm problems, and the hydrogen gas in solution in the liquid metal is partitioned uniquely strongly during solidification, inflating bifilms to become porosity. Thus there have been important and unsuspected problems with attempts to make castings in Al and its alloys. The metallurgical aspects of these issues will be dealt with in this section, and the practical engineering aspects of metal handling and casting later in Chapter 10 onwards.

Finally, these issues should not take our focus away from the central problem with the casting of all Al alloys; during the turbulence of pouring, the incorporation of the surface oxide into the bulk of the alloy, where it constitutes a bifilm crack that impairs properties. Al alloys have therefore gained a reputation for unreliability. It is necessary to understand this problem to ensure that it is effectively avoided. This is the reason for making an early start in this chapter with the oxidation of the melt immediately following the short introduction of the main alloy types.

The strengthening of the surprisingly weak and soft pure aluminium metal has led to the development of three families of casting alloys.

1. The Al-Si alloys form the mainstream of the industrially important engineering cast alloys. The presence of hard, metallic Si particles in the alloys gives the alloys the character of an in situ metal/matrix composite (MMC). Additions of Mg promote the precipitation of Mg_2Si in precipitation hardening heat treatments, creating one of the most widely used structural engineering alloys: Al-7Si-0.4Mg. Cu addition to Al-Si alloys similarly will precipitate and harden by the precipitation of $CuAl_2$ although the Cu addition impairs the corrosion resistance of the alloy. The Al-Si alloys, with perhaps Mg and/or Cu additions will feature significantly in this section.

In contrast to the two-phase Al-Si alloys, two alloys very different to the Al-Si types, essentially single-phase solid solution alloys, are also important:

2. The Al-Mg series, in which the Mg content may rise to around 10 wt% are reasonably strong ductile alloys with significant corrosion resistance, and capable of taking a bright white anodising coat much loved by the food industry. The choice of Mg levels is typically 5–6 wt%. The strongest Al-10 Mg alloy has been generally avoided over recent years as a result of people fearing rumours of it becoming brittle with age, particularly in warm conditions such as in the tropics or in engine compartments of cars. This is a travesty of the facts, and it is a pity that this excellent alloy has suffered such unjust criticism; it is true that it loses a little of its huge ductility after some years, possibly as a result of the precipitation of Mg_5Al_8 (or perhaps Mg_2Al_3) in grain boundaries (probably on bifilms—suggesting a remedy by eliminating bifilms). Even so, its generous reserves of ductility ensure that it always remains ductile, in fact, usually more ductile than most of the Al-Si alloys even when at their best. Much of its ductility arises as a result of its uniform single-phase solid solution structure (contrasting

sharply with the MMC structure of the Al-Si alloys). When liquid, the Al-Mg alloys react strongly with their environment, and usually require to have inhibitors in the mould to reduce metal/mould reactions if cast into sand.

3. The second important series of single phase solid solution alloys is that based on Al-Cu. Copper contents up to about 5.0 wt% are possible, with many alloys with a nominal composition around 4.5%. These alloys, when subjected to solution treatment and ageing can achieve high strength and ductility. A famous alloy, A201, contains approximately 0.7%Ag and is one of the strongest and toughest of all the cast Al alloys. Unfortunately, having silver lying about in the foundry introduces a security headache, adding to the daily problems the founder would rather avoid.

6.3.1 OXIDE FILMS ON Al ALLOYS

Considering first the reaction of liquid aluminium with oxygen, the equilibrium solubility of oxygen in aluminium is extremely small; less than one atom in about 10^{35} or 10^{40} atoms. This corresponds to less than one atom of oxygen in the whole world supply of the metal since Al was discovered and extraction began. Even allowing for the dynamic effects described in Chapter 1 in which the metal can have much higher levels of oxygen in solution, we can safely approximate its solubility to zero. Yet everyone knows that aluminium and its alloys are full of oxides. How is this possible? The oxides certainly cannot have been precipitated by reaction with any oxygen in solution. Oxygen can only reach and react with the metal surface. Furthermore, the surface can only access the interior of the metal if it is entrained or folded in. This is a mechanical, not a chemical process. The presence of oxygen in aluminium thereby has to be understood not in terms of chemistry but in terms of mechanical entrainment.

It makes sense therefore to start with an understanding of the reactions at the surface of the metal. For the process and mechanisms of oxidation, the reader is referred to the few original sources concerned with the oxidation of liquid aluminium alloys (Theile, 1962; Drouzy and Mascre, 1969). This short review of the oxidation of Al-based alloys is based on these works.

For the case of pure liquid aluminium, the oxide film forms initially as an amorphous variety of alumina that quickly transforms into a crystalline variety, gamma-alumina. These thin films, probably only a few nanometres thick, their thickness consisting of only a few molecules, inhibit further oxidation. However, after an incubation period the gamma-alumina in turn transforms to alpha-alumina which allows oxidation at a faster rate. Although many alloying elements in aluminium, including iron, copper, zinc and manganese have little effect on the oxidation process (Wightman and Fray, 1983), other alloys mentioned later exert important changes.

Figure 6.3 shows, approximately, the rate of thickening of films on aluminium and some of its alloys based on weight gain data by Theile (1962). The extremes are illustrated by the rate of thickening of Al-1.0 atomic %Mg at about 5×10^{-9} ms^{-1} that is more than 1000 times faster than the 1 atomic %Be alloy. Interestingly, Theile found that water vapour in an oxidising environment inhibited the rate of oxidation of Al (Figure 1.3). This finding seems to be in accord with shop floor experience of operating furnaces in which the incoming Al alloy charge is preheated by the spent furnace gas (flue gas necessarily containing much water vapour). Although Al melts would be normally expected to increase their hydrogen content in proportion to the water vapour in the environment, over a certain concentration of moisture the oxide film appears to become protective, inhibiting further uptake (but regrettably Cochran (1977) found this benefit does not extend to high Mg melts) so that high levels of hydrogen are not experienced in the melted metal from such furnaces.

The Al-Mg system is probably typical of many alloy systems that change their behaviour as the percentage of alloying element increases. For instance, where the aluminium alloy contains less than approximately 0.005%Mg the surface oxide is pure alumina. Above this limit, the alumina can convert to spinel, Al_2MgO_4. It is important to note for later reference that the spinel crystal structure is quite different from any of the alumina crystal structures. Finally, when the alloy content is raised to above approximately 2%Mg, then the oxide film on Al converts to pure magnesia, MgO (Ransley and Neufeld, 1948). These critical compositions change somewhat in the presence of other alloying elements.

In fact, the majority of aluminium alloys have some magnesium in the intermediate range so that although a pure alumina film forms almost immediately on a newly created surface, given time, it will usually be expected to convert to a spinel film.

The films have characteristic forms under the microscope. The newly formed alumina films are smooth and thin (Figure 2.10(a)). If they are distorted or stretched they show fine creases and folds that confirm the thickness of the film to be typically in the 20–50 nm range. The magnesia films are corrugated, as a concertina, and typically 10 times thicker (Figure 2.10(b)). The spinel films are different again, resembling a jumble of crystals that look rather like coarse sand paper (Figure 2.10(c)).

Rough measurements of the rate of thickening of the spinel film on holding furnaces show its growth to be impressively fast, approximately 10^{-9} to 10^{-10} kgm^{-2}s^{-1}. Although these speeds appear to be small, they are orders of magnitude faster than the rates of growth of protective films on solid metals. Because the oxide itself is fairly impervious, its rate of growth expected to be controlled by the rate of diffusion of ions through the oxide lattice, how can further growth occur quickly after the first layer of molecules is laid down?

It seems that this happens because the film is permeated with liquid metal. Fresh supplies of metal arrive at the surface of the film not by diffusion, which is slow, but by flow of the liquid along capillary channels, which is, of course, far faster. The structure of the spinel film as a porous assembly of oxide crystals percolated through with liquid metal, as coffee percolates through ground beans, may be an essential concept for the understanding of its behaviour.

We have already seen that progressive Mg additions to Al change the oxide from alumina, to spinel and finally to magnesia. A cursory study of the periodic table to gain clues of similar behaviour that might be expected from other additives quickly indicates a number of likely candidates. These include the other group IIA elements, the alkaline earth metals including beryllium, calcium, strontium and barium. The Ellingham diagram (Figure 1.5) confirms that these elements have similarly stable oxides, so stable in fact that in sufficient concentration, alumina can be reduced back to aluminium and the new oxide take its place on the surface of the aluminium alloy. The disruption or wholesale replacement of the protective alumina or spinel film can have important consequences for the melt.

In the case of additions of beryllium at levels of only 0.005%, the protective qualities of the film on Al-Mg alloy melts is improved, with the result that oxidation losses are reduced as Figure 6.3 indicates. Low-level additions of Be have been found to be important for the successful production of wrought alloys by continuous DC (direct chill) casting possibly because of a side effect of the strengthening of the film, causing it to 'hang up' on the mould or ladle, and so avoid entering the casting, as discussed earlier.

Attempts to measure the strength of films on Al alloys have been made from time to time (Kahl and Fromm, 1984, 1984; Syvertsen, 2006), but the measurements are not easy to make and the results seem of uncertain value at this time.

Strontium is added to Al-Si alloys to refine the structure of the eutectic in an attempt to confer additional ductility to the alloy. However, strontium has a significant effect on the oxidation behaviour of an Al-7Si-0.4Mg alloy as determined by Dennis et al. (2000). Strontium, as with magnesium, seems also to form a spinel, its oxide combining with that of aluminium to form $Al_2O_3 \cdot SrO$ alternatively written Al_2SrO_4. In addition, the resistance to tearing of the film is probably also increased, affecting the entrainment process. Because of this additional powerful effect on the oxide film, the action of Sr as an addition to Al alloys is complicated. It is therefore dealt with separately in Section 6.3.5.

Sodium is also added to modify the microstructure of the eutectic silicon in Al-Si alloys. In this case, the effect on the existing oxide film is not clear and requires further research. Sodium will have much less of an effect in sensitising the melt to the effect of moisture because it is less reactive than strontium. In addition, sodium is lost from the melt by evaporation because the melt temperatures used with aluminium alloys, typically in the range 650–750°C, approach its boiling point of 883°C. The wind of sodium vapour issuing from the surface of the melt will act to sweep moisture vapour and hydrogen away from its environment. Both the reduced reactivity and the vapourisation would be expected to reduce any hydrogen problems associated with Na treatment compared to Sr treatment, in agreement with general foundry experience.

However, Wightman and Fray (1983) found that all alloys that vapourise disrupt the film and increase the rate of oxidation. The additions they tested included sodium, selenium and (>900°C) zinc. The disruption of the film acts in opposition to the benefit of the wind of vapour purging the environment in the vicinity of the melt. Thus the total effect of these opposing influences is not clear. It may be that at these low concentrations of solute any beneficial wind of vapour is too weak to be useful, allowing the disruption of the film to be the major effect. However, the overall rate is in any case likely to be dominated by the reduced reactivity of Na with moisture compared with Sr.

Experience of handling liquid aluminium alloys in industrial furnaces indicates that the character of the oxide film is visibly changed when sodium, strontium or magnesium is added. For instance, as magnesium metal is added to an Al-Si alloy, the surface oxide on the melt is seen to take on a glowing red hue that spreads out from the point that the addition is made. This appears to be an effect of emissivity, not of temperature. Also, the oxide appears to become thicker and stronger. The beneficial effects found for the improved ductility of Al-Si alloys treated with sodium or strontium may be due not only therefore to the refined silicon particle size, as discussed in Section 6.3.5.

6.3.2 ENTRAINED INCLUSIONS

In aluminium alloy castings, the standard well known inclusions are isolated fragments of aluminium oxide. The fragments are too thick and chunky to be newly formed oxide, but are almost certainly particles carried over from the melting furnace. To be included in the alloy, they will have necessarily undergone an entrainment event so that they will themselves be wrapped in an envelope of new, thin oxide film. (Figure 2.3) If this entrainment was some minutes or hours before casting, there is a good chance that the defect can be regarded as extensively if not completely 'old' oxide, with its new envelope more or less welded in place over much of its surface as a result of continuing reaction with the entrained layer of air separating the wrapping from the granular oxide. Some sintering may also have occurred to supplement the welding process. Thus it will probably remain permanently compact, its welded patches inhibiting its relatively new envelope from re-opening in the mould cavity.

Such alumina inclusions are extremely hard. During machining, they are often pulled out of the surface, forming long tears on the machined face. The ease of pulling out is almost certainly the direct result of the inclusion being unbonded to the matrix as a result of its surrounding envelope acquired during its entrainment into the liquid. These rock-like inclusions cause the machinist additional trouble by chipping the edge of the cutting tool.

Carbides, and sometimes borides, are also reported for some Al alloys, but these appear to have been little researched. It seems likely in any case that the precipitation of such phases may occur on oxide bifilms, so, in a way, it is perhaps more of a bifilm problem than a carbon contamination problem. Only further research will clarify this situation.

Turning now to the subject of entrainment of the liquid surface, together with its oxide film, during surface turbulence: when the surface of the melt becomes folded in, the doubled-over films take on a new life, setting out on their journey as bifilms. The scenario has been discussed in some detail in Section 6.2 in the description of liquid metal as a slurry of defects.

As an overview of these complicated effects, Figure 6.5 gives an example of the kinds of populations of defects that may be present. This figure is based on a few measurements by Simensen (1993) and on some shop floor experiences of the author. Thus it is not intended to be any kind of accurate record; it is merely one example. Some melts could be orders of magnitude better or worse than the figures shown here. However, what is overwhelmingly impressive, and clearly shown, are the vast differences that can be experienced. Melts can be very clean (1 inclusion per litre) or dirty (1000 inclusions per cubic millimetre). This difference is a factor of a 1000 million. It is little wonder that the problem of securing clean melts has presented the industry with a practically insoluble problem for so many years. These problems are only now being resolved in some semi-continuous casting plants, and even in this case, many of these plants are not operating particularly well. It is hardly surprising therefore that most foundries for shaped castings have much to achieve. There is much to be gained in terms of increased casting performance and reliability. The mechanics of cleaning the melt, eliminating the oxides, and then ensuring that the oxides are not re-introduced before casting, is dealt with in detail in Chapter 10 and the following chapters.

6.3.3 GRAIN REFINEMENT (NUCLEATION AND GROWTH OF THE SOLID)

There are many reasons given for refining the grain size of castings. Yield strength and toughness are increased; microsegregated phases are more evenly distributed at grain boundaries—the susceptibility to hot tearing is reduced. The Al alloys are well behaved in this respect, so that the grain refining action is now thought to be relatively well understood and appears to be under good control. Even here, however, everything is not what it seems because the immense importance of bifilms, strongly influencing these processes, has been mainly overlooked.

The addition of titanium in various forms into aluminium alloys has been found to have a strong effect in nucleating the primary aluminium phase. It is instructive to consider the way in which this happens.

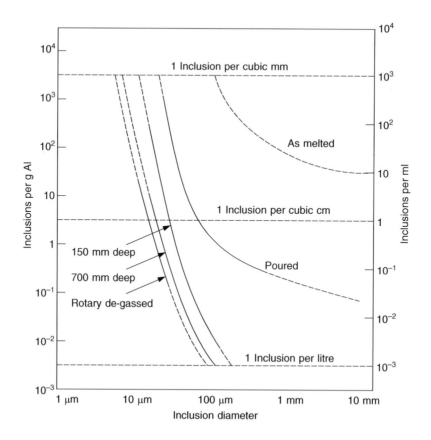

FIGURE 6.5

An example of inclusion content in an Al alloy.

Sigworth and Kuhn (2007) described a well-accepted theory in which the addition of Ti to Al alloys raises the melting point by 5°C. At the same time, the first phase to form is rich in Ti as a result of the relatively rare condition in which the partition coefficient is more than unity, causing solute to be preferentially absorbed by the growing solid rather than being segregated ahead. They point to elegant micrographic studies revealing Ti-rich dendritic shapes in the centres of refined grains as evidence of the correctness of the theory. All this experimental evidence is certainly in line with expectations from the phase diagram (Figure 6.6), but does not explain *refinement*.

The fundamental reason for the nucleation of grains appears to be the prior existence of $TiAl_3$ nuclei. There is no doubt that $TiAl_3$ is an active nucleus for aluminium because $TiAl_3$ is found at the centres of aluminium grains, and there is a well-established orientation relationship between the lattices of the two phases (Davies et al., 1970). These nuclei are stabilised by the huge undercooling of hundreds of degrees Celsius (Figure 6.6). At such high undercoolings, the $TiAl_3$ phase might even form by homogeneous nucleation. It seems quite probable that this could happen in the melt during the Ti addition, when the local Ti concentrations around the dissolving Ti addition start as extremely high Ti levels, well into the $TiAl_3$ undercooled region of the phase diagram before dispersing and falling to lower levels at which the $TiAl_3$ would be ineffective. Equally probably, the $TiAl_3$ particles are likely to be added ready-made in the grain refining addition.

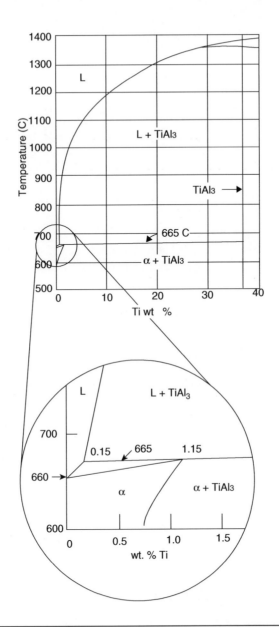

FIGURE 6.6

Binary Al–Ti phase diagram.

TiAl₃ in the liquid metal at a Ti concentration above about 0.15 wt% would be expected to be stable (Figure 6.6). However, many Al alloys contain Ti but at levels lower than this, so there is the danger that the TiAl₃ can disappear into solution in the melt. This is probably the mechanism for the 'fade' of the grain refinement effect with time.

Results of several researchers shown in Figure 6.7 illustrate that the effect of titanium in the grain refinement of aluminium starts at concentrations well below 0.10Ti. How titanium can be effective at concentrations anything lower

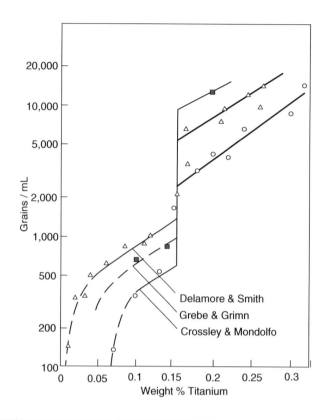

FIGURE 6.7

Increase in grain refinement with increasing titanium addition, especially at the peritectic 0.15Ti as found by several researchers.

than 0.15% has remained a mystery until the epoch-making research by Schumacher and Greer (1993, 1994). These researchers carried out their studies on an amorphous aluminium alloy as an analogue of the liquid state. Nucleation is more easily observed because the kinetics of reaction are 10^{16} times slower. Using TEM (transmission electron microscopy) they observed that $TiAl_3$ was present as adsorbed layers on titanium boride (perhaps more accurately, titanium diboride) (TiB_2) crystals, and so its existence was stabilised at lower levels of Ti than would be expected from the phase diagram, and it was thereby effective in nucleating aluminium for far longer periods as a result of the near-insolubility of TiB_2.

Effective grain refinement, however, seems to require more than simply the nucleation of new grains. A second important factor is the suppression of their growth. If growth is fast, grains can grow large before others have chance to nucleate. Conversely, if growth rates can be suppressed, greater opportunity exists for more grains to nucleate. For complex systems, where many solutes are present, the rate of growth of grains is assumed by Greer and colleagues (2001) to be controlled by the rate at which solute can diffuse through the segregated region ahead to the advancing front. They use a *growth restriction parameter* Q based on the concentration head of the front, defined as

$$Q = m(k-1)C_0 \tag{6.2}$$

Although they note that this relation should be modified by the rate of diffusion D of the solute, these factors are not well known for many solutes. They therefore assume that the rates of diffusion are fairly constant for all solutes of interest in aluminium (this is not far from the truth for the substitutional solutes shown in Figure 1.4(a)) Thus the

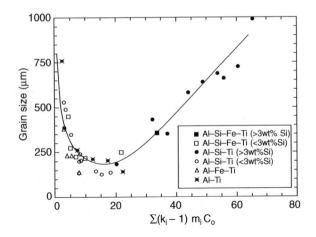

FIGURE 6.8

Effect of the growth restriction factor on the grain size of various Al alloys (Greer, 2001).

potentially more accurate summation of the effects of different solutes in solution, weighted inversely by their diffusivities as proposed by Hodaj and Durand (1997), is neglected in favour of a simple summation of Q values. The effect is shown in Figure 6.8. Initially, grain size clearly decreases with increasing total values of Q. With negligible values of growth restraint grains are seen to grow to nearly 1 mm in diameter. As restraint is increased grain size reduces to nearly one tenth of this, but can get no smaller. After this minimum, increasing Q now results in grain size returning to 1 mm.

The subsequent apparent growth of grains with increasing Q above about 20 is thought by Greer (2002) to be the result of the special effect of Si in 'poisoning' the grain refinement action of Ti at Si contents over 3%. The higher Q data are defined only by alloys with Si contents above 3%. For other solutes at high Q, particularly Cu, it is thought that the grain size remains small. Anyway, it seems that the attainment of fine grain size in Al-Si casting alloys has fundamental limitations, the attainable sizes being 5 to 10 times larger than those in some other casting alloys and in most wrought alloys.

Sigworth and Kuhn (2007) reported practical tests in Al-Si, Al-Si-Cu, Al-Cu, Al-Mg and Al-Zn-Mg alloys. They find a more confusing overall picture in which the grain refining response is different for each alloy system. With today's powerful Al-Ti-B refiners, they suggest there is no reason for large additions of soluble titanium in most alloys. In fact, they suggest it is more accurate to consider that we grain refine with boron, not titanium. The recommended addition is 10–20 ppm of boron, preferably in the form of Al-5Ti-1B or Al-3Ti-1B rod. Lower dissolved titanium levels provide better grain refinement.

However, copper-containing Al-Si casting alloys are the exception in their work. In alloys such as 319 or 355, they find it is best to have a minimum of about 0.1% Ti, which is of course in line with our earlier discussion in this section, but Chen and Fortier (2010) recommended that Ti should be near but not exceed 0.10% for all Al-Si alloys (see also Section 5.2.2). Clearly, we still have much to learn about optimising grain refinement.

Melt cleaning by grain refinement

We cannot leave the subject of grain refinement without adding a major caution to the reader. It is certain that the grain refining additives act to precipitate and grow on oxide bifilms which then become heavy and sediment to the bottom of the melt. The newly cleaned melt will then result in castings with improved properties. If the sediment is stirred in an (unfortunately misguided) effort to maximise the amount of grain refiner transferred to the casting then the bifilms with their precipitates of titanium-rich compounds will enter the casting, and properties are likely to be impaired. Thus it must be kept in mind that the major benefits of grain refinement are usually only marginally or even negligibly associated with the refining of grains, but mainly the result of the cleaning of the melt from oxide bifilms. This important aspect is discussed more fully in Section 9 on properties. The effect was elegantly demonstrated by Nadella, Eskin and Katgerman (2007) in which billets of continuously cast high strength 7075 alloy were found to crack on cooling. In an effort to reduce cracking, grain refiner was added to the metal flowing in the launder on its way into the mould, resulting in excellently grain refined ingots but still cracked. However, when the grain refiner was added to the ladle, permitting bifilms time to sediment, the ingots were grain refined but crack-free.

Other nucleation and growth effects are happening during the solidification of many Al alloys as a result of the many solutes that are present, both intended and unintentional.

6.3.4 DENDRITE ARM SPACING (DAS) AND GRAIN SIZE

In Figure 5.36, we can see the close link between freezing time and dendrite arm spacing (DAS), and in Figure 2.47 the close link between DAS and improved tensile properties. This is the reason, of course, for the strong motivation to chill castings to increase their strength and ductility.

In general, DAS is controlled by freezing rate alone (alloying has a minor effect that can usually be ignored). However, there is an interesting benefit to DAS from the action of grain refinement in removing bifilms from suspension. This occurs because the thermal conductivity of the melt increases when the bifilm barriers to heat flow are removed. Thus heat can be extracted more quickly from the casting, and DAS is consequently reduced.

In contrast, grain size is controlled mainly by the chemistry of grain refining additions, plus any mechanical action that will fragment grains, leading to an effect sometimes called grain multiplication. The combined and contrasting effects of DAS and grain size on mechanical properties are dealt with in Chapter 9.

6.3.5 MODIFICATION OF EUTECTIC Si IN Al-Si ALLOYS

The *modification* of Al-Si alloys is the refinement of the Si eutectic phase by the addition of a small amount of a *modifier* such as Na or Sr. Figure 6.9 illustrates the dramatic change in structure which can be achieved.

When Al-Si alloys are solidified the eutectic silicon is seen on polished sections to consist of coarse, sharp-edged plates (Figure 6.9(a)). These have usually been thought to be detrimental to mechanical properties, being assumed to act as crack initiators. For this reason, for alloys containing more than about 5–7% Si, the addition of sodium or strontium to the melt has been favoured to refine the eutectic silicon phase, converting, as if by magic, the coarse Si flakes into an attractively fine eutectic (Figure 6.9(b)). Modification usually benefits the mechanical properties. However, life is rarely so simple. This straightforward scenario is beset with contradictions and complications. We shall try to unravel some of these problems here as a salutary and instructive exercise in the influence of bifilms on behaviour of melts and cast products. The patience of the reader is requested while we plough through what appears to be an essentially simple mechanism, but clouded by a long history of misconceptions.

Hyper-eutectic Al-Si alloys

In *hyper-eutectic* Al-Si alloys, the primary silicon forms as compact polyhedral shapes. Here the nucleant is most definitely an aluminium phosphide, most probably AlP (although there is the possibility that the nucleus is actually AlP_3 as suggested by Dahle and co-workers, 2008). The particle has an epitaxial relation with Si as is clear from the images recorded by Arnold and Presley in 1961 (Figure 6.10).

The study by Cho (2008) confirmed that an aluminium phosphide (assumed here for simplicity to be AlP) is the preferred nucleant for Si, suggesting that when P is added to be present in sufficiently high concentration the AlP probably nucleates homogeneously in the melt. As small particles, the precipitating Si can wrap completely around them, therefore taking up a compact form.

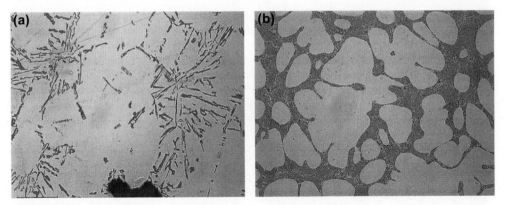

FIGURE 6.9

(a) An unmodified Al-Si eutectic in Al-7Si alloy; (b) the alloy modified by Sr addition.

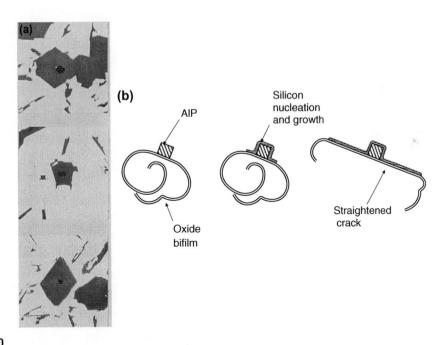

FIGURE 6.10

(a) Primary Si particles nucleated on AlP showing the clear crystallographic relationship (Arnold and Presley, 1961) and (b) the aluminium phosphide particle nucleating on an oxide bifilm. The subsequent nucleation and growth of Si straightens the bifilm, effectively extending and flattening the crack, thus reducing properties in this unmodified condition.

However, of course, at the same time the presence of a copious quantity of oxide bifilms will ensure that plenty of AlP particles will also precipitate on bifilms as favoured substrates for most intermetallics. The Si that precipitates on the nuclei attached to the oxides will not be able to wrap around completely because at least one side of the AlP will be firmly in contact with the oxide bifilm. Thus the Si will wrap around as far as it can, but subsequently proceed to grow out along the bifilm as a second favoured substrate, straightening the bifilm crack somewhat during its progress, and thereby progressively reducing mechanical properties. It seems that the oxide bifilm is not a sufficiently good substrate to encourage *nucleation* of the Si phase (although it seems just possible that this may occur if P concentration is sufficiently low), but contact with the Si reduces the overall energy of the combination sufficiently to encourage *growth*. The final alloy microstructure is now mixed, with some Si forming as particles wrapped around free-swimming AlP particles but much Si continuing to grow as plates on the substantial population of bifilms. This is typical of a hyper-eutectic Al-Si structure. Figure 6.11(c) shows the compact primary Si particles with surrounding Si primary platelets.

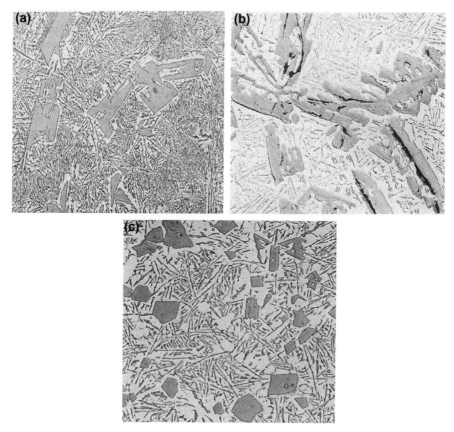

FIGURE 6.11

(a) The microstructure of a hyper-eutectic alloy, showing a mix of primary particles and unmodified eutectic silicon with relatively few AlP nuclei. (b) The cracks associated with growth on oxide bifilms can be clear, especially if solidification occurs unfed or with high hydrogen. (c) The hyper-eutectic alloy modified with additional phosphorus, as can be seen from the AlP nuclei in the centres of refined primary particles.

Please take note of this new observation that appears to be an important universal principle for new phases that grow freely in the liquid: *the new phase takes on the morphology of its substrate.* This disarmingly simple principle helps to explain many features of cast microstructures. Thus primary Si can adopt two quite distinct forms.

1. a *compact polyhedral particle* when it grows on AlP *particles*, and
2. a *plate-like form* when it grows on planar features such as *bifilms*.

(We shall see similar behaviour in other intermetallics and in the Fe-C eutectics.)

Hypoeutectic Al-Si alloys

Part of the action of modification was first explained by Flood and Hunt in 1981. They interrupted the solidification of unmodified and Na-modified Al-Si eutectic alloy by quenching partway through solidification to study the form of the growth front. Sections of their castings are seen in Figure 6.12. In the case of the unmodified alloy, they found the growth front appeared ragged, with some nucleation of eutectic phase apparently ahead of the freezing front, but after the addition of sodium, a smooth planar growth of the eutectic front was found. The effective length of solidification front was now a factor of 17 times shorter than the irregular front. Thus for the same given quantity of heat extracted by the mould over a given area, Flood and Hunt suggested that the interface in the case of the planar front would advance 17 times faster, and therefore have a much finer structure than the unmodified alloy.

The smoothing action of Na on the eutectic front has been confirmed by Crossley and Mondolfo (1966). In their excellent thoughtful, classic paper, they conclude that AlP as a nucleant is neutralised, and observe that instead of Si crystals taking the lead, growing ahead of the freezing front of the eutectic in unmodified alloys, but after modification,

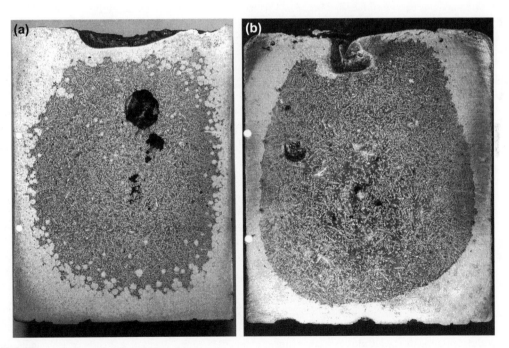

FIGURE 6.12

(a) A laboratory sample of unmodified Al-12Si alloy showing nucleation of eutectic ahead of the general solidification front; (b) the alloy modified with Na, showing suppression of nuclei in the melt, resulting in the advance of a planar solidification front. (Flood and Hunt, 1981).

FIGURE 6.13

Thin casting of Al-7Si alloy modified with Sr, its unfed condition reveals its cellular growth form.

Al dendrites take the lead, implying that the eutectic now grows at a lower temperature. This clear thinking and accurate reporting was well ahead of its time.

Smoothing of the solidification front by Na to give a uniform and strong casting skin is used to good effect in most gravity die casting (permanent mould) shops. As the operator takes each casting from the die, he will check it for any sign of the drawing-in of the casting surface at a hot spot. This local collapse occurs because of the sucking-in of the surface that in places will have liquid near to the surface as a result of the ragged form of the freezing front, thinning and weakening the outer surface of the casting. On finding such a 'draw' or 'sink', the operator will add sodium to the melt. The local collapsing of the surface is instantly cured, only to return some time later, after the sodium has evaporated from the melt. However, the operator is watching for this event, and when he judges that a sink has become almost unacceptable, he adds sodium once again, straightening the freezing front and thus strengthening the solidifying wall of the casting, and the surface perfection is renewed.

Interestingly, in contrast with the *planar* front resulting from the Na addition, the form of the freezing front for the Sr-modified eutectic alloy is usually *cellular* in the freezing conditions commonly found in castings. This fairly smooth modified front has been studied by Anson et al. (2000). The intermediate (cellular) condition between extremely ragged (roughly dendritic) and perfectly smooth explains the intermediate performance of Sr. The cellular pattern is seen occasionally on the surface of Sr-modified castings that have suffered some degree of poor feeding, encouraging the loss of some residual liquid from around the cell, and so giving slight depressions on a cast surface that outline the cell boundaries (Figure 6.13).

The cellular structure of the front may lead to problems of feeding Sr-modified castings as suggested by McDonald and co-workers (2004). The cellular growth creating possible feeding problems is contrasted with the smooth front resulting from modification by Na. Surface-initiated porosity is favoured by modification with Sr because of weak patches of the casting surface between the cells, whereas modification with Na creates a smooth, uniformly strong cast surface.

Nucleation of Si

Flood and Hunt appreciated that the ragged interface in the unmodified condition could be explained by nucleation of silicon ahead of the general solidification front, but they admitted that the nature of the nucleus was at that time unknown.

After the realisation that oxide bifilms were common in liquid Al alloys, it was clear that Si crystals could form on oxide films as is seen in Figure 5.28, and reported by Jorstad (2008). The detailed mechanism of their formation is crucial to our understanding of Al-Si alloys. Current evidence indicates that the bifilms may not provide a sufficiently favoured substrate to *nucleate* Si, but they are sufficiently effective to provide a favoured substrate for its *growth*.

The paper by Cho and co-workers (2008) on the effect of Sr and P on Al-Si alloys presented excellent experimental data, confirming the nucleation of Si on AlP nuclei. However, these authors overlooked the probability that the phosphides will themselves nucleate on oxide bifilms (in common with many other intermetallics that appear to favour precipitation on oxide bifilms). In this case, if Si particles now form on the phosphides as illustrated in Figures 6.10 and 6.11(c), then the subsequent continued growth of the Si will naturally follow the oxide substrate on which the phosphide particle is sitting, illustrated schematically in Figure 6.10(b).

Although at this time these authors dismissed the possibility of a double oxide substrate because no central crack can be discerned in the particles (as in Figure 6.11(b) by Ghosh and Mott, 1964), it has to be kept in mind that the central crack will only obligingly reveal itself if opened by shrinkage or gas. In a well fed and well degassed sample, the crack will remain firmly closed, and essentially invisible, as in Figure 6.11(a). It may appear mischievous to make an apparently perverse recommendation for the use of deliberately poorly fed castings for such research to make bifilms clearly identifiable, but it would help to clarify many issues. (The alternative use of higher gas levels to open and reveal bifilms will be less reliable because of the difficulty of control of hydrogen levels, and the rapidity of the changes of hydrogen content as a result of its rapid rate of diffusion either in or out of the sample. The use of shrinkage pressure, equivalent to a tension in the liquid, to open bifilms is subject to fewer uncertainties.)

A potentially important observation can be discerned in Figure 6.14(a) in the paper by Cho and colleagues (2008). A transmission electron microscope (TEM) image of P-rich particles surrounded by Al_2Si_2Sr are significantly, perhaps, arranged on an impressively straight line. This has all the hallmarks of being an oxide bifilm. This is an interesting possibility because bifilms are rarely imaged successfully in the TEM. The line is also characterised by fine pores (probably fragments of the interior unbonded interface of the bifilm) and by clear discontinuities in the strain bands, as would be expected across the plane of a crack. Because the original AlP precipitates occur before the general solidification of dendrites, it suggests that the alignment is the result of the AlP particles having formed on the planar substrate of an oxide bifilm (corroborating the increasing evidence that nearly all intermetallics seem to favour formation on oxide bifilm substrates).

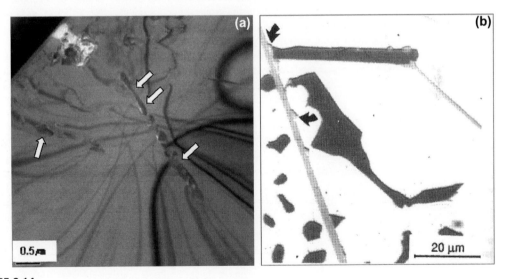

FIGURE 6.14

(a) TEM image by Cho (2008) of Al2Si2Sr phase containing P-rich particles (arrowed) and (b) SEM image by Pennors (1998) showing a row of AlP particles (arrowed) apparently on a β-Fe particle.

Courtesy F H Samuel (2009).

An optical image by Pennors (1998) from a quite different study shows a lineup of AlP particles on a bifilm, implied by Figure 6.14(b). The presence of the beta-Fe particle implies the prior location of an oxide bifilm. One can infer that the AlP particles arrived first, occupying one side of the bifilm and the beta-Fe particles arrived later, occupying the other side and straightening the bifilm, which all the time remains invisible. The alpha aluminium phase grows around the whole assembly later.

In many Al-Si alloys, P is already present in small concentrations. Because of the relatively low concentration of P, no free particles of AlP would be expected apart from those already precipitated on bifilms, favouring the subsequent growth of Si on bifilms, thus forming the typical plate-like primary Si particles that we know as the *unmodified* Al-Si structure (Figure 6.9(a)).

Ludwig and colleagues (2013) quantify the P addition and finds outcomes in line with our expectations. At low levels of P addition (0.2 to 0.5 ppm), the melt undercools 4.5°C and recalesces, achieving a fine, modified Si eutectic. At intermediate levels (2 to 3 ppm), the planar freezing front transitions to nucleation ahead of the front, creating equiaxed grains. At 20 ppm the undercooling is only 1.5°C, and the freezing is characterised by coarse flakes of Si and a great increase in nucleation rate of eutectic grains ahead of the front. Growth now takes place close to the normal 577°C temperature.

The formation of the *modified* Al-Si eutectic occurs as a result of the primary Si platelet formation being prevented. Cho (2008) found that AlP can be deactivated (i.e. poisoned) by the addition of Sr to the melt; the mechanism of deactivation being the precipitation of Al_2Si_2Sr on its surface. Thus all the microscopic nuclei of AlP attached to bifilms would then be deactivated, preventing any formation of Si on bifilms. Pandee and co-workers (2014) found exactly similar action by Sc in Al-Si alloys; the modification effect being the elimination of nucleant AlP by its conversion to the inactive ScP. (Note the closeness of Sr and Sc in the periodic table of the elements.)

(I had originally thought that the bifilms themselves acted as nucleating substrates, and the addition of Sr or Na caused modification by deactivating the bifilms as nuclei. However, the finding by Cho that the precipitation of Al_2Si_2Sr on AlP causes the deactivation of the AlP nucleant now appears to me to be convincing. Thus the concept of a hierarchy of nucleating substrates, first AlP, then oxide bifilms, then some lower temperature nucleus for the eutectic, now requires to be modified because bifilms appear to act only weakly or not at all as nucleation sites. Nevertheless, of course, the bifilms remain active as important growth substrates.)

When sufficient Sr has been added so there are no longer phosphide nuclei or oxide growth substrates that can operate to form Si particles or plates, a third process now comes into play. Si is forced to grow as a eutectic, probably nucleated on a boride such as CrB as suggested by Felberbaum and Dahle (2011), in the regions of significant undercoolings on or near the mould walls or near the melt surface. Because the easy nucleation mechanisms are now denied to Si, it is now forced to grow at a significantly lower temperature. At these greater 'undercoolings', the Si now takes the form of a regular, classical, fine, coupled eutectic, the *modified* Al-Si eutectic, as is usually observed (for instance, Glenister and Elliott, 1981).

Shamsuzzoha et al. (2012) researched Al-17Si alloy, finding that an addition of 3%Ba converts the eutectic to a modified form. Although the authors go to some lengths to invoke the effect of the solution of Ba in Si as some kind of explanation for the effect, it seems far more likely that Ba, an adjacent neighbour in the Periodic Table of the Elements, is simply acting precisely analogously as Sr. The expectation therefore is that the 3% level of Ba could be well over the top. Ultimately, however, the fascinating result is an alloy of excellent hardness, strength and toughness (160 HV, UTS 475 MPa and 5% elongation) and unusually high stiffness (85 GPa). The high Si alloys in the totally modified form would also be expected to be easily machinable. They clearly represent an untapped but impressive commercial opportunity.

Al-Si phase diagram

The *modified* or *non-modified* scenario can be described in terms of the Al-Si phase diagram in Figure 6.15. For normal melts of Al-Si alloys, which would all be expected to contain a generous population of bifilms, the eutectic freezing point would be 577°C, and the growth of the Si particles would follow the bifilms, forming large irregular plates. Interestingly, the eutectic composition appears to be not well defined, as might be expected with the melts freezing on rather variable, uncontrolled substrates. The diagram is based on the composition being 11.5%Si, although other works (Brandes and Brook, 1992) suggest 12.6%Si. This unmodified eutectic is marked 'U' in the figure.

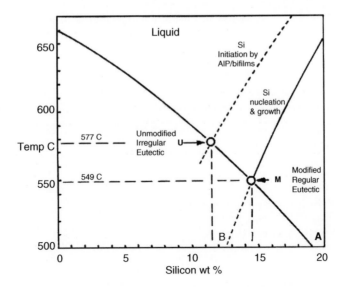

FIGURE 6.15

The Al-Si phase diagram showing the two eutectics, one at high temperature forming Si platelets on bifilms in suspension in the liquid and the other at low temperature as a classic coupled eutectic.

On addition of Sr, deactivating the AlP nuclei on the bifilms by coating them with Al_2Si_2Sr prevents Si nucleation so that the melt continues to undercool to the point at which the modified eutectic forms at point M. This point is located at $14.5 \pm 0.5\%Si$, as identified from the maximum fluidity clearly recorded by many studies of fluidity of modified Al-Si alloys (see Chapter 3), indicating an undercooling in the region of 28°C. This figure is subject to significant uncertainty because of the uncertainties in the compositions, and might therefore have a value somewhere between extremes of about 12–33°C. Whatever its exact value, it is a significant undercooling. In thermal analysis techniques, the undercooling is used to monitor the effectiveness of a modifying addition.

As an aside, it is probable that non-modified Al-Si alloys would also record a maximum fluidity in the region of 14.5% Si because the bifilms will be compacted during this period of rapid flow and will have little time to unfurl in the fairly rapid freezing associated with a fluidity test casting (Figure 3.18).

Although the reader with scientific interests will know that eutectic growth has been assumed to occur in a regime below the liquidus, or the extrapolation of the liquidus, at some point in a region skewed to the right under point U, in view of the new form of the phase diagram in which two eutectics are now clearly separated, it seems likely that eutectic growth actually takes place in the region MAB which is simply under both extrapolated liquidus lines. No complex argument to explain a skewed region appears to be necessary.

The consistent picture that emerges is of a succession of nucleation and growth processes operating at progressively lower temperatures. Summarising:

1. With adequate P addition, AlP particles may form freely in the melt and will nucleate Si particles (Figures 6.10 and 6.11(c)).
2. At the same time (and at more normal lower P concentrations), AlP particles will nucleate on oxide bifilms if present in suspension in the melt. These will form a favoured growth substrate for Si (Figures 6.10(b) and 6.11(c)), the growth starting at the AlP particle on the film; the eutectic freezing at U.
3. The addition of Sr coats the AlP nucleus with Al_2Si_2Sr and deactivates its nucleating potential. Such a surface mechanism, concentrating the action of the modifier onto microscopic areas, would explain why only parts per million of Sr in solution can be effective.

4. Thus neither AlP nor bifilms can operate to nucleate Si particles or grow Si plates. The melt will continue to cool without precipitation of Si until a new nucleus finally becomes effective to nucleate the eutectic phase at M. Felberbaum and Dahle (2011) found that borides such as CrB can nucleate the eutectic. The eutectic grows cooperatively with the Al phase on a substantially planar front (if modified with Na as in Figure 6.16(b)) or on a cellular front (if modified by Sr). The Si now takes on the form of a fine, fibrous eutectic seen deeply etched in Figure 6.17 and in its more familiar optical micrographic structure as a fine, rod-like or coral-like eutectic phase (Figure 6.9) that we know as a *modified* structure.

For (4), after modification the Si grows as a fine, fibrous eutectic, the inter-phase distance being controlled by diffusion in the liquid. A feature of all such eutectics that contain a phase that grows in a crystalline, faceted fashion is that this phase wishes to grow in a single direction, but is forced to continuously modify its growth direction because of the close proximity of its neighbours. The result is a fibrous, 'coral' structure resulting from the necessity to continuously generate growth defects, usually twins, to accommodate the enforced changes in direction (Figure 6.17). Thus such eutectic phases will naturally be expected to contain a high density of twin and other defects. This has been often reported, for instance as described by Steen and Hellawell (1975) and Song and Hellawell (1989), and in the absence of other explanations was once thought to explain the phenomenon of modification, even though no one could quite understand why. Contrary to all previous authors, we can conclude that the high density of growth twinning is not *a cause*

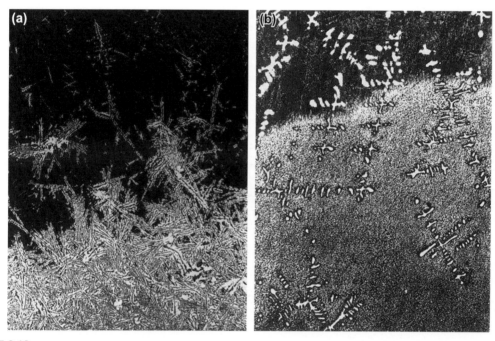

FIGURE 6.16

(a) Growth front of the high temperature unmodified Al-Si eutectic showing Si growth as platelets in the liquid phase; (b) the relatively compact planar front of the modified alloy growing as a classical coupled eutectic. (The eutectic front is seen growing through an open mesh of Al dendrites because this composition near the commonly accepted 12%Si is far from what appears to be a real location of the eutectic at nearer 14.5%Si.)

Courtesy Shu-Zu Lu and John Hunt (2010).

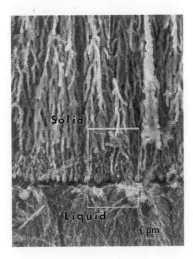

FIGURE 6.17

A magnified image of a quenched interface of the coupled eutectic (growing downwards in this image) illustrating the typical coral form of the silicon.

of the coral morphology but a natural *consequence*. The heavily twinned coral structure contrasts with the internal perfection of the primary crystals that grow freely, unrestricted in liquid ahead of the front.

Note that in contrast to the unmodified eutectic in which the Si is in the form of discontinuous particles which have had to repeatedly re-nucleate on separately swimming oxides, large regions of the modified eutectic nucleate only once in some undercooled region usually near to the mould wall, and becomes subsequently a continuously growing phase. Its main distinguishing feature is therefore its *continuous growth* leading to the development of *continuous lengths* of branching coral of Si. It grows as a 'classic' eutectic.

Formation mechanisms (1) to (2) explain the plate-like morphology and comparative crystallographic perfection of plate-like Si because at this stage of growth Si follows the extensive planar substrates which are sufficiently flexible to provide minimum mechanical constraint, so that the crystal grows essentially freely in the liquid. The growth of hypereutectic Si on AlP nuclei is another example of separated growth on particulate substrates. The eutectic morphologies that grew separated and necessarily not crystallographically related have been traditionally known as *irregular* or *anomalous* eutectics in contrast to *regular* or *normal* eutectics. The first has second phase particles that are irregular, whereas the normal eutectic, enjoying the coupled growth of both Si and Al controlled by diffusion, is highly ordered and predictable.

Practical application of Sr modification

In practice, the choice between the use of sodium or strontium for modification depends on the circumstances. The sodium is usually lost fairly quickly from a melt as a result of evaporation because the sodium is close to its boiling point. For instance, for a 200 kg crucible furnace a Na addition is lost within a time of the order of 15–30 min. Strontium, on the other hand, is a normal, stable, alloy in liquid Al alloys. Although it is slowly lost by oxidation at the surface, and sometimes can soak away into refractory furnace linings until they saturate, the alloy will usually survive several re-meltings.

The addition of strontium to the melt is an expensive option, but is taken in an effort to improve the ductility of the alloy. However, the results are not always straightforward to understand, and are accompanied by a number of problematical factors.

Effect of modification on mechanical properties

The precipitation and growth of Si on the originally crumpled and compact bifilm will force the straightening of the bifilm because the Si, as a diamond cubic lattice, has a distinctly favoured growth morphology, taking on the form of a plate. This is the reason that unmodified alloys are generally associated with poor properties—because the Si has become attached to the bifilm—so that its planar growth necessarily straightens the bifilm, forcing them to unravel, unfolding them to become serious planar cracks that can lower properties. The Si particle has always been assumed to be brittle, but the observed cracking behaviour is certainly associated with the presence of their internal bifilm cracks incorporated during their growth.

After modification the bifilms in the liquid no longer act as substrates, so they are no longer straightened, and remain in their original, crumpled, compact state. The mechanical properties now remain reasonably high. This appears to be the mechanism whereby modification tends in most cases to improve mechanical properties. To re-state this conclusion in different words, modification appears only to raise the properties of those alloys whose properties are already low because of the presence of compact oxides, but are lowered further by the growth of unmodified Si. Modification is a technique for avoiding the worsening of poor properties. We shall discuss the considerable evidence for this later.

Sr modification and porosity

Much confusion has existed over whether strontium can be successfully added without the deleterious effect of hydrogen pickup. Amid the confusion, the hydrogen already in solution in the strontium master alloy addition has often been blamed for this problem. However, we can, with some certainty, dismiss this minute source as negligible. What alternative possibilities remain?

Some have doubted that Sr leads to any increase the gas content of Al-Si melts. However, there are good fundamental reasons why Sr would be expected to lead to an increase of hydrogen in the melt, given the correct conditions of water vapour or other sources of hydrogen in the immediate environment. The enhanced reactivity of Sr will lead to enhanced reaction with environmental water vapour, leading to the oxidation of Sr and the release of hydrogen into the melt (Eqn (1.2)).

Even so, it is certainly the case that some foundries experience no problems because they happen to enjoy conditions for minimal uptake of hydrogen because of multiple factors:

1. low hydrogen content, usually associated with low humidity, of the local environment of the melt;
2. protection of the melt from its environment (as for instance in an enclosed low-pressure casting unit);
3. when undisturbed, the melt may continue to be protected from the environment by its oxide and create little additional surface oxide by reaction with water vapour or hydrocarbon gases with the associated release of hydrogen. Dennis and colleagues (2000) found that Al-7Si-0.4Mg alloy with 250 ppm Sr exhibited a short-term increase in the rate of oxidation, but over several hours exhibited a longer term dramatic decrease in rate. Thus melts subjected to repeated disturbance of the surface will react rapidly, whereas undisturbed melts will become practically inert. The stabilisation of the oxide by moisture as seen in Figure 1.3 is a further factor to strengthen this behaviour. However, disturbance of the melt surface by continuous bubble de-gassing would be expected to undermine such benefits.

The elegant research led by Gruzleski (1986) and Dinayuga (1988) employed minimal disturbance of the surface of the melt, and so minimised pickup of hydrogen, whereas Zhang et al. (2001) find a major uptake of hydrogen when regularly disturbing the surface by repeated sampling of their melt after Sr addition.

In an industrial furnace the stirring and ladling actions would be expected to fracture the protective oxide and so allow further reaction, encouraging the ingress of hydrogen in a furnace open to the atmosphere. In these conditions, the addition of strontium will usually be accompanied by an increase of hydrogen content of the melt. The hydrogen naturally increases with increasing strontium content, temperature, time and the presence of environmental water. This makes it practically impossible to add strontium without an increase in porosity in most foundry melting systems. In the experience of the author a single 0.05 wt% addition to a 1000 kg holding furnace caused the gas level to rise to such a high level that gas porosity caused all the castings during that shift to swell, solidifying oversize because of their generous pickup of gas, and had to be scrapped. It took 3 days of waiting for the melt to return to a castable quality. (The arrival of rotary de-gassing has, thankfully, eliminated such lengthy waits nowadays.) The absorption of hydrogen can, of course, be reduced by reducing the time available for absorption, for instance, by the casting of the whole melt immediately after the strontium treatment. Treating the strontium as a late addition in this way has been adopted successfully (Valtierra, 2001).

Strontium is, however, generally used with success in low-pressure casting furnaces where the melt is transferred immediately after treatment into an enclosed furnace that excludes any environmental moisture. It may then be held indefinitely, provided that the pressurising gas that is introduced into the furnace from time to time is dry or inert, and that the furnace lining is already saturated with strontium from previous additions.

So far we have examined the role of Sr in the increase of hydrogen content of the melt, leading to increased porosity. There appear to be additional factors at work.

A mechanism for the additional enhancement of porosity because of Sr has been put forward by Campbell and Tiryakioglu (2009). Briefly, their argument is as follows: After an Sr addition, the bifilms are no longer favoured substrates for the precipitation of Si. Thus the bifilms are no longer encapsulated inside the Si plates, but remain floating freely in suspension in the melt. These are now available for:

1. blocking interdendritic channels, reducing the permeability of the dendrite mesh practically to zero, as convincingly demonstrated by Fuoco and co-workers (1997, 1998) Figure 6.18(a). It is easily envisaged that piles of bifilms will be sucked into the dendrite mesh, where they will accumulate to block interdendritic channels. Furthermore, the bifilm at the base of the heap, experiencing the reduced pressure because of shrinkage, will open, effectively decohering from its other half because of its lack of bond, to create the start of a shrinkage pore (Figure 6.18(b)).

2. The precipitation of hydrogen into the central crack, expanding the bifilm into a pore (recall that pores are volume defects and cannot therefore be initiated by solidification alone; they have to be initiated from an entrainment defect such as a bubble or bifilm). Thus the freely floating bifilms can initiate the creation of hydrogen pores. Lui et al. (2003) affirm that oxides are responsible for pore formation in Sr modified alloys. Zhang (2001) describes how fissure-like pores gradually swell to become rounded after Sr addition, effectively describing the expansion of relatively large bifilms by the absorption of gas. Other authors only report the development of rounded pores immediately after the addition of Sr, indicating that their population of bifilms had a distribution of much smaller sizes, and thus less evidently initiating as cracks.

These bifilm actions leading to impairment of feeding, leading in turn to shrinkage problems, and their effectiveness as initiation sites for both shrinkage and gas pores, simply add to the already serious uptake of hydrogen possible under some conditions as we have seen. Thus the association of porosity with Sr modification has many aspects which have confused and resisted attempts at simplistic explanations for many years. The previous discussion, like much of foundry science, is seen to be non-trivial, and may yet be far from complete. It is hardly piety, but simply an admission, that at times it is necessary to respect the natural complexity of the real world.

Outside of the Al-Si system, the mechanism whereby Sr deactivates bifilm substrates for Si appears to be a standard mechanism applicable to other precipitates in other alloy systems. The approach explains for instance how the refinement of the structure of the Al alloys containing Mg_2Si precipitates is modified by the addition of lithium metal (Hadian, 2009).

Mixed experience with Sr

It is possible that modification, with its fine structure of the silicon phase in the Al-Si eutectic, may have better properties than the coarse unmodified structure. There are many results of published work that cite the benefits to strength and ductility following modification which have been thought to be the result of the finer Al-Si eutectic. However, despite eutectic refinement, some researchers have reported only mediocre improvements. Some have reported no benefit. A few have reported reductions in properties. This confusion of experience requires some examination.

Different foundries require massively different levels of Sr addition to achieve a useful measure of modification. This variable performance between foundries can be understood in terms of the deactivation by Sr of AlP and possibly oxide bifilms as favoured substrates for Si. Those with higher residual P contents, or possibly greater surface area of oxide per volume of melt, will require more Sr for deactivation.

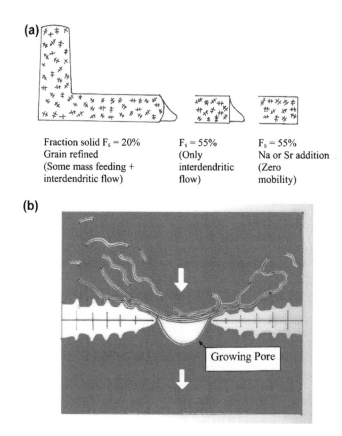

(a)

Fraction solid F_s = 20%
Grain refined
(Some mass feeding +
interdendritic flow)

F_s = 55%
(Only
interdendritic
flow)

F_s = 55%
Na or Sr addition
(Zero
mobility)

(b)

Growing Pore

FIGURE 6.18

(a) Results by Fuoco (1997/8) revealing that the small amount of interdendritic flow at 55% fraction solid was completely stopped by eutectic modification; (b) the probable reason being the release of bifilms that would choke flow and have the potential to initiate porosity.

Conventional foundries might typically require 300–500 ppm Sr to achieve modification. This is the situation in many small foundries and research laboratories where the processing of the melt in small batches makes melt quality practically impossible to control. Furthermore, probably all castings and test pieces will have been cast under gravity, mostly using relatively poor filling system designs, and thus their material will be expected to be impaired with new oxide films. For such poor melt quality, the action of Sr in preventing the straightening of a large proportion of large oxide bifilms is a major advantage, and benefits are likely to be experienced.

On the other hand, a good casting operation may achieve relatively low levels of oxides and so require treatment with much lower levels of Sr, perhaps in the region of 50 ppm.

For a more ideal casting operation, taking for instance the counter-gravity filling of sand moulds to cast engine blocks, such a system may have very few new films because of the relatively quiescent handling, and the possibility of gentle counter-gravity filling. Using such a process, Hetke and Gundlach (1994) found their castings suffered a reduction in strength and ductility after Sr addition, and porosity was significantly increased. Similarly, Byczynski and Cusinato (2001) re-confirmed this finding in the Cosworth Process foundry operated by Ford/Nemak in Windsor, Ontario. Almost certainly these near-ideal operations do not need to discourage the straightening of bifilms because there are few bifilms to straighten. In fact, these better foundries find that Sr impairs the mechanical properties of their castings, almost certainly because, with no favourable effects of Sr, only its unfavourable effects are now experienced. Any increase in the hydrogen content would cause the few bifilms to open by hydrogen precipitation, thus suffering an impairment mechanism because of inflation (instead of the other impairment mechanism of crystallographic straightening) and so increasing microporosity and reducing properties.

The pickup of hydrogen is more serious in sand foundries, as described previously, because there is a significant influx of hydrogen from the sand binder during mould filling, and the slower solidification gives greater time for the inflation of the bifilms by hydrogen diffusion. The hydrogen level will rise during the flow of melt into the mould because of the reaction with the sand binder in the same way that hydrogen increases in greensand moulds as observed by Chen (1994). In heavy sections, there will be several minutes for the hydrogen to diffuse into the casting sections. From Figure 1.4(a) the coefficient of diffusion of hydrogen is seen to be close to 10^{-6} m^{2}s^{-1}. From Eqn (1.5), for a typical time around 100 s, the average diffusion distance for hydrogen in aluminium is close to 10 mm. Thus the gas will easily penetrate the thickest sections of castings such as automotive cylinder blocks. The degradation of the sand binder raises the hydrogen level as the melt enters the mould, so increasing the damaging effect of the heightened gas level with maximum effectiveness.

In permanent mould foundries, the hydrogen influx from the mould is reduced (although it remains from any cores of course) and the time for expansion of the bifilms is reduced. Thus the significant disadvantages of Sr are greatly reduced, allowing the beneficial aspects to dominate. Thus Sr may now show a net benefit, and will particularly show a benefit if the melt has a poor quality (i.e. has a high residual P and high density of oxide bifilms). In this case, the amount of the addition will have to be high to be effective for such a poor melt, but the overall effect should be positive.

In summary, the most likely explanation of the action of Sr is that most operators using poorly designed gravity filling systems will benefit because the new bifilm defects introduced by pouring are prevented from forming plate-like silicon particles. Some will enjoy a useful net benefit if the increase of hydrogen can be controlled. In contrast, those operators using a process such as Cosworth Process will have few new films, and so will automatically achieve a high level of mechanical properties despite the automatically higher hydrogen level and longer freezing time in the sand mould. This high level of properties cannot be raised much further because there are practically no defects to deactivate, and will only suffer negatively from the problem of extra porosity. The refinement of the eutectic, much sought after by metallurgists and assumed to be the main mechanism of property enhancement, appears to have little effect either way. As we have noted before, the properties are not controlled so much by the microstructure, but are mainly under the influence of the defect population.

It seems inescapable therefore that the really important quality requirement should perhaps be the absence of bifilms. This means really clean metal and excellent designs of melt handling. If such conditions were achieved expensive additions of Sr would achieve nothing, and could be abandoned.

Summary of the modification mechanism

We can summarise the evidence, laying out the steps of the logic for the central role played by oxide bifilms; this is a long and detailed list of explanations of the different facets of the modification phenomena.

1. Bifilms are to be expected in our Al alloys because of the way our metals are handled in our cast houses and foundries.
2. The bifilm has two important morphologies: (a) a compact original state in which the bifilm has been crumpled by internal turbulence in the liquid, becoming a fairly harmless defect and (b) a second condition in which it may become straightened to become a serious planar crack.
3. All intermetallics studied so far (including Si, alpha-Fe, beta-Fe and Mg$_2$Si etc.) appear to precipitate and grow on bifilms. The cracked morphology of the intermetallic cannot be explained by solidification (which is incapable of creating such volume defects because of the effectively unbreakably powerful interatomic forces that hold the melt, the solid, and the interfaces in close atomic contact). The cracks can only originate by entrainment mechanisms (i.e. introduced externally into the liquid), explaining their necessary association with bifilms.
4. The known nucleation sites for Si are AlP particles. The AlP nuclei are most probably formed on oxide bifilms.

 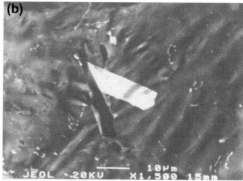

FIGURE 6.19

SEM image of an iron-rich particle nucleated on the underside of a thin alumina film imaged by (a) secondary electrons and (b) back-scattered electrons (Cao, 2000).

5. The Si particles are intermetallic with a diamond cubic crystal structure that forces its growth morphology to take up a plate-like form, attaching to the bifilm and therefore straightening the crumpled substrate into a flat planar bifilm. Thus the central bifilm crack now takes on the form of a large flat crack that has a serious effect on properties. The rather straight cracks sometimes observed in the centrescentres of primary Si and beta-Fe particles are thereby explained, together with the *apparent* 'weakness and brittleness' of these particles which in reality are expected to be strong and crack resistant (see also later in Figure 6.19).

6. The dilute addition of Sr (or Na) deactivates the AlP as a favourable nucleant for Si, and may consequently also deactivate oxide bifilms as favoured substrates.

7. After modification, early and easy nucleation of Si particles on bifilms can no longer occur. Thus the bifilms remain in suspension in the melt in their crumpled form (they are no longer straightened out by the plate-like growth of Si) so that properties are not degraded.

8. Having removed the oxide bifilms as nuclei, the next available opportunity for Si to form is as a eutectic at a significantly lower growth temperature, possibly nucleating on particles such as CrB. The eutectic spacing is now controlled by diffusion and so is significantly finer and controlled by the rate of heat extraction. Its coral-type morphology is a consequence of dense twinning, forced by the continuous change of direction of the Si fibres as a result of the constraint of neighbours.

9. The non-straightening of the bifilms (the retention of the bifilms in their original fairly harmless compact form) by the addition of Sr is expected to be the major benefit to mechanical properties, particularly increased strength and ductility. (The finer spacing of the eutectic is expected to be a minor contribution to any increased strength, as illustrated by the experiments of Cho and Loper (2000) who demonstrate with Bi additions how properties are unaffected by the modification of the structure of an A356 alloy.)

10. The addition of Sr has the potential disadvantage of increasing porosity in the casting. This effect is the result of (a) enhanced the reactivity of the melt, resulting in the danger of increased hydrogen pick-up in some circumstances and (b) the bifilms, no longer encased in Si, remain free-floating in the melt, where they can block interdendritic feeding and initiate porosity, although it is predicted that faster freezing and lower gas-content moulds can reduce these disadvantages.

Non-chemical modification

The bifilm hypothesis would predict that Al-Si melts completely free from oxide bifilms would be automatically modified without any necessity for additions. This is an interesting and testable proposition that appears in fact to have good support. Jian and colleagues (2005) used ultrasonics to drive bifilms out of suspension, causing modification without the aid of any chemical additions. Similar observations at the University of Windsor have been noted using electromagnetic stirring to centrifuge Al-Si melts clear of bifilms, producing a beautifully modified eutectic (Sokolovski, 2002). Wang et al. (2007), using what appears to be extremely clean Al-Si produced by direct electrolysis, found that a fully modified structure can be retained up to 16%Si. Above this level of Si, these authors find that primary Si is formed, but this does not seem to be a fundamental limitation, but was probably the result of the introduction of oxides because the additional Si was added using a master alloy containing 50%Si which would be certain to contain generous quantities of oxide.

It seems true that for very clean Al-Si alloys primary Si will be unable to precipitate, and the structure will therefore be automatically modified without the intervention of Na or Sr.

The experiments by Li and Xia (2014) are also fascinating. These researchers investigated the effects of high temperature melt treatment on Al-20%Si alloy in a scanning calorimeter. When increasing the temperature from 750 to 1000°C, they noticed a small exothermic peak at 951°C. Subsequently, they found the *eutectic* Si had been modified from a coarse flake to a fine coral form (although some finer but blocky *primary* Si is also present as would be expected from the equilibrium phase diagram but perhaps not expected compared with other non-equilibrium research results in this field). The authors suggested the mechanism was a heredity effect from the breakup of Si–Si atomic clusters in the melt when subject to high temperature. However, no evidence is presented in corroboration. It is tempting to assume that the effect is due to the change of oxide constituting the bifilms from gamma to alpha alumina, thus changing the effectiveness of the substrate, either discouraging the precipitation of AlP or discouraging the subsequent growth of Si. Similar effects of high temperature treatment of the melt are reported by Wang et al. (March 2014) and have been commented on by me (Campbell, September 2014).

6.3.6 IRON-RICH INTERMETALLICS

In addition to Si, it seems that most intermetallics also find oxides to be favoured substrates for growth. The iron-rich intermetallics alpha-iron ($Al_{15}Fe_3Si_2$) and beta-iron (Al_5FeSi) are included in this list.

Several authors have concluded that beta-iron first nucleates on AlP particles (Cho et al., 2008). If so, it would be expected, as in the case of Si nucleation, that the AlP particles would be already sited on oxide bifilms. Pennors and colleagues (1998) used elegant metallography to show AlP in a straight line, apparently attached to a beta-iron platelet (Figure 6.14(b)). Because the AlP appears to precipitate first, it seems reasonable to conclude that AlP must have aligned itself against a bifilm because no other extensive planar solids are known to be present at that stage of solidification. The beta-iron will arrive later, sited on the other side of the bifilm, and so, perhaps, in no way connected to the AlP. It seems most probable therefore that both AlP and beta-iron form on bifilms, but these are independent and unrelated events. They only appear to be related when the AlP forms on one side of the bifilm and the beta-Fe forms on the opposing side; because the bifilm is usually not easily seen, it appears that the AlP particles are sitting on the beta-Fe.

Cao and Campbell (2000) discovered that βFe plates in Al-Si alloys precipitated on the wetted outside surfaces of bifilms. Later work by these authors in 2003 indicated that a wide variety of oxide substrates might be effective, including $αAl_2O_3$, $γAl_2O_3$, $MgAl_2O_4$ and MgO. During the early stages of the growth of the precipitate, the βFe particle is sufficiently thin that it can follow the folds of the bifilm. On a fracture surface, the iron-rich phase can be clearly seen through the thin oxide film that represents one half of the bifilm (Figure 6.19). At this early stage, it is faithfully following the undulations of the oxide film. However, as the βFe particle thickens, the particle becomes increasingly rigid, taking on its preferred crystalline plate-like form, and so forcing the film to straighten. Finally, the bifilm is often seen as a crack aligned along the centre of the βFe particle (Figure 6.20(b)), or along the matrix/particle interface if the βFe happened to nucleate only on one side of the bifilm (Figure 6.21). The cracks are not always seen because the individual halves of the bifilm are usually extremely thin and are often tightly closed together. It usually requires some opening mechanism to come into play before the cracks are easily seen. Thus gas can inflate the bifilm, or shrinkage can pull it apart, as can stress applied during or after freezing.

The weakening effect of the straightened cracks seriously reduces the mechanical properties of the metal. *This* is the mechanism whereby iron is the most damaging impurity in Al alloys; it weakens the alloy as a result of its effectiveness in straightening oxide bifilms. It is important to remember that such inclusions as the βFe particle are probably extremely strong and resistant to fracture, in common with many intermetallics. As bears repeating, the observed cracks are not the

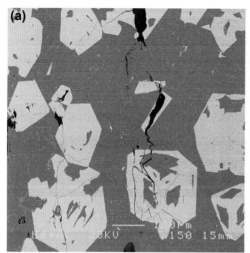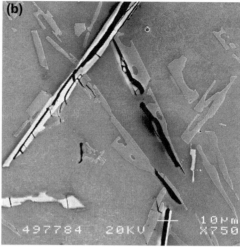

FIGURE 6.20

Fe-rich particles in an Al-12Si alloy growing on oxide bifilms, showing (a) alpha-Fe with its cubic symmetry wrapping around and sealing in situ compact bifilms, thereby maintaining properties; and (b) beta-Fe with its monoclinic crystal structure, and silicon particles with their diamond cubic structure, growing as platelets, flattening bifilms, thereby straightening cracks, and reducing properties. (Cao, 2000).

result of intrinsic brittleness but the result of the presence of the oxide bifilm. The strong effect of Fe in reducing the properties of Al alloys is additionally heightened because although the iron content in the alloy is low, the size of both alpha- and beta-iron inclusions is disproportionately large because their volume is increased fivefold by the content of aluminium, the formula having a 5:1 ratio in both $Al_{15}Fe_3Si_2$ and Al_5FeSi, respectively.

Wang and colleagues (2009) observed the formation of beta-Fe particles directly in a solidifying Al alloy using synchrotron radiation. They noted that the beta-Fe nucleates at a temperature between 550° and 570°C. This is a surprisingly large range for a well-characterised crystal, and strongly suggests that the problem lies not with the crystal but with the variability of its messy and fragmented oxide substrates, each one of which will be different.

Figure 6.20(b) demonstrates one of the convincing predictive successes of bifilm theory. When two oxide films originally impinge to form a bifilm, the film sizes will be random, never perfectly matched, so one film will always have a larger area than the other. Thus one film, which we call side 'A', will have to settle against its smaller partner 'B' by forming a series of small transverse folds on its side of the bifilm. When the βFe particles nucleate on A, they will only be able to grow as far as the nearest transverse fold which will form a barrier to further progress. Thus this side will have a series of short crystals separated by transverse cracks. Side B, the originally smaller side, will have no transverse folds and therefore will permit a continuous straight growth across its surface. The βFe particle is predicted therefore to be split by a central crack, but with transverse cracks on one side only, and a continuous crystal on the opposite side, exactly as seen in Figure 6.20(b).

Figure 6.21 illustrates different views of a bifilm, its presence inferred by the presence of a βFe plate. One side (note only one side) of the βFe plate, as a result of the plate having formed on only one side of the bifilm, has been inflated, growing away by gas precipitation or pulled away by shrinkage forces, opening a pore on one side of the βFe plate. Because it has been common to observe an association between pores and βFe particles, it has in the past been assumed that the βFe particles blocked the movement of feed liquid along interdendritic channels, and so caused shrinkage porosity. However, in view of the three-dimensional access routes for feed liquid and in view of the strong probability that pores probably cannot be formed without bifilms, any restriction of feeding is almost certainly less important than the presence of bifilms. An observation by Miresmaeili (2006) in which platelets of βFe were scattered and randomly

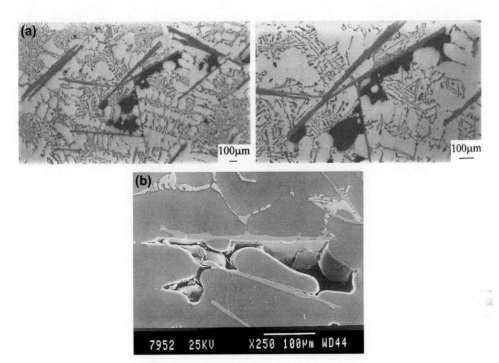

FIGURE 6.21

Platelets of βFe in an Al-Si alloy showing (a) pores opened by shrinkage or gas on only one side of each platelet; and (b) a 3-dimensional insight into the plate and pore geometry.

(a) Courtesy Cao (2000); (b) courtesy Samuel et al. (2001), Internat J Cast Metals Research.

oriented over a wide array of dendrites illustrated that the associated porosity at the sides of many of the plates were randomly oriented with respect to the dendrite direction and feeding direction. If the plates had been blocking the flow the pores should all have been sited 'down-stream' of the plates. Their random siting indicated that the association was not the result of blockage of feed metal, but more likely the result of the growth of a pore from the decoherence of the bifilm from the side of those plates formed on only one side of the bifilm.

Intermetallics with close-packed cubic symmetries (fcc, body-centred cubic) have such a high degree of symmetry that they often show little or no strong growth morphology; any direction of growth is as favourable as any other. The αFe phase has a hexagonal lattice which is close to cubic, so it can grow in practically any direction. For this reason, the αFe phase is usually seen to grow in a meandering fashion, following around the irregular form of the bifilm in its original crumpled state. In this way the compact, convoluted bifilm, with its associated central crack, is sealed in place, achieving a minimal impact on properties. This is why αFe as Chinese script is generally encouraged by metallurgists. On polished microsections the convoluted forms of αFe are often seen to contain a skeleton of convoluted cracks—the entombed trace of the originating crumpled bifilm (Figure 6.20(a)).

We might speculate that the form of Chinese script may arise, at least partially, from the original convoluted form of the bifilm, the αFe depositing on, and faithfully following, the original convolutions.

The conversion of the extended planar crack of the βFe particles (Figure 6.20(b)) into convoluted compact cracks inside the Chinese script particles (Figure 6.20(a)) is one of the key techniques for reducing the deleterious impact of iron impurity in Al alloys. There have been a very large number of research programmes to determine the best metallurgical solution to this problem. The common approach has been to use 0.5Mn addition for every 1.0Fe content (all in wt%).

Many other alloying solutions have been explored: Mahta and colleagues (2005) find the 1.0Fe level can be converted with 1.0Co or 0.33Cr. (Interestingly, this work indicates that primary Si in the form of compact nodules appears also to be stabilised by these additions, indicating that in addition to AlP there may be other effective nuclei for nodular Si based on transition metals.)

All of these alloying approaches to the control of βFe come with the built-in disadvantage that the Mn addition increases the total volume of unwanted intermetallic phases as pointed out by Crepeau in his 1995 review. Other elements have been used as listed by Crepeau. These include Be, Mo, Ni and S. This list is highly suspect; it suggests wildly different mechanisms. For instance, Ni may act as an alloying element similarly to the other transition metals. Mo seems unlikely to fit into this category and remains a mystery. Be will almost certainly not act in solution but as a significant strengthener of the oxide film on the melt, reducing entrainment of bifilms by a 'hanging up' mode during pouring, so reducing the total oxide bifilm content in the melt. Sulphur seems likely to be analogous to phosphorus and may provide or modify a nucleating substrate. Clearly, much more work remains to clarify this messy subject.

An alternative approach based on a further finding by Mahta may be important. The occurrence of the βFe phase appears to be particularly sensitive to the rate of solidification. Although more rapid freezing makes most features of the microstructure finer, there may be a reason for the βFe to be particularly sensitive. This is because for the particle to grow, the bifilm has to be mechanically unfurled as an extendable substrate against the viscous drag of the liquid. This means that the βFe can only form at relatively slow cooling rates. The αFe phase can also be suppressed by faster freezing as would be expected, but is less sensitive, requiring higher rates of freezing to reduce its formation.

It is not just Fe-rich intermetallics which are considered as exhibiting poor structure and properties. As just one instance of many, very many, Chen and colleagues (2013) investigated the intermetallics Al_7Cu_2Fe and T phase (AlZnMgCu) in high-strength Al alloys ascribed the loss of properties to these 'weak and brittle' phases leading to intergranular fracture along high angle boundaries and transgranular microvoid-induced fracture. The fallacy of such standard reasoning is examined critically in Section 6.3.7.

Si and Fe competition for sites

Caceres (2004) made a series of alloys of increasing Si and Fe contents. He found that as Si was increased the βFe particles were refined. Conversely, Liu (2003) noted that for some common Al-based casting alloys, higher Fe resulted in finer Si particles. This reciprocal relation between the formation of Si and βFe particles reflects that they are both competing for the same limited number of suitable substrates, as is clearly seen from their formation on bifilms in Figure 6.20(b).

Shabestari and Gruzleski (1995) found that Sr suppressed the precipitation of Si (presumably on bifilms) but that βFe particles still straightened and grew, but were slightly reduced in average length. This finding indicates that Sr is extremely effective in deactivating the precipitation of Si on the oxide substrate, possibly because of its action on the tiny AlP nuclei sitting on the oxide bifilms. The action of Sr is similar, but somewhat less effective for βFe, possibly because the βFe is nucleated directly on the oxide substrate, and this more extensive area may require deactivation by correspondingly more strontium. Mulazimoglu (1993) noted that Sr promotes the formation of more fragmented αFe particles in the form of a kind of Chinese script and coarsens the $CuAl_2$ phase in many copper-containing common Al alloys. His observations are again explained by the partial deactivation of the oxide substrates, resulting in fewer active substrates and consequently coarser precipitates.

Finally, it is significant that Caceres (2000) concluded that ductility of Al-Si alloys is controlled by the cracking of both Si and βFe particles. Both phases nucleate and grow on the oxide bifilms in suspension in the alloy. Thus both naturally contain similar cracks.

6.3.7 OTHER INTERMETALLICS

Miresmaeili and co-workers (2005), studying Al-7Si-0.4 Mg alloy with Sr additions, found Al_2Si_2Sr precipitated on both γAl_2O_2 and MgO films. In addition, alignments of the precipitates indicated an epitaxial relationship. The experimental technique was an admirable model of simplicity. The melts were poured into a small steel cup the size of an eggcup. The outside surface of the casting was then studied by scanning electron microscope (SEM). The intermetallics could easily be seen through the thickness of the oxide skin of the casting. Simple!

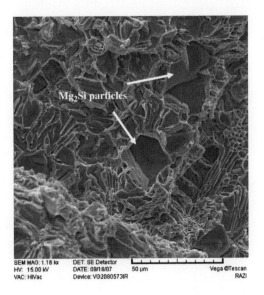

FIGURE 6.22

Fracture surface of an in situ Al-Mg$_2$Si intermetallic MMC showing the ductile failure of the upper intermetallic particle (Hadian et al., 2009).

There is some evidence that the CuAl$_2$ intermetallic in Cu-containing Al alloys also forms on oxide bifilms (Campbell, 2009). Thus the strong and tough Al-4.5Cu series of alloys, of which there are many, exhibit the CuAl$_2$ phase along grain boundaries. On careful examination, although it seems clear that the Cu segregates to the interdendritic regions around every dendrite arm, the CuAl$_2$ phase occurs only in some locations. It seems that the phase only precipitates when a bifilm is present; otherwise, the Cu merely remains in solution (Figure 6.23). (Although it is possible that the absence of the CuAl$_2$ phase is the result of a locally lower Cu segregation, this seems unlikely because the interdendritic segregation appears to be especially uniform.) A further important observation is that, probably as a result of its cubic symmetry, the CuAl$_2$ phase does not appear to have a strong straightening effect on the bifilm. Thus in contrast to Al-Si alloys in which the main alloying agent, Si, straightens the bifilms to reduce properties, this does not happen in Al-Cu alloys, explaining an important contribution to their excellent behaviour. (Another important contribution is its high Cu content creating the high precipitate density after heat treatment of course.)

Excess Ti in Al-Si alloys, often specified to be in the range 0.13–0.18%Ti, leads to the formation of TiAl$_3$-type intermetallics that form at temperatures well above the liquidus temperature of the alloy. These crystals can take up to 37%Si in solution, making a mixed intermetallic, usually written Ti(AlSi)$_3$. They take either a block or flake morphology (possibly as a result of the different sizes of bifilm on which they may nucleate). For optimum effect with modern grain refiners in most Al-Si alloys, including the popular Al-7Si types, Chen and Fortier (2010) recommended that 10–20 ppm B should be targeted, whereas the Ti level should be near, but not exceed 0.10%. The reduction of Ti reduces the content of the Ti-rich intermetallics.

The fracture resistance of intermetallics

The conventional description of intermetallics as brittle and weak does not stand up to critical appraisal. Although mentioned several times previously, it is worth devoting some space to some fundamental facts. It should be common knowledge that intermetallics are generally strong; in fact micrometre-sized intermetallic particles will be expected to be extremely strong. The great strength, hardness and crack resistance of carbides and borides are used worldwide for cutting tools for machining hard metals. It is also common knowledge that intermetallic alloys are currently being developed for turbine blade production and other ultra-high performance uses which require at least some toughness.

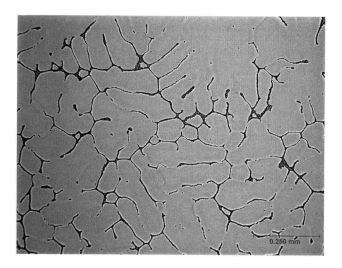

FIGURE 6.23

The as-cast Al-4.5Cu alloy showing CuAl$_2$ phase at many but not all interdendritic boundaries even though Cu segregation can be visually discerned in all boundaries.

Furthermore intermetallics are even likely to be ductile. There is evidence for this in occasional observations of intermetallics that have been plastically deformed (for instance, the Mg$_2$Si intermetallic particle observed by Hadian, 2009, Figure 6.22). Although carbides are renowned for their apparent hardness and brittleness, Yoshimi and co-workers (2014) studied two carbides of quite different structures, (Mo,Ti)C and (Mo,Ti)$_2$C, during high temperature deformation of Mo alloys. They observed significant plastic deformation of both carbides without any formation of cracks.

Ductile behaviour in hard crystals is well-known in physics; we need to divert to describe a simple but revealing experiment which deserves to be more universally known. A crystal of common salt is usually brittle when crushed in air, but when deformed in water becomes strong and ductile as a result of the dissolving away of surface defects that can lead to failure (Mendelson, 1962). All strong crystals deposited and grown in a solvent, and stressed while remaining in their solvent, are expected to be similarly strong and ductile. This is the case for all intermetallics grown in situ in metal matrices. It is to be expected therefore in Si and βFe particles formed during the solidification of Al alloys. Their apparent brittleness is therefore a deception arising from their formation on bifilms and therefore appearing to be cracked. We have all lived with this deception for years, and clearly it will take much unlearning.

Incidentally, we need to emphasise this behaviour is not to be expected of MMCs made by *mechanical mixing*; i.e. stirring intermetallic particles into a melt of the matrix. Such introduced particles naturally entrain the surface film on the melt during entry to the melt, as though stepping into a paper bag when entering. They are therefore effectively isolated from the matrix by the surrounding film and its associated layer of air. The entrained particle does not contact the melt (Figure 2.3) and effectively does not know that it is submerged in a metallic melt because it only experiences the layer of surrounding air. Such particles are thus always associated with, if not actually surrounded by, bifilm cracks. The attainment of ductility in such mechanically mixed MMCs is thereby fundamentally limited as seen in the work of Emamy et al. (2009). These submerged particles are equivalent to our grains of salt in air and are consequently truly brittle.

6.3.8 THERMAL ANALYSIS OF Al ALLOYS

The cooling curve of a solidifying alloy can be informative, usually in several ways. This is a potentially large subject, and although useful for many alloys (particularly cast irons), we shall deal only with Al alloys as an example of features that can be observed. For a variety of cooling curves the reader is recommended the excellent book by Backerud, Chai and Tamminen (1990) which gives clear curves for all common Al casting alloys. Readers are also recommended to the excellent review by Stefanescu (2015). In this short section, we shall only consider an Al-7Si alloy as a particularly simple example. This is shown in Figure 6.24(a).

In the first edition of this book it was assumed, in common with standard interpretations, that the cooling curve of temperature T versus time t shows the undercooling that is necessary to give the driving force to nucleate the alpha-Al dendrites. Strictly, this is an error. Figure 6.25(a) illustrates that the equilibrium freezing point is, of course, not shown on the cooling curve, but when inserted will often be found to lie above the dip in the curve. It is immediately clear therefore that undercooling will have occurred down to the nucleation start as found by construction from the first inflection of the first derivative curve and by extension of the alpha phase portion of the cooling curve. Only a single nucleation might occur at this point, so the dip in the cooling curve might only, in principle, correspond to the undercooling required to accelerate the growth of the nucleated solid. Thus, again emphasising in principle, the dip normally attributed to the undercooling required for nucleation may only correspond in fact to a condition of delayed heat evolution because of

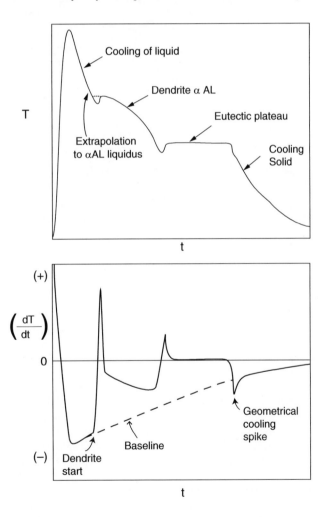

FIGURE 6.24

The thermal analysis of a simple Al-7Si alloy.

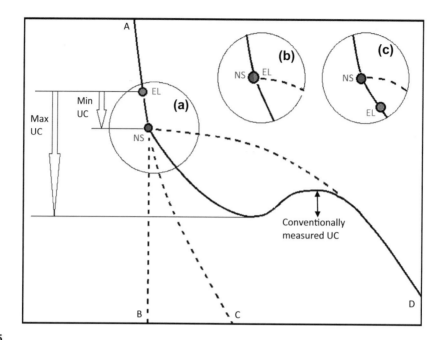

FIGURE 6.25

A cooling curve showing how the key measurement is sometimes incorrectly assumed to be the obvious dip in the curve. (a) The correct undercooling required below the equilibrium liquidus (EL) temperature to the nucleation start (NS) temperature; (b) illustrates zero undercooling condition; (c) shows nucleation on a highly favoured substrate before EL. The latter is common, for instance, for Si precipitation in Al-Si alloys and graphite in cast iron, where bifilms are present in the liquid.

slow growth. Perhaps more probably, however, is that additional nucleation is occurring during the downward portion of the dip, only stopping after the bottom of the dip when the rate of release of heat from the growth of the nucleated phase overcomes the rate of cooling, so that the temperature recalesces (literally 'reheats').

By the provision of effective grain refiners the initial minimum and dip undercoolings can both be reduced practically to zero (Figure 6.25(b)), and so can be used to assess the state of grain refinement of the melt prior to casting. Similarly, the undercooling necessary to initiate the eutectic phase, and the steady-state growth temperature can monitored to assess the state of modification of the eutectic. This 'plateau' is usually not so level in more complex alloys as a result of continued segregation during the course of the eutectic solidification, resulting in the 'plateau' taking on an increasing downward slope.

There has been occasional speculation among researchers that a sufficiently favoured substrate might cause the alpha phase to appear even above its equilibrium temperature. This seems to me to be quite possible, and is shown in Figure 6.25(c). It would be useful to identify such events.

The differential curve dT/dt versus time t is used by Backerud and others to show how the area between the cooling curve and the baseline can yield the heat evolved during the formation of that particular phase that is solidifying at the time. Thus if the heats of formation of the phases is known, the relative volumes of phases in the solidifying structure can be found. In this way, for instance, the quantity of Fe-rich phases in Al-Si alloys can be fairly accurately assessed.

Later developments of this technique to use the cooling curve to determine the quantity of the different phases occurring during freezing has been carried out by workers at the University of Windsor, Ontario, using a Newtonian method (Emadi, Whiting et al., 2004). More recently still, Chinese workers (Xu and Liu et al., 2012) have developed a clever technique employing extrapolations of the cooling curve itself to construct a baseline. These recent quantitative techniques agree excellently and are to be recommended.

A curious feature often observed at the end of freezing on a cooling curve is a 'spike' of cooling mysteriously diving down below the baseline. This does not appear to have been ever explained, but is simply the result of the rapid acceleration of freezing of the sample because of the geometrical effect of the residual melt dwindling to zero. In contrast with the rest of the thermal analysis record, it has nothing to do with any metallurgical reaction. It is the same effect as causes the 'reverse chill' in grey cast iron. It is quantified in Section 5.1.2 and Figure 5.10.

6.3.9 HYDROGEN IN Al ALLOYS

We turn now to the presence of hydrogen in aluminium. Hydrogen behaves quite differently to other solutes.

The aluminium-hydrogen system is a classic model of simplicity. The only gas that is soluble in aluminium in any significant amounts is hydrogen. (The magnesium-hydrogen system is similar, but rather less important in the sense that the hydrogen concentration increases only by a factor of about 1.4 on freezing, compared with a factor of about 20 for aluminium, so that dissolved gas in liquid magnesium is usually much less troublesome. Other systems are in general more complicated as we shall see later.)

Figure 6.26 is calculated from Eqn (1.4) illustrating the case for hydrogen solubility in liquid aluminium. It demonstrates that on a normal day with 30% relative humidity, the melt at 750°C should approach about 1 mLkg^{-1} (0.1 mL 100 g^{-1}) of dissolved hydrogen. This is respectably low for most commercial castings (although perhaps just uncomfortably high for aerospace standards). Even at 100% humidity, the hydrogen level will continue to be tolerable for most applications. This is

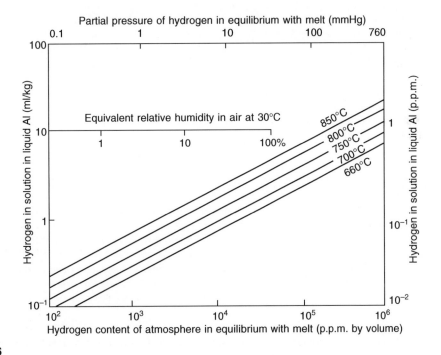

FIGURE 6.26

Hydrogen content of liquid Al shown as increasing with temperature and the hydrogen content of the environment as hydrogen gas or as water vapour.

the rationale for de-gassing aluminium alloys by doing nothing other than waiting. If originally high in gas, the melt will equilibrate by losing gas to its environment (as is also illustrated by the copper-based alloy in Figure 1.2).

Further consideration of Figure 6.26 indicates that where the liquid aluminium is in contact with wet refractories or wet gases, the environment will effectively be close to 1 atm pressure of water vapour, causing the concentration of gas in solution to rise to nearer 10 mLkg^{-1}. This spells disaster for most normal castings. Such metal has been preferred, however, for the production many non-critical parts, where the precipitation of hydrogen as dispersed porosity can compensate to some extent for the shrinkage on freezing, and thus avoid the problem and expense of the addition of feeders to the casting. Traditional users of high levels of hydrogen in this way are the permanent mould casters of automobile inlet manifolds and rainwater goods such as pipes and gutters. Both cost and the practicalities of the great length to thickness ratio of these parts prevent any effective feeding.

Raising the temperature of the melt increases exponentially the solubility of hydrogen in liquid aluminium. At a temperature of 1000°C, the solubility is over 40 mLkg^{-1}. However, of course, if there is no hydrogen available in its environment, the melt will not be able to increase its gas content, no matter what its temperature is. This self-evident fact is easy to overlook in practice because there is nearly always some source of moisture or hydrogen, so that, usually, high temperatures are best avoided if gas levels are to be kept under good control. Most aluminium alloy castings can be made successfully at casting temperatures of 700–750°C. Rarely are temperatures in the 750–850°C range actually required, especially if the running system is good.

A low gas content is only attained under conditions of a low partial pressure of hydrogen. This is why some melting and holding furnaces introduce only dried air, or a dry gas such as bottled nitrogen, into the furnace as a protective blanket. Occasionally, the ultimate solution of treating the melt in vacuum is employed (Venturelli, 1981). This dramatically expensive solution does have the benefit that the other aspects of the environment of the melt, such as the refractories, are also properly dried. From Figure 6.26, it is clear that gas levels in the melt of less than 0.1 mL/kg are attainable. However, the rate of de-gassing is slow, requiring 30–60 min because hydrogen can only escape from the surface of the melt, and takes time to stir by convection, and finally diffuse out. The time can be reduced to a few minutes if the melt is simultaneously flushed with an inert gas such as nitrogen.

For normal melting in air, the widespread practice of flushing the melt with an inert gas from the immersed end of a lance of internal diameter of 20 mm or more is only poorly effective. The useful flushing action of the inert gas can be negated at the free surface because the fresh surface of the liquid continuously freshly exposed by the breaking bubbles represents ideal conditions for the melt to equilibrate with the atmosphere above it. If the weather is humid the rate of re-gassing can exceed the rate of de-gassing.

Systems designed to provide numerous fine bubbles are far more effective. The free surface at the top of the melt is less disturbed by their arrival. Also, there is a greatly increased surface area, exposing the melt to a flushing gas of low partial pressure of hydrogen. Thus the hydrogen in solution in the melt equilibrates with the bubbles with maximum speed. The bubbles are carried to the surface and allowed to escape, taking the hydrogen with them. Such systems have the potential to degas at a rate that greatly exceeds the rate of uptake of hydrogen.

Rotary de-gassing systems can act in this way. However, their use demands some caution. On the first use after a weekend, the rotary head and its shaft will introduce considerable hydrogen from their absorbed moisture. Thus it is to be expected that the melt will get worse before it gets better. Thus de-gassing to a constant (short) time is a sure recipe for disaster when the refractories of the rotor are damp. In addition, there is the danger that if a vortex is formed at the surface of the melt, it may carry down air, thus degrading the melt by manufacturing oxides faster than they can be floated out. This is a common and disappointing mode of operation of a technique that has reasonable potential when used properly. The simple provision of a baffle board to prevent the rotation of the surface will suppress the vortex formation at reasonable rates of rotation. However, it is common to see such high rates of stirring that additional vortices are generated by the baffle board, making a bad situation worse.

High rates of loss of hydrogen during rotary de-gassing are entertaining because of the mass of flashes of blue flames of hydrogen burning as it is released from the melt surface. If there are no blue flames there is probably no hydrogen in the melt.

Despite these proper considerations surrounding the de-gassing of melts, a major effect of rotary de-gassing may not be the de-gassing. The action of greatest importance is most probably the floating out of the major oxide films resulting from the skins of charge materials. These oxides are fractions of square metres in area. It is essential to eliminate them if castings of any integrity are to be achieved. It seems likely that these large area films will be eliminated within the first few minutes. At that point the major benefit of rotary de-gassing may be over.

The danger that subsequently arises in continued operation is the generation of millions of small oxide bifilms from the bursting of small de-gassing bubbles as they arrive at the surface of the melt. As they release their contents of inert gas plus any flushed out hydrogen, the waft of fresh air across the newly opened surface will oxidise the surface. Thus as the sides of the opened bubble collapse together, they will create a new oxide bifilm. Thus the original massive bifilms are replaced by millions of small bifilms. The compositions are likely to be different too; the old films will be mainly spinels but the new films will be relatively pure alumina. Only later will the alumina films convert to spinel if some Mg is available in the alloy. This behaviour of rotary de-gassing is not fully researched and understood, so an optimum practical procedure is not easily recommended at this time.

One technique that has yielded good quality metal is the use of rotary de-gassing together with the use of a flux. Once again, rather than any significant effect on hydrogen removal, the action seems to be important for the wetting and elimination of oxide bifilms, or possibly the gluing shut of bifilms (keeping them permanently closed and therefore resistant to forming pores or cracks). Here too the recommended operational details are sketchy. The big Al producers of the world are known to have carried out their research, but keep the details a closely guarded secret.

The rate of pickup of hydrogen

When dealing with the rate of attainment of equilibrium in melting furnaces, the times are typically 30–60 min. This slow rate is a consequence of the large volume to surface area ratio. We call this ratio the 'modulus'. Notice that it has dimensions of length. For instance, a 10-tonne holding furnace would have a volume of approximately $4 \, m^3$, and a surface area in contact with the atmosphere of perhaps $10 \, m^2$, giving a modulus of $4/10 \, m = 0.4 \, m = 400 \, mm$. A crucible furnace of 200 kg capacity would have a modulus nearer 200 mm.

The values around 300 mm for large bodies of metal contrast with those for the pouring stream and the running system. If these streams are considered to be cylinders of liquid metal approximately 20 mm in diameter, then their effective modulus is close to 5 mm. Thus their reaction time would be expected to be as much as $300/5 = 60$ times faster, resulting in the approach towards equilibrium within times of the order of 1 min. This is the order of time in which many castings are cast and solidified. We have to conclude, therefore, that the melt will continue to equilibrate actively with its environment at all stages of its progress from furnace to mould.

There are methods available of protecting the liquid by an inert gas during melting and pouring which are claimed to reduce the inclusion and pore content of many alloys that have been tested, including aluminium alloys, and carbon and stainless steels (Anderson et al., 1989). It requires to be kept in mind that such techniques only protect the free surface of the melt. Hydrogen may continue to rise if moisture is present in the furnace refractories.

Ultimately, however, the gas in solution is normally not a problem to the castings of most metals and alloys (steels might be an exception because the exact mechanism of hydrogen embrittlement is not yet understood). If hydrogen can be persuaded to remain in solution, it appears to be normally perfectly harmless; it simply acts as an ordinary solute in solution, and therefore might provide some strengthening. Retaining the hydrogen in solution, but reducing oxide bifilms by such techniques as appropriate handling and treatment, and the elimination of pouring by, say, counter-gravity filling of moulds, hydrogen control becomes largely irrelevant.

6.4 COPPER ALLOYS

The ease of reducing copper from its ore has ensured copper and its alloys have enjoyed a long history, lasting thousands of years. Even so, copper and its alloys continue to be important at the present day for corrosion resistant cast products from domestic water fitting to high-strength marine applications in ships.

Pure copper castings are used for applications in which maximum thermal conductivity is required; a demanding application being blast furnace tuyeres. Similarly, pure copper is used in applications where maximum electrical conductivity is required, such as heavy electrical switch gear and bus bars (although dilute additions of cadmium or chromium are often used to increase strength without reducing conductivity too much). 'Pure' copper and its dilute alloys have a reputation for being particularly difficult to cast, the castings often suffering hot tears and porosity. This is almost certainly the direct result of poor casting technique, leading to the entrainment of surface oxides. The reader is recommended the sections of this work on casting manufacture for solutions to this common problem. The answer, as most often for castings, lies in the correct casting practice (for instance, the alloy composition and other metallurgical aspects being relatively unimportant).

The common brasses are red brass (70Cu30Zn) and yellow brass (60Cu40Zn), although many more complex and stronger brasses based on these alloys are also common.

Bronze alloys were traditionally solely alloys with tin (for instance bell metal is almost universally 77Cu23Sn), but nowadays aluminium bronze (5–10Al) and silicon bronze are important and widely used. Al bronze with additions of Fe, Ni and Mn is used for ships' propellers.

Gunmetals are alloys in which Zn has been added to a tin bronze. For instance 'Navy Gun Metal' was 88Cu8Sn4Zn. If lead is also added the alloy is known as 'leaded gunmetal'. One of the most famous of these is *ounce metal*, so-called because it was made up of a pound of copper, an ounce of zinc, an ounce of tin and an ounce of lead, also known as 85-5-5-5 metal. The lead was useful because of its action to seal pores in these very long freezing range alloys. Leaded brasses and bronzes are used for plain bearings, the lead- and tin-rich phases probably acting as high pressure lubricants. However, lead has been phased out of domestic water fittings, and is being steadily eliminated from nearly all foundry alloys as a result of the toxicity of its fumes and contamination of sand moulds during casting. Much exploratory work has been expended to investigate Si-containing alloys as a replacement in domestic water fittings (Fasoyinu, 1998).

6.4.1 SURFACE FILMS

Pure liquid copper in a moist, oxidising environment, causes water molecules to break down on its surface, releasing hydrogen to diffuse away rapidly into its interior. This behaviour is, of course, common to many liquid metals. The oxygen released in the same reaction (Eqn (1.2)), and copper oxide, Cu_2O, may be formed as a temporary intermediate product, but is also soluble, at least up to 0.14% oxygen. The oxygen diffuses and dissipates more slowly in the metal so no permanent film is created under oxidising conditions unless the solubility limit is exceeded. If the solubility limit is exceeded at the melt surface if the rate of arrival exceeds the rate of solution, a surface film of Cu_2O will build up. However, this will dissolve away if the rate of input of oxygen falls below the rate of dissolution of the film. In reducing conditions no film of any kind is expected. Thus pure liquid copper can be free from film problems in many circumstances. (Unfortunately, this may not be true or in the presence of certain carbonaceous atmospheres that may create a carbonaceous film, as we shall see later. Carbon-based films on copper and its alloys do not appear to have been researched so far.)

Certain Cu alloys, particularly the various aluminium bronzes that contain typically 5–10 wt% Al are quite a different matter. The great reactivity of Al with oxygen in the atmosphere and the high melting temperatures of these alloys combine to create conditions for the growth of a thick, tough and tenacious oxide that can give major problems if allowed to enter the casting.

6.4.2 GASES IN COPPER-BASED ALLOYS

Copper-based alloys have a variety of dissolved gases and thus a variety of possible reactions. In addition to hydrogen, oxygen is also soluble. These elements in solution (denoted by square brackets) can produce water vapour according to the reversible reaction:

$$2[H] + [O] = H_2O \tag{6.3}$$

Thus water vapour in the environment of molten copper alloys will increase both hydrogen and oxygen contents of the melt. Conversely, on rejection of stoichiometric amounts of the two gases to form porosity, the principal content of the pores will not be hydrogen and oxygen but their reaction product, water vapour. An excess of hydrogen in solution will naturally result in an admixture of hydrogen in the gas in equilibrium with the melt. An excess of oxygen in solution will result in the precipitation of copper oxide.

Much importance is often given to the so-called *steam reaction*:

$$2[H] + Cu_2O = 2Cu + H_2O \qquad (6.4)$$

This is, of course, a nearly equivalent statement of Eqn (6.3). The generation of steam by this reaction has been considered to be the most significant contribution to the generation of porosity in copper alloys that contain little or no deoxidising elements. This seems a curious conclusion because the two atoms of hydrogen are seen to produce one molecule of water. If there had been no oxygen present, the two hydrogen atoms would have produced one molecule of hydrogen, as indicated by Eqn (1.3). Thus the same volume of gases is produced in either case. It is clear therefore that the real problem for the maximum potential of gas porosity in copper is simply hydrogen. Depending on how much oxygen is present in solution, dissolved hydrogen will produce either a molecule of water vapour or a molecule of hydrogen. The volumes of gas are the same in either case.

(However, as we shall see in later sections, the presence of oxygen will be important in the nucleation of pores in copper, but only if oxygen is present in solution in the liquid copper, not just present as oxide. The distribution of pores as sub-surface porosity in many situations is probably good evidence that this is true. We shall return to consideration of this phenomenon later.)

Proceeding now to yet more possibilities in copper-based materials, if sulphur is present in solution, then a further reaction is possible:

$$[S] + 2[O] = SO_2 \qquad (6.5)$$

and for the copper-nickel alloys, such as the monel series of alloys, the presence of nickel introduces an important impurity, carbon, giving rise to an additional possibility:

$$[C] + [O] = CO \qquad (6.6)$$

Systematic work over the past decade at the University of Michigan (see, for instance, Ostrom et al. (1981)) on the composition of gases that are evolved from copper alloys on solidification confirms that pure copper with a trace of residual deoxidiser evolves mainly hydrogen. Brasses (Cu-Zn alloys) are similar, but because zinc is only a weak deoxidant the residual activity of oxygen in solution gives rise to some evolution of water vapour. Interestingly, the main constituent of evolved gas in brasses is zinc vapour because these alloys have a melting point above the boiling point of zinc (Figure 1.6). Pure copper and the tin bronzes evolve mainly water vapour with some hydrogen. Copper-nickel monels with nickel above 1% have an increasing contribution from carbon monoxide as a result of the promotion of carbon solubility by nickel.

Thus when calculating the total gas pressure in equilibrium with melts of copper-based alloys, for instance inside an embryonic bubble, we need to add all the separate contributions from each of the contributing gases.

The brasses represent an interesting special case. The continuous vapourisation of zinc from the free surface of a brass melt carries away other gases from the immediate vicinity of the surface. This continuous out-flowing wind of metal vapour creates a constantly renewed clean environment, sweeping away gases which diffuse out of the melt, carrying them from the alloy surface and preventing contamination of the local environment of the metal surface with furnace gases or other sources of pollution. For this reason cast brass is usually found to be remarkably free from gas porosity.

For alloys with more than 20%Zn, there is sufficient zinc vapour to burn in the air with a brilliant flame known as zinc flare. Flaring may be suppressed by a covering of flux. Similarly, aluminium is often added to brasses to reduce the loss of zinc, probably as a result of the formation of a dense alumina film. However, the beneficial de-gassing action is thereby suppressed, raising the danger of porosity, mainly from hydrogen.

The boiling point of pure zinc is 907°C. But the presence of zinc in copper alloys does not cause boiling until higher temperatures because, of course, the zinc is diluted (strictly, its activity is reduced). Figure 6.27 shows the effects of increasing dilution on raising the temperature at which the vapour pressure reaches 1 atm, and boiling occurs. The onset of vigorous flaring at that point is sufficiently marked that in the years before the wider use of thermocouples foundrymen used it as an indication of casting temperature. The accuracy of this piece of folklore can be appreciated from Figure 6.27,

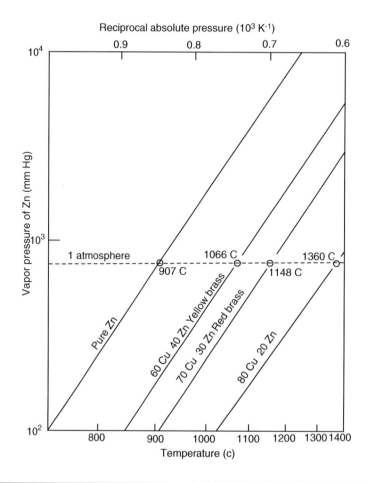

FIGURE 6.27

The vapour pressure of Zn and some brasses.

in which it is clear that the flaring temperatures increase in step with the increasing copper contents (i.e. at greater dilutions of zinc), and thus with the increasing casting temperatures of the alloys.

Around 1% of zinc is commonly lost by flaring and may need to be replaced to keep within the alloy composition specification. In addition, workers in brass foundries have to be monitored for the ingestion of zinc fumes.

Melting practice for the other copper alloys to keep their gas content under proper control is not straightforward. Following are some of the pitfalls.

One traditional method has been to melt under oxidising conditions, thereby raising the oxygen in solution in the melt in an attempt to reduce gradually the hydrogen level. Before casting, the artificially raised oxygen in solution is removed by the addition of a deoxidiser such as phosphorous, lithium or aluminium. The problem with this technique is that even under good conditions the rate of attainment of equilibrium is slow because of the limited surface areas across which the elements have to diffuse. Thus in fact little hydrogen may have been removed. Worse still, the original oxidation has often been carried out in the presence of furnace gases, so raising oxygen and (unwittingly) hydrogen levels simultaneously (Eqn (1.2)) high above the values to be expected if the two dissolved gases were in equilibrium. The addition of deoxidiser therefore may leave hydrogen at near saturation.

The further problem with this approach is that the deoxidiser precipitates out the oxygen as a suspension of solid oxide particles in the melt, or as surface oxide films. Either way, these by-products are likely to give problems later as non-metallic inclusions in the casting, and, worse still, as bifilm initiation sites to assist the precipitation of the remaining gases in solution, thus promoting the very porosity that the technique was intended to avoid. In conclusion, there is little to commend this approach.

An additional problem should be noted with respect to the relatively common practice of deoxidation with phosphorus. Gunmetals typically require only 0.01%P and Sn bronze typically 0.2%P (French, 1957). If excess is used a liquid Cu_3P film forms on the top of the melt. If this is poured with the alloy, it penetrates the mould leading to impaired surface finish and possibly entrained phosphide liquid.

A second reported method is melting under reducing conditions to decrease losses by oxidation. Hydrogen removal is then attempted just before casting by adding copper oxide or by blowing dry air through the melt. Normal deoxidation is then carried out. The problem with this technique is that the hydrogen-removal step requires time and requires the creation of free surfaces, such as bubbles, for the elimination of the reaction product, water vapour. Waiting for the products to emerge from the quiescent surface of a melt sitting in a crucible would probably take 30–60 min. Fumes from the fuel-fired furnace would be ever-present to help to reverse any useful de-gassing. Clearly, therefore, this technique cannot be recommended either!

Marin and Utigard (2010) described a variant of this technique in which they directed a flame of mixed O_2 and CH_4 (methane) on to the surface of the melt. Higher ratios of CH_4/O_2 gave higher reducing conditions that raised the rate of deoxidation of the copper. However, they noted that hydrogen is simultaneously increased by this practice, eventually becoming so high that the melt 'boils' with the evolution of water vapour by reactions as in Eqn (6.3). Boiling may now flush the melt to low levels of hydrogen, but the surface turbulence of the boil may now increase reaction with the air, increasing oxygen once again. All this resounds of lack of control.

A less dramatic and more straightforward technique involves a simple cover of granulated charcoal over the melt to provide the reducing conditions. This is a genuinely useful way of reducing the formation of drosses (dross is a mixture of oxide and metal, so intimately mixed that it is difficult to separate) as can be demonstrated from the Ellingham diagram (Figure 1.5), the traditional free energy/temperature graph. The oxides of the major alloying elements copper, zinc and tin are all reduced back to their metals by carbon, which preferentially oxidises to carbon monoxide (CO) at this high temperature. (The temperature at which the metal oxide is reduced, and carbon is oxidised to CO, is that at which the free energies for the formation of CO exceed that of the metal oxide, i.e. CO becomes more stable. This is where the lines cross on the Ellingham diagram.)

However, even here, it is as well to remember that charcoal contains more than just carbon. In fact, the major impurity is moisture, even in well-dried material that appears to be quite dry. An addition of charcoal to the charge at an early stage in melting is therefore relatively harmless because the release of moisture, and the contamination of the charge with hydrogen and oxygen, will have time to be reversed. In contrast, an addition of charcoal at a late stage of melting will flood the melt with fresh supplies of hydrogen and oxygen that will almost certainly not have time to evaporate out before casting. Any late additions of anything, even alloying additions, introduce the risk of unwanted gases.

Reliable routes for melting copper alloys with low gas content include:

1. Electric melting in furnaces that are never allowed to go cold.
2. Controlled use of flaring for zinc-containing alloys.
3. Controlled dry environment of the melt. Addition of charcoal is recommended if added at an early stage, preferably before melting. (Late additions of charcoal or other sources of moisture are to be avoided.)

In summary, the gases and vapours which can be present in the various copper-based alloys are:

Pure copper	H_2, H_2O
Brasses, gunmetals	H_2, H_2O, Zn, Pb
Cupro-nickels	H_2, H_2O, CO, (N_2?)

Various tests have been proposed from time to time for copper-based alloys. Dion (1979) recommended a test for oxygen by simply dipping a carbon rod into the melt. If the oxygen is high, the rod 'sings' with the rapid evolution of microscopic bubbles of CO, producing a vibration that can be sensed by hand. This test can detect oxygen in solution down to a limit of about 0.01% at which the rod stops singing. The rod needs to be at the temperature of the melt because a cold graphite rod evolves gas. This raises the potential problem, of course, of the probability of moisture in the graphite, although it is possible that the test itself will assist to eliminate this providing the rod is reasonably dry. If the rod is dry it will also contribute to the flushing out of any hydrogen.

The total sum of gases has often been assumed to be testable using the Reduced Pressure Test exactly similar to that used for Al alloys. Studies of the test are reported by numerous workers, including Matsubara (1972) and Ostrom (1974, 1976). The test is also described in the ASM Handbook 2008 vol. 15. As with experience in Al alloys, the test has proved controversial and confusing for the same reason: the test is also highly sensitive to the presence of oxide bifilms. In reality, the test would be a better test of oxides than a test of gas, but with intelligent use, should give a useful indication of both.

Sub-surface pinholes in copper base alloys have been widely researched. One of the more thorough reports is by Fischer (1988). The phenomenon appears to be exactly analogous to that observed in Al alloys. The pores arise as a result of a combination of factors, naturally leading to widespread confusion in the literature. The key factors include the following.

1. The gases in solution in the melt (of which, of course, there are several in copper alloys).
2. The gases released by the mould binder material and which diffuse into the casting surface.
3. The build-up of gases ahead of the advancing freezing front, raising local concentration of the total gas content after a distance of 1–2 mm (i.e. just sub-surface).
4. The presence of a population of oxide bifilms in suspension in the melt as a result of poor melting practice or poor casting practice. The bifilms act as initiation sites and are themselves pushed ahead of the front, so being in precisely the correct location for maximum effect of pore generation. The variation of difficulty of unfurling of the bifilms will give a spectrum of pore initiation times, giving a mix of round and dendritic pore morphologies corresponding to early and late arrivals respectively (see Figures 7.40 and 7.41).

6.4.3 GRAIN REFINEMENT

There has been much practical work carried out on the grain refinement of Cu-based alloys, but, it seems, little fundamental research so far.

It was Cibula in 1955 who used iron and cobalt borides, plus zirconium carbide and nitride, to grain refine bronzes and gunmetals. As is usual with such work, hot tears were found to be reduced although porosity became more evenly dispersed or at times concentrated as layer porosity. Finer grains also appeared to be associated with reduced ductility and a lack of pressure tightness. These apparently perverse results are almost certainly the result of a high oxide bifilm population in his cast material which is only to be expected from the awful casting techniques in general use at that time. Techniques such as those described later in this book should greatly assist to reduce these defect levels, and reverse such effects of loss of ductility and pressure tightness.

With similar poor casting methods, Couture and Edwards (1973) used Zr master alloys to successfully grain refine a variety of gun metals, finding improved strength and hot tear resistance, but confirmed Cibula's finding of a drastic reduction in elongation and pressure tightness.

Gould (1960) studied the grain refinement of pure copper, and found additions of Li and Bi to be effective, but many additions that were tried were of no use.

Clearly, copper alloys would benefit from additional work to understand the fundamentals of grain refinement. In addition, casting with improved filling system designs should help to remove the current confusion relating to loss of ductility and pressure tightness that has accompanied early work.

6.5 CAST IRON

Cast irons are nature's gift to foundrymen. They melt at accessible temperatures, requiring relatively low energy, they run like water, are strong enough for many engineering applications and are relatively insensitive to poor casting techniques and so yield adequate castings despite sloppy and out-dated casting practices.

The commercial exploitation of simple grey irons was a major factor triggering the start of the industrial revolution, but the current spectrum of different cast irons is now dauntingly broad. It stretches from grey irons (so-called because of their fracture surface is coloured grey from the graphite flakes that provide the fracture path) through compacted graphite irons, spheroidal graphite (known as ductile) irons, to white irons (the apparently brittle iron carbides providing a bright, white fracture surface) for wear resistance. Highly alloyed irons are commonly used for heat and oxidation resistance. Special heat treatments, particularly austempering, can provide high-performance irons with strengths approaching 1 GPa—the traditional reserve of high-strength steels.

For cast irons, carbon is the key alloying element. In pure binary Fe-C alloys, the minimum carbon content for a cast iron to distinguish it from a steel is close to 2%, but the liquidus temperature at this carbon content is high at 1290°C and the long freezing range of about 150°C and limited graphite content would make feeding difficult. At a carbon content of approximately 4.3%, the alloy is of eutectic composition and the freezing temperature is comfortably low at 1150°C with negligible freezing range. This is also close to the composition at which the expansion—because of graphite precipitation—almost exactly counters the contraction resulting from austenite solidification, conferring on eutectic irons the enormous benefit of requiring little if any feeding. Beyond the eutectic composition, the liquidus temperature rises once again and the precipitation of primary graphite prior to general solidification at the eutectic temperature means that graphite flotation in these hyper-eutectic irons can become a problem, limiting carbon contents usually to a maximum close to 4.5%.

With silicon as the next most common element in grey irons, equivalent structures can be achieved with reduced carbon, trading off C against increased Si. It is usual to quantify this relation by an *effective* carbon content, known as the carbon equivalent (C_E) or sometimes the carbon equivalent value (CEV). This is approximately

$$C_E = \%C + \%Si/3$$

A typical engineering quality of grey iron would consist of around 3.2C and 2.0Si, to give a C_E close to 3.9%C, just slightly hypoeutectic. In general, grey irons contrast with ductile irons which have a C_E commonly of 4.2–4.4%.

In the days of making elaborate ornamental cast iron ware, or for certain wear-resistance applications, 1–2% of phosphorus was often added. This addition increased fluidity of the melt (Figure 3.13) and the hard phosphide phases added wear resistance to such applications as brakes. The %P acted similarly to the Si in its replacement of effective carbon, so that their combined effect gave approximately:

$$C_E = \%C + (\%Si + \%P)/3$$

The reasonable cost, huge availability and attractive machinability of the graphitic irons, especially grey irons, guarantees the prosperous future of these alloys (despite all criticisms of its weight compared to Al alloy castings for instance).

Having said all this, it is true that cast irons have always been and continue to be the most used casting alloys, the most researched, but the least understood. This is why this Section 6.5, perhaps unfortunately, is effectively pressurised into being the most speculative section in this book. For want of any other coherent explanation, I present my own approach to an understanding of the microstructures of irons in terms of the mechanisms of inoculation and nodularisation. Until disproven, it is offered as no more than a potentially useful working hypothesis. In the meantime, we can look forward to the outcome of researches that might confirm the approach, or suggest an improved hypothesis.

6.5.1 REACTIONS WITH GASES

Before moving on to describe the formation of porosity from gases in solution or reacting with its environment, it is worth emphasising that most porosity I have seen in most castings, including cast irons, is mostly air bubbles entrained by the poor design of filling system. Such clumps of porosity, rather than uniformly dispersed pores, are easily identified as filling system defects.

Moving on now to reactions leading to porosity, as for copper-based alloys, the production of iron-based alloys is also complicated by the number of gases that can react with the melt and that can cause porosity by subsequent evolution on solidification. Again, it must be remembered that all the gases present can add their separate contributions to the total pressure in equilibrium with the melt. We shall deal with the gases in turn.

Oxygen is soluble and reacts with the high carbon content of cast irons. CO is the product, following Eqn (6.6). Carbon dioxide is seen to be blamed by many experimenters who note bubbles in cast irons usually in association with oxide (usually silicate) slags. It has been assumed that the high carbon content of the iron has reacted with the oxygen in the slag to create bubbles of CO. However, this is almost certainly an error. The slag is more usually not the result of carryover from the ladle, but is created in situ in the mould by surface turbulence. The bubbles are much larger than could be generated by reaction because the rate of reaction is limited by the rate of diffusion which is necessarily rather slow, producing, as for the dispersed microporosity of Al alloys, bubbles of maximum diameter perhaps 0.5 mm. It would require the coalescence of 1000 of these bubbles to make the 5 mm diameter bubbles typical of those seen in irons. Thus it is certain that the bubbles are not the product of a diffusion-controlled reaction but are simply *air bubbles* introduced by the turbulence involved with the poor filling systems usually employed. Naturally, they will contain some traces of carbon monoxide. However, with a well designed naturally pressurised filling system, both slag and so-called CO bubbles will disappear.

CO is still expected to be an important gas for the creation of porosity observed from time to time in cast irons, simply because of the high availability of both carbon and oxygen. However, it needs to be kept in mind that such bubbles will be expected to be fine, usually submillimetre in size, and evenly distributed.

CO can be encouraged to evolve from cast iron as a froth of bubbles, known as a carbon boil. Such an evolution of CO can be induced in molten cast iron, providing the silicon is low, simply by blowing air onto the surface of the melt (Heine, 1951), thus reinforcing that, if not already obvious, oxygen can be taken into solution in cast iron even though the iron already contains high levels of carbon. During subsequent solidification, in the region ahead of the solidification front, carbon and oxygen are concentrated still further. It is easy to envisage how, therefore, from relatively low initial contents of C and O, they can increase together so as to exceed a critical product $[C] \cdot [O]$ to cause CO bubbles to form in the casting. The equilibrium equation, known as the solubility product, relating to Eqn (6.7) is:

$$[C] \cdot [O] = kP_{CO} \tag{6.7}$$

We shall return to this important equation later. It is worth noting that the equation could be stated more accurately as the product of the activities of carbon and oxygen. However, for the moment we shall leave it as the product of concentrations, as being accurate enough to convey the concepts that we wish to discuss.

Hydrogen is soluble, as in Eqn (1.3), and exists in equilibrium with the melt, as indicated in Eqn (1.4). Although a vigorous carbon boil would reduce any hydrogen in solution to negligible levels by flushing it from the melt, such techniques are not usual in iron melting. Thus any hydrogen will tend to remain in the melt. Even so, although some hydrogen might be present, it does not generally appear to lead to any significant problems in irons. This is probably the result of the hydrogen solubility being rather low, in contrast to steels, where higher melting temperatures lead to the possibility of high hydrogen levels in solution.

In addition, of course, the rate of diffusion of hydrogen is astonishingly high. Thus hydrogen might be quickly lost from a melt, probably only in a few minutes, if the melt has the benefit of a dry environment. Similarly, in the presence of moisture or hydrogen, the hydrogen content of a melt could rise quickly. It clearly pays to work with furnaces and their backup refractories that are always kept hot.

Nitrogen is also soluble in liquid iron. The reaction follows the normal law for a diatomic gas:

$$N_2 = 2[N] \tag{6.8}$$

and the corresponding equation to relate the concentration in the melt $[N]$ with its equilibrium pressure P_{N_2} is simply:

$$[N]^2 = kP_{N_2} \tag{6.9}$$

As before, the equilibrium constant k is a function of temperature and composition. It is normally determined by careful experiment. As for hydrogen, in general, it seems that nitrogen is probably not an important source of porosity in irons. Exceptional circumstances might include the reactions involving high-nitrogen binders because the decomposition of amines they contain appear to lead to such problems as nitrogen fissures (see the Nitride Films section in this chapter). However, these seem to be as much surface nitride problems as nitrogen solution problems.

6.5.2 SURFACE FILMS ON LIQUID CAST IRONS

When cast iron is held at a high temperature (e.g. 1500°C) in a furnace or ladle lined with a traditional refractory material such as ganister, a fascinating sight can be witnessed. The surface of the liquid iron is seen to be continuously punctuated by the silent and mysterious arrival of bright circular patches. These suddenly appear and spread from nothing to their full size of several centimetres within about a second. The patches drift around, coalesce with other patches, and finally attach themselves to the wall of the vessel where they cool and add to the solidified rim of slag. These patches are droplets of liquid refractory, melted from the walls and bottom of the vessel. As the vessel is tips and empties, upstanding 'stalactites' on the base, and upward runs and drips on the walls can usually be clearly seen, marking the sites where the drops detached.

In common with all the components of molten metal systems, the slag will be changing its composition rapidly as it interacts with the molten metal. At high temperature its contents of iron, manganese and silicon will be reduced from their respective oxides and taken into solution in the liquid iron, whereas the remaining stable oxides, such as those of aluminium and calcium, will remain to accumulate as a dry slag, sometimes called a dross. These reactions will be explained further in the next section.

Such layers of slag on the surface of molten iron can be anything from 0.1 mm thickness upwards. (In the cupola, of course, the thickness is often around 100 mm or more.) It is not intended to consider such macroscopic surface-layer problems in this section. The following section considers only the microscopically thin surface film that, under certain conditions, will form automatically on the surface of the melt (no matter how good is the melting resistance of the lining material of the holding vessel).

Oxide films

Work by Heine and Loper at the University of Wisconsin, dating from 1951, has done much to explain the complex formation of surface films on cast irons. A slightly later study by Merz and Marincek (1954) is also illuminating. Based on these studies, we can explain the changes that occur as the temperature falls.

When the iron is at a high temperature, 1550°C, the Ellingham diagram indicates that CO is a more stable oxide than SiO_2. Thus carbon oxidises preferentially and is therefore lost at a higher rate than silicon, as is seen in Figure 6.28. Here the blowing of air on to the surface of a small crucible of molten metal serves to accentuate the effect. Silicon is observed to fall only after all the carbon has been used up. At this high temperature, no film is present on the melt—any silicon oxide, SiO_2, would be immediately reduced to silicon metal which would be dissolved in the melt, simultaneously forming CO which would escape to atmosphere.

At around 1420°C the stability of the carbon and silicon oxides is reversed. The exact temperature of this inversion seems to be dependent on the composition of the iron as pointed out by Merz and Marincek; de Sy (1967) reported a range of 1410–1450°C for the irons that he investigated, whereas the Ellingham diagram (Figure 1.5) predicts an inversion temperature for pure Fe-C alloys of about 1500°C. The agreement is, perhaps, as good as can be expected because of compositional uncertainties. Below approximately 1400°C, therefore, SiO_2 appears on the surface as a dry, solid film, rather grey in colour. This film cannot be removed by wiping the surface because it constantly reforms.

At a temperature of 1300°C, and in alloys that contain some manganese, it is clear from the Ellingham diagram that MnO is the least stable, SiO_2 is intermediate, and CO the most stable. Thus manganese is oxidised away preferentially, followed by silicon, and finally by carbon. The contribution of MnO to the film at this stage may reduce the melting point of the film, causing it to become liquid.

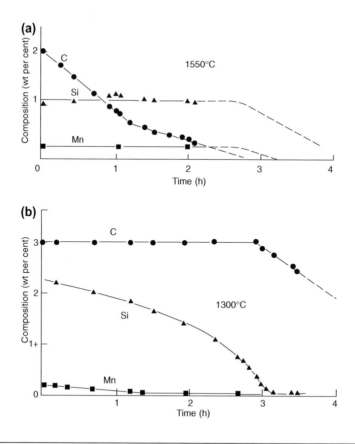

FIGURE 6.28

Change in composition of 3.6 kg of molten grey iron held in a silica crucible, whilst air was directed over its surface at the rate of 22 mL/s (a) melt at 1550°C; and (b) melt at 1300°C.

Data from Heine (1951).

At around 1200°C, iron oxide, FeO, contributes to the further lowering of the melting point at the ternary eutectic between FeO, MnO and SiO_2. If sulphur is also present in the iron, then MnS will contribute to a complex eutectic of melting point 1066°C (Heine and Loper, 1966).

The author finds that, in general terms, the previous considerations nicely explain his observations in an iron foundry where he once worked. For a common grade of grey iron, the surface of the iron was seen to be clear at 1420°C. As the temperature fell, patches of solid grey film were first observed at about 1390°C. These grew to cover the surface completely at 1350°C. The grey film remained in place until about 1280°C, at which temperature it started to break up by melting, finally becoming completely liquid at 1150°C.

When casting grey iron in an oxidising environment, the falling temperature during the pour will ensure that the surface film will be liquid at the most critical late stage of the filling of the mould. If the film becomes entrained in the molten metal, it will therefore quickly spherodise into compact droplets. The droplets are of much lower density than the iron and so will float out rapidly. On meeting the surface of the casting they will mutually assimilate, and be assimilated by, the existing surface liquid film, and so spread over the casting surface. The glassy sheen of some grey iron castings may be this solidified skin. The harmless dispersal of the oxide film in this way is the reason for the good natured behaviour of cast iron when cast into greensand moulds; it is one of the very few metal-mould combinations capable of exhibiting tolerance towards surface turbulence. Even so, there appears to be some experience indicating the irons cast without turbulence exhibit improved properties. This confusing situation requires to be resolved by future research.

Only on one occasion has the accumulation of liquid oxide at a casting surface given the author some problems. This was in a grey iron casting where a small amount of surface turbulence was known to be present just inside the ingate, because it was not easy to lower the velocity below 0.4 ms^{-1} at this point and was judged to be a negligible risk of any kind of internal defect. However, so much liquid surface was created at that location, and so many droplets of the entrained slag floated out at a point just down-stream, that the layer of surface slag accumulated at the down-stream location exceeded the machining allowance, scrapping the casting.

In **ductile irons**, in contrast to grey irons, the entrainment of the surface is nearly always a serious matter. The small percentage of magnesium that is required to convert the iron from flake to the spheroidal graphite type dramatically alters the nature of the oxide film.

Above 1454°C, Heine and Loper (1966) found that the surface of liquid ductile iron remains clear of any film. Below this temperature, a film starts to form, increasing in thickness to 1350°C, at which point the surface exhibits solidified crusty particles. By the time the temperature has reached 1290°C, the entire surface is covered with a dry dross. Magnesium vapour distils off through the dross because the molten iron is above the boiling point of magnesium. Presumably, the oxidation of the vapour to powdery MgO at the upper surface of the dross is a major contributor, causing the dross to grow quickly and copiously. The dross makes life difficult for the ductile iron foundryman, forming films, and agglomerating into dry, non-wetting heaps, that, if entrained, spoil otherwise excellent castings. Ductile iron is renowned for being difficult to cast cleanly, without unsightly dross defects. Surface films something like those on Al alloys are to be expected as in Figure 6.29.

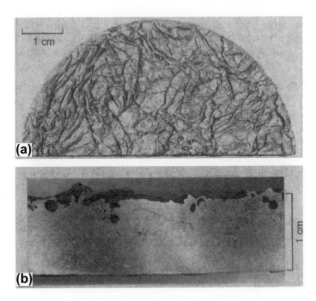

FIGURE 6.29

(a) Oxide skin on liquid Al-9Si-4Mg alloy wrinkled by repeated disturbance of the surface: and (b) a cross-section of the solidified metal. The appearance in both cases is extremely similar to graphite films on grey iron.

Courtesy of Agema and Fray (1990).

Oxide bifilms naturally in suspension in grey irons

De Sy (1967) has shown that liquid cast iron generally contains significant quantities of oxygen in solution in excess of its solubility. He concluded, on the basis of careful and rigorous experiments, that the undissolved fraction of oxygen was present as SiO_2 particles. Interestingly, by heating to 1550°C he confirmed the expectation that the SiO_2 solids dissolved because they became less stable than CO, but reappeared on cooling once again. Hartman and Stets report for irons containing Mg not only the presence of SiO_2 in suspension, but also the more complex iron-magnesium-silicate, olivine, $2(Mg,Fe)O \cdot SiO_2$. Hoffman and Wolf (2001) find a variety of oxides including SiO_2, FeO and MnO among others, but superheating and holding at high temperature eliminated many of these. In their elegant study of the thermodynamics Mampaey and Beghyn (2006) show how mainly SiO_2 together with some FeO forms in a typical melt when cooling from 1480° to 1350°C.

It seems reasonable to speculate that these oxides almost certainly would not be compact spheres, cubes or rods etc., but would most likely be in the form of *films*. Only films would have a sufficiently low Stokes velocity to remain in suspension for long periods of time associated with these experiments, and the long periods during which irons are held molten in holding furnaces.

In any case, of course, the *film* morphology is to be expected. During melting in the cupola as droplets of iron rain down, the natural folding-in of the surface film of SiO_2 of each droplet would ensure a natural population of SiO_2-rich double films (bifilms). Additional treatments or handling such as pouring actions, stirring in induction furnaces, and the oxide introduced from the surface of the charge (whether steel, pig or foundry returns) would increase this already large, natural population of oxide bifilms. Furthermore, it is well known that iron from electric furnaces is more liable to chill formation problems in thin sections than cupola iron, and this effect has been widely accepted as the loss of nuclei (in agreement with proposals made here), especially during extended time in holding furnaces or pouring systems.

Even if no silica bifilms are formed naturally in the iron as described previously, they will certainly be introduced during the process of inoculation, when Si-rich particles are added through the silica surface, entraining the surface silica as they penetrate and become submerged, as illustrated in Figure 2.3.

Thus there appear to be at least two quite different populations of oxide bifilms in suspension in liquid iron. (1) In grey irons, the silica-rich bifilms are a natural population in equilibrium with the melt, the amount of silica-rich phase being predictable by thermodynamics. (2) In ductile irons, the magnesia-rich bifilms are the result of mechanical accidents involving the turbulent entrainment of the surface film into the bulk liquid, resulting in non-reversible damage. We expand on this problem in ductile irons in this next section.

Bifilms in ductile iron

The population of oxide bifilms observed in ductile irons arises from the trauma of a turbulent filling system. In this case, the presence of Mg stabilises the magnesia, MgO, in the surface film, although Si might also contribute, thus forming a magnesium silicate $MgO \cdot SiO_2$ (also written as $MgSiO_3$). Both magnesium oxide and magnesium silicate are extremely stable, and when entrained, represent permanent damage to the liquid metal and subsequently to the casting. Although the bifilms are known to have an initially compact morphology as a result of the turbulence during their formation, and so are relatively harmless as cracks, the subsequent straightening of these bifilms by various natural processes such as the growth of dendrites, creating extensive planar cracks, is common, resulting in the development of brittleness in the form of plate fracture (Figures 6.30 and 6.31), as discussed in more detail in the next section.

The ductile casting industry has referred to entrained surface films observed on polished microstructural samples as 'dross stringers' (Figure 6.32). This name, based on their one-dimensional appearance on a polished two-dimensional section, has led to a comforting self-deception, concealing their real nature as extensive planar defects in three-dimensional space in the form of films floating about in a sea of metal. The occasional appearance of clusters of graphite nodules that have floated up and been trapped under such 'stringers' corroborates their real nature as films; nodules would not be trapped under one-dimensional 'strings' but naturally collect under two-dimensional films. Also, as we are now aware, if the film is solid (as it clearly is in this case), the entrainment process will fold them in dry side to dry side, thus forming a bifilm, a crack.

FIGURE 6.30

Plate fracture in the feeder neck of a ductile iron casting (Karsay, 1980).

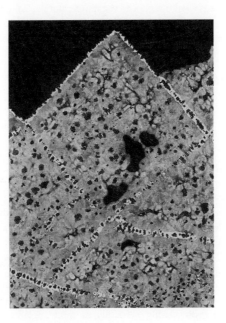

FIGURE 6.31

Polished microsection through the fracture (Barton, 1985).

Courtesy Casting Technology International.

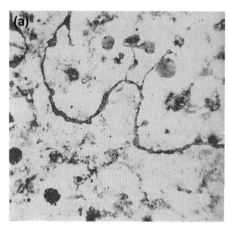 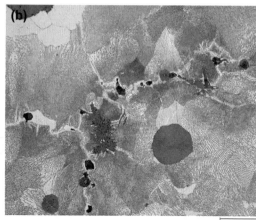

FIGURE 6.32

(a) So-called silicate 'stringer' in ductile iron (actually a visible silicate bifilm; (b) an alignment of mis-shapen nodules at a grain boundary in a pearlitic ductile iron, indicating the probable presence of an invisible bifilm.

The films appear on the liquid metal only at low temperature as we have seen, and seem to be mainly magnesium silicate, probably with a thick upper layer of solid MgO. If the ductile iron is cast at a low temperature, and if the surface is entrained, the creation of seriously damaging bifilms is guaranteed. Naturally, as the hot liquid iron cools during its passage through the running system it is likely to cool to the temperature at which the solid film starts to form, so that defects will be expected in most filling systems in which surface turbulence is not controlled. Once entrained, the defects can, of course, lead to a variety of additional problems. One of these serious problems is discussed later.

The magnesium silicate dross 'stringers' (actually, we should immediately stop calling them 'stringers' and call them 'bifilms') are particularly thick bifilms, and very clear when seen on a polished microsection as seen in Figure 6.32(a). Figure 6.32(b) probably also shows a bifilm in ductile iron, but the bifilm in this case is too thin to reveal its presence directly. We can be fairly sure a bifilm is present from the long line of particles of flaky graphite and mis-shapen spheroids, all typical of features that prefer to form on an oxide bifilm, and all apparently following a grain boundary, a common site for a bifilm. Elsewhere in this alloy, away from contact with the bifilm, the spheroids are beautifully formed.

Plate fracture defect in ductile iron

As with nitrogen fissures in grey irons, plate fracture in ductile irons has also never been satisfactorily explained. *Ductile irons*, should, of course, always exhibit a *ductile* mode of failure. Sometimes, however, a casting will exhibit poor strength and poor elongation to failure, with the fracture surface exhibiting large planar facets, the alloy appearing to consist of large embrittled grains. These unpredictable events give rise to serious concern that the material is not under the proper control that either the foundry or the customer would like to see. Everyone's faith is shaken. The question naturally arises, 'Is ductile iron a reliable engineering material?' These are the questions that should never arise, and that no one wishes to hear.

Following the description given by Karsay (1980) and Gagne and Goller (1983), the features of the plate fracture are large, flat, apparently brittle fracture planes, in ductile irons that in normal circumstances would exhibit only ductile failure. The planar facets appear to grow mainly vertically at right angles to the bottom surface of the casting (Figures 6.30). When viewed closely in cross-section (Figure 6.31), the planes are seen to be studded with small, irregularly shaped graphite spheroids, arranged with an accuracy almost resembling a crystal lattice. When polished and etched the planes are characterised by a matrix that is somewhat lighter than the rest of the casting. Karsay suggests that the colour difference may be the result of a higher Si content in this region, stabilising the ferrite. Finally, in this region, there is a high incidence of small inclusions that appear to be mainly magnesium silicates.

All these features are consistent with the defect being an oxide bifilm, probably a magnesium silicate, explaining the high Si content, the high inclusion content, and possibly the malformed spheroids as a result of local loss of Mg together with the natural tendency of graphite to form preferentially on bifilms. The planar form of the failure surface arises from the bifilm being pushed by the raft of austenite dendrites and organised into an interdendritic sheet, similar to that commonly seen in other alloy systems (Figures 2.43, 2.44(a) and 2.46). The vertical orientation is also understandable because of the greater rate of heat transfer from the base of the casting where gravity retains its contact with the mould, enhancing cooling, so that grains growing vertically from the base grow fastest and furthest. In addition, the magnesium silicate bifilm will necessarily be trapped at the mould wall (otherwise, dendritic straightening without some part of the film being anchored is not easily envisaged—bunches of film will only be pushed ahead if the film is not anchored) and buoyancy will encourage its vertical orientation, and so assist the advancing dendrite to straighten the film. Spheroids in interdendritic regions would then be revealed at the regular spacing dictated by the dendrite arm size. During solidification and cooling at the high temperature the bifilm probably disintegrates to some extent because of its surface energy tending to spherodise it. What remains are the changes in chemistry and numerous silicate fragments as inclusions plus remnants of the original central bifilm crack to encourage the direction of growth of the crack that finally causes failure.

Other features of plate fracture are its occurrence in slowly cooled regions, such as in a feeder neck. This may be the result of the lower rate of growth allowing the dendrites to straighten films more successfully (at high growth velocity, the drag resistance of films would resist dendrite growth, and resist film straightening).

Loper and Heine (1968) found that '*spiking*' observed on fracture surfaces of both white and ductile irons are similar and sometimes appears as oxidised facets. The occasional oxidation is easily understood if the bifilm connects to the surface and allows the ingress of air deep into the casting. Similar *internal oxidation* during heat treatment of irons is described in Section 9.10.

The less common appearance of plate fracture in irons of higher carbon equivalent and its reduction in resin-bonded sand moulds reported by Barton (1985) is probably not so much the result of a more rigid mould as he suggests, but an indication that the entrainment of the oxide film is less damaging in this more carbonaceous environment.

Heine and Heine (1968) described some fascinating observations and use telling language: 'when iron is damaged by melting and pouring so that it solidifies "spikey" it will not permit feeding even though proper sized feeders are provided'. This description acknowledges the possibility of deterioration of the melt by improper melting and handling and the consequent facetted nature of the freezing pattern. Furthermore, the presence of extensive oxides, probably from wall to wall in the casting, and bridging the necks of feeders, act to prevent the flow of feed metal. I am always impressed by intuition of good foundrymen for understanding the underlying scientific mechanisms.

In passing, it is worth reminding ourselves of the possibility that other non-oxide bifilms are to be expected in cast irons.

Nitride Films

Nitride bifilms probably form in cast irons giving rise to the 'nitrogen fissure' defects associated in the past with high nitrogen binders. Nitrogen fissures in grey iron castings are large cracks, often measured in centimetres, which appear to have been associated with the use of sand binders that contain high levels of nitrogen. They are an enigma that has never been satisfactorily explained. The blame is normally given to high-nitrogen binders that contain amines, whose breakdown probably contributes both nitrogen and hydrogen to the liquid iron. However, although such binders are known to be associated with fissure defects, their use does not always result in fissures.

It seems most likely that entrained bifilms, perhaps consisting of nitride films, are also required, so that the filling system may also be highly influential. Any involvement of the filling system has not previously been suspected, but would explain the current confusion in the results of studies carried out so far. Once entrained, the combination of high hydrogen and nitrogen pressure in the iron might be sufficient to inflate any nitride bifilms to some extent, opening the bifilms to revealing their presence as crack-like features. Further discussion is given in Section 7.2.3 under Nitrogen Porosity.

Carbon films (lustrous carbon)

The liquid film present on cast iron at low temperature in an oxidising environment has made iron easy to cast free from serious defects. This marvellous natural benefit of cast iron when cast into moulds made from sand bonded with clay and water must have played an important part in the success of the industrial revolution. In general (although acknowledging a number of infamous and tragic exceptions) the bridges did not fall into the river, and the steam engines continued to power machinery. Later, this benefit was to be extended to moulds made using one of the first widely used chemical binders: sodium silicate. This environmentally friendly chemical is still widely used today as a low cost sand binder for the production of strong moulds (despite a number of significant disadvantages that some foundries have been prepared to live with, but which are now being successfully overcome).

However, it is one of those ironies of history that the arrival of modern chemical binders based on resins was to change all of this.

Binders based on various kinds of resins; furan, phenolic, acrylic, polyurethane, etc. were heralded as the breakthrough of the twentieth century. Indeed, the new binders had many desirable properties, making accurate and stable moulds, with excellent surface finish, at good rates of production from simple low-cost equipment, and with good breakdown after casting.

However, when iron was poured into some of these early resin-bonded moulds, especially those based on polyurethane, a new defect was discovered. It became known as lustrous carbon. It had been occasionally seen, especially if the volatile additions to greensand had been high, but it was never so common nor so damaging. This shiny, black film resulted in casting skins wrinkled like elephant hide. Studies deduced from X-ray diffraction patterns (Draper, 1976) concluded that it was pyrolytic carbon (a microcrystalline form) that was deposited from the gas phase on to hot surfaces in the temperature range of at least 650–1000°C. Being a form of pyrolytic carbon, similar to carbon black, it was unlike the nicely crystallographic regularity of graphite.

The hot surface was originally assumed to be the sand grains of the mould, and somehow the deposit was pushed ahead of the advancing liquid front, to become incorporated into the surface as folds (Naro and Tanaglio, 1977). The explanation is clearly problematical on several fronts: in most instances, the sand surface is rather cold, and thus incapable of chemically 'cracking' (i.e. breaking down) the polymeric gases to precipitate carbon. Also, it is difficult to imagine how a film deposited on the complex and rough sand grains could be detached from its grip on these three-dimensional shapes before the arrival of the liquid metal and so be pushed into rucks and folds. After the arrival of the metal the film would be assisted in keeping its place by being held against the surface of the sand grains by the pressure of the liquid. Clearly, this explanation cannot be correct.

The only explanation that fits all the facts is that the graphitic film forms on the surface of the molten metal itself. Photographs of lustrous-carbon defects, particularly those seen on the fracture surfaces of parts that have suffered brittle failure, beautifully reveal their origin as the surface film on the liquid metal (Bindernagel, 1975; Naro and Tanaglio, 1977). The caption to the photograph of the oxide skin on an aluminium alloy (Figure 6.29) could be changed to read that it was a carbonaceous skin on a grey iron; to the unaided eye, the appearance of the two types of film is practically identical.

Part of the confusion that has surrounded the lustrous carbon film, claiming that it deposited on sand grains, dates from a misreading of the brilliant original work by Petrzela (1968). This Czech foundry researcher devised a test in which he demonstrated that the vapours released from coal tar and other hydrocarbon additions to moulding sands would decompose, depositing carbon as a shiny, silvery film on a metal strip, resistance heated to at least 1300°C. In his test, it happened that the sand was also heated to this temperature. Thus he observed carbon to be deposited directly onto the sand grains in addition to that deposited on the heated metal strip. The mistake of subsequent generations of researchers has been to assume that lustrous carbon always deposits onto sand grains, even though, at the instant that the metal is filling the mould, the sand is usually nowhere near the temperature at which a hydrocarbon vapour could be decomposed.

The reader is recommended to Petrzela's engaging, chatty and candid account. He was clearly one of our great foundry characters. His writing contains other fascinating asides to some of his observations on the release of carbon from hydrocarbons. He was possibly the first to describe a sooty deposit among sand grains that had a fibrous, woolly appearance. It has been observed many times since, although not previously explained. It is considered later, for instance in Figure 6.35.

Other work has studied the generation of lustrous carbon in greensand moulds. It is clear that the mould atmosphere can provide a hydrocarbon environment for the liquid metal if sufficiently high concentrations of hydrocarbons are added to the sand mixture. Such additives help the mould to resist wetting by the metal, and so improve surface finish, as appreciated in the original work of Petrzela (1968) and later by Bindernagel et al. (1975). Excess additions have sometimes been claimed to give lustrous carbon defects. One such defect is shown in Figure 6.33. However, it is certain that the defects form only if the surface turbulence can cause the film to be entrained. Otherwise, if the film is retained on the surface by careful uphill filling as in Figure 6.34, it is a valuable effect, significantly enhancing the surface finish of the casting. It is a pleasure to see iron castings that shine like new shoes.

The mechanism for the improvement of surface finish by the addition of hydrocarbons to the mould repays examination in some detail. The carbon film forms on the front of the advancing liquid. There is a suggestion by me (Campbell and Naro, 2009) that the carbon deposits on a precursor oxide film (Figure 6.34), on the grounds that the surface graphite, Kish graphite, appears to do the same (as noted by Liu and Loper, 1990) and flake graphite seems likely to form on oxides. Whether or not the film forms on a prior oxide film, it becomes trapped between the melt and the mould, and is held there by friction. Thus, as the meniscus advances, the film bridging the meniscus is forced to tear, splitting apart, but of course, immediately re-forming, as illustrated for the films on the advancing melts shown in Figures 2.2 and 6.34. The film is therefore continuously formed and laid down between the melt and the mould by the advancing metal, as though the advancing metal were rolling out its own track like a track-laying vehicle. The film forms a mechanical barrier between the metal and the mould. It is the mechanical rigidity of this barrier, helping to bridge the sand grains, that confers the improved smoothness to the cast surface. Thus, over the years, although there has been much talk about the action of surface tension bridging the gaps between sand grains (and this is clearly true to some extent) the main action in many alloys appears to be the result of the presence of the mechanically rigid surface film. The paper by Campbell and Naro (2009) beautifully illustrated the lustrous carbon film resembling a plate of steel, easily bridging sand grains and thus smoothing the surface of the casting (Figure 6.35(a)–(d)).

History now appears to have turned full circle because some resin binders for sands have recently been developed to yield iron castings with reduced incidence of lustrous carbon defects. At this stage, it is not clear whether the surface finish of the castings has suffered as a result. If better filling systems using bottom gating had been employed, the lustrous carbon would have been given its best chance to enhance the surface finish of the casting with no danger of an entrainment defect. Thus the new reduced lustrous carbon binders need never have been developed.

In lost foam castings using polystyrene foam, lustrous-carbon films cause troublesome defects. In this situation, the vapourisation of the polystyrene to styrene, and the subsequent decomposition of the styrene to lower hydrocarbons and eventually to carbon, deposits thick carbonaceous films on the advancing surface of the iron (Figure 4.9 shows the decomposition products). Gallois et al. (1987) found the film to consist of three main layers: (1) an upper lustrous multilayered structure of amorphous carbon; (2) an intermediate layer of sooty fibres consisting of strings of crystallites; and (3) a layer adhering strongly to the surface of the iron consisting of polycrystalline graphite enriched in manganese, silicon and sulphur. Clearly there has been some exchange of solutes from the iron into the film.

In sand castings the often-observed apparent decohesion of the carbon film from the metal surface and adhering to the surface of the mould (Figure 6.35) has previously led to much confused thinking, concluding that the lustrous carbon had formed on the mould. The decohesion of the carbon film from its original substrate had been difficult to explain because most if not all other films such as oxides strongly adhere to their originating matrix, as is demonstrated by anodised layers on Al alloys.

The presence of the woolly, fibrous carbon behind the film and among the sand grains in Figure 6.35(a)–(d) is illuminating and probably holds the explanation of this behaviour. Clearly the fibrous carbon forms after the film is put in place against the mould wall. Thus it seems that the fibrous material forms at perhaps lower temperatures or at lower concentrations of hydrocarbons. The dense array of fibres, interconnected between the film and the grains, explains how the film, originally on the metal, can become rather firmly attached to the mould. Further encouragement for the film to detach from the casting will arise from the enormously different thermal contractions of the substrate casting and the carbon surface film.

Graphite films that have grown on molten iron have been studied in the form of crystals formed on the surface of Fe-C alloys held in graphite crucibles, and so saturated with carbon before being allowed to cool (Sumiyoshi, 1968).

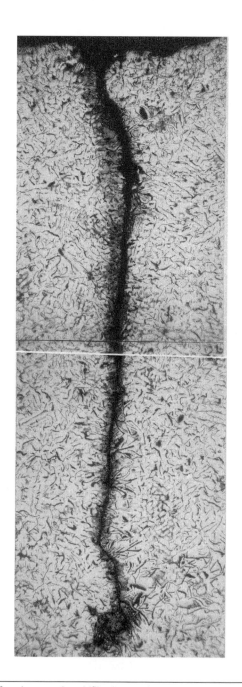

FIGURE 6.33

A folded-in lustrous carbon film, forming a carbon bifilm in grey iron.

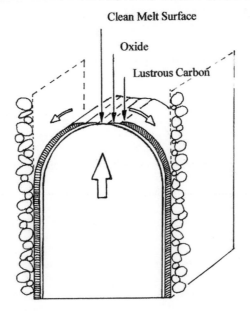

FIGURE 6.34

A schematic view of the advancing cast iron front causing the deposited films of oxide and carbon to be trapped between the mould and the metal, leading to the necessary splitting of the films at the advancing front.

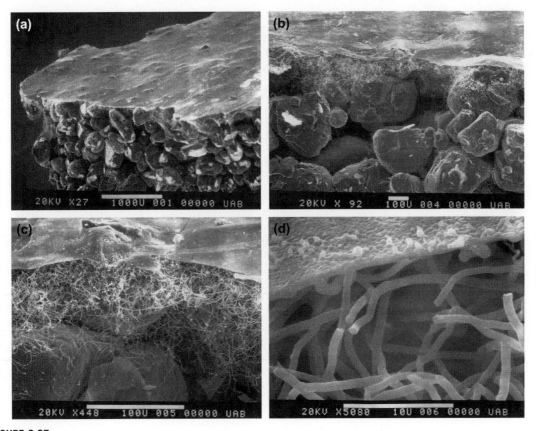

FIGURE 6.35

SEM images (a-d) illustrating the lustrous carbon film bridging sand grains.

These shiny black sheets that float on the surface are analogous to the Kish graphite that separates from hyper-eutectic cast irons during cooling, and are likely therefore to be growing on the underside of the surface oxide film. (For the scientifically minded, the graphite films in this research had interesting features. They were single crystals with numerous cracks along certain crystal directions, and hexagonal growth steps on the underside that showed how the film grew by gradual deposition of carbon atoms, probably onto ledges from emergent screw dislocations.)

The growth of graphite on melts saturated with carbon, as discussed previously, is easy to understand; but how does graphite grow in the case of lustrous-carbon films where the composition of the iron is far from saturated? Such films should go into solution in the iron! Every student of metallurgy knows that at the eutectic temperature the carbon in solution in iron has to exceed 4.3 wt% before free (Kish) graphite will precipitate. At higher temperatures, the carbon concentration for saturation increases, following the liquidus line for the solidification of Kish graphite in hyper-eutectic irons, as is clear from the Fe-C phase diagram. It seems unlikely that a graphite film can result from an equilibrium reaction. The explanation that follows is an example first mentioned in Section 2.5 relating to soluble films. It shows the effect relies on an interesting dynamic condition.

In an atmosphere containing hydrocarbons, if the rate of arrival of reactants at the free liquid surface is low, then both carbon and hydrogen can diffuse away from the surface into the bulk liquid. The free surface of the melt therefore remains clean.

However, in a highly concentrated environment of hydrocarbon gases the rate of arrival of reactants may exceed the rate of diffusion away into the bulk. Thus carbon will become concentrated on the surface (hydrogen less so, because its rate of diffusion is much higher) and may exceed saturation, allowing carbon to build up at the surface as a solid in equilibrium with the local high levels of carbon. Once formed, it would then take time to go into solution again, even if the conditions for growth and stability were removed. Thus it would appear to have a pseudo stability, with a life just long enough so that in some conditions the film could be frozen into the casting if a chance event of surface turbulence were to enfold the surface into the melt.

The folding in of the graphitic film is known to result in the familiar lustrous carbon defect (Figure 6.33). This is, of course, simply a bifilm crack lined with the lustrous carbon films. In heavy sections, such defects are not seen, probably because they will have time to go into solution before being frozen into the casting. In thinner sections of a ferritic matrix or mixed ferrite/pearlite matrix, the films can be seen to be partly dissolved as indicated by the layers of the higher carbon content pearlite on either side of the defect.

More speculatively, many irons are poured so turbulently that it is to be expected that huge numbers of graphitic films will be entrained. This raises a distinct possibility that many of the graphite flakes seen on a polished section of grey iron will not be formed as a result of a metallurgical precipitation reaction, but may be the remnants of entrained graphitic bifilms. The occasional appearance on microsections of what seem to be isolated large flakes amid uniform smaller flakes is suggestive of the bi-modal distribution to be expected if such a mixed source of graphite were present.

Recent research has indicated that the conditions for the growth of the graphite film on liquid metals are similar to the conditions required for the growth of diamond films. Reviews by Bachmann and Messier (1984) and Yarborough and Messier (1990) listed conditions for the growth of diamond as the breakdown of hydrocarbons and the presence of hydrogen. In the case of iron, the temperature is a little too high for diamond, and so would tend to stabilise the formation of graphite films. But for metals such as aluminium in a hydrocarbon environment, the conditions seem optimum for the creation of diamond on the metal surface. Prospectors and investors will be disappointed to note, however, that the rate of growth is slow, only 1 μm/h. Thus in the time that most liquid metal fronts exist while pouring a casting, the diamond layers, if any, will be so thin as to be a disappointing investment.

6.5.3 CAST IRON MICROSTRUCTURES

The forms of graphite in cast irons have been the subject of intense interest and huge research efforts over many decades, but a full understanding has been elusive. Readers are referred to the review by Loper (1999) for a wide-ranging synopsis covering many details not included in this study. Here, a rather different review is made of the literature, exploring the possibility of a unifying approach based on the hypothesis that oxide bifilms are present in liquid irons.

We have seen how a comprehensive understanding of the microstructure of Al-Si alloys has been proposed in terms of bifilms, explaining both the mechanism of modification and the structures of hypo- and hyper-eutectic alloys. Nakae and Shin (1999), among many others, have drawn attention to the analogous features of Al-Si and Fe-C alloys. This section of the book is an extension of the bifilm hypothesis, apparently valuable to an understanding of the Al-Si system, to a possible understanding of the various morphologies of carbon in the form of graphite and carbides in the Fe-C alloy system.

Adopting an analogous phase diagram to that shown in Figure 6.15 for Al-Si, the equivalent for the Fe-C system is shown in Figure 6.36(a). The eutectic at 4.3%C and 1130°C seems fairly well established, although freezing point values up to 1150°C have also been commonly assumed. This seems to correspond to the formation of graphite as flakes on the silica bifilms (in detail, most probably nucleated not directly on the bifilms but on nuclei already formed on the bifilms). If easy growth on silica is avoided by elimination of the silica, formation of graphite then occurs at some lower temperature, growing as compacted or nodular forms as discussed later. The precise value of the lower temperature is not known at this time, but is probably in the region of 1100–1125°C, and will certainly be affected by the growth form as is clear from thermal analysis curves.

The eutectic composition is also moved to higher levels of carbon, so that the higher carbon levels typical of ductile irons may be at least partly the result of this effect, and not solely the result of the founder seeking greater graphite content to reduce feeding problems. Interestingly, the peak in fluidity, normally indicating the composition of a eutectic, are seen from the work by Porter and Rosenthal (1953) to be close to 4.5%C (Figure 3.11), even though this work was carried out with grey iron. This result indicates that rapidly flowing iron, which would contain be turbulently ravelling and retaining bifilms in their compact form, flows similarly to ductile iron which contains no bifilms, as might be expected.

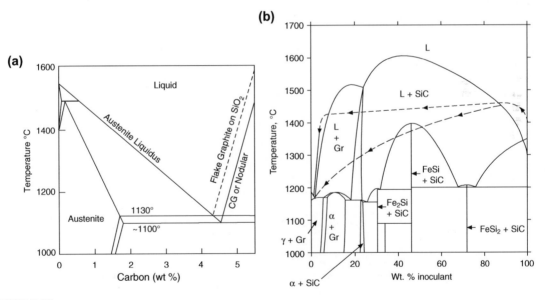

FIGURE 6.36

(a) Fe-C phase diagram showing (i) the high temperature eutectic in which graphite forms on silica-rich bifilms in suspension in the melt and (ii) the lower temperature eutectic of Mg-treated iron in which the eutectic can take different forms in the partial or complete absence of bifilms. (b) The Fe-FeSi phase diagram showing possible melting and mixing routes for a dissolving FeSi inoculant particle (Harding et al., 1997).

Graphite nuclei

Mizoguchi and co-workers (1997) have demonstrated that austenite is ineffective in nucleating graphite. In fact, they find that undercoolings below the liquidus of between 200 and 400°C are required to trigger nucleation by austenite. A more unfavourable nucleus than austenite would be difficult to imagine. The question therefore arises, 'what does nucleate graphite?' This question is all the more intriguing following the work by Mampaey and Xu (1997) in which they found that a single population of nuclei (although the nature of the nuclei remained unidentified) could explain both grey and ductile irons.

There is now a growing consensus that both flake and spheroidal graphite nucleate on similar if not identical nuclei (for instance, Warrick, 1966) composed of particles of complex oxides and sulphides. This was the conclusion reached in the first study following the development in the United Kingdom of the microprobe analyser (Jacobs, 1974). Jacobs and colleagues were the first to carry out an elegant study suggesting that within graphite nodules there is a central seed of a mixed sulphide (they suggested Ca and Mg sulphide) surrounded by a mixed (Mg,Al,Si,Ti) oxide spinel. They found matching crystal planes between the central sulphide, the spinel shell, and the graphite nodule, indicating a succession of nucleating reactions. This exemplary work has been confirmed a number of times, most recently by Solberg and Onsoien (2001).

Many confirmations of this general conclusion have because been made (for instance, Skaland, 2001) suggesting that the oxy-sulphide mix of the various elements will have a spectrum of lattice spacings ensuring that at least part of the compound will match graphite, and therefore possibly constitute a favoured substrate. As an example of a recent study, while working on preconditioning treatments for grey irons (treatments involving small additions of elements such as Al before inoculation—possibly to enhance the population of naturally occurring nuclei in uninoculated irons), Riposan et al. (2008) defined a three-stage model for the nucleation of graphite that differs in some details from that by Jacobs:

1. Small oxides (<2 μm) are formed in the melt (from preconditioner)
2. Complex sulphides (<5 μm) nucleate on and wrap around the oxides (the presence of Mn and S is necessary; many authors have reported the beneficial effects of S, for instance Chisamera et al., 1996)
3. Graphite nucleates on the sulphides (Ca, Sr, Ba in the sulphide assists nucleation), and, we shall suppose, will attempt to wrap itself around the sulphide particle.

Whether the details of the successive steps of nucleation follow exactly that described by Jacobs or by Riposan are not important for our understanding of the overall mechanism of inoculation. In the following discussion, the mechanisms proposed to explain the various morphologies of graphite are based on the possibility of

1. *nucleation* on specific oxy-sulphide nucleating particles which are effective for all types of graphite and
2. *growth* morphology of the graphite depends on the presence or absence of favourable growth substrates such as oxide bifilms for flake graphite and oxysulphide particles for nodular graphite (it is proposed in this work that the initial spherical growth on a nucleating particle appears to require stabilisation by plastic deformation of austenite at a later stage of growth.)

For a proper understanding of inoculation, it will be essential for the reader to keep in mind the separate actions of nucleation and growth.

6.5.4 FLAKE GRAPHITE IRON AND INOCULATION

The treatment of cast irons by the deliberate addition of material to aid the formation of graphite is generally called inoculation. Inoculation of cast irons is important to achieve a reproducible type and distribution of graphite to give reproducible mechanical properties and machinability.

Uninoculated iron is characterised by poor control of the graphite flake morphology. Flakes occur, but are relatively few in number, and uncontrolled in size. The relatively few opportunities for the carbon to precipitate lead to relatively large regions of the iron elsewhere being supersaturated with carbon. Thus iron carbide (Fe_3C) precipitation is likely in regions that are deficient in graphite nuclei. The mechanical properties of the iron are generally poor. In general, as will become clear during the progress of this account, it seems that some nuclei exist prior to inoculation, but their number and effectiveness cannot be relied on.

The history of inoculation

I am indebted to my good friend, Reginald Forrest, for an account of the history of the inoculation process.

Ross and Meehan were co-owners of the Ross Meehan Foundry based in Chattanooga, Tennessee, USA (a city later to be immortalised by Glenn Miller with his swing band hit 'Chattanooga Choo Choo'). Gus Meehan was an inquisitive and observant foundryman. He was intrigued by the known benefits of 'treating' the melt with floor sweepings and put the treatment on a more controlled basis in his foundry. The practice became 'standard' in the Ross Meehan foundry and customers were happy. By chance an English foundryman working in Newcastle, UK, learned about this development and corresponded with Meehan to exchange experiences. This was Oliver Smalley, another observant, smart and practical foundryman. They experimented with various mixtures and materials and finally decided on calcium silicide as a major component of their mix, together with ferrosilicon fines and graphite. The mixtures were a closely guarded secret. All this time, the consistency of the Ross Meehan cast iron castings was becoming widely known.

They decided to name their iron 'Meehanite' and later 'Meehanite Metal'. They patented their invention in 1922 and began a system of licencing foundries to produce iron castings from controlled (inoculated) liquid cast iron. This system, spearheaded by Smalley, who had the marketing flair, became progressively better controlled with measured response (wedge controls) and producing a spectrum of grades of grey iron known variously as Meehanite irons.

Smalley, the genius of marketing Meehanite, developed the concept of Meehanite as bridging the gap between steel and cast iron. Meehanite licencee foundries mushroomed in the United States and Europe; the issue of licenses being limited so far as possible to good, disciplined foundries who would follow the prescribed practices. Smalley realised that by reaching the end user (the engineers and designers) and convincing them that Meehanite was a high quality, consistent and reliable material and that castings produced by Meehanite licenced foundries were to be trusted—he tapped into a rich vein. This was perhaps one of the first examples of technological foundry marketing.

Gus Meehan was the origin of the Meehanite name that became so famous in the cast iron industry—but Smalley was the real driving force in its commercial exploitation. The Americans and Smalley eventually went their separate ways, splitting the operation: Meehanite Metal Corporation controlled the American and Far East operations and International Meehanite Metal Company based in the UK controlled European Licensee operations. Materials & Methods Limited was the birth child of International Meehanite Metal Company and produced and marketed all of the inoculants and, later, nodularisers used by the licenced Meehanite foundries (and later sold openly to any foundry).

The gradual introduction of the inoculation process from 1922 onwards, and its continued development to the present day (for instance, Skaland, 2001 and Hartung et al., 2008), was found to greatly increase the number of nuclei available, giving a copious crop of graphite flakes of good uniformity of size, with a reduced tendency to carbide formation, and a consequent benefit to the mechanical properties and machinability of the iron. These benefits are now freely available to all cast iron foundries.

The mechanism of inoculation

Clearly, inoculation was some kind of process to provide the nucleation of graphite particles. However, the detailed mechanism remained unknown.

What was known was that successful inoculants include ferrosilicon (an alloy of Fe and Si, usually denoted as an inaccurate shorthand, FeSi, but usually containing approximately 75 wt% silicon, so sometimes written Fe75Si), calcium silicide and graphite. These were added to the melt as late additions, just before casting. Additions designed to work over 15–20 min were used in a granular form, of size around 5 mm diameter, whereas very late additions (made to the pouring stream) were generally close to 1 mm. Late inoculation was favoured because the inoculation effect gradually disappeared; a process known as 'fade'.

Ferrosilicon is the normally preferred addition, and is known as a 'clean' inoculant. Calcium silicide is known to be a rather 'dirty' addition, almost certainly because the calcium will react with air to give solid CaO surface films (in contrast to FeSi that will cause liquid silicate films). The folding-in of CaO films during turbulent filling would create particularly drossy bifilm inclusions together with entrained air as porosity. The calcium silicide addition would probably have been much more acceptable with better-designed filling systems that reduce surface turbulence. (Ductile iron casters experience similar problems when using Mg as a noduliser.)

It is immediately clear that the common inoculant, FeSi, does not perform any nucleating role itself. This is because liquid iron at its casting temperature (perhaps 1350–1400°C) is well above the melting point of the FeSi intermetallic compound (1210°C), so that the whole FeSi particle melts (Figure 6.36(b)).

The evidence now suggests that the inoculants, originally a particle, continue to exist for some time as a molten high Si region in the liquid iron. Although the Si-rich region is liquid, and the iron is liquid, and the two liquids are completely miscible, the two nevertheless take time to inter-diffuse. This time is probably the fade time. The Si-rich region slowly dissipates in the melt, eventually disappearing completely. However, during its short lifetime, it provides a local environment with a high effective CEV. To get some idea of the scale and importance of this effect it is instructive (although admittedly not really justified, as we shall see) to calculate the carbon equivalent in one of these regions. For an iron of carbon content about 3%, assuming CEV = (%C) + (%Si/3), we have CEV = 3 + 75/3 = 28%C. Extrapolating the carbon liquidus line on the equilibrium diagram to an iron alloy with the huge level of 28%C predicts a liquidus temperature in the region of several thousand degrees Celsius. (This is actually not surprising in view that graphite itself has an effective melting point of more than 10,000°C.) Clearly, therefore, there seems good reasons for believing that the carbon in solution in the Si-rich regions is, in effect, enormously undercooled. It is a form of artificial *constitutional undercooling* (because the graphite is effectively *undercooled* as a result of a change in the *constitution* of the alloy).

Now, in reality, it is not appropriate to extrapolate the CEV beyond the eutectic value of 4.3%C. In fact, when this part of the equilibrium phase diagram is calculated, the liquidus surface is nothing like linear, as seen in Figure 6.36(b) (Harding, Campbell and Saunders, 1997). Even so, this figure shows the liquidus in the hyper-eutectic region to be very high, so that the essential concept is not far wrong. The path of the dissolving particle is marked on the figure, showing its gradual loss of silicon as the melted liquid region makes its way from right to left across the figure. Different paths can be envisaged for different rates of loss of temperature as are seen in the Figure. This slowly disappearing liquid 'package' has to pass through regions where it will experience undercooling of several hundred degrees Celsius, providing a massive driving force for the precipitation of graphite.

The large undercooling, creating such a substantial driving force in these liquid regions is almost certainly the reason why, over the years, so many different nuclei have been identified for the initiation of graphite. It seems that even nuclei that would hardly be expected to work at all are still coaxed into effectiveness by the extraordinary and powerful undercooling conditions that it experiences. Studies have shown that many particles that are found in the centres of graphite spherules, and thus appear to have acted as nuclei, whereas identical particles are also seen to be floating freely in the melt of the same casting, having nucleated nothing (Harding, Campbell and Saunders, 1997). This is understandable if the nuclei are not particularly effective. They will only be forced to act as nuclei if they happen to float through a region that is highly constitutionally undercooled.

Studies by quenching irons just after inoculation have revealed a complex series of shells around the dissolving FeSi particle. Hurum (1952, 1965) was the first to draw attention to this phenomenon, but it has been studied by several others since (for instance, Fredriksson, 1984). Although FeSi itself contains almost no carbon, the carbon in the cast iron diffuses into the liquid FeSi region quickly. Data from Figure 1.4(c) and Eqn (1.5) indicate a time of 1 s for an average diffusion distance d = 0.1 mm, but 100 s for d = 1 mm. The flow resulting from the buoyancy of the high Si melt, and the internal flows of metal in the mould cavity, will smear the liquid Si-rich region into streamers, reducing the diffusion distance to give shorter estimated times for the homogenisation of carbon. Thus the shell of silicon carbide (SiC) particles around a dissolving FeSi particle (Figure 6.37) appears logical as a result of the high undercooling in the part of the phase diagram where SiC should be stable (Figure 6.36(b)). It seems likely that the SiC nucleates homogeneously because of the high constitutional undercooling. In a shell further out from the centre of the dissolving inoculant particle, graphite starts to form. It may be that graphite does not simply nucleate homogeneously as a result of the generous undercooling but can also form in this region by the decomposition of some of the SiC particles.

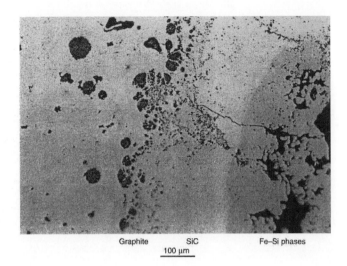

Graphite   SiC   Fe–Si phases
100 µm

FIGURE 6.37

Microsection of a dissolving FeSi particle in a ductile iron, quenched from the liquid state (Bachelot, 1997).

If all of this were not already complicated enough, there is even more complexity. In addition to the dissolving FeSi particle providing (1) a local solute enrichment of Si there will also be (2) a release of sundry complex inclusions including oxides and sulphides. Commercially available inoculants contain various impurities, and various deliberate additions that supplement the natural nucleating action in this way. These additions include group 1A elements of the periodic table, Mg, Ca, Sr and Ba, and often some rare earths such as La and Ce, that will react to create oxides and sulphides. At least some of these may be good heterogeneous nuclei for the formation of new graphite crystals (or perhaps new SiC crystals that may subsequently transform to graphite particles). Also, of course, these particles are provided exactly where they are needed, in the heart of the highly undercooled liquid region. These intentionally added particles will augment the naturally occurring population of nuclei already present in the melt. The overwhelming driving force explains the wide variation of successful nuclei which, in other circumstances, would be expected to be of only mediocre, if any, effectiveness.

This action of the inoculating material in providing a combination of copious heterogeneous nuclei together with good growth conditions explains the action of graphitisers such as ferrosilicon, and the importance of the traces of impurities such as aluminium, rare earths and sulphur that raise the efficiency of inoculation.

Ferrosilicon and calcium silicide are not, of course, the only materials that can act as inoculants. SiC is also effective, as is graphite itself. Both of these materials can be seen to provide similar transient conditions consisting of pockets of liquid in suspension in the melt in which high constitutional undercooling promotes the nucleation of graphite.

The chain of nucleating effects, oxides-sulphides-graphite, and only in the effectively supercooled regions, has the outcome that graphite particles exist in the melt at temperatures well above the eutectic. The prior existence of graphite particles in the liquid at high temperature, well above the temperature at which austenite starts to form is quite contrary to normal expectations based on the equilibrium phase diagram, but explains many features of cast iron solidification. The expansion of graphitic irons prior to freezing (the so-called 'pre-shrinkage expansion') has in the past been difficult to explain (Girshovich, 1963). The existence of graphite spheroids growing freely in the melt above the eutectic temperature has been a similar problem, seemingly widely known, and seemingly widely ignored, but now provided with an explanation, even though it would be highly desirable to have some additional confirmation as soon as possible.

Harding and co-workers (1997) pointed out that once nucleated in the regions of high driving force for initiation, the graphite particles attached to their nuclei now will emerge into the general melt where they will become unstable and

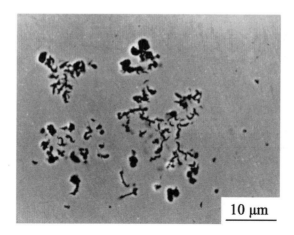

10 μm

FIGURE 6.38

Coarsening of tips of graphite particles on emerging from the undercooled FeSi region (Benaily, 1998).

start to re-dissolve. Feest et al. (1980) found that although the Si-rich inoculant regions disperse relatively rapidly, the graphite which formed rapidly in these regions is slow to re-dissolve.

Although they were in the undercooled region, the graphite particles would have been expected to grow with extreme speed, thus adopting a thin and branching dendritic morphology. However, on leaving this rapid growth environment and entering a region where growth will suddenly be arrested, and re-solution starts, the dramatic change will be expected to result not only in the arrest of growth but the coarsening of the graphite dendrite tips. This effect is seen in material quenched from this region (Figure 6.38) by Benaily (1998).

The observations by Loper and Heine (1961) confirmed that graphite can form and survive in both hypo- and hyper-eutectic irons at 1400°C, well into the liquid range, high above the expected liquidus temperatures. Mampaey (1999) also confirmed that graphite forms in the melt before the appearance of austenite. (These observations are quite contrary to expectations based on the equilibrium diagram. However, of course, the equilibrium diagram is based on the assumptions not only of (1) equilibrium behaviour but also (2) perfectly uniform composition, neither of which applies during the inoculation of cast irons.)

Thus nuclei will have initiated graphite nucleation in the *constitutionally undercooled* pockets of liquid, but will emerge and start to re-dissolve in the open melt. However, if they happen to pass through other undercooled regions the graphite will experience sudden bursts of growth, followed by slow dissolution in the bulk of the melt. Finally, the graphite particle will approach the eutectic front, which will be *thermally undercooled*, and so enjoy stability and a final spurt of growth before being frozen into the advancing eutectic.

Given sufficient time, all the inoculant particles will have melted and dispersed, leaving no pockets of undercooling floating about in the melt. This is almost certainly the phenomenon known to all foundry personnel as 'fade' of the inoculation effect, occurring within a time of approximately 5–20 min.

Assuming that the nucleated graphite particles survive, whether their subsequent growth occurs in the form of flakes or spheroids is a completely separate issue, unrelated to the nucleation/inoculation treatment. This is a growth problem. We shall deal with growth separately.

Growth of graphite

Bearing in mind that many second phases and intermetallics precipitate on bifilms as preferred substrates, it seems reasonable to assume that these new graphite nuclei would also preferentially form on substrates provided by oxide bifilms. Having nucleated on the bifilm, the nucleus would in turn nucleate graphite. Figure 6.39 schematically shows a graphite nucleus formed on an oxide bifilm. As the bifilm by chance enters an undercooled region provided by a

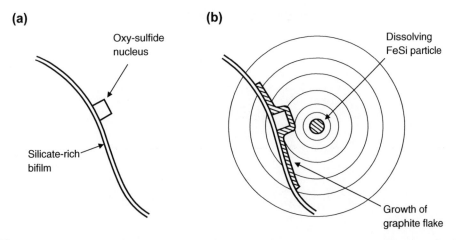

FIGURE 6.39

The mechanism of inoculation: (a) a bifilm in suspension in the melt, with oxy-sulphide nucleus (from trace impurities or preconditioners); (b) the bifilm floating into a constitutionally supercooled region, causing graphite to nucleate and grow on the bifilm.

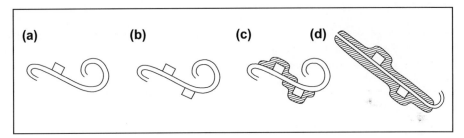

FIGURE 6.40

(a) A bifilm possibly supporting some initial nuclei; (b) additional nuclei provided by inoculant; (c) graphite forms on nuclei in supercooled regions; (d) graphite grows as fairly straight flake, straightening the bifilm to create a central crack.

dissolving FeSi inoculant particle, the nucleus experiences a massive driving force as a result of hundreds of degrees of effective undercooling, forcing graphite to nucleate around the nucleus, and forcing rapid growth.

The newly forming graphite is not able to grow completely around the nucleating particle because the particle itself has itself grown on the planar bifilm substrate so that at least one of its faces is inaccessible (Figure 6.40). The silica-rich bifilm will form a 'next best' substrate for graphite, so although insufficiently favoured to cause nucleation, it is sufficiently favoured to support the further growth of the graphite. Thus in grey irons the graphite extends across the bifilm, leading to the fairly flat morphology of flakes in grey iron. The flakes grow in regions ahead of the solidification front (i.e. slightly above the general eutectic freezing temperatures) because of the energetically favoured growth of graphite on the oxide substrates in suspension (Figure 6.41). The growth morphology of graphite, extending in the directions in its basal plane, would favour the straightening of the bifilm (Figure 6.40(d)). The bifilm would be expected to be extremely thin, possibly measure in nanometres, its minimal rigidity exerting

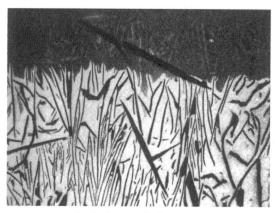

(10a) flake graphite; R = 1.2 μ/s

FIGURE 6.41

A straight graphite flake formed on a bifilm freely floating in the melt is overtaken by the coupled growth of the eutectic, and incorporated into the solid (Li, Liu and Loper, 1990).

negligible constraint of the advancing graphite crystal. The freedom from restraint would explain the development of relatively perfect crystals of graphite as observed growing in the liquid ahead of the coupled eutectic graphite (Figure 6.41).

The mechanism proposed previous explains the growth of flake graphite from nucleating particles introduced by inoculation. Particles that appear to be nuclei for the initiation of flakes have often been observed and seem likely to be a universal phenomenon in both flake and nodular irons (Rong and Xiang, 1991).

Although several experimenters have concluded that graphite nucleates on silica (Tyberg and Granehult, 1970; Gadd and Bennett, 1984; Nakae et al., 1991), this is probably an understandable error. In fact, it is far more probable that they were observing *growth* of the graphite on silica. Nakea (1993) goes further to identify the particular structural form of silica as cristobalite. It seems certain in fact that graphite nucleates on compact oxy-sulphide particles, but subsequently grows on silicon oxide bifilms, having the structure of cristobalite if Nakea is correct.

Eventually, the advancing solidification front will overtake those flakes growing on bifilms floating freely in suspension in the liquid. Thus eventually, these freely floating flakes will become incorporated into the solid as seen in the centre of the solidified coupled eutectic in Figure 6.41 by Li, Liu and Loper (1990). Thus it would be expected to be common to see grey irons with two separate populations of flakes: (1) those formed as primary particles by free growth in the liquid and (2) those formed by coupled growth with austenite at lower temperatures. The coupled growth mode is discussed later. A bi-modal distribution of graphite flakes is therefore to be expected in many microstructures. A bi-modal distribution is suggested in Figure 6.41 but is more clearly seen in Figure 6.42. Less obvious but important bi-modal distributions are probably common, as may be inferred from the work of Enright and colleagues (2000) using automated fractal analysis of microstructures. If lustrous carbon surface films on the liquid iron are also incorporated by turbulent entrainment events, it is conceivable that trimodal graphite forms can be present. No one has yet looked for such curious features so it is not known whether they exist.

Practical experience with inoculation

Goodrich (2008) attributes the *type C* iron (ASTM A247), characterised by large, very straight flakes, with some branching, to the result of the growth of the flakes in the liquid, unencumbered by the presence of austenite. He calls these 'proeutectic' flakes. They originate in suspension in the melt and are therefore capable of flotation to the upper regions of

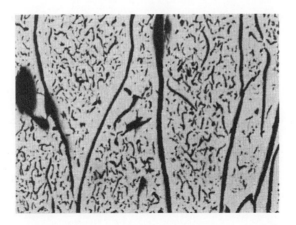

FIGURE 6.42

Two populations of flakes; one formed on bifilms and the other formed as a classical couple eutectic appearing as undercooled or coral form. (Hillert and Rao, 1967).

a casting. The more common *type A* graphite flakes are similar, displaying only minimal irregularity, suggesting a similar origin and behaviour in the melt. Loper and Fang (2008) use deep etching to reveal what they call 'pre-eutectic' flakes with elegant hexagonal symmetry, and apparently largely free from defects. For many other irons, the presence of a dense mesh of austenite dendrites constrains the size and shapes of flakes and prevents any significant buoyancy problems (Loper, 1999).

More usually in castings, the graphite flakes are seen to branch relatively frequently. In terms of the bifilm substrate, this is straightforwardly understood from the irregular structure of the bifilms. During their entrainment from the liquid surface into the bulk melt they tend not to entrain as nicely parallel double films, but as randomly folded, messy structures. Thus folds leading to parts of the double film at irregular angles to the main bifilm fold are to be expected, and would account for the branching of growing flakes.

Experience of variable performance is also to be expected. For instance, on a Monday morning, after melt has been held for the weekend, operators commonly find the iron has poor graphite structure, despite attempts to provide nuclei by inoculation. We may speculate that this is the result of the gradual floating out of the bifilm substrates. Similarly, iron heated to a high temperature suffers a similar degradation of graphite structure, almost certainly as a result of the dissolution of the bifilms because of the instability of SiO_2 above about 1450°C in the presence of carbon. It would be interesting to know whether the melt, after losing its silica-rich bifilms at high temperature, would regain its good solidified structure when cooled once again, because although de Sy reports that the silica reappears in the melt on cooling, without some kind of surface turbulence the form of the silica may not be a bifilm, nor even a film, but may be a compact particle. As such, it is not likely to be a good substrate for the development of a good flake structure. Even so, the process of inoculation, entraining the surface oxide during the act of submerging the inoculants particles, and the turbulence of the final pouring of the melt into the mould may address this problem, making the problem essentially invisible to those attempting to study the effect.

In agreement with the prediction that graphite grows on oxide bifilms, the effect of oxygen addition to the melt during the pouring of iron into the mould is demonstrated by a number of authors. For instance, Basdogan et al. (1980) and Chisamera et al. (1996) found oxygen to be highly effective in converting carbidic irons into beautifully 'inoculated' flake graphite irons. Liu and Loper (1990) found that oxygen was necessary to nucleate Kish graphite on the surface of grey iron melts. Larger quantities of Kish were formed at temperatures below 1400°C, below which SiO_2 is stable, but Kish was not observed in Si-free melts. Moreover, Johnson and Smart (1977) described a critical experiment in which they use sophisticated Auger analysis to prove that two or three atomic layers of oxygen (and interestingly, sulphur) are present on fracture surfaces of the matrix adjacent to graphite flakes (fractured and observed in high vacuum), but the surfaces of the hollows left by spheroidal graphite nodules exhibited no oxygen. This is behaviour consistent with a bifilm hypothesis, and with the assumption that the graphite itself may be strong, the presence of the bifilm gives it the appearance of weakness in tension. Briefly, if flake graphite formed on one side of an oxide bifilm, the fracture surface would necessarily travel along the centre of the oxide bifilm, revealing the oxide on both the graphite and the matrix, but oxides would be absent in the case of spheroidal graphite iron (as discussed later). Interestingly, if the graphite flake had nucleated on both sides of the oxide the fracture path would have passed through the flake, but would still have revealed the presence of oxygen, this time on both of the graphite faces. Johnson and Smart do not appear to have tested this condition.

Although all the previous discussion relates to silica-based bifilms, there is evidence that alumina-based, or possibly Al-containing Si-based bifilms (for instance based on mullite or other stable alumino-silicate compound) exist. Carlberg and Fredriksson (1977) found that cast irons based on Fe-C-Si exhibit fine graphite structures, whereas those based mainly on Fe-C-Al display coarse graphite flakes. Chisamera (1996) confirmed that conventional grey irons containing Al develop coarse graphite flakes.

The relatively poor mechanical properties of grey iron seems likely to be more to do with the presence of bifilm cracks down the centers of graphite flakes (or the sides of graphite flakes if graphite grows on only one side of the bifilm—the impression given that the flake has decohered from the matrix) rather than any intrinsic weakness of the graphite itself. A crack down the centrecentre of a flake is seen in Figure 6.43. In their studies of crack initiation and propagation in irons, Voigt and Holmgren (1990) reported many centreline cracks in graphite flakes plus some decoherence from the matrix. As is well known, graphite exhibits easy slip parallel to the basal plane. However, what is only recently uncovered from molecular dynamics simulations is that at right angles to its basal plane van der Waal forces result in a reasonable tensile strength approximately 0.65 GPa (Okamoto, 2013).

The previous discussion relates to those graphite flakes growing freely in the melt giving rise to large, randomly oriented flakes which tend to float or settle irregularly, creating what has been called in the past an 'anomalous' or 'irregular' eutectic; i.e. a eutectic whose phases are not regularly sized and spaced, in contrast to a classical eutectic whose spacing is regularly ordered and controlled by diffusion.

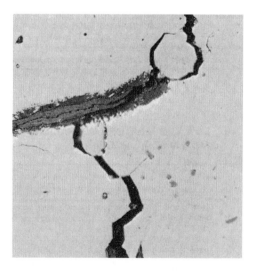

FIGURE 6.43

Graphite flake exhibiting a central crack (the solid state precipitation of surrounding temper graphite is also fractured off) (Karsay, 1971).

6.5.5 NUCLEATION AND GROWTH OF THE AUSTENITE MATRIX

The nucleation of the austenitic matrix of cast irons has, to the author's knowledge, never been systematically researched, although it is interesting that in irons with high Ti and S levels Moumeni and colleagues (2013) found a mixed Mn + Ti sulphide in the form of stars and ribs in the middle of austenite dendrites, strongly suggesting a nucleating role. Even so, it is not especially clear that the problem is at all important. For instance, if a fine austenite grain size could be obtained, would it be beneficial? The answer to this question appears to be not known. Moreover, in the section on steels the grain refinement of austenite is seen to be unsolved. Thus in all this disappointing ignorance, we shall turn to other matters about which at least something is known.

Only recently have two different teams of researchers revealed for the first time the growth morphology of the austenite matrix in which the graphite spherulites are embedded. Ruxanda et al. (2001) studied dendrites that they found in a shrinkage cavity, finding them to be irregular, each dendrite being locally swollen and mis-shapen from many spherulites beneath their surfaces (Figure 6.44). Rivera et al. (2002) developed an austempering treatment directly from the as-cast state that revealed the austenite grains clearly. The grains were large, about 1 mm across, clearly composed of many irregular dendrites, several hundred eutectic cells, and tens of thousands of spherulites. The dendrites from both these studies have some resemblance to the aluminium dendrite shown in Figure 5.33.

It seems fairly certain therefore that the growth of the austenite dendrites occurs into the melt in which there exists a suspension of graphitic particles. The particles hover with what seems to be neutral buoyancy despite their very different density. This arises because their small size confers on them such a low Stokes Velocity that they are carried about by the flow of the liquid: using Stokes relation it is quickly shown that a 1 μm diameter particle has a rate of flotation of only about 1 μms^{-1}, corresponding to a movement of the order of one dendrite arm spacing in a minute. Particles of 10 μm

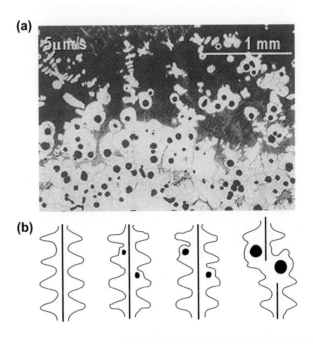

(a)

(b)

FIGURE 6.44

(a) The growth front of ductile iron (Li, Liu and Loper, 1990); (b) the distortion of dendrites as a result of the internal expansion of nodules.

After Hillert (1968).

diameter might have a more irregular form as in Figure 6.38. Thus despite their larger size, this would reduce their overall average density difference, and increase their viscous drag, so their flotation rate would hardly be higher. Thus many particles would have plenty of time to be incorporated into the dendrite structure.

Once trapped, the surrounding dendrite would be expanded and distorted by the continued growth of the graphite particle as in Figure 6.44, because, at these temperatures, the surrounding solid will be no barrier to the rapid diffusion of carbon to feed its growth. This micro-expansion of the dendrites to accommodate its content of graphite nodules translates of course to the macroscopic expansion of the whole casting, the expansion of the mould, and even the expansion of the surrounding steel moulding box, if any. Sub-microscopic rearrangements of atoms can accumulate to become irresistible forces in the macroscopic world. This is *mould dilation* leading to an increased demand from feeders, or, if not fed, to increased porosity in the casting.

6.5.6 COUPLED EUTECTIC GROWTH OF GRAPHITE AND AUSTENITE

So far, we have considered the separate nucleation and growth modes of graphite and austenite, as though these were unrelated. In fact the relation between the two is sometimes so poor to make any eutectic relation hardly discernible. This rather loose relation between the two major phases in grey iron castings has sometimes resulted in such eutectics being called 'anomalous' or 'irregular' as mentioned previously.

In this section, we move the focus from 'anomalous' to the truly regular, 'classical' eutectic form, in which austenite and graphite grow in a closely cooperative mode, known as coupled growth.

In the absence of suitable nuclei that have formed on oxide substrates in suspension in the melt, the carbon in solution will be unable to precipitate. Thus the melt will continue to undercool until the undercooling finally becomes sufficient to provoke precipitation on some other (less favourable) substrate. Only relatively few such nuclei will operate, activating in those parts of the melt that are especially cool, such as those regions close to the mould walls. The subsequent evolution of heat will inhibit other nuclei from becoming active.

Muhmond and Fredriksson (2013) found that if the Mn or S is too low to aid the formation of nuclei for flakes, then only an undercooled classical eutectic forms, outlining the austenite dendrites. Similarly, Moumeni et al. (2013) found that an addition of 0.35Ti suppresses the formation of graphite as flakes, and causes it to grow as a superfine 10 μm interdendritic graphite/austenite eutectic. Such behaviour is also clearly seen when the melt is cooled quickly.

At modest undercoolings, the coupled growth takes the form of rosettes, often called 'cells' (Figure 6.45). Thus it seems likely that a single initiation event on a nucleus, often sited on the mould wall, expands the coupled growth front as a hemisphere to form the rosettes, or cells (Figures 6.45 and 6.46(a)). The cells are beautifully regular structures, with inter-flake distances now strictly controlled by diffusion in the boundary layer immediately ahead of the advancing front. The rosette form seems to be a strictly coupled growth, not requiring the presence of bifilms. It seems probable that it forms in the pockets of undercooled liquid that would impinge from time to time on an outside surface, thus the constitutional undercooling and the thermal undercooling near to the wall of the mould would be additive to promote nucleation on some marginally favourable particle. The strongly undercooled (i.e. fine) graphite at the centres of some rosettes seen in Figure 6.46(a) seems to confirm this suggestion for some conditions, but Figure 6.46(b) suggests conditions more gently undercooled on this occasion. The even spacing is easily understood if nucleation is prolific, because those nucleation events that occur too near to a neighbour will be less favoured and may even re-melt as its neighbours give off their heat of formation as they grow.

At still lower undercoolings, after initiation, the growth of the coupled eutectic is probably so fast that it will spread sideways to cover large undercooled regions at the mould wall. It will subsequently proceed on a substantially planar growth front, growing away from the wall. Thus only *growth* can now occur. It is the growth phenomenon that dominates the structure we call coral graphite (Figure 6.17). This fine, more highly undercooled eutectic, sometimes designated Type D or E according to ASTM specification A247, seems in general to have been avoided for general engineering castings. This is possibly because the inter-flake diffusion distance is now so small that only ferrite can be formed, limiting the strength of such irons in the as-cast or heat treated conditions.

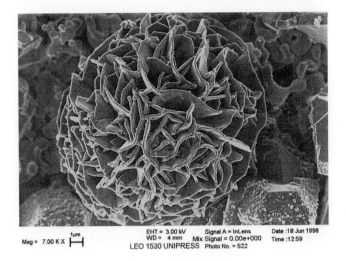

FIGURE 6.45

A rosette of flake graphite, expanding to form a 'cell' or 'eutectic grain'.

Courtesy of Fraz, Gorney and Lopez (2007).

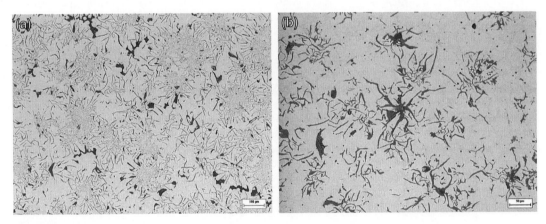

FIGURE 6.46

An array of eutectic graphite grains (cells) in a grey iron (a) rapidly solidified at a high undercooling to give a fine graphite spacing at the center of the cell; (b) slower freezing showing a coarser growth.

Courtesy Serge Grenier QIT 2010.

During coupled growth, flakes have to continually realign their growth direction because of the intrusion of their neighbours into their growth space. Because the growth direction of graphite is mainly parallel to the basal (0001) plane, this means that the crystal has to develop faults to allow it to change direction. This explains the 'coral' type of graphite morphology which is highly faulted (Zhu et al., 1985). We would expect therefore that types D and E graphite would be highly faulted, containing high defect densities, whereas rosette (or cell) graphite would represent an intermediate case as a result of its larger spacing. Flake graphite, as mentioned previously, would be expected to contain the least faults, growing while floating in the liquid, experiencing no significant restraint, and resulting in a nearly perfect crystal.

For interested readers, Nakae and Shin (1999) presented beautiful micrographs to illustrate in detail the close similarity between the coral forms of Fe-C and Al-Si alloys (even though they fail to mention, and appear possibly not even to have noticed, the coral growth of Fe-C shown by their work).

6.5.7 SPHEROIDAL GRAPHITE IRON (DUCTILE IRON)

When excess magnesium is added to the melt, the oxide bifilms are completely eliminated. In the case of silica bifilms the silica will be reduced by magnesium to (1) silicon metal which will dissipate into solution in the matrix and (2) solid magnesium oxide that will precipitate probably on the pre-existing nuclei that originally sat on the films, possibly augmenting these original particles. The reaction is simply:

$$SiO_2 + 2Mg = Si + 2MgO$$

The sudden and total loss of bifilms means that only solids remaining in suspension in the melt now are the original particulate nuclei, possibly augmented by additional MgO. If sulphur is also present in the melt, the MgO likely to contain a component of MgS. These compact nuclei are now the only nucleation sites available for the precipitation of graphite. The precipitating graphite grows over the compact nucleus, wrapping completely around it so as to form a compact initiating morphology.

The disappearance of the bifilms and the initiation of spheroids are shown schematically in Figure 6.47. The 'wrapping around' process (Figure 6.47(d)) may consist of re-nucleation of many separate microscopic grains of graphite on favourable fragments of the oxy-sulfide surface. The growth mode was originally proposed by Hillert and Lindblom (1954) to be an addition of carbon atoms to spiral growth steps generated by <0001> oriented screw dislocations (Figure 6.48). In this way, the radial structure of graphite nodules develops from the graphite grains growing radially out from the compact nucleus to form the familiar approximately spherical nodule (Figure 6.49). Lacaze and co-researchers (Theuwissen et al., 2012) found recent evidence that growth might occur instead by a repeated nucleation from the radial grain boundaries, so that growth occurs by rapid propagation in the 'a' direction.

Johnson and Smart (1977) used the sophisticated and respected perturbation analysis by Mullins and Sekerka to suggest that interfacial energies are of importance in spherodising graphite nodules up to a diameter of perhaps 50 nm, after which the spherical form can no longer be stabilised. Thus, much speculation by earlier authors that interfacial energies may be important in defining the shape of spheroidal graphite seems irrelevant.

In his review, Stefanescu (2007) concluded that all the evidence points to nodules initially growing freely in the liquid, subsequently developing a shell of austenite, and finally contacting and becoming incorporated into an austenite dendrite. A minor modification of this development may be envisaged in which the graphite nodule does not grow a shell of austenite until it contacts an austenite dendrite. At that moment, a shell of austenite would be expected to wrap itself

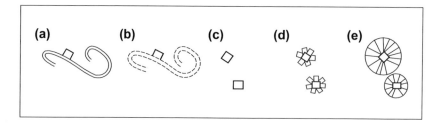

FIGURE 6.47

(a) The melt with a population of bifilms with sundry attached nuclei from impurities or preconditioning additions; (b) the elimination of the silica-rich bifilms by Mg; (c) the survival of the existing nuclei plus possible additional nuclei from inoculation; (d) the nucleation of graphite, wrapping completely around nuclei, particularly if they happen to pass through supercooled regions; and (e) growth of spheroids.

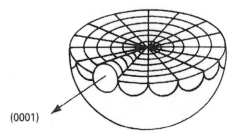

(0001)

FIGURE 6.48

The probable structure of a graphite nodule (Stefanescu, 1988).

FIGURE 6.49

Graphite nodules in an austempered iron indicating nucleation on a small central inclusion followed by radial growth (Hughes, 1988).

rapidly around the nodule. Painstaking metallography would be required to clarify this detail. Anyway, whatever the finer details of the encapsulation process, the shell of austenite seems a key feature associated with the growth of spheroids.

The separate nature of the growth problem can be appreciated from a close look at the graphite structure around some central nucleating particles. The structure in graphite spheroids close to the nucleating particle is sometimes seen to be highly irregular (Figure 6.50). The graphite form in this region appears chaotic, as perhaps might be expected if the effective undercooling leading to dramatically fast initial growth were to be in some kind of dendritic form (Figure 6.38). Clearly, after a very short growth distance, whatever original crystallographic orientation the graphite might have enjoyed with the oxy-sulphide nucleus is quickly lost in the rapid, chaotic growth. However, after a further small distance, the graphite organises itself, and develops its nicely ordered radial grains typical of a good spheroid. Thus the organisation of the growth takes time to develop and is a macroscopic phenomenon. There is a strong analogy with the planar growth condition of a metal under conditions of low constitutional undercooling.

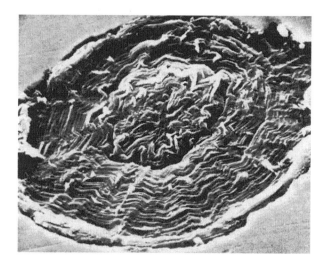

FIGURE 6.50

The chaotic growth structure of a graphite spherule, cathodically etched in vacuum, and viewed at a tilt of 45° in the SEM (Karsay, 1985, 1992).

Reprinted with permission of the American Foundry Society.

The spherical growth form almost certainly has at least some contribution from macroscopic influences. To influence the roundness of the growth form, a mechanism cannot be on an atomic scale, but must act on the scale of the spheroid itself. Such mechanisms might include (1) a low constitutional undercooling condition in the surrounding liquid when in the free-floating state, encouraging a smooth interface analogous to the planar growth of the metal at low driving force for growth, or (2) a mechanical constraint imposed on the expanding sphere when surrounded by solid, but plastically deforming, austenite. It is just possible that (3) some adsorption on the surfaces of the growing crystal, limiting growth directions, may be important. There are no shortages of theories on this issue, and facts are hard to establish.

We shall attempt to evaluate here what seems to me to be perhaps the most likely mechanism: the mechanical constraint by the surrounding solid.

The spherical morphology of the graphite nodules may be encouraged by the mechanical constraint provided by the nodule having to force its growth against the resistance provided by its surrounding shell of austenite (Roviglione and Hermida, 2004). Several studies have clearly revealed the deformation of austenite dendrites by the growth of internal nodules (Figure 6.44). This lumpy dendrite morphology has been attributed to various mechanisms, all of which are likely to contribute to some degree:

1. Ruxanda and colleagues (2001) and Stefanescu (2008) assumed the protrusions to be the natural growth shapes arising from cooperative growth of austenite and graphite by diffusion from the liquid.
2. Buhrig-Polackzed and Santos (2008) indicated in a schematic illustration that the contact between nodules already surrounded by austenite shells and the austenite dendrites results in the mutual assimilation of the two sources of austenite, to create a local bump on the dendrite. (Some subsequent surface smoothing driven by surface energy would be expected to occur rapidly.)
3. Deformation of the dendrite by plastic flow, locally expanding the surrounding solid to accommodate the increasing volume occupied by the graphite has to be important. This effect appears to have been generally overlooked, but appears to be important and worthy of examination, as discussed later.

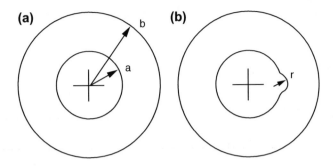

FIGURE 6.51

(a) A thick spherical shell expanding plastically because of internal pressure; (b) a perturbation radius *r* is not favoured because higher local pressure is required, so that sphericity is encouraged.

The pressure developed in a thick spherical shell (Figure 6.51) deforming plastically because of internal pressure is quantitatively expressed by (Chadwick, 1963)

$$P = 2Y \ln b/a \qquad (6.10)$$

Where *P* is the internal pressure, *Y* is the yield stress and *a* and *b* are the internal and external radii of the shell. The logic is as follows: if a perturbation to the spherical graphite shape were to occur, having a necessarily smaller radius, *r*, the pressure to plastically extend the growth at this location would (according to Eqn (6.1)) be increased (Figure 6.51(b)). Thus growth of the extension of smaller radius would be discouraged because additional pressure would be required to stabilise the perturbation. The easier spherical growth mode, simply expanding the uniform radius *a* would therefore be encouraged. It seems therefore that there is some *qualitative* justification for believing that mechanical forces stabilise the spherical growth mode of the nodule.

However, it is useful to ascertain whether there is *quantitative* justification for this mechanism. If we take *Y* to be approximately 6 MPa (Campbell, 1967) for austenite at the melting point of iron, and *a* = 2 nm and *b* = 20 nm, we find *P* = 30 MPa approximately. Even at values of *a* = 20 μm and *b* = 200 μm, *P* is of course unchanged at 30 MPa because the ratio *a*/*b* is the same, indicating that there is a substantial restraining pressure, approximately five times the yield stress, on the growth of the nodule during most of its life.

With regard to the possible asymmetric effect of a perturbation of radius *r*, taking *r* = *a*/2 to *a*/10 locally increases *P* to approximately between 35 and 60 MPa, respectively. Thus a rounding effect from mechanical smoothing of the forces to expand the austenite shell appears to be important. Although a creep model rather than the above plastic model might give a somewhat more accurate result, the previous result can be relied upon to give us an order-of-magnitude estimate of the effect. I have found creep models and plastic flow models to give very similar answers provided the rates of deformation are similar (Campbell, 1968a). Even so, clearly, more work is required to confirm this preliminary indication.

In general agreement with this conclusion, Jiyang et al. (1990) used colour etching to reveal the austenite shells around graphite. They found that if the shell formed quickly and completely, the nodule developed as a sphere, whereas slow-developing or non-enveloping shells led to mis-shapen nodules, in agreement with our previous logic.

As part of their work in the metals treatment industry, Lalich and Hitchings (1976) observed that some nuclei for nodules were far from round. They noted that the graphite originally wrapped completely around such curious shapes (in agreement with our assumptions in this book) but as growth of the nodule advanced, so the nodule quickly became progressively more round as would be expected from the effect of mechanical constraint encouraging smoothing described here.

In passing, it may be significant that on the addition of Mg causing dissolution of the silica-rich oxide bifilms, any residual gas trapped between the films is expected to be released. In this way, it seems possible that clouds of fine bubbles, consisting mainly of argon, will be released into the melt. It seems likely that some Mg vapour will also diffuse into the bubbles. It is not easy to define the sizes of such bubbles with any accuracy. For instance, a bifilm of 100 μm square and average 1 μm gas gap would yield a pore of approximately 20 μm diameter. A similar bifilm of average 0.01 μm gas gap would form a pore approximately 5 μm diameter. A very small bifilm of 10 μm square and 0.01 μm average spacing would create a 1 μm diameter pore. Thus it seems that a fog of bubbles in the range of approximately 1–20 μm is to be expected.

It is intriguing, but not perhaps relevant, that there is a theory proposing that nodules nucleate from Mg bubbles in suspension in the melt (Gorshov, 1964; Itofugi, 1996). However attractive this hypothesis might be to explain graphite coatings inside pores in solidified castings, as a result of the reduction in strain energy involved, any strain energy relief in the liquid state is zero, and the reduction of surface energy to encourage such precipitation in the liquid state seems negligible. Furthermore, the successful incorporation of solid particulate oxide nuclei into the bubble depends critically on further reduction in interface energies. This is unlikely for particles of oxide formed by precipitation in situ in the liquid which will be in atomic contact with the melt (i.e. will be well 'wetted', being a necessary condition for the particle to be a nucleating agent). Such particles will be energetically rejected by bubbles. Thus although a mechanism for the presence of extremely fine bubbles may be provided by the present analysis, it appears to be irrelevant to the formation of graphite nodules. It does not offer support to the gas bubble nucleation hypothesis.

A further interesting aside can be noted. The transformation of the planar cracks sandwiched inside the bifilms into clouds of fine bubbles that may float and escape from the alloy is the essence of the process by which apparently brittle grey iron becomes ductile. (Ductile iron only becomes embrittled once again, as noted later, if oxide bifilms are re-introduced by turbulence by handling of the melt or poor filling system design of the casting.)

Finally, curious observations such as those reported by Yamamoto and colleagues (1975) in which flake iron is converted into nodular iron by simply purging the melt with fine bubbles of nitrogen, argon or carbon dioxide become explicable. From experience in the light metals industries, it is known that purging with gases can eliminate bifilms from melts. Thus spherodisation of the graphite appears to be achievable via a purely mechanical route, replicating the condition achieved chemically by the addition of Mg. However, the observations by Aylen et al. (1965), in which the graphite structure of pig iron is refined by bubbling nitrogen through the melt, is less easily explained. Perhaps nitride bifilms are active, or possibly sufficient oxygen contaminates the nitrogen to create thin silica films. Such observations are tantalising, and will not be understood without incisive research.

Mis-shapen spheroids

The presence of ill-formed spheroids, particularly if present in large numbers, is widely known to be associated with the reduction in mechanical properties of nodular irons. Hughes (1988) describes how a good ductile iron can achieve at least 90% nodularity but less good irons can fall to as low as 50% or less and suffer reduced properties. The 50% or so component of flakes or other non-nodular shapes does not particularly affect properties such as proof strength but greatly reduces those properties related to failure, such as tensile strength and ductility. Hughes attributes this loss of fracture resistance to the sharp notches at the root of flakes. However, this widely held belief presupposes that the flakes act as cracks. This is probably only true if the flakes are formed on bifilms; the bifilm providing the crack. It is the presence of the crack that has to be viewed as the principal cause of failure.

In terms of the bifilm hypothesis, graphite would be expected to grow on oxide bifilms. Thus, if Mg treatment were carried out to eliminate silica bifilms, spheroids would be created in suspension, floating upwards only very slowly as a result of their small size. However, many Mg addition techniques are extremely turbulent, so that, unfortunately, large quantities of MgO and Mg silicates are expected to be created by the mechanical entrainment of the liquid surface. These new bifilms will be permanent defects formed from highly stable magnesia (MgO) or magnesium silicate (MgSiO$_3$). Before pouring into the mould, some of these will fortunately float out in the ladle, adding to the Mg-rich slag. Thus, given a reasonable time for separation (an interesting and clearly important process variable that seems not well researched), not all of these Mg-rich bifilms find their way into the castings to impair the structure and properties.

Unfortunately again, on pouring into the mould, additional large quantities of Mg-rich oxide bifilms are likely to be re-introduced, particularly if the mould filling system is a poor design.

On contact with an oxide bifilm, a spheroid would tend to attach firmly to the film, thereby reducing the overall energy of the system. Further growth of the spheroid would then necessarily spread over the plane of the film. The symmetrical spherical constraint previously provided by the surrounding austenite is now also destroyed, aiding the non-spherical development. Thus the spheroid will grow to become significantly mis-shapen.

This effect can be seen in Figures 6.31 and 6.32(b). In Figure 6.31, several oxide bifilms have been straightened by the growth of dendrites to lie along 100 planes. The nodules attached to the bifilms are clearly poorly shaped. Additional mis-shapen nodules elsewhere in the structure are expected to be lying on random areas of bifilm not straightened by dendrite growth. In Figure 6.32(b), it seems likely that the bifilm lies along the grain boundary, decorated with mis-shaped nodules, ferrite and (probably) some scattering of porosity or cracks.

The phenomenon of mis-shapen nodules has up to now appeared a mystery. The effect is therefore predicted to be associated with either (1) insufficient dwell time for the damage introduced during Mg addition to float out or (2) poor casting practice, in which an otherwise nicely inoculated and spherodised melt is re-contaminated with bifilms.

It is interesting to predict that more time after the spherodising treatment to allow the melt to clear, together with a properly designed filling system, or counter-gravity filling system, should completely solve this problem. A step in the right direction was taken by Takita and colleagues (1999) who observe that nodules are converted from mis-shapes to spheroids by the use of a filter to take out the 'inclusions produced by inoculation'. This positive step contrasts with that taken by Liu and co-workers (1992) who, after the Mg addition, added 'postinoculants'. The postinoculants were highly successful to increase the nodule count, but led to a disastrous fall in nodularity. This was almost certainly a result of adding the inoculants through the melt surface. The inoculants would, of course, have their own oxide skins, which would have doubled up with the entrained the surface oxide of the melt to create an asymmetrical bifilm with an alloy oxide on one side and a Mg-rich silicate on the other. On contact with growing graphite particles, this major source of fresh bifilm contamination would lead to the growth of non-spheroids.

Furthermore, of course, the significant reduction of properties associate with malformed spheroids cannot be the direct result of the shape of the spheroids because they occupy such a small volume fraction of the alloy. The loss of properties is predicted to be the result of the presence of the bifilms in the melt, occupying a vastly greater cross-sectional area than the spheroids. These extensive bifilms will act as cracks in the casting, significantly reducing properties.

Exploded nodular graphite

'Exploded' spheroids (Figure 6.52) are commonly seen in irons subject to graphite flotation, and especially if the composition of the iron is sufficiently hyper-eutectic (Sun and Loper, 1983; Druschiz and Chaput, 1993).

This undesirable morphology is not easily explained at this stage as a result of relatively little experimental work to clarify the problem. Cole (1972) suggested they had suffered re-melting as a result of being carried by convection in and out of hot zones of the liquid. This seems most unlikely, however, because if the nodule had grown uniformly in a compact morphology the uniform graphite would be expected to have a substantially uniform rate of dissolution; solidification and re-melting would be expected to be reversible.

Because exploded nodules appear exclusively in the flotation region of hyper-eutectic irons two far more likely factors are.

1. Nodules growing in a sufficiently hyper-eutectic melt (Sun and Loper, 1983) will experience an enhanced driving force for growth because of the carbon supersaturation that develops as the melt cools. This will encourage growth instabilities leading to 'dendritic' rather than 'planar' growth, leading to exploded rather than smooth spheroid surfaces.
2. The nodules may have nucleated early in the liquid phase and grown without the benefit of the mechanical constraint of the austenite. When not mechanically pressurised to remain spherical, the nodule will be free to grow more like a dendrite, developing instabilities that grow into projections to its growth front, finally developing the characteristic exploded forms. Evidence for mechanical restraint as a powerful effect is presented in the section on nodular graphite previous.

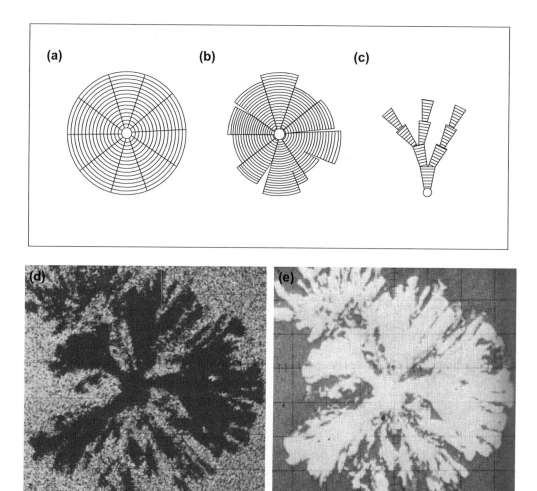

FIGURE 6.52

(a) Spheroid and (b) malformed spheroid; (c) chunky graphite (after Liu et al., 1983; (d) SEM iron image of an exploded spheroid; (e) electron image (Cole, 1972).

6.5.8 COMPACTED GRAPHITE IRON

If the addition of magnesium is more carefully controlled to some level intermediate between spheroidal and flake iron, compacted ('vermicular' based on the Italian for 'worm-like') graphite is the result (Figure 6.53).

In our bifilm model, it is clear that most of the oxide bifilms will be quickly dissolved by the addition of Mg. However, small patches may remain if the Mg addition is not too high; the tiny patches on which the original nuclei sat will be resistant to dissolution because they will be stabilised by their attachment to the nuclei. (Naturally, it will have been energetically favourable for the nuclei to attach to the bifilm, so that the combination of nucleus and film will enjoy a reduction in overall energy, stabilising the combination.) Only half of the bifilm will be retained in this way, its distant 'twin' half not enjoying the protective influence of the nucleus will dissolve and disappear because it is separated by an

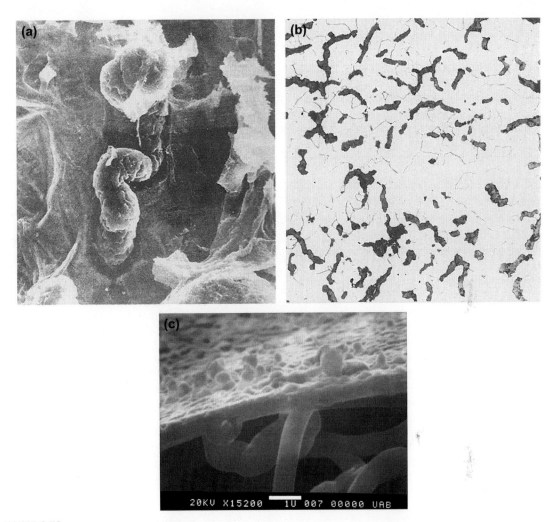

FIGURE 6.53

CGI viewed by (a) SEM deep etching, (b) optical metallography (Stefanescu et al., 1988) and (c) carbon wool fibre (growing off a lustrous carbon surface film on a grey iron) as a possible vapour phase–equivalent of compacted graphite (Campbell and Naro, 2010).

air gap. Only the small part of one half of the bifilm together with its unbonded interface, the remnants of the layer of air, will remain (Figure 6.54(b)).

The subsequent nucleation of graphite on the nucleus will result in rapid spreading of growth around the nucleus. On arrival at the non-wetted interface belonging to the residual patch of bifilm this spreading will be arrested (Figure 6.54(c)). The graphite has now grown to reach the residual layer of air on the remaining patch of half of the bifilm. The further growth of graphite is forced to occur not radially, but in general unidirectionally away from the bifilm residue (Figure 6.54(d)). Clearly, the growth cannot now be spherical, but its exact form is not possible to predict. Cole (1972) had observed a fine, unidirectional spiral structure similar to the worm-like growth mode clearly seen in Figure 6.53. Liu and

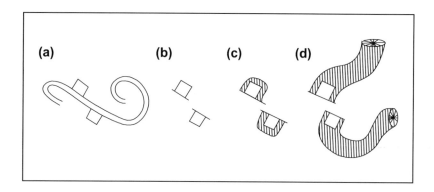

FIGURE 6.54

Formation of CGI by addition to (a) nuclei on bifilms of just sufficient Mg to (b) eliminate most of the silica-rich bifilms except for the remnant attached to the nuclei. (c) Inoculation promotes graphite initiation on the nuclei; (d) growth cannot occur by wrapping completely around the nucleus, so that initial growth cannot be spherically symmetrical, possibly favouring unidirectional growth.

colleagues (1980) found that the growth direction is along the C axis (0001 direction perpendicular to the basal planes) and appears to develop by a spiral dislocation mechanism as witnessed by the coarse and irregular spirals that they observe.

The great sensitivity of the compacted graphite morphology to magnesium concentration is corroborated by the proposed bifilm mechanism. If the Mg level is too low, residual bifilms will encourage flake graphite, whereas if the Mg level is too high, the limited stability enjoyed by the residual bifilm patches will be overcome, and the last remaining patches of bifilm will be dissolved, allowing the growth of totally spherical grains.

After nucleation of the compacted graphite form, a number of workers (for instance, Su, 1982 and Mampaey, 2000) found that the continued growth of the graphite appears to occur solely in the liquid. Because the graphite stays in contact with the liquid, it transfers its expansion directly to the liquid, reducing feeding requirements (Altstetter and Nowicki, 1982). This welcome and valuable good behaviour is in contrast to ductile iron in which the graphite transfers its expansion to its surrounding solidified shell, expanding the casting, and the mould, thereby increasing feeding requirements. Alonso et al. (2014) appeared to find this behaviour in their study of graphite expansion of flake and nodular irons.

As a fascinating final thought concerning compacted graphite iron (CGI), the fibrous graphite seen growing from the vapour phase in Figure 6.35, and shown in close up in Figure 6.53(c) appears to be identical in form to the graphite grown from the melt as CGI or 'worm-like' graphite, and might therefore be viewed as a three-dimensional model, giving an insight into the internal structure of CGI. These free-growing vapour growth phase and liquid growth phase forms of graphite appear to be identical. It is possible that the carbon concentration in each of these environments is similar, resulting in similar kinetics of deposition, and similar growth modes.

6.5.9 CHUNKY GRAPHITE

Chunky graphite is often observed concentrated in the centers of heavy sections of nodular iron castings. 'Chunky' is not a particularly helpful descriptive adjective for this variety of graphite. Its 'chunkiness' is only apparent under the microscope at high magnification; otherwise, it simply appears to be fine, irregular, branched and interconnected fragments (Figures 6.52(c) and 6.55). Once again, the properties of nodular iron are reduced. However, it seems the loss of properties may, once again, be at least partly associated with the short diffusion distances between branches of the graphite filaments, promoting the development of ferrite instead of the stronger pearlite eutectoid phase (Liu et al., 1983).

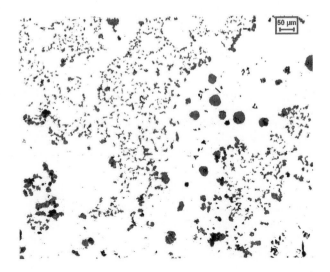

FIGURE 6.55

Graphite nodules and areas of fine, chunky graphite in the thermal centre of a 200 mm cube casting (Kallbom, 2006).

Liu and co-workers (1980, 1983) found evidence that chunky graphite grows along the C-axis direction, as does both nodular and compacted graphite. Furthermore, they reported observations on spheroids that exhibit gradual degeneration, gradually taking on the growth forms of chunky graphite. Thus they concluded that chunky graphite is a degenerate form of spheroidal graphite, and their work implied that chunky graphite grows out from spheroids. Itofugi and Uchikawa (1990) confirmed the identical growth modes of spheroidal, compacted and chunky graphites as illustrated schematically in Figure 6.52.

All these workers observed the characteristic form of chunky graphite, as an apparently 'stop/start' growth in the C-direction consisting of nearly separate pyramidal 'chunks' linked by a narrow neck, like beads on a string (Figure 6.52(c)). The individual chunk sections comprise layers parallel to the basal plane, but only nanometres thick. This characteristically lumpy growth may be the result of a pulsating or irregular advance of the growth front, with the austenite advancing to nearly grow over the top of the graphite, forming the nearly pinched-off neck of the graphite, only to be overtaken again because the carbon in solution will now buildup in the liquid ahead of the front, accelerating the next phase of growth of the graphite until the local carbon concentration is depleted once again.

Observations by Kallbom and co-workers (2006) are consistent with an origin associated with bifilms. They observe the chunky graphite to be concentrated in the center of heavy sections, explained by the growth of the freezing front pushing bifilms ahead, and explaining their observations of 'stringers' of graphite nodules. These features are almost certainly sheets of oxides decorated with graphite nodules that have been nucleated on the oxide (analogously to those seen in Figure 6.32). The earlier paper by these authors (Kallbom, 2005) described how the outer several centimetres of the casting can be perfect, with good nodularity, good strength and ductility, but the structure can change abruptly, over a short distance equivalent to only a nodule diameter, to chunky graphite. Thus the central volume of the casting is weak and brittle. Because the problem is buried in the centre of thick section castings it can be difficult to find by non-destructive methods.

All previous evidence that suggests chunky graphite requires both the presence of bifilms combined with an absence of nuclei. It is possible to imagine a mechanism in which following a turbulent pour, many bifilms will be pushed ahead of the forest of growing dendrites, forming, at times, a distinct and abrupt separation of the outer dendritic region from the inner residual liquid. The presence of the concentration of bifilms in the centre will suppress the normal pattern of free circulation that would ensure a good supply of nodules from the outer, cooler regions into the hot central region. Furthermore, because so much time is available, particles such as nuclei and nodules already in suspension in the centre will have time to float out, or existing nodules to dissolve (because they will be unstable at these higher temperatures) depleting the centres of nuclei so that spheroids cannot form.

For those nuclei now floated out to the edge of this region of higher temperature and enhanced segregation, any nodules formed on the nuclei will not enjoy the benefit of a surrounding austenite shell, so that their growth mechanism will more nearly resemble that of an exploded spheroid. Because these will be at the boundary of the central region, their growth is most likely to be an extension of the nodules along the C-axis (Liu et al., 1980) in the direction of the gradually advancing solidification front. In the absence of nuclei, the whole region would be expected to fill with this continuous growth form. The extended size of chunky graphite regions, much larger than cells of other types of graphite (Itofugi and Uchikawa, 1990), corroborate the absence of nuclei in these regions.

The presence of bifilm cracks concentrated in the chunky graphite regions would explain the poor properties that are observed; it is difficult to see how otherwise a continuous graphite phase could reduce properties, particularly because in other irons (such as CGI) a continuous graphite phase is associated with excellent and reproducible tensile properties.

It is hoped that in the near future the explanation for the origin of chunky graphite might be clarified and confirmed by further careful experiments. The key word here is 'careful'. For instance, the experiment by Asenjo and co-workers (2009) involving the placement of inoculants in different branches of a runner system to simulate the casting of separate moulds from one melt. In this way, it should have been possible to compare the effects of different mould inoculation in separate heavy castings. The idea was clever, but regrettably experimentally flawed because, in common with most iron casting, the runners were not designed to be pressurised and fill on a single pass. Thus a reverse flow is likely to have contaminated the mould cavities, and all the cast material would have suffered from turbulence and air entrainment. All the cavities would therefore have been contaminated with varying amounts of inoculants from neighbouring cavities, and all would have contained unknown quantities of oxide bifilms. Clearly, in the future, much greater sophistication of melting and casting will be required for experiments designed to clarify the solidification mechanisms for cast irons.

6.5.10 WHITE IRON (IRON CARBIDE)

Work by Rashid and Campbell (2004) demonstrated the nucleation and growth of carbides on oxide bifilms in vacuum-cast Ni-base superalloys (seen later in Figures 6.62 and 6.63). It would be expected therefore that an analogous reaction would occur in Fe-C alloys, because the austenite forming during solidification also possesses a closely similar fcc structure.

Carbides in irons appear to form preferentially at grain boundaries, and appear often to be associated both with residual graphite (sometimes as nodules, malformed nodules or flakes aligned with the boundary) and pores all forming on the same boundaries (Figure 6.32(b)). These are all clues to the possible presence of an otherwise invisible bifilm.

Faubert et al. (1990) have studied carbides in heavy section austenitic ductile iron. Towards the top of their castings they find degradation of properties more serious than they would have expected from the carbides themselves. They suspected that the real impairment was caused by the presence of films that had floated into this region. The 'films' would, of course, have been 'bifilms', in line with their profound effect on properties. Bifilms would segregate to grain boundaries, and possibly actually constitute the boundary. The presence of the bifilm is not only inferred from (1) the cracked carbides but also (2) from the linear rows of nodules seen in micrographs from this work, (3) from the pores as the residues of air bubbles trapped between the films, (4) from the graphite flakes sitting in the boundary (called by the authors, unflatteringly, 'degenerate' graphite). Both graphite and carbides are expected to form on the wetted, outer surfaces of the bifilms. The presence of the central unbonded region (including the pores) between the films, constituting the crack through the interiors of the carbides, explains the *apparent* brittleness of the carbides. These intermetallic compounds would otherwise be expected to be strong, resistant to failure by cracking at the modest stresses that can be induced by solidification and cooling. The plasticity and crack resistance of intermetallic phases has been discussed in Section 6.3.7.

Stefanescu (1988) quoted the work of Hillert and Steinhauser (1960) in which the growth of iron carbide eutectic (ledeburite) occurs by the spreading of carbide (cementite) across a plane, followed by the development of a rod type of eutectic at right angles. It is tempting to consider that the original planar expansion would have been facilitated by growth across the surface of a bifilm. The bifilm would originally have been randomly crumpled, but would have been straightened by the progress of the carbide across its face, thus creating an essentially planar crack that would constitute a serious defect in the carbide. Associated branching cracks would have arisen from irregular folds in the bifilm.

Although carbides in grey irons are feared for their disastrous effect on machinability, they are of course desirable in those components required to withstand wear and abrasion. Furthermore, a complete white iron structure is required before subjecting the iron to malleabilising treatments. Although the role of malleable iron has been greatly eroded by the rise of ductile iron, for some products and some foundries the economics are not so different, so that at this time a respectable tonnage of malleable iron is still produced. It will be interesting to see the eventual outcome of this competition.

6.5.11 GENERAL

Several workers have noted that the initial stages of formation and growth of the graphite particles start with a spheroidal shape, regardless of whether these are destined to grow into spheroids or flakes, or any other type of graphite (for instance, Jianzhong, 1989; Itofugi et al., 1983). This observation is consistent with the scenarios portrayed in Figures 6.39, 6.40 and 6.47(d). The spread of the graphite around the originating nucleus is always nearly spherical, or partly spherical. Even the partly spherical forms will appear spherical in some sections.

Overall, it seems clear that both nucleation and growth mechanisms influence graphite morphology in both flake and spheroidal forms, although these mechanisms dominate to different extents in different circumstances. For compacted, undercooled coral and chunky graphites nucleation occurs once to initiate each cell, after which continuous growth dominates the development of the structure, leading to continuous branching morphologies. Indeed, the finer, fairly continuous forms, coral, chunky and compacted graphites, are so similar that they often seem to be confused in the casting literature (see Figure 6.56). They do seem similar in the sense that they all appear to be more or less coupled growth forms, advancing together with the austenite.

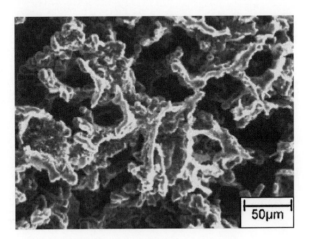

FIGURE 6.56

SEM image of deeply etched CGI closely resembling coupled eutectic coral structure, illustrating the overall similarity of these structures.

Courtesy T Prucha, AFS 2009.

If, as proposed here, the various forms of graphite are significantly influenced by the presence or absence of bifilms, it would explain the historical reluctance of cast irons to give up their secrets and allow an understanding. This seems typical of bifilm phenomena.

Furthermore, it seems likely that the principle cause of reduced mechanical properties in all cases of non-spheroidal forms of cast irons is the presence of various kinds of oxide bifilms acting as cracks. Ductile iron owes its ductility to graphite that forms in the absence of oxide bifilms, but when oxide bifilms (mainly magnesia-rich) are entrained by poor casting technology, even ductile iron can be seriously reduced in ductility, possibly failing brittlely by the 'plate fracture' mechanism as we have seen.

6.5.12 SUMMARY OF STRUCTURE HYPOTHESIS

1. Cast iron melts normally contain oxide double films (mainly SiO_2 bifilms) in suspension.
2. Inoculation produces oxy-sulfide particles which nucleate on silica-rich oxide bifilms (and unintentionally introduces entrained oxide bifilms to further aid the growth of flake graphite).
3. Graphite nucleates on the oxy-sulfide particles, and grows primary graphite, spreading over the bifilms, straightening the bifilms and forming flakes of crystallographically near-perfect graphite; the presence of the bifilms, as cracks, trapped inside or alongside graphite flakes accounts for the relatively poor tensile properties of flake irons.
4. Compacted graphite nucleates on oxy-sulfide nuclei occupying bifilm residues. Growth occurs by fibrous filaments extending freely into the liquid.
5. Spheroidal graphite nucleates on oxy-sulphide nuclei free from bifilms. The spherical growth morphology is initiated by the freedom that the graphite has to wrap completely around the nucleus, but is later importantly encouraged by mechanical constraint of the austenite matrix; the absence of bifilms, eliminated by the Mg addition, explains the high mechanical properties.
6. Eutectic coral graphite nucleates on (currently unknown) nuclei at low temperatures, expanding to form cells of coupled growth with austenite, consisting of highly faulted continuous branching filaments of graphite in the austenite matrix; bifilms play no part in this growth mode.
7. Mis-shaped spheroids appear to be spheroids that have encountered an oxide (probably Mg-rich) bifilm, subsequently growing along the bifilm and losing sphericity. The presence of the bifilm also destroys the symmetrical mechanical constraint of the austenite that favours sphericity.
8. Chunky graphite occurs in heavy section ductile iron in regions. The central regions are isolated by, and probably full of, bifilms created by the poor casting technique. Thus convection and the redistribution of nuclei and nodules are suppressed, and existing nuclei can float out, so that the central region becomes devoid of nuclei. Graphite therefore develops akin to coral morphology. As such, it may be a coupled eutectic form, and its 'beads on a string' morphology may result from an unstably advancing growth front.
9. Exploded spheroids may be the result of growth in the liquid, without the benefit of the mechanical constraint of an austenite shell. They are favoured by (1) high CEV and (2) the presence of oxide bifilm which interfere with the sphericity of growth.
10. Carbides form at very low temperatures on oxide bifilms. The presence of bifilms in the carbides explains the apparently brittle behaviour of these strong, crack-resistant intermetallics.

6.6 STEELS

There are such a wide variety of steels of widely differing properties that it is possible only to generalise with extreme caution. In general, we shall consider the simplest of steels; the carbon steels (often more accurately referred to as carbon/manganese steels, because Mn is such a common additional alloying element) and stainless steels. Stainless steels fall into main groups: ferritic, austenitic and duplex (i.e. mixed ferrite and austenite).

Other steels are commonly known from their microstructures, including ferritic, austenitic, pearlitic, martensitic and bainitic. Tool steels (another wide category) encompass forming and cutting tools often with high Mo or W additions to form hard carbides.

Although astonishing strengths available in the final worked and heat-treated products, sometimes approaching or even in the GPa range, from the point of view of the casting technologist, the key differences between the steels and the light alloys are as follows.

1. The high melting and casting temperatures encouraging more severe and faster reaction with the environment.
2. The possibility of benefiting from a partly or completely melted surface oxide film.
3. The higher strengths of the steels result in higher internal stress during freezing, leading to higher driving forces for the initiation of such defects as shrinkage porosity, hot tearing and cracking.
4. The higher density of steels requires stronger, reinforced moulds and cores to resist flotation and heavy weights to prevent the lifting of copes.
5. The greater difference in density between the metal and its oxides encourages the faster flotation of entrained defects after pouring events.

6.6.1 CARBON STEELS

Steelmaking practice for the production of carbon steels traditionally starts from pig iron produced from the blast furnace. The high carbon in the iron, in the region of 3–4%, is the result of the liquid iron percolating down through the coke in the furnace stack. (A similar situation exists in the cupola furnace used in the melting of cast iron used by iron foundries.) Oxygen is added to oxidise the carbon down to levels more normally in the range of a few tenths of a percent. The bonus from the burning of the carbon is the huge and valuable increase of temperature that is needed to keep steels molten. Oxygen to initiate the CO reaction is added in various forms, traditionally as shovelfuls of granular FeO thrown onto the slag, but in modern steelmaking practice by spectacular jets of supersonic oxygen. The stage of the process in which the CO is evolved as millions of bubbles is so vigorous that it is aptly called a 'carbon boil'.

After the carbon is brought down into specification, the excess oxygen that remains in the steel must be lowered by deoxidising additions to prevent a 'carbon boil' as a result of the positive segregation of carbon and oxygen during freezing. Common deoxidisers have been manganese, silicon and/or aluminium. In modern practice, a more complex cocktail of deoxidising elements is added as an alternative or in addition. These often contain small percentages of rare earths to control the shape of the non-metallic inclusions in the steel. It seems likely that this control of shape is the result of reducing the melting point of the inclusions so that they become at least partially liquid, adopting a more rounded form that is less damaging to the properties of the steel. Also, of course, such compact spherical inclusions float out more rapidly than solid film-type deoxidation products, so that the steel is much cleaner, and properties are higher and more uniform as a result of the absence of bifilm-type cracks.

In most steel foundries, only *steel melting* is carried out from scrap steel (not made from pig iron, as in *steelmaking*). Because the carbon is therefore already low, there is often no requirement for a carbon boil.

The potentially significant problem that remains in the absence of a carbon boil is that hydrogen may remain in the melt. In contrast to oxygen in the melt that can quickly be reduced by the use of a deoxidiser, there is no quick chemical fix for hydrogen. Hydrogen can only be encouraged to leave the metal by providing a dry and hydrogen-free environment. Hydrogen then will gradually evaporate off from the melt, tending to equilibrate with its surroundings. If a carbon boil can be induced, the loss of hydrogen will be rapid. However, if this cannot be induced, possibly in some artificial way, and if environmental control is insufficiently good, or is too slow, then the comparatively expensive last resort is vacuum de-gassing, although the use of argon-oxygen de-gassing treatment is now more common which can efficiently flush hydrogen down to very low levels sometimes required for large castings whose dimensions (measured in large fractions of metres) exceed the distance that hydrogen can diffuse out during heat treatments.

6.6.2 STAINLESS STEELS

The percentage ferrite in cast stainless steels at room temperature depends mainly on the composition of the alloy. This is summarised in the Schoefer diagram presented in Figure 6.58. The equivalent Cr and Ni contributions are estimated from the weight percentage of alloying elements:

$$Cr_{equiv} = \%Cr + 1.5(\%Si) + 1.4(\%Mo) + \%Nb - 4.99$$

$$Ni_{equiv} = \%Ni + 30(\%C) + 0.5(\%Mn) + 26(\%N - 0.02) + 2.77$$

Fully austenitic and fully ferritic stainless steels suffer a number of cracking and intergranular corrosion problems that seem to me to be the result mainly of oxide bifilm problems, most probably as a result of dendrite straightening, leading to a population of cracks throughout the alloy.

However, most cast austenitic steels do not have the single phase structure that their name implies; their structure is usually duplex. Ferrite in the range 5–25 volume percent is found to be valuable for (1) improved strength; (2) improved weldability and castability (reducing cracking problems); and (3) improved resistance to corrosion, stress corrosion cracking and intergranular corrosion attack in certain corrosive environments. It seems possible that the formation of the austenite during the growth of the primary ferrite dendrites might interrupt and hamper bifilm straightening in some way.

The advantages of the duplex structure are maximised at about 50/50 volume percent mixture (actually encompassing the small range of approximately 40–60 volume percent). It is *only* these steels that are known in the industry as '*duplex stainless steels*'.

An additional method of classifying stainless steels is according to their stainlessness, or ability to resist corrosion as quantified by their pitting resistance equivalent. A commonly used equation for ranking stainless compositions is given by Davidson (1990):

$$PRE = \%Cr + 3.3(\%Mo) + 16(\%N)$$

Four categories of improving levels of pitting resistance equivalent can be listed (together with typical examples):

	Pitting Resistance Equivalent	**Example**
1.	Approximately 25	(23Cr steels with zero Mo)
2.	30–36	(22Cr steels)
3.	32–40	(High-alloy 25Cr steels)
4.	>40	(Super duplex steels 25Cr-7Ni-4Mo)

FIGURE 6.57

Bifilm cracks revealed by red dye in a 20Cr-20Ni-6Mo steel casting. The alignment of the cracks by the growth of columnar grains is evident. The deep red indications are most probably bubble trails.

Courtesy S Scholes.

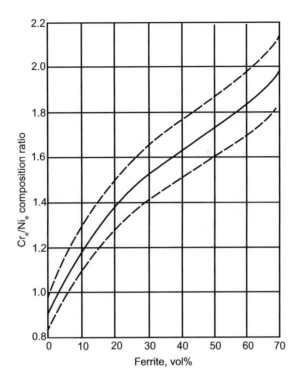

FIGURE 6.58

Schoefer diagram for estimating the ferrite content of steel castings in the composition range 16-26Cr, 6-14Ni, 0-4Mo, 0-2Mn, 0-2Si, 0-1Nb, 0-0.2C, 0-0.19N. Broken lines denote limits of uncertainty.

One of the standard problems with many cast stainless steels is the formation of the infamous sigma phase (σ) usually at grain boundaries. As with most intermetallics, the phase appears to be brittle, exhibiting cracks and leading to failure by cracking. Once again, this maligned intermetallic has all the familiar signs of having nucleated and grown on an oxide bifilm, thereby naturally exhibiting cracks in a material that would normally be expected to be strong and crack-free (see Section 6.3.7). Once again, the avoidance of this problem lies not necessarily in metallurgical control but in the use of appropriate casting technology. Contact pouring and a naturally pressurised filling system is predicted to largely eliminate bifilms, removing these favoured substrates, and thereby suppressing the formation of sigma phase. A large benefit to tensile properties would result from the elimination of the bifilm cracks, and a perhaps smaller benefit from the presence of additional solutes in solution (because the sigma phase may not precipitate in the absence of an attractive substrate) that would enhance strength. The pitting resistance equivalent might also improve, and possibly the resistance of some of the stainless steels to such other imponderables such as stress corrosion cracking.

6.6.3 INCLUSIONS IN CARBON AND LOW-ALLOY STEELS: GENERAL BACKGROUND

Svoboda and colleagues (1987) reported on a large programme carried out in the United States, in which more than 500 macro-inclusions were analysed from 14 steel foundries. This valuable piece of work has given a definitive description of the types of inclusions to be found in cast steel, and the ways in which they can be identified. A summary of the findings is presented in Figure 6.59 and is discussed later.

Each inclusion type can be identified by (1) its appearance under the microscope and (2) its composition.

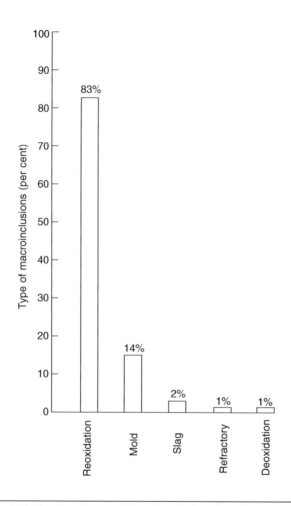

FIGURE 6.59

Distribution of types of macro-inclusions in carbon and low-alloy steel castings, from a sample of 500 inclusions in castings from 14 foundries (Svoboda et al., 1987).

1. Acid slags can be identified by their high FeO content (typically 10–25%), and glass-like microstructure.
2. Basic slags and furnace slags from high-alloy melts can be traced by the calcia (lime), alumina, and/or magnesia that they contain.
3. Refractories from furnace walls and/or ladles have characteristic layering, flow lines and a pressed and sintered appearance including sintered microporosity. Their compositions are reminiscent of those of the refractories from which they originated (e.g. pure alumina, pure magnesia, phosphate bonded aggregates etc.).
4. Moulding sand is identified from the shape of residual sand grains and from its composition high in silica.
5. Mould coat material is normally easily distinguished because of its composition (e.g. alumina, zircon).
6. Deoxidation products are always extremely small in size (typically less than 15 μm) and are composed of the strongest deoxidisers. These inclusions are likely to have formed at two distinct stages. (1) During the initial addition of strong deoxidiser to the liquid steel, when small inclusions will be nucleated in large numbers as a result of the high supersaturation of reactive elements in that locality of the melt. Any larger inclusions will have some opportunity to float out at this time. (2) During solidification and cooling. These stages will be discussed further later.

7. Re-oxidation products are large in size, usually 2–10 mm in diameter, and consist of a complex mixture of weak and strong deoxidisers. In carbon steels, the mixture contains aluminium, manganese and silicon oxides. In high-alloy steels the mixture often contains a dark silica-rich phase, and a lighter coloured Mn + Cr oxide-rich phase. Entrapped metal shot is found inside most of these inclusions. The shot is probably incorporated by turbulence (rather than by chemical reduction of the FeO by the strong deoxidisers). These larger inclusions have been previously known as *ceroxide* defects, not as a result of their content of cerium and other powerful rare earth deoxidisers as I had originally expected; the name was simply made up from '*ceramic oxides*'.

Most research on the formation of inclusions in steels is directed, disappointingly, only at the diffusion reactions between the raw metallic deoxidiser, such as aluminium and liquid steel. The inclusions generated in the supersaturation surrounding the deoxidants during its first seconds of melting and dispersing are interesting, including mixes of dendritic, star-like and spherical forms (Tiekink, 2010). Van Ende and co-workers (2010) did not comment on the presence of the crack that separated the liquid Al sitting upon their sample of liquid steel. This is clearly present in all their work. Even when bifilms are clear they are nearly always ignored. The inter-diffusion between deoxidisers and liquid steel elucidate the processes occurring during *deoxidation* but the much more important processes involved in *re-oxidation*, including entrainment during pouring, have mainly been overlooked so far.

It seems clear that the re-oxidation products are bifilms of various sorts, somewhat scrambled, and often unable to re-open as a result of their formation from partially molten oxide mixtures. They confirm their identity by their content of droplets of metal and bubbles of residual air.

The re-oxidation products are formed during the tapping of the furnace into the ladle. In general they will largely have floated out of the metal, forming the slag layer on the top of the melt in the ladle during the several minutes that it takes to lift the ladle out of the pit and travel to the casting station. Only the pour into the mould is expected to create the re-oxidation products that are observed. There is excellent evidence for this expectation; Melendez, Carlson and Beckermann (2010) developed a computer model that predicts the locations of re-oxidation inclusions in the mould. They achieve good descriptions of the awful mess castings become when they adopt a filling system design which contains a conical funnel basin, uniform internal diameter refractory tubes for the sprue, runners and ingates, and finally gates into parts of the casting that are not at the lowest points of the mould cavity. This filling system was clearly designed to break all the rules, and so be as bad as possible, to illustrate the accuracy of the computer model in predicting castings destined for the scrap heap.

These efforts to predict the distribution of re-oxidation defects seem curiously misplaced when it is reasonably widely known (Kang, 2005; Puhakka, 2010), and certainly predictable, that if a properly designed filling system is applied to the casting the re-oxidation inclusions disappear! Clearly, the so-called re-oxidation problems during casting are entrainment problems.

6.6.4 ENTRAINED INCLUSIONS

Previously, most inclusions introduced from outside sources have been called exogenous inclusions, but this name, besides being ugly, is unhelpful because it is not descriptive. 'Entrained' indicates the mechanism of incorporation as an impact of opposed surfaces, as occurs in the incorporation of a droplet, or the folding-in of the surface. Also, the word entrained draws attention to the fact that as a necessary consequence of their introduction to the melt, such inclusions have passed through its surface, and so will be wrapped in a film of its surface oxide. Depending on the dry or sticky qualities of the oxide, and the rate at which the wrapping may react with the particle, the entrained inclusion and its wrapping can act later as an initiation site for porosity or cracks. Metal too, may become entrapped in the entraining action, and thus form the observed shot-like particles.

Svoboda has determined the distribution of types of macro-inclusion in carbon and low alloy cast steels from the survey (Figure 6.59). The results are surprising. He finds that re-oxidation defects comprise nearly 83% of the total macro-inclusions. These are our familiar bifilms created by the surface turbulence during the transfer of the melt from the furnace into the ladle, and from the ladle, through the filling system and into the mould. In addition, he found nearly 14% of macro-inclusions were found to be mould materials. Because we know that mould materials are also introduced to the melt as part of an entrainment process, it follows that approximately 96% of all inclusions in this exercise were entrainment defects resulting from damaging pouring actions.

Only approximately 4% of inclusions were due to truly extraneous sources; the carryover of slag, refractory particles and deoxidation products.

This sobering result underlines the importance of the reaction of the metal with its environment after it leaves the furnace or ladle. The pouring, and the journey through the running system and into the mould, are all opportunities for reaction of those elements that were added to reduce the original oxygen content of the steel in the furnace. The unreacted, residual deoxidiser remains to react with the air and mould gases. Such observations confirm the overwhelming influence of reactions during pouring or in the running system as a result of surface turbulence; these effects are capable of ruining the quality of the casting.

However, good running systems are not usually a problem for small steel castings. I used to think that large steel castings were a separate matter because of the high velocities that the melt necessarily suffers. This is partly a consequence of the use of bottom-pour ladles and partly the result of the long fall of the melt down tall sprues. However, even these challenges are now known to be solvable.

The historical use of rather poor filling system designs has given steel the reputation for a high rate of attack on the mould refractories. Unfortunately, the attempt at a solution to this problem has resulted in the use of pre-formed refractory tubes for the running system. The joining of these standard pipe shapes means that nicely tapered sprues cannot easily be provided, with the result that much air goes through the running system with the metal. The chaos of surface turbulence in the runner, and the splashing, foaming and bursting of bubbles rising through the metal in the mould cavity, will mean that re-oxidation product problems are an automatic, unavoidable penalty.

It follows that a common feature of steel foundries is that the foundry often employs more welders in the 'upgrading' department than people in the foundry making castings. Common black humour among steel foundry workers includes 'More weld metal goes out of the foundry door than cast metal' and 'the drag is made in the foundry and the cope in the welding shop'. These regrettable jibes follow from the surface turbulence caused during pouring. The difficulties are addressed in 'casting manufacture', in which pouring basins are recommended to be eliminated and the filling system is recommended to be a naturally pressurised design moulded in sand. In the meantime, we shall examine the problems caused by the current poor filling systems.

Solid oxide surface films

Some liquid steels have strong, solid oxide films covering their surface. The high melting point of these oxides ensures that the surfaces of these liquid metals appear to be perfectly dry. They occur on many stainless steels rich in chromium and molybdenum, especially the super duplex stainless steels. In casting above about 250 kg in weight, the filling systems are sufficiently large to pass bifilms up to 100 mm across or more. Entrained air bubbles and surface turbulence in the mould cavity create even more films in situ in the forming casting. Clusters of bifilms are identified on radiographs as resembling faint, dispersed microshrinkage porosity. When attempting to grind away such regions, hollowing out deep cavities in the casting in an effort to eliminate such apparently porous metal, it is common to check periodically with the red penetrant dye to ascertain whether the region is yet free from defects. The bifilms in these regions appear as an irregular red-coloured spider's web. Figure 6.57 shows a typical valve casting in a 20Cr-20Ni-6Mo steel which is notorious for its problems. The bifilm cracks are mainly at grain boundaries, of course, and are noticeably aligned by the growth of the columnar grains at the outer edge of the casting. The casting has received some attempt at cleaning up and grinding out of defects. The deep red indications are almost certainly bubble trails because these tend to hold a lot of dye.

The grain boundary bifilms are often somewhat opened up by cooling strains. When viewed under the optical microscope they have given rise to the description the 'loose grain effect' in some stainless steel foundries because the grains appear to be separated from each other by deep cracks at the grain boundaries, as though they might rattle if shaken. This is, of course, nearly true. The bifilms are segregated by the growing grains, being pushed ahead, and thus usually come to rest in the boundaries between grains. Thermal strains during cooling open the bifilms to give the crazed appearance of the microstructure. This maze of thin, deep cracks often has to be excavated completely through walls sometimes 100 mm in thickness and sometimes greater before these regions can be re-built by welding.

However, in small castings of these particular steels, such as many lost wax castings, there now seems to be evidence that the ingates can be sufficiently narrow (measured in millimetres) so that strong, rigid plates of oxide cannot pass through (Cox et al., 1999). Thus, paradoxically, alloy steels which are notoriously difficult for the pouring of large castings can be used to make small castings that are relatively free from defects.

Low carbon/manganese and low alloy steels are typically deoxidised with Si, Mn and Al in that order. They can suffer from a stable alumina film on the liquid if the final deoxidation with Al has been carried out too enthusiastically. This causes similar problems to those described previous. In addition, I have known ingots of high alloy steels and Ni-base alloys to break up on the first stroke of a forging press after high levels of Al deoxidation have been used before casting. Those ingots that survived the forge usually cracked later during rolling or extrusion. When Al was reduced and Ca added (the effect of the Ca addition is explained later) the problem was solved for some alloys; the ingot forged and rolled like butter. For other alloys, the solution has not worked. We need to do far more research into the achievement of liquid oxidation production on molten steels.

Partly liquid surface oxide films

When adding the usual level of final deoxidation with Al, approximately only 1 kg or less Al per 1000 kg steel, most low carbon/manganese and low alloy steels do not usually suffer severe internal defects. Because of the high melting temperatures of such steels, the surface oxides contain a mix of SiO_2, MnO and Al_2O_3, amongst other oxide components. The mix is usually partially molten. On being entrained during pouring, the internal turbulence in the melt tumbles the films into sticky agglomerates. Because of the presence of the liquid phases that act as an adhesive, the bifilms cannot re-open, and grow by agglomeration. The matrix becomes therefore relatively free from defects in this way. Also, the oxide conglomerates are now rather compact. Their compactness and their low density cause them to float out rapidly, gluing themselves to the surface of the cope as a 'ceroxide' defect. Cope defects are common *surface defects* in these steels. In castings weighing a 10,000 kg or more the defects can sometimes grow to the size of a fist. They are, of course, labour-intensive to dig out and repair by welding. However, their compact form makes this job somewhat easier and not quite in the league of the extensive webs of bifilms bridging the walls of some stainless castings.

Liquid surface oxide films

Over recent years, it has become popular to give a final deoxidation treatment with calcium in the form of calcium silicide or ferro-silicon-calcium because the steel has been found to be much cleaner. This is quickly understood from the following. Alumina and calcia both have melting points in excess of 2000°C. The dry Al_2O_3 surface oxide (if deoxidising with only Al, plus possibly Si and Mn) is converted by addition of Ca into a liquid oxide of approximate composition $Al_2O_3 \cdot CaO$ that has a melting point at or lower than 1400°C. Any folding-in of the liquid film that may now occur will quickly be followed by agglomeration of the impacting liquid films into droplets. The compact form and low density of the droplets will ensure that they float out quickly and will be assimilated into the original liquid eutectic film at the surface, leaving the steel without defects. Any inclusions that remain will be small and spherical, and thus having a minimal impact on properties.

The previous example of the substantial benefits of targeting a liquid condition for the surface film on the steel is a critical factor which has been largely overlooked in the focus for the development of deoxidation reactions. Over recent years, I have revolutionised my approach to quality issues in steel castings by emphasising the importance of attaining a liquid surface film on the steel before and during pouring.

With a liquid surface film, entrainment of the surface no longer results in the creation of bifilms. Thus, Lind and Holappa (2010) in their review of the effects of calcium additions to alumina in steels, record the benefits to ductility and toughness of high-strength low-alloy steels and high quality structural steels by Al + Ca deoxidation.

The reviewers go on to describe how MnS inclusions in flat rolled plate and sheet grades of steels elongate to form stringers or platelets, lowering the through-thickness ductility and toughness. This behaviour seems most likely to be the result of alumina bifilms forming elongated stringers or platelets, onto which manganese sulphide precipitates. The central alumina bifilms are too thin to see easily, but form the cracks that reduce the through-thickness properties.

The boron containing steels might be an important class of steels with liquid surface oxides because of the role of the very low melting point oxide BO_2. The high strength and toughness of these steels may owe much to bifilm reduction, or possibly bifilm adhesion as a result of the BO_2 forming a liquid sandwich inside higher melting point and stable oxide films such as Al_2O_3 or Cr_2O_3. The glue formed by the boron oxide will not be strong, but it will be enormously better than nothing, and would be expected to be particularly effective in preventing the bifilm from shearing. The filled bifilm, or possibly absence of bifilms, may also explain the extraordinary effect which tiny boron additions make to the hardenability of steels. The cleaner metal (or metal with filled bifilms) would have significantly higher thermal conductivity, conferring a much improved penetration of the action of a quench, thus aiding the depth of hardening heat treatments.

Hadfield manganese steels appear to be another example of an alloy which benefits from the oxidation of its major constituents, creating a liquid MnO_2-rich surface film which is probably something like a manganese silicate MnO_2SiO_2 or $MnSiO_4$. The absence of bifilms in this cast steel explains its formidable toughness, despite poor filling system designs that would ensure poor properties in any ordinary steel. It is used for such punishing applications as rail track crossings and points, and the only (rare) failures tend not to be related to cracking but to improper bolting and support problems as a result of mechanical maintenance issues.

Aluminium nitride

In passing, it seems worth mentioning a class of defect that has been the subject of huge amounts of research, but which has never been satisfactorily explained. A tentative explanation is presented here. The phenomenon is the so-called 'rock candy fracture'. This type of defect was seen when the ductility of the casting was especially low, despite the metal appearing to have precisely the correct chemistry and heat treatment. The fracture surface was characterised by intergranular facets that on examination in the scanning electron microscope were found to contain aluminium nitride. Naturally, the aluminium nitride was concluded to be brittle.

This defect seems most likely to be an entrained surface film. The film would probably originally consist of alumina, but would also contain some enfolded air. The nitrogen in the entrained air would be gradually consumed to form aluminium nitride as a facing to the crack. The defect would, of course, be pushed by the growing dendrites into the interdendritic spaces, particularly to grain boundaries. The central crack in the bifilm would give the appearance of the nitride being brittle. On examination, only the nitrogen is likely to be detected, constituting four-fifths of the air, and the oxygen being in any case not easily analysed. The defect is analogous to the plate fracture defect in ductile irons, and the planar fracture seen in Al alloys and other alloy systems (Figures 2.44, 2.46 and 6.30).

Thus in this case, despite the chemistry of the steel being maintained perfectly within specification, the defect would come and go depending on chance entrainment effects. Such chance effects could arise because of slight changes in the running system, or the state of fullness of the bottom-pour ladle, or the skill of the caster, etc. It is not surprising that the defect remained baffling to metallurgists and casters for so long.

6.6.5 PRIMARY INCLUSIONS

When the liquid alloy is cooling, new phases may appear in the liquid that precede the appearance of the bulk alloy. We shall deal with the formation of the primary metallic matrix phase in Section 6.6.7. Whether any newly forming dense phase gets called a phase or an inclusion largely depends on whether it is wanted or not: keen gardeners will appreciate the similar distinction between 'plants' and 'weeds'!

New phases that precede the appearance of the bulk alloy are especially likely following the additions to the melt of such materials as deoxidisers or grain refiners, but may also occur because of the presence of other impurities or dilute alloying elements.

For instance, in the case of steel that has a sufficiently high content of vanadium and nitrogen, vanadium nitride, VN, may be precipitated according to the simple equation:

$$V + N \leftrightarrow VN$$

Whether the VN phase will be able to exist or not depends on whether the concentrations of V and N exceed the solubility product for the formation of VN. To a reasonable approximation the solubility product is defined as:

$$K = [\%V] \cdot [\%N]$$

where the concentrations of V and N are written as their wt%. More accurately, a general relation is given by using, instead of wt%, the activities a_V and a_N, in the form of a product of activities:

$$K^{'} = a_V \cdot a_N$$

It is clear then that VN may be precipitated when V and N are present, where sometimes V is high and N low, and vice versa, providing that the product %V × %N (or more accurately, a_V × a_N) exceeds the critical value K (or K'). It is interesting to speculate that [N] may be high very close to the surface where the melt may be dissolving air. Thus the formation of a surface film of VN may be more likely. It is also necessary to bear in mind that even though all thermodynamic conditions for the formation of VN are met (meaning the solubility product is exceeded), the compound may still not form. This is because there is often a substantial nucleation problem that has to be overcome. This is either achieved by extremely high supersaturations of either V or N or both, or by the provision of a favourable nucleus. As we have seen repeatedly, oxide films are common substrates for the formation of many precipitating intermetallics. The film itself may not be the nucleant, but the nucleant will almost certainly have already nucleated on the film. Thus the film is the starting point for growth, starting at the site of a suitable incumbent nucleus, but spreading out over the film as its favoured growth substrate.

In the case of deoxidation of steel with aluminium, the reaction is somewhat more complicated:

$$2Al + 3O = Al_2O_3$$

and the solubility product now takes the form:

$$K^{''} = [a_{Al}]^2 \cdot [ao]^3$$

where the value of K'' increases with temperature. Again, the surface conditions are likely to be different from those in the bulk, with the result that a surface film of AlN or Al_2O_3 is to be expected, even if concentration for precipitation in the bulk are not met.

We have considered examples of nucleation at various points in the book, especially in Section 5.2.2. At this stage, we shall simply note that any primary inclusions form before the arrival of the matrix primary phase. Thus they appear in a sea of liquid. During this 'free-swimming' phase, primary inclusions have been thought to grow by collision and agglomeration (Iyengar and Philbrook, 1972). Whether this is true or not probably depends on the nature of the inclusions.

For instance, it is not clear whether all liquid inclusions will coalesce even if they impinge. It seems probable that coalescence may be hindered by the presence of a surface film. However, if coalescence does occur, droplets would be expected to result in large spherical inclusions whose compact shape will enable them to float rapidly to the surface and become incorporated into a slag or dross layer which can be removed by mechanically raking off or can be diverted from incorporation into the casting by the use of bottom-pouring ladles.

For solid inclusions, any agglomeration process that may occur might take the form of loosely adhering aggregates or clouds. However, agglomerations apply to particles. Entrained solid alumina films or other entrained solid films will not, in general, be particles but will be messy crumpled masses of double films that on a polished section might easily appear as a cloud of particles, particularly if parts of the alumina bifilm are so thin as to avoid detection, so that only thicker fragments are visible. Furthermore, at steel-casting temperatures, an extremely thin entrained and ravelled alumina film may condense into arrays of compact particles, analogous to the way in which a sheet of liquid metal breaks up into droplets (a spectacular example is given in Figure 2.13), an effect driven by the reduction of surface energy. Subsequent working by rolling or extrusion will elongate films or arrays of separated particles thus explaining the observed alumina 'stringers' (an extremely poor name because these are clearly not one-dimensional linear features but two-dimensional arrays of planar features) often seen in wrought steels. The occasional cracks and pores associated with alumina inclusions will almost certainly be the residue of the central unbonded interface of the original alumina bifilm (it will not be expected to be the result of so-called brittleness of the alumina phase, nor its effect to initiate cracks in the matrix). Work to clarify these suppositions would be welcome.

Hutchinson and Sutherland (1965) have studied the formation of open-structured solids. They found that flocs can form by the random addition of particles. If these particles are spherical and adhere precisely at the point at which they first happen to encounter the floc, then the floc builds up as a roughly spherical assembly, with maximum radius R, and about half the number of spheres within a region $R/2$ from the centroid. The central core has an almost constant density of 64% by volume of spheres. Occasional added spheres will penetrate right into the heart of the floc. Graphite nodules in ductile iron appear to be a good example of this kind of flocculation; melts of hyper-eutectic ductile irons suffer a loss of graphite by the floating out of loose flocs of spherulites (Rauch et al., 1959).

We have only touched on examples of oxides and nitrides as inclusions in cast metals. Other inclusions are expected to follow similar rules and include borides, carbides, sulphides and many complex mixtures of many of these materials. For instance, carbo-nitrides are common, as are oxy-sulphides. In C-Mn steels, the oxide inclusions are typically mixtures of MnO, SiO_2, and Al_2O_3 (Franklin et al., 1969) and in more complex steels deoxidised with ever more complex deoxidisers the inclusions similarly grow more complex (Kiessling, 1978).

In his substantial review of inclusions in steels, Kiessling points out that steel that contains only as little as 1 ppm oxygen and sulphur will contain more than 1000 inclusions/g. Thus it is necessary to keep in mind that steel is a composite product, and probably better named 'steel with inclusions'. Even so, steels are often much cleaner than light alloy castings that might contain 10 or 100 times more inclusions, partly helping to explain the relatively poor ductility and absence of a fatigue limit exhibited by Al-based casting alloys compared with steels.

Finally, of course, not all inclusions will be formed during the liquid phase of the metal alloy. Many, if not most, will be formed later as the metal freezes. These are termed secondary inclusions, or second phases, and are dealt with in the following section.

6.6.6 SECONDARY INCLUSIONS AND SECOND PHASES

After the primary alloy phase has started to freeze, usually in the form of an array of dendrites, the remaining liquid trapped between the dendrite arms progressively concentrates in various solutes as these are rejected by the advancing solid. Because the concentration ahead of the front is increased by a factor $1/k$, where k, the partition coefficient, can often be a rather small number, greatly enhancing the segregation effect, the number of inclusions can be greatly increased compared to those that occurred in the free-floating stage in the liquid. However, the size population is usually different, being somewhat finer and more uniform as a result of the smoothing action of diffusion in the tiny volumes between the dendrites.

The secondary inclusions or second phases form at or close to the freezing front. One of the most common and important second phases is a eutectic. We have already seen how microsegregation can lead to the formation of eutectic at bulk compositions that are much below those expected from the equilibrium phase diagram.

However, the nucleation of a first phase is likely to prevent the subsequent nucleation of any other phase that might also require one of the same elements for its composition. The availability of solute is clearly limited by a naturally occurring 'first come first served' principle.

In the subsequent observation of inclusions in cast steels, those that have formed in the melt before any solidification are, in general, rather larger than those formed on solidification within the dendrite mesh. The possible exceptions to this pattern are those inclusions that have formed in channel segregates, where their growth has been fed by the flow of solute-enriched liquid. Similarly, in the cone of negative segregation in the base of ingots, the flow of liquid through the mesh of crystals would be expected to feed the growth of inclusions trapped in the mesh, like sponges growing on a coral reef feeding on material carried by in the current. In Figure 5.44, the peak in inclusions in the zone of negative segregation is composed of macro-inclusions that may have grown by such a mechanism. Elsewhere, particularly in the region of dendritic segregation around the edge of the ingot, there are only fine alumina inclusions.

Sulphides

It would not be right to leave the subject of inclusions without mentioning the special importance of the role of sulphide inclusions in cast steel. The ductility of plain carbon steel castings is sensitive to the type of sulphide inclusions that form.

Type 1 sulphides have a globular form. They are produced by deoxidation with silicon.

Type 2 sulphides take the form of thin grain boundary films that seriously embrittle the steel. They usually form when deoxidising with aluminium, zirconium or titanium.

Type 3 sulphides have a compact form and do not seriously impair the properties of the steel. They form when an excess of aluminium or zirconium (but not apparently titanium!) is used for deoxidation.

Mohla and Beech (1968) investigated the relation between these sulphide types and concluded that the change from type 1 to type 2 is brought about by a lowering of the oxygen content. Additionally, it seems that the new mixed sulphide/oxide phase has a low interfacial energy with the solid, allowing it to spread along the grain boundaries. Also, it might constitute a eutectic phase. Type 3 sulphides were thought by Mohla and Beech to be a primary phase. In common with all other investigators of sulphide embrittlement of steels, it is clear that these authors were less than happy with their tentative findings because doubts and confusion still existed.

However, type 2 inclusions have all the hallmarks of an entrained bifilm. It is significant that this type of inclusion forms only when the melt is deoxidised with Al or other powerful deoxidisers that are known to create solid films on the melt. The surface film might originally have been enriched with the other highly surface-active element, sulphur. The entrainment of an oxide film would in any case be expected to form a favourable substrate for the precipitation of sulphides. The film would naturally be pushed into the interdendritic regions by the growing dendrites, so that it would automatically sit at grain boundaries.

Even so, an explanation of type 3 sulphides remains elusive. These results illustrate the complexity of the form of inclusions, and the problems to understand their formation. Much additional research is required to elucidate the mechanism of formation of these defects.

A final question we should ask is 'How do inclusions in the liquid become incorporated into the freezing solid?'

It seems that for small inclusions, especially those that are in the relatively quiet region of the dendrite mesh, the particles are pushed ahead of the front, concentrating in interdendritic spaces.

For larger inclusions, generally larger about 10 μm in diameter, trapping between dendrite arms is only likely if the inclusion is carried directly into the mesh by an inward-flowing current. This may be the mechanism by which inclusions are originally trapped within the cone of negative segregation, where they subsequently grow to large size, feeding on solute-rich liquid percolating through the dendrite mesh (Figures 5.44 and 5.45).

Where the front is relatively planar and strong currents stir the melt, the larger inclusions are not frozen in to the advancing solid as a consequence of the velocity gradient at the front. Delamore (1971) found that those particles that do approach the interface cannot be totally contained within the boundary layer, and as a result spin or roll along it because of the torque produced by the velocity gradient. In this way the larger particles finally come to rest in the centre of castings. Rimming steels benefited for the same reason from an absence of large inclusions in their pure rim.

Steel inclusion summary

The liquid metal in the melting furnace is probably fairly clean at a late stage of melting because of the large density difference and plenty of time for flotation. However, before casting, deoxidation by Si, Mn and Al etc. creates a large population of fine inclusions in situ in the melt. A proportion of the inclusions generated from this action separate quickly by gravity. There has been much research on these in situ–generated inclusions (see, for instance, Tiekink et al., 2010).

Much less research has been carried out on the entrainment inclusions created during the two major pouring events. These are as follows.

1. The transfer of the melt from the furnace into the transfer ladle creates a fresh, dense crop of inclusions, the air reacting with the exposed melt surface, oxidising the residual deoxidiser additions that remain in solution in the steel. These surface inclusions, often in the form of a surface film, become entrained into the bulk melt by the action of the pour. They are known as re-oxidation inclusions. Fortunately, a large proportion of this contribution towards the loss of quality of the steel will float out in the time taken to lift the ladle out of the pit and transfer it to the casting station. This huge volume of defects constitutes most of the layer of slag on the surface of the ladle by the time the pouring of castings begins.

2. A second crop of re-oxidation inclusions are now entrained by the pouring action into the mould. Depending on the freezing time and geometry of the casting, this late crop of inclusions may little or no time to float out.

Huge numbers of inclusions, generally known in the trade as re-oxidation products, result from these turbulent pouring processes, although, clearly, the second pour is far more damaging to the product than the first.

However, in addition to these sources, steels are also noted for the number of very fine additional inclusions that form later, during solidification. The remaining unreacted deoxidising elements are concentrated in the interdendritic liquid, where more inclusions pop into being by a nucleation and growth process. The interdendritic regions are small, limiting the size to which such inclusions can grow. Svoboda et al. (1987) observe that these inclusions are often also associated with small amounts of MnS. This is to be expected because both manganese and sulphur will also be concentrated in the interdendritic spaces. It is possible that, because of their small size, some of the deoxidation products may be pushed ahead of the growing dendrites and so become the nuclei for precipitates that arrive later.

As solidification proceeds, other inclusions may form by this concentration of segregated solutes in the interdendritic spaces. These may include nitrides such as TiN, carbides such as TiC, sulphides such as MnS and oxy-sulphides etc. In general, they will be most concentrated and largest in size in the regions between grains, and in regions of the cast structure where segregation is highest, such as in channel segregates and the tops of feeders. They may also enjoy late excursions into the casting as this remaining enriched metal from the feeder is sucked into the casting during the very last moments of feeding. The region under the feeder is known for its segregation problems resulting from the concentration of light elements, particularly carbon, but it is probable that the residual liquid will also carry plenty of additional unwanted solutes and solid debris.

Later still, further precipitation of inclusions will occur in the solid state. In general, these are up to 10 times finer as a result of the much lower coefficient of diffusion in the solid. The driving force for their appearance is the decreasing solubility of the elements as the temperature falls. Many hardening reactions are driven in this way; for instance, the formation of aluminium and vanadium carbides and nitrides in steels and the precipitation of $CuAl_2$ phase in Al-4.5Cu alloy. The hardening is the consequence of the very fine size and spacing of the precipitate, making the inclusions effective as impediments to the movement of dislocations.

The precipitation of inclusions in the liquid and solid states creates populations of phases that are entirely different from entrained inclusions. Such intrinsically formed (in situ) inclusions grow atom by atom from the matrix and so are in perfect atomic contact with the matrix. They would *never* be expected therefore to initiate volume defects such as pores and cracks.

In contrast, the entrained inclusion is characterised by its unbonded wrapping of oxide, so it can easily nucleate pores and cracks from this lightly adherent coat; it easily peels away during subsequent plastic working or creep. Alumina-rich inclusions in rolled steel are often seen to be associated with cavities. These are usually assumed to be the result of the brittle breakup of the inclusion during working, or the tearing away of the matrix from sharp corners acting as stress raisers. It seems likely to me that neither of these explanations is correct because alumina is strong, and volume defects are difficult if not impossible to initiate in the solid state. The real explanation is most probably the fact that cracks and pores that are evident in such strings of inclusions are the fragments of the original alumina bifilm. This will never completely close despite much plastic working, partly as a result of the inert and stable nature of the ceramic phase, alumina and partly as a result of the residual argon gas in the entrained layer of air. The residual argon is highly insoluble in metals.

6.6.7 NUCLEATION AND GROWTH OF THE SOLID

During the cooling of the liquid steel, a number of particles may pre-exist in suspension, or may precipitate as primary inclusions. The primary iron-rich dendrites will nucleate in turn on some of these particles. The work by Bramfitt (1970) illustrated how only specific inclusions act as nuclei for iron.

Bramfitt carried out a series of elegant experiments to investigate the effect of a variety of nitrides and carbides on the nucleation of solid pure iron from the liquid state. In this case, of course, the solid phase is the body centred cubic delta-iron (δFe). In his work he found that his particular sample of iron froze at approximately 39°C undercooling (i.e. 39°C below the equilibrium freezing point). Of the 20 carbides and nitrides that were investigated, 14 had no effect and the remaining six had varying degrees of success in reducing the undercooling required for nucleation.

The results are shown in Figure 6.60. They give clear evidence that the best nuclei are those with a lattice plane giving a good atomic match with a lattice plane in the nucleating solid. Extrapolating Bramfitt's theoretical curve to the value for the supercooling of his pure liquid iron indicates that any disregistry between the lattices beyond approximately 23% means that the foreign material is of no help in nucleating solid iron from liquid iron.

Another interesting detail of Bramfitt's work was that a number of additions were ineffective because they either melted or dissolved in the liquid iron as it was cooled to promote freezing. This consequent lack of effectiveness was despite, in some cases, quite low values of disregistry. This underlines the perhaps self-evident point (but often forgotten) that any addition has to be present in solid form for it to nucleate another solid. In addition, all of Bramfitt's work was concerned with the nucleation of δFe, the body-centred cubic form of iron.

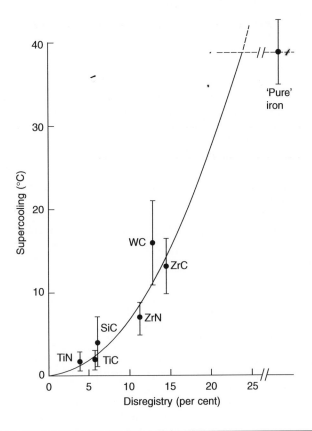

FIGURE 6.60

Supercooling of liquid iron in the presence of various nucleating agents.

Data from Bramfitt (1970).

The fcc form of iron, γFe, or austenite, has been considerably more resistant to past attempts to nucleate it. Until recently no one had succeeded to identify any previously existing solid that acted as a nucleus for the γFe phase. In 2013, Moumeni and colleagues found a mixed sulphide of Mn and Ti (Mn,Ti)S, in the centres of austenite dendrites in cast irons, indicating that they may have acted as nuclei. This has yet to be confirmed. Japanese workers led by Mizumoto (2008) claimed to have successfully grain refined an austenitic stainless steel (12Cr 18Ni with $Cr_{eq}/Ni_{eq} = 0.83$; the values of this ratio are explained later). Even here, the work was carried out only on a laboratory scale (melt weight 90 g) and had to be performed in seconds otherwise the niobium carbide contained in their master alloy dissolved so quickly as to become ineffective. Again, no mechanical testing was carried out to ascertain whether there appeared to be any benefit to this procedure.

Jackson (1972) listed a large number of additives that were unsuccessful in attempts to refine austenitic steels. His first success was the addition of FeCr powder together with floor sweepings! This impressively economic but hardly commendable formula caused him to persevere, searching for more scientifically chosen additions. He later found that calcium cyanamide was quite useful, but required the nitrogen content in the alloy to be raised to 0.3% to be successful in 18:8 stainless steels. At this level of gas content, severe nitrogen porosity is the unwelcome product. Jackson was able to define satisfactory conditions for 18/10/3Mo stainless steel, again providing that the nitrogen content was above 0.08 wt%. The modest improvement that he reports in mechanical properties he attributes not so much to the reduction in grain size as to the increase in the alloying effect of nitrogen! This work is further complicated by the expectation that the compound calcium cyanamide will probably have decomposed at steel casting temperatures.

The grain refinement of austenitic stainless steels therefore remains a challenge to future metallurgists. It seems likely that only those cast austenitic stainless steels that solidify first to δFe, before their subsequent transformation in the solid state to γFe, are definitely able to benefit from grain refinement resulting from a nucleation process.

Suutala (1983) proposed a factor that allows the prediction of whether the steel will solidify to austenite or primary ferrite (delta iron); this is the ratio of the chromium equivalent to the nickel equivalent, Cr_{eq}/Ni_{eq}, where the chromium and nickel equivalents are calculated from (elements in wt%):

$$Cr_{eq} = \%Cr + 1.5Si + 1.37Mo + 2Nb + 3Ti$$

$$Ni_{eq} = \%Ni + 22C + 0.31Mn + 14.2N + Cu$$

(Readers will note a close similarity with the definition of the Schoefer equations given earlier that define the structures of stainless steels at room temperatures.) Suutala proposes the ratio $Cr_{eq}/Ni_{eq} = 1.55$ is the critical value; at values higher than this, the solidification changes from primary austenite to ferrite. (This value applies for shaped castings and ingots. The equivalent value for welds because of their much faster rates of freezing is 1.43.) However, careful work by Ray and co-workers (1996) suggest that the division between austenite and ferrite is not sharp, but is better defined by a parameter the ferrite potential (FP) that assesses the fraction of primary ferrite in austenitic stainless steels

$$FP = 5.26\left(0.74 - Cr_{eq}/Ni_{eq}\right)$$

Barbe and co-workers (2002) found that an FP value below 3.5 is useful for mainly ferritic steels, corresponding to a maximum $Cr_{eq}/Ni_{eq} = 1.40$. The retention of some austenite limits the grain growth of ferrite, and is suggested to reduce the susceptibility that ferritic steels have to 'clinking', i.e. cracking during continuous casting. When slabs are cooled to room temperature cracks are found on the slab edges. Sometimes the slabs crack into two. This problem is reminiscent of problems in continuously cast direct chill aluminium alloys, particularly the strong 7000 series alloys. These appear to suffer from oxide bifilms entrained during the first few seconds of casting in which the metal falls and splashes turbulently during the first seconds of the filling of the mould. The bifilm cracks created in this moment float randomly into higher regions of the ingot during the casting operation, causing it to crack catastrophically, sometimes weeks after being cast. It is dangerous to be near such an event. The high Cr ferritic steels might be expected to behave analogously as a result of entrained Cr-rich films.

Roberts et al. (1979) confirmed that only ferritic material was refinable with titanium additions, and confirm that TiC and TiN have lattices that are good fits with ferrite, but poor fits with austenite. Baliktay and Nickel (1988) reported that titanium additions can also refine the grain size of the widely used high-strength stainless steel 17-4-PH. The ratio $Cr_{eq}/Ni_{eq} = 2$ approximately for this material, confirming that it solidifies to ferrite, in agreement with the finding by Roberts. More recent work (Wang et al., 2010) has demonstrated good refinement under laboratory conditions in a 12Cr steel ($Cr_{eq}/Ni_{eq} = 25$) using an addition of Fe-Ti-N.

Now, let us stop this account of the grain refinement of cast steels and stand back a little to take stock of the situation in which we find ourselves. Whereas it is common knowledge that the grain refinement of wrought steels is a valuable feature, the grain refinement of cast steels does not seem necessarily beneficial. Having recorded these struggles to achieve significant grain refinement in cast steels, the question arises, 'Is it worth it?' The answer is not clear as the following reports indicate.

Doubt is inferred from the work by Campbell and Bannister (1975) on the ferritic alloy Fe-3Si. They showed that the best refinement was obtained by the addition of TiB_2 to the melt. However, on metallographic examination the grain boundaries were found to be surrounded by a phase that appeared to be iron boride which would have done little for the mechanical properties. These were not tested, but were likely to have been impaired. It would be valuable to explore further whether conditions could be found in which grain refinement would improve properties.

Encouragement that useful results might be gained is given by Church et al. (1966), but even these results are not all good news. Their work on a high-strength steel, 0.33C-0.7Mn-0.3Si-0.8Cr-1.8Ni-0.25Mo-0.040S-0.040P, revealed that although grain refinement was successfully accomplished with 0.60Ti, the benefit was negated by the presence of interdendritic films of titanium sulphide, causing severe embrittlement. However, toughness and ductility could be improved by smaller additions of titanium in the range 0.1–0.2%, which was still successful in achieving grain refinement. The doubt remains that much of this research was undermined by poor casting technique, introducing quantities of deleterious bifilm cracks, particularly at grain boundaries. Sulphur would be expected to precipitate preferentially on such substrates, giving the impression of brittle sulphide films at these locations.

In contrast to the limited success from attempts to nucleate grains, other approaches involving 'seeding', using granular metal additions of the same composition as the melt, have been repeatedly confirmed by different workers over the past half-century as potentially successful (Jackson's addition of FeCr powder is an instance, although the floor sweepings are probably not; other genuine instances include Campbell and Caton, 1977).

Part of the reason for the overall disappointing progress on the grain refinement of cast steels, that might appear to be a potentially important metallurgical advance, lies in the special properties of steels. First, many steels transform and recrystallize more than once to different crystal structures as they cool, thereby automatically providing a fine grain size at room temperature (even though, of course, the original segregated boundaries may still be in place). Secondly, unlike Al alloys with long-lived bifilm populations, they do not benefit from being 'cleaned up' by sedimentation of the bifilms by the grain refining addition, because the high density of steels ensures they are already generally clean by flotation at the time of the grain refining addition. No significant additional cleaning is possible, thus the major advantage of a cleaning action plus grain refinement enjoyed by Al alloys does not happen and is not necessary for steels.

6.6.8 STRUCTURE DEVELOPMENT IN THE SOLID

The grain structure of the steel that forms on solidification may turn out to be the same as that seen in the finished casting. However, this would be somewhat unusual. It happens only in those cases in which the metal is a single phase from the freezing point down to room temperature. Examples include some austenitic stainless, and some ferritic stainless. However, even the ferritic stainless may undergo a transformation to martensite or bainite depending mainly on its carbon content. The transformer steel, Fe-3.25%Si, is a common ferritic steel that, on a polished and etched section, clearly displays at room temperature a structure that has changed little from that originating during solidification. For that reason, it is a useful model alloy for research.

Even in these single-phase materials, there is opportunity for grain boundary migration, possibly grain growth, and possibly recrystallisation. Complete recrystallisation would be expected in those parts of castings that had been subject to considerable plastic deformation during cooling. This would be expected, for instance, at junctions of flanges that restrain the contraction of the casting.

In materials that change phase during cooling to room temperature, the situation can be very much more complicated. Low-carbon and low-alloy steels are a good example, illustrating the problems of understanding a structure that, after freezing, has undergone at least two further phase changes during cooling to room temperature. Figure 6.61 lists the changes.

(a) The liquid solidifies to delta iron dendrites.

(b) When solidification is complete, the principal grain boundaries (shown as full lines) have their positions delineated and to some extent fixed in position by segregates, particulate inclusions and bifilms. The slight misalignments between parts of the dendrite raft result in a network of less important subgrain boundaries (shown as broken lines).

(c) During cooling and differential contraction of the casting, the plastic strains will create dislocations that will migrate to form an additional network of new subgrain boundaries. These are, of course, all low angle boundaries and may not be readily visible.

(d) On reaching the temperature for the formation of the gamma-iron phase, austenite grains will nucleate on the original grain boundaries or other discontinuities, particularly bifilms because the residual 'air gap' between the films will reduce the strain energy required for the nucleation of a new phase which requires to change both its volume and shape. Their growth into the delta-grains will sweep away most traces of the subgrain network.

(e) When the conversion to austenite is complete the original delta-grain boundaries (shown as broken lines) will still usually be discernible as ghost boundaries because of the fragmentary lines of segregates, particularly second phases decorating bifilms.

(f) Further cooling strains will generate a new subgrain structure.

(g) Austenite will start to convert to ferrite, usually nucleating at grain corners and boundaries, once again likely to be associated with bifilms as a result of the reduction in strain energy, sweeping away the substructure once again.

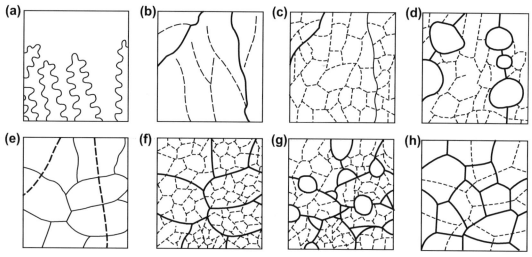

FIGURE 6.61

Successive stages of grain evolution in a low carbon steel, from its freezing point to room temperature (see text for full explanation).

(h) The final ferrite grains will again show the ghost boundaries of the previous austenite grains because these will have experienced sufficient time at temperature to have gathered some segregates by diffusion to the boundary.

(i) Subsequently, a further series of subgrains may be created, although by now the temperature is sufficiently low that any strains will generate fewer dislocations, and that such dislocations will not be sufficiently mobile at lower temperature to migrate into low-energy positions, forming low-angle boundaries; thus, the alloy will have become sufficiently strong to retain any further strain as elastic strain. The structure of the alloy will no longer be affected during further cooling, but elastic stresses will build up.

The final structure on a polished section will be a grain size that has been refined by two successive phase changes (but possibly coarsened a little by intervening grain growth) and may still retain ghost boundaries of delta iron and austenite. The underlying structure of the original delta iron dendrites will probably still be present, as can be revealed by etching to highlight the differences in chemical composition.

This sequence of events neglects the other many phase changes that can occur in some steels. Thus transformation of austenite to pearlite is usual for carbon/manganese steels, or transformations to martensite or bainite is also possible at higher cooling rates.

For a formal review of the development of structure in castings, see Rappaz (1989). Further detailed work on cast structures has been carried out during extensive work on the structures of welds in steels. For a review of this work, see Sugden and Bhadeshia (1987). This work draws attention to the complicating effects of the formation of Widmanstatten and acicular ferritic structures, and the presence of martensite, bainite, pearlite and retained austenite. The solidification morphology of the steel in this review of welding seems to be principally cellular or possibly cellular/dendritic (i.e. dendrites without side branches). Also, of course, in successive weld deposits there are the additional effects of the subsequent heat treatment of the previous runs in the laying down of the subsequent deposits.

6.7 NICKEL-BASE ALLOYS

The reader does not have to spend much time in a foundry casting steels and Ni-base alloys to realise that even though many steels are not easy, there are special difficulties with casting good Ni-base castings. This has always been a mystery to those involved with casting these high-temperature metals; a mystery we shall explore and seek to solve during the course of this chapter.

There are of course a wide range of Ni-base alloys, together with their cousins, the Co-base alloys. When alloyed with Cr, the alloys can achieve great resistance to heat and oxidation. For a special class of highly creep-resistant alloys, strong and oxidation resistant at high temperatures, the name '*high-temperature alloy*' is often used.

The class of alloys known as '*superalloys*' has, additionally, several wt% of Al as the major strengthening element, precipitating Ni_3Al (the so-called gamma prime phase, written as the Greek letter γ'). The '*gamma prime*' phase is an extremely fine and stable second phase. Ti similarly contributes to some extent as Ti_3Al. Many superalloys appear to contain most of the elements of the periodic table, making them probably more complex than any other metallic alloys. Recent developments have seen the incorporation of many rare and expensive elements into high temperature alloys. In particular, hafnium (Hf) has been used and noted for its action to reduce porosity. This somehow has been attributed to its lower coefficient of expansion (Chen and Knowles, 2003), but seems much more likely to be the result of a stronger oxide on the surface of the liquid, thus holding up on the lip of the crucible or in the mouth of the mould, and so avoiding incorporation as bifilms into the casting.

The high quantity of highly oxidisable elements, Al, Cr, Ti, Hf etc. in the high-temperature Ni-base alloys is the reason for the problems suffered by these alloys during casting. These elements all react rapidly with oxygen (and possibly with nitrogen) in the environment, creating solid films on the liquid surface. The entrainment of the solid surface film leads to bifilm creation and hence pore and crack defects. In their common wall-to-wall form, bifilms can also be expected to be a barrier to the growth of single crystals. As for steels, the conversion of the solid alumina-based surface film into a liquid surface film with the aid of Ca addition is a great help, avoiding bifilm formation, and thus raising properties, but does not seem to be effective for all alloys. For those alloys for which this phenomenon is effective the technique is important and is mentioned repeatedly later. Even here, it is possible that for some alloys where this would be helpful, Ca may be considered a harmful impurity for other reasons. The superalloys are complicated!

6.7.1 AIR MELTING AND CASTING

Many Ni-base alloys are melted and cast in air, both into ingots for subsequent working into plate, sheet, bar, tube and wire etc. and into shaped castings. Such castings behave similarly to steels, so that deoxidisers such as Al require to be added to avoid a carbon boil reaction during freezing. Alternatively, Niu and colleagues (2003) described how carbon is useful to deoxidise, eliminating more than 50% of the oxygen during the melting stage, even though Al is useful later to reduce the oxygen content further. Once again, as for steels, if Al would normally be used for deoxidation, a 50:50 mixture of Al and Ca is recommended to be used instead. For some Ni-base alloys, this technique is valuable to avoid the embrittling effects of alumina bifilms, replacing these with liquid oxide eutectic that, even if entrained by poor casting technique, still has the chance to spherodise and float out of the casting. At this time, it is not clear which alloys respond successfully to this deoxidation technique. More research would be so welcome.

For a high-temperature alloy, its alloy content of Al means that no additional Al will be required for deoxidation because the alloy will naturally have a low oxygen content (although sometimes not so low to prevent a carbon boil occurring during freezing). However, the alloy will definitely form a strong aluminium oxide bifilm on pouring; although there is only a few percentage of Al in the alloy, the high temperature encourages its extremely rapid growth. Thus a small addition of Ca immediately before pouring is valuable to liquefy the surface oxide, and thereby reduces cracking on solidification or during subsequent working.

Two factors currently work against success in the Ni casting industry.

1. For some reason lost in the mists of the early history of Ni-base alloy casting, it seems that most if not all melts are treated with Mg. A more inappropriate alloy addition can hardly be imagined: any residual Mg will ensure the development of a strong, solid MgO film on the liquid which would be highly damaging to the liquid metal after a turbulent pour. For confirmation, a study of Inconel 718 by Chen and colleagues (2012) found all the oxides were MgO or spinel $MgAl_2O_4$. Work by Ren (2014) shows that Mg deoxidation of steels can form globular

liquid oxides if Al is present together with Ti in sufficient concentration: if no Ti is present the oxides are solid at 1600°C but if 0.2Ti is present the melting point lowers to 1450°C. Thus the highly dangerous addition of Mg might be made safe by the presence of other alloy additions. The once-sacrosanct Mg addition requires to be critically reassessed immediately; it may be permissible for some alloys, whereas others will be made even more difficult to cast.

2. Ni-base alloys have a melting point sometimes 100°C below that of steels, so that Ni-base casters generally pour at temperatures well below those used for steels. In addition, of course, as is usual in foundries, casters understandably work with concepts such as minimising superheat, providing just sufficient to keep the alloy liquid during casting. These circumstances cooperate to ensure that the surface oxide on the liquid metal is solid, making Ni-base alloys difficult to cast and forge without cracking. This cracking behaviour has always been a mystery for such a malleable metal. These alloys may require much higher casting temperatures to help liquefy their surface oxide if turbulent filling systems are retained. Alternatively, *of course*, non-turbulent filling could also eliminate cracking.

6.7.2 VACUUM MELTING AND CASTING

One of the principal castings made from Ni-base (and some Co-base) alloys are turbine blades mainly for aero engines, but also nowadays for power generating turbines. As usual, these alloys contain aluminium and titanium as the principal hardening elements together with high levels of chromium for high temperature oxidation resistance. All of these elements can assist to create oxide films. Because such castings are produced by investment (lost wax) techniques, the running systems have been traditionally poor. It is usual for such castings to be top poured, introducing severe surface turbulence, and creating high scrap levels. Even most bottom-gated filling systems, still filled by pouring from a highly placed furnace, designed to fill the mould cavity in an uphill direction, are still poor; the defects introduced in the metal during the turbulence of the pour are simply carried over into the mould cavity to spoil the casting, or interfere with the progress of directional solidification or single crystal growth. When studying faults in single crystals, Carney and Beech (1997) found oxides at the root of most stray grains.

Turbine blades that have failed from major bifilm defects have caused planes to crash and have cost lives. Clearly, the problem is not trivial and the consequences are tragic. It is not satisfactory that we continue to live with gravity-poured blades fitted in aero engines. Such gravity-poured castings necessarily contain bifilm defects, many of which are not detectable by either X-ray radiography or by dye penetrant inspection (DPI). In contrast, blades produced by counter-gravity filling could probably be guaranteed free from filling defects. Furthermore, perhaps we would neither have to check them by radiography nor by DPI. If we *have* to live with a defective production technique for parts for aircraft engines, more discriminating testing by multi-resonance techniques would ensure greater safety.

When melting Ni-base superalloys in vacuum, a slag collar usually builds up from the accretion of oxides in suspension in the alloy. Some of these oxides are from the skins of the charge materials, but many are from the population of bifilms generated during the prealloying production process. Here the alloy maker blends his alloy in a large melting furnace. The melt is poured into launders and poured again through drops of several metres, falling into long steel tubes in which it solidifies into cast rods (again this occurs in vacuum, but the vacuum is not sufficiently good to prevent the creation of oxide bifilms of course). It is known that the larger diameter tubes (for the preparation of larger logs to be melted for the production of larger castings) have higher oxide content than the slimmer tubes in which turbulence would have been better suppressed.

Casting in vacuum is probably essential for such products as turbine blades. The wall thickness of these products is so small, becoming measured in micrometres at the trailing edges of blades, that any backpressure of gas will inhibit filling.

It is quite clear, however, that casting in vacuum is not a complete solution from the point of view of eliminating solid surface films. A good industrial vacuum is around 10^{-4} torr (approximately 10^{-2} Pa). This is not good enough to prevent the formation of bifilm defects (from entrainment of the surface oxide or nitride film) during pouring.

Equilibrium theory predicts that not even the vacuum of 10^{-18} torr (10^{-20} Pa) that exists in the space of near-Earth orbit is good enough to prevent the formation of Al_2O_3 (the most likely and most usual oxide) because 10^{-40} torr

(10^{-42} Pa) is required to dissociate this particular oxide. However, recent research on pure liquid aluminium has revealed that the equilibrium prediction is hugely inaccurate as a result of the alumina dissociating to a sub-oxide Al_2O. This occurs at around 10^{-8} Pa, reducing to 10^{-6} Pa at approximately 1000°C, causing a 'wind' of Al_2O to flow away from any Al_2O_3 on the surface of liquid Al, decomposing the Al_2O_3 and conveying away its oxygen in the form of the sub-oxide. The wind of sub-oxide vapour prevents oxygen from arriving at the surface of the liquid metal (Giuranno, 2006; Molina, 2007). The aluminium surface can then slowly become perfectly clean, free from oxide (Aguilar-Santillan, 2009). Even without the action of the evaporation of the sub-oxide, Zemcik (2015) reminds us that alumina can be reduced by carbon in only modest vacuum conditions at the temperatures used for casting. Unfortunately, he omits to check whether other oxides can be similarly reduced in casting conditions.

Despite this beneficial evaporation of the aluminium sub-oxide, and possible reduction by carbon, it remains a fact that films are found entrained in current vacuum cast alloys. Perhaps they are not oxides, but nitrides? Perhaps they are oxides of Cr, Ti, Nb, Hf or any one of the long list of elements sometimes present in the alloy which carbon is unable to reduce at these pressures and temperatures.

We need to give some further consideration to the fact that the surface films on high-temperature Ni-based alloys might in some cases be AlN. The Ni-base superalloys are well known for their susceptibility to react with nitrogen from the air and so become permanently contaminated, especially when the casting is cooled by opening the furnace door to air immediately after pouring. It seems more likely that the contamination is actually a nitride film problem rather than a problem caused by nitrogen in solution. In any case, in air, the reaction to the nitride may be favoured even if the rates of formation of the oxide and nitride are equal, simply because air is 80% nitrogen. Niu et al. (2002) reported in many superalloys that there are an order of magnitude more nitride than oxide inclusions. These authors reported higher porosity and loss of rupture life with higher nitrogen, lending some support to the concern that the nitrogen is in the form of nitride bifilms.

These complications and uncertainties about the films on the liquid metal emphasise, if emphasis is needed, that the real solution to entrainment problems is not to attempt to prevent the formation of the surface film by, for instance the quest for better vacuums, but simply to avoid *entrainment* of films.

Gravity pouring using a well-designed bottom-gated filling system might therefore be a significant improvement on most current filling system designs. However, the ultimate answer would be the complete avoidance of any type of pouring, using a counter-gravity system of filling. It would make beautiful castings that would probably not require expensive testing (the multi-resonance technique is fast and low cost). Furthermore, the easy automation of counter-gravity would greatly lower costs and enhance quality.

Whatever the films are, oxides or nitrides, an example is seen in Figure 6.62 that happens in this case to be an oxide. Other examples of film-like defects in vacuum cast superalloys can be seen in the work of Ocampo (1999) and Malzer (2007). Ocampo described a Co-base alloy with continuous grain boundary films decorated with porosity and carbides which are the typical signatures of the presence of bifilms at the boundaries. Figure 6.63 also shows a bifilm crack surrounded by carbides.

The precipitation of topologically close-packed (TCP) phases in some single crystal alloys seems to be associated with porosity and with preferred fracture planes. Work by Graverend (2012) and Shi et al. (2014) clearly showed TCP phases on {111} planes thought to be the habit plane of the precipitate, but more likely the preferred close packed plane favouring the straightening of a bifilm, on which the TCP has formed (Figure 6.64). This interpretation is strongly confirmed by the observation of decohesion only on one side of the TCP phase, and transverse cracks in the TCP phase on the opposing side of the crack, exactly as is typical of bifilm structure. It is interesting to compare the exactly analogous structure in an Al alloy (Figure 6.20(b)).

There is an interesting situation with the Ni-base alloys containing Hf. It seems that when these alloys are cast, a curious unexplained glassy surface defect on the blades that appears to run down the casting surface like a congealed river, but if a filter is placed in the filling system the defect does not occur. It seems likely that the defect represents an attack of the ceramic mould by HfO surface films generated by turbulence. The HfO will most likely form a low-melting-point oxide mixture with the components of the mould, often mainly Al_2O_3. If the filter is present, any HfO already present will be reduced by filtering out, and the reduced turbulence after the filter will not generate more HfO.

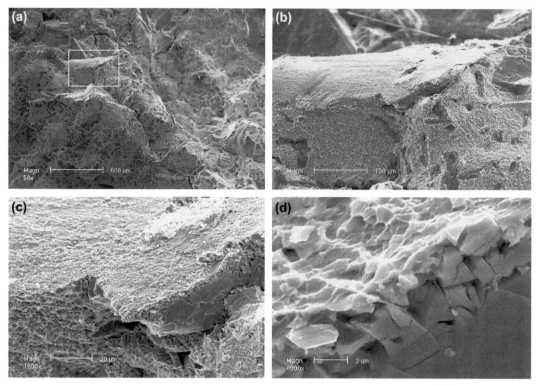

FIGURE 6.62

(a) to (d) show successively closer SEM images of the fracture surface of a Ni-based superalloy melted and cast in vacuum. The light-grey oxide clearly occupies more than 10 mm^2 of surface. The selected region, like a glacier, is only about 1 µm thick, but has a 20 µm deep underlay of carbides. At its high temperature of formation, the oxide film has recrystallised (Rashid and Campbell, 2004).

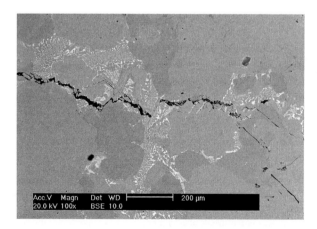

FIGURE 6.63

SEM image of a Ni-base alloy casting polished through a freckle containing segregated carbides precipitated on a bifilm crack (Rashid and Campbell, 2004).

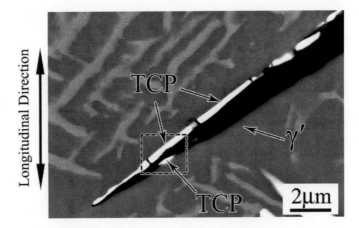

FIGURE 6.64

A bifilm crack in a Ni-based superalloy, revealed by its precipitate of a topologically close packed (TCP) phase (only a fragment of the phase is on one side of the bifilm crack) with characteristic short transverse cracks as a result of folds in one of the constituent oxide films of the bifilm (Figure 6.20(b) shows an analogous defect in an Al-Si alloy. Courtesy Professor Qiang Feng (Shi et al., 2014).

The larger bifilm defects, clearly seen on fracture surfaces, may be less common than those that give rise to the 'stars at night' symptoms created by clouds of bifilm fragments that intersect the surface of vacuum investment superalloy castings (for instance, reported by Wauby, 1986) when viewed in ultraviolet light after treatment by fluorescent dye penetrant. However, it is difficult to be sure. The fluorescent dye can penetrate the folded bifilms, and so exude again during inspection under ultraviolet light. Some indications are large and bright, indicating large internal cavities. Others can appear on both front and back of blades, possibly indicating that they completely penetrate the wall. A casting made with a good filling system (probably of lower cost than a conventional filling system, for those who assert that quality has to be paid for) can display zero indications. In the presence of a group of investment foundry personnel, I recall casting a blade that was put through the DPI process. It was then presented to an experienced inspector who, after a long and puzzled inspection, concluded it had not undergone immersion in the dye.

Further important aspects of Ni-base alloys, particularly for turbine blades, are discussed in Chapter 17 (Controlled Solidification Techniques).

6.8 TITANIUM

Titanium is a curious and special metal. In its liquid state, it is practically impossible to handle without it reacting with everything. However, in its solid state, it can be wonderfully resistant to oxidation and corrosion, whilst being impressively strong and having a density occupying an interesting 'halfway house', partway between the Al alloys and steels.

Its chemically inert behaviour makes it favoured for use for valves and fittings in the chemical industry, marine applications and in prostheses for the human body, including replacement hip and knee joints. Its unique combination of lightness and strength make it ideal for aircraft, particularly undercarriage structures, and blades for some of the cooler parts of turbines.

Despite these glowing advantages, the whole technology of melting and casting of Ti and its alloys is, in my view, deeply disappointing as a result of poor melting and casting technology. Parallels with the casting of superalloy turbine blades are clear. For this reason, I am risking unpopularity to state my views, because I believe that the industry has the capability to develop what I believe to be its true, very much greater potential which has so far not been attained.

6.8.1 Ti ALLOYS

Naturally, the metallurgy of Ti alloys is an extensive subject that cannot be covered in any detail here. The reader is recommended to Polmear's book *Light Alloys* (2006). However, for a quick overview in a few sentences, the metallurgy of Ti is divided into (1) the low-temperature α-alloys generally stabilised by O, N, C and slightly by Sn and Zr and (2) the higher temperature β-alloys, stabilised at room temperatures by Mo, V, Cr and Fe. Common α-alloys include commercially pure Ti (actually a Ti-O alloy) and Ti-5Al-2.5Sn. Probably the most common of all Ti alloys is the mixed αβ-alloy Ti-6Al-4V (often known simply as 6-4 alloy).

Other important alloys include the titanium aluminides, TiAl and Ti_3Al both of which exist over a range of compositions and usually contain significant additional alloys. They have excellent specific stiffness, oxidation resistance and good creep resistance up to 700°C. Disadvantages include significant microsegregation as a result of the 'double cascade' effect of the two significant peritectic reactions that occur in series during its solidification. The alloy also suffers from all the typical symptoms of high bifilm content, including brittleness, low thermal conductivity and difficulty to machine, creating chipping, cracking and grain pullout.

Despite these drawbacks, much development work is in progress with the aluminides because of their huge potential for weight reduction in gas turbines for aircraft.

6.8.2 MELTING AND CASTING Ti ALLOYS

There is a fundamental challenge with the melting and casting of Ti and its alloys. Ti reacts with and dissolves practically everything, so there are few materials in which liquid Ti can be contained. In addition, the high melting temperature creates other technical problems, particularly rapid loss of temperature during casting.

Much Ti is therefore melted in a water-cooled copper crucible. Roberts (1996) describes the early history of the process in which melting was carried out in an inert atmosphere or vacuum by electric arc, electron beam or by plasma in a hemispherical copper crucible. After the liquid metal had been poured out, a hemispherical shell of frozen metal remained. When this was lifted out of the crucible, its shape was remarked to be like a skull. Hence early workers referred to the process as 'melting in a skull' which became shortened to 'skull melting'.

One of the serious issues that arises as a result of melting in a cold crucible is that relatively little average superheat can be achieved in a melt. With modern induction heating typical average superheats in vacuum are around 15°C, whereas an argon atmosphere seems to 'keep the melt warm' achieving around 35°C (Rishel, 1999). These very limited values have promoted workers to adopt (erroneously, in my view) techniques for extremely rapid filling of moulds, such as centrifugal casting. The melt is poured down a central downsprue and radially out to the moulds along spoke-like runners. Moulds are arranged around the outside of a rotating platform, carefully balanced before spinning. All this happens in a large chamber providing an environment of vacuum or inert gas.

The disastrous turbulence generated by this approach ensures that castings are full of defects (Cotton, 2006). Nevertheless, this problem seems to be accepted as normal in the Ti casting industry. Castings are repaired by grinding out defects and refilling by gas-tungsten arc welding. Finally, all Ti alloy castings are subjected to hot isostatic pressing (hipping) as standard procedure. Hipping appears to be especially effective for Ti alloys because pores are not only closed, but also appear to be effectively welded, because any oxide or nitride film on the original pore is taken into solution. Thus the casting defects are mainly repaired by these additional (expensive) procedures.

Harding (2006) has demonstrated the casting of TiAl by tilt pouring. Compared with centrifugal casting, this is a vast improvement in terms of reduced damage to the product. Even so, tilt is not an easy process at the low superheats available from a cold crucible. In addition, many castings have a geometry for which a tilt-filling solution is impossible without a free fall at some point in the mould and consequent damage to the liquid.

My approach would be to use counter-gravity to fill the mould. This is a powerful filling technique that can apply to practically all geometries of castings. The melt is transferred vertically into the mould by dipping a riser tube into the centre of the melt in the crucible and applying a differential pressure. This is a simple technique. In addition, a gentle uphill fill would make good use of the limited superheat because most moulds even up to a metre or so high could be filled within a very few seconds without exceeding 0.5 or perhaps 1.0 m/s. Such a technique would provide an essentially defect-free casting. Furthermore, of course, the castings would not require hipping, or, if hipping were insisted upon (as would be likely with many quality specifications apparently formulated without regard to logic) would not benefit further. Counter-gravity would ensure that the most frequent inclusions, solidified droplets of tungsten metal (W) from the splatter of weld electrodes in scrap and returned material, would remain sitting firmly on the bottom of the crucible, and thus not appear in the castings. The other major inclusion type, hard alpha particles, again mainly from foundry returns, are less easily dealt with, and largely invisible to X-rays. They can be detected by sophisticated ultrasound techniques.

It is doubtful if a low melt temperature would be a problem when filling moulds by counter-gravity. If temperature were a problem, I would opt to trial a CaO crucible. This material (lime) is one of the most available and low-cost materials, in contrast to the more usual ceramic material for melting Ti, cerium oxide, which is a potentially highly suitable material, but rare and costly. A calcia crucible allows significantly greater temperatures, probably 100°C or more, to be reached in the melt. Perhaps more importantly, these temperatures are reached at lower power densities so the melt is subjected to much less turbulence. In a cold crucible, the melt jumps about in the crucible and may be a problem to prevent it jumping out of the crucible when using very high power in an effort to achieve a tolerable casting temperature. Thus much entrainment of the liquid surface will occur with all this violent activity of the surface. Although researchers are concerned for oxygen that will be picked up from the CaO by the melt, the real problem may be the pickup not of oxygen itself, but of oxide films entrained by the turbulence in the cold crucible. Thus the action of CaO to introduce oxygen into solution in the alloy may be harmless, or relatively harmless, whereas oxide films introduced by turbulent melting will act as cracks and will therefore seriously reduce properties, particularly ductility. As usual, there is no shortage of avenues to explore.

A further potential benefit of a ceramic crucible is that there is some chance of its accretion of sundry debris in suspension in the melt (the 'fly paper' principle). Thus a melt would be expected to gradually become clean. This action occurs with practically all other ceramic and refractory containment of liquid metals; in fact the buildup of a collar of 'slag' is a common complaint with a melting process, but in reality is probably an unappreciated benefit. Such a scavenging action might be enhanced by the use of a very small amount of flux or slag. Such a liquid second phase afloat the main melt, occupying the meniscus valley at the crucible wall, is a common technique to encourage suspended material to stick to and be absorbed by the second liquid.

One of the most exciting developments in the history of titanium has to be the work of Schwandt and Fray (2014) who demonstrate the melting of a titanium alloy in an alumina or magnesia crucible, in air. The melt is protected against the absorption of oxygen by melting under a CaF_2-rich flux which transfers cathodic protection to the liquid metal from an iridium anode and titanium cathode. The technique might be most useful in the use of ceramic crucibles in vacuum, thereby gaining the benefit of adequate casting temperatures without turbulence-inducing energy inputs needed for the cold crucible technique.

Chandley (2000) described a quite different approach to reducing the contamination problems during melting. Titanium is heated under vacuum in a graphite-lined, induction-heated crucible. The vacuum is then replaced by argon and liquid aluminium is poured into the heated solid titanium and the power turned on. The exothermic reaction between the metals forms the TiAl molten alloy which is then transferred by suction counter-gravity into the investment shell mould.

Mould materials for Ti and its alloys are also problematic because of the high reactivity of Ti. Machined graphite moulds were the first to be used for Ti in 1954 (Cotton, 2006), soon to be followed by rammed graphite (Antes, 1958). More recently, oxide-based materials that can be investment cast are becoming the norm. Cheng et al. (2014) found some alumina is allowable to dilute the expense of yttria face coats. In all graphitic and oxide-based moulds, reaction with the mould stabilises the Ti alpha phase, thus converting the surface of the casting to this hard, brittle phase, known as the 'alpha skin'. The alpha skin has to be removed from the casting by machining or chemical dissolution. Unfortunately, this high-cost operation adds to the high cost of the alloy, the high cost of processing and the high cost of shipping. The selection of Ti and its alloys for a casting requires a dedicated and determined buyer, usually one who has no choice but to opt for a titanium product.

Permanent metal moulds for simple shapes have the great advantage of avoiding the creation of an alpha skin, but such moulds are often themselves expensive, consisting of alloys based on either Mo, W or Nb (Choudry, 1999).

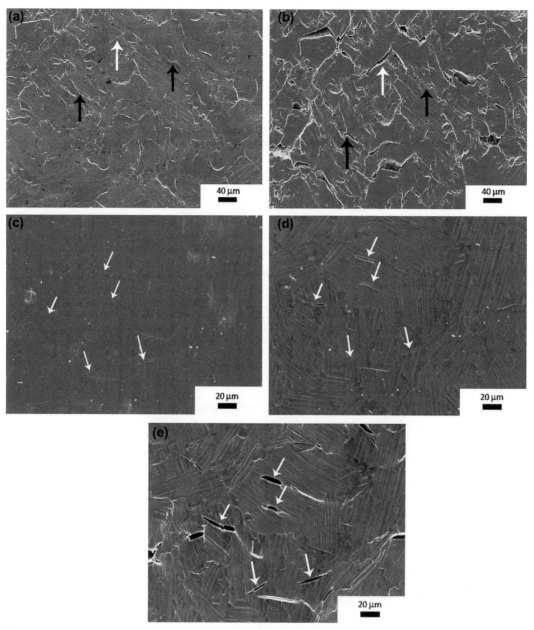

FIGURE 6.65

Cracks opening with time during creep at 700°C in Ti-45Al-TiB$_2$ alloy (a, b) at 300MPa showing development of cracks at B phase (white arrow) and colony boundaries (black arrows); (c,d,e) evolution of cracks with time at 400MPa only at B phase.

Courtesy Munoz-Moreno (2013)

6.8.3 SURFACE FILMS ON Ti ALLOYS

Mi (2002) has produced evidence that films do occur and can occasionally be seen by careful examination under the scanning electron microscope in castings.

In general, however, titanium alloys may not be troubled by a surface film. Certainly during the hot isostatic pressing (hipping) of these alloys, any oxide seems to go into solution. Careful studies have indicated that a cut (and, at room temperature, presumably oxidised) surface of the intermetallic alloy, TiAl, can be diffusion bonded to full strength across the joint, and with no detectable discontinuity when observed by transmission electron microscopy (Hu and Loretto, 2000). It seems likely, however, that the liquid alloy may exhibit a *transient film* in some conditions, such as the oxide on copper and silver and the graphite film on cast iron. Transient films are to be expected where the film-forming element is arriving from the environment faster than it can diffuse away into the bulk. This is expected to be a relatively common phenomenon because the rates of arrival, rates of surface reaction and rates of dissolution are not likely to be matched in most situations.

In conditions for the formation of a transient film, if the surface happens to be entrained by folding over, although the bifilm formed in this way is continuously dissolving, it may survive sufficiently long to create a legacy of permanent problems. These could include the initiation of porosity, tearing or cracking, prior to the complete disappearance of the bifilm. In this case, the culprit responsible for the problem would have vanished without trace.

Other consequences of the transient presence of bifilms in Ti alloys may be the morphology of the borides in some alloys of titanium aluminide. These are of extremely variable lengths in castings, short from some melts, but long from others, their lengths appearing to be bafflingly independent of casting parameters, and wildly out of the control of the metallurgist. This is typical of bifilm problems. In addition, on close examination, the boride particles are often seen to be associated with crack-like porosity along their lengths (too thin to be easily seen in Figure 6.64), typical of that which would occur if a bifilm had started to separate a little during the growth of the boride over its surfaces. Munoz-Moreno and colleagues (2013) studied centrifugally cast and hipped Ti-45Al alloy, finding, significantly, cracks associated with both boride particles and colony boundaries in creep tests at 700°C (Figure 6.65). It is expected that centrifugally cast material will have a high density of bifilms from this turbulent casting process, some of which will be favoured substrates

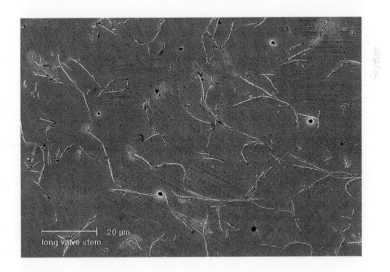

FIGURE 6.66

A titanium aluminide casting poured in argon atmosphere. Round pores may be argon bubbles, but all fine, crack-like pores are associated with platelet growth of borides (white linear features in image) and may indicate the bifilm substrates for the borides.

for the precipitation of borides, whose growth will then straighten the bifilms, and some will be pushed into grain (colony) boundaries. Chandravanshi (2013) studied creep fracture surfaces of an Alpha Ti alloy, finding clear decohesion from borides. Similarly, Cowen and Boehlert (2007) observed borides on the polished surfaces of specimens of Ti-22Al-26Nb alloy tested in creep. Immediately after the stress was applied, they saw the borides opening like cracks. As is often assumed for graphite in cast iron, the borides may also be weak, but their strengths are not known and may actually be high across transverse planes, in which case tensile failure would only be expected if the borides were formed on bifilms. Research is needed to clarify this (Figure 6.66).

POROSITY

7.1 SHRINKAGE POROSITY

There are so many ways to obtain porosity in a casting. Mainly, the porosity in castings seems to be entrained air bubbles, although sometimes is a more uniform dispersion of bifilms partly or fully inflated with a gas from solution in the melt. At other times, what appears to be shrinkage porosity is most often not shrinkage at all, but is a mass of oxides generated by the entrainment of the oxide surface in the turbulence of pouring. Finally some porosity is the result of mould or core blows, or even 'micro-blows' fired into the casting from microscopic pockets of volatile binder trapped in recesses in mould particles.

As should become clear, as a result of founders generally knowing how to feed castings, shrinkage porosity is rather rare in castings. On my travels around the foundries of the world, I rarely see shrinkage problems. Foundry people are aghast to hear such blasphemies because everyone knows that most casting scrap is assigned as 'shrinkage'. This is a mistake; a mis-identification. When viewing regions of apparent shrinkage porosity on a radiograph, I recommend that all foundry people should never say the words 'it is shrinkage' but should learn by heart and always use the phrase 'it *appears* to be shrinkage' adding for good measure 'but probably is not'. The mistaken identity arises from the scrambled mess of oxide bifilms and air bubbles which has a convincing appearance of shrinkage, but is not the result of lack of feeding, but the result of a poor filling technique. Entrainment defects constitute the world's most common foundry defect, perhaps accounting for 80–90% of all scrap.

Nevertheless, we need to discuss shrinkage here to highlight some of the conceptual problems of shrinkage conditions which can lead to reduced or negative pressures that can initiate pores. These concepts are, frankly, complicated to explain. I shall do my best.

7.1.1 GENERAL SHRINKAGE BEHAVIOUR

The molten metal in the furnace occupies considerably more volume than the solidified castings that are eventually produced, giving rise to several problems for the founder. The metal contracts at three quite different rates when cooling from the liquid state to room temperature, as Figure 7.1 illustrates.

1. As the temperature reduces, the first contraction to be experienced is that in the liquid state. This is the normal thermal contraction observed by everyone as a mercury thermometer cools; the volume of the liquid metal reduces almost exactly linearly with falling temperature.

   When making castings, the shrinkage of the liquid metal is usually not troublesome; the extra liquid metal required to compensate for this small reduction in volume is provided without difficulty. It is usually not even noticed, being merely a slight extension to the pouring time if freezing is occurring while the mould is being filled. Alternatively, it is met by a slight fall in level in the feeder.

2. The contraction on solidification is quite another matter, however. This contraction occurs at the freezing point, because (in general) of the greater density of the solid compared with that of the liquid. Contractions associated with freezing for several metals are given in Table 7.1. The contraction causes several problems. These include (1) the requirement for 'feeding', which is defined here as any process that will allow for the compensation

Complete Casting Handbook. http://dx.doi.org/10.1016/B978-0-444-63509-9.00007-8

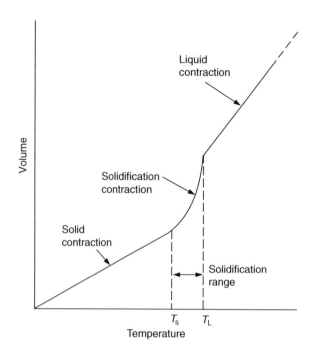

FIGURE 7.1

Schematic illustration of three shrinkage regimes: (i) in the liquid; (ii) during freezing and (iii) in the solid.

of solidification contraction by the movement of either liquid or solid, and (2) 'shrinkage porosity', which is the result of failure of feeding to operate effectively. These issues are dealt with at length in this chapter.

3. The final stages of shrinkage in the solid state can cause a separate series of problems. As cooling progresses, and the solidified casting attempts to reduce its size in consequence, it is rarely free to contract as it wishes. It usually finds itself constrained to some extent either by the mould, or often by other parts of the casting that have solidified and cooled already. These constraints always lead to the casting being somewhat larger than would be expected from free contraction alone. This is because of a certain amount of plastic stretching that the casting necessarily suffers. It leads to difficulties in predicting the size of the pattern because the degree to which the pattern is made oversize (the 'contraction allowance' or 'patternmaker's allowance') is not easy to quantify. The mould constraint during the solid-state contraction can also lead to more localised problems such as hot tearing or cracking of the casting. The conditions for the generation of these defects are discussed in Chapter 8.

7.1.2 SOLIDIFICATION SHRINKAGE

In general, liquids contract on freezing because of the rearrangement of atoms in the liquid structure changes from a rather open 'random close-packed' arrangement to a regular crystalline array of significantly denser packing in the solid.

The densest solids are those that have close-packed (face centred cubic (fcc) and hexagonal close packed (hcp)) symmetry. Thus the greatest values for contraction on solidification are seen for these metals. Table 7.1 shows the contractions to be in the range of 3.2–7.2%. The solidification shrinkage for the less closely packed body-centred cubic (bcc) lattice is in the range 2–3.2%. Other materials that are less dense in the solid state contract by even smaller amounts on freezing.

Table 7.1 Solidification Shrinkage for Some Metals

Metal	Crystal Structure	Melting Point °C	Liquid Density (kgm^{-3})	Solid Density (kgm^{-3})	Volume Change (%)	References
Al	fcc	660	2368	2550	7.14	1
Au	fcc	1063	17,380	18,280	5.47	1
Co	fcc	1495	7750	8180	5.26	1
Cu	fcc	1083	7938	8382	5.30	1
Ni	fcc	1453	7790	8210	5.11	1
Pb	fcc	327	10,665	11,020	3.22	1
Fe	bcc	1536	7035	7265	3.16	1
Li	bcc	181	528	—	2.74	4, 5
Na	bcc	97	927	—	2.6	4, 5
K	bcc	64	827	—	2.54	4, 5
Rb	bcc	303	11,200	—	2.2	2
Cd	hcp	321	7998	—	4.00	2
Mg	hcp	651	1590	1655	4.10	3
Zn	hcp	420	6577	—	4.08	2
Ce	bcc	787	6668	6646	−0.33	1
In	tetrag	156	7017	—	1.98	2
Sn	tetrag	232	6986	7166	2.51	1
Bi	rhomb	271	10,034	9701	−3.32	1
Sb	rhomb	631	6493	6535	0.64	1
Si	diamond	1410	2525	—	−2.9	2

fcc, face centered cubic; bcc, body centered cubic; hcp, hexagonal close packed; tetrag, tetragonal; rhomb, rhombic.
References: 1, Wray (1976); 2, Lucas (quoted by Wray, 1976); 3, This book; 4, Lida and Guthrie (1988); 5, Critical review by J. Campbell in Brandes (1983).

The exceptions to this general pattern are those materials that expand on freezing. These include water, silicon and bismuth (Table 7.1), and, of course, perhaps the most important alloy of all, cast iron.

Remember the ice cubes are always stuck fast, having expanded in the ice tray. Analogously, the success of 'type metal', a bismuth alloy used for casting type for printing, derives some of its ability to take up fine detail required for lettering because of its expansion on freezing. Graphitic cast irons with carbon equivalent above approximately 3.6 (Figure 7.2) similarly expand because of the precipitation of the low-density phase, graphite. The non-feeding requirements of grey irons are in sharp contrast to other cast irons that solidify 'white', i.e. containing the non-equilibrium carbide phase in place of graphite. The white irons in general have solidification contractions similar to those of steels.

For the majority of materials that do contract on freezing, it is important to have a clear idea of what happens in a poorly fed casting. As an ideal case of an unfed casting, it is instructive to consider the freezing of a sphere. We shall assume that the sphere has been fed via an ingate of negligibly small size up to the stage at which a solid shell has formed of thickness x (Figure 7.3). The source of supply of feed metal is then frozen off. Now as solidification continues with the freezing of the following onion layer of thickness dx, the reduced volume occupied by the layer dx compared with that of the original liquid means that either a pore has to form, or the liquid has to expand a little, and the surrounding solid correspondingly has to contract a little. If we assume for the moment that there is no favourable nucleus available for the creation of a pore, then the liquid has to accommodate this by expanding, creating a state of tension, or negative pressure. At the same time, of course, the liquid is in mechanical equilibrium with the enclosing solid shell, effectively sucking it

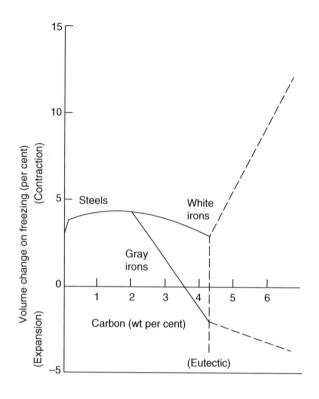

FIGURE 7.2

Volume change on freezing of Fe-C alloys. The relations up to 4.3% carbon are due to Wray (1976); higher carbon values are estimated by the author.

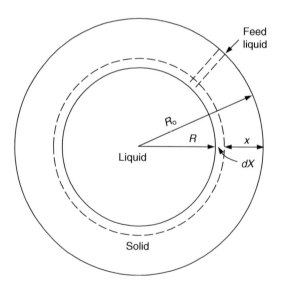

FIGURE 7.3

Solidification model for an unfed sphere.

inwards. As more onion layers form, so the tension in the liquid increases, the liquid expands and the solid shell is drawn inwards, initially suffering only elastic deformation, but as the stress rises, finally suffers inward plastic contraction.

The reader may find the solidifying sphere model easier to visualise by thinking of the exact converse situation of a material that expands on solidification: in this case, the formation of solid squeezes the remaining liquid into a smaller volume. The compressed liquid therefore experiences a positive pressure. The internal positive pressure then clearly acts on the liquid to compress it. In turn, the liquid acts on the solid shell to expand it. It is an instructive exercise to go back and reverse this logic to ensure that the concept of negative pressure, or hydrostatic tension, in the liquid is fully understood before proceeding further with this chapter. However, if the reader finds this challenging, further examples to clarify the concept of negative pressures in liquids are presented in Section 7.1.2.1.

Calculations based on a solidifying sphere model were first carried out by the author (Campbell, 1967, 1968) that show the high tensile stresses in the residual liquid cause the solid to collapse plastically by a creep process. This inward movement of the solid greatly reduces the tension in the liquid, but negative pressures of −100 to −1000 atm (10–100 MPa tensile stress) may be expected under ideal conditions. However, in conditions in which rapid cooling leads to temperature gradients which cause the outside of the casting to shrink, Susumu Oki (1969) estimates that these maximum values can be greatly reduced. Later researchers (Forgac, 1981; Ohsasa et al., 1988) provide further refinements in confirmation of the development of tensile stresses in the residual liquid of a spherical casting. The liquid itself is easily able to withstand such stresses because the tensile strength of liquid metals is in the range of −50,000 to −500,000 atm (5–50 GPa). Incidentally, the liquid–solid interface (whether planar or dendritic is immaterial) is also easily able to withstand the stress, because it is not a favourable site for failure by the nucleation of a cavity (see Section 7.1.5.3).

Even in situations that remain connected to feed metal, it can be shown that the resistance to viscous flow of the residual liquid through a pasty zone can cause the hydrostatic tension to build up, eventually becoming large enough for the casting to collapse by plastic or creep flow. This condition is described more fully later in the sections on inter-dendritic feeding and shrinkage porosity.

Thus, in many casting situations, the pressure in the solidifying metal can fall to low, zero or negative values. The negative values correspond to a hydrostatic stress that powers the formation of shrinkage porosity. However, at the same time, of course, the pressure gradient between the outside and inside of the casting is also the driving force for the various feeding mechanisms that help to reduce porosity.

Whether the driving force for pore formation wins over the driving force for feeding will depend on whether nuclei for pore formation exist. If not (i.e. if the metal is clean), then no pores will nucleate, and feeding is forced to continue until the casting is completely frozen and completely sound. If favourable nuclei are present, then pores will be created at an early stage before the development of any significant hydrostatic stress, with the result that little feeding will occur, and the casting will develop its full percentage of porosity as defined by the physics of the phase change (as given, for instance, in Table 7.1). In most cases, the real situation is somewhere between these extremes, with castings displaying some evidence for partially successful liquid feeding in the form of feeder heads which have drawn down somewhat, and parts of the casting displaying evidence of some solid feeding (Figure 7.4) because of some surface sinks, but the interior exhibiting some porosity. Clearly, in such cases, feeding has continued under increasing pressure differentials until the development of a critical internal stress at which some particular nuclei, or surface puncture, can be activated at one or more points in the casting. Feeding is then stopped at such a locality and pore growth starts.

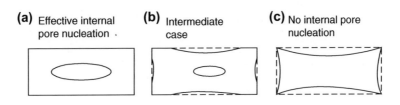

(a) Effective internal pore nucleation

(b) Intermediate case

(c) No internal pore nucleation

FIGURE 7.4

The three forms of shrinkage porosity: (a) internal; (b) mixed; and (c) external shrinkage porosity.

In those cases in which a pore does form at an early stage of solidification, a freezing contraction of perhaps 3% does not at first sight appear to be of any great significance. However, the reader can easily confirm that in a 100 mm diameter sphere the corresponding cavity will be 31 mm in diameter. For an aluminium casting at 7% contraction, the cavity would be 41 mm in diameter. These considerable cavities require a dedicated effort to ensure that they do not appear in castings. This is especially difficult when it realised that on occasions a substantial casting might be scrapped if it is found to contain a defect of only approximately 1 mm in size. (The scrapping of castings because of the imposition of unreasonably high inspection standards is a widespread injustice that would benefit from the injection of common sense.)

For the vast majority of cast materials, therefore, shrinkage porosity is one of the most important threats to a casting. Paradoxically, this even includes grey cast irons because of the effect of mould dilation. The problem of mould dilation occurs if the mould is weak, allowing the benefits of the expansion of the graphite phase to be wasted on outwards expansion of the casting, rather than being used on inwards compression of cavities.

The serious consequences of the porosity occurring on the inside of the casting, where it causes a disproportionally large hole, contrasts with the occurrence of porosity on the outside of the casting. If our 100 mm diameter sphere had a common solidification contraction in the region of 3% and had no internal nuclei for the initiation of porosity, no pores would form here and the solid would therefore be forced to collapse. The final casting would be slightly smaller by approximately 0.5 mm over its whole surface. This 1% reduction in diameter is sufficiently small to be not noticeable in most situations. If it were to be important for a casting requiring special accuracy, this small adjustment could be added to the patternmaker's shrinkage allowance. Thus the elimination of internal porosity and its displacement to the outside of the casting is a powerful technique that is strongly recommended.

Although shrinkage porosity can be reduced by improvements to the cleanness of the metal as mentioned previously, this chapter deals later with the other major approach to this problem, the design and provision of good feeding, ensuring good conditions for easy accommodation of the volume change during freezing. The special case of provision of good liquid feeding, the usual way to deal with a feeding problem, is dealt with as a 'methoding' issue in Chapter 10 rule 6 'Avoid shrinkage porosity'.

In the meantime, understanding negative pressure (hydrostatic tension) in castings is sufficiently critical to an appreciation of casting behaviour that it will be valuable to divert to discuss examples before proceeding further.

7.1.2.1 Hydrostatic tensions in liquids

In the original book 'Castings 1991', this section was to have been an appendix. However, it proved to be so interesting and so central to the understanding of defect generation in castings that the subject was included as part of the main text. The reader will understand why the subject is included here even though few of the examples described in this section are metal castings. Most relate to water or other organic systems but the principles will be seen to apply perfectly to metals.

Liquids have been known to be able to withstand tension for many years. Mercury is an interesting case. In a clean glass barometer freshly filled with clean mercury, the liquid would adhere to the glass, 'hanging up', supporting the weight of the long length of the heavy liquid like a rope in tension. If the glass was tapped sharply, breaking the liquid away from the glass, the liquid would fall, opening up a vacuum above it. It would settle at the height at which it could be supported by the pressure of 1 atm acting at its base. This height is approximately 760 mm: the barometric height. The so-called vacuum is of course not perfect. A tiny pressure from the vapour pressure of mercury would operate, pushing the upper surface down about 1 μm at room temperature which we shall ignore. Actually, if the glass and the metal had been really clean, in principle the mercury column could have hung from a height somewhere in the stratosphere, because the tensile strength of the liquid is extremely high.

The very high tensile strengths of liquid metals should not be a surprise. The interatomic distances in liquid metals, as random close-packed atomic structures, are only a few percent larger than the distances in solid metals, which have more regular close-packed structures. Thus the strengths would be expected to be similar, which is the case.

In 1850, Berthelot cooled various organic liquids in sealed glass tubes, watching until cavitation occurred as a shower of small bubbles. The number of bubbles was subsequently explained by Lewis (1961), who observed that the rupture was always associated with the sudden appearance and collapse of a single bubble, which was immediately followed by the formation of a large number of tiny bubbles along the length of the tube. The asymmetric collapse of the primary bubble would form an impacting jet and create a series of shock waves. These would be ideal conditions for the formation of additional bubbles. This simple experiment is of special interest because it is analogous to the solidifying casting, with the internal solidifying liquid progressively occupying less volume, and with the surrounding solid gradually being pulled inwards, its elastic and plastic resistance to deformation steadily increasing, building up the tensile stress in the liquid, to the point at which the liquid fractures.

Berthelot's experiment has been repeated many times in different ways. For instance, Vincent and Simmons (1943) sealed their tube by freezing the liquid in the mouth of the capillary. They found that water would withstand -157 atm and a mineral oil would withstand -119 atm. This is a fascinating observation because it indicates that the freezing interface will withstand high tensions; it is not necessarily a good nucleation site for pores, as has been confirmed theoretically (Section 7.1.5). The facts reveal the error of those assumptions, widely presumed in the solidification literature, of ease of pore nucleation at the solidification front.

For a liquid sealed in a tube, Scott et al. (1948) calculated the build-up of tensile stress ΔP as the liquid cools an amount ΔT, using the coefficient of thermal expansion of the liquid α and the glass σ, together with the modulus of rigidity G and the coefficient of compressibility β of the glass tube:

$$\Delta P = [(\alpha - \sigma)/(\beta - (1/G))]\Delta T$$

They also checked their results by measuring the length of the liquid columns after fracture. This work emphasises the elastic strains built up in the liquid and the surrounding solid during the experiment.

Lewis (1961) found tensile strengths in water and carbon tetrachloride in excess of -60 atm. Also, when water and carbon tetrachloride were introduced into the same tube, the two liquids separated at a clear interface. Fracture was not observed to occur at the interface. The liquid–liquid interface seems as poor a nucleation site as the liquid–solid interface discussed previously. These observations confirm the expectations from classical nucleation theory as presented in Section 7.1.5.

Tyndall (1872) carried out experiments on the internal melting of ice by shining light through the ice. The light was absorbed by particles frozen inside. The heated particles then melted a small surrounding region of the shape of a snowflake (known now as a Tyndall figure, or, occasionally, a negative crystal). However, because ice reduces in volume on melting, the liquid volume created in this way was under tension. As melting progressed, the tensile stress became so great that the liquid fractured with an audible click. The melted region was then observed to contain a small bubble. Kaiser (1966) demonstrated that the critical pressure for fracture was in excess of -150 atm. Tyndall proved that the bubble was a vapour cavity containing practically no air by carefully melting the ice towards the figure using warm water. As soon as a link was established with the bubble, it collapsed and disappeared; no air bubble rose to the surface. The Tyndall figures are analogous to the isolated regions of metal castings, where vapour cavities are expected to be formed at critical stresses during solidification.

Briggs (1950) centrifuged water in a capillary tube, spinning it at increasing speed to stretch the water until it was observed to snap at a tensile stress of -280 atm.

The early work on the determination of the tensile strengths of liquids held in various containers such as a Berthelot tube gave notoriously scattered results. This was probably to be expected from the different degrees of cleanness of the container, and any microscopic imperfections it may have had on its walls. Experiments to test the liquid away from the influence of walls have been carried out using ultrasonic waves focused in the centre of a volume of liquid. The rarefaction (tensile) half of the vibration cycle can cause cavitation in the liquid. These experiments have been moderately successful in increasing the limits, but impurities in the liquid have tended to obscure results.

More recently, Carlson (1975) has reflected shock waves through a thin film of mercury and recorded strengths of up to $-30,000$ atm. This result is one of the first to give a strength value of the same order as that predicted by various theoretical approaches. Despite the dynamic nature of the test method, the result is not particularly sensitive to the length of the duration of the stress (varying only from $-25,000$ to $-30,000$ atm as the time decreased from 1 s to 10^{-7} s). This again is to be expected because nucleation is a process involving the rearrangement of atoms, where the atoms are vibrating about their mean positions at a rate of approximately 10^{13} times per second.

An attempt to measure negative pressures in solidifying aluminium ingots directly was made by Ohsasa et al. (1988a,b). They immersed a stainless steel disc into the top of a solidifying metal to almost fill the top of the crucible in which the metal was held. When the solidification front reached the disc, a volume of liquid was effectively trapped

Table 7.2 Fracture Pressures of Liquids

Liquid	Surface Tension, Nm^{-1}	Atom Diameter, nm	ΔP				Critical Embryo (No. of Vacancies)	ΔP	
			From Eqn (7.6)		From Fisher			Complex Inclusion	
			GPa	atm	GPa	atm		MPa	atm
Water	0.072	—	—	—	0.13	1320	—	1.6	16
Mercury	0.5	0.30	1.67	16,700	2.23	22,300	14	20	200
Aluminium	0.9	0.29	3.10	31,000	3.00	30,000	52	36	360
Copper	1.3	0.26	5.00	50,000	5.00	50,000	72	60	600
Iron	1.9	0.25	7.6	76,000	7.00	70,000	72	85	850

underneath. As this volume solidified, the stress in the liquid was measured by a transducer connected to the disc. Stresses needed to cause pore formation were typically only about −0.1 atm. However, occasionally stresses of up to −2.5 atm were recorded. On these occasions, the walls of the ingot were observed to suffer some inward collapse. Clearly, in this work, the stresses were limited by the nucleation of cavities against the surface of the stainless steel disc. This would be poorly wetted as a result of the presence of the Cr-rich oxide on the steel. At the higher stresses that were measured, these were in turn limited by the inward plastic collapse of the solidified casting.

In general, therefore, it seems that the various attempts to measure the strengths of mercury and the organic liquids have achieved results intermediate between the minimum predicted for a complex inclusion (potentially the weakest point in a liquid metal as explained in Section 7.1.5.3) and the maximum set for the prediction of homogeneous nucleation (Table 7.2). This is probably as good a result as can be expected. However, the one high result by Carlson is reassuring, suggesting that the physics seems basically correct, and really does apply to liquid metals.

7.1.3 FEEDING CRITERIA

Because the first half of this book is concerned with the metallurgy of cast metals, the practical aspects of feeding is taken up later, becoming an important feature from Chapter 10 onwards. However, it is desirable to indicate some central concepts, and some small amount of repetition will do no harm.

To allow for the fact that extra metal needs to be fed to the solidifying casting to compensate for the contraction on freezing, it is normal to provide a separate reservoir of metal. We shall call this reservoir a *feeder* because its action is to *feed* the casting (i.e. to compensate for the solidification shrinkage). The American term *riser* is not recommended. It is not descriptive in a helpful way and could lead to confusion with other features of the *rigging* such as vents, up which metal also rises. However, the American term *rigging* is helpful. It is a general word for the various appendages of runners, gates and feeders etc. No equivalent term exists in UK English.

The provision of a feeder can be complicated to get right. There are seven rules that the author has used to help in the systematic approach to a solution. Readers of the book 'Castings 1991' will note this is one more rule than before. The additional rule was cited in 'Castings' but not elevated to the status of a rule. The reader will appreciate that the new Rule 1 'Do not feed' has a valuable place in the new listing. Chapter 10, Section 6.4, explains that the rule is not as perverse as it first seems.

More recently, Tiryakioglu (2001) has further simplified the Seven Rules as is noted in at the end of this section, but is described in more detail in Chapter 10, Section 6.5.

7.1.3.1 The seven feeding rules

Rule 1. The main question relating to the provision of a feeder on a casting is 'Should we have a feeder at all?' This is a question well worth asking, and constitutes Rule 1 '*Do not feed (unless absolutely necessary)*'.

The avoidance of feeding is to be greatly encouraged, in contrast to the teaching in most traditional foundry texts. There are several reasons to avoid the placing of feeders on castings. The obvious one is cost. Feeders cost money to put on and money to take off. In addition, many castings are actually impaired by the inappropriate placing of a feeder. This is especially true for thin-walled castings, where the filling of the feeder diverts metal from the filling of the casting, with the creation of a misrun casting. Probably half of the small and medium-sized Al alloy castings made today do not need to be fed. Sometimes the casting suffers delayed cooling, impaired properties and even segregation problems as a result of the presence of a feeder. Finally, it is easy to make an error in the estimation of the appropriate feeder size, with the result that the casting can be more defective than if no feeder were used at all.

However, there is one potential major problem if a feeder is not used, but where a feeding problem remains in the casting. The feeding problem will show itself as a region of reduced pressure in the casting at a late stage of freezing. The reduced pressure will act to open bifilms as in the reduced pressure test (RPT) used for nonferrous alloys. Thus in an aluminium alloy casting with medium thickness sections, enhanced size and number of internal microshrinkage pores will be experienced if the sections are unfed. Whether this is important or not depends on the specification of microstructure and properties that the casting is required to meet. If the casting is required to have good elongation properties in a tensile test it is likely that feeding to maintain pressure during solidification (and thus keep the bifilms closed) will be needed to achieve this.

Just for the moment, we shall assume that a feeder is required. The next question is 'How large should it be?'

There is of course an optimum size. Figure 7.5 shows the results of Rao et al. (1975), who investigated an increasing feeder size on the feeding of a plate casting in the Al-12Si alloy. Interestingly, when the data are extrapolated to zero feeder size, the porosity is indicated to be approximately 8%, which is close to the theoretical 7.14% solidification

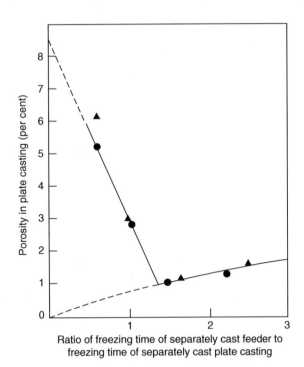

FIGURE 7.5

Effect of increasing feeder solidification time on the soundness of a plate casting in Al-12Si alloy.

Data from Rao et al. (1975).

shrinkage for pure aluminium (Table 7.1). At a feeder modulus of around 1.2 times the modulus of the casting, the casting is at its most sound. The residual 1% porosity is probably dispersed gas porosity. As the feeder size is increased, further the solidification of the casting is now progressively delayed by the nearby mass of metal in the feeder. Thus, although this excessive feeder is no disadvantage in itself, the delay to solidification of the whole casting increases the time available for further precipitation of hydrogen as gas porosity and the unfolding of bifilms. However, it is clear from this work that an undersized feeder will result in very serious porosity, whilst an oversize feeder causes less of a problem. Considering the economics, a slightly oversize feeder may be a good investment to reduce the threat of the total loss of the casting because of porosity.

The previous work illustrates that it is more important to deal with the shrinkage problems than with gas problems in castings. (This conclusion might raise the eyebrows of practised foundry people. It needs to be kept in mind that most of what previously has been generally described as 'gas' in castings has actually been turbulently entrained *air bubbles* introduced by our poor filling systems. In general, good eyesight is needed to see true gas porosity. We shall deal with entrained air bubbles by the provision of non-entraining filling system designs.)

There is a vast amount of literature on the subject of providing adequate-sized feeders for the feeding of castings. It is mainly concerned with two feeding rules. Numbering onwards from Rule 1 mentioned previously, the conventional rules are as follows.

Rule 2. The feeder must solidify at the same time as, or later than, the casting. This is the heat-transfer criterion, attributed to Chvorinov.

Rule 3. The feeder must contain sufficient liquid to meet the volume-contraction requirements of the casting. This is usually known as the volume criterion.

However, there are additional rules which are also often overlooked, but which define additional thermal, geometrical and pressure criteria that are absolutely necessary conditions for the casting to freeze soundly:

Rule 4. The junction between the casting and the feeder should not create a hot spot, i.e. have a freezing time greater than either the feeder or the casting. This is a problem that, if not avoided, leads to '*underfeeder shrinkage porosity*'. The junction problem is a widely overlooked requirement. The use of Chvorinov's equation systematically gives the wrong answer for this reason, so the junction requirement is often found to override Chvorinov (Rule 2).

Rule 5. There must be a path to allow feed metal to reach those regions that require it. The reader can see why this criterion has been often overlooked as a separate rule; the communication criterion appears self-evident! It is, nevertheless, often violated.

Rule 6. There must be sufficient pressure differential to cause the feed material to flow, and the flow needs to be in the correct direction, towards the region requiring the additional liquid (no apologies for spelling out the glaringly obvious—it is surprising how often these rules are broken and the casting scrapped).

Rule 7. There must be sufficient pressure at all points in the casting to suppress the formation and growth of porosity. In Section 9, we shall see that the action of pressure also retains the good mechanical properties of the casting by suppressing the opening of bifilms. The bifilms often constitute a low background level of porosity (that may or may not be just visible) that is highly effective in reducing strength and ductility of castings.

It is essential to understand that *all* the rules must be fulfilled if truly sound castings are to be produced. The reader must not underestimate the scale of this problem. The breaking of only one of the rules may result in ineffective feeding, and a porous casting. The wide prevalence of porosity in castings is a sobering reminder that solutions are often not straightforward.

Because the calculation of the optimum feeder size is therefore so fraught with complications, is dangerous if calculated wrongly, and costs money and effort, the casting engineer is strongly recommended to consider whether a feeder is really necessary at all. This is Rule 1. You can see how valuable it is.

Of the remainder of castings that do suffer feeding demand, many could avoid the use of a feeder by the judicious application of chills and/or cooling fins or pins.

This leaves only those castings that have heavy sections, isolated heavy bosses or other features that *do* need to be fed with the correct size of feeder.

Finally, in a recent development of the concepts of feeding, it is worth drawing attention to the work of Tiryakioglu (2001). He has demonstrated that Rules 2, 3 and 4 can be gathered together under only one new criterion 'The thermal centre of the total casting (i.e. the casting + feeder combination) should be in the feeder'. This deceptively straight-forward statement has been shown to allow the calculation of optimum feeders with unprecedented accuracy. Discussion of this approach is detailed in Chapter 10 under Casting Rule 6.

7.1.3.2 Criteria functions

The development of computer software to predict the solidification of castings is not yet developed to predict the occurrence of porosity from first principles, i.e. calculating the pressure drop in the various parts of the casting, and thereby assessing the potential for nucleation and growth of cavities. This represents a Herculean task. As a useful shortcut, therefore, many workers have searched for parameters that can be relatively easily calculated and enable an assessment of the potential for pore formation.

Niyama et al. (1982) was perhaps the first to develop a useful criterion function. Dantzig and Rappaz (2009) give a rigorous and elegant account of his derivation. Based on a simple model assuming Darcy's law for the flow of residual liquid through the dendrite mesh, he proposed the parameter $G/R^{1/2}$ to assess the difficulty of providing feed liquid, where G is the temperature gradient at the solidus temperature and R the cooling rate. This is probably the most widely used criteria function. It has been found useful to predict the propensity for porosity in steels, often in the form assuming that porosity will occur when $G/R^{1/2}$ falls below some critical value in the region of 1 $K^{1/2}min^{1/2}cm^{-1}$. These awful units of the critical Niyama criterion (quoted shamelessly by Dantzig and Rappaz) translate with sufficient accuracy into the respectable SI equivalent 1 $K^{1/2}s^{1/2}mm^{-1}$ (for the purist $1.00\ K^{1/2}s^{1/2}mm^{-1} = 1.29\ K^{1/2}min^{1/2}cm^{-1}$).

In their study of the formation of porosity in steel plates of thickness 5–50 mm, with and without end chills, Minakawa et al. (1985) confirmed that usefulness of the Niyama criterion in assessing the conditions for the onset of porosity in their castings. They also looked at G the temperature gradient along the centreline of the casting at the solidification front, and the fraction solid along the centreline. Neither of these alone was satisfactory.

In another theoretical study, Hansen and Sahm (1988) support the usefulness of $G/V_S^{1/2}$ for steel castings. However, in addition they go on to argue the case for the use of a more complex function $G/V_S^{1/4}V_L^{1/2}$ where V_L is the velocity of flow of the residual liquid. They proposed this relation because they noticed that the velocity of flow in bars was 5 to 10 times the velocity in plates of the same thickness which, they suggest, contributes to the additional feeding difficulty of bars compared with plates. (A further contributor will be the comparatively high resistance to the plastic collapse of bars, compared with the efficiency of solid feeding in plates as will be discussed later.)

They found that $G/V_S^{1/4}V_L^{1/2}$ is less than a critical value, which for steel plates and bars is approximately 1 $Ks^{3/4}\ mm^{-7/4}$. Their parameter is, of course, less easy to use than that of Niyama because it needs flow velocities. The Niyama approach only requires data obtainable from temperature measurement in the casting.

Carlson and Beckermann (2010) proposed a dimensionless form of the Niyama criterion (especially useful in view of the difficult units $K^{1/2}s^{1/2}mm^{-1}$) and find it to be useful for predicting shrinkage porosity in a steel, an Al alloy and a Mg alloy, demonstrating impressively wide applicability.

Unfortunately, however, there are many limitations of most of the other criteria functions that are widely overlooked. They include the following.

1. Strictly, they assess only the difficulty of interdendritic feeding.
2. They apply best to strong materials such as steels in relatively cold moulds. Here, interdendritic feeding can become impossibly difficult, generating high internal tension in the casting, a condition that the computer would interpret as leading to porosity. However, in conditions in which the solid is softer and can yield plastically, the mould can collapse slightly to give an internally sound casting. Such conditions apply, for instance, in steels cast into hot investment moulds and light alloys in cold moulds where significant solid feeding can occur, causing the criteria function to become inaccurate.
3. Criteria functions cannot predict conditions for porosity arising as a result of the many other mechanisms for the production of porosity in castings. These include the major shrinkage pores as a result of the isolation of major liquid regions, the creation of surface-initiated porosity and the mechanically entrained porosity originating from bubbles of air and mould gases introduced by poor filling systems. Experience shows that in general entrainment defects cause most of the porosity found in casting. Shrinkage porosity is actually quite rare because, as we have

noted previously, founders understand shrinkage and can avoid it. In fact, the ability to avoid shrinkage problems is continuously improving at this time because of the continuous development of prediction using computer models.

4. Sigworth et al. (1994) point out an obvious but often overlooked limitation of criteria functions that they cannot take account of the important effects of melt treatment, such as, in particular, its bifilm content.

7.1.4 FEEDING: THE FIVE MECHANISMS

During the solidification of a casting, the gradual spread and growth of the solid, often in the form of a tangled mass of dendrites, presents increasing difficulties for the passage of feeding liquid. In fact, as the freezing liquid contracts to form the solid, the pressure in the liquid falls, causing an increasing pressure difference between the inside and the outside of the casting. The internal pressure might actually fall far enough to become negative, as a hydrostatic tension.

The generation of reduced pressure, and even actual hydrostatic tension, is undesirable in castings. Such phenomena provide the driving force for the unfurling and inflation of bifilms, leading to reduced properties and the initiation and growth of volume defects such as porosity. Where internal defects do not open up, the reduced pressure can lead to surface sinks on castings.

There appear to be at least five mechanisms by which hydrostatic tension can be reduced in a solidifying material, although, of course, not all five processes are likely to operate in any single case. Adequate feeding by one or more of these feeding processes will relieve the stress in the solidifying liquid and thus reduce the possibility of the formation of defects.

I first identified and described the five feeding mechanisms as set out in Figure 7.27 in 1969. Since then, I see my drawing everywhere (I could draw better when I was young). In the original publication, solid-state diffusion was added as a sixth effect that would cause pore shape to change somewhat after solidification was complete. Shape changes in pores would occur because the forces of surface tension and mechanical stress are sufficient to cause material flow at temperatures near the melting point of the solid. Such changes to the pore shape and size are detectable under a microscope. However, these considerations are the reserve of the research scientist, and reflect the author's early interests, having been trained as a physicist. Nowadays, as a somewhat more practical foundryman trying to make good castings, the first five mechanisms are all that matter.

The mechanisms are dealt with in the order in which they might occur during freezing. The order coincides with a progressive but ill-defined transition from what might usefully be termed 'open' to 'closed' feeding systems.

7.1.4.1 Liquid feeding

Liquid feeding is the most 'open' feeding mechanism and generally precedes other forms of feeding (Figure 7.6). It should be noted that in skin-freezing materials it is normally the only method of feeding. The liquid has low viscosity, and for most of the freezing process the feed path is wide, so that the pressure difference required to cause the process to operate is negligibly small. Results of theoretical model of a small cylindrical casting 20 mm diameter (Figure 7.7) indicate that pressures of the order of only 1 Pa are generated in the early stages. By the time the 10 mm radius casting has a liquid core of radius 1 mm, corresponding to 99% solid, the pressure difference has increased only to 100 Pa which is approximately one-thousandth of an atmosphere and smaller than about one-tenth of the hydrostatic pressure due to depth. (It is worth emphasising that the theoretical model represented in Figure 7.7 and elsewhere in this book represents a worst case. This is because the temperature gradient in the solidified shell has been neglected. The lower temperature of the outer layers of the shell will cause the shell to contract, compressing the internal layers of the casting and thus reducing the internal hydrostatic tension. In some cases, the effect is so large that the internal pressure can become positive, as shown in the excellent treatment by Forgac et al., 1979.)

For all practical purposes, therefore, liquid feeding occurs at pressure gradients that are so low that these gentle stresses will never lead to problems.

The rules for adequate liquid feeding are the seven feeding rules listed in Section 7.1.3.

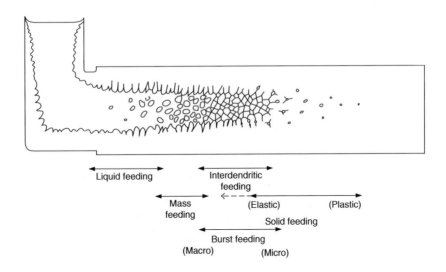

FIGURE 7.6

Schematic representation of the five feeding mechanisms in a solidifying casting (Campbell, 1969a,b).

Inadequate liquid feeding is often seen to occur when the feeder has inadequate volume. Thus liquid flow from the feeder terminates early, and subsequently only air is drawn into the casting. Depending on the mode of solidification of the casting, the resulting porosity can take two forms:

1. Skin-freezing alloys will have a smooth solidification front that will therefore result in a smooth shrinkage pipe extending from the feeder into the castings as a long funnel-shaped hole. In very short freezing range metals, the surface of the pipe can be as smooth and silvery as a mirror.
2. Long freezing range alloys will be filled with a mesh of dendrites in a sea of residual liquid. In this case liquid feeding effectively becomes interdendritic feeding, of course. In the case of an inadequate supply of liquid in the feeder, the liquid level falls, draining out to feed more distant regions of the casting, and sucking in air to replace it. The progressively falling level of liquid will define the spread of the porosity, decreasing as it advances because of the decreasing volume fraction of residual liquid as freezing proceeds. The resulting effect is that of a partially drained sponge, as shown in the tin bronze casting in Figure 7.8. Sponge porosity is a good name for this defect.

When sectioned, the porosity resembles a mass of separate pores in regions separated by dendrites. It is therefore often mistaken for isolated interdendritic porosity. However, it is, of course, only another form of a primary shrinkage pipe, practically every part of which is connected to the atmosphere through the feeder. It is a particularly injurious form of porosity in castings that are required to be leak-tight, especially because it can be extensive throughout the casting, as Figure 7.8 illustrates. Furthermore this type of porosity is commonly found. It is an indictment of our feeding practice.

The author recalls an investigation into the nature of the porosity in the centre of a balanced steel ingot. To ascertain whether the so-called secondary porosity was connected to the atmosphere via the shrinkage cavity in the top of the ingot, water was poured onto the top of the ingot. It created a never-forgotten drenching from the shower that issued from the so-called secondary pores. The lesson that the pores were perfectly well connected was not forgotten.

7.1.4.2 Mass feeding

Mass feeding is the term coined by Baker (1945) to denote the movement of a slurry of solidified metal and residual liquid. This movement is arrested when the volume fraction of solid reaches anywhere between 0 and 50%, depending on

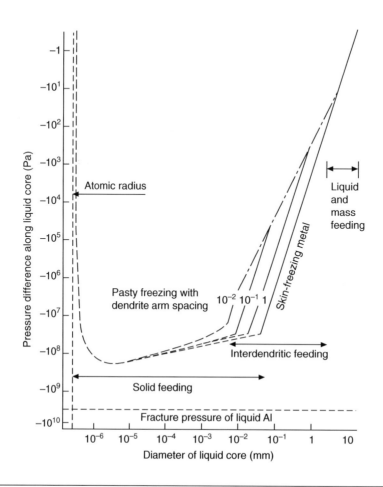

FIGURE 7.7

Hydrostatic tensions in the residual liquid calculated for the various feeding regimes during the freezing of a 20 mm diameter aluminium alloy cylinder (Campbell, 1969a,b).

FIGURE 7.8

Porosity in the long freezing range alloy Cu-10Sn bronze, cast with an inadequate feeder, resulting in a spongy shrinkage pipe.

the pressure differential driving the flow, and depending on what percentage of dendrites are free from points of attachment to the wall of the casting. However, it seems that smaller amounts of movement can continue to occur up to about 68% solid which is the level at which the dendrites start to become a coherent network, like a plastic three-dimensional space frame (Campbell, 1969a,b).

In thin sections, where there may be only two or three grains across the wall section, mass feeding will not be able to occur. The grains are pinned in place by their contacts with the wall. However, as the number of grains across the section increases to between 5 and 10, the central grains are definitely free to move to some extent. In larger sections, or where grains have been refined, there may be 20 to 100 grains or more, so that the flow of the slurry can become an important mechanism to reduce pressure differential along the flow direction. Clearly, the important criterion to assess whether mass flow will occur is the ratio (casting section thickness)/(average grain diameter). This is probably one of the main reasons why grain refinement is useful in reducing porosity in castings (other important reasons being the greater dispersion of gases in solution, their reduced segregation, and the cleaning effect resulting from the sedimentation of bifilms if the grain refiner addition is correctly carried out in the ladle and allowed to settle).

(In passing, we may note that in some instances mass feeding may cause difficulties. There is some evidence that the flow of the liquid/solid mass into the entrance of a narrow section can lead to the premature blocking of the entrance with the solid phase. Thus the feed path to more distant regions of the casting may become choked.)

As mass feeding progresses, the point at which the grains finally impinge strongly and stop is the point at which feeding starts to become appreciably more difficult. This is the regime of the next feeding mechanism, interdendritic feeding.

7.1.4.3 Interdendritic feeding

Allen (1932) was one of the first to use the term 'interdendritic feeding' to describe the flow of residual liquid through the pasty zone. He also made the first serious attempt to provide a quantitative theory. However, we can obtain an improved estimate of the pressure gradient involved simply by use of the famous equation by Poisseuille that describes the pressure gradient dP/dx required to cause a fluid to flow along a capillary:

$$\frac{dP}{dx} = \frac{8v\eta}{\pi R^4} \tag{7.1}$$

where v is the volume flowing per second, η is the viscosity and R is the radius of the capillary. It is clear without going further that the resistance to flow is critically dependent on the size of the capillary. For a bunch of N capillaries which we can take as a rough model of the pasty zone, the pressure gradient is correspondingly reduced:

$$\frac{dP}{dx} = \frac{8v\eta}{\pi R^4 N} \tag{7.2}$$

For the sake of completeness, it is worth developing this relation to evaluate a more realistic channel that includes the effect of simultaneous solidification so as to close the channel by slow degrees. The treatment is based on that by Piwonka and Flemings (1966) (Figure 7.9). Given that the average velocity V is $v/\pi R^2$, and, by conservation of volume, equating the volume flow through element dx with the volume deficit as a result of solidification on the surface of the tube beyond element dx, we have, all in unit time:

$$\pi R^2 V = 2\pi R(L - x)\left(\frac{\alpha}{1 - \alpha}\right)\frac{dR}{dt} \tag{7.3}$$

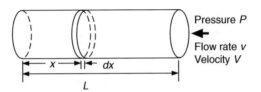

FIGURE 7.9

A tube of liquid, solidifying inwards, whilst being fed with extra liquid from the right.

By substituting and integrating, it follows directly that:

$$\Delta P = \frac{16\eta}{N} \left(\frac{\alpha}{1-\alpha}\right) \frac{dR}{dt} \cdot \frac{1}{R^3} \left(Lx - \frac{x^2}{2}\right)$$

(7.4)

We can find the maximum pressure drop ΔP at the far end of the pasty zone by substituting $x = 0$. At the same time, we can substitute the relation for freezing rate used by Piwonka and Flemings, $dR/dt = -4\lambda^2/R$ approximately, where λ is their heat-flow constant. Also using the relation $Nd^2 = D^2$ where d is the dendrite arm spacing and D^2 is the area of the pasty zone of interest, we obtain at last:

$$\Delta P = 32\eta \left(\frac{\alpha}{1-\alpha}\right) \frac{\lambda^2 L^2 d^2}{R^4 D^2}$$

(7.5)

This final solution reveals that the pressure drop by viscous flow through the pasty zone is controlled by several important factors such as viscosity, solidification shrinkage, the rate of freezing, the dendrite arm spacing and the length of the pasty zone. However, in confirmation of our original conclusion, the pressure drop is most sensitive to the size of the flow channels.

Additional refinements to this equation, such as the inclusion of a tortuosity factor to allow for the non-straightness of the flow do not affect the result significantly. However, more recent improvements have resulted in an allowance for the different resistance to flow depending on whether the flow direction is aligned with or across the main dendrite stems (Poirier, 1987).

The overriding effect of the radius of the flow channel leads to ΔP becoming extremely high as R diminishes. In fact, in the absence of nuclei that would allow pore formation to release the stress, the high hydrostatic stress near the end of freezing will be limited by the inward collapse of the solidified outer parts of the casting, as indicated in Figure 7.7. This plastic flow of the solid denotes the onset of 'solid feeding', the last of the feeding mechanisms. The natural progression of interdendritic feeding followed by solid feeding is confirmed by more recent models (Ohsasa et al., 1988a,b).

Effect of eutectic presence

The rapid increase of stress as R becomes very small explains the profound effect of a small percentage of eutectic in reducing the stress by orders of magnitude (Campbell, 1969a,b). This is because the eutectic freezes at a specific temperature, and the progress of this specific isothermal plane through the mesh corresponds to a specific planar freezing front for the eutectic. The front occurs ahead of the roots of the dendrites, so that the interdendritic flow paths no longer continue to taper to zero, but finish, abruptly truncated as shown in Figure 7.10. Thus the part of the dendrite mesh most difficult to feed is eliminated.

Larger amounts of eutectic liquid in the alloy reduce ΔP even further, because of the increased size of channel at the point of final solidification. As the percentage eutectic increases towards 100%, the alloy feeds only by liquid feeding, of course, which makes such materials easy to feed to complete soundness.

Because most long freezing range alloys exhibit poor pressure tightness, the use of the extremely long freezing range alloy 85Cu-5Sn-5Zn-5Pb for valves and pipe fittings seems inexplicable. However, the 5% lead is practically insoluble in the remainder of the alloy, and thus freezes as practically pure lead at 326°C, considerably easing feeding by acting analogously to a eutectic, as discussed previously.

The appearance of non-equilibrium eutectic in pure Fe-C alloys is predicted to be rather close to the equilibrium condition of 2%C (Clyne and Kurz, 1981) because carbon is an interstitial atom in iron, and therefore diffuses rapidly, reducing the effect of segregation during freezing. However, in the presence of carbide-stabilising alloys such as manganese, the segregation of carbon is enhanced to some extent, causing eutectic to appear not at the equilibrium 2%C but probably only somewhere near 1.0%C as seen in Figure 5.41. A similar result for 1%C1.5%Cr steels is shown in Figure 5.42.

In Al-Mg alloys, layer porosity is observed in increasing amounts as magnesium is increased, illustrating the growing problem of interdendritic feeding as the freezing range increases. However, at a critical composition close to 10.5%Mg, the eutectic beta-phase is first seen in the microstructure and the porosity suddenly disappears (Jay and Cibula, 1956).

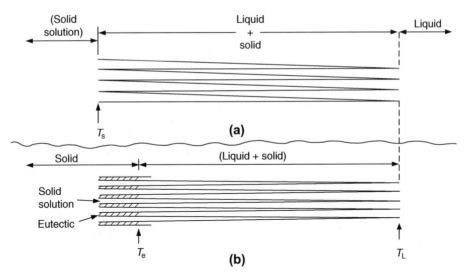

FIGURE 7.10

A diagrammatic illustration of (a) how the tapering interdendritic path increases the difficulty of the final stage of interdendritic feeding and (b) how a small percentage of eutectic will eliminate this final and narrowest portion of the path, thereby greatly easing the last stages of feeding.

The arrival of eutectic at 10.5%Mg confirms the non-equilibrium conditions and compares with the prediction of 17.5%Mg for equilibrium conditions. Lagowski and Meier (1964) found a similar critical transition in Mg-Zn alloys as zinc is progressively increased. Their results are presented in Figure 9.6.

However, one of the most spectacular displays of segregation of a solute element in a common alloy system is that of copper in aluminium. In the equilibrium condition, eutectic would not appear unless the copper content exceeded 5.7%. However, in experimental castings of increasing copper content, eutectic has been found to occur at concentrations in the 0.25–0.5%Cu range. This concentration corresponds to a peak in porosity, and the predicted peak in hydrostatic tension in the pasty zone (Figure 7.11).

Many property–composition curves are of the cuspoid, sharp peaked type (note that they are not merely a rounded, hump-like maximum). Examples are to be found throughout the foundry research literature (although the results are most often interpreted as mere humps!). For instance, in a series of bronzes of increasing tin contents, Pell-Walpole (1946) was probably the first to conclude that the peak in porosity at 5%Sn, not 14% as expected from the equilibrium phase diagram, was the result of the maximum in the effective freezing range. Spittle and Cushway (1983) find a sharp maximum in the hot-tearing behaviour of Al-Cu alloys at approximately 0.5–0.8%Cu (Figure 8.9). The analogous results by Warrington and McCartney (1989) can be extrapolated to show that their peak is nearer 0.5%Cu (Figure 8.6), close to the peak in porosity, as described previously.

7.1.4.4 Burst feeding

Where hydrostatic tension is increasing in a poorly fed region of the casting, it seems reasonable to expect that any barrier might suddenly yield, like the bursting of a dam. Feed metal would then suddenly flood into the poorly fed region. This feed mechanism was proposed by the author simply as a logical possibility (Campbell, 1969a,b). As solidification proceeds, both the stress and the strength of the barrier will be increasing together, but at different rates. Failure will be expected if the stress grows to exceed the strength of the barrier. The barrier may be only a partial barrier, i.e. a restriction to flow, and failure may or may not be sudden.

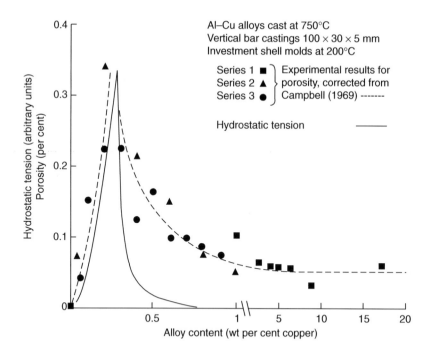

FIGURE 7.11

Predicted peak in hydrostatic tension in the pasty zone, and the measured porosity in test bars, as a function of composition in Al-Cu alloys (Campbell, 1969a,b).

In terms of Figure 7.12, the nucleation threshold diagram (explained later in Section 7.1.5.3; see Figure 7.21), the build-up of stress in the casting might be rapid, leading to the creation of a pore at P. However, if the casting is hotter, or is a weaker alloy, the development of stress will be somewhat slower, and might result in a burst of feed liquid by the collapse of a barrier to flow at point A. The internal stress will then be relieved, but after some additional build-up of gas in solution, the subsequent fall of pressure to B will result in a second burst event, once again relieving the internal stress, and returning the internal pressure to 1 atm. The effect of these bursts is to prevent conditions reaching the nucleation threshold, so that porosity is avoided.

If the feeding barrier is substantial, it may never burst, causing the resulting stress to increase and eventually exceed the nucleation threshold at P. This time, the release of stress, returning the internal pressure to 1 atm, corresponds to the creation and growth of a pore. There can be no further feeding of any kind in that region of the casting after a pore creation event; the driving force for feeding is suddenly eliminated.

Previously, the author has quoted the following observation as a possible instance of a kind of microscopic type of burst feeding. During observation of the late stages of solidification of the feeder heads of many aluminium alloy castings, it is clearly seen that the level of the last portion of interdendritic liquid sinks into the dendrite mesh not smoothly, but in a series of abrupt, discontinuous jumps. It was thought that the jumps may be bursts of feeding into interdendritic regions. However, it now seems more likely that the jumps are the result of the repeated, sudden failure of the surface oxide film, caught and stretched between supporting dendrites at the surface. The liquid draining down into the dendrite mesh will attempt to drag down its surface film, that will repeatedly burst and repair, resisting failure again for a time. The phenomenon is an illustration of the strength of the film, its capacity for stretching to some extent elastically and the capacity of the solidified metal in the feeder head at its freezing point to exhibit a certain amount of elastic recoil behaviour.

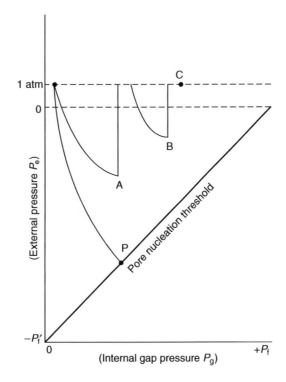

FIGURE 7.12

Gas-shrinkage map showing the path of development to early pore nucleation at P. In a contrasting case, slow mechanical collapse of the casting delays the build-up of internal tension, culminating in complete plastic collapse in the form of burst feeding processes at A and B. Such delay can be successful in avoiding pore nucleation because conditions never reach the pore threshold if, for instance, freezing is complete at C.

A macroscopic type of barrier can be envisaged for those parts of castings where mass flow has occurred, causing equiaxed crystals to block the entrance to a section of casting.

Macroscopic blockages have been observed directly in waxes, where the flow of liquid wax along a glass tube was seen to be halted by the formation of a solidified plug, only to be restarted as the plug was burst. This behaviour was repeated several times along the length of the channel (Scott and Smith, 1985).

In iron castings, such behaviour was intentionally encouraged in the early twentieth century. Nearly all large castings were subjected to 'rodding'—one or two men would stand on the mould and ram an iron rod up and down through the feeder top. Extra feed metal might be called for and topped up from time to time. This procedure would last for many hours until the casting had solidified. Nowadays, it is more common to provide a feeder of adequate size so that feeding occurs automatically without such strenuous human intervention!

On a microscale, a type of burst feeding is the rupture of the casting skin, allowing an inrush of air or mould gases (Figure 7.19). However, this is, of course, a gaseous burst that corresponds to the growth of a cavity, not a feeding process. Pellini (1953) drew attention to this possibility in bronze castings. It is expected that such surface-initiated porosity will be relatively common in castings of many long freezing range alloys.

In conclusion, it has to be admitted that whilst burst feeding might be an important feeding mechanism, it is not easy to quantify its effects by modelling. Despite some interest in using the concept of burst feeding as an explanation of some casting experiments, the explanations remain speculative. The existence of burst feeding has never been unambiguously demonstrated. It therefore seems difficult to understand it and difficult to control it. At this stage, we probably have to content ourselves with the conclusion that logic suggests that it does exist.

7.1.4.5 Solid feeding

At a late stage in freezing, it is possible that sections of the casting may become isolated from feed liquid by premature solidification of an intervening region.

In this condition, the solidification of the isolated region will be accompanied by the development of high hydrostatic stress in the trapped liquid—sometimes high enough to cause the surrounding solidified shell to deform, sucking it inwards by plastic or creep flow. This inward flow of the solid relieves the internal tension, like any other feeding mechanism. In analogy with 'liquid feeding', the author called it 'solid feeding'. An equally good name would have been 'self-feeding'.

When solid feeding starts to operate, the stress in the liquid becomes limited by the plastic yielding of the solid and so is a function of the yield stress Y and the geometrical shape of the solid. (The yield stress Y is, of course, a function of the strain rate at these temperatures when assuming an elastic/plastic model.) The procedure is practically equivalent to the assumption of a creep stress model, and results in similar order-of-magnitude predictions for stress (Campbell, 1968a,b). For instance, for a sphere of radius R_0, with internal liquid radius R (Figure 7.3), the onset of complete plasticity is defined by:

$$P = 2Y \ln(R_o/R)$$

Mechanical engineers will recognise this relation as the classical formula for the failure of a thick-shell pressure vessel stressed by internal pressure to the point at which it is in a completely plastic state. This equation is expected to give maximum estimates of the hydrostatic tensions in castings because: (1) the shape is the most difficult to collapse inwardly and (2) the equation neglects the opposing contribution of the thermal contraction of the solidified shell which will tend to reduce internal tension (Forgac et al., 1979). Nevertheless, it is still interesting to set an upper bound to the hydrostatic tensions that might arise in castings.

This early model (Campbell, 1967) used the concept that the liquid radius R had to be expanded to some intermediate radius R', and the solid had to be shrunk inwards from its original internal radius $R + dR$ to the new common radius R'. At this new radius, the stress in the liquid equals the stress applied at the inner surface of the solid.

The working out of this simple model indicated that for a solidifying iron sphere of diameter 20 mm, the elastic limit at the inner surface of the shell was reached at an internal tensile stress of about 4 MPa (-40 atm); and by the time the residual volume of liquid was only 0.5 mm in diameter a plastic zone had spread out from the centre to encompass the whole shell. At this point, the internal pressure was in the range of approximately 20–40 MPa (-200 to -400 atm) and the casting was 99.998% solid. Solidification of the remaining drop of liquid increased the pressure in the liquid to approximately 100 MPa (-1000 atm). Later estimates using a creep model and cylindrical geometry confirmed similar figures for iron, nickel, copper and aluminium (Campbell, 1968a,b).

A minute theoretical point of interest to those of a scientific disposition is the effect of the solid–liquid interfacial tension. Although this is small, it starts to become important when the liquid region is only a few hundred atoms in diameter. The interfacial tension causes an inward pressure $2\gamma_{LS}/R$ that starts to compress the residual liquid. This is the explanation for the theoretical curves to take an upward turn in Figure 7.7 as freezing nears completion, creating a limit to the maximum internal tension.

We have to bear in mind that these estimates of the internal tension are upper bounds, likely to be reduced by thermal contraction of the shell, and reduced by geometries that are easier to collapse, such as a cylinder or a plate. Also the predictions are in any case lower for smaller trapped volumes of liquid, as might occur, for instance, in interdendritic spaces. Figure 7.13 shows schematically the effect of plastic zones spreading from isolated unfed regions of the casting.

For an infinite, flat plate-shaped casting in a skin-freezing metal, the internal stress developed is zero, which is an obvious solution because there can be no restraint to the inward movement of infinite flat plates separated by a solidifying liquid; the plates simply move closer together to follow the reduction in volume. For real plates, their surfaces are held apart to some extent by the rigidity of the edges of the casting, so the development of internal stress would be expected to be intermediate between the two extreme cases. The ease of collapse of the central regions of flat plates emphasises the importance of geometry.

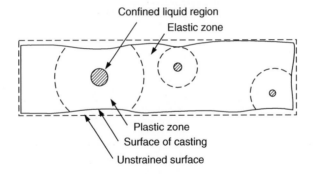

Confined liquid region
Elastic zone
Plastic zone
Surface of casting
Unstrained surface

FIGURE 7.13

Plastic zones spreading from isolated volumes of residual liquid in a casting, showing localised solid feeding in action (Campbell, 1969a,b).

Figures 7.14 and 7.15 show results of some of my early experiments concerning porosity in small plates of an investment-cast nickel-based alloy. It happens to be an excellent example of solid feeding in action. At low mould temperatures, the solid gains strength rapidly during freezing and therefore retains the rectangular outer shape of the casting, and the steep temperature gradient concentrates the porosity in the centre of the casting. As mould temperature is

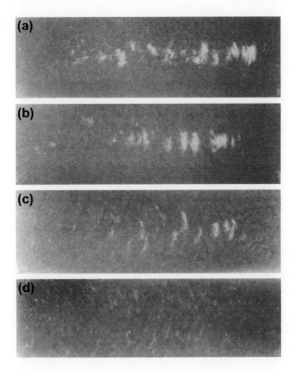

FIGURE 7.14

(a) Radiographs of bar castings $100 \times 30 \times 5$ mm in nickel-based alloy cast at 1620°C in vacuum 15 μmHg into moulds at: (a) 250°C; (b) 500°C; (c) 800°C; and (d) 1000°C (Campbell, 1969a,b). Centre-line macroporosity is seen to blend into layer porosity and finally into dispersed microporosity.

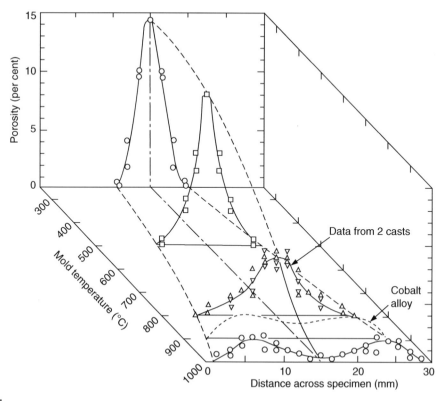

FIGURE 7.15

Porosity across an average transverse section of vacuum-cast nickel-based alloy as a function of mould temperature, quantifying the effect shown in this figure. The effect of solid feeding by the plastic collapse of the section is
clear from the shape of the porosity distribution at high mould temperatures.

increased, the falling yield stress of the solidified metal allows progressively more collapse of the centre, solid feeding
acting to reduce the total level of porosity. However, some residual porosity remains noticeable nearer the side walls,
where geometrical constraint prevents full collapse. Note that these results were obtained in vacuum, with zero
contribution from exterior positive atmospheric pressure. It follows, therefore, that all of the solid feeding in this case is
the result of internal negative pressure. In fact, surface sinks are commonly seen in vacuum casting. Sinks are therefore
not solely the consequence of the action of atmospheric pressure, as generally supposed.

Figure 7.16 shows solid-feeding behaviour in spherical wax castings. The example is interesting because it is evident
that sound castings can, in principle, be produced without any feeding in the classical sense. In this case, it is clear that
feeding can be successfully accomplished by skilful choice of mould temperature to facilitate uniform solid feeding.

Figure 7.17 shows a similar effect in unfed Al-12Si alloy as a function of increasing casting temperature. The full
6–7% of internal shrinkage porosity at low casting temperature is gradually reduced by external collapse of the casting as
casting temperature increases (Harinath et al., 1979).

If solid feeding is controlled so that it spreads itself uniformly in this way, then the accompanying movement of the
outer surface of the casting becomes negligible for most purposes. For instance, the high volume shrinkage of about 6%
suffered by Al-Si alloys corresponds to a linear shrinkage of only 2% in each of the three perpendicular directions. For a

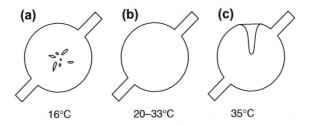

FIGURE 7.16

Cross-section of 25 mm diameter wax castings injected into an aluminium die at various temperatures. (a) a cold, strong solidified skin, creating internal shrinkage pores; (b) optimum conditions; and (c) a warm, soft solidified skin allowing an exterior collapse to form an externally connected shrinkage pipe.

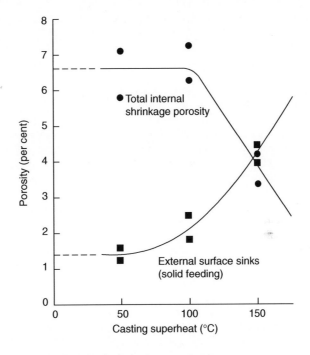

FIGURE 7.17

Al-12Si alloy cast into unfed shell moulds showing the full 6.6% internal shrinkage porosity at low casting temperature, giving way to solid feeding at higher casting temperature.

Data from Harinath et al. (1979).

datum in the centre of the casting, this means an inward wall movement of only 1% from each of the opposite surfaces. Thus a 25 mm diameter boss would be 250 µm small on radius if it were entirely unfed by liquid.

In practice, of the 6% volume contraction in aluminium alloy castings, usually about 5% is relatively easily fed by liquid and interdendritic modes, leaving only about 1% as a feeding problem, and thus available to be provided by solid feeding. This corresponds to an inward movement of a surface of only 0.16%, which is only about 40 µm for our 25 mm diameter casting. Thus dimensional errors resulting from solid feeding reduce to the point at which they become not measurable.

In contrast to the 0.25 mm worst case reduction in radius for the 25 mm diameter feature, if all the shrinkage were concentrated at the centre of the casting, the internal pore would have a diameter of 10 mm. This point has been made before, but is worth repeating. The difference between the extreme seriousness of internal porosity, compared with its harmless dispersion over the exterior surfaces of the casting is a key factor to encourage the development of casting processes that would automatically yield such benefits. The elimination of internal defects in the liquid that could nucleate a cavity is a major beneficial feature of a good casting process.

It is also worth emphasising that solid feeding will occur at a late stage of freezing even if the liquid is not entirely isolated. The case has been discussed in the section on interdendritic feeding, and is part of the meaning of Figure 7.7. It is also clear in Figure 7.15. The effect is the result of the gradual build-up of tension along the length of the pasty zone because of viscous resistance to flow. At the point at which the tension reaches a level where it starts to cause the collapse of the casting the region is effectively isolated from the feeder. Although liquid channels still connect this region to the feeder, they are by this time so small they become ineffective in delivering feed liquid.

An experimental result by Jackson (1956) illustrates an attempt to reduce solid feeding by increasing the internal pressure within the casting by raising the height of the feeder. Jackson was casting vertical cylinders 100 mm in diameter and 150 mm high in Cu 85-5-5-5 alloy in greensand. He employed a plaster-lined feeder of only 50 mm in diameter (incidentally, failing Feeding Rules 2 and 3, which explains why he observed such high porosity in his castings). Nevertheless, the beneficial effect of increasing the feeder height is clear in Figure 7.18. His data indicate that, despite the unfavourable geometry, if he had raised his feeder height to 250 mm, all exterior shrinkage would have been eliminated. The interior porosity would have fallen to about 2.0%, almost certainly being the residual effects from the combination of gas porosity, and the residual shrinkage from his poorly sized feeder.

In a study of two small shaped castings in three different Al-Si alloys, of short, medium and long freezing ranges, Li et al. (1998) measured the internal porosity of the castings by density, and the external porosity (the total surface sink effect) by measuring the volume of the casting in water. We found that the internal porosity in the castings in all three alloys was about the same at approximately 1 volume%. However, the external sinks grew from an average of 3.1, to 6.4 to 7.5 volume% for the short-, medium- and long freezing range alloys. This significant increase in solid feeding for the long freezing range material probably reflects the easier collapsibility of the thinner solidified shell and its internal mesh

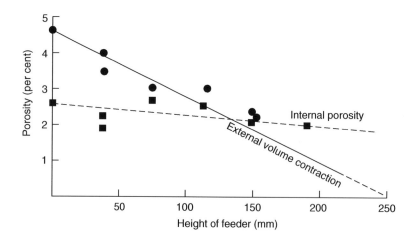

FIGURE 7.18

Gunmetal casting showing the reduction in solid feeding as liquid feeding is enhanced by extra height and volume of feed metal.

Data from Jackson (1956).

of dendrites. The more severe internal stress because of the greater difficulty in interdendritic feeding may also be a significant contributor. Conversely, of course, the absence of any corresponding increase in internal porosity confirms that feeding of the castings in the shorter freezing range alloys occurred by the simpler and easier more open liquid feeding mechanisms.

Amid the significant value of useful solid feeding, we need to keep in mind the possible dangers accompanying this feeding mechanism; hence, the brief summary that follows. Clearly, for the condition that the liquid is free from bifilms, the casting will not contain internally initiated pores. However it may generate:

1. Surface-initiated pores or even
2. Surface sinks.
   In the presence of one or more easily opened bifilms, the situation changes significantly. The casting will now exhibit:
3. A large interior shrinkage pore in the presence of a bifilm in the stressed region, if the hydrostatic stress becomes sufficiently high and if the stressed volume is large.
4. A population of internal microscopic cracks. This is the subtle danger arising from the usual presence of a population of bifilms in the stressed liquid. In this situation, the compact bifilms are subjected to a tensile stress in the liquid, providing a driving force to unfurl. The mechanical properties, especially the ductility and ultimate tensile strength (UTS), of the casting are thereby impaired in this region, even though there may be no visible indication if the bifilms are not opened sufficiently to reveal their presence. In a nearby region of the casting that had enjoyed better feeding, bifilms would have remained fairly furled, so the ductility and UTS would have been higher.

A final personal remark concerning solid feeding that is a source of mystery to the author is the widespread inability of many to comprehend that it is a fact. This lack of comprehension is difficult to understand in view of the obvious evidence for all to see as surface sinks (even in castings solidified in vacuum) and that isolated bosses can be cast sound provided that the metal quality is good (i.e. few nuclei to initiate internal pores). Foundries that convert poor filling systems to well-designed filling systems suddenly find that internal porosity and hot tears vanish, but the castings now require extra feeding to counter surface sinks (Tiryakioglu, 2001). The increased solid feeding at higher mould temperatures is widely seen in investment castings. The easy collapse of flat plates, especially of alloys weak at their freezing points like Al alloys, explain their long and difficult-to-define feeding distances. The better-defined feeding distances of steels are the consequence of their better-defined resistance to collapse; their greater strength resisting solid feeding. Additionally, of course, hot isostatic pressing (hipping) is a good analogy of an enforced plastic collapse of the casting, as is also direct squeeze casting.

Who needs good imagination in the presence of all these facts?

7.1.5 INITIATION OF SHRINKAGE POROSITY

In the absence of gas, and if feeding is adequate, then no porosity will be found in the casting.

Unfortunately, however, in the real world, many castings are sufficiently complex that one or more regions of the casting are not well fed, with the result that the internal hydrostatic tension will increase, reaching a level at which an internal pore may 'pop' into existence in several ways. Conversely, if the internal tension is kept sufficiently low by effective solid feeding, the mechanisms for internal pore formation are not triggered; the solidification shrinkage then appears on the outside of the casting. All this is discussed in more detail later.

7.1.5.1 Internal porosity by surface initiation

If liquid from the feeder becomes cut off, the pressure inside the casting falls. Liquid that is still connected to the outside surface may then be drawn from the surface, causing the growth of porosity connected to the surface (Figure 7.19). The sucking of liquid from the surface in this way naturally draws in air, following interdendritic channels, spreading along these routes into the interior of the casting. The phenomenon is a kind of feeding by a fluid, where the fluid in this case is air. However, of course, this is hardly a feeding process but a pore growth process. This porosity in the interior of the casting is usually hard to distinguish from microporosity caused in other ways: on a polished section it appears to be a

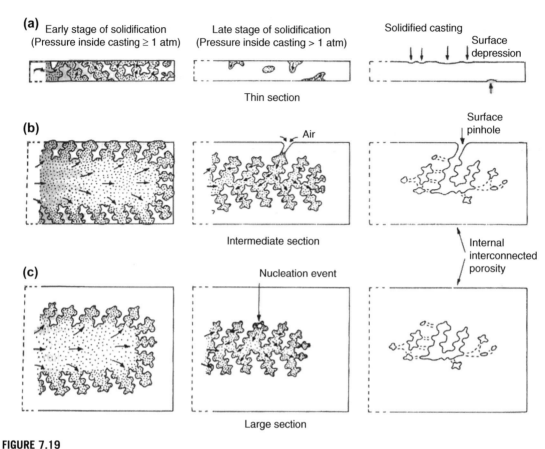

(a) Early stage of solidification (Pressure inside casting ≥ 1 atm)   Late stage of solidification (Pressure inside casting > 1 atm)   Solidified casting   Surface depression

Thin section

(b)   Air   Surface pinhole

Intermediate section   Internal interconnected porosity

(c)   Nucleation event

Large section

FIGURE 7.19

(a) Thin sections require little or no feeding because they can contain zero shrinkage porosity; (b) intermediate sections can develop surface connected porosity; (c) thick sections contain nucleated internal porosity (Campbell, 1969a,b).

series of separate interdendritic pores, whereas in reality it is a single highly complex shaped interconnected pore, linked to the surface. It is sometimes possible to identify its origins as surface-initiated porosity by its oxidised internal surface near to the surface of the casting. Deeper into the casting, away from the surface, the oxygen is totally consumed, so the surface film changes from an oxide to a nitride.

Figure 7.19 illustrates how the withdrawal of surface liquid is negligible in thin-section castings, explaining why thin sections require little feeding, or even no apparent feeding, but automatically exhibit good soundness. The effect is easily seen in gravity die castings because of their shiny surface when lifted directly from the die. In a section of intermediate thickness, the experienced caster will often notice a local frosting of the surface. This dull patch is a warning that interdendritic liquid is being drained away from the surface indicating an internal feeding problem that requires attention.

Pericleous and Bounds et al. (1997) was the first to predict this form of surface-initiated porosity using a computer model of the freezing of a long freezing range alloy. His result is shown in Figure 7.20.

This pore formation mechanism from the surface seems to be much more common than is generally recognised. It is especially likely to occur in long freezing range alloys at a late stage in freezing, when the development of the dendrite

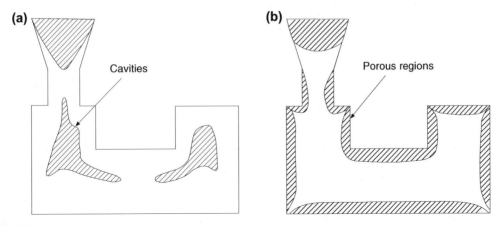

FIGURE 7.20

Regions of computer-simulated shrinkage porosity, (a) internally in a short freezing range alloy, and (b) externally (surface-initiated) in a long freezing range alloy. The latter was the first prediction of surface-initiated porosity by computer simulation.

After Pericleous and Bounds et al. (1998).

mesh means that drawing liquid from the nearby surface becomes easier than drawing liquid from the more distant feeder. The point at which liquid may be drawn from the surface may be anywhere (or practically everywhere) for an alloy of sufficiently wide freezing range.

However, in an alloy of intermediate freezing range, the initiation site is often a hot spot such as an internal corner or re-entrant angle. As has been mentioned before, the gravity die caster pouring an Al-Si alloy looks for such defects on each casting as it is taken from the die. If such a 'draw' or cavity or frost appearance is noticed in a re-entrant angle, he immediately doses the melt with sodium. The straightening of the solidification front (Figure 6.12(b)) strengthens the alloy at the corner so that it can better resist local collapse. The outcome is a pore hidden inside the casting if the melt quality is poor so that nucleation is easy. Alternatively, if the melt quality is good, no internal pore can easily form, so that the rise in internal tension will cause more general collapse of the casting. Solid feeding will have been encouraged.

The connection of two opposite surfaces of the casting by pores that are extensively connected internally is one of the major reasons why long freezing range alloys cannot easily be used for pressure-tight applications such as hydraulic valves or automobile cylinder heads. In such complex castings, it is often difficult to meet the essential requirement that the interior of the casting has a positive pressure at all locations so as to prevent surface-connected internal porosity.

7.1.5.2 External porosity (surface sinks)

If internal porosity is not formed (either by surface-linked initiation or nucleation events) then the lowering of the internal pressure will lead to an inward movement of the external surface of the casting (Figure 7.13). If the movement is severe and localised, then it constitutes a defect known as a '*sink*' or a '*draw*'. The feeding of the internal shrinkage by the inward flow of solid is, of course, '*solid feeding*' or '*self-feeding*'.

Adequate internal pressure within the casting will reduce or eliminate solid feeding, so maintaining the shape of the casting and keeping it sound. In such favourable feeding conditions, neither internal nor external porosity will occur.

If the remedy of the application of internal pressure is carried out too enthusiastically, the natural inward movement of solid feeding will be reversed, causing the casting to swell. Such growth is common in grey iron castings and castings that have a high head of metal. Swelling of cast metal is also commonly seen in pressure die casting if the casting is removed from the die before it is fully solid. This is because the gas bubbles entrained by the extreme turbulence are under extremely high pressure. The technique is useful for identifying hot spots in the casting (i.e. regions last to freeze and which therefore require additional local cooling in the die to raise productivity). Gas entrained in hot spots can cause the casting to locally explode if released very early from the constraint of the die.

Returning to conventional gravity castings, in real situations it sometimes happens that a certain amount of external collapse of the casting will occur before the internal pressure falls to the level required to nucleate an internal pore. Once the pore is formed, the internal stress will be eliminated so that further solid feeding is arrested. The action of the remaining solidification shrinkage is simply to grow the pore. The final effect of solidification shrinkage on the casting is usually found to be partly external and partly internal as illustrated in Figure 7.4. The balance between external and internal porosity can be widely seen in foundries. Examples are seen in Al-12Si (Figure 7.18) and in aluminium-based metal/matrix composites (Emamy and Campbell, 1997).

7.1.5.3 Nucleation of internal porosity

Although the problem of the nucleation of cavities and bubbles is in principle similar to that of the nucleation of condensed phases such as the metal matrix and non-metallic inclusions, there are differences that make it worthwhile to look at non-condensed phases such as vacuums, vapours and gases separately in some detail. In particular, we shall find that there are special difficulties with the nucleation of gas and void phases, forcing us to adopt new concepts. We shall deal first with shrinkage cavities, then gas pores, not forgetting that both can cooperate in both the formation and growth of such volume defects.

Short-freezing-range alloys, such as aluminium bronze and Al-Si eutectic, do not normally exhibit surface-connected porosity. They form a sound, solid skin at an early stage of freezing, and liquid feeding continues unhindered through widely open channels. Any final lowering of the internal pressure from poor feeding towards the end of freezing may then create a pore by nucleation in the interior liquid. In this case, there is clearly no connection to the outside surface of the casting, as illustrated in the larger section shown in Figure 7.19(c). After nucleation, further solidification shrinkage will provide the driving force for growth of the pore, which, on sectioning or radiography, may be more or less indistinguishable from the surface-initiated type.

In alloys of short freezing range, therefore, porosity is probably normally nucleated, and is concentrated near the centre of the casting, usually well clear of the casting surface. In castings of large length to thickness ratio, this is widely referred to as centreline porosity. Thus unless subsequent machining operations cut into the porosity, castings in such alloys are normally leak-tight. (The leak paths commonly provided by folded oxide films or bubble trails generated during a turbulent fill are a separate problem requiring solution by other means such as improved filling, and/or the use of filters.)

Tiwara et al. (1985) has suggested a way of initiating internal porosity in specific regions of castings by the addition of nuclei in the form of fragments of refractory. These foreign particles contain much porosity, so that the growth of pores from such sites proceeds without difficulty. The result is a large internal pore, which, to some extent, can be sited in a chosen location in the casting. Additional feeders or chills are therefore not required, and internal porosity in an unwanted location is avoided. External porosity is also successfully avoided because internal pressure is prevented from falling to negative values. However, as inventive as this technique is, for the majority of castings that are required to be sound throughout, and are required to be free of pieces of refractory, it is, unfortunately, only of academic interest.

Homogeneous nucleation

Following the beautifully elegant approach by Fisher (1948), we can quantify the conditions required for the formation of porosity in liquid metals. A quantity of work is associated with the reversible formation of a bubble in a liquid. If the local pressure in the liquid is P_e, we need to carry out an amount of work $P_e V$ to push back the liquid far enough to create a bubble of volume V.

The formation and stretching out of the new liquid–gas interface of area A requires additional work γA, where γ is the interfacial energy per unit area.

The work required to fill the bubble with vapour or gas at pressure P_i is negative and equal to $-P_iV$. (The negative sign arises because the pressure inside the bubble clearly helps the formation of the bubble, as opposed to the other work requirements, which tend to oppose bubble formation.) Thus the total work is:

$$\Delta G = \gamma A + P_e V - P_i V$$
$$= 4\gamma\pi r^2 + (4/3)\pi r^3 (P_e - P_i)$$

where clearly $(P_e - P_i)$ is the pressure difference between the exterior and the interior of the bubble which we may write as ΔP for convenience. Similarly to dense phase nucleation, a plot of ΔG versus bubble radius r shows a maximum that constitutes an energy barrier to nucleation, as in Figure 5.27. The critical radius $r*$ in this case is:

$$r^* = 2\gamma/\Delta P^* \tag{7.6}$$

Because bubbles growing from the bulk liquid will grow an atom at a time as the result of statistical thermal fluctuations, it is evident that small bubbles with radii less than $r*$ will tend to disappear. Only exceptionally will a long chain of favourable energy fluctuations produce a bubble exceeding the critical radius $r*$. When this rare event does happen, the embryonic bubble will then have the potential to grow to an observable size.

Fisher goes on to apply some delightfully elegant rate theory to derive values for the critical pressure difference $\Delta P*$ at which nucleation will occur. The reader is strongly recommended to consult Fisher's original paper. However, for our purposes, we can obtain a sufficiently good estimate very easily and quickly using Eqn (7.6). Using experimentally determined values of atomic sizes and surface energy γ for liquid metals, if we assume that the critical radius is perhaps in the region of a couple of atomic diameters (in the first edition of *Complete Casting Handbook*, the critical radius was erroneously assumed to be only one atomic diameter), we obtain Table 7.2.

The reasonable agreement between the calculated critical pressures is corroboration that the critical embryo is actually about four atoms across and therefore occupies the volume of approximately 50 vacancies. However, whether or not these figures really are accurate is a detail that need not concern us here. The important message is that the pressures that are required for nucleation are *extremely* high and reflect the real difficulty of homogeneous nucleation of pores in liquid metals. It is clear that the strengths of liquid metals are almost as high as those of solid metals (for liquid iron the fracture strength corresponds to nearly 8 GPa). As noted previously, this is hardly surprising because the atomic structure is similar, liquid metals being close packed random structures, compared with solid metals being close packed regular structures. In either case, the atoms are about the same distance apart, and it is similarly difficult to separate them; they are resistant to be forced apart to create a void.

The problem of nucleation is reduced by the presence of surface-active impurities in the liquid. The non-metals oxygen, sulphur and phosphorus are particularly active in iron melts: the presence of only 0.2 wt% of oxygen reduces the surface tension of liquid iron from 1.9 to approximately 1.0 Nm^{-1}. This approximately halves the estimates of the pressure required for nucleation as shown by Eqn (7.6). Similar reductions in surface tension (and therefore in fracture pressures) are to be found in liquid copper when contaminated with high levels of the non-metals O, S and P.

Such high concentrations of oxygen (and the other non-metals) are probably often found on solidification because of the concentration of solutes ahead of the freezing front. Once again, for the case of liquid iron, the partition coefficient for oxygen is approximately 0.05, giving a factor of 20 increase in concentration at the advancing front. Thus an average of only 0.01 wt% oxygen in the bulk melt can produce 0.2 wt% at the front.

If the levels of oxygen rise sufficiently to precipitate FeO as a liquid inclusion at the front, then the nucleation problem is reduced yet further because FeO has a surface tension of between 0.6 and 0.5 Nm^{-1}, depending on its oxygen content (Popel and Esin, 1956). Thus a gas pore will preferentially nucleate in such a liquid inclusion, where the critical pressure is easily shown to be reduced to around 1.7 GPa. Effectively, this is still homogeneous nucleation in a pure liquid where, in this case, the pure liquid in the liquid Fe is in the form of regions, possibly minute droplets, of FeO.

Even this pressure is still so high as to be probably unattainable. What other possibilities are there?

It is possible that nucleation might occur on a solid impurity particle. A solid foreign substrate, if a poorly wetted surface, might make a location for nucleation. This is known as heterogeneous nucleation. If this poorly wetted solid surface happened to be inside the liquid FeO inclusion, we shall see how we can reduce the critical fracture pressure 1.7 GPa yet further in the following section.

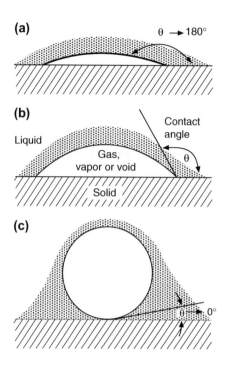

FIGURE 7.21

Geometry of a bubble in contact with a solid substrate showing (a) poor wetting and easy decohesion of liquid from the solid; (b) medium wetting; and (c) good wetting in which liquid cohesion to the solid is high, and the bubble is displaced out of contact with the solid.

Heterogeneous nucleation

Fisher considers the case of the nucleation of a bubble against the surface of a solid substrate. The liquid is considered to make an angle θ with the solid. This contact angle defines the extent of wetting; $\theta = 0°$ means complete wetting, whereas $\theta = 180°$ is complete non-wetting. The geometry is shown in Figure 7.21. Fisher shows that nucleation is easier by a factor:

$$P_{het}^*/P_{hom}^* = 1.12\left\{(2 - Cos\theta)(1 + Cos\theta)^2/4\right\}^{1/2} \qquad (7.7)$$

The factor 1.12 arises because of the fewer sites for nucleation on a plane surface of atoms, compared with the greatly increased number of possibilities in the bulk liquid. This factor alone makes the nucleation of pores on wetted surfaces unfavourable. In fact, nucleation on solid surfaces does not become favourable until the contact angle exceeds 60–70°, as shown in Figure 7.22. For this reason the nucleation of pores against the growing solid such as a dendrite is *not* favoured (because a melt wets the solid formed from itself).

This factor is contrary to those other factors that favour the nucleation of a pore close to a front. The favourable factors include the high gas contents and low surface tension usually present in the highly segregated liquid. Additionally, there are likely to be inclusions present, pushed and concentrated by the advancing front into the residual liquid. Again, the inclusions that are pushed by the front are likely to be the non-wetted variety, and so will constitute good nuclei for pores. Contrary, therefore, to many published opinions elsewhere, the interface is theoretically *not* a favoured site. Yet, paradoxically, in practice, pores will nucleate there because of all the other favourable conditions that prevail adjacent to an advancing solidification interface. The received wisdom turns out to be correct for the wrong reasons!

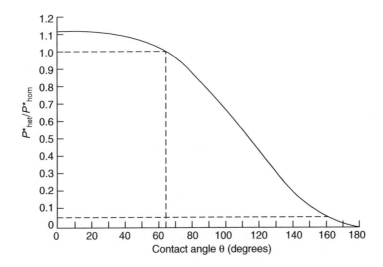

FIGURE 7.22

Relative difficulty of nucleating a pore as the contact angle with the solid changes from wetting to non-wetting. Only when the angle exceeds about 65° does heterogeneous nucleation on the solid become favourable.

It is also important to note that not all inclusions are good nucleation sites for porosity. Those that are well wetted will not be favoured. These include the rather more metallic inclusions such as borides, carbides and nitrides. (However, being well wetted, they are mostly good nuclei for the solid phase, and so can assist with grain refinement, as we have seen at several points in Section 5.)

The wetting requirements for the nucleation of pores are completely opposite: good nuclei must be not wetted. Such substrates include the non-metals such as oxides. However, the situation is especially bad for *entrained* oxides as will be described later in this chapter.

The reader should be aware that there is widespread misunderstanding of the important fact outlined previously: that the nuclei required for the formation of a solid (i.e. for grain refinement) are quite different to the nuclei required for the formation of pores and voids.

A nice experiment was carried out by Gernez in 1867 in which he demonstrated that crystalline solids which had been grown in the liquid and had never been allowed to come into contact with air were incapable of inducing effervescence in a liquid supersaturated with gas. Otherwise identical solids that had surfaces which had been allowed to dry always caused effervescence.

Oelsen describes a related experiment that he carried out in 1936 in which he isolated a sample of liquid iron on all sides by a liquid slag. Because the iron contained 2% carbon and 0.035% oxygen, it was supersaturated considerably in excess of equilibrium. In fact, Oelsen estimated that the internal pressure of carbon monoxide would be approximately 40 atm. When an iron rod was immersed in the melt to destroy the perfection of the containment, a violent eruption of gas immediately occurred, which ceased once again when the rod was withdrawn.

Figure 7.22 indicates that as the contact angle increases to 180° any difficulty of heterogeneous nucleation should fall to zero. In fact, there are good reasons to believe that such perfect non-wetting is probably not possible, and that the maximum contact angle attainable in practice is perhaps close to 160°. Certainly, no contact angle greater than this appears ever to have been observed (see, for instance, the work by Livingston and Swingley (1971)). Assuming this to be true, then from Figure 7.22, we see that heterogeneous nucleation on the most non-wetted solid known, having a contact angle of 160°, requires only about one-twentieth of the pressure required for nucleation in the bulk liquid.

Returning to our liquid FeO inclusion in solidifying iron: if a highly non-wetting inclusion were present inside the liquid FeO inclusion, then the lowest pressure for nucleation of a pore in this complex inclusion would be approximately 1.7 GPa/20 = 85 MPa. Although this pressure is still high, it might now (just) be attainable in iron and steel castings. In Al and Mg alloys, such a nucleation condition from an equivalent complex inclusion seems unlikely to be attained because these weaker materials would collapse plastically under the internal tension.

Similar reasoning can almost certainly be applied to other alloy systems: complex inclusions are likely to be present in all alloys. In fact, it seems likely that most if not all inclusions are complex. The apparent simplicity of many inclusions may be an illusion; microscopic regions of impurities, perhaps only an atom or so thick, may be distributed in patches. It would be difficult to find such patches, and, conversely, difficult to prove that they did not exist.

Perhaps therefore the large fracture pressures predicted by the classical nucleation theory are to be expected to be always reduced by the presence of low surface energy liquid inclusions that in turn contain very poorly wetted solid particles. If this is the case, then assuming that all the values are reduced from their homogeneous nucleation values by the same total factor approximately 83 found for pure iron, the final column in Table 7.2 is estimated.

There is a natural concern that although these stresses are greatly reduced from those required for nucleation in the pure liquid, they are still probably too high to be attainable. Thus, in a nutshell, even fairly dirty liquid metals do not meet a reasonable criterion for fracture considering the classical nucleation theory.

Thus, although the formation of pores by random, thermal atomic fluctuations within the liquid and against plane solid surfaces, has been a problem in classical physics that has fascinated many scientists since the first attempt by Fisher in 1948, all of the solutions that have been found so far (see, for instance, the short review by Campbell (1968)) have shown that pores are difficult, and actually, probably impossible, to nucleate, despite invoking the most active of heterogeneous substrates.

If liquid metals do not fracture, the internal initiation of porosity in castings is impossible. Nevertheless, the fact is that pores in castings are the norm rather than the exception. Clearly, there is a major mismatch between theory and fact. The failure of classical nucleation theory to account for porosity is not widely appreciated. In this work, the fundamental inability of the classical theory has driven the search for other pore initiation processes as outlined later.

This conclusion is so important and so surprising that it is worth emphasising by stating it another way. The student will usually have been persuaded that because classical homogeneous nucleation of pores in liquid metals is difficult, classical heterogeneous nucleation on a solid particle must therefore occur. The key conclusion from our previously discussion is that this is almost certainly untrue. Classical nucleation of a cap-shaped bubble cannot occur even on the most non-wetted solid. This alarming fact forces a re-evaluation of initiation processes for porosity in castings.

Other potential processes are examined later in Section 7.1.5.4. However, it will become clear that the most likely candidate mechanism for pore initiation are entrainment defects, as will be described later in the Section 7.1.5.4.

Nucleation conditions for shrinkage pores

The problem of the nucleation of shrinkage cavities has been widely overlooked. Somehow it has been assumed that they are fundamentally different to gas pores, and that they 'just arrive'. After all, it is argued, they *must* occur in an unfed isolated volume of liquid because the concept of shrinkage means that there is a volume deficit. It is assumed that this volume deficit *must* result in a cavity.

However, as we have seen previously, there really is a difficulty in the initiation of a cavity in a liquid. The various analogous systems described in Section 7.1.2.1 demonstrate that the liquid can withstand hydrostatic tension, and may never fail. If we accept this, then it follows that the liquid is stretched elastically, and the surrounding solid drawn inwards, first elastically, then plastically as the stress in the liquid increases (Campbell, 1967). These predictions explain many common observations in the foundry, as will be referred to repeatedly in this work. Only if the stress in the liquid reaches the critical fracture pressure P_f will a pore appear, growing in milliseconds to a size which will dispel the stress. This is the instant at which the pore grows explosively, releasing the tension in the liquid.

Thus, in case of any doubt, all the conclusions reached previously for nucleation of pores in liquids, estimated by Fisher's formula, apply to the nucleation of shrinkage cavities. These are briefly summarised later.

Fracture strengths are, of course, reduced by the presence of weakly bonded surfaces in the liquid. Thus the previous discussion on heterogeneous nucleation also applies. Shrinkage cavities are therefore expected to nucleate only on non-wetted interfaces.

Good nuclei for shrinkage cavities include oxides. Complex inclusions that consist of low-surface-tension liquid phases containing non-wetted solids might be especially efficient nuclei, as have also been discussed.

Unfavourable nuclei on which the initiation of a shrinkage cavity will *not* occur include wetted surfaces such as carbides, nitrides and borides, and other metal surfaces such as the surfaces of dendrites that constitute the solidification front. (Readers need to beware that many authors assume, incorrectly, that dendrites are good nuclei for pores although, perhaps somewhat perversely, pores often do nucleate at a dendritic front for other reasons as listed in Section 7.1.5.3). All these substrates are unfavourable for decohesion simply because the bonding between the atoms across the interface is so strong. This is reflected in the good wetting (i.e. small contact angle) of the liquid on these solids.

Interestingly, although oxides are included previously as good potential nuclei for pores, this is only true of their non-wetted surfaces. Those surfaces that have grown off the melt, and thereby in perfect atomic contact with the melt, are not expected to be good nuclei. This illustrates the important distinction between wetting defined by contact angle, and wetting defined as being in perfect atomic contact with the liquid. Ultimately, it is the atom-to-atom contact, and the strong interatomic bonding that is important.

Nucleation conditions for combined gas and shrinkage pores

Following Fisher's analysis through once again, considering now that there is gas at pressure Pg on the inside of the pore effectively pushing and a negative pressure Ps in the bulk liquid effectively pulling the embryonic pore into existence, the final result for the critical fracture pressure Pf is:

$$Pf = Pg + Ps \qquad (7.8)$$

This equation illustrates how gas and shrinkage cooperate to exceed the critical pressure for nucleation. The significance of this equation can perhaps be better appreciated by deriving an analogous relation as follows.

If we write the condition for simple mechanical equilibrium of a bubble of radius r in a liquid of surface tension γ, in which the bubble has internal pressure Pi and external pressure in the liquid Pe, we have:

$$2\gamma/r = Pi - Pe \qquad (7.9)$$

When the pore is of critical size, radius r^*, and when the internal pressure is the pressure of gas Pg in equilibrium with the liquid, and the external pressure Pe is the (negative) pressure because of shrinkage $-P_S$, then Eqn (7.8) becomes the analogue of Fisher's equation:

$$2\gamma/r^* = Pg + Ps \qquad (7.10)$$

The fracture pressures of various liquid metals can then be estimated from this relation assuming that r is approximately one or two atomic diameters, giving the values presented in Table 7.2.

The cooperative action of gas and shrinkage quantified previously in Eqns (7.8) and (7.10) was predicted by Whittenberger and Rhines (1952). Their ground-breaking concept was enshrined by them in a nucleation diagram as shown in Figure 7.23. We shall devote some space here to a consideration of this insightful map and show how it can be developed to a fascinating degree of sophistication, greatly assisting the description of pore-forming conditions within a casting.

Turning now to a detailed consideration of Figure 7.23: for a well-fed casting, $Ps = 0$, and as freezing proceeds the gas is progressively concentrated in the residual liquid, progressively raising the equilibrium gas pressure. Thus conditions in the casting progress along the line ADCE. At point E, the conditions for heterogeneous nucleation of a gas pore on nucleus one are met, so that a gas pore will pop into existence at that instant. The initial rapid growth of the gas bubble will deplete its surroundings of excess gas in solution, so that conditions in the locality of the bubble will reverse off the nucleation threshold, back towards D. Thus a second pore will be unable to nucleate in the immediate neighbourhood of the first pore. Other gas pores may nucleate elsewhere, beyond the diffusion catchment area of pore number 1.

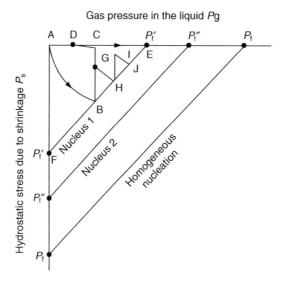

FIGURE 7.23

Gas-shrinkage map showing the path of conditions within the residual liquid in the casting in relation to the nucleation threshold for pore formation.

Notice also that other heterogeneous nuclei are also present, floating about in the liquid (threshold two in Figure 7.23) but, being less favourable, are not activated.

We turn now to the quite different situation where the casting is free from gas, but is poorly fed. The internal pressure P_s in the casting falls because of the action of shrinkage, progressing along the line AF. At F, the fracture pressure for nucleation on heterogeneous nucleus one is met, and a cavity forms. The hydrostatic tension is explosively released and conditions in the casting shoot back to point A.

In practice, both gas and shrinkage will be present to some degree in the average casting, and both will cooperate, causing the conditions to progress along a curve AB. In the absence of any foreign nuclei, it is unlikely that the condition Pf would be met before the casting had solidified completely. This is because the homogeneous nucleation threshold is so far away. (Figure 7.23 is not to scale. If it had been, then the heterogeneous threshold would have been a minute triangle only a millimetre or two in size up at the top left hand corner of the page, making all the action undecipherable!) Thus the casting would have been sound if no favourable inclusions had been present. However, in the presence of nuclei one and two, the combined gas and shrinkage pore will form at B on nucleus one. Nucleus two will never be needed.

On the formation of a mixed pore at B, the pressure in the liquid immediately reverts to point C. Subsequent slower diffusion of gas into the pore will deplete the immediate surroundings of the pore, causing the local environment to progress to D. Outside this diffusion distance, conditions elsewhere in the casting continue to be free of stress and may progress to point E, at which many more gas pores may nucleate. These will add to the original mixed pore that will in the meantime have continued to grow under the combined action of shrinkage and gas.

The area of influence of the release of the hydrostatic tension from shrinkage has been assumed in the previously discussion to apply to the whole casting. In practice, it is true that its influence is vastly greater than that of the depletion of gas. The distance over which shrinkage is depleted is probably a factor of 1000–10,000 times greater than that of gas. This is evident from the fact that diffusion distances of solutes in the liquid are usually measured in micrometres, whereas the distances between layers in typical layer porosity can be measured in millimetres or centimetres. As will be explained in Section 7.1.7.2, each layer in layer porosity represents a separate nucleation event, effectively isolated from its neighbouring layer, because the threads of liquid that connect the two have become so fine as to be practically impassable because of viscous restraint.

Thus in terms of our nucleation diagram, Figure 7.23, the nucleation event at B may cause the reduction in tension in its own locality to drop to nearly zero at point C. However, at a distant location elsewhere, it may drop to only to about one-quarter of the original tension at G (as is explained in the case of the formation of layer porosity to be discussed in Section 7.1.7.2). A second nucleation event may then occur at H, leading via I to a third pore formation event at J, and so on.

The nucleation diagram is useful for visualising the effect of such variables as casting under an applied pressure, so that the starting point A is raised, effectively making the nucleation threshold more distant. If it is sufficiently distant then there is a chance that it may not be reached before the casting has solidified.

The diagram also illustrates how it is of little benefit to the soundness of a casting to melt in vacuum and cast in vacuum (although it may be better for other reasons to cast in vacuum, to avoid any back pressure of gas that might resist the filling of very thin sections for instance. It may also be useful to avoid the entrainment of a protective gas such as argon, which cannot be subsequently diffused out, nor fully closed by hipping). This is because both the gas pressure and the pressure in the liquid are both shifted by 1 atm, in the same direction, meaning that the pore nucleation threshold remains the same distance away, cancelling any advantage. Conversely, melting under vacuum but solidifying under atmospheric pressure is seen to be a benefit, pushing the threshold away by 1 atm pressure.

7.1.5.4 Non-classical initiation of pores
High-energy radiation
The radioactive decay of naturally occurring isotopes, and, unfortunately, contaminating artificial radioactive elements, occurs around us all at every point of our lives. Naturally occurring radioactive materials are relatively common in metals and alloys, and general radioactive contamination is presently increasing annually, arising from industrial and medical sources and general fall-out since the first nuclear explosions in 1945. This is a sad outcome, but even if all future contamination of the environment were to be prevented, we will still have to come to terms with accepting the historical legacy of a radioactive environment as a fact of life.

Liquid metals are, in common with all other present-day materials, subjected to a constant barrage of high-energy particles from these internal radioactive decay processes. The passage of these high-energy particles through the liquid causes thermal or displacement spikes, the name given to regions of intense heating, or actual displacement of atoms, effectively raising the local temperature of the liquid to well above its boiling point. It is possible, therefore, that these transient heated regions might become vapour bubbles sufficiently large to satisfy Eqn (7.1), thus constituting effective nucleation sites for gas or shrinkage pores.

Johnson and Orlov, in their review (1986), describe defect regions in solid metals of up to 100 atoms in diameter. Energy can be channelled away from such events along crystallographic directions in ordered solids, reducing the local damage. However, the lack of any long range order in liquids means that no such safety valve is possible, so that energy deposition will be much more localised. It follows that the production of a bubble of 100 atoms diameter in liquid iron should be easy, giving an equivalent fracture pressure of approximately 1500 atm, much lower than that required for classical homogeneous nucleation. In liquid FeO, the fracture pressure would be 400 atm. If the event occurred close to a non-wetted surface, then the fracture pressure would be only 20 atm. Such pressures would be much more easily met in castings.

Analogous events are actually observed directly in the bubble chamber, a device full of a transparent liquid that can be vapourised by a high-energy particle, defining its path by a string of nucleated vapour bubbles. It is sobering to note that a bubble chamber can only be constructed using steel made before 1945, in Europe commonly sourced from the German battleships on the seabed at Scapa Flow in the north of Scotland. Later steel introduces too much spurious background radiation.

Claxton (1967) has carried out a detailed study of the nucleation of vapour bubbles in liquid sodium subjected to a wide variety of different high-energy particles, including photons, electrons, protons, neutrons, alpha-particles, xenon and strontium as fission fragments and alpha-recoils. His preliminary analysis suggests that the only interactions capable of initiating nucleation of bubbles are 'knocked-on' atoms of the liquid produced by fast neutrons. Claxton (1969) suggests that for heavy recoils arising from alpha-particle decay the rate of energy transfer in liquid aluminium would be about three times higher than that in sodium, and in liquid iron should be about 10 times, giving rise to the possibility of nucleation, depending on the isotope responsible.

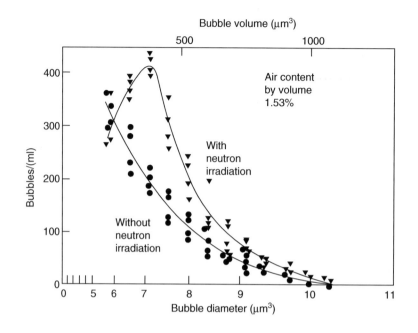

FIGURE 7.24

Micro-bubble spectrum in tap water with and without neutron irradiation (Hammitt, 1973).

Significantly, the micro-bubble spectrum in water is seen to be augmented when the water is irradiated with neutrons (Figure 7.24).

Additional factors that are likely to enhance any effects of the presence of radioactive isotopes in metals will be the concentration of such elements ahead of the freezing front. This is precisely where such events can be most effective. The region has high gas content, low surface tension and high density of assorted solid debris, some of which may be effective nuclei.

Furthermore, if a particularly troublesome radioactive isotope happened to be present in the melt, as an alloy in solution, it would be fundamentally different in character from suspensions of bubbles or inclusions in that it could not simply be eliminated by sedimentation or filtration. These uncertainties have never been investigated.

Differences in the levels of contamination by trace isotopes may be just one of many possible reasons for the occasional different behaviour from one batch of metal to the next that is often experienced in the foundry, and that often seems inexplicable. Kato (1999) is one of the first to record a check for the presence of alpha-emitters in his high-purity copper castings. It is a concern that one day such checks may have to become a routine.

Pre-existing suspension of bubbles

Several studies have indicated that there may be a micro-bubble spectrum in most liquids.

Studies on tap water have demonstrated that there seem to be approximately 300 bubbles of around 5 μm diameter in each millilitre (Figure 7.24). Hammitt (1973) has carried out work that implies similar distributions of bubbles in liquid sodium circulating as coolant in atomic reactors. Even higher densities of bubbles have been measured by Outlaw et al. (1981) in vacuum-cast pure aluminium (Figure 7.25). Although, of course, we have to be careful not to assume that this pore distribution in the solid reflects that originally present in the liquid, the result does underline that there are distributions of fine pores in circumstances in which we may not have any reason to suspect their presence.

Chen and Engler (1994) examine the very old proposal that pores exist in irregular crevices of solid inclusions. They propose that for a conical cavity a gas pocket would have an indefinite existence (they neglect the complications of

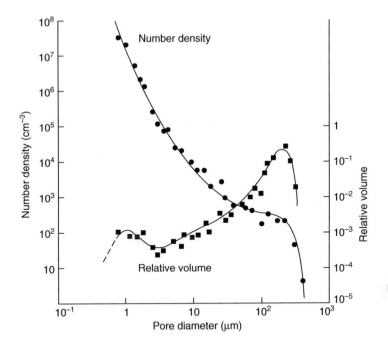

FIGURE 7.25

Hydrogen porosity in vacuum melted and cast 99.99Al. Total porosity is 0.71% (Outlaw et al., 1981).

chemical reactions and dissolution that we shall consider later). In the case of old oxides, from the surfaces of melting or holding furnaces, that have been relatively recently entrained, this model is probably accurate. Even so, it seems likely that in most current situations, this source of inclusions will be of less importance than new films from more recent entrainment events. Bifilms are expected to exist in many liquid metals and alloys. They are not envisaged as a distribution of spherical pores, but as a wide spectrum of sizes and shapes of films of air trapped between folded entrained surface films (the crack-like pores illustrated by Chen and Engler are good examples of bifilms in the early stages of opening). The films will usually be oxides, but may be several other non-metallic phases such as graphite or nitrides etc. The bifilms are, of course, a kind of crevice, so that, to some extent, the theory elaborated by Chen and Engler remains appropriate. Even so, we shall see how the bifilm model as a somewhat flexible folded film, exhibiting some rigidity and enclosing films and bubbles of entrained gas, in general, fits the facts more closely than that of a model of a solid particle with a gas-filled crevice.

For an entrained oxide film, the oxygen in the air entrapped in the folded film will be quickly consumed by reaction with the metal to form more oxide. The nitrogen may subsequently be consumed more slowly to form nitrides. Ultimately, however, there will be a tiny residue of only about 1% of the original volume of entrapped air, consisting mainly of argon. The inert gases are practically insoluble in liquid metals (Boom et al., 2000). Thus a spectrum of very fine volumes of inert gas, trapped within oxide fragments, will be expected to be rather stable over long periods of time.

Figure 7.26 shows the order of magnitude relation showing the size and number of bubbles equivalent to that volume percentage of porosity. For instance, 1% porosity can correspond to either a mere 10 pores per millilitre when the pores are 1 mm in diameter, or 10 million pores per millilitre when the pores are only 10 μm in diameter. (In Figure 7.26, the scale of gas content of the melt, assuming the melt to be aluminium, is of course only accurate at larger bubble sizes.

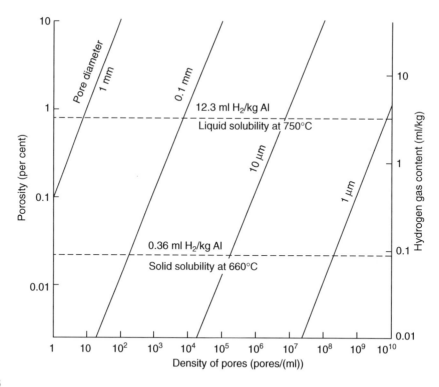

FIGURE 7.26

General relation between volume % porosity and number and size of pores. The solubility of hydrogen in aluminium is superimposed.

It becomes increasingly inaccurate as sizes fall below 0.1 mm in diameter because the internal gas becomes increasingly compressed by the action of surface tension.)

In other more reactive metals, such as liquid titanium, both oxygen and nitrogen have high solubility, leaving only the inert gases that may be insoluble. A micro-bubble distribution introduced on the surface oxide (if any) may be less stable in this material after the oxide has dissolved because the bubbles would be more free to float out. Similarly, in other high-temperature liquids such as irons and steels the greater speed of reactions, the higher density of the melt and its higher surface tension will all tend to limit the lifetime and spread of sizes of any micro-bubble population introduced from this source. Nevertheless, enough of the population may still survive for long enough to cause problems in the short time needed to achieve solidification.

Pore initiation on bifilms

In contrast with all other pore initiation mechanisms, apart from the pre-existing population of pores with which this section has much in common, the bifilm is seen to possess the potential to initiate pores with negligible difficulty. It simply opens by the separation of its unbonded halves. Surface tension is not involved (as is usually assumed when nucleating a pore in a liquid). This is simply a mechanical action between parallel films of (effectively) vanishingly thin oxide separated by a vanishingly thin layer of gas. The acid test of any theory of pore initiation is its capability to explain the experimental data that has previously appeared to be inexplicable. Such data are described later.

The definitive research on the growth of porosity in aluminium alloy castings was that carried out at Alcan Kingston Laboratories by a team led by Fred Major (Tynelius et al., 1993). In this exemplary work, small tapered plates and end-chilled plates of Al-7Si-0.4Mg alloy were cast under varying conditions to separate the effects of gas content, alloy composition, freezing time, and solidus velocity on the growth of porosity in the castings. The reader is recommended to consult this impressively logical piece of work; the first of its kind, and not since repeated at the time of writing.

Normally of course, the effects of solidification time, temperature gradient, and solidification rate are all so closely linked that they are effectively inseparable in most practical casting experiments. However, the experiment was cleverly designed so that the effect of casting geometry was separated out, revealing that the most appropriate thermal parameters to predict porosity were the solidification time and solidus velocity. These gave better results than any of the various temperature gradient terms. They also quantified the dominant effect of hydrogen and the important contribution of strontium.

The outstanding mystery from this work was the parameter 'areal pore density' (i.e. the number of pores per unit area). The results are shown in Figure 7.27. It was found that at short solidification times the pore density *increased* with increasing hydrogen content. However, at long solidification times, the pore density *decreased* as hydrogen content increased. The authors correctly surmised that this curious and baffling result must depend somehow on the nucleation processes at work. However, the finding has remained unexplained ever since. We shall see how the action of bifilms explains this enigma in a natural way.

We assume that conditions at the lower left corner of Figure 7.27 will be characterised by convoluted bifilms of which only the longest and outermost fold is inflated with sufficient gas to be seen as a pore on a polished section. With time (going vertically up the figure), or with additional concentration of gas (travelling along the bottom towards the right side of the figure), the bifilm will partly inflate additional sections, thus creating what would appear to be a cluster of small pores. Eventually, as the remaining sides of Figure 7.27(a) are traversed in progress towards the top right corner, the separate micro-inflated bifilm sections will become sufficiently inflated that the whole bifilm will unfurl, blowing up like a single spherical balloon.

The interpretation of Figure 7.27(a) is outlined in Figure 7.27(b). In the lower left hand corner, gas content is so low and time available so short that any bifilms will be mostly still be closed, having been ravelled up by the action of bulk turbulence. If anything, only the longest section of the bifilm at the outside of the compacted inclusion will be observable as a pore on a polished section because this part of the bifilm will open first, having the largest area to gain gas from solution, and being unshielded from the arrival of supplies of diffusing gas. (Internal folds within the convoluted film will be shielded from outside supply of gas until the defect unfolds.)

As we move from the lower left hand corner up to the top left hand corner, the increased solidification time will have allowed additional gas to diffuse to the bifilm and will begin to open many of the folds. We can call this the micro-inflated stage. A polished section through the bifilm will give the appearance of about 10 times the number of pores, in agreement with observations (Figure 7.27(a)). On the other hand, a move from the lower left corner directly to the lower right corner will similarly micro-inflate the bifilm. Although the time is still short, the amount of gas available is now sufficient to achieve this.

Finally, moving to the top right gives sufficient time and sufficient gas to power the complete inflation of the film. This is expected to occur by successive unfolding actions, as fold after fold opens out. Finally, the bifilm is fully inflated like a balloon. Thus on a polished section only one pore is seen. The main features of Figure 7.27(a) are thereby explained.

Keen readers can check the observed pore densities from the original publication, allowing a rough calculation of the number of bifilms in the original cast metal and the size distribution of the final pores. A further check is the total fraction of porosity. Such checks confirm the consistency of the whole scenario.

The assumption that bifilms are present allows a description of porosity in terms of initiation and growth characterised by considerable sophistication and complexity; the important experimental work by Fred Major would probably otherwise remain unexplained and inexplicable.

As an additional related exercise, we can gain some idea of the rate of unfurling of bifilms from a simple mechanical model as illustrated in Figure 7.28. We shall assume that the unfolding of the bifilm is resisted by a force F of the same type as that resisting the motion of a sphere in a viscous liquid (as in the derivation of Stoke's law). Thus the force would be $3\pi\eta RV$ if it were evenly distributed over the square face of area $R \times R$. Because the velocity V is that at the tip of the

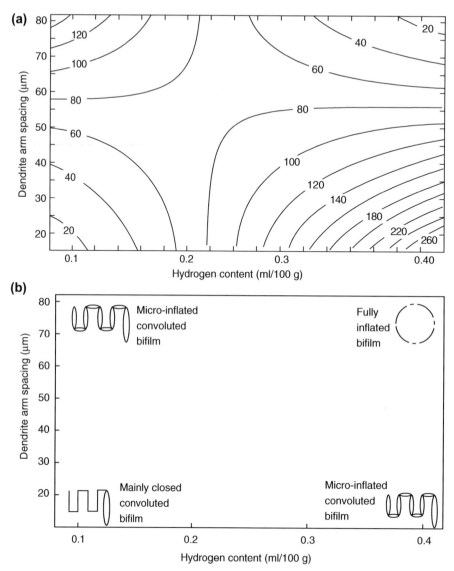

FIGURE 7.27

(a) Experimental results by Tynelius et al., (1993) showing that pore density per cm^2 (the numbers in the figure) increases at fast freezing rates (small dendrite arm spacing, DAS) but decreases at slow freezing rates (large DAS); (b) interpretation by a bifilm model.

bifilm and the pivot of the bifilm is fixed, on average, the resisting force is $3\pi\eta RV/2$. The opening force is that due to the pressure P in the gas phase of the bifilm, acting over the area hR. Equating moments we have:

$$2PRh \cdot (h/2) = 3\pi\eta RV \cdot (R/2)$$

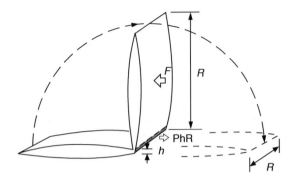

FIGURE 7.28

Model of the unfolding of a bifilm.

so that we can find the opening time t from the speed V and the distance travelled πR:

$$
\begin{aligned}
t &= \left(3\pi^2/2\right)(\eta/P)(R/h)^2 \\
&= 15(\eta/P)(R/h)^2
\end{aligned}
$$
(7.11)

For viscosity $\eta = 1.4 \times 10^{-3}$ Nsm^{-2}, and reasonable figures for P of about 0.2 atm (0.2×10^5 Pa) above ambient, and for $R = 5$ mm and $h = 5$ μm, t is only approximately 1 s. Whereas if $R = 10$ mm and $h = 1$ μm, t is approximately 2 min. Thus the rate of opening is seen to be highly dependent on the geometry of the bifilm, as might be expected. Nevertheless, the rates are of the correct order of magnitude to explain the rate of loss of properties in most ordinary castings as the freezing time increases. This wide scatter in the performance of bifilms supports the interpretation of variable nucleation conditions surmised as a result of the sub-surface porosity described in Section 7.2.

7.1.6 GROWTH OF SHRINKAGE PORES

For internal shrinkage pores that are nucleated within a stressed liquid, the initial growth is extremely fast; in fact, it is explosive! The elastic stress in the liquid and the surrounding solid can be dissipated at the speed of sound. The tensile failure of a liquid is like the tensile failure of a strong solid; it goes with a bang.

On a more gentle scale, the audible clicks that are heard as pores nucleate within the melted Tyndall volumes in ice are an exact analogy, and a reminder of the smaller but not negligible tensile stress that is supported in the liquid water before a cavity is nucleated (see Section 7.2.2.1).

In the direct observation of the freezing of Pb-Sn alloys under a glass cover, Davies (1963) recorded the solidification of isolated regions of alloy on cine film at 16 frames per second. It was observed that as these regions shrank, a pore would suddenly appear in between frames. Thus the growth time for this initial growth phase had to be shorter than approximately 60 ms. Almost certainly it was very much faster than this. If in fact the rate of expansion is close to the speed of sound, close to 6 kms^{-1} in liquid aluminium, a pore 1 mm in diameter will form 1000 times faster, in 0.1 μs.

After this explosive growth phase, the subsequent growth of the pore observed by Davies was more leisurely, occurring at a rate controlled by the rate of solidification. Thus this second phase of growth is controlled by the rate of heat extraction by the mould.

For pores that are surface-initiated, the initial stress is probably lower, and the puncture of the surface will occur relatively slowly as the surface collapses plastically into the forming hole. Thus the initial rapid growth phase will be less dramatic.

For shrinkage porosity that grows like a pipe from the free surface of the melt, there is no initial fast growth phase at all. The cavity grows at all stages simply in response to the solidification shrinkage, the rate being dictated by the rate of extraction of heat from the casting.

7.1.7 SHRINKAGE PORE STRUCTURE

7.1.7.1 Shrinkage cavity or pipe

During liquid feeding, the gradual progress of the solidification front towards the centre of the casting is accompanied by the steady fall in liquid level in the feeder. These linked advances of the solid and liquid fronts generate a smooth conical funnel as shown in Figure 7.29. This is a shrinkage pipe.

It is sometimes called a primary shrinkage pipe to distinguish it from so-called secondary shrinkage seen on sections cut through the casting. Secondary shrinkage appears to take the form of a scattering of disconnected pores below the primary pipe. Generally, however, it is easy to demonstrate that water, or a dye, poured in the top of the primary pipe will find its way into most, if not all, of the so-called secondary cavities. Thus in reality the secondary shrinkage cavities are merely the extension of the primary pipe. Thus the primary shrinkage cavity penetrates further into the casting than might be apparent at first sight, as Figure 7.29 makes clear.

In the situation where the shrinkage problem is in an isolated central region of the casting, a narrow freezing range material will give a smooth single cavity. This is occasionally called a macropore to distinguish it from microporosity. There has been much written to emphasise the differences between these two forms of porosity. However, as will be clear from evidence presented in the next section, there seems to be no fundamental difference between them; one gradually changes into the other as conditions change from micro-volumes to macro-volumes of unfed regions, and from skin freezing to pasty freezing. Figure 7.8 illustrates a macropore in a bronze casting of very long freezing range, resulting in a single tapering cavity that has numerous branches, and appearing on a cut section to be thousands of separate cavities constituting a spongy structure.

In the case of the single isolated area of macroporosity, it is important to note that its final location will not be in the thermal centre of the isolated region, as might at first be supposed. This is because the shrinkage pore could be nucleated anywhere within the volume of isolated, stressed liquid. However, immediately after it is formed, it will float to the top of the isolated liquid region. Conversely of course, the situation can be viewed from the point of view of the liquid. This phase, being heavier, will sink, finding its own level in the bottom of the isolated region. The final position of the shrinkage cavity in this case will be as shown in Figure 7.30. Also, of course, the shape and the position of the porosity can be altered by changing the angle of the casting, because the pore floats (or the residual liquid finds its lowest level). An analogy is a stoppered bottle, partly filled with liquid, which is turned through different angles. The bottle corresponds to the outline of the isolated liquid volume in Figure 7.30.

Note once more that the long parallel walls of the casting give a corresponding long tapering extension of the shrinkage cavity. On a cut section, any slight out-of-straightness of this tubular cavity can once again be easily misinterpreted as dispersed porosity, so-called 'secondary pipe', as it weaves its way in and out of the plane of sectioning Figure 7.30 (stage 4). In a casting of more complicated shape the shape of the shrinkage pore will take on a corresponding complexity.

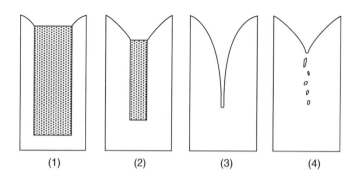

(1)   (2)   (3)   (4)

FIGURE 7.29

Stages in the development of a primary shrinkage pipe. Stage 4 is the appearance of stage 3 on a planar cut section if the central pipe is not exactly straight (i.e. it is not a series of separate pores).

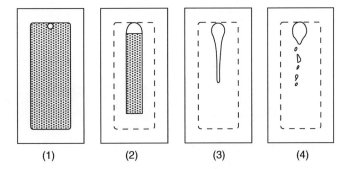

FIGURE 7.30

Stages in the development of an internal shrinkage cavity. Stage 4 is again the equivalent cut section to stage 3. Note that the porosity is not concentrated in the thermal centre, but is offset from the centre of the trapped liquid region, outlined by the broken line, by gravity.

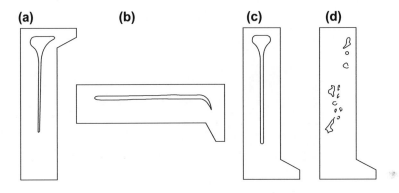

FIGURE 7.31

((a)–(c)) Shrinkage cavity in a short freezing range alloy as a function or orientation. Porosity shown in (d) illustrates some other source of porosity (it cannot be a shrinkage type because of its random form, not linked to the casting geometry).

The effect of gravity on the final form and distribution of porosity is illustrated in Figure 7.31. Clearly, the porosity can be moved from one end of the casting to the other simply by making the casting in a different orientations, a, b and c. (Figure 7.31(d) is simply slipped in to emphasise that porosity having no relation to the casting geometry cannot be shrinkage porosity.)

7.1.7.2 Layer porosity

Alloys of long freezing range are particularly susceptible to a type of porosity that is observed to form in layers parallel to the supposed positions of the isotherms in the solidifying casting. It is known as layer porosity.

Conditions favourable to the formation of layer porosity appear to be a wide pasty zone arising from long freezing range and/or poor temperature gradients. Poor gradients are typical of alloys of high thermal conductivity (such as the light alloys and copper-based alloys) and of moulds having low rate of heat extraction either because of their high temperature (such as in investment casting) or low thermal conductivity (as in sand or plaster castings).

Given these favourable conditions for layer porosity, this variety of porosity has been observed in practically all types of casting alloys, including those based on magnesium (Lagowski and Meier, 1964), aluminium (Pollard, 1964), copper (Ruddle, 1960), steels and high-temperature alloys based on nickel and cobalt (Campbell, 1969a,b). An example in steel is shown in Figure 7.32 and in a nickel-based vacuum cast alloy in Figure 7.14.

It has been argued (Liddiard and Baker, 1945; Cibula, 1955) that layer porosity is the result of thermal contraction, in a manner analogous to hot tearing. Briefly, the theory goes that after the establishment of a coherent dendrite network by the impingement of the dendrite tips, subsequent cooling causes the structure to shrink, imposing tensile stresses on the network, causing the network to tear perpendicularly to the stress, i.e. parallel to the isotherms. If the tears are not filled by the inflow of residual liquid, then layers of porosity are frozen into the structure. Baker (1945) attempted to test whether the porosity was the result of thermal contraction on cooling by casting test pieces in a mould designed to accentuate hot tearing. However, no significant increase in layer porosity was found.

The negative result of Baker's critical experiment is substantiated by numerous observations, particularly the observations by Lagowski and Meier (1964) on Mg-Zn alloys covering the range of zinc contents up to 30% zinc. These clearly reveal that hot tearing peaks at 1%Zn, corresponding to the maximum of the thermal contraction, and well separated from layer porosity that peaks at 6%Zn (Figures 8.8 and 9.11).

Finally, the thermal contraction model is seen in any case to be fundamentally flawed; it is not easy to envisage how such differential cooling could arise to pull apart the centre, when the whole casting is solidifying in the absence of significant temperature differences. In any case, close inspection of the pore structure reveals that it is not the dendrite mesh that is pulled apart; the mesh remains in place, and only the residual interdendritic liquid is missing.

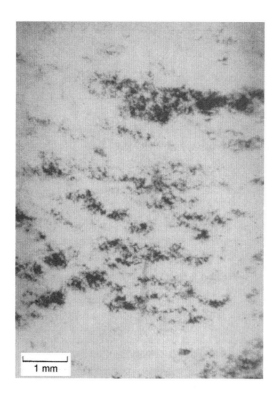

1 mm

FIGURE 7.32

Radiograph of interdendritic porosity in a carbon steel (Campbell, 1969a,b).

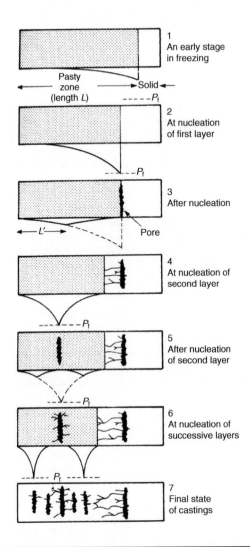

FIGURE 7.33

Schematic representation of the formation of layer porosity (Campbell, 1968a,b).

I therefore felt it necessary to forward a new explanation of layer porosity (JC, 1968c). The new approach avoided the difficulties mentioned previously because it was based not on thermal contraction in the solid as a driving force, but on the contraction of the liquid on solidification; the problem was interpreted as a feeding problem.

The sequence of events in the solidifying casting is shown in Figure 7.33. The stress in the liquid is assumed to be zero near to the feeder on the extreme left. From Eqn (7.5), it is clear that the hydrostatic tension increases parabolically with distance x through the pasty zone of length L, as shown in stage 1, Figure 7.33. The stresses continue to increase with advancing solidification (as the radius R of interdendritic channels continues to decrease) until the local stress at some point along the parabola exceeds the threshold at which a pore will form (by either internal nucleation on an inclusion or by surface puncture). This threshold is labelled the fracture pressure P_f in stage 2.

As soon as a pore is created by some mechanism, it will immediately spread along the isobaric surface (this surface of constant pressure also probably coincides with an isothermal surface and an isosolid surface), forming a layer and instantly dissipating the local hydrostatic tension. The elastic energy that is available for the initial explosive growth stage is proportional to the difference in areas under the pressure-distance curves before and after this growth. The energy is clearly proportional to the area $L \times Pf$ under the curve. As discussed in the previous section, this first stage of growth would last probably only microseconds, or at the most, milliseconds.

As solidification proceeds further, the solidification contraction in the centre of the remaining liquid region is now fed both from the feeder and by fluid (whether residual liquid, gas or vapour) from the newly created pore. This is a slower growth phase for the new pore, extending via channels towards the region requiring feed metal. The new layer-shaped pore effectively provides a free liquid surface, adjacent to which no large stresses can occur in the liquid.

The maximum stress in the liquid has at this stage fallen by approximately a factor of 4 because the effective length L has now approximately halved. However, because of the progressive decrease in radius R of the interdendritic channels, stress once again gradually increases with time until another pore formation event occurs as at stage 4.

Further nucleation and growth events produce successive layers until the whole casting is solidified. The final state consists of layers of porosity that have considerable interlinking.

Although these arguments have been presented for the case of porosity being formed only by the action of solidification shrinkage, the reader is reminded that the action of gas and shrinkage in combination also fits the facts well as discussed in Section 7.1.5.3 and illustrated in Figure 7.23.

It is important to observe that layer porosity is quite different to a hot tear. A hot tear is formed by the linear contraction of the casting pulling the grains and/or dendrites apart. When they are sufficiently separated, there is insufficient residual liquid to fill the increasing volume, so that a true crack opens up. The crack, of course, supports no load, and represents a serious defect.

In contrast, layer porosity is formed by the nucleation of a pore in the residual liquid. The liquid is in a state of hydrostatic tension, so that the pore spreads along the surface of maximum hydrostatic tension, through the liquid phase. The dendrites stay fixed in place. The final defect is a layer-like pore threaded through with dendrites. It is akin to a crack spot-welded at closely spaced points, and so has considerable tensile strength. On a polished microsection therefore, a hot tear is clear, whereas porosity as part of a layer may or may not be easily identified. It only becomes really clear on a radiograph when the radiation is aligned with the plane of the layers. The layer porosity illustrated in this book (Figures 7.14 and 7.32) are taken from radiographs.

Figures 7.14 and 7.15 show layer porosity in an investment casting as a function of conditions that vary progressively from skin freezing to pasty freezing. It reveals that macroscopic centreline porosity, layer porosity and microscopic dispersed porosity transform imperceptibly from one to the other as mould temperature alone is increased, effectively reducing the temperature gradient during solidification. It is clear, therefore, contrary to many widely held views, that there are no fundamental differences between the various types of macroporosity and microporosity. They are simply different growth forms of shrinkage porosity under different solidification conditions.

Similarly, Jay and Cibula (1956) carried out interesting work on Al-Mg alloys, in which they showed that as the gas content of the alloy was increased, the porosity changed gradually from layer porosity to dispersed pinhole porosity (Figure 7.34). Thus these two extreme categories of shrinkage and gas microporosity were demonstrated to be capable of being mixed, allowing a complete spectrum of possibilities from pure shrinkage layer type to pure dispersed gas type.

This merging and overlap of all the different types of porosity makes diagnosis of the cause of porosity sometimes difficult in any particular case. However, it is better to know of this difficulty, and thus be better able to guard against falling into the trap of being dogmatic. Section 7.3 gives some guidance on the diagnosis of the various types of porosity.

The previous classical description of the formation of layer porosity was conceived before the discovery of bifilms. Thus the theory, based on the nucleation of pores by the accumulated build up tension because of the difficulty of interdendritic flow, and their subsequent growth along isobaric surfaces, seems to me probably still essentially correct. However, now that the presence of bifilms has been realised, it is necessary to take account of the effect they may have in such situations where interdendritic feeding is difficult. It seems likely that bifilms would interfere with the flow of residual liquid through the dendrite mesh. The close spacing of dendrites, providing support for the films, would ensure

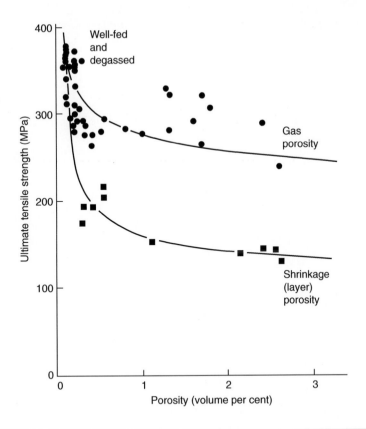

FIGURE 7.34

Reduction in UTS of an Al-11.5Mg alloy by dispersed porosity and by layer porosity.

Data from Jay and Cibula (1956).

that they would be capable of resisting large pressure differences across their surface. Thus bifilms transverse to the flow would halt the upstream flow, but be sucked open by the downstream demand, creating a series of shrinkage cavities arranged generally transverse to the flow direction. Such a scenario is not greatly different from that of the classical theory presented previously and might be difficult to separate experimentally because the nucleation events in both explanations would be assumed to be bifilms. Ultimately, cleaning the melt will be a certain technique to eliminate layer porosity irrespective of its formation mechanism.

7.2 GAS POROSITY

Sub-surface porosity in casting is a common fault, if not more common than any other type of gas porosity from gases in solution in the metal. Sometimes, the sub-surface pores are broken into by the loss of surface scale during heat treatment, as is common for steel castings, or are broken into during shot blasting to clean the castings. However, sometimes the problem is not seen until the first machining cut. Pores can originate by completely different processes (1) from entrained air bubbles during turbulent pouring; (2) from core blows or condensation on chills; and (3) by the precipitation of gas from solution in the melt. The processes are considered in order later.

7.2.1 ENTRAINED PORES (AIR BUBBLES)

Sub-surface pores can form as a result of the flotation of air bubbles that have been entrained during the pouring of a casting. These bubbles are about 5 mm in diameter or smaller because larger bubbles will in general escape, but at 5 mm and below they do not have sufficient buoyancy to break through the double oxide (their own oxide skin and the oxide skin on the surface of the casting) when they reach the surface of the casting, and so remain trapped. They sit immediately under the surface of the casting, only two oxide thicknesses deep, and so extremely near to the surface and extremely easily broken into. These bubbles are always clustered in specific sites, having floated into position, often above ingates, and have a spread of sizes, typically 1–5 mm in diameter.

This 5 mm limit on size is the result of a mechanical limitation and is a certain identification of these pores as being air bubbles entrained during the pouring of the casting. This size of bubble is too large to be generated by a diffusion reaction, as described later, thus they are definitely not, for instance, hydrogen pores nucleated and grown in situ during the freezing of the casting. It is a waste of time therefore to attempt to remove them by degassing the melt!

There is a large amount of research literature claiming that such bubbles in cast irons, associated with oxide slags, are the result of a reaction between the carbon in the iron and the oxygen in the slag, to create a CO bubble. This is clearly a mistake for a variety of reasons.

1. The bubbles are nearly always too large to be produced by a diffusion reaction.
2. The size range of bubbles corresponds to bubbles entrained by turbulent filling systems.
3. The filling systems are always turbulent.
4. The slag is nearly always generated in the filling system by the turbulent mixing of the melt and air, and so naturally occurs in association with the entrained air (although it is recognised that after swimming through the melt, the bubble will have acquired a contribution of CO, and probably H_2).

It is important to recognise the sub-surface bubbles generated by a turbulent filling system. They are quite different to those bubbles formed in situ during solidification which are characterised by their uniform size (i.e. typically 0.5 mm in diameter or smaller for castings up to a few hundred kilograms or so, hence normally smaller than one tenth of the size of entrained air bubbles) and uniform distribution over much of the outer surface of the casting (in contrast with the clustering of entrained air bubbles). Sometimes in situ formed bubbles are uniformly distributed only the upper parts of the casting because the pressure at greater depth suppresses the formation of bubbles.

A widely accepted theory of the origin of the in situ grown sub-surface porosity has been summarised by Turkdogan (1986). He describes how sub-surface porosity occurring in cast irons and steels poured into greensand moulds is a consequence of metal-mould interaction. Gas bubbles form in crevices of the mould in contact with the metal, and bubble into the metal, where they become trapped during the early stages of solidification. The action of alloying elements on the process is discussed in terms of their effect on the surface tension of the liquid metal; a lower surface tension allows bubbles to enter the metal more easily, thereby increasing the sub-surface porosity.

The theory is similar to the micro-blow theory outlined later in Chapter 10, Rule 5. However, micro-blows are probably effectively suppressed by the presence of a strong surface film, such as is normally found on many Al alloys, irons and steels. Thus there are several difficulties with the theory as put forward by Turkdogan.

1. The metal does not in general enter the crevices of the mould. The high surface tension (plus the effect of a strong, fairly rigid surface film) causes the liquid to bridge between high spots, leaving the crevices empty. Mould washes that promote the wetting of the mould by the metal, such as those that are based on sodium silicate, actually reduce sub-surface porosity.
2. The pressure required to force small bubbles (radius of 1 mm or less) into the liquid metal against the resistance of surface tension is high. Conversely, it is known that the pressures attainable at the surface of a mould, from which gas can easily migrate away through the mould to the atmosphere, are very low. Thus it seems

unlikely that small bubbles could be forced into the metal in this way. (Large bubbles, with radii measured in centimetres, and therefore small pressure requirements, can be forced in to the metal from outgassing cores; Chapter 10, Rule 5.)

3. Tellurium additions to cast iron reduce surface tension and should, according to the penetration theory, increase porosity. In fact, tellurium additions are found to decrease porosity.
4. The theory is not capable of explaining the occurrence of sub-surface porosity in inert moulds such as investment moulds, which are free from gas-forming materials such as moisture and hydrocarbons.

Features 1 and 2 of the list may not always be relevant, because the micro-blow conditions may apply as described in Chapter 10, Rule 5. However, features 3 and 4 remain important damning evidence.

We shall therefore assume that in general the formation of sub-surface porosity does not occur by mechanical penetration of the liquid surface by bubbles from the mould.

7.2.2 BLOW HOLES

When sand cores are surrounded by liquid metal, the heating of the core will expand its content of air contained between sand grains and will start to degrade its binder causing gas to be generated. Normally, the core will be designed to permit gas to escape through the core prints and so be dissipated in the mould. In this way, high gas pressure inside the core can be avoided.

In some circumstances, however, the gas in the core may be generated at such a high rate, as in a thin core in a steel casting, faster than it can escape. Its internal pressure may then rise to a level higher than that in the liquid, forcing a bubble out into the melt. This forcing of gas into the liquid can be nearly explosive. The gas bubble is *blown* into existence. *Blow defect* is therefore a good name for this type of gas pore.

The reader needs to be aware that the name '*blow-hole*' is, unfortunately, widely misused to describe almost any kind of hole in a casting. In this work the name '*blow-hole*', '*blowhole*' or simply '*blow*', is strictly reserved solely for those defects which are forced (actually '*blown*') into the liquid metal via the forced mechanical penetration of the liquid surface. The term is therefore quite specific and accurately descriptive. (The term excludes pores nucleated internally by the precipitation of gas dissolved in the liquid, or diffusing in from the surface, and also excludes bubbles entrained by surface turbulence.) The reader is encouraged to use the name '*blowhole*' with accuracy.

The contribution of surface tension to increasing the pressure required is practically negligible in the case of *core blows* because of the large size of the bubbles that are formed in this process (Figure 7.35). Core blow bubbles are large, lazy, wobbling bubbles, containing gas at low pressure. They are quite unlike micro-blows, if these exist, to be discussed later. The effect of surface tension at such small radii causes these tiny bubbles to behave like hard spheres resembling mini ball bearings.

Core blow bubbles create trails as they rise through a melt. The trails are, of course, a similar form of bubble damage to that of turbulence-entrained bubbles discussed under Rule 4. However, they are sufficiently distinct that they benefit from separate consideration.

For instance, bubble damage arising from surface turbulence in the filling system takes quite a different form. The bubbles are generated by the high velocities in the front end of the system (in the basin, sprue or runner). The high shear stresses in the melt, especially in tall castings, ensure that the bubbles are chopped mainly into small sizes, in the range 1–10 mm diameter. Some of the smaller bubbles have been observed in video radiographic studies to coalesce in the gate. These coalesced bubbles float quickly, before any significant solidification has taken place, and so burst at the liquid surface and escape. A bubble smaller than about 5 mm in diameter has only a tenth of the buoyancy of a 10 mm bubble, and cannot force the splitting of the oxide films that bars its escape. Bubble 'A' in Figure 7.36(a) illustrates such a case. If such a bubble succeeds to reach the top of the casting it therefore remains trapped at a distance only a double oxide skin depth beneath the surface of the casting.

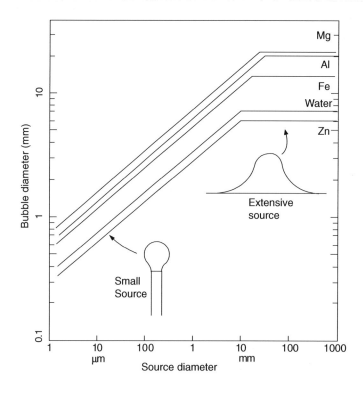

FIGURE 7.35

Size of bubbles detaching from a range of sizes of source in a variety of liquids.

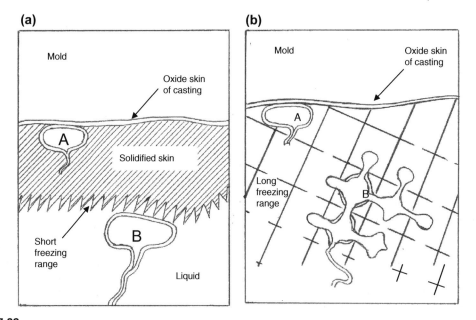

FIGURE 7.36

Bubbles arriving at the cope in (a) a skin-freezing alloy and (b) a long freezing range alloy. Bubbles 'A' arrive early, probably originating from entrainment during pouring. Bubbles 'B' arrive late; in the dendritically freezing alloy the bubble now grows interdendritically, but can, of course, be grown either by additional gas or by shrinkage.

The bubble given off by the outgassing of a core is quite different. These bubbles are large. In irons and steels, the single core blow bubble is about 13 mm in diameter. In light alloys, the effective bubble diameter is approximately 20 mm (Figure 7.36(b)). Although these large bubbles have high buoyancy, they are not produced immediately. The timing of their eruption into the melt determines the kind of defect that is formed in the casting.

If, in relatively thick sections, the bubble detaches before any freezing, the repeated arrival of bubbles at the surface of the casting can result in repeated build-up of bubble skins, forming multiple leaves of oxide, well-named as an *exfoliation defect* (Figure 7.37(a)). Exfoliated structures are those that have been expanded and delaminated, like puff pastry, or like the conversion of slate rock into vermiculite or like leaves from a book.

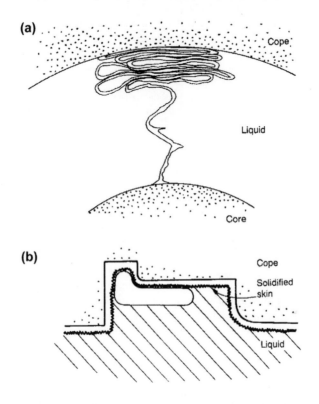

FIGURE 7.37

(a) An exfoliated dross defect produced by copious gas from a core blow before any solidification; (b) trapped gas from a core blow evolved sometime after solidification.

Exfoliated dross defects are recorded for steels containing 0.13C, 1Cr, 0.5Mo and 0.4A1 (Frawley et al., 1974), and for ductile iron (Loper and Saig, 1976). The content of film-forming elements in these alloys is of central significance. For ductile iron (containing magnesium) the defect is described as a wrinkled or pastry-like area, and when chipped out appears to be a mass of tangled flakes resembling mica that can be easily peeled away. Polished microsections of the defect confirmed its structure as composed of tangled films, and contrasted with the structure of the rest of the casting which was reported as excellent. Exfoliated defects are observed in irons and steels where the earlier outgassing of the core precedes any significant freezing of the skin of the casting (this contrasts of course with the light alloys where the leisurely development of pressure in the core usually allows time for some skin freezing of the casting, with the consequent development of a characteristic massive bubble under the solidified skin) (Figure 7.37(b)).

Once a core has blown its first bubble, additional bubbles are easily formed because the bubble trail seems usually to remain intact and keeps re-inflating to pass additional bubbles along its length. The bubbles contain a variety of gases, including water vapour, that are aggressively oxidising to metals such as aluminium and higher melting point metals. Bubble trails from core blows are usually particularly noteworthy for their characteristically thick and leathery double oxide skin, built up from the passage of many bubbles. This thick skin is part of the reason why core blows result in such efficient leak defects through the upper sections of castings. Bubble trails from core blows are particularly damaging because they are, of course, automatically connected from a cored volume of the casting, and often penetrate to the adjacent core (because little solidification will usually have occurred between cores to stop it). Alternatively, in thicker sections, they travel from the core to the very top of the casting.

After the emergence of the first blow into the melt, the passage of additional bubbles contributes to the huge growth of some blow defects. Often the whole of the top of a casting can be hollow. The size of the defect can sometimes be measured in fractions of metres. An accumulating bubble will spread under the dendrites, as, inversely, a large limp balloon would float down to sag on the tips of conifer trees.

These trapped bubbles are recognisable in castings by their size (often measured in centimetres or even decimetres!), their characteristic flattened shape and their position several millimetres under the uppermost surfaces of local regions of the casting (Figure 7.37(b)). In addition, of course, they will faithfully follow the contours of the upper surface of the casting. Like the bubble trail mentioned previously, the core blow defect is sometimes so impressively extensive as to be difficult to perceive on a radiograph.

As an example, the author remembers making a delightful compact aluminium alloy cylinder block for a small pump engine. At first sight, the first casting appeared excellent and unusually light. It seemed that the designer had done a good job. A close study of the radiograph revealed it to be clear of any defects. However, on standing back to take a more general view it was noticed that a curious line effect, as though from a wall thickness, could not be found on the designer's drawing. It took some time to realise the casting was completely hollow; the smooth and extensive form of the core blow faithfully followed the contours of the casting. The prolific outgassing of the water jacket core around the bores of the block was found to be the source. The water jacket core was stood upright on its core prints (the only vents) that located in the drag. The problem arose because of the very slow filling velocity that had been chosen. Vapours were driven ahead of the slowly rising metal, concentrating in the tip of the water jacket core (unfortunately unvented at its top). Thus, by the time the melt arrived at the top, it was unable to cover the core because of the rate of gas evolution. One unhelpful suggestion was that the technique seemed an excellent method for making ultra-light-weight hollow castings. The problem was solved by simply increasing the fill rate, covering the core by the melt before outgassing started. Core gasses were then effectively sealed into the core, and were vented in the normal way via the prints at the base of the core.

A lesson is clearly drawn from this experience: it is far better if possible to vent cores from their top. A somewhat slow filling rate is then not such a danger.

In the case of core blows completely through thin walls, the bubble has no chance to grow to a large size before it has penetrated the wall. Thus the 'bubble trail' becomes merely a hole in the wall of a thin wall casting. Figure 7.38(a) shows a fracture surface made through a core blow defect in a grey iron casting that provided a leak path through the wall thickness (the left and right hand edges of the wall can be seen in the figure). The other images in this figure illustrate progressively closer views of this defect, revealing its astonishing detailed structure. The bubble trails are formed from an oxide in the form of a glassy iron silicate. Figure 7.38(b–d) shows the unusual form of these trails; they appear to reveal clusters of paths made by individual bubbles. Cast iron appears unique in that its bubble trails appear to detach from the matrix. (Similar behaviour has been noted for the lustrous carbon film that forms on the liquid iron meniscus, but becomes trapped against the mould during mould filling, and finally detaches from the casting and adheres to the mould surface after the casting is shaken out; see Section 6.5.)

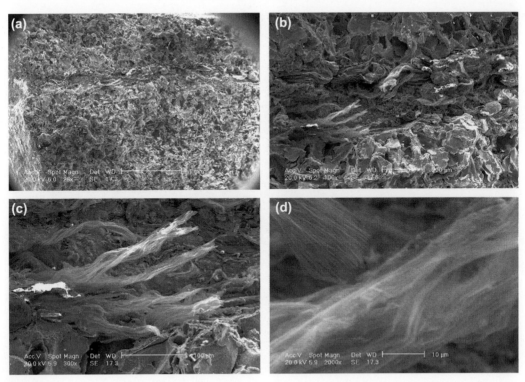

FIGURE 7.38

A leakage defect from a core blow through a 4 mm wall in grey iron. (a) Secondary electron image of a fracture through the defect, showing the leakage channel only about 300 μm diameter; (b) an irregular group of glassy silicate bubble trails near the centre of the leak path; (c) closer view of the twisted form of some silicate trails; (d) close-up of translucent glassy fluted trails (foreground trail is necessarily out of focus).

At other times, the bubbles trails are formed from a lustrous carbon film (Figure 7.39). The carbon film appears to be somewhat more rigid than most oxide films, as suggested by its characteristic smooth surface shown in Figure 7.39(b) and (c), resisting the complete collapse of the trail, and retaining a more open centre. In effect, the bubbles punch holes through the cope surfaces of the casting, so that their more open trails, as irregular tubes, form highly efficient leak paths.

The core blow seen in Figure 7.39(a) has caused a severe leak through the wall of an iron casting. The detail is revealing: the region A is a carbon based bubble trail; C is a portion of matrix iron, whereas B is an irregular mass of iron silicates that was probably the original series of bubble trails when the emerging gases from the core blow were mainly expanding air and water vapour, both highly oxidising, and thus creating glassy silicate trails at that early stage. While the water vapour is evaporating from the core, the temperature of the core is held down at close to 100°C. When the core is finally dry, its temperature now increases sufficiently to drive off carbonaceous volatiles. These would now react on contact with the matrix to create carbonaceous bubble trails. However, in the meantime, the original glassy silicates would partially melt and be churned by the continuing and possibly accelerating maelstrom of gases into irregular masses of slag (the iron silicates). The finding of slag in castings, especially in bubble trails, is usually the result of the local

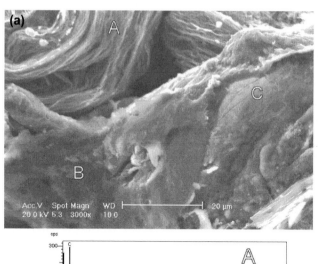

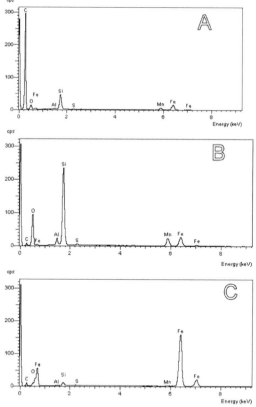

FIGURE 7.39

Bubble trail that had caused a leak, probably formed later by core blow in grey iron. (a) General view inside trail showing 'A' carbon films, 'B' oxides (silicates) and 'C' matrix alloy; (b) other areas of carbon film; (c) close-up of plate-like carbon film.

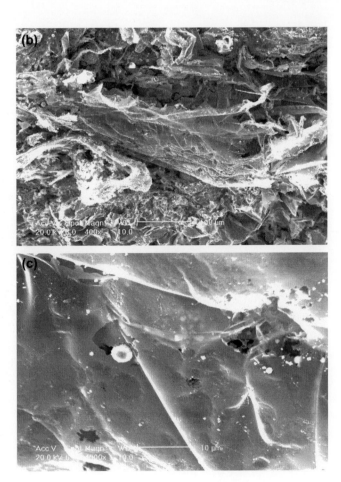

FIGURE 7.39 Cont'd

generation of slag by local oxidation of the melt; in my experience, especially when using a good offset step basin, the appearance of slag is rarely the result of the ingress of slag from the poured metal; it is the result of in situ creation of oxides by entrained air or core gases.

Blows can sometimes form from moulds. Whereas founders are familiar with the problem of blows from cores, blows from moulds are rarely considered. In fact, this is a relatively common problem (even though this section remains entitled 'Core Blows' as a result of common usage). The huge volumes of gas that are generated inside the mould have to be considered. They need room to expand and escape. Any visitor to an iron or steel foundry will be impressed with the jets of burning gasses issuing from the joints of moulding boxes. Effectively the gases and volatiles will be fighting to get out. It is prudent therefore to provide them with escape routes because the alternative escape route via the liquid metal in the mould cavity can spell disaster for the casting.

Mould or core blows usually emerge from the highest point of the core, especially if this is a sharply upward-pointed shape, like a steeple of a church (Figure 7.40). The bubbles will collect and detach preferentially upward from an upward-pointing shape. (The detaching bubble is the upside-down equivalent of the drop of water falling preferentially from the

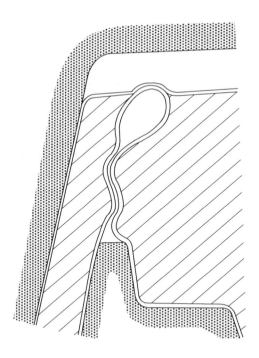

FIGURE 7.40

Detachment of a bubble from the top of a core, bequeathing a bubble trail as a permanent legacy of its journey. This bubble, unusually, may be early enough to escape at the free surface of the rising metal.

tip of a downward-pointing stalactite.) Such shapes are therefore best avoided or the whole mould assembly turned upside down for casting. Core bubbles will, of course, detach from flat surfaces but with much greater reluctance.

Theory indicates (Campbell, 1971) that the size of a detaching bubble is linked to the size of the source. The results of calculated bubble diameters are shown in Figure 7.35 and reveal that even though a bubble might detach from a horizontal surface, and so be of maximum size, its size is limited to 10–15 mm or so, depending on the liquid. However, final sizes of core blow defects vary in size typically from 10 to 100 mm in diameter, showing that they are usually the accumulation of many bubbles, or possibly even the result of a constant stream of gas funnelled along a single bubble trail, arriving and coalescing under the solidifying skins of castings.

Blows from cores surrounded by light alloys are relatively gentle; leading to the generation of one or usually a succession of large bubbles. The minimum diameter of bubble that can form on a core is perhaps 10–20 mm (Figure 7.35). If the core contains gas at sufficient pressure to overcome the hydrostatic pressure and cause a blow defect, then it will usually have enough volume to continue to blow for some time. In this way, strings of bubbles from the core can accumulate to create a massive single gas cavity near the top of the casting. The diameter of this cavity can easily reach 100 mm. The whole top half of a casting can be made hollow by this process. (It suggests an interesting manufacturing technique for hollow ware!)

For cores surrounded by liquid cast iron, or even more especially, steel, the effects are explosive. The very much higher temperature ensures that the core outgassing occurs with much greater energy and much earlier, before much if any freezing of the metal has taken place. Instead of bubbles issuing from the pressurised source, gas can escape as high velocity jets. Metal can therefore be violently ejected from the open tops of moulds and the resulting casting is usually found to contain a mass of confusing defects; the quality of the casting can be effectively destroyed. There are usually no remains of the single large void or heavily oxidised bubble trail as is found for the gentler reactions in the light alloys.

In the light alloys, the core will usually take time to absorb heat, and build up its pressure before blowing; during this time, the top of the casting will usually have become partly frozen. Thus because of its late arrival, a core blow in these metals is often some distance beneath the top surface, with its upper surface faithfully following the contours of the top of the casting. (Note that this behaviour contrasts sharply with (1) 'real' gas porosity that starts only 1 or 2 mm sub-surface and contrasts with (2) turbulence-entrained gases that arrive early, together with entry of the liquid metal into the mould, and which are prevented from escape only by the thickness of the surface oxide film.)

In the event of large cavities in the top of the casting, an unvented core should always be viewed with suspicion. All cores benefit from being internally vented to reduce the build-up of pressure that might lead to blows. It is also important to be aware of blows from sources other than cores and moulds. The practical and quantitative aspects of this subject are addressed in Section 10.5.

Micro-blows

It is conceivable that small pockets of volatiles on the surface of moulds might cause a small localised explosive release of vapour that would cause gas to be forced through the (oxidised) liquid surface of the casting to form an internal bubble. Poorly mixed sand containing pockets of pure resin binder might perform this. Alternatively, sand particles with small cavities filled with a volatile binder material might act as cannons to blast pockets of vapour into the metal. On arrival of the hot metal, the small pocket of binder would be expected to heat to a temperature probably well above its boiling point. Thus after a few seconds the superheated binder would explosively vapourise, like a cannon, firing its vapour jet into the casting. There seems strong evidence that the synthetic spherical particle aggregates made from powdered refractories are in this category. Close examination of these under the scanning electron microscope might reveal evidence of a neck connecting to the external atmosphere.

Other evidence for micro-blows comes from the strip casting of steel between twin rolls (Ha et al., 2008). 'Dent' defects are observed that are 0.3–1.5 mm in diameter and up to 0.5 mm deep. Sometimes taking the form of a simple depression, but sometimes a characteristic form of a blow defect, connected to the surface with a narrow neck linked to an internally spherical shape. These miniature gas pockets were found to be the result of the roughness of the roll surface, so that gases entrapped in microscopic depressions in the roll surface expanded on arrival of the melt, causing a gas bubble to be driven into the liquid.

Figure 7.35 indicates that for the blowing of bubbles of less than 1 mm diameter in some common liquid metals, the source of the high-pressure gas needs to be only a few micrometres in diameter. The pressures inside such minute bubbles are easily shown to be high, of the order of 1 MPa (10 atm).

For sand castings, there is no direct evidence for this mechanical mechanism of sub-surface pore formation. The mechanism for the creation of sub-surface porosity remains obscure. It awaits creative and discriminating experiments to separate the problem from classical processes such as diffusion, in which the gases travel from the mould, across the mould/metal interface into the casting as separate atoms, albeit in swarms, the interface being mechanically undisturbed by this inward flow of elements. Such diffusive transfer is assumed in most of the remainder of this book. However, the reader should be aware of the possibility of mechanical transfer of gases, possibly as eruptive events, disrupting the casting interface as minute explosions.

An observation of a surface interaction that is worth reporting, but whose cause is not yet known, is described below. It may therefore have been allocated to the wrong section of the book. It is hoped that future research will provide an answer.

In the study of interactions between chemically bonded sand moulds and liquid metals, a novel and simple test was devised in the author's laboratory (Mazed and Campbell, 1992). It was prosaically named the 'wait and see' test. It consisted simply of a flat horizontal surface of a bonded aggregate onto which a small quantity of metal was poured to give a puddle of 50–100 mm in diameter. The investigator then waited to see what happened. If anything was going to happen it was not necessary to wait long.

When Al-7Si alloy was poured on to the flat surface of either a phenolic-urethane (PF) or furane/sulphonic acid (FS) bonded silica sand the puddle of liquid alloy maintained its mirror smooth upper surface. The same was true when Al-7Si-0.4Mg alloy was poured on PF-bonded sand. When the puddles of liquid metal had finally solidified, they had retained their top mirror-like surfaces and after sectioning were found to be sound.

However, when Al-7Si-0.4Mg alloy was poured on the FS-bonded sand, the effect was startlingly different. After several seconds, a few bubbles not exceeding 1 mm in diameter were observed to arrive at the surface, underneath the surface oxide film, raising up minute bumps on the mirror surface. After a few more seconds, the number increased, finally becoming a storm of arriving and bursting bubbles, destroying the mirror. The final sessile drop was found to be quite porous throughout its depth, corroborating the expectation that the bubbles had entered the drop at its base because of a reaction with the sand binder.

That this reaction was only observed with certain binders and with certain alloys originally suggested to the author a chemical mechanism. At the time of the work, this was thought unlikely because the diffusive transfer of gas into the melt would result in gas in solution. Having an equilibrium pressure of only about 1 atm (because the pressure at the mould surface was originally thought to be limited to this level), it would have been difficult to nucleate small bubbles. However, it has been more recently realised that the presence of bifilms in the alloy practically eliminates the nucleation problem, and so might make the process feasible. This will be especially true because some melts will, partly by accident and partly as a result of alloy susceptibility, have different quantities and different qualities of bifilms. The 0.4Mg-containing alloy would have been expected to contain a different concentration and a different type of bifilm compared to the Mg-free Al-Si alloy.

Even so, a mechanical process cannot be ruled out at this stage. For instance the presence of Mg in the Al-7Si alloy may cause MgO formation at the metal/mould interface leading to a convoluted and microscopically fractured surface oxide film that may be more easily penetrated by high-pressure gas. High local pressures might arise particularly if the FS binder was not as well mixed as the PF binder, leaving minute pockets of pure binder in the sand mixture of the plate core. Thus the quality of mixing may be important.

Clearly, there is no shortage of important factors to be researched. Probably it will always be so. In the meantime, the best we can do is be aware of the possible effects and possible dangers, and have patience to live with uncertainty until the truth is finally uncovered.

7.2.3 GAS POROSITY INITIATED IN SITU

General

I hesitate to call this 'true gas porosity', but that would convey the correct sense. It is porosity that arises as a result of gas, in solution in the liquid metal, which precipitates out during solidification.

It is necessary to re-emphasise that the formation of pores is not necessarily easy. The presumption that pores form immediately on solidification or because the solubility limit has been reached is, unfortunately, a widespread delusion. The gas in solution in the liquid, rejected ahead of the advancing solidification front, can reach high levels of supersaturation without the formation of a single pore if no convenient nucleation sites are present. The gas would simply remain in supersaturated solution. For aluminium alloys, the presence of supersaturated hydrogen in the solid does not appear to be harmful. For hydrogen in ferritic steels, however, the hydrogen may lead to the serious problem of hydrogen embrittlement. (Although even this does not seem to be certain. Even hydrogen embrittlement in steels may be a bifilm problem. If it is, this would explain why it has proven so difficult to understand over the many years that the phenomenon has been the subject of intensive research.)

In this section, we shall assume that there are plenty of nuclei on which gases can precipitate during freezing. The only viable nuclei will normally be bifilms because they offer such an easy precipitation site and minimal (if somewhat variable) resistance to unfurling.

In metals that solidify dendritically, the precipitated gas appears in the interdendritic spaces, and so is a well-dispersed, fine precipitate. This is its usual form. For castings of a few kilograms in weight the pores are usually in the range of 0.01–0.5 mm in diameter, and uniformly distributed throughout the whole casting. They can be up to a millimetre or so in diameter for larger castings. In general, however, these pores are so small that they are often invisible on a machined surface of an Al alloy because a cutting tool with a carbide tip will smear the metal over, concealing such microscopic defects. On the other hand, a diamond-tipped tool will cut cleanly, and thus render the defect visible, perhaps with the aid of a magnifying glass.

Al alloy automotive pistons in the United Kingdom used to be machined with carbide-tipped tools. On the introduction of polycrystalline diamond–tipped tools in the 1970s, the piston foundries were brought to an alarming stop. They were suddenly confronted with the appearance of porosity in their castings as a result of the cleaner cut of the sharper diamond tools. In reality, of course, the pistons were unchanged. However, no customer was prepared to buy pistons that were clearly peppered with holes. The industry was forced to make a traumatic leap in the quality of its liquid metal, and lived to fight another day.

It has to be kept in mind that the source of gas in solution can be that in the metal from the melting conditions. For a sand casting, additional gas in solution will have been acquired during its journey through the runner, and from continuing reaction with the mould during and after mould filling. All of these sources cooperate, and may raise the gas content of the metal over the threshold at which pores will initiate and grow.

We shall see how it is to be expected that gas pores will form as the consequence of normal segregation ahead of the solidification front, and the normal processes of nucleation and growth of pores from gases in solution in the metal. A supply of gas diffusing from the mould or core into the casting surface will enhance the effect. Also, assuming the presence of bifilms completes a convincing model as we shall see.

We have seen in Section 5.3.1 how the early growth of the freezing front from the mould wall is planar because of the high temperature gradient. Thus the 'snow plough' build-up of solute occurs, starting from the original solute content C_0, and increasing over the first millimetre or so of travel of the front, until the solute reaches a level at which a pore may nucleate (Figure 5.39). Thus the gas content concentrated in this way is too low in the first 1 mm or so to create significant porosity, whereas across all the central parts of the casting pores are possible. A typical form of this porosity is shown in Figure 7.41. It should perhaps be better named surface-layer-free porosity because, if the gas content of the melt is sufficiently high, the porosity in such cases is rarely simply sub-surface, but is distributed everywhere *except* in the 1 mm surface layer as Figure 7.41 illustrates. (The 1 mm beneath the surface contrasts with entrained air bubbles which are only a couple of oxide film thicknesses away from the surface.)

This microsegregation process is common in many alloy systems where the solute has a low partition coefficient k, resulting in a high concentration of solute ahead of the advancing front. (Recall from Section 5.3.1 that the maximum concentration of solute is C_0/k.) For these systems, sub-surface porosity is the standard form of gas porosity. It may or may not be the result of the release of gas from a metal-mould reaction that can subsequently diffuse into the casting. Such a source will certainly increase the gas available, and may in fact be the sole source of gas. However, it must not be forgotten that sub-surface porosity is the normal appearance of gas porosity whether metal-mould reactions contribute or not.

For instance, Klemp (1989) describes sub-surface porosity in a low-alloy steel cast into an investment mould. Such moulds are fired at high temperature (commonly about 1000°C) and can be safely assumed to be dry and free from outgassing materials. In this case, the sub-surface porosity is definitely *not* the result of surface reaction; the gases were already in solution in the cast metal at the moment of pouring.

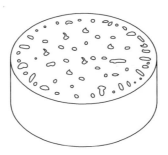

FIGURE 7.41

Sub-surface porosity revealed on a cut section of a bar casting.

In contrast, but not in conflict, Turkdogan (1986) reports subsurface porosity in cast irons cast in greensand, but had no reports of any such defects in irons cast into iron moulds. In this case, the porosity in the greensand moulds *is* the result of surface reaction; little gas was in solution in the cast metal.

We have discussed the action of bifilms to initiate porosity in liquid metals in Sections 7.1.5.3 and 7.1.5.4. The interesting feature of the mechanical model for the opening of bifilms in relation to the growth of pores is illustrated in Eqn (7.11). If the gas in solution in the liquid is assumed to be hydrogen, and assuming it to be approximately in equilibrium with the entrapped gas in the bifilm, the internal pressure will be proportional to $[H]^2$ (other diatomic gases will act similarly of course, although their approach to equilibrium may be slower because of their slower rates of diffusion to the bifilm). The rate of unfurling is therefore especially sensitive to the amount of gas in solution in the alloy.

In the case of iron and steel where an important contributor to the internal pressure will be expected to be carbon monoxide, CO, the internal pressure will approach that dictated by the product of the activities of carbon and oxygen in the melt, approximately $[C].[O]$. In addition, of course, nitrogen and hydrogen will also contribute to the total pressure.

In one of the most exciting pieces of research ever published in this field, Tiberg (1960) describes the growth of CO bubbles in liquid steel whilst actually observing the inside surface of the growing bubbles. He achieves this miracle by using high-speed cine-film to record the nucleation and growth of bubbles on the inside wall of a fused silica tube that contained the steel. (One day, in the 1970s, I was introduced in passing to a pleasant young Scandinavian. His name turned out to be Tiberg. I enquired whether it his work on observing bubble growth inside liquid steel because this was the most exciting research I had ever read. He affirmed it was him, agreeing that it was the most exciting research he had ever carried out. We shook hands in on such a warm and unexpected discovery of mutual interests. We had only time to bid each other farewell. We have never met since.)

The classical theories of pore growth assume that the geometry of the pore and its collection volume are spherical, and that growth is steady. This seems to be far from true in the experiment in which Tiberg tested these assumptions.

At high rates of growth, he found that the speed of expansion of the bubble surface *dr/dt* was indeed constant from the time the bubble was first observed at a size of 30 µm. However, after the addition of the deoxidisers, aluminium or silicon, the growth rate was slower and varied considerably from one bubble to another. In some bubbles growth suddenly halted and then continued at a slower rate. In fast-growing bubbles a small bright spot was observed.

The observation of the bright spot is interesting. It is most likely to have been an inclusion of alumina or silica (perhaps actually in the form of a bifilm) because this behaviour was only observed after the addition of the corresponding deoxidiser. The transparency or translucency (or even its hollowness if in the form of a partially opened bifilm) of the inclusion would have allowed the interior of the inclusion to be visible, giving an observer a view into the interior of the melt. This would appear as a bright enclosure, the classical 'black body cavity' of the physicist, radiating a full spectrum corresponding to the temperature of the interior of the steel, and therefore appearing as a bright spot. (The remainder of the bubble surface radiating its heat away to the outside world via the transparent silica vessel, and partially reflecting the cooler outside environment from its surface, would therefore appear cooler). We may speculate that the enhanced rate of transfer of gas into the bubble may have resulted from either (1) the short-circuiting of a surface layer that was hindering the transfer of gas into the bubble or (2) the attached inclusion having a large surface area and a high rate of diffusion for gas. Its surface area would then act as a collecting zone, funnelling the gas into the growing bubble through the small window of contact. A bifilm would have been expected to be especially effective in this way.

Tiberg's extraordinary observations link with work by the Japanese researcher, Kita (1979), who studied the insides of pores in steels by scanning electron microscope. He found that the inner surfaces of pores had curious markings, with radial 'petals', like flowers, suggesting that this may have been the site of gas leaking into the pores via some special route. The details of the markings were different for pores created by hydrogen, nitrogen and carbon monoxide. In addition, some markings were clearly oxide inclusions.

It seems therefore that the growth of pores in steels, and perhaps in most liquid metals, is complicated, and far removed from the ideal spherical model assumed in all the theoretical approaches to the problem. This subject would greatly benefit from a more thorough study with modern research equipment.

In the absence of more detailed information, we shall continue to assume that the growth of gas pores in liquid metals is controlled mainly by the rate of diffusion of gases through the liquid metal. There are many data in support of this, especially in simple systems such as the Al-H system. In general metal systems, the rate of growth is probably dominated by the rate of arrival of the fastest diffusing gas. From Figure 1.4(c), it is clear that in liquid iron, hydrogen has a diffusion coefficient approximately 10 times higher than that of any other element in solution. Thus the average diffusion distance d is approximately $(Dt)^{1/2}$ so that in comparison with other diffusing species, the radius over which hydrogen can diffuse into the bubble is $(10/1)^{1/2} = 3$ times greater. Thus the spherical volume over which hydrogen can be collected by the bubble, in comparison with other diffusing species, is therefore $3^3 = 30$ times greater. Thus it is clear that hydrogen has a dominant influence over the *growth* of the bubble.

It should be remembered that hydrogen makes a comparatively small contribution to the *nucleation* of the bubble because it concentrates relatively little ahead of the advancing freezing front, in comparison with the combined effects of oxygen and carbon to form CO in liquid iron and steel. The situation is closely paralleled in liquid copper alloys, where oxygen controls the nucleation of pores because of the snow plough mechanism, whereas hydrogen contributes disproportionately to growth because of its greater rate of diffusion.

This clarification of the different roles of oxygen and hydrogen in copper and steel explains much early confusion in the literature concerning which of these two gases was responsible for subsurface pores. Zuithoff (1964, 1965) published the first evidence that confirmed the present hypothesis for steels. He succeeded in showing that aluminium deoxidation would control the appearance of pores. Clearly, if the oxygen was high, then pores could nucleate, but they would not necessarily grow unless sufficient hydrogen was present. Conversely, if hydrogen was high, pores might not form at all if no oxygen was present to facilitate nucleation. The hydrogen would therefore simply remain in solution in the casting. The same arguments apply, of course, to the roles of hydrogen and oxygen in copper-based alloys.

A useful simple test for steels that deserves wider use is proposed by Denisov and Manakin (1965): a sample test piece was developed 110 mm high, and 30×15 mm at the top, tapering down to 25×12 mm at the base. A metal pattern of the sample quickly creates the shaped cavity in the sand, into which the metal is poured. Immediately after casting, the sample is knocked out and quenched in water. It is then broken into three pieces in a special tup. The entire process takes 1–2 min. It was found that the tapered test piece gave an accurate prediction of the risk of subsurface porosity; if such problems were seen in the sample, they were seen in the castings and vice versa. The test therefore warned of danger, and avoiding action could be taken, such as the addition of extra deoxidiser to the ladle. This test for steel castings cast in greensand moulds should be applicable to other alloy and sand systems prone to this problem. Although the simplicity and apparent usefulness of this test is commendable, perhaps the 'wait and see' test by the author, described in Section 10.5.4 might be even simpler and quicker. Quick, reliable tests for use on the shop floor are very much needed. The reader is recommended to try these techniques.

In some alloy systems, the rate of growth of pores is not expected to be simply dependent on the rate of diffusion in the matrix. The rate can also be limited by a surface film.

Ultimately however, the maximum amount of gas porosity in a casting depends partly on simple mechanics, as illustrated by the well-known **general gas law**. The use of this law assumes that the gas in the pore behaves as a perfect gas, which is an excellent approximation for our purposes. We shall also assume that all the gas precipitates (which is a less good approximation of course).

$$PV = nRT \tag{7.12}$$

where n is the amount of gas in gram-moles (in much use of this equation, n is somewhat misleadingly assumed to be unity), R is the gas constant $8.314 \ JK^{-1}mol^{-1}$, and P is the applied pressure in $N/m^2 = $ Pa.

The equation can be restated to give the volume V explicitly as:

$$V = nRT/P \tag{7.13}$$

It follows as a piece of rather obvious logic that the volume of the porosity is directly proportional to n, the amount of gas present in solution. This is graphically shown in Figure 7.42. The effect appears to have been well understood by Percy Riley, a well-known and irrepressible character in the early days of the development of the Al casting industry in the United Kingdom. He was an ex-Foseco employee working as a representative for Frankel Alloys Limited. He is said to have travelled around the country with a trunkload of potatoes in his car. All casting problems at that time were solved

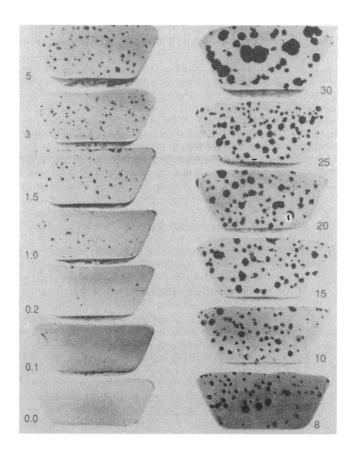

FIGURE 7.42

Gas porosity at various percentage levels in sectioned samples from the reduced pressure test.

Courtesy Stahl Speciality Co. (1990).

by adding potatoes to the melt. Clearly, the generous supply of hydrogen expanded bifilms giving copious hydrogen porosity, but shrinkage problems were solved. Percy's approach is not necessarily recommended at this time.

The relation of hydrogen porosity to hydrogen content is revealed in Figure 7.42. This test shows cut sections of a small sample of melt cast into a metal cup about the size of an egg cup and then solidified in vacuum. The test is sometimes known as the reduced pressure test (RPT) or the *Straube-Pfeiffer test*. The solidification under reduced pressure expands the pores, making the test more sensitive and easier to use than the old foundry trick of pouring a small pancake of liquid onto a metal plate, and watching closely for the evolution of tiny bubbles (you need good eyes for this).

The general gas equation also shows that the volume of a gas is inversely proportional to the pressure applied to it. For instance, in the RPT, to determine the amount of hydrogen in a liquid aluminium alloy, the percentage porosity is commonly expanded by a factor of 10 by freezing at 0.1 atm (76 mmHg) residual pressure rather than at normal atmospheric pressure (760 mmHg).

This sensitivity to pressure needs to be kept in mind when using the test. For instance, if the vacuum pump is overhauled and starts to apply not 76 mmHg (0.1 atm) but only 38 mmHg (0.05 atm) as a residual pressure, then the porosity in the test samples will be doubled, although, of course, the gas content of the liquid metal will be unchanged. Rooy and Fischer (1968) recommend that for the most sensitive tests the applied pressure should be reduced to

2–5 mmHg (approximately 0.003–0.006 atm). Clearly this will yield about a further 10-fold increase in porosity in the sample for any given gas content. However, care needs to be taken because these simple numerical factors are reduced by the additional loss of hydrogen from the surface of the test sample during the extra time taken to pump down to these especially low pressures, plus the losses of bubbles by bursting at the surface of the melt. In general, my experience is that a residual pressure of 76 mmHg (0.1 atm) is about right for all practical purposes, and most RPT kits operate at this value.

As has been mentioned before in the case of vacuum casting, the effect of pressure on pore growth is an excellent reason to melt and pour under vacuum, but to solidify under atmospheric pressure. It makes no sense to solidify under vacuum from the point of view of controlling porosity because pore expansion will act to negate the benefits of lower gas content (although solidification under vacuum may be important to prevent attack of the surface of the casting by nitrogen and oxygen if air is admitted to the vacuum chamber at an early stage). In terms of the general gas law, the pore volume V will be decreased by lower n, but increased correspondingly by low P. Whether the effects will exactly cancel will depend, among other things, on whether the melt has had time to equilibrate with the applied vacuum so as to reduce its gas content n. Taylor (1960) gives a further reason for not freezing under vacuum: For a nickel-based alloy containing 6% aluminium, the vapour pressure of aluminium at 1230°C is sufficient to form vapour bubbles at the working pressure of the vacuum chamber. He correctly concludes that the only remedy is to increase the pressure in the chamber immediately after casting. During the melting of TiAl intermetallic alloys at temperatures close to 1600°C, the evaporation of Al causes a loss of Al from the alloy, and a messy build-up of deposits in the vacuum chamber. Melting under an atmosphere of argon greatly reduces these problems. However, pouring under argon cannot be recommended if the pouring is turbulent, because of the danger of the entrainment of argon bubbles (another reason for the adoption of counter-gravity filling of moulds).

If the rate of diffusion of the gas in the casting is slow, the volume of the final pore will be less than that indicated by the general gas law and be controlled by the time available for gas to diffuse into the pore. In Figure 7.5, the benefits of increasing feeder size are seen to be enjoyed up to a critical size. After that, any further increase in the feeder merely delays solidification of the casting so that gas porosity increases.

Although these general laws for the volume of a gas-filled cavity are well known and nicely applied in various models of pore growth (see for instance the elegant work by Kubo and Pehlke (1985), Poirier (1987) and Atwood and Lee (2000)), some researches have shown that the detailed mechanism of the growth of pores can be very different in some cases.

A direct observation of pore growth has been carried out for air bubbles in ice. At a growth rate of 40 μms^{-1}, Carte (1960) found that the concentration of gas built up to form a concentrated layer approximately 0.1 mm thick. He deduced this from observing the impingement of freezing fronts. When the bubbles nucleated in this layer, their subsequent rapid growth so much depleted the solution in the vicinity of the front that growth stopped and clear ice followed. The concentration of gas built up again and the pattern was repeated, forming alternate layers of opaque and clear ice.

On examination of the front under the microscope, Carte saw that the bubbles seemed to originate behind the front; the first 0.1 mm deep layer of solid appeared to be in constant activity; threads of air approximately 10 μm in diameter spurted along what seemed to be water-filled channels, and were squeezed out of the ice. Sometimes bubbles arrived in quick succession, the first being pushed away and floating to the surface. Those bubbles that remained attached to the front would then expand, but finally be overtaken and frozen into the solid. It seems that pore growth might involve more turmoil than we first thought! Much of this activity arises, of course, from the expansion of the ice on freezing, and so forcing liquid back out of interdendritic channels. The opposite motion will occur in most metallic alloys as a result of the contraction on freezing. Also, it is to be expected that the movement in metals will be somewhat less frenetic. Nevertheless, no matter how the pore might grow in detail, we can reach some conclusions about the final limits to its growth.

Poirier (1987) uses the fact that the pore deep in a dendrite mesh will grow until it impinges on the surrounding dendrites. The radius of curvature of the pore is therefore defined by the remaining space between the dendrite arms. However, of course, although the smallest radius defining the internal pressure is now limited, the pore can continue to grow, forcing its way between the dendrite arms. Again, as has been mentioned before, an interdendritic morphology should not be taken as the definition of a shrinkage pore. Whether grown by gas or shrinkage, its morphology as spherical or interdendritic is merely an indication of the timing of its growth relative to the timing of the growth of the dendrites. Thus round pores have grown early. Interdendritic pores have grown late. Either can have been driven by gas or shrinkage, or both.

Fang and Granger (1989) found that hydrogen porosity in Al-7Si-0.3Mg alloy was reduced in size and volume percentage, and was more uniformly distributed, when the alloy was grain refined. In this case, the growth of bubbles will be limited by their impingement on grains. It may be that a certain amount of mass feeding may also occur, compressing the mass of grains and pores. In later work Poirier et al. (2001) confirm by a theoretical model that finer grains do reduce porosity to some degree. If grain refinement were made in the ladle, inadvertently cleaning the melt, the reduced bifilm content would probably have resulted in fewer but larger pores. This was not observed and so, with perhaps some surprise, can be concluded to be unimportant in this work.

Another limitation to growth occurs when the bubble can escape from the freezing front. This will normally happen when the front is relatively planar, and is typified by the example of the rimming steel ingot. In general, however, escape from a mesh of dendrites is likely to be rare.

A final limitation to growth is seen in those cases where the porosity reaches such levels that it cannot be contained by the casting. This occurs at around 20–30% porosity. During the freezing of the sample in the reduced pressure test, gas can be seen to escape by the bursting of bubbles at the surface of the sample. The effect is suggested in Figure 7.42. Also, Figure 7.43 shows that as gas in increased, measurements of porosity above 20% in such samples show a lower rate of increase of porosity than would be expected because of this loss from bubbles bursting at the surface, releasing their gas to the environment. (The theoretical curve in Figure 7.43 is based on 1% porosity in the solid being equal to 10 mL hydrogen in 1000 mL aluminium; this is equivalent to 10 mL hydrogen in 2.76 kg aluminium, or 0.362 mL hydrogen in 1 kg aluminium. Incidentally, Figure 7.43 also shows the enormous reduction in porosity as a result of increasing the difficulty of nucleation, because of the removal of nuclei by filtration. Clean metal makes better castings.)

It is worth drawing attention to the considerable volume of work over many years in which people have drilled into steel castings immersed in mercury or oil and have collected and analysed the gases in pores. In almost every case, the dominant gas was hydrogen. This led early workers to conclude that the bubbles were caused by hydrogen (see, for instance, the review of early work by Hultgren and Phragmen in 1939). This was despite calculations by Muller in 1879

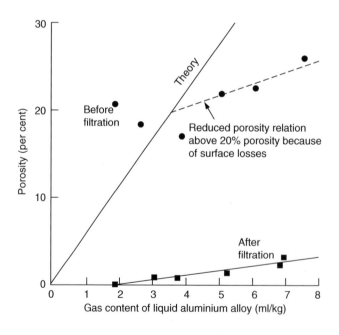

FIGURE 7.43

Porosity of RPT samples frozen at 0.005 atm as a function of gas content.

Data from Rooy and Fischer (1968).

that the CO pressure in pores in steel castings was up to 40 atm and the consequent correct deduction by Ledebur in 1882 that the hydrogen content of pores in steel castings at room temperature was the result of the continued accumulation of hydrogen after solidification was complete.

We can see that the high final hydrogen content of pores at room temperature is a natural consequence of the high rate of diffusion of hydrogen in both the liquid and solid states. Thus hydrogen is the dominant gas contributing to the growth of pores, and continuing to contribute additional gas during the cooling to room temperature.

Even after the casting reaches room temperature the growth of gas pores may still not be complete! Talbot and Granger (1962) showed that hydrogen in cast aluminium could continue to diffuse into pores in the solid state during a heat treatment at 550°C. Porosity was found to increase during this treatment, with pores becoming larger and fewer. With very long annealing in vacuum, the hydrogen could be removed from the sample and the porosity could be observed to fall or even disappear.

We have so far in this section discussed mainly porosity which can be uniformly distributed throughout the section of a casting, even though subsurface porosity, or perhaps more accurately, if more clumsily termed, 'subsurface porosity-free zone', is a product of the accumulation of segregated gas ahead of an advancing planar solidification front. It is a condition partway between the completely uniform distribution of gas pores as discussed earlier, and the highly directional form that is to be discussed later.

In metals that solidify on a planar front the gas pores may grow, keeping pace with the growth of the front, causing long, tunnel-like defects that are sometimes called wormholes (Figure 7.44). They are often observed in ice cubes because tap water generally freezes on a fairly planar front. The pores may be up to 1 mm or so in diameter and of various lengths of up to 100 mm or more. An example of pores nearly 20 mm long is seen in Figure 7.45. This radiograph of an aluminium-bronze plate cast in a sand mould using non-degassed metal contrasts with a similar radiograph of a plate (not shown) cast from properly degassed melt, in which no defects at all could be seen.

Beech (1974) was perhaps the first to point out that the environment of a long bubble is not necessarily homogeneous. In other words, what may be happening at one end of the bubble may be quite different from what is happening at the other. This is almost certainly the case with the kind of subsurface porosity that continues to grow into an array of wormholes. The metal-mould reaction will continue to feed the base of the bubble, which of course remains within the diffusion distance of the surface. If the gas content of the melt is high the front of the bubble may also be gaining gas from the melt. However, if the melt has a low gas content, the front part of the bubble may lose gas. The bubble effectively acts as a diffusion short-circuit for the transfer of gas from the surface reaction to the centre of the casting. The effect is shown diagrammatically in Figure 7.44.

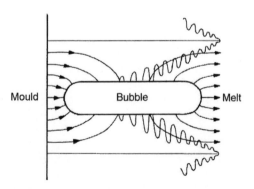

FIGURE 7.44

Subsurface pore growing in competition with dendrites, into a melt of low gas content. The pore gains gas by diffusion form the surface reaction, and loses it from its growing front.

After Beech (1974).

FIGURE 7.45

Radiograph of a Cu-10Al casting $200 \times 100 \times 10$ mm with a high hydrogen content poured at 1285°C with gate velocity 0.85 ms^{-1} into a sand mould (Helvaee and Campbell, 1997).

Depending on the relative rates of gain and loss, the pore may grow or even give off bubbles from the growing front. Alternatively, it may stop growing and thus be overtaken by the dendrites and frozen into the thickening solid shell. The swellings and narrowings of these long bubbles probably reflect the variation in growth conditions, such as sudden variations in gap between the casting and the mould, or variations from time to time of convection currents within the casting.

The long CO bubbles in rimming steel ingots, cast into cast iron moulds, were a clear case where the growing bubble was fed with gas from the growing front, and not from the mould surface.

Conversely, in aluminium bronze castings made in greensand moulds Matsubara et al. (1972) provides an elegant and clear demonstration of the feeding of the pore with gas from the mould, together with the combined effect of residual gas in the metal. Radiographs show wormhole porosity approximately 100 mm long. The amount of porosity was shown to increase as the gas content of the metal was increased, and as the water content of the moulds was increased from 0 to 8%. Halvaee and Campbell (1997) illustrates a similar structure for aluminium bronze cast into a sand mould bonded with a phenolic urethane resin (Figure 7.45). A proper melting procedure to give a low gas content in the melt before casting eliminated the porosity completely.

In this section on gas porosity, we have so far discussed gases in general. In many cases, it is not easy to make a clear separation in this account of the various gases because all cooperate in some alloy systems and many systems are analogous. However, a rough division particular to each gas will now be attempted, the reader being requested to bear in mind the necessarily ragged edges between sections.

Hydrogen porosity

It is important to remember that both water and hydrocarbons (that are available in abundance from the sand binders employed in most sand castings) can decompose at the metal surface, both releasing hydrogen. The surface of the melt and the casting will therefore have no shortage of hydrogen; in fact, from Section 4.4, it is seen that in general the mould atmosphere often contains up to 50% hydrogen, and may be practically 100% hydrogen in many cases.

What happens to this hydrogen?

Although much is clearly lost by convection to the general atmosphere in the mould, some will diffuse into the metal if not prevented by some kind of barrier (see later). If the hydrogen does manage to penetrate the surface of the casting, how far will it diffuse?

We can quickly estimate an average diffusion distance d from the useful approximate relation Eqn (1.5), $d = (Dt)^{1/2}$. Some researchers increase the right hand side of this relation by a factor of two in a token attempt to achieve a little more accuracy. We shall neglect such niceties, and treat this equation merely as an order of magnitude estimate. Taking the diffusion rate D of hydrogen as approximately 10^{-7} $m^2 s^{-1}$ for all three liquid metals, aluminium, copper and iron (see Figures 1.4 (a), (b) and (c)) then for a time of 10 s d works out to be approximately 1 mm. For a time of 10 min, d grows to approximately 10 mm.

Clearly, hydrogen from a surface reaction can diffuse sufficiently far in the time available during the solidification of an average casting to contribute to the formation and growth of subsurface porosity in most of our engineering metals.

The distance that the front has to travel before the solute peak reaches its maximum is actually identical to the figures we have just derived, as explained in Section 5.3.1. Thus conditions are exactly optimum for the creation of the maximum gas pressure in the melt at a point a millimetre or so under the surface of the casting. The high peak will favour conditions for the nucleation of pores whilst the closeness to the surface will favour the transport of additional gas, if present, from a surface reaction. Naturally, if there is enough gas already present in the melt, then contributions from any surface reaction will only add to the already existing porosity.

In an investigation of a wide variety of different binders for the moulding sand, Fischer (1988) finds that subsurface porosity in copper based castings is highly sensitive to the type of binder, although degassing and deoxidising of the metal did help to reduce the problem. These observations are all in line with our expectations of contributing agencies.

In aluminium alloys where hydrogen gas in solution in the melt segregates strongly on freezing, the partition coefficient is approximately 0.05, corresponding to a concentrating effect of 20 times. Figure 7.46 shows subsurface porosity in an Al-7Si-0.4Mg alloy solidified against a sand core bonded with a phenolic urethane resin. The general gas content of the casting is low, so that pores are only seen close to the surface, within reach by diffusion of the hydrogen from the breakdown of the resin binder. Some of the pores in Figure 7.46(a) are clearly crack-like, and seem likely therefore to have formed on bifilms.

Close-up views of another casting showing sub-surface porosity in the same alloy and same bonded sand (Figure 7.47(a)) confirms that the pores are of widely different form, some being perfectly round (Figure 7.47(b)), some dendritic (Figure 7.47(c)) and some of intermediate form (Figure 7.47(d)). It seems reasonable to assume that all the pores experienced the same environment, consisting of a uniform field of hydrogen diffusing into the melt from the degeneration of the core binder. Their growth conditions would therefore have been expected to be identical. Their very different forms cannot therefore be the result of growth effects (i.e. some are not shrinkage and others gas pores). The differences therefore must be a result of differences in ease of nucleation. The simple explanation is that the round pores nucleated early because of easy nucleation, and thus grew freely in the liquid. The dendrite-lined pores are assumed to have nucleated late, as a result of a greater difficulty to nucleate, so that they expanded when the matrix dendrites were already well advanced. The differences in ease of nucleation can be easily understood in terms of the randomly different conditions in which the nuclei, bifilms, are created. Some bifilms will come apart easily, whereas others will be ravelled tightly, or may be partially bonded as a result of being older or being contaminated with traces of liquid salts from the surface of the melt.

The case of some subsurface pores initiating their growth very late, when freezing must have been 80% or more complete, raises an interesting extrapolation. If the bifilms had been even more difficult to open, or if not even present at all, then *no* pores would have nucleated. This situation may explain the well-known industrial experience, in which subsurface porosity comes and goes, is present one day, but not the next, and is more typical of some foundries than others. It is a metal quality problem.

Note that both round and dendritic pores can be both gas pores. They could also both be shrinkage pores, or both generated by combined gas + shrinkage contributions. Whether grown by gas or shrinkage, or both, the shape difference merely happens because of the timing of the pore growth in relation to the dendrite growth. Although it common for gas pores to form early and so be rounded, and shrinkage pores to form late and so take on an interdendritic morphology, it is not necessary. It is very important not to fall into the standard trap of assuming that round pores result from gas and dendritic forms result from shrinkage.

The work by Anson and Gruzleski (1999) describes particularly careful work in an attempt to distinguish between gas and shrinkage pores. Their study concentrated on the appearance and spacing or pores. They pointed out that, on a

(a)

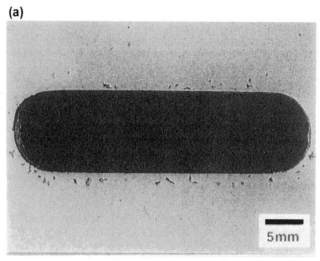

Macrosection (Aqua blasted)

(b)

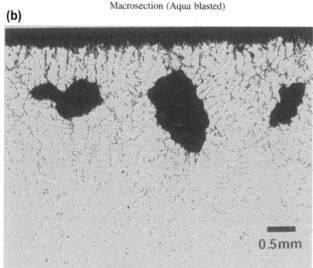

Macrosection (Polished)

FIGURE 7.46

(a) Polished section, lightly blasted with fine grit, showing subsurface porosity around a sand core in an Al-7Si-0.4Mg alloy casting of low overall gas content; and (b) an enlarged view of some pores on a polished section.

polished section, groups of apparently separate, small interdendritic pores were almost certainly a single pore of irregular shape (Figure 7.48). Despite apparently clear differences in shape and spacing, it is finally evident in their case that all pores were gas pores because they all grow at the same rate as hydrogen is increased. In this case the pores that were assumed to be shrinkage pores were almost certainly partially opened and/or late opened bifilms. Their irregular cuspoid outlines probably derived partly from the irregular, crumpled form of the bifilms, together with their late opening in the

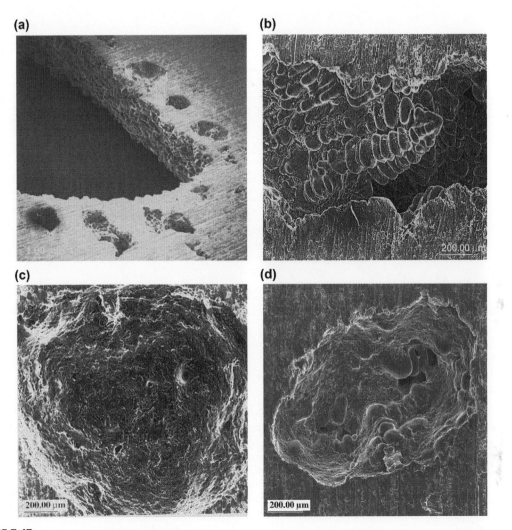

FIGURE 7.47

Thin slice of an Al-7Si-0.4Mg alloy casting taken from around a phenolic urethane bonded sand core. (a) A general view, showing the sand-cast surface made by the core and several subsurface pores; and close-ups of (b) a dendritic pore; (c) a spherical pore; and (d) a mixed pore.

interdendritic spaces. Such mis-identification of pore shapes is easily understood, and is common. We need to be on our guard against such mistakes.

Carbon, oxygen and nitrogen

For the case of the casting of aluminium alloys, we have only to concern ourselves about hydrogen, the only known gas in solution.

For the case of copper-based alloys, several additional gases complicate this simple picture. The rate of diffusion of oxygen in the liquid is not known, but is probably not less than 10^{-8} m^2s^{-1}. In the case of liquid iron-based alloys,

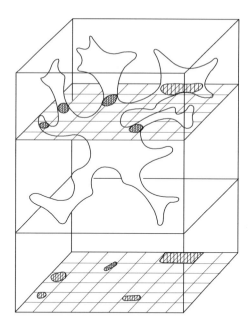

FIGURE 7.48

Complex interdendritic pore, appearing as a group of pores on a polished section.

After Anson and Gruzleski (1999).

oxygen and nitrogen, diffuse at similar rates (Figure 1.4). Thus for all of these liquids the average diffusion distance d is 1–2.5 mm for the time span of 1–10 min. It seems therefore that all these gases can enter and travel sufficiently far into castings of all of these alloy systems to contribute to the formation of porosity.

In copper-based alloys, the effect is widely seen and attributed to the so-called 'steam reaction' (Eqn 6.3). It seems certain, however, that SO_2 and CO will also contribute to the total pressure available for nucleation in copper alloys that contain the impurities sulphur and carbon. Carbon is an important impurity in Cu-Ni alloys such as the monels. Zinc vapour is also an important contributing gas in the many varieties of brasses and gunmetals.

From the point of view of nucleation, the action of oxygen is likely to be central. This is because it is probably the most strongly segregating of all these solutes (with the possible exception of sulphur). Thus deoxidation practice for copper-based alloys is critical.

When gunmetal has been deoxidised with phosphorus, Townsend (1984) reports that an optimum addition is required, as illustrated in Figure 7.49. Too little phosphorus allows too much oxygen to remain in solution in the melt to be concentrated to a level at which precipitation of water vapour will occur as freezing progresses. Too much phosphorus will reduce the internal oxygen to negligible levels, suppressing this source of porosity. However, the melt will then have enhanced reactivity with its environment, the excess phosphorus picking up oxygen and hydrogen from a reaction at the metal surface with water vapour from the mould. The porosity in the cast metal is the result of the sum of these internal and external reactions, resulting in a minimum at approximately 0.015% phosphorus for the case of this particular sample of alloy.

A similar reaction occurs in grey iron in the presence of 0.005–0.02% aluminium or 0.04% titanium. The reaction is characterised by subsurface pores that have a shiny internal surface covered with a continuous graphite film. (The

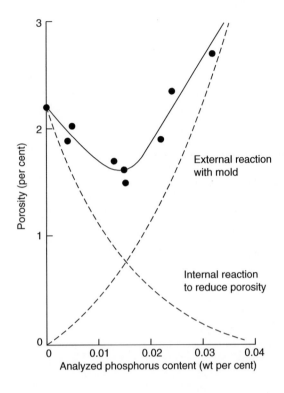

FIGURE 7.49

Effect of phosphorus on the porosity in 75 mm thick plates of leaded gunmetal LG2 cast at 1100°C.

Data from Townsend (1984).

graphite film is present simply because the free surface provided by the pore allows the graphite to accommodate its volume expansion on precipitation most easily. Similarly, these pores are often seen to be filled with a frozen droplet of iron, again simply because the pore is an available volume into which liquid can be exuded during the period when the graphite expansion is occurring. Such droplets would be expected to be more common in casting made in rigid moulds, where the expansion could not be easily accommodated by the expansion of the mould.)

Carter et al. (1979) describes the analogous problem caused by the presence of magnesium in **ductile iron**. Clearly, this double effect of the addition of a strong deoxidiser, resulting in an optimum concentration of the addition, is a general phenomenon.

In much of the work on *subsurface* pores in **irons and steels**, the phenomenon is called *surface* pinhole porosity. This is almost certainly the result of the loss of the surface of the casting by a combination of oxidation and/or severe grit blasting, allowing the pores to be broken into, or simply uncovered. The *surface pinholes* almost certainly originated as *subsurface pinholes*.

In the case of low-carbon equivalent grey irons, it is found that small surface pinholes occur that have an internal surface lined with iron oxide, and whose surrounding metal is decarburised, as witnessed by a reduction in the carbide content of the metal. Although Dawson et al. (1965) and others make out a case for these defects to be the result of a reaction with slag, it seems more reasonable to suppose that once again the pores were originally subsurface, but the high oxygen content of the metal promoted early nucleation, with the result that the pores were extremely close to the surface of the casting. The thin skin of metal quickly oxidised, opening the pore to the air at an early stage, and allowing plenty of time for oxidation and decarburisation whilst the casting was still at a high temperature. Tests to check whether the pores have been connected to the atmosphere do not appear to have ever been carried out.

Dawson reports that an addition of 0.02% aluminium usually eliminates the problem. This relatively high addition of aluminium is probably to be expected because the oxygen in solution in these low-carbon equivalent irons will be higher than that found in normal grey irons. However, if even higher levels of aluminium were added, the problem would be expected to return because of the increased rate of reaction with moisture in the mould, following the example in Figure 7.49.

Oxygen and carbon are important when CO is the major contributing gas, although, of course, in cast irons, where carbon is present in excess, the CO pressure is effectively controlled solely by the amount of oxygen present. This deduction is nicely confirmed for malleable cast irons by the Italian workers Molaroni and Pozzesi (1963), who found a strong correlation between their proposed '*oxidation index*', I, defined as:

$$I = C + 4Mn + 1.5Si - 0.42FeO - 5.3$$

where the symbols for the elements carbon C, manganese Mn, etc., represent the weight percentage of the alloying elements in the iron. Compositions of irons that gave a positive index were largely free from pores, whereas those with an increasingly negative index were, on average, more highly porous.

In steels, there are several gases that can be important in different circumstances. The most important are CO (Eqn 6.5), N_2 (Eqn 6.7) and H_2 (Eqn 1.3). Because, at the melting point of iron, hydrogen has a solubility in the liquid of approximately 245 mlkg^{-1} and in the solid of 69 mlkg^{-1} (extrapolating slightly from Brandes and Brook 1992), its partition coefficient is $69/245 = 0.28$, with the result that it is concentrated ahead of the solidification front by a factor of $1.0/0.28 = 3.55$. It therefore makes a modest contribution to the gas pressure for nucleation of pores in iron alloys.

Nitrogen seems to have a similar importance in nucleation. Its solubility at the melting point of iron is 0.0129 wt% in the solid and 0.044 wt% in the liquid (Brandes, 1983) giving a partition coefficient 0.29, and a concentration effect for nitrogen ahead of the freezing front of approximately 3.4 times. This situation has already been discussed in Section 4.6.3 on metal-mould reactions, and is discussed further in the following section.

Because $k = 0.05$ for oxygen in iron, and $k = 0.2$ for carbon in iron, the concentration factors are 20 and 5, respectively, so that when combined, the equilibrium CO pressure at the solidification front is $20 \times 5 = 100$ times higher than in the bulk melt (The distribution coefficients refer to bcc delta-iron; those for fcc gamma-iron would be nearer to unity, implying much less concentration ahead of the solidification front for solidification to austenite). Because of the multiplying factor 100, oxygen in solution in the iron is the major contributing gas in the nucleation of CO gas pores during the solidification of many irons and steels.

Nitrogen porosity

There has been a massive effort to understand the metal-mould reactions in which nitrogen is released. This gives problems in both iron and steel castings as subsurface pores. A review for steel castings is given by Middleton (1970).

Subsurface porosity is common for steels cast into moulds bonded with urea formaldehyde resin (Middleton, 1970), or bonded with other amines that release ammonia, NH_3, on heating. These include hexamine in Croning shell moulds (Middleton and Canwood, 1967). The ammonia breaks down at casting temperatures to release both nitrogen and hydrogen.

The nitrogen problem in ferrous castings has resulted in the production of a whole new class of sand binders known as 'low nitrogen' binders. However, later work (Graham et al., 1987) investigating the relation between total nitrogen content of the binder and the subsurface porosity and fissures in iron castings found no direct correlation. However, Graham did find a good correlation with the ammonia content of the binder. Ammonia is released during the pyrolysis of important components of many binders, such as urea, amines (including hexamine used in shell moulds) and ammonium salts. The ammonia in turn will decompose at high temperature as follows

$$NH_3 \Leftrightarrow N + 3H$$

thus nascent nitrogen and nascent hydrogen are released (the word *nascent* meaning *in the act of being born,* implying high reactivity of the atomic forms of these elements rather than their molecular forms). Both will contribute to the formation of pores in the metal. Both nitrogen and hydrogen will have a similar influence in the nucleation of a pore, concentrating reasonably strongly ahead of the freezing front. For the subsequent growth, however, hydrogen will be the major influence because of its much faster rate of diffusion. That both gases are released simultaneously by ammonia explains the extreme effectiveness of ammonia in creating porosity. Nitrogen alone would not have been particularly effective. Even if it may have been successful in nucleating pores, without the additional help from hydrogen any subsequent growth would have been limited. Hydrogen is supplied by gas in solution in the liquid metal, and any fresh supply through the surface from a surface reaction.

It seems that ammonia can build up in greensand systems as the clays and carbons absorb the decomposition products of cores. Lee (1987) confirms that an ammoniacal nitrogen test on the moulding sand was found to be a useful indicator of the pore-forming potential of the sand, even though the test was not a measure of total nitrogen.

The action of gases working in combination is illustrated by the work of Naro (1974) in his work on phenolic urethane-isocyanate binders. He showed that, from a range of irons, ductile iron (high carbon and low oxygen) was least susceptible and low-carbon equivalent irons (high oxygen) were most susceptible to porosity from the binder. Once again, it seems logical that the oxygen remaining in solution in the iron plays a key role encouraging nucleation, contributes modestly to both nucleation and growth, whereas hydrogen overwhelmingly dominates growth.

Barriers to diffusion

In some unusual conditions, hydrogen appears to be prevented, or at least inhibited, from diffusing into some metals.

For magnesium alloys, potassium borofluoride (KBF_4) has been known for many years to be an effective suppressant of metal/mould reactions for Mg alloys. In fact, if not added to the sand moulds of some Mg castings, both mould and casting will be consumed by fire—the ultimate metal/mould reaction! However, Al-5Mg and Al-10Mg casting alloys, and even Al-7Si-0.4Mg alloy, also benefit from KBF_4 or K_2TiF_6 additions to suppress reactions with the mould. We might speculate that liquid oxyfluorides, produced by the dissolution of the alumina film in the flux, assist to seal the surface of the liquid metal.

Naro (1974) confirms the widely reported fact that the addition of 0.25% iron oxide to phenolic urethane-isocyanate–bonded sands reduces subsurface pores in a wide range of cast irons. This is a curious fact and difficult to explain at this time. One suggestion is that the oxide creates a surface flux, possibly an iron silicate. This glassy liquid phase is likely to reduce the rate at which gases can diffuse into the casting.

The rate of uptake of nitrogen in stainless steel is inhibited by the presence of silicon in the steel that, at certain oxidation potentials, forms SiO_2 on the surface in preference to Cr_2O_3 (Kirner et al., 1988).

Even when the surface film consists only of a layer or so of adsorbed surface-active atoms, the presence of the layer reduces the rate at which gases can transfer across the surface. This happens, for instance, in the case of carbon steels: sulphur and other surface-active impurities hinder the rate at which nitrogen can be transferred. An excellent review of this phenomenon is given by Hua and Parlee (1982).

However, the precise mechanisms and effectiveness of many of these inhibition reactions are not clear at this time.

7.3 POROSITY DIAGNOSIS
7.3.1 GAS POROSITY

To be able to take effective action to cure a porosity problem, clearly it has to be correctly identified. Table 7.3 is designed to assist with diagnosis of the wide variety of types of gas pores. Some care is needed where gas and shrinkage effects overlap, but these situations can usually be identified with some certainty.

Table 7.3 Porosity Diagnosis Summary

	Intrinsic (Metallurgical Defects)		Extrinsic (Entrained Casting Defects)	
	Gas from Solution in Melt	**Gas from Mould Reaction**	**Entrained Air Bubbles**	**Core Blows**
Spatial distribution	1–2 mm subsurface clear from pores. Bulk of casting has a uniform distribution of pores	1–2 mm under surface	Clustered above ingates	Uniform depth under cope
Size (mm)	0.05–0.5	0.05–0.5	1–5	5–500
Features	Associated with bifilms in suspension in melt	Associated with bifilms in suspension in melt	Associated with oxides (solid oxides include bifilms + bubble trails) (liquid oxides include slags)	Connected by thick bubble trail to originating core
Examples	Figures 7.41, 7.43	Figure 7.41		Figures 7.35, 7.36, 10.10 and 10.11

7.3.2 SHRINKAGE POROSITY

The various forms of shrinkage cavity are summarised in Figure 7.50. Without exception, all the morphologies are dictated by (1) the *geometry* of the casting and (2) *gravity*. These two key features allow shrinkage-dominated porosity to be clearly differentiated from other sources of porosity. The various forms that shrinkage porosity can take include the following.

a. Centreline porosity is formed in a skin-freezing alloy that has suffered an inadequate supply of liquid from the feeder. The geometry dictates that the pore is closely parallel to the thermal axis of the casting.
b. Sponge porosity. Formed in a long freezing range alloy, with adequate temperature gradient, but inadequate feed liquid from the feeder (e.g. Figure 7.8).
c. Layer porosity; the result of inadequate interdendritic feeding in a poor temperature gradient. The nucleation of internal porosity indicates a poor cleanness of the liquid metal. Geometry dictates that the pores are closely at right angles to the thermal axis of the casting.
d. Surface-initiated porosity generated in a long freezing range alloy in conditions of poor temperature gradient, but adequate cleanness of the melt.
e. Surface sink (external shrinkage porosity) formed in conditions of no liquid available from the feeder, but good cleanness of melt, in a relatively skin-freezing alloy, resulting in good solid feeding. Gravity dictates that the sink is usually sited on the cope surface of the casting.

Notice that the shrinkage pores are always influenced by the geometry of the casting, whether sometimes parallel to the axis of the casting as in centreline shrinkage, or (perhaps seemingly perversely) sometimes at right angles to the axis of the casting, as in layer porosity. Fortunately, the other major influence to the distribution of shrinkage porosity, gravity, is kinder to our intuition, acting only downwards.

If the porosity is clearly *not* strongly influenced by the shape of the casting and by gravity (for instance, porosity in random locations or well away from the thermal axis of the casting as in Figure 7.31(d)), we can conclude it is *not* shrinkage porosity. Very often, in fact most often, it will be porosity associated with masses of bifilms and bubble damage. This is common variety of porosity that only 'appears' to be shrinkage porosity.

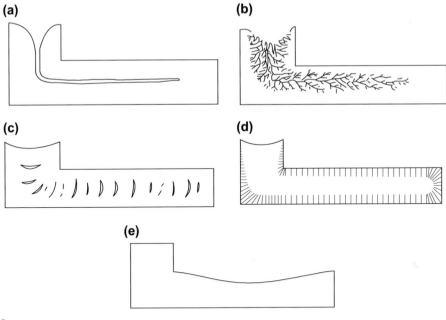

FIGURE 7.50

Summary of the various types of shrinkage porosity.

The casting engineer needs to keep at the forefront of the mind that in *most* cases the pores are *not* shrinkage pores. Nearly all porosity I see identified on micrographs and labelled 'shrinkage' are actually oxides. It is useful to repeat here the valuable mantra learned earlier, learning to say not 'it is shrinkage' but 'it *appears* to be shrinkage', bearing in mind that the pores are in nearly every case clusters of oxides together with entrained air, or bubble trails, introduced during the turbulence of the pour. These unwelcome features disappear if the filling system design is good.

Readers might find helpful the short summary of the different types of porosity and how they might be identified presented in Table 7.3.

CRACKS AND TEARS

8.1 HOT TEARING

8.1.1 GENERAL

A hot tear is one of the most serious defects that a casting can suffer. Although it has been widely researched, and is understood in a general way, it has remained a major problem in the foundries, particularly with certain alloys that are especially prone.

Over recent years, the work by Rappaz (1999–2009) has provided illuminating details of the microscopic and mesoscopic behaviour of metals leading to conditions in which hot tearing occurs. More will be said about this elegant work later in this chapter.

Even so, it has to be admitted that important insights into hot tearing behaviour have emerged not only from scientific experiments in the laboratory, but from experience on the shop floor of foundries. However, these important shop floor findings have not been published in the scientific journals.

Briefly, for those who wish to read no further, the key result from the shop floor is that hot tearing can usually be eliminated in castings by simply improving the filling system. The improved systems are described in Chapter 10 onwards.

In the experience of the author, every hot tear, wherever it appears in a casting, can be eliminated by addressing the problems of the filling system. Once oxide bifilms are eliminated from the melt, the solidifying alloy cannot tear, reinforcing my conclusion that there is no such thing as a solidification defect; there are only casting defects. When subjected to a tensile strain by contraction of the casting, material free from bifilms simply stretches because metals are normally extremely soft and ductile at temperatures close to their melting points. It is essential to keep this basic piece of experience in mind while ploughing through the details of this section. Thus although traditional foundrymen cluster around the hot tear, discussing how the mould or core can be reduced in strength to reduce the stress on the contracting casting, my approach is to ignore the hot tear, simply taking it as the evidence of a poor filling design. Thus the key action is to fix the filling design. The hot tear then disappears like magic. No action to reduce the strength of the mould or core is necessary.

This piece of practical experience appears to be disbelieved, or at least systematically overlooked, by every researcher in this field. This is a pity because there is much useful background that has been clarified by careful, systematic research over the years. This interesting background is reported here. It will become clear that it is consistent with the view that bifilms (as prior cracks suspended in the liquid) introduced by a poor running system are implicated as the fundamental cause, actually *becoming* the hot tear. Nevertheless, the patient reader will find it illuminating to review the experimental data, keeping in mind that bifilms are the major underlying cause.

The reader should also note the large number of terms used for hot tearing, such as hot cracking, hot shortness or hot brittleness. We shall use them interchangeably here.

Characteristics of a hot tear

The defect is easily recognised from one or more of several characteristics:

1. Its form is that of a ragged, branching crack, generally follow intergranular paths. This is particularly clear on a polished section viewed under the microscope (Figure 8.1).
2. The failure surface reveals a dendritic morphology (Figure 8.1).

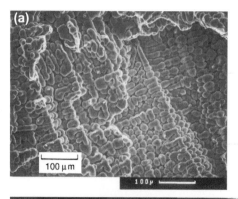

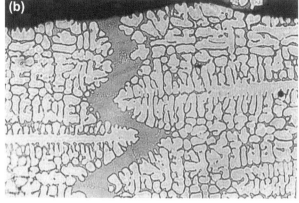

FIGURE 8.1

(a) Scanning electron microscope image of the surface of a hot tear in Al-7Si-0.4Mg alloy sand casting. (b) A filled hot tear in an Al-10Cu alloy (Spittle and Cushway, 1983).

3. The failure surface is often heavily oxidised (prior, of course, to any subsequent heat treatment). This is more particularly true of higher temperature alloys such as steels.
4. Its location is often at a hot spot where contraction strain from adjoining extensive thinner sections may be concentrated.
5. It does not always appear under apparently identical conditions; in fact, it seems subject to a considerable degree of randomness in relation to its appearance or non-appearance and to its extent.
6. The defect is highly specific to certain alloys. Other alloys are virtually free from this problem.

Before we go on to discuss the reasons for all this behaviour, it is worth bearing in mind the most simple and basic observation:

The defect has the characteristics of a tear.

This disarmingly obvious characteristic immediately alerts us to a powerful clue about its nature and its origin. We can conclude that:

A hot tear is almost certainly a uniaxial tensile failure in a weak material.

This may appear at first sight to be a trivial conclusion. However, it is fundamental. For instance, it allows us to make some important deductions immediately:

1. Those theories that assume hot tearing is the result of feeding difficulties can almost certainly be dismissed instantly. This is because feeding problems result in hydrostatic (i.e. a triaxial) stress in the residual liquid, causing pores or even layer porosity in the liquid phase. If the triaxial stress does increase to a level at which a defect nucleates, then the liquid separates and expands (triaxially) to create a pore among the dendrites. The dendrites themselves are not affected, and are not pulled apart. They continue to interlace and bridge the newly formed volume defect, as was discussed for layer porosity (Section 7.1.7.2).

This is in contrast to the hot tear, where it is clear from micrographs and radiographs (Figure 8.1(b)) that the dendrites open up a pathway *first*. The opened gap drains free of liquid *later*.

The pulling apart of the dendrites, separating to form a eutectic-rich path through the structure (Figure 8.1(b)) probably brings with it no significant problems of loss of strength or other properties of the casting despite its alarming and unwelcome appearance on radiographs or cut sections. However, if the path had subsequently drained of eutectic liquid, a hot tear would have been formed, which would, of course, have seriously affected strength. This formation of an open tear might have been avoided if feed metal had been locally available to keep the interdendritic regions full of residual liquid.

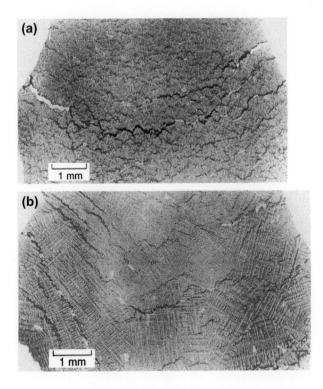

FIGURE 8.2

(a) Radiograph of a hot tear in an Al-6.6Cu grain-refined alloy. Dark regions are Cu-rich eutectic; white areas are open tears. (b) Radiograph of hot tears in Al-10Cu alloy not grain refined.

Rosenberg et al. (1960). Courtesy of Merton C Flemings.

A second confusion arises from the linking of hot tearing in general with the defect formed on the surface of the casting above a grossly underfed region. The severe local collapse of the casting surface above this hot spot will deform the surface considerably, concentrating much strain in this local region. Thus a hot tear may form over this badly fed volume. The experimental arrangement described by Paray et al. (2000) is of this type. This special type of hot tear can also be solved by improving the local feeding, even though the tear itself is the result of the intense local strain; only indirectly the result of feeding.

2. If the defect is a tear, then it has to be understood in terms of its initiation by exceeding a certain critical tensile stress, in common with all types of tensile failure. Rappaz et al. (1999) find a critical rate of tensile strain that will initiate a pore, finding that this gives a remarkably accurate estimate of the susceptibility to hot tearing. (The important work by Rappaz and his team will feature repeatedly in this review of the fundamentals of hot tearing.)

Keeping these general thoughts in mind will help us to keep the important features of the process in perspective while we deal with a host of other aspects. It is not surprising that the research literature on this subject is confusing; the subject is genuinely complicated.

However, before proceeding with the details of the mechanisms associated with this defect, it is probably necessary to dismiss two other contenders for explanations of these hot failures. Dickhaus (1993) has proposed that the effect of surface tension, in the phenomenon of viscous adhesion, might explain hot strengths and the failure mechanisms of solidifying metals. However, simple order-of-magnitude estimates indicate that surface tension is perhaps capable of generating only one hundredth, or even one thousandth, of the stresses involved in hot tearing. In a quite different approach, Fredriksson et al., (2005) has proposed that hot crack formation might occur because of the condensation of vacancies, expected to be present at concentrations of 0.01–0.1% in the metal lattice at these high temperatures. Vacancies are certainly present at maximum equilibrium concentrations at the melting point of metals and certainly condense during cooling. However, in practice, vacancies have never been observed to condense into volume defects such as cracks and pores. In every case examined, and in every metal studied so far, vacancies clusters collapse into either dislocation rings or stacking fault tetrahedra. These studies include direct observation in the electron microscope and computer simulations of the behaviour of large assemblies of atoms (known as molecular dynamics simulations). Basically, the forces between atoms are so great that pores or cracks cannot be opened up without the imposition of huge stresses in the GPa range. Solidifying metals are in general soft and ductile near their melting points, and so cannot generate and sustain such stresses (Campbell, 2010).

This leaves us clear to study the more likely contenders for the generation of these defects.

8.1.2 GRAIN BOUNDARY WETTING BY THE LIQUID

It was C.S. Smith who, during 1949–1952, first formulated the concept of the wettability of grain boundaries by the presence of a liquid phase in the boundary. Figure 8.3 summarises his concept. The shape of the grain boundary particles is largely controlled by the relative surface energies of the grain-to-grain interface itself, γ_{gg} and the grain-to-liquid interface γ_{gL}. The balance of forces is:

$$\gamma_{gg} = \gamma_{gL} \text{Cos}\theta \qquad (8.1)$$

It is clear that for most values of the equilibrium dihedral angle 2θ, the grain boundary liquid assumes compact shapes. However, it will, of course, occupy a greater area of the boundary as its volume fraction increases. The relation between (1) the area of the boundary which is occupied by liquid, (2) the dihedral angle and (3) the volume fraction of liquid present is a complicated geometrical calculation which the author was proud to identify and tackle (Campbell, 1971), setting off the subsequent improved treatment by Tucker and Hochgraf (1973), and, finally, the comprehensively solution by Wray (1976). Hochgraf (1976) went on later to develop a fascinating study of the conditions for the spread of the liquid phase under non-equilibrium conditions, where the dihedral angle becomes effectively less than zero.

The importance of the dihedral angle being zero for complete wetting is illustrated in the work of Fredriksson and Lehtinen (1977). They observed the growth of hot tears in Al-Sn alloys using the scanning electron microscope. The

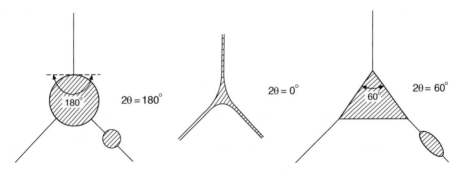

FIGURE 8.3

Shapes of the liquid phase at grain corners as a function of the dihedral angle (Smith, 1948, 1952).

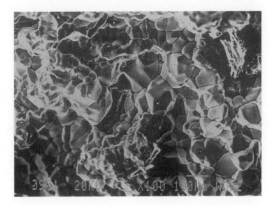

FIGURE 8.4

Hot tear surface of an Al-1%Sn alloy.

Courtesy Chakrabarti (2000).

liquid tin wetted the grain boundaries of the aluminium, leading to intergranular brittle failure when subjected to tension (Figure 8.4). In Al-Cd alloys, the liquid cadmium at the grain boundaries did not wet and therefore did not spread over the boundaries, but remained as compact pools, so that when subjected to tension these alloys failed by ductile fracture.

There have been several observations of failure by hot tearing where, on subsequent observation under the microscope, the fracture surface has been found to exhibit separate, nearly spherical droplets that appear to be non-wetting towards the fracture surface. This has been seen in systems as different as Al-Pb (Roth et al., 1980) and Fe-S (Brimacombe and Sorimachi, 1977; Davies and Shin, 1980). It seems certain that the liquid phase would originally wet a normal grain boundary. It is not clear therefore whether the observation is to be explained by the subsequent de-wetting of the liquid phase after the crack is exposed to the air, or, perhaps more probably, because the boundary consists of a remaining half of a poorly wetted oxide bifilm.

8.1.3 PRE-TEAR EXTENSION

Whilst the casting is cooling under conditions in which liquid and mass feeding continue to operate, if the casting is contracting the solid grains swim about, manoeuvring into new positions. In these conditions clearly no tearing can occur.

The problem starts when grains grow to the point at which they finally collide firmly against each other, but are still largely surrounded by residual liquid.

Patterson et al. (1967) were among the first to consider a simple geometrical model of cubes. We shall develop this concept further as illustrated in Figure 8.5. It is clear that for grains of average diameter a separated at first by a liquid film of thickness b, the pre-tear extension ε can reach approximately:

$$\varepsilon = b/a \tag{8.2}$$

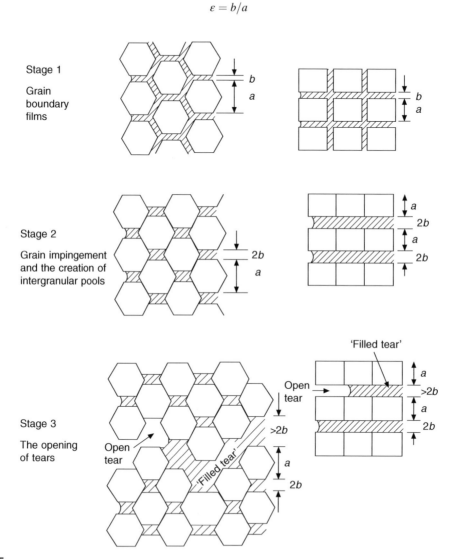

FIGURE 8.5

Stages of hot tearing using two-dimensional square and hexagonal grain models. Grains size 'a' is surrounded by a liquid film of thickness 'b'. Stage 1: unstrained structure; stage 2 shows isolated regions of segregate; stage 3 shows open tears plus so-called 'filled' hot tears. Continued strain eventually will drain the liquid film completely, completing the tear.

For both the cube and the hexagon models in two dimensions, the relation $b/a = f_L/2$ can quickly be seen to be true, where f_L is the volume fraction of liquid. For a three-dimensional cube model, the reader can easily confirm the further relation:

$$b/a = f_L/3 \qquad (8.3)$$

Thus for between 3 and 6% residual liquid phase, we have between 1 and 2% extension before the impingement of the grains. The pre-tear extension being proportional to the amount of liquid present is an observation confirmed many times by experiment. Furthermore, those alloys with large amounts of eutectic liquid during freezing, such as the Al-7Si and higher Si alloys, are usually free from hot tear problems probably for this reason; there is plenty of extension that can be accommodated before any danger of the initiation of a tear.

Also, for a given amount of liquid present, the extension is inversely proportional to the grain size. Thus for finer grains, more strain can be accommodated by easy slip along the lubricated boundaries without the danger of cracking.

After the grains have impinged, a certain amount of grain boundary sliding may continue, as we shall discuss later, although this later phase may contribute only a very limited amount of further extension.

Even in the case of the solidification of pure metals, the grain boundaries are known to have a freezing point well below that of the bulk crystalline material (see, for example, Ho et al. (1985) and Stoltze et al. (1988)). The presence of liquid at the grain boundaries even in pure metals, but perhaps only a few atoms thick, may help to explain why some workers have found tearing behaviour at temperatures apparently below the solidus temperature. However, many of the observations are also explainable simply by the presence of minute traces of impurities that have segregated to the grain boundaries. The two effects are clearly additive.

Because of the presence of the grain boundary film of liquid, bulk deformation of the solid will occur preferentially in the grain boundaries, so long as the strain is below a critical value (Burton and Greenwood, 1970). This explains why the extension of the solid during fracture can be accounted for completely by the sum of the effects of (1) grain boundary sliding plus (2) the extension resulting from the opening up of cracks (Williams and Singer, 1968).

Later, during grain boundary sliding where the grains are now in contact over their complete surfaces, there has to be some deformation of the grains themselves. Novikov et al., (1966) found by careful X-ray investigation that the deformation is confined to the surface of the sliding grains. In addition, at a temperature close to the melting point, recovery of the grains is so fast that they do not work harden. Because they remain in a relatively soft condition, the general flow of the bulk material can continue relatively easily. Thus although the flow is now actually controlled by bulk deformation of the grains, the appearance under the optical microscope is simply that of the sliding of the grains along their boundaries.

It is necessary to keep in mind that the total extension resulting from the various kinds of grain boundary sliding (whether 'lubricated' or not) amount to only perhaps 1–2% strain. Further strain of at least this magnitude arises during the extension of the crack itself, as is discussed later.

8.1.4 STRAIN CONCENTRATION

It was Pellini in 1952 that drew attention to the concentration of strain that could occur at a hot spot in a casting. It is instructive to quantify Pellini's theory by the following simple steps.

If the length of the casting is L, and if it has a coefficient of thermal expansion α, during its cooling by ΔT from the liquidus temperature it will contract by an amount $\alpha \Delta TL$. If all this contraction is concentrated in a hot spot of length l, then the strain in the hot spot is given by:

$$\varepsilon = \alpha \Delta TL/l \qquad (8.4)$$

Clearly, in the hot spot, the casting contraction strain is increased by the factor L/l.

For a casting 300 mm long and a hot spot of approximately 30 mm length at its end, the strain in the hot spot is concentrated 10 times. This would be expected to be a fairly typical result—although it seems possible that strain concentrations of up to 100 or more may sometimes occur.

It is interesting to note that the problem in the hot spot depends on the amount of strain concentrated in it, and this depends on the size of the adjacent casting and the temperature to which it has cooled while the hot spot remains hot and in a weak state.

We can clarify the size of the problem by evaluating an example of an aluminium casting. Assume that $\alpha = 20 \times 10^{-6} \, C^{-1}$ and that the casting has cooled 100°C. If its contraction is hindered, the strain that will result is, of course, $20 \times 10^{-6} \times 100 = 0.002 = 0.2\%$. This level of strain puts the material as a whole above the elastic limit even at room temperature. (In materials that do not show clear yield points, the yield stress is often approximated to the so-called proof stress, at which 0.1–0.2% permanent strain remains after unloading.) In the hot spot, therefore, if the strain concentration factor lies between 10 and 100, then the strain will be between 2 and 20%. These are strains giving an amount of permanent plastic extension that is relatively easily withstood by sound material. However, in material that is weakened by the presence of bifilms at the grain boundaries, and which can withstand typically only 1–2% of strain before the start of failure, as we shall see later, it is no wonder that the casting fails.

In addition to the consideration of the amount of strain concentrated into the hot spot, it is also necessary to consider how many grain boundaries the hot spot will contain. If the grain size is coarse, the hot spot may contain only one boundary, with almost certain disastrous consequences, because all the strain will be concentrated in that one liquid film. If the hot spot contains fine grains, and thus many boundaries, then the strain per boundary is reduced. We may quantify this because the number of grains in the length l of the hot spot is l/a for grains of diameter a. Hence if we divide the strain in the hot spot (Eqn (8.4)) by the number of boundaries in it, then we have the strain per boundary ε_b:

$$\varepsilon_b = \alpha \Delta TLa/l^2 \tag{8.5}$$

It is clear that to reduce the strain that is trying to open up the individual grain boundaries, the beneficial factors include (1) reduced temperature differences, (2) smaller overall lengths between hot spots and (3) finer grain size. However, Eqn (8.5) reveals for the first time that the most sensitive parameter is the length l of the hot spot; if this is halved, the grain boundary strain is increased four times.

8.1.5 STRESS CONCENTRATION

The problem of how sufficient stress arises during cooling to initiate and grow the hot tear may not be relevant because the forces available during cooling are massive, greatly exceeding what is necessary to create a failure in the rather weak casting. Thus we may consider the forces available as being irresistible, forcing the casting to deform. Because this deformation will always occur, the question as to whether a hot tear will arise is clearly *not* controlled by stress, but must depend on other factors, as we shall discuss in this section.

Nevertheless, although overwhelmed, the forces of resistance offered by the casting are not quite negligible. Guven and Hunt (1988) have measured the stress in solidifying Al-Cu alloys. Although the stresses are small, they are real, and show a release of stress each time a crack forms. The loads at which failure occur are approximately 50 N in a section 20×20 mm. Thus the stresses are approximately 0.1 MPa (compared with a strength of >100 MPa at room temperature). Also, as an interesting detail, a simultaneous change in the rate of heat transfer across the casting-mould interface was detected each time the force holding the casting against the mould was relaxed.

In rough agreement with Guven and Hunts' results, Forest and Berovici (1980) carried out careful tensile tests and found that an Al-4.2Cu alloy has a strength of more than 200 MPa at 20°C, that falls to 12 MPa at 500°C, 2 MPa at the solidus temperature, and finally to zero at a liquid fraction of about 20%.

As we have mentioned before, the other stress that may be present could be a hydrostatic tensile stress in the liquid phase. Although this may contribute to the nucleation of a pore, which in turn might assist the nucleation of a tear, the presence of a *hydrostatic* stress is clearly not a necessary condition for the formation of a tear, as we have discussed earlier. We need a *uniaxial* tensile stress to create a tear.

One final point should be emphasised about stresses at these high temperatures. Because of the creep of the solid at high temperature, any stress will depend on the rate of strain. The faster the solid is strained, the higher will be the stress with which it resists the deformation.

Zhao et al. (2000) have determined the rheological behaviour of Al-4.5Cu alloy, and thereby have determined the stress leading to the critical strain at which hot tearing will cause failure. This novel approach may require the densities of bifilms to be checked for similarity between their rheological sample and their hot tearing test piece, which is clearly poured rather badly. The elegant test piece devised by Rappaz and his team (Mathier, 2009) has no defined filling technique, and one can only assume that the filling is poor, manufacturing entrainment defects that will confuse the results. It would be valuable to redefine this attractive-looking test to ensure that it could be filled without compromising the quality of the metal being tested.

8.1.6 TEAR INITIATION

Probably the most important insight into the problem of tear initiation was provided by Hunt (1980) and Durrans (1981). Until this time, the nucleation of a tear was not widely appreciated as a problem. The tear was assumed simply to form! The experiment by these authors is an education in profound insight provided by a simple technique.

These researchers constructed a transparent cell on a microscope slide that enabled them to study the solidification of a transparent analogue of a metal. The cell was shaped to provide a sharp corner around which the solidifying material could be stretched by the turning of a screw. The idea was to watch the formation of the hot tear at the sharp corner.

The amazing outcome of this study was that no matter how much the solidifying material was stretched against the corner it was not possible to start a hot tear in clean material: the freezing mixture continued to stretch indefinitely, the dendrites continuing simply to move about and rearrange themselves.

However, on the rare occasion of the arrival of a small inclusion or bubble near the corner, then a tear opened up immediately, spreading between the dendrites and away from the corner. In their system, therefore, hot tearing was demonstrated to be a process dependent on nucleation. In the absence of a nucleus, a hot tear did not occur no matter how much strain was applied. This fact immediately explains much of the scattered nature of the results of hot-tearing work in castings: apparently identical experimental conditions do not give identical tears, or at times even any tears at all.

It is necessary to remember that in the Hunt and Durrans' work, the liquid would wet the mould, adhering to the sharp corner, and so would require a volume defect to be nucleated; the defect would not be created easily. In the case of a casting, however, a sharp re-entrant corner may have liquid present at the casting surface, but the liquid will not be expected to wet the mould. In fact there is the complication that the liquid will be retained inside the surface oxide of the casting. The drawing of liquid away from this surface, analogous to the case of surface-initiated porosity, would represent the growth of the crack from the surface, and may involve a nucleation difficulty. Thus in the case of cracks initiated at the surface, the Hunt and Durrans observation may not apply universally. However, the concept is still of value, as we shall discuss next. In the case of internal hot tears, their observation remains crucial. However, in the case of alloys that form strong oxide skins, there may still be a difficulty in drawing inwards a rather rigid surface film, or of nucleating a tear on the underside of the surface oxide film; either way, surface initiation might be difficult in many casting alloys.

However, even at the surface of a casting in an alloy that does not form a surface film, the initiation of a tear may not be straightforward. It is likely that the tear will only be able to start at grain boundaries, not within grains, because the dendrites composing the grain itself will be interconnected, all having grown from a single nucleation site. Dendrites from neighbouring grains will, however, have no such links, and in fact the growing together and touching of dendrite arms has not been observed in studies of the freezing of transparent models. The arms are seen to approach, but final contact seems to be prevented by the flow of residual liquid through the gap. Thus if a grain boundary is not sited conveniently at a hot spot, and where strain is concentrated, then a tear may be difficult to start. This will be more common in large-grained equiaxed castings, as suggested by Warrington and McCartney (1989).

If a grain boundary is favourably sited, it may open along its length. However, on meeting the next grain, which in general will have a different orientation, further progress may be arrested, at least temporarily. Thus a tear may be limited to the depth of a single grain. The effect can be visualised as the first stage of the spread of the tear in the hexagon grain model shown in Figure 8.5. Considerably more strain will be required to assist the tear to overcome the sticking point in its further advance beyond the first grain.

For the case of columnar grains, the boundaries at right angles to the tensile stress direction will provide conditions for easy initiation of a tear along such favourably oriented grain boundaries. The effect is analogous to the rectangular grain model in Figure 8.5.

For fine-grained equiaxed material where the grain diameter can be as small as 0.1–0.2 mm, the dispersion of the problem as a large number of fine tears, all one grain deep, is effectively to say that the problem has been solved. This is because the crack depth would then be only approximately 0.1 mm. This is commensurate with the scale of surface roughness because average foundry sands also have a grain size in the range 0.1–0.2 mm. The fine-scale cracking would have effectively disappeared into the surface roughness of the casting.

Nevertheless, it is fair to emphasise that the problem of the nucleation of tears has been very much overlooked in most previous studies. Nucleation difficulties would help to explain much of the apparent scatter in the experimental observations. A chance positioning of a suitable grain boundary containing, by chance, a suitable nucleus, such as a folded oxide film, would allow a tear to open easily. Its chance absence from the hot spot would allow the casting to freeze without defect; the hot spot would simply deform, elongating to accommodate the imposed strain.

In practice, there is much evidence to support the assertion that most hot tears initiate from entrained bifilms. As has been mentioned previously, the author has personally solved every hot tearing problem he has encountered in foundries by simply improving the design of the casting filling system. The proposal to use such an approach has generated almost universal scorn and disbelief. However, when good filling system designs were implemented systematically in an aerospace foundry, all hot tearing problems in the difficult Al-4.5Cu-0.7Ag (A201) alloy disappeared, to be replaced by surface sink problems. In comparison to the hot tears, the surface sinks were welcomed and easily dealt with by improved feeding techniques (Tiryakioglu, 2001).

The study by Chadwick and Campbell (1997) of A201 alloy poured by hand into a ring mould containing a central steel core showed that failure by hot tearing in such a constrained mould was almost guaranteed. Conversely, when the metal was passed through a filter, and caused to enter the mould uphill at a speed of less than 0.5 ms^{-1} to ensure the avoidance of defects, no rings exhibited hot tears. This was an amazing result, showing no failures in one of the world's most hot tear prone alloys, in a test designed to maximise hot tearing conditions. (Memorably, when Chadwick arrived in my office to report the results of the hot tearing test, I was amazed, shocked into silence to see the cast rings so tightly contracted onto the steel cores that they could not be extracted, but completely without tears. While I was gathering my thoughts to comment appropriately on this immensely important and exciting result, Chadwick leaned back in his chair. "Yes" he said, "The test was a complete failure." I was aghast at this remark and could only manage to get out the word "What!" He went on with resignation "What was the use of a hot tear test that did not hot tear?")

A similar study was repeated for Al-1%Sn alloys by Chakrabarti and Campbell (2000). Surfaces of hot torn alloys illustrate the brittle nature of the failure in this alloy (Figure 8.4). The alloy has such a large freezing range, close to 430°C (extending from close to pure Al at 660°C down to nearly pure Sn at 232°C) that in the hot tear ring test the alloy appears even more susceptible to failure by hot tearing than A201 alloy. When subjected to the uphill filled version of the ring test, most castings continued to fail. However, about 10% of the castings solidified without cracks. Once again, the existence of even one sound casting would be nothing short of amazing.

Further evidence can be cited from the work of Sadayappan et al. (2001), who demonstrated that their as-melted Al alloy gave many and large hot tears, whereas after cleaning the metal by degassing they observed only a few small tears. Dion et al. (1995) found that in their castings of yellow brass, the addition of aluminium to the alloy promoted hot tearing, as would be expected from the presence of the entrained alumina film resulting from their very turbulent filling system.

8.1.7 TEAR GROWTH

We have touched on the problem of tear growth in the previous section. However, it bears some repeating that (1) the birth of the hot tear and (2) its growth, sometimes to awesome maturity, are quite separate phenomena.

The evidence is growing that tears are closely associated with bifilms. It remains to clarify the nature of the link. For instance, (1) do tears initiate on bifilms and subsequently extend into the matrix alloy? Or (2) do bifilms constitute the tears, so that the growth of the tear is merely the opening of the bifilm, so that the defect is revealed, in fact becoming obvious. The evidence is accumulating that the important mechanism is (2).

The easy growth in columnar grains where the direction of tensile stress is at right angles to the grain boundaries has been mentioned. Spittle and Cushway (1983) observed that the linear boundary formed between columnar crystals growing together from two different directions was an especially easy growth route for a spreading crack. This is confirmed by experience in the rolling industry, where the diagonal plane issuing from the corners of rectangular ingots, defining the joint plane of the two sets of columnar grains from the two adjacent sides, is a common failure plane during the early reduction passes. The problem is reduced by rounding the corners, or reducing the levels of critical impurities. In steel ingots the significant impurities are usually sulphur, and the so-called tramp elements such as lead and tin.

The explanation in terms of bifilms is that columnar grains both align the bifilm cracks along their length, but also push other bifilms ahead into intergranular spaces, so that failure along these surfaces is to be expected. The presence of an invisibly thin film organised into position between dendrite arms explains the fracture surface seen in Figure 8.1(a). The fracture follows the bifilm, the fracture surface exhibiting steps at integral numbers of dendrite arms, as explained in Figure 2.43. Naturally, the hot tear morphology is also seen in the room temperature fracture seen in Figures 2.44 and 6.29. Clearly, after being extended and flattened by the growth of dendrites, the bifilm is simply a hot tear waiting to be opened and so be revealed. If is not opened during solidification, then it can wait until opened later in a tensile test, or, more worryingly, lying in wait to open later still, causing a failure in service.

A radiograph of an Al alloy (Figure 2.43(c)) by Fox and Campbell J (2000) in which the bifilms have been opened up by the action of reduced pressure show the bifilms near to the mould surface to be organised at right angles to the surface by the pushing action of the growing columnar grains.

In general, the bifilms and the consequential porosity are sited at grain boundaries, leading to intergranular failures, but as we have noted previously, bifilms can be incorporated into growing grains, leading to transgranular failures. In all cases, however, the individual dendrites that constitute a growing grain never cross a bifilm. They cannot grow through air.

Warrington and McCartney (1989) confirm the findings of Spittle and Cushway (1983) when they find that fine equiaxed grains also promote easy growth conditions for a hot tear. This seems to be because the tear can propagate intergranularly along a path that, because of the fine grain size, can remain almost perpendicular to the applied stress on a macroscopic scale. In terms of bifilms, the result is simply the observation of the opening of that particular bifilm favourably oriented with respect to the stress direction, out of the many bifilms usually present between small grains.

Conversely, coarse equiaxed grains gave increased resistance to the spread of the crack. In this case the bifilms would be segregated to planes sometimes well away from the stress direction, causing greater plastic deformation in the attempted return of the crack to its average growth direction. In addition, of course, the distance travelled by the crack would be significantly increased.

The question of the amount of plastic work that is expended during the propagation of a tear is interesting. The work of deformation is easily shown to be of the order of at least 10^4 times greater than the work required to create the newly formed surfaces of the tear. Thus arguments based on the effect of the surface energy of the crack limiting its growth (as in the case of a classical Griffiths crack in a brittle solid such as glass) are clearly not relevant in the case of the failure of plastic solids such as metals at their melting points.

Novikov and Portnoi (1966) draw attention to the fact that, despite the rather brittle appearance of the fracture, the hot torn surfaces cannot usually be fitted back together again, confirming the expectation that considerable plastic deformation occurs during hot tearing. Furthermore, they found that the gap between the poorly fitting surfaces corresponded almost to the total elongation, indicating that the elongation was associated almost entirely with crack propagation in their work. The further implication was that (1) nucleation of the crack was easy because high stresses would have given high elongations before cracking and (2) the pre-tear extension resulting mainly to grain boundary sliding was limited in their experiments. These observations are again consistent with propagation along a bifilm. The brittle, intergranular appearance of the fracture surface is typical of a crack that has followed the central unbonded interface of a double oxide film.

Although high stresses cannot be envisaged at the stage of the nucleation of a crack (because crack nucleation will find easy start sites such as the opening up of bifilms), when the crack has started to grow, the sharpness of its tip ensures that high stress is available locally at this point. For large-grained material, therefore, the occasional absence of a favourably oriented grain boundary can be expected to result in further propagation by two means:

1. The continuation of the crack across the grain that attempts to block its path. Such occasional transgranular growth has been observed by Davies et al., (1970) in models using low-melting-point Sn-Pb alloys. He saw that in such cases the crack followed the cell boundaries, which were clearly the next best route for the crack. Bifilms would be expected in both grain and cell interfaces of course, and present a similarly cogent explanation.
2. The crack either renucleates a short distance ahead in a favourably oriented grain boundary, or more likely, travels around the grain by a path out of the plane of the section, appearing once again a little ahead. Fredriksson and Lehtinen (1977) have directly observed such behaviour by pulling specimens of Al-Sn alloys in the scanning electron microscope. As the crack continues to open, the intervening region of obstructing grain is seen to deform plastically, like a collapsing barrier. The stepwise propagation of the crack, linked by plastic barriers at the steps, explains the irregular, branching appearance that is a characteristic feature of hot tears. The failure of plastic bridges between randomly sited, disconnected bifilms explains the observations similarly.

Returning to our geometrical model, at the end of the pre-tear extension period, the residual liquid is separated into pools between grains, as is seen in stage 2 of Figure 8.5. The corresponding pools in the real casting are seen in Figure 8.2. These compact segregates must pose a problem for subsequent solution heat treatments, even though their existence does not seem to have been previously recognised. At this point, further strain must cause some concentration of failure at a weak point in the structure of the solidifying casting. If a grain near the surface is separated from its neighbour by the presence of a bifilm, the growth of the crack can then occur with little further hindrance. The necessarily irregular nature of its progress mirrors that of the real crack; once again, compare Figures 8.2 and 8.5. Straighter portions of the crack resemble the cube model in Figure 8.5.

A potentially important feature of the hot tearing literature needs to be raised. From Figures 8.1, 8.2 and 8.5, it is quite clear that the portions of the cast structure in which the dendrites have separated but which still contain residual liquid have *always* contained residual liquid. This may seem self-evident. However, the casting literature is full of references to 'healed' hot tears, meaning tears containing residual liquid, but implying that the tears were once empty, and, fortuitously, were somehow subsequently filled by an inflow of liquid. Whether the term 'healed hot tears' is really intended to imply original emptiness and subsequent refilling is not clear, but the term is misleading and would be better discontinued. The term 'filled tear' is more explicit. 'Un-emptied' tear would be even more accurate, but hardly an attractive name! The *filled tear* is simply the region between grains that have been separated by uniaxial strain, but which still remain full of intergranular liquid. If it solidifies while still full, as in Figure 8.1(b), it will constitute a region of segregate, but will almost certainly still be as strong as the bulk of the casting, and so not constituting a defect that might impair its serviceability. It is *not* a hot tear and has *never* been a hot tear.

Only if it becomes empty does it become a major defect meriting the name of hot tear. There are two quite separate mechanisms that could provide an empty tear.

1. The separation of the grains by continued strain to the point at which the residual liquid is no longer capable of keeping the tear filled. This is the mechanism displayed in our simple model (Figure 8.5), and as seems to be shown in Figure 8.2(a). The available liquid has been insufficient to keep the intergranular regions filled, and so, with increasing demand from surrounding liquid regions as these regions are pulled apart, has simply sucked the main crack region dry.
2. The previous mechanism contrasts with the hot tear whose conditions for formation are identical (i.e. grains are separated by contraction strain) but the region between the grains now contains one or more bifilms that, on separation of the grains, also separates the halves of the bifilms. This creates a crack instantly and easily.

Both of these mechanisms seem possible. Both seem likely to operate. However, the first is expected to be more difficult to grow because the grains deform plastically as the crack attempts to propagate from grain to grain. The presence of the bifilm in the second mechanism is expected to create a defect of a serious size with ease.

8.1.8 PREDICTION OF HOT TEARING SUSCEPTIBILITY

Over the years, there have been many attempts to provide a useful working theory of hot tearing. Recently the attempts have narrowed to a few serious contenders. The exercise that has been found to be most useful to discriminate between them has been the attempt to predict susceptibility to hot tearing as a function of composition for binary alloys.

This is a useful test in alloy systems that display a eutectic. At zero solute content the theory has to contend with a pure metal; at low solute contents, only solid solution dendrites are present; above a critical solute content eutectic liquid appears for the first time, steadily increasing to towards 100% as the solute content increases towards the eutectic composition. The ability to deal with all of these aspects across a single alloy system constitutes a searching test of any theory, and covers the majority of solidification conditions in real castings.

A typical experimental result is shown in Figure 8.6. It reveals a steeply peaked curve that Feurer (1976) has called a lambda curve, after the shape of the Greek capital letter Λ. The problem is to find a theoretical description that will allow the lambda curves to be simulated for different alloy systems.

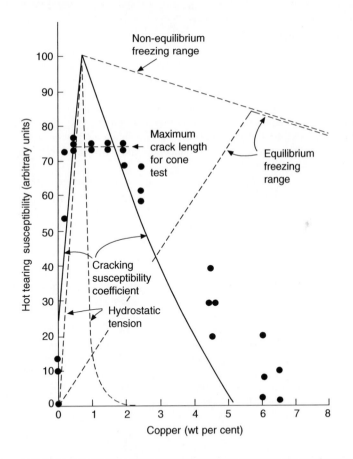

FIGURE 8.6

Hot tearing response of Al-Cu alloys showing a peak (necessarily extrapolated somewhat) at approximately 0.7Cu using the conical ring die test by Warring ton and McCartney (1989) compared with various theoretical models. Freezing ranges and hydrostatic tension by JC (1989); CSC by Clyne and Davies (1977).

It is salutary to note that for the Al-Cu alloy system any prediction based on the equilibrium diagram is completely wrong. Here the maximum freezing range would be predicted to be at 5.7Cu, which might lead the unwary to believe that the maximum problem in porosity and hot tearing should be at this copper content. From Figure 8.6 the problem in hot tearing is clearly centred on a rather dilute alloy of approximately only 0.5Cu. Any problems of hot shortness have disappeared on reaching 5.7Cu!

It is interesting to look at my early effort at prediction (Campbell, 1969a,b) that deals with the analogous problem of porosity in a spectrum of binary alloy compositions. Here the relative hydrostatic tension developed by the flow of feed metal through the dendrite mesh was calculated. The form of the relationship calculated for the Al-Cu system is shown in the figure. This particular result is based on the assumption that the residual liquid is 1% by volume. The peak is almost exactly in the correctly predicted location, confirming the fundamental importance of the arrival of eutectic liquid at that critical concentration of solute. The result closely agrees with the model by Rappaz et al., (1999), as would be expected because both models are based on the development of hydrostatic tension at the root of the dendrites.

The remainder of the predictions based on hydrostatic tension follows this particular experimental data by Warrington and McCartney (1989) poorly, but follows the data by Spittle and Cushway (1983) rather more closely. This intermediate agreement reflects the general capability of the models to achieve fair agreement with experimental data that are in themselves of rather variable quality.

It is surprising that models based on hydrostatic tension agree as well as they do with hot-tear initiation because they are not expected to be necessarily closely related, as we have mentioned before. The hydrostatic tension falls steeply with the arrival of eutectic liquid, dramatically reducing shrinkage porosity as seen in the original work, but not reducing hot tearing so steeply as found experimentally by Warrington and McCartney.

We also need to note that hot tearing is closely related to the peak of the non-equilibrium freezing range, but bears no relation to the subsequent slope of the non-equilibrium freezing range curve, as is clear in Figure 8.6.

The theoretical approach by Feurer (1976) that appears to explain the form of the lambda curves can be similarly discounted because it is also based on the modelling of liquid flow and hence the development of hydrostatic stress, not uniaxial tension (Campbell and Clyne, 1991).

An alternative theoretical approach to hot tearing was proposed by Clyne and Davies (1979). They implicitly assume that the failure is the result of uniaxial tension, but point out that strain applied during the stage of liquid and mass feeding is accommodated without problem by the casting. The problem of accommodating strain only occurs during the last stage of freezing, when the grains are no longer free to move easily. They define a cracking-susceptibility coefficient (CSC):

$$CSC = t_V/t_R \tag{8.6}$$

Where t_R is the time available for stress-relaxation processes such as liquid and mass flow and t_V is the vulnerable period when cracks can propagate between grains. The concept is clear, but defining the limits of applicability of these various regimes for different alloy systems is not easy. However, as a first attempt the authors assume that the stress-relaxation period spans a fraction liquid f_L of approximately 0.6–0.1, and the vulnerable period spans f_V values 0.1–0.01 (Figure 8.7). The predictions of the scheme for the Al-Cu system are shown in Figure 8.6 to fit the Warrington and McCarthy data better. For the Al-Si system, they correctly predict a lambda curve with the correct form and peak at approximately 0.3Si, as found by their experiments. For the Al-Mg and Al-Zn systems, they found the agreement less good. The agreement for Al-Mg was improved later by Katgerman (1982), who modified the CSC limits. For magnesium-based alloys Clyne and Davies (1981) use their model to predict a peak at 2.0Zn for the Mg-Zn system and 3.0Al for the Mg-Al system. The poorer tearing resistance of the zinc-containing alloy is the result of its considerably greater freezing range. However, from the ring test results shown in Figure 8.8, the peaks are actually observed to be at approximately 1.0Zn and 1.0Al. It is necessary to keep in mind that the experimental data will be significantly affected by the presence of bifilms, which will, in general, reduce the composition of the peak susceptibility, bringing agreement between theory and experiment a little closer in line.

The approach therefore seems basically sound and useful, even though not always especially accurate in its predictions, as we have seen. Thus, in common with most theories, it invites further development and refinement.

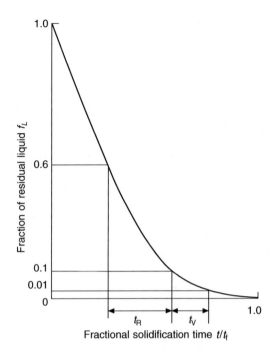

FIGURE 8.7

Model by Clyne and Davies (1977) for the regimes during which either stress relaxation or vulnerability to hot tearing occur.

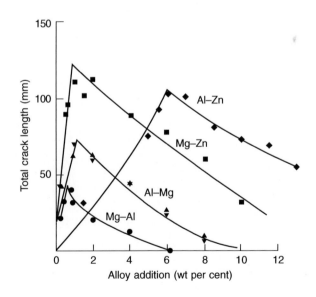

FIGURE 8.8

Hot tearing behaviour of various alloys subjected to the ring die test.

Data from Dodd (1955), Dodd et al. (1957), Pumphrey and Lyons (1948) and Pumphrey and Moore (1949).

As a start it would seem useful to combine the CSC with Eqn (8.5) derived previously for the strain per boundary in the hot spot. This gives a modified CSC as:

$$CSC_b = \frac{\alpha \Delta TLa}{l^2} \cdot \frac{t_V}{t_R} \tag{8.7}$$

There is already considerable evidence to suggest that this more general equation is at least approximately accurate. Figure 8.9 shows how grain size is significant. Also, several researchers with different diameter ring tests confirm that the cracking susceptibility is proportional to the circumference of the ring (Isobe et al., 1978). This proportionality to length is implicit in the design of the various I-beam tests using graded lengths of beam (see later). Pekguleryuz (2010) compared several CSC parameters for Al-Si alloys and found Eqn (8.7) to give the best correlation.

The CSC model has been extended to the cracking of steels (Clyne et al., 1982). In Fe-S alloys, the prediction of a CSC peak at 0.1%S was observed to be accurately fulfilled in experiments by Davies and Shin (1980). Here, as a consequence of the complexity of iron-based alloys, the CSC model is extremely useful in providing a framework to understand the phenomena. For instance, Rogberg (1980) found that for stainless steels that solidified to delta-iron the alloys were insensitive to the impurities As, Bi, Pb Sn, P and Cu, whereas those that solidified to gamma iron suffered serious loss of hot ductility. Kujanpaa and Moisio (1980) confirmed that S and P embrittled gamma iron, but not delta ferrite, but the best resistance to embrittlement was provided by a mixture of gamma and delta irons.

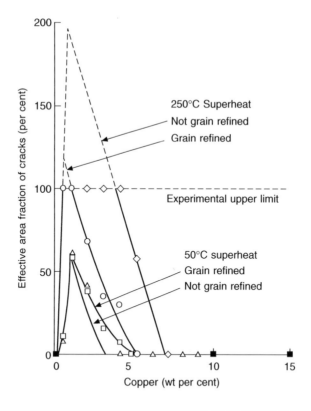

FIGURE 8.9

Hot tearing behaviour of Al-Cu alloys using an I-beam-type test showing the benefits of low casting temperature and grain refinement (Spittle and Cushway, 1983). Peaks are extrapolated to illustrate the close agreement with Figure 8.6.

The most sophisticated development of attempts to predict the susceptibility to hot tearing has been the series of developments since 1999 by Rappaz and colleagues. Initially, their model assessed the linear contraction and volumetric shrinkage contributions to the hydrostatic tension, admitting that the model at that stage only predicted the formation of the initial pore, and did not extend to the development of a crack. A series of publications has followed since 2006 expanding the concept of a granular model, but one in which the grains were randomly distributed. This has been a great advance on the simplistic geometrical models discussed earlier, allowing the details of the nature of the pasty zone to be simulated during the progressive stages of freezing. The network of channels of residual liquid are seen in Figure 8.10. A result for an Al-1Cu alloy pasty zone is illustrated in Figure 8.11, and the highly non-uniform flow through the zone is shown in Figure 8.12. The authors summarise their work in the creation of morphological maps (Figure 8.13). Although at this stage these simulations are only two-dimensional, an extension to three-dimensional is claimed to be relatively easy. It will represent a major step forward in our understanding of the processes within the pasty zone. Even so, the presence of bifilms will complicate these already complex flow patterns still further. It is not surprising that the pasty zone sometimes will allow no flow whatever when jammed with bifilms, which appears to be the only explanation of the results shown in Figure 6.18.

8.1.9 METHODS OF TESTING

A complete survey of the methods used to assess hot tearing is beyond the scope of this book. The interested reader is recommended to various reviews, including historical accounts by Middleton (1953), Dodd (1955) and Hansen (1975), and more recently several more brief accounts by Guven and Hunt (1988) and Warrington and McCartney (1989).

The methods fall into two main groups: (1) the various tests of straight castings solidifying in moulds in which the tensioned test piece is restrained at each end as an I-beam (sometimes known as 'dog bone') tests and (2) the ring or conical mould tests. Both these types of tests use the occurrence of cracks as a measure of hot tearing propensity. Sundry other tests include the pulling of tensile test pieces during their solidification in a tensile testing machine as a direct measure of the resistance of the casting to failure.

The I-beam has many variants. One of the most common is the arrangement of various lengths of rod castings from a single runner. Each of the rods has a T-shaped end to provide a restriction to its contraction. When metal is poured, the contraction of the rods will take place with various degrees of constraint, those with rods greater than a critical length failing by tearing at the hot spot, which is the joint between the runner and the rod. From such a test, therefore, only one result is obtained, and its accuracy is limited by the increments by which the rods increase in length. The limited discrimination provided by the limited number of bars of different lengths in this test is further confounded by the fundamentally limited reproducibility of hot shortness tests.

Other sorts of I-beam test gain a potentially more discriminating result by measuring the lengths of cracks in the hot-spot region. However, even in this case, the actual test volume of material is limited and the stress and strain distribution in the hot spot is far from uniform.

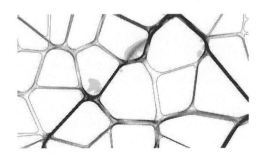

FIGURE 8.10

Preferred flow paths through a simulated intergranular mesh of equiaxed grains (Phillion et al., 2009).

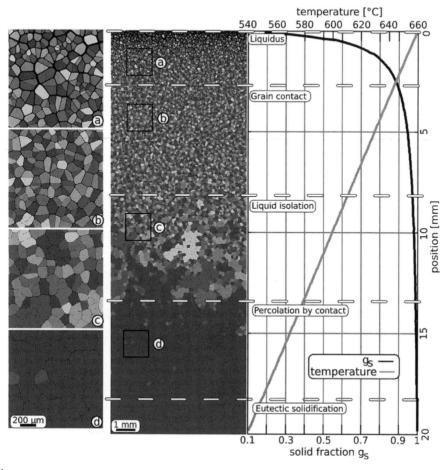

FIGURE 8.11

Calculated pasty zone of equiaxed grains for an Al-1%Cu alloy cooled at −1 K/s in a gradient of 6 K/mm. Grains in mechanical contact are shaded the same grey level (Vernede et al., 2006).

The ring test is a constrained geometry that could hardly be simpler. It is normally carried out in a steel die. Metal is poured into the open annulus between the inner and outer parts of the die. As it cools, it contracts onto the inner steel core. The core also expands slightly as it heats up. The resulting constraint on the casting is severe, opening up transverse tears all around the ring in a susceptible alloy. In this respect, the technique is useful in that it tests a large volume of material, subjecting it all to a uniform constraint condition. The method of assessment is by measuring the total length of cracks. This gives a reasonably large number that can be assessed accurately, making the test more discriminating. Thus, despite the criticisms normally levelled at the ring test, it has many advantages, one of the most important of which is simplicity. In general, the many different researches that have been carried out over the years by different workers using this approach are seen to agree tolerably well. Pumphrey (1955) carried out a series of repeated checks with different workers over a period of months. These showed an impressive consistency. This is more than can be said for most other tests for hot tearing.

Another aspect of the ring test die is worth commenting on: the test is unusual in that its rigid restraint is sufficient to cause hot tearing in the absence of any real hot spot and in the absence of strain concentration.

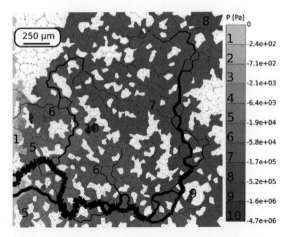

FIGURE 8.12

A model of the fluid flow and pressure distribution in an equiaxed pasty zone as a result of solidification shrinkage in Al-1%Cu alloy at 98.4% solid. The width of each channel is magnified in proportion to the local flow; local pressure is indicated by the grey scale (Vernede et al., 2006).

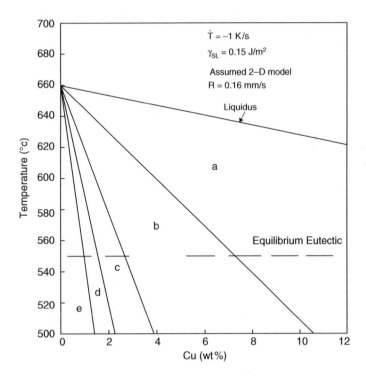

FIGURE 8.13

Map of the pasty zone based on the granular model by Vernede et al. (2006). Region (a) contains mostly isolated grains; (b) some grain contact leading to isolated clusters; (c) larger clusters containing some isolated liquid; (d) a continuous solid network but continuous films no longer exist; (e) solidification complete.

Gruznykh and Nekhendzi (1961) and DiSylvestro and Faist (1977) describe similar ring tests for steels using sand moulds.

The recent variant of the ring test by Warrington and McCartney (1989) provides a tapered centre core that gives a constraint condition increasing linearly along the axis of the cone. This seems a useful test, but is clearly limited in very hot short materials to a maximum tear of the length of the side of the cone. This artificial cutoff is seen in Figure 8.6. More work would be helpful to confirm the limitations of reproducibility of the test. For instance, from Eqn (8.7), it is to be expected that the width of the hot spot that the authors place along the cone by painting a strip of insulating wash down its length will be critical. It is not clear whether such a hot spot is necessary or even desirable in the test, but its presence is certain to have a major effect on results.

Trikha and Bates (1994) employ a nicely designed hot tearing test for steels that consists of opposed conical cores (Figure 8.14). The conical castings are gated at the horizontal joint defining a longitudinal hot spot. Any cracks in the cores can be measured, giving a pair of values to assess the reproducibility of the results. The tearing can be assessed in terms of density of cores and binder materials.

A new approach to measure the strength of Al alloys in the mushy state (but, note, not necessarily a hot tearing test) is described by Mathier et al. (2008). A specially shaped tensile specimen is contained in a hot stainless steel mould and gradually cooled. The precise dimensions of the mould define a precisely shaped specimen which in turn can be closely studied by carrying out, in parallel, a numerical simulation. In this way, the information that can be generated by such a test is maximised. The only disappointment to the reporting of this interesting work is that the filling of the test mould is not specified and has clearly been overlooked as being of critical importance to the integrity of the material under test.

8.1.10 METHODS OF CONTROL

The casting engineer can be reassured that there are several different approaches to tackling hot-tearing problems in castings, or even, preferably, preventing such defects by appropriate precautions.

8.1.10.1 Improved mould filling

In practice, as bears repeating any number of times, the author has never failed to deal with hot tearing by simply up-grading the filling system, ensuring that air is not entrained into the metal at any point. Whole foundries have been revolutionised in this way. Thus this is probably the single most important technique for dealing with hot tearing problems. It is the most convincing evidence that hot tears are strongly linked to the presence of bifilms. To the author's knowledge, sadly, this overwhelming evidence has not yet found its way into the scientific casting literature.

8.1.10.2 Casting design

Much can be achieved at the design stage of the casting. A publication by Kearney and Raffin (1987) concerns itself almost exclusively with the prevention of hot tears by adjustments to the casting design. In general it can be summarised by the following.

1. Do not design sharp re-entrant corners;
2. Do not provide a straight join between two potential hot spots; curve such members;
3. Provide curves in gates so that deformation can be accommodated easily;
4. Angle and off-set stiffeners and ribs to allow easier accommodation of strain; and
5. For gravity die (permanent mould) casting the rapid removal of any internal steel core is recommended to reduce constraint.

The reader will appreciate the general philosophy, although this severe condensation hardly does justice to the original work of these authors.

It is also worth bearing in mind that flash on castings as the result of the poor fitting of moulds and cores can be a major source of casting constraint. To identify the real sources of constraint is necessary, therefore, to check the casting straight out of the mould (not after it has been nicely dressed to appear on the manager's desk!).

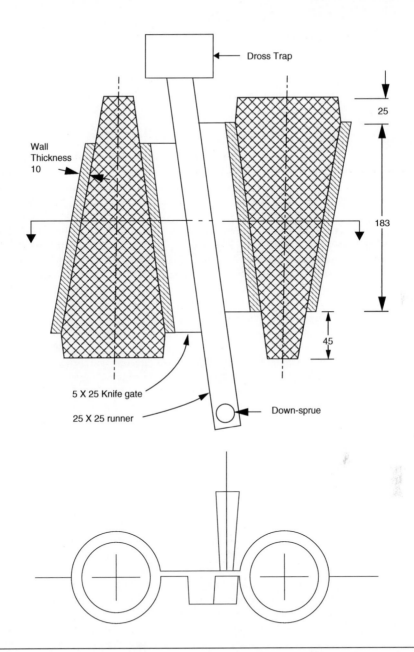

FIGURE 8.14

The hot tearing test of Trikha and Bates (1994) consisting of opposed cones, giving a pair of results that can be compared for reproducibility.

This list of dire warnings needs to be set into context. In the experience of the author, the provision of sharp corners and other so-called geometrical dangers have never given a problem providing the liquid metal is of good quality, and provided the metal has not been impaired by a poor filling system design.

In any case, it is usual to find that the design is fixed, and any changes will involve significant negotiation with the customer and/or designer, and may eventually not be agreed. In such situations, the casting engineer has to fall back on other options. These include the following.

8.1.10.3 Chilling

The chilling of the hot spot is a useful technique. This reduces the temperature locally, thus strengthening the metal by taking it outside of its susceptible temperature range before any significant strain and stress is applied. By reducing the temperature locally nearer to that of the casting as a whole, the temperature differential that drives the process is reduced, and any strain concentration is redistributed over a larger region of the casting. Local chilling is therefore usually extremely effective.

In addition to this conventional explanation, recent research has revealed that the action of chills is more complex. Chills cause the solidification front to move away from the chill in a rapid, unidirectional movement. Bifilms, with their enclosed layer of air, are the ultimate barrier to dendrites because the solidification cannot progress through the air. The result is that the bifilms are pushed ahead of the front, and so pushed away from the vulnerable hot spot.

This is the reason that many entrainment defects, as opposed to genuine shrinkage problems, appear to be cured by the placing of a chill. The usual (and usually quite wrong) interpretation being that the chill has cured the 'shrinkage' porosity.

Some oxide bifilms appear, however, to be attached to the casting surface (possibly at the point at which they were originally folded in) so that they are not completely free to be pushed ahead. Also, some are oriented so that some of the tips of the dendrites in the advancing dendrite raft pass either side of the bifilm, as seen in Figure 2.43. Either way, the bifilm is organised into a planar sheet parallel to the dendrite growth direction and then pinned in this position until solidified in place. Cracks at right angles to the chill surface can be seen in the radiograph shown in Figure 2.43. Such planes easily separate along the air layer as the result of a tensile strain. This may occur during freezing, in the form of a hot tear, as shown in Figure 8.1. Alternatively, the planes and the characteristic steps can be seen in room temperature tensile fracture surfaces as in Figures 2.44 and 2.46. Clearly, once again, the message is clear: it is far more effective to remove the bifilms than to impose chills.

8.1.10.4 Reduced constraint

A reduction in the stress on the contracting casting can be achieved by reducing the mould strength. But as we have noted before, this is more easily said than done. The options are: (1) reducing the level of binder in the sand core, although there is normally little scope for this, because those foundries not already operating at minimum strengths to reduce costs and ease shake-out are using strength as an insurance against core breakages and defects as a result of mishandling; and (2) weakening the core by using it in a less dense form (such as produced by blowing rather than by hand ramming), or modifying its design by making it hollow. Earlier shake-out from the mould, and more rapid decoring, may also be useful. However, Twitty (1960) found the opposite effect; his white iron castings suffered more hot tears when shaken out earlier as a result of the extra stresses put on the casting during the removal of the mould!

El-Mahallawi and Beeley (1965) show how appropriate tests can be carried out on sands containing different binders. Their deformation/time test for sand under a gradient heating condition would be expected to provide a good assessment of the constraint imposed by different types of sand/binder systems. DiSylvestro and Faist (1977) use their sand moulded ring test to check the effect of different sand binders on the hot tearing in carbon steels. In later work on steel, SCRATA (1981) list the sand binders in increasing hot-tearing tendency for steel casting sections of less than 30 mm. These are:

Greensand (*least hot-tearing*)
Dry sand (clay bonded)
Sodium silicate bonded (CO_2 and ester hardened types)
Resin bonded shell sand
Alkyd resin/oil (perborate or isocyanate types)

Oil sand

Phenol formaldehyde resin/isocyanate

Furan resin (*worst tearing*)

In the case of the worst sand binder, furan resin, it is difficult to believe that the thermal and mechanical behaviour of the binder is responsible for the increased incidence of hot tearing. It seems likely that the sulphur (or phosphorus) contained in the mixed binder will contaminate the surface of the steel casting, promoting grain boundary films of sulphides or phosphides, and thus rendering the metal more susceptible to tearing.

In thicker sections, the greater amount of heat available causes more burning out of the binder, with better collapse of the sand mould. Organic binders benefit from this effect. Paradoxically, the inorganic system based on sodium silicate in this list is also seen to benefit, probably as a result of the extra heat leading to greater general softening and/or melting of the binder at high temperatures. As the binder cools, however, it is well known to become a fused, glassy mass which can be difficult to remove from the finished casting. Even so, modern silicate binders contain breakdown agents that appear to have overcome this problem.

For ferrous castings, the inclusion in the sand of material that will burn away rapidly, leaving spaces into which the sand can move, allows faster and greater accommodation of the movement of the casting. Common additions to greensand are wood flour, cellulose and polystyrene granules. Slabs of polystyrene foam of 25–50 mm thickness have been inserted in the mould or core at approximately 6–12 mm away from the casting/mould interface, depending on the thickness of the casting. Conversely, reinforcing rods or bars in sand moulds or cores greatly reduce the collapsibility of that part of the mould in which they are placed, and may cause local tearing.

In the case of gravity die casting, in which the core may be made from cast iron or steel, it is common to construct the core so that it can be withdrawn or can be collapsed inwards as soon as possible after casting. Almost all aluminium alloy pistons are made in this way, with complex collapsible five-piece internal cores.

Ultimately, however, it has to be emphasised that removal of constraints after casting is not always a reliable technique because the timing is difficult to control precisely; too early an action may result in a breakout of liquid metal, and too late will cause cracking. It is really better to rely on passive systems.

8.1.10.5 Brackets

The planting of brackets across a corner or hot spot can sometimes be useful. The brackets probably serve not only for strengthening but also as cooling fins. Perhaps predictably (because of the poorer conductivity of steel castings compared with those in aluminium, for instance), for steel castings they are generally less good at preventing tearing than are chills (SCRATA, 1981). Even so, some large steel castings are sometimes seen to be covered in 'tear brackets', the casting bristling like a porcupine. Such features confirm the existence of their poor filling system designs.

8.1.10.6 Grain refinement

In general, fine grains reduce the susceptibility to hot tears, as predicted by Eqn (8.7). Figure 8.9 shows the improved resistance to tearing by grain refinement of Al-Cu alloys. Davies (1970) finds a similar result for other Al alloys. Novikov and Grushko report the beneficial effects of refinement by Sc and Zr in Al-Cu-Li alloys. Twitty (1960) confirmed that his white cast iron alloyed with 30% chromium was severely hot short when not grain-refined, whereas 0.10–0.25Ti addition reduced the grain size and eliminated hot tearing.

However, in a really illuminating experiment to reduce hot cracking of a direct chill ('continuous') Al alloy, Nadella et al. (2007) introduce grain refiner in the ladle before casting, successfully grain refining and eliminating cracking. However, when they added the same grain refiner into the metal stream during pouring into the mould, the casting was successfully grain refined, but cracked. This is clearly an instance of the Ti-rich grain refiner precipitating on bifilms in the ladle and sedimenting to the bottom of the ladle, so that the cast metal was clean. When the grain refiner was added to the launder bifilms were clearly unable to separate, and thus were carried over into the casting, creating excellent grain refinement, but also creating excellent conditions for cracking in the presence of cooling strains. This experiment seems an excellent vindication of the critical action of bifilms in hot tearing and cracking.

8.1.10.7 Reduced casting temperature

Reduction of the casting temperature can sometimes help, as is seen clearly for Al-Cu alloys (Figure 8.9). This effect is likely to be the result of the achievement of a finer grain size. If, however, the effect also relies on the reorganisation of bifilms, then the effect may be sensitive to geometry: some directions may become more prone to failure if bifilms are caused to lie across directions of tensile strain.

8.1.10.8 Alloying

A variation of the alloy constituents, often within the limits of the chemical specification of the alloy, can sometimes help.

The addition of elements to increase the volume fraction of eutectic liquid can be seen to help by (1) increasing the pre-tear extension by lubricated grain boundary sliding and (2) decreasing the CSC. Couture and Edwards (1966) confirm that copper-based alloys benefit from increased amounts of liquid during the final stages of freezing.

Manganese in steels is well known for reacting with sulphur to form MnS, so that the formation of the deleterious FeS liquid at the grain boundaries is reduced.

For other more complicated alloy systems the answers are not so straightforward. For instance, in early work on the Al-Cu-Mg system by Pumphrey and Lyons (1948), the relation between hot tearing and composition is complex, as was confirmed by Novikov (1962) for various Al-Cu-X systems. Ramseyer et al. (1982) investigated the Al-Cu-Fe system and found that for certain ranges of composition increasing levels of iron were desirable to control tearing. This is a most surprising conclusion which most metallurgists would not have predicted; iron is usually an embrittling impurity in most high-strength aluminium alloys when assessed by room temperature tests of ductility. Later work by Chadwick (1991) revealed that the effect of the iron is to provide a network of iron-rich crystals around the primary aluminium dendrites, as a scaffold framework that appears to support and reinforce the weaker dendrite array. When attempting to reinterpret this fascinating result in terms of bifilm theory, it seems that the scaffold framework would consist of beta-Fe particles which had grown on and straightened bifilms. However, the beta-Fe framework appeared to have impressive rigidity and strength, despite, we assume, its associated central cracks.

8.1.10.9 Reduced contracting length

Shortening the length over which the strain is accumulated is sometimes conveniently achieved by placing a feeder in the centre of the length. Such a large concentration of heat in the centre of the cooling section will allow the strain to be accommodated in the plastic region close to the feeder. Any opening up of intergranular pathways is likely to be easily fed from the nearby feeder. The careful siting of feeders in this way effectively splits up a casting into a series of short lengths. If each length is sufficiently short, then strains that can cause a tear will be avoided. The siting of insulating coatings can also be used in the same way. In terms of Eqn (8.7), the technique is equivalent to a method of reducing the strain concentration by multiplying the number of hot spots and thus increasing the total length l of the hot spot whilst reducing the contracting length L.

8.1.11 SUMMARY OF THE CONDITIONS FOR HOT TEARING AND POROSITY

The findings of research such as that by Spittle and Cushway (1983) are summarised schematically in Figure 8.15(a). The benefits of normal metallurgical controls during casting such as fine grain size and reduced casting temperatures are clear. However, a liberty is taken to add to the figure the newly discovered but as yet relatively little known overriding benefits from clean metal.

The differences between the conditions for the occurrence of porosity and the occurrence of hot tearing have been mentioned several times. Figure 8.15(b) represents a summary of these conditions.

Clearly, porosity forms in the hydrostatic tension as a result of poor feeding, especially in the kind of conditions found in a pasty freezing alloy. The interdendritic feeding leads to a reduction in pressure described by the so-called Darcy relation, in which laminar flow through interdendritic channels suffers viscous drag, causing the pressure to fall. Thus these conditions exist in those alloys that solidify as a solid solution, growing as dendrites whose interdendritic

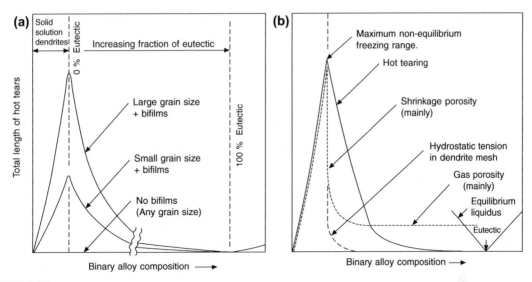

FIGURE 8.15

Summary of (a) the modest effects of grain size and the powerful effects of bifilms on hot tearing and (b) the close and potentially confusing relation between the conditions for porosity (dependent on hydrostatic pressure) and hot tearing (mainly dependent on uniaxial strain).

channels taper down to nothing, therefore maximising viscous drag, and maximising the potential for the creation of porosity.

Increasing the solute content of the alloy sufficiently to form some eutectic, particularly in non-equilibrium freezing conditions, the hydrostatic (triaxial) tension is immediately greatly reduced because the dendrite channels now no longer taper to a point, but are now truncated with a planar front of eutectic. The potential for porosity now falls precipitously as is clear in Figure 8.15(b).

If any uniaxial tension exists, the potential for hot tearing at first increases similarly to the potential for pore formation, but the steep drop at the first appearance of the non-equilibrium eutectic liquid may or may not correspond to the hot tearing susceptibility depending on how this is measured experimentally.

The different regimes of triaxial and uniaxial tension correspond to the regimes for the incidence of porosity and hot tearing in alloy systems as illustrated in Figure 8.15. The continuing low level of porosity at higher solute contents approaching the eutectic is the result of a residual amount of gas porosity. This too, of course, usually only occurs because of the presence of a population of (usually) relatively small bifilms.

The maximum in the freezing range corresponds to the maximum potential for porosity and hot tearing (Campbell, 1969a,b; Rappaz 1999). Interestingly, this peak also corresponds closely to the compositions of many wrought alloys. This is because the wrought alloys have been optimised for maximum alloy content in solution, avoiding the presence of hard eutectic phases which contribute little additional strength and reducing other properties such as resistance to corrosion. Thus these alloys should suffer maximum hot tearing and cracking problems during casting. The only reason such compositions are successful in practice is because of the great precautions taken by the wrought industry to cast metal as clean as possible, and under the highest temperature gradient possible provided by intense water cooling. Even so, the continuous casting industry is troubled by cracking of ingots, particularly of some of the stronger alloys, and particularly at the start of casting. This is because after all the cleaning of the melt, the cast is started by the melt being dropped several hundred millimetres into the mould where, of course, it re-creates huge bifilms. These cause the cracking of the cast material around the dovetail key at the top of the starter bar. Although, once the mould is filled, the remainder of the pour creates no further bifilms; the cracking troubles sometimes extend a long way up the length of the casting because of the mixing and progressive dilution of the original melt pool as the cast proceeds, spreading the original bifilms that are free to float far up the length of the product.

8.1.12 HOT TEARING IN STAINLESS STEELS

The failure of stainless steel castings during solidification and cooling can be complicated by the phase changes that can occur during freezing. The critical parameter is the Cr_{eq}/Ni_{eq} ratio as discussed in some detail in Section 6.6. Nayal (1986) does not hesitate to make the useful, if obvious point, that it is the structure of the steel during freezing that is important, rather than the structure at room temperature. He categorises the steels according to their structural reactions during freezing:

A.	$L \rightarrow L + \gamma \rightarrow \gamma$
	$L \rightarrow L + \gamma \rightarrow L + \gamma + \delta \rightarrow \gamma + \delta$
B.	$L \rightarrow L + \delta \rightarrow L + \gamma + \delta \rightarrow \gamma + \delta$
C.	$L \rightarrow L + \delta \rightarrow \delta \rightarrow \gamma + \delta$

The range of solidification mode B occurs over values of Cr_{eq}/Ni_{eq} from 1.49 to 2.0. In later work, the researchers find in the case of freezing entirely in mode B that no cracks were observed, whereas the presence of some component of A and C always led to some cracking.

The researchers Kujanpaa and Moisio (1980) find that P and S are particularly harmful in encouraging tearing in steels that solidify purely to austenite, but do not affect those that solidify with some delta ferrite. The division between the two steels is sharply defined at $Cr_{eq}/Ni_{eq} = 1.5$. Another Scandinavian worker, Rogberg (1980), confirms this general rule for the particular case of two steels pulled at high temperature in a tensile testing machine. The steel solidifying to austenite was sensitive to a wide range of impurities, but those containing ferrite seemed insensitive to these problems.

8.1.13 PREDICTIVE TECHNIQUES

The ability to predict the occurrence of tearing in a casting would be valuable. This quest has driven several different approaches. One of the early attempts assuming the rheological behaviour of solidifying alloys has been made by Qingchun et al. (1991), with some success. A not dissimilar viscoplastic model has benefited from the power of computers by Pokorny, Monroe and Beckermann (2009), who have shown that the sites and severity of hot tears appears to be predictable in simulations of relatively complex light alloy castings. These forerunners of examples of early computing approaches are likely to become routine provisions of commercial computer simulations in due course. But for more scientific insights into the mechanisms of hot tearing, the highly descriptive granular model under development by Rappaz et al. (1999–2009) is a likely contender as a future useful model to understand and predict tensile failure in the partly solidified state. This model describes the intergranular and interdendritic flow through the pasty zone, defining a condition in which viscous drag grows to reach a nucleation threshold for a pore, which is then assume to grow into a crack. The model may one day be tested for bifilm-free melts, but in the meantime appears not to apply as a result of bifilms blocking all intergranular and interdendritic flow of liquid as demonstrated by the experiments of Fuoco (Figure 6.18).

8.2 COLD CRACKING

8.2.1 GENERAL

Cold cracking is a general term used to emphasise the different nature of the failure from that of hot tearing. Whereas the word 'hot' in hot tearing implies a failure occurring at temperatures above the solidus, 'cold' simply means lower than the solidus temperature; in some cases, therefore, it can be rather warm! Solidus here means, of course, the real non-equilibrium solidus (not necessarily the value picked off an equilibrium phase diagram!).

The term 'cracking' is also to be contrasted with 'tearing'. Whereas a tear is a ragged failure in a weak material, a crack is more straight and smooth and occurs in strong materials. Because it represents the failure of a strong material, the stress required to nucleate and propagate the failure is high (in hot tearing stress was generally insignificant, whereas strain was important).

Occasionally, a failure appears to fall somewhere between the tear and crack categories. Such borderline problems include the cracks that form in steels because of the presence of low melting point residual elements, such as copper-rich phases, at the grain boundaries of steels. The copper-bearing phase is liquid between 1000 and 1100°C, with its dihedral angle falling to zero in this temperature range, thus wetting the grains. Over this range of temperature, therefore, steel is particularly susceptible to tensile failure (Wieser, 1983). The temperature is well below the solidus (in terms of the Fe-C system), and cracking will not occur easily at higher temperatures because the liquid phase does not wet so well and therefore does not cover such a large proportion of the grain boundary area.

Returning to the 'cold' crack, the driving force for the nucleation and spread of a cold crack is stress. The various ways in which stress can arise and be concentrated in castings have been dealt with previously and are not repeated here. However, one mechanism not previously discussed is phase change in the solid state. In steels transforming from a δ- to a γ structure, the large volume reduction has been suggested as a potential source of stress because of the large strains involved (Gelperin, 1946; Grill and Brimacombe, 1976). In fact, the volume change occurring during the collapse of the body centred cubic ferrite phase to the close packed austenite phase is close to 1.14%, corresponding to a linear contraction 0.38% (being one third of the volume contraction). This is a massive linear strain, well over the yield point (in the region of a 0.05–0.1% strain), and so would be expected to open up any fragments of bifilms to form the nucleation sites of larger cracks that would propagate as the steel cooled, generating high stresses to propagate the cracks to observable size. This mechanism seems to be the cause of the widely experienced cracking of low carbon and low alloy steel continuously cast billets in the carbon range 0.05–0.20 with a peak at 0.10°C. Figure 8.16 is given as an example.

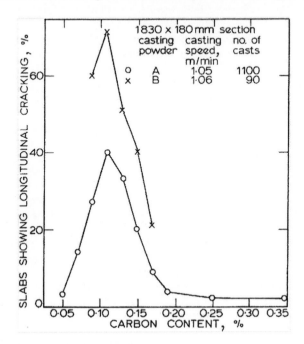

FIGURE 8.16

Effect of carbon content on the longitudinal cracking of continuously cast steels.

The peak corresponds nearly exactly with the peak extent of the ferrite plus austenite transformation range. The reduced cracking at less than 0.05°C and higher than 0.20°C reflects the solidification of the steel directly to delta ferrite or to austenite respectively, avoiding the strains associated with the solid state transformation between the two. These considerations are in agreement with Nayal's list.

8.2.2 CRACK INITIATION

Cracks start from stress raisers. A stress raiser can be an abrupt change of section in a casting. However, this problem is well known to designers, and in any case is not likely to cause increases of stress by much more than a factor of two.

More severe stresses are raised by sharper natural defects which are assumed to be present in all metals, known as Griffith cracks. I have had problems acknowledging the existence of Griffith cracks because such features cannot be generated by solidification, and recent computer studies (molecular dynamics) appear to show that cracks cannot exist in solid metal until the theoretical strength of the metal is exceeded. Such stresses are in the realm of many tens of GPa. The molecular dynamic studies indicated that volume defects such as pores and cracks cannot exist in the metal lattice because the bonds between atoms are so strong; any such volume defect collapses to form dislocation rings and stacking fault tetrahedra and other similar atomically dense, coherent structures. The logical outcome is the conclusion that bifilms, whose presence *is* to be expected, actually *are* the Griffith cracks (JC, 2011).

These cracks which are cast into place at the time of the filling of the casting are all the more dangerous because they can occupy a large portion of the section of a casting, but at the same time are difficult to detect. Oxide bifilms are probably the most important initiators of cracks in castings of all types. This is clear because from daily foundry experience, castings made with good running and gating systems are usually not sensitive to problems of cracking; the heads of large castings can be flame cut off without the generation of any cracks.

It is important to keep in mind that if it is true that all cracks initiate from bifilms, which I think is not only possible but probable, then cracking of metals can be controlled, and can be eliminated. The prospects of eliminating failure by cracking would revolutionise metallurgy and engineering. Such thoughts are so exciting to be beyond belief. Time will tell whether this will come to pass.

Turning back once again to more pedestrian issues, it is interesting to consider briefly the welding of steels. Stresses involved during and after the solidification of a weld can be extremely high because of the constraint of the surrounding solid metal. This surrounding material is near to room temperature and thus very strong. In such a case, Dixon (1987) has observed that cracks start from such innocuous sites as micropores. However, such small and relatively rounded defects are unlikely initiators despite the very high driving force. It is also most likely that the micropore is sited on a bifilm, and it is the bifilm that causes the initiation of the crack. This is just one illustration of the possible universality of cracks only forming from bifilms. Only very careful observation using the scanning electron microscope would be capable of resolving such issues.

To clarify further the issues relating to welding: it seems probable that all welding cracks originate from bifilms introduced into the parent plate material when originally cast and which have survived processing into plate and sheet. Thus the huge amount of research on ways to reduce weld cracking almost certainly require to be redirected from the weld conditions and turned to address the casting conditions of their parent plate materials.

In the case of the strong 7000 series Al alloys, Lalpoor et al. (2009) used finite element analysis to estimate the stresses in semi-continuously direct chill cast billets and slabs. Knowing the levels of stress, they estimated the critical defect sizes to initiate catastrophic failure. The diameters of such critical penny-shaped cracks need to be between 3 and 10 mm to initiate failure. Oxide bifilms are to be expected easily exceeding this size as a result of the turbulent initial fill of the mould. In fact, bifilms up to 100 mm across would not be surprising. Thus sudden brittle failures are to be expected.

8.2.3 CRACK GROWTH

As the casting cools, stress-relaxation by creep becomes progressively slower and eventually stops altogether. Thus from this stage onwards, further contraction only builds up elastic strain and consequent stress. The accumulating stress is available to grow the crack with great speed to large size. Lees (1946) reports the casting of a high-strength aluminium alloy into a sand mould designed to produce hot tearing by constraining the ends of the cast bars. If the restraint was not released before the castings cooled to 200°C, then a loud crack was heard, corresponding to the complete fracture of the bar. Similar failures during the cooling of steel castings are also well known, as was, for instance, reported by Steiger as long ago as 1913.

In steels, if the crack is open to the atmosphere, the colour of its surface is a useful guide to when it formed: an uncoloured metallic surface will indicate that the crack occurred at a temperature near to room temperature; the normal 'temper colours' (the light interference colours reflecting the thickness of the oxide) range from light straw, formed at somewhere near 300°C, through yellow, blue and finally to brown indicate greater exposure to time at temperature, with temperatures probably approaching 600–700°C for the darker colours.

After the strains of delta to austenite transformation between about 1400 and 1500°C, depending on the carbon content, during the further cooling of steel castings there are a succession of particularly vulnerable temperature ranges. The following list is not expected to be exhaustive.

1. Carbon steel castings will embrittle if they dwell for an excessive period between approximately 950 and 600°C (Harsem et al., 1968). Butakov et al. (1968) cites hydrogen, sulphur and phosphorus as increasing embrittlement in the 900–650°C range, and the intergranular fracture surface as exhibiting various forms of sulphides, particularly MnS and FeS. Harsem et al., (1968) adds carbides and AlN to this list. From our privileged gift of the wisdom of hindsight, it seems probable that many of these researches were influenced by the presence of bifilms. At least some of these problems are likely to be associated with the strains of the austenite to ferrite transition. New, clarifying researches are now needed.

2. Impure low-alloy steels suffer similarly in the range 550–350°C (Low, 1969).

3. Low-carbon steels are susceptible to brittle failure if deformation occurs in the range 300–150°C, the so-called blue brittleness, or temper embrittlement range (Sherby, 1962). In his review, Wieser (1983) lists the principal contributors to temper embrittlement as antimony, arsenic, tin and phosphorus. These elements have been found to segregate to the prior austenite grain boundaries. Because these boundaries will in general coincide with the ferrite boundaries, then the crack will usually be intergranular once again. However, the effect of these and other elements in the various kinds of steels is complicated, not being the same in every case; for instance, tin at concentrations up to 0.4% embrittles Ni-Cr-Mo-V steel but not 2.25Cr-1Mo steel. Copper, manganese and silicon are also deleterious in some steels, whereas the effect of molybdenum is generally beneficial.

4. Most body-centred cubic iron alloys become brittle at subzero temperatures, −150 to −250°C. This is often known as the cold brittleness range. The ductile to brittle transition is one of the most important parameters which many steels are required to meet before being accepted for service. It is generally determined from a series of impact tests at successively lower temperature.

PROPERTIES OF CASTINGS

Having made our casting, it may appear visually perfect, but the warm hopes encouraged by its comforting appearance can quickly be dashed when its properties are checked.

Very often, the tensile properties reveal that the casting may be strong, passing the requirement for strength specified by the customer, but its ductility is below the specified requirement. This is common. Alternatively, everything appears to be fine, but the casting leaks. This is also common.

The customer can be impressively intransigent when it comes to demanding his specified properties. After all, when accepting the order from the customer, the foundry had agreed to provide the specified properties.

This section reviews some of the potential problems.

9.1 TEST BARS

Test bars are used to test the tensile properties of the cast metal. They are used in various ways:

1. To test a melt before or at the same time as the pouring of the casting (even though the test bar will probably require heat treatment, and so often be tested days after the pouring of the casting);
2. To test the casting as a cast-on feature of the casting, accompanying the casting through heat treatment, and finally cut off and tested, sometimes in the presence of the customer's inspector;
3. To test the casting as a specimen excised (i.e. cut out) from the casting as part of a destructive test of the casting.

The cast-on test bars and excised bars both will be tolerably representative of the material of the casting and tend not to cause any serious problems of interpretation, even though exact simulation of conditions throughout the main body of casting are clearly not achieved and are probably not achievable. These bars are generally accepted as being sufficiently representative.

The main problems with test bars are with the separately cast bars. This section focuses on these.

Invariably, at the present day, separately cast test bars are poorly designed, so they give poor representations of the properties of the castings. These unrepresentative low properties make the task of the foundry, already hard, even harder. To reduce their problems with obtaining reasonably representative bars, foundries will give the job of pouring to one particular person who is known to have a good pouring 'technique' and who can therefore achieve properties that are believed to be representative. Stories abound of those foundries who should pour test bars for each melt from day to day, but who have poured a week's supply on Monday morning because they happened to know it was a good melt and their expert pourer was available at the time. Such stories probably show my age; I have not heard of such practices in recent years.

The design of all current test bars pre-dates the concepts of surface turbulence, air entrainment and critical velocity and pre-dates experimental verification of their filling action by such techniques as video X-ray and computer simulation. Thus at the time of writing it is regrettable that all our world standards of test bar designs are unsatisfactory. Some are seriously poor. None can faithfully reproduce the actual quality of the melt; all impair the melt during the filling of the mould, explaining the scatter in properties that are usually observed. The scatter can be important, possibly causing a perfectly good batch of castings to be scrapped if properties appear (misleadingly) to be low.

Because all current national and international test bar moulds known to the author appear to be designed to maximise defects, there seems to be no point in describing any of them. I am taking the initiative to describe the features of one test bar mould below that has been properly developed and is known to give representative results.

The 10 Test Bar Mould

The concept is shown in Figure 9.1. The design has been proven by video X-ray to be free from air entrainment.

This mould casting 10 bars at a time was specifically designed for research. The design was optimised by Dispinar and Campbell (2005) using Al-7Si-Mg alloy poured into resin-bonded sand moulds viewed by X-ray radiographic video. All 10 bars fill beautifully, so that their tensile properties have been found to be indistinguishable and to exhibit minimal scatter. This makes the mould ideal for carrying out studies on the optimisation of heat treatment for instance, in which from only 10 moulds, 100 test bars are quickly produced, all with practically identical as-cast properties. Other studies requiring large numbers of bars for statistical purposes are also quickly generated from relatively few moulds.

However, of course, even this design does not deliver a test bar free from damage. This is because the act of pouring metal into the pouring basin is somewhat damaging, and the initial stages of the priming of the down-sprue will be likely to be messy and additionally damaging. The damage in the down-sprue might be reduced by redesigning the assembly so that the bars were horizontal, thus much smaller fall heights would be suffered, giving much lower filling velocities and

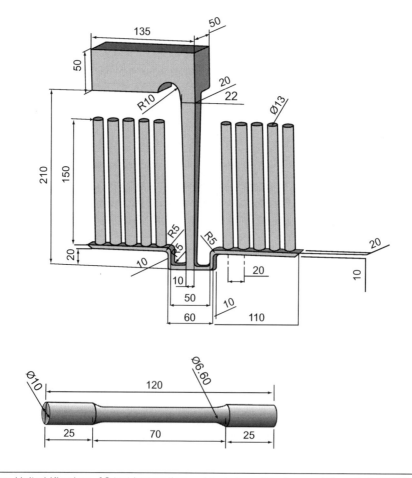

FIGURE 9.1

The Birmingham, United Kingdom, 10 test bar casting and test piece (Dispinar and Campbell, 2005).

thus reduced damage to the metal. It would be useful to compare the results of vertical and horizontal moulded systems. Ultimately, the pouring basin could be eliminated by contact pouring, and the down-sprue damage could be totally eliminated if the mould were filled from a counter-gravity filling system. The adoption of counter-gravity filling systems is to be strongly encouraged; only then would it be possible to achieve the filling of a test bar with metal that would be truly representative of the melt and the casting.

The mould works quite well for the Al-7Si-Mg alloys in a static pour mode (i.e. the mould is simply poured under gravity and left to cool.) This design fulfils my rule 1 for feeding 'Do not feed'. It seems to work adequately for most alloys, and especially light alloys in which solid feeding is relatively efficient, and where the alloys are of a reasonably good quality—i.e. a small population of small bifilms that are relatively resistant to opening under the reduction of pressure suffered without a feeder.

Early trials targeted at the production of a shaped test bar were abandoned because of the difficulty to ensure that the grip length furthest from the feeder was sound. The provision of a feeder to ensure soundness was probably possible but was not straightforward without introducing other dangers of parallel filling systems and possible overspill between the different filling routes. The penalty that the bars all required to be machined was accepted. The development of a well-shaped test bar capable of being pulled without machining would be welcome.

The offset stepped basin and tapered sprue is designed to ensure that minimal air enters the mould (which appears to be one of the most serious criticisms of current designs). The minimum depth in the basin used for the calculation of the filling system is 25 mm; thus, the sprue is likely to start filling before the basin is filled to its minimum operational height unless a stopper is used in the sprue entrance. Preferably a stopper is employed and only raised when the basin has been filled above the minimum level line. The basin is continued to be topped up after the removal of the stopper.

The mould works well as a sand mould. It would also be appropriate as a permanent metal mould, although the rather narrow channels would require to be increased in size to allow for the thickness of a mould coat.

Naturally, the design could be adapted for the casting of perhaps only two test bars at a time, making it more universally useful. Furthermore, the development at Birmingham showed, in line with expectations, that the design worked well for all types of metals from light alloys via cast irons to steels.

However, for some alloys that are more difficult to feed, and probably as a prudent design where metal quality might be less than optimum, the bars would benefit from the enlargement of the horizontal runners from 20×10 mm to 20×20 mm, plus the provision of a feeder of at least 20×20 mm standing on each arm of the runner (a test bar on each side may need to be sacrificed to make room for the feeder). The feeder on each arm can then pressurise the runner and in turn maintain pressure in the 13 mm diameter castings during freezing. An additional feature that is easily provided to aid feeding from an enhanced temperature gradient is the provision of a cooling fin at the top of each test bar. This is laid on the parting line, 2 mm thick with dimensions approximately 20×30 mm and possibly open at the top of the mould to eliminate any back pressure (metal rarely fills all of the fin).

Alternatively, for alloys that are difficult to feed, a second variant of the 10 test bar mould is designed to be rolled over through 180° immediately after casting. In this case, the runner bar is increased in section to 25×25 mm to act as a feeder after the mould has been inverted. This is somewhat better for alloys which require much feeding, and, of course, reproduces more faithfully the conditions of those processes that use a roll-over mechanism such as the Cosworth operation. Most of the fatigue research from the Birmingham, UK, department has been performed with this casting technique.

The roll-over technique requires the use of a small pivoted cradle and hand crank (it is impressively primitive to look at!) for clamping the mould halves together and inverting the complete mould immediately after casting.

A practical inconvenience experienced with roll-over was that the sprue emptied and poured over the floor. Originally we had suffered this, thinking that the ensuing mess was simply to be tolerated, and assuming that the casting would be unharmed because of the swan neck link to the runner/feeder bar which would have prevented this vital feed metal from the runner bar being lost. However, this was not always true. The metal would sometimes siphon around the swan neck and empty from the runner/feeder which would then collapse inside its oxide skin like a punctured balloon.

Thus a refinement of technique which we found necessary to introduce was to take the sprue cut from the previous casting and use this to plug and freeze the sprue of the current casting immediately before the roll-over. This was a slight inconvenience, but saved the mess on the floor and retained the runner/feeder filled so that feeding could occur efficiently.

Once again, such practical inconveniences would be avoided by the adoption of a side-filled counter-gravity filling system with an integral roll-over facility. Counter-gravity is such a sensible, civilised and economic technology!

K-mould test (Evaluation of melt cleanness)

The K-mould test (Figure 9.2) is used not for the evaluation of mechanical properties but for the evaluation of the cleanness of molten aluminium. It was invented in Japan in 1973 by Kitaoka but not widely published until 2001. A description is given by Wannasin and colleagues (2007). The test is sensitive principally to large oxide bifilms and is valuable for assessing the quality of incoming metal to the foundry and in optimising melt treatments to raise melt quality.

FIGURE 9.2

K-bar test mould.

Courtesy N-Tec (2009).

The test is used in different ways by users. For instance, one user defines a K-value (the number expressing the cleanness) in terms of the total number of oxide films observed on the fracture surfaces. Another user defines K as the number of fracture surfaces (made by breaking at the five V-grooves along the length of the bar) that contain oxides. If all five contain oxides, then the rating is '5'. Most users, understandably, tend to ignore oxide defects smaller than some critical size, usually about 1 mm.

The fundamental feature of the K-bar test is that the fracture is forced to occur at the V-groove; thus, the material is sampled at that point whether there is a defect present or not. Thus the test checks random samples of the material. This differs from a test like the tensile test or a bend test in which failure occurs at the largest defect, giving a pessimistic indication of the general quality of the material.

It will be interesting in the future if material is made without a substantial population of bifilms. The familiar brittle fractures at the five V-notches should then become impossible. The material will simply bend, refusing to fracture.

9.2 THE STATISTICS OF FAILURE

Unfortunately, the properties of materials are not accurately reproducible. For castings in particular, scatter of tensile and fatigue properties has been a traditional concern for all casting users.

This is recognised by many of the world's standards for the testing of castings. If a tensile test bar fails below the strength value required by the specification, the specified procedure is to record the low properties with a record of any visible anomaly on the fracture surface and then break two additional test bars from the same production batch to show that the low property break was due to an inclusion or other anomaly and did not reflect the yield, elongation and ultimate measurements of the metal.

Effectively, the failed result is discarded. The discarding of low values is a practice that has crept in to both foundries and laboratories. Somehow the low results are viewed as a mistake. We turn a blind eye to them. This approach is, of course, less than honest. Furthermore, it deprives us of valuable information on the real properties of the products we are making. It is a real result, illustrating how the properties reduce as a result of the presence of defects. Because all of the test bar results, 'good' or 'bad', are influenced by the population of defects, the 'outliers' are a good indication of the real spread of results.

When dealing with scattered property results, the common approach is simply to take an average. Occasionally, a standard deviation might also be calculated. The attempt to assess scatter by the use of a standard deviation, usually denoted sigma, σ or s, is of course much better than nothing, but does make the implicit assumption that the distribution of results is Gaussian, or normal. The familiar bell-shaped distribution is shown in Figure 9.3. The value $\pm\,\sigma$ encloses approximately 68% of the scattered results. The value $\pm 2\sigma$ includes 95% and $\pm 2.5\sigma$ includes 99%. Higher multiples of sigma, for instance 3σ (99.73%) and 4σ (99.994%), are less commonly used.

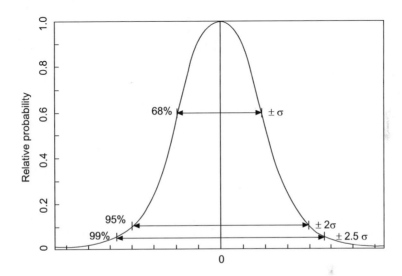

FIGURE 9.3

Gaussian distribution of strengths, showing assessment of scatter using different multiples of the standard mean deviation σ.

If instead we retain all results, good and bad, and plot them to reveal their distribution, we can obtain a histogram such as that in Figure 9.4(a). The shape of the distribution curve is usually not the symmetrical Gaussian form but is skewed, as are many strength distributions (usually negatively skewed) and elongation and fatigue life distributions (usually positively skewed). A close approximation to the shape of the curve was derived theoretically by Weibull (1951). The curve (Figure 9.4(a)) can be plotted as the familiar cumulative distribution (Figure 9.4(b)) and by further mathematical manipulation the curved ends of this plot can be straightened in what is known as a Weibull probability plot (Figure 9.4(c)). In other words, a Weibull probability plot is simply another way of presenting data and checking whether the data follow the Weibull distribution.

In general, the failure properties of metals are accurately described by Weibull (pronounced 'Vybl' where 'y' is as in 'why') distribution. There are good fundamental reasons for this. Weibull derived his distribution in 1951 based on the 'weakest link' concept developed by Pierce in 1926.

In materials research, Weibull analysis was originally used almost exclusively for ceramics and glasses. Although Weibull used tensile strength data from castings as a case study in his historic paper, it was after the work by Green and Campbell (1993, 1994); this approach was applied much more commonly to castings.

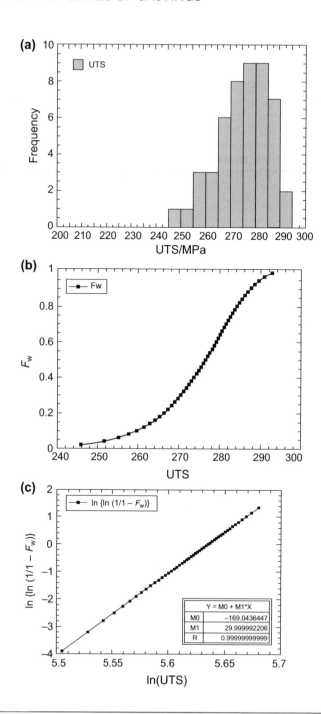

FIGURE 9.4

(a) Typical skewed distribution of tensile strengths as might be obtained from a cast Al-Si alloy *(synthetic data by Green, 1995)*. (b) Distribution identical to that in (a), but replotted as a cumulative distribution (Green, 1995).
(c) Plot of the data in (b) showing the simple straight line form of the two-parameter Weibull cumulative distribution (Weibull modulus $m = 30$ and position parameter $\sigma = 280$ MPa) (Green, 1995).

9.2.1 BACKGROUND OF USING WEIBULL ANALYSIS

I am grateful to Professor Murat Tiryakioglu for his major contribution to this section. The Weibull cumulative distribution function is given by the expression

$$F_w = 1 - \exp\left\{-\left(\frac{x - x_T}{x_0}\right)^m\right\}$$

(9.1)

where F_w is the cumulative fraction of *failures* up to a given value of x, say strength or fatigue cycles, x_0 is a position (or scale) parameter, x_T is a lower strength threshold below which no specimen fails and m is the shape parameter, referred to as the Weibull modulus. The average and standard deviation of the three-parameter Weibull distribution are given by

$$\bar{x} = x_T + x_0 \Gamma\left(1 + \frac{1}{m}\right)$$

and

$$s_x = x_0 \sqrt{\Gamma\left(1 + \frac{2}{m}\right) - \left(\Gamma\left(1 + \frac{1}{m}\right)\right)^2}$$

where Γ represents the gamma function. The skewness of the Weibull distribution is a function of only the Weibull modulus. When the Weibull modulus is less than 3.6, then the distribution is positively skewed. For $m > 3.6$, the distribution is negatively skewed.

The threshold is that value below which no failures are expected; even though thousands of tensile tests might be performed, for a finite threshold no test would fail at a lower stress. The presence of a threshold is comforting. However, it can only be derived using the three-parameter Weibull analysis. If it is assumed that there is no threshold (i.e. it is assumed that strength can fall to zero on occasions), then the threshold can be neglected, allowing a simpler analysis using two-parameter Weibull. We shall present examples of both these analyses later.

The previous expression is frequently used with the lower strength x_T set to zero, giving the so-called two-parameter Weibull approximation. Equation (9.1) becomes

$$F_w = 1 - \exp\left\{-\left(\frac{x}{x_0}\right)^m\right\}$$

(9.2)

In general, the three-parameter relation should be used, or at least explored, before making the assumption that a lower threshold of value x_T does not exist. Forcing the Weibull fit to assume no threshold can give a misleading interpretation of the data. For instance, Figure 9.7 showing top-poured material versus bottom-gated material shows the same data interpreted (1) as a two-parameter straight line plot or (2) a three-parameter curve. Clearly the three-parameter result is a more faithful representation of the data. Furthermore, the shape parameter (m) and position parameter (scale parameter) x_0 are quite different, so that extreme caution needs to be exercised when attempting to compare the parameters; it is essential to know whether the parameter refer to two- or three-parameter interpretations. An additional fundamental difference between the two plots is that the two-parameter interpretation predicts that strengths can scatter to zero, whereas the three-parameter fit confirms that the results extrapolate to a minimum threshold value, below which no results will be expected. The threshold value is, of course, of enormous importance.

An example of a distribution of Weibullian failures described by Eqn (9.2) is shown in Figure 9.4(a). Unlike the normal distribution there is no reflective symmetry about the mean. To obtain the two-parameter Weibull modulus and position parameter from such a plot requires mathematical curve fitting. Alternatively, these can be found by reducing the cumulative distribution (Figure 9.4(b)) to a straight line plot (Figure 9.4(c)).

After rearranging and taking natural logarithms twice gives

$$\ln\{\ln(1/(1 - F_w))\} = m \ln(x) - m \ln(x_0)$$

(9.3)

This can now be presented as a straight line by plotting $\ln(-\ln(1 - F_w))$ as a function of $\ln(x)$ giving the slope m and intercept $-m \ln(x_0)$, from which can be deduced the scale parameter σ. The slope m and the position parameter σ are the two parameters to emerge from this approach, giving the name 'two-parameter Weibull' to this approach. The data of Figure 9.4(b) are replotted in Figure 9.4(c) with a straight line fitted to the data by simple regression analysis.

The slope of the Weibull plot in the two-parameter analysis is a measure of the average to standard deviation (signal-to-noise) ratio and is known as the Weibull modulus, m.

In two-parameter Weibull distributions, the ratio of average to standard deviation is written as

$$\frac{\bar{x}}{s_x} = \frac{\Gamma\left(1 + \frac{1}{m}\right)}{\sqrt{\Gamma\left(1 + \frac{2}{m}\right) - \left(\Gamma\left(1 + \frac{1}{m}\right)\right)^2}}$$

where $\bar{x}$ is average fracture stress, s_x is the standard deviation and Γ represents the gamma function. When the previous equation is plotted, we obtain the following figure, which yields a surprisingly linear relationship (Tiryakioğlu, 2009)

$$m = 1.279\frac{\bar{x}}{s_x} - 0.5$$

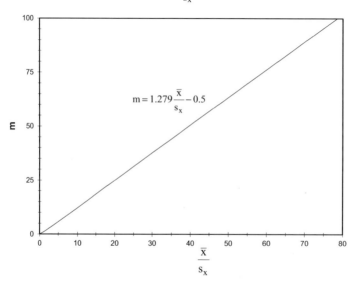

The other important parameters that are derived from a Weibull approach are the position parameter (a measure of the average strength) and the minimum threshold value.

The Weibull analysis can, in principle, estimate the probable value of strength of the weakest casting in a sample of 100 typical castings, or perhaps the weakest one in 10,000. However, of course, when using extrapolations from limited data, the accuracy of the extrapolation needs to be carefully assessed. In addition, as with all extrapolations, uniformity of conditions when extending into the unknown is never certain so that extreme caution, if not outright scepticism, is required. The reader is advised to use the percentile estimation methods described by Hudak and Tiryakioğlu (2009).

In much work using two-parameter Weibull, it has been found that for pressure die castings m is often between 1 and 10, whereas for many gravity-filled castings, it is between 10 and 30. For good quality aerospace castings, a value between 50 and 100 is more usual. Values of 150–250 are probably somewhere near a maximum limit defined by the limits of accuracy of strength measurements (i.e. even if all strength results were perfectly identical, the expected modulus of infinity could not be demonstrated because the tensile testing machine would record slightly different strengths because of errors in the machine). It should be noted that all these Weibull parameters are based on the two-parameter interpretation of the Weibull analysis. The more widely applicable three-parameter analysis is dealt with later in this chapter.

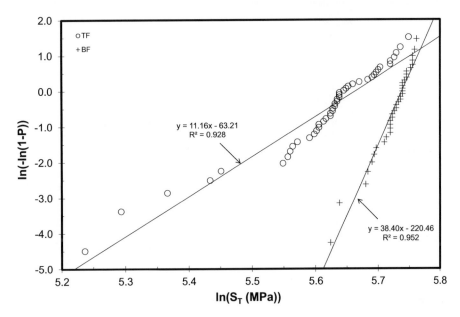

FIGURE 9.5

Two-parameter Weibull plot of strength results for Al-7Si-0.4Mg alloy (i) top filled and (ii) bottom-gated with filter. (Green and Campbell (1994) replotted by Tiryakioglu, Hudak and Okten, 2009).

Figure 9.5 shows data for a single alloy cast in a two ways, displaying the difference in reproducibility between the two filling processes. The data are from sand castings poured variously 'badly' and 'well', giving a Weibull modulus varying between about 10 and 40. These kind of data are some of the simplest demonstrations of the importance of filling techniques on the reliability of cast products. It is also, of course, a clear demonstration of the action of entrainment defects.

9.2.2 PROCEDURE FOR TWO-PARAMETER WEIBULL ANALYSIS

It is implicit in the Weibull distribution that the failure strength of a material is determined by the distribution of defects resulting from its manufacture. The Weibull distribution can be of great use because it is possible to extrapolate back to the very low probability of failure (high reliability) region, and to select a design stress to give a desired failure rate. The following procedure can be used when the volume effect is small. Tiryakioglu and Hudak (2011) give an exemplary guide to the use of the two-parameter approach.

1. Rank the data in ascending order of strength (fatigue life etc.) and assign corresponding ascending failure rank j.
2. Make the extrapolation from a sample of specimens to the population they are drawn from. This is required because the first specimen to fail in the sample tested has a probability of survival of 100% at a stress below its failure stress. However, there is a finite probability that if more specimens were tested, one of them would fail at a lower stress. The most probable unbiased estimate of the probability of survival of the jth specimen in the ranked data is given by Tiryakioglu and Hudak (2008) as

$$F_j = \frac{j-a}{N+b}$$

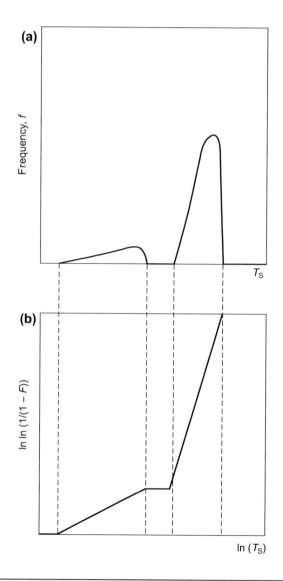

FIGURE 9.6

(a) Bimodal distribution and (b) its corresponding plot in terms of two separate two-parameter Weibull distributions.

where N is the total number of specimens tested and the parameters a and b are listed in Table 9.1.

3. Calculate $\ln\{\ln 1/(1 - F)\}$ for each F_j.
4. Calculate $\ln(x)$ for each x.
5. Plot $\ln\{\ln 1/(1 - F)\}$ as a function of $\ln x$ and regress through this a best-fit straight line. The Weibull modulus is the slope m of the line.

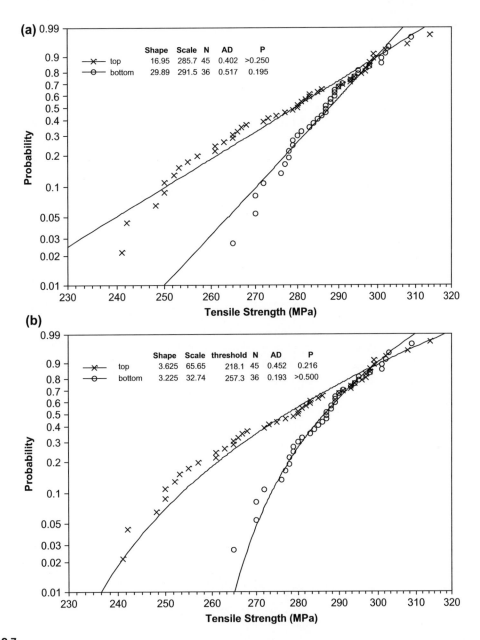

FIGURE 9.7

(a) A pair of strength distributions from top-poured versus bottom-gated castings, ill-fitted by a two-parameter Weibull analysis; (b) a three-parameter Weibull plot illustrating that both distributions have thresholds (lower limit values below which no results are expected).

Courtesy Tiryakioglu (2009).

Table 9.1 Unbiased Probability Estimators (Tiryakioglu and Hudak, 2008)

N	a	b
5	0.173	0.500
6	0.243	0.390
7	0.280	0.310
8	0.309	0.251
9	0.322	0.210
10	0.348	0.190
11	0.367	0.160
12	0.371	0.130
13	0.382	0.110
14	0.388	0.100
15	0.394	0.080
17	0.407	0.050
20	0.417	0.030
22	0.430	0.000
25	0.443	0.000
27	0.448	0.000
30	0.455	0.000
32	0.460	0.000
35	0.465	0.000
40	0.472	0.000
45	0.481	0.000
50	0.486	0.000
55	0.499	0.000
60	0.503	0.000
65	0.509	0.000
70	0.518	0.000
75	0.522	0.000
80	0.516	0.000
90	0.518	0.000
100	0.519	0.000

6. Conduct a goodness-of-fit test by using the coefficient of determination, R^2 of the best fit line. For this test, first calculate the critical value, $R^2_{0.05}$ (Tiryakioğlu et al., 2009)

$$R^2_{0.05} = 1.0637 - \frac{0.4174}{n^{0.3}}$$

If the R^2 of the linear regression from the Weibull probability plot is less than the $R^2_{0.05}$ value, then it can be concluded that the data do not come from a Weibull distribution. If $R^2 \geq R^2_{0.05}$, then the distribution of the mechanical testing data is indeed Weibull.

7. Calculate position parameter x_0

Tiryakioğlu and Hudak (2011) recommend several other steps in the two-parameter Weibull analysis, such as determining confidence intervals for the shape and scale parameters. If a percentile needs to be estimated, such as the design strength, determine the percentile and the confidence level, e.g. 99% confident that 95% of castings have a higher tensile strength. For estimation of Weibull percentiles, the reader is referred to the paper by Hudak and Tiryakioğlu (2009).

9.2.3 THREE-PARAMETER WEIBULL ANALYSIS

Although the two-parameter Weibull approach has been widely used for the analysis of scatter in the mechanical test results of castings, there are plenty of instances where this approach is clearly not valid. If the two-parameter Weibull plot is not a straight line, the distribution is clearly not described by two-parameter Weibull assumptions. The real distribution might be one of several, but the most useful rapid check is to ascertain whether the original assumption making the minimum strength limit x_T equal to zero is correct. If in fact there is a minimum limit to strengths, Eqn (9.3) becomes

$$\ln\{-\ln(1 - F)\} = m \ln(x - x_T) - m \ln x_0 \tag{9.4}$$

The Weibull plot then becomes a plot of the left hand side of Eqn (9.4) as the vertical axis, versus $\ln(x - x_T)$ along the horizontal axis. Alternatively, when $\ln(x)$ is used as the x-axis, the distribution is three-parameter type if the data suggest a curve, rather than a line. The data then have to be fitted to give optimised values for the three parameters—m, x_0 and x_T—giving the appropriate appellation 'three-parameter Weibull' analysis.

Those casting processes and conditions that can deliver a distribution that is terminated by a minimum threshold are of enormous importance. It is not easy to predict at this early stage what feature of the casting process would lead to this key advantage. We may find the threshold is reproducible if the processing is carried out with attention to some key detail, such as, for instance, settling time before casting, or turbulence during the pour. A sharp cutoff to the distribution scatter may be an independent factor related to melt quality and turbulence, rather than some metallurgical factor such as cooling rate, chemistry or heat treatment etc.

Conversely, finding that the data indicate that x_T is actually zero, or even negative (i.e. the distribution of strengths or ductilities extrapolate to zero), is a salutary piece of information, which, if correct, warns of enormous danger for safety critical components, some of which will be expected to have zero strength or zero toughness.

Figure 9.8(a) shows strength distributions of two sets of castings. The castings with lower average strength follow a three-parameter Weibull distribution as confirmed by its p value for the Anderson-Darling goodness of fit test being in excess of 0.05. Their threshold strength is 281 MPa. Thus, although the average strength is low, heat treatment, the castings exhibit a minimum strength below which no failures are expected. In contrast, the higher strength castings are members of a population at risk of some with zero strength.

9.2.4 bi-WEIBULL DISTRIBUTIONS

Figure 9.6 shows the case of a bimodal distribution i.e. two separate modes or distributions. Such multiple distributions are relatively common, and an indication of such an effect is to be seen in the poorest data shown in Figure 9.5. The various slopes are usually found to correspond to different distributions of defects. Several slopes can sometimes be distinguished corresponding to pores, new bifilms, old bifilms and sound material. Populations of different varieties of bifilms are to be expected in castings as described in Section 2.2. In addition, Tiryakioğlu has drawn attention to the fact that tensile and fatigue life data should be interpreted differently. For tensile data, Weibull mixtures are the result of distinctly different defects, usually 'old' and 'new' bifilms. For fatigue data, *surface* or *interior* initiation of cracks limit fatigue (Tiryakioğlu, 2015).

For instance, a bimodal distribution of tensile strength results in many castings is to be expected. The melt will already contain a background population of bifilms, usually sub-millimetre size but in large number density, but if the melt enters the mould cavity turbulently the bifilms created at that late stage of filling are usually of large size, at least in the centimetre size range, giving quite different failure modes. A typical example is seen in Figure 2.44.

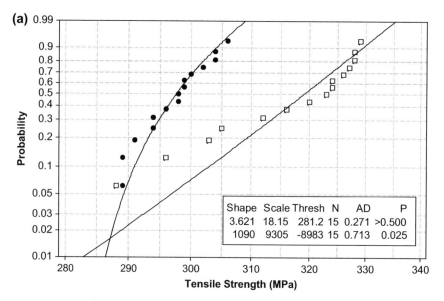

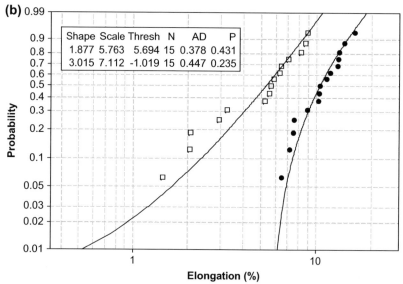

FIGURE 9.8

(a) A comparison of the strength distribution of two sets of castings by three-parameter Weibull. Those with the lower average strength had a minimum threshold of 281 MPa, whereas those with the higher average strength had a strength distribution that extrapolated to zero (actually indicating negative in the curve-fitted result). (b) Elongation results conveyed the message that one set of results, this time higher, were more reliable than those which extrapolated to zero.

Courtesy Alotech Limited and M Tiryakioglu (2009).

In fatigue life data, Weibull mixtures are due to two failure mechanisms being active: failure from cracks initiating from surface defects and interior defects. Surface defects are predominant and interior defects lead to fatigue failure only when there are no cracks initiated by surface defects.

An attempt has been made to fit the castings with higher average strength in Figure 9.8(a) to a three-parameter Weibull plot without success, as indicated by the very low p value. The poor fit suggests a negative (i.e. zero in practical terms) threshold. However, the data strongly suggest that the data may follow a bi-Weibull distribution, as indicated by the two separate regions of the curve. A bi-Weibull distribution is characterised by two different slopes for each part of the Weibull curve; this indicates the presence of two separate failure mechanisms, implying two different defect distributions or two different failure mechanisms.

Figure 9.8(b) shows the ductility results for the same two sets of castings. Both datasets follow three-parameter Weibull distribution as indicated by p values above 0.05. One set has a threshold value of 5.7%, confirming that no castings from this population are expected to fail at an elongation below 5.7%. For the other castings, the threshold elongation value is negative. (The negative result is a mathematical impossibility but effectively equivalent to a result of zero.) Hence some castings from this source can be expected to fail without any elongation. Even here, although the p value indicates a reliable result, the appearance of the data in Figure 9.8(b) may indicate that a bi-Weibull distribution would be an even better description of the data. If so, even for this relatively poor data, a minimum threshold might be found to apply.

The bi-Weibull approach is too complex to present here. The interested reader is recommended to the guidelines set out by Tiryakioglu and me (2010). However, the power and flexibility of this technique to describe the complexity of property distributions in castings is illustrated by Figure 9.9. Sets of samples subjected to various hot isostatic pressing (HIP) routines were subjected to tensile tests. The distributions of the results illustrate that at least two defect populations appeared to be present in all the samples. In some results from this work, there is even a suggestion that three populations were present, but more data would have been required to confirm this.

9.2.5 LIMITS OF ACCURACY

Reliance on the precision of the Weibull moduli derived from a limited number of tests has to be viewed with some caution. For instance, Tiryakioglu and Hudak (2008) have shown (Table 9.1) that for a simulated result of 30 tensile tests that should have a true modulus $m = 26.5$, the actual experimental results to within 95% confidence limits (looking up ± 0.025 in the Table 9.2 giving the fractional errors) will vary from 0.667 to 1.405 times the true result, equivalent to $m = 17.7$ to 37.2. This is a soberingly widespread, warning that the value of m is often less accurately estimated than we might like.

The use of about 30 samples has often been regarded as sufficient to obtain a valid Weibull plot. However, Tiryakioglu's cautionary analysis would indicate that an extrapolation of strength data to a failure rate of one in a 100 will give a corresponding expansion of the uncertainty in the extrapolated result. Despite these legitimate concerns, the result does not further deteriorate too drastically. For instance, for our population of 30 tensile results with true $m = 26.5$, the 0.01 and 0.99 columns of Table 9.1 give factors of 0.611 and 1.493, corresponding to a possible range of $m = 16.2$ to 39.6.

However, further extrapolation to a rate of, for instance, one in a 1000 seems likely to expand the uncertainty of the result to the point of it being worthless. To retrieve some precision many more tests, perhaps a hundred or more, will be needed depending on what accuracy is required.

9.2.6 EXTREME VALUE DISTRIBUTIONS

There is often keen interest in the extrapolation of results of a limited number of tests to assess the failure rate of a casting at a probability, for instance, of one in 10,000 or one in 10 million. The design of safety-critical components on automobiles and aircraft falls into this category. Interestingly, there are techniques for dealing with such improbable extrapolations.

Beretta and Murakami (2001) developed a technique for the quantification of the probability of finding inclusion defects in very clean steels. This awesome task can be likened to the search for the occasional needle in the haystack,

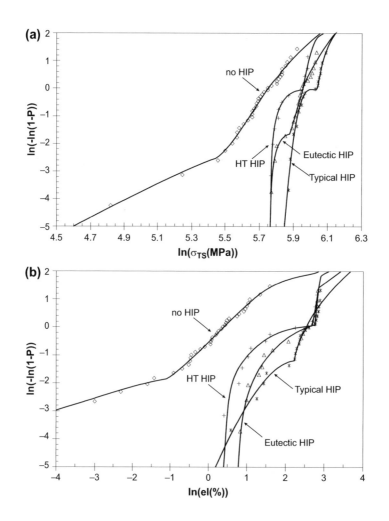

FIGURE 9.9

Three-parameter bi-Weibull plots of (a) strength and (b) elongation for non-HIPped and castings with three different HIPping techniques showing at least two distributions of defects in each set. (Staley et al., 2007).

followed by the necessity of predicting precisely how many needles may be in other random haystacks. Their simple solution is the counting only of the maximum defects in any studied area of the sample, and analysing these data with a largest-extreme-value distribution. They selected the Gumbel distribution (named after its inventor) for their study. They successfully used the technique for the study of pores and cracks.

Tiryakioglu (2008, 2009) has developed the science of extreme value distributions for application to the properties of castings. In particular, in an exercise (2008) to find the best description of the size of fatigue-initiating defects in cast Al alloys, he assesses a wide variety of different distributions including Weibull, lognormal, Frechet, General Pareto, Gumbel and General Extreme Value (GEV). He concludes that Gumbel and GEV descriptions are accurate and appropriate. At this time, these descriptions of distributions are hardly known in the casting community, but clearly deserve wider appreciation and use.

Table 9.2 Percentage Points of the Distribution of Modulus _m_ obtained by using the Unbiased Probability Estimators in Table 9.1

N	0.005	0.01	0.025	0.05	0.1	0.9	0.95	0.975	0.99	0.995
5	0.292	0.325	0.381	0.434	0.504	1.630	2.014	2.458	3.169	3.653
6	0.320	0.358	0.418	0.475	0.549	1.555	1.852	2.189	2.694	3.052
7	0.356	0.391	0.445	0.501	0.579	1.502	1.747	2.031	2.462	2.753
8	0.374	0.409	0.470	0.530	0.602	1.480	1.705	1.950	2.293	2.579
9	0.392	0.427	0.488	0.545	0.618	1.444	1.650	1.836	2.126	2.371
10	0.404	0.447	0.504	0.563	0.638	1.419	1.605	1.791	2.051	2.298
11	0.429	0.465	0.529	0.582	0.652	1.399	1.574	1.746	2.003	2.194
12	0.436	0.484	0.539	0.593	0.662	1.380	1.543	1.690	1.892	2.042
13	0.450	0.488	0.550	0.610	0.677	1.362	1.510	1.644	1.837	1.982
14	0.459	0.503	0.560	0.618	0.687	1.350	1.493	1.631	1.792	1.930
15	0.462	0.503	0.568	0.625	0.695	1.334	1.465	1.603	1.769	1.900
17	0.491	0.528	0.594	0.643	0.708	1.324	1.449	1.574	1.735	1.859
20	0.523	0.557	0.611	0.664	0.728	1.299	1.412	1.514	1.644	1.760
22	0.536	0.575	0.635	0.683	0.746	1.282	1.384	1.480	1.610	1.714
25	0.565	0.597	0.648	0.697	0.754	1.266	1.363	1.447	1.568	1.643
27	0.559	0.601	0.652	0.702	0.762	1.254	1.341	1.430	1.538	1.613
30	0.576	0.611	0.667	0.716	0.773	1.241	1.328	1.405	1.493	1.565
32	0.592	0.633	0.684	0.729	0.781	1.234	1.311	1.384	1.480	1.537
35	0.612	0.645	0.690	0.738	0.790	1.226	1.300	1.369	1.453	1.523
40	0.632	0.662	0.708	0.749	0.798	1.207	1.276	1.342	1.424	1.486
45	0.633	0.670	0.723	0.764	0.813	1.196	1.260	1.320	1.388	1.440
50	0.654	0.684	0.734	0.773	0.820	1.190	1.250	1.303	1.374	1.420
55	0.670	0.700	0.748	0.785	0.830	1.183	1.239	1.291	1.355	1.406
60	0.676	0.709	0.749	0.790	0.833	1.174	1.229	1.277	1.332	1.380
65	0.691	0.720	0.762	0.798	0.841	1.168	1.218	1.266	1.321	1.361
70	0.703	0.731	0.769	0.804	0.845	1.159	1.210	1.259	1.312	1.348
75	0.704	0.735	0.776	0.810	0.850	1.158	1.207	1.252	1.304	1.347
80	0.713	0.740	0.780	0.817	0.855	1.150	1.197	1.238	1.289	1.327
90	0.723	0.755	0.790	0.824	0.862	1.141	1.184	1.222	1.273	1.304
100	0.742	0.769	0.805	0.834	0.870	1.132	1.173	1.211	1.250	1.287

9.3 EFFECT OF DEFECTS

In general, any non-metallic discontinuity in the structure, whether a hard, soft, brittle, gas or vapour phase, will probably impair the properties to some degree. Figure 9.10 is an excellent illustration.

However, of course, the more compact the defect, the less damage will be suffered. At the point at which a defect is smaller than the micro-structural features of the alloy (i.e. smaller than the dendrite arm spacing [DAS]), it probably ceases to be important. It is important therefore to make a distinction between the less harmful defects considered in this section and the more extensive planar defects considered in the next section. (Very large planar defects can measure 10–10,000 times the linear dimensions of the DAS.)

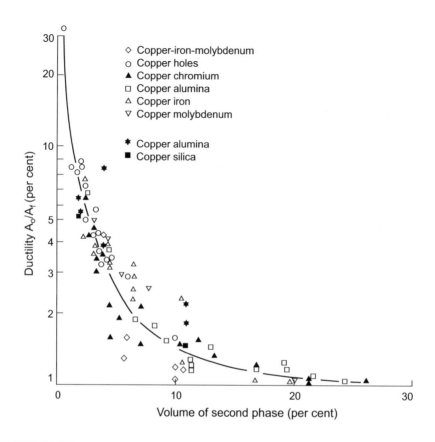

FIGURE 9.10

Ductility of copper containing a dispersion of second phases.

Data from Edelson and Baldwin (1962).

However, the density of defects (the number of defects present in a given volume) is also important, and in some cases will over-ride the effects of the compactness of the defect. We need to keep this in mind because we shall look into the twin aspects of defect size and defect density in the section on properties.

It is noteworthy that defects considered here are not solidification defects. So far as most shaped castings are concerned, apart from a few obvious features as shrinkage porosity and some segregation issues, which are comparatively rare, there is probably no such feature as a *solidification defect*. The only serious defects in castings are *casting defects*. These are all entrainment defects arising from surface turbulence before or during the filling of the mould.

9.3.1 INCLUSION TYPES

Inclusions fall into two major categories: (1) those formed in situ in the liquid or solid matrix by precipitation from solution and (2) those introduced from the outside the liquid metal via an entrainment process, and which therefore have to penetrate the oxide surface.

Those formed atom by atom from the matrix will, of course, enjoy an interface with the matrix which is atomically perfect. This will be an immensely strong interface. Traditional, classical continuum models and the latest molecular dynamics simulations of aggregates of atoms both agree that such interfaces cannot be decohered; their strength corresponds closely to the ultimate theoretical strength of the matrix, and is therefore in the region of 10–100 GPa. Such inclusions, usually present as desirable second phases, are usually beneficial as strengthening features for the alloy.

The contrast with those inclusions entrained as surface oxide could hardly be more dramatic. Entrained oxide decoheres easily from itself. Similarly, when foreign inclusions become entrained, they arrive in the matrix naturally wrapped in surface oxide, as illustrated in Figure 2.3. As a result of their unbonded wrapping, these inclusions also decohere easily from the matrix. This is the reason for the interesting uniform behaviour of pores and inclusions in copper seen in Figure 9.10. The results for pores and entrained inclusions all fall on the same curve; the higher the defect density, the worse the properties.

Even so, notice that the ductility only approaches disaster levels of 1% at defect densities of more than 10% volume fraction. This is a huge volume fraction. Soberingly, Al alloys are known to have ductilities of 1% or lower with film densities which are so low as to be difficult to estimate, but perhaps in the region of only 10^{-3} vol% (assuming a 1 mm square film 10 nm thick in one cubic mm.).

Whatever the exact numbers, it is clear that films are enormously more damaging than volume defects by a margin of at least several orders of magnitude. In view of this simple deduction, it is difficult to understand the preoccupation which researchers have in identifying compact, chunky inclusions and assuming that these are the initiation sites for cracks, when even a tiny film nearby or in association will effortlessly outweigh its effects. The message is clear: bifilm inclusions (because they act as unbonded cracks) are important and, in densities commonly found in many metals, effectively control properties.

9.3.2 GAS POROSITY

There are several different types of gas porosity, all of which are common in castings. They have to be diagnosed correctly, otherwise a wrong, and consequently ineffectual treatment, may be attempted. The main categories are listed here.

1. Gas precipitated from solution
2. Entrained air bubbles
3. Core blows
4. Micro-blows
5. Layer porosity

Layer porosity is that type of shrinkage porosity that forms in long freezing range alloys under conditions intermediate between those for concentrated macroporosity, and dispersed microporosity. This important intermediate condition is clear from Figure 7.15. Whereas macroshrinkage in the form of a shrinkage pipe, or large central cavity, can be viewed as a compact defect, and dispersed micropores can be viewed as individual compact defects, the intermediate condition is more severe, in the sense that a relatively small volume of porosity has formed itself into an extended defect by concentrating into a sheet, or layer.

To form an opinion of the seriousness of this defect, it is helpful to recall the way in which layer porosity is formed. The pore spreads across a surface of constant tension that exists in the residual liquid in the dendrite mesh. This formation mechanism results in the unique structure of the defect. The layer is not exactly equivalent to a crack because the dendrites are hardly disturbed during its growth; it is only the residual liquid that moves, opening up the planar defect. In fact the dendrites continue to bridge the pore as a dense array, effectively stitching the layer together with closely spaced threads. For this reason, its effect on tensile properties seems to be intermediate between that of individual pores and a fully open hot tear or crack. This intermediate position is seen in its rather mild effects on mechanical properties illustrated in Figure 9.11.

The presence of layers is usually not easily discerned optically on a polished section because there are so many bridging dendrites that the overall pattern is almost hidden, causing the porosity to appear as normal dispersed microporosity. However, it is clearly revealed on radiographs if the view direction is oriented along the plane of the porosity.

A summary of gas pore diagnostic information is given in Table 7.3.

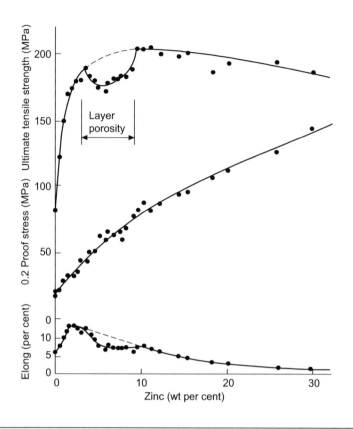

FIGURE 9.11

Effect of layer porosity on the mechanical properties of Mg-Zn alloys.

Data from Lagowski and Meier (1964).

9.3.3 SHRINKAGE POROSITY

Microshrinkage

The effect of shrinkage porosity on mechanical properties is probably practically identical to that of any other kind of porosity. On fracture surfaces, it is common to identify the pore as the origin of the failure. However, we have to keep in mind that, although not obvious to a casual viewer, a bifilm will almost certainly be associated with the pore, probably having initiated the formation of the pore, and will almost certainly be the starting point of the crack leading to failure. The relative likelihood of initiation by the pore or its associated bifilm is easily assessed by comparing the radius of curvature that each provides as a stress raiser. The radius of curvature of a shrinkage cavity will correspond to the curvature dictated by the curvature possible between dendrites of the order of 10 μm for a typical small casting. The curvature formed by the tip of a bifilm would be expected to be in the region between 10 and 1000 times smaller, thus being a significantly more threatening initiation site for a crack propagated by stress.

A further pitfall to bear in mind is that practically all shrinkage defects identified by radiography are actually not shrinkage defects but oxide film defects. A compact, convoluted bifilm, or tangled mass of many bifilms, containing some residual entrapped sprinkling of gas will appear identical to shrinkage porosity on a radiograph. Such attempted identifications should *not* be labelled 'shrinkage' but instead '*appears* to be shrinkage' (with the proviso 'but probably not' added for good measure). In my experience, most so-called 'shrinkage defects' turn out to be oxide defects introduced by poor filling systems.

True shrinkage defects are, in my opinion, rather rare. This is easily understood, because most foundry people understand shrinkage and the necessity for the provision of feeders. Thus most castings are reasonably well fed and are unlikely to show 'true' shrinkage defects, only 'apparent' shrinkage defects which are in fact filling system defects.

For those castings that do suffer some reduced internal pressure as a result of a poor feeding condition, if any part of the liquid is still contact, or nearly in contact, with the outside skin of the casting, the skin might be sucked in, and a shrinkage pore grown by extension of this microscopic pipe drawn in from the outside surface. I have previously called this *externally initiated shrinkage porosity* (Figures 7.13 and 7.17). On a section cut through the casting, it is indistinguishable from an internally initiated porosity. However of course, because it is externally linked, it constitutes a leak to the outside world. It would perform badly as a vacuum component, or would internally corrode, or harbour contamination in a food processing or chemical plant. If machined into from the far side, it would form a through-wall leak.

For other genuine shrinkage conditions in a long freezing range alloy, the microshrinkage pores can take the form of layer porosity (Figure 7.15)—as has already been described in the previous section—as also possibly formed by gas precipitation. As has been repeated mentioned throughout the book, both shrinkage and gas may have cooperated to form the porosity. It is probably not possible to detect the driving source of the porosity, both shrinkage and gas varieties, and the combined variety, would probably all appear identical if formed at the same stage of dendrite development in the casting.

Macroshrinkage (Shrinkage Pipe)

A macroshrinkage pipe, usually from a feeder head, can take the form of a smooth walled cone for a eutectic alloy, or an extensive region of sponge porosity if a long freezing range alloy (Figure 7.8). The sponge appears to be separate interdendritic pores on a section, but the pouring of a coloured liquid through the pores quickly illustrates how interconnected they are.

The various shapes of macro pores are very much influenced by gravity (effectively they float to the top of trapped liquid volumes—or perhaps we should say the residual liquid in the trapped liquid region naturally sits at the bottom of these volumes). Typical shapes are seen in Figures 7.30 and 7.31.

Such macroporosity is simply a sign of inadequate feeding. The rules for good liquid feeding are presented in 'Casting Manufacture'.

9.3.4 TEARS, CRACKS AND BIFILMS

All planar defects such as cracks are capable of constituting extensive faults in the casting, and so are of great concern with regard to its strength, toughness and fatigue resistance. Not only do such features reduce the load bearing area, but increase stress on the residual area even further because of the effect of stress intensification at the crack tip. The overall effect is significant as is clear in Figure 9.12.

Interestingly, the effect of these defects on the yield stress is practically zero. Figure 9.11 illustrates a clear result for Mg-Zn alloys. The effect arises as a result of the reduction in area and notch effects leading to plastic flow around the crack tip which raises the strength by work hardening. The loss of area and the gain in strength match sufficiently well to yield a negligible net change.

A fundamental, but not yet resolved issue, needs to be raised at this point. It is possible that tears, and cracks and bifilms are all the same defect, all being simply bifilms. The difference is merely that tears and cracks are bifilms that are open; tears have opened early, while the casting is still partly liquid, whereas cracks have opened late, after full solidification. Tears open with little or no stress, merely responding to strain, whereas cracks in the solid tend to only open after significant stress has accumulated, and possibly with a loud audible 'crack' as elastic energy is released. Bifilms are of course effectively cracks, but if they remain substantially closed, they can remain undetected and unsuspected. It is frustrating that insufficient research has been carried out to ascertain whether the statement 'all tears and cracks are originally bifilms' is true. However, on balance, it seems to me to be likely to be true.

Hot tears and cold cracks open during cooling as a result of tensile strain and stress respectively in the casting. They typically occur, therefore, across sections and in radii, where they will have an especially damaging effect on the serviceability of the casting. Their diagnostic features have already been discussed in Chapter 8. As open failures, they are naturally more serious defects than a closed bifilm. Although the bifilm can act as a crack when subjected to a tensile stress at right angles to its plane, it may not have developed significant planar morphology by unfurling, and even if

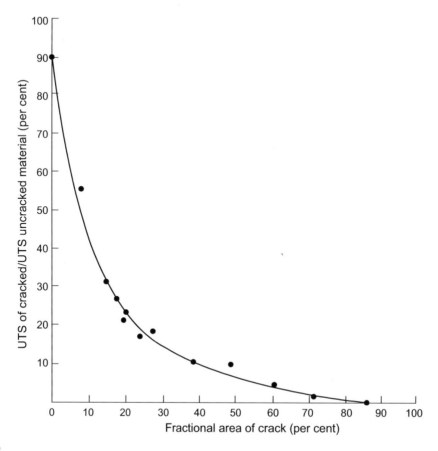

FIGURE 9.12

UTS of a casting as a function of the area of the crack.

From Clyne and Davies (1975).

completely unfurled, and thus not capable of withstanding a tensile stress normal to itself, it will be able to withstand significant shear stress because of friction between the oxide halves, and because of jogs and creases that will provide mechanical obstruction to shear. Thus a closed crack, or closed bifilm, is a crack-like defect, but exhibits some strength under some loading conditions. The open crack or opened bifilm is a seriously weak feature.

Such defects, especially if open, can extend to the whole casting cross-section, in which case, of course, the casting is already broken, and therefore perhaps fortunately identifiable as defective before being put into service!

However, a crack connected to the surface may have formed and opened at high temperature may close again at the surface as the casting cools because of the normal development of compressive stress in the casting surface during the final stages of cooling, as discussed in Section 8.2. This is a standard cooling regime for castings. Thus cracks that are closed, and effectively invisible at the surface, should be expected as the norm. Such cracks, closed tightly under pressure, would be extremely difficult to detect by normal methods such as dye-penetrant testing and radiography. It is possible, therefore, that this is a common but rarely detected defect. It represents a serious concern in safety-critical components. Its detection might require pre-stressing to open the crack whilst being viewed by real-time radiography, or while being investigated by a penetrant dye. The writer is not aware that such testing techniques are ever used. However, for the most important of applications they would be well worth careful consideration. Sophisticated resonance testing would be another discriminating testing technique.

It is worth restating that the extensive tear or crack-like defects are probably not so much initiated by bifilms, but actually *are* the bifilm. These entrainment defects can easily extend over the whole casting cross-section, as we have seen. The visible presence of one of these defects will almost certainly warn of the presence of many more, less easily seen neighbours, and is the reason why it is sometimes futile to attempt to repair such defects by welding for instance.

The lesson remains clear. Intensive inspection and repair is no substitute for correct casting procedures to achieve good metal, quiescently filled.

Bifilms

The double film defect, the bifilm, is formed during the re-aggregation or re-assimilation of a dispersed or fragmented liquid. Thus actions such as the folding-in of the liquid surface, or the mutual assimilation of droplets, all result in the surface oxide film meeting as an impingement of dry face to dry face. Whereas liquid metals during turbulent casting are re-aggregation events of a fragmented liquid, close analogies exist with several solid-state processes such as powder metallurgy and spray forming which are similarly re-aggregation processes, re-forming bulk solids from oxide-covered particulates.

The composition of the surface film is usually an oxide, nitride or pure carbon. The bifilm defect can be extensive, occupying the whole of a cross-section. Alternatively may be small and even negligible, but no one knows how small bifilms can be and still exert some important effect on properties. Similarly, placing and orientation are random, so that either its chance placing, or its orientation, or its size, or all factors together, in a highly stressed region, may or may not be damaging. It is the unpredictable nature of the double film defect that is such a concern. In this way, it contrasts with the bifilms which must exist in powder metallurgy products because here the films are uniformly thin and of a controlled size and dispersion. This is all perfectly predictable.

I am often questioned why a sporadic defect such as a crack always appears at the same place in a particular casting. Clearly, the pattern of stress concentration in the component is fixed. The random distribution of bifilms means that in the case of relatively few bifilms sometimes a bifilm will happen to be sited in the stressed region, and thus be opened by the strain. However, if there is a high concentration of bifilms, as often happens, there will then always be at least one in this region, with the result that all castings will exhibit a similar crack.

Figure 9.13 shows two simple plates 10 mm thick cast in 99.5% Al in the author's laboratory (Runyoro, 1992). They were cast in a vertically parted mould via a bottom gate into the centre of the long side at ingate speeds of above and below 0.5 ms^{-1}, respectively, and were afterwards subjected to a three-point bend test (later work progressed to use the

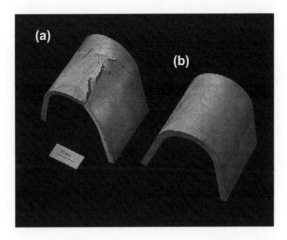

FIGURE 9.13

Plates of 10 mm thick cast in 99.5Al subjected to three-point bending (a) filled at an ingate speed greater than 0.5 m/s and (b) less than 0.5 m/s.

Courtesy Runyoro (1995).

much improved four-point bend test). The turbulently filled casting exhibited several cracks, none of which was along the line of the maximum stress. In fact, the cracks were aligned in *random* orientations. The *randomness* was a strong clue to their origin as the products of surface turbulence because turbulence means randomness, chaos, and essentially unpredictability. The material between the cracks is clearly highly ductile, as would be expected for rather pure cast aluminium. Also, interestingly, under the microscope, the tips of the cracks are found to be blunt. They have been unable to propagate in the ductile aluminium, which constitutes a crack-blunting matrix. It follows that the cracks have not propagated but must have pre-existed as cast, and have merely opened when subjected to bending. All these features are to be expected from entrained bifilms. The test plate (Figure 9.13(b)) cast at a speed at which entrainment is not possible is clearly free from bifilm defects.

If the bifilm is sufficiently large, and if it breaks the surface, then it *may* be detected by normal dye-penetrant methods. However, as a result of the bifilm being naturally tightly closed or forced closed at the surface by surface compressive stress, the dye penetrant approach is not reliable, indicating perhaps only minute pinpoint contact in places. The 'stars at night' appearance of defects on vacuum-cast Ni base alloys when the fluorescent penetrant dye is viewed under ultraviolet light are mostly defects of this kind. This is easily demonstrated by putting on a well-designed filling system, after which the 'stars' all disappear! Bifilm defects are also difficult to detect by radiography unless the incidence of radiation happens to be in or near the plane of the defect, or unless the central region is sufficiently open to constitute an open crack, or the enclosing films sufficiently thick to appear as inclusions.

In contrast to the difficulty of observing bifilms non-destructively, they can sometimes be clearly seen on fracture surfaces. Very thin films require careful study of the fracture surface to reveal their minute characteristic signatures such as the tracery of fine folds. LaVelle (1962) studied the fracture surfaces of aluminium alloy pressure die castings and found layer upon layer of oxide films throughout the structure, giving the fracture a fibrous appearance. In some cases, the surface contact of these extensive defects was minimal, indicating how unreliable their detection by a surface penetrant dye technique could be.

Because film defects are potentially so damaging, and because they are practically impossible to detect by non-destructive methods with any certainty, the only suitable method to deal with this problem is not to attempt to inspect for them, but to set in place manufacturing techniques that guarantee their avoidance. At the present time, no specification calls for specific manufacturing techniques that would eliminate film defects. This unsatisfactory situation has to change.

Only control of the quality of the melt and of the filling of the casting will prevent film defects entering the mould. Once good quality metal is in the mould, measures will be required to prevent the reintroduction of defects by such processes as the outgassing of cores or other forms of surface turbulence after the filter (if any) in the runner. Only by carefully planned and monitored processing will bifilm creation be overcome with confidence.

Again, it bears plentiful repeats: the achievement of good castings lies not in inspection, but in process control.

9.4 TENSILE PROPERTIES

The fracture of metals is a complex subject that we can only touch on with respect to its castings aspects. Even so, the outline of knowledge presented here will be helpful to guide the casting engineer to achieve the best properties for his castings.

9.4.1 MICRO-STRUCTURAL FAILURE

Three main varieties of micro-structures and their mechanical behaviour can be distinguished:

1. Where the matrix of the alloy is ductile, and the second phase, perhaps as hard particles, introduced by mixing (i.e. entrained), the added particles are not bonded to the matrix. When subjected to tensile strain the particles act like pores as seen in Figure 9.10 as a result of their wrapping of unbonded oxide film, as seen in Figure 2.3. This is easily understood; the mix has isotropic properties, and the microscopic failure associated with each particle has no preferred directionality.

2. Where the second phase particles have formed in situ in the alloy, if their formation was not associated with entrained particles, they would be perfectly bonded to the matrix. No fracture within the particles or decoherence at the interface could occur. This is perhaps more clearly seen in the precipitates formed in the solid, such as the GP zones in Al alloys, where debonding is inconceivable. Gall et al. (2000) use atomic simulations to estimate that the force to separate Si from the Al matrix would probably exceed 30 GPa. This is 100 times a typical stress to cause tensile failure of an Al-Si alloy. Thus debonding (and we might also assume, cracking) will be impossible for Si particles.

3. For those second-phase particles that nucleate and grow on entrainment defects such as bifilms, the associated unbonded interface in the bifilm causes interesting specific failure modes. As always, Al-Si alloys are the most intensively studied system comprising an essentially non-metallic second phase, silicon metal, in the soft metallic Al matrix. As such, the findings are instructive because it exhibits behaviour typical of many in other such alloy systems, particularly the cast irons.

When subjected to a tensile load, the silicon particles appear to either fracture or decohere from the matrix. In terms of bifilm theory, this behaviour occurs if the silicon happens to have precipitated on both sides of the bifilm, with the crack in the centre (or only on one side, with the crack between the particle and the matrix).

Some admirable evidence supporting this hypothesis comes from the work of Lopez and his team (Fras, 2007) who painstakingly measured individual particles, noting their orientation to the tensile axis and whether they fractured internally or decohered. Their fascinating results are presented in Figure 9.14.

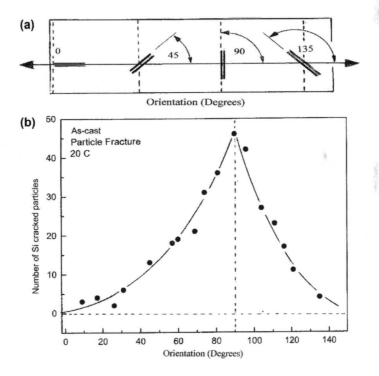

FIGURE 9.14

(a) Silicon particles in Al-Si alloy A319 subject to tensile stress. (b) At room temperature, the particles mainly fracture when at 90° to the stress axis. (c) At 300°C ductile shear at 45° promotes decohesion, in addition to tensile failure continuing near 90°; (d) for alpha-Fe particles at room temperature, their convoluted bifilms lead to fracture at random angles (Fras, Wiencek, Gorny, Lopez, 2007).

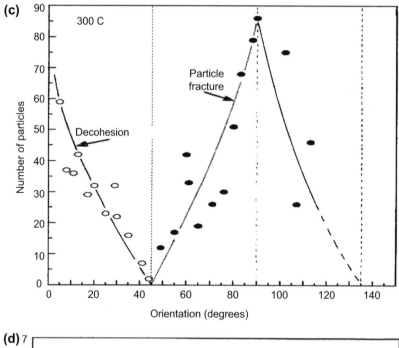

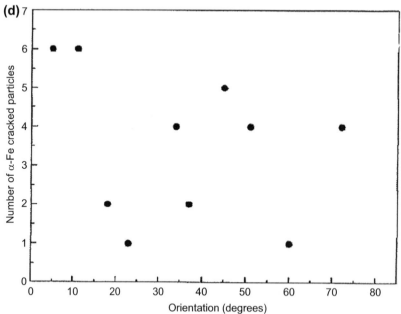

FIGURE 9.14 Cont'd

Clearly, at room temperature, most particles fractured when the stress was at right angles to the plane of the particle. This is no surprise. However, it is interesting that no decohesion is reported for this orientation. In the case of the silicon particles, perhaps the tensile force operates on the extreme edges of the two particles to separate them successfully, whereas when operating on the edges of a vanishingly thin and weak half of the bifilm the strain might tweak the edges a little, but is not expected to pull the bifilm completely clear as in the case of the strong, rigid particle.

At 300°C, the situation becomes more complex. The matrix now softens and can flow plastically under tensile load. Those particles aligned parallel to the stress direction and which have formed on only one side of a bifilm will now be subject to material shearing across and at 45° away from the particle, opening up the unbonded interface. When oriented at 45° to the stress, the maximum resolved shear will be aligned parallel to the length of the particle, so that decohesion is no longer favoured. Fracture of particles starts to occur at increasing angles of orientation, maximising at 90° once again as would be expected.

As an interesting comparison, alpha-Fe particles were also checked for their sensitivity to orientation with respect to the tensile stress direction. As is clear from Figure 9.14(d), there is no favoured direction. This is to be expected in view of the ability of this cubic phase to wrap around convoluted bifilms during its growth, sealing in the convolutions, rather than straightening them as happens with Si particles and beta-Fe particles. The marked difference in behaviour is clearly shown in Figure 6.20.

Turning now to a common feature of fracture surfaces, Figure 2.44(d) shows a typical so-called ductile fracture consisting of a dense array of ductile dimples. At the base of each dimple is a fractured Si particle. If the proposition is that Si particles are strong and resistant to fracture, then surely there must be a bifilm present in every ductile dimple? This reasoning indicates that huge, possibly unreasonably high, numbers of bifilms are present.

In practice, this appears to be only partly true. It seems there is a reasonable explanation of how such numerous fractures could occur from a much smaller population of bifilms as follows. Each bifilm will consist of one smooth side and one side with numerous transverse folds, constituting short transverse cracks. An illustration is seen in Figure 6.20(b) which shows a long beta-Fe particle with one continuous length on one side of the main crack, and separate short lengths of crystal on the other side separated by transverse cracks. This structure is to be expected in all bifilm formation because, during the turbulence of folding in the surface, the impinging of two areas of surface are not likely to each have the same area; the larger surface will be forced to adopt rucks, as transverse folds, while the opposite surface film will remain essentially unfolded. Thus the growth of the beta-Fe crystal will be continuous along one side of the continuously smooth side of the bifilm and will be a series of short lengths on the other side of the bifilm as a result of a natural presence of short transverse cracks.

During the solidification of the Al-Si–coupled eutectic, the bifilms in suspension will tend to be dragged down into the hollows between the advancing Al and Si phases (Figure 9.15). This will occur because the bifilms will be pushed ahead by the Al phase because the oxide is not preferentially 'wetted' by the alpha phase, but will preferentially attach to the Si phase, lowering the overall energy of the system. The bifilms will have one smooth side and one with transverse folds (i.e. subsidiary transverse cracks connected to the main longitudinal crack). This is shown in Figure 9.15. If it happens that the transverse cracks are on the side of the Al phase, they will be pushed ahead, effectively flattened against the advancing Si particles as a result of the Al phase tending to 'push' the bifilm. Thus the Al phase will not be cracked, explaining the perfection of the plastic shear between Si particles, forming razor-sharp cusps of necked-down material. On the other hand, if the transverse cracks happen to be on the side of the Si particles, the Si will grow around these, incorporating the transverse cracks into the advancing Si crystal. Thus the advancing Si phase will be expected to be cracked repeatedly along its length. If Figure 6.20 is typical, we might expect to see transverse cracks every 10–20 μm which appears reasonably consistent with the observed fractures.

9.4.2 DUCTILITY

Figure 9.10 is a famous result showing the ductility (in terms of reduction of area, RA) of a basically highly ductile material, pure copper, being reduced by the addition of various kinds of second-phase particles, including pores. It is clear that the clean material has a ductility of more than 30%. An RA of 100% (meaning necking down to zero residual area) would be expected for extremely clean copper, but there is a large deleterious effect of the second phases, more or

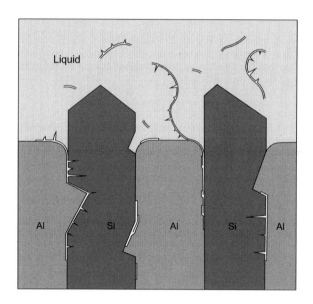

FIGURE 9.15

Bifilms with subsidiary transverse branches are entrained between the Al and Si phases during the growth of the Al-Si eutectic. Branches are pushed back and rejected by the Al phase, but are incorporated in the Si phase, leading to easy Si fracture and decoherence.

less irrespective of their nature. The lack of sensitivity to the nature of the particles or holes is almost certainly the result of the relatively easy decohesion of the particles from the matrix when deformation starts. Thus all particles act as holes.

This result is predictable if the particles are introduced into the melt by some kind of stirring-in process. As the particles penetrate the surface, they necessarily take on the mantle of oxide that covers the liquid metal (Figure 2.3(b)). Thus all immersed particles will be expected to be coated with a layer of the surface oxide, with the dry side of the oxide adjacent to the particle. The absence of any bonding across this interface will ensure the easy decohesion that is observed. In practice, the submerged particles will often remain in clumps despite intense and prolonged stirring (Figure 2.3(c)). This seems to be most probably the consequence of the particles entering the liquid in groups and being enclosed inside a packet of oxide. With time, the enclosing wrapping will gather strength as it thickens by additional oxidation, using up the enclosed air, and so gradually improve its resistance to being broken apart.

In castings, the volume of pores rarely exceeds 1%. (Only occasionally is 2–3% porosity found.) For 1% porosity, Figure 9.10 indicates that the ductility will have fallen from the theoretical maximum (which will be 100% reduction in area for a perfectly ductile material) to approximately 10%, an order-of-magnitude reduction! However, 10% elongation (note we are neglecting that the elongation is not accurately equivalent to RA) is fairly typical, if rather generous, for most light alloys, indicating the possibility of at least 1% by volume (or area) of defects normally present whether detected or not.

Why should an assembly of holes in the matrix affect the ductility so profoundly?

Figure 9.16 shows a simple model of ductile failure. For the sound material, the extension to failure is of the order of the width l of the specimen, because of the deformation of the specimen along 45° planes of maximum shear stress. For the test piece with the single pore of size d, the elongation to failure is now approximately $(l - d)/2$. In the general case for a spacing s in an array of micropores, we have

$$\text{Elongation} = s - d$$
$$= 1/n^{1/2} - (f/n)^{1/2} \qquad (9.5)$$
$$= \left(1 - f^{1/2}\right)/n^{1/2}$$

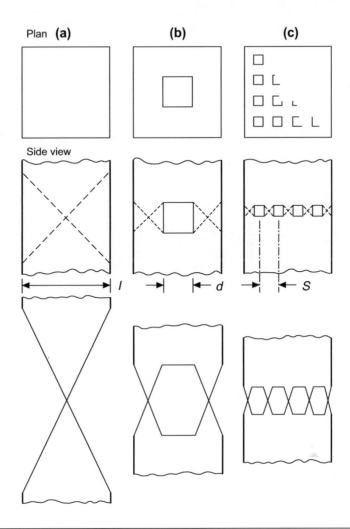

FIGURE 9.16

Simple ductile failure model, representing (a) sound specimen that necks down 100% reduction in area; (b) a single central pore reducing RA to about 50% with a single cup-and-cone fracture; and (c) multiple pores leading to poor elongation with ductile dimple fracture surface.

where n is the number of pores per unit area, equal to $1/s^2$ and f is the area fraction of pores on the fracture surface, equal to nd^2.

Equation (9.5) is necessarily very approximate because of the rough two-dimensional model on which it is based. (For a more rigorous treatment, the reader is recommended to the pioneering work by Thomason (1968) and the more recent work by Huber (2005).) Nevertheless, our order-of-magnitude relation indicates the relative importance of the variables involved. It is useful, for instance, in interpreting the work of Hedjazi (1976), who measured the effect of different types of inclusions on the strength and ductility of a continuously cast and rolled Al-4.5Cu-1.5Mg alloy. From measurements of the areas of inclusions on the fracture surface, Hedjazi reached the surprising conclusion that the film defects were less important than an equal area fraction of small but numerous inclusions. His results are seen in Figure 9.17. One can see that for a given elongation, the microinclusions are about 10 times more effective in lowering ductility. However, he reports that there were between 100 and 1000 times the number of microinclusions compared with

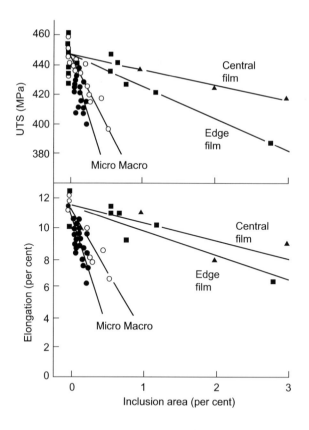

FIGURE 9.17

Strength and ductility of an Al-4.5Cu-1.5Mg alloy as a function of total area of different types of inclusions in the fracture surface.

Data from Hedhazi et al. (1976).

film-type defects in a given area of fracture surface. From Eqn (9.5), an increase in number of inclusions per unit area by a factor of 100 would reduce the elongation by a factor of 10, approximately in line with the observations.

The other observation to be made from Eqn (9.5) is that ductility falls to zero when $f = 1$, for instance in the case of films which occupy the whole of the cross-section of the test piece. This self-evident result can easily happen for certain regions of castings where the turbulence during filling has been high and large films have been entrained. This is precisely the case for the example for the ductile alloy Al-4.5Cu that failed with 0.3% elongation, nearly zero ductility, seen in Figure 2.44(a). This part of the casting was observed to suffer a large entrainment effect that had clearly created extensive bifilms. Elsewhere, other parts of the same casting had filled quietly, and therefore contained no new bifilms, but only the background scatter of old bifilms inherited from the distribution already present in the melt. In this condition the ductility of the cast material rose to a rather mediocre 3.5% (Figure 2.44(b)).

Pure aluminium is so soft and ductile that it is almost possible to tie a length of bar into a knot and pull the knot tight. However, Figure 9.13 illustrates how the presence of bifilms has caused even this ductile material to crack when subjected to a three-point bend test. Notice the material close to the tips of the cracks is highly ductile, so the cracks could not have propagated as normal stress cracks, because the crack tips would have blunted rather than propagated (as confirmed under the microscope). Thus the only way for such cracks to appear in a ductile material like pure aluminium is for the cracks to have been introduced by a non-stress mechanism. The random accidents of the folding-in of the surface from surface turbulence seems the only likely mechanism, corroborated by the random directions of the cracks, not necessarily aligned along the direction of maximum strain. The *randomness* of their *direction* and *size* is strong confirmation of their origin as the result of a chaotic process such as flow suffering surface turbulence.

Regardless of the inclusion content of a melt, one of the standard ways to increase the ductility of a casting is to freeze it rapidly. This is usually a powerful effect. Figure 2.47 illustrates an approximate 10-fold improvement as the DAS drops from 90 to 30 μm. The effect follows directly from the freezing-in of bifilms in their compact form, reducing the time available for the operation of the various unfurling mechanisms. (There may also be some a contribution from the dendrites pushing the bifilms away from surface regions, effectively sweeping the surface regions clear, and concentrating the bifilms in the centre of the casting where they will be somewhat less damaging to properties as is clearly seen to happen in the channels, mainly the runner and gates, of filling systems of castings. This effect has not been investigated, but may be significantly affected if the bifilms are not quite cleared from the surface regions, but are organised into planar sheets by the columnar grains, as in Figure 2.43(c), and whether therefore the benefits are now dependent on the direction of stress.)

The converse aspect of this benefit is that if the ductility of a casting from a particular melt quality is improved by chilling, this can be probably taken as proof that oxides are still present in the melt. (A simple quality control test can be envisaged based on this proposition.)

The great sensitivity of ductility to the presence of defects in the form of pores or cracks (or any other easily decohered interface such as those obligingly provided by bifilms) is the most powerful limitation to the attainment of good ductility. The awesome spread of ductilities of even aerospace castings is graphically illustrated in Figure 9.18(a) for Al-7Si-0.4Mg alloy by Tiryakioglu (2009a). Because strength is increased by heat treatment, so the elongation of the alloy falls. However, the major effect can be seen to be not heat treatment, but some other arbitrary factor, causing many results to fall lamentably short of their potential. This scatter is attributed to the presence of bifilms. No other conclusion seems possible because aerospace foundries pride themselves on the control of all casting variables. Unfortunately, the one variable which has been unsuspected and has therefore escaped control so far has been turbulence during pouring. Tiryakioglu (2009b,c) produces similar results for the higher strength Al-4.5Cu-type casting alloys A206 and A201, the latter containing 0.7Ag (Figure 9.18(b)). It is clear that excellent properties could be achieved with simple, low-cost Al-Si alloys if they were sufficiently free from bifilms.

Furthermore, the limit shown in Figure 9.18 is only the limit of *uniform* elongation. After the initial extension of the test piece in a uniform manner, it then starts to *neck down*, beginning the *non-uniform* elongation regime. Alexopoulos and Tiryakioglu (2009) show that the total elongation is nearly doubled by the addition of the non-uniform elongation. The elongation associated with necking-down is expected to be an important feature of metals free from bifilms. Zhong and colleagues (2013) suggest it is the work hardening and the strain rate hardening close to the onset of necking that matter for ductility and formability, rather than the overall values of these characteristics. Thus when metals and alloys are produced free from bifilms, they will have a high but as yet unknown elongation, but be expected to achieve routinely 100% RA as a demonstration of their perfection. The RA might become then a more common and discriminating measure of ductility.

Finally, there is the practical challenge of achieving the best ductility from a given batch of alloy. For instance, the ductility falls with aging, but having reached a minimum at the peak hardening response, the ductility then rises once again. Thus occasionally over-aging is recommended equally with under-aging to achieve good ductility. Figure 9.19 illustrates that this is unfortunately not always true for the precipitation hardening of Al alloys. The under-aging treatment optimises ductility for Al-Si-Mg alloys. However, the opposite appears to be true for an Al-Zn-Mg-Cu alloy (Figure 9.20).

9.4.3 YIELD STRENGTH

Steels have a well-defined yield point separating elastic and plastic flow regions so the yield stress can be defined precisely. Al alloys show no such definitive behaviour. The smooth and gradual transition between elastic and plastic behaviour is so difficult to define that it forces us to define an effective yield stress in terms of a 'proof stress'. We shall

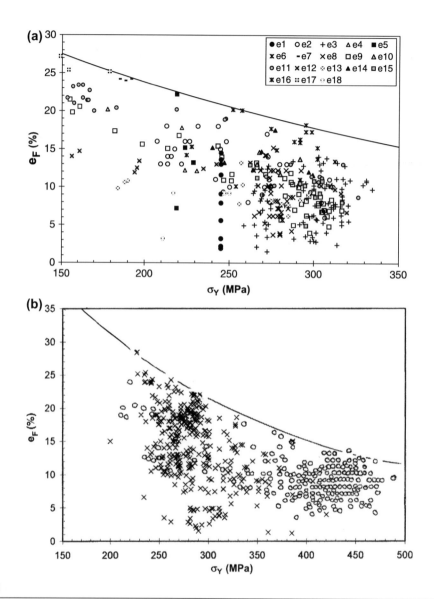

FIGURE 9.18

The scatter of uniform elongation to fracture results for (a) Al-7Si-Mg alloys (356 and 357 types) and (b) Al-4.5Cu alloys (201 and 206 types), reported mostly from aerospace foundries. (Tiryakioglu et al., 2009). Clearly, the attainments of many fall far short of the potential available.

assume that the 0.2% proof stress (0.2 PS) is sufficiently equivalent for our purposes. (The 0.2 PS is that stress at which reversible elastic deformation has just exceeded its limit, and a small amount, 0.2%, of permanent plastic deformation remains on unloading. Some define yield at 0.05 PS and others at 0.1 PS. All these definitions are closely similar to within a few MPa.)

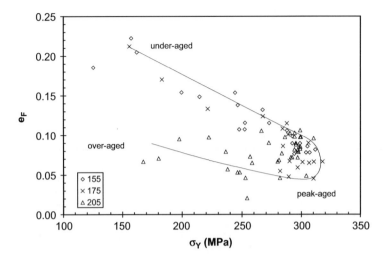

FIGURE 9.19

Elongation versus yield strength for Al-7Si-0.4Mg alloy showing under-aged strengths to be superior to those developed by over-aging (Alexopoulos and Tiryakioglu, 2009).

Whereas both the ultimate strength and the ductility can be greatly damaged by defects, the yield strength is interestingly insensitive to their presence. This behaviour is explained by the load bearing area being reduced by the presence of the defect, thus increasing the stress in the matrix and thereby causing work hardening because of local plastic flow. The increased strength of the matrix as a result of the work hardening to some extent counters the effect of the reduction in load bearing area, keeping the yield point reasonably constant.

This insensitivity of yield to layer porosity is seen in Figure 9.11. The few percent or so loss of area from layer porosity, clearly reducing the ultimate tensile strength (UTS) and elongation, cannot be detected in the scatter of the results for the 0.2 PS. A similar fairly constant yield stress result is seen in Figure 2.47.

In those cases where the yield stress *is* found to be significantly reduced, it is a warning that the material had lost a significant fraction of its load bearing area. In the past it has not been easy to accept that the reduction of 0.2 PS by 90% is the result of a crack occupying 90% of the area, especially when the fault is not easily seen, even on the fracture surface! (Figure 2.44(a) which we have discussed several times before is a clear example showing a fracture of a tensile specimen expected to show 10% elongation but in fact achieved close to zero elongation, corresponding to the 100% loss of area provided by large bifilms. The 0.3% elongation recorded for this sample was found on close examination to have resulted from the interlocking of dendrites either side of the main bifilm crack, giving an extension to failure approximately equal to the DAS.)

Micro-structural Size Effects: The Hall–Petch Relation

As the grain size d of a metal is reduced, its yield strength σ_y increases. The widely quoted formula to explain this result is that due to Hall and Petch (see, for instance, the derivation by Cottrell (1964)):

$$\sigma_y = a + b \cdot d^{-1/2} \tag{9.6}$$

where a and b are constants. The equation is based on the assumption that a slip plane can operate with low resistance across a grain, allowing one part of the grain to slide over the other, and so concentrating stress on the point where the slip plane impinges on the next grain. With the further spread of yielding temporarily blocked, the stress on the neighbouring grain increases until it exceeds a critical value. Slip then starts in the next grain, and so on. The process is exactly analogous to the spreading of a crack, stepwise, halting at each grain boundary.

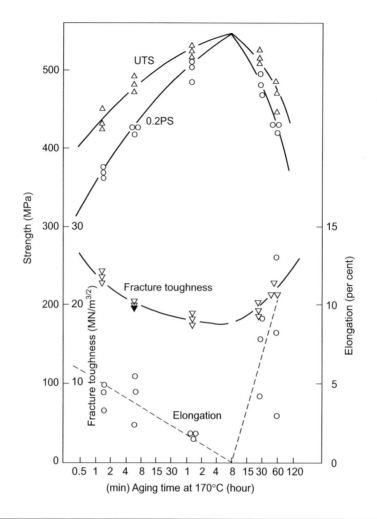

FIGURE 9.20

Mechanical properties of a cast high-strength alloy Al-6Zn-2.7Mg-1.7Cu as a function of effective aging time at 170°C.

Data from Chen et al. (1984).

Although the Hall–Petch relation was originally conceived to explain the effect of grain size, its fundamental power is not limited to this particular mechanism for the hindering of slip planes. For instance, slip can be halted at strong precipitates that might reside in interdendritic regions or inside grains. For instance, Gurland (1970) uses data from the literature to show for several steels in which their strength increases linearly with the inverse of the square root of the grain size for ferritic steels or the inter-particle spacing for pearlitic steels. Similarly, for Al-Si alloys, he shows the relationship between spacing and strength is the same for alloys containing 0–25% Si where the definition of spacing changes from (1) the grain size in pure aluminium, to (2) the dendrite cell size in 1.6–2.5% Si alloys and (3) the planar centre-to-centre nearest neighbour distance between Si particles in 5.3–25% Si alloys.

Both grain size and DAS are important factors affecting the strength of castings, and both rely in different ways on the Hall–Petch relation. We shall deal with these factors separately.

9.4.3.1 Grain size
General

The development of small grains during the solidification of the casting is generally thought to be an advantage. When the grain size is small, the area of grain boundary is large, leading to a lower concentration of impurities in the boundaries. The practical consequences that generally appear to follow from a finer grain size are

1. Improved resistance to hot tearing during solidification.
2. Improved resistance to cracking when welding or when removing feeders of steel castings by flame or arc cutting.
3. Reduced scattering of ultrasonic waves and X-rays, allowing better non-destructive inspection.
4. Improved resistance to grain boundary corrosion.
5. Higher yield strength (because of Hall-Petch relationship).
6. Higher ductility and toughness.
7. Improved fatigue resistance (including thermal fatigue resistance).
8. Reduced porosity and reduced size of pores. This effect has been shown by computer simulation by Conley et al. (1999) and is the consequence of the improved inter-granular feeding and better distributed gas emerging from solution. Improved mass feeding will also help as described in Section 7.1.4.2.
9. Improved hot workability of material cast as ingots.

However, it would not be wise to assume that all these benefits are true for all alloy systems. Some types of alloys are especially resistant to attempts to reduce their grain size, whilst others show impaired properties after grain refinement. Furthermore, this impressive list is perhaps not so impressive when the effects are quantified to assess their real importance. These apparent inconsistencies will be explained as we go.

In addition, important exceptions include the desirability of large grains in castings that require creep resistance at high temperature. Applications include, in particular, ferritic stainless steel for furnace furniture, and high temperature nickel-based alloy castings. Single-crystal turbine blades are, of course, an ultimate development of this concept. These applications, although important, are the exception, however. Because of the limitations of space, this section neglects those specialised applications that require large grains or single crystals and is devoted to the more usual pursuit of fine grains.

Some of these benefits are explained satisfactorily by classical physical metallurgy. However, it is vital to take account of the presence of bifilms. These will be concentrated in the grain boundaries. The influence of bifilm defects is, on occasions, so important as to over-ride the conventional metallurgical considerations. For instance, with the exception of item 5, every other item on the list is controlled by bifilms! For a traditionally educated metallurgist, this is a sobering if not shocking realisation.

For instance, in the case of the propagation of ultrasonic waves through aluminium alloy castings, this was long thought to be impossible. Aluminium alloys were declared to be too difficult. They were thought to prevent ultrasonic inspection because of scatter of the waves from large as-cast grains; no back-wall echo could be seen amid the fog of scattered reflections. However, in the early days of Cosworth Process, with long settling time of the liquid metal, and quiescent transfer into the mould, suddenly back-wall echoes could be seen without difficulty despite the absence of any grain refining action. The clear conclusion is that the scatter was from the gas film between the oxide layers of the bifilms at the grain boundaries.

By extrapolation, it may be that the so-called 'diffraction mottle' that confuses the interpretation of X-ray radiographs, and usually attributed to the large grain size, is actually the result of the multitude of thin-section pores, or the glancing angle reflections from the air layer of bifilms at grain boundaries. It would be interesting to compare radiographs from material of similar grain size, but different content of bifilms to confirm this prediction.

The strong link between bifilms and micro-structure, particularly grain size, is illustrated particularly well in Figure 2.44. Images (a) and (b) are the fracture surfaces of test bars taken from different parts of a single casting whose filling was observed by X-ray video radiography. The large grain size in the turbulently filled test bar (a) contrasts with the fine grain size in the quietly cast bar (b). The large grains are probably the result of reduced thermal convection in the casting because of the presence of wall-to-wall bifilms which obstructed local thermal convection, so that dendrites could grow without thermal and mechanical disturbance that is needed to melt off dendrite arms, leading to grain multiplication. In (b), the presence of numerous pockets of porosity suggests the presence of many smaller bifilms (that cannot be seen directly). These are older bifilms already present in the melt before pouring. The small bifilms will not be a hindrance to the flow of the melt, so that the small grain size is the result of grain multiplication because of convection during freezing.

Unfortunately, nearly all the experimental evidence that we shall cite regarding the structure and mechanical properties of castings is influenced by the necessary but unsuspected presence of bifilms. We shall do our best to sort out the effects so far as we can, although, clearly, it is not always possible. Only new, carefully controlled experiments will provide certain answers. At this time, we shall be compelled to make our best guesses.

Grain refinement

The Hall–Petch equation has been impressively successful in explaining the increase of the strength of wrought steels with a reduction in grain size, and has been the driving force behind the development of high-strength constructional steels based on manufacturing processes, especially controlled rolling or other thermomechanical treatment, to control the grain size. This is cheaper than increasing strength by alloying, and has the further benefit that the steels are also tougher—an advantage not usually gained by alloying.

The extension of these benefits to cast steels has been much less successful as has been related in Section 6.6.

The development of higher-strength magnesium casting alloys with zirconium as the principal alloying element has also been driven by such thinking. The action of the zirconium is to refine the grain size, with a useful gain in strength and toughness. The sharp increase in yield strength as grain size is reduced is seen in Figure 9.21. The zirconium is almost insoluble in both liquid and solid magnesium, so that any benefit from other alloying mechanisms (for instance, solid solution strengthening) is negligible; all the benefit is from the reduced grain size. In fact, the test to ensure that the zirconium has successfully entered the alloy is simply a check of the grain size. Yue and Chadwick (1991) used squeeze casting to demonstrate the impressive benefits of the grain refinement of magnesium castings. This is especially clear work that is not clouded by other effects such the influence of porosity. It appears that magnesium benefits significantly from the effect of small grain size because factor b in the Hall–Petch equation is high. This is the consequence of the grain boundaries being particularly effective in preventing slip, because hexagonal close packed lattices have relatively few slip systems, and only on the basal plane, so that slip is not easily activated in a randomly oriented neighbour.

This behaviour contrasts with that of face-centred-cubic materials such as aluminium, where the slip systems are numerous, so that there is always a slip system close to a favourable slip orientation in a neighbouring grain. Thus although grain refinement of aluminium alloys is widely practised, Flemings (1974) draws attention to the fact that its effects are generally over-rated. In Figure 9.21, for normally attainable fine grain sizes in the range reducing from 1 mm down to 0.1 mm all the Al alloys have small or zero slope, showing small or zero benefit of grain refinement. (Only the Al-Si has a modest slope for reasons explained in the next section.)

The usual method of grain refinement of aluminium alloys is by the addition of titanium and/or boron. The effect has been discussed in Section 6.3.4. The practical difficulties of controlling the addition of grain-refining materials are discussed by Loper and Kotschi (1974), who were among the first to draw attention to the problem of fade of the grain-refinement effect. Sicha and Boehm (1948) investigated the effect of grain size on Al-4.5Cu alloy, and Pan et al. (1989) duplicated this for Al-7Si-0.4Mg alloy. However, both these pieces of work confirm the useful but relatively unspectacular benefits of refinement on strength. These authors suggest that the effects are complicated by the alloying effects of titanium, particularly the precipitation of large TiAl$_3$ crystals at higher titanium levels because the expected linear increase in yield strength with $d^{-1/2}$ is not achieved. It seems likely, once again, that the presence of bifilms is complicating behaviour. The large crystals of TiAl$_3$ suggest that these crystals had precipitated on bifilms, and further suggest that the authors had stirred the melt in an effort to transfer as much Ti into the casting as possible. This counter-productive

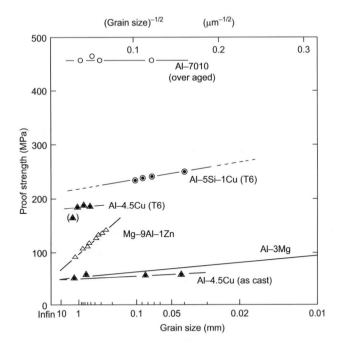

FIGURE 9.21

Hall-Petch plots of yield strength versus grain size$^{-1/2}$ illustrating (a) the relatively poor response of Al alloys contrasted with data for Mg alloys.

Data for 7070, Mg alloy, Al-4.5Cu from T M Yue and G A Chadwick (1991); Al-5Si-1Cu from Wierzbinska (2004); Al-3Mg from Lloyd and Court (2003).

technique eliminates the benefits of the settling of bifilms by the weight of Ti-rich precipitates attached to them as discussed in Section 6.3.3 'Melt cleaning by grain refinement' and Section 14.4.2 'Cleaning by Detrainment'.

It should not be assumed that the advantages, even if small, of fine grain size in wrought steels and cast light alloys automatically extend to other alloy systems. It is worth devoting some space to the difficulties and imponderables elsewhere.

For instance, steels that solidify to the body-centred-cubic form of the iron lattice are successfully grain refined by several additives, particularly compounds of titanium and similar metals, as discussed in Section 6.6 (although not necessarily with benefits to the mechanical properties, as we shall see later). In contrast, steels that solidify to the face-centred-cubic lattice do not appear to respond to attempts to grain refine with titanium and are resistant to attempts to grain refine with most materials that have been tried to date.

The grain-refinement work carried out by Cibula (1955) on sand-cast bronzes and gunmetals showed that these alloys could be grain refined by the addition of 0.06% of zirconium. This was found to reduce the tendency to open hot tears. However, this was the only benefit. The effect on strength was mixed, ductility was reduced, and although porosity was reduced in total, it was redistributed as layer porosity, leading to increased leakage in pressure-tightness tests. Although this was at the time viewed as a disappointing result, an examination of the tests that were employed makes the results less surprising. The test castings were grossly underfed, leading to greatly enhanced porosity. Had the castings been better poured and better fed, the result might have been greatly improved.

Remarkably similar results on a very different casting (but that also appears to have been underfed) were obtained by Couture and Edwards (1973). They found that various bronzes treated with 0.02Zr exhibited a nicely refined grain structure, and had improved density, hot-tear resistance, yield and ultimate strengths. However, ductility and pressure-tightness were drastically reduced. It is possible to conjecture that if the alloy had been better supplied with feed metal during solidification then pressure-tightness would not have been such a problem. The presence of copious supplies of bifilm defects are to be expected to complicate the results as a consequence of the poor casting technique.

The poor results by Cibula in 1955 have been repeatedly confirmed in Canadian research; Sahoo and Worth (1990), Fasoyinu et al. (1998), Popescu et al. (1998) and Sadayappan et al. (1999) using, mainly, permanent mould test bars. These castings are not badly fed, so that the disappointing results by Cibula cannot be entirely ascribed to poor feeding. It seems likely that bifilms are at work once again.

There may be additional fundamental reasons why the copper-based alloys show poor ductility after grain refinement. Couture and Edwards noted that the lead- and tin-rich phases in coarse-grained alloys are distributed within the dendrites that constitute the grains. In grain-refined material the lead- and tin-rich phases occur exclusively in the grain boundaries. Thus it is to be expected that the grain boundaries are weak, reducing the strength of the alloy by: (1) offering little resistance to the spread of slip from grain to grain, and so effectively lowering the yield point and (2) allowing deformation in their own right, as grain boundary shear, like freshly applied mortar between bricks. However, this seems unlikely to be the whole story because some of the poor results are found in copper-based alloys that contain no lead or tin. A further confusion probably exists because of the probability that the lead and tin phases may precipitate preferentially on bifilms, so that the observed grain boundary weakening effect may not be directly the result of the metallic alloy additions.

Many of the previous studies of copper-based alloys have used Zr for grain refinement. Thus is seems possible that they may have been seriously affected by the sporadic presence of zirconium oxide bifilms at the grain boundaries. Thus the loss of strength and ductility and the variability in the results would be expected as a result of the overriding damage caused by surface turbulence during the casting of the alloys.

The work by Nadella, Eskin and Katgerman (2007), which is referred to elsewhere in this book, beautifully illustrates the current contradictions which arise when the presence of bifilms is overlooked. These workers aimed to use grain refinement reduce the cracking in direct chill (DC) continuously cast Al alloy ingot. When they added Ti-rich grain refiner to the ladle, and then transferred this to the continuous caster, the ingot was grain refined and free from cracks. However, when they added grain refiner to the launder, immediately before the melt entering the continuous casting mould, the ingot was beautifully grain refined, but cracked. This result, baffling to the researchers, is easily understood as the settlement of bifilms by the precipitation of Ti-rich compounds on them, so the melt was cleaned from bifilms before casting. When added to the launder, the sedimentation action was no longer possible, so the bifilms were allowed to enter the casting and initiate cracks.

As the reader will now appreciate, much of our research work on the strengths of cast alloys is, frankly, in a mess. Much of this work will require to be repeated with relatively bifilm-free material. We shall then benefit from understandable and reproducible data for the first time.

9.4.3.2 Dendrite arm spacing (DAS)

DAS can refer to the spacing of the primary dendrite stems in those rather rare cases where no secondary arms exist. More usually, however, the DAS refers to the spacing between the secondary arms of dendrites. However, if tertiary arms are present at a smaller spacing, then it would refer to this.

If the DAS is reduced, then the mechanical properties of the cast alloy are usually improved. A typical result by Miguelucci (1985) is shown in Figure 2.47. Near the chill, the strength of the alloy is high and the toughness is good. As the cooling rate decreases (and DAS grows), the ultimate strength falls somewhat. Although the decrease in itself would not perhaps be disastrous, the fall continues until it reaches the yield stress (taken as the proof stress in this case). Thus, on reaching the yield stress, failure is now sudden, without prior yield. This is disastrous. The alloy is now brittle, as is confirmed by elongation results close to zero.

Because of the effect of DAS, the effect of section size on mechanical properties is seen to be important even in alloys of aluminium that do not undergo any phase change during cooling. For ferrous materials, and especially cast irons, the effect of section size can be even more dramatic, because of the appearance of hard and possibly brittle non-equilibrium phases such as martensite and cementite in sections that cool quickly.

The improvement of strength and toughness by a reduction in DAS is such a similar response to that given by grain refinement that it is easy to see how these two separate processes have often been confused. However, the benefits of the refinement of grains and DAS cannot be the result of the same mechanisms. This is because when considering DAS no grain boundary exists between the arms of a single dendrite to stop the slide of a slip plane. A dislocation will be able to run more or less without hindrance across arm after arm, since all will be part of the same crystal lattice. Thus, in general, it seems that the Hall–Petch equation should not apply. Also, of course, Hall–Petch can only explain an increase in yield with an improved refinement, whereas refined DAS offers also offers an improved ductility.

Why then does a reduction in DAS increase both strength and toughness?

This question appears never to have been answered.

Classical physical metallurgy has been unable to explain the effect of DAS on mechanical properties. Curiously, this important failure of metallurgical science to explain an issue of central importance in the metallurgy of cast materials has been consistently and studiously overlooked for years.

In the first edition of *Castings* the author suggested that the answer seemed to be complicated, and to be the result of the sum of several separate effects, all of which seem to operate beneficially. These beneficial processes are listed and discussed in the following sections. However, after these effects have been reviewed and assessed, it will become clear that the major benefit from a refinement of DAS remains largely unexplained.

We shall arrive at what seems to be an inescapable conclusion: the action of bifilms will be presented as the dominant effect, explaining for the first time the widely appreciated benefits of small DAS in castings, as we shall see.

Residual Hall–Petch hardening

It might be supposed that there was some residual Hall–Petch effect affecting strength. A perfect atomic lattice between adjacent dendrite arms would allow slip to progress from arm to arm without problem. This seems largely true for relatively pure Al and its solid solution alloys such as the low magnesium Al-Mg alloys and many of the wrought alloys that are relatively low in solute. Thus there are fundamental reasons why the effect of Hall–Petch hardening should be negligible in these Al alloys, and in many other solid solution, single-phase alloys (also including of course many stainless steels). The small or negligible benefit to the single-phase Al alloys is seen in Figure 9.21 for 7010, Al-4.5Cu and Al-3Mg alloys.

Slight faults during growth will cause the dendrite arms within a grain to become slightly misoriented. This will result in a low-angle grain boundary between the arms. The higher the degree of misorientation, the greater the resistance will be to the passage of a slip plane.

When studying the structure of a cast alloy under the microscope, most dendrite arms are seen to be, so far as one can tell by unaided observation, fairly true to their proper growth direction. Thus any boundary between the arms will have an almost vanishingly small misorientation, presenting a minimal impediment to slip across the boundary. However, it is also usually possible to see a proportion of arms at slight deviations of several degrees, perhaps as a result of mechanical damage. If mechanical disturbance during freezing is increased, for instance by stirring or vibration, then the number of misaligned arms, and their degree of misalignment, would be expected to increase.

Thus one might expect some small resistance to slip even from rather well aligned dendrites because of the lack of perfection; the result of the existence of subgrains within the grains. Some contribution from Hall–Petch hardening might therefore be expected to be present at all times.

For those alloys that have interdendritic precipitates, however, there may be a positive action to arrest slip between dendrite arms. This is true in Al-Si alloys, as is evident from the modest increase in strength noted in the impressive early work by Gurland (1970) and later by Wierzbinska and Sieniawski (2004) noted in Figures 9.22 and 9.23. Interestingly, for the Al-5Si-1Cu alloy, there is a maximum potential benefit of about 15 MPa if very efficiently grain refined, plus a further maximum benefit from the refinement of DAS by chilling that might yield about 30 MPa. Thus by taking maximum advantage of refinement of grains and dendrites a total addition to strength could be 45 MPa. In practice, one would expect to achieve something rather less than this in most casting conditions, so that benefits to strength of Al-Si higher than 5–7%Si are not especially impressive, but worth having.

Comparing the benefits achievable in a single-phase solid solution alloy such as Al-3Mg, Figure 9.22 illustrates that only about 10 MPa might be achieved by rapid solidification, but no benefit from grain refinement would be expected as a result of grain boundaries being easily traversed by slip planes in such materials. For most practical purposes, this benefit is negligible.

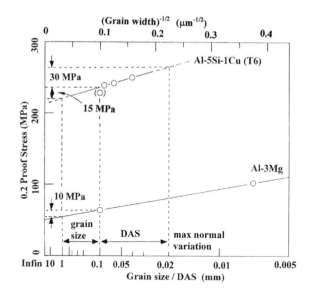

FIGURE 9.22

The poor strengthening effect of up to approximately 10 MPa for the best grain size reduction of Al-3Mg, with nothing for DAS refinement for this solid solution alloy compared with 15 MPa for DAS plus 30 MPa for grain refinement of Al-Si alloys.

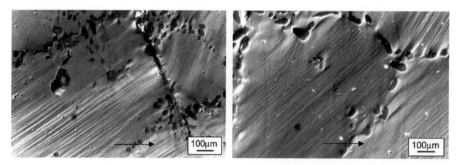

FIGURE 9.23

Change of slip band direction and intensity at interdendritic boundaries in Al-5Si-1Cu alloy (Wierbinska and Sieniawski, 2004).

We can conclude that in general, although the Hall–Petch mechanism is likely to be a contributor to increased strength, in most castings it will be negligibly small.

The final fact that eliminates the Hall–Petch effect as a major contributor to the DAS effect is the fundamental fact that Hall–Petch strengthening affects only the yield strength. Figure 2.47 and many similar results in the literature indicate that yield strength is hardly affected by DAS. The main effect of changes in DAS is seen in the ductility and ultimate strength values which Hall–Petch is not able to explain.

Restricted nucleation of interdendritic phases

As the DAS becomes smaller, the residual liquid is split up into progressively smaller regions. Although in fact these interdendritic spaces remain for the most part interconnected, the narrowness of the connecting channels does make them behave in many ways as though they were isolated.

Thus as solutes build up in these regions, the presence of foreign nuclei to aid the appearance of a new phase becomes increasingly less probable as the number of regions is increased. As DAS decreases, the multiplication of sites exceeds the number of available nuclei, so that an increasing proportion of sites will not contain a second phase. Thus, unless the concentration of segregated solute reaches a value at which homogeneous nucleation can occur, the new phase will not appear.

Where the second phase is a gas pore, Poirier et al. (1987) have drawn attention to the fact that the pressure due to surface tension, becomes increasingly high as the curvature of the bubble surface is caused to be squeezed into progressively smaller interdendritic spaces. The result is that it becomes impossible to nucleate a gas pore when the surface tension pressure exceeds the available gas pressure. Thus as DAS decreases, there becomes a cutoff point at which gas pores cannot appear. Effectively, there is simply insufficient room for the bubble! The model by Poirier suggests that this is at least part of the reason for the extra soundness of chill castings compared with sand castings. Later work by Poirier (2001) and the theoretical model by Huang and Conley, assuming no difficulty for the nucleation of pores, confirm the improvement of soundness with increasing fineness of the structure.

In summary, therefore, we can see that as DAS is reduced, the interdendritic structure becomes, on average, cleaner and sounder. These qualities are probably significant contributors to improved properties.

Restricted growth of interdendritic phases

Meyers (1986) found that for unmodified alloys of the Al-7Si system the strength and elongation were controlled by the average size of the silicon particles, but where the particles were uniformly rounded as in structures modified with sodium, the strength and elongation were controlled by the number of silicon particles per unit volume. These conclusions were verified by Saigal and Berry (1984), using a computer model. This important conclusion may have general validity for other systems containing hard, plate-like particles in a ductile matrix.

The highly deleterious effect of iron impurities in these alloys is attributed to the extensive plate-like morphology of the iron-rich phases. Vorren et al. (1984) have measured the length of the iron-rich plates as a function of DAS, and find, as might be expected, the two are closely related; as DAS reduces, the plates become smaller. From the work of Meyers, Saigal and Berry, we can therefore conclude that the strength and toughness should be correspondingly increased, as was in fact confirmed by Vorren. Ma and coworkers (2008) measure the lengths of the β-Fe particles in Al-7Si-0.4Mg alloy as a combined assessment of the increase of both Fe content and cooling time, plotting strength properties as a function of this length (Figure 9.24). They show how ductility is reduced dramatically, tensile strength is somewhat reduced, but yield strength is hardly affected.

It seems likely that although there may be an element of cause and effect in the restriction of the growth of second phases by the dendrite arms, the major reason for the close relation between the size of secondary phases and DAS is that both are dependent on the same key factor: the time available for growth. Thus local solidification time controls the size of both dendrite arms and interdendritic phases.

The reason underlying the importance of the large iron-rich plates in Al alloys is proposed in this work to be the result of the presence of oxide bifilms inside the plates, as has been stated in Section 6.3, and will be referred to later.

Improved response to heat treatment

To summarise the effect of DAS on heat-treatment response: as DAS is reduced, so (1) the speed of homogenisation is increased, allowing more complete homogenisation, giving more solute in solution and so greater strength from the subsequent precipitation reaction. (2) The speed of solution is also increased, allowing a greater proportion of the non-equilibrium second phase to be dissolved. The smaller numbers and sizes of remaining particles, if any, and the extra solute usefully in solution, will bring additional benefit to strength and toughness.

Even when the second-phase particles are equilibrium phases, the high temperature homogenisation and solution treatments have a beneficial effect, even though the total volume of such phases is probably not altered. This is because

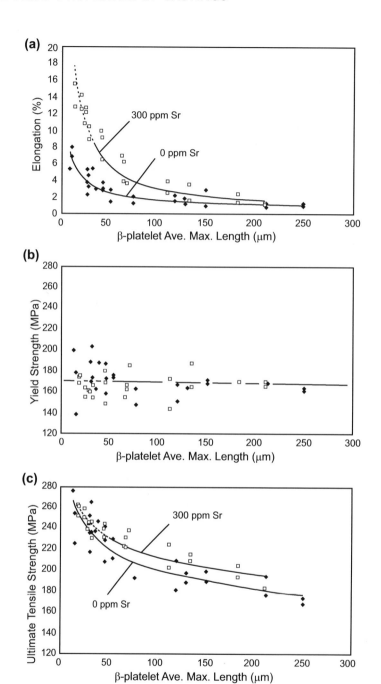

FIGURE 9.24

(a) Elongation; (b) yield strength; and (c) tensile strength of Al-7Si-0.4Mg alloy with 0 ppm and 300 ppm Sr as a function of average maximum length of β-Fe particles (Ma, Samuels, Dot and Valtierra, 2008).

the inclusions tend to spherodise; their reduced aspect ratio favours improved toughness as discussed previously. Any remaining pores will also tend to spherodise, with similar benefit.

There is no doubt, therefore, the improved response to heat treatment is a valuable benefit from refinement to DAS arising from the reduced presence of interdendritic phases (although, ironically, this effect is seen to reduce any Hall–Petch contribution) and possibly the additional solute now in solution. These real improvements to properties if heat treatment is carried out are a separate and additional factor to the important factor described later.

We need to keep in mind, however, that the improvements in properties as a result of refined DAS are enjoyed whether heat treatment is carried out or not. Thus there remains some other major factor at work, so far not considered. We consider this now.

Effect of bifilms

The following examination of the effect of bifilms in the liquid alloy suggests that their effect on properties, as indicated by the DAS, is expected to be dominant.

Rather than consider that refined DAS improves properties, it is more accurate to start with the view that the properties of castings should initially be good, but are degraded as DAS coarsens. This revised viewpoint is helpful as is explained later.

The importance of the presence of bifilms only becomes seriously damaging as solidification time is extended. This is due to the fact that the bifilms arrive in the casting in a compact form, tumbled into compactness by the bulk turbulence during the filling of the mould cavity. In this form, their deleterious effects are minimised. They are small and rounded. Their effectiveness as defects grows as they slowly unfurl, gradually enlarging to take on the form of planar cracks up to approximately 10 times larger diameter than the diameter of their original compact shape. As we have seen in Chapter 2, there are several driving forces for this straightening phenomenon. These include the precipitation of gas inside, or second phases (particularly iron-rich phases in Al alloys) outside, or the action of shrinkage to aid inflation, or the action of growing grains to push and thereby organize the films into interdendritic or inter-granular planes. All these mechanisms extend and flatten the bifilm cracks so as to reduce properties.

The bifilms are seen therefore to evolve into serious defects, given time. Time is a crucial factor. Thus, if there is little time, the defects will be frozen into the casting in a compact form of diameter in the range of 0.1–1 mm or less in smaller castings. However, given time, the bifilms will open to their full size. This might be as large as 10–15 mm across in an Al alloy casting of 10–100 kg weight. In Al-bronze castings and some stainless steel castings weighing several tonnes bifilms in the size range of 50–100 mm are not uncommon. These constitute massive cracks that seriously degrade properties.

The fall of ductility with increasing DAS becomes clear therefore. It is not the DAS itself that is important. *The DAS is merely the indicator of the time available* for the opening of bifilms. The DAS is our independent clock. It is the opening of the bifilms into extensive planar cracks that is important in degrading properties.

The time required for the complete opening of bifilms depends, of course, on the rate at which they can open. This in turn depends on how ravelled the bifilm is, and depends on the various driving forces available for opening. Thus the defects will open at variable rates depending (1) somewhat on their geometry, but faster for (2) higher concentration of gas in solution, (3) poorer conditions for feeding, and, in aluminium alloys, (4) higher iron contents.

A practical example is given in Figure 9.25. Sketch (a) shows the opening of bifilms in a poorly fed region whose solidification is delayed by its heavy section. In other regions of the casting the bifilms have no time to open, and are therefore frozen-in as compact and relatively harmless defects. The mechanical properties are good in the rapidly so-lidified regions, and poor in the slowly solidified and poorly fed regions. Sketch (b) illustrates the benefit of keeping some pressure on the solidifying liquid, in this case by the planting of a feeder onto the casting. The opening of the bifilms is thereby resisted to some extent, and improved properties are retained by the solidification under pressure, even though the pressure is quite modest. Sketch (c) illustrates the case where the metal is superbly clean. The properties are good everywhere regardless of cooling conditions and regardless of pressurisation from a feeder. (The reader will notice some solid feeding, however, which would be avoided by the pressurisation provided by a good-sized feeder.)

We can therefore make an interesting prediction for results such as those presented in Figure 2.47. The prediction is: if there were no bifilms in the liquid metal, the curves of UTS and Elongation to failure versus DAS would both be nearly horizontal lines. It is possible that a slight downward slope might be expected as a result of other factors, such as interdendritic regions becoming less clean, as discussed earlier. However, apart from such minor effects, properties

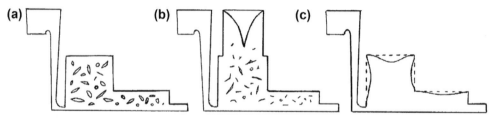

FIGURE 9.25

(a) Opened bifilms resulting from a lack of feeding leading to poor ductility; (b) improved ductility from maintenance of pressure by a feeder; and (c) excellent ductility irrespective of feeding or pressure in the absence of bifilms (however, possible external sinks in heavy sections as shown).

would become essentially *insensitive* to changes in DAS. In other words, in the absence of bifilms, properties would remain high regardless of how slowly or quickly the casting solidified.

Good evidence that this is true, that properties of clean metals are substantially independent of DAS, comes from widely different sources:

1. For steels, Polich and Flemings (1965) showed that low-alloy steels that had been subjected to directional solidification in an upward direction, encouraging the floating out and pushing ahead of bifilms, exhibited a constant elongation despite DAS increasing from 20 to 150 μm. Conversely, conventionally solidified equiaxed versions of the same steel, in which bifilms would not have separated, showed a fall in ductility as DAS increased (Figure 9.26).

2. From early experience with the Cosworth Process, when Cosworth Engineering made its racing car engines from castings poured in conventional foundries, at least 50% of all the cylinder heads (four-valves-per-cylinder designs) failed by thermal fatigue between the neighbouring exhaust valve seats. In 1980, when cylinder heads became available for the first time from the new process, characterised by quiescent transfer and counter-gravity filling, the failures fell immediately to zero. The DAS was hardly changed, and was in any case considerably larger than that available from permanent mould casting, for which resistance to thermal fatigue at that time was poor and variable. Thus DAS was not controlling in this case. It is difficult to avoid the conclusion that the action of bifilms was crucial in this experience.

   The action of some casting customers specifying DAS to avoid thermal fatigue in critical locations like the exhaust valve bridge in cylinder heads is seen therefore to miss a valuable potential benefit. Certainly control of DAS will be important for casting systems that provide poor filling control and thus contain a high density of bifilms. However, in a process, such as Cosworth Process, specifically designed to provide good quality metal, it is almost certainly an unnecessary complication and expense. In general, for casting processes as a whole, it would be more constructive to specify the reduction in bifilms by good processing techniques. This would achieve benefits throughout the whole casting, not merely in those designated regions where a limited DAS has been specified.

3. The new ablation process (Section 17.14) completely separates any observable effect of DAS from properties. DAS can be huge if the application of the cooling water is delayed, allowing the dendrites to grow and coarsen. However, the properties can still be remain high if the residual eutectic liquid is frozen quickly, thus retaining the bifilms small size, the Fe-rich particles having no time to grow and straighten the bifilms and so reduce properties. Properties could remain adequate even when working with melts that were known to contain large populations of bifilms. Figure 9.27 is an example for an Al-7Si-0.4Mg-0.4Fe alloy showing the eutectic phase so fine as to be difficult to resolve, and Fe-rich platelets so small, approximately 1 μm or less long, that they are difficult to find even at the highest optical magnifications (Tiryakioglu, 2009).

9.4.4 TENSILE STRENGTH

UTS is a somewhat dated mouthful, yet nicely descriptive, for its equivalent, the tersely modern, economic, tensile strength (TS). It is a composite property composed of the total of (1) the yield stress plus (2) additional strengthening

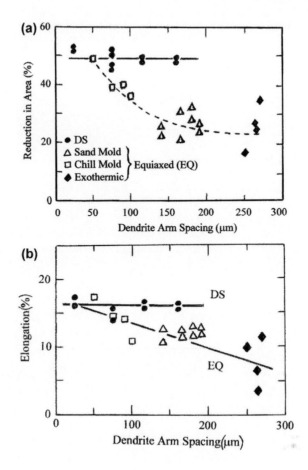

FIGURE 9.26

Vertically oriented DS low-alloy steel is expected to be relatively free from bifilms, and thus DAS has no influence on ductility; conventional equiaxed solidification traps bifilms, permitting DAS to control ductility measured as RA (a) and Elongation (b) by Polich and Flemings (1965).

from work hardening during the plastic yielding of the material before failure. These two components make its behaviour more complicated to understand than the behaviour of yield stress or ductility alone.

TS equals the yield, or proof, stress when (1) there is no ductility, as is seen in Figure 2.47 and Figure 9.20, and (2) when the work hardening is zero. The zero work hardening condition is less commonly met, but occurs often at high temperatures when the rate of recovery equals or exceeds the rate of hardening.

The problem of determining the TS of a cast material is that the results are often scattered. The problems of dealing with this scatter are important, and are dealt with at length in Section 9.2 'The statistics of failure'. Section 9.2 is strongly recommended reading.

Generally, for a given alloy in a given heat treated condition, proof strength is fixed. Thus as ductility is increased (by, for instance, the use of cleaner metal, or faster solidification) so TS will usually increase, because with the additional plastic extension, work hardening now has the chance to accumulate and so raise strength. The effect is again clear in Figure 2.47. For a cast aluminium alloy, Hedjazi et al. (1975) show that TS is increased by a

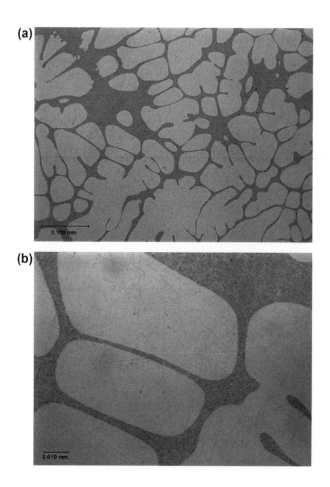

FIGURE 9.27

Ablation cooled Al-7Si-0.4Mg-0.4Fe alloys at nominal (a) 200 × and (b) 1000 × magnification, showing Si particle size and spacing approximately 1 μm, and β-Fe particles too small to be easily found.

Courtesy Alotech.

reduction in defects, as shown in Figure 9.28. The response of the TS is mainly due to the increase in ductility, as is clear from the linear response, and the shift of the properties mainly to the right, rather than simply upwards, for cleaner material.

The rather larger effect that layer porosity is expected to have on ductility will supplement the smaller effect because of loss of area on the overall response of TS. Figure 9.11 shows the reduction in TS and elongation in Mg-Zn alloy systems where the reduction in properties is modest. In Figure 7.34, the TS of an Al-11.5 Mg alloy shows more serious reductions, especially when the porosity is in the form of layers perpendicular to the applied stress. Even so, the reductions are not as serious as would be expected if the layers had been cracks, a result emphasising their nature as 'stitched', or 'tack welded' cracks, as discussed in Section 9.3.2.4.

When the layers are oriented parallel to the direction of the applied stress, then, as might be expected, Pollard (1965) has shown that layer porosity totalling even as high as 3% by volume has a practically undetectable effect on properties.

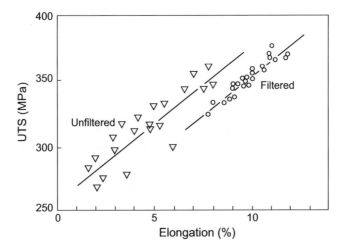

FIGURE 9.28

Mechanical properties of Al-4.5Cu-1.5Mg alloy in the unfiltered and filtered conditions, illustrating the strong response of ductility.

Data from Hedjazi et al. (1975).

Finally, it is clear that cracks or films occupying the majority of the cross-section of the casting will be highly injurious for TS as they are for ductility. The self-evident general understanding that the TS falls to zero as the crack occupies progressively more of the area under test is quantified by Clyne and Davies (1975) in the Figure 9.12.

9.5 FRACTURE TOUGHNESS

Fracture toughness is a material property that is generally independent of the presence of gross defects (although could be affected by a dense population of small defects, as will become clear). This is because it is assessed by the force required to extend a crack that has been artificially introduced in the material usually by extending a machined notch by fatigue. Thus fracture toughness measures the properties of the matrix at the point at which the notch is placed, i.e. it is a material property. It is not a property like tensile strength or ductility in which the crack finds its own start location, ensuring failure from the largest defect. When preparing the toughness assessment test piece, the machining of a notch into the specimen at a pre-fixed location would be unlikely to encounter a major defect by chance.

Fracture toughness is the material property that allows the prediction of the shapes and sizes of defects that might lead to failure. It is a basic tenet of fracture mechanics that fracture may begin when the stress-intensity factor K exceeds a critical value, the fracture toughness K_{1C}. A detailed presentation of the concept of fracture toughness is beyond the scope of this book. The interested reader is recommended to an introductory text like that by Knott and Elliott (1979). Here we shall simply assume some basic equations and the experimental results.

Stress intensity K as well as fracture toughness has the dimensions of stress times the square root of length, and is most appropriately measured in units of $MNm^{-3/2} = MPa\, m^{1/2}$. (Care is needed with units of this property. Fracture toughness is sometimes measured less conveniently in $Nmm^{-3/2} = MPa\, mm^{1/2}$ that differ by a factor of $(1000)^{1/2} = 31.6$.) For a penny-shaped crack of diameter d in the interior of a large casting, the critical crack diameter at which failure will occur is approximately:

$$d = 2K_1 C^2 / \pi \sigma^2 \tag{9.7}$$

from which, for an aluminium alloy of fracture toughness 32 MPa m$^{1/2}$ at its yield point of 240 MPa, the critical defect size d is 11 mm. For an edge crack, Eqn (9.7) is modified by a factor of 1.25, giving a corresponding critical defect size of 9 mm, indicating that edge cracks are somewhat more serious than centre cracks, but in any event defects of about a centimetre across would be required, even at a stress at which the casting is on the point of plastic failure. These are comparatively large defects, which is reassuring in the sense that such large cracks may have a chance to be found by non-destructive tests before the casting going into service. However, they underline the conclusion that most current radiography standards which state 'no linear defects' of any size, or which reject aluminium alloy castings for flaws of only approximately 1 mm in size, may not be logical.

The important message to be learned from the concept of fracture toughness is that the maximum defect size that can be tolerated can be precisely calculated if the fracture toughness and the applied stress are both known.

Equation (9.7) can be used to indicate that at the limit, when the applied stress σ equals the yield stress σ_y, the greatest resistance to crack extension that can be offered by a material is controlled by the value of K_{1C}/σ_y (units most conveniently in m$^{1/2}$). This parameter is a valuable measure of the defect tolerance of a material.

Figure 9.29 shows how the permissible defect size in ductile irons is large for low-strength irons, but diminishes with increasing strength because of the fall of fracture toughness. Thus for the stronger irons the permissible defect size is below 1 mm. This is particularly difficult to detect and sets a limit to the stress at which a strong ductile iron casting may be used with confidence.

Steels can have high fracture toughness, with correspondingly good tolerance to large defects at low applied stress. However, because they are used in highly stressed applications, the permissible defect size is reduced, as Eqn (9.10) indicates. Jackson and Wright (1977) discuss the serious problem of detecting flaws when the permissible size is small. At the time of writing, most non-destructive testing methods have been found wanting in some respect. However, the technology of non-destructive techniques has improved significantly and is expected to continue to improve, allowing the detection of smaller defects with greater certainty, and so extending the range over which castings can be stressed whilst

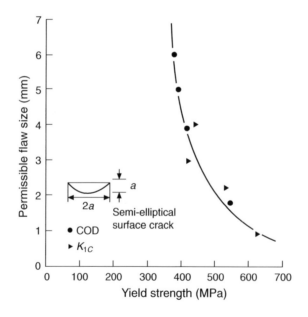

FIGURE 9.29

Permissible edge defect sizes in ductile irons of increasing strength.

Data from Seetharamu and Srinivasan (1985).

reducing the risk of failure. In this respect, the most significant advance is probably the development of sophisticated resonance testing described in Section 19.8.

Nevertheless, Jackson and Wright's conclusion is good advice: choose an alloy with a large critical flaw size, which can be detected, and measured accurately. Also, if the working load is fluctuating, it will be necessary to calculate from crack propagation rate curves the time for a flaw to grow to the critical size. This can then be compared with the design life of the component.

To be safe, they recommend the following assumptions:

1. A string of cracks is equivalent to one long crack.
2. A group of flaws is one large flaw of size comparable with the envelope that circumscribes them.
3. Unless clearly seen to be otherwise, all flaws have a sharp aspect ratio.
4. All flaws reside on the surface.

A casting designed on this basis is likely to be somewhat overdesigned, but perhaps not so much as if there had been no use of the principles of fracture mechanics.

There has been relatively little work carried out on the fracture toughness of cast Al alloys. Tiryakioglu and Hudak (2008) have therefore developed a relation between round notched tensile strength σ_{NT} and fracture toughness K_{1C}, both normalised by dividing by the yield stress σ_y, as shown in Figure 9.30(a) for Al-7Si-0.6Mg alloy. The direct comparison between σ_{NT} and K_{1C} is given in Figure 9.30(b) showing limits of accuracy at 5% and 1% error to within 95% confidence. The best fit line is given by

$$K_{Ic} = \beta \sigma_{NT} \tag{9.8}$$

where $\beta = 6.33 \times 10^{-2} \, \text{m}^{0.5}$ with $R^2 = 0.72$ and standard mean deviation 1.48 MPa^2 m. Round tensile test specimens with a circumferential sharp notch are easy and quick to produce, so the important parameter, K_{1C}, is suddenly made much more accessible.

So far all the discussion has related to linear elastic fracture mechanics (LEFM), in which the fracture resistance of a material is defined in terms of the elastic stress field intensity near the tip of the crack. In fact, the fracture toughness parameter K_{1C} is valid only when the region of plastic yielding around the tip is small.

For lower-strength materials, plastic deformation at the tip of the crack becomes dominant, requiring the application of yielding fracture mechanics (YFM) and the concept of either (1) crack opening displacement (COD or δ) or (2) the J integral, as a measure of toughness.

The critical COD is the actual distance by which the opposite faces of the crack separate before unstable fracture occurs. This critical opening is known as δ_c. The critical flaw size is given by:

$$d_c = c\left(\delta_c / \varepsilon_y\right) \tag{9.9}$$

where c is a material constant and ε_y is the strain at yield. For materials that are on the borderline for treatment by LEFM or YFM, Jackson and Wright (1977) indicate that a unified test technique can be used to determine K_{1C} or δ_c from a single test piece, and that the following useful relation may be employed

$$\delta_c / e_y = \left(K_{1C} / \sigma_y\right)^2 \tag{9.10}$$

Hence the ratio δ_c / e_y in YFM is comparable with $(K_{1C}/\sigma_y)^2$ in LEFM, i.e. it is a measure of the defect tolerance of a material. It deserves to be much more widely used in the design and specification of castings.

The transition between K_{1C} and COD, when properly applied, is seen to give consistent predictions of the critical defect size. The results in Figure 9.29 for ductile irons show a region of smooth overlap between the two approaches.

Although the COD has a clear physical basis, in both its formulation and its application, there are empirical assumptions that are difficult to justify by the rigorous application of mechanics at this time. For this reason, more attention is being given to the use of the J integral in structural-integrity assessment, especially in the United States. J is a crack-tip driving force that, under certain conditions, can be applied to elastic–plastic situations to describe the onset

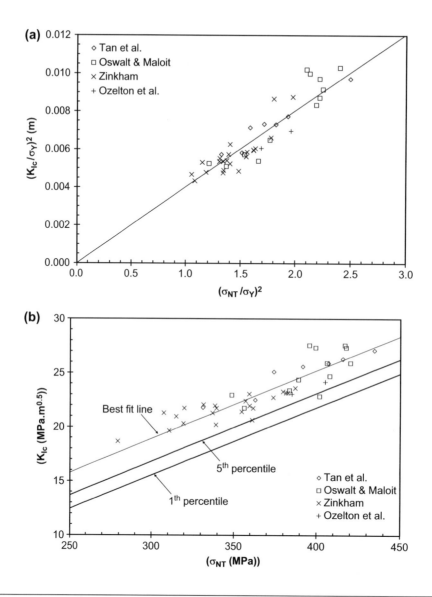

FIGURE 9.30

(a) Relation between $(K_{1C}/\sigma_Y)^2$ and the square of notch tensile strength σ_{NT}/yield strength σ_Y. The best fit line goes through the origin, suggesting a convincing relationship. (b) K1C values from all studies is plotted versus notch-tensile strength (lower bounds at 95% and 99% confidence limits are also shown). (Tiryakioglu, 2008.)

of stable crack growth. For a given geometry of loading, size of test piece and size of defect, J is related directly to the area under the load–displacement curve obtained for the cracked body, where displacements are measured at the loading points.

A general overview of the fracture toughness and strength of cast alloys is given in Figure 9.31. This indicates the rather poor properties of flake irons, and the excellent properties of some steels, nickel alloys and titanium alloys. Ductile

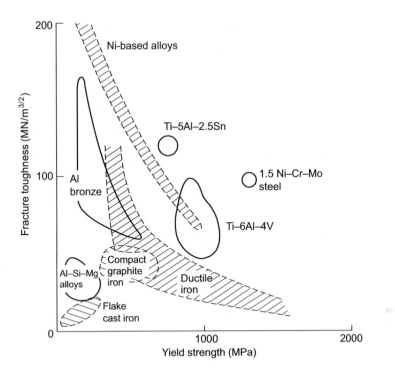

FIGURE 9.31

Map of fracture toughness versus yield strength for various cast alloys.

Data from Speidel (1982) and Jackson and Wright (1977).

irons occupy an interesting middle ground. In general, it seems that most groups of alloys can exhibit either high strength and low toughness or high toughness and low strength. The result is the hyperbolic shape of the curve as seen for ductile irons. As more results for other alloys are determined, Figure 9.31 will be expected to develop into a series of overlapping hyperbolic regimes containing the various alloy systems.

Figure 9.32 shows how the fracture toughness values can be translated into permissible flaw diameters, assuming a central penny-shaped crack, and a given level of applied tensile stress. Lalpoor and colleagues (2009) work out from computed stresses the sizes of critical flaws in a continuously cast high-strength Al alloy 7050, finding that the diameters of critical penny shaped cracks are in the region of 20–50 mm. Because spontaneous failures of these products are relatively common (and expensive of course) it is clear that such large cracks do exist. It seems inescapable that these are bifilms introduced during the initial fall of the melt into the mould; the rest of the mould filling action is practically perfect and not capable of introducing such defects.

Figure 9.20 presents interesting results for the mechanical properties of a high-strength aluminium alloy that is not normally poured to make shaped castings. In this work, it was cast and given a range of heat treatments covering the peak response condition. In the optimally hard state, it had an elongation close to zero and would not normally be used in this condition because a designer or user would be nervous about brittle failure. It is in the nature of brittle failure (i.e. failure without general plastic deformation) that there is no benefit from either (1) the redistribution of stress or (2) any prior warning of impending failure by the general plastic yielding of the casting. Nevertheless, the fracture toughness is clearly little influenced by aging and retains a respectably high value regardless of the low ductility. This means that any crack or other defect will always have to exceed a certain critical size before failure occurs, even if such failure is eventually of a brittle character, and therefore probably catastrophic.

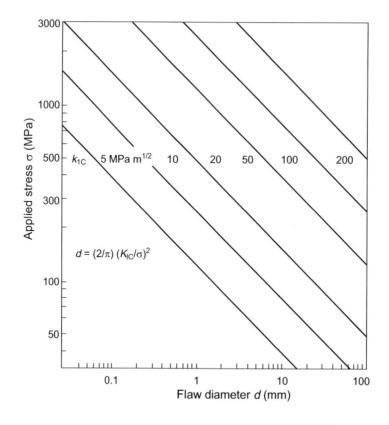

FIGURE 9.32

Relation between fracture toughness, applied stress and the critical defect size for failure in a material containing a central circular flaw.

This general behaviour is confirmed by Vorren and colleagues (1984) for the more common casting alloy Al-7Si-0.5Mg: although these researchers find the fracture toughness falls with increasing iron content in the alloy, the fall is significantly less than the fall in ductility, assessed by the reduction in area.

In those cases where general deformation of the casting cannot be tolerated, therefore, the use of fracture toughness is a more appropriate measure of the reliability of the casting than ductility. It seems that for many years, we have been using an inappropriate parameter to assess the reliability of many of our castings.

9.6 FATIGUE

9.6.1 HIGH CYCLE FATIGUE

A huge volume of work exists on the fatigue of castings. Therefore this lightweight and shallow condensation hardly scratches the surface of a profound and increasingly mature discipline. We shall endeavour to mention some of the more important issues.

Initiation

A historically classical study of fatigue in a cast alloy Al-7Si-0.5 Mg was carried out by Pitcher and Forsyth (1982). These workers were able to show that in general the fatigue performance of cast alloys was poor because the initiation of the fatigue crack during stage 1 of the fatigue process was short. This observation appears to have been confirmed for cast Al alloys many times, for instance recently by Wang, Apelian and Lados (2001), who found *zero* time was required for crack initiation and that the fatigue life was the time merely for propagation of the crack. With the wisdom of hindsight, we can conclude that this was almost certainly a consequence of the presence of bifilms. Thus the cast specimens were effectively pre-cracked. This conclusion is strongly indicated by the rather poor casting technique used for the preparation of test specimens and additionally confirmed by the appearance of pores that appeared to be associated with films in some of their micrographs. Thus the size of initiating defects in their castings effectively eliminated stage 1.

Stage 2 of the fatigue process is a small forward movement of the crack for each tensile part of the stress cycle. This stepwise propagation across the majority of the section leaves a characteristic pattern of 'beach marks' as seen in Figure 9.35(f). By counting back from the thousands of beach marks from the final mark at which the remaining cross-section was too small to withstand the load, and so failed catastrophically, it can be deduced that the initiation of the fatigue crack probably started on the first cycle.

However, Pitcher and Forsyth found that stage 2 of the growth of the crack was remarkably slow compared with high-strength wrought alloys. The slow rate of crack propagation seemed to be the result of the irregular branching nature of the crack, which appeared to have to take a tortuous path through the as-cast structure. This contrasts with wrought alloys, where the uniformly even micro-structure allows the crack to spread unchecked along a straight path. With the wisdom of hindsight, we can speculate that the crack followed randomly oriented and branching bifilms, as would have occurred if the sample shown in Figure 2.40(b) had been subjected to fatigue.

The results by Polich and Flemings (1965) on the solidification of steels are highly informative. They found that steels directionally solidified (DS), in which the growth direction was upwards, had properties significantly better than those solidified in a non-directional equiaxed mode. We can speculate that the equiaxed mode would have trapped bifilms between the growing grains, whereas the vertical solidification would have allowed bifilms to float clear, and even if encountered by the front, would have a chance to be pushed ahead and thus avoid incorporation in the solid. The absence of bifilms in the DS material is corroborated by the insensitivity of its ductility to wide changes in DAS (in contrast to equiaxed material) as seen in Figure 9.26. Furthermore, these authors found the properties of the DS material were also insensitive to the direction of grain growth, so that grain direction and grain boundary direction was, it seems, unimportant to the properties and the development of cracks. This important counter-intuitive finding further confirms that only the presence of bifilms at grain boundaries influences properties, and that otherwise, boundaries are practically as strong as the matrix so that cracks are hardly influenced by the presence of boundaries. (Despite widespread glib assertions of the weakness of grain boundaries, they are clearly strong; otherwise, all our bridges would be in the rivers.)

One conclusion to be drawn from the work indicating that initiation takes no time but crack propagation is slow, is that if initiation can be delayed, then fatigue lives of cast metals might be extended considerably, perhaps in excess of those of wrought alloys.

An indication of the general validity of this conclusion is confirmed by Pitcher and Forsyth, who shot-blasted their specimens to generate residual compressive stresses in the surface. This did help to slow initiation, and total fatigue lives were improved. For sand castings, fatigue life was increased to that expected of chill cast material as shown in Figure 9.33. (It is a pity that the important high-stress/low-cycle part of the fatigue curve is often not investigated. It is interpolated here between the high-cycle results and the TS values; i.e. the single cycle to failure result.) Chill-cast material is subject to a similar substantial improvement when shot blasted. For steel castings, Naro and Wallace (1967) also found that shot peening considerably improved the fatigue resistance.

Although the improvement of fatigue resistance as a result of the compressive surface stresses introduced by shot peening is perhaps to be expected, the effect of grit blasting is less easy to predict, because the fine notch effect from the indentations of the grit particles would impair, whereas the induced compressive stresses would enhance fatigue life. From tests on aluminium alloys, Myllymaki (1987) finds that these opposing effects are in fact tolerably balanced, so that grit blasting has little net effect on fatigue behaviour.

Effect of defects
Improvement is expected if cast defects are eliminated, so that the fatigue crack now needs to be initiated, thus introducing a lengthy stage 1. Figure 9.33 illustrates the case of a pearlitic ductile iron where consistently improved

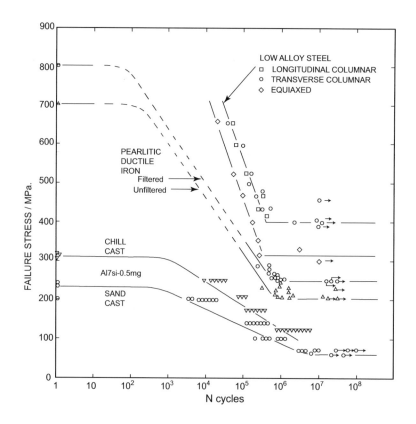

FIGURE 9.33

Fatigue data for an Al alloy (Pitcher and Forsyth, 1982), a pearlitic ductile iron (Simmons and Trinkl, 1987) and a low-alloy steel (Polich and Flemings, 1965).

performance was obtained from the use of material with a reduced density of defects produced by filtering in the mould. The elimination of thermal fatigue cracking in the early Cosworth Process Al-7Si-0.4Mg alloy cylinder heads is another instance of the importance of the elimination of defects, so as to introduce a long stage 1 to the fatigue crack formation.

This industrial experience has now been confirmed by a careful laboratory study by Nyahumwa, Green and Campbell (1998, 2000) on Al-7Si-0.4Mg alloy castings. To vary the number density and size of oxide film defects in the castings, test bars were cast using bottom gated filling systems with and without filtration. Test pieces were machined from the castings and were fatigue tested in pull–pull sinusoidal loading at maximum stresses of 150 and 240 MPa under stress ratio $R = +0.1$.

The use of the pull–pull mode of testing (positive R ratio) is important for two reasons. (1) The fracture surface is undamaged (in contrast to the hammering that normal reversing load fatigue specimens suffer) and so can be studied under the scanning electron microscope (SEM) in great detail. This is most important when searching for elusive features such as oxide films, and probably has been a significant factor explaining how such extensive defects have been overlooked until recently. (2) In service conditions in many castings, the existence of residual tensile stress is a common feature because of inappropriate quenching practice after solution heat treatment. Some residual stress is to be expected even in carefully quenched castings. Thus fatigue failure is most likely to occur in regions already experiencing significant tensile loads. Thus pull–pull testing conditions will represent such regions more accurately. It follows that fatigue testing in laboratories using reversed loading (push–pull, with negative R ratios) should be reconsidered in favour of pull–pull (positive R ratio).

Test bars of an Al-7Si-Mg alloy (2L99) were cast in chemically bonded silica sand moulds. Two batches of test bars were cast using (1) a bottom-filling unfiltered system through which a liquid metal entered the ingate at a velocity greater than 0.5 ms^{-1}, and (2) a bottom-filling filtered system through a ceramic foam filter with 20 pores per inch (corresponding to an average pore diameter of approximately 1 mm); in this system, a liquid metal entered the ingate at a velocity less than 0.5 ms^{-1}. An SEM examination was carried out on the fracture surfaces of every one of the 64 failed specimens to ascertain the crack initiator in every case. This was a huge exercise, probably never attempted previously.

Fatigue life data obtained from the filtered and unfiltered castings tested at 150 and 240 MPa are plotted in Figure 9.34. The probability of failure for each of the specimens in a sample was defined as the rank position divided by the sample size (i.e. the test results were ranked worst to best in ascending order; so, for instance, the 20th sample from the bottom in a total of 50 samples exhibited a 40% probability of failure).

Figure 9.35 shows the three main types of fracture surface observed: (a) oxide film defects which were categorised either as young, thin film or older, thicker film, based on the thickness of folds; (b) slip mechanisms indicated by a typical faceted transgranular appearance; and (c) fatigue striations, often called beach marks, denoting the step by step advance of the crack. Later examination of the so-called slip planes indicated a mix of what appeared to be a true slip plane, and a fairly flat plane formed from a straightened oxide film.

There was much to learn from this work. Figure 9.34 shows that for the unfiltered castings, most failures initiated from defects, sometimes pores, but usually oxides. The oxides were a mix of young and old. Only three specimens were fortunate to contain no defects. These exhibited 10 times longer lives, and finally failed from cracks that had initiated by the action of slip bands that created interesting facets on the fracture surface. Thus these specimens had lives limited by the metallurgical properties of the alloy, although even in this case, an 'ultimate' attainment of life could not be assumed since it seems likely that entrainment defects had initiated the slip bands (Tiryakioglu, Nyahuwa and Campbell, 2011)

The filtered castings (bold symbols in Figure 9.34) were expected to be largely free from defects, but even these were discovered to have failed mainly as a result of defects, the defects consisting solely of old oxides that must have passed through the ceramic foam filter. There were no young oxides and no pores. A check of the equivalent initial flaw size (determined from the square root of the projected area of fatigue defect initiators) showed that 90% were in the range 0.1–1.0 mm, with a few as large as 1.6 mm. Thus most would have been able to pass through the filter without difficulty.

Again, those few samples of cast material free from oxides displayed an order of magnitude improved life, the best of which agreed closely with the best of the results from the unfiltered tests. This agreement confirms the relatively defect-free status of these few results.

The detrimental effect of mixed oxide films in the unfiltered castings is reflected by the lower fatigue performance compared with that of the filtered castings containing old oxide films. This indicates that the old oxide films, which were observed to act as fatigue crack initiators in the filtered castings, were less damaging than mixed oxide films. Clearly, we can conclude that the young films are more damaging, almost certainly as a result of their lack of bonding. This contrasts with the old oxide films that have probably benefited from the closing and partial re-bonding of the interface. This observation is consistent with the expectation that pores would be associated with new films, whose bifilm halves could separate to form pores, but not with old films, whose bifilm halves were (at least partly) welded closed.

These findings are confirmed in general by the results at the higher stress 240 MPa (Figure 9.34(b)) The fatigue failures of unfiltered castings are initiated from a mixture of pores, and old and new oxides; the 33 filtered castings initiated from 29 old oxides, three pores and one slip plane; and the 33 unfiltered and HIPped tests initiated from 32 old oxides and one slip plane. The various types of fracture surface are shown in Figure 9.35.

The presence of facets on the fracture surface is interesting. Whereas it has been usual to ascribe the formation of facets only to a slip plane mechanism, Tiryakioglu et al. (2011) have suggested that some facets appear to have formed from straightened bifilms. There appear therefore to be two forms of facets which are quite distinct: (1) the slip plane–generated facets have nearly flawless mirror-smooth planes with sharply defined edges and steps as seen in

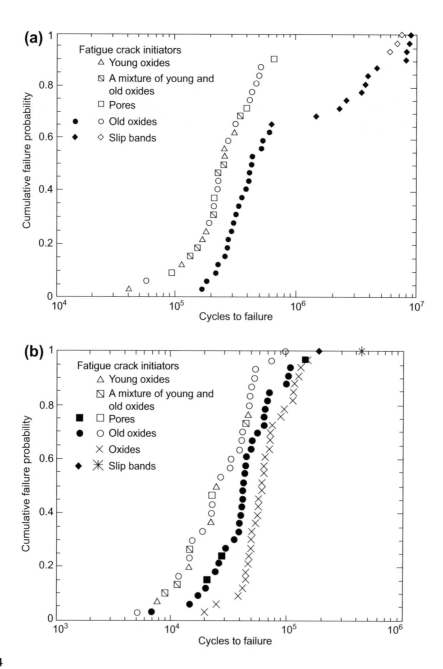

FIGURE 9.34

Fatigue life of unfiltered (open symbols), filtered (solid symbols) and unfiltered and HIPped (cross and star symbols) of Al-7Si-0.4Mg alloy castings tested in pull–pull at R = +0.1 and (a) 150 MPa and (b) 240 MPa (Nyahumwa et al., 1998, 2001).

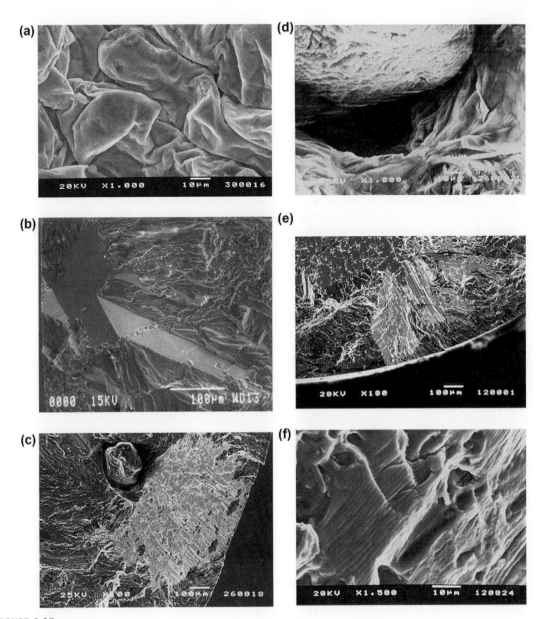

FIGURE 9.35

Fatigue fracture surfaces illustrating (a) oxide initiator; (b) slip plane–generated facets; (c) bifilm facet with failure probably initiated from entrained sand particle; (d) close-up of sand particle showing it to be enveloped by an oxide film as a result of its entrainment; (e) mixed slip plane and bifilm facets; (f) 'beach mark' striations from the stepwise advance of the fatigue crack (Nyahumwa et al., 1998, 2001).

Figure 9.35(b). (2) The straightened bifilm originated plane is less smooth, with no sharp steps or edges, but edges in places extending out from the facet to become contiguous with a crumpled oxide film, and its surface covered with small pores as would be expected to have been entrained together with the bifilm as is shown in Figure 9.35(c) and 9.35(e). Figure 9.35(c) and the higher magnification Figure 9.35(d) illustrate that the oxide wrapping of the entrained grain of silica sand appears to be the contiguous with the planar film, strongly suggesting these two oxide films are one and the same film, the area beyond the sand grain having been straightened by the advance of dendrites. Interestingly, both forms (1) and (2) act as cracks, and both are of size dictated by the grain size of the casting, thus these authors found that their effect on fatigue life was effectively identical. Clearly, when these are observed the fatigue life is usually long, but may not represent an 'ultimate limit' as has previously been supposed.

Other workers (for instance Wang et al., 2001) using Al-7Si-0.4Mg alloy have in general confirmed that for a defect of a given area, pores are the most serious defects, followed by films of various types. This is to be understood in terms of pores having zero tensile strength, but bifilms may be partly welded, but in any case always have some degree of geometrical interlocking as a result of their convoluted form as is clear from Figure 2.41.

However, the pre-eminence of pores in fatigue failure is not to be taken for granted. Nyahumwa found films to be most important in his work on this alloy. He had the advantage of having the latest techniques to search for and identify the full extent of films, whereas it is not known how thorough previous studies have been, and films are easy to overlook even with the best SEM equipment. In contrast, Byczynski (2002) working with the same techniques in the same laboratory as Nyahumwa found that pores were the most damaging defects in the more brittle A319 alloy used for automotive cylinder blocks. This difference may be the explicable by the differences in ductility of the two alloys that were studied. Nyahumwa's alloy was ductile, and so required the stress concentration of bifilms acting as cracks. Byczynski's alloy was brittle, so that cracks could more easily occur from pores.

Thus we may tentatively summarise the hierarchy of defects that initiate fatigue in order of importance: these are pores and/or young bifilms, followed by old bifilms. It would be expected that in the absence of larger defects, progressively finer features of the micro-structure would take their place in the hierarchy of initiators. However, it seems that such an attractively simple conceptual framework remains without strong evidence, even doubtful, at this time as is described later.

The considerable work now available on Al alloys illustrates the present uncertainties. In Al alloys, it has been thought that silicon particles may become active in the absence of other initiators. However, the action of bifilms is almost certainly involved in the occasional observations of the nucleation of cracks from these sources. For instance, the observed decohesion of silicon particles from the matrix (Wang et al., 2001, part I) is difficult to accept unless a bifilm is present, as seems likely. The initiation of cracks from iron-rich phases occurs often if not exclusively from bifilms hidden in these intermetallics (Cao and Campbell, 2002). The initiation of fatigue from eutectic areas, and reported many times (for instance Yamamoto and Kawagoishi, 2000 and Wang et al., 2002, part II) is understandable if bifilms are pushed by growing dendrites into these regions. The fascinating fact that Yamamoto and Kawagoishi observe silicon particles sometimes initiating fatigue cracks and sometimes acting as barriers to crack propagation strongly suggests bifilms are present sometimes and not at other times, as would be expected. It is not easy to think of other explanations for this curious observation.

Performance

Figure 9.36 is intended to illustrate the panorama of performance that can be seen in castings. The poor results can sometimes be seen in pressure die castings, in which the density and size of defects can cause failure to occur on or even before the first cycle (some test pieces from a poor pressure die casting fell to pieces in the hand before being loaded into the testing machine). However, it is to be noted that the unpredictability of this process sometimes will yield excellent results if the defects are, by chance, or by manipulation, in an insensitive part of the casting, or where perhaps the bifilms are aligned parallel to the stress axis. The uncertainty of the effects of the certainly present bifilms in pressure die castings is the key reason why it is never easy to select such castings for safety critical applications.

In terms of this panorama in Figure 9.36, the results of Nyahumwa are presented as 'fair' and 'good', respectively, showing a few tests that exhibit outstandingly good lives, reaching 10^7 cycles. Clearly, a series of ideal castings, free from defects, would display identical lives all at 10^7 cycles. The important lesson to draw from Figure 9.36 is that most engineering designs have to be based on the minimum performance. It is clear, therefore, that even the castings

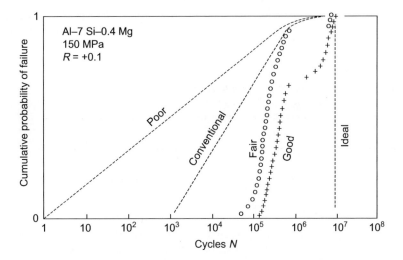

FIGURE 9.36

Schematic overview of the extreme range of fatigue performance found in structural Al alloy castings. There is recent evidence that the 'ideal' performance should be shifted substantially to the right (Tiryakioglu, Campbell and Nyahumwa, 2011).

designated 'good' in this figure have a potential to increase their lowest results by at least two orders of magnitude, i.e. 100 times. It means that for most of the aluminium alloy castings in use today, we are probably using a maximum of only 1% of their potential fatigue life. To gain this hundredfold leap in performance, we do not need different or more costly alloys, we merely need to eliminate defects.

The sobering density of defects in many special steels arises as a result of the appalling way in which the steels are cast. Whereas large tonnages of steels are cast by continuous casting, and so tolerably well cast, smaller quantities of special steels are cast into ingots. The process is shown in Figure 12.1. The dreadful mix of liquid metal and air creates masses of bifilm in steels with solid oxides such as the 1%C1.5%Cr steels for roller-bearing applications. The bifilm networks created during the pouring of the ingots have been widely interpreted as fatigue cracks (Figure 9.37) and certainly lead to fatigue failures by parts of the bearing surface spalling away. The contact pouring of ingots would solve this problem as seen in Figure 12.1.

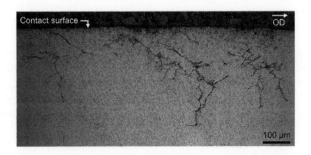

FIGURE 9.37

Bifilm cracks leading to the fatigue failure of the main steel roller bearings in a windmill.

Courtesy Martin Evans (2012).

As a closing update on the current understanding of the key aspects of fatigue, Tiryakioglu (2009) finds that the size distribution for fatigue-initiating defects in Al-7%Si-Mg alloys is described by an extreme value distribution known as the Gumbel distribution (and not Weibull or lognormal as has been commonly assumed). In addition, the fatigue life model based on the Paris–Erdogan equation with no crack initiation stage provides respectable fits to the data he has been able to test so far. A new distribution function combining the size distribution of fatigue-initiating defects and fatigue life model provides fits to data that could not be rejected by goodness-of-fit tests, in contrast to two-parameter Weibull. Tiryakioglu has gone on to explain the use of Weibull and other distributions, providing helpful examples and illustrations of comparisons between different distribution models (Tiryakioglu, 2010, 2011).

9.6.2 LOW CYCLE, HIGH STRAIN AND THERMAL FATIGUE

Thermal fatigue is a dramatically severe form of fatigue. Whereas normal *high-cycle fatigue* occurs at *stresses* comfortably in the elastic range (i.e. usually well below the yield point) *thermal fatigue* is driven by thermal *strains* that force deformation well into the plastic flow regime. The maximum stresses, as a consequence, are therefore well above the yield point.

Thermal fatigue is common in castings in which part of the casting experiences a fluctuating high temperature whilst other parts of the casting remain at a lower temperature. The phenomenon is seen in grey iron disc brakes and aluminium alloy cylinder heads and pistons for internal combustion engines, particularly diesel engines and air-cooled internal combustion engines. It is also common in the casting industry with the crazing and sometimes catastrophic failure of high-pressure die-casting dies made from steels and gravity dies (permanent moulds) and ingot moulds made from grey cast iron.

The valve bridge between the exhaust valves in a four-valve per cylinder diesel engine is an excellent example of the problem, and has been examined in detail by Wu and Campbell (1998). In brief, the majority of the Al alloy casting remains fairly cool, its temperature controlled by water cooling. However, the small section of casting that forms a bridge, separating the exhaust valves, can become extremely hot, reaching a temperature in excess of 300°C. The bridge therefore attempts to expand by $\alpha \Delta T$, where α is the coefficient of thermal expansion and ΔT is the increase in temperature. For a value of α about 20×10^{-6} K^{-1} for an Al alloy, we can predict an expansion of $300 \times 20 \times 10^{-6} = 0.6\%$. This is a large value when it is considered that the strain to cause yielding is only about 0.1%. Furthermore, because the casting as a whole is cool, strong and rigid, the bridge region is prevented from expanding. It therefore suffers a plastic compression of about 0.6%. If it remains at this temperature for a sufficient time (an hour or so) stress relief will occur, so that the stress will fall from above the yield point to somewhere near zero.

However, when the engine is switched off, the valve bridge cools to the temperature of the rest of the casting, and so now suffers the same problem in reverse, undergoing a tensile test, plastically stretching by up to 0.6%.

The starting and stopping of the engine therefore causes the imposition of an extremely high strain and consequent stress on the exhaust valve bridge. For those materials, such as a poorly cast Al alloy, that has perhaps only 0.5% elongation to failure available, it is not surprising that failure can occur in the first cycle. What perhaps is more surprising is that any metallic materials survive this punishing treatment at all. It is clear that modern, well cast cylinder heads can undergo thousands of such cycles into the plastic range without failure.

The experience from the early days of setting up the Cosworth Process provided an illustration of the problem as described earlier. In brief, before the new process became available, the Cosworth cylinder heads intended for racing were cast conventionally via running systems that were probably well designed by the standards of the day. However, approximately 50% of all the heads failed by thermal fatigue of the valve bridge when run on the test bed. These engines were, of course, highly stressed, and experienced few cycles before failure. From the day of the arrival of the castings, otherwise substantially identical in every way, but made by the new counter-gravity process, no cylinder head ever failed again. The presence of defects is seen therefore to be critical to performance, particularly when the metal is subjected to such extreme strains as are imposed by thermal fatigue conditions.

Thermal fatigue tests can be carried out in the laboratory on nicely machined test pieces resembling a tensile test piece. One of the interesting observations is that for some ductile Al alloys, the repeated plastic cycling for those

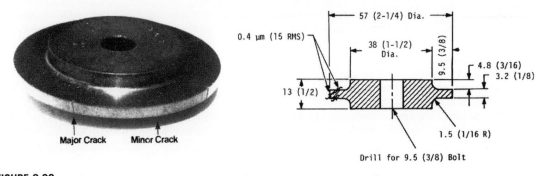

FIGURE 9.38

A simple and low-cost thermal fatigue specimen (Roehrig, 1978).

specimens that survive hundreds or thousands of cycles causes them to deform into shapeless masses. This gross deformation appears to be resisted more successfully for higher strength alloys (Grundlach, 1995). However, such tests require sophisticated and costly equipment which limit the number of investigations that can be carried out and rate at which results can be achieved.

There are, however, other forms of test pieces that are to be recommended for their extreme simplicity. Roehrig (1978) described three such tests, the simplest of which is shown in Figure 9.38. Such simple test pieces provide their own strain and stress development without the need for costly equipment. A typical testing scenario might be a stack of finned disc samples, bolted together through the central hole. The stack is immersed in a fluidised bed of alumina grit at near 1000°C for 5 s followed by transferring into a second bed at near 100°C for 120 s. These conditions may be selected for the simulation of stresses in a particular design or new alloy intended for an automotive brake disc. Some samples and some alloys fail after only a few cycles, whereas others survive for 1000 cycles or more. Other tests described by Roehrig are designed to simulate conditions specifically in cylinder heads, and some raise the severity of the cooling by water quenching. Such tests are excellent to optimise the selection of alloy for a particular casting requiring a specific thermal cycle resistance.

9.7 ELASTIC (YOUNG'S) MODULUS AND DAMPING CAPACITY

In an engineering structure, the *elastic modulus* is the key parameter that determines the *rigidity* of the design. Thus steel with an elastic modulus of 210 GPa is very much preferred to the light alloys aluminium (70 GPa) and magnesium (45 GPa). However, of course, the specific modulus (the rigidity modulus divided by density) gives a somewhat more favourable comparison because they are all closely similar at about 26 MPam^{3}kg^{-1}.

Because the elastic properties of metals arise directly as a result of the interatomic forces between the atoms, there is little that can normally be done to increase this performance. Alloying, for instance has practically no effect.

The one exception worth noting to this general rule is the alloy system Al-Li, which creates such a large volume fraction of Al-Li intermetallics that the elastic modulus is raised by about 10%. At the same time, the density is reduced by about 10%, resulting in an overall increase in specific elastic modulus of 20%. This relatively modest increase in stiffness seems to be an example of the highest that can be expected for most engineering metals.

It was all the greater surprise to the author, therefore, when he arranged for a computer simulation to be carried out on the largest bell in the United Kingdom. This was Great Paul, the 17,000 kg bell in St Paul's Cathedral, London. The investigators, Hall and Shippen (1994), found that agreement with the fundamental note and the harmonics could only be obtained by assuming that the elastic modulus was only 87 GPa, not 130 GPa expected for alloys of copper. This alarming 33% reduction was almost certainly the consequence of Great Paul being full of defects, mainly porosity and bifilms. This is true for most bells because, unfortunately, they are top poured via the crown of the bell. (The Liberty Bell in the United States is a famous example of a crack opened from the massive bifilm constituted by the oxide flow tube that can be seen to have formed around the sinuous flow of the falling metal stream, starting high on the shoulder of the bell, and continuing down the side to its lip.) At the time of writing, the one bell foundry in the world known to the author (Nauen, Tonsberg, Norway) that uses bottom gating for its bells, placed mouth upwards, claims that it has never suffered a cracked bell, and its bells sound for twice as long as conventional bells. The longevity of the sound is understandable in view of the reduced quantity of bifilms to be expected, so reducing the loss of energy by internal friction between the rubbing surfaces of the bifilms as the bell vibrates.

It is clear, therefore, that many castings will suffer from a lower elastic modulus than expected because of their high content of bifilms. Furthermore, because elastic modulus is very rarely checked, this fact is not widely known. For the future, the full stiffness of the material will be gained only if the metal quality is good, and if the casting is poured with a minimum of surface turbulence.

The effect of a high density of bifilms in castings reducing the elastic modulus of the cast material is an interesting hypothesis that has a widely known and accepted metallurgical analogy that lends convincing comparison. The example is that of flake cast irons compared with steels. The graphite flakes in cast irons act as cracks because they are unable to withstand significant tensile stress (they may actually have formed around cracks if the bifilm theory presented in Chapter 6 is correct, which I think it is!). The iron, behaving like a steel matrix laced with a multitude of cracks, finds its modulus reduced from 210 GPa to only 152 GPa, a reduction of 28%, a value comparable to the reduction expected for many defective cast materials. In addition, of course, flake graphite irons are renowned for their excellent damping capacity, its absorption of vibrations and deadening of noise making it an ideal choice for the beds of machine tools.

Sand cast plates of Mg alloy ZE41A-T5 showed variations in elastic modulus in the as-cast condition that could be plotted on a Weibull distribution, with measurements generally in the 6–37% range below the theoretical 44.7 GPa (Xinjin Cao, 2009).

Further Examples Cannot Be Resisted

Zildjian cymbals, much prized by percussionists and drummers, own much of their unique sound and mechanical resilience (lesser cymbals can turn inside out when struck) to the fact that their original cast preform is tested by striking to see if the material has a ring in its cast state. If it does not, it is remelted. Clearly, this example confirms our expectations that the effects of the presence of bifilm defects survive in some alloys, even after the extensive working, applied in this case by forging and spinning into a thin disc.

The writer reveals his age when he admits to recalling the wheel tapper, moving along the side of the rail track, tapping the wheels of the train with his long handled hammer. He was checking that the 'ring' was correct. A cracked wheel would give a different note. Less romantically, but perhaps more reliably, the wheel tapper has now been superseded by the use of automated checking by ultrasonics while the wheels are rolled in a special checking station.

Ultimately, therefore, although there are fundamental reasons why the elastic modulus of many castings and their alloys cannot be improved, the reader may be able to raise the elastic modulus of a badly poured casting by 20–30% with negligible cost, exceeding the achievements of the famous Al-Li alloy which cost a fortune to develop.

Elastic properties of castings can now be measured quickly and with great precision using such modern tools as vibration frequency spectrum analysis, as described in Section 19.8.

9.8 RESIDUAL STRESS

Unseen and often unsuspected, residual stress can be the most damaging defect of all. This is because the stress can be so large, outweighing all the benefits of heat treatment and expensive alloying, and outweighing even the effects of all other defects.

Residual stress is usually never specified to be low. This is a grave indictment of the quality of component specifications and of standards in general. It is also practically impossible to measure in a non-destructive way in the interior of a complicated casting. However, it can be controlled by correct processing—another vindication of prudent, intelligent manufacture compared with costly, difficult and potentially unreliable inspection.

Also, of course, as will be noted in due course, in rare instances, residual stress can be manipulated to advantage. However, in the general case it should be assumed for the sake of safety that somewhere in the finished part, the retained stress will be in opposition to the strength of the casting. It can be viewed as subtracting from the casting's strength. Alternatively, it can be viewed as adding to the applied stress. Either way, a high residual stress means that the casting can be subjected to a total stress near to its point of failure even when relatively trivial loads are applied. Unfortunately, a conservative assessment of residual stress would have to assume that it reached the yield stress. This condition is expected to be common.

The remedy is, of course, either (1) the avoidance of stress-raising treatments such as quenching castings into water following solution treatment (and so accepting the somewhat reduced strengths available from safer quenchants such as forced air) or (2) the application of stress relief as already discussed (remembering, of course, that stress relief will effectively negate much or all of the strength gained from heat treatment). Either way, again, the use of the casting with lower internal stress and lower strength is vastly safer, more predictable and more reliable than a high strength casting with high internal stress.

As an example of a part that suffered from internal stress, a compressor housing for a roadside compressor in Al-5Si-3Cu alloy (the UK specification LM4) was thought to require maximum strength and was therefore subjected to a full solution treatment, water quench and age (T6 condition). Two housings exploded in service. Fortunately, on these occasions, no one was near at the time so no one was hurt. The manufacturer was persuaded to carry out a heat treatment that would reduce the internal stress, but also, of course, reduce the strength. This was strenuously resisted, but finally very reluctantly agreed. However, on testing the parts to destruction, the implementation of a heat treatment not to increase strength but merely to provide stress relief (a TB7 treatment) after casting, followed by air cooling, gave a part with only half the strength of the fully heat treated 'strong' casting, but double the burst pressure. No failures have occurred since.

This sobering lesson is expanded as rule 9 in the 10 *Rules for Casting* dealt with in Chapter 10.

9.9 HIGH TEMPERATURE TENSILE PROPERTIES

Creep

The gradual deformation of metals under load at high temperatures has been traditionally viewed as taking place in three stages:

1. Primary creep is the rapid early phase that gradually reduces in rate, eventually leading into stage two.
2. Secondary creep is the steady-state regime, in which creep rate is constant.
3. Tertiary creep is the final stage in which the rate of strain increases because of the growth of microscopic internal pores and tears that gradually link to cause the fracture of the whole component.

However, the reader should be aware that this traditional view is strongly criticised by Wilshire (see Wilshire and Scharning, 2008; Williams et al., 2010), who proposes a revolutionary single mechanism and single equation to fit all parts of the extension versus time creep curve. Wilshire's argument is that primary creep is subject to a decreasing rate, whereas tertiary creep is subject to an accelerating rate, so that the so-called secondary creep is merely a region in which these two mechanisms overlap to give a minimum creep rate. Wilshire's approach allows an accurate interpolation and extrapolation from relatively few experimental data points for several quite different metals and alloys, suggesting it is a valuable new fundamental interpretation of creep behaviour. We shall not delve deeper into the creep theory, but for our purposes, it is simply important to know that creep happens, and what we as founders might do to improve creep performance.

Components are said to have failed by creep if, during this slow deformation process, they exceed some critical size or shape. An example might be a turbine blade, whose length grows under the centrifugal stress so that it eventually scrapes the outer casing of the engine (although nowadays both blades and casings are designed to accept such problems). Alternatively, creep failure under tension might mean the fracture of the component at the end of tertiary creep.

Bifilms at grain boundaries are prime sites for the initiation of such catastrophic failures. Failure of the Ni-base alloy 718 described by Jo and colleagues (2001) appears to be precisely such a process initiated by oxide or nitride bifilms, judging from the micrographs published by these authors, although confirmatory evidence from the appearance and chemistry of the fracture surfaces was not carried out.

During the early development of the Pegasus engine for the Harrier Jump Jet 25, polycrystalline Ni-based alloy turbine blades that had previously been scrapped because of their content of porosity were subjected to HIPping, and were fitted to a test engine alongside sound blades to evaluate whether hipping might be a satisfactory reclamation technique for blades that otherwise would be scrapped. The HIPped blades failed within a few hours, damaging the engine and forcing a rapid shutdown of the test. The failures had occurred by creep cavitation at the grain boundaries of recrystallised regions in the centre of the castings. Almost certainly the original porosity would be caused by films rich in aluminium oxide entrained by the severe turbulence that is usual during the vacuum casting process. (The vacuum is known to contain plenty of residual air to ensure the formation of an oxide or nitride film on the liquid metal.) The great stability of the films, formed at the high casting temperature, would ensure that they were resistant to any re-bonding action during HIPping. The recrystallisation would have happened because of the large plastic strains that were a necessary feature of the collapse of the porosity. The grains would grow, expanding until they reached local barriers such as bifilms. Thus the bifilms, effectively unbonded, and so acting as efficient cracks, were automatically located at the grain boundaries from where the failures were seen to occur.

Superplastic Forming

In a related phenomenon of superplastic deformation, creep is deliberately induced at a high rate as a deformation process often involving several 100% of elongation. It is mainly useful for the shaping of sheet metals. The forming process is limited by the opening of microscopic voids, usually at grain boundaries, a process referred to in this field as 'cavitation'. A typical example in Mg alloys is shown in Figure 9.39. The voids grow, causing the mechanical properties of the material to deteriorate, and progressively link, eventually leading to complete failure.

Several Al alloys can be produced in a superplastic condition. Al-Zn alloys are commonly used, although superplasticity has been developed in many Al alloys including Al-Li and Al-Mg varieties.

Chang (2009) describes how his Al-4.5Mg alloy cavitated at grain boundary precipitates of $Al_6(Mn,Fe)$ and Mg-Si particles, strongly suggesting these particles occupied sites on bifilms that happened to be in grain boundaries.

Even more direct evidence of a bifilm-controlled failure mechanism in this alloy is provided by the work of Kulas and coresearchers (2006). Their SEM observations confirmed that voids opened at grain boundary intermetallics during grain boundary sliding. The intermetallics seem to have formed on oxide bifilms, because close examination of the voids by these authors indicated the presence of 'filaments' aligned with the deformation direction, stretching across the voids. The filaments were not analysed (indeed, this would have been a problem because they were clearly extremely thin, perhaps only 100 nm or less) and did not receive further comment by the authors, but can hardly be anything other than alumina films; the remnants of the bifilms causing the failure.

Similar observations can be listed for steels. During creep at 540°C of a power plant steel (Perets, 2004) containing 2.25Cr, 1Mo, 0.15C, carbides form on grain boundaries, where cracks and pores are observed, together with decohesion of the matrix from the carbides. A 12Cr steel studied by Wu and Sandstrom (1995) showed cavity initiation so early in their creep tests that they concluded the cavities were probably already present before testing. A super-austenitic stainless steel studied by Fonda (2007) using micro-tomography revealed that every sigma particle was associated with at least one void.

Intermetallic compounds are generally infamous for their brittle behaviour. However, in his studies of Fe_3Al Frommeyer (2002) reports amazing superplastic behaviour with elongations of 350%, indicating there seems to be nothing fundamentally wrong with the formability of this intermetallic. His work could not contrast more with that reported later for the same intermetallic by Yu and Sun (2004). They produced this material by vacuum melting and vacuum pouring into a sand mould, but experienced the ingots cracking and sometimes falling into pieces during hot forging which they attributed to the large, coarse grains. It seems certain that the fractures were the result of alumina bifilms from the turbulence of the pour because pour heights are high in typical vacuum furnaces. Figure 16.30 illustrates the problem. The large bifilms produced during the filling of the mould would have suppressed convection in the

FIGURE 9.39

Castings, solution treated, extruded 30:1 and finally subjected to tensile strain at 10^{-3} s^{-1} at 200°C (a) Mg and (b) Mg-0.9Al alloy showing so-called cavitation; pore growth from the decoherence of grain boundaries.

Courtesy Stanford et al. (2010).

solidifying ingot and lead to the undisturbed growth of large grains, instead of the fine grains produced from grain multiplication in normal convection currents.

The implication of these creep and superplastic failures by the gradual accretion of grain boundary voids is that they would not occur if bifilms were not present. The failure mechanism would then probably be 100% ductile failure leading to 100% RA. If true, the prospects for improvement in these materials are awe inspiring.

9.10 **OXIDATION AND CORROSION RESISTANCE**

During the cooling of the casting in the mould, and to some extent after it has been extracted from the mould, and especially during high temperature heat treatment, air can gain access to and can attack the interior of the casting via surface-connected bifilms. Similarly, if the casting experiences a corrosive environment, internal access of the corrodant can lead to a variety of problems. The role of the bifilm is central to the understanding of these important problems.

9.10.1 INTERNAL OXIDATION

When a metal such as an aluminium alloy is heat treated, the external surface naturally develops a thick oxide skin. However, in addition, some researchers have noted the development of internal oxides.

How is this possible? Oxygen is insoluble in aluminium and its alloys, and so, in principle, is not able to penetrate the alloy.

Clearly, such internal oxidation can only occur if the oxidising environment has access to the interior through some kind of hole. Surface-connected shrinkage porosity might create such a hole linked to extensive internal cavities. However, reasonably adequate pressurisation of the solidifying metal in the casting by proper feeding will prevent the formation of surface-connected porosity. Much more probable is access via bifilms connected to the outside surface. In the case of machined castings, particularly test bars, the machined surface will certainly cut into many bifilms, opening up many points of access for penetration by gas or liquid and attack deep into the interior of the casting.

It is easy to understand therefore that when working with an Al-4Mg alloy subjected to a solution treatment at 520°C for several hours Samuel and coworkers (2002) expected to find the usual benefits including not only the taking of solutes into solution, but fragmentation and spherodising of intermetallics, resulting in an improvement in mechanical properties. However, they reported the development of spinel (the $Al_2O_3 \cdot MgO$ mixed oxide) in the interior of the casting, and the reduction in tensile properties.

In the author's laboratory, Fenn and Harding (2002) treated some as-cast ductile iron test bars at 950°C for 1 h to eliminate carbides to allow them to be machined. The tensile properties turned out to be unexpectedly low, especially the ductility. Elongations in the region of 20% or more for this grade were to be expected, because ductile iron, as its name implies, is expected to be reliably ductile. It was a shock therefore to discover that elongations were highly scattered in the range of only 1–10%. Many specimens appeared to fail brittlely. SEM studies revealed thick uniform carpets of oxide on the fracture surfaces (Figure 9.40). These oxides clearly did not have the appearance of entrained films, and did not contain the content of Mg that would be expected if the oxides were bifilms entrained in the liquid metal during casting. It seems that the oxygen had penetrated the core of the bifilm and was added to the original magnesium silicate from which the original bifilm was constituted. The additional oxide appeared to be nearly pure iron oxide, as would be expected from a solid-state reaction. At the relatively low temperature of 950°C (low compared with the casting temperature of about 1450°C), the magnesium in solution in the solid alloy would be fixed, being unable to diffuse to the surface of the bifilm to enhance the magnesium concentration in the oxide. A further confirmation of the solid-state thickening of these oxides is that the graphite nodules appear through holes in the carpet. Clearly, the iron oxide cannot thicken above the nodules, giving the curiously dimpled or quilted appearance of the surface of the oxide seen in Figure 9.40(a). In Figure 9.40(b), the nodules themselves appear to be disintegrating as might be expected as their graphite oxidised away.

These internal oxidation problems are almost certainly common in cast alloys that have been subjected to heat treatment.

The same effect will occur, of course, automatically during the cooling of the casting in the mould. Additional oxidation of the originating bifilm will occur for surface-connected bifilms, thickening the film and possibly masking its original form. The original composition of the oxide may also be diluted and/or hidden by overgrowth of new oxide resulting from the solid-state reaction.

The solution to the avoidance of such internal oxidation is the avoidance of bifilms. Although this would be a complete solution, it may not be always practical. A next-best solution might be one of the many techniques to keep the bifilms closed, because, clearly, any action to open the bifilm, for instance by shrinkage, will enhance the access routes to the interior. It follows that a well-fed casting (i.e. pressurised by the atmosphere) or a casting artificially pressurised internally during solidification (as provided by some casting processes) will be less susceptible to the ingress of air during high temperature treatment. It will therefore retain its mechanical properties relatively unchanged (whether originally good or bad of course).

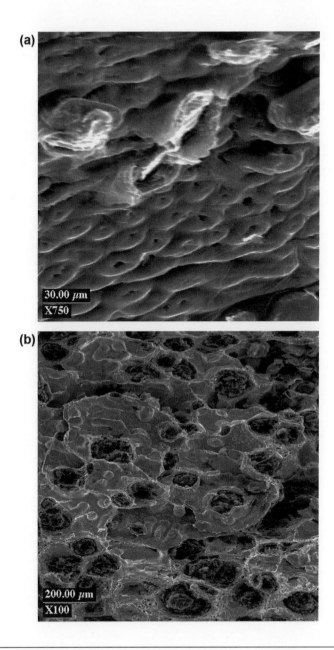

FIGURE 9.40

Fracture surface of a ductile iron damaged during annealing at 950°C for 1 h by internal oxidation of a surface-connected bifilm, showing (a) the formation of a carpet of solid-state-grown oxide and (b) oxidation of graphite nodules (Fenn and Harding, 2002).

9.10.2 **CORROSION**

Corrosion of metals, particularly aluminium alloy castings, and wrought products such as alloy plate and sheet, is a troublesome feature that has attracted much research in an effort to understand and control the phenomenon. Naturally, no comprehensive review of such a vast discipline can be undertaken here. The reader is referred to some recent reviews (Leth-Olsen and Nisancioglu, 1998). The purpose of this section is to present the evidence that a large number, possibly most corrosion problems, not only in shaped castings, but also in wrought alloys, arise from casting defects.

All corrosion pits that the author has seen are the sites where bifilms happen to meet the surface. In the absence of bifilms, it is proposed that there would probably be no corrosion of metals from surface pits. Corrosion might still be expected, but would probably be vastly reduced, having to occur uniformly over the whole surface of the metal. Alternatively, if localised sites were favoured, localised corrosion might continue to occur on a reduced scale from other inclusions, grain boundaries, or from dislocations that intersect the surface.

Many of the current theories of the corrosion of metals have been principally concerned with environmental attack on an essentially continuous unbroken planar substrate, regarding the surface of the metal as a uniform re-active layer (Leth-Olsen and Nisancioglu, 1998). The result has been that theories of filiform and inter-granular corrosion of aluminium alloys are at a loss to explain many of the observed features of these phenomena because these corrosion processes clearly do not exhibit uniformity of attack; the attack is extremely localised and specific in form.

The presence of bifilms generated in the pouring process has not, of course, been considered up to now as a factor contributing to the severity of corrosion. It will become clear in this section how bifilms help to explain many of the observed features of metallic corrosion. The link occurs because bifilms are, of course, often connected to the surface, allowing them to be detected sometimes by dye penetrant techniques. Similarly, in a corrosive environment, such bifilms will allow the local ingress of corrosive fluids between their unbonded inner surfaces (Figure 9.41). The presence of different varieties of second phases precipitated on the outer surfaces of bifilms will be expected to act as a further enhancement of the corrosion process by the creation of an electrochemical corrosion couple, explaining the major differences observed, for instance, between Al alloys of different Fe, Mn and Cu contents.

The stages of corrosion, leading to some pits formed by cathodic action involving alkaline dissolution, and other pits formed by anodic action leading to acidic dissolution, are illustrated schematically in Figure 9.42(a) and (b).

Direct and clear observations of oxide film tangles associated with corrosion sites have been made by Nordlien and Davenport (2000) and Afseth and coworkers (2000). Figures 9.43 and 9.44 are typical examples of sea water corrosion of a wrought Al alloy.

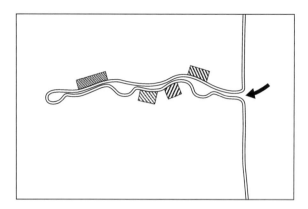

FIGURE 9.41

An image of a surface-connected bifilm, decorated with pockets of air and precipitated compounds of various kinds, is effectively open to the environment and makes an effective and complex corrosion cell if any aqueous corrodant such as sea water or rain water enters.

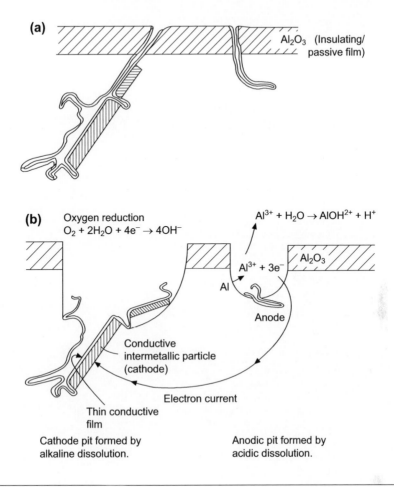

FIGURE 9.42

Mechanism of pitting corrosion from surface-connected bifilms (a) before and (b) during corrosion.

Adapted from Bailey and Davenport (2002).

9.10.3 PITTING CORROSION

Although there are many instances in which the corrosion of metals occurs uniformly across the whole surface, the special case of concentrated corrosion at highly localised sites, generating deep pits, is sometimes a serious concern. Most of the studies of pitting corrosion have been carried out on steels. Because we cannot in this short work survey this vast subject, we shall take only Al and its alloys as an example, following the review by Szklarska-Smialowka (1999), and see how pitting corrosion relates to the cast structure.

The main message of this section is that, in general, the familiar corrosion pit is not, originally, the product of corrosion. It pre-exists as a bifilm which emerges at the free surface of the metal. This pre-existence appears to have been generally overlooked until now because the originating defect is usually practically, if not actually, invisible. The corrosion process develops the pit into a highly visible and deleterious feature.

The corrosion proceeds as illustrated in Figure 9.42. The intermetallic particle precipitated on the bifilm acts as a cathode, with the electrical current passing through the electrolyte to anodic areas of the surface. It has been generally

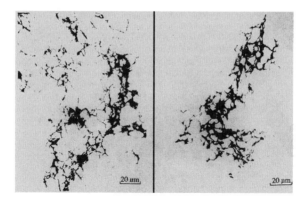

FIGURE 9.43

Two views of forged 7010-T736 alloy subjected to seawater corrosion.

Courtesy Forsyth (2000).

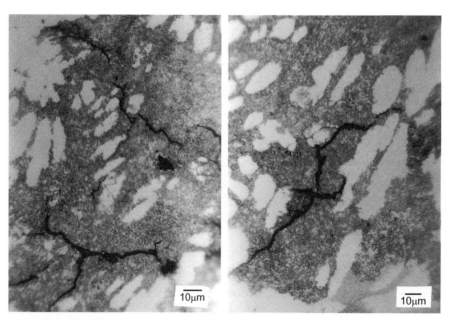

FIGURE 9.44

Typical polished and etched 7010 alloy in the solutionised condition subjected to seawater corrosion illustrating the interdendritic nature of the cracks.

Courtesy Forsyth (2000).

thought that the intermetallic particles provide the conductive path through the insulating alumina film. However, it is probable that the bifilm itself is sufficiently thin to be conductive, or will have been fractured during the cooling of the casting, so aiding this effect. The cathodic pit is the bifilm pit containing the intermetallic, whereas the anodic pits may be part of the same bifilm pits but distant from the intermetallic, or may be quite separate surface-intersecting bifilms that do not happen to contain intermetallics.

Oxygen is reduced at the cathode, demanding electrons, and so forming hydroxyl ions according to

$$O_2 + 2H_2O + 4e^- = 4OH^-$$

The alkaline conditions created by the hydroxyl ions assist to dissolve the material around the intermetallic, enlarging the pit. Conversely, at the anodic pit, conditions are acidic because of the generation of hydrogen ions as follows

$$Al = Al^{3+} + 3e^-$$

$$Al^{3+} + H_2O = AlOH^{2+} + H^+$$

Thus this pit also enlarges as matrix material is dissolved. The electrical circuit is, of course, completed by electrons travelling though the aluminium matrix from the anode pit to the cathode pit.

The random nature of the creation of such defects, being linked to the action of surface turbulence at several stages of manufacture of the sheet, explains why the corrosion behaviour is so variable, changing in severity from one supplier of metal to another and from one production batch of alloy to the next. Also, of course, every pit will be different because of the random nature of the oxide tangles. The tangled geometry is indicated in Figures 9.43 and 9.44. This randomness has been a major problem to investigators.

The bifilms are expected to survive, and may even grow during plastic deformation, with their residual entrained air continuing to oxidise any newly created elongation of the defect until supplies of air are exhausted. Further extensions of the defect may then weld. Thus surface-linked cracks, possibly plated with intermetallics, will be expected not only characteristic of castings but also of wrought products.

In their studies of the effect of corrosion on the wrought alloy Al 6061 in the T6 heat treated condition Almaraz et al. (2014) notice that fatigue crack initiation was frequently associated with two or more corrosion pits, which greatly aided crack nucleation. These authors admitted that they could not understand why the effect was so severe. When researching etch pit formation in DC ingots, Jaradeh and Carlberg (2011) observed oxides associates with porosity (the clear signatures of bifilms) which developed into circular etch pits in NaOH solution. Interestingly, the numbers of pits corresponded reasonably well with the numbers of oxides in the liquid metal counted by Prefil.

9.10.4 FILIFORM CORROSION

In a standard corrosion scratch test, filiform corrosion takes the form of a high surface density of superficial corrosion paths, called filaments, which propagate extensively over the next few days from a scribe mark on a test plate. The corrosion proceeds away from the scratch along filamentary lines aligned with the original rolling direction. They travel under any protective layer such as paint, occasionally tunnelling beneath the metal surface, only to break out at the metal surface once again after a few millimetres or so. The lengthwise growth and subsequent sideways spreading of the filaments eventually causes any protective coating, such as a paint layer, to exfoliate. The length of filaments has been found to be generally in the 1–10 mm range. However, reviewers confirm (Leth-Olsen and Nisancioglu, 1998) that quantification of the phenomenon suffers from significant scatter that has hampered these studies.

The concentration of corrosion at strictly localised sites (the filaments) is clear. However, it is important to observe that the great majority of the metal surface remains completely free from attack (despite the long and deep breach of the protective coating by the scratch). Also clear is the different behaviour of different casting batches of nominally identical material, on different occasions giving filaments shallow or deep, or short (1 mm) or long (10 mm).

Growth of filaments stops when the length reaches some value between 1 and 10 mm. This has been suggested to be the result of chloride depletion in the head of the filament (Leth-Olsen and Nisancioglu, 1998), but is clearly more likely that the corrosion has reached the end of that particular bifilm.

In his review of the subject, Nordlien (1999) describes how the filaments of corrosion can grow at up to 5 mm per day. They occur on all families of aluminium alloys (1000, 2000, 3000, 5000, 6000, 7000 and 8000 series) and on all product forms (sheet, foil, extrusions).

Interestingly, a surface of rolled aluminium alloy sheet can be sensitised to the formation of filiform corrosion (in corrosion jargon it is 'activated') by annealing at 400°C. This effect can be understood as the growth of oxidation products on the internal surfaces of cracks that will assist to open the cracks (see Section 9.10.1). The deactivation by etching probably corresponds to the preferential attack and removal of surface cracks and laminations. Re-activation by subsequent annealing seems likely to be the result of the opening of slightly deeper defects by oxidation. The removal of defects by etching removes only a few micrometres of depth of the surface. Considering the defects are commonly 1–10 mm in size, there will be no shortage of new defects to open on subsequent re-activation cycles.

In severe cases of surface corrosion, the frequent observations of delamination (Leth-Olsen and Nisancioglu, 1998) can be understood as the lifting of irregular fragments of bifilm that lie just under the metal surface. Other related observations of blistering (see Chapter 19.2) can also be understood as the inflation of just-subsurface bifilms by hydrogen evolved from the chemical reaction between the corrodant and the intermetallic compounds associated with the bifilm.

9.10.5 INTER-GRANULAR CORROSION

Inter-granular corrosion in its various forms is also proposed here to be associated in some cases with the newly identified bifilm defects, as a result of the natural siting of bifilms at grain boundaries in the cast structure.

Metcalfe (1945) records studies of the intercrystalline corrosion of the heads of rivets in an Al-Mg alloy from an aircraft that has been flown near marine environments. He concludes that the effect is one of stress corrosion cracking. Undoubtedly, there would be residual stress that may have played a part in the failures that are described. More especially so because the cracks were observed to follow grain boundaries sensitised by prolonged in-service aging, and the convoluted form of the crevices was due to the fact that the flattened grains themselves were distorted in this fashion by the complex flow pattern of the worked metal. Even so, a look at a section of one of the decapitated rivets in his work reveals a convoluted crack that can hardly have been propagated by stress. The stress would have been reduced to near zero after the spread of the first crack across the neck of the rivet. In fact, there is the trace of a crack which has repeatedly turned, spreading back and forth across the neck of the rivet at least five or six times. This type of crack is typical of a folded oxide defect. Its presence would ensure the stability of the convoluted form of the grain boundaries, which it would pin. Furthermore, in this vintage of alloy, a high density of entrainment defects would be the norm. The defect has provided an easy path for the attack of corrodant.

Forsyth (1995, 1999) describes seawater corrosion leading to intergranular cracking in 7010Al alloy. Corroded surfaces that have been polished back through the worst of the surface layer are presented in Figure 9.44. The inter-granular and transgranular cracks were, once again, typical of the localised tangled arrays of films that are normal in aluminium alloys produced via the melting and casting route. The cracks exhibit the typical irregular branching and changes of direction on several different size scales, often unrelated to the general size of the grain size of the matrix. Alloy material between such damaged regions was recorded to be completely free from attack. These observations are difficult to explain without the existence of random entrainment defects from the original casting.

When etching to reveal the dendrite structure, the cracks were seen (Forsyth, 1999) to be confined to the inter-dendritic regions (Figure 9.44) as is expected from the dendrite pushing of oxide bifilms. The defects are therefore concentrated in the residual liquid in the interdendritic regions, and in grain boundaries.

Forsyth (1999) also investigated the corrosion of 7010 alloy in seawater as a result of machining or bruising of the surface. In the case of bruising, the deformation of the surface would be expected to open any entrained defects at or near the surface, creating highly localised and deeply penetrating inter-granular pathways for attack.

Forsyth also draws attention to the especially damaging nature of the attack, in that despite rather little dissolution of material, complete blocks of material could be removed simply by the penetration of the attack along narrow planes in different directions. This observation corroborates his earlier report (Forsyth, 1995) in which subsequent anodising of the surface led to the incorporation of unanodised grains of metal in the corrosion debris remaining from such localised attack. The metal grains remained unanodised because they were found to be electrically isolated from their surroundings. This would not be surprising if double oxide films, separated by their inter-layer of air, surrounded the grains.

9.10.6 STRESS CORROSION CRACKING

Stress corrosion cracking (SCC) is a particularly serious form of failure. It seems to occur in conditions in which both stress and corrosive environment combine. It is often unexpected, involving the corrosion of minute amounts of material, but producing extensive and sometimes disastrous cracks. There remains much research but few conclusive results. The problem remains a mystery.

Once again, the presence of bifilms might prove to be the unsuspected major influence in most of the research conducted so far. Two very different examples are presented later, one for a Mg alloy and the other for an austenitic stainless steel.

Winzer and Cross (2009) describe the stress corrosion cracking of Mg alloys containing β particles ($Mg_{17}Al_{12}$). They describe how the consensus of opinion is growing that β particles are associated with the transgranular crack propagation and suggest several mechanisms that might be involved without reaching any conclusions. However, the simplest explanation seems to be the association between β particles and bifilms. Bifilms will certainly be present, and β particles will be expected to form on them, and the bifilm crack will provide a route for the corrodants. It all seems so straightforward. However, it has yet to be proven.

Andresen and colleagues (2009) carried out work on an austenitic stainless steel intended for application as a core component in a nuclear light water reactor. The corrodant in this case was deaerated, demineralised water. Once again, the authors admit to inconclusive results. However, the technique for the production of their samples (as roundly criticised by me in a published letter – Campbell, 2010), including vacuum induction melting and casting followed even more inappropriately by vacuum arc remelting guarantees a generous population of oxide bifilm defects. Several of the SEM images of the fracture surfaces showed clear examples of oxide films that had been formed on the melt. On occasions, these filled the field of view of the image and so could be deduced to be at least approximately 0.1–0.2 mm in size.

Further evidence of SCC initiated by bifilms was found by Lu and coworkers (2014) for an austenitic stainless steel in water at temperatures in the 250–320°C range. The SCC crack formed preferentially at the bottom of a corrosion pit (indicating that the initiator was a bifilm) and propagated along the phase boundary between austenite and ferrite (a typical location for a bifilm, arising from the ferrite initiating preferentially on bifilms to reduce its strain energy of formation).

In conclusion, it seems there is considerable evidence that in the absence of bifilms, some types of corrosion and possibly stress corrosion cracking might be reduced or eliminated. The elimination of bifilms would revolutionise metals and improve the quality of our lives in many ways.

9.11 LEAK-TIGHTNESS

Leak-tightness has usually been dismissed as a property hardly worthy of consideration, being merely the result of 'porosity'.

However, of all properties specified that a casting must possess, such as strength, ductility, fatigue resistance, chemical conformity etc., leak-tightness is probably the most common and the most important. This might seem a trivial requirement to an expert trained in the metallurgy and mechanical strengths of materials. However, for the foundry engineer, it is a critical requirement not to be underestimated.

A cylinder head for an internal combustion engine is one of the most demanding examples, requiring it to be free from leaks across narrow walls separating pressurised water above its normal boiling point, very hot gas, hot oil at high pressure, and all kept separate from the outside environment. A failure at a single point is likely to spell failure for the whole engine. In this instance, as is common, leakage usually means 'through leaks', in which containment is lost because of a leak path completely through the containing wall.

However, leakage sometimes refers to surface pores that connect to an enclosed internal cavity inside a wall or boss. Such closed pores give problems in applications such as vacuum equipment, where outgassing from surfaces limits the attainment of a hard vacuum. Problems also arise in instances of castings used for the containment of liquids, where capillary action will assist the liquid to penetrate the pore. If the pore is deep or voluminous, the penetrated liquid may be impossible to extract. This is a particular problem for the food processing industry where bacterial contamination residing in surface-connected porosity is a concern. Similarly, in the decontamination of products used in the chemical, pharmaceutical or nuclear industries, aggressive mechanical and chemical processes fail to achieve 100% decontamination almost certainly as a result of the surface contact with bifilms and possibly with shrinkage cavities. Such industries require castings made from clean metal, transferred into moulds with zero surface entraining conditions. Only then would performance be satisfactory.

It is true that leaks are sometimes the result of shrinkage porosity, especially if the alloy has a long freezing range, so that the porosity adopts a sponge or layer morphology. Clearly, any form of porous metal resulting from poorly fed shrinkage will produce a leak, especially after machining into such a region.

Leaks are seldom caused by gas porosity i.e. bubbles of gas precipitated from solution in the liquid metal. The following logic provides an explanation.

Gurland (1966) studied the connections between random mixtures of conducting and non-conducting phases by measuring the electrical resistance of the mixture. He used silver particles in Bakelite, gradually adding more silver to the mix. He found the transition from insulating to conducting to be quite abrupt, in agreement with stochastic (i.e. random) models. The results are summarised as follows.

% Ag	% Conducting
1	0
1.73	50
2.5	100

In the case of about 1–2% gas porosity in cast metals, the metal must surely therefore be permeable to gas. Why is this untrue? It is untrue because the distribution of gas pores is not random as in Gurland's mixtures. Gas pores are distributed at specific distances, dictated by the diffusion distance for gas. In addition, the pores are kept apart by the presence of the dendrite arms. Thus leakage from connections between gas pores cannot occur until there are impossibly high porosity contents in the region of 20–30% by volume (see Figure 7.36).

The only possible exception to this rule is the relatively rare occurrence of wormhole-type bubbles formed by the simultaneous growth of gas bubbles and a planar solidification front. Such long tunnels through the cast structure naturally constitute highly effective leak paths (see Figures 7.38 and 7.39). Fortunately they are rare, and easily identified, so that corrective action can be taken.

In the author's experience, most leaks in light-alloy and aluminium bronze castings are the result of oxide inclusions. These fall into two main categories:

1. Some are the result of fragments of old, thick oxide films or plates which are introduced from the melting furnace or ladle, in suspension in the melt, and which become jammed, bridging between the walls of the mould as the metal rises. The leak path occurs because the old oxide itself suffered an entrainment event; as it passed through the surface, it would take in with it some new surface oxide as a thin, non-wetting film covering. The leak path is the path between the rigid old oxide fragment and its new thin wrapping.
2. The majority of leaks are the consequence of new bifilms introduced into the metal by the turbulent filling of the mould. A clear example in an iron casting is shown in Figure 19.7. These tangled layers of poorly wetted surface films, folded over dry side to dry side, constitute major leak paths through the walls of castings. The leaks are mainly concentrated in regions of surface turbulence. Such regions are easily identified in Al alloys as areas of frosted or grey striations down the walls of top-gated gravity castings outlining the path of the falling metal. The remaining areas of walls, away from the spilling stream are usually bright and clear, free from any visible oxide striations and are free from leaks. The reader should be able to confirm, and take pride in, the identification of an aluminium alloy casting which has been top-poured from a distance of at least 100 m! Unfortunately, this is not a difficult exercise, and plenty of opportunity exists to keep oneself in training in most light alloy foundries! It is to be hoped that this regrettable situation will improve.

An example of a sump (oil pan) casting, top poured into a gravity die (permanent mould), is shown in Figure 9.45. The leakage defects in this casting are concentrated in the areas that have suffered the direct fall of the melt. The surface oxide markings are seen on both the outside and inside surfaces of these parts of the casting walls (Figure 9.45(a) and (b)). Other distant areas where the melt has filled the mould in a substantially uphill mode are seen to be clear of oxide markings and free from leaks. The precise points of leakage are found by the operator who inverts the casting, pressurises it with air and immerses it under water. He is guided by the stream of air bubbles emerging from leaks, and deals a rapid

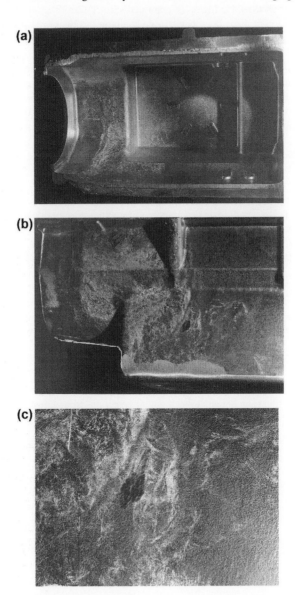

(a)

(b)

(c)

FIGURE 9.45

Views of (a) the inside and (b) outside of a top-poured ail pan (sump) casting showing the light traces of entrained oxides and (c) the corresponding leak defects repaired by peening, seen in close up.

series of blows from a peening gun. This hammering action deforms the surface locally to close the leak. The peening marks are seen in close up in Figure 9.45(c).

The linkage between oxide films and leakage problems was noted by Burchell (1969), when he attempted to raise the hydrogen gas content of aluminium alloys by stirring with wood poles, dipped in water. The porosity of the castings increased as was intended to counter feeding problems, but so did the number of leaks. Burchell identified the presence of oxide films on the fracture surfaces of tensile test bars that were cast at the same time.

It is unfortunate therefore that the folded form of entrained surface films creates ideal opportunities for leak paths through the casting. This source of leakage is probably more common than leaks resulting from other kinds of porosity. Although shrinkage porosity is often blamed for leaks, this seems to be the result of the common mis-diagnosis of bifilm tangles as shrinkage.

As an instance of the seriousness of leaks in castings that are required to be lead-tight, many foundries have been historically reluctant to cast aluminium manifolds and cylinder heads with sections less than 5 mm. This is because of the increased incidence of leaks that require the casting to be repaired or scrapped. The lack of pressure tightness relates directly to the presence of oxides whose size exceeds 5 mm (an interesting confirmation of the non-trivial size and widespread nature of these defects), and that can therefore bridge wall to wall across the mould cavity, connecting the surfaces by a leak path.

Leaks are often associated with bubble-damaged regions in castings. This is because all bubbles will have been originally connected to a surface as a necessary feature of their entrainment process. Some bubbles will have retained their bubble trail links to the outside world, whereas others will have broken away during the turmoil of filling. Bubble trails are particularly troublesome with respect to leak-tightness because they necessarily start at one casting surface and connect to the surface above, and as part of their structure, have a continuous pipe-like hollow centre. The inflated bubble trails characteristic of high-pressure die castings (Figure 2.39) make excellent leak paths. A core blow also leaves a serious defect in the form of a collapsed bubble trail (Figure 2.32). Despite its collapsed form, the thickness and residual rigidity of its oxide will ensure that the trail does not completely close, so that a leak path is almost guaranteed.

In general, the identity of a leakage defect in a casting can be made with certainty by sawing the casting to within a short distance of the defect, and then breaking it open and studying the fracture surface under the microscope. A new oxide film (probably from surface turbulence during pouring, or from a bubble trail) is easily identified from its folded and wrinkled appearance; an old oxide fragment (perhaps from the melting furnace or crucible) from its craggy form, like a piece of rock; and shrinkage porosity by its arrays of exposed dendrites. Fracture studies are a quick and valuable test and are recommended as one of the most powerful of diagnostic techniques. The reader is recommended to practice this often – despite its unpopularity with the production manager; the destruction of one casting will often save many.

Leak Detection

We turn now from the nature of leakage defects to methods of detection. Bubble testing has already been mentioned in which the inside of the casting is simply pressurised with air, the casting immersed in water and any stream of bubbles observed.

For castings which require a demonstration of leak-tightness against very high pressures they cannot be pressurised safely with a gas, but require to be pressurised with a liquid such as water. Thus the geometry of the usual bubble technique is reversed; leakage is detected by the emergence of a coloured dye in the water.

Even these time-honoured and apparently simple techniques are not to be underestimated because providing effective and rapid sealing of all the openings from the interior of the casting may not be easy in itself. It will almost certainly require careful planning to ensure that the correct amount of dressing or machining has been carried out to eliminate troublesome flash and gating systems etc. so that the sealing surfaces can be easily accessed and effectively sealed. Also, of course, the techniques are slow, not quantifiable, and both demand the constant attention of skilled personnel. After testing, the part often requires to be dried.

Hoffmann (2001) describes three basic methods of dry air leak testing suitable for production line applications: they measure (1) the rate of decay of gauge pressure, (2) the rate of decay of differential pressure and (3) leakage rate directly in terms of mass flow. For highly specialised applications, helium mass spectrometry offers testing capability beyond the limits attainable by dry air methods.

1. The first technique, the rate of decay of gauge pressure, is the simplest and lowest cost, and is generally suitable where pressures do not exceed 2 bar and volumes do not exceed 100 mL.
2. The differential pressure method pressurises a non-leaking reference volume along with the test part. A transducer reads any difference in pressure that occurs over time. The differential technique reduces errors from temperature changes, and is more accurate and faster than the direct pressure decay method. The technique is also well-suited to applications specifying higher test pressures, exceeding 10 bar, and where relatively small cavities must be tested to a very low leak rate.
3. The mass flow method pressurises the test cavity, then allows any leakage to be compensated by actively flowing air into the cavity. The in-flowing air is measured directly by a mass flow meter in terms of volume per second. The method involves a single measurement, usually less than 1 s. (It avoids the taking of two measurements over a time interval—during which temperature may change for instance—that is required for the first two techniques, thus halving errors and increasing speed of response.) The method can tackle a wider range of volumes, and is accurate down to 0.001 mLs^{-1}.

For castings that are required to be leak tight to even greater standards, helium mass spectrometry can measure down to rates that are 10,000 times lower. This is principally because the helium atom is much smaller than molecules of nitrogen and oxygen, and so can penetrate much smaller pores.

Krypton gas has been used for the detection of very fine leaks, because of its content of 5% radioactive Krypton 85 (Glatz, 1996). As before, the part to be tested is placed in an evacuated chamber to suck air out of the cavities. Krypton gas is then introduced to the chamber and allowed time to penetrate the surface pores. The Krypton is then pumped out, ready for re-use and air is admitted. The rate at which Kr is slowly released can be monitored to assess the volume of surface-connected internal pores. In addition, the spraying of the surface with a liquid emulsion of silver halide particles makes the surface sensitive to the low energy beta particles given off by the radioactive decay of Kr85. After the emulsion is developed by conventional photographic techniques, the part reveals the site and shapes of surface pores and cracks. The beta particles can penetrate approximately 1 mm of metal, revealing subsurface cracks (if connected to the surface elsewhere of course) and magnifying the width of pores and cracks that otherwise would be too small to see. The technique is more sensitive than dye penetrant testing because of the viscosity of gases is typically only 1/100 of that of liquids, making the test extremely searching.

Finally, it is worth emphasising that a good melt quality combined with a good filling system will usually eliminate most of the leaks found in castings (providing core blows can be avoided by careful design or venting of cores). This conclusion is confirmed by casting operations using intrinsically quiescent melt handling processes such as Cosworth Process. These foundries are so confident of the quality of their products that they do not even bother to test for leak tightness; none is ever found to leak.

9.12 SURFACE FINISH

There are two major aspects to the achievement of a good surface finish for castings; the first, applying to some metals, is surface tension of the liquid, and the second, important for many important alloys, the solid surface film on the liquid.

9.12.1 EFFECT OF SURFACE TENSION

It has been well understood for many years (Hoar, 1953) that when in contact with many non-metals, liquid metals enjoy the benefit of capillary repulsion; the resistance felt by the metal when attempting to enter a small hole or channel.

(In passing, we should note that if the liquid metal wetted the mould materials it would experience capillary attraction, effectively being sucked into the mould surface as water is sucked up a capillary tube. Mould materials are therefore selected however for their non-wettability. The founder goes to some lengths to avoid wetting conditions which cause penetration of metal into the mould.)

Assuming non-wetting conditions, when the pressure P in the liquid metal becomes sufficiently high, surface tension γ is no longer able to resist the penetration of the metal into the spaces between the sand grains of the mould. The size of the holes between the sand grains can be roughly estimated assuming that the radius of the inter-granular spaces r is only approximately 15.4% of the radius of the sand grains.

The following arguments already presented in the development of Eqn 4.1, the resistance offered by surface tension is $P = 2\gamma/r$. If the pressure in the melt, density ρ, arises simply from its depth h, then $P = \rho gh$ where g is the acceleration due to gravity. From these equations the critical depth at which penetration first occurs can easily be estimated. For instance for liquid steel where $\gamma = 1.5$ N/m and $\rho = 7000$ kg/m^3, a grain size of 500 µm corresponding to a pore size of 75 µm and radius 37.5 µm indicates a critical depth at which penetration will occur as approximately 1.5 m.

This static estimate neglects the important dynamic problems created by a poor filling system in which the melt may violently impact the sand, this hammering effect by the liquid metal causing the mould to suffer additional transient penetration pressures.

The penetration of the mould in this way produces a 'furry' casting that may be quite unsaleable. The penetration may be only one grain deep, giving effectively an excessively rough surface. However penetrations of 20–50 mm are not uncommon in large castings. In a classical series of experiments, Hoar and Atterton (1950–1956) demonstrate that once the critical pressure difference to force the metal into the sand is exceeded, then penetration occurs rapidly, within a second or so. Figure 4.10 illustrates that once the surface grains are penetrated, the penetration is a run-away effect. Ultimately, the depth of penetration is controlled by the freezing of the leading edge of the advancing metal when it reaches the freezing isotherm in the sand. Clearly, this distance is greater for larger castings. The mix of solid metal and sand is difficult to remove from castings.

Attempts are made to resist such mould penetration by reducing the size of the pores by:

1. The use of finer sand for the mould or core. For cores, this approach is limited by the requirement to maintain the permeability of the core material so that core gases can escape during casting. Fine grain sand is, however, widely used for vacuum (V) process moulding, where the vacuum that is applied to maintain the rigidity of the mould assists in drawing the liquid metal into the pores between the sand grains. Thus, whereas normal sand castings have an average grain size in the range 250–500 µm, sands for the V process are approximately 50–150 µm. This can result in a dust nuisance when casting hotter metals such as iron, which is a pity because the vacuum moulding process is otherwise excellent in its environmental benefits. Newer plants are improving their designs to tackle this problem. Croning shell process benefits surface finish by its fine grains, usually less than 200 µm, and high percentage of binder, but permeability is maintained in this case by the thinness of the shell mould, usually never more than 10 mm.

2. The application of a mould wash—a ceramic slurry applied as a paint to fill the spaces between the sand grains. The pores in the dried coating are one or two orders of magnitude smaller than the pores between the grains, thus, in line with Eqn 2.2, enabling the mould surface to withstand 10–100 times greater pressures before metal penetration. Although the metal will often still succeed in penetrating the sand coating through cracks, the penetrated sand only adheres to the casting at the isolated points of failure of the coat, and so is relatively easy to remove. This action is generally used to counter metal penetration at the base of tall moulds, which can sometimes be more than a metre high.

The application of a core wash is only marginally successful, however, in resisting metal penetration with the use of sand cores in low-pressure die casting. This seems to be the result of the rather poor pressure control on most low-pressure machines, and the additive effect of the momentum of the metal, giving a pressure peak at the instant when the liquid hits the top of the mould (see Section 3.1.2). Sand cores are only rarely used, therefore, in low-pressure die casting. For similar reasons, sand cores have proved impractical for high-pressure die casting. In this case other solutions such as water-soluble salt, glass, or ceramic cores have sometimes achieved success. In general, only withdrawable steel cores continue to be used successfully.

Filling pressures are high in pressure die casting. This is widely alleged to be for the purpose of increasing surface finish and definition, i.e. the ability of the metal to fill small radii so as to reproduce fine detail. Pressure die-casting machines commonly operate at metal pressures of 1000 atm (100 MPa). Equation 2.1 indicates that such pressures will force the liquid into radii of only 10 nm, approaching atomic dimensions! This is, of course, a vast overkill. The student should therefore be on his guard against such loose thinking. The high pressures are actually needed mainly to reduce the bubble defects produced by the turbulently entrained mould gases – the bubbles are simply squashed to acceptable dimensions. Consideration of Eqn 2.1 reveals that a mere 10 atm (1 MPa) would reproduce a radius of about 1 μm, which would be more than good enough for most engineering purposes.

Filling pressures are enhanced for the genuine purpose of reproducing detail in the centrifugal casting of jewellery. A casting travelling at 10 m/s on an arm of radius 1 m will experience an acceleration of 100 m/s^2. This is close to 10g. In the absence of mould gases, and replacing the acceleration g due to gravity with the total acceleration $(g + 10g) = 11g$ as the arm goes from the vertically up through the vertically down part of its stroke, Eqn 2.3 predicts that an improvement in the fineness of detail by a factor of 11 should be achievable. These predictions are likely to be all the more accurate for jewellery alloys based on noble metals such gold and platinum since, in principle, they will not be troubled by surface oxides (alloy components may assist to detract from this benefit of course).

In engineering, uses of centrifugal casting much higher accelerations are normally used, typically 50–100g. The high pressures in the liquid are, however, not normally needed for filling because the moulds are usually of simple shapes. The technique is valuable for totally unrelated reasons: (1) for pipes and cylinders and the like, because the centrifugal action avoids the requirement for a central cylindrical core to make a hollow shape; (2) for enhancing the pressure in the casting during freezing to reduce porosity; and (3) centrifuging the oxides and similar less dense inclusions into the centre of the bore, where, if necessary, they can be machined off.

As explained in more detail in Section 16.4, the production of shaped castings by this route involves pouring the liquid metal down a central down-runner, and accelerating it out along radial runners, to arrive in the mould at such a high speed that considerable damage is done to both mould and metal. The high centrifugal pressures are then needed to help to repair this damage constituted by entrained porosity and inclusions in the casting. Shaped castings would probably be cheaper and better if not centrifuged at all, but simply produced with a properly designed gravity-running system.

9.12.2 EFFECTS OF A SOLID SURFACE FILM

The previous considerations of the control of surface finish by surface tension do not apply to many metals and alloys that have strong, solid surface films, particularly if they are filled uphill by bottom gating of gravity poured castings, or by counter-gravity filling. The metal rises upwards, its surface oxide thickening during the time it moves across the top of the melt, arriving at the mould wall to become the skin of the casting (Figure 2.2). This leisurely formation and sideways sliding of the surface film allows it time to thicken. It seems possible that the quality of surface finish may be a function of the rate of rise of the liquid in the mould because this would be expected to be directly related to the thickness and rigidity of the surface film.

Having travelled across the meniscus and arrived at the wall of the casting, the metal effectively rolls out its solid skin like a track-laying vehicle against the mould surface (Figure 2.2). No liquid metal contacts the mould; the two are separated by a rigid, solid, plate-like film. The solid film has sufficient rigidity to bridge the gaps between grains of aggregate, thereby creating a smooth surface. The smoothness of the casting surface reflects the smoothness of the surface of the original liquid meniscus at the time the original surface film formed.

This mechanical process, allowing time for thickening and strengthening the oxide film on the liquid, can be supplemented by various chemical effects:

1. Aluminium alloys enjoy this benefit because of its strong alumina film. These include aluminium bronzes (Cu-10Al etc.) because the higher temperatures of these alloys give an even thicker and stronger alumina skin.
2. An Alcoa patent by DeYoung and Dunlay in 2002 describes how the addition of strontium (Sr) to aluminium alloys strengthens the surface oxide and consequently improves the surface of DC ingots.
3. Similarly, in 1989, Sare found that when melting the surface of cast irons by gas tungsten arc welding, grey iron produced a rough surface but ductile iron was smooth as a result of its small content of Mg that promoted a strong magnesia-rich surface film.

Perhaps the most impressive demonstration of the benefits of a strong skin is shown by cast iron in moulds that contain volatile carbonaceous additions such as coal dust or certain resin binders. The hydrocarbons chemically decompose ('crack') on the surface of the rising hot metal, releasing hydrogen which escapes somewhere (it would be interesting to know where it goes; whether into the metal or the mould atmosphere), but certainly depositing carbon on the surface of the liquid in the form of a strong film. This is known as lustrous carbon (Figures 6.33 and 6.34). Iron castings that are nicely bottom-gated can emerge from the mould with a black, glossy surface almost like a mirror. Such a good finish is the sign of a casting filled nicely.

For steel castings with poor filling systems, Puhakka (2011) reports a poor finish with generous amounts of so-called 'burn-on'. The poor finish almost certainly arises from impacting the mould surface with high momentum, with little or no time for the development of a useful thickness of protective film barrier. Conversely, the surface of his bottom-gated castings with good filling systems exhibit zero burn-on and an overall finish so smooth to the touch as to be almost strokable.

9.13 QUALITY INDICES

The confusion of the interplay of properties among yield strength, tensile strength and elongation, the strengths going up as the elongation reduces with different heat treatments and other factors, is solved by the concept of a quality index. This brilliant innovation introduces only a single value to indicate the quality of a cast material. So far, quality indices appear to be exclusively used in the Al alloy casting industry, and in particular for the widely used Al-7Si-Mg alloys.

The concept was invented by French researchers (Drouzy, Jacob and Richard, 1980) in the following form

$$Q_{DJR} = \sigma_{TS} + 150 \text{Log E}$$

where the tensile strength σ_{TS} was in units of MPa and elongation E was in units of percent. The formulation as a single parameter is undoubtedly useful, and has become popular in the industry. Those castings achieving an index above 500 MPa were regarded as outstanding, but it is probably only a matter of time before techniques improve to yield values of 600 MPa or more.

Even though the single parameter quality index concept has been recognised as excellent, the Drouzy formulation of the parameter has been criticised by several subsequent researchers. In particular, the use of TS and elongation has elements of double counting and was not based on the really essential engineering design parameter, yield strength σ_Y (or near equivalent such as 0.2% proof stress). Finally, the incorporation of the log term was inconvenient and in fact hardly necessary. The author's laboratory (Din, Rashid and Campbell, 1996) came up with the simpler

$$Q_{DRC} = \sigma_Y + kE$$

where k is a constant determined from experiment to be 50 for both 356 and 357 alloys (respectively the low and high Mg versions of the Al-7Si alloys).

Both of these quality measures were based on empirical approaches which had no fundamental significance. In practice, a reader may wonder why there is even a specification on elongation because parts are designed to handle stresses well below the yield strength.

The answer on a legalistic level is put forward by the automotive companies who wish their components to bend rather than fracture in a crash, so that the component cannot be cited as a potential cause of the crash.

On a more general engineering level, it is clearly more desirable for a component to fail by plastic deformation than fracture because plastic deformation absorbs more energy. This statement points to the assessment of energy (or toughness) as described by the energy under the true stress–strain curve as a measure of quality. However, we all know that almost all component failures happen by fatigue fracture, although both fracture toughness and fatigue life are both strongly affected by defects. Therefore, quality indices giving a measure of the severity of defects would be expected to be useful in giving an indication of fatigue life.

Tiryakioglu (2009) initially suggested a more rational assessment based on the energy to fracture. This could be closely approximated to the average $(\sigma_{TS} + \sigma_Y)$ multiplied by E (giving the area under the stress/strain curve). Because σ_Y is relatively fixed, and σ_{TS} is a function of σ_Y and E, it follows that E is a dominant factor in this proposed assessment of quality. Finally, therefore, if E is measured at a known value of σ_Y, E alone can give an excellent measure of the quality of the material.

Tiryakioglu and colleagues (2009) proposed from fundamental grounds there should be an upper bound to the uniform elongation to fracture of Al alloys, because necking before fracture was generally never observed. Their theoretical predictions were revealed to be closely modelled by experimental data; the upper bound to elongation was revealed by plotting all the available data, mainly from aerospace alloy foundries. The results for the 356 and 357 alloys are shown in Figure 9.18(a) and the results for the Al-4.5Cu alloys A206 and A201 in Figure 9.18(b). A measure of quality at any value of yield strength and experimentally measured elongation E can now be defined immediately as

$$Q_T = E/E_m \%$$

The quality index now becomes the percentage of the maximum elongation E_m attainable at that level of yield strength (for instance as resulting from a particular heat treatment). This formulation of the quality concept by Tiryakioglu using Q_T is strongly recommended as giving a direct measure of the level of attainment of potential quality. The new index is applicable to all Al alloys and should be applicable to all cast alloys.

In fact, it is sobering to see the masses of mediocre or actually bad results in Figure 9.18 compared with the maximum elongation that is possible in the alloy at each particular yield strength. Clearly, a few results are close to $\sigma_T = 100\%$, but many achieve only 50% and a few only 5% or less of the potential value. There is plenty of scope for improvement in the industry.

9.14 **BIFILM-FREE PROPERTIES**

It is fascinating to look ahead, extrapolating a little from our current knowledge, to envisage the properties of metals if manufactured free from bifilms.

It is essential to realise that all the evidence currently available points strongly to the conclusion that metals would never fail by cracking (Campbell, 2011). This arresting prediction follows from a survey of molecular dynamics studies which investigate the atomic structures of metals by computer simulation. When the atomic lattices of various metals are subjected to tensile stress, the metals are seen to stretch, but are never observed to crack until the stress reaches the ultimate tensile strength of the metal, usually many tens of GPa. Thus failure mechanisms which were supposed to operated, such as a dislocation pile-up against a barrier such as an inclusion, leading to the formation of a crack, do not seem to occur in practice. This follows simply as a result of the enormous strength of the interatomic bonding; metal atoms are highly resistant to being pulled apart.

This profound conclusion is perfectly in line with the knowledge widely known for decades from the simple formula, $P = 2\gamma/r$, for the strength of liquid metals by the formation of an atomic sized bubble, which effectively cannot be solved at normally available stresses. Once again, tensile failure only occurs at stresses close to the ultimate failure strength measured in many GPa.

There appear therefore to be no intrinsic mechanisms which can trigger the failure of metals by cracking. This amazing conclusion reinforces the role of bifilms. These unsuspected and largely undetectable defects appear to be the only features capable of initiating cracks. They *are* the Griffith cracks required by all the theories of failure. This is an exciting discovery, and one which gives this book a significance I could not have dreamed of. Casting defects and their lingering existence in wrought metals constitute the only sources of failure by cracking. This massively important conclusion predicts that metals which cannot fail by cracking are possible by appropriate casting techniques. The processes to achieve this are easy, low-cost processes.

In the absence of failure by cracking, failure in tension would occur by plastic flow. Thus elongation to failure and RA would be expected to increase greatly achieving 100% for metals completely free from bifilms. Experimental evidence for the truth of this prediction is seen in Al alloy castings from aerospace foundries in Figure 9.18 in which results from 1% to 30% elongation are found even in those critical operations which would claim to have all casting parameters under control, whereas unfortunately overlooking the major parameter: the turbulence during pouring. Notice that these results are for elongation rather than RA; the elongation is sensitive to the rate of work hardening and is therefore not straightforwardly predictable. Even so, the values of typically 50% for steels reflects their better freedom from bifilms compared with Al alloys as a result of the density difference which assists steels to reduce its bifilm content rapidly by flotation. For both steels and Al alloys, values of hundreds of percent can be achieved for conditions of superplasticity; such values would be expected to be more common as metal purities increased.

The benefits of directional solidification (DS) of castings in an upward direction are explainable for the first time. Such slow freezing conditions provide sufficient time to favour the flotation of bifilms, particularly in dense alloys such as Ni-base alloys and steels, and the pushing of bifilms ahead of the freezing front, so that the casting becomes reasonably free from bifilms. Polich and Flemings (1965) show for their DS steel castings (Figure 9.26) that tensile properties are independent of DAS compared with equiaxed casting which suffered from entrapped bifilms in which DAS is used as an indicator of tensile properties (see Section 9.4.3.2).

Ni-based single crystals are the ultimate example of bifilm-free behaviour, or possibly low bifilm content castings. Figure 17.1 illustrates how these products achieve a factor of 10 improvement in properties compared with equiaxed equivalents (in which their bifilm population would have no chance to escape, but would be trapped by rapid crystal growth from all directions).

Some steels, by virtue of their composition and high melting temperatures giving them the benefit of liquid surface films, display huge toughness. An excellent example is Hadfield Manganese steel containing approximately 13%Mn so that its surface oxide is rich in the low melting point MnO_2 (probably containing also SiO_2). Kim (2014) confirms the inclusions in this steel are beautifully spherical typically consisting of MnS patches in a glassy amorphous phase formed from a supercooled liquid, $MnO-Al_2O_3-SiO_2$. This steel is used for crossing points on rail track, and despite its severe service conditions, is highly resistant to failure by cracking.

The addition of boron to steels is widely known to increase toughness and hardenability of steels. These may be the result of the surface oxide on the liquid steel becoming a liquid boride, with a melting point of 1000°C or less, so that bifilms cannot form during pouring. Toughness is easily understood as the result of fewer crack-like defects. The benefits to hardenability might be the result of the reduction of bifilms leading to an increase in thermal conductivity, so that a quenching action will reach deeper into the steel because of the direct flow of heat rather than the tortuous flow around arrays of bifilms. Although the boride is certain to be liquid on the steel surface during casting, the liquid boride may not take into solution the stable oxides such as alumina and chromia which are also likely to be present, but will merely float on top of the solid film. If this occurs, the clearly beneficial action of the liquid boride may be the result of merely forming the centre of a liquid sandwich, bonding the stable bifilms together like a glue. Yu's investigation (2013) of a superalloy with added boron reports 'improved grain boundary adhesion' increased rupture life and eliminated notch sensitivity. All these observations would be expected from reduced bifilms or bonded bifilms at grain boundaries. Among the multitude of research which confirm the beneficial effect of B additions to steels, we can cite Mejia (2013) who finds improved hot ductility, Das (2013) who finds fewer ductile dimples on fracture surfaces (fewer inclusions on unbonded bifilms) and Song and Cooman (2013) who find the bainite formation at grain boundaries delayed by boron additions to a lean alloy steel. This latter observation seems associated with the ease of nucleation on a bifilm of those phases which are accompanied by a volume and/or shape change—this is easy into the 'air gap' at the centre of the bifilm. In the absence of a bifilm, the grain boundary will not constitute a favoured nucleation site for such transformations.

Turning finally to the effect of the absence of bifilms on corrosion: It cannot be claimed to be a piece of rigorous scientific evidence, but in the absence of any nicely researched information, I perhaps should report that I did once cast an aluminium alloy wheel with a good filling system. The wheel was subjected to the standard salt spray corrosion test used for wheels but was not found to corrode. This was a surprise at the time, but is now understandable from the viewpoint that bifilms are almost certainly the mechanism for pitting and other localised forms of corrosive attack as described in Section 9.10. Properly controlled experiments to demonstrate the role of bifilms in corrosion are, regrettably, not yet available but would be welcome.

CASTING MANUFACTURE

INTRODUCTION TO THE CASTING MANUFACTURING INDUSTRY

The foundry world never ceases to amaze me with its kaleidoscopic mix of metals and processes. It has to be one of the most, if not the most, complex industry. Its complexity makes it a challenge to describe, but all the more interesting.

The main metal divisions are, perhaps naturally, ferrous and nonferrous. This split is presented in some detail in the accompanying Figure 10.01. It is somewhat artificial in the sense that Ni and Co base alloys share much similar technology and grade smoothly into the various varieties of stainless steels. Even so, tradition has it that Ni and Co are not ferrous, and it has to be admitted that as face-centred cubic metals, they share many properties in common with Cu and Al.

The alternative common approach to categorising the casting industry is in terms of its mould making, with greensand easily holding its magnificent first place in terms of productivity, but a wide variety of alternative moulding processes each designed for their own niches, and the niches sometime of huge size, such as pressure die casting, and the manufacture of Al alloy wheels by low-pressure permanent mould (Figure 10.02).

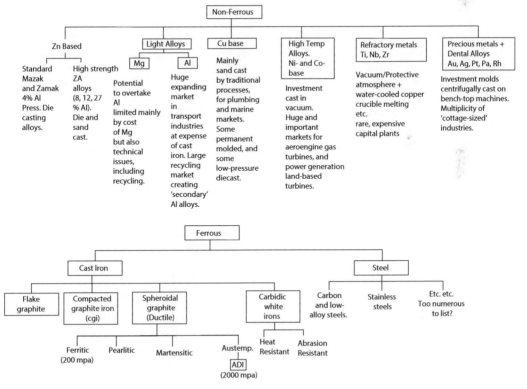

FIGURE 10.01

The structure of the casting industry (1) by metal.

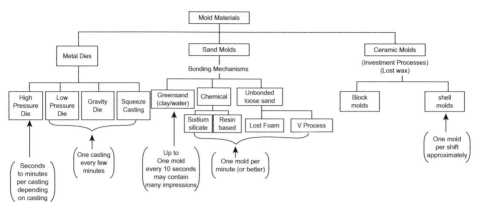

FIGURE 10.02

The structure of the casting industry (2) by molding technology.

The main process divisions for manufacturing castings include moulding and casting. It is an interesting exercise to set up a matrix with a vertical list of moulding options and a horizontal list of casting options. Correlating these can be seen to fill nearly every box of the matrix, although there are interesting gaps that the reader can quickly discover.

However, the description of casting manufacture is not that simple, because in addition to the main processing steps of moulding and casting, there are very many other process steps including melting and solidification which add complexity, so that the two-dimensional matrix quickly multiplies into a multidimensional array. Thus it has not been possible in this account to tackle a description of the industry as a matrix. The processes are, so far as possible, discussed as separate topics.

The processes include (1) melting, (2) moulding, (3) casting, (4) solidification control (sometimes) and (5) many postcasting processing operations. Each processing step has yet more options, making the attempt to summarise the industry even more problematic. Many of these processing steps have generally been selected for such laudable features as their rate of production, acceptable efficiency or low cost.

However, with regret, the previous choices of (1) melting and (3) casting processes have usually been selected badly, only in terms of productivity and apparent cost; these are the weakest links in the production sequence. In fact, it has to be admitted that most of our melting operations and most casting operations in the casting industry are awful. Few are selected for the *quality* of the product, even though all will admit that a system that consistently produces nearly zero or actually zero scrap would constitute a major advantage. This major feature has been consistently overlooked. We shall go out of our way not to overlook it here.

Finally, before the processing is even considered, the casting technique has to be designed correctly. This is a nontrivial task. Thus the next three chapters are devoted to getting this right. The new rules listed in Chapter 10, and the procedures for the design of the filling systems for the castings, are, with regret, often in conflict with much traditional practice. New research has revealed many of the reasons for the poor performance of tradition systems and has pointed the way to substantial improvements.

The central problem within the metal casting industry is the *pouring* of our liquid metal.

Pouring occurs at multiple stages of our production processes. We have been lulled into complacency from the innocent images in our minds of pouring a glass of water or a cup of tea, but have failed to appreciate that liquid metals are greatly damaged by such actions. The use of pouring under gravity as an aid to making castings is a two-edged sword: gravity pouring is easy, but gravity accelerates our metals to unwanted and damaging speeds. As an interim solution, we can improve our gravity filling systems to reduce (but not eliminate) the damage associated with pouring. Ultimately, however, for most castings, we have to rethink our processing to eliminate pouring at every stage. This may sound alarming, but in fact is not such a daunting engineering challenge. An increasing number of foundries are now successful to eliminate their reliance on pouring.

In the meantime, referring back to an interim possibility, there are some strategies for reducing the damage inflicted by pouring that will be helpful while more revolutionary systems can be put in place. The improved systems are repaying careful application in a number of foundries. I am glad to report that the new systems recommended in this book come with warm recommendations from both their users and their accountants.

JC
Ledbury
25 November 2010

THE 10 RULES FOR GOOD CASTINGS

The 10 Rules are my personal checklist (Campbell, 2004), ensuring that I have not forgotten any essential aspect of casting manufacture. It cannot be emphasised too strongly that the failure of only one of the rules can result in total failure of the casting. This is not meant to be alarmist, but simply practical. No one has ever promised that making castings would be easy. However, following the rules is a great help.

We start off with a quick summary, followed by a detailed assessment of each rule in turn in the remainder of this section.

Rule 1. Start with a good quality melt

Immediately before casting, the melt shall be prepared, checked, and treated, if necessary, to bring it into conformance with an acceptable minimum standard. Prepare and use so far as possible only near-defect-free melt.

Rule 2. Avoid turbulent entrainment of the surface film on the liquid

This is the requirement that the liquid metal front (the meniscus) should not go too fast. Maximum meniscus velocity is approximately 0.5 m/s for most liquid metals. This requirement also implies that the liquid metal must not be allowed to fall more than the critical height corresponding to the height of a sessile drop of the liquid metal. The maximum velocity may be raised to 1.0 m/s or even higher, and the critical fall height might be correspondingly raised to approximately 50 mm, in sufficiently constrained running systems or thin section castings.

Rule 3. Avoid laminar entrainment of the surface film on the liquid

This is the requirement that no part of the liquid metal front should come to a stop before the complete filling of the mould cavity. The advancing liquid metal meniscus must be kept "alive" (i.e. moving) and therefore free from thickened surface film that may be incorporated into the casting. This is achieved by the liquid front being designed to expand continuously. In practice, this means progress only *uphill* in a continuous *uninterrupted* upward advance (i.e. in the case of gravity-poured casting processes, from the base of the sprue onwards). This implies

- Only bottom-gating is permissible.
- No falling or sliding downhill of liquid metal is allowed.
- No horizontal flow of significant extent.
- No stopping of the advance of the front due to arrest of pouring or waterfall effects, etc.

Rule 4. Avoid bubble entrainment

No bubbles of air entrained by the filling system should pass through the liquid metal in the mould cavity. This may be achieved by:

- Properly designed off-set step pouring basin; fast back-fill of properly designed sprue; preferred use of stopper; avoidance of the use of wells and all other volume-increasing features of filling systems (such as expanding channels sometimes known as 'diffusers'); small volume runner and/or use of ceramic filter close to sprue/runner junction; possible use of bubble traps. A naturally pressurised filling system fulfils most of these criteria.
- No interruptions to pouring.

Rule 5. Avoid core blows

- No bubbles from the outgassing of cores or moulds should pass through the liquid metal in the mould cavity. Cores to be demonstrated to be of sufficiently low gas content and/or adequately vented to prevent bubbles from core blows.
- No use of impermeable clay-based core or mould repair paste.

Rule 6. Avoid shrinkage
- No feeding uphill in larger section thickness castings. Feeding against gravity is unreliable because of (1) adverse pressure gradient and (2) complications introduced by convection.
- Demonstrate good feeding design by following all seven Feeding Rules, by an approved computer solidification model, and by test castings.
- Once good feeding is attained, fix the temperature regime by controlling (1) the level of flash at mould and core joints, (2) mould coat thickness (if any), and (3) temperatures of metal and mould.

Rule 7. Avoid convection
Assess the freezing time in relation to the time for convection to cause damage. Thin- and thick-section castings automatically avoid convection problems. For intermediate sections, either (1) reduce the problem by avoiding convective loops in the geometry of the casting and rigging, (2) avoid feeding uphill, or (3) eliminate convection by rollover after filling.

Rule 8. Reduce segregation
Predict segregation to be within limits of the specification, or agree out-of-specification compositional regions with customer. Avoid channel segregation formation if possible.

Rule 9. Reduce residual stress
No quenching into water (cold or hot) following solution treatment of light alloys. (Polymer quenchant or forced air quench may be acceptable if casting stress can be shown to be acceptable.)

Rule 10. Provide location points
All castings to be provided with location points for pickup for dimensional checking and machining. Proposals are to be agreed on with quality auditor, machinist, etc.

10.1 RULE 1: USE A GOOD-QUALITY MELT
10.1.1 BACKGROUND
The melt needs to be demonstrated to be of good quality. A good-quality liquid metal is one that is defined as follows.

1. Substantially free from suspensions of non-metallic inclusions in general and bifilms in particular.
2. Relative freedom from bifilm-straightening and bifilm-opening agents. These include certain alloy impurities in solution such as Fe in Al alloys and hydrogen or other gases. However, it should be noted that Si can precipitate in its primary form as a bifilm straightening element and is, of course, inescapable when using unmodified Al-Si alloys, but is avoided in modified alloys.

Please note that the good quality of the melt should not be taken for granted, and, without proper treatment, most often fails this requirement.

There are a few exceptions in which good quality might be assumed. Such metals may include pure liquid gold, iridium, platinum, perhaps mercury and possibly some liquid steels whilst in the melting furnace at a late stage of melting. These instances are, however, either rare or tantalisingly inaccessible, such as the steel in the melting furnace; in the process of getting it out of the furnace, much damage is done to the liquid by the pouring that is currently an integral feature of our conventional steel foundries.

An important distinction is useful to identify four major oxide inclusion types as listed in Section 2.9. All of these different oxide films will necessarily have the structure of a bifilm as a result of them becoming submerged, having penetrated the oxidised surface of the liquid metal. Clearly, the major task is to eliminate the macroscopic bifilms. However, for reliable properties, it is also essential to eliminate the majority of the larger fraction of mesoscopic and microscopic bifilms.

Regrettably, many liquid metals are actually so full of sundry solid phases floating about that they begin to more closely resemble slurries than liquids. In the absence of information to the contrary, this condition of a liquid metal should be assumed. Over recent years, the evidence for the real internal structure of liquid metals being crammed with defects has been growing as investigation techniques have improved. Some of this evidence is described later. Much of the evidence applies to aluminium and its alloys where the greatest research effort has been. Evidence for other materials is presented elsewhere in this book.

It is sobering to realise that many of the strength-related properties of metals can only be explained by assuming that the initial melt is full of defects. Many of our theoretical models of liquid metals and solidification that are formulated to explain the occurrence of defects neglect to address this critical fact. Classical physical metallurgy and solidification science has been unable to explain the important properties of cast materials such as the effect of dendrite arm spacing and the ease of formation of pores and cracks. During the early stage of fatigue, the problem of the anomalously rapid growth of short cracks despite the low stress intensity is easily explained if the cracks pre-exist. We shall see that in general the behaviour of cast metals arises naturally from the population of defects.

However, it has to be admitted that is often not easy to confirm the presence of non-metallic inclusions in liquid metals, and even more difficult to quantify their number and average size or spread of sizes. McClain et al. (2001) and Godlewski and Zindel (2001) have drawn attention to the unreliability of the standard approach studying polished sections of castings. A technique for liquid aluminium involves the collection of inclusions by pressurising up to 2 kg of melt, forcing it through a fine filter, as in the porous disc filtration analysis and Prefil tests. Pressure is required because the filter is so fine. The method overcomes the sampling problem by concentrating the inclusions by a factor of about 10,000 times (Enright and Hughes, 1996 and Simard et al., 2001). The layer of inclusions remaining on the filter can be studied on a polished section. (The total quantity of inclusions is assessed as the area of the layer as seen in section under the microscope, divided by the quantity of melt that has passed through the filter. The unit is therefore the curious quantity mm^2kg^{-1}. We can hope that at some future date this unhelpful unit will, by universal agreement, be converted into some more meaningful quantity such as volume of inclusions per volume of melt. In the meantime, the standard provision of the diameter of the filter in reported results would at least allow readers the option to convert the values for themselves.)

To gain some idea of the range of inclusion contents an impressively dirty melt might reach $10 \ mm^2kg^{-1}$, an alloy destined for a commercial extrusion might be in the range 0.1–1, foil stock might reach 0.001 and computer discs $0.0001 \ mm^2kg^{-1}$. For a filter of 30 mm diameter, these figures approximately encompass the range 10^{-3} (0.1%) down to 10^{-7} (0.1 part per million) volume fraction.

Other techniques for the monitoring of inclusions in Al alloy melts in the past included Liquid Metal Cleanness Analyser (LiMCA), in which the melt is drawn through a narrow tube while measuring the voltage drop applied along the length of the tube. The entry of an inclusion of different electrical conductivity (usually non-conducting) into the tube causes the voltage differential to rise by an amount that is assumed to be proportional to the size of the inclusion. The technique is generally thought to be limited to inclusions approximately in the range 10–100 µm, presuming the inclusions to be particles. Although once widely used for the casting of wrought alloys, the author regrets that that technique has to be viewed with great reservation. As we have mentioned, the key inclusions in light alloys are not particles but (double) films, and although often extremely thin, can be up to 10 mm in diameter. Such inclusions sometimes succeed to find their way into the LiMCA tube, where they tend to hang in the metal stream, caught up at the mouth of the tube and rotate into spirals like a flag tied to the mast by only one corner. Asbjornsonn (2001) has reported piles of helical oxides in the bottom of the LiMCA crucible. It is to be regretted that most workers using LiMCA have been unaware of these serious problems. However, the general disquiet about the appropriateness of this technique has finally caused the LiMCA device to be dropped by the manufacturer. It seems unlikely to be used significantly in the future.

Ultrasonic reflections have been used from time to time to investigate the quality of melts. The early work by Mountford and Calvert (1959) is admirable, making recommended reading, and has been followed up by considerable development efforts in Al alloys (Mansfield, 1984), Ni alloys and steels (Mountford et al., 1992). Ultrasound is efficiently reflected from oxide films because the films are double, and the elastic wave cannot cross the intermediate layer of air and thus is reflected with mirror-like efficiency. However, the reflections may not give an accurate idea of the size of the defects because the irregular, crumpled form of such defects, and their tumbling action in the melt. The tiny mirror-like facets of large defects reflect back to the source only when the facets happen to rotate to face the beam. The result is a general scintillation effect, apparently from many minute and separate particles. It is not easy to discern whether the images correspond to many small or a few large bifilms.

Neither LiMCA nor the various ultrasonic probes can distinguish any information on the types of inclusions that they detect. In contrast, the inclusions collected by pressurised filtration can be studied in some detail. In aluminium alloys many different inclusions can be found. Table 1.1 lists some of the principal types.

Nearly all of these foreign materials will be deleterious to products intended for such products as thin foil or computer discs. However, for engineering castings, those inclusions such as carbides and borides are probably not harmful at all (although they would be unpopular with machinists). This is because, having been precipitated from elements in solution in the melt, they would be expected to be in excellent atomic contact with the matrix alloy. These non-metallic phases enjoy well-bonded interfaces and are thereby unable to act as void-type initiators or volume defects such as pores and cracks. On the contrary, they may act as grain refiners. Furthermore, their continued good bonding with the solid matrix is expected to confer on them a minor or negligible deleterious influence on mechanical properties. They may strengthen the matrix to some degree.

Generally, therefore, this book concentrates on those inclusions that have a major detrimental influence on mechanical properties, because of their action to initiate other serious problems such as pores, cracks and localised corrosion. Thus the attention will centre on *entrained surface films, known as bifilms*. Usually, these inclusions will be oxides. However, carbon films are also common, and occasionally nitrides, sulphides and other substances. These entrained films exhibit a unique structure, with outer faces that are in atomic contact with the melt, but have an inner interface that is unbonded, and thus lead to the spectrum of problems of pores and cracks all too familiar to foundry people.

The pressurised filtration tests can find some of these entrained solids, and the analysis of the inclusions present on the filter can help to identify the source of many inclusions in a melting and casting operation. However, films that are newly entrained into the melt as a result of surface turbulence remain undetectable but can be enormously important. These films are commonly entrained during the pouring of castings, and so, perhaps, are not normally subjected to detection or assessment in a melting and distribution operation at this late stage. They are typically only 20 nm thick, and so remain invisible under an optical microscope, especially in a ceramic filter because they will be difficult to discern when draped around a piece of the filter that, when sectioned, appears many thousands of times thicker.

The only fairly reliable detection technique for such inclusions is the lowly reduced pressure test (RPT). This test opens the films (because they are always double and contain residual entrapped air or other gases such as hydrogen) so that they can be seen by eye on a polished section or on a radiograph. The radiography of the cast test pieces reveals the size, shape and numbers of such important inclusions, as has been shown in studies by Fox and Campbell (2000) and Dispinar and Campbell (2006). A central slice of the small cylindrical RPT casting to yield a parallel section gives an improved radiographic result (Figure 10.1). Viewing this work in retrospect the approach using X-rays now appears somewhat over the top, and probably unnecessary. Simply viewing a polished section by unaided eye has subsequently proved almost equally effective, and vastly cheaper and quicker.

Figure 10.2 shows how a quality map can be constructed to show the total length and total number of bifilms on a polished RPT section. The large central square might form a suitable quality window for an operation making 'space filling' components for which filling and feeding are not critical as a result of relatively low requirements from the casting. The '50 × 50' regime might be a minimum requirement for parts requiring some strength and toughness. The Cosworth foundry used a window '3 × 3' during my time and quite often achieved '0 × 0', giving castings of exceptional soundness and high properties.

However, even the RPT technique probably only reveals the more extensive bifilms, between perhaps 0.1 and 10 mm diameter. This is because the effective tensile opening stress applied to the bifilm can only be a maximum of 1 atm (0.1 MPa). In a tensile test of a solidified metal, the stress can easily reach 1000 times greater than this and so is capable of opening much smaller bifilms. For instance, in some Al-Si alloys cracked Si particles at the base of many ductile dimples indicate the failure of the particle by cracking. If we assume this failure can only occur because of the presence of a bifilm, then a true density of defects might approach 10^{15} m^{-3} (corresponding to 10^6 mm^{-3}) at a diameter of between 1 and 10 μm. This is a huge density of defects, but is consistent with the direct ultrasonic observation of a melt by Mountford and Calvert (1959), in which they describe a dirty melt appearing as a fog. After the inclusions were allowed to settle (aided by precipitating heavy grain-refining titanium-rich compounds on them), the melt cleared

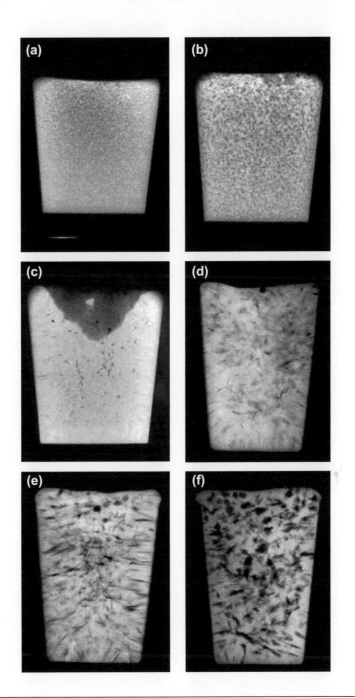

FIGURE 10.1

Radiographs of RPT samples of Al-7Si-0.4Mg alloy illustrating different bifilm populations. (For description of each part (a to f) see text).

Courtesy S. Fox.

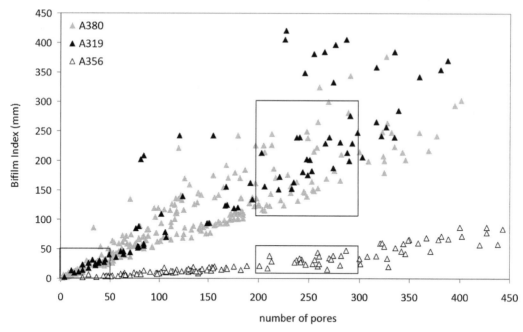

FIGURE 10.2

RPT results for three casting alloys, showing number and total length of bifilms on the polished section (Dispinar and Campbell, 2011).

completely so that a clear back-wall echo could be seen. They were able to repeat this phenomenon simply by stirring up the melt to recreate the fog, and watch the weighed-down oxides clearing once again to build up a layer of bottom sediment.

It is unfortunate that many melts start life with poor, sometimes grossly poor, quality in terms of its content of suspended bifilms. The 'fog' persists in most alloys because the bifilms are usually neutrally buoyant. Figure 10.1 gives several examples of different poor qualities of liquid aluminium alloy. The figures show results from RPT samples observed (somewhat unnecessarily as has already been noted!) by X-ray radiography. Because the samples are solidified under only one tenth of an atmosphere (76 mm residual pressure compared with the 760 mm of full atmospheric pressure), any gas-containing defects, such as bubbles, or bifilms with air occluded in the centres of their sandwich structures, will be expanded by 10 times. Thus rather small defects can become visible for the first time.

We shall assume that pores are always initiated by bifilms, giving initially crack-like or irregular pores. The formation of rounded pores simply occurs as a result of the bifilm being opened beyond this initial condition by excess precipitation of gas, finally achieving a pore diameter greater than the original dimensions of the bifilm. Thus the RPT is an admirably simple device for assessing (1) the number of bifilms; (2) their average size (even though this might be somewhat of an overestimate if much hydrogen is present); and (3) gas content is assessed by the degree of lowering the average density of the cast sample by the opening of the bifilms from thin crack-like forms to fairly spherical pores.

If the melt contained no gas-containing defects, the cut and polished section (or in this case the radiographs) of the RPT would be clear.

However, as we can see immediately, and without any benefit of complex or expensive equipment, the melts recorded in Figure 10.1 are far from this desirable condition. Figure 10.1(a) shows a melt with small rounded pores that indicates that the bifilms that initiated these defects were particularly small, of the order of 0.1 mm or less. The density of these defects, however, was high, between 10 and 100 defects per cubic cm. Figure 10.1(b) has a similar defect distribution, but with slightly higher hydrogen content. Figure 10.1(c) illustrates a melt that displayed a deep shrinkage pipe, normally interpreted to mean good quality, but showing that it contained a scattering of larger pores, probably as a result of fewer bifilms opened mainly by shrinkage rather than by gas. Figure 10.1(d) has considerably larger bifilms, of size in the region of 5 mm in length, and in a concentration of approximately $1/cm^3$. Figure 10.1(e) and (f) show similar samples but with increasing gas contents that have inflated these larger bifilms to reasonably equiaxed pores.

Naturally, it would be of little use for the casting engineer to go to great lengths to adopt the best designs of filling and feeding systems if the original melt was so poor that a good casting could not be made from it.

Thus this section deals with some of the aspects of obtaining a good-quality melt.

In some circumstances, it may not be necessary to reduce both bifilms and bifilm-opening agents. An interesting possibility for future specifications for aluminium alloy castings (where residual gas in supersaturated solution does not appear to be harmful) is that a double requirement may be made consisting of (1) the content of dissolved gas in the melt to be high, but (2) the percentage of gas porosity to be low. The meeting of this interesting double requirement will ensure to the customer that bifilms are not present. Thus these damaging but undetectable defects will, if present, be effectively labelled and made visible on X-ray radiographs and polished sections by the precipitation of dissolved gas. It is appreciated that such a stringent specification might be viewed with dismay by present suppliers. However, at the present time we have mainly only rather poor technology, making such quality levels out of reach. We shall not necessarily suffer such backward processing for ever.

The possible future production of Al alloys for aerospace, with high hydrogen content but low porosity, is a fascinating challenge. As our technology improves, such castings may be found not only to be manufacturable, but offer a guaranteed reliability of fatigue life, and therefore command a premium price.

The prospect of producing ultra-clean Al alloys that can be *demonstrated* in this way to be actually extremely clean, raises the issue of contamination of the liquid alloy from the normal metallurgical additions such as the various master alloys, and grain refiners, modifiers, etc. It may be that for superclean material, normal metallurgical additions to achieve refinement of various kinds will be found unnecessary, and possibly even counter-productive because of the unavoidable contamination by primary oxide skins from the outer surfaces of the master alloy additions.

The improvement of Al alloys by melting with dry hearth furnaces to eliminate the primary oxide skins of the charge, simply by scraping these off the hearth through a side door, is to be strongly recommended. The subsequent treatment of the melt with Ti-rich grain refiners is also recommended, not necessarily to grain refine, but to precipitate heavy Ti-rich inclusions on oxide bifilms to sediment these from the melt. It seems likely that extremely clean Al might be obtained in this way. These and other sedimentation techniques are discussed in more detail under the section on melting.

For steels, the content of hydrogen may be a more serious matter, especially if the section thickness of the casting is large. In some steel castings of section thickness above about 100 mm up to 1000 mm or more, the hydrogen cannot escape by diffusion during the time available for cooling or during the time of any subsequent heat treatments. Thus the high hydrogen content retained in these heavy sections may lead to hydrogen embrittlement and catastrophic failure of the section by cracking. Although Ren et al. (2008) claim that hydrogen might nucleate voids with the aid of vacancies, it is difficult to avoid what appears to be the more probable speculation that the hydrogen cracking might initiate by hydrogen precipitating into, and forcing open, bifilms. To avert hydrogen cracking problems when making very large steel castings, such as backup rolls for rolling mills, it was traditional practice to induce a carbon boil to reduce the hydrogen content but now is reduced by modern, rather costly techniques such as treatment by argon oxygen degassing. Having done this, it is necessary to work quickly to cast the metal because the hydrogen content tends to rise rapidly once again, re-establishing its equilibrium with its environment such as damp atmospheres and/ or damp refractories in ladles etc.

For ductile iron production, the massive amounts of turbulence that accompanies the addition of magnesium in some form, such as magnesium ferro-silicon, are almost certainly highly damaging to the liquid metal. It is expected that immediately after such nodularisation treatment, the melt will be massively dirty. It will be useful therefore to ensure that the melt can dwell for sufficient time for the entrained magnesium oxide–rich bifilms to float out. The situation is analogous to the treatment of cast iron with CaSi to effect inoculation (i.e. to achieve a uniform distribution of graphite of desirable form). In this case the volumes of calcium-oxide-rich films are well known, so that the CaSi treatment is known as a 'dirty' treatment compared to FeSi inoculation. The author is unsure about in-mould treatments therefore with such oxidisable elements as Ca and Mg. Clearly these treatments spherodise the graphite as they are designed to do, but it is not clear whether the matrix suffers from additional bifilms which would lower properties? Work to clarify this question would be valuable.

For nickel-based superalloys melted and cast in vacuum, it is with regret that the material is, despite its apparently clean melting environment, found to be sometimes as crammed with oxides (and/or nitrides) as an aluminium alloy (Rashid and Campbell, 2004). This is because the main alloying element in such alloys is aluminium, and the high temperature favours rapid formation and thickening of the surface film on the liquid. This problem occurs even when casting in vacuum, because the vacuum is, of course, only dilute air. If the film becomes entrained, the liquid and any subsequent casting is damaged. The process for the production of the alloys designed for remelting to make castings in foundries involves, unfortunately, melting and pouring processes that leave much to be desired:

The melting and alloying process in the manufacture of Ni-based alloys starts in an induction crucible furnace, and is poured under gravity via a series of sloping launders, falling several times, and finally falling 1–2 m or more into steel tubes that act as moulds. This awfully turbulent primary production process for the alloy impairs all the downstream cast products. More recent moves to increase productivity using larger diameter tubes to make larger diameter alloy billets have only made the bad situation worse. An alternative recent move towards the production of Ni-base alloy bar by horizontal continuous casting is to be welcomed as the first step towards a more appropriate production technique for these key ingredients of our modern aircraft turbines. Production of improved material is currently limited, but deserves to be the subject of demand from customers.

Even so, the subsequent melting and gravity casting operations in the investment casting foundries to produce Ni- and Co-base turbine blade castings are also currently extremely disappointing, so that good quality starting material would at this time probably be a waste. The aircraft industry has been trapped within its rigid procedures designed to ensure safety, but which have inhibited the rational engineering development of improved casting techniques that could deliver far safer products at reduced costs.

Melting systems and melt treatments designed to provide improved melt quality are dealt with in more detail in Volume 3 'Melting'.

10.2 RULE 2: AVOID TURBULENT ENTRAINMENT (THE CRITICAL VELOCITY REQUIREMENT)

10.2.1 INTRODUCTION

The avoidance of surface turbulence is probably the most complex and difficult rule to fulfil when dealing with gravity pouring systems.

The requirement is all the more difficult to appreciate by many in the industry because everyone working in this field has always emphasised the importance of working with 'turbulence-free' filling systems for castings. Unfortunately, despite all the worthy intentions, all the textbooks, all the systems and all the talk, so far as the author can discover, it seems that no one appears to have achieved this target so far. In fact, in travelling around the casting industry, it is quite clear that the majority (at least 80%) of all defects are directly caused by turbulence. Thus the problem is massive; far more serious than suspected by most of us in the industry.

In fact, Johnson and Baker concluded from their experiments in 1948 that 'gating systems did not function as commonly supposed' and 'no gating system prevented turbulence'. It seems we have all been warned for a long time.

To understand the fundamental root of the problem, it is clear in Section 2.1 that any fall greater than the height of the sessile drop (of the order of 10 mm) causes the metal to exceed its critical velocity, and so introduces the danger of defects in the casting. Because most falls are in fact at least 10 or 100 times greater than this and because the damage is likely to be proportional to the energy involved (i.e. proportional to the square of the velocity), the damage so created will usually be expected to be 100–10,000 times greater. Thus in the great majority of castings that are poured simply under the influence of gravity, there is a major problem to ensure its integrity. In fact, the situation is so bad that the best outcome of many of the filling system design solutions proposed in this book are merely damage limitation exercises. Effectively, it has to be admitted that at this time it seem impossible to guarantee the avoidance of some damage when pouring liquid metals.

This somewhat depressing conclusion needs to be tempered by several factors.

1. Expectations. The world has come to accept castings as they are. Thus any improvement will be welcome. This book described techniques that will create very encouraging improvements.
2. Continuing development. This book is merely a summary of what has been discovered so far in the development of filling system design. Better designs are to be expected now that the design parameters and filling system concepts (such as critical velocity, critical fall height, entrainment, bifilms, etc.) are defined and understood.
3. There *are* filling systems that can yield, in principle, perfect results.

Of necessity, such perfection is achieved by fulfilling rule 2 by avoiding the transfer of the melt by pouring. Thus consider the three modes of filling a mould:

1. Downhill pouring under gravity;
2. Horizontal transfer into the mould achieved by (a) tilt casting in which the tilt conditions are accurately controlled to achieve horizontal transfer, or (b) by the 'level pour' or 'level transfer' techniques;
3. Uphill (counter-gravity) casting in which the melt is never poured, but caused to fill the mould in only an uphill mode.

Only the last two processes have the potential to deliver castings of near perfect quality. In my experience, I have found that in practice it is often difficult to make a good casting by gravity, whereas by a good counter-gravity process (i.e. a process observing all the 10 rules) it has been difficult to make a bad casting. The jury is still out on horizontal transfer by tilting. This approach has great potential, but is applicable only to certain limited shapes of castings, and requires a dedicated effort to achieve the correct conditions. Horizontal transfer has great potential but unfortunately has fallen into disuse.

Thus in summary, filling of moulds can be carried out down, along, or up. Only the 'along' and 'up' modes totally fulfil the non-surface-turbulence condition.

However, despite all its problems, it seems more than likely that the *downward* mode, *gravity casting*, will continue to be with us for the foreseeable future. Thus in this book we shall devote some considerable length to the damage limitation exercises that can offer considerably improved products, even if, unfortunately, those products cannot be ultimately claimed as perfect.

Most will shed no tears over this conclusion. Although potential perfection in the *along* and *up* modes is attractive, the casting business is all about making *adequate* products; products that meet a specification and at a price a buyer can afford.

The question of cost is interesting; perhaps the most interesting. Of course the costs have to be right, and often gravity casting is acceptable and sufficiently economical. However, more often than might be expected, high quality and low cost can go together. An improved gravity system, and certainly one of the better counter-gravity systems, can be surprisingly economical and productive. Such opportunities are often overlooked. It is useful to be alert to such benefits.

10.2.2 MAXIMUM VELOCITY REQUIREMENT

Some years ago, I was seated in the X-ray radiograph viewing room using an illuminated viewing screen to study a series of radiographic films of cylinder head castings made by our recently developed counter-gravity casting system at the Cosworth foundry. Each radiograph in turn was beautiful, having a clear, 'wine glass' perfection of which every founder dreams. I was at peace with the world. However, suddenly, a radiograph appeared on the screen that was a total disaster.

It had gas bubbles, shrinkage porosity, hot tears, cracks and sand inclusions. I was shocked, but sensed immediately what had happened. I shot out of the viewing room into the foundry to query Trevor, our man on the casting station. 'What happened to this casting?' He admitted instantly, 'Sorry governor, I put the metal in too fast'.

This was a lesson that remained with me for years. This chance experiment by counter-gravity, using an electro-magnetic pump allowing independent control of the ingate velocity, had kept all the other casting variables constant (including temperature, metal quality, alloy content, mould geometry, mould aggregate type, binder type etc.), showing them to be of negligible importance. Clearly, the ingate velocity was dominant. By only changing the speed of entry of metal we could move from total success to total failure. This was a fundamental lesson. Clearly, it applies to all casting processes.

Thinking further about this lesson, common sense tells us all that there is an optimum velocity at which a liquid metal should enter a mould. The concept is outlined in Figure 2.18. At a velocity of zero, the melt is particularly safe (Figure 2.18(a)), being free from any danger of damage. Regrettably, this stationary condition for the melt is not helpful for the filling of moulds. In contrast, at extremely high velocities the melt will enter the mould like a jet of water from a fire-fighter's hosepipe (Figure 2.18(c)), and is clearly damaging to both metal and mould. At a certain intermediate velocity, the melt rises to just that height that can be supported by surface tension around the periphery of the spreading drop (Figure 2.18(b)). The theoretical background to these concepts is dealt with at length in Chapter 2. For nearly all liquid metals, this critical velocity is close to 0.5 ms^{-1}. This value is of central importance in the casting of liquid metals, and will be referred to repeatedly in this section and when designing filling systems for castings.

Japanese workers optimising the filling of their design of vertical stroke pressure die-casting machine using both experiment and computer simulation (Itamura et al., 1995) confirm the critical velocity of 0.5 ms^{-1} for their Al alloy, finding at velocities above this value that air bubbles have a chance to become entrained. (These workers go further to define the amount of liquid that needs to be in the mould cavity above the ingate to suppress entrainment at higher ingate velocities.)

Looking a little more closely at the detail of critical velocities for different liquids, it is close to 0.4 ms^{-1} for dense alloys such as irons, steels and bronzes and about 0.5 ms^{-1} for liquid aluminium alloys. The value is $0.55-0.6 \text{ ms}^{-1}$ for Mg and its alloys. Taking an average of about 0.5 ms^{-1} for all liquid metals is usually good enough for most purposes related to the design of filling systems for castings, and will be generally used throughout this book.

Returning to Figure 2.18(b) showing the liquid metal emerging close to its critical velocity, and spreading slowly from the ingate. The shape of the drop is closely in equilibrium, its surface tension holding its compact shape, and just balancing the head of pressure due to its density that would tend to spread the drop out infinitely thin. As the ingate steadily supplies metal at close to 0.5 ms^{-1} the steadily growing drop is closely resembles the shape of a *sessile drop* (from the Latin word for 'sitting'. The word contrasts with *glissile drop*, meaning a gliding or sliding drop). A sessile drop of Al sitting on a non-wetted substrate is always approximately 12.5 mm high. Corresponding values for other liquids are Fe 10 mm, Cu 8 mm, Zn 7 mm, Pb 4 and water 5 mm.

Recent research has demonstrated that if the liquid velocity exceeds the critical velocity, there is a danger that melt will overshoot the height supportable by surface tension, so that on falling back again the surface of the liquid metal may be folded over. This *entrainment* of the liquid surface I have called *surface turbulence*. At risk of overly repeating this important phenomenon, this entrainment of the surface can occur if there is sufficient energy in the form of velocity in the bulk liquid to perturb the surface against the smoothing action of surface tension. In addition, notice that damage is not *necessarily* created by the falling back of the metal. The falling is likely to be chaotic, so that any folding action may or may not occur. The significance of the critical velocity is clear therefore: below the critical velocity the melt is safe from entrainment problems; above the critical velocity there is the danger, but not the necessity, of surface entrainment leading to defect creation.

To be more precise about the entrainment action, actually any disturbance of the surface of the liquid irreversibly extends the area of oxide on the surface, with the result that some entrainment of the additional oxide area is unavoidable. This is because the surface oxide film forms almost instantly, but once formed, cannot reduce its area without crumpling in some way, leading to the folding in of the excess surface oxide (see, for instance, Figure 2.2(b)). Thus even below the critical velocity some entrainment may occur, but, clearly, above the critical velocity the rate of entrainment suddenly increases.

Thus we see there is a chance that if the speed of the liquid exceeds about 0.5 ms^{-1}, its surface film may be folded into the bulk of the liquid. This folding action is an *entrainment* event. It leads to a variety of problems in the liquid that we can collectively call *entrainment defects*. The major entrainment defects are.

1. Bubbles as, of course, air bubbles (but widely misinterpreted as reaction products of solidification such as hydrogen porosity, or reaction products between oxide slag and graphite in cast iron to create CO bubbles).
2. Bifilms, as doubled-over oxide films. The author has named these folded-in films '*bifilms*' to emphasise their double, folded-over nature. Because the films are necessarily folded dry side to dry side, there is little or no bonding between their dry interfaces, so that the double films act as cracks. The cracks (alias bifilms) become frozen into the casting, lowering the strength and fatigue resistance of the metal. Bifilms may also create leak paths, causing leakage failures, and provide the ingress of corrodants leading to pitting and other varieties of corrosion.

The folding-in of the oxide is a random process, leading to scatter and unreliability in the properties on a casting-to-casting, day-to-day and month-to-month basis during a production run.

The different qualities of metals arriving in the foundry will also be expected to contain populations of bifilms that will differ and type and quantity from batch to batch. Thus the performance of the foundry will suffer additional variation. The foundry needs to have procedures in place to smooth variations of its incoming raw material so far as possible so as to fulfil rule 1.

The maximum velocity condition effectively forbids top gating of castings (i.e. the planting of a gate in the top of the mould cavity, causing the metal to fall freely inside the mould cavity). This is because liquid aluminium reaches its critical velocity of about 0.5 ms^{-1} after falling only 12.5 mm under gravity. The critical velocity of liquid iron or steel is exceeded after a fall of only about 10 mm. (These are, of course, the heights of the sessile drops.) Naturally, such short fall distances are always exceeded in practice in top gated castings, leading to the danger of the incorporation of the bifilms, and consequent porosity, leakage and crack defects.

Castings that are made in which velocities everywhere in the mould never exceed the critical velocity are consistently strong, with high fatigue resistance, and are leak-tight (we shall assume they are properly fed, of course, so as to be free from shrinkage porosity).

Experiments on the casting of aluminium have demonstrated that the strength of castings may be reduced by as much as 90% or more if the critical velocity is exceeded (Figure 2.19). The corresponding defects in the castings are not always detected by conventional non-destructive testing such as X-ray radiography or dye penetrant because, despite their large area, the folded oxide films are sometimes extremely thin, and do not necessarily give rise to any significant surface or X-ray indications.

The speed requirement automatically excludes conventional pressure die-castings as having significant potential for reliability because the filling speeds are usually 10–100 times greater than the critical velocity. Even so, over recent years there have been welcome moves, introducing some special developments of high-pressure technology that are capable of meeting this requirement. These include the vertical injection squeeze casting machine, and the shot control techniques. Such techniques can, in principle, be operated to fill the cavity through large gates at low speeds, and without the ingress of air into the liquid metal. Such castings require to be sawn, rather than broken, from their filling systems of course. Unfortunately, the castings remain somewhat impaired by the action of pouring into the shot sleeve. Even here, these problems are now being addressed by some manufacturers, with consequent benefits to the integrity of the castings, but involving technical, maintenance and cost challenges that are not easily overcome.

Other uphill filling techniques such as *low-pressure* filling systems are capable of meeting rule 2. Even so, it is regrettable that the critical velocity is practically always exceeded during the filling of the low-pressure furnace itself because of the severe fall of the metal as it is transferred into the pressure vessel, damaging the melt prior to casting. In addition, many low-pressure die casting machines are in fact so poorly controlled on flow rate that the speed of entry into the die can greatly exceed the critical velocity, thus negating one of the most important potential benefits of the low-pressure system. Processes such as Cosworth Process avoid these problems by never allowing the melt to fall at any stage of processing, and control its upward speed into the mould by electromagnetic pump.

These processes are to be discussed in greater detail in the chapter 16 'Casting'. The conventional rammed lining in such low-pressure furnaces is a further source of major defects as a result of the bubbles which emerge from the lining each time the furnace is depressurised. This serious but little-known problem is dealt with in detail in Section 16.31.

Metals that can also suffer from entrained surface films include the zinc-aluminium alloys and ductile irons. For a few materials, particularly alloys based on the Cu-10Al types (aluminium and manganese bronzes), the critical velocities were originally thought to be much lower, in the region of only 0.075 ms^{-1}. However, from recent work at Birmingham (Halvaee and Campbell, 1997), this low velocity seems to have been a mistake probably resulting from the confusion caused by bubbles entrained in the early part of the filling system. With well-designed filling systems, the aluminium bronzes accurately fulfil the theoretically predicted 0.4 ms^{-1} value for a critical ingate velocity.

Some ferrous alloys have similarly dry oxide films that give classical, unbonded bifilm defects, although in certain irons and steels the entrained bifilms agglomerate as a result of being partially molten and therefore somewhat sticky. These sticky masses of oxide therefore remain more compact and float out more easily to form surface imperfections in the form of slag macroinclusions on the surface. The subject of entrainment is described more extensively in the metallurgy of cast steels (Section 6.6).

Because of the central importance of the concept of critical velocity, the reader will forgive a re-statement of some aspects in this summary.

1. Even if the melt does jump higher than the height of a sessile drop, when it falls back into the surface there is no certainty that it will enfold its surface film. These tumbling motions in the liquid can be chaotic, random events. Sometimes the surface will fold badly and sometimes not at all. This is the character of surface turbulence; it is not predictable in detail. The key aspect of the critical velocity is that at velocities less than the critical velocity the surface is always safe. Above the critical velocity, there is always the *danger* of entrainment damage. The critical velocity criterion is therefore seen to be a necessary but not sufficient condition for entrainment damage.

2. If the whole, extensive surface of a liquid were moving upwards at a uniform speed, but exceeding the critical velocity, clearly no entrainment would occur. Thus the surface disturbance that can lead to entrainment should more accurately described not merely as a velocity but in reality a velocity difference. It might therefore be defined more accurately as a critical velocity gradient measured across the liquid surface. For those of a theoretical bent, the critical gradient might be defined as the velocity difference achieving the critical velocity along a distance in the surface of the order of the sessile drop radius (approximately half its height) in the liquid surface. To achieve reasonable accuracy, this approach requires allowing for the reduction in drop height with velocity. Hirt (2003) solves this problem with a delightful and novel approach, modelling the surface disturbances as arrays of turbulent eddies, and achieves convincing solutions for the simulation of entrainment at hydraulic jumps and plunging jets. Such niceties are neglected here. The problem does not arise when considering the velocity of the melt when emerging from a vertical ingate into a mould cavity. In that situation, the ingate velocity and its relation to the critical velocity is clear.

3. If the melt is travelling at a high speed, but is constrained between narrowly enclosing walls, it does not have the room to fold-over its advancing meniscus. Thus no damage is suffered by the liquid despite its high speed and despite the high risk involved. This is one of the basic reasons underlying the design of extremely narrow channels for filling systems that are proposed in this book. Very narrow filling channels have the great advantage of filling the running system in *one pass*, the melt filling the channel with its meniscus operating as a piston to push the air ahead. In this way, despite its high speed, it retains its quality. (This contrasts with oversize channels in which there is sufficient room for the melt to jet and splash, rebounding from the end of the channels to flow counter to the incoming melt, churning with the air. The high shearing speeds fragment the bubbles and oxides into numerous relatively small defects prior to entering the mould cavity. However, on those occasions when the melt does enter the mould cavity too quickly, any resulting damage tends to be even more serious. This is because such defects are then formed in the casting itself, and at the somewhat lower speeds than in the runner, so tending to avoid fragmentation, resulting in relatively few but massive defects.)

10.2.3 **THE NO-FALL REQUIREMENT**

It is quickly shown that if liquid aluminium is allowed to fall more than 12.5 mm, then it exceeds the critical speed 0.5 m/s. The critical fall height can be seen to be a kind of re-statement of the critical velocity condition. Similar critical velocities and critical fall heights can be defined for other liquid metals. The critical fall heights for all liquid metals are in the 3–15 mm range .

It follows immediately that practically all pouring of metals is bad. Figure 10.3 shows an all too common and lamentably efficient destruction of the melt quality. Clearly, neither the employee, manager nor designer of the plant knew they were engaged in such a destruction of quality. In addition of course, this destruction of metal simply forms oxide in the form of dross. The fall shown in Figure 10.3 would probably have caused the loss of 2–3% of metal; a large and avoidable cost.

It also follows similarly that *top gating* of moulds, almost without exception, will lead to a violation of the critical velocity requirement. In addition, for many forms of gating that enter the mould cavity at the mould joint, if any significant part of the cavity is below the joint, these will also violate this requirement.

In fact, for conventional sand and gravity die casting, it has to be accepted that some fall of the metal is necessary. Thus it has been accepted that the best option is for a single fall, concentrating the total fall of the liquid at the very beginning of the filling system, and causing the fall to be surrounded by a close-fitting, tapered conduit known as a *sprue*, or *down-runner*. This conduit brings the melt to the lowest point of the mould. The damage potential of the fall

FIGURE 10.3

A typical melt transfer, seriously degrading the melt.

is limited if the sprue is as narrow as possible, so as to constrain the melt, giving it little chance to damage itself by folding in its oxide surface. Thus the sprue is designed to have a cross-section only just large enough to take the required flow rate to fill the casting in the target fill time. The distribution system from the point that the melt hits the base of the sprue, consisting of runners and gates, should progress only *horizontally* or preferably *uphill*. The metal should never suffer a further drop.

Considering the mould cavity itself, the no-fall requirement effectively rules that all gates into the mould cavity enter at the bottom level, known as *bottom-gating*. The siting of gates into the mould cavity at the top *(top-gating)* or at the joint *(gating at the joint line)* are not options if safety from surface turbulence is required.

Also excluded are any filling methods that cause waterfall effects in the mould cavity (Figure 2.30). This requirement dictates the siting of a separate ingate at every isolated low point on the casting.

Even so, the concept of the critical fall distance does require some qualification. If the critical limit is exceeded, it does not mean that defects will necessarily occur. It simply means that there is a risk that they may occur. This is because for falls greater than the critical fall distance the energy of the liquid is sufficiently high that the melt is potentially able to enfold in its own surface. Whether a defect occurs or not is now a matter of chance.

There is, however, further qualification that needs to be applied to the critical fall distance. This is because the critical value quoted previously has been worked out for a liquid, neglecting the presence of any oxide film. In practice, it seems that for some liquid alloys, the surface oxide has a certain amount of strength and rigidity, so that the falling stream is contained in its oxide tube and so is enabled to better resist the conditions that might enfold its surface. This behaviour has been investigated for aluminium alloys (Din et al., 2003). It seems that although the original fall distance limit of 12.5 mm continues to be the safest option, fall heights of up to about 100 mm might be allowable in some instances, corresponding to a speed of 1.4 m/s. In the unusual circumstances of a ceramic foam filter being inserted into the flow, the fall distance downstream of the filter might be increased to approximately 200 mm without undue damage to the melt as explained in Section 11.2. However, falls greater than 200 mm definitely entrain defects; the velocity of the melt in this case is over 2 ms^{-1}. Entrainment seems unavoidable at this and higher speeds. Also, of course, other alloys may not enjoy the benefits of the support of a tube of oxide around the falling jet. This benefit requires to be investigated in other alloy systems to test what values beyond the theoretical limits may be used in practice.

The initial fall down the sprue in gravity-filled systems does necessarily introduce some oxide damage into the metal. The damage is concentrated during the first few seconds during which the sprue is priming. This period of fall of the metal can be very messy. The period can be shortened by contact pouring or by the use of a stopper in an offset step-pouring basin to ensure the basin is filled before the sprue being allowed to fill. Even so, it seems reasonable to conclude that gravity poured castings will never attain the degree of reliability that can be provided by counter-gravity and other systems that can totally avoid surface turbulence.

Of necessity therefore, it has to be accepted that the no-fall requirement applies to the design of the filling system beyond the base of the sprue. The damage encountered in the fall down the sprue has to be accepted; although with a good sprue, good contact pouring or good pouring basin and stopper this initial fall damage can be reduced to a minimum.

It is a matter of good luck that it seems that for some alloys much of the oxide introduced by the initial turbulence in the filling system does not appear to find its way through and into the mould cavity. It seems that much of it remains attached to the walls of the running system. This fortunate effect is clearly seen in many top-gated castings, where most of the oxide damage (and particularly any random leakage problem) is confined to the area of the casting under the point of pouring, where the metal is falling, as is seen in Figure 9.45. Severe damage does not seem to extend into those regions of the casting where the speed of the metal front decreases, and where the front travels uphill, but there does appear to be some carryover of defects. Thus the provision of a filter immediately after the completion of the fall is valuable. It is to be noted however that the filter will not take out all of the damage.

The requirement that the filling system should cause the melt to progress only uphill after the base of the sprue appears to force the decision that the runner must be in the drag and the gates must be in the cope for a horizontal single jointed mould. However, it has to be admitted that this preferred arrangement is not always possible, and in any case the velocities dictated by the acceleration from gravity ensures that the melt takes little notice of such niceties; the metal simply overrides this small difference in height.

The no-fall requirement may also exclude some of those filling methods in which the metal slides down a face inside the mould cavity, such as some tilt-casting type operations. This undesirable effect is discussed in more detail in the section devoted to tilt casting. As we shall see, sliding downhill is not a necessary filling mode for tilt casting; if feasible, horizontal transfer of the melt is strongly recommended for the tilt-casting process as explained in Section 16.2.2.

It is noteworthy that these precautions to avoid the entrainment of oxide films also apply to casting in inert gas or even in vacuum. This is because the oxides of Al and Mg (as in Al alloys, ductile irons and high temperature Ni-base alloys for instance) form so readily that they effectively 'getter' the residual oxygen in any conventional industrial vacuum, and form strong oxide films on the surface of the liquid.

Rule 2 in its no-fall embodiment applies to 'normal' castings with walls of thickness over 3–4 mm. For channels that are sufficiently narrow, having dimensions approaching a millimetre, the curvature of the meniscus at the liquid front should keep the liquid front from disintegration, the effect of surface tension becoming ever more powerful. Thus the action of surface tension in narrow filling system geometries is valuable to conserve the liquid as a coherent mass, and so acting to push the air out of the system ahead of the liquid. The filling systems therefore fill in one pass. (Unfortunately, in some conditions, this highly desirable mode of filling degenerates into micro-jetting as mentioned again later. Possibly this damaging form of flow applies only to certain alloys or certain environmental conditions as mentioned in Section 2.2.7.)

A good filling action, pushing the air ahead of the liquid front as a piston in a cylinder, is a critically valuable action. Such systems deserve a special name such as perhaps '*One pass filling designs*'. Although I do not usually care for such jargon, the special name emphasises the special action. It contrasts with the turbulent and scattered filling often observed in systems that are over-generously designed, in which the melt can be travelling in two directions at once along a single channel; a fast jet travels under the return wave that rolls over its top, rolling in air and oxides at the boundary between the opposing streams.

For a wide, shallow horizontal channel, any effect of surface tension is clearly limited to channels that have dimensions smaller than the sessile drop height for that alloy. Thus for Al alloys, the maximum channel height would be 12.5 mm, although even this height would exert little influence on the melt because the roof would just touch the liquid, exerting no pressure on it. Similarly, taking account that the effect of surface tension is doubled if the curvature of the liquid front is doubled by a second component of the curvature at right angles, a channel of square section could be 25 mm square, and be just contained by surface tension. In practice, however, for any useful restraint from the walls of channels, these dimensions require to be at least halved, effectively compressing the meniscus, as a mechanical spring, into the tight confines of the channel.

For very thin-walled castings, of section thickness less than 2 mm, the effect of surface tension in controlling filling becomes dominant. The walls are so much closer than the natural curvature of a sessile drop that the meniscus is tightly compressed and requires the application of pressure to force it into such narrow gaps. The liquid surface is now so constrained that it is not easy to break the surface i.e. once again there is no longer room for splashing or droplet formation. Thus the critical velocity is higher, and metal speeds can be raised by approximately a factor of two without danger.

In very thin-walled castings, with walls less than 2 mm thickness, the tight curvature of the meniscus becomes so important that filling can sometimes be without regard to gravity (i.e. can be uphill or downhill) because the effect of gravity is overpowered by surface tension. This makes even the uphill filling of such thin sections problematic because the effective surface tension exceeds the effect of gravity. Instabilities therefore occur, whereby the moving parts of the meniscus continue to move ahead in spite of gravity because of the reduced thickness of the oxide skin at that point. Conversely, other parts of the meniscus that drag back are further suppressed in their advance by the thickening oxide, so that a runaway instability condition occurs. This dendritic advance of the liquid front is no longer controlled by gravity in very thin castings, making the filling of extensive sections, whether horizontal or vertical, a major filling problem.

The problem of the filling of thin walls occurs because the flow happens, by chance, to avoid filling some areas because of random meandering. Such chance avoidance, if prolonged, leads to the development of strong oxide films or even freezing of the liquid front. Thus the final advance of the liquid to fill such regions is hindered or prevented altogether.

The dangers of a random filling pattern problem are relieved by the presence of regularly spaced *ribs* or other geometrical features that assist to organise the distribution of liquid. Random meandering is thereby discouraged and replaced by regular and frequent penetration of the area, so that the liquid front has a better chance to remain 'live' i.e. it keeps moving so that a thick restraining oxide is given less chance to form.

The further complicating effect of the microscopic break-up of the front known as micro-jetting observed in casting sections of 2 mm and less in sand and plaster moulds is not yet understood. The effect has not yet been systematically researched, and may not occur at all in the dry mould conditions such as are found in gravity die (permanent mould) casting. What little is known about this mysterious phenomenon is discussed briefly in Chapter 2.

10.3 RULE 3: AVOID LAMINAR ENTRAINMENT OF THE SURFACE FILM (THE NON-STOPPING, NON-REVERSING CONDITION)

10.3.1 CONTINUOUS EXPANSION OF THE MENISCUS

If the liquid metal front continues to *advance* at all points on its surface, effectively, continuing to *expand* at all points on its surface, like a progressively inflating balloon, then all will be well. This is the ideal mode of advance of the liquid front. In general, this is the mode of advance of a counter-gravity filling system, which is why counter-gravity filling of castings is such a powerful technique for the production of good castings.

In fact, we can go further with this interesting concept of the requirement for continuous surface expansion. There is a sense in that if surface is lost (i.e. if any part of the surface experiences contraction) then, because the surface oxide cannot contract, some entrainment of the surface oxide necessarily occurs. Thus this can be seen to be an all-embracing and powerful definition of the condition for entrainment of defects, simply that surface area must not be lost. Clearly, surface area is effectively lost by being *enfolded* (in the sense that the fold now disappears inside the liquid, as has been the central issue described under rule 2), or by simply *shrinking* i.e. being lost (necessarily leading to folding) as described later. Thus, in a way, this condition 'avoid loss of surface' can be seen to supersede the conditions of critical velocity or Weber number. It promises to be a useful condition that could be recognised in numerical simulation and thus be useful for computer prediction of entrainment.

In practice, however, an uphill advance of the liquid front, if it can be arranged in a mould cavity, is usually a great help to keep the liquid front as 'alive' as possible, i.e. keeping the meniscus moving, and so expanding and continuously creating new oxide.

It seems possible that the condition for exceeding the critical velocity may lead to more severe entrainment than simple surface contractions, and that surface contractions might lead to entrained film that remains closely attached to the oxide on the surface, instead of becoming displaced into the deep interior of the casting as happens with surface turbulence. These distinctions, if they are real, remain to be clarified by research.

Although the surface is being continuously expanded when filling the mould uphill, the casting has the benefit that the older, thicker oxide is continuously being displaced to the walls of the mould where it becomes the skin of the casting. Thus very old and very thick oxide does not normally have chance to form and become entrained. In fact, one of the great benefits of a good filling system is to ensure that the older oxides on the surface of the ladle or pouring basin etc. do not enter the mould cavity. For instance, in the tilt pouring of aluminium alloy castings, the filling of a new casting can be checked by dropping a fragment of paper on the metal surface as the tilt commences. This marker should stay in place, indicating that the old skin on the metal was being retained by the runner, so that only clean metal underneath could flow into the mould cavity. If the paper disappears into the runner, the runner is not doing its job. In a way, the use of the teapot pouring ladle and bottom-pour ladles common in the steel casting industry are in response to the acute form of these same problems, in which the high rates of reaction with the environment at such high temperatures encourages the surface oxide to grow from microscopic to macroscopic thickness, to constitute the familiar slag layer.

10.3.2 ARREST OF FORWARD MOTION OF THE MENISCUS

Problems arise if the front becomes pinned by the rate of advance of the metal front being too slow, or if it stops or reverses.

If the liquid front stops, a thick surface film has chance to form. Figure 10.4(a) shows a common way in which this can happen—by the abrupt enlargement of area at some stage during the filling of the mould. The stationary surface film may become so thick and strong that if pressure increases later to encourage flow to re-start, the front finds itself pinned in place, no longer able to move. As the advance re-starts elsewhere in the mould, the melt may overflow and submerge the thick stationary film, forming an entrainment defect (Figure 2.25). As the new metal rolls over the old film, a new fresh oxide is laid down over the old thick oxide, forming our familiar double film, with dry side facing dry side so that the double layer forms an unbonded interface, as a crack. On this occasion, the bifilm is a thick/thin asymmetric variety. This mode of creation of a bifilm can constitute a large geometrical defect, sometimes in the form of a horizontal crack extending across the whole casting, or as a horizontal lap around its complete perimeter.

A good name for this defect is an *oxide lap*. (It is to be distinguished from the solidification defect that often has a superficially similar appearance, called here a *cold lap* caused not necessarily by the strength of the oxide film but by the freezing in place of the liquid front. The method of treating these defects is quite different. A cold lap can be cured by increasing the casting temperature, whereas increasing the temperature of an oxide lap is likely to make it worse.)

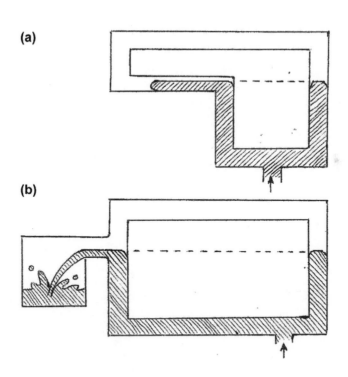

(a)

(b)

FIGURE 10.4

Two common filling situations in which the general advance of the melt is stopped, introducing the danger of a lap defect. (a) An enlarged area, and (b) a 'waterfall' effect to halt the advance of the liquid front.

10.3.3 WATERFALL FLOW: THE OXIDE FLOW TUBE

Instead of a large horizontal defect, a curious but major geometrical defect in the form of a cylindrical tube that I call an *oxide flow tube* can form in several ways.

If the liquid falls vertically, as a plunging jet, the falling stream is surrounded by a tube of oxide (Figures 2.30 and 10.4(b)). Despite the high velocity of the falling metal inside, the oxide tube remains stationary, thickening with time, until finally surrounded by the rising level of the metal in the mould cavity. This rising metal rolls up against the oxide tube, forming a double oxide crack (once again of a highly asymmetrical thick/thin form). Notice the curious cylindrical form of this crack and its largely vertical orientation. The arrest of the advance of the front in this case occurred by the curious phenomenon that although the metal was travelling at a high speed parallel to the jet, its transverse velocity, i.e. its velocity at right angles to its surface, was zero. It is the zero velocity component of a front that allows the opportunity for a thick oxide skin to develop.

These oxide flow tubes are often seen around the falling streams of many liquid metals and alloys as they are poured. The defects are also commonly seen in castings. Although occasionally located deep inside the casting where they are not easily found, they are often clearly visible if formed against the casting surface.

A patent dating from 1928 (Beck) describes how liquid magnesium can be transferred from a ladle into a mould by arranging for the pouring lip of the ladle to be as close as possible to the pouring cup of the mould, and to be in a relatively fixed position so that the semi-rigid oxide flow tube which forms automatically around the jet is maintained unbroken, thus protecting the metal from contact with the air (Figure 2.23(a)). A similar phenomenon is seen in the pouring of aluminium alloys and other metals such as aluminium bronze.

A large Ni-base superalloy turbine blade, cast in vacuum, and destined for a land-based turbine, was placed upside down and top-poured through its root. The root had almost certainly created a cluster of vertically oriented oxide flow tubes as the melt poured in a series of separate streams through the root, to start its long fall towards the blade tip. When the transverse grooves of the fir tree root were machined by grinding, vertical cracks were observed which were argued to be grinding cracks from over-enthusiastic grinding. However, their origin as oxide flow tubes was corroborated by the observation under the microscope of a series of sulphide precipitates with an alignment that was characteristic of occupying one side of a vertical oxide film. Micrographs could not be released for security reasons, so a sketch to illustrate the effect is seen in Figure 10.5(a). The interpretation of the micrograph is shown in Figure 10.5(b). The presence of the bifilm in the form of an oxide flow tube explains both the vertical cracking

(a) **(b)**

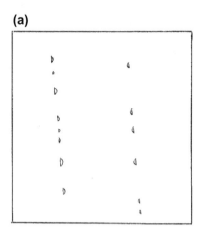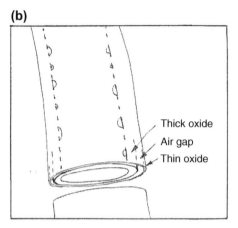

FIGURE 10.5

(a) Vertical lines of sulphide inclusions seen on a polished section of a vacuum top-poured Ni-base superalloy. (b) Interpretation in terms of precipitation on an oxide flow tube (oxide thicknesses and air gap are greatly exaggerated for clarity).

during grinding, and the vertical alignment of the sulphides which have clearly grown only on one side of an interface, which in this case appears to be the inner wetted interface of the original flow tube. Precipitates have been reported to form on only once side of an asymmetric bifilm, as seen in a confluence weld (Garcia-Garcia et al., 2007). (The air gap between the two oxide films shown in Figure 10.5(b) is exaggerated for clarity. Clearly in practice the two films will be in contact, and the air gap will be comprised only of the air trapped in the microscopic roughness of the two contacting surfaces.)

The stream does not need to fall vertically. Streams can be seen that have slid-down gradients in such processes as tilt casting when carried out under poor control. Part of the associated flow tube is often visible on the surface of the casting, as a witness to the original presence of the metal stream. Alternatively, a wandering horizontal stream can define the flow tube, as is commonly seen in the spread of liquid across a horizontal surface. Figure 2.24 shows how, in a thin horizontal section, the banks of the flowing stream remain stationary whilst the melt continues to flow. When the flow finally fills the section, coming to rest against the now-rigid oxide forming the banks of the stream, the banks will constitute long meandering asymmetrical bifilms as cracks, following the original line of the flow.

The jets of flow in pressure die castings can sometimes be seen to be leave permanent legacies as oxide tubes as seen in Figure 2.31.

Even with the best design of gravity-poured system, the rate of fill of the mould may be far from optimum at certain stages during the fill. For instance, Figure 10.4 shows two common geometrical features in castings that cause the advance of the liquid metal to come to a stop. The heavy section filled downhill will cause the metal front to stop, possibly causing a lap-type defect at points on the casting well away from the real cause of the problem. Until the recess is filled by the pouring metal, the remainder of the liquid front cannot advance. There are several reasons for avoiding any 'waterfall' action of the metal during the filling of the mould.

1. A cylindrical oxide flow tube forms around the falling stream. If the fall is from a reasonable height, the tube is shed from time to time, and plunges into the melt where it will certainly contribute to severe random defects. The periodic shedding of oxide flow tubes into the melt is a common sight during the pouring of castings. Several square metres of oxide area can be clearly seen to be introduced in this way during the filling of the mould.
2. The plunging jet is likely to exceed the critical velocity. Thus the metal that has suffered the fall is likely to be impaired by the addition of randomly entrained bifilms generated near the point of impact.
3. As the melt rises around the tube, supporting it to some degree and reducing the height of the fall, the flow tube remains in place and simply thickens. As the general level of the melt rises around the tube, the new oxide rolls up against the surface of the cylinder, forming the curious cylindrical bifilm that acts as a major cylindrical crack around a substantially vertical axis.
4. During the period of the waterfall action, the general rise of the metal in the rest of the mould will be interrupted, causing an oxide to form across the whole of the stationary level surface. If this has time to thicken significantly, it may be too strong to move when the levels equalise. Thus rising melt will overflow the film, as the flooding over an ice sheet, rolling a second thin film into place on top of the thick stationary layer. Thus a major horizontal lap defect may be created as a horizontal bifilm. The bifilm will be highly asymmetrical, with one thick and one thin film facing each other, as described earlier.

Defects (1) and (2) are the usual fragmented and chaotic type of bifilm. Defects from (3) and (4) are the major geometrical bifilms.

Waterfall problems are usually easily avoided by the provision of a gate into the mould cavity at every low point in the cavity. Occasionally, deep recesses can be linked by channels through the mould or core assembly; the links being removed during the subsequent dressing of the casting.

10.3.4 HORIZONTAL STREAM FLOW

If the melt is allowed to spread without constraint across a horizontal thin-walled surface, gravity can play no part in persuading the flow to propagate on a broad stable front, as would happen naturally in a vertical or sloping plate (Figure 2.24). The front propagates unstably in the form of a river bounded by river banks composed of thickening oxide.

A fascinating example of a flow tube can be quoted from observations of flow up a sloping open channel driven by a travelling magnetic field from a linear motor sited under the channel. When used to drive liquid aluminium alloy uphill, out of a furnace and into a higher-level receiver, the travelling melt is seen to flow inside its oxide tube. When the magnetic field is switched off, the melt drains out of the oxide tube and back into the furnace, with the tube collapsing flat on the bottom of the channel. However, when the field is switched on again, the same oxide tube magically refills and continues to pass metal as before. Clearly, the tube has considerable strength and resilience. It is sobering to think that such features can be built into our castings, but remain unsuspected and almost certainly undetectable. Clearly, the casting methods engineer requires vigilance to ensure that such defects will not form.

The vertical oxide flow tube is probably more common than any of us suspect. The example given later is simply one of many that could be described.

Figure 10.6 illustrates the bronze bell, sometimes known as the 'Freedom Bell' as a larger-sized replica of the Liberty Bell, hung outside the railway station in Washington, DC. Horizontal weld repairs, etched for enhanced clarity by rain water, record for all time the fatal hesitations in the pouring process that led to the horizontal oxide bifilms that would have appeared as horizontal laps. Perhaps these were the points at which the pouring ladles were changed. The vertical weld repairs record the passage of the falling stream that created the vertical flow tube that led to cracks through the thickness of the casting. These vertical cracks were opened by the hoop stress created by the casting contracting onto its core. This is a common source of failure for bells, nearly all of which are top-poured through the crown.

The renowned Liberty Bell (the only survivor of three attempts, all of which cracked) reveals a magnificent example of a flow tube defect that starts at the crown, curves sinuously around and over the shoulder and finally falls vertically down the skirt to the mouth of the bell. Although there are many examples of bells that exhibit these long cracks, it is perhaps all the more surprising that any bells survive the top-pouring process. It seems likely that in the majority of cases of bells of thicker section the oxide flow tube is not trapped between the walls of the mould to create a through-thickness pair of parallel cracks. In such thicker sections, the tube is more likely to be detached and carried away, crumpling into a somewhat smaller defect that can be accommodated elsewhere. It is to be hoped that the new resting place of the defect will not pose any serious future threat to the product. Clearly, the top pouring of castings is a risky manufacturing technique.

FIGURE 10.6

Inspection of the Freedom Bell outside the railway station in Washington, DC, unfortunately spoiled by welds from an attempt to repair cracks caused by vertical oxide flow tubes from top pouring and horizontal oxide bifilms from filling hesitations.

Photograph by Sheila.

Oxide flow tubes are common defects seen in a wide variety of castings that have been filled across horizontal sections or down sloping downhill sections. The deleterious oxide flow tube structures described previously that form when filling *downwards* or *horizontally* cannot form when filling *vertically upwards* i.e. in a *counter-gravity* mode. The requirement for reliable castings that the meniscus only travels uphill is sacred.

Although moulds can be filled substantially without risk only by *counter-gravity filling,* even in this case vigilance is still required to avoid other filling defects. Even in this most favourable mode of filling, a related *oxide lap* defect, or even a *cold lap* defect, can still occur if the advance of the meniscus is stopped at any time as we have seen.

In all cases, it will be noticed that such interruptions to flow, where, for any reason, the surface of the liquid locally stops its advance, a large, asymmetric double film defect is created. These defects are always large, and always have a recognisable, predictable geometrical form (i.e. they are cylinders, planes, meandering streams etc.). They are quite different to the double films formed by surface turbulence, which are random in size and shape, often extremely thin or of variable thickness, but essentially completely unpredictable as a result of their chaotic origin.

10.3.5 HESITATION AND REVERSAL

If the meniscus stops at any time, it is common for it to undergo a slight reversal. Minor reversals to the front occur for a variety of reasons.

1. A reversal will practically always occur when a waterfall is initiated. This occurs because at the point of overflow, the liquid will be at a level slightly above the overflow, dictated by the curvature of its meniscus, i.e. for a liquid Al alloy it will be about 12.5 mm above the height of the overflow because this is the height of the sessile drop. However, immediately after the overflow starts, the general liquid level drops, no longer supported by the surface tension of the meniscus. In the case of liquid aluminium alloy this fall in general level of the liquid will be perhaps about 6 mm, just enough to flatten the more distant parts of the meniscus against the rest of the mould walls.

2. Hesitations to an advancing flow will often be accompanied by slight reversals because of inertial effects of the flow. Momentum perturbations during filling will cause slight gravity waves, the surface therefore experiencing minor slopping and surging motion, oscillating gently up and down.

These minute reversals of flow flatten the oxidised surface of the meniscus. When advancing, the meniscus adopts a rounded form, but when flattened, the oxidised surface now occupies a smaller area. A fold necessarily develops, wrinkling the surface, endangering the melt with the possibility that the entrainment of this excess oxide is permanent; the entrainment folding action is not reversed once the melt is able to continue its advance. The folding in of a small crack attached to the surface of the casting is illustrated in Figure 10.7. Such shallow surface cracks occurring as a result of hesitation and/or reversal of the front are common in aluminium alloys, and are revealed by dye penetrant testing.

It is instructive to estimate the maximum depth that such oxide folds might have. In Figure 10.7, if the front of the liquid in Figure 10.7(a) is a cylinder of radius r, the perimeter of the quarter of a cylinder is $\pi r/2$, so that the maximum length of excess surface if the melt level now drops a distance r to become horizontal is approximately $\pi r/2 - r = r((\pi/2) - 1) = r/2$. The radius of the meniscus r is approximately 6 mm for liquid aluminium (as a result of the total height of a sessile drop being approximately 12 mm), giving the excess length 3 mm. If this is folded just once to create a bifilm, its potential depth is therefore found to be approximately 1.5 mm. This may not seem a lot, but an appearance of such a crack during the NDT examination of a casting by dye penetrant will almost certainly fail any safety critical casting. Clearly, the depth of the crack cannot easily be known during the dye penetrant inspection, and in any case might genuinely be serious for a highly stressed part experiencing fatigue.

If the melt continues its downward oscillation, the defect can be straightened out as shown in Figure 10.7(c). Alternatively, if the bifilm created in this way holds itself closed as in Figure 10.7(b), possibly because of viscous adhesion (i.e. the trapped liquid metal takes time to escape from between the films) or possibly as the result of other forces such as Van der Waals forces, then there is the danger that the fold is not reversible, and additional folds may be created on each oscillation cycle.

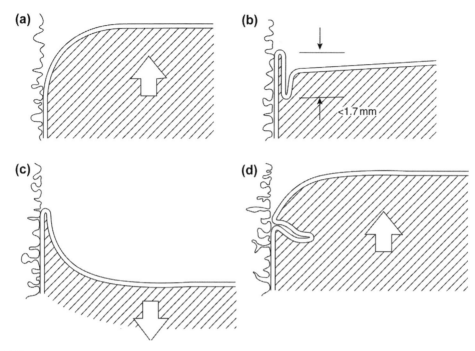

FIGURE 10.7

The creation of bifilm cracks of the order of millimetres deep by the reversal of the front, causing the meniscus to flatten and enfold the excess surface area of film. If (a) the advancing front suffers (b) a small reversal, or (c) a somewhat larger reversal, then (d) the re-starting of the flow may generate a bifilm crack from the enfolding of the excess (stretched) area of oxide film.

In fact, many of these defects are not as deep as the maximum estimate of 1.5 mm for several reasons. (1) The melt surface may not drop the full distance r; (2) the film may be folded more than once, creating a greater number of shallower folds; and (3) the fold-like crack may hinge to lie flat against the surface of the casting. The action of internal forces as a result of flow of the liquid, surging through the mould sections, may be helpful in this respect. For these reasons, such defects are usually only a fraction of a millimetre deep, so that they can often be removed by grit blasting. Only relatively rarely do they reach the maximum possible depth approaching 1.5 mm. Even so, for castings requiring total integrity, requiring resistance to high stress or fatigue, these minor oscillations of the front are very real threats that are best avoided.

The ultimate solution, as we have emphasised here, is that the melt should be designed to be kept on the move, advancing steadily forwards at all times. All we foundry people should be deeply grateful for the simplicity of this solution.

10.3.6 OXIDE LAP DEFECTS

The flooding of the melt over a large oxide film that has grown as a result of a hesitation in the vertical rise of the liquid, and the problem of joining of flows if one has stopped, are essentially equivalent. The first is usually a horizontal double film, where the second is usually vertical. Both are highly asymmetrical, consisting of a thick film (grown on the stationary interface) and a thin film (grown on the moving front).

These very deleterious defects, as major cracks, can be avoided by increasing the rate of filling of the mould. Care is needed of course to avoid casting at too high a rate at which surface turbulence may become an issue. However, providentially, there is usually a comfortably wide operational window in which the fill rate can meet all the requirements to avoid defects. This is another simple solution for which we are all grateful.

10.4 RULE 4: AVOID BUBBLE DAMAGE

Entrainment defects are caused by the folding action of the (oxidised) liquid surface or by the impact of splashes or droplets, all of which meet dry side to dry side. Sometimes only oxides are entrained, as doubled-over film defects, called bifilms. Sometimes the bifilms themselves contain small pockets of accidentally enfolded air, so that the bifilm is decorated by arrays of trapped bubbles. Much, if not all, of the microporosity observed in castings either is, or has originated from, a bifilm.

Sometimes, however, the folded-in packet of air is so large that the buoyancy of the newly created bubble confers on it a life of its own. This oxide-wrapped piece of air that we call a bubble is a massive entrainment defect. It can be sufficiently buoyant to power its way through the liquid and sometimes through the dendrites. In this way it develops its own distinctive damage pattern in the casting.

10.4.1 THE BUBBLE TRAIL

The passage of a single bubble through an oxidisable melt is likely to result in the creation of a bubble trail as a double oxide crack, a kind of long bifilm, in the liquid (Figure 2.32). A bubble trail is the name coined (JC, 1991) to describe the defect that was predicted to remain in a film-forming alloy after the passage of a bubble through the melt. Thus, even though the bubble may be lost by bursting at the liquid surface, the trail remains as permanent damage in the casting.

The bubble trail occurs because the trail is nearly always attached to the point where the bubble was first entrained in the liquid. The enclosing shroud of oxide film covering the crown of the bubble attempts to hinder its motion. This restrain is so large that bubbles having relatively little buoyancy, of diameter less than about 5 mm, are held back, unable to rise. However, if the bubble is larger, its upward buoyancy force will split this restraining cover. Immediately, of course, the oxide on the crown re-forms, and splits and re-forms repeatedly. The expanding region of oxide film on the crown effectively slides around the surface of the bubble, continuing to expand until the equator of the bubble is reached. At this point the area of the film is a maximum. Figure 10.8 shows residual fragments on such a bubble in a Ti alloy (also shown in this figure is some shrinkage porosity that has grown from the underside of the bubble presumably because the casting had been poorly fed. The bubble trail is just visible below this region). Although the film was able to expand by splitting and re-forming, it is not able to contract because it is a solid. Further progress of the bubble causes the film to continue sliding around the bubble, gathering together in a mass of longitudinal pleats under the bubble, forming a trail that leads back to the point at which the bubble was first entrained as a packet of gas. Any spiralling motion of the bubble will twist and additionally tighten this rope-like tether that continues to lengthen as the bubble rises.

The structure of the trail is a kind of collapsed tube. It is star-like in section but with a central portion that has resisted complete collapse because of the small residual rigidity of the oxide film (Figure 2.32). This is expected to form an excellent leak path if it joins opposing surfaces of the casting or if cut into by machining. In addition, of course, the coming together of the opposite skins of the bubble during the formation of the trail ensure that the films make contact dry side to dry side, and so constitute our familiar classical bifilm crack.

Poor designs of filling systems can result in the entrainment of much air into the liquid stream during its travel through the filling basin, during its fall down the sprue and during its journey along the runner. In this way, dozens or even hundreds of bubbles can be introduced into the mould cavity.

10.4.2 BUBBLE DAMAGE

Bubble damage is usually a mixture of the residue of bubbles and bubble trails. It is a typical entrainment mess.

When many bubbles are involved, the later bubbles have problems to rise through the morass of bubble trails that remain after the passage of the first bubbles. Figure 10.9 shows counts of the number of bubbles per cubic centimetre

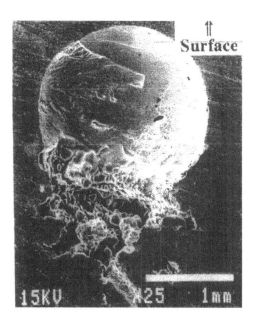

FIGURE 10.8

A bubble in a centrifugally cast Ti-6Al-4V alloy by Suzuki et al. (1996), showing oxide slipping around its surface, and some shrinkage porosity clustered around the start of its bubble trail in the bottom of the image.

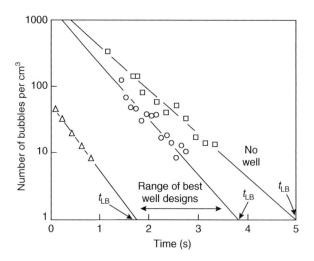

FIGURE 10.9

Water model of bubbles entrained by surface turbulence in a well (Isawa and Campbell, 1994) for runners twice the area of the sprue, showing their decrease with time for different well designs, extrapolated to the time for the last bubble t_{LB}.

arising in a small experimental casting as a result of the provision of a 'well' at the base of the sprue. Clearly, hundreds of bubbles are generated, and only die away after several seconds. Larger castings would be expected to generate many times this number, and experience bubble production persisting for proportionally longer. Furthermore, if this were not bad enough, even more bubbles are created in the pouring basin (requiring the offset step basin to keep many of these out of the filling system) and many more are generated at the base of the sprue and in the running system unless these are designed to be naturally pressurised.

It is easy to understand that the free flotation and escape of late-arriving bubbles is hampered by the accumulation of the tangle of residual bifilms. If the density of films is sufficiently great, fragments of bubbles remain entrapped as permanent features of the casting. This mass of fragments of bifilm trails and bubbles is collectively called 'bubble damage' (Figure 2.38). In the experience of the author, bubble damage is probably the most common defect in castings, but up to now has been almost universally unrecognised.

Bubble damage in the form of masses of trails is nearly always mistaken for shrinkage porosity as a result of the irregular form of the air porosity that it encloses, usually with characteristic cusp-like morphology. Another common feature of bubble damage is its non-uniform distribution. Unlike shrinkage porosity, it is often not in areas where shrinkage porosity might be expected. For instance, shrinkage porosity would be expected in hot spots, but oxide damage resembling shrinkage porosity can be found almost anywhere in the casting.

In some stainless steels, the phenomenon is seen under the microscope as a mixture of bubbles and cracks. (A remarkable combination! Without the concept of the bifilm, such a combination would be extremely difficult to explain; with bubbles that would naturally expand to relieve and eliminate stresses, and cracks that therefore could not form because their formation would need stresses.) In these strong materials, the high cooling strain leads to high stresses that open up the double-oxide bubble trails.

10.4.3 BUBBLE DAMAGE IN GRAVITY FILLING SYSTEMS

In gravity-filled running systems, the requirement to reduce bubbles in the liquid stream during the filling of the casting calls for offset stepped basins or another non-entraining filling system such as contact pouring, for instance. (The conventional conical or funnel-shaped pouring basin is the worst possible design of filling device; it concentrates at least 50%, but sometimes hundreds of percent, of air in the metal.)

The sprue is required to be tapered, the taper calculated to match, or very slightly compress, the natural form of the falling stream; the stream naturally narrows during its fall because of its acceleration under the action of gravity. By tailoring the shape of the sprue to the natural shape of the stream, the melt has the best chance to avoid the entrainment of air.

Parallel or reversed taper sprues are not recommended. Elliott and Mezoff (1948) show that their parallel slot sprues entrain air bubbles, finding that at low casting temperatures, their Mg alloy castings exhibit bubbles, whereas at high casting temperatures, the castings exhibit mainly oxide films. Although the authors conclude that the air is consumed by the Mg alloy, which will certainly be partly if not wholly true, it seems also likely that many larger bubbles would have time to escape at high temperatures before the formation of a solidified skin. The authors are clearly not aware of the seriousness of these residual oxides.

Special precautions are needed to reduce the damage introduced if a parallel or reverse tapered sprue design has to be used. Such features include the radial choke or a filter at the entrance to the sprue.

At one time, it was mandatory that each sprue had a sprue well at its base. The designs of filling systems recommended in this book explicitly avoid wells. This departure from tradition is possible only because the new filling systems are characterised by runners of approximately the same area as that of the sprue exit. It is to be noted that the traditional choice of sprue exit/runner/gate ratios of 1:2:2 and 1:2:4 etc. are automatically bad. The runner is too large to fill completely, regrettably entraining air, and so ensuring bubble damage problems.

An additional beneficial consequence of the avoidance of a well is the addition friction to the liquid provided by the additional solid surface of the mould at the point impacted by the metal as it turns the corner, therefore slowing the velocity of the liquid. This reduction in velocity is usually not great, in the region of 20%, but any reduction in speed is to be welcomed. If the sprue/runner junction is nicely formed, bubbles are formed for precisely 0 s. This awkward way of making a simple statement that no bubbles are formed is deliberate. It emphasises the contrast with filling systems that have been accepted as conventional up to now (for instance as in the study for Figure 10.9). Nowadays it is not necessary to accept a design that introduces any bubbles at all.

It is mandatory that no interruption to the pour occurs that leads to the lowering of the melt in the pouring basin below the minimum design level. If the sprue entrance is unpressurised in this way air will enter the running system. In the worst instance of this kind, if the basin level drops to the point that the sprue entrance becomes uncovered this has to be viewed as a disaster. A provision must be made for the foundry to reject automatically any castings that have suffered an interrupted pour, or slow pour that has allowed the basin to empty to a level below the designed minimum level.

To be safe, if a basin is to be used, it is worth ensuring that it has up to twice the required minimum depth to keep the sprue filled, and ensuring that the pourer keeps the basin topped up as highly as possible, well above the minimum level. In this way the casting may run a little faster, but air will be excluded and bubble damage avoided.

The best option is the avoidance of any kind of pouring basin by using contact pouring. This not only eliminates bubbles but bifilms too (basins have been designed to eliminate bubbles, but remain great manufacturers of bifilms).

10.4.4 BUBBLE DAMAGE IN COUNTER-GRAVITY SYSTEMS

Pumped systems such as Cosworth Process, or low pressure castings systems into sand moulds or dies, are highly favoured as having the potential to avoid the entrainment of bubbles if, and only if, the processes are carried out under proper control. The reader needs to be aware that good control of a *potentially* good process should not be assumed; it requires to be demonstrated.

Although low pressure filling systems can in principle satisfy the requirement for the complete avoidance of bubbles in the metal, a leaking riser tube in a low-pressure casting machine can lead to a serious violation (Figure 10.10(a)). The

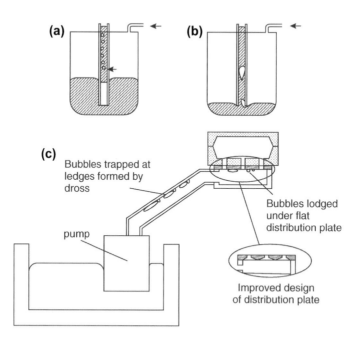

FIGURE 10.10

Bubbles introduced by defective counter-gravity systems. (a) Leak in a riser tube of a low pressure casting unit; (b) the dangerous ingestion of massive bubbles if the melt level is too low; (c) bubbles entrapped by dross and poor design features of some pumped systems, creating bubbles that are released erratically.

stream of bubbles from a leak in a defective riser tube will float directly up the tube and enter the casting. Unfortunately, this problem is not rare. Thus regular checks for such leakage, and the rejection of castings subjected to such consequent bubble damage, will be required. If the melt in the pressurised vessel becomes too low a huge amount of gas can erupt up the riser tube (Figure 10.10(b)) ejecting the liquid metal ahead.

For pumped systems, bubbles can be released erratically from the interior wall of a sloping tube launder system, especially if it is not cleaned with regular maintenance. Bubbles can become trapped and escape from time to time from the underside of a badly designed distribution plate used in some counter-gravity systems such as the early variants of the low-pressure sand system (Figure 10.10(c)).

A particularly insidious and little known source of a copious supply of bubbles in a low pressure casting system can come from a refractory lining in the pressurised furnace. This extremely important problem is dealt with in detail in Section 16.3.1.

All counter-gravity filling systems need to be designed to avoid the ingress of air or the trapping of air bubbles that might be randomly released.

10.5 RULE 5: AVOID CORE BLOWS
10.5.1 BACKGROUND

Gas bubbles can be forced into the liquid metal from a mould of core if their internal pressure from outgassing reactions exceeds the pressure in the melt. This section deals with an attempt to quantify the problem and the ways in which this problem can be avoided. However, there are additional dangers from gases 'blown' or 'fired' into the liquid metal from other sources which can be dealt with first.

Blows from the explosive volatilisation of core adhesives are well known if excess adhesive has been squeezed out from a sand joint so that it is brought into direct contact with the liquid metal.

Less well known is that considerable volumes of water vapour are given off from clay-based core repair and mould repair pastes. This is because the clay contains water of crystallisation, so that even after thoroughly drying the core repair at 100 or 200°C, the water bound in the structure of the clay remains unchanged, only being released at a high temperature, in the region of perhaps 600°C. Thus the water is released only when the clay contacts the liquid metal. This is particularly unfortunate, because the clay is composed of such fine particles that it is substantially impermeable, preventing the escape of the water into the core or mould, so that the water is forced to boil off through the metal. Repair of cores with clay-based pastes therefore generally leads automatically to blow defects. The wide use of core repair pastes illustrates that this danger is little known.

Unless you know from independent tests that your core repair technique does not result in core blows, the clear lesson is 'Do not repair cores.'

The generation of blows off metal chills is the result of an almost identical process. When a block metal chill is placed in a bonded aggregate mould, the pouring of the metal causes a rapid outgassing of the volatiles in the aggregate/binder mixture. The volatiles, particularly water vapour, are driven ahead of the spreading liquid metal and condense on any cold surface, such as a metal chill. When the liquid metal finally arrives and overruns the chill the condensates boil off. Because the chill is impermeable, the vapour is forced to bubble through the melt (Figure 10.11).

The prevention of blows from condensation on chills is widely known and generally well applied. The chill should normally be coated with a ceramic wash or spray that is afterwards thoroughly dried to give an inert, permeable and non-wetted surface layer. If this coat has residual surface roughness, so much the better. The effective permeability of the surface can be further enhanced by providing deep V grooves in a criss-cross pattern. The grooves are bridged to some degree by the action of the surface tension of the melt, so that the bottoms of the grooves act as surface vents, tunnelling the expanding vapours to freedom ahead of the advancing melt. Additionally, the V grooves are thought to enhance the effectiveness of the chilling action by increasing the contact between the casting and the chill, particularly during the cooling and contraction of the surface of the casting, forcing the grooved cast surface sideways against the walls of the grooves on the chill.

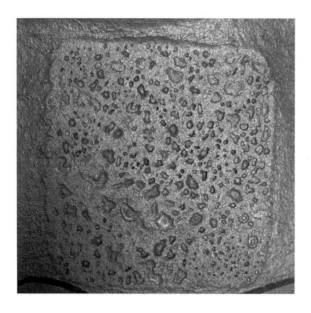

FIGURE 10.11

Blows off a damp steel chill in a steel sand casting. (In addition, it is to be noted that bubble damage would be expected in a vertical zone above the chill).

Courtesy S. Scholes.

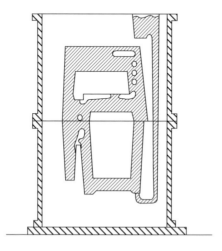

FIGURE 10.12

Casting in a close-fitting steel mould box on an unvented flat steel plate, showing blows from an upwardly oriented feature on the lower part of the mould.

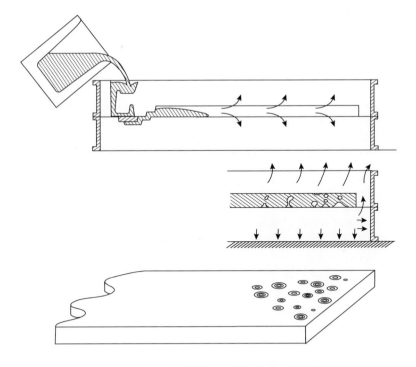

FIGURE 10.13

A large flat plate casting with an enclosed drag. Volatiles are driven ahead and condense in the cooler distant mould, exacerbating defects at the far end.

To demonstrate that a chill, a core, or assembly of cores, does not produce blows may require a procedure such as the removal of all or part of the cope or overlying cores, and taking a video recording of the filling of the mould. A small (probably sacrificial!) videocamera built into a completed mould might provide a better and economical test. If there are any such problems, the eruption of core gases will be clearly observable and will be seen to result in a boiling action, creating a froth of surface dross that would of course normally be entrapped inside the upper walls of the casting. A series of video recordings might be found to be necessary, showing the steady development of solutions to a core-blowing problem and recording how individual remedies resulted in progressive elimination of the problem. The video recording requires being retained by the foundry for inspection by the customer for the life of the component. Any change to the filling rate of the casting, or core design or the core repair procedure should necessitate a repeat of this exercise.

It should be noted that the danger of forming blows (bubbles in the melt) only arises if the liquid metal is subjected to the pressure in the mould (or core) and the pressurised gas has no option but to escape via the melt.

For instance, the general buildup of pressure in the mould cavity during filling is not usually a problem from the point of view of creating bubbles in the melt. In fact, for a fast-filling mould, the buildup of backpressure inside the mould cavity can be useful in slowing the final stages of filling, reducing the mould penetration that can occur because of the impact at the final instant of filling. Naturally, the converse is true for some difficult-to-fill castings, where mould backpressure can be sufficient to prevent the complete filling of moulds. The escape of gases entrapped in the mould cavity is made more difficult by the application of mould coats, so that pressures can easily be doubled (Ohnaka, 2003). Thus the internal pressure in the mould cavity may affect the filling of the mould positively or negatively, but will never usually lead to the creation of bubbles.

The buildup of pressure within the body of the mould *can* lead to problems. It can be severe in moulds that are enclosed in steel boxes and that are placed on a steel plate or on a concrete floor. The gases are relatively free to escape from the cope, but gases attempting to escape from the drag are sealed in by the overlying liquid in the mould cavity (Figures 10.12 and 10.13). The problem is enhanced if the casting is a tight fit in the moulding box, as is usually the case of course because the casting engineer is always trying to get as much value as possible out of each box. In fact the buildup of pressure inside sand moulds crammed into tight-fitting steel boxes has, in the author's experience, contributed to several spectacularly defective castings, and in one instance, to a casting that persistently refused to fill its mould because the backpressure of gases rose so high.

Fortunately, mould backpressure is easily dealt with by the provision of one or more whistlers. These are narrow, pencil-shaped vents connecting the mould cavity to the outside world via the cope.

Finally, it is worth commenting on the curiously misguided but common provision of whistler vents through the top of the mould in an effort to eliminate 'gas' porosity in the form of bubbles from some source. Clearly, this action is of no use. A moment's reflection reveals the self-evident fact that the pores (i.e. the entrained air bubbles) are already in the metal, and the metal itself would have to rise up the vent to eliminate the porosity from the casting. The error in thinking arises because of the confusion between gas entrapped in the mould cavity and gas entrained in the melt. When these are separated into their logically separate categories, confusion disappears and the correct remedy can be identified.

10.5.2 OUTGASSING PRESSURE IN CORES

The sudden heat from the liquid metal causes the volatile materials in the mould to evaporate fiercely. In greensand moulds and many other binder systems, the main component of this volatilisation is water. Even in so-called dry-binder systems, there is usually enough water to constitute a major contribution to the total volume of liberated gas. On contact with the hot metal, much of the water is decomposed to surface oxide films and to hydrogen. The high hydrogen contents of analysed mould gases are clearly seen in Figure 4.2.

In the case of the mould, the generation of copious volumes of gas is usually not a problem. The gas has plenty of opportunity to diffuse away through the bulk of the mould. The pressure buildup in a greensand cavity during mould filling is normally only of the order of a 100 mm water gauge (0.01 atm) according to measurements by Locke and Ashbrook (1972). This corresponds to merely 10 mm or so head pressure of liquid iron or steel. However, even this rather modest pressure might have been overestimated because their experimental arrangement corresponded to a closely fitting steel moulding box, and escape for mould gases only occurs via the cope. Even so, in greensand systems where the percentage of fines and clay and other constituents is high, the permeability of the mould falls to levels at which the ability of the mould volatiles to escape becomes a source of concern. The venting of the cope by needling with wires of about 3 mm diameter is a time-honoured method of re-introducing useful levels of permeability.

Chemically bonded moulds are usually of no concern from the point of view of generating a backpressure during the filling of the mould. This is because the sand is usually bought in as ready washed, cleaned and graded into closely similar sizes (a 'three pan sand'). In addition, usually only a 1 or 2 volume percent of binder is used, leaving an open, highly permeable bonded mould. A single measurement by the author using a water manometer showed a pressure rise during the filling of a cylinder head mould of less than 1 mm water gauge. Even this completely negligible rise seemed to decay to nothing within a second or less.

In the case of cores, however, once the core is covered by liquid metal, the escape of the core gases is limited to the area of the core prints, if the metal is not to be damaged by the passage of bubbles through it. Furthermore, the rate of heating of the core is often greater than that of the mould because it is usually surrounded on all sides by hot metal, and the volume of the core is, of course, much less. All these factors contribute to the internal pressure within the core rising rapidly to high values.

Many authors have attempted to provide solutions to the pressure generated within cores. However, there has until recently been no agreed method for monitoring the rate or quantity of evolved gases that corresponds with any accuracy to the conditions of casting. A result of one method by Naro and Pelfrey (1983) is shown in Figure 10.14. (This method is an improvement on earlier methods in which the water and other volatiles would condense in the pipework of the measuring apparatus, reducing the apparent volume of gases, and thereby invalidating the experimental results.)

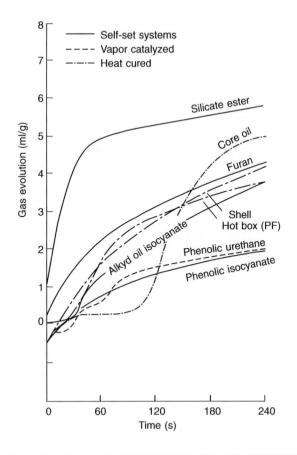

FIGURE 10.14

Gas evolution from carious binder systems using an improved test procedure that includes the contribution from water and other volatiles (Naro and Pelfrey, 1983).

The really important quantity given by these curves is the *rate* of evolution of gas. The rates, of course, are equal to the slope of the curves in Figure 10.14, and are presented in Figure 10.15. Only a few of their results are presented for clarity.

It is sufficient to note that the rates of outgassing are very different for different chemical binder systems. The initial high rate for the silicate is the result of its water content evolved at relatively low temperature during the early stages of the heating of the core. Other organic binders evolve their carbonaceous gases at a variety of temperatures, and apart from core oil, generally show much gentler rates that are more easily vented.

Anyway, taking these recent estimates of Q, the rate of volume of gas generated from a given weight of core in $mLg^{-1}s^{-1}$ (or preferably the identical-sized value in the equivalent unit $Lkg^{-1}s^{-1}$) as being of tolerable relevance to the real situation in castings, we can construct a core outgassing model. We shall roughly follow the method originally pioneered by Worman and Nieman (1973).

We first need to define the concept of permeability. This is a measure of the ease with which a fluid (the core gas in our case) can flow through a porous material (the core in our case). Permeability P_e is defined as the rate of gas flow Q (as a volume per unit time) through a permeable material of area A and length L and driven by a pressure difference ΔP:

$$P_e = QL/A\Delta P$$

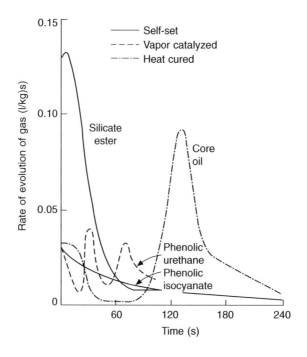

FIGURE 10.15

Rates of gas evolution from various sand binders based on the slopes of the curves shown in Figure 10.13.

The SI units of P_e are quickly seen to be:

$$[P_e \text{ units}] = [\text{litre}/s][m]/[m^2] \ [Pa]$$

$$= Ls^{-1}m^{-1}Pa^{-1}$$

Consider now our simple model of a core shown in Figure 10.16. The measured volume of gas evolved per second from a kilogram of core material is Q. If we allow for that this will have been measured at temperature T_1, usually above 100°C (373 K) to avoid condensation of moisture, and the temperature in the core at the point of generation is T_2, then the volume of gas produced in the core is actually QT_2/T_1 where T is measured in degrees K. For the casting of light alloys, the temperature ratio T_2/T_1 (in K remember) is about 3, whereas for steels it is nearly 6.

If we multiply this by the weight of sand heated by the liquid metal, then we obtain the total volume of gas evolved per second from the core. Thus if the heated layer is depth d, the core area A_c and density ρ, then the volume of gas evolved per second is $QdA_c\rho T_2/T_1$. If the core is surrounded by hot metal, this volume of gas has to diffuse to the print and force its way through the length L of the print of area A_p. We shall assume that the pressure drop experienced by the gas in diffusing through the bulk of the core is negligible in comparison with the difficulty of diffusing through the print (this is of course not always accurate depending on the shape of the core). Considering then the permeability definition for only the pressure drop along the print, we obtain the pressure in the core (above the ambient pressure at the outside tip of the print):

$$P = QdA_c\rho \cdot LT_2/A_pT_1P_e$$

This simple model emphasises the direct role of permeability P_e and of Q, the rate of gas evolution on the generation of pressure in the core. It is to be noted that the high casting temperature for steels is seen to result in values for P

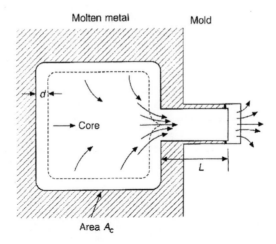

FIGURE 10.16

Core model, showing heated layer thickness d outgassing via its print. In this particular case, metal flash along the sides of the print force gas to exit only from the end area A.

approximately twice that for aluminium alloys. Thus cores in steel castings will be twice as likely to create blows than cores in aluminium alloy castings. For this reason, an enclosed core that would give no problems in an aluminium alloy casting may cause blows when the same pattern is used to make an equivalent bronze or iron casting.

Our model also highlights the various geometrical factors of importance. In particular, the area ratio of the core and the print, A_c/A_p, is a powerful multiplier effect, and might multiply the pressure by anything between 10 and 100 times for different core shapes. Also emphasised is the length L of the core print. If the print is a poor fit, then L may be unnecessarily lengthened by the flashing of the metal into the print so as to enclose the flow path in an even longer tunnel. If the liquid metal completely surrounds the end of the print too, then, of course, all venting of gases is prevented. Gases are then forced to escape through the molten metal, with consequential bubble damage to the casting.

The provision of a vent such as a drilled hole along the length of the print will effectively reduce L to zero; the model predicts that the internal pressure in the core will then be eliminated (the only remaining pressure will, of course, be the much smaller pressure to overcome the resistance to flow through the core itself). The value of vents in reducing blowing from cores has been emphasised by many workers. Caine and Toepke (1966), in particular, estimate that a vent will reduce the pressure inside a core by a large factor, perhaps 5 or 10. This is an important effect, far outweighing all other methods of reducing outgassing pressure in cores.

Vents can be moulded into the core, formed from waxed string. The core is heated to melt out the wax, and the string can then be withdrawn prior to casting. This traditional practice was often questioned as possibly being counter-productive, because of the extra volatiles from the wax that, on melting, soaks into the core. Such fears are seen to be happily unfounded. The technique is completely satisfactory because the presence of the vent completely overrides the effect of the extra volatile content of the core.

A final prediction from the model is the effect of temperature. In theory, a lowering of the casting temperature will lower the internal core pressure. However, this is quickly seen to be a negligible effect within the normal practical limits of casting temperatures. For instance, a large change of 100 K in the casting temperature of an aluminium alloy will change the pressure by a factor of approximately 100/900. This is only 11%. For irons and steels, the effect is smaller still. It and can therefore be abandoned as a useful control measure.

We shall now move on to some further general points.

Cores are almost never made from greensand because the volatile content (particularly water, of course) is too high and the permeability is too low (because of its 5–10% clay and other fines). In addition, the cores would be weak and unable to support themselves on small prints; they would simply sag. If greensand is used at all, then it is usually dried in an oven, producing 'dry sand' cores (their name should perhaps use the past tense and so be more accurately 'dried sand' cores). These are relatively free of volatiles and are mechanically strong. But because they retain the poor permeability of the original greensand they therefore usually require good venting. This is usually time-consuming and labour-intensive.

Sand cores are therefore not nowadays made from dried sand, but generally chemically bonded from clean, washed and dried silica sand that is closely graded in size to maintain as high a level of permeability as is possible. The limit to the size of sand grains and the permeability is set by the requirements of the casting to avoid (1) penetration by the metal and (2) the production of internal surfaces of the casting that are unacceptably rough.

These cores are bonded with a chemical binder that is cured by heat or chemical reaction to produce a rigid, easily handled shape. The numerous different systems in use all have different responses to the heat of the casting process, and produce gases of different kinds, in different amounts, at different times and at different rates, as illustrated in Figures 10.14 and 10.15. These results are not to be taken as absolute in any sense. Manufacturers' products are changing all the time for a variety of reasons: health and safety; economics; commercial; changes in world markets and supplies of raw materials etc. Thus binder formulations change and new systems are being developed all the time. At present, the phenolic urethane systems are among the lowest overall producers of volatiles, which explains their current wide use for intricate cores, for instance in the case of water jackets for automobile cylinder heads and blocks.

Part of the reason for the historical success enjoyed by the phenolic urethane binders is their high strength, which means that the addition levels needed to achieve an easily handled core are low. This is one of the important factors in explaining their position near the bottom of Figure 10.14; the volume of gas evolved is, of course, proportional to the amount of binder present. This self-evident fact is clearly substantiated in the work of Scott et al. (1978), shown in Figure 10.17. (If allowance is made for the fact that these workers used a core sample size of 150 ml, corresponding to a weight of approximately 225 g, then the rate of evolution measurements converted to $1 \text{ kg}^{-1}\text{s}^{-1}$ agree closely with those presented in Figure 10.15. This is despite the significant differences in the techniques. The data in Figure 10.15 may therefore be of more universal application than is apparent at first sight.)

Renewed interest these days is being taken in inorganic binders based on salts. Preferably the salts should contain no water, particularly water bound up in the crystalline structure that could be evolved at the high temperature required for casting. It will repay us all to monitor these developments closely.

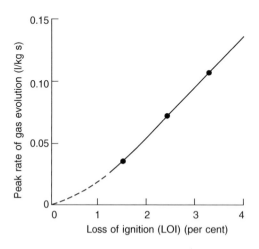

FIGURE 10.17

Increase in the peak rate of outgassing as loss on ignition (LOI) increases.

Data recalculated from Scott et al. (1978).

10.5.3 **CORE BLOW MODEL STUDY**

Campbell (1950) (not the author) devised a useful test that assessed the pressure in cores compared with the pressure in the liquid metal. A modified version of his test is shown in Figure 10.18; the following is a modification and further development of his explanation.

If the mould is filled quickly, then the hydrostatic pressure resulting from depth in the liquid metal is built up more rapidly than the pressure of gas in the core. This dominance of metallostatic pressure can persist throughout filling and solidification. The higher metal pressure effectively suppresses any bubbling of gas through the core at point A as illustrated in Figure 10.18(a). Bubbles will form at A only if the core pressure reaches the metal pressure because either the venting of the core is poor (Figure 10.18(b)) or the core is covered too slowly (Figure 10.18(c)).

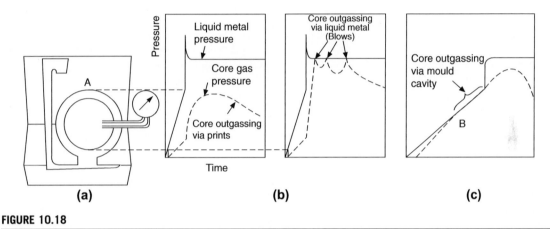

FIGURE 10.18

Effect of fill rate on core outgassing. (a) Fast fill with well-vented core; (b) fast fill but badly vented core; (c) slow fill.

Adapted from Campbell (1950).

If the mould is filled slowly, the gas pressure in the core exceeds the metal pressure at B and remains higher during most of the filling of the mould. Gas escapes through the top of the core during the late stages of the filling, so that, although the metal pressure in Figure 10.18(c) finally exceeds the gas pressure in this case, it is doubtful whether the gas will stop flowing. This is because, in practice, when the liquid metal reaches the top of the casting at A, it is prevented from closing and joining together by the constant evolution of gas. The migration of volatiles ahead of the slowly rising metal front also concentrates the gas source near A, and the creation of a well-established, well-oxidised bubble trail will further hinder the welding and closing together of the meniscus at this point. Gas issuing from the top of a submerged core can be seen to cause the melt to tremble and flutter against the surface of the core, the melt attempting to close together to stop the flow, but repeatedly failing in this attempt as drossy surface films thicken.

Figure 10.18(b) illustrates the case in which the pressure in the core rises only just above the metal pressure, with the result that a limited number of bubbles escape. If the casting has started to solidify, then although the bubble may be forced into the melt, it is unlikely to be able to escape from the upper surface of the casting.

10.5.4 **PREVENTION OF BLOWS**

By far the best solution to the evolution of gases from cores is the use of a sand binder for the core that has little or no evolution of gas as the core becomes hot. This would represent a perfect solution. The best hopes here are the inorganic binders that contain no water of crystallisation. However, the few binders that have so far been developed to meet this criterion are usually not satisfactory in other ways. At the time of writing the perfect core binder has yet to be developed!

In the meantime, one of the best actions to avoid blows from cores (or more occasionally from moulds) is to increase the permeability of the core by the use of a coarser aggregate and/or by the use of venting.

Because the core print is usually the area where all the escaping gas has to concentrate, a simple hole through the length of the print makes a huge impact on the problem, as has been shown previously. For some castings, this can be a complete solution. However, of course, if the vent hole can be continued to the centre of the core this is even better. The further provision of easy escape for gases through the mould and out to the atmosphere is necessary for copper-based and iron and steel castings where the outgassing problem becomes severe because of the higher temperatures. In iron and steel foundries, many readers will have seen the vigorous jets of burning carbon monoxide issuing from vents in moulds for many minutes after pouring. Figure 10.19 illustrates a succession of improved venting techniques.

For low-volume production involving the making of cores by hand, a vent can be provided along a curved path through a core by laying a waxed rope inside whilst it is being made. The core is subsequently heated to melt the wax so that the rope can be pulled out. As we have noted earlier, the additional volatiles from the wax remaining in the core are easily accommodated by the provision of the vent; the vent is an overwhelming benefit.

For more delicate low volume work, the author has witnessed long curved cores for aerospace castings being drilled by hand, using a drill bit fashioned from a length of piano wire held in a three-jaw rotary chuck driven by a small motor. The tip of the high-carbon-steel wire is hammered flat and ground to a sharp point shaped like an arrowhead. The core is drilled by hand in a series of straight lengths, the piano wire drill buzzing quickly through the core. Each hole is targeted to intersect the previous hole, the straight holes emerging on the bends, where the openings are subsequently plugged by a minute wipe of refractory cement (Figure 10.20). The complete vent is checked to ensure that it is continuous, and free from leaks, by blowing smoke in to one of the vent openings, and watching for the smoke to emerge from the far opening. Only when the smoke emerges freely at the far end, and from no other location, is the core accepted for use. It is then stored in readiness for mould assembly.

Occasionally, instead of an opening to the atmosphere, it is necessary to link the outside opening to a vacuum line. This is relatively common practice in gravity die (permanent mould) casting, increasing the efficiency of the extraction of gases from a resin-bonded sand core. However, the evolution of volatiles from the binder creates problems by condensing as sticky resins and tars in the vacuum line, so that, for long production runs, regular attention is required to avoid blockage, often dictating the timing of the withdrawal of the tooling for maintenance.

The reader is advised caution with regard to the application of a vacuum line to aid venting. The author once tried this on an extensive thin-section core with a single small area print around which was poured liquid stainless steel at 1600°C. The resulting rapid buildup of pressure was so dramatic that it blew off the vacuum connection with a bang! However, the discerning reader will notice the extreme circumstances described here, and rightly conclude that in this case the author was testing the patience of Providence.

The application of vacuum connections to core prints is widely practised in the gravity die casting of Al alloy automotive cylinder heads. The water jacket core is particularly difficult to vent without such extreme measures. However, because of the wide use of resin-based binders for this core, the vacuum line quickly becomes blocked by condensing resins and tars. Thus after perhaps only a dozen castings the die may have to be withdrawn from service to clean out the vacuum lines.

If there is an option, it is far better to arrange that the core vents through prints that are directed vertically upwards (Figure 10.21). This is because as the melt rises in the mould, the volatiles migrate through the core ahead of the metal, concentrating in the last part of the core. If the core is vented at its base, this is a potential disaster. The volatiles are too far from the print and will continue to be pushed ahead, finally being pushed into the form of an eruption of bubbles from the top of the core. This problem can be reduced by covering the core with liquid metal as quickly as possible. Venting from the base is then given its best chance.

Even so, a print allowing outgassing from the top of the core is ideal. If a vent cannot be provided up the centre of the top print, a top print is still valuable, even though it may contain no central vent because the volatiles will travel up the core surface. This can be seen on core prints that emerge from the tops of aluminium alloy castings. The melt is seen to

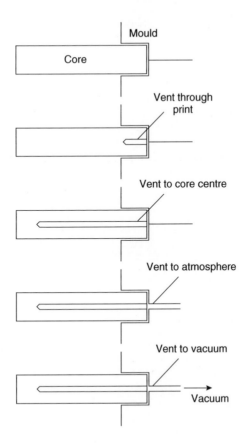

FIGURE 10.19

Venting of a core, illustrating progressively more effective techniques.

flutter, trembling against the side of the core print as gas rushes up in the form of minute, high-frequency waves, causing ripples to radiate out across the surface of the melt.

The provision of soft ceramic paper gaskets, preferably with a central hole, shaped like a washer, placed on the end of core prints is an excellent provision for the escape of gases. This simple remedy prevents the melt from flashing over the end of the print to block the vent (Figure 10.21). The compressible washer allows for the sealing of the core print against ingress by liquid metal, but allows closure of the mould without danger of the crushing of the core.

Finally, if the core can be covered quickly with liquid metal, and the pressure in the metal quickly raised to be at all times greater than the internal pressure generated inside the core, then bubble formation will be suppressed. Thus simply filling the mould faster is often a quick and complete solution to a core blow problem. The provision of an additional top feeder to increase hydrostatic pressure needs care because if the feeder has large volume the delay in the rise of pressure to fill it may be counter-productive. If feeding of the casting is not really required, the sprue and pouring basin can provide the early pressurisation that is needed, it would be better to leave well alone and not be tempted to add the top feeder.

Eventually, it is hoped, we can look forward to the day when computer simulation will provide an accurate description of each core and mould, allowing in detail for the effect of outgassing of cores of intricate geometries and the complicating effect of rate of filling. A welcome start has been made by Maeda et al. (2002), who demonstrated a computer simulation of the flow of gas through an aggregate core, and Starobin et al. (2010), who investigated water

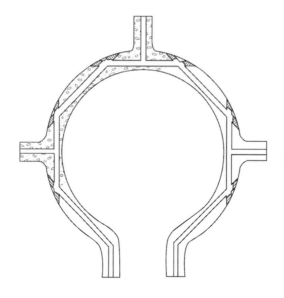

FIGURE 10.20

Drilled and plugged holes to vent a narrow ring-shaped core.

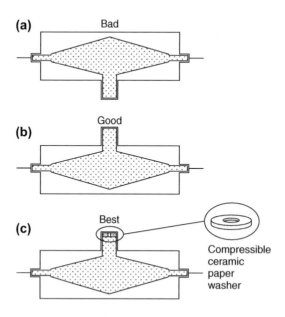

FIGURE 10.21

(a) Downwards venting core vents poorly, compared with (b) an upward vented core. (c) The cored compressible washer prevents the closure of the vent by ingress of metal. A further benefit would be a central hole though the core print, preferably extending into the centre of the core.

jacket cores for cylinder blocks. Perhaps we can now look forward to such studies becoming a commonplace feature of the design of a new casting.

10.5.5 SUMMARY OF BLOW PREVENTION

The various actions that can be taken to eliminate blow holes are:

1. Use a core binder that has minimal outgassing (i.e. minimise the binder so far as possible) or delayed outgassing (change to a late outgassing material).
2. Improve the permeability of the core aggregate.
3. Provide vents, particularly through the print.
4. If possible, vent the core from its top-most point.
5. Apply a vacuum to the core prints.
6. Redesign the core (or invert the casting) to avoid upward-pointing features.
7. Fill quickly to cover the core with liquid metal as soon as possible, before its internal pressure rises high enough to force a bubble into the melt.
8. Use a high hydrostatic head of liquid metal to suppress the evolution of gas from the surface of the core.
9. Raise the casting temperature. This is a last resort. The technique is quite commonly applied. It has the effect of giving more time for the core to outgas through the liquid metal before a solidified shell can form, so that bubbles will not be trapped by the growth of the solid skin of the casting. The approach can only be recommended with some reluctance, perhaps justifiable in an emergency, or perhaps only for grey iron cast with dry sand cores. In most situations, the passage of the bubbles through the melt will, even if the bubbles escape, result in damage to the casting in the form of the creation of bubble trails, leading possibly to both leakage and dross defects.

10.6 RULE 6: AVOID SHRINKAGE DAMAGE
10.6.1 DEFINITIONS AND BACKGROUND

Before getting launched into this section, we need to define some terms.

There is widespread confusion in parts of the casting industry, particularly in investment casting, between the concepts of *filling* and *feeding* of castings. It is essential to separate these two concepts.

Filling is self-evidently the short period during the pour and refers to the filling of the filling channels themselves (sometimes called the *priming* of the running system) and the filling of the mould cavity. This may only last seconds or minutes.

Feeding is the long, slow process that is required during the contraction of the liquid that takes place on freezing. This process takes minutes or hours depending on the size of the casting. It is made necessary as a result of the solid occupying less volume than the liquid, so the difference has to be provided from somewhere. This contraction on solidification is a necessary consequence of the liquid being a structure resembling a random close-packed array of atoms, compared to the solid, which has the denser regular close packing in a structure known as a crystal lattice.

Chapter 7 is required reading. It reminds the reader that three separate stages of contraction accompany the cooling of a liquid metal down to room temperature, but the linear contraction of the liquid and of the solid are of little consequence for feeding; we shall concentrate here on the solidification contraction. This contraction gives a volume deficit that can result in problems for the casting.

The volume contraction during freezing can be fed by several mechanisms, but mainly by liquid feeding from a feeder. If for some reason liquid metal to feed the contraction cannot easily be supplied, the contraction starts to act on the surrounding solid, drawing it in by plastic deformation (solid feeding) and at the same time stretching the liquid metal. This elastic expansion of the liquid is naturally accompanied by the development of a hydrostatic tension, a negative pressure which can provide the driving force for the nucleation and growth of porosity. These concepts of negative pressure are not easily understood. The reader is strongly recommended to read Chapter 7, where this phenomenon is described in detail.

10.6.2 FEEDING TO AVOID SHRINKAGE PROBLEMS

To allow ourselves the luxury of some repeated emphasis and further definitions: To provide for the fact that extra metal needs to be fed to the solidifying casting to compensate for the contraction on freezing, it is normal to provide a separate reservoir of metal. We shall call this reservoir a *feeder* because its action is to *feed* the casting i.e. to compensate for the solidification shrinkage (obvious really!).

In much casting literature the reservoir is known, less than obviously, as a *riser*, and worse still, may be confused with other channels that communicate with the top of the mould, such as vents, or whistlers because metal *rises* up these openings too. The author reserves the name *riser* for (1) the special kind of feeder that is connected to the side of the casting via a slot gate, and in which metal rises up at the same time as it rises in the mould cavity, and (2) the 'surge riser' designed to avoid the surge of first metal into the ingate.

If a feeder is *not* placed on a chunky casting that definitely requires feed metal, then we can expect a significant shrinkage problem. Figure 10.22 illustrates for an Al casting the absence of a feeder would result in a pore volume of up to 7%. That is a big porosity problem. However, as the feeder size increases in proportion to the casting, the resulting porosity reduces, eventually reaching a minimum at a feeder-to-casting modulus ratio of about 1.25. The porosity in the casting at this minimum depends on the gas content; at zero gas, the casting will be sound. At higher gas levels, the porosity rises, but it is clear that gas porosity does not raise the total porosity level so significantly as an undersized feeder. Although there is much discussion concerning the dangers of overfeeding of castings, the figure illustrates that an oversize feeder is far less damaging than an undersized feeder.

It is most important to be clear that the filling system (sometimes called the running system) is not normally required to provide any significant feeding. The filling system and the feeding system have two quite distinct roles: one fills the casting so its role is completed quickly, and the other feeds the shrinkage during solidification, whose role takes the much

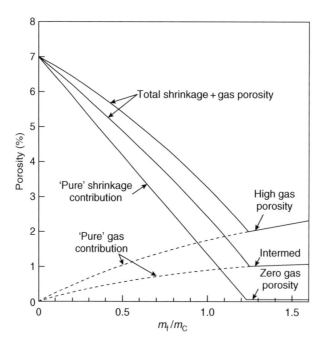

FIGURE 10.22

Generalised relation between gas and shrinkage as feeder size is increased in terms of the modulus ratio, based on particular results by Rao et al. (1975), shown in Figure 7.5.

longer time that the casting requires to freeze. (On occasions it is possible and valuable to carry out some feeding via the filling system, but this is somewhat unusual and requires special precautions. We shall deal with this later.)

Where computer modelling is not carried out, the following of the *seven feeding rules* by the author is strongly recommended. Even when computer simulation is available, the seven rules will be found to be good guidelines. For the computer itself, the following of Tiryakioglu's reduced rules constitutes the most powerful logic and is recommended, although the same rules also constitute a useful check for those of us who determine feeder sizes by pen and paper.

In addition to observing all the requirements of the rules for feeding, the use of all *five mechanisms* for feeding (as opposed to only liquid feeding) should also be used to advantage. These, once again, are described at length in Chapter 7. They will be found to be especially useful when attempting to achieve soundness in an isolated boss or heavy section where the provision of feed metal by conventional techniques may be impossible. However, a reminder of the attendant dangers of the use of solid feeding is presented later.

10.6.3 THE SEVEN FEEDING RULES

The *feeding rules* have already been briefly outlined in volume one, but are presented here in greater detail.

It is essential to understand that all of the rules must be fulfilled if castings are to be produced that require soundness, accuracy and high mechanical properties. The reader must not underestimate the scale of this problem. The breaking of only one rule may result in ineffective feeding and a defective casting. The wide prevalence of porosity in castings is a sobering reminder that solutions are often not straightforward. Because the calculation of the optimum feeder size is therefore so fraught with complications, it is dangerous if calculated wrongly, and the casting engineer is strongly recommended to consider whether a feeder is really necessary at all. (For instance, many modest feeding problems are easily dealt with by the provision of a chill or cooling fin rather than a feeder. Many thin-walled castings do not even require this because thin sections in general act good naturedly as though self-feeding.)

Feeding rule 1: Do not feed (unless necessary)

This first and most important question relating to the provision of a feeder on a casting is 'Should we have a feeder at all?' It is a question well worth asking.

For instance, most castings I see being produced in foundries are covered in feeders when it is evident there is no real shrinkage problem. The so-called 'shrinkage' being actually a mass of bifilms and bubbles because of a poor filling system.

Another practice which deserves a separate feeding rule of its own is the topping up of feeders with hot metal after the mould has been filled. This is a seriously damaging practice; the minimal aid to feeding temperature gradient by this addition of hot metal is vastly countered by the serious damage done to the casting by the surface oxide on the feeder being dragged down by the falling metal. This entrained oxide is carried down through the feeder and enters the casting, producing serious defects that *appear* to be shrinkage just below the feeder. Thus if a feeder is to be provided to a casting, we should adopt the additional feeding rule 'Never top up feeders'.

The avoidance of feeding is to be greatly encouraged, in contrast to the teaching in most traditional foundry texts. There are several reasons to avoid the placing of feeders on castings. The obvious one is cost. They cost money to put on and money to take off. In addition, many castings are actually impaired by the inappropriate placing of a feeder. This is especially true for thin-walled castings, where the filling of the feeder diverts metal from the filling of the casting, with the creation of a misrun casting.

Probably half of the small- and medium-sized castings made today do not need to be fed. This is especially true as modern castings are being designed with progressively thinner walls. In fact, as we have already mentioned, the siting of heavy feeders on the top of thin-walled castings is positively unhelpful for the filling of the casting because the slow filling of the feeder delays the filling of the thin sections at the top of the casting, with consequent misruns. Sometimes, as a result of the presence of a feeder, the casting suffers delayed cooling, impaired properties, and even segregation problems. Finally, it is easy to make an error in the estimation of the appropriate feeder size, with the result that the casting can be more defective than if no feeder were used at all.

As a general rule, therefore, it is best to avoid the placing of feeders on thin-walled castings. The low feed requirement of thin walls can be partly understood by assuming that of the total of 7% solidification shrinkage in an aluminium casting, 6% is easily provided along the relatively open pathway through the growing dendrites. Only about the last 1% of the volume deficit is difficult to provide. Thus if this final percentage of contraction on freezing has to be provided by solid feeding, moving the walls of the casting inwards, this becomes, at worst, 0.5% per face, which on a 4 mm thick wall is only 20 μm. This small movement is effectively not measurable because it is less than the surface roughness. If this deficit does appear as internal porosity, then it is in any case rather limited, and normally of little consequence in commercial castings. (It may require some attention in castings for safety-critical and aerospace applications.)

The other feature of thin-walled castings is that considerable solidification will often take place during pouring. Thus the casting is effectively being fed via the filling system. The extent to which this occurs will, of course, vary considerably with section thickness and pouring rate. If the section thickness (or rather, modulus; see later) of the filling system is similar to that of the casting, then feeding via the filling system might be a valuable simplification and cost saving. This important and welcome benefit to cost reduction is strongly recommended.

Because the calculation of the optimum feeder size is so fraught with complications, is dangerous if calculated wrongly, and costs money and effort, the casting engineer is strongly recommended to consider whether a feeder is really necessary at all. Rule 1 is probably the most valuable rule.

Of the remainder of castings that do suffer some feeding demand, many could avoid the use of a feeder by the judicious application of chills or cooling fins. The general faster freezing of the casting might then allow the provision of sufficient residual feeding via the filling system as indicated previously. Minor revisions, opening up restrictions to the feed path along the length of the filling system may provide valuable (and effectively 'free') feeding from the pouring basin.

However, this still leaves a reasonable number of castings that have heavy sections, isolated heavy bosses, or other features which cannot easily be chilled and thus *do* need to be fed with the correct size of feeder. The remainder of the rules are devoted to getting these castings right.

Feeding rule 2: The heat-transfer requirement

Chvorinov's heat transfer requirement for successful feeding can be stated as follows: *The feeder must solidify at the same time as, or later than, the casting.*

Nowadays, this problem can be solved by computer simulation of solidification of the casting. Nevertheless, it is useful for the reader to have a good understanding of the physics of feeding, so that computer predictions can be checked. We need to keep in mind that the computer simulation may not be especially accurate because much of the basic input data are sometimes not well known or wrongly inputted. Also, of course, computer time could be usefully avoided in sufficiently simple cases. In this chapter we shall concentrate on approaches which do not require a computer.

The freezing time of any solidifying body is approximately controlled by its ratio (volume)/(cooling surface area) known as its modulus, m. Thus the problem of ensuring that the feeder has a longer solidification time than that of the casting is simply to ensure that the feeder modulus m_f is larger than the casting modulus m_c. To allow a factor of safety, particularly in view of the potential for errors of nearly 20% when converting from modulus to freezing time, it is normal to increase the freezing time of the feeder by 20% i.e. by a factor of 1.2. Thus the heat-transfer condition becomes simply

$$m_f > 1.2 \, m_c \tag{10.1}$$

It is important to notice that the modulus has dimensions of length. Using SI units, it is appropriate to use millimetres. (Take care to note that in French literature, the normal units are centimetres, and in the United States at the present time, to the despair of all those promoters of the welcome logic of SI units (the *Systeme International*), a confusing mixture of millimetres, centimetres, inches and feet. It is essential therefore to quote the length units in which you are working.)

Figure 10.22 confirms the optimum feeder modulus of factor approximately 1.2 times the casting modulus. The danger of insufficient feeding is clear, whereas overfeeding is clearly preferable. The contribution of gas porosity to the total unsoundness of the casting is also seen superimposed on the feeding response.

The modulus of a feeder can be artificially increased by the use of an insulating or exothermic sleeve. It can be further increased by an insulating or exothermic powder applied to its open top surface after casting. Recent developments in such exothermic additions have attempted to ensure that after the exothermic reaction is over, the spent exothermic material continues in place as a reasonable thermal insulator. These products are constantly being further developed, so the manufacturer's catalogue should be consulted when working out minimum feeder sizes when using such aids. However, as a guide as to what can be achieved at the present time, a cylindrical feeder in an insulating material is only 0.63D in diameter compared with the diameter D in sand. This particular insulated feeder therefore has only 40% of the volume of the sand feeder. Useful saving can therefore be made, but have, of course, to be weighed against the cost of the insulating sleeve and the organisational effort to purchase, store and schedule it etc. However, a further benefit that is easily overlooked from the use of a more compact feeder is the faster pressurisation (1) of thin sections that may aid filling and so reduce losses from occasional incomplete filling of mould cavities and (2) of cores to reduce the chance of blows.

When working out the modulus of the casting, it is necessary to consider which parts are in good thermal communication. These regions should then be treated as a whole, characterised by a single modulus value. Parts of the casting that are not in good thermal communication can be treated as separate castings.

For instance, castings of high thermal conductivity such as those of aluminium- and copper-based alloys can nearly always be treated as a whole because when extensive thin sections cool attached thicker sections and bosses, the thin sections act as cooling fins for the thicker sections. Conversely, of course, the thick sections help to maintain the temperature of thinner sections. The effect of thin sections acting as cooling fins extends for up to approximately 10 times the thickness of the thin section.

However, for castings of low thermal conductivity materials such as steel and nickel-based alloys (and surprisingly, the copper-based Al-bronze), practically every part of the casting can be treated as separate from every other. Thus a complex product can be dealt with as an assembly of primitive shapes: plates, cubes, cylinders etc. (making allowance, of course, for their common mating faces, which do not count as cooling area in the modulus estimate).

Table 5.3 lists some common primitive shapes. Familiarisation with these will greatly assist the estimation of appropriate feeder modulus requirements.

Feeder rule 3: Mass transfer (volume) requirement

The second and widely understood and well-used rule, usually known as the volume criterion is as follows: *The feeder must contain sufficient liquid to meet the volume-contraction requirements of the casting.*

At first sight, it may seem surprising that when the heat transfer requirement discussed previously is satisfied then the volume requirement is not automatically satisfied also. However, this is definitely not the case. Although we may have provided a feeder of such a size that it would theoretically contain liquid until after the casting is solid, in fact it may still be too small to deliver the volume of feed liquid that the casting demands. Thus it will be prematurely sucked dry, and the resulting shrinkage cavity will extend into the casting.

Different metals and alloys need significantly different amounts of feed metal. Figure 10.23(a) illustrates a section of a feeder on a plate casting in which the required shrinkage volume is just nicely concentrated in the feeder. This is the success we all hope for. However, success is not always easily achieved, and Figure 10.23(b),(c) and (d) show the complication posed by the different shrinkage behaviour of different alloys. The pure Al and the Al-12Si alloy are both short freezing range, and contrast with the Al-5Mg alloy which is a long freezing range material.

Some additional points of complexity in the operation of feeders in real life need to be emphasised.

1. The Mg-containing alloy in Figure 10.23(d) will almost certainly contain some fine, scattered microporosity that will have acted to reduce the apparent shrinkage cavity. Thus in practice the demand from the feeder will vary with the gas content of the metal.
2. The complicated form of the pipe in Al-12Si alloy (Figure 10.23(c) and (e)) almost certainly reflects the presence of large oxide films that were introduced by the pouring of the castings. These large planar defects fragment both the heat flow and the mass flow in the feeder, and the short freezing range and surface tension conspire to round off the cavities in the separated volumes of liquid isolated by the planes of oxide. In addition, the oxide, together with the solidifying crust on the top surface of the feeder also has some strength and rigidity, again complicating the collapse of the feeder top and influencing the shape of the shrinkage pipe as it, and its associated oxide skin, gradually expands downwards. These effects are additional reasons for the 20% safety factor often used for the calculation of feeder sizes. Feeders often do not have the simple carrot-shaped shrinkage pipe predicted by the

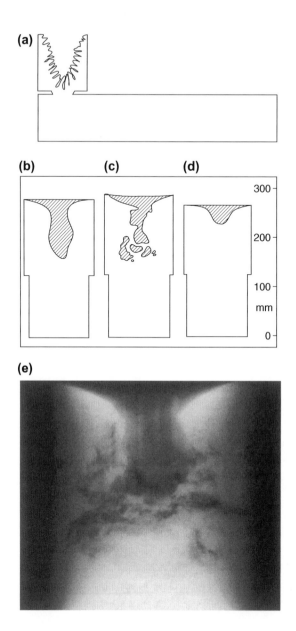

FIGURE 10.23

Cross-section of (a) simple plate casting, nicely fed, with all its shrinkage porosity concentrated in the feeder; (b) 99.5Al; (c) Al-12Si; (d) Al-5Mg; (e) radiograph of Al-12Si alloy feeder.

Courtesy Bird (1989).

computer. The radiograph of the feeder in Figure 10.23(e) is an excellent example of the action of large bifilms fracturing and diverting the melt draining from a feeder.

Figure 10.23 illustrates that normal feeders are relatively inefficient in the amount of feed metal that they are able to provide. This is because they are themselves freezing at the same time as the casting, depleting the liquid reserves of the reservoir. Effectively, the feeder has to feed both itself and the casting. We can allow for this in the following way. If we denote the efficiency e of the feeder as the ratio (volume of available feed metal)/(volume of feeder, V_f) then the volume of feed metal is, of course, eV_f. Because the liquid contracts by an amount α during freezing, then the feed demand from both the feeder and casting together is $\alpha(V_f + V_c)$, and hence:

$$eV_f = \alpha(V_f + V_c) \tag{10.2}$$

or:

$$V_f = (e - \alpha)V_c \tag{10.3}$$

For aluminium where $\alpha = 7\%$ approximately (see Table 7.2 for values of α for other metals), and for a normal cylindrical feeder of $H = 1.5D$ where $e = 14\%$, we find:

$$V_f = V_c \tag{10.4}$$

That is, there is as much metal in the feeder as in the casting! This is partly why the yield (measured as the weight of metal going into a foundry divided by the weight of good castings delivered) in many aluminium foundries is rarely above 50%. Metal in the running system, and scrap allowance will reduce the overall yield of good castings even further so that overall yields are often closer to 45%. (In comparison, the economic benefits of higher-yield casting processes such as counter-gravity casting, in which metallic yields of 80–90% are common, appear compellingly attractive, especially for high-volume foundries.)

For steels, the value of α lies between 3 and 4%, depending on whether solidification is to the body-centred cubic or face-centred cubic structures. For pure Fe-C steels the face-centred cubic structure applies above 0.1% carbon where the melt solidifies to austenite. For $\alpha = 4$ and $e = 14\%$, Eqn (10.3) gives:

$$V_f = 0.40V_c$$

and for steel that freezes to the body-centred cubic structure (delta ferrite) with $\alpha = 3$, and using a feeder of 14% efficiency we have:

$$V_f = 0.27V_c$$

Thus, compared with Al alloys, the smaller solidification shrinkage of ferrous metals reduces the volume requirement of the feeder considerably. For graphitic cast irons, the value reduces even further of course, becoming approximately zero in the region of 3.6–4.0% carbon equivalent. Curiously, a feeder may still be required because of the difference in *timing* between feed demand and graphite expansion, as will be described later.

The interesting reverse tapered feeder (Figure 10.24(c)) has been promoted for many years (Heine, 1982; Creese and Xia, 1991) and is currently widely used for ductile iron castings. Even so, the reader needs to be aware that in the opinion of the author, Figure 10.24 may not be as accurate we would like. At this time, the extent of the uncertainties is not known following the recent work of Sun and Campbell (2003). This investigation of the effect of positive and negative tapers on the efficiencies of feeders, found that the reverse tapered feeder (Figure 10.24(c)) appeared to be less efficient than parallel sided cylindrical feeders, or even feeders with a slight positive taper. These doubts are an unwelcome sign of the extent of our ignorance of the best feeder designs at this time.

Whether the size of the feeder is dictated by the thermal or volume requirement is related to the geometry of the casting. Figure 10.25 shows a theoretical example, calculated neglecting non-cooling interfaces for simplicity. Curve A is the minimum feeder volume needed to satisfy the thermal condition $m_f = 1.2m_c$; and curve B is the minimum feeder volume needed to satisfy the feed demand criterion based on 4% volume shrinkage and 14% metal utilisation from the feeder. Figure 10.25 reveals that chunky steel plates up to an aspect ratio of about 6 or 7 length to thickness are properly

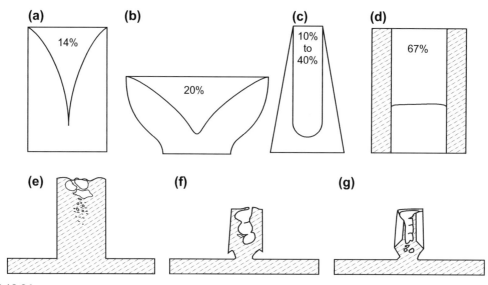

FIGURE 10.24

Metal utilisation of feeders of various forms moulded in sand. The (a) cylindrical and (b) hemispherical heads have been treated with normal feeding compounds; (c) efficiency of reverse tapered feeders depends on detailed geometry (Heine, 1982; Heine and Uicker 1983); (d) exothermic sleeve (Beeley, 1972); metal utilisation for ductile iron plates with (e) cylindrical sand feeder; (f) insulating feeder; and (g) cruciform exothermic feeder. (*After Foseco, 1988.*)

fed by a feeder dictated by freezing time requirements. However, thin section steel plates above this critical ratio always freeze first, and so require a feeder size dictated by volume requirements.

In fact the shape of the shrinkage pipe in the feeder is likely to be different for each of these conditions. For instance, the feeder efficiencies shown in Figure 10.24 are appropriate for the feeding of chunky castings because the continuing demand of feed metal from the casting until the feeder itself is almost solid naturally creates a long, tapering shrinkage pipe, resembling a carrot.

In the case of the more rapid solidification of thin castings, the relatively large diameter feeder needed to provide the volume requirement will give a shallowly dished shape in the top of the feeder because the feed metal is provided early, before the feeder itself has solidified to any great extent. The efficiency of utilisation of the feeder will therefore be expected to be significantly higher, as confirmed by Figure 10.25.

Research is needed to clarify this point. In the meantime, the casting engineer needs to treat the present data with caution, and conclusions from Figures 10.30 and 10.31, for instance, have to be viewed as illustrative of general principles rather than numerically accurate. Clearly, it is desirable to achieve smaller, more cost-effective feeders. The change of feeder efficiency depending on whether freezing or volume requirements are operating requires more work to clarify this uncertainty. In the meantime, this problem illustrates the power of a good computer simulation to avoid the necessity for simplifying assumptions.

A further use of feeders where the casting engineer requires care is the use of a blind feeder sited low down on the casting. The problems are compounded if such a low-sited blind feeder is used together with an open feeder placed higher. It must be remembered that during the early stages of freezing the top feeder is supplying metal to the blind feeder as well as the casting. The blind feeder has to be treated as though it is an integral part of the casting. The size of the top feeder needs to be enlarged accordingly.

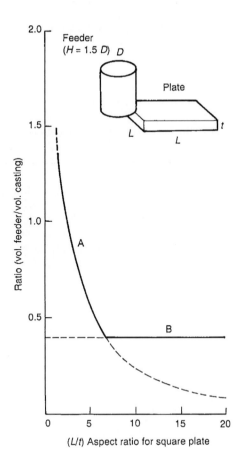

FIGURE 10.25

Feeder volume based on a feeder moulded in sand, and calculated neglecting non-cooling interfaces for simplicity. Curve A is the minimum feeder volume to satisfy $m_f = 1.2m_c$ and curve B is the minimum volume to satisfy the feed demand of 4% volume shrinkage and 14% utilisation of the feeder.

The blind feeder only starts to operate independently when the feed path from the top feeder freezes off. This point occurs when the solidification front has progressed a distance $d/2$, where d is the thickness of the thickest casting section between the top and the blind feeders. Thus the volume of the blind feeder is now reduced by the $d/2$ thickness layer of solid that has already frozen around its inner walls.

If this caution were not already enough, a further pitfall is that the thickness of solid shell inside the blind feeder may now exceed the length of the atmospheric vent core, creeping over its end and sealing it from the atmosphere. The blind feeder is therefore prevented from breathing, and is unable to provide any feed metal. In fact it now works in reverse, sucking metal out of the casting.

There are therefore subtleties in the operation of blind feeders that make success illusive. It is easy to make a mistake in their application, and the correct operation of the atmospheric vent is not always guaranteed, so it is difficult to recommend their use on smaller castings. For larger castings, where the size of the blind feeder is large, the collapse of the top of the feeder is more predictable, so that the action of the blind feeder becomes more reliable.

Whereas the size of feeders for alloys such as those based on alloys that shrink in a conventional fashion on freezing are straightforward to understand and work with, graphitic cast irons are considerably more complicated in their behaviour. They are therefore more complicated to feed, and the estimate of feeder sizes subject to more uncertainty.

The amount of graphite that is precipitated depends strongly on factors that are not easy to control, particularly the efficiency of inoculation. In addition, the expansion of the graphite can lead to an expansion of the casting if the mould and its container are not rigid. Because a mould and a moulding box can never be perfectly rigid, this leads to a larger volume casting that requires more metal to feed it, with the danger that the feeder may now be insufficient to provide this additional volume. Shrinkage porosity as a result of mould dilation is a common feature of iron castings.

One of the ways to reduce (but of course, never to eliminate) this problem is to use very dense, rigid mould in rigid, well-engineered boxes. Furthermore, the expansion of the graphite can be accommodated without swelling the casting by allowing the excess melt to exude out of the casting and back into the feeder. The provision of a small feeder is therefore essential to the production of many geometries of small iron castings. This small feeder provides essential feed metal during the time that normal shrinkage occurs, but also acts as an overflow during graphite expansion. It follows therefore that subsequent examination of the feeder indicates, mysteriously, that no net volume of liquid has flowed out from the feeder. If, as a result of appearing to have provided no feed metal, and concluding that the feeder had served no purpose, the feeder was removed; the result would have been a porous casting!

Ductile cast irons commonly use a reverse-tapered feeder such as that shown in Figure 10.24(c). If a feeder remains full for too long, its top would freeze over, isolating it from the atmosphere and so preventing the delivery of any liquid. It is logical therefore to encourage the feeder to start feeding almost immediately, by tapering the feeder so as to concentrate the action of shrinkage throughout the casting all on the small area at the top of the feeder. In this way, the level of liquid in the top of the feeder falls quickly, becoming surrounded by hot metal, so very soon there is no danger of it freezing over. For this to happen, it is essential that no feeding continues to be provided via the running system. This would keep the feeder full for too long. Thus it is necessary to design the ingate to freeze quickly. The feeder then works well.

This feeding technique, although at this time used exclusively in the ductile iron industry so far as the author is aware, would be expected to be applicable to a wide variety of metals and alloys.

Feeding rule 4: The junction requirement

The junction problem is a pitfall awaiting the unwary. It occurs because the simple act of placing a feeder on a casting creates a junction. As we shall see in Section 12.4.8, a junction with an inappropriate geometry will lead to a hot spot, with the danger that the hot spot may cause a shrinkage cavity that extends into the casting.

We shall see later (in the section on filling system design, dealing with the design of ingates) that is it easy to create a junction, planting either a feeder or a gate on to a casting, only to find that it has created a hot spot, and so leads to a localised feeding problem at the junction. After the feeder or gate is cut off, a shrinkage cavity is found extending into the casting. This occurs when the ratio of modulus of the addition (the feeder or gate) to the modulus of the casting is close to 1:1.

The simplest example illustrating the problem clearly is that of the feeding of a cube. The cube casting has the reputation of being notoriously difficult to feed. This is because the casting technologist, carefully following rule 2, calculates a feeder of 1.0 or perhaps 1.2 times the modulus of the cube. If the cube has side length D, then the feeder of 1:1 height to diameter ratio works out to have a diameter of $1.2D$. Thus the cube appears to require a feeder of rather similar volume sitting on top. However, the cube and its feeder are now a single compact shape that solidifies as a whole, with its thermal centre in the centre of the new total cast shape i.e. approximately in the centre of the junction. The combination therefore develops a shrinkage cavity at the junction, the hot spot between the casting and feeder. When the casting is cut off from the feeder, the porosity that is found is generally called '*under-feeder shrinkage porosity.*' This rather pompous pseudo-technical jargon clouds the clear conclusion that *the feeder is too small.*

The junction rules, developed later in the section on design of ingates, indicate that to avoid creating a hot spot we need to ensure that the feeder actually has twice the modulus of the casting. Thus the cube should have had a feeder of side length $2D$. The shrinkage cavity would then have been concentrated only in the feeder.

The junction problem is a widely overlooked requirement. The use of Chvorinov's equation systematically gives the wrong answer for this reason, so the junction requirement is often found to override Chvorinov. This is an important result that has caused much trouble for methoding engineers over the years and requires us all to update our thinking.

However, in some cases, the junction problem can be avoided. The simplest solution is not to place the feeder directly on the casting so as to create a junction. It happens that this rule is not easily applied to a cube because there is no alternative site for it.

However, in the case of a plate casting, there are options. The feeder should not be placed directly on the plate, but should be placed on an extension of the plate.

The general rules to solve the junction problem are therefore as follows:

1. Appendages such as feeders and ingates should not be planted on the casting so as to create a T- (or an L-) junction (although the L-junction is rather less detrimental than the T-junction). They are best added as extensions to a section, as an elongation to a wall or plate, effectively moving the junction off the casting.
2. If there is no alternative to the placing of the feeder directly on the casting, then to avoid the hot spot in the middle of the junction, the additional requirement that the feeder must meet is, if a T-junction,

$$m_f > 2m_c \qquad\qquad (10.5)$$

or if an L-junction

$$m_f > 1.33m_c \qquad\qquad (10.6)$$

The value of the constants is taken from Sciama (1974).

Note that no safety factor of 1.2 has been applied to these feeder sizes. This is because the shrinkage cavity does not occur exactly at the geometric centre of the freezing volume trapped at the thermal centre; the cavity 'floats' to the top, and the feed liquid finds its level at the base of the isolated region. Thus the final shrinkage cavity is naturally displaced above the junction interface, giving a natural 'built-in' safety factor.

Note that we have assumed that the feeder is above the casting to feed downwards under gravity. This is the recommended safe way to use feeders. If the feeder (or large gate) were placed *below* the casting, gravity would now act in reverse, so that any shrinkage cavity caused by the junction would float into the casting (in other words, the residual liquid metal in the casting drains into the feeder). This action illustrates one of the dangers of attempting to feed uphill.

Although it is not a good idea to make the feeder any larger than is really required, if it is only marginally adequate, the root tip of porosity seen in Figures 10.29 and 10.30 may on occasions just enter the casting, and may therefore be unacceptable. This necessitates the application of a safety factor, giving a feeder of larger size on average, but still just acceptable even when all the variables are loaded against it. It is common to use the safety factor 1.2 for the estimation of its modulus.

Feeding rule 5: Feed path requirement (the communication requirement)

There must be a path to allow feed metal to reach those regions that require it. It is clearly no use having feed metal available at one point on the casting, unable to reach a more distant point where it is needed. Clearly there has to be a way through. The reader can see why this criterion has been often overlooked as a separate rule; the communication criterion appears self-evident! Nevertheless, it does have several geometrical implications which are not so self-evident and which will be discussed.

In a valuable insight, Heine (1968) has drawn attention to the highest-modulus regions in a casting are either potential regions for shrinkage porosity if left unfed, or may be feeding paths if connected to feed metal. He recommends the identification of feed paths that will transport feed metal through castings of complex geometry, such as the hot spots at the T-junctions between plates (Figure 10.26). (He also draws attention to the fact that certain locations are never feed paths. These include corners or edges of plates or the ends of bars and cylinders.)

The various ways to help to ensure that feed paths remain open are considered in this section.

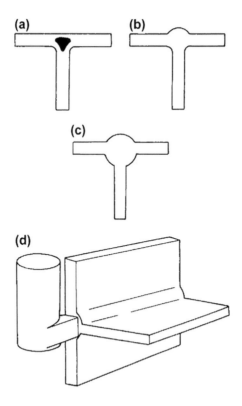

FIGURE 10.26

(a) T-junction with normal concave fillet radius; (b) a marginal improvement to the feed path along the junction; (c) convex fillets plus pad that doubles feeding distance along the junction; and (d) practical utilisation of a T-junction as a feed path (Mertz and Heine, 1973).

Directional solidification towards the feeder

If the feeder can be placed on the thickest section of the casting, with progressively thinner sections extending away, then the condition of progressive solidification towards the feeder can usually be achieved.

A classical method of checking this according to Heuvers can be used in which circles are inscribed inside the casting sections. If the diameters of the circles increase progressively towards the feeder, then the condition is met (Figure 10.27). Lewis and Ransing (1998) draw attention to the fact that Heuvers' technique is only two-dimensional, and that the condition would be more accurately represented in three dimensions by a progressive change in the radius of a sphere, effectively equivalent to the progressive increase in casting modulus towards the feeder. The fundamental reason for tapering the casting in this way is to achieve taper in the liquid flow path (Sullivan et al., 1957). For convenience, we shall call this the modulus gradient technique.

Failure to provide sufficient modulus gradient towards the feeder can be countered in various ways by (1) either re-siting the feeder (2) providing additional feeder(s) or (3) modifying the modulus of the casting. Ransing describes a further option, (4) in which he proposes a change in heat transfer coefficient. The latter technique is a valuable insight because it is easily and economically computed by a geometrical technique, and so contrasts with the considerable computing reserves and effort required by finite element and finite difference methods. If R is the radius of the inscribed sphere, the local solidification time t is proportional to R^2/h where h is the heat transfer coefficient at the metal/mould

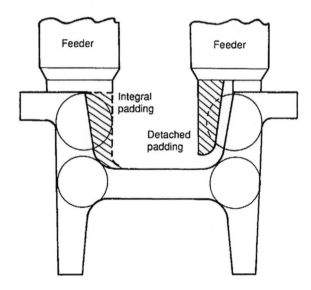

FIGURE 10.27

Use of Heuvers circles to determine the amount of added padding (Beeley, 1972) and the use of detached (or indirect) padding described by Daybell (1953).

interface. Thus at locations 1 and 2 we have $h_2 = h_1(R_1/R_2)^2$. This relation allows an estimate of the change of h that is required to ensure that the freezing time increases steadily towards the feeder. Ransing uses a change of 10% increase of solidification time for each different geometrical section of the casting (i.e. for feeding via a thin section to a thick distant section he increases the freezing time of the thin section over that of the thick by 10%). The technique can be usefully employed in reverse, in the sense that known values of h produced by a chill can quickly be checked for their effectiveness in dealing with an isolated heavy section. In every case, the target is to eliminate the local hot spot and ensure a continuous feed path back to the feeder. This simple technique has elegance, economy and power and is strongly recommended.

The casting modulus can be modified by providing either a chill or cooling fin or pin to speed solidification locally, or by providing extra metal to thicken the section and delay solidification locally. The provision of extra metal on the casting is known as padding. The addition of padding is most usefully carried out with the customer's consent, so that it can be left in position as a permanent feature of the casting design. If consent cannot be obtained, then the caster has to accept the cost penalty of dressing off the padding as an additional operation after casting.

Occasionally, this problem can be avoided by the provision of detached, or indirect, padding as shown in Figure 10.27. Daybell (1953) was probably the first to describe the use of this technique. The author has found it useful in the placement of feeders close to thin adjacent sections of casting, with a view to feeding through the thin section into a remote thicker section.

The principle of progressive increase of modulus towards the feeder, although generally accurate and useful, is occasionally seen to be not quite true. Depending on the conditions, this failure of the principle can be either a problem or a benefit, as shown later. (Even so, the Ransing technique described previously is unusual because it successfully takes this problem into account.)

In a re-entrant section of a casting, the confluence of heat flow into the mould can cause a hot spot, leading to delayed solidification at this point and the danger of local shrinkage porosity in an alloy that shrinks during freezing, such as an Al alloy. Alternatively, in an alloy that expands such as a high carbon-equivalent cast iron, the exudation of residual liquid into the mould as a result of the high internal pressure created during the precipitation of graphite can cause penetration of the aggregate mould material, with unwelcome so-called *burned-on* sand in a particularly inaccessible region. Such a hot spot can occur despite an apparent unbroken increase in modulus through that region towards the feeder. This is because the simple estimation of modulus takes no account of the geometry of heat flow away from cooling surfaces; all surfaces are assumed to be equally effective in cooling the casting. Such a hot spot requires the normal attention such as extra local cooling by chill or fin or additional feed via extra padding or feeders.

The failure of the modulus gradient technique can be used to advantage in the case of feeder necks to reduce the subsequent cutoff problem. Feeders are commonly joined to the main casting via a feeder neck, with the modulus of the neck commonly controlled to be intermediate between that of the casting and the feeder; the moduli of casting, neck and feeder are in the ratio 1.0:1.1:1.2 (Beeley, 1972). However, the neck can be reduced considerably below this apparently logical lower limit because of the hot spot effect and because of the conduction of heat from the neighbouring casting and feeder that helps to keep the neck molten for a longer period than its modulus alone would suggest. This point is well illustrated by Sciama (1975); his results are summarised in Figure 10.28. The results clearly demonstrate that for steel, feeder necks can be reduced to half of the diameter D of the feeder, providing that they are not longer than $0.1D$. The higher thermal conductivity copper- and aluminium-based materials can have necks almost twice as long without problems.

By extrapolation of these results towards smaller neck sizes, it seems that a feeder neck in steel can be only $0.25D$ in diameter, providing it is no more than approximately $0.03D$ in length. Similarly, for copper and aluminium alloys the $0.25D$ diameter neck can be up to $0.06D$ long. These results explain the action of the Washburn core, or breaker core, which is a wafer-thin core with a narrow central hole, and which is placed at the base of a feeder, allowing it to be removed after casting by simply breaking it off. In separate work the dimensions of typical Washburn cores is recommended to be a thickness of $0.1D$ and a central hole diameter of $0.4D$ (work by Wlodawer summarised by Beeley, 1972). The hole size and thickness appear to be very conservative in relation to Sciama's work. However, Sciama may predict optimistic results because he uses a feeder of nearly 1.5 times the modulus of casting which would tend to keep the

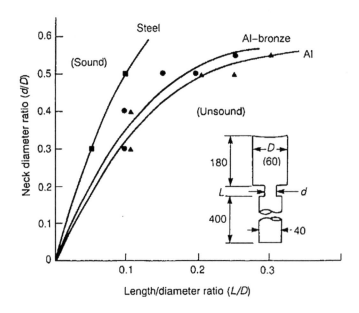

FIGURE 10.28

Effect of a constricted feeder neck on soundness of steel, aluminium bronze, and 99.5Al castings. The experimental points by Sciama (1975) denote marginal conditions.

junction rather hotter than a feeder with a modulus of only 1.2 times that of the casting (it would be valuable to repeat this work using a more economical feeder). Also, of course, conservatism may be justified where feeding conditions are less than optimum for other reasons in the real foundry environment.

The aspect of conservatism because of the real foundry environment is an interesting issue. For instance, the feeder neck could, in the extreme theoretical case, be of zero diameter when the thickness of the feeder neck core was zero. The reductio ad absurdum argument illustrates that an extreme is not worth targeting, especially when sundry debris in suspension, such as rather rigid bifilms, could close off a narrow aperture.

In fact, this danger appears to be very real for metals that commonly suffer a high population of bifilms, such as the light alloys. During the rapid filling of the mould, the bifilms swarm into the feeder, propelled by the momentum of the flow. However, during feeding, the liquid in the feeder drifts into the feeder neck, causing the gradual accumulation of bifilms, piling them up at a narrowed exit. Feeding may then be prevented by the simple mechanical action of the lowest bifilm opening, its low half expanding away into the casting as the metal in the casting attempts to suck down feed liquid. At the same time, the pileup of bifilms above the feeder exit may be sufficiently rigid to support the weight of metal in the feeder, thus preventing feeding. The eventual outcome is the mysterious case of the full feeder with a large shrinkage cavity immediately underneath. The pompous 'underriser shrinkage porosity' jargon now no longer means 'the feeder was too small', but 'the melt was too dirty to allow feeding'.

Minimum temperature gradient requirement

Experiments on cast steels have found that when the temperature gradient at the solidus (i.e. the temperature gradient at which the final residual liquid freezes) falls to below approximately $0.1–1$ Kmm^{-1} then porosity is observed even in well-degassed material. Although there is much scatter in other experimental determinations, it seems in general that the corresponding gradients for copper alloys are around 1 Kmm^{-1} and those for aluminium alloys are around 2 Kmm^{-1} (Pillai et al., 1976). It seems therefore that the temperature gradient defines a critical threshold of a non-feeding condition. As the flow channel nears its furthest extent and becomes vanishingly narrow, it will become subject to small random fluctuations in temperature along its length. This kind of temperature 'noise' will occur as a result of small variations in casting thickness, or of density of the mould, thickness of mould coating, blockage or diversion of heat flow direction by random entrained films etc. Thus the channel will not reduce steadily to infinite thinness, but will terminate when its diameter becomes close to the size of the random perturbations.

There has been some discussion about the absolute value of the critical gradient for feeding on the grounds that the degree of degassing, or the standard of soundness, to which the casting was judged, will affect the result. These are certainly very real problems, and do help to explain some of the wide scatter in the results.

Hansen and Sahm (1988) draw attention to a more fundamental objection to the use of temperature gradients as a parameter that might correlate with feeding problems. They indicate that the critical gradient required to avoid shrinkage porosity in a steel bar is five to 10 times higher than that required for a plate, and point to other work in which the critical gradient in a cylindrical steel casting is a function of its diameter. Thus the concept of a single gradient which applies in all conditions seems too simplistic. It seems the rate of flow of the residual feeding liquid may also be important.

Feeding distance

It is easy to appreciate that in normal conditions it is to be expected that there will be a limit to how far feed liquid can be provided along a flow path. Up to this distance from the feeder the casting will be sound. Beyond this distance the casting will be expected to exhibit porosity.

This arises because, along the length of a flow channel, the pressure will fall progressively because of the viscous resistance to flow (Section 7). When the pressure falls to a critical level, which might actually be negative (simply becoming a tensile stress in the liquid), then porosity may form. Such porosity may occur from an internal initiation event (such as the opening of a bifilm) or from the drawing inwards of feed metal from the surface of the casting because this may now represent a shorter and easier flow path than supply from the more distant feeder.

There has been much experimental effort to determine feeding distances. The early work by Pellini and coworkers (summarised by Beeley (1972)) at the US Naval Materials Laboratory is a classic investigation that has influenced the thinking on the concept of feeding distance ever since. They discovered that the feeding distance L_d of plates of carbon steels cast into greensand moulds depended on the section thickness T of the casting: castings could be made sound for a distance from the feeder edge of 4.5T. Of this total distance, 2.5T resulted from the chilling effect of the casting edge; the remaining 2.0T was made sound by the feeder. The addition of a chill was found to increase the feeding distance by a fixed 50 mm (Figure 10.29). They found that increasing the feeder size above the optimum required to obtain this feeding distance had no beneficial effects in promoting soundness. The feeding distance rule for their findings is simply:

$$L_d = 4.5T \tag{10.7}$$

Pellini and colleagues went on to speculate that it should be possible to ensure the soundness of a large plate casting by taking care that every point on the casting is within a distance of 2.5T from an edge, or 2.0T from a feeder.

Note that all the semi-empirical computer programs written since have used this and the associated family of rules, as illustrated in Figure 10.29, to define the spacing of feeders and chills on castings. However, the original data relate only to steel in greensand moulds, and only to rather heavy sections ranging from 50 to 200 mm. Johnson and Loper (1969) have extended the range of the experiments down to section a thickness of 12.5 mm and have re-analysed all the data. They found that for plates, the data, all in units of millimetres, appeared to be more accurately described by the equation:

$$L_d = 72m^{1/2} - 140 \tag{10.8}$$

and for bars:

$$L_d = 80m^{1/2} - 84 \tag{10.9}$$

where m is the modulus of the cast section in millimetres. The revised equations by Johnson and Loper have usually been overlooked in much subsequent work. What is also overlooked is that all the relations apply to cast mild steels in greensand moulds, not necessarily to any other casting alloys in any other kinds of mould.

In their nice theoretical model, Kubo and Pehlke (1985) find support for Pellini's feeding distance rules for steel castings, but it is a concern that no equivalent rule emerged for Al-4.5Cu alloy that they also investigated.

In fact, colleagues of Flinn (1964) found that whereas the short-freezing-range alloys manganese bronze, aluminium bronze and 70/30 cupro-nickel all had feeding distances that behaved like steel, increasing with section thickness, the long-freezing-range alloy tin bronze appeared to react in the opposite sense, giving a reduced feeding distance as section thickness increased. (The nominal composition of this classical long-freezing-range material is 85Cu, 5Sn, 5Zn and 5Pb. It was known among traditional foundrymen as 'ounce metal' because, to make this alloy, they needed to take 1 lb of copper to which 1 oz of tin, 1 oz of zinc and 1 oz of lead was added. This gives, allowing for small losses on addition, the ratios 85:5:5:5.)

Kuyucak (2002) reviews the relations for estimating feeding distance in steel castings, and finds considerable variation in their predictions. This makes sobering reading.

Jacob and Drouzy (1974) found long feeding distances, greater than 15T, for the relatively long-freezing-range aluminium alloys Al-4Cu and Al-75i-0.5Mg, providing the feeder is correctly sized.

All this confusion regarding feeding distances remains a source of concern. We can surmise that the opposite behaviour of short- and long-freezing-range materials might be understood in terms of the ratio pasty zone/casting section. For short-freezing-range alloys, this ratio is less than 1, so the solidified skin of the alloy is complete, dictating feeding from the feeder, and thus normal feeding distance concepts apply.

For the case of long-freezing-range materials where the pasty zone/casting section ratio is greater than 1, and in fact might be 10 or more, the outer solid portions of the casting are far from solid for much of the period of solidification. The connections of liquid through to the outer surface will allow flow of liquid from the surface to feed solidification shrinkage if flow from the more distant feeder becomes more difficult.

In addition, the higher temperature and lower strength of the liquid/solid mass will allow general collapse of the walls of the casting inwards, making an important contribution to the feeding of the inner regions of the casting by the 'solid feeding' mechanism. It is for this reason that the higher conductivity and lower strength alloys of Al and Cu can be characterised by practically infinite feeding distances, particularly if the alloys are relatively free from bifilms. For clean metal, internal porosity simply does not nucleate, no matter how distant the casting happens to be from the feeder; the outer walls of the casting simply move inwards very slightly.

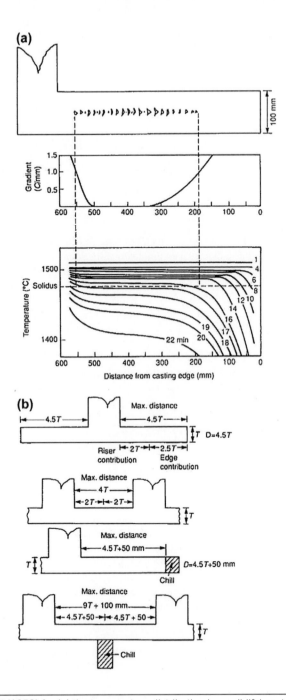

FIGURE 10.29

The famous results by Pellini (1953) for (a) the temperature distribution in a solidifying steel bar and (b) the feeding distances for steel plates cast in greensand.

Thus, although the general concept of feeding distance is probably substantially correct, at least for short-freezing-range alloys, and particularly for stronger materials such as steels, it should be used, if at all, with great caution for non-ferrous metals until it is better understood and quantified. In summary, it is worth noting the following:

1. The data on feeding distances have been derived from extensive work on carbon steels cast in greensand moulds. Relatively little work has been carried out on other metals in other moulds.
2. The definition of feeding distance is sensitive to the level of porosity that can be detected and/or tolerated.
3. It is curious that the feeding distance is defined from the edge of a feeder (not its centreline).
4. The quality of the cast metal in terms of its gas and oxide content would be expected to be crucial. For instance, good quality metal achieved by the use of good degassing and casting technique, possibly using filters (i.e. having the benefit of a low bifilm content) would be expected to yield massive improvements in feeding distance. This has been demonstrated by Romero et al. (1991) for Al-bronze. Berry and Taylor (1999) report a related effect, whilst reviewing the benefit to the feeding distance of pressurising the feeder. This work is straightforwardly understood in terms of the pressure on the liquid acting to suppress the opening of bifilms.

A final note of caution relates to the situation in which the concept of feeding distance probably does apply to an alloy, but has been exceeded. When this happens, it is reported that the sound length is considerably less than it would have been if the feeding distance criterion had just been satisfied. If true, this behaviour may result from the spread of porosity, once initiated, into adjacent regions. The lengths of sound casting in Figure 10.29(a) are considerably shorter than the maximum lengths given by Eqns (10.7) to (10.9), possibly because the feeding distance predicted by these equations has been exceeded and the porosity has spread. Mikkola and Heine (1970) confirm this unwelcome effect in white iron castings.

Criteria functions

For a discussion of the use of criteria functions to assess the difficulty to feed and the propensity of the metal to develop porosity, please refer to the earlier Section 7.1.3.2.

Feeding rule 6: Pressure gradient requirement

Although all of the previous feeding rules may be met, including the provision of feed liquid and a suitable flow path, if the pressure gradient needed to cause the liquid to flow along the path is not available, feed liquid will *not* flow to where it is needed. Internal porosity may therefore occur.

One of the most usual causes of failure to feed is not taking advantage of gravity. As opposed to *filling uphill* (which is, of course, quite correct), *feeding* should only be carried out *downhill* (using the assistance of gravity).

Attempts to feed uphill, although possible in principle, are unreliable in practice, and may lead to randomly occurring defects that have all the appearance of shrinkage porosity. In castings of a modest size, feeding uphill appears in general to be successful using the technique 'active feeding' as will be discussed later. In many castings, particularly larger castings, problems occur when attempting to feed uphill because of the difficulties caused by various effects: (1) adverse pressure gradient as discussed later; (2) danger of upward floating of gas or air bubbles into the casting and (3) adverse density gradient leading to convection as dealt with in rule 7 of the 10 Rules.

A positive pressure gradient from the outside to the inside of the casting will help to ensure that the feed material (either solid or liquid) travels along the flow path into those parts of the casting experiencing a shrinkage condition. The various feeding mechanisms are seen to be driven by the positive pressure such as atmospheric pressure and/or the pressure due to the hydrostatic head of metal in the feeder. The other contributor to the pressure gradient, the driving force for flow, is the reduced or even negative pressure generated within poorly fed regions of some castings. All of these driving forces happen to be additive; the flow of feed metal is caused by being pushed from the outside and by being pulled from the inside.

Figure 10.30 illustrates the feeding problems in a complicated casting. The casting divides effectively into two parts either side of the broken line. The left-hand side has been designed to be fed by an open feeder Fl and a blind feeder F2. The right-hand side was intended to be fed by blind feeder F3.

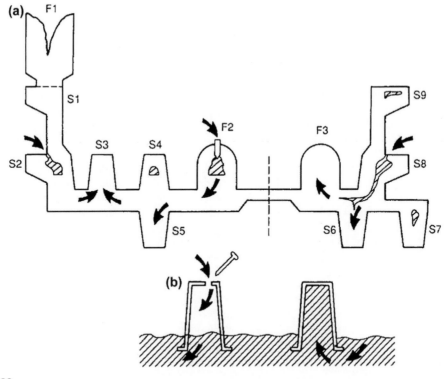

FIGURE 10.30

(a) Castings with blind feeders; F2 is correctly vented but has mixed results on section S3 and S4. Feeder F3 is not vented and therefore does not feed. The unfavourable pressure gradient draws liquid from a fortuitous skin puncture in section S8. See text for further explanation. (b) The plastic coffee cup analogue: the liquid is held in the upturned cup and cannot be released until air is admitted via a puncture. The liquid it is holding is then immediately released.

Feeder Fl successfully feeds the heavy section S1. This feeder is seen to be comparatively large. This is because it is required to provide feed metal to the whole casting during the early stages of freezing, whereas the connecting sections remain open. At this early stage, the top feeder is feeding the whole casting and both blind feeders.

Feeder F2 feeds S5 because it is provided with an atmospheric vent, allowing the liquid to be pressurised by the atmosphere as in the plastic coffee cup experiment illustrated in Figure 10.30, so forcing the metal through into the casting. (The reader is encouraged to try the coffee cup experiment.)

The identical heavy sections S3 and S4 show the unreliability of attempting to feed uphill. In S4, a chance initiation of a pore has created a free liquid surface, and the internal gas pressure within the casting happens to be close to 1 atm. Thus the liquid level in S4 falls, finding its level equal to that in the feeder F2. The surface-initiated pore in S2 has grown from the hot spot that has weakened and broken through the corner of the casting. The pore has grown, equalising its level exactly with that in the feeder F2 because both surfaces are subject to the same atmospheric pressure. In section S3, by good fortune, no pore initiation site is present, so no pore has occurred, with the result that atmospheric pressure via F2 (and unfortunately also via the puncture by the atmosphere at the hot spot in the re-entrant section S2) will feed solidification shrinkage here, causing the section to be perfectly sound.

Turning now to the right-hand part of the casting, although feeder F3 is of adequate size to feed the heavy sections S6, S7 and S8; unfortunately, its atmospheric vent has been forgotten. This is a serious mistake. The plastic coffee cup experiment shows that such an inverted airtight container cannot deliver its liquid contents. The pressure gradient is now reversed, causing the flow to be in the wrong direction, from the casting to the feeder! The detailed reasoning for this is as follows. The pressure in the casting and feeder continues to fall as freezing occurs until a pore initiates, either under the hydrostatic tension, or because of a buildup of gas in solution, or because of the inward rupture of the surface at a weak point such as the re-entrant angle in section S8. The pressure in section S8 is now raised to atmospheric pressure whilst the pressure in the feeder F3 is still low, or even negative. Thus feed liquid is now forced to flow from the casting into the feeder as freezing progresses. A massive pore then develops because feeder F3 has a large feed requirement, and drains section S8 and the surrounding casting. The defect size is worse than that which would have occurred if no feeder had been used at all!

Section S6 remains reasonably sound because it has the advantage of natural drainage of residual liquid into it. Effectively it has been fed from the heavy section S8. The pressure gradient resulting from the combined actions of gravity, shrinkage and the atmosphere from S8 to S6 is positive. The only reason why S6 may display any residual porosity may be that S8 is a rather inadequate feeder in terms of either its thermal requirement or its volume, or because the feed path may be interrupted at a late stage.

Section S7 cannot be fed because there is no continuous feed path to it. S9 is similarly disadvantaged. This has been an oversight in the design of the feeding of this casting. In a sand casting, it is likely that S7 and S9 will therefore suffer porosity. This will be almost certainly true for a steel casting, but less certain if the casting is a medium-freezing-range aluminium alloy. The reason becomes clear when we consider an investment casting, where, if a high mould temperature is chosen, and if the metal is clean, will allow solid feeding to operate, allowing the sections the opportunity to collapse plastically, and so become internally sound, provided that no pore-initiation event interrupts this action. Solid feeding is often seen in aluminium alloy sand castings, but more rarely in steel sand castings because of the greater rigidity of the solidified steel, which successfully resists plastic collapse in cold moulds.

If feeding had taken place with the assistance of gravity, feeder F2 in Figure 10.30 would have successfully fed sections S2 and S4 either if it was taller, or if it had been placed at a higher location, for instance on the top of S4.

It is clear that F3 may not have fed section S8 if the corner puncture occurred, even if it had been provided with an effective vent, because the pressure gradient for flow would have been removed. A provision of an effective vent, and the re-siting of the base of the feeder F3 to the side of S8, would have maintained the soundness of both S6 and S8 and would have prevented the surface puncture at S8. S7 and S9 would still have required separate treatment.

The exercise with the plastic coffee cup shows that the water will hold up indefinitely in the upturned cup until released by the pin causing a hole. The cup will then deliver its contents immediately (but not before!). Blind feeders are therefore often unreliable in practice because the atmospheric vent may not open reliably. Such feeders then act to suck feed metal from the casting, making any porosity worse.

If a blind feeder is provided with an effective atmospheric vent, then the available atmospheric pressure may help it to feed uphill. The atmosphere is capable of holding up several metres of head of metal. For liquid mercury the height is approximately 760 mm, being the height of the old-fashioned atmospheric barometer of course. Equivalent heights for other liquid metals are easily estimated allowing for the density difference. Thus for liquid aluminium of specific gravity 2.4 compared with liquid mercury of 13.9, the atmosphere will hold up about $(13.9/2.4) \times 0.76 = 4$ m of liquid aluminium or $(13.9/7.0) \times 0.76 = 1.5$ m liquid iron.

Summarising the maximum heights supportable by 1 atm for various pure liquids near their melting points:

Mercury	0.760 m (the barometric height 760 mm = 1 atm)
Steel	1.48 m
Zinc	1.58 m
Aluminium	4.36 m
Magnesium	6.54 m
Water	10.40 m

Whilst no pore exists, the tensile strength of the liquid will in fact allow the metal, in principle, to feed to heights of kilometres because in the absence of defects the liquid can withstand tensile stresses of thousands of atmospheres. The liquid can, in principle, hang up in a tube, its great weight stretching its length somewhat. However, the random initiation of a single minute pore will instantly cause the liquid to 'fracture', causing the liquid in the tube to fall, finally stabilising at the level at which atmospheric pressure can support the liquid. Thus any height above that supportable by 1 atm is clearly at high risk.

Moreover, there is even worse risk. If when attempting to feed against gravity, there is a leak path to atmosphere, allowing atmospheric pressure to be applied in the liquid metal inside the solidifying casting, the melt will then fall further, the action of gravity tending to equalise levels in the mould and feeder. Thus, if the feeder is sited below the casting, the casting will completely empty of residual liquid. Regrettably, this is an efficient way to cast feeders of excellent soundness, but seriously porous castings.

Clearly, the initiation of a leak path to atmosphere (via a double oxide film or a liquid region in contact with the surface at a hot spot) is rather easy in many castings, making the whole principle of uphill feeding so risky that it should not be attempted in circumstances where porosity cannot be tolerated. It is a pity that the comforting theories of pushing liquid uphill by atmospheric pressure or even hanging it from vast heights using the huge tensile strength of the liquid cannot be relied upon in practice.

For all practical purposes, the only really reliable way to feed is *downhill*, using gravity.

The conclusion to these considerations is: *place feeders high to feed downhill*. This is a general principle of great importance. It is of similar weight to the general principle discussed previously, *place ingates low to fill uphill*. These are fundamental concepts in the production of good castings.

Feeding rule 7: Pressure requirement

The final rule for effective feeding is a necessary requirement like all the others. Sufficient pressure in the residual liquid within the casting is required to suppress both the initiation and the growth of cavities from both internal and external sources.

This is a *hydrostatic* requirement relating to the suppression of porosity and contrasts with the previous pressure gradient requirement that relates to the *hydrodynamic* requirements for flow (especially flow in the *correct* direction!)

A fall in internal pressure may cause a variety of problems:

1. Liquid may be sucked from the surface. This is particularly likely in long-freezing-range alloys, or from re-entrant angles in shorter-freezing-range alloys, resulting in internal porosity initiated from, and connected to, the outside.
2. The internal pressure may fall just sufficiently to unfurl, but not fully open the population of bifilms. The population of bifilms will therefore remain effectively invisible. The result will be an apparently sound casting but, inexplicably, the mechanical properties will be poor, particularly the elongation and fatigue performance. (It is possible that some so-called '*diffraction mottle*' may be noted on X-ray radiographs as a result of the newly appearing population of slightly opened cracks.)
3. The internal pressure may fall sufficiently to open bifilms, so that a distribution of fine and dispersed microporosity will appear. The mechanical properties will be even lower. (For the technically punctilious, the mechanical opening of bifilms is driven not so much by pressure reduction, a *stress* phenomenon, but by the progression of shrinkage, a *strain* phenomenon experienced by the liquid.)
4. The internal shrinkage may cause macro-shrinkage porosity to occur, especially if there are large bifilms present as a result of poor filling of the casting. Properties may now be in disaster mode and/or large holes may appear in the casting. Figure 10.31 illustrates pressure loss situations in castings that can result in shrinkage porosity.
   Figure 10.32 illustrates the common observation in Al alloy castings in which a glass cloth is placed under the feeder to assist the break-off of the feeder after solidification. Bafflingly, it sometimes appears that the cloth prevents the feeder from supplying liquid, so that a large cavity appears under the feeder. The truth is that double oxide films plaster themselves against the underside of the cloth as the feeder fills, keeping the bifilms in the casting, and ensuring that the feeder contains clean metal. The half of the bifilm against the cloth tends to weld to the cloth, possible by a chemical fusing action, or possibly by mechanical wrapping around the fibres of the cloth. Whatever the mechanism, the action is to hold back the liquid above, whereas the lower half of the (unbonded) double film is

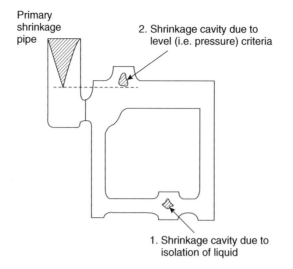

FIGURE 10.31

Pressure loss situations in castings leading to the possibility of shrinkage porosity.

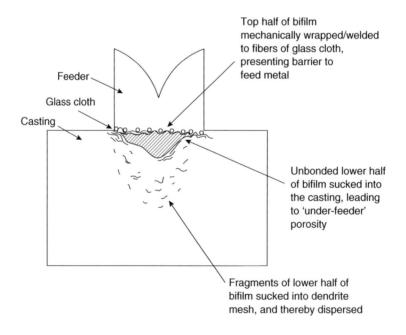

FIGURE 10.32

Apparent blocking of feed metal by a glass cloth strainer in an Al alloy casting because of bifilms collected on the underside of the cloth.

easily pulled away by the contracting liquid, opening the void that was originally the microscopically thin interface of air inside the bifilm. Once again, bifilms are seen to interfere with the action of a feeder, and create the curious phenomenon of shrinkage porosity immediately under a feeder which remains completely full of (clean) metal.

5. If there is insufficient opportunity to open internal defects, the external surface of the casting may sink to accommodate the internal shrinkage. (The occurrence of *surface sinks* is occasionally referred to elsewhere in the literature as '*cavitation*'; a misuse of language to be deplored. Cavitation properly refers to the creation and collapse of minute bubbles, and the consequent erosion of solid surfaces such as those of ships' propellers, although, I admit, it is beginning to have a tolerably respectable use to describe the opening of pores during super plastic forming.)

Often, of course, the distribution of defects observed in practice is a mixture of the previous list. The internal pressure needs to be maintained sufficiently high to avoid all of these defects.

Finally, however, it is worth pointing out that over-zealous application of pressure to reduce the previous problems can result in a new crop of different problems.

For instance, in the case of long-freezing-range materials cast in a sand mould, a high internal pressure, applied for instance to the feeder, will force liquid out of the surface-linked capillaries, making a casting having a 'furry' appearance. Overpressures are not easy to control in low-pressure sand casting processes, and are the reason why these processes often struggle to meet surface-finish requirements. Work at the University of Alabama (Stefanescu et al., 1996) on cast iron castings has shown that the generation of excess internal pressure by graphite precipitation can lead to exudation of the residual liquid via hot spots at the surface of the casting, leading penetration of the sand mould. Later work on steels has shown analogous effects (Hayes et al., 1998).

In short-freezing-range materials the inward flow of solid can be reversed with sufficient internal pressure. Too great a pressure will expand the casting, blowing it up like a balloon, producing unsightly swells on flat surfaces (Figure 10.33).

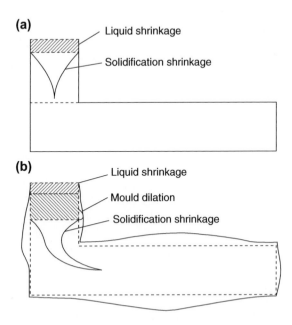

FIGURE 10.33

Comparison between the external size and internal shrinkage porosity in a casting as a result of (a) moderate pressure in the liquid, and adequately rigid mould and (b) too much pressure and/or a weak mould.

Feeding of cast iron castings by control of pressure

The successful feeding of cast irons is perhaps the most complex and challenging feeding task compared to all other casting alloys as a result of the curious and complicating effect of the expansion of graphite during freezing. The effects are most dramatically seen for ductile irons.

The great prophet of the scientific feeding of ductile irons was Stephen Karsay. In a succession of engaging and chattily written books, he outlined the principles that applied to this difficult metal (see, for instance, Karsay, 1992, 2000). He drew attention to the problem of the swelling of the casting in a weak mould as shown in Figure 10.33, in which the valuable expansion of the graphite was lost by enlarging the casting, causing the feeder to be inadequate to feed the increased volume. He promoted the approach of making the mould more rigid, and so better withstanding stress, and at the same time reducing the internal pressure by providing feeders that acted as pressure relief valves. The feeder, after some initial provision of feed metal during the solidification of austenite, would reverse its action, back-filling with residual liquid during the expansion of the solidification of the eutectic graphite, relieving the pressure and thus preventing mould dilation. The final state was that *both* the casting and feeder were substantially sound. (Occasionally, one hears stories that such sound feeders have been declared to be evidently useless, having apparently provided no feed metal. However, their removal would immediately cause all subsequent castings to become porous!)

The reproducibility of the achievement of soundness in ductile iron castings is, of course, highly sensitive to the efficiency of the inoculation treatment because the degree of expansion of graphite is directly affected. This is notoriously difficult to keep under good control, and makes for one of the greatest challenges to the iron founder.

Roedter (1986) introduced a refinement of Karsay's pressure relief technique in which the pressure relief was limited in extent. Some relief was allowed, but total relief was prevented by the premature freezing of the feeder neck. In this way, the casting was slightly pressurised, elastically deforming the very hard sand mould, and the surrounding steel moulding box (if any). The elastic deformation of the mould and its box would store the strain energy. The subsequent relaxation of this deformation would continue to apply pressure to the solidifying casting during the remainder of solidification. Thus soundness of the casting could be achieved, but without the danger of unacceptable swells on extensive flat surfaces. Even so, the use of strain energy is clearly seen to be vulnerable, because it can only reverse only a very small amount of deformation as is explained later.

For somewhat heavier ductile iron castings, however, it has now become common practice to cast completely without feeders. This has been achieved by the use of rigid moulds, now more routinely available from modern greensand moulding units. Naturally, the swelling of the casting still occurs because, ultimately, solids are incompressible. However, as before, the expansion is restrained to the minimum by the elastic yielding of the mould and its container and distributed more uniformly. Thus the whole casting is a few percentages larger. If the total net expansion was 3 volume %, this corresponds to 1 linear % along the three orthogonal axes, so that from a central datum, each point on the surface of the casting would be approximately 0.5% oversize. This uniform and very reproducible degree of oversize is usually negligible. However, of course, it can be compensated, if necessary, by making the pattern 0.5% undersize.

The use of the elastic strains to re-apply pressure is strictly limited because such strains are usually limited to only 0.1 linear % or so. Thus only a total of perhaps 0.3 volume % can be compensated by this means. This is, as we have seen previously, only a fraction of the total volume change that is usual in a graphitic iron, and which permanently affects the size and dimensions of the casting. The judgement of feeder neck sizes to take advantage of such small margins is not easy.

With the steady accumulation of experience in a well-controlled casting facility, the casting engineer can often achieve such an accuracy of feeding that even such a modest gain is considered a valuable asset. Even so, the reader will appreciate that the feeding graphitic irons is still not as exact a science and still not as clearly understood as we all might wish.

10.6.4 THE NEW FEEDING LOGIC

Much of the formal calculation of feeders has been of poor accuracy because of several simplifying assumptions that have been widely used. Tiryakioglu has pioneered a new way of analysing the physics of feeding, having, in addition, the good fortune to have as a critical test the exemplary experimental data on optimum feeder sizes determined by his late father, Ergin Tiryakioglu, many years earlier (1964). This was carried out at the University of

Birmingham, UK, where Ergin and I shared an office for a time while we were both researching for our doctorates. The reader is recommended to the original papers by Murat Tiryakioğlu (1997–2002) for a complete description of his admirable logic. We shall summarise his approach only briefly here, following closely his excellent description (Tiryakioğlu et al., 2002).

As we have seen in rules 2–4, an efficient feeder should (1) remain molten until the portion of the casting being fed has solidified (i.e. the solidification time of the feeder has to be equal to, or exceed, that of the casting), (2) contain sufficient volume of molten metal to meet the feeding demand of that same portion of the casting and (3) not create a hot spot at the junction between feeder and casting. An optimum feeder is then defined as the one with the smallest volume, for its particular shape, to meet these criteria. A feeder that is less compact or that has less volume than the optimum feeder will result in an unsound casting.

The standard approaches to solve these problems have usually been based on the famous rule by Chvorinov (1940) for the solidification time of a casting is:

$$t = (V/A)^n \tag{10.10}$$

where B is the mould constant, V is the casting volume, A is the surface area through which heat is lost, and n is a constant ($n = 2$ in Chvorinov's work for simple shaped castings in silica sand moulds). The V/A ratio is known as the modulus m, and has been used as the basis for several approaches to determining the size of feeders for the production of sound castings, as described in feeding rule 2.

Despite its wide acceptance, Chvorinov's rule has limitations because of the underlying assumptions used in deriving the equation. As a result of these limitations, the exponent, n, fluctuates between 1 and 2, depending on the shape and size of the casting, and the mould and pouring conditions. One of the reasons for this anomaly is that Chvorinov's rule originally did not take the shape of the casting into consideration. A new geometry-based model (Tiryakioğlu et al., 1997a) proved that the modulus includes the effect of both casting shape and size. These two independent factors were separated from each other by the use of a shape factor k where

$$t = B'k^{1.31}V^{0.67} \tag{10.11}$$

and B' is the mould constant. The shape factor, k, is the ratio (the surface area of a sphere of same volume as the casting)/ (the surface area of the casting). In Eqn (10.11), V assesses the amount of heat that needs to be dissipated for complete solidification, and k assesses the relative ability of the casting shape to dissipate the heat under the given mould conditions.

$$k = A_s/A = 4.837V^{2/3}/A \tag{10.12}$$

where A_s is the surface area of the sphere.

Adams and Taylor (1953) were the first to consider *mass transfer* from feeder to casting. They realised that during solidification, a mass of αV_c needs to be transferred from the feeder to the casting (α is fraction shrinkage of the metal). However, as Tiryakioğlu (2002) explains, their development of the concept unfortunately introduced errors so that the final solution was not accurate.

Moreover, the lack of knowledge about the effect of *heat transfer* between a feeder and a casting has led the researchers to the mindset of considering the feeder and casting separately. In other words, almost all feeder models have been based on calculation of solidification times for the feeder and casting independently, and then assuming that the same solidification characteristics will be followed when they are combined. However, it should be remembered that the feeder is also a portion of the mould cavity, and the solidifying metal does not know (or care) which portion is the casting and which is the feeder.

The objective of the foundry engineer when designing a feeding system is to have the thermal centre of the total casting (the feeder-casting combination) in the feeder. In fact, all three requirements for an efficient feeder listed (1–3) previously can be summarised as a single requirement: *the thermal centre of the feeder-casting combination will be in the feeder.* This new approach, which treats the casting-feeder combination as a single, total casting constitutes the foundation of Tiryakioğlu's new approach to characterise heat and mass transfer between feeder and casting.

The new approach

Let us consider a plate casting that is fed effectively by a feeder. Knowing that the solidification contraction of the casting is αV_c, this volume is transferred from the feeder to the casting, resulting in the final volume of the feeder being $(V_f - \alpha V_c)$. Solidification contraction of the feeder is ignored sincebecause it does not change the heat content of the feeder. If the feeder has been designed according to the rules for efficient feeding, the last part to solidify in this combination is the feeder. In other words, the thermal centre of the casting-feeder combination (the total casting) is in the feeder. Therefore the solidification time of the total casting is exactly the same as the feeder, and both have the same thermal centre.

This is not true for the casting, however. The thermal centre of the casting is also in the feeder, but its solidification time may or may not be equal to that of the total casting. Hence

$$k_t^{1.31} V_t^{0.67} = k_f^{1.31}(V_f - \alpha V_c)^{0.67} \tag{10.13}$$

where subscript t refers to the total casting. So far, we have ignored the heat transfer between the casting and the feeder. Using optimum feeder data obtained by systematic changes in feeder size for an Al-12wt%Si alloy (Tiryakioğlu, 1964), the solidification times of casting and feeder are compared in Figure 10.34(a) and total casting and feeder in Figure 10.34(b). Figure 10.34(a) shows the $(t_c - t_f)$ relationship when mass transfer is taken into account and heat transfer is ignored. It should be kept in mind that the scatter in Figure 10.34(a) is not due to experimental error because all values were calculated. Figure 10.34(b) shows the relationship between the solidification times of feeder and total casting (feeder + casting combination). Although the agreement in Figure 10.34(b) is encouraging, at low values the error is up to 30%. However, solidification time should be identical for the feeder and the total casting. The error is due to the neglect of the heat exchange between feeder and casting. Mass is transferred from the feeder to the casting throughout the solidification process. Solidification takes place over a temperature range, so that subtracting αV_c from V_f adjusts for mass exchange completely, but for heat exchange this treatment assumes isothermal conditions, and therefore is not sufficient. Hence the feeder solidification time needs to be adjusted for the heat exchange with the casting. We can treat this heat exchange as if it were superheat extracted from/given to the feeder. For the superheat model, we will use the model by E. Tiryakioğlu (1964) for its simplicity and its independence from actual pouring temperature. Equation (10.13) can now be rewritten as

$$k_t^{1.31} V_t^{0.67} = k_f^{1.31}(V_f - \alpha V_c)^{0.67} e^{\xi \Delta T_f} \tag{10.14}$$

where ξ is a constant dependent on the alloy ($0.0028°\text{C}^{-1}$ for Al-Si eutectic alloy (Tiryakioğlu, 1964) and $0.0033°\text{C}^{-1}$ for Al-7%Si (Tiryakioğlu et al., 1997b)), ΔT_f is the temperature change (rise or fall) in feeder because of the heat exchange, and can be easily calculated using Eqn (10.14). The sum of change in heat content of the feeder and casting is zero (heat lost by one is gained by the other). Therefore

$$C(V_f - \alpha V_c)\Delta T_f + C(V_f + \alpha V_c)\Delta T_c = 0 \tag{10.15}$$

where C is the specific heat of the metal. Hence

$$\Delta T_c = (V_f - \alpha V_c)\Delta T_f/(V_f + \alpha V_c) \tag{10.16}$$

The total solidification time of the casting can now be written as

$$t_c = k_c^{1.31}(V_f + \alpha V_c)^{0.67} e^{\xi \Delta T_c} \tag{10.17}$$

The solidification times of feeder and casting can now be compared. This comparison is presented in Figure 10.34(c) which shows a practically perfect fit and a relationship that can be expressed as;

$$t_t = t_f = \alpha t_c \tag{10.18}$$

The data for Al-Si alloy shown in Figure 10.34(c) give $\alpha = 1.046$.

In a separate exercise, using the data for steel by Bishop et al. (1955) assuming ξ of $0.0036°\text{C}^{-1}$ for steel (Tiryakioğlu, 1964), a similar excellent relationship is obtained where α is found to be 1.005 (Tiryakioğlu, 2002).

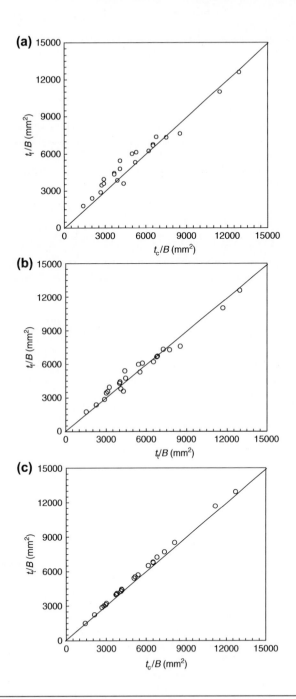

FIGURE 10.34

Comparison of calculated solidification times of (a) casting and feeder; (b) total casting and feeder; (c) total casting (or feeder) versus casting after adjustment to account for heat transfer between feeder and casting (Tiryakiolğu et al., 2002).

Thus the solidification time of optimum-sized feeders in the feeder-casting combination was found to be only a few percent longer than that of castings both for Al-Si alloy and steel castings.

We can conclude that, for an accurate description of the action of a feeder, both mass and heat transfer from feeder to casting during solidification have be taken into account simultaneously. Previous feeder models that account for mass transfer assume that the transfer takes place isothermally and at the pouring temperature. This previous assumption overestimates the additional heat brought into the casting from the feeder. The new model incorporates the effect of superheat and is based on the equality of solidification times of feeder and total casting.

The requirements for efficient feeders: (1) solidification time; (2) feed metal availability and (3) prevention of hot-spot at the junction; can be combined into a single requirement when the casting-feeder combination is treated as a single, total casting. The three criteria reduce to the simple requirement: 'The thermal centre of the total casting should be in the feeder'.

The disarming simplicity of this conclusion conceals its powerful logic. It represents the ideal criterion for judging the success of a computer model of a casting and feeder combination.

10.6.5 FREEZING SYSTEMS DESIGN (CHILLS, FINS AND PINS)

In this section on feeding, we are of course mainly concerned with the action of chills and fins to provide localised cooling of the casting. This is my first choice for avoiding the planting of a feeder on a hot spot. I always try to avoid feeding.

Chills (or fins or pins) can assist directional solidification of the casting towards the feeder, thus assisting in the achievement of soundness. Details of this thermal action are presented in Section 5.1.2 on increased heat transfer.

Chills (or fins or pins) also act to increase the ductility and strength of that locality of the casting. For some alloys, this benefit to mechanical properties results from the well-known Hall–Petch effect associated with shortening slip distances, and the difficulty of re-initiating slip in the adjacent grain. However, it is more likely that most alloys will benefit most from the freezing of bifilms in their compact state. This justification of this conclusion is outlined in Section 9.4 on the mechanical properties of castings.

10.6.6 FEEDING: THE FIVE MECHANISMS

The five feeding mechanisms—liquid, mass, interdendritic, burst and solid feeding—are different ways in which cast material can flow to feed the solidification contraction. The mechanisms are dealt with in detail in Chapter 7. In general, it is fair to state that only liquid, interdendritic and solid are important for our purposes of making sound castings. However, when relying on solid feeding for soundness there are precautions that it is wise to observe.

Dangers of solid feeding

The great benefit of solid feeding is that a casting can be made sound, even though it demands feed metal but cannot be fed by liquid from a conventional feeder. In favourable conditions, the casting can collapse plastically; the shrinkage volume is merely transferred from the inside to the outside of the casting. Here, if the volume is distributed reasonably evenly over a large area of the casting, the shrinkage will cause only a negligible and probably undetectable reduction in the size or shape of the casting.

This feeding benefit is commonly gained in light alloy castings which are relatively weak at their freezing points and so collapse rather easily; no large internal tensile stresses are generated so that pores are not necessarily created internally. Similarly, the easy collapse of weak solid is found in many ferrous and higher temperature alloys in investment casting, where the high mould temperature retains the plasticity of the solidified skin of the casting.

There are, however, dangers in relying on solid feeding for the soundness of a casting, as are illustrated in Figure 9.25.

1. If the outside shrinkage is not distributed so favourably, but remains concentrated in a local region, a surface sink is the result. This may, of course, result in a portion of the casting being out of tolerance, possibly no longer cleaning up on machining, so that the casting is scrapped.

2. When operating without feeders, a second possibility is the formation of shrinkage pores, grown from initiation sites (almost certainly bifilms), so that solid feeding immediately fails. It seems that such events tend to be triggered by rather large bifilms such as arise from random variations in pouring techniques. The achievement of soundness using solid feeding then becomes a hit-and-miss affair.
3. A further possibility will be commonly, but not easily perceived. If the melt has a distribution of small, possibly microscopic, bifilms, these will be unfurled to some extent by the reduced pressure in the unfed region thus being converted from crumpled compact features of negligible size to flat thin extensive cracks. Thus although the casting may continue to appear perfectly sound in the unfed region, and solid feeding declared to be a complete success, the mechanical properties of this part of the casting will be reduced. In particular, although the yield strength of the region will be hardly affected, that part of the casting will exhibit reduced strength, ductility and fatigue resistance.
4. If the localised shrinkage problems are even more severe, the distribution of small bifilms will develop further. After unfurling to become flat cracks, additional reduction of pressure in the liquid will open them further still to become visible microporosity. The pores may even grow to such a size that they become visible on radiographs. This situation is common in steel castings where the high strength of the solidified skin creates large internal stresses, providing irresistible forces to open bifilms and thus bring solid feeding to a halt.

In contrast to these dangers, the action of a feeder to pressurise the melt and so help to resist the unfurling and even the inflating of bifilms, is seen to be a benefit. The bifilms although still present, remain out of sight. Using a domestic analogy from home decoration, there is a very real sense in which adding feeders to castings is almost literally 'papering over the cracks'.

However, this rather severe censure is not completely valid, in the sense that closed bifilms in a casting do exhibit some strength because they can withstand shear stress (although, of course, not tensile stress at right angles to the bifilm). The pressure provided by the (liquid) feeder has a valuable role in keeping the bifilms closed.

10.6.7 COMPUTER MODELLING OF FEEDING

Good computer models have demonstrated their usefulness in being able to predict shrinkage porosity with accuracy and have therefore brought about a revolution within the industry. A simulation using a reliable modelling software package should now be specified as a prior requirement to be carried out before work is started on making the tooling for a new casting. This minor delay has considerable benefits in shortening the overall development time of a new casting, and greatly increases the chance of being 'right first time'.

However, the limitations of some computer simulations require to be recognised. For instance, these include

1. no allowance for the effect of thermal conduction in the cast metal (rare nowadays);
2. no allowance for the important effects due to convection in the liquid (common);
3. neglect of, or only crude allowance for, the effect of the heating of the mould by the flow of metal during filling (now commonly addressed); and
4. no capability of any design input. Thus gating and feeding designs will be required as inputs (just beginning to be addressed at this time).

For the future, it is to be expected that software packages will evolve to provide intelligent solutions to all these requirements. Examples of a good start in this direction are shown by Dantzig and coworkers (Morthland et al., 1995 and McDavid and Dantzig, 1998). In the meantime, it remains necessary to use computer models with some discretion. For instance, in the work by the Morthland team, they warn that the results are specific to the feeding criterion used. If a more stringent temperature gradient criterion were used (for instance $2 Kcm^{-1}$ instead of $1 Kcm^{-1}$) the feeder would have been larger.

The previous approaches to the optimisation of the feeding requirements of castings have involved the use of numerical techniques such as finite element and finite difference methods. Ransing et al. (2004) propose a geometrical method based on an elegant extension of the Heuvers circle technique. This technique was described previously in the section describing the feed path requirements for feeding.

10.6.8 RANDOM PERTURBATIONS TO FEEDING

It is helpful if the computer simulation can provide a robust casting solution. This is because on the foundry floor, several practical issues conspire to change the casting conditions, and thus threaten the quality of the casting.

In aluminium castings, flash of approximately 1 mm thickness and only 10 mm wide has been demonstrated to have a powerful effect on the cooling of local thin sections up to 10 mm thick, speeding up local solidification rates by up to 10 times. The effect is so much reduced in ferrous castings because of their much lower thermal conductivity that any effect of cooling by fins can usually be neglected.

Thus for high-conductivity alloys flash has to be controlled, or used deliberately because, in moderately thick sections, it has to potential to cut off feeding to more distant sections. The erratic appearance of flash in a production run may therefore introduce uncertainty in the reproducibility of feeding, and the consequent variability of the soundness of the casting. Flash on very thick sections is usually less serious because convection in the liquid in thick sections conveys the local cooled metal away, effectively spreading the cooling effect over other parts of the casting, giving an averaging effect over large areas of the casting. In general, however, it is desirable that these uncertainties are reduced by good control over mould and core dimensions.

The other known major variable affecting casting soundness in sand and investment castings is the ability of the mould to resist deformation. This effect is well established in the case of cast irons, where high mould hardness is a condition for soundness. However, there is evidence that such a problem exists in castings of copper-based alloys and steels. A standard system such as statistical process control or other technique should be seen to be in place to monitor and facilitate control of such changes. Permanent moulds such as metal or graphite dies are relatively free from such problems. Similarly many other aggregate moulding materials are available that possess much lower thermal expansion rates, and so produce castings of greater accuracy and reproducibility. Many of these are little, if any, more expensive than silica sand. A move away from silica sand is already under way in the industry, and is strongly recommended.

The solidification pattern of castings produced from permanent moulds such as gravity dies and low-pressure dies may be considerably affected by the thickness and type of the die coat which is applied.

For some permanent moulds, pressure die-casting and some types of squeeze casting the feeding pattern is particularly sensitive to mould cooling. After the development and acceptance of the casting, any further changes to cooling channels in the die, or to the cooling spray during die opening, will have to be checked to ensure that corresponding deleterious changes have not been imposed on the casting. The quality of the water used for cooling also requires to be seen to be under good control if deposits inside the system are not to be allowed to build up and so cause changes in the effectiveness of the cooling system with time.

10.6.9 THE NON-FEEDING ROLES OF FEEDERS

Feeders are sometimes important in other ways than merely providing a reservoir to feed the solidification shrinkage during freezing.

We have already touched on the effect that feeders can have on the metallurgical quality of cast metal by assisting to restrain the unfurling and opening of bifilms by maintaining a pressure on the melt. This action of the feeder to pressurise the casting therefore helps to maintain mechanical properties, particularly ductility and fatigue strength.

A further key role of many feeders, however, is merely as a flow-off or kind of dump. Many filling system designs are so poor that the first metal entering the mould arrives in a highly damaged condition. The presence of a generous feeder allows some of this metal to be floated out of the casting. This role is expected to be hindered, however, in highly cored castings where the bifilms will tend to attach to cores in their journey through the mould.

In general, experience with the elimination of feeders from Al alloy castings has resulted in the casting 'tearing itself apart'. This is a clear sign of the poor quality of metal probably resulting mainly from the action of the poor running system. The inference is that the casting is full of serious bifilm cracks. These remain closed, and so invisible, whilst the feeder acts to pressurise the metal. If the pressurisation from the feeder is removed, the bifilms will be allowed to open, becoming visible as cracks. This phenomenon has been seen repeatedly in X-ray video radiography of freezing castings. It is observed that good filling systems do not lead to the casting tearing itself apart, even though the absence of a feeder has created severe shrinkage conditions. In this situation the casting shrinks a little more (under the action of solid feeding) to accommodate the volume difference.

Summarising and thinking further we have.

1. As we have seen, pressurisation raises mechanical properties, particularly ductility.
2. Pressurisation together with some feeding helps to maintain the dimensions of the casting. Although the changes in dimensions by solid feeding are usually small, and can often be neglected, on occasions the changes may be outside the dimensional tolerance. A feeder to ensure the provision of liquid metal under some modest pressure is then required.
3. Pressurisation can delay or completely prevent blow defects from cores.

In summary, providing the filling system design is good so as to avoid creating large bifilms, and provided the solidification rate is sufficiently fast to retain the inherited population of bifilms compact, castings that do not require feeders for feeding should not be provided with feeders.

10.7 RULE 7: AVOID CONVECTION DAMAGE

10.7.1 THE ACADEMIC BACKGROUND

Convection is the flow phenomenon that arises as a result of density differences in a fluid.

In a solidifying casting, the density differences in the residual liquid can be the result of differences in solute content as a consequence of segregation. This is a significant driving force for the development of channel defects known as the 'A' and 'V' segregates in steel ingots and as freckle trails in nickel- and cobalt-based investment castings. The name 'freckles' comes from the appearance of the etched components that shows channels containing randomly oriented grains that have been partly remelted in the convecting flow and detached from their original dendrites. Although, for many reasons, channel defects are unwelcome, they are usually not life threatening to the product. These defects are discussed earlier in Section 5.3.4 and are not discussed further here.

Convection can also arise as a result of density differences that result from temperature differences in the melt. There have been numerous theoretical studies of this phenomenon taking examples such as the solidification of low melting point materials in simple cubical moulds, of which one side is cooled and the other not. The resulting gentle drift of liquid around the cavity, down the cool face and up the non-cooled face, changes the form of the solidifying front. A schematic example is shown in Figure 10.35. These are interesting exercises, but give relatively little assistance to the understanding of the problems of convective flow in engineering systems.

The results from Mampaey and Xu (1999), who studied the natural convection in an upright cylinder of solidifying cast iron, showed that the thermal centre of the liquid mass was shifted upwards, and graphite nodules in spheroidal graphite irons were transported by the flow. Such studies reflect the gentle action of convection in small, simple shaped, closed systems; the kind of action one would expect to see in a cooling cup of tea. These facts have lulled us into a state of false security, assuming convection to be essentially harmless and irrelevant. We need to think again.

10.7.2 THE ENGINEERING IMPERATIVES

Convection was practically unknown as an important factor in shaped castings until the early 1980s. Even now, it is not widely known nor understood. However, it can be life and death to a casting, and has been the death of several attempts to develop counter-gravity casting systems around the world. Most workers in this endeavour still do not know why they failed. The Cosworth Casting Process nearly foundered on this problem in its early days, only solving the problem by its famous (infamous?) rollover system.

Thus convection is not merely a textbook curiosity. The casting engineer is required to understand and come to terms with convection as a matter of urgency. The problem can be of awesome importance and lead to major difficulties, if not impossibilities, to achieve a sound and saleable casting.

Convection enhances the difficulty of uphill feeding in medium section castings, making them extremely resistant to solution. In fact increasing the amount of (uphill) feeding by increasing the diameter of the feeder neck, for instance,

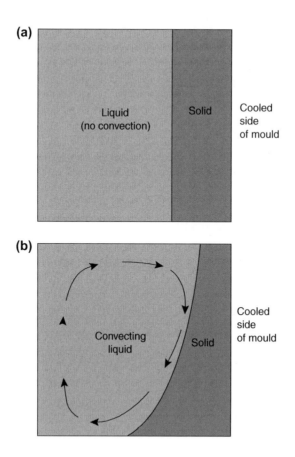

FIGURE 10.35

Solidification in a two-dimensional box of which only the right-hand side is cooled (a) planar front growth in the case of no convection; (b) the distortion caused by convective flow.

makes the feeding problem worse by increasing the opportunity for convection. Many of the current problems of low-pressure casting systems derive from this source.

In contrast, having feeders at the top of the casting, and feeding downwards under gravity is completely stable and predictable, and gives reliable results.

The instability of convective problems is worth emphasising. Because the heavy, cool liquid overlays the hotter less dense liquid, the situation is metastable. If the stratified layers of liquid are not disturbed there is a chance that the heavy liquid will remain wobbling around on the top and may solidify in place without incident. However, a small disturbance may upset the delicate balancing act. Once started, the cold melt will slip sideways, plunging downwards to the bottom, and the hot liquid surge upwards, so that a convective circulation will quickly establish. In practice therefore, several castings may be made successfully if the metastable equilibrium is not disturbed, but, inexplicably, the next may exhibit massive remelted and unfed regions.

Triggers to initiate the unstable flow could arise from many different kinds of uncontrolled events. A significant trigger could be an event such as the rising of a bubble from a core blow, as a result of an occasionally ill-fitting of a core print, leading to the chance sealing of the core vent by liquid metal.

Momchilov (1993) gives one of the very few accounts of the exasperating randomness of convection problems. He found that with the use of two riser tubes from a furnace containing liquid metal into one die cavity, successive castings could be observed to have completely different internal temperature histories. The first casting might be fine. However, the subsequent casting would suffer a die temperature inexplicably overheating by 120°C and the temperature in the furnace simultaneously dropping by 65°C. These are powerful and important exchanges of heat between the die and the crucible below. These changes caused the second casting to be partially remelted.

The use of twin riser tubes by Momchilov raises an important feature of convection. Convective flows are required to be continuous, as in a circulation. Thus in the case of two riser tubes into one die cavity, the conditions for a circular flow, up one tube and down the other, are ideal. It is likely that Momchilov would have solved his problem, or at least greatly reduced it, simply by blocking off one of the tubes.

The elimination of ingates in this way to solve convection problems in counter-gravity *fed* castings should be considered as a standard *first* step. This was found to be a useful measure in the early days of the Cosworth Process when it operated merely as a static low-pressure casting process. (The later development of the rollover concept represented a welcome total solution.)

The only other description of the problems of convection ever discovered by the author comes from a patent by Rogers and Heathcock (1990). They fall foul of convection during the attempt to make an aluminium alloy cylinder block casting in a counter-gravity filled permanent mould. They found that as the mould heated up the problem became worse, and the rate of flow of the convection currents increased. The microstructure of the casting was unacceptable in the area affected by convection. They dealt with the problem by providing strong cooling just above the ingates. This solution clearly threatened the provision of feed metal whilst the casting was solidifying, and so was a risky strategy. There is no record that the patent was ever implemented in production. Perhaps convection secured another victim.

Castings that employ a third mould part to site the running system under the casting are at risk of convective effects causing the melt to circulate up some ingates and down others via the casting above and the runner underneath. This is especially dangerous if the runner is a heavy section. Pressurising the runner with an adequate feeder is a way of maintaining the net upward movement of metal required for the feeding of the casting, thus reducing the deleterious effects of the convection to merely that of delaying freezing. In this case, the worst that happens is the development of a locally coarser structure.

Investment casting often provides numerous convective loops in wax assemblies as a result of attaching the wax patterns at more than one point to increase the strength of the complete wax assembly. A typical wax assembly for the casting of polycrystalline Ni-base turbine blades is illustrated in Figure 10.36(a). The central upright is surrounded by six blades (only two are shown in the section), so that in addition to its heavy section designed to act as a feeder, it is kept even hotter by the presence of the surrounding blade castings that prevents loss of heat by radiation. Conversely, of course, the blades cool quickly because they can radiate heat freely to the cool surroundings. A convective loop is therefore set up with hot metal rising up the central feeder and falling through the cooling castings. The final grain structure seen on the etched component reveals the path of the flow (Figure 10.36(b)). The casting is designed to have fine surface grains nucleated by the cobalt aluminate addition to the primary coat of the mould. However, because hot metal enters the mould cavity from the top, sweeping down through the casting after the chill grains are formed, the original chill grains are remelted in those patches where the flow brushes against the walls. The flow becomes a concentrated channel as it exits the base of the blade. The very narrow section of the trailing edge of the casting is not penetrated, and so escapes remelting, as does the large region in the bottom right that the flow has missed.

Very large blades for the massive land-based turbines for power generation are sometimes cast horizontally. In this case, each end of the casting is subject to convective problems as is seen in Figure 10.37.

The cutting of convective links in wax assemblies is recommended and cries out for wide attention in most current investment casting operations. The strengthening of wax assemblies by wax links inadvertently provides convective links and should be avoided. Ceramic rods can provide strengthening, or, if wax connections are used, they should be plugged with a ceramic disc to avoid metal flow. These simple modifications to the wax assembly will completely change the mode of solidification of the castings, allowing for the first time an accurate understanding of filling and feeding effects.

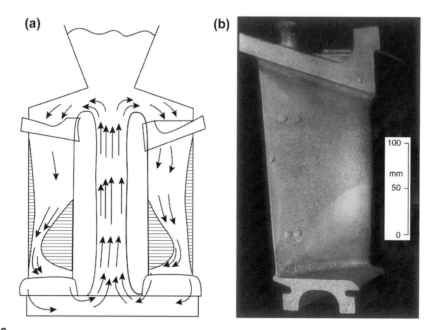

FIGURE 10.36

(a) Lost wax assembly of six Ni-base turbine blades around a central feeder, showing the expected convective loops; (b) an etched blade showing the remelting of the fine surface grains created by the cobalt-aluminate nucleant in the mould surface, and the subsequent growth of coarse grains that define the patches where the flow paths impinged on the casting surface.

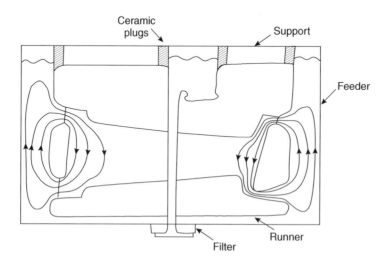

FIGURE 10.37

Horizontal orientation of a large investment-cast turbine blade, illustrating convective loops in the root and shroud. The flows convey heat from the cylindrical feeders, remelting regions of the casting.

Other problems in sand castings are illustrated in Figure 10.38. Gravity die (permanent mould) castings are less prone to these problems because of their more rapid rate of heat extraction by the metal mould. For castings in metal moulds, the sections have to be considerably larger before convection starts to be a threat. Figure 10.39 shows the convection effects from side or bottom feeding compared with the relative stability of top feeding.

It is evident that many computer predictions of heat flow and the feeding of castings will be quite inadequate to deal with convection problems because it is usual to consider the loss of heat from castings simply by conduction. Clearly, thicker sections in a loop will cool more quickly than the computer would predict because convection allows them to export their heat. Conversely, of course, thin sections in the same loop will suffer the arrival of additional heat that will greatly delay their solidification. In fact, if the hot section has an independent source of heating, such as the electrical heating provided in many counter-gravity systems, the sections in the loop can circulate for ever. The computer would have particular difficulty with this.

Even so, the greater speed and sophistication of computing will eventually provide the predictions containing the contribution of convection that are so badly needed. It is hoped that future writers and founders will not need to lament our poor abilities in this area.

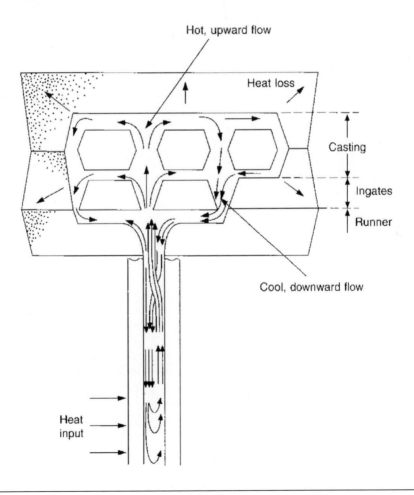

FIGURE 10.38

Convection driven flow within a solidifying low-pressure casting.

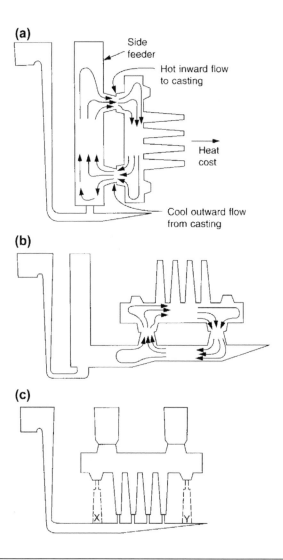

FIGURE 10.39

Encouragement of thermal convection by (a) side feeding; (b) bottom feeding; (c) its near elimination by top feeding.

10.7.3 CONVECTION DAMAGE AND CASTING SECTION THICKNESS

If the solidification time of the casting is similar to the time taken for convection to become established, extensive remelting can be caused by convective flows. Serious damage to the micro- and macro-structure of the casting can then occur. The convective flow takes about one or two minutes to gather pace and organise itself into rapidly flowing plumes. This is occurring at the same time as the casting is attempting to solidify. The flows cut channels through the newly solidified material, remelting volumes of the casting.

Castings that freeze in a time either shorter than 1 min or longer than perhaps 10 min are expected to be largely free from convection problems as indicated below.

Thin section castings are largely free from convection difficulties. They can therefore be fed uphill simply because the thin section gives (1) the viscous restraint of its nearby walls makes any convective tendency more difficult and (2) more rapid freezing allows convection less time to develop and so less time to wreak damage in the casting. Thus instability is (1) suppressed and (2) given insufficient time, respectively, so that satisfactory castings can be made.

Conversely, thick section castings taking perhaps 10 min to 10 h to freeze are also relatively free from convection problems, because the long time available before freezing allows the metal plenty of time to convect, re-organising itself so that the hot metal floats gently into the feeders at the top of the casting, and the cold metal slips to the bottom. All this activity occurs and is complete before any significant amount of solidification has occurred. Thus the system reaches a stable condition before damage can be caused. Once again, castings are predictable.

In what can only be described as a perverse act of fate, convection does its worst in the most common sizes of castings, the problem emerging in a serious way in the wide range of intermediate section castings. These include the important structural castings such as automotive cylinder heads, cylinder blocks and wheels, and the larger investment cast turbine blades in nickel-based alloys amongst many others. Convection can explain many of the current problems with difficult and apparently intractable feeding problems with such common products.

Channels cut through the structure of the casting which is attempting to solidify will contain a coarse microstructure because of their greatly delayed solidification and in addition may contain shrinkage porosity if unconnected to feed metal. This situation is likely if the feeders solidify before the channels as undoubtedly happens on occasions, because the channels derive their energy for flow from some other heat source, such as a very heavy section low down on the casting, or the ingate attached to the riser tube of a counter-gravity furnace for instance.

For conventional gravity castings that require a lot of feed metal, such as cylinder heads and blocks, and that are bottom-gated, but top-fed, this will dictate large top feeders, because of their inefficiency as a result of being furthest from the ingates, and so containing cold metal. This is in contrast to the ingate sections at the base of the casting that will be nicely pre-heated. The unfavourable temperature regime is of course unstable because of the inverted density gradient in the liquid, and thus leads to convective flow, and consequent poor predictability of the final temperature distribution and effectiveness of feeding. It is the standard legacy of bottom filling: the favourable filling conditions leading to the worst feeding conditions. Life was never easy for the casting engineer.

The upwardly convecting liquid within the flow channels usually has a freezing time close to that of the pre-heated section beneath, which is providing the heat to drive the flow. In the case of many low-pressure systems, the metal supply system is artificially heated, leading to a constant heat input, so that the convecting streams rising out of these regions never solidify. This is what happened to the Cosworth system in the early days of its development. When the mould and casting (which should by now have been fully solidified) were hoisted from the casting unit, liquid poured from the base of the mould, emerging from remelted channels to the amazement of onlookers. Everyone had assumed that after the appropriate length of time for solidification, no liquid could possibly still be present, and present in such quantity.

When removing a convecting casting from a counter-gravity filling system in this way, the draining of liquid from the interdendritic regions leaves regions in the casting that appear convincingly like shrinkage defects, and are usually confused as such.

The convection of hot metal up and the simultaneous movement of cold metal down the riser tube of a low-pressure casting unit delays the freezing of the casting in the mould above and can lead to a significant reduction in productivity. A thermocouple in the riser tube reveals the chaotically varying temperature of the hot and cold eddies as they swirl past each other.

The author is aware of a series of castings being made on a low pressure machine whose freezing time kept increasing as the melt was subjected to increasingly thorough rotary degassing treatment. It seems that each rotary degassing treatment reduced the amount of bifilms in suspension. Because the effective viscosity of the melt was progressively reduced in this way the convection increased, extending the time taken for the casting to solidify. Thus clean metal is free to convect, whereas melt with an internal semi-rigid lattice of bifilms will be more resistant to flow. It is perhaps superfluous to add that reliance on the inhibition of convection by filling the liquid with oxides to increase viscosity cannot be recommended.

10.7.4 COUNTERING CONVECTION

Solutions to the problems of convection are summarised as follows.

1. Rollover.

The inversion of the mould after casting effectively converts the pre-heated bottom ingate filling system into a top feeding system, thus gaining a really efficient feeding system.

Furthermore, of course, the massive technical benefit of the inversion of the system to take the hot metal to the top, and the cold at the bottom, confers stability on the thermal regime. Convection is eliminated. For the first time, castings can be made reliably without shrinkage porosity.

Batty (1935) was well ahead of his time, using the roll over technique for steel castings. He showed by careful measurement of the temperature gradient the reasons for his success to convert 'troublesome' steel castings to 'reliable' steel castings.

The massive productivity and economic benefit of this technique when applied to Al aggregate moulded castings follows because the mould now contains its liquid metal all below the entry point in the mould, allowing the mould to be removed from the casting station without waiting for the metal to freeze (which is of course the standard productivity delay suffered by most counter-gravity casting processes). In this way, cycle times can be reduced from about 5 to less than 1 min. This is a powerful and reliable system used by such operations as Cosworth Process which achieved the casting a mould containing two cylinder blocks every 45 s. The technique has made this sand casting process the fastest automotive block production process in the world. In addition, the castings are of a superb integrity.

2. Tilt casting

Those tilt casting processes in which the rollover is used *during* casting actually to effect the filling process can also satisfy the top feeding requirement.

However, in practice care is needed because many geometries are accompanied by waterfall problems, if only by the action of the sliding of the metal in the form of a stable, narrow stream down the sloping side of the mould. Thus meniscus control is, unfortunately, often poor. Where the control of the meniscus can be improved to eliminate entrainment problems, tilt casting techniques are valuable. Ultimately, if the tilt is controlled to perfection, a kind of horizontal transfer of the melt can be achieved. This system does not seem difficult or costly to attain, and is commended strongly.

3. Cut convective loops.

Explore the elimination of ingates when counter-gravity feeding of castings. The widespread use of convective loops in investment castings wax assemblies should be reduced by the wider use of ceramic supports and stops.

10.8 RULE 8: REDUCE SEGREGATION DAMAGE

Section 5.3 describes the many mechanisms whereby concentration of solutes can be enhanced or depleted in some regions of the casting during the short time of solidification.

Microsegregation between dendrite arms is, as we have seen, a necessary consequence of normal freezing of an alloy, but the fine scale of the effect allows a significant amount of redistribution during subsequent homogenising and solutionising heat treatments.

The various forms of macrosegregation, on the other hand, are irreversible and have to be lived with. They can sometimes be sufficiently serious to threaten the serviceability of the casting, and in any case will threaten to scrap the casting if the composition is found to be outside the specification limits in places, whether or not this is important for the function of the casting. (At times, the founder has to live with the injustice of castings rejected for illogical or unreasonable objections, even though destructive testing to simulate service conditions would in principle provide evidence of the complete harmlessness of most deviations from over-tight specifications.)

One of the most common of macrosegregation problems is that arising from dendritic freezing. This perfectly normal mode of segregation occurs in practically all castings to some degree: the positively partitioning solute is concentrated against a cooling surface. In sand castings, the effect is rarely a problem because of the relatively small temperature gradients, but can be serious in metal moulds and when using metallic chills. Thus the use of chills requires caution in those alloys prone to segregation. In addition, the use of a fin or a pin can simulate the action of a chill so as to give a local segregation effect.

For Al alloys, segregation is not usually a problem, especially in alloys such as those based on the Al-Si-Mg system. However, for those Al alloys that contain Cu, particularly if the Cu content is in the region of 4.5–5.5%Cu, the variations in Cu content throughout a casting can be serious.

Figure 10.40(a) gives an example. Dendritic segregation will lead to a concentration of Cu at a chill face, but an opposite pattern adjacent to the feeder. Both regions can easily be well outside specification. A reduced section thickness can act as a cooling fin and so simulate the action of a chill, causing a peak in Cu segregation adjacent to the fin as shown in Figure 10.40(b). However, just inside the fin, the conditions are those of a thin-section casting joined on to a heavy section, thus delaying freezing, so that the opposite pattern of segregation is to be expected. This steep change in composition will, of course, result in a correspondingly large variation in mechanical properties, particularly after heat treatment. It is a concern that these large changes in strength, possibly from brittle to ductile behaviour, are concentrated at the change in section at which externally applied stress will also be concentrated. These effects do not appear to have been investigated so far, but clearly should be explored. Design rules for avoiding or reducing these problems would be welcome.

The other macrosegregation effect that is troublesome in slowly cooled castings is the channel defect. This arises from the effect of gravity on the solute in the dendrite mesh, causing liquid high in solute to migrate, organising its flow into channels that remelt a pathway through the mesh. The effect has been countered in some alloy developments by the design of more neutral buoyancy into the segregated melt. However, this luxury is not normally open to the majority of castings that have to conform to a chemical specification, even though the specification is often rather arbitrary and rarely of any consequence for the satisfactory operation of the product in service.

Channel segregates appear also to concentrate bifilms, as evidenced by the numbers of cracks that can occur in these lengthy pencil-like channels. The bifilms will be pushed into these regions. Later, when most of the casting is solid, the highly segregated liquid in the channel will remain liquid as a result of its low melting point. The local high gas content in solution, and the local reduction in pressure resulting from the problems of feeding down such long channels, partly blocked by tumbling melted-off grains (the 'freckles' when viewed by sectioning and etching), will combine to unfurl and expand the bifilms, creating planar cracks across the width of the channel. The cracks are, of course, much more serious than the segregates themselves and are probably the sole reason why channels are considered to be deleterious. If the mould could be filled with good-quality metal, channel defects would probably be harmless.

The vigorous operation of channel segregates in heavy steel castings funnels low-density solutes, mainly high in carbon, into the feeder head of the casting. Unfortunately, this segregated liquid accumulating in the feeder is not harmlessly put away, but awaits its turn to degrade the casting at a later stage, when the feeder comes into operation to

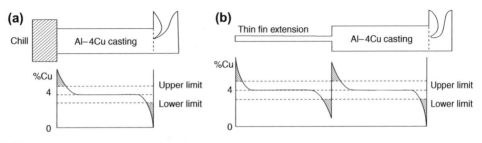

FIGURE 10.40

(a) Dendritic segregation pattern, concentrating solute against a chilled face; (b) analogous pattern produced by a reduction in section thickness acting as a cooling fin.

feed the solidification shrinkage in the body of the casting. At this late stage, the high carbon liquid is drawn down into the casting, giving a carbon-rich region immediately below the feeder (Figure 10.41). If this action is especially severe, as occurs in heavy steel ingots, channels are melted from the feeder into the casting by this sucked-down low melting point liquid. We call these channels 'V segregates'.

The authors Li et al. (2014) describe an excellent solution to the huge segregation problems of very large ingots, especially those intended for power plant forgings. Instead of filling up a single ingot mould with a single pour of metal, the mould is filled in stages, using small pours from a succession of small ladles. Each pour is allowed to solidify to some degree, and when approximately 20% residual liquid remains, the next ladle is poured, and so on until the mould is completely full. Thus a 300 tonne ingot might require 10 pours of 30 tons, each waiting until sufficient solidification had taken place, but with sufficient remaining liquid to ensure a problem-free joint. The final ingot is impressively uniform.

In shaped castings, the carbon segregation under the feeder is controlled to some degree by the diameter of the feeder. Clearly, in the extreme, a feeder of the same diameter of the casting would cause minimal segregation. However, this is clearly not a particularly useful solution but may indicate that a modest increase in feeder diameter might be helpful, and a narrow feeder neck breaker core may do more harm than good. Faster freezing of the casting would help most. The founder needs to be aware of this solute pattern and may ultimately have to target the fine balance between too little carbon in the main body of the casting and too much beneath the feeder. Once again, the foundry engineer would greatly benefit from some quantitative guidance from a program of carefully conducted research.

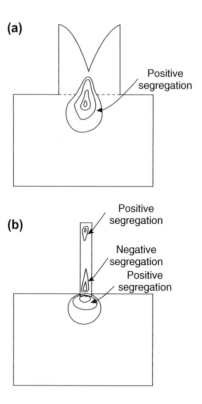

FIGURE 10.41

(a) Positive segregation under a feeder; (b) positive segregation under a cooling fin, but negative closely adjacent inside the fin. These extremes of concentration of solute are both close to a vulnerable change in section, where stress may be concentrated.

The under-feeder segregation is compared and contrasted with the opposite thermal condition: an under-fin condition giving some more complexity overall, but similar results, if on a smaller scale, in the casting (Figure 10.41(b)). Naturally, an under-chill condition is similar. The fin and pin actions are not likely to be noticeable in steel castings as a result of the poor thermal conductivity, but the situation is likely to be non-trivial in Al, Mg and many Cu alloys.

A final observation on segregation is important. All segregations take time to build up. Thus reducing the time available by almost any method is a valuable and powerful general strategy to reduce macrosegregation. The provision of additional chills arranged to reduce the temperature gradients that would have been set up by a single chill, especially using internal chills if possible, is strongly recommended. If the customer would agree, a move to a less segregating alloy would also be an effective strategy.

10.9 **RULE 9: REDUCE RESIDUAL STRESS**
10.9.1 **INTRODUCTION**

It seems that, in general, the engineering community has not been made aware of the potentially awesome importance of residual stress in the manufacturing of engineering components. All manufactured components contain internal stress, often high. The problem is that this very real danger is invisible.

Most foundries and machining operations have stories about the casting that flew into pieces with a bang when being machined or even when simply standing on the floor! Benson (1946) describes one such event. The author and several colleagues narrowly missed the shrapnel from an aluminium alloy compressor housing that exploded in their midst while being cut on a band saw. It happened late one evening after the foundry had stopped work for the day. The building was quiet and in darkness. After the initial bang and the return to silence, the sound of fragments bouncing off the roof and clattering echoes of them distantly falling to the floor was awesome.

It is easy to disregard such stories. However, they should be viewed as warnings. They warn that, in certain conditions, castings can have such high stresses locked inside that they are dangerous and unfit for service. In fact, they can act like small bombs. We are unaware that the casting may be on the brink of catastrophic failure because, of course, the problem is invisible; the casting looks perfect.

There are those metallurgists within the industry, some eminent, and whose opinions on other matters I respect, that have taken issue with me. They have argued that the presence of residual stresses, particularly those from quenching, are actually irrelevant because the whole component is in balance with its own stresses. The question of overall balance is certainly true. However, this argument overlooks that the *distribution* of stress is to be expected to be far from uniform, and parts of the component may be near to their failure stress even before the application of any stress in service. Usually, as we shall see, the major tensile stress is in the centre, and it is this part of the component that fails first under tensile load.

Admittedly, not all components are necessarily endangered by internal stress. Indeed, in some cases the stress can be beneficial (some examples are given later). However, in general, the very real risk exists that the stress may not be beneficial. The residual stress may add to the service stress and so promote premature failure at only low service stress, to the bewilderment of the designer who imagines the material of his component to be uniform and who has completely overlooked the possibility that the part contains invisible threats from stresses. He responds by increasing the thickness of the section—a response which is likely to be counter-productive.

Because of the complexity of some castings, and the complexity of the state of stress, it is usually not easy to estimate the magnitude of either the internal residual stress or its precise action. Often however, the stress is at least equal to or exceeds the yield stress. Thus it is not trivial. In fact at this level it will dominate all other designed loads. If the casting is in a fatigue condition, it will certainly lead to early failure. It is ignored at our peril.

This section takes a look at the wide spectrum of stresses in castings, and attempts to clarify those that are important and which should be controlled, in contrast to those that can be safely neglected.

10.9.2 **RESIDUAL STRESS FROM CASTING**

There have been several test pieces that have been used over the years to help assess the parameters affecting the residual stress in castings. Most of these are based on the form of the three-bar frame casting shown in Figure 10.42.

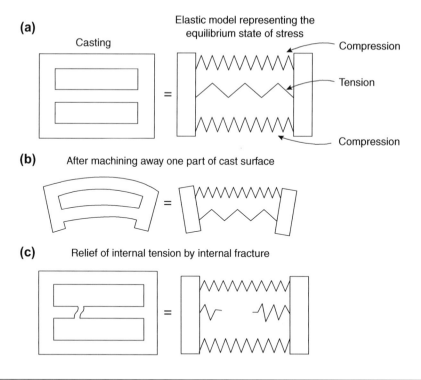

FIGURE 10.42

Heyn's (1914) model of the balance of internal stresses after rapid cooling: (a) quenched casting showing high internal tensile stress and relatively low external compressive stress; (b) the distortion of the casting after one side is machined away; and (c) the common condition of internal tensile failure.

In practice the stress remaining in the casting is usually assessed by scribing two lines on the central bar and accurately measuring their spacing. The bar, of length L, is then cut between the lines, and, usually, the cut ends spring apart as the cut is completed. The distance between the lines is then measured again and the difference ΔL is found. The strain ε is therefore $\Delta L/L$ and the stress σ is simply found, assuming the elastic (Young's) modulus E, by the definition:

$$E = \sigma/\varepsilon \qquad (10.19)$$

Such studies have revealed that the residual stress in castings is a function of the cooling rate in the mould, as shown for aluminium alloy castings from the effect of water content of the mould in Figure 10.43. Dodd (1950) cleverly illustrated that this effect is not the result of the change of mould strength by preparing greensand moulds with various water contents, then drying each carefully so that they all had the same water content. This gave a series of moulds with greatly differing strengths. When these were cast and tested, there was found to be no difference in the residual stress in the castings. This result was further confirmed by testing castings made in moulds rammed to various levels of hardness. Again, no significant difference in residual stress was found. Dodd also checked the effect of casting temperature and noted a small increase in residual stress as casting temperature was increased.

As with cases of the constraint of the casting by the mould, removing the casting from the mould at an early stage would be expected to be normally beneficial in reducing residual stress. Figure 10.44 shows the result for iron and a high-strength aluminium alloy. The higher residual stress for cast iron reflects its greater rigidity and strength. The effects of percentage water in the sand binder, and of stripping time and casting temperature, have been confirmed in other work on high strength aluminium alloys and grey iron using a rather different three-bar frame (IBF Technical Subcommittee,

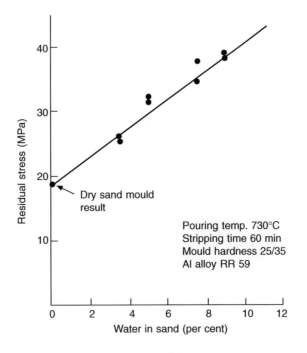

FIGURE 10.43

Internal stress in the centre member of a three-bar frame as a function of water content of the greensand mould (Dodd, 1950).

1949, 1952). It is noteworthy that all these direct measurements show casting stresses to be modest; only approximately 10% of the yield strength.

All these observations appear to be explainable assuming that the main cause of the development of residual stress is the interaction of different members of the casting cooling at different rates. Beeley (1972) presents a neat solution to the problem.

The strain $\Delta L/L$ resulting from differential contraction is determined by the temperature difference ΔT and the coefficient of thermal expansion of the alloy α. Thus, we have the strain

$$\varepsilon = \Delta L/L = \alpha\Delta T \tag{10.20}$$

and so from Eqn (10.19), the stress is

$$\sigma = \alpha E\Delta T \tag{10.21}$$

The stress therefore depends on the temperature difference between members. It is also worth noting that the stress is independent of casting length L.

A rather different result illustrating the influence of the geometry of the three-bar frame casting was found by Steiger (1913). He measured the increase of stress in the centre bar of grey iron castings after increasing the rigidity of the end cross-members. He found that a centre bar of more than twice the diameter of the outer bars would suffer a residual stress of more than 200 MPa, sufficient to fracture the bar during cooling. Working Eqn (10.21) backwards, it is quickly shown that the temperature difference was only about 100°C to produce this failure stress. Clearly, such temperature differences will be common in castings, and often exceeded. Thus, high stresses may be expected for some castings. Even so, this conclusion was reached from very early work (1913), and should really be confirmed with modern techniques and equipment before we place too much faith in this result.

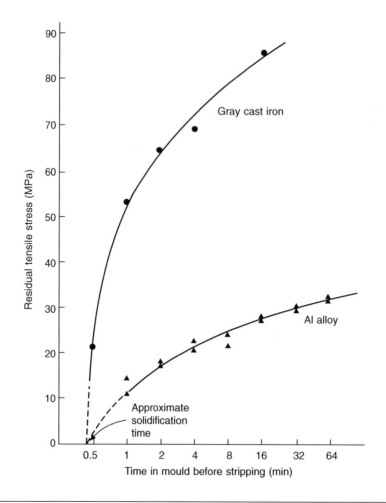

FIGURE 10.44

Residual stress in Al alloy and grey iron castings as a function of stripping time.

Data from Dodd (1950) and IBF Technical Committee (1949).

In ferrous castings that experience a gamma to alpha (austenite to ferrite) phase change during cooling, any stresses that are built up before this event are probably reduced, their memory diluted by the uniform plastic flow that the transformation causes throughout the metal. It seems probable therefore that the temperature differences and cooling rates applying below the gamma-alpha transformation temperature that are the most important for the final remaining levels of stress in these steels.

This fact prompts Kotsyubinskii (1961) to recommend that heavy sections of ferrous castings be cooled by forced air or chills to equalise their cooling rates with those of the thinner sections, up to the point at which the pearlite reaction occurs. Below this temperature, little can be done to avoid the buildup of stress. This is because the metal is largely elastic, and plastic relaxation, occurring only slowly by creep, becomes progressively less effective; thus cooling should at that stage be slow and even, so as to take advantage of as much natural stress relief as possible.

In an aggregate mould, castings are cooled relatively slowly, so that the final internal stress in the product will normally be relatively low and can often be neglected (although the possible exception warned by Steiger previously is noted). It is true that the dimensions of the casting will often be changed by stress during cooling, but on shaking out from the mould the final, residual, stress will not normally be high. In addition, the distortions that have arisen during cooling in the mould are usually extremely reproducible. This is a consequence of the reproducible conditions of production, in which the mould is the same temperature each time, and the metal is the same temperature each time, so that the final shape is closely similar each time. This reproducibility of sand castings is probably greater than for any other casting process.

This repeatable regime is not quite so well enjoyed by the various kinds of die casting, particularly gravity die (permanent mould) casting, as a result of many factors, but in particular the variability of mould size and shape as a result of variation of mould temperature. The somewhat faster cooling, particularly because of the earlier extraction of the casting from the mould, is an additional factor that does not favour low final stress.

In general, internal stress remaining from the casting process is rarely high enough to be troublesome, but we cannot always be complacent about this. The ability to predict stresses using computer simulation will be invaluable to maintain a cautious watch for such dangers.

Ultimately, however, particularly for aluminium alloys, the stresses from casting are usually eliminated by any subsequent high temperature solution heat treatment which acts as an excellent stress-relieving treatment.

In confirmation, Benson (1938) made the important commonsense generalisation that it is the last thermal treatment, and the rate of cooling from this final treatment, that is important so far as residual stresses are concerned. Thus the last treatment might simply be casting, or annealing, or quenching from a solution treatment. Having considered casting, we shall now turn our attention to these subsequent treatments.

10.9.3 RESIDUAL STRESS FROM QUENCHING

When quenching castings from a high temperature heat treatment, the time for cooling the outer sections of castings is shorter than the time required for heat to diffuse out from interior sections. The outer parts of the casting therefore cool first to form a rigid, strong frame that will contract to squash the hot, weak, plastic central features to shorter lengths. Later, the inner sections, now somewhat smaller in length, will cool, gathering strength and contracting further, but this late contraction will suffer the restraint of the outside sections that have by now have become cold and rigid. Thus, the interior sections, now also becoming strong and rigid, go into tension, forcing in turn the outer parts into compression. Thus the major problems of internal tensile stress and distortion of the casting are usually created in quench operations.

Furthermore, the stresses are not significantly reduced by a subsequent ageing treatment. The temperatures and times for ageing treatments are too low to lead to stress relief.

It is unfortunate that many heat treatments require a quenching stage, intended to cool the casting sufficiently quickly to freeze solutes in to a solid solution, thereby preventing them from precipitating too early, and thus still being available for controlled precipitation during the ageing treatment. If the quench is slow some solute may be lost by precipitation from solution, thus making it unavailable for subsequent hardening reactions, so that the final strength of the casting is reduced. This reasoning has driven metallurgists to seek quenching rates as fast as possible.

The problem has been that all such research by metallurgists to optimise heat treatments has been carried out on test bars of a few millimetres in diameter that represent no problem to cool quickly. The outside and inside of the bars are in excellent thermal communication, and the high thermal conductivity of most metals ensures that the cooling throughout the section is essentially uniform. Thus, the world's standards on heat treatment often dictate water quenching to obtain the highest material properties.

Quite clearly, the problem of larger components, or certain components of special geometrical complexity in which uniform cooling is an impossibility, has been overlooked. This is a most serious oversight. The performance of the whole component may therefore be undermined by the application of these techniques that have been optimised by research on small test bars, and which therefore are inappropriate, if not actually dangerous, for many large and complex components.

This is such a common problem that when a troubled casting user telephones me to say words to the effect 'My aluminium alloy casting has broken. What is wrong with it?' This is such a regular question that my standard, and rather tired, reply now is 'Do not bring the casting to me. I will tell you *now* over the telephone why it has failed. It has failed because it has been poured badly and therefore contains bifilms that reduce its strength. However, in addition, you have carried out a solution heat treatment accompanied by a water quench'. The caller is usually stunned, incredulous that I know that he has water quenched his casting, and asks how I know. My experience is this: in all my life investigating the causes of failure of perhaps hundreds of Al alloy castings, only one failed because of embrittlement caused as a result of the alloy being wildly outside chemical specification. All the rest failed for only two reasons: (1) weakening by bifilms, together with (2) massive internal stresses that have loaded the already weakened casting close to its failure stress even before any service stress was applied.

I have to record, with some sadness, that all the standard metallurgical investigations into casting failures that I see appear completely irrelevant; they include such costly and time consuming activities as the determination of the chemical specification, the metallurgical structure, the mechanical properties such as hardness and other standard metallurgical tests. It underlines the importance of understanding the new metallurgy of cast metals in which the residual stresses and bifilms together play the dominating roles in the performance of engineering components, particularly cast engineering components. It is necessary to sympathise with the metallurgist attempting to carry out the traditional failure investigation because both these dominating macroscopic defects are hard to detect: the stress is perfectly invisible, and the bifilms are mostly invisible, but these two invisible factors are in control.

Equation (10.21) explains why not all shapes and sizes of castings necessarily suffer a problem. Compact or small castings, and those for which the quenchant can easily reach all parts, are often not seriously affected, because ΔT is necessarily small. It is essential to remember this: not all castings are seriously affected by quench stresses.

However, for those castings that are affected by uneven temperatures during cooling, it is salutary to estimate the actual magnitude of the strain ε. For aluminium the coefficient of expansion α is about 20×10^{-6} K^{-1} and the temperature fall during a quench is approximately 500 K. The strain works out to be therefore approximately as $20 \times 10^{-6} \times 500 = 10^{-2} = 1\%$. For steels, α is approximately 14×10^{-6}, and the temperature change for quench from many heat treatments in the region of 900 K, again giving a strain close to 1%. Because the yield (or proof) strain is only approximately 0.1%, these imposed quench strains are about 10 *times* the yield strain. These dramatic strains can therefore be seen to take the component well into the plastic deformation range.

For steels with elongations to failure of perhaps 30–50%, the imposition of 1% strain is usually not a serious threat to the serviceability of the casting, even though the corresponding stress is very high.

However, for many Al alloys, with only 1–2% elongation to failure, the imposition of 1% strain can be highly threatening. The comforting thought for the future is that as elongations for Al alloys approach those of steels because of the steady improvement of cleanness, the danger of residual stress in such castings will decrease.

In the meantime, casting geometries that are particularly susceptible include large, thick section castings, where the heat of the interior takes time to reach the outside of the casting, giving high ΔT. Ingots or other block-type products can be seriously stressed for this reason. Direct chill continuously cast ingots of aluminium alloys are severely cooled by water, but are often larger than 300 mm diameter. Whilst sitting on the shop floor waiting further processing the strong 7000 series alloy ingots have sometimes been known to explode like bombs.

As an aside, the length of time taken before the ingot decides to fail is curious and interesting. It seems likely that the failure under the high internal stresses is initiated from one of the large bifilms that is expected to be entrained during the turbulent start of casting. The gradual precipitation of hydrogen into the bifilm will gradually increase the pressure in the bifilm crack, encouraging it to extend as a stress crack. The hydrogen may be already in solution in the metal or may be gradually accrued by reaction with water vapour in the atmosphere during storage, especially if part of an extensive bifilm is near to the ingot surface where it can receive hydrogen by diffusion from the outside surface, but is therefore able to distribute the pressurised gas over the whole area of the bifilm to create significant stress at the sharp edges of the defect. Other penetrating contaminants may include air to cause additional internal oxidation, or fluxes, or traces of chlorine gas, or sulphides from greases, to act as surface active additions to reduce the surface energy of the metal and so further encourage crack growth. Research to clarify these possibilities would be valuable.

Other varieties of castings that are susceptible to damaging levels of residual stress include those that are hollow, with limited access for the quenchant into the interior parts of the casting, and which also have interior geometrical features such as dividing walls and strengthening ribs (Figure 10.45). This latter series of geometrical requirements might seem to be an unlikely combination of features that would eliminate most castings. Perhaps surprisingly therefore, the list of castings that fulfils these requirements is rather long and includes such excellent examples as automotive cylinder heads

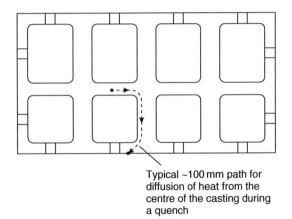

Typical ~100 mm path for
diffusion of heat from the
centre of the casting during
a quench

FIGURE 10.45

Schematic representation of a hollow casting with internal walls and small ports to the outside world, such as an automotive cylinder head, illustrating the long diffusion path for heat from its centre during a quench, leading to internal residual tension.

and blocks and housings for components such as compressors and pumps. When immersed in the water quench, the water attempts to penetrate the entrances into the hollow interior of the casting. However, because the casting is originally above 500°C, any water that succeeds in entering will convert almost instantaneously to steam, blowing out any additional water that is attempting to enter. The result is that the outside of the casting in contact with the quenchant cools rapidly, whereas the interior can cool only at the rate that thermal conduction will conduct the heat along the tortuous path via interior walls of the casting to the outer surfaces.

The rate of conduction of heat from the interior to the exterior of the casting can be estimated from the order of magnitude relation

$$x = (Dt)^{1/2} \tag{10.22}$$

where x is the average diffusion distance, D is the thermal diffusivity of the alloy and t is the time taken. The thermal diffusivity is defined as

$$D = K/\rho C \tag{10.23}$$

where K is the thermal conductivity, about 200 $Wm^{-1}K^{-1}$ for aluminium, the density ρ is about 2700 kgm^{-3} and the specific heat C is approximately 1000 $Jkg^{-1}K^{-1}$. These values yield a value for the thermal diffusivity D close to 10^{-4} m^2s^{-1}. (The corresponding value for steel is approximately 10^{-5} m^2s^{-1}.) Equation (10.22) is used to generate Figure 10.46 in which the distance for diffusion of heat out of a product indicates the approximate boundaries of safe regimes constituting conditions in which sufficient time is available for the diffusion of heat from the interior during the quench. The time of cooling in different quenchants is provided by results such as that shown in Figure 10.47. These results were obtained by siting a thermocouple in the centre of a 10mm wall of an Al-7Si-0.4Mg alloy casting. Comparative results would be valuable for ferrous materials.

For Al alloys, the temperature range of approximately 500°C down to 250°C appears to be the critically important range during cooling because any beneficial effects of solute retention occur in this range and little can be affected at temperatures below 250°C.

For a solid aluminium bar of 20 mm diameter (approximately equivalent to a solid plate of 10 mm thickness), Figure 10.47 indicates that quenching in water will reduce the temperature in its centre from 500 to 250°C in about 5 s. Substituting 5 s in Eqn (10.22) shows that on average heat will have travelled 20 mm in this time. The 20 mm bar or 10 mm plate will both therefore enjoy a reasonably uniform temperature so that minimal stress will be generated.

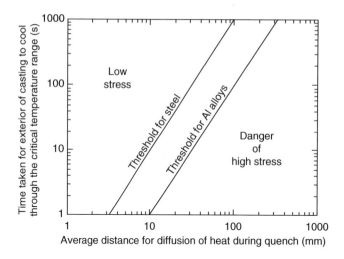

FIGURE 10.46

Regime for low stress in terms of quench rate and distance for heat flow.

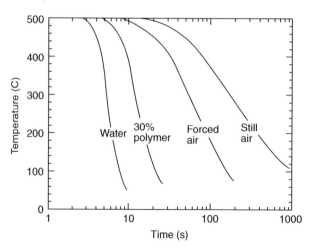

FIGURE 10.47

Quench rates in a 10 mm thick Al plate casting in a variety of quench media.

In such castings as automotive cylinder heads, whose links between the internal sections and the outside world are via small holes in the outer walls and tortuous routes around the water jackets the distance that heat has to diffuse from the centre of the casting is of the order of 100 mm compared with the outside walls which are 10 mm or less. Figure 10.47 indicates that the cooling times are approximately 100 and 5 s, respectively. This large difference in the cooling and contraction results in internal tensile stress over the yield point and well into the plastic flow regime.

10.9.4 CONTROLLED QUENCHING USING POLYMER AND OTHER QUENCHANTS

Figure 10.47 shows that there are other intermediate quenching rate options available. The use of water, whilst being cheap and environmentally pleasant, causes problems for the quenching of most alloys, whether light alloys or ferrous.

The rapidity of the quench is not suitable for larger parts, especially hollow parts with internal structures, as we have seen previously. In addition to this problem, water gives an uneven and non-reproducible quench because of its boiling action. When the parts are immersed in the water, they are at a temperature hundreds of degrees above the boiling point. Thus the water in contact with the hot surface boils, coating it with a layer of vapour that conducts heat poorly, temporarily insulating that area whilst surrounding areas that may happen to remain in contact with the water continue to cool rapidly via a nucleate boiling regime (Figure 10.48; this figure is the result of an investigation by Xiao et al., 2010 with a window-frame shaped casting of 122 mm square with sections of 10–25 mm). The pattern of contact varies rapidly and irregularly as the vapour film forms and collapses in the turbulent water. Agitation of the water is often used to even the cooling effect, so far as possible. However, Figure 10.48 shows that agitation has an effect only on the vapour blanket stage but is not a major effect, and has no effect at all on the nucleate boiling stage which is the process by which heat transfer occurs at maximum rate.

Thus the resulting stress pattern in the casting is complex and different from casting to casting. This is expected to be especially true when the castings are stacked closely in a heat treatment basket because those in the centre of the basket will experience quite different cooling conditions to those on the outside.

To overcome the blanketing action of the vapour, liquids with higher boiling point such as oils have been used. However, the flammability hazard and the smoke and fumes have caused such quenchants to become increasingly unacceptable. Cleaning of the casting after the quench is also an environmental problem. Water-based solutions of polymers have therefore become widely used. They are safer and somewhat less unpleasant in use. Fletcher (1989) reviews their action in detail. We shall simply consider a few general points.

Some polymers are used in solution in water and appear to act simply by the large molecular weight and length of their molecules increasing the viscosity and the boiling point of the water. Such viscous liquids are resistant to boiling and so provide a more even quench, with the quenchant remaining in better contact with the surface of the casting.

Sodium polyacrylate solution in water produces cooling rates similar to those of oils. However, its action is quite different. It seems to stabilise the vapour blanket stage by enclosing the casting in a gel-like casing (Fletcher and Griffiths, 1995).

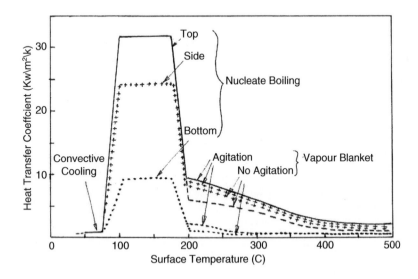

FIGURE 10.48

Heat transfer coefficient (HTC) measured for a small frame casting quenched into water at 50°C. HTC is a strong function of temperature and orientation, but is relatively little influenced by agitation except at high temperature.

Smoothed data from Xiao et al. (2010).

Other polymers have a so-called reverse temperature coefficient of solubility. This lengthy technical description means that the polymer becomes less soluble as the temperature of the solution is raised. Many, but by no means all, of the polymers are based on glycol. One widely used polymer is polyalkylene glycol. This material becomes insoluble in water above about 70°C. The commercial mixtures are usually sold already diluted with water because the product in its pure form would be intractably sticky, like rather solid grease, and would therefore present practical difficulties on getting it into solution. It is usually available containing other chemicals such as antifoaming agents, corrosion inhibitors, and possibly anti-bacteriological components.

Such polymers have an active role during the quench. When the quenchant contacts the hot casting, the pure polymer becomes insoluble. It separates from the solution and precipitates both on the surface of the casting, and in the hot surrounding liquid as clouds of immiscible droplets. The sticky, viscous layer on the casting, and the surrounding viscous mixture, inhibit boiling and aid the uniform cooling condition that is required. When the casting has cooled to below 70°C, the polymer becomes soluble once again in the bulk liquid, and can be taken back into solution. Re-solution is unfortunately rather slow, but the agitation of the quench tank with, for instance, bubbles of air rising from a submerged manifold to scrub the sticky residue off the castings, reduces the time required.

Polymer quenchants have been highly successful in reducing stresses in those castings that are required to be quenched as part of their heat treatment. The properties developed by the heat treatment are also found to be, in general, more reproducible. Capello and Carosso (1989) has shown that the elongation to failure of sand-cast Al-7Si-0.5Mg alloy, using 2.5 times the standard deviation to include 99% of expected results, exhibits greater reliability, as shown in Table 10.1. Thus the average properties that are achieved may be somewhat less than those that would have been achieved by a cold water quench, but the products have the following advantages:

1. The properties of heat treated castings are more uniform and reproducible.
2. The minimum values of the random distribution of strength values are raised.
3. With castings nearly free from stress, the user has the confidence of knowing that all of the strength is available and that an unknown level of internal stress is not detracting from the strength indicated (misleadingly) by a test bar.
4. The castings will have significantly reduced distortion.

Table 10.1 Elongation to Failure Results from Different Quenching Media

	Elongation (%)	
	Minimum	**Mean ± 2.5σ**
Hot-water quench (70°C)	2.01	4.73 ± 2.72
Cold-water quench	4.80	6.47 ± 1.67
Water-glycol quench	4.85	5.81 ± 0.96

These authors carried out quenching tests on an aluminium plate $150 \times 100 \times 1.5$ mm, and found that, taking the distortion in cold water as 100%, a quench with water temperature raised to 80°C reduced the distortion only marginally to 86% of its previous value. Quenching in a mixture of 20% glycol in water gave a distortion of only 3.5%.

Ramesh and Prabhu (2013) introduce a dimensionless cooling performance parameter for the characterisation of quench media. For cylindrical metal bars

$$\text{Cooling parameter} = D^2 R / \alpha \Delta T$$

Where D is the diameter of the bar (m), R is cooling rate (K/s), α is its diffusivity (m²/s) and ΔT is the difference between the initial temperature of the bar and the temperature of the quench medium (K). This quantifying parameter might help to clarify future work in this field. In the meantime, we continue to have to live with some uncertainties as described later.

For many castings, the use of a boiling water quench seems to be of limited help in reducing quench stress, even though some workers claim it to be useful (for instance, Godlewski et al., 2013). Thus, although the rate of quench may be reduced by the use of hot or boiling water, the results are not always reliable. This is almost certainly the consequence of the variability of the vapour blanket that forms around the hot casting. The blanket forms and disappears irregularly, depending on many factors including the precise geometry of the part, its inclination during the quench, and the proximity of other hot castings etc. In addition, from a practical point of view, a hot water quench is costly to install, run and maintain.

For steels, other quenching routes to achieve a low stress casting have been developed involving the use of an intermediate quench into a molten salt at some intermediate temperature of approximately 300°C for approximately 20 s before the final quench into water (Maidment et al., 1984). Despite the advantages claimed by the authors, the expense and complexity of this double quench are likely to keep the technique reserved for aerospace components, if anything.

Quenches into fluid beds at different temperatures almost certainly deserve more attention for the control of quench rates (Ragab et al., 2013). Hard and wear-resistant synthetic aggregates would be expected to make good fluid media to limit the formation of dust. There may be some castings which may not be suitable if their geometry has large horizontal planes which suppress the fluidisation, but for the majority of castings the method achieves excellent heat transfer and good environmental advantages.

However, it is with regret they I conclude that the polymer quench is not currently a long-term practical solution for production for an automotive product. The polymer solution has to be cleaned out of internal cavities where it could lodge; otherwise, it would become concentrated and carbonised during the subsequent ageing treatment, creating copious fumes that pour from the ageing furnace and spread throughout the foundry. In addition, any residual core sand in such locations would be effectively bonded into place by this concentrated polymer binder, and would be practically impossible to dislodge quickly. Perversely, such bonded-in sand is likely to cause damage later in the life of the engine when it finally decides to free itself and starts to clog oil passages and destroy bearings. Finally, the use of polymer in the foundry environment is notoriously messy; splashes from the quench or from the rinse tank cause the surrounding equipment to become black and sticky, and your feet tend to stick to the floor.

In contrast to its problems with automotive castings, polymer is excellent for aerospace castings where the extra trouble to clean each casting individually does not outweigh the benefit of superb heat treatment response and reduced internal stress.

There seems to be little information on another quenchant that holds significant promise: a water/carbon dioxide solution. The mixture is of course highly environmentally friendly. Different concentrations of CO_2 provide different degrees of control over the rate and uniformity of the quench. It seems likely that at least some of the action of the solution is to break up the vapour blanket, making the quench more uniform. More information on this interesting and attractive technique would be welcome.

In a further development, the gas supply companies have been investigating such CO_2 quenchants, combining them with inert nanoparticles such as alumina to form liquid muds. These are claimed to give sufficiently slow quenches to avoid the well-known problems of the quench-cracking of highly alloyed steels (Stratton, 2010).

10.9.5 CONTROLLED QUENCHING USING AIR

The author recalls with pain memories of quenching complex cylinder heads into water, requiring the consequential banana-shaped products to be straightened with a huge 50,000 kg press specially bought-in to rectify the damage. The castings subjected to straightening were those that appeared to have survived failure from cracking in the quench itself (although internal cracks inside the water jackets were often difficult to detect, being usually only found on subsequent return of service failures). In addition, it was found, the castings failed by fatigue in service after only short lives. (These were highly stressed racing cylinder heads so that the disappointment of failure was to some extent compensated by the rare, if not unique, benefit of rapid feed-back of service life data.)

It was this experience that drove the author to the air quenching of cylinder head castings. When the cylinder head casting, considered previously, was subjected to quenching in a blast of air, Figure 10.47 indicates that cooling was now a leisurely 100 s or so. Thus sufficient time was available for the internal sections to lose their heat to the outside so that the casting maintained a reasonably uniform temperature during the quench. The generation of high internal stress and resulting distortion was avoided.

Air quenching proved to be a complete solution for automotive castings. It was clean, quick, environmentally friendly, low cost and quickly and easily implemented in a series production environment. The castings retained their accuracy, and quench failures and fatigue failures disappeared. We were able to restore productivity and profitability (and sell the 50,000 kg straightening press to get our money back).

10.9.6 STRENGTH REDUCTION BY HEAT TREATMENT

Figure 10.49 illustrates how the overall effective strength of a casting can be reduced by a heat treatment designed to strengthen the alloy.

Figure 10.49(a) shows the stress-strain curve for the alloy, and the imposition of 1% tensile strain on the inner parts of the casting as a result of a water quench. This quench strain results in a quench stress close to the failure stress of the material. If no ageing treatment is carried out, this stress is locked into the component for the rest of its life. Naturally, it has little residual strength left, and is likely to fail on the first application of a stress in service.

However, after an ageing treatment designed to double the yield strength of the alloy, the situation is shown in Figure 10.49(b). Assuming the benefit of a small amount of stress relief (the amount indicated in the figure may be rather generous), the residual quench stress is only slightly lower; substantially unchanged. If additional service stress in tension is applied to the central parts of the casting, the residual tensile stress in these parts is effectively a starting point for the additional loading. Thus, effectively, the new stress-strain curve for the component is shown in Figure 10.49(c). It is clear that the new overall stress-strain capability of the casting has been reduced compared to the original unheat-treated material; thus, tragically, as a result of our lengthy, complex and costly heat treatment the component is effectively weaker.

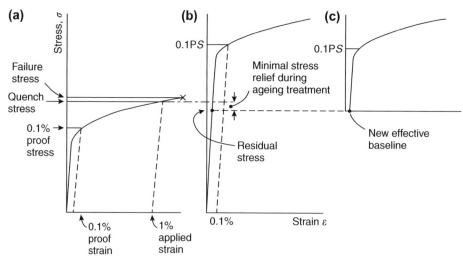

FIGURE 10.49

Evolution of the stress/strain properties of a precipitation-hardened Al alloy as heat treatment progresses: (a) after quench; (b) after ageing to double the yield strength; (c) the final effective stress/strain relation after allowing for the presence of residual stress. Although clearly the alloy is stronger, the load bearing ability of the casting has been reduced.

In summary, for certain sensitive castings such as automotive cylinder heads, the residual stress in aluminium alloy castings quenched into water in this way are well above the yield point of the alloy. Even after the strengthening during the ageing treatment, the stress remains at between 30 and 70% of the yield stress, with a useful working approximation being 50%. Thus the useful strength of the alloy is reduced from its unstressed state of 100%, down to around 50%. This massive loss of effective strength makes it inevitable that residual tensile stresses are a significant cause of casting failure in service, particularly fatigue failure because the residual stress is always generously above the fatigue limit of the alloy.

Turning now to steels: in contrast to the behaviour of Al alloys, the thermal diffusivity D is approximately 10 times lower, of the order of only 10^{-5} m^{2}s^{-1}. The reader can quickly show that the corresponding distances to which heat can flow are 7 mm in 5 s but only 30 mm in 100 s. For a given rate of quench, therefore, steels will suffer a higher residual stress (Figure 10.46). Nevertheless, they are much more able to withstand such disadvantages, having not only higher strength, but more particularly, higher elongation to failure (Figure 10.50). Thus, although the final internal stress is high, the steel product is nowhere near the failure condition experienced by the aluminium alloy casting. The aluminium alloy casting also experiences about 1% imposed elongation but has only a few percent, perhaps on occasions even less than 1% elongation before failure. Thus it can fail actually in the quench, or early in service. In contrast, the steel casting has 10–20 times greater elongation (as a result primarily of its reduced bifilm content). Thus although the 1% or so of imposed quench strain resulting from unequal cooling may result in 1% or so of distortion of the steel casting, its condition is far from any dangerous condition that might result in complete failure because enormously greater strain has to be imposed to reach a failure condition (Figure 10.50).

The previous statements are so important they are worth repeating in different words for additional clarity. The rapid quenching of steels for metallurgical purposes (such as the stabilisation of austenite for Hadfield Manganese steel) is not usually a problem. The reason is that most steels are particularly clean because of the rapidity with which entrainment defects are deactivated and/or detrained after pouring with the result that they typically have

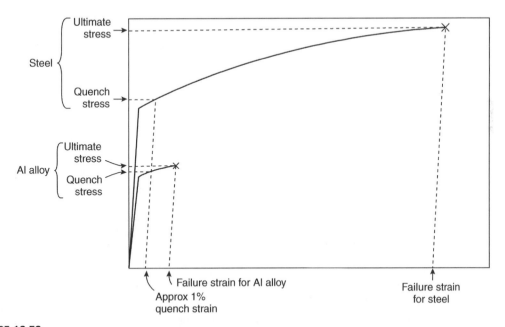

FIGURE 10.50

Comparison of the stress/strain relations for an Al alloy and a steel illustrating the relatively dangerous condition of the Al alloy after a quench.

elongations to failure of 40–50%. In contrast, most Al alloys (and probably most Mg alloys) do not enjoy this benefit; suffering from a high density of neutrally buoyant bifilms they typically achieve less than a tenth of this ductility. Thus, the application of 1% strain takes the aluminium alloy close to, or even sometimes in excess of its failure strain. For steels, even though the 1% strain applied by the quench will take the part into the plastic region, causing huge stresses, the steel remains safe; its greater freedom from bifilms permits it to endure enormously greater extension before it will fail (Figure 10.50).

For the future, the production of Al alloys with low bifilm concentration promises to offer ductilities in the range of that of steels. Already, good foundries know that high strengths together with elongations of 10–20% are achievable if good care is taken.

Slower quenching techniques are safer, although, of course, the strength attained by the heat treatment is somewhat reduced. Even so, the reduced mechanical strength when using slower and more controlled quenches such as a polymer or a forced air quench is more than compensated by the benefit of increased reliability from putting unstressed (or more accurately, low-stressed) castings into service. Thus, the casting designer and/or customer needs to accept somewhat reduced mechanical strength and hardness requirements in order to gain a superior performance from the product. The reductions of strength and hardness are expected to be in the 5–10% range, but the improvement in casting performance can be expected to be approximately 100% or more (as a result of avoiding the loss on the order of 50%). These are huge benefits to be gained at no extra cost.

The proper development of quenching techniques to give maximum properties with minimum residual stress is a technique known as quench factor analysis. It is also much used to optimise the corrosion behaviour of aluminium alloys. The method is based on the integration of the effects of precipitation of solute during the time of the quench. In this way, any loss of properties caused by slow quenching or stepped quenching can be predicted accurately. The interested reader is recommended to the nice introduction by Staley (1981), and his later more advanced treatment (Staley, 1986).

10.9.7 DISTORTION

Residual stresses in castings are not only serious for parts that require the ability to withstand stress in service: they are also of considerable inconvenience for parts that are required to retain a high degree of dimensional stability. This problem was understood many years ago, being first described as early as 1914 in a model capable of quantitative development by Heyn. His model was the three-bar casting shown in Figure 10.42. The internal stresses were represented by two outer springs in compression, each carrying half of the total load of internal compressive stress and an inner spring in tension carrying all of the internal tensile stress. If one of the surfaces of the casting was machined away, one of the external stresses would be eliminated. It is predictable therefore that the casting would deform to give a concave curvature on the machined side as illustrated in the figure.

The distortion of castings both before and after machining is a common fault and typical of castings that have suffered a water quench. Once again, it is a problem so frequently encountered that I have, I regret, wearied of answering the telephone to these enquires too. After all, it is difficult to understand how a casting could avoid distortion if parts of it are stressed up to or above its yield point.

For light alloy castings in particular, a more gentle quench, avoiding water (either hot or cold) and choosing polymer or air will usually solve the problem instantly. As mentioned briefly previously, polymer performs well for aerospace castings but is expensive and messy, whereas air is recommended as being clean, economical and practical for high volume automotive work. Otherwise, stress relieving of castings by heat treatment before machining is strongly recommended. In all cases, of course, some fraction of the *apparent* strength of the product has to be sacrificed.

10.9.8 HEAT TREATMENT DEVELOPMENTS

Although not strictly relevant to the question of reducing residual stresses, it is worth emphasising the newer developments in heat treatments that give approximately 90% or more of total attainable strength, but with much reduced stress and greatly reduced cost. The reduced cost is always an attention-grabbing topic and materially helps the introduction of technology that can deliver an improved product.

Figure 10.51 illustrates the progression of recent developments in heat treatment of Al alloys where the problem of stress is central.

The traditional full heat treatment of a precipitation-hardened alloy, that constitute the bulk of cast structural components at this time, consists of a solution treatment, water quench and age as illustrated in Figure 10.51(a). The treatment results in excellent apparent strength for the material, but is energy intensive in view of the long total times, and the water quench may be severely deleterious (depending on the geometry of the casting) as we have seen.

Illustration Figure 10.51(b) shows how the traditional treatment can be reduced significantly in modern furnaces that enjoy accurate control over temperature, thereby reducing the risk of overheating of the charge because of random thermal excursions. An increase in temperature by 10°C will allow, to a close approximation, an increase in the rate of treatment by a factor of 2. Thus times at temperature can be halved. These benefits are cumulative, such that a rise of 20°C will allow a reduction in time by a factor of $2 \times 2 = 4$, or a rise of 30°C a reduction in time of a factor $2 \times 2 \times 2 = 8$, although, of course, there are clearly limits to how far this simple relation can be pushed. Even so, it seems to hold within acceptable accuracy over many decades of temperature change. The reader will appreciate that the tiny additional energy required by the higher temperature is of course generously offset by the savings in overall time at temperature.

In Figure 10.51(a) and (b) require separate furnaces for solution and ageing treatments (if long delays waiting for the solution furnace to cool to the ageing temperature are to be avoided). Thus floor space requirement is high. Floor space requirement is increased further by the quench station, and, if a polymer quench is used, by a rinsing tank station.

If an air quench is used to gain the benefits of reduced residual stress, the additional benefits to the overall cycle time are seen in Figure 10.51(c); the quench is now easily interrupted and the product transferred to the ageing furnace already

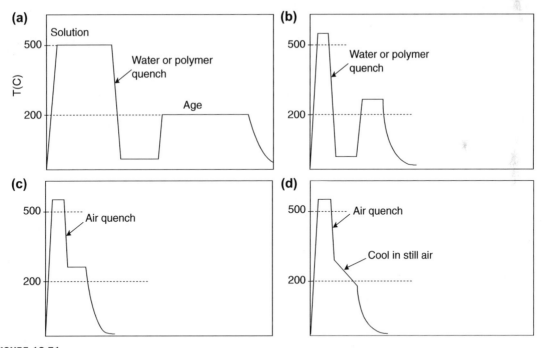

FIGURE 10.51

A progression of developments for the precipitation heat treatment of Al alloys. (a) Traditional full T6 heat treatment giving excellent properties but taking 12–24 h; (b) shortened treatment giving nearly equivalent results; (c) use of air quench to reduce time, energy and residual stress; (d) a possible ultimate short and simple cycle, ageing during the high temperature cool in still air.

at the correct temperature for ageing, saving time and reheat energy. Additional benefits include the fact that the air quench is environmentally friendly; the castings are not stained by the often less-than-clean water; the conveyor is straightforward to build and maintain; there is no mechanism required for lowering into water that normally results in complex and rusting plant. As we have repeatedly emphasised, the products from this type of furnace have somewhat lower apparent strengths and hardness, but greatly improved performance in service.

Figure 10.51(d) shows an ultimate system that might be acceptable for some products. The ageing treatment is simply carried out by interrupting the air quench slightly above the normal ageing temperature, and allowing the part to cool in air (before final rapid cooling by fans if necessary). This represents a kind of natural ageing process in which no ageing furnace is required. Strengths will suffer somewhat, but the lower costs and simplicity of the process may be attractive, making the process suitable for some applications.

10.9.9 BENEFICIAL RESIDUAL STRESS

Not all residual stress need be bad.

Bean and Marsh (1969) describe a rare example in which the stress remaining after quenching was used to enhance the service capability of a component. They were developing the air intake casing for the front of a turbojet engine. The casting has the general form of a wheel, with a centre hub, spokes, and an outer shroud. In service the spokes reached 150°C and the shroud cooled to −40°C. The expansion of the spokes and contraction of the shroud gave problems in service. With additional high loads from accelerations up to 7g and other forces, some casings were deformed out of round, and some even cracked. To counter this problem, the casting was produced with tensile stress in the spokes and compressive loading in the shroud. This was achieved by wrapping the spokes in glass fibre insulation, whilst allowing the shroud to cool quickly at the full quenching rate, but the spokes to cool and contract later. By this means, approximately 40 MPa tensile stress was introduced into the spokes. This was tested by cutting a spoke on each fifteenth casting, and measuring the gap opening of approximately 2 mm.

Another method of equalising quenching rates in castings is by the clamping of shielding plates around thinner sections to effectively increase their section. The method is described by Avey et al. (1989) for a large circular clutch housing in a high-strength aluminium alloy. The technique improved the fatigue life of the part by more than 400%.

It may be significant that both of these descriptions of the positive use of residual stress relate to rather simple rotationally symmetric castings.

10.9.10 STRESS RELIEF

For traditional grey iron foundries of the past century, the original method of providing some stress relief in iron castings was simply to leave the castings in the foundry yard. Here, the long passage of time, of weeks or months, and the changeable weather, including rain, snow, frost and sun, would gradually do its work. It was well known that the natural ageing outdoors was more rapid and complete than ageing done indoors, presumably because of the more rapid and larger temperature changes.

Nowadays, the more usual method of reducing internal stress is both faster and more reliable (although somewhat more energy intensive!). The casting is reheated to a temperature at which sufficient plastic flow can occur by creep to reduce the strain and hence reduce the stress. This is designed to take place within a reasonable time, of the order of an hour or so. As pointed out earlier, it is then most important that stress is not re-introduced by cooling too quickly from the stress-relieving treatment.

An apparently perverse and quite exasperating feature of internal tensile stress in castings is that the casting will often crack while it is being reheated as part of the stress-relieving process to avoid the danger of cracks! This happens if the reheating furnace is already at a high temperature when the castings are loaded. The reason for this is that if the casting already has a high internal tensile stress, on placing the casting in a hot reheating furnace the outside will then be heated first and expand, before the centre becomes warm. Thus the centre, already suffering a high stress, will be placed under additional tensile load, the total being sometimes sufficient to exceed the tensile strength.

Table 10.2 Approximate 1-h Stress-Relieving Temperatures

Alloy	Stress-Relief Temperature (°C)
Al-2.2Cu-1Ni-1Mg-1Fe-1Si-0.1Ti	300
Brass Cu-35Zn-1.5Fe-3.7Mn	400
Bronze Cu-10Sn-2Zn	500
Grey iron 3.4C-2Si-0.38Mn-0.1S-0.64P	600
Steel (C-Mn types)	700

The problem is avoided by reheating sufficiently slowly that the temperature in the centre is able, within tolerable limits, to keep pace with that at the outside. Consideration of the thermal diffusivity using Eqn (10.22) will give some guidance of the times required.

Figure 10.52 shows the temperatures required for stress relief of various alloys. (Strictly, the figure shows results for 3 h, but the results are fairly insensitive to time, a factor of two reduction in time corresponding to an increase in temperature of 10°C, hardly moving the curves on the scale used in the figure.) It indicates that nearly 100% of the stress can be eliminated by an hour at the temperatures shown in Table 10.2.

Bhaumik et al. (2010) describes experiments to prove an elegant and simple technique to determine the times and temperatures needed to achieve specific degrees of stress reduction. The technique is known as a stress relaxation test. A tensile test bar of the material is mounted in a tensile testing machine, and the test bar is enclosed by a heater. The test bar is loaded to near its yield stress while the heater is raised to the test temperature. The tensile test machine will then automatically record the fall in stress with time. At low temperatures, the times to achieve a 90% drop in stress might take hours or days, whereas at high temperatures this degree of stress relief might take only minutes. This simple, accurate, quantitative test is strongly recommended.

There are numerous examples of the use of such heat treatments to effect a valuable degree of stress relief. One example is the work by Pope (1965) on cast iron diesel cylinder heads that were found to crack between the exhaust valve seats in service, despite a stress-relief treatment for 2 h at 580°C. A modification of the treatment to 4 h at 600°C cured the problem. From Figure 10.52 and Table 10.2, we can see that even 1 h at 600°C would probably have been sufficient.

The work by Kotsyubinskii et al. (1968) highlights the fact that during the thermal stress relief the casting will distort. They carried out measurements on box-section castings in grey iron, intended as the beds of large machine tools, for which stress-relieving treatment is carried out after some machining of the top and base of the box section. He suggests that the degree of movement of the castings is approximately assessed by the factor $(w_1 - w_2)/w_c$ where w_1 and w_2 are the weights of metal machined from the top and base of the casting, respectively, and w_c is the weight of the machined casting. This interesting observation has, to the author's knowledge, never been confirmed.

Moving on now from heat treatment, there are other methods of stress relief that are sometimes useful. In simple castings and welds, it is sometimes possible to effect relief by mechanical overstrain as described in the excellent review by Spraragen and Claussen (1937).

Kotsyubinskii (1962) describes a further related method for grey iron in which the castings are subjected to rapid heating and cooling between 300°C and room temperature at least three times. The differential rates of heating within the thick and thin sections produce the overstrain required for stress relief by plastic flow.

More drastic heating rates are required to effect stress relief by differential heating in aluminium alloys because of the thermal smoothing provided by the high thermal conductivity. Hill et al. (1960) describe an 'up-quenching' technique in which the casting is taken from cryogenic temperatures, having been cooled in liquid nitrogen, and is reheated in jets of steam. This thermo-mechanical treatment introduces a pattern of stresses into the casting that are opposite to those introduced by normal quenching. One of the benefits of this method is that it is all carried out at temperatures below normal ageing temperatures, so that the effects of the final heat treatment and the resulting mechanical properties are not affected. One of the

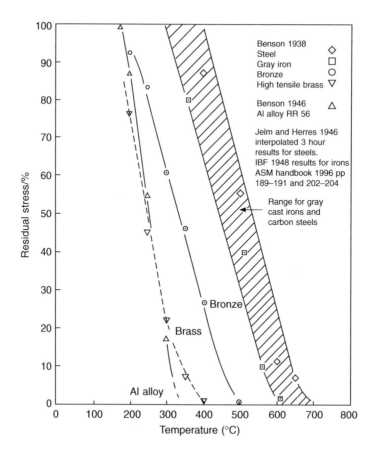

FIGURE 10.52

Stress relief of a selection of alloys treated for 3 h at temperature.

Data from Benson (1938), Jelm and Herres (1946) and Institute of British Foundrymen (1948).

possible disadvantages that the authors do not mention is the enhanced tensile stress in the centre of the casting during the early stages of the up-quench. Some castings would not be expected to survive this dangerous moment.

A variant of these approaches is stress relief by vibration. This is probably effective in some shapes, but it is difficult to see how the technique can apply to all parts of all shapes. This is particularly true if the component is treated at a resonant frequency. In this condition, some parts of the casting will be at nodes (will not move) and some parts at antinodes (will vibrate with maximum amplitude). Thus the distribution of energy in the casting will be expected to be highly heterogeneous. Some investigators have reported the danger of fatigue cracks if vibrational stresses over the fatigue limit are employed (Kotsyubinskii, 1961). The technique may require some skill in its application because null results are easily achieved (IBF Technical Subcommittee, 1960).

There may be greater certainty of a valid result with sub-resonant treatment. This technique emerged later as a possible method of stress relief (Hebel, 1989). In this technique, the casting is vibrated not on the peak of the frequency–amplitude curve, but low on the flank of the curve. At this off-peak condition the casting is said to absorb energy more efficiently. Furthermore, it is claimed that the progress towards complete stress relief can be monitored by the gradual change in the resonant frequency of the casting. When the resonant frequency ceases to change, the casting is said to be fully stress relieved. If this technique could be verified, then it would deserve to be widely used.

10.9.11 EPILOGUE

Although the *strength of the material* will therefore be lowered by a slower quench, the *strength of the component* (i.e. the failure resistance of the complete casting acting as a load bearing part) in service will be increased.

If water quench is avoided with a view to avoiding the dangers of internal residual stress, it is common for the customer to complain about the 5% or so loss of apparent properties. In answer to such understandable questions, an appropriate reply to focus attention on the real issue might be 'Mr. Customer, with respect, do you wish to lose 5% or 50% of your properties?'

In the experience of the author, several examples of castings that have been slowly quenched, losing 5–10% of their strength, are demonstrated to double their performance in service. This was the case for a roadside compressor casting subjected to a full T6 heat treatment. The compressor housing exploded in service, fortunately not resulting in any injury to passersby. After the second explosion the manufacturer requested help. I recommended a stress relief treatment. This was declared to be impossible because it was claimed the compressor housing needed maximum strength. Despite the reservations of the manufacturer, stress relieved castings (that had only half the strength of the fully heat treated product) were tested to destruction in comparison with fully heat-treated castings. The stress-relieved products proved twice as resistant to failure under pressure compared with the fully heat-treated parts with the metallurgically 'strong' alloy.

Finally, therefore, it remains deeply regrettable, actually a scandal, that many national standards for heat treatment continue to specify water quenching without any warning of the dangers for certain geometries of casting. This disgraceful situation requires to be remedied. In the meantime, the author deeply regrets to have to recommend that such national standards be set aside. It is easy for the casting supplier to take refuge in the fact that our international and national standards on heat treatment often demand quenching into water and thereby avoid the issue that such a production practice is risky for many components—and in any case provides the user with a casting of inferior performance. However, the ethics of the situation are clear. We are not doing our duty as responsible engineers and as members of society if we continue to ignore these crucial questions. We threaten the performance of the whole component merely to fulfil a piece of metallurgical technology that from the first has been woefully misguided.

Our inappropriate heat treatments have been costly to carry out and have resulted in costly failures. It has to be admitted that this has been nothing short of a catastrophe for the engineering world for the past half-century, and particularly for the reputation of light-alloy castings, not to mention the misfortune of users. As a result of the unsuspected presence of bifilms, they have suffered poor reliability so far, but as a result of the unsuspected presence of residual stress, this has been made considerably worse by an unthinking quest for material strength regardless of component performance.

10.10 RULE 10: PROVIDE LOCATION POINTS

This rule is added simply because the foundry can accomplish all the other nine rules successfully, and so produce beautiful castings, only to have them scrapped by the machinist. This can create real-life drama if the castings have been promised in a just-in-time delivery system. This rule is added to help to avoid such misfortunes and allow all parties to sleep more soundly in their beds.

Before describing location points, their logical precursors are datum planes. We need to decide on our datums first (for language purists the plural of datum is data, but this would introduce unwelcome confusion with the conventional meaning of the word, thus we all need to grit our teeth with the inelegant plural 'datums').

10.10.1 DATUMS

A datum is simply a plane defining the zero from which all dimensions are measured.

For a casting design, it is normal to choose three datum planes at right angles to each other. In this way, all dimensions in all three orthogonal directions can be uniquely defined without ambiguity.

In practice, it is not uncommon to find a casting design devoid of any datum, there being simply a sprinkling of dimensions over the drawing, none of the dimensions being necessarily related to each other. On other designs, the dimensions relate with great rigour to each other and to all machined features such as drilled holes etc., but not to the casting. In yet other instances that the author has suffered, datums on one face have not been related to datums on other views of the same casting. Thus the raft of features on one face of the casting shifts and rotates independently of the raft of features on the opposite face.

It is fortunate today that such poor dimensioning of castings is not easy, and is perhaps impossible, with computer-aided design. Even so, there are good reasons to be cautious.

Figure 10.53(a) shows a sump (oil pan) for a diesel engine designed for gravity die casting in an aluminium alloy. The variations in die temperature and ejection time result in variability of the length of the casting that are well-known with this process, and not easily controlled. The figure shows how the dimensioning of this part was very logically laid out, but has made the part nearly unmanufacturable. Three fundamental criticisms can be made:

1. The datum is at one end of the product. If the datum had been defined somewhere near the centre of the part, then the variability produced by the length changes of the casting would have been approximately halved.
2. There is only one feature on the component whose location is critical; this is the dipstick boss. If the boss is slightly misplaced then it fouls other components on and around the engine. It will be noticed that the dipstick boss is at the far end of the casting from the datum. Thus variability in length of the casting will ensure that a large proportion of castings will be deemed to have a misplaced boss. If the datum had been located at the other end of the casting, near to the boss, the problem would have been reduced to negligible proportions. If the datum had been chosen as the boss itself, the problem would have disappeared altogether, as in Figure 10.53(b).
3. The datum is not defined with respect to the casting. It is centred on a row of machined holes which clearly do not exist at the time the casting is first made and when it is first required to be checked and before any machining of the casting has taken place. Depending on whether the machinist decides to fix the holes in the centre of the flange, or relate them by measurement to the more distant dipstick boss, or to the centre of the casting averaged from its two ends, or any number of alternative strategies, the drilled holes could be almost anywhere, including partly off the flange or even completely off the flange!

Figure 10.53(b) shows how these difficulties are easily resolved. The datum is located against the side of the dipstick boss and hence is fixed in its relation to the casting and goes some way to halving errors in the two directions from this

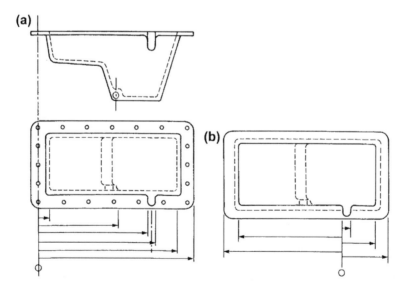

FIGURE 10.53

(a) A badly dimensioned sump casting, resulting in a casting manufactured only with difficulty and high scrap losses.
(b) A change of datum to the most critical feature of the casting results in easy and efficient manufacture.

plane. It also means that the dipstick boss itself is now impossible to misplace, no matter how the casting size varies; all the other dimensions are allowed to float somewhat because movement of other points on the casting is not a problem in service. This part can then be produced easily and efficiently, without trauma to either producer or customer!

In summary, the rules for the use of datums (partly from Swing (1962)) are:

1. Choose three orthogonal datum planes.
2. Ensure the planes are parallel to the axes of motion of the machine tools that will be used to machine the part (otherwise unnecessary computation and opportunity for error is introduced).
3. Fix the planes on real casting features, such as the edges of a boss, or the face of a wall (i.e. not on a centreline or other abstract constructional feature). Choose casting features that are
   a. critical in terms of their location and
   b. as near the centre of the part as possible.

10.10.2 LOCATION POINTS

Location points are those tiny patches on the casting that are used to locate the casting precisely and unambiguously in three dimensions. They are required by all parties involved in the manufacture of the product including.

1. the toolmaker because he can construct the tooling with reference to them;
2. the founder, to check the core and mould assembly if necessary and the casting once it has been made, and
3. the machinist, who uses them to locate the casting before the first machining operation.

These features therefore integrate the manufacture of the product, ensuring its smooth transmission as it progresses from toolmaker to founder to machinist.

Whereas the casting datums are invisible planes that define the concept of a zero in the dimensional space in and around a casting, the tooling points are real bits of the casting. The datums are the software, whereas the tooling points are the hardware, of the dimensioning system.

It is useful, although not essential, for the datums to be defined coincident with the tooling points.

The location points are required to be *actual cast-on features* of the casting. This point cannot be over-emphasised. It is not helpful, for instance, to define a location feature as a centre-line of a bore. This invisible feature only exists in space (perhaps we should say 'free air'). Virtual features such as centrelines have to be found by locating several (at least three) points on the internal as-cast solid surface of the bore, and its centre thereby calculated. Clearly, these 'virtual' or 'free-air' so-called location points necessarily rely on their definition from other nearby as-cast surfaces. These ambiguities are avoided by the direct choice of as-cast location features.

The location features need to be cast nicely, without obscuring flash, or burned-on sand, and definitely should not be attacked by an enthusiastic finisher wielding an abrasive wheel.

It is essential that the location points are *not* machined. If they are machined, the circumstance poses the infinitely circular question 'What prior datums were used to locate and position the casting accurately to ensure that the machining of the machining locations (from which the casting would be picked up for machining) were correctly machined?' Unfortunately, such indefensible nonsense has its committed devotees.

Location points are known by several different names, such as machining locations (which is a rather limiting name) or pickup points. On drawings, tooling point, called TP, is used as the common abbreviation for the drawing symbol. Although in practice I tend to use all the names interchangeably it is proposed that 'location points' describes their function most accurately and will be used here.

Before the day of the introduction of location points in the Cosworth engine-building operation, I was accustomed to a complex cylinder head casting taking a skilled man at least 2 h to measure to assess how to pick up the casting for machining. The casting was repeatedly re-checked, re-orienting it slightly with shims to test whether wall thicknesses were adequate and whether all the surfaces required to be machined would in fact clean up on machining. After the 2 h, it was common to witness our most expensive casting dumped unceremoniously on the scrap heap; no orientation could be found to ensure that it was dimensionally satisfactory and could be completely cleaned up on machining. All this changed on the day when the new foundry came on stream. A formal system of six location points to define the position of the casting was introduced. After this date, no casting was subjected to dimensional checking. All castings were received from the foundry and on entering the machine shop, and were immediately thrown on to a machine tool, pushed up against their location points, clamped, and machined. No tedious measurement time was subsequently lost, and no casting was ever again scrapped for machining pickup problems.

It is essential that every casting has defined locations that will be agreed with the machinist and all other parties who require to pick up the casting accurately.

For instance, it is common for an accurate casting to be picked up by the machinist using what appear to be useful features, but which may be formed by a difficult-to-place core, or a part of the casting that requires some dressing by hand. Thus although the whole casting has excellent accuracy, this particular local feature is somewhat variable in location. The result is a casting that is picked up inaccurately, and does not therefore clean up on machining. As a result it is, perhaps rather unjustly, declared to be dimensionally inaccurate.

The author suffered precisely this fate after the production of a complex pump body casting for an aerospace application that achieved excellent accuracy in all respects, except for a small region of the body that was the site where three cores met. The small amount of flash at this junction required dressing with a hand grinder, and so, naturally, was locally ground to a flat, but at various slightly different depths beneath the curved surface of the pump body. This hand-ground location was the very site that the machinist chose to locate the casting. The result was disaster. Furthermore, it was not easily solved because of the loss of face to the machinist who then claimed that the location options suggested by the foundry were inconveniently awkward. The fault was not his of course. The fundamental error lay in not obtaining agreement between all parties before the part was made. If the location point used by the machinist really was the only sensible option for him, the casting engineer and toolmaker needed to ensure that the design of the core package would allow this.

Ultimately, this rule is designed to ensure that all castings are picked up accurately and conveniently if possible, so that unnecessary scrap is avoided.

Different arrangements of location points are required for different geometries of casting. Some of the most important systems are listed later.

Rectilinear systems

1. Six points are required to define the position of a component with orthogonal datum planes designed for essentially rectilinear machining, as for an automotive cylinder head or block. (Any fewer points than six are insufficient to define the position of the casting, and any more than six will ensure that one or more points are potentially in conflict.)

On questioning a student on how to use a six-point system to locate a brick-shaped casting, the reply was 'Oh easy! Use four points around the outside faces and one top and one bottom.' This shows how easy it is to get such concepts wildly wrong!

In fact, the six points are used in a 3, 2, 1 arrangement as shown in Figure 10.54. The system works as follows: three points define plane A, two define the orthogonal plane B and one defines the remaining mutually orthogonal plane C. The casting is then picked up on a jig or machine tool that locates against these six points. Example (a) shows the basic use of the system: points 1, 2 and 3 locate plane A; points 4 and 5 define plane B; and point 6 defines plane C. Planes A, B and C may be the datum planes. Alternatively, it is often just as convenient for them to be parallel to the datum planes, but at accurately specified distances away.

Clearly, to maximise accuracy, points 1, 2 and 3 need to define a widely based triangle, and points 4 and 5 similarly need to be as widely spaced as possible. A close grouping of the locations will result in poor reproducibility of the pickup of the casting; tiny errors in the position or surface roughness of the tooling points will be magnified if they are not widely spaced.

Example (b) shows an improved arrangement whereby the use of a tooling lug on the longitudinal centreline of the casting allows the dimensions along the length of the casting to be halved. The largest dimension of the casting is usually subject to the largest variability, so halving its effect is a useful action.

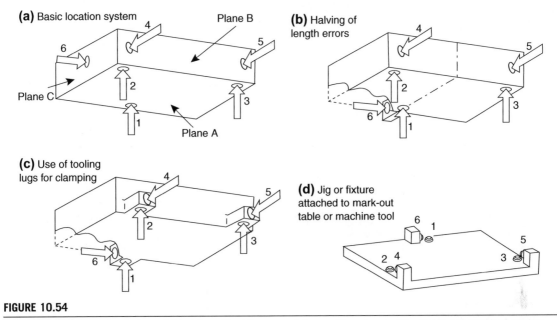

FIGURE 10.54

(a) The six-point location system; (b) halving of length errors; (c) use of lugs to combine tool location with clamping points; (d) a jig to cradle the casting during dimensional checking or machining (clamps are omitted for clarity).

Example (c) is a further development of this idea, creating lugs that serve the additional useful purpose of allowing the part to be clamped immediately over the points of support and off the faces that require to be machined. In practice the lugs may be existing features of the casting, or they may be additions for the purpose of allowing the casting to be picked up for inspection or machining.

A final development of the concept (not shown in our illustration) uses all the lugs arranged on all the centrelines of the casting so as to halve the errors in all directions.

For maximum internal consistency between the tooling points, all points should be arranged to be in one half of the mould, usually the fixed or lower half, although sometimes all in the cope. The separation of points between mould halves, or having some defined from the mould and some from cores, will compromise accuracy. However, it is sometimes convenient and correct to have all tooling points in one internal core, or even one half of an internal core (defined from one half of the core box) if the machining of the part requires to be defined in terms of its internal features.

It is noteworthy that this introduces a small error of principle. The faces of tooling points 1, 2 and 3 are formed in the drag, whereas 4, 5 and 6 are formed in the cope. This is a small error in this case, because the cope-drag joint is among the most reliable of all the mould joints, introducing only a small error in the definition of the location of plane A. It is a good rule to place all six points in one piece of mould, often in the drag, so that they all relate to each other accurately, but because of the small potential for cope/drag error, an exception may be allowable in this case.

In general, it is useful if the tooling locations on the casting can remain in position for the life of the part. It is reassuring to have the tooling points always in place, if only to resolve disputes between the foundry and machine shop concerning failure of the part to clean up on machining. It is therefore good practice to try to avoid placing them where they will be eventually machined off. Using existing casting features wherever possible avoids the cost of additional lugs, and avoids the cost of subsequent removal if necessary.

The definition of the six-point locations, preferably on the drawing (nowadays the computer model), before the manufacture of the casting, is the only method of guaranteeing the manufacturability of the part.

Cylindrical systems

Most cylindrical parts do not fall nicely into the classical six-point rectilinear system as described previously. The errors of eccentricity and diameter both contribute to a rather poor location of the centre using this approach. The unsuitability of the orthogonal pick-up system is analysed nicely by Swing (1962).

In fact, the obvious way to pick up a cylinder is in a three-jaw chuck. The self-centring action of the chuck gives a useful averaging effect on any out-of-roundness and surface roughness, and is of course insensitive to any error of diameter.

In classical terms, the three-jaw chuck is equivalent to a two-point pickup because it defines an axis. We therefore need four more points to define the location of the part absolutely. Three points longitudinally abutting the jaws will define the plane at right angles to the central axis, and one final point will provide a 'clock' location. Figure 10.55 shows the general scheme.

Another location method that is occasionally useful is the use of a V block. This is a way of ensuring that a cylindrical part, or the round edge of a boss, is picked up centrally, averaging errors in the size and, to some extent, the shape of the part. The method has the disadvantage that errors in diameter of the part will cause the whole part to be shifted either nearer to or away from the block, depending on whether the diameter is smaller or larger. (The reader can quickly confirm these shifts in position with rough sketches.)

A widely used but poor location technique is the use of conical plugs to find the centre of a cored hole. Even if the hole is formed by the mould, and so relatively accurately located, any imperfection in its internal surface is difficult to dress out, and will therefore result in mis-location. If a separate core forms the hole then the core-positioning error will add to the overall inaccuracy of location. Location from holes is not recommended.

It is far better to use external features such as the sides of bosses or walls, as previously discussed. These can be more easily maintained cleanly cast.

Triangular systems

For some suitable parts of triangular form, such as a steering gear housing, a useful and fundamentally accurate system is the cone, groove and plane method (Figure 10.56).

Thin-walled boxes

For prismatic shapes, comprising hollow, box-like parts such as sumps (oil pans), the pickup may be made by averaging locations defined on opposite internal or external walls. This is a more lengthy and expensive system of location often tackled by a sensitive probe on the machine tool that then calculates the averaged datum planes of the component and

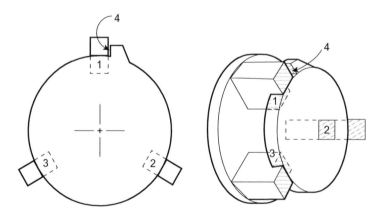

FIGURE 10.55

Use of a three-jaw self-centring chuck for casting location and clamping (note that location 4 is occasionally required as a 'clock' location).

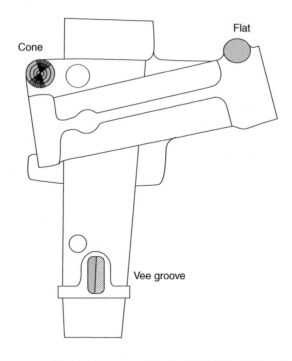

FIGURE 10.56

Plan view of a steering housing for an automobile, showing a flat, a groove and a cone location system.

orients the cutter paths accordingly. This technique is especially useful where an average location is definitely desirable, as a result of the casting suffering different degrees of distortion of its relatively thin walls.

10.10.3 LOCATION JIGS

Figure 10.54(d) illustrates a basic jig that is designed to accept a casting with a six-point location system. The jig is simply a steel plate with a series of small pegs and blocks. It contrasts with many casting jigs, which are a nightmare of constructional and operational complexity.

Our simple jig is also simple to operate. When placing the casting on the jig, the casting can be slid about on locations 1, 2 and 3 to define plane A, then pushed up against locators 4 and 5 to define plane B, and finally slid along locators 4 and 5 until locator 6 is contacted. The casting is then fixed uniquely in space in relation to the steel jig plate. It can then be clamped, and the casting measured or machined. The six locations can, of course, be set up and fixed directly onto, effectively integrated into the machine tool that will carry out the first machining operation.

After the first machining operations, it is normal to remove the casting from the as-cast locations and proceed with subsequent machining using the freshly machined surfaces as the new location surfaces. McKim and Livingstone (1977) go on to define the use of functional datums which may become useful at this stage. They are machined surfaces that normally relate to features locating the part in its intended final application.

Other jigs can easily be envisaged for cylindrical and other shaped parts.

10.10.4 CLAMPING POINTS

During machining, the forces on the casting can be high, requiring large clamping loads to reduce the risk of movement of the casting. Clamping points require to be planned and designed into the casting at the same time as the location points.

This is because the application of high clamping loads to the casting involves the risk of the distortion of the casting, and of spring-back after the release of the clamps at the end of machining. Surfaces machined flat are apt to become curved after unclamping because of this effect.

The great benefit of using tooling lugs as shown in Figure 10.54(c) can therefore be appreciated. The location point and the clamping point are exactly opposed on either side of the lug. In this way, the clamping loads can be high, without introducing the risk of the overall distortion of the casting.

Further essential details of the design of the clamping action include the requirement for the action to move the part on to, and hold it against, the location point.

For softer alloys that are easily indented, the clamp face needs to be 5–10 mm in diameter, similar to the working area of the tooling point. Even so, a high clamping load will typically produce an indentation of 0.5 mm in a soft Al alloy, decreasing to 0.2 mm in an Al alloy hardened by heat treatment, and correspondingly less still in irons and steels.

10.10.5 POTENTIAL FOR INTEGRATED MANUFACTURE

As we have already emphasised, the tooling points should be defined on the drawing of the part (i.e. more usually on the computer model), and should be agreed by (1) the manufacturer of the tooling, (2) the caster and (3) the machinist. It is essential that all parties work from *only* these points when checking dimensions and when picking the part up for machining.

The method allows an integrated approach right from the start of the creation of the tooling, because the pattern or tool maker can use the tooling points as the critical features of the tooling in relation to which all measurements will be defined. The foundry engineer can check his core and mould assembly, and will know how to pick up the casting to check dimensions after the production of the first sample castings. The machinist will use the same points to pick up the casting for machining. They all work from the same reference points. It is a common language and understanding between design, manufacture, and inspection of products. Disputes about dimensions then rarely occur, or if they do occur, are easily settled. Casting scrap apparently resulting from dimensioning faults, or faulty pick up for machining, usually disappears.

This integrated manufacturing approach is relatively easily managed within a single integrated manufacturing operation. However, where the pattern shop, foundry and machinist are all separate businesses, all appointed separately by the customer, then integration can be difficult to achieve. It is sad to see a well-designed six-point pickup system ignored because of apparent cussedness by one member of the production chain.

The industry and its customers very much need purchasing and manufacturing policies based on teamwork and cooperation, together with the adoption of integrated and fundamentally correct systems. We all hope to arrive in this utopia one day.

FILLING SYSTEM DESIGN

The heart of achieving a good casting lies in achieving a good filling system. We discuss here (1) the basic principles, followed by (2) a description of all the various component parts of a filling system and finally, (3) the challenge of putting them all together in a logical and practical design.

FILLING SYSTEM DESIGN FUNDAMENTALS

Getting the liquid metal out of the crucible or melting furnace and into the mould is a critical step when making a casting: it is likely that most casting scrap arises during this few seconds of pouring of the casting.

The scrap arises because of the incorporation of entrainment defects. As the reader will now know, the two most important entrainment defects are bubbles and bifilms. One is highly visible, the other is often invisible. However, the highly visible air bubble creates more damage en route through the casting by leaving a bubble trail. Bubble trails are like long bifilms with a central channel which gives a deeply coloured dye-penetrant test result, reflecting the extensive depth of the defect.

Recent work observing the liquid metal by video X-ray radiography as it travels through the filling system confirms that most of the damage is done to castings by poor filling system design, and most of the damage occurs in the filling system. It is also worth reflecting on the fact that every gram of metal in the casting has, of necessity, travelled through the filling system. Leaving the design of the filling system to chance, or even to the patternmaker (with all due respect to all our invaluable and highly skilled patternmakers), is a risk not to be recommended.

It is fair to say that the avoidance of surface turbulence is probably the most complex and difficult requirement to fulfil when dealing with a gravity pouring system. The requirement is all the more difficult to appreciate by many in the industry because everyone working in this field has always emphasised the importance of working with 'turbulence-free' filling systems for castings. Unfortunately, despite all the worthy intentions, all the textbooks, all the systems and all the talk, so far as the author can discover, it seems that no one appears to have achieved anywhere near this target. In fact, in travelling around the casting industry, it is quite clear that the majority (I used to think 80%, but I now think this is far too conservative) of all defects are directly caused by turbulence. Thus the problem is massive; far more serious than suspected by most of us in the industry.

To understand the fundamental root of the problem, it will become clear from casting rule 2 that any fall greater than the height of the sessile drop (on the order of 10 mm) causes gravity to accelerate the metal, far exceeding its critical velocity, and so introduces the danger of defects in the casting. Because most falls are in fact in the range 10 or 100 times greater than this, and because the damage is likely to be proportional to the energy involved (i.e. proportional to the square of the velocity), the damage so created will usually be expected to be 100–10,000 times greater. Thus, in the great majority of castings that are poured simply under the influence of gravity, there is a major problem to ensure its integrity. In fact, the situation is so bad that the best outcome of many of the solutions proposed in this book are merely damage limitation exercises. Effectively, it has to be admitted that at this time it seems impossible to guarantee the avoidance of some turbulence with the consequence of some damage when pouring liquid metals.

This somewhat depressing conclusion needs to be tempered by several factors.

First: the world has come to accept castings as they are; thus, any improvement will be welcome. This book describes techniques that can create very encouraging improvements.

Second: this book is merely a summary of what has been discovered so far in the development of filling system design. Better designs are to be expected now that the design parameters (such as critical velocity, critical fall height, entrainment, etc.) are defined and understood for the first time.

Third: there *are* filling systems that can yield, in principle, perfect results.

Of necessity, such perfection is achieved by fulfilling casting rule 2 by avoiding the transfer of the melt into the mould by pouring downhill under gravity. Considering the three directions of filling a mould,

1. Downhill pouring under gravity;
2. Horizontal transfer into the mould (achieved by such techniques as 'level pour', and achieved by tilt casting in which the tilt conditions are accurately controlled to avoid any downhill flow);
3. Uphill (counter-gravity) casting in which the melt is caused to fill the mould in *only* an uphill mode.

The counter-gravity mode is not to be confused with a bottom-gating technique, in which melt is poured, falling down the sprue under gravity but filling the mould cavity from the bottom upwards as in Figure 11.2(b). True counter-gravity does not suffer this initial or any fall.

Only processes 2 and 3 have the *potential* to deliver castings of near-perfect quality. In my experience, I have found that in practice it is often difficult to make a good casting by gravity (mode 1), whereas by a good counter-gravity process (i.e. process 3 which automatically observes all the 10 casting rules) it has been difficult to make a bad casting. The jury is still out on horizontal transfer by tilting. The horizontal transfer approach has great potential, but not all castings have a suitable geometry to achieve the benefits, and those that do require a dedicated effort to achieve success.

Thus, in summary, filling of moulds can be carried out down, along or up. Only the 'along' and 'up' modes have the potential to fulfil completely the non-surface turbulence condition.

However, with regret, and despite all its problems, it seems more than likely that the *downward* mode, *gravity casting*, will continue to be with us for the foreseeable future. Thus we shall devote some considerable length to the damage limitation exercises that can offer considerably improved products, even if, unfortunately, those products cannot be ultimately claimed as perfect. Most readers will shed no tears over this conclusion. Although potential perfection in the *along* and *up* modes is attractive, the casting business is all about making *adequate* products; products that meet a specification and at a price a buyer can afford.

The question of cost is interesting; perhaps the most interesting. Of course, the costs have to be right, and often gravity casting is acceptable and sufficiently economical. However, more often than might be expected, high quality and low cost can go together. An improved gravity system, or even one of the better counter gravity systems, can be surprisingly economical and effective. Such opportunities are often overlooked. It is useful to have an open mind to take advantage of such benefits.

11.1 THE MAXIMUM VELOCITY REQUIREMENT

Casting rule 2 described in some detail how a melt is safe from entrainment problems if its velocity is below the critical velocity, approximately 0.5 ms^{-1}. The danger of entrainment increases as its velocity increases to about 1.2 ms^{-1}, after which entrainment damage is probably unavoidable. Acknowledging these limits, I find myself always targeting ingate speeds between 0.5 and 1.0 m/s. The situation is described further in the following section.

11.2 GRAVITY POURING: THE 'NO-FALL' CONFLICT

Although we have a rule forbidding the fall of the metal, gravity pouring involves pouring, which necessarily means the metal falls. The obviousness of this pedantic statement emphasises the fundamental problem faced when pouring castings under gravity. The whole design problem for filling systems revolves around the solution to this conflict.

When melts are transferred by pouring from heights less than the critical heights predicted in Table 2.2 (the heights of the sessile drop), there is no danger of the formation of entrainment defects. Surface tension is dominant in such circumstances and can prevent the inward folding of the surface, thus preventing the creation of entrainment defects. Critical velocities and critical fall heights can be defined for all liquid metals. The critical fall heights for all liquid metals are in the range 3–15 mm.

It is just possible to meet this stringent requirement for open top moulds in the absence of a filling system (Figure 11.1) because the pouring ladle can be lowered to within a few millimetres of the bottom of the mould. Thus, although the world's first metal engineering structure, the Iron Bridge in the United Kingdom, with impressive 23 m long spans, was cast in this way, most moulds nowadays are closed and therefore require specially designed filling channels to lead the metal into the mould cavity.

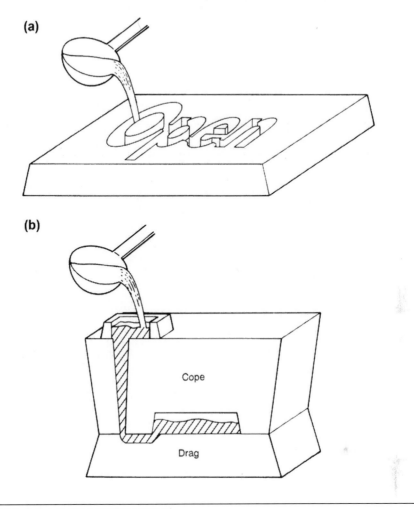

FIGURE 11.1

(a) An open and (b) closed mould partially sectioned.

It is unfortunate for founders that the critical fall height is such a minute distance. Most falls that an engineer might wish to design into a melt handling system, or running system, are nearly always greater, if not vastly greater. The small value of the critical fall height is one of those extremely inconvenient distances that we casting engineers have to learn to live with.

It follows immediately that *top-gating* of a casting (Figure 11.2(a)), almost without exception, will lead to a violation of the critical velocity requirement. Many forms of gating that enter the mould cavity at the mould joint, if any significant part of the cavity is below the joint, will also violate this requirement.

In fact, for conventional sand and gravity die (permanent mould) casting, it has to be accepted that some fall of the metal is necessary. Thus it has been accepted that the best option is for a single fall, concentrating the loss of height of the liquid at the very beginning of the filling system. The fall takes place down a conduit known as a *sprue*, or *down-runner*. This conduit brings the melt to the lowest point of the mould. It is important that the sprue be designed to be as slim as possible so that the melt has minimum opportunity to fold over its meniscus to fold in air or do itself damage by folding in

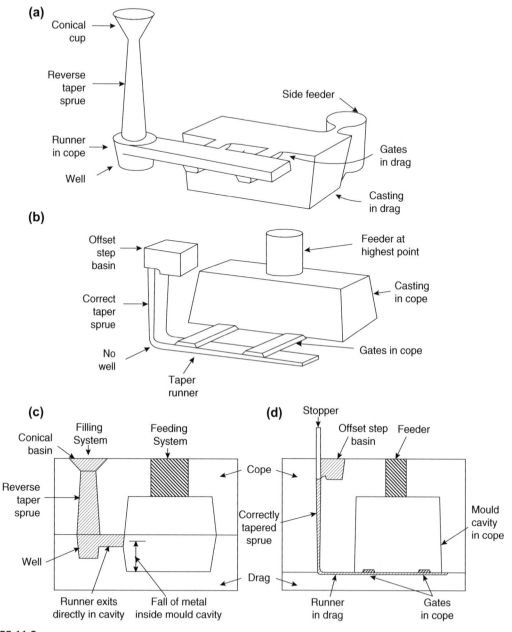

FIGURE 11.2

(a) Poor top gates and side-fed running system, compared with (b) a more satisfactory bottom-gated and top-fed system; (c) poor system gated at joint and (d) recommended economical and effective system.

its own oxide. The melt needs maximum constraint by the walls of the sprue during this fall. When exiting the base of the sprue, the distribution system from that lowest point, consisting of runners and gates, should progress only *uphill*.

Considering the mould cavity itself, the requirement effectively rules that all gates into the mould cavity enter at the lowest possible level, known as *bottom-gating*. The siting of gates into the mould cavity at the top *(top-gating)* or at the joint *(gating at the joint line)* are not options if freedom from defects is required.

Also excluded are any filling methods that cause waterfall effects in the mould cavity. This requirement dictates the siting of a separate ingate at every isolated low point on the casting.

In Figure 2.23(a), a fall of a few millimetres takes place inside a stationary tube of oxide. From greater fall heights, and depending on the alloy and the flexibility of the oxide film, oxide may concertina to form a dross ring (Figure 2.23(b)). Although this represents an important loss of metal on transferring liquid aluminium and other dross-forming alloys, it is not clear at this stage whether defects are also dragged beneath the surface and thereby entrained.

At higher speeds still, the dross is definitely carried under the surface of the liquid, together with entrained air, as shown in Figure 2.23(c). Turner (1965) has reported that, above a pouring height of 90 mm, air begins to be taken into the melt with the stream, to reappear as bubbles on the surface.

Falls in excess of the critical heights have been investigated for aluminium alloys (Din, Kendrick and Campbell, 2003). The author's own tendency is often to allow himself up to 1.0 m/s gate velocity, but never beyond 1.2 m/s which appears to be the velocity at which problems first occur, corresponding to a 70 mm fall. Falls of 100 mm, corresponding to a falling velocity of 1.4 ms^{-1}, appear always to entrain serious defects unless the melt is smoothed (the flow is 'laminised') by the use of a ceramic foam filter. In the case of a flow smoothed by a filter, even a fall of even 200 mm (2 ms^{-1}) appears to be tolerable as clearly shown later in Section 12.8 on filtration. This is in general agreement with the water model results of Goklu and Lange (1986) who found that the internal turbulence within a plunging jet influenced its surface smoothness which in turn influenced the amount of air entrainment. Practical maximum pour heights require to be investigated for other alloy systems to test what values beyond the theoretical limits may be used in practice.

Clearly, the benefits of defect-free pouring are easily lost if the pouring speed at the entry point into the mould is too high. This is often observed when filling the pouring basins of castings from an unnecessary height. In aluminium foundries, this is usually by robot. In iron foundries, it is commonly via automatic pouring systems from fixed launders sited over the line of moulds. In steel foundries, it is common to pour from bottom poured ladles that can contain more than a metre depth of steel above the exit nozzle (the pouring conditions for steel from bottom-teemed ladles are further complicated by the depth of metal in the ladle decreasing progressively from pour to pour). These three systems are all at risk of entraining defects in the pouring basin. However, the effect becomes sharply greater as we progress from filling the pouring basin by

1. lip pouring to
2. robot pouring to
3. automatic pouring from overhead launders to
4. bottom-poured ladles.

The situation is so bad for bottom-poured ladles, or bottom teeming as it is sometimes known, that it seems to me that a pouring basin concept will never work satisfactorily for this mode of filling. Entrainment will be occurring at a rate faster than detrainment from the basin can occur. This is a serious situation that is not widely appreciated and causes widespread problems in steel foundries. We shall deal with this problem later, and note that one of the few solutions to this serious issue is the complete elimination of the pouring basin and the implementation of contact pouring.

When the melt finally enters the down-sprue, the initial fall down the sprue in gravity-filled systems is nearly always a ragged and messy affaire, necessarily introducing oxide damage into the metal. Some of this raggedness is reduced by the use of a stopper to pre-fill the pouring basin. However, it seems reasonable to conclude that gravity-poured castings will never attain the degree of reliability that can be provided by counter gravity and other systems that can avoid surface turbulence. Of necessity therefore, it has to be accepted that the no-fall requirement applies to the design of the filling system downstream of the base of the sprue. The damage encountered in the fall down the sprue has to be accepted; although with a good sprue and pouring basin design, or better still, the adoption of contact pouring, this initial fall damage can be reduced to a minimum as we shall see.

11.3 REDUCTION OR ELIMINATION OF GRAVITY PROBLEMS

Over the years, several valuable techniques have been developed to overcome the problems that gravity pouring involves. These will be dealt with as technological issues later in the Processing Section under 'Casting'. Essentially, they can be summarised as follows,

1. Accepting as inevitable the unwanted high speeds as a result of the acceleration from gravity, and thus designing filling systems to cope with the necessarily high velocities but, despite the great danger to the melt, attempting to reduce the opportunities for damage by entrainment. This is achieved mainly by constraining the melt in narrow channels, too slim to allow the melt to fold over on itself and thus enclose its own surface, and only just wide enough to carry the volume of liquid at the rate to achieve the target fill time. This is the basis behind the 'naturally pressurised' filling system design described at length in this book.
2. Using essentially horizontal transfer of various types, including so-called 'level pour' techniques, or a special approach to tilt casting controlled in such a way to encourage only horizontal transfer while the mould rotates.
3. True counter-gravity filling, in which the melt is transferred only uphill against gravity, using either pressurisation from below by pump system or gas pressure, or vacuum from above.
4. Filling moulds with such narrow walls and sections that the melt is constrained by surface tension which now controls the filling instead of gravity. This phenomenon is discussed later below.

At the time of writing, most castings are made by pouring the liquid metal into the opening of the running system, as in (1) previously, using the action of gravity to effect the filling action of the mould. This is a simple and quick way to make a casting. Thus *gravity sand casting* and *gravity die-casting* (*permanent mould casting* in the United States) are important casting processes at the present time. Gravity castings have, however, gained a poor reputation for reliability and quality, simply because their running systems have in general been badly designed. Surface turbulence has led to porosity and cracks and unreliability in leak-tightness and mechanical properties.

Nevertheless, as emphasised in this book, there are rules for the design of gravity-running systems that, although admittedly far from perfect, are *much* better than nothing. Such rules were originally empirical, based sometimes on water modelling work and some confirmatory tests on real castings. We are now a little better informed by access to real-time video radiography of moulds during filling and sophisticated computer simulation, so that liquid aluminium or liquid steel can be observed or at least simulated as it tumbles through the mould. Despite this, many uncertainties still remain. The rules for the design of filling systems are still not the mature science for which we all might wish. Even so, some rules are now evident, and their intelligent use allows castings of the highest quality to be made. They are therefore described at length (but not, we hope, ad nauseam) and constitute essential reading!

It is hoped to answer the questions 'Why is the running system so complicated?' and 'Why are there so many different features?' It is a salutary fact that the apparent complexity has led to much confused thinking.

An invaluable clarifying rule that I recommend to all those struggling with the understanding of running and gating systems is 'If in doubt, visualise water'. Most of us have clear perceptions about the mobility and general flow behaviour of water in the gentle pouring of a cup of tea, the splat as it is spilled on the floor, the flow of a river over a weir or the spray from a high-pressure hose pipe. A general feeling for this behaviour can sometimes allow us to cut through the mystique, and sometimes even through the calculations! In addition, the application of this simple criterion can often result in the instant dismissal of many existing filling systems intended for the production of a reliable quality of casting as being quite clearly useless!

There are, of course, numerous ways to get the metal into the mould, but some are disastrously bad, some tolerable, some good. To appreciate the good, we shall have to devote some space to the bad. If this reads like a sermon, then so be it. A *good running system* is a *good cause* that deserves the passionate concern of the casting engineer. Too many castings with hastily rigged running systems have appeared to be satisfactory in limited prototype trials, but have proved to have disastrous levels of scrap when put into production. This is normally the result of surface turbulence during filling that produces non-reproducible castings, some apparently good, some definitely bad. This result confirms the nature of turbulence. Turbulence implies chaos; chaos implies unpredictability. When using a running system that generates surface turbulence a typical scrap rate for a commercial vehicle casting might be 15%, whereas a turbine blade subjected to much more stringent inspection, rejections can easily reach 75%.

In general, however, experience shows that foundries that use exclusively turbulent filling methods such as most investment foundries, experience on average about 20–25% scrap, of which 5–10% is the total of miscellaneous minor processing problems such as broken moulds, castings damaged during cutoff etc. The remaining 15–20% is composed of random inclusions including mould fragments, random porosity and misruns – the standard legacy of turbulent running systems: the inclusions are created by the battering of the mould and the folding of the liquid surface, as are the random pockets of porosity; and the misruns by the unpredictable ebb and flow in different parts of the casting during filling. In sand casting foundries, most of the so-called mould problems leading to sand inclusions are actually the result of the poor filling system designs. With good filling systems, problems such as mould erosion and sand inclusions disappear. Widespread problems in steel and Ni alloy foundries are cracks. These again disappear with the good filling system.

In a foundry making a variety of castings, the 15–20% running system scrap is made up of difficult castings which might run at 85–95% scrap (almost never 100%!) and easy castings which run at 5% scrap (almost never zero!). The non-repeatable results continuously raise the characteristic false hope that the problems are solved, only to have the hopes dashed again by the next few castings. The variability is baffling because the foundry engineer will often go to extreme lengths to ensure that all the variables believed to be under control are held constant.

Only a carefully worked out running system will give filling that is characterised by low surface turbulence, and which is therefore reproducible every time.

Interestingly, this can mean 100% scrap. However, this is not such a bad result in practice because the defect will be reproducibly repeated in every casting. It is therefore easy to identify and correct, and when corrected, stays corrected. After the first trials, the good running system should yield reliable, repeatable castings, and be characterised by a scrap rate close to or actually zero.

A good running system, perhaps something like that shown in Figure 11.2(b), will also be tolerant of wide variations in foundry practice in contrast with the normal experience accompanying turbulent filling, in which pouring conditions are critical. Many foundries will have the experience that certain castings can only be poured successfully by certain operators.

In contrast, a good running system will ensure that filling of the casting will now be under the control of the running system, not the pourer, and casting temperature will no longer be dictated by the avoidance of misruns, but might be set independently to control, for instance, grain size without the addition of grain refiners. It is clear, therefore, that a good running system is a good ally in the creation of economical products of high quality.

The requirements of a good system are:

1. Economy of size. A lightweight system will increase yield (the ratio of finished casting weight to total cast weight), allowing the foundry to make more castings from the existing melt supply. It may also help to get more castings into a given mould box. This has a big effect on productivity and economy. Also, critically, a slim system works better than an oversized system, delivering better castings.

2. Filling of the mould at the required speed. In the method proposed in this book, the whole running system is designed so that the velocity of the metal in the gates is below the critical value. This value varies negligibly from one alloy to another, being generally close to 0.5 ms^{-1}. There is now much experimental and theoretical data to support this value (Boutorabi, 1991). Data on the density of castings produced by gating uphill have shown that air entrapment can occur above approximately 0.5 ms^{-1} (Suzuki, 1989). In computer simulations of flow, Lin and Hwang (1988) show that when liquid aluminium enters the mould horizontally at 1.1 m/s, it hits the far wall with such force that the reflected wave breaks, causing surface turbulence. These figures confirm the safety of 0.5 ms^{-1}, and the danger of exceeding 1 ms^{-1}. Even so, I often allow myself to work at velocities up to 1 ms^{-1} because the 0.5 ms^{-1} is often difficult to meet, and the 1 ms^{-1} is usually just sufficiently slow to avoid any serious problems.

3. The delivery of only liquid metal into the mould cavity, i.e. not other phases such as slag, oxide and sand. However, in most cases, the overwhelmingly important and unwelcome phase is air (probably mixed with other mould gases of course). The design of filling systems to achieve the exclusion of air is a major preoccupation in this book. The main weapon here is pressurisation of the metal in the filling system. The naturally pressurised filling system proposed in this work automatically fulfils this essential requirement.

4. The elimination of surface turbulence, preferably at an early stage in the runner system, but certainly by the time that the metal arrives in the mould cavity. The problem here is that, by the time the metal has fallen the length of the sprue to reach the lowest level of the casting, its velocity is well above the critical velocity for surface turbulence. Despite this danger, the running system should, so far as possible, prevent the resulting fragmentation of the stream. Any fragmentation will result in permanent damage to the casting in most alloys. However, if fragmentation occurs, the best that can now happen is that it should be followed by an action to gather the stream together again. In this way, the melt enters the mould as a coherent, compact spreading front, preferably at a velocity sufficiently low that the danger of any further break-up of the front is eliminated.

5. Ease of removal. Preferably the system should break off. As a next best option, it should be removable with a single stroke of a clipping press or a straight cut on a saw. Curved cuts take more time and are more difficult to dress to finished size by grinding or linishing. Internal or shielded gates may need to be machined off, in which case the expense of setting up the casting for machining might be avoidable by carrying out this task later, during the general machining of the casting.

(Note that in general practice it is usually best to assume that there is no requirement for the filling system to act as a feeder i.e. to compensate for the contraction on solidification. We should ensure that the feeding function if necessary at all, is carried out by a separate feeder placed elsewhere, preferably high up, on the casting (Figure 11.2(b)). Occasionally, it is possible to use a running system that can also provide some feed metal. This especially favourable option should be exploited whenever possible. It is worth noting that in investment casting, the almost universal confusion between filling and feeding systems is deeply regrettable. In this book, the two functions are treated totally separately.)

Because the previous list of criteria has been so difficult to meet in practice, there has been a move away from gravity casting as a result of what have been believed to be insoluble barriers to the attainment of high quality and reliability.

Uphill filling, against gravity, known as *counter-gravity casting*, has provided a solution for the elimination of surface turbulence. (Counter-gravity should not be confused with a variety of gravity pouring known as *bottom-gating* as in Figures 11.2 and 11.3.) This development has therefore provided the impetus for the growth of low-pressure die casting, low-pressure sand casting and various forms of counter-gravity filling of investment castings. A form of high-pressure die-casting has also been developed to take advantage of the quality benefits associated with counter-gravity filling followed by high-pressure consolidation. These different techniques of getting the metal into the mould will all be discussed later.

Although counter-gravity filling fulfils all the previous requirements, our task when faced with gravity pouring is to optimise filling conditions so far as possible to meet the previous difficult set of criteria in spite of the difficulties.

Requirement 3 discussed previously for good gating is important: only liquid metal should enter the casting. Thus all bubbles entrained by the surface turbulence characterising the early part (the priming stage) of the running system should have been eliminated by this stage. If the running system is poor and bubbles are still present, their rise and bursting at the liquid surface in the mould violates casting rules 2 and 4. These violations result in several problems, including bubble trails, splash defects and the retention of the scattering of smaller bubbles that remain trapped under the oxide skin of the rising metal. These cause concentrations of medium-sized pores (0.5–5 mm diameter) at specific locations in the casting, usually at upper surfaces of the casting above some of the ingates.

The other point in requirement 3, that dross or slag not enter the mould cavity, is interesting. In the production of iron castings, it is normal for the runner to be placed in the cope and the gates in the drag, as is illustrated in Figure 11.2(a). The thinking behind this design of system is that slag will float to the top of the runner and thus will not enter the gates. Such thinking is at fault because it is clear that at least some of the first metal to enter the runner will fall down the first gate that it meets, taking with it not only the first slag but also air. This premature delivery of metal into the mould before the runner is full is clearly unsatisfactory. The metal has had insufficient time to settle down, to organise itself free from dross, oxide and bubbles. That such systems are widely used and are found in practice to reduce bubble and slag defects

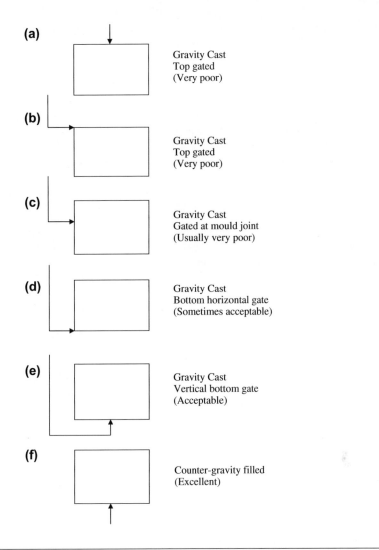

(a)   Gravity Cast
Top gated
(Very poor)

(b)   Gravity Cast
Top gated
(Very poor)

(c)   Gravity Cast
Gated at mould joint
(Usually very poor)

(d)   Gravity Cast
Bottom horizontal gate
(Sometimes acceptable)

(e)   Gravity Cast
Vertical bottom gate
(Acceptable)

(f)   Counter-gravity filled
(Excellent)

FIGURE 11.3

Various direct gating systems applied to a box shaped casting. (Note that all of the gravity systems shown here are poor: the sprue base connects directly with the ingate into the casting. All need mechanisms (not shown for clarity) to reduce the velocity of the melt.)

in the casting actually reveals how poor the front end of the running system is. Clearly, both bubbles and slag are being generated throughout the pour, so the off-take of gates at the base of the runner is valuable in this case. Nearly all of the slag problems I find in iron castings are not the result of carryover from the ladle, but are generated by turbulence in the filling system.

A traditional solution is illustrated in Figure 11.2(b). Here the runner is in the drag and the gates in the cope. The purpose in this system is to fill the runner first before the gates are reached. Thus the metal has a short time to rid itself of bubbles and dross, most of which can be trapped in a dross trap or against the upper surface of the runner. Only a limited amount of slag or dross will be unfortunately placed to enter the gate. Provided the mould height is low, the velocity of the metal may not be too high, giving the slag a good chance of being held against the ceiling of the runner or gate, and so not entering the casting. Figure 11.2(d) illustrates an optimum system designed to resist the entrainment of air at all stages of the system. However, in practice, this system requires further sophistication to avoid bubbles and slag entering the mould cavity. The reason for the non-optimum performance is that the metal velocity is usually far too high, so the small height difference between runner and gate is of no use to prevent premature filling of early gates; we have failed requirement 4 discussed previously.

Requirement 4 is deceptively simple. However, the task of achieving no surface turbulence is so important, and so central to the quest for good castings, that we have to consider it at length.

Texts elsewhere often refer to turbulence-free filling as laminar filling. The implication here is that turbulence as defined by Reynold's number is involved, and that the desirable criterion is that of laminar flow of the bulk. However, it is not bulk turbulence that is relevant because turbulent flow in the bulk liquid can still be accompanied by the desirable smooth flow of the surface. Our attention requires concentrating on the behaviour of the liquid surface. Thus provided we ensure that by 'laminar fill' we mean 'surface laminar fill,' then we shall have our concepts correct, and our thinking accurate.

Requirement 4 is clearly violated by splashing during filling. It can be seen immediately that top gating will probably therefore always introduce some defects (the exception is very thin wall castings where surface tension becomes dominant). Figure 11.3 illustrates the spectrum of gating options from very poor to excellent. The systems can be judged in the degree to which the metal is allowed to suffer a free fall into the mould cavity. Bottom-gated systems are always required if surface turbulence is to be eliminated. Counter-gravity is, of course, even more powerful than any bottom-gated gravity system because the velocity of the metal can be controlled *at all times* to be below critical.

However, although bottom-gating is necessary, it is not a sufficient criterion. It is easy to design a bad bottom-gating system! In fact, it is possible to state the case more forcefully: a bad bottom-gated system is usually worse than most top-gated systems.

For instance, it is common to see bottom-gated systems proudly displayed with the base of the runner turned so that metal directly enters the mould (Figures 11.3(d) and 11.4). Such systems are compact and appear economical until the percentage scrap figures are inspected. The sequence of events is clear if we consider the fall of the first liquid down the length of the sprue. The high velocity of the metal on its impact at its base is not contained. The resulting splash may be likened to an explosion of a high-velocity jet fired like a projectile directly into the mould. The jet ricochets from the far wall, causing more surface turbulence as the rebounding wave washes back, breaking and rolling over to entrain yet more surface oxide and more gas. The bulk of the metal follows in an untidy fashion as fragments of jets mixed with air and mould gases. The elimination of the entrained bubbles by bursting as they rise to the surface of the melt causes additional droplets to be created by splashing inside the mould cavity.

It is important, therefore, to design the down-sprue with care so that it will fill quickly, excluding air as quickly as possible, and to design the runner and gate to constrain the metal, avoiding any provision of room for splashing (Figure 11.5(a)). Further improvements might be allowable as in Figure 11.5(b) and (c) in which the fall heights down the sprue are progressively reduced, reducing velocities in the mould, by simply re-orienting the casting. This is a valuable technique.

The base of the sprue should be the lowest point in the whole system: having reached here, all subsequent flow of the liquid should be uphill, displacing the air ahead in a controlled and progressive advance. So far as possible, the liquid should be slowed as it goes, experiencing as much opportunity as possible to become quiescent before entering the mould. It should finally enter the mould at a velocity between 0.5 and 1.0 m/s. In this way, a good and reproducible casting is favoured.

It is important to note that these precautions to avoid the entrainment of oxide films also apply to casting in inert gas or even in vacuum. This is because the oxides of Al and Mg (as in Al alloys, ductile irons or high temperature Ni-base alloys, for instance) form so readily that they effectively 'getter' the residual oxygen in any conventional industrial vacuum, and form strong films on the surface of the liquid.

The reader might notice that there is no emphasis in this book regarding the huge amount of research that has been carried out on the filling systems of castings, treating the flow of the metal as an exercise in hydraulics. Using this scientific method, the friction factors and loss coefficients at bends and orifices etc. can be summed to find overall rates of

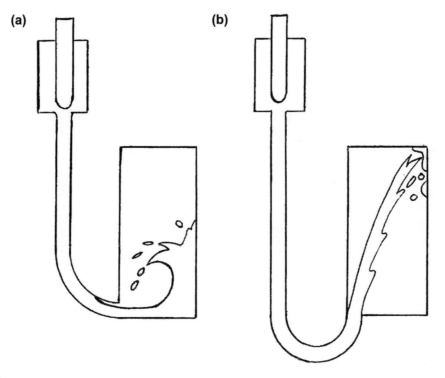

FIGURE 11.4

(a, b) Two bottom-gated systems that behave poorly because of direct entry of metal into the mould cavity at an uncontrolled velocity.

After Grube and Kura (1955).

flow (for instance see Bradley et al., 1992). In general, only perfectly filled channels are considered. This approach is neglected here only because the major problem for the filling systems of castings is flow in (at least initially) unfilled channels and unfilled moulds, where surface turbulence can wreak havoc. Even with a well-designed filling system, the problem to obtain a good casting may be controlled by the priming of the system, in which the 100% of air has to be supplanted by, it is hoped, 100% of metal, without in the meantime harming the quality of the liquid metal. Thus the technology for casting is essentially different, and, it seems, significantly more difficult to solve. That is what this book is about.

11.4 SURFACE TENSION CONTROLLED FILLING

Casting rule 2 – the avoidance of pouring and therefore exceeding the critical velocity, applies to 'normal' castings with walls of thickness of 3–4 mm or more. It is important to consider the different circumstances for extremely thin-walled products.

For channels that are sufficiently narrow, having dimensions of only a few millimetres, the curvature of the meniscus at the liquid front can help to preserve the liquid front from disintegration. Thus, narrow filling system geometries are valuable in their action to conserve the liquid as a coherent body, and so acting to push the air out of the system ahead of the liquid. The filling systems therefore fill in one pass. Such a filling action, pushing the air ahead of the liquid front as a piston in a cylinder, is a critically valuable feature. These systems deserve a special name such as perhaps 'one pass

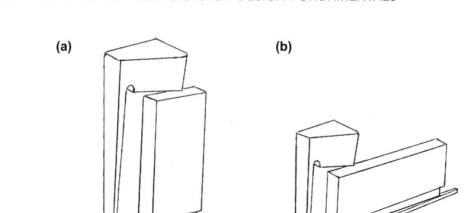

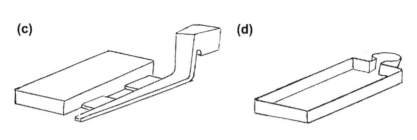

FIGURE 11.5

Improved bottom gated system; (a–c) progressively improved filling systems by height reductions down to (d) zero height by tilt casting.

filling designs'. Although I do not usually care for such jargon, the special name emphasises the special action. It contrasts with the turbulent and scattered filling often observed in systems that are over-generously designed, in which the melt travels in opposite directions at the same time along a single channel; a fast jet travels under the return wave that rolls over its top, rolling in air and oxides.

For a thin but wide horizontal channel, any effect of surface tension is clearly limited to channels that have dimensions smaller than the sessile drop height for that alloy. Thus the maximum channel height would be 12.5 mm for Al and only 8 mm for steel, although even these heights would exert little influence on the melt, because the roof would just touch the liquid, exerting no pressure on it. Similarly, taking account that the effect of surface tension is doubled if the curvature of the liquid front is doubled by a second component of the curvature at right angles, a channel of square section 25 mm square for Al and 16 mm for steel, would just constrain the melt by surface tension. In practice, however, for any useful restraint from the walls of channels in the absence of dynamic effects, these dimensions require to be at least halved, effectively compressing the liquid into the channel. If, however, the liquid advances through these limited sized channels at high speed, the channels would fill as a result of the dynamics of friction along the walls, causing drag against the walls and the building up of a back pressure, pressurising the melt against the walls.

In some thin-walled castings, the liquid may not be able to enter the mould at all. This is to be expected if the pressure is too low to force melt into the narrow section against the action of capillary repulsion. It is an effect resulting from surface tension γ. If the liquid surface is forced to take up a sharp curvature to enter a non-wetted mould, then it will be subject to a repulsive force that will resist the entry of the metal. Even if the metal manages to enter, it will still be subject to the continuing resistance of surface tension, which will tend to reverse the flow of metal, causing it to empty out of the mould if there is a loss of filling pressure. These are important effects in narrow-section moulds (i.e. thin-section castings) and have to be taken into account.

We may usefully quantify our formulation of this problem with the well-known equation (stated in an earlier chapter as Eqn (3.10) but reintroduced here in a revised form for our new purpose)

$$Pi - Pe = \gamma(1/r_1 + 1/r_2) \tag{11.1}$$

where Pi is the pressure inside the metal and Pe the external pressure (i.e. referring to the local gaseous environment in the mould). The two radii r_1 and r_2 define the curvature of the meniscus in two planes at right angles. The equation applies to the condition when the pressure difference across the interface is exactly in balance with the effective pressure resulting from surface tension. To describe the situation for a circular-section tube of radius r (where both radii are now identical), the relation becomes:

$$Pi - Pe = 2\gamma/r \tag{11.2}$$

For the case of filling a narrow plate of thickness $2r$, one radius is, of course, r, but the radius at right angles becomes infinite, so the reciprocal of the infinite radius equates to zero (i.e. if there is no curvature there is no pressure difference). The relation then reduces to the effect of only the one component of the curvature, r:

$$Pi - Pe = \gamma/r \tag{11.3}$$

We have so far assumed that the liquid metal does not wet the mould, leading to the effect of *capillary repulsion*. If the mould is wetted, then the curvature term γ/r becomes negative, allowing surface tension to assist the metal to enter the mould. This is, of course, the familiar phenomenon of *capillary attraction*. The pores in blotting paper attract the ink into them; the capillary channels in the wick of a candle suck up the molten wax; and the water is drawn up the walls of a glass capillary. In general, however, the casting technologist attempts to avoid the wetting of the mould by the liquid metal. Despite all efforts to prevent it, wetting sometimes occurs, leading to the penetration of the melt into sand cores and moulds.

Continuing now in our assumption that the metal-mould combination is non-wetting, we shall estimate what head of metal will be necessary to force it into a mould to make a wall section of thickness $2r$ for a gravity casting made under normal atmospheric pressure. If the head of liquid is h, the hydrostatic pressure at this depth is ρgh, where ρ is the density of the liquid and g the acceleration from gravity. The total pressure inside the metal is therefore the sum of the head pressure and the atmospheric pressure, Pa. The external pressure is simply the pressure in the mould due to the atmospheric pressure Pa plus the pressure contributed by mould gases Pm. The equation now is

$$(Pa + \rho gh) - (Pa + Pm) > \gamma/r \tag{11.4}$$

giving immediately (reflecting the fact that atmospheric pressure acts uniformly everywhere, so effectively cancelling)

$$\rho gh - Pm > \gamma/r \tag{11.5}$$

The back-pressure resulting from outgassing in the mould lowers the effective head driving the filling of the mould. It is good practice, therefore, to vent narrow sections, reducing this resistance to zero if possible.

It is also clear from the previous result, provided the mould is permeable and/or well vented, that atmospheric pressure plays no part in helping or resisting the filling of thin sections in air because it acts equally on both sides of the liquid front, cancelling any effect. Interestingly, the same equation and reasoning applies to casting in vacuum, which, of course, can be regarded as casting under a reduced atmospheric pressure. Clearly, a *vacuum casting* is therefore not helpful in overcoming the resistance to filling provided by surface tension (although, to be fair, it may help by reducing Pm by outgassing the mould to some extent before casting, and it will help where the permeability of the mould is low, where residual gases may be compressed ahead of the advancing stream. Vacuum casting may also help to fill the mould by reducing—but not eliminating – the effect of the surface film of oxide or nitride).

The case of *vacuum-assisted filling* (not vacuum casting) is quite different because the vacuum is applied only to the advancing front as illustrated in Figure 11.6. This application of a reduced pressure to only one side of the meniscus creates a differential pressure that drives the flow. The differential pressure is created by atmospheric pressure continuing

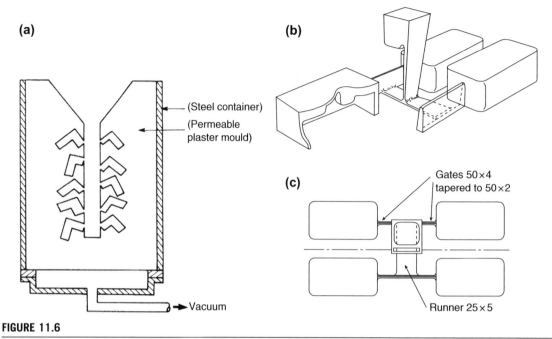

FIGURE 11.6

(a) A plaster mould encased in a steel box using vacuum-assisted filling through the base of the mould. No formal running system is required for such small thin-walled castings. (b) A perspective and (c) plan view of a sand mould to make four cover castings, using narrow slot filling system to maximise benefits from surface tension and wall friction.

to apply to the liquid metal via the running system, but the atmospheric pressure in the mould is reduced by applying a (partial) vacuum in the mould cavity. This is achieved by drawing the air out either through the permeable mould or through fine channels cut through to the section required to be filled (as is commonly applied to the trailing edge of an aerofoil blade section). In this way, Pm is guaranteed to be zero or negligible and Pa remains a powerful pressure to assist in the overcoming of surface tension as the equation indicates:

$$Pa + \rho gh > \gamma/r \qquad (11.6)$$

It is useful to evaluate the terms of this equation to gain a feel for the size of the effects involved. Taking, roughly, g as $10 \ ms^{-2}$ and the liquid aluminium density ρ as $2500 \ kgm^{-3}$ and γ as $1.0 \ Nm^{-1}$, the resistance term γ/r works out to be 2 kPa for a 1 mm section (0.5 mm radius) and 10 kPa for a 0.1 mm radius trailing edge on an air-conditioning fan blade.

For a head of liquid aluminium h 100 mm, the head pressure ρgh is 2.5 kPa, showing that the 1 mm section might just fill. However, the 0.1 mm trailing edge has no chance; the head pressure being insufficient to overcome the repulsion of surface tension. However, if vacuum assistance were applied (note below: not vacuum casting, remember) then the additional 100 kPa of atmospheric pressure normally ensures filling. In practice it should be noted that the full value of atmospheric pressure is not easily obtained in vacuum-assisted casting; in most cases, a value nearer half an atmosphere is more usual. Even so, the effect is still important: 1 atm pressure corresponds to 4 m head of liquid aluminium and approximately 1.5 m head of denser metals such as irons, steels and high-temperature alloys. In modest-sized castings of overall heights around 100 mm or so, these powerful pressures to assist filling are not easily obtainable by other means, but may in many instances be far too great, leading to high and damaging velocities. The pressure delivered by a feeder placed on top of the casting may only apply the additional head corresponding to its height of perhaps 0.1–0.4 m; only one tenth of the pressures that can be applied by the atmosphere.

For those castings that have sections of only 1–2 mm or less, the surface tension wields strong control over the tight radius of the front. Filling is only possible by the operation of additional pressure, such that provided by vacuum assistance or such mechanical devices as the jeweller's centrifuge. Filling can occur upwards or downwards, ignoring gravity without problems, being always under the control of the surface tension, which is effectively so strong in such thin sections that it keeps the surface intact. Surface turbulence is thereby suppressed. The liquid has insufficient room to break up into drops, or to jet or splash. The integrity of the front remains under the control of surface tension at all times. This special feature of the filling of very thin-walled castings means that they do not require formal running systems. In fact, such thin-walled plaster investment castings are made successfully by simply attaching wax patterns in any orientation directly to a sprue (Figure 11.6(a)). The metal flows similarly either with gravity or counter to gravity, and no 'runner' or 'gate' is necessary.

To gain an idea of the head of metal required to force the liquid metal into small sections, from Eqn (2.6) we have:

$$\rho g h = \gamma / r$$
$$h = \gamma / r \rho g$$

(11.7)

Using the values for aluminium and steel given previously, we can now quickly show that to penetrate a 1 mm section we require heads of at least 80 and 60 mm, respectively, for these two liquids.

If the section is halved, the required head for penetration is, of course, doubled. Similarly, if the mould shape is not a flat section that imposes only one curvature on the meniscus, but is a circular hole of diameter 1 mm, the surface then has an additional curvature at right angles to the first curvature. Equation (11.2) shows the head is doubled again.

In general, because of the difficulty to predict the shape of the liquid surface in complex and delicate castings, the author has found that a safety factor of two is not excessive when calculating the head height required to fill thin sections. This safety factor is quickly used up when allowances for errors in the wall thickness, and the likely presence of surface films are taken into account.

For steels and high-temperature alloys with a density of approximately 7000 kgm^{-3} and surface tension of 2.0 Nm^{-1}, the reader can quickly show that the depths of metal required to fill the 1 mm radius and the 0.1 mm radius are approximately 30 and 300 mm, respectively. For many small turbine blades, the trailing edge hardly has such a depth of metal above the upper sections of its trailing edge, with the result that non-filling of the edge is a constant problem.

In Figure 11.6(a), the castings have a wall thickness of only 0.4 mm and because the sprue is only about 100 mm high, providing too little head pressure, the castings will not normally fill. This is the reason for connecting the base of the mould on to a vacuum manifold that can deliver perhaps 0.5 bar reduction in pressure through the permeable mould. The 0.5 atm suction is equivalent to nearly 2 m head of liquid Al alloy that encourages the castings to fill practically instantaneously.

The resistance to flow provided by surface tension can be put to good effect in the use of slot-shaped filling systems. In this case, the slots are required to be a maximum thickness of only 1 or 2 (perhaps three at the most) mm thickness for aluminium alloy engineering castings (although, clearly, jewellery and other widget type products might require even thinner filling systems). Figure 11.6(b) shows a good example of such a system. A similar filling system for a test casting designed by the author, but using a conical basin (not part of the author's original design!), was found to perform tolerably well, filling without the creation of significant defects (Groteke, 2002). It is quite evident, however, when filling is complete such narrow filling channels offer no possibility of significant feeding. This is an important issue that should not be forgotten. In fact, in these trials, this casting never received the proper attention to feeding, and as a consequence suffered surface sinks and internal microporosity (the liquid alloy was clearly full of bifilms that were subsequently opened by the action of solidification shrinkage).

Finally, however, in some circumstances there may be fundamental limitations to the integrity of the liquid front in very thin sections.

1. There is a little-researched effect that the author has termed microjetting (Castings, 2003). This phenomenon has been observed during the filling of liquid Al-7Si-0.4Mg alloy into plaster moulds of sections between 1 and 3 mm thickness (Evans et al., 1997). It seems that the oxide on such small liquid areas temporarily restrains the flow, but

repeatedly splits open, allowing jets of liquid to be propelled ahead of the front. When viewed directly towards the advancing front, the result resembles advancing spaghetti. The mechanical properties are impaired by the oxide films around the jets that become entrained in the maelstrom of progress of the front. Whether this unwelcome effect is common in thin-walled castings is unknown, and the conditions for its formation and control are also unknown. It is a concern that this turbulent advance in very thin sections will undermine the benefits we have discussed previously. The reality is not known. Very thin-walled castings remain to be researched.

2. In pressure die castings, a high velocity v of the metal through the gate is necessary to fill the mould before too much heat is lost to the die. Speeds of between 25 and 50 ms^{-1} are common, greatly exceeding the critical velocity of approximately 0.5 ms^{-1} that represents the threshold between surface tension control and inertial control of the liquid surface. The result is that entrainment of the surface necessarily occurs on a large scale. The character of the flow is now dictated by inertial pressure, proportional to v^2, vastly exceeding the restraining influences of gravity or surface tension. This behaviour is the underlying reason for the use of PQ^2 diagrams as an attempt to understand the filling of pressure die castings. In this approach, a diagram is constructed with a vertical axis denoting pressure P and the horizontal axis denoting the flow rate Q. The parabolic curves are linearised by squaring the scale of the Q values on the horizontal axis. The approach is described in detail in much of the pressure die casting literature (see, for instance, Wall and Cocks, 1980). In practice, it is not certain how valuable this technique is now that computer simulation is beginning to be accepted as an accurate tool for the understanding of the process.

FILLING SYSTEM COMPONENTS

12

12.1 POURING BASIN
12.1.1 THE CONICAL BASIN

The in-line conical basin (Figure 11.2(a)), used almost everywhere in the casting industry, appears to be the worst system that anyone could imagine. It is not recommended. It is probably responsible for the production of more casting scrap in the casting industry than any other single feature. It has cost the industry dearly. It requires to be eliminated from our foundries as soon as possible.

The problems with the use of the conical basin arise as a result of several factors:

1. The device works as an air pump. This is the worst possible feature of an entrance to a filling system designed to receive only liquid metal. The pump action concentrates air into the flow (the action is analogous to other funnel-shaped pumps in which a fast stream of fluid directed down the centre of the funnel is designed to entrain a second surrounding fluid – good examples are steam ejectors and the vacuum suction device that can be driven from a compressed air supply). The mix of air and metal in terms of volume is often about 50/50, corresponding to the widely quoted finding of 0.5 discharge coefficient for a normal sprue. Because air is the single most important contaminant in running systems, this is the most severe disadvantage, yet, regrettably, is not widely appreciated to be a problem.

   An example that the author has witnessed many times can be quoted. Bottom-teemed steel ingots produced by the conventional arrangement that consisted of pouring the steel into a central conical cup, affixed to the top of a spider distribution system of ceramic tubes connected to the centre of the base of a group of up to six or more surrounding ingot moulds (Figure 12.1). Because the top of the ingot mould remains open during filling, the upwelling cascade of air bubbles in the centre of the rising metal is clear for all to see. This boiling action of emerging air almost certain undermines the quality of most present-day ingot casting. It is fortunate that much steel is now continuously cast, thereby avoiding the problem, but contact pouring would eliminate most of the problems for ingot casting. Furthermore, it is no more expensive, and significantly safer for operators.

2. The small volume of the basin makes it difficult for the pourer to keep full (its *response time* is too short, as explained later), so that air is automatically entrained as the basin becomes partially empty from time to time during pouring. The pourer can be completely unaware of any problem because, although it seems to the pourer that the basin remains full, the aspiration of air usually takes place under the surface at the basin/sprue junction.

3. The metal enters at an unknown velocity, and so making the estimation of the design of the remainder of the running system problematical.

4. The metal enters with a high, unchecked velocity. Because the main problem with running systems is to reduce the velocity, this adds to the difficulty of reducing surface turbulence.

5. Any contaminants such as dross or slag that enter with the melt are necessarily taken directly down the sprue.

6. The mould cavity fills differently depending on precisely where in the basin the pourer directs the pouring stream, whether at the far side of the cone, the centre, or the near side. Computer simulations of these three filling modes clearly show that three castings are produced with quite different defect patterns. Thus the castings are intrinsically not reproducible.

Complete Casting Handbook. http://dx.doi.org/10.1016/B978-0-444-63509-9.00012-1

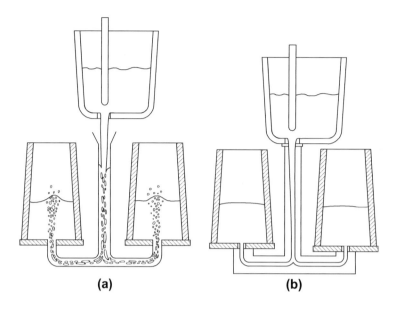

FIGURE 12.1

The pouring of ingots from a bottom teemed ladle (a) traditional trumpet entrance to refractory pipe filling system showing massive entrainment problems; (b) a contact pour, with ladle sealed against sprue entrance with ceramic felt, and filling system preferably moulded in sand. A vortex or tangential filter ingate could preferentially form the end gate into the casting to avoid initial jetting.

7. This type of basin is most susceptible to the formation of a vortex because any slight off-axis pouring will tend to start a rotation of the pool. There has been much written about the dire dangers of a vortex, and some basins are provided with a flat side to discourage its formation. In fact, however, this so-called disadvantage would only have substance if the vortex continued down the length of the sprue, along the runner and into the mould cavity. This is unlikely. Usually, a vortex will 'bottom out', giving an air-free flow into the remaining runner system as will be discussed later. This imagined problem is almost certainly the least of the difficulties introduced by the conical basin. In any case, the use of the offset step basin recommended later avoids any trace of vortex action.

All the previous disadvantages are present if the conical basin is well made, fits the sprue entrance accurately and is accurately placed, as shown in Figure 12.2(a). This is the best situation which is already unacceptably bad.

However, the already long list of damning faults is made even worse for a variety of reasons. A basin that is too large for the sprue entrance (Figure 12.2(b)) jets metal horizontally off the exposed ledge formed by the top of the mould, creating much turbulence and preventing the filling of the sprue. The problem is unseen by the caster, who, because he is keeping the basin full, imagines he is doing a good job. The cup shape of basin (Figure 12.2(c)) is bad for the same reason. The basin that is too small (Figure 12.2(d)) has painful memories for the writer: a casting with an otherwise excellent running system was repeatedly wrecked by such a simple oversight! Again, the caster thought he was doing a good job. However, the aspirated air caused a staggering amount of bubble damage in an aluminium sump casting; it was possible to push a finger through the papery wall into the bubble damaged areas.

The expansion of the sprue entrance to act as a basin (Figure 12.2(e)) may hold the record for air entrainment (however, the author has no plans to expend effort investigating this black claim). Worse still, the top of this awful device is usually not sufficiently wide that the pourer can fill it because it is too small to hit with the stream of metal without the danger of much metal splashed all over the top of the mould and surrounding personnel. Thus, this

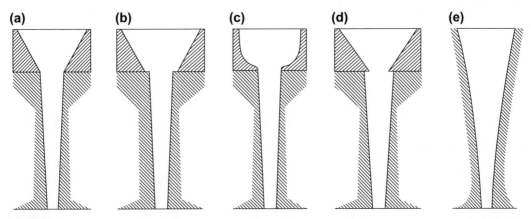

FIGURE 12.2

A rogues gallery of non-recommended scrap-generating conical basin and sprue combinations, showing (a) perhaps least damaging; (b) basin too large creating a ledge; (c) cup form creating a right angle edge; (d) basin too small creating an undercut; (e) enlarged sprue to act as a combined basin and sprue, greatly accelerating the flow.

combined 'basin/sprue' necessarily runs partly empty for most of the time. Furthermore, the velocity of the melt is increased as the jet is compressed into the narrow exit from the sprue (this important point is discussed in detail later). The elongated tapered basin system has been misguidedly chosen for its ease of moulding. There could hardly be a worse way to introduce metal to the mould.

For very small castings weighing only a few grams and where the sprue is only a few millimetres diameter, there is a strong element of control of the filling of the sprue by surface tension. For such small castings the conical pouring cup probably works tolerably well. It is simple and economical, and, probably fills well enough. This is as much good as can be said about the conical basin. Probably even this praise is too much.

Where the conical cup is filled with a hand ladle held just above the cone, the fall distance of about 50 mm above the entrance to the sprue results in a speed of entry into the sprue of approximately 1 ms^{-1}. At such speeds, the basin is probably least harmful. On the other hand, where the conical cup is used to funnel metal into the running system when poured directly from a furnace, or from many automatic pouring systems, the distance of fall is usually much greater, often 200–500 mm. In such situations, the rate of entry of the metal into the system is between 2 and 4 m per second. From the bottom-poured ladles in steel foundries, the metal head is usually over 1 m giving an entry velocity of 5 ms^{-1}. This situation highlights one of the drawbacks of the conical pouring basin; it contains no mechanism to control the speed of entry of liquid.

The pouring cup needs to be kept full of metal during the whole duration of the pour. If it is allowed to empty at any stage, then air and dross will enter the system. Many castings have been spoiled by a slow pour, where the pouring is carried out too slowly, allowing the stream to dribble down the sprue, or simply poured down the centre without touching the sides of the sprue, and without filling the basin at all (which is the trouble with the expanded sprue type, Figure 12.2(e)). Alternatively, harm can be done by inattention, so that the pour is interrupted, allowing the bush to empty and air to enter the down-sprue before pouring is restarted. Because of the small volume of the basin, it is not easily kept full so that these dangers are a constant threat to the quality of the casting.

Unfortunately, even keeping the pouring cup full during the pour is no guarantee of good castings if the cup exit and the sprue entrance are not well matched, as we have seen previously. This is the most important reason for moulding the cup and the sprue entrance integral with the mould if possible.

The operator also has the task to aim the pour with great precision because it is known that hitting the far side of the cone will give a different filling condition and a different quality of casting from a pour whose stream hits the near side of the cone, or falls directly down the axis of the sprue. Variations between these targets during the course of a pour will deliver a variety of different defects in successive castings, giving the foundry the familiar problem of generating scrap castings from time to time even though 'nothing has been changed'.

Finally, even if the pour is carried out as well as possible, any witness of the filling of a conical basin will need no convincing that the high velocity of filling, aimed straight into the top of the sprue, will cause oxides and air to be carried directly into the running system, and so into the casting. For castings where quality is at a premium, or where castings are simply required to be adequate but repeatable, the conical basin can never be recommended and should never be used.

12.1.2 INERT GAS SHROUD

A shroud is the cloth draped as a traditional covering over a coffin. In the foundry the word is used to describe the draping of an inert gas, often contained within a metal enclosure, to shield and thereby protect the metal stream from oxidation by air.

It is commonly used during the pouring of steel castings and is described in more detail in Section 14.4. Although undeniably useful to some extent, I have to admit to viewing the use of a shroud as a vulnerable and incomplete solution for the exclusion of air. Thus I have preferred to put in place systems that offer more robust solutions to the avoidance of ingestion of gases into the filling system. These various systems are described later.

12.1.3 CONTACT POURING

This is a brilliant solution, easy to apply, with no extra cost, and with dramatic benefits to the casting.

The attempt to exclude air during the pouring of castings is carried to its ultimate logical solution in the concept of contact pouring. In this system, the metal delivery system and the mould are brought into contact so that air is effectively sealed out. Reports of the use of contact pouring date back to 1962 (Jeancolas et al., and Bracale).

The direct contact system is of course necessary, and taken for granted in the case of counter-gravity systems, in which the mould is placed directly over a source of metal. The metal is then displaced upwards by pump or differential pressure.

Contact pouring is practised in the Al casting industry by the use of bottom pour ladles (Figure 12.3). These are generally rather small ladles containing 10–20 kg of metal. Their great advantage is that a robot can dip them vertically into the metal in a furnace, with the stopper open so that the ladle fills from underneath the metal surface. The stopper can be closed and the ladle lifted out and transferred to the mould. It is then set in place aligning with the sprue and the stopper raised to fill the mould. As the melt level in the ladle falls, the level in the mould rises, so that the sprue does not lose its pressurisation, thus keeping out air. The ladle can be used to retain metal in the ladle as a heel that can remain in the ladle between casts. Alternatively, if the speed of emptying is sufficient, the surface oxide skin in the ladle rolls down the ladle walls and can be sucked out with the melt so that the ladle remains clean, requiring no attention or cleaning between pours (the discarded oxide is the last to exit the ladle, and so sits harmlessly at the top of the sprue). Contact pour using a bottom stoppered ladle handled by a robot deserves wider use, and would be far preferable to the current widespread use of dipping duck type ladles and the messy filling and emptying operations they perform.

A high volume operation (VAW, now Hydro Aluminium Limited, Dilligem, Germany) casting aluminium alloy transports the mould up to the underside of a launder, the base of which is a nozzle closed by a stopper. When the mould is presented to and pressurised against the nozzle, the stopper is opened. After the mould is filled, the stopper is closed and the mould can be removed. This system works reliably and well.

The author has used contact pouring to make a 50,000 kg 4%Cr steel casting using the stopper in the base of the 50,000 kg bottom-poured ladle to deliver directly into the mouth of a sprue to make a highly successful backup roll casting (Kang, 2001).

A standard challenge of contact pouring is the achievement of an accurate alignment of the ladle nozzle with the sprue entrance. Usually, a white ceramic fibre blanket gasket is provided around the sprue to form a seal. When casting steel, the white surround is valuable because the light radiated from the glowing stopper in the ladle shines downwards through the nozzle, illuminating a circular patch on the white surround. The ladle can be easily manoeuvred into place

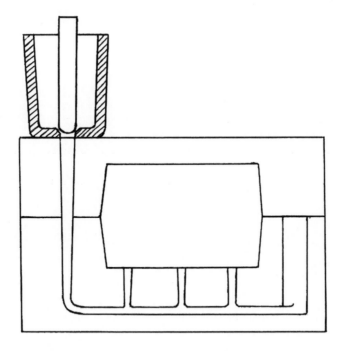

FIGURE 12.3

Contact pour, avoiding basin problems.

using the guiding light. For lower temperature metals such as aluminium alloys, a mechanical location system or some form of laser triangulation and feedback to the crane may be required.

Contact pouring has been used on automatic moulding lines at hundreds of moulds per hour for the pouring of cast iron (Disa, 2002). The iron is poured into an intermediate pouring box fitted with a nozzle and stopper in its base. The box is planted on the mould and the stopper is raised to fill the mould. The system offers the lowest possible head pressure to enhance surface finish and reduce speeds in the mould. When filling is complete, the stopper is lowered and the box is lifted clear. The small amount of metal that drains is collected in a small indentation in the mould surface, generally shrinking back into the sprue as the metal solidifies. The top of the mould remains clean, instead of swimming with pools of metal, making a contribution to overall yield. The box is then refilled to its original level. The absence of a pouring basin in the mould further increases yield. Why anyone should want to use any other kind of gravity filling system is a mystery to me.

12.1.4 THE OFFSET BASIN

Another design of *basin* (sometimes called a *bush*) that has been recommended from time to time is the offset basin (Figure 12.4(a)).

The floor of this basin is usually arranged to be horizontal (but sometimes sloping). The intention is that the falling stream is brought to rest before entering the sprue. This, unfortunately, is not true. The vertical component of flow is of course zero, but the horizontal component is practically unchecked. This sideways jet across the entrance to the sprue prevents approximately half of the sprue from filling properly, so that air is entrained once again. The horizontal component of velocity continues beneath the surface of the liquid throughout the pour, even though the basin may be filled.

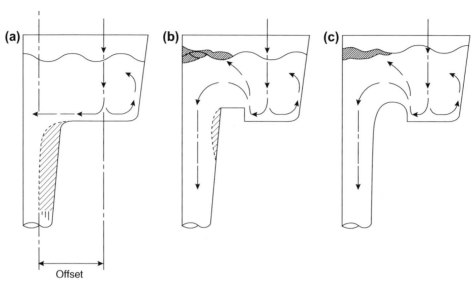

FIGURE 12.4

Offset pouring basins (a) without step (definitely not recommended); (b) sharp step (not recommended);
(c) radiused step (recommended).

There has been research using this type of basin over the years in which the discharge coefficients from sprues have been measured and found to be in the region of 50% or less. These low figures confirm that the sprue is only 50% or less filled, so that the major fluid being discharged is air. The quality of any castings produced from such devices must have been lamentable.

This type of basin is definitely not recommended.

12.1.5 THE OFFSET STEP (WEIR) BASIN

The provision of a vertical step, or weir, in the basin (Figure 12.4(b) and (c)) brings the horizontal jet across the top of the sprue to a stop. It is an essential feature of a well-designed basin.

Interestingly, this basin has a long history. West describes the first use of a basin conforming almost exactly to the offset step concept even though the step is not quite as sharp as one would like. This was in 1882. Since then, the principle has become progressively lost. Sexton and Primrose described a closely similar design (but once again without a well-formed step) in their textbook on iron founding published in 1911. If this basin is really valuable (as is recommended here), the reader will be curious why it has been known for so long, but has been extremely unpopular in foundries, whose experience of it has been discouraging. There are several reasons for this bad experience. Sometimes the basin has been made incorrectly, neglecting the important design features listed later. However, more serious than this, it has been common practice to place this excellent design of basin on a filling system that completely undoes all the benefits provided by the basin. Thus the benefits of the basin are never realised, and the basin is unjustly blamed.

Despite the revered age of this basin design, the precise function and importance of each feature of the design had not been investigated until recent computer studies by Yang and Campbell (1998). These studies make it clear that:

1. The *offset* blind end of the basin is important to bring the vertical downward velocity to a stop. The *offset* also avoids the direct *inline* type of basin, such as the conical basin, where the incoming liquid goes straight down the sprue, its velocity unchecked, and taking with it unwanted components such as air, dross and slag.

In older designs of this device, the blind end of the basin was often moulded as a hemispherical cup. This was not helpful because metal could easily be returned, ejected upwards and out of the basin by the sloping sides. The spraying of melt all around the foundry is bad for yield and therefore unpopular with the foundry accountant. It is also not recommended for operators. The flat floor and near-vertical sides of the basin were therefore significant advantages. In fact, the use of sharp internal corners to the offset side of the basin is positively helpful to avoid metal being ejected by the basin.

2. The *step (or weir)* is essential to eliminate the fast horizontal component of flow over the top of the sprue which would prevent it from filling fully. Basins without this feature commonly only about half fill the sprue, giving an effective so-called discharge coefficient of therefore only approximately 0.5 (how could it be higher if the sprue is only half full?). The provision of the step yields a further bonus because it reverses the downward velocity to make an upward curving flow in the basin, giving some opportunity for lighter phases such as slag and bubbles to separate out to the top surface of the melt in the basin before entering the sprue. Floating debris that has detrained in this way is shown schematically in Figure 12.4(b) and (c). Early basin designs were less than ideal because the step was not vertical (Swift, 1949) so that its effect was compromised. The step needs a vertical height at least equal to the height of the stream at that point to ensure that it brings the horizontal component of flow to a complete stop. Commonly, this height will be at least a few millimetres for a small casting, and might be 10–20 mm for a casting weighing several tonnes.

3. Finally, the provision of a generous *radius* over the top of the step (Figure 12.4(c)), smoothing the entrance into the sprue, further aids the smooth, laminar flow of metal. Swift and coworkers (1949) illustrated this effect clearly in their water models of various basins. The effect is also confirmed by the computer study by Yang and Campbell (1998).

The action of the step in preventing the entrainment of bubbles into the sprue, and encouraging detrainment is seen clearly in the water model by Kuyukak (2008). This model shows the great benefit of the step in detraining hundreds of buoyant bubbles. It needs to be kept in mind, of course, that non-buoyant or slowly buoyant defects such as oxide bifilms will not be detrained, but will be carried over into the casting. Thus, a good fraction of the damage caused by the turbulence of filling the pouring basin will degrade the casting; it is sobering to conclude that the damage suffered by gravity pouring starts even before any metal enters the sprue. (This situation contrasts with that of the contact pour ladle which can fill and discharge beautifully from its bottom nozzle with practically zero entrainment of air bubbles or bifilms. We shall repeated return to the benefits of contact pouring without apology. This is a technology we should all adopt instantly!)

There are plenty of opportunities for getting important details of this basin design incorrect, so that it functions poorly. Figure 12.5 shows some of these: (a) shows a common fault in which an extended flat between the step and the sprue allows the melt to regain some of its horizontal component of velocity, and thus overshoots the sprue entrance and causing air entrainment. The use of basins from a store, and simply dropped onto all types of moulds leads often to (b) in which there is a mis-match, leading to serious air entrainment. Figure 12.5(c) shows a less serious but unnecessary shape fault that will cause some entrainment. The practice of placing a boom or dam across the top of the basin as in Figure 12.5(d) to hold back floating debris is almost certainly counter-productive. It is seen to interfere with the natural circulation in the basin that would favour the automatic separation of buoyant phases. For this reason, a dam is not recommended. The use of a completely open base (e) is not a perfect solution, but may be acceptable because the velocities at the sprue entrance are not high if the basin is not too deep. The optimum design is seen in (f) together with the bringing forward of the back wall to be flush with the sprue entrance, constraining the flow of the melt to enter the sprue exactly parallel to its axis. The close wrapping around of the back wall of the basin in this way is not helpful if a stopper is used; the thin flash of metal around the back of the stopper can freeze and prevent the lifting of the stopper.

In practice, compared with the conical basin, the off-set step design is so easy to keep full it becomes immediately popular with both caster and quality technologist alike. And, naturally, when teamed up with a well-designed filling system, the basin can demonstrate its full potential for quality improvement of the casting, even though it is not perfect because it cannot hold back non-buoyant (i.e. bifilm) defects. The feature that makes the carryover of bifilms acceptable

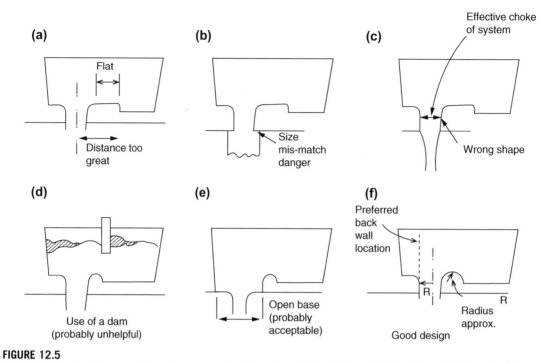

FIGURE 12.5

Offset basins (a) offset too large; (b) mis-match causing an undercut; (c) incorrect matching of tapers; (d) a dam prevents beneficial upward flow to encourage detrainment of bubbles and slag; (e) open base is acceptable if sprue entrance is radiused; (f) correctly designed basin, with further benefit if basin walls more closely wrapped around sprue entrance.

is that they arrive in the mould cavity in a compact, convoluted form, and are therefore a minimal threat to properties, providing they are retained compact by either rapid freezing and/or good pressurisation from a feeder.

An understandable criticism of offset basins is that they are so voluminous that they reduce yield and are therefore costly. The usual design is shown in Figure 12.6(a). Clearly, the yield criticism can be completely met by ensuring that the basin drains as completely as possible by arranging it to be sufficiently higher than the casting. However, of those cases in which the basin has to be placed lower and will not drain, the problem is to some extent addressed by the design variant shown in Figure 12.6(b) and (c). In addition to saving money, these basins work even better because they constrain the melt more effectively; they encourage the funnelling of the melt into the sprue with excellent laminar directional guidance. Even so, there is a limit to which these designs should be pushed in terms of reduced volume because the dwell time in the basin is useful for detrainment of defects.

These offset step basins can be made as separate cores, stored, and planted on moulds, matching up with the sprue entrance when required. However, because they will be required for many different castings, and so will need to mate up with different sprue entrance diameters, there is a concern about any mis-match of the basin exit and the sprue entrance. However, the problem is much less acute than mis-match of conical basins, because the speed of the falling stream at this point is considerably lower, in fact only at about its critical velocity. In these circumstances, surface tension is able to bridge modest outstanding ledges without significant entrainment of the liquid surface. An overhanging ledge is probably more serious and to be avoided. Thus a selection of stored basins with excess exit diameter is to be preferred. In fact, it may be preferable to arrange the bush to have its base completely removed on the sprue side (Figure 12.5(e)). The bush

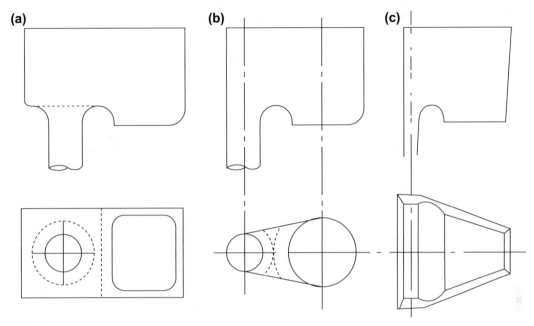

FIGURE 12.6

Side and plan views of offset basins (a) conventional rectangular; (b) slimmed shape to streamline flow and improve metal yield; (c) 'reverse delta' for high aspect ratio rectangular sprues.

Recommended by Puhakka (2010).

will then fit practically any mould. Provided the entrance to the sprue on the top surface of the cope is nicely radiused, the metal will probably be adequately funnelled into the sprue.

Ultimately however, the author prefers to mould the basin integral with the sprue, so avoiding the link-up and alignment problems. This is easily achieved with a vertical mould joint, but less easy, but still possible, with a horizontally jointed mould.

The basin is easier to use, and in general works sufficiently effectively, if its *response time* is approximately 1 s. To the author's knowledge there is no definition of response time. I therefore adopt a convenient measure as the time for the basin to empty completely if the pourer stops pouring. In practice, of course, the pourer does not usually stop pouring, so that the actual rate of change of level of the basin is relatively leisurely, allowing the pourer to maintain a consistent level of melt in the basin. Different pourers or pouring systems may require times shorter or longer than this.

The volume of the basin V_b (m^3) to give a response time t_r (in seconds) at a pouring rate Q (m^3 per second) is given simply by

$$V_b = Q/t_r.$$

Clearly, when $t_r = 1$ s, $V_b = Q$ numerically when using the recommended SI units. This is the volume of a '1-s basin'. Occasionally, if the quality of the casting demands it, I use a larger '2-s basin'.

Overall, as we have seen, the use of a basin is not altogether satisfactory from the point of view of the metal quality that is delivered into the casting, and is technically complicated, requiring such extras as pouring control using laser triangulation (Secom, 2003). Once again, contact pouring is far preferable.

Use of offset stepped basin with a bottom-pour ladle

Ladles equipped with a nozzle in the base are common for the production of large steel castings. Pouring from the base is sometimes known as 'teeming'. The benefits are generally described to be:

1. for large castings the tipping of a ladle to effect a lip pour becomes impractical, so that the bottom-teemed ladle is the only practical option;
2. the accuracy of the placing and the vertical direction of the pour is valuable;
3. the metal is delivered from beneath the surface of the melt, so avoiding the transfer of slag.

Even so, it is widely known in the trade that steel foundries using bottom pour ladles suffer dirtier castings than those steel foundries that use lip pour ladles. This follows as a natural consequence of the great difference in pouring speeds into the conical basin, with the consequent great difference in the rate of entrainment of air by the intake cone acting as an air pump.

Although it might appear that the use of bottom-pour ladles with an offset stepped basin at the entry to the mould has the potential to avoid this central problem, the offset step basin has a problem to cope with the huge entrance velocity. Even more seriously, even if the offset step basin were used optimally, it is likely that steel casting free from oxides would still be unachievable. This serious condition arises because the high speeds of entry from the ladle into the basin entrain so much air and oxide that these defects cannot be detrained in the time available before entering the sprue. This is a massive disadvantage suffered by steel casters.

Kuyucak (2008) describes a water model illustrating the use of a nozzle extension applied to the nozzle exit on the ladle. The extension is introduced deep into the basin so as to remain submerged during the pour. This virtually eliminates all entrainment. It seems likely that a good seal would be required between the nozzle and the extension tube; otherwise, air could be drawn in to the stream at this point as experienced on occasions by the long silica tube extension used by Harrison Castings.

A problem with the use of the nozzle extension is the initial splash when the nozzle is first opened, causing metal to jet vertically out of the basin and fountain liberally around. Kuyucak illustrates how this can be reasonably controlled by sinking the tube exit into a 'well' sunk into the base of the basin. Interestingly, he does not use an undercut form around the base of the basin walls. His water model indicates some residual entrainment action of any concave feature, whereas a convex surface, as an upturned cup, placed directly under the falling stream appeared to be helpful to control this initial transient.

For a successful use of the nozzle extension concept, it is necessary to examine how the offset step basin can be optimally filled. This in itself is not straightforward.

The common problem when using an offset stepped basin is that although a pourer using a lip-pour ladle can continue to adjust the rate of pour to maintain the level of liquid at the required height in the basin, this is easier said than done if the melt is being supplied from a bottom-pour ladle whose rate of delivery often cannot be controlled, the stopper may be either open or closed. Any attempt to adjust the rate of delivery may result in sprays of steel in all directions.

In addition to this problem, as the bottom-teemed ladle gradually empties, it reduces its rate of delivery. In the case of pouring a single casting from a ladle, it is fortunate that the filling system for the casting actually requires a falling rate of delivery as the net head (the level in the basin minus the level of metal in the mould) of metal driving the flow around the filling system gradually falls to zero. Even so, it is clear that the two rates are independently changing and may be poorly matched at times. The match of speeds might be so bad that the basin runs empty, but even if this disaster is avoided, well before this condition occurs, filling conditions can be expected to be bad. At a filling level beneath the designed fill level in the basin, the top of the liquid will appear to be covering the entrance to the sprue, but underneath, the sprue will not be completely filled, and so will be aspirating air. It is essential therefore to ensure, somehow, that the level in the basin remains at least up to its designed level. At this time, the problem of satisfactorily matching speeds can only be solved in detail by computer. Most software designed to simulate the filling of castings should be able to tackle this problem. However, it is perhaps more easily solved by simply having a basin with greatly increased depth, for instance perhaps up to four times the design depth. The ladle nozzle size is then chosen to deliver at a higher rate, causing the basin to overfill its design level, and so effectively running the casting at an increased speed. This increased speed is far preferable to the danger of under-filling of the basin with the consequential ingestion of air into the melt.

In general, therefore, a greatly increased depth to the basin is very much to be recommended. The problem of overfilling and increased speed of running may not be as serious as it might first appear. The reason is quickly appreciated. If the height of metal in the pouring basin rises to a level twice the design depth, the rate of delivery from the ladle would be approximately 40% higher (a factor of $2^{1/2} = 1.414$). A basin four times the minimum height will accommodate delivery from the ladle at up to twice as fast as the running system was designed for. The increase in pressure that this provides will drive the filling system to meet the higher rate. (Notice that the narrow sprue exit is not acting as a so-called choke, illustrating how wrong this concept is.) Thus the system is, within limits, automatically self-compensating if the basin has been provided with sufficient freeboard. It is important therefore to make sure that offset stepped basins in collaboration with a bottom-poured ladle do have sufficient additional height.

The preferred option to overfill the basin in terms of height is valuable in the other common experience of using a large bottom-pour ladle to fill a succession of castings. Let us take as an example a 20,000 kg ladle that is required to pour nine castings each of 2000 kg. (The final 2000 kg in the ladle will probably be discarded because it will pour too slowly, contain too much slag and be too low in temperature). The first castings will be poured extremely rapidly because the head of metal in the ladle will be high. However, the most serious problem is that the final castings in the sequence will be poured slowly, perhaps too slowly.

The important precaution therefore is to ensure that the minimum height in the pouring basin is still met so the final casting is still poured sufficiently quickly. This is a key requirement, and will ensure that the final casting is good. Thus all of the filling design should be based on the filling conditions for the last casting. Clearly, all the preceding castings will all be over-pressurised by increased heights of metal in their pouring basins, and so will fill correspondingly faster, with correspondingly higher velocities entering the mould. This should be checked to ensure that the velocities are not so very high as to cause unacceptable damage. Usually, this approach can be made to work out well.

In some cases, the first castings may have their pouring basins filled high, but the metal has not yet arrived in the feeders to give a signal to the operator to stop pouring. In this case, the only option seems to monitor the progress of the pour by some other factor, such as precise timing, or better still, a direct read-out load cell on the overhead hoist carrying the ladle if such a cell can provide the required accuracy.

The matching of the speed of delivery from the ladle with the speed of flow out of the pouring basin is greatly assisted if the rate of delivery from the ladle is known. This is a complex problem dependent on the height of metal in the ladle, its diameter and the diameter of the nozzle. The interaction of all these factors can be assessed using the nomogram provided in the Appendix. However, much of the problem of matching the rates out of the ladle and the rate out of the pouring basin is eliminated by eliminating the pouring basin; contact pouring has significant advantages.

12.1.6 THE SHARP-CORNERED OR UNDERCUT BASIN

In addition to the matching of the rate of flow between the ladle and the casting, there are additional problems with the application of offset weir basins for use with bottom-poured ladles.

As we have discussed previously, the velocity of the melt exiting the base of the bottom-poured ladle when the stopper is first opened is awesomely high. This is because the melt at the base of a full ladle is highly pressurised. Effectively it has fallen from the upper surface of the melt in the ladle; often as much as a metre or more. Thus the exit speed is often in the region of 5 ms^{-1}. This is so high that if this powerful jet is directed into the blind end of a step basin, the liquid metal flashes outwards over the base, hits the radii in the corners of the vertical sides, where it is turned upwards to spray all over the foundry (Figure 2.12(a)). Such spectacular pyrotechnic displays are not recommended; little metal enters the mould.

The radii around the four sides of the off-axis well of the basin are extremely effective in redirecting the flow upwards and out of the basin. One solution to this problem is therefore simply the removal of the radii. The provision of sharp corners to all four sides reduces the splashing tendency to a minimum (the top of the weir step leading over to the sprue entrance should still be nicely radiused of course).

The sharp-cornered basin is a useful design. However, an ultimate solution to the splashing problem is provided by a simple re-entrant undercut at the base of the basin (Figure 12.7(b)). (I recall demonstrating such a basin in a steel foundry whilst foundry personnel hid behind pillars and doors. On the opening of the ladle stopper the stream gushed into the

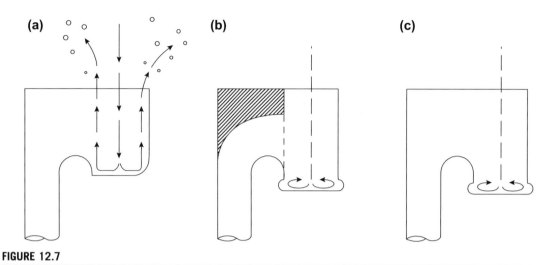

FIGURE 12.7

Offset basins for high velocity input. (a) No undercut empties spectacularly upwards (not recommended). (b) The provision of an undercut gives a basin that resists splashing. The shaded area can be moulded in a vertical jointed mould to further improve flow and metal yield, and prevents any risk of pouring directly down the sprue. However, its use may be questionable because it leaves little room for detrainment action of buoyant phases. (c) a recommended design.

basin, but not a drop emerged. The pouring process was quiet; its intense energy tamed for the first time. The foundry personnel emerged from their hiding places to gaze in wonder.)

The undercut is, of course, a problem for many greensand moulding operations making horizontally parted moulds. This is why the sharp-edged basin is so useful. Even so, where extreme incoming velocities are involved, an undercut edge to all four sides of the filling well of the basin may be the only solution.

The undercut may be difficult to mould, but it can be machined. The upgrading of a sprue cutter on a high production greensand line to three-dimensional machining unit equipped with a ball-ended high speed cutter would make short work of the basin, complete with its undercut and sprue entrance, and providing all this within the moulding cycle time. Such a unit would be an expensive sprue cutter, but would be a good investment.

The undercut is not a problem for vertically jointed moulds. Its use on machines such as Disamatics is popular and welcomed by the foundry operators. Its quiet filling is easily controlled and there is complete absence of splashed metal (splashed metal is commonly seen as pools, sometimes nearly lakes, swimming around on the tops of moulds). The reduction of pouring overspill is a significant contribution to the raising of metal yield in the foundry and saves the narrow bars of the shakeout grid from becoming jammed with spills.

The precise vertical depth of the undercut is important. It is required to be twice as deep as the incoming jet flowing over the base to allow for the depth of the outgoing reversed flow. If the depth is less than this, the reversal cannot occur efficiently. If the depth is much greater, the undercut will not fill and is likely to entrain air. Its depth can be estimated as follows.

Assuming the metal maintains its speed approximately after impacting on the floor of the basin, if the infalling jet of metal has diameter d, its area $\pi d^2/4$ must equal the area of the stream after it has expanded and flattened to thickness t on reaching the perimeter of the basin. If the floor diameter is D, the area of the perimeter flow at thickness t is πDt. Equating these areas, assuming they correspond to approximately equal flow rates, we have $\pi d^2/4 = \pi Dt$. The thickness of the undercut 2t is therefore simply $d^2/2D$. (The author apologises for the factor of two oversight in the 2011 edition of this book.) This does require an estimate of the diameter of the metal stream exiting the ladle.

A note of caution surrounds the use of the undercut. More research is needed to ensure that the quality of the melt is not degraded by entrainment because the design is based on the re-entrainment of the high-speed flow. However, entrainment below the surface seems likely to be an improvement on the massive entrainment suffered by a conical basin. It would be useful to see this expectation verified by careful practical measurements.

The moulding of the sprue cover (the shaded region in Figure 12.7(b)) ensures that metal is never poured in error directly down the sprue, and saves a little metal. However, it is almost certainly counter-productive because the buoyant phases (slags and bubbles) now have practically no opportunity to detrain. Overall, therefore, it is not recommended. The additional metal cast in Figure 12.7(c) will be worth the small additional cost, and the larger area of the basin at the pouring end (seen in Figure 12.6(b)) will encourage the pourer to target the pour correctly.

If, having fulfilled all the previous concerns, the offset stepped basin is successfully maintained full, the head of metal provided by the height of the down-runner will be steady, and the rate of flow will be controlled by the sprue. The filling rate will be no longer at the mercy of the human operator on that day. The running system will have the best chance to work in accord with the casting engineer's calculations.

12.1.7 STOPPER

As a further sophistication of the use of the offset step basin, some foundries place a small sand core in the entrance to the sprue. The core floats only after the bush is full and therefore ensures that only clean metal is allowed to enter the sprue. Alternatively, a wire attached to the core, or a long stopper rod lifted by hand accomplishes the same task. For a large casting, the raising of the stopper will require a more ruggedly engineered solution, involving a the benefit of the action of a long lever to add to the mechanical advantage and keep the operator well away from sparks and splashes. However it is achieved, the delayed opening of the down-sprue until the basin is filled to its designed depth is valuable in many foundry situations.

The early work on the development of filling systems at Birmingham concentrated on the use of the offset step basin. A stopper was not used because it was considered to be too much trouble. However, after about the first 12 months, as a gesture to scientific diligence, it was felt that the action of a stopper should be checked, if only once, by observing the filling of a sprue using the video X-ray radiographic unit, comparing filling conditions with and without a stopper. A stopper was placed in the sprue entrance, sealing the sprue. The metal was poured into the basin. When the basin was filled to the correct level the stopper was raised. The pouring action to keep the basin full was then continued until the mould was filled. The results were unequivocal. The use of a stopper improved the filling of the sprue to a degree that was amazing. It was with some resignation that the author affirmed this result. For every casting after that day, a stopper was always used.

Latimer (1976) demonstrated that the use of a stopper reduced the fill time by 60%. This is further proof that the system runs fuller (and indicates indirectly that without the stopper the system must be entraining more than 50% air).

There seems little doubt therefore that, despite the inconvenience, when the best quality castings are required, a stopper is advisable. Thus the author always recommends its use for aerospace castings.

In addition, the use of stoppers is particularly useful for very large castings where different levels of the filling system are activated by the progressive opening of stoppers as the melt level rises in the mould, so bringing into action new sources of metal to raise the filling speed.

12.2 SPRUE (DOWN-RUNNER)

The sprue has the difficult job of getting the melt down to the lowest level of the mould whilst introducing a minimum of defects despite the velocity of the stream being accelerated to well over its safe velocity, 0.5 m/s.

The fundamental problem with the design of sprues is that the length of fall down the sprue greatly exceeds the critical fall height; the height at which the critical velocity is reached corresponds to the height of the sessile drop for that liquid metal. Thus, for aluminium, this is about 13 mm, whereas for iron and steel it is only about 8 mm. Because sprues are typically 100–1000 mm long, the critical velocity is vastly exceeded. How then is it possible to prevent damage to the liquid? This question is not easily answered and illustrates the central problem to the design of filling systems that work using gravity. (Conversely, of course, counter-gravity systems can solve the problem at a stroke, which is their massive technical advantage.)

For the sprue at least, the problem is soluble. The secret of designing a good sprue is to make it as narrow as possible, so that the metal has minimal opportunity to break up and entrain its surface during the fall. The concept of protecting the liquid from damage is either (1) to prevent it from going over its critical velocity which in this case is not possible or (2) if the critical velocity has to be exceeded, to protect it by *constraining* its flow in channels as narrow as possible so that it is not able to jump and splash.

Theoretically, a design of the sprue can be seen to be achieved by tailoring a funnel in the mould of exactly the right size to fit snugly around a freely falling stream of metal, carrying just the right quantity of metal per second (Figure 12.8). We call the funnel the down-runner, or sprue for short. Many old founders I have worked with call it the spue, or spew (which, incidentally, does not appear to be a joke).

Most sprues are oversized. This is bad for metallic yield, and thus bad for economy. However, it is much worse for the metal quality, which is damaged in important ways:

1. The sprue takes more time to fill, extending the damaging processes listed later.
2. Air is therefore taken down with the metal, causing severe surface turbulence in the sprue because of the high velocity of the fall. This, of course, leads to a buildup of oxide in the sprue itself, and much consequential damage downstream in the running system and the casting from entrained oxide and air. The amount of damage to the metal caused by a poor basin and sprue can be quickly appreciated from the common observation of the blockage of filters. Even with good quality liquid metal, a poor basin and sprue will create so much oxide that a filter is quickly overloaded and blocked. Such poor front ends to filling systems are so common that filter manufacturers give standard recommendations of how much metal a filter can be expected to take before becoming choked. However, in contrast to what the manufacturers say, with a good basin and sprue (and providing, of course, the quality of the melt is not too bad) a filter seems capable of passing any amount of liquid metal without problem.
3. The free fall of the melt in an oversized sprue, together with air to oxidise away the binder in the sand, is a potent combined assault that is highly successful in destroying moulds. The hot liquid ricochets and sloshes about, its high speed and agitation punishing the mould surface with a hammering and scouring action. At the same time, the pockets of air in the rebounds of the flow will displace air backwards and forwards through the sand like blasts from

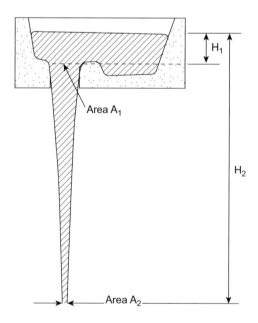

FIGURE 12.8

The geometry of the stream falling freely from a basin.

a blacksmith's bellows, causing the organic matter in the binder to glow, and, literally, to disappear in a puff of smoke! When the binder is burned away, reclaiming the sand back to clean, unbonded grains, the result is, of course, severe sand erosion. Figure 12.9 shows a typical result for an aluminium alloy casting in a resin-bonded sand mould. An oversize sprue is a liability.

Conversely, if the sprue is correctly sized the metal fills quickly, excluding air before any substantial oxidation of the binder has a chance to occur. The small amount of oxygen in the surface region of the mould is used up quickly by the burning of a small percentage of binder, but further oxidation has to proceed at the rate at which new supplies of air can arrive by diffusion or convection through the body of the mould. This is, of course, slow, and is therefore not important for those parts of the mould, such as the sprue, that are required to survive for only the relatively short duration of the pour. Furthermore, because the liquid metal now fills the volume of the sprue, the oxide film forming the metal–mould interface is stationary, protecting the mould material in contact with the sprue, and transmitting the gentle pressure of the steady head of metal to keep the mould surface intact. The result is a perfectly cast sprue (Figure 12.9), free from sand erosion and oxide laps. A correct-sized sprue for an aluminium alloy casting will shine like a new pin. (But beware, an undersized sprue will too!) Figure 12.2 illustrates some examples of good and bad systems. A test of a good filling system design in any metal is how well the running system has cast. It should be perfectly formed.

How then is it possible to be sure that the sprue is exactly the right size? The practical method of calculating the dimensions of the sprue is explained in Chapter 13. Basically, the sprue is designed to mimic the taper that the falling stream adopts naturally as a result of it acceleration because of gravity (Figure 12.10). The shape is complex, its area as a function of height is given by an equation for a hyperbola (interestingly, not a parabola as widely stated). Because most sprues have been traditionally formed by a straight taper, the curved sides of the stream cause the metal to become detached from the walls at about halfway down as shown in this figure. For modest-sized castings this (together with other errors, mainly from the geometry and friction of the flow in the basin) is simply corrected by making the sprue entrance about 20% larger in area (corresponding of course to about 10% increase in diameter). Thus, straight-tapered sprues, especially if the entrance area is enlarged by 20%, appear to work reasonably satisfactorily.

For very tall castings, the straight tapered approximation to the sprue shape is definitely not satisfactory. In this case, it is necessary to calculate the true diameter of the sprue at close intervals along its length. The correct form of the falling stream can then be followed with sufficient accuracy, keeping the stream in contact with the mould during the whole course of its fall, and air entrainment during the fall can be avoided.

Using this detailed approach the author has successfully used sand sprues for very large castings. An interesting example is a 50,000 kg steel casting 7 m high. The sprue was assembled from a stack of tubular sand cores, accurately located by an annular stepped joint. (Only one core box was required, but the central hole required a pile of separately

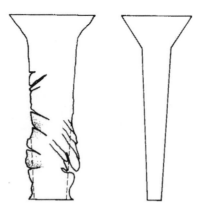

FIGURE 12.9

An oversize sprue that has suffered severe erosion damage because of air entrainment during the pour. A correctly sized sprue is usually well formed.

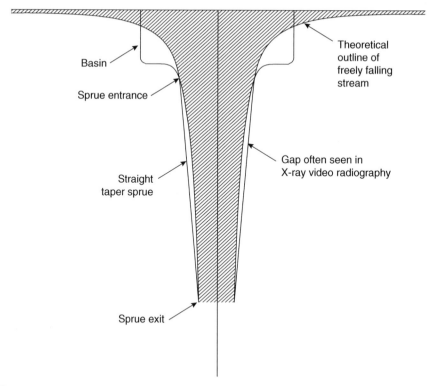

Basin

Sprue entrance

Theoretical outline of freely falling stream

Straight taper sprue

Gap often seen in X-ray video radiography

Sprue exit

FIGURE 12.10

Theoretical hyperbola shape of the falling stream, illustrating the complicating effects of the basin and sprue entrance.

turned loose patterns to make a gradually curved taper to the sprue.) The conventional use of ceramic tubes for the building of filling systems for steel castings was thereby avoided, with advantage to the quality of the casting. As an interesting aside, the appearance of this sprue after being broken from the mould was at first sight disappointing. It seemed that considerable sand erosion had occurred, causing the sprue to increase in diameter by more than 10 mm (about 10%). On closer examination, however, it became clear that no erosion had occurred, but the chromite sand cores had softened and been compressed under the high hydrostatic pressure of the liquid steel, losing the air spaces between the grains to become a solid mass. It had partially softened, probably as a result of the use of a silicate binder system; the silicate had reacted with the chromite to form a lower melting point phase. Because such a growth in diameter would necessarily have occurred by a kind of creep process, in which pressure, temperature and time would be involved, it follows that much of this expansion would have happened after the casting had filled because pressure was then highest, the sand fully up to temperature and more time would be available because the time for the solidification of the sprue would be as much as 10 times longer than the pouring time. Thus during the pour the sand-moulded sprue would almost certainly have retained a satisfactory shape, as corroborated by the predicted fill time being fulfilled, the surface chromite wash coat was still intact, and the cleanness of the metal rising in the mould cavity was clearly seen to be excellent.

Sand-moulded filling systems for steel castings are, of course, prone to erosion if the system design is bad, and particularly if, as is usual, the system is oversized. In this case. however, the sand-moulded hyperbolic sprue worked considerably better than the conventional refractory tube system. However, there would be no doubt that the refractory tubes might be satisfactory if they could be specifically designed and produced for the sprues for each individual steel casting. Naturally, at the present time, this is not easily arranged. Also, there is the doubt about their suitability because of their lack of permeability.

An interesting option might be the creation of sand cores to replace the refractory hollowware. Thus the foundry would be free to construct its own hollowware to fit each casting, using one core box with a variety of turned, central loose pieces to form the tapered internal form.

The cross-section of the sprue can be round or rectangular. Some authorities have strongly recommended square in the interests of reducing the tendency of the metal to rotate, forming a vortex, and so aspirating air. This may have been important in castings using conical pouring basins because any out-of-line pouring would induce the rotation of the melt in the pouring cup. However, the author has never seen any vortex formation with an offset step basin. The problem seems not to exist with good basin design.

In addition, of course, the vortex appears to be unjustifiably maligned. The central cone of air will only act to introduce air to the casting if the central core extends into the mould cavity. This is unlikely, and in its use with the vortex sprue and other benign uses of vortices, the design is specifically arranged to suppress the possibility of air entering the casting. The vortex can be a powerful friend, as we shall see.

To summarise: for ease and safety of design at this time, the sprue should be a simple, smooth, nearly vertical, tapering channel, containing no connections or interruptions of any kind. The rate of filling of the mould cavity should be under the absolute control of its cross-section area. If, therefore, the casting is found in practice to be filling a little too fast or too slow, then the rate can be modified without difficulty by slight adjustment of the size of the sprue.

Significantly, it is not simply the sprue exit that requires modification in this case. If correctly designed, the whole length of the sprue acts to control the rate of flow. This is what is meant by a naturally pressurised system. We can get the design absolutely correct for the sprue along its complete length. Although methoding engineers have been carrying out such calculations correctly for many years, somehow only the sprue exit has been considered to act as the choke. We need to take careful note of this widespread error and perhaps take time to re-think our filling system concepts. The whole length of a filling system, including the sprue, runner and gates, should each be slightly pressurised, and so each be making its contribution to the overall 'choking' of the flow. The choke is localised in no particular place. We are now dealing with a *distributed* choke effect.

Turning now to a common problem with many automatic moulding units for the manufacture of horizontally parted moulds: we have already lamented that these units do not provide for a properly moulded basin. (Despite such a satisfactory offset step basin being rapidly machined as mentioned previous, cutting by machining is usually never actioned. With a sufficient stroke and correct radius of cutter, even the sprue could be fully machined to its proper hyperbolic taper.)

Worse still, it is regrettable that a reverse-taper sprue is usually the only practical option. The sprue pattern needs to be permanently fixed to the pattern plate, and therefore has to be mouldable (i.e. the mould has to be able to be withdrawn off the sprue when stripping the mould off the pattern) as seen in Figure 11.2(c).

Because of the previous basin and sprue problems, there is a strong case for believing that all our high production moulding lines are especially designed to make only scrap castings. However, for the dedicated engineer, even when pitted against such odds, all is not yet lost. The strategies to reduce the problems that are suffered are detailed later.

The sprue entrance should be maintained at its correct calculated size, and the taper (now the wrong sign, remember) down the length of the sprue should be kept to a minimum. (A polished stainless steel sprue pattern can often work perfectly well with zero taper providing the stripping action is accurately square.) The sprue exit is now very much oversize of course, so that much turbulence is to be expected at the base of the sprue.

Even though all precautions are taken in this way to reduce the surface turbulence to a minimum, the consequential damage to the melt by a reverse or zero tapered sprue is preferably reduced further by the provision of:

1. A filter as soon as possible after the base of the sprue. The friction provided by the filter acts to hold back the flow, and thus assists the poorly shaped sprue to backfill as completely and as quickly as possible, and so reduce the total damage. The filter will also act to filter out some of the damage, although it has to be realised that this filtering action is not particularly efficient. The use of filters is dealt with in detail later in Section 12.8.
2. A radial choke. This is a clever 'quick fix' suggested by Groteke (2005) described in Section 12.2.5.

We need to dwell a little longer on the importance of the use of the correct taper, so far as possible, for sprues.

The effect of too little, or even negative, taper has been seen previously to be detrimental to casting quality. Surely, one might expect that the opposite condition of too much taper would not be a problem because it seems reasonable to assume that the velocity of the metal depends only on the distance of fall. However, this is not true. The head of metal in the pouring basin provides the driving force experienced by the melt entering the sprue. If the sprue taper matches the natural taper of the falling stream, the only acceleration experienced by the melt *is* the acceleration from gravity. If, however, the taper of the sprue is greater than this, the melt is correspondingly speeded up as the sprue constricts its area. This extra speed is unwelcome because the task of the filling system designer is to reduce the speed. The effect of varying taper has been studied by video X-ray techniques. In experiments in which the sprue exit area was maintained at a constant, a doubling of the sprue entrance area was seen to nearly double the exit speed, with the generation of additional turbulence in the runner. Three times greater entrance area led to such increased velocities in the runner that severe bubble entrainment was created (Sirrell and Campbell, 1994). This is one of the reasons why the elongated basin/sprue (Figure 12.2(e)) is so bad.

This effect is illustrated in Figure 12.11. In Figure 12.11(a), the negatively tapered sprue and (b) zero taper the velocity at the sprue exit is merely that from the fall of metal. The rate of arrival is of course controlled by the area of the sprue top, and the falling metal does not touch the walls of either sprue. For the correctly sized sprue (c), the velocity and rate of delivery are substantially unchanged, although it will be noticed that the whole of the length of the sprue is now contacting and controlling the stream, to the benefit of the melt quality. Those sprues with too much taper (d,e) continue to deliver metal at nearly the same rate (in kgs^{-1} for instance), but at much higher speed (in ms^{-1} for instance) in proportion to the reduction in area of the exit. Far from acting as an effective restraint, the narrow sprue exit merely increases problems.

These effects were studied pouring liquid aluminium in a sand mould but viewing the fall of metal with real-time X-ray radiography (Sirrell et al., 1995). The aim was to optimise the taper, measuring the time for the sprue to back-fill and measuring the speed of the exiting melt (Figure 12.12). This work confirmed that the long-used 20% increase of the area of the sprue entrance was a valuable correction for a straight tapered sprue. The consequential 20% increase in

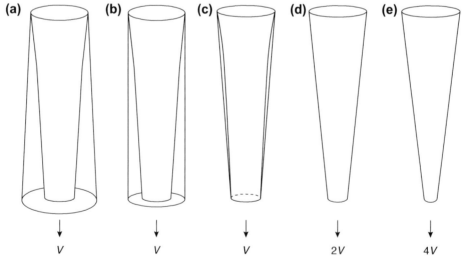

(a)   (b)   (c)   (d)   (e)

V V V $2V$ $4V$

FIGURE 12.11

A variety of straight tapered sprues. (a) negatively tapered; (b) zero taper; (c) correct taper giving optimum filling and minimal exit velocity, but improved by adding 20% entrance area or, better still, a true hyperbolic taper; (d) showing an effective choke of half exit area, approximately doubling exit speed; (e) quartered exit giving approximately four times exit speed.

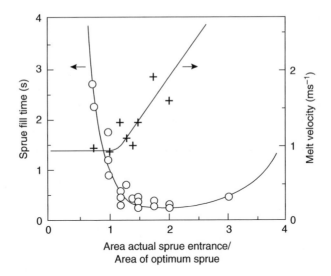

FIGURE 12.12

Experimental data from video radiographic observations of sprue filling time and velocity of discharge. A taper of 20% greater than the optimum taper (corresponding to 1.2 on the horizontal axis) is shown to be close to an optimum choice (Sirrell, 1995).

velocity into the runner was an acceptable penalty to ensure that the sprue primed faster and more completely despite its straight-wall approximate shape.

Thus to summarise the effect of sprue taper; the taper has to be correct (within the 20% outlined previously). Too little or too much taper both lead to damage of the melt.

12.2.1 MULTIPLE SPRUES

In magnesium alloy casting, the widespread use of a parallel pair of rectangular slots to act as the sprue seems to be due to the desire for the reduction in vortex formation (especially, as we have noted, if poor designs of pouring basin are employed). Swift et al. (1949) have used three parallel slots in their studies of the gating of aluminium alloys. However, the really useful benefit of a slot shape is probably associated with the reduction in stream velocity by the effect of friction on the increased surface area. The slots can be tapered to tailor their shapes to that of the falling stream. Such sprues would probably benefit the wider casting industry. A study to confirm this extent of this expected benefit would be valuable.

A really important benefit from using a single-slot sprue appears to have been widely overlooked. This is the accuracy with which it can be attached to a slot runner of the same area and aspect ratio to give an excellent geometry of the sprue–runner junction. This benefit is described in detail in a section later concerning the design problems of this junction. It seems that we should perhaps be making much more regular use of slot sprues.

When pouring a large casting whose volume is greater than can be provided from the ladle, it is common to use more than one ladle. The sequential pouring of one ladle after the other into a single basin has to be carried out smoothly because any loss of metal level to below the design minimum level is almost certain to create defects in the casting. Simultaneous pouring is often carried out. Occasionally, this can be accomplished with a single sprue, but using an enlarged pouring basin, often with a double end, either side of the sprue, allowing the ladles access from either side (Figure 12.13(b)). Single pouring can still be conveniently used with a double sprue (Figure 12.13(a)).

(a)   **(b)**

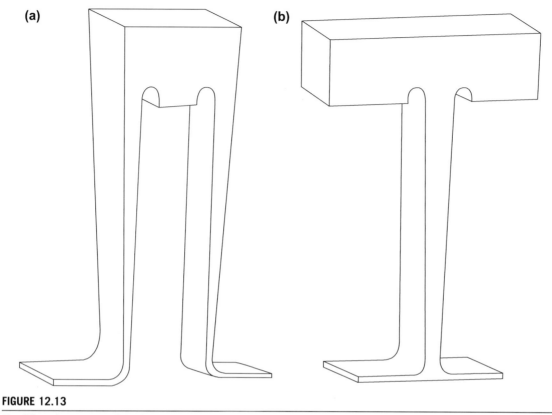

FIGURE 12.13

The filling of a runner splitting into two directions by (a) two sprues or (b) a single sprue with a double-ended basin to allow two pourers either together or pouring in series without a break.

Often, however, two or more sprues are used, sited at opposite ends of the mould, to give plenty of accessibility for ladles and cranes and reduce the travel distance for the melt in the filling system. When using more than one sprue, the correspondingly smaller area of each channel is an advantage because they fill more easily and quickly, excluding their air more rapidly. Multiple sprues for larger castings are recommended and should be considered more often.

For very large castings, an interesting technique can be adopted. Several sprues can connect to runners that are arranged around the mould cavity at different heights. For instance in the pour of a 3 m high iron casting weighing 37,000 kg described by Bromfield (1991), four sprues were arranged to exit from two pouring basins. The sprues were initially closed with graphite stoppers. The trough was filled first. The stoppers to the lowest level runner were then opened. The progress of the filling was signalled by the making of an electrical contact at a critical height of metal in the mould. In other instances witnessed by the author, the progress of filling could be observed by looking down risers or sighting holes placed on the runners. When the next level of runner was reached, announced by the bright glow of metal at the base of the sighting hole, the next level of sprues was brought into action to deliver the metal to this level of runner. In the case of the casting that was witnessed, three levels of runners were provisioned by six sprues. The technique had the great advantage that the rate of pouring did not start too fast, and then slowly decrease to zero during the course of the very long duration of the pour. The rate could be maintained at a more consistent level by the action of bringing in additional sprues as required. In addition, the temperature of the advancing front of the melt could also be maintained by the fresh supplies of hot metal arriving at the different levels, thus reducing the need for excessive casting temperatures to avoid mis-runs. Again, the significant advantages of multiple sprues are clear.

12.2.2 DIVISION OF SPRUES

The attempt to provide gating or feeding off various parts of the sprue at various heights is almost always a mistake and is to be avoided. Examples of poor junctions affecting the sprue are shown in Figure 12.14. Overflow from such channels can introduce metal into parts of the mould prematurely, where it can fall, splashing and damaging the casting and mould before the general arrival of the melt via the intended bottom gate. Small puddles and droplets often solidify and oxidise and thus create lap-type defects in the casting because they cannot easily be re-melted and assimilated. Even if the channels are carefully angled backwards to avoid premature filling, they then act to aspirate air into the metal stream. Thus divided sprues usually either act to let metal out or let air in.

However, of course, sprues can be divided if this is carried out with extreme care. The extreme care is needed because of the extreme velocity in the sprue. Any slight error can create disproportionate amounts of entrainment damage. Division of sprues into separate sources of metal is required in vertically parted moulds, especially those requiring to fill all cavities at the same time to reduce the total filling time of the mould to a minimum figure corresponding to the filling time of a single cavity.

Figure 12.15(a) shows a technique for dividing up the flow of a sprue so that equal quantities reach all the branches. The figure shows a sprue design for six cavities. The area at A_0 carries the total flow of course and is notionally divided into six equal areas. The areas A_1, A_2 and A_3 are calculated assuming the free fall of the melt, taking off the appropriate fractions of the area of the flow at each level.

The question nearly always arises, 'What about the sharp corners that are revealed after the taking off of each of the runners at the various levels?' Fortunately, these seem to survive well because the whole sprue volume fills quickly, after which all portions of the sprue wall are pressurised and therefore protected by the melt. Any attempt to round the corners to protect (beyond the millimetre or so radius to permit moulding and stripping) is counter-productive, because such a region cannot be properly filled and pressurised, with the result that the mould is destroyed at this location, and sand inclusions will appear in the casting.

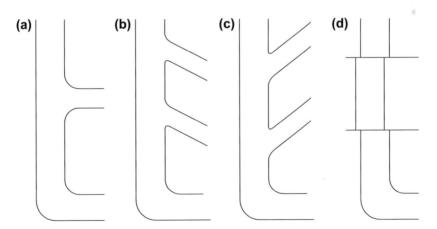

FIGURE 12.14

A variety of sprue junctions; non are recommended: (a) a horizontal connection allowing some metal spillage and air entrainment; (b) downward angled connections giving uncontrolled division of the flow; (c) upward angled junctions avoiding metal spillage but encouraging air entrainment; (d) mis-match leading to gas entrainment.

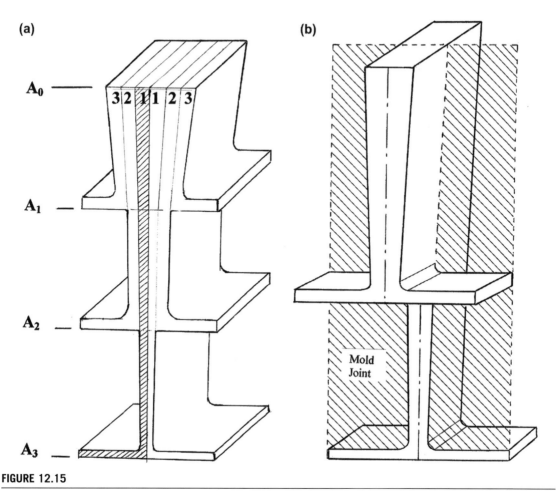

FIGURE 12.15

(a) A technique for splitting flow off from a sprue to fill multiple cavities in a vertically split mould at the same time. (The sharply moulded corners survive well in the absence of entrained air.) (b) A single sprue split to fill cavities at the same time can be conveniently made across the mould joint.

If only one division is required, for instance for four cavities, the sprue can be conveniently divided across the mould joint (Figure 12.15(b)). This technique works well, once again despite the sharp edges of sand which appear so vulnerable.

12.2.3 SPRUE BASE

The point at which the falling liquid emerges from the exit of the sprue and executes a right-angle turn along the runner requires special attention. The design of this part of the liquid metal plumbing system has received much attention by researchers over the years, but with mixed results that the reader should note with caution.

Figure 12.16(a) illustrates a common round sprue to flat runner junction. This well known and apparently innocuous system gives unbelievably bad results as shown in the simulations of Gebelin and colleagues (2005, 2006) and confirmed by video radiographic studies of the filling of castings. The ricochet behaviour of the melt as it zigzags along the runner

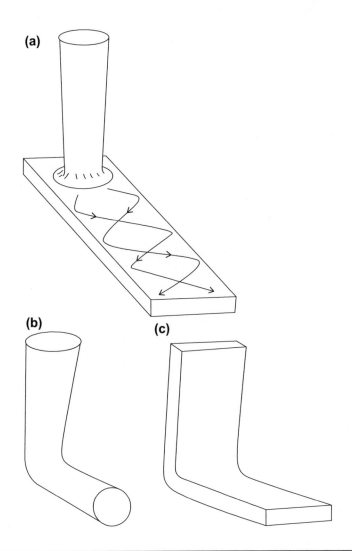

FIGURE 12.16

(a) There is no solution to the problem of joining a round sprue to a rectangular runner; the flow ricochets from wall to wall, never properly filling the runner. Simple sprue–runner junctions at (b) and (c) can work perfectly.

ensures that the runner never properly fills and the melt is subject to severe oxidation. Despite much research, there seems no way to optimise this particular type of junction. Clearly, it needs to be avoided.

The simple geometries shown in Figure 12.16(b) and (c) do not require a computer simulation to convince a reader that they have the potential deliver metal essentially without turbulence. These simple designs are recommended.

12.2.4 THE WELL

One of the widely used designs for a sprue base is a *well*. This is shown in Figure 12.17(a). Its general size and shape has received enormous research effort, striving to provide optimum efficiency in the reduction of air entrainment in the

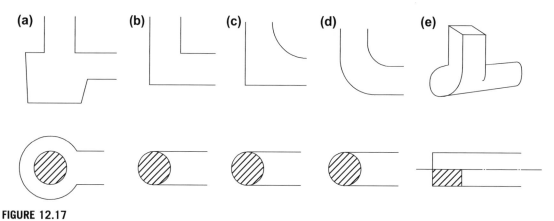

FIGURE 12.17

A variety of sprue–runner junctions in side and plan views from poorest (a) to best (d). The offset junction at (e) forms a vortex flow along the cylindrical runner. Early results suggest it performs well, but more research is probably needed (Section 12.6.3).

runner. The final optimisation from early researches was a well of double the diameter of the sprue exit and double the depth of the runner. This optimisation was confirmed in an elegant study by Isawa (1994), who found that the elimination of the hundreds and thousands of bubbles that were generated initially reduced exponentially with time. The exponential relationship gave a problem to define a finite time for the elimination of bubbles because the data could not be extrapolated to zero bubbles; clearly, the extrapolation predicted an infinite time! He therefore cleverly extrapolated back to the time required to arrive at the last bubble and used 'the time to the last bubble' to compare different well designs.

However, it should be noticed that both this and all the research into wells had been carried out on water models, and all had used runners of large cross-section that were not easy to fill. The result was a well design that, at best, cleared the liquid of bubbles after about 2 s.

For small castings that fill in only a few seconds, we have to conclude that such well designs are counter-productive. In these cases, it is clear that much of the filling time will be taken up conveying highly damaged metal into the mould cavity. Thus the comforting and widely held image of the well as being a 'cushion' to soften the fall of the melt is seen to be an illusion. In reality, the well was an opportunity for the melt to churn, entraining quantities of oxide and bubble defects.

Clearly, an abrupt right angle (Figure 12.17(b)) was not a satisfactory solution because it could be certain that plenty of bubbles would be entrained from the downstream portion of the vena contracta (Figure 12.18). This region of flow, at the trailing extremity of the air and vapour pocket, is particularly unstable and sheds bubbles into the flowing stream.

The systematic X-ray radiographic studies started in 1992 have been revealing. They have shown that in a sufficiently narrow filling channel with a good radius at the sprue–runner junction, the high surface tension of the liquid metal assists to retain the integrity of a compact liquid front, constraining the melt. These investigative studies on dramatically narrow channels in real moulds with real metals quickly confirmed that the sprue–runner junction was best designed as a simple turn (Figure 12.17(c) and (d)), provided that the channels were of minimum area.

The minimum area condition is important. For instance, if the inside radius seen in Figure 12.17(c) is approximately the same as the thickness of the sprue exit, the junction has a modest volume that is somewhere close to optimum. If the inside radius becomes much greater than twice this value, the additional volume starts to introduce entrainment behaviour. Furthermore, the liquid in the corner of the junction revolves like a ball bearing, reducing the friction experienced by the melt undertaking the turn. Thus the velocity of the melt out of the junction is closely equal to the entering velocity; very little energy is lost.

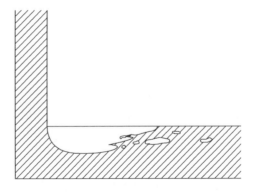

FIGURE 12.18

The vena contracta problem at a right angle with inadequate radius. The downstream part of this decohered region of flow is unstable, shedding bubbles into the flow.

From this point of view the junction (Figure 12.17(d)) is a great improvement because the redundant volume in the sharp angle of the bend is eliminated, reducing entrainment problems. In addition, the outside wall of the bend, being stationary and microscopically rather rough, introduces friction into the turn, so that the exiting melt is somewhat reduced in speed. Reductions of 20% or so in velocity have been recorded. Any such speed reductions, even though small, are welcome. Thus it is useful, if possible, to also radius the outside of the bend to maintain a parallel channel.

In summary, despite what was recommended by the author in *Castings* 1991, more recent research confirms that wells are no longer recommended; even the best designs generate bubbles. The simple turn, maintaining a uniform cross-section area as closely as possible, generates no bubbles.

12.2.5 THE RADIAL CHOKE

This is a clever 'quick fix' for the evils of a reverse tapered sprue suggested by Groteke (see Sikovski and Groteke, 2005) in which a small cylinder can be planted on the drag pattern at the base of the sprue. It forms a narrow annulus through which the melt is constrained to flow in a radial direction (Figure 12.19). No other changes to the pattern are usually made. The corrective action to the sprue, in getting it to fill in about half the time, is immediate and valuable. (As an interesting aside, the provision of a reduced volume at the base of the sprue is diametrically opposed to the concept of providing an enlarged volume in the form of a well: one works, the other is a disaster.)

Computer simulations of the detailed filling of the radial choke shows it to be somewhat messy in practice, indicating that improvements to the concept can surely be made and would be desirable. An optimisation study would be welcome. Even so, even in its current relatively undeveloped form, it is quick and easy to fit and acts as an immediate and significant improvement over the unaided reverse tapered sprue. It can help a lot until an overall better system can be put in place.

12.2.6 THE RADIUS OF THE SPRUE–RUNNER JUNCTION

It has been shown that for small castings, generally up to a few kilograms in weight, the melt can be turned through the right angle at the base of the sprue simply by putting a right angle bend into the channel. However, if no radius is provided, the melt cannot follow the bend, so that a *vena contracta* is created (Figure 12.18). The trailing edge of this cavitated region is unstable, so that its fluttering and flapping action sheds bubbles into the stream.

The vena contracta is a widely observed phenomenon in flowing liquids. It occurs wherever a rapid flow is caused to turn through any sharp change of direction. An important example has already been met in the offset pouring basin if

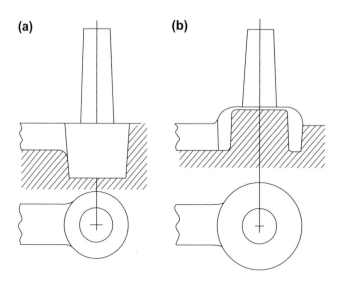

FIGURE 12.19

(a) Traditional well, nearly as deleterious as a conical basin; (b) radial choke, a substantial benefit compared with a 'well' but may benefit further from additional optimisation.

no step is provided (Figure 12.4(a)). This creates a vena contracta that showers bubbles down the sprue. However, the base of the sprue is probably an even more vulnerable location because the speeds are much higher here. The loss of contact of the stream from the top of the runner immediately after the turn has been shown to be the source of much air in the metal. Experiments with water have modelled the low-pressure effect here, demonstrating the sucking of copious volumes of air into the liquid as streams and clouds of bubbles (Webster, 1967). This is expected to be particularly severe for sand moulds, where the permeability will allow a good supply of air to the region of reduced pressure.

In fact, when pouring castings late at night, when the foundry is quiet, the sucking of air into the liquid metal can be clearly heard, a haunting sound in the darkness like bathwater down the plug hole! Such castings always reveal oxides, sand inclusions and porosity above the gates which are the tell-tale signs of air bubbles aspirated into the running system.

In contrast, provided that the internal corner of the bend is given a sufficiently large radius, the melt will turn the corner without cavitation or turbulence (Figure 12.17(c)). In fact, the action of the advancing metal is like a piston in a cylinder: the air is simply pushed ahead of the advancing front, never becoming mixed. To be effective, the internal radius of the bend needs to be at least equal to the diameter of the sprue exit, and possibly twice this amount, but probably not more. Too large a radius introduces too much redundant volume into the filling channel, allowing the development of surface turbulence and entrainment of air and oxides. The precise radius might benefit from further research. The action of the internal radius is significantly improved if the outside of the bend is also provided with a radius to maintain the parallel form of the channel (Figure 12.17(d)).

For larger casting where surface tension becomes progressively less important, the channels are filled only by the available volume of flow. Initially, during the first critical period as the filling system is priming, there is considerable danger of damage to the metal.

To limit such damage, it is helpful to take all steps to prime the front end of the filling system quickly. This is assisted by the use of a stopper in the pouring basin. However, in *Castings* 1991, the author has considered the use of various kinds of chokes at the entrance to the runner as a possible solution to these problems. Again, recent research has not upheld these recommendations. It seems that any such constriction merely results in the jetting of the flow into the more distant expanded part of the runner.

This finding emphasises the value of the concept of the *naturally pressurised* system. It is clearly of no use to expand the running system to fulfil some arbitrary formula of ratios, in the hope that the additional area will persuade the flow velocity to reduce. The flow tends to maintain its speed and direction (steadfastly ignoring any enlargement of area we may provide in the hope that the melt might fill this additional area and slow down).

The use of a *vortex sprue*, or even simply a *vortex base* or *vortex runner* (Figure 12.17(e)), in conjunction with a conventional sprue represent exciting and potentially important new developments in running system design. These concepts are described more fully in Section 12.6.

12.3 RUNNER

The runner is that part of the filling system that acts to distribute the melt horizontally either around or under the mould, reaching distant parts of the mould cavity quickly to reduce heat loss problems.

The runner is usually necessarily horizontal because it simply follows the normal mould joint in conventional horizontally parted moulds. In other types of moulds, particularly vertically jointed moulds, or investment moulds where there is little geometrical constraint, the runner would often benefit from being inclined uphill.

It is especially useful if the runner can be arranged under the casting, so that the runner is connected to the mould cavity by vertical gates. All the lowest parts of the mould cavity can then be reached easily this way. The technique is normally achieved only in a three-part mould in which the joint between the cope and the drag contains the mould cavity, and the joint between the lower mould parts (the drag and the base) contains the running channels (Figure 12.20(a)). The three-part mould is often seen as a costly option. Sometimes the three-level requirement can be achieved by use of a large core (Figure 12.20(b)). The distribution system can be assembled from refractory or sand core sections and built into the mould as the moulding box is filled with sand (Figure 12.20(c)), but the author has never achieved good results with this particular technique. Designing all the various parts to be moulded in sand, and have exactly the correct size and area can give excellent results. Similarly, of course, these benefits translate into other moulding systems such as investment castings and permanent mould castings.

More usually, however, a two-part mould requires both casting and running system to be moulded in the same joint between cope and drag. To avoid any falls in the filling system, the runner has to be moulded in the drag, and the casting and its gates in the cope (Figure 12.21).

The usual practice, especially in iron and steel foundries, of moulding the casting in the drag (Figure 11.2(a)) is understandable from the point of view of minimising the danger of run-outs. A leak at the joint, or a burst mould is a possible danger and a definite economic loss. This was an important consideration for hand-moulded greensand, where the moulds were rather weak (and was of course part of the reason for the use of the steel moulding box or flask). However, the placing of the mould cavity below the runner causes an uncontrolled fall into the mould cavity, creating the risk of imperfect castings. There is no longer such a danger for the dense, strong greensand moulds produced from modern automatic moulding machines, nor for the extremely rigid and strong moulds created in chemically bonded sands. For products whose reliability needs to be guaranteed, the arrangement of the runner at the lowest level of the mould cavity, causing the metal to spread through the running system and the mould cavity only in an uphill direction is a challenge that needs to be met (Figure 12.21). Techniques to achieve this include the clever use of a core (Figure 12.20(b)) or for some open-centred castings the use of central gating (Figures 12.22(b) and 12.23).

Webster (1964) carried out some early exploratory experiments to determine optimum runner sizes. We can summarise his results in terms of the comparative areas of the runner/sprue exit. He has found that a runner that has only the same area as the sprue exit (ratio one of 1) will have a metal velocity that is high. A ratio of 2, he claims, is close to optimum because the runner fills rapidly and excludes air bubbles reasonably efficiently. A ratio of 3 starts to be difficult to fill, and a ratio of 4 is usually simply wasteful for most castings. Webster's work was a prophesy, foretelling the dangers of large runners that foundries have, despite all this good advice, continued to use.

For the best results, however, recent careful studies have made clear that even the expansion of the area of flow by a factor of 2 is not achieved without a serious amount of surface turbulence. This is now known from actual observations of metal flow by video X-ray radiographic studies, and from detailed examination of the scatter of mechanical properties of castings using highly sensitive and quantifying technique, Weibull analysis.

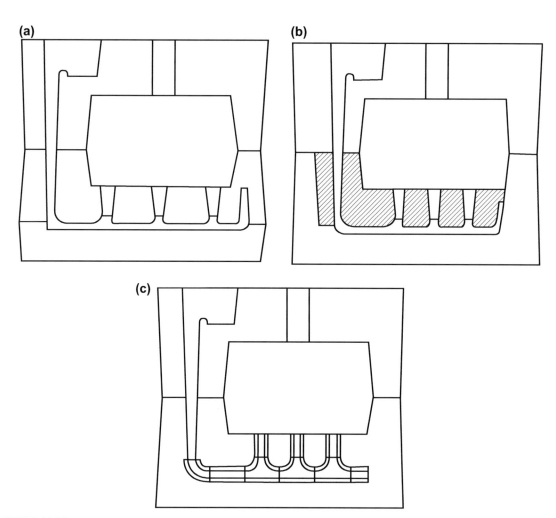

FIGURE 12.20

Bottom-gated systems achieved by (a) a three-part mould with accurately moulded running system; (b) making use of a core and (c) a two part mould with pre-formed refractory channel sections moulded into drag.

The best that can easily be achieved without damage is merely the reduction of about 20% in velocity by the friction of the sprue/runner bend, necessitating a 20% increase in area of the runner as has been discussed previously. Any greater expansion of the runner will cause the runner to be incompletely filled, and so permit conditions for damage. Even this is probably not to be recommended on the grounds that the transition to the increased area around the bend may require a precision difficult to meet. A computer model could in principle provide digital information for a carefully created accurate pattern. However, there is no simple rule of thumb to guide the patternmaker to avoid positioning the expansion too early or too late, both of which would lead to turbulent flow.

Greater speed reductions, and thus greater opportunities for expansion of the runner occur if the number of right angle bends is increased, because the factor of 0.8 reduction in speed is cumulative from one bend to the next. After three such bends, the speed is reduced by half ($0.8 \times 0.8 \times 0.8 = 0.5$). Right-angle bends were anathema in filling system designs when large cross-sections were the norm. However, with very narrow systems, there is less room for surface

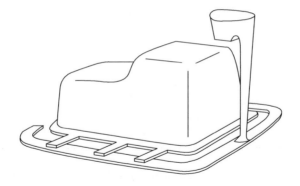

FIGURE 12.21

An external running system arranged around an automotive sump (oil pan).

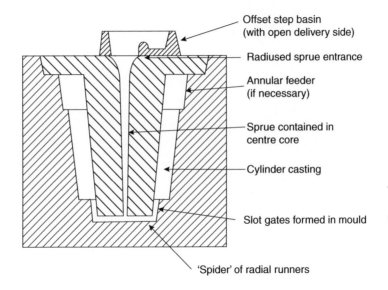

FIGURE 12.22

Cross-section of an internal running system for the casting of a cylinder.

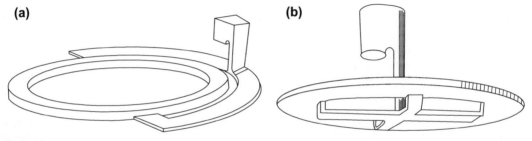

FIGURE 12.23

Ring casting produced using (a) an external and (b) an internal filling system.

turbulence. Even so, great care has to be taken. For instance, video X-ray studies have confirmed that the bends operate best if their internal and external radii provide a parallel channel. The lack of an external radius can cause a reflected wave in larger channels.

One of the most effective devices to reduce the speed of flow in the runner is the use of a filter. The close spacing of the walls of its capillaries ensures a high degree of viscous drag. Flow rate can often be reduced by a factor of 4 or 5. This is a really valuable feature, and actually explains nearly all of the beneficial action of the filter (i.e. when using reasonable quality metal in a well-designed filling system the filter does very little filtering. It is really important action in improving the quality of castings is its reduction of velocity.) The use of filters is considered later (Section 12.8).

There has, over the years, been a considerable interest in the concept of the separation of second phases in the runner. Jeancolas et al. (1969) carried out experiments on ferrous metals to show that at Reynold's numbers below the range 7000–12,000, suspended particles of alumina could be deposited in the runner but at values in excess of 15,000, they could not precipitate. Although these findings underline the importance of working with the minimum flow velocities wherever possible, it is quickly shown that for a steel casting of height 1 m, giving a velocity of flow of 4.5 ms^{-1}, for $\eta = 5.5 \times 10^3$ Nsm^{-2}, and for a runner of 80 mm square, Re is higher than 100,000. Thus it seems that conditions for the deposition of solid materials such as sand and refractory particles in runners will not be met. In general therefore, it seems misguided to attempt to design filling systems that might be thought to encourage flotation of inclusions.

Even so, a worker in a foundry casting iron may protest that slag accumulates on the tops of runners, where it is much to be preferred than in the casting. He will point out that separation in this case happens because of (1) the great difference in density between the slag and the metal; (2) the large size of the slag droplets; and (3) his large section runner that promotes slower flow. Although there may be a modicum of truth in all of this, it is certain that in most cases, most of the slag does not arrive with the metal because this is often via bottom-poured launder, but is generated in situ in the running system, mainly because the filling system channels are too large, and the provision of special slag traps above the runner merely provide opportunity for the high speed metal to generate slag locally. In my experience, most slag in running systems and in castings is generated by turbulence in the running system.

At this point, the reader might, among the confusing detail, be starting to appreciate the basic scenario and some of the major targets of the filling design:

1. The metal arrives in some chaos at the bottom of the sprue.
2. After this initial trauma, it may be gathered together once again by the integrating action of a feature such as a filter or radial choke to provide some delay and back-pressure;
3. Rises steadily against gravity, filling section after section of the running system;
4. Finally arrives in the mould in fair condition, not having suffered too much damage, and at a speed below the critical velocity.

It should be noted that such a logical system and its consequential orderly fill is not to be taken for granted. For instance, it is usually a problem if the runner is moulded in the cope. This is mainly because the gates, which are in either the drag or the cope, will inevitably start to fill and allow metal into the mould cavity before the runner is full, as is clear from Figure 11.2(a). The traditional running of cast iron in this way fails to achieve its potential in its intended separation of metal and slag. This is because the first metal and its load of slag (mainly generated in the sprue and early part of the runner) enters the gates immediately, before the filling of the runner, and thus before the chance that the slag can be trapped against the upper surface of the runner. In short, the runner in the cope results in the violation of the fundamental "no fall" criterion. The runner in the cope is not recommended for any type of casting – not even grey iron! Despite all, the reader needs to be aware that regardless of all these good reasons and good intentions, it is sometimes highly inconvenient or sometimes impossible, to site the runner in the drag.

The complexity of behaviour of some filling system designs is illustrated by a runner in a gravity die, positioned in the cope, which acted to reduce the bubble damage in the casting (Figure 12.24). This result, apparently in complete contradiction to all the behaviour warned previously, arose because of the exceptionally tall aspect ratio of the runner, which was shaped like a vertical slot. This shape retained bubbles high above the exits to the gates moulded below. In fact, it seems that the reduction in bubbles into the casting by placing the gates low in this way only really resulted because of the extremely poor front end of the filling system. The filling system was a bubble-manufacturing design, so that almost any remedy had a chance to produce a better result. However, there is a real benefit to be noted (running

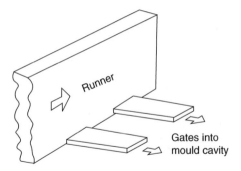

FIGURE 12.24

Tall slot runner with bottom gates.

systems are perversely complicated) because the gates would prime slowly as a head of metal in the runner was built up, thus avoiding any early jetting through into the mould cavity. This is a benefit not to be underestimated, and highlights the problem of generalising for complex geometries of castings and their filling systems that can sometimes contain not just liquid metal but sometimes emulsions of slag and/or air.

In gravity die (permanent mould) castings, the placing of the runner in the cope, and taking off gates on the die joint (Figure 12.2(a)), is especially bad. This is because the impermeable nature of the die prevents the escape of air and mould gases from the top of the runner. Thus, the runner never properly fills. The entrapped gas floating on the surface of the metal will occasionally dislodge, as waves race backwards and forwards along the runner and as the gasses heat up and expand. Large bubbles will therefore continue to migrate through the gates from time to time throughout the pour. Some bubbles might dislodge from the runner and into the casting at a late stage, causing not only bubble trails and splash problems, but also the advancing solidification front may trap whole bubbles.

This scenario is tempered if a die joint is provided along the top of the runner to allow the escape of air. Alternatively, a sand core sited above the runner can help to allow bubbles to diffuse away.

12.3.1 THE TAPERED RUNNER

It is salutary to consider the case where the runner has two or more gates and where the stepping or tapering of the runner has been unfortunately overlooked. The situation is shown for three gates in Figure 12.25(a). Clearly, the momentum of the flowing liquid causes the furthest gate, number 3, to be favoured. The rapid flow past the opening of gate 1 will create a reduced-pressure region in the adjacent gate at this point, drawing liquid out of the casting! The flow may be either in or out of gate 2, but at such a reduced amount as to probably be negligible. In the case of a non-tapered runner, it would have been best to have only gate 3.

Where more than one gate is attached to the runner, the runner has been traditionally reduced in cross-section as each gate is passed. In the past, such reductions have usually been carried out as a series of steps, producing the well-known stepped runner designs. For three ingates, the runner would be reduced in section area by a step of one third the height of the runner as each gate was passed. However, real-time X-ray studies have noted that during the priming of such systems, the high velocity of the stream causes the flow to be deflected, leaping into the air when impacting the step, and ricocheting off the roof of the runner. Needless to say, the resulting flow is highly aerated and does not achieve its intended even distribution. It has been found that simply reducing the cross-section of the runner gradually, usually linearly, giving a smooth, straight taper geometry does a reasonable job of distributing the flow evenly (Figure 12.25(b)).

Kotschi and Kleist (1979) allow a reduction in the runner area of just 10% more than the area of the gate to give a slight pressurisation bias to help to balance the filling of the gates. However, they used a highly turbulent non-pressurised system that will not have encouraged results of general applicability. In contrast, computer simulation of the narrow runners recommended in this work has shown that the last gate suffers some starvation as a result of the accumulation of friction along the length of the runner. Thus, for slim systems, the final gates require some additional area, not less. Thus, instead of

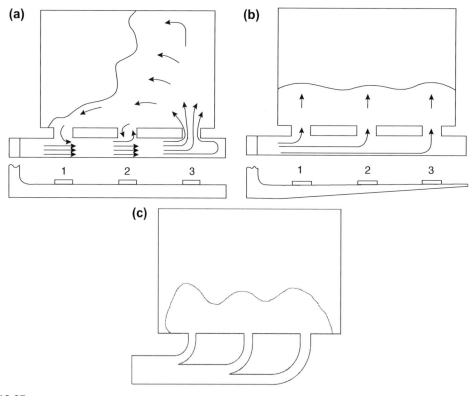

FIGURE 12.25

(a) An unbalanced delivery of melt into the mould as a result of an incorrect runner design; (b) a tolerably balanced system. (c) Cuspoid gates giving excellent distribution for low entry speeds, but jetting from the gates will limit their use for high speeds on large castings unless some prior speed reduction is achieved.

tapering the runner to zero thickness, which is an overcorrection favouring most of the filling via gate 1, some compromise, such as tapering to 50% of the thickness will usually be found to give a more even favouring of gates (Figure 12.25(b)). The taper can, of course, be provided horizontally or vertically (an important freedom of choice often forgotten).

In the interests of health and safety, at all times avoid tapering the runner to zero. The thinning section adds no advantage but to provide a pointed tip on which people in the foundry keep stabbing themselves. Tapering to approximately 50% of thickness usually gives a satisfactory distribution and sufficiently blunt tip, preferably more than a reasonably safe 5 mm thick.

A technique for dividing flow from a runner almost perfectly is recommended; the cusp-shaped junctions in Figure 12.25(c), illustrating the runner being divided equally into three gates. The only provision is that the filling system requires to be designed to exclude air (air entrained in the flowing metal will destroy cusps moulded in sand, but cusps will survive perfectly satisfactorily if only subjected to liquid metal.)

12.3.2 THE EXPANDING RUNNER

In an effort to slow the metal in its early progress in the runner, several methods of expanding the area of the runner have been tried. Such expanding sections of runners have often been graced by the name of 'diffuser'. It has to be admitted that none of these designs is successful. However, it is worth describing these attempts for our education.

The simple expansion of the runner at an arbitrary location along the runner is of no use at all (Figure 12.26(a)). The melt progresses without noticing the expansion, progressing as a central jet, and finally hitting the far end of the runner

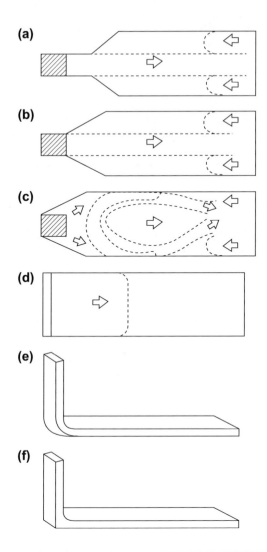

FIGURE 12.26

Plan views of a square section sprue connected to a shallow rectangular runner showing attempts to expand the runner (a,b) that fail completely. Attempt (c) is better, but flow ricochets off the walls generates a central starved, low-pressure region; (d) a slot sprue and slot runner produce a uniform flow distribution in the runner shown in (e) (recommended) and (f) (usually acceptable).

and creating back waves, which in their reverse flow entrain bubbles and oxides where the flows shear side by side. Even expanding the runner directly from the near side of the sprue (shown as having a square section for clarity) is not helpful (Figure 12.26(b)). However, expanding the runner from the far side of the sprue (Figure 12.26(c)) does seem to work considerably better. But even here the front tends to progress in two main streams on either side of the central axis of the runner, leaving the centre empty, or relatively empty, forming a low-pressure region some distance downstream in the runner, similar to the pattern seen previously for a round sprue on a rectangular runner (Figure 12.16). This development of this double jet flow seems to be the result of the attempted radial expansion of the flow as it impacts on the runner, but finds itself constrained by and reflected from the walls of the runner. This situation for high temperature liquids such as irons and steels leads to the downward collapse of the centre of the runner in sand moulds because this becomes heated by radiation, and so expands (especially if mould consists of silica sand and its temperature is raised above its alpha/beta transition), but is unsupported by the pressure of metal. The closing down of the runner in this way can be avoided by a central moulded support, effectively separating the runner into two separate, parallel runners. In practice, I have found that a slot runner about 100 mm wide for irons and steels is close to the maximum that can resist collapse. It may be that larger dimensions would be safe if the flow were parallel (Figure 12.26(d), (e) and (f)). Also, of course, a cylindrical-shaped channel would be expected to avoid collapse more reliably.

Despite this possible benefit of the cylindrical form, the use of slot sprues linked to slot runners promises to be a complete solution to the problem of the sprue–runner junction and deserves wider exploitation.

What is certain that most normal efforts to expand the runner to reduce speed not only fail, but as a result of the melt becoming depressurised and entraining gas the failure is a disaster for the casting quality. Much more research is needed to test what can be achieved. Successful and reliable speed reduction by expanding runners may prove impossible. To be safe in the meantime, and perhaps for ever, diffusers should probably be laid to rest.

12.4 GATES

The gates, or 'ingates', are those last features of the filling system that introduce the metal into the mould cavity. They can have a profound effect on the melt flow into the cavity, for good or evil. Every part of the filling system has its own special purpose, and is critical to success. Every part has to be correct, including the gates.

12.4.1 SITING

When setting out the requirements for the site of a good gate, it is usual to start with the questions:

"Where can we get it on?"
and   "Where can we get it off?"
Other practical considerations include:
"Gate on a straight side if possible"
and   "Locate at the shortest flow distance to the key parts of the casting".
This is a good start, but, of course, just the start. There are many other aspects to the design of a good gate.

12.4.2 DIRECT AND INDIRECT

In general, it is important that the liquid metal flows through the gates at a speed lower than the critical velocity so as to enter the mould cavity smoothly. If the rate of entry is too high, causing the metal to fountain or splash inside the mould cavity, then the battle for quality is probably lost. Turbulence inside the mould cavity is the most serious turbulence of all. Turbulence occurring early in the running system may or may not produce defects that find their way into the casting because many bifilms remain attached to the walls of the runners and many bubbles escape. However, any creation of defects in the mould cavity ensures unavoidable damage to the casting.

One important rule therefore follows very simply:

Do not place the gate at the base of the down-runner so that the high velocity of the falling stream is redirected straight into the mould, as shown in Figure 11.4. In effect, this *direct* gating is *too direct*. An improved, somewhat *indirect* system is shown in Figure 11.2(b) illustrating the provision of a separate runner and gate, and thus incorporating several right-angle changes of direction of the stream before it enters the mould. These provisions are all used to good effect in reorganising the metal from a chaotic mix of liquid and gases into a coherent moving mass of liquid. Thus although we may not be reducing the entrainment of bifilms, we may at least be preventing bubble damage in the mould cavity.

As we have mentioned previously, all of the oxides created in the early turbulence of the priming of the running system do not necessarily find their way into the mould cavity. Many appear to 'hang up' in the running system itself. This seems especially true when the oxide is strong as is known to be the case for Al alloys containing beryllium. In this case, the film attached to the wall of the running system resists being torn away, so that such castings enjoy greater freedom from filling defects. The wisdom of lengthening the running system, increasing friction, especially by the use of right angle bends, adds back-pressure for improved back-filling and reduces velocity. It also provides more surface to contain and hold the oxides generated during priming.

12.4.3 TOTAL AREA OF GATE(S)

A second important rule concerns the sizing of the gates. They should be provided with sufficient area to reduce the velocity of the melt into the range 0.5–1.0 ms^{-1}. The concept is illustrated in Figure 2.18. In the author's experience, velocities above 1.2 ms^{-1} for Al alloys always seem to give problems. Velocities of 2 ms^{-1} in film-forming alloys, unless onto a core as explained later, would be expected to have consequences sufficiently serious that they could not be overlooked. With even higher velocities the problems simply increase.

Occasionally, there is a problem to obtain a sufficient size of gate to reduce the melt speed to safe levels before it enters the mould cavity. In such cases, it is valuable if the gate opens at right angles onto a thin (thickness a few millimetres) wall. This is because the melt is now forced to spread sideways from the gate and suffers no splashing problems because the local section thickness of the casting is too small. As it spreads away from the gate, it increases the area of the advancing front, thereby reducing its velocity. Thus by the time the melt arrives in a thicker section of the casting it is likely to be moving at a speed below critical. Effectively, the technique uses a thin wall section of the casting as an extension of the filling system.

This is a good reason for gating direct onto a core, but, once again, is contrary to conventional wisdom. In the past, gating onto a core was definitely bad. This was because of the amount of air entrained in the flow. The air assisted to hammer and oxidise the surface of the core and thus led to sand erosion and the local destruction of the core. With a good design of filling system, however, in which air is largely excluded, the action of the hot metal is safe. Little or no damage is done to the core despite the high velocity of the stream, because the melt merely heats the core whilst exerting a steady pressure that holds the core material in place. Thus with a good filling system design, gating directly onto a core is recommended.

Returning to the usual gating problem whereby the gate opens into a large section casting. If the area of the gate is too small, then the metal will be accelerated through, jetting into the cavity as though from a hosepipe. Figure 2.18 shows the effect. In many castings, the jet speed can be so high that the metal bounces around the mould cavity, scouring roof and walls of the cavity. Historically, many castings have been gated in this way, including most steels and grey cast iron castings. The approach has enjoyed mediocre success whilst greensand moulding has been employed, but it seems certain that better castings and lower scrap rates would have been achieved with less turbulent filling. In the case of cores and moulds made with resin binders that cause graphitic films on the liquid iron, the pressurised delivery system is usually unacceptable. The same conclusion is true for ductile irons in all types of moulds.

We may define some useful quick rules for determining the total gate area that is needed. For an Al alloy cast at 1 kgs^{-1} assuming a density of approximately 2500 kgm^{-3} and assuming that we wish the metal to enter the gate at its critical speed of approximately 0.5 ms^{-1} it means we need approximately 1000 mm^2 of gate area. The elegant way to describe this interesting ingate parameter is in the form of the units of area per mass per second; thus for instance "1000 mm^{2}kg^{-1}s".

Clearly we may pro-rata this figure in different ways. If we wished to fill the casting in half the time, but at the same low ingate velocity, we would require 2000 mm^2 and so on. It can also be seen that the area is quickly adjusted if it is decided that the metal can be allowed to enter at twice the speed, thus the 1 kgs^{-1} would require only 500 mm^2, or if directly onto a core in a thin section casting, perhaps twice the rate once again, giving only 250 mm^2.

Allowing for the fact that denser alloys such as irons, steels and copper-based alloys, have a density approximately three times that of aluminium, but the critical velocity is slightly smaller at 0.4 ms^{-1} the ingate parameter becomes, with sufficient precision, 500 mm^{2}kg^{-1}s.

The values of approximately 1000 mm^{2}kg^{-1}s for light alloys and 500 mm^{2}kg^{-1}s for dense alloys are useful parameters to commit to memory.

12.4.4 GATING RATIO

In its progress through the running system, the metal is at its highest velocity as it exits the sprue. If possible, we aim to reduce this in the runner, and further reduce as it is caused to expand once again into the gates. The aim is to reduce the velocity to below the entrainment threshold (the 0.4–0.5 ms^{-1} if possible, but up to 1.0 ms^{-1} if necessary) at the point of entry into the mould cavity.

It is worth spending some time describing the traditional ratio method of defining running systems which is widely used. It is to be noted that it is not recommended!

It has been common to describe running systems in terms of ratios based on the area of the exit of the sprue. For instance, a widely used area ratio of the sprue/runner/gates has been 1:2:4. Note that in this abbreviated notation, the ratios given for both the runner and the gates refer back to the sprue, so that for a sprue exit of 'A', the runner area is 2A and the total area of the gates is 4A. It is clear that such ratios cannot always be appropriate, and that the real parameter that requires control is the velocity of metal entering the mould. Thus on occasions this will result in ratios of 1:1:10 and other unexpected values. The design of running systems based on ratios is therefore a mistake.

Having said this, I do allow myself to use the ratio of the area of sprue exit to the (total) area of the gates. Thus if the sprue is 200 mm tall (measured of course from the top of the metal level in the pouring basin) the velocity at its base will be close to 2 ms^{-1}. Thus a gate of 4 times this area will be required to get to below 0.5 ms^{-1}. (Note therefore that the old 1:2:4 and 1:4:4 ratios can be seen therefore to be applicable only up to 200 mm sprue height. Beyond this sprue height the ratios are insufficient to reduce the speed below 0.5 ms^{-1}.) (The reader should also notice the damaging increase in area of the runner which is implied in these figures.)

I am often asked what about the problem that occurs when the mould cross-sectional area reduces abruptly at some higher level in the mould cavity. The rate of rise of the metal will also therefore be increased suddenly, perhaps becoming temporarily too fast, causing jetting or fountaining as the flow squeezes through the constriction. Fortunately, and perhaps surprisingly, this is extremely rare in casting geometries. In 40 years dealing with thousands of castings, I have difficulty to recall whether this has ever happened. The narrowest area is usually the gate, so the casting engineer can devote attention to ensuring that the critical velocity is not exceeded at this critical location, not forgetting the danger just inside the mould because of the transverse velocity – the sideways spreading flow (see later). If the velocities through the gate and across the floor of the mould cavity are satisfactory, it usually follows that the velocity will be satisfactory at all other levels in the casting.

Even in a rare situation where a narrowing of the mould is severe, it would still be surprising if the critical velocity were exceeded, because the velocity of filling is at its highest at the ingate, and usually decreases as the metal level rises, finally becoming zero when the net head is zero, as the metal reaches the top of the mould.

Once again, of course, counter-gravity filling wins outright. In principle, and usually with sufficient accuracy in practice, the velocities can be controlled at every level of filling.

12.4.5 MULTIPLE GATES

As will become clear, there are several reasons why more than one gate may be desirable. This might be to (1) distribute heat more evenly in the mould, (2) reduce horizontal velocities in the mould, (3) reduce contact hot spots, (4) reduce distortion of the casting, (5) provide liquid at all the lowest points in the mould cavity to avoid waterfall effects, (6) provide greater gate area for heavier parts of the casting and (7) ease cutoff. These factors will be discussed.

First it is necessary to consider how to divide up the runner to deliver metal to the gates. The various systems are shown in Figures 12.27 and 12.28 showing the benefit of working with cusp-shaped junctions where this is possible. The traditional foundryman will be surprised by what appear to be such fragile features which would normally be washed away in any conventional filling system. However, with a system designed to keep out air, the battering effect of the mix of bubbles and metal is avoided, and the smooth pressure of the metal is not a problem. The feather edge of sand is avoided by using a minimum angle of at least 30° at the tip.

12.4.6 PREMATURE FILLING PROBLEM VIA EARLY GATES

Sutton (2002) applied Bernoulli's theorem to draw attention to the possibility that a melt travelling along a horizontal runner will partly enter vertical gates placed along the length of the runner, despite the fact that the runner may not yet have completely filled and pressurised (Figure 12.27). This arises as a result of the pressure gradient along the flow and is proportional to the velocity of flow. In real casting conditions, the melt may rise sufficiently high in such gates that cavities attached to the gates might be partially filled with a slow dribble of upwelling metal before the filling of the

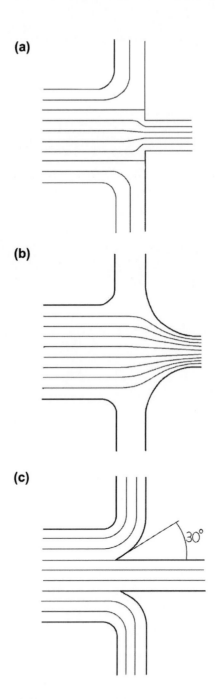

FIGURE 12.27

Three way junctions, illustrating how junction with three equal exits divide the flow (a) with no corner radii the flow is somewhat concentrated forward; (b) with tradition large radii the flow is almost exclusively forward; (c) with cusped division the flow can be very nearly equal in all three directions. The cusps are limited to 30° for practicality.

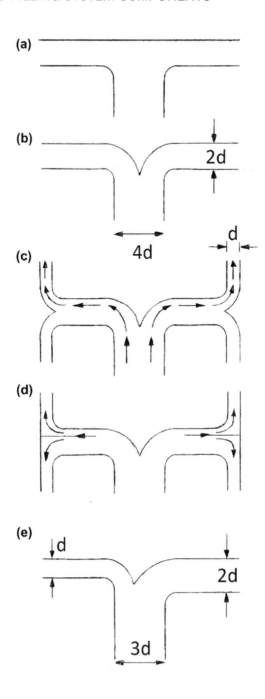

FIGURE 12.28

The division of a runner into equal branches has several options. (a) The traditional split is improved at (b) in which excess metal, with opportunity for turbulence, is avoided. Additional friction also provides a minor benefit. (c) Cautions that a second cusp split is wrong because flow favours only one direction of the second split. (d) The plane second split works better. (e) Unequal splits are easily and accurately provided when using cusp divisions.

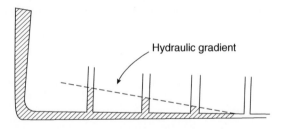

FIGURE 12.29

The partial filling of vertical ingates along the length of a runner as a result of the pressure gradient formed naturally by friction losses.

runner, and therefore before the main flow up the vertical gates. These upward dribbles of metal in the cavity are poorly assimilated by the arrival of the main metal supply, and so usually constitute a lap defect resembling a mis-run or partly filled casting.

This same effect would be expected to be even more noticeable in horizontal gates moulded in the cope, sited above a runner moulded in the drag (Figure 11.2(b)). The head pressure required to simply cross the parting line and start an unwanted early filling of part of the mould cavity would be relatively small, and easily exceeded.

12.4.7 HORIZONTAL VELOCITY IN THE MOULD

When calculating the entry velocity of the metal through the gates, it is easy to overlook what happens to the melt once it starts to spread sideways into the mould cavity. The horizontal sideways velocity away from the gate can sometimes be high. In many castings where the ingate enters a vertical wall, the transverse spreading speed inside the mould is higher than the speed through the gate, and causes a damaging splash as the liquid hits the far walls (Figure 12.30). We can make an estimate of this lateral velocity V_L in the following way.

The lateral travel of the melt will normally be at about the height h of a sessile drop. We shall assume the section thickness t, for a symmetrical ingate, area A_i. The melt enters the ingate at the velocity V_i, and spreads in both directions away from the gate. Equating the volume flow rates through the gate and along the base of the casting gives

$$V_i \cdot A_i = 2V_L \cdot h \cdot t$$

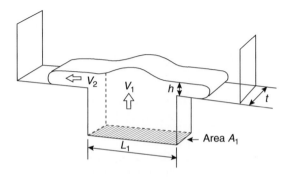

FIGURE 12.30

Sideways flow inside mould cavity.

If we limit the gate velocity and the transverse velocity to the same critical velocity V_C (for instance 0.5 ms^{-1}) and adopt a gate thickness t the same as that of the casting wall, the relation simplifies to the fairly self-evident geometrical relation in terms of the length of the ingate L_i

$$L_i = 2h$$

The message from this simple formula is that if the length of the gate exceeds twice the height of the sessile drop, even if the gate velocity is below the critical velocity, the transverse velocity may still be too high, and surface turbulence will result from the impact of the transverse flow on the end walls of the mould cavity.

To be sure of meeting this condition, therefore, for aluminium alloys where $h = 13$ mm, gates must always be less than 26 mm wide (remembering that this applies only to gates that have the same thickness as the wall of the casting). For irons and steels gates should not exceed 16 mm wide. If we allow ourselves 1 ms^{-1} for both ingate and transverse flow inside the mould cavity, the previous figures remain unchanged.

Clearly, it is a concern that in practice, gate lengths are often longer than these limits and may be causing unsuspected problems inside the mould cavity.

Because we often require areas considerably greater than can be provided by such short gates, it follows that multiple gates are required to achieve the total ingate area to bring the transverse velocity below critical. Two equally spaced gates of half the length will halve the problem, and so on. In this way the individual gates lengths can be reduced, reducing the problem correspondingly. Our relation becomes simply for N ingates of total ingate length L_i

$$L_i/N = 2h$$

Or directly giving the number of ingates N that will be required to limit velocities to 0.5 ms^{-1}

$$N = L_i/2h \tag{12.1}$$

These considerations based on velocity through the gate or in the casting take no account of other factors that may be important in some circumstances. For instance, the number of ingates might require to be increased for a variety of reasons as listed at the beginning of this multiple gate section.

12.4.8 JUNCTION EFFECT

When the gates are planted on the casting, they create a junction. This self-evident statement requires explanation.

Some geometries of junction create the danger of a hot spot. The result is that a shrinkage defect forms in the pocket of liquid that remains trapped here at a late stage of freezing. Thus when the gate is cut off, a shrinkage cavity is revealed in the casting. This defect is widely seen in foundries. In fact, it is almost certainly the reason why most traditional moulders cut such narrow gates, causing the metal to jet into the mould cavity with consequent poor results to casting quality.

The magnitude of the problem depends strongly on what kind of junction is created. Figure 12.31 shows the different kinds of junctions. An in-line junction (c) is hardly more than an extension of the wall of the casting. Very little thermal problem is to be expected here. The T-junction (a) is the most serious problem. It is discussed later. The L-junction (b) is an intermediate case and is not further discussed. Readers can make their own allowances assuming conditions intermediate between the zero (in-line junction) and T-junction cases.

To help understand the hot spot risk associated with the formation of a T-junction, it is instructive to examine the freezing patterns of T-sections. In the 1970s, Kotschi and Loper carried out some admirable theoretical studies of T-junctions using only simple calculations based on modulus. These studies pointed the way for experimental work by Hodjat and Mobley in 1984 that broadly confirmed the predictions. The data are interpreted in Figures 12.32 and 12.33 simply as a set of straight lines of slopes 2, 1 and 1/2. (A careful study of the scatter in the data shows that the predictions are not infallibly correct in the transitional areas, so that some caution is required.)

Figure 12.34 presents a simplified summary of these findings. The upright leg of the 'T' is the gate and the horizontal arm is the casting section. It is clear that a gate of section equal to the casting section (the 1:1 geometry) has a hot spot in the junction, and so is undesirable. In fact, Figure 12.34 makes it clear that any medium-sized gate less than twice as

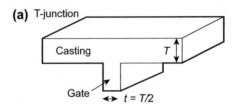

(a) T-junction

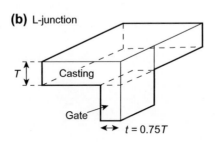

(b) L-junction

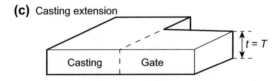

(c) Casting extension

FIGURE 12.31

Maximum allowable gate thickness to avoid a hot spot at the junction with the casting for (a) a T-junction; (b) an L-junction; and (c) a casting extension type junction.

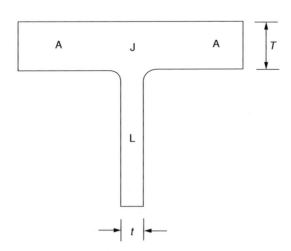

FIGURE 12.32

Geometry of a T-shaped junction.

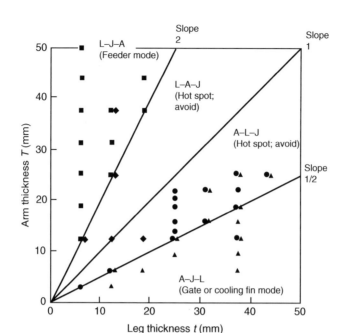

FIGURE 12.33

Solidification sequence for T-shaped castings (A = arm, J = junction, L = leg).

Experimental data from Hodjat and Mobley (1984).

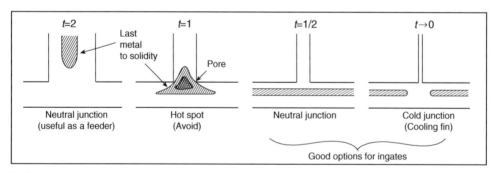

FIGURE 12.34

Array of different T-junctions.

thick, or more than half as thick, will give a troublesome hot spot. It is only when the gate is reduced to half or less of the casting thickness that the hot spot problem is eliminated.

In the case of gates forming T-junctions with the casting, the requirement to make the gate only half of the casting thickness ensures that under most circumstances no localised shrinkage defect will occur, and almost no feeding of the casting will take place through the gate.

Slot gates much less than half the section of the casting act as cooling fins. This effect can be put to good use in setting up a favourable temperature gradient in the casting, encouraging solidification from a bottom gate towards a top feeder. Such cooling fin gates have been used to good effect in the production of aluminium and copper-based alloys because of their high thermal conductivity (Wen et al., 1997). (The effect is much less useful in irons and steels, to the point at which any benefit is practically unnoticed.) Also, the common doubt that the cooling effect would be countered by the pre-heating because of the flow of metal into the casting is easily demonstrated to be, in most cases, a negligible problem. The pre-heating only occurs for the relatively short filling time compared with the time of freezing of the casting, and the thin gate itself has little thermal capacity. Thus following the completion of its role as a rather hot gate it quickly cools and converts to acting as a cooling fin.

Where a gate cannot conveniently be made to act as a cooling fin, the author has planted a cooling fin on the sides of the gate. (This is simple if the slot gate is on a joint line.) By this means, the gate is strongly cooled, and in turn, cools the local part of the casting.

The current junction rules have been stated only in terms of thickness of section. For gates and casting sections of more complex geometry, it is more convenient to extend the rules, replacing 'section thickness' by 'equivalent modulus'. The more general rule that is inferred is 'the gate modulus should be half or less than the local casting modulus'.

It is worth drawing attention to the fact that not all gates form T-junctions with the casting. For instance, those that are effectively merely extensions of a casting wall may clearly be continued on at the full wall thickness without any hot-spot effect (Figure 12.31(c)).

Gates which form an L-junction with the wall of the casting are an intermediate case (Sciama, 1974), where a gate thickness of 0.75 times the thickness of the wall is the maximum allowable before a hot spot is created at the junction (Figure 12.31(b)).

It is possible that these simple rules may be modified to some degree if much metal flows through the gate, locally pre-heating this region. In the absence of quantitative guidelines on this point it is wise to provide several gates, well distributed over the casting to reduce such local overheating of the mould. The Cosworth system devised by the author for the ingating of cylinder heads used 10 ingates, one for every bolt boss. It contrasted with the two or three gates that had been used previously, and at least partly accounts for the immediate success of the gating design (although it did not help cutoff costs, of course!).

If the casting contains heavy sections that will require feeding then this feed metal will have to be provided from elsewhere. It is necessary to emphasise the separate roles of (1) the filling system and (2) the feeding system. The two have quite different functions. In the author's experience, attempts to feed the casting through the gate are to be welcomed if really possible; however, there are in practice many reasons why the two systems often work better when completely separate. They can then be separately optimised for their individual roles.

It is necessary to make mention of some approaches to gating that attempt to evaluate the action of running systems with gates that operate only partly filled (Davis, 1977). The reader will confirm that such logic only applies if the gates empty downhill into the mould, like water spilling over a weir. This is a violation of one of our most important filling rules. Thus, approaches designed for partially filled gates are not relevant to the technique recommended in this book. The placing of gates at the lowest point of the casting, and the runner below that, ensures that the runner fills completely, then the gates completely, and only then can the casting start to fill. The complete prior filling of the running system is an essential requirement; it avoids the carrying through of pockets of air as waves slop about in partially filled systems. Complete filling eliminates waves.

As in most foundry work, curious prejudices creep into even the most logical approaches. In their otherwise praiseworthy attempt to formalise gating theory, Kotschi and Kleist (1979) omit to limit the thickness of their gates to reduce the junction hot spot, but curiously equalise the areas of the gates so as to equalise the flow into the casting. In practice, making the gates the same is rarely desirable because most castings are not uniform. For instance, a double flow rate might be required into part of the casting that is locally twice as heavy.

The design of gates may be summarised in these concluding paragraphs.

The requirement for gates to be limited to a maximum thickness naturally dictates that the gates may have to become elongated into a slot-type shape if the gate area is also required to be large. Limitations to the length of the slots to limit the lateral velocities in the mould may be required, dictating more than one gate. The limitation of lateral velocities in addition to ingate velocities is a vital feature.

The slot form of the gates is sometimes exasperating when designing the gating system because it frequently happens that there is not sufficient length of casting for the required length of slot! In such situations, the casting engineer has to settle for the best compromise possible. In practice, the author has found that if the gate area is within a factor of 2 of the area required to give 0.5 ms^{-1}, then an aluminium alloy casting is usually satisfactory. Any further deviation than this would be cause for concern. Grey irons and carbon steels are somewhat more tolerant of higher ingate velocities.

As a final part of this section on gating, it is worth examining some traditional gating designs.

12.4.9 THE TOUCH GATE

The touch gate, or kiss gate, is shown in Figure 12.35(a). As its name suggests, it only just makes contact between the source of metal and the mould cavity. In fact, there is no gate as such at all. The casting is simply placed so as to overlap the runner. The overlap is typically 0.8 mm for brass and bronze castings (Schmidt and Jacobson, 1970; Ward and Jacobs, 1962) although up to 1.2 mm is used. More than a 2.5 mm overlap causes the castings to be difficult to break off, negating the most important advantage. The elimination of a gate in the case described by the authors was claimed to allow between 20% and 50% more castings in a mould. Furthermore, the castings are simply broken off the runner, speeding production and avoiding cut-off costs and metal losses from sawing. The broken edge is so small that for most purposes dressing by grinding is not necessary; if anything, only shot blasting is required (Figure 12.35(a)).

A further benefit of the touch gate is that a certain amount of feeding can be carried out through the gate. This happens because (1) the gate is pre-heated by the flow of metal through it, and (2) the gate is so close to the runner and casting that it effectively has no separate existence of its own; its modulus is not that of a tiny slot, but some average between that of the runner and that of the local part of the casting to which it connects. Investigations of touch gate geometry have overlooked this point, with confusing results. More work is needed to assess how much feeding can actually be carried out. The result is likely to be highly sensitive to alloy type so that any study would benefit from the inclusion of short- and long-freezing-range alloys and high and low conductivity metals.

Ward and Jacobs report a reduced incidence of mis-run castings when using touch gating. This observation is almost certainly the result of the beneficial effect of surface tension control in preventing the penetration of the gates before the runner is fully filled and at least partly pressurised. Only when the critical pressure to force the metal surface into a single curvature of 0.4 mm is reached (in the case of the 0.8 mm overlap) will the metal enter the mould cavity. This pressure corresponds to a head of 30–40 mm for copper-based alloys.

The use of such narrow slots for gates might give a user some concern about the melt jetting through such a narrow constriction. In practice, surface tension probably acts to avoid this because the meniscus has to be compressed to a radius of only 400 μm when passing through a 0.8 mm overlap. Immediately after the narrowest point is passed, the surface tension acts to spring the meniscus back to a larger radius, closer to its sessile drop radius in the region of 5 mm or more. Thus the speed of advance of the front immediately reduces. The velocity is only high in the very centre of the gap, but becomes gentle once again after passing this point. Thus the mould fills relatively quiescently.

The only concern about jetting relates to the little-known phenomenon of 'microjetting'. This unknown effect occurs repeatedly in our consideration of flow of metals in narrow spaces. It is possible that for certain strongly oxidising alloys such a mode of advance might occur in the narrow overlap, the microjets causing oxidation damage in the cavity before the arrival of the main body of liquid. This effect could easily be tested for each alloy with a partly open cavity. Any microjetting would be quickly seen.

A further concern may arise as the result of the bifilm content of the melt. If the bifilms are large and strong, they may tend to block the gate. In any case, occasional large bifilms will almost certainly be prevented from entering the casting; the narrow gate operating as a filter. This effect might be an important benefit in some situations, and anyone researching this gate will be required to take this effect into account.

Finally, with such a thin gate, variations of only 0.1 mm in thickness have been found to change performance drastically. With the runner in one half of the mould and the casting in the other, this is clearly seen to be a problem from small variations in mis-match between the mould halves. This problem is solved later.

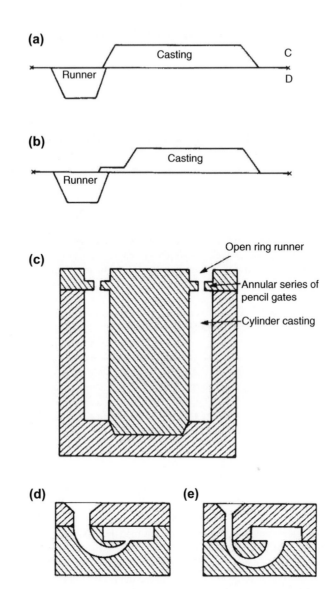

FIGURE 12.35

(a) Touch gate, (b) knife gate, (c) pencil gate, (d) normal and (e) reversed horn gates.

12.4.10 **KNIFE GATE**

The previous capillary repulsion problem arising as a result of the variable thickness of the touch gate can be countered in practice by providing a small gate attached to the casting, i.e. in the same mould half as the casting cavity, so that the gate geometry is fixed regardless of mis-match. This is sometimes called a *knife gate* (Figure 12.35(b)).

Although it is perhaps self-evident, touch and knife gates are not viable as knock-off gates on the modern designs of accurate, thin-walled, aluminium alloy castings. This is simply because the gate has a thickness similar to the casting, so that on trying to break it off, the casting itself bends! The breaking off technique only works for strong, chunky castings or for relatively brittle alloys (Figure 12.35(b)).

The system was said to be unsuitable for aluminium bronze and manganese bronze, both of which are strong film-forming alloys (Schmidt and Jacobson, 1970), although this discouraging conclusion was probably the result of the runner being usually moulded in the cope and the castings in the drag and a consequence of their poor filling system, generating quantities of oxide films that would threaten to choke gates. The unfortunate fall into the mould cavity would further damage quality, as was confirmed by Ward and Jacobs (1962). These authors that uphill filling of the mould was essential to providing a casting quality that would produce a perfect cosmetic polish.

The system has been studied for several aluminium alloys (Askeland and Holt, 1975), although the poor gating and downhill filling used in this work appears to have clouded the results. Even so, the study implies that a better quality of filling system with runner in the drag and casting impressions in the cope could be important and rewarding.

The fundamental fear that the liquid may jet through the narrow gate may be unfounded, as has been discussed previously. It is worthwhile dwelling a little more on this issue. As mentioned earlier, there may actually be no jetting problem at all as a result of the high surface tension of liquid metals. Whereas water might be expected to jet through such a narrow constriction, liquid aluminium is effectively compressed when forced in to any section less than its natural sessile drop height of 12.5 mm. The action of a melt progressing through a thin gate, equipped with an even thinner section formed by a sharp notch was observed for aluminium alloys in the author's laboratory by Cunliffe (1994). The gate was 4 mm thick and the thickness under the various notches was only 1–2 mm. The progress of the melt along the section was observed via a glass window. The metal was seen to approach, cross the notch constriction and continue on its way without hindrance, as though the notch constriction did not exist! This can only be explained if the melt immediately re-expands to fill the channel after passing the notch. It seems the liquid meniscus, acting like a compressed, powerful spring, immediately expands back to fill the channel when the point of highest compression is passed.

If the surface turbulence through touch and knife gates is tolerable, or minimal, as expected, it deserves to be much more widely used. It would be so welcome to be able to end the drudgery of sawing castings off running systems, together with the noise and the waste. With good quality metal provided by a good front end to the filling system, and uphill filling of the mould cavity after the gate, it seems likely that this device could work well. It would probably not require much work to establish a proper design code for such a practice.

12.4.11 THE PENCIL GATE

Many large rolls for a variety of industries are made from grey cast iron poured in greensand moulds. They often containing a massive proportion of grey iron chills around the roll barrel to develop the white iron wear surface of the roll. It is less common nowadays to cast rolls in *loam moulds* produced by strickling. (*Loam* is a sand mixture containing high percentages of clay and water, like a mud, which allow it to be formed by striking or strickling and *sleeking* into place. It needs to be thoroughly dried before casting.) Steel rolls are cast similarly.

Rolls, whether hollow or solid, can be beneficially bottom-gated tangentially into their base. If the front end of the filling system design is good, excluding all air, this is an excellent technique for roll production which deserves wider use. The author has made excellent rolls this way.

Where the roll or cylinder is hollow, it may be centrifugally cast, or it may be produced by a special kind of top gating technique using pencil gates. Neither of these techniques is expected to be as satisfactory or as simple as the tangential bottom gating mode.

Figure 12.35(c) represents a cross-section through a mould for a roll casting. Such a casting might weigh more than 60,000 kg, and have dimensions up to a 5 m diameter by a 5 m face length, with a wall thickness 80 mm (Turner and Owen, 1964). It is cast by pouring into an open circular runner, and the metal is metered into the mould by a series of pencil gates. The metal falls freely through the complete height of the mould cavity, gradually building up the casting. The metal-mould combination of grey iron in green sand with iron chills is reasonably tolerant of surface turbulence. In addition, the heavy-section thickness gives a solidification time in excess of 30 min, allowing a useful time for the floating out and separation of much of the oxide entrained by splashing. The depth of splashing is limited by the slimness of the falling streams from the narrow pencil gates.

The solidification geometry is akin to continuous casting. The slow, controlled buildup of the casting ensures that the temperature gradient is high, thus favouring good feeding. The feeder head on top of the casting is therefore only minimal because much of the casting will have solidified by the time the feeder is filled. This beneficial temperature gradient is encouraged by the use of pencil gates: the narrow falling streams have limited energy and so do not disturb the pool of liquid to any great depth (a single massive stream would be a disaster for this reason).

Top gating in this fashion using pencil gates is expected to be useful only for the particular conditions of: (1) grey iron; (2) heavy sections; and (3) greensand or inert moulds. It is not expected to be appropriate for any metal-mould combinations in which the metal is sensitive to the entrainment of oxide films, especially in thin sections where entrained material has limited opportunity to escape.

Even so, this top pouring, although occurring in the most favourable way possible as discussed previously, still results in occasional surface defect in products that are required to be nearly defect-free. The use of bottom gating via an excellent filling system, entering the mould at a tangent to centrifuge defects away from the outer surface of the roll would be expected to yield a superior product. No matter what the casting method, there is no substitute for a good filling system.

12.4.12 THE HORN GATE

The horn gate is an ancient device used by a traditional greensand moulder to make a quick and easy connection from the sprue into the base of the mould cavity without the need to make and fit a core or provide an additional joint line (Figure 12.35(d)). The horn pattern could be withdrawn by carefully easing it out of the mould, following its curved shape.

Although the ingenuity of the device can be admired, in practice, it cannot be recommended. It breaks one of our fundamental rules for filling system design by allowing the metal to fall downhill. In addition, there are other problems.

When used with its narrow end at the mould cavity, it causes jetting of the metal into the mould. This effect has been photographed using an open-top mould, revealing liquid iron emerging from the exit of the gate, and executing a graceful arc through the air, before splashing into a messy, turbulent pool at the far side of the cavity (Subcommittee T535, 1960).

It has occasionally been used in reverse in an attempt to reduce this problem (Figure 12.35(e)). However, the irregular filling of the first half of the gate by the metal running downhill in an uncontrolled fashion and slopping about in the valley of the gate is similarly unsatisfactory. Furthermore, the large end junction with the casting now poses the additional problem of a large hot spot that requires to be fed to avoid shrinkage porosity. Also, of course, it is now a non-trivial effort to cut off.

The horn gate might be tolerable for a rough grey iron casting in greensand. Otherwise, it is definitely to be avoided.

12.4.13 VERTICAL END GATE

Sometimes it is convenient to place a vertical gate at the end of a runner. Whereas the slowing of the flow by expanding the channel was largely unsuccessful for the horizontal runner, an upward-oriented expanding fan-shaped gate can be extremely beneficial because of the aid of gravity. As always, the application is not completely straightforward (Figure 12.36).

Figure 12.36(a) shows that if the fan gate is sited directly on top of a rectangular runner, the flow is constrained by the vertical sides of the runner, so that the liquid jets vertically, falling back to fill the fan gate from previously.

Figure 12.36(b) shows that if the expansion of the fan is started from the bottom of the runner, the flow expands nicely, filling the expanding volume and so reducing in speed before it enters the mould cavity. This result is valuable because it is one of the very few successful ways in which the speed of the metal entering the mould cavity can be reduced.

(a)

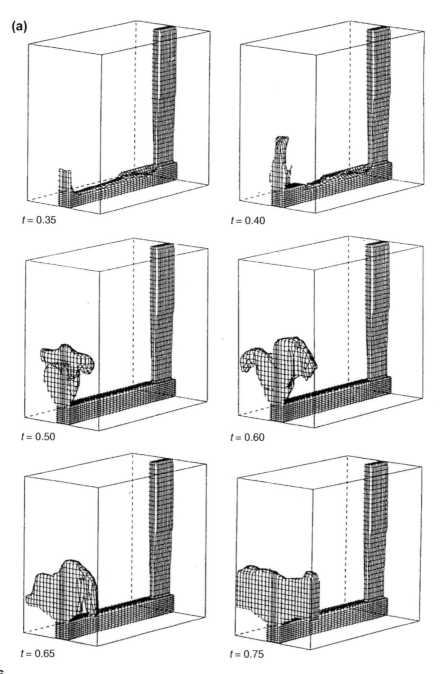

$t = 0.35$ $\qquad\qquad$ $t = 0.40$

$t = 0.50$ $\qquad\qquad$ $t = 0.60$

$t = 0.65$ $\qquad\qquad$ $t = 0.75$

FIGURE 12.36

A vertical fan gate at the end of a runner showing the difference in flow as a result of a (a) top connection and (b) bottom connection to the runner.

(b)

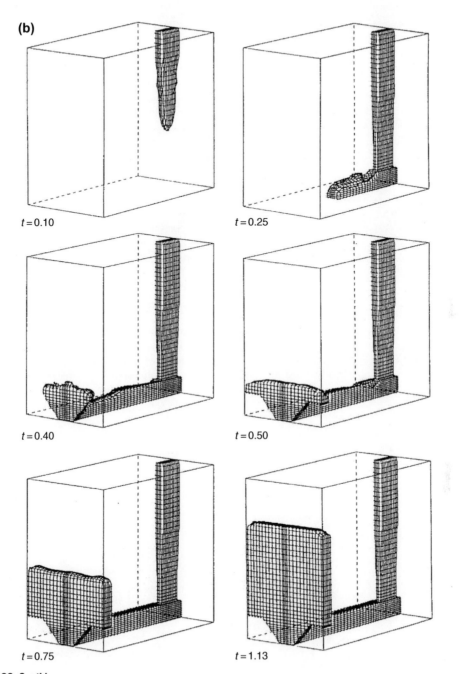

$t = 0.10$ $t = 0.25$

$t = 0.40$ $t = 0.50$

$t = 0.75$ $t = 1.13$

FIGURE 12.36 Cont'd

The work in the author's laboratory (Rezvani et al., 1999) illustrates that this form of gate produces castings of excellent reliability. Compared with conventional slot gates, the Weibull modulus of tensile test bars filled with the nicely diverging fan gate was raised nearly four times, indicating the production of castings of four times greater reliability.

Itamura (2002) and coworkers have shown by computer simulation that the limiting 0.5 ms^{-1} velocity is safe for simple vertical gates, but can be raised to 1.0 ms^{-1} if the gate is expanded as a fan. However, expansion does not continue to work at velocities of 2 ms^{-1} where the flow becomes a fountain. Similar results have been confirmed in the author's laboratory by Lai and Griffiths (2003), who used computer simulation to study the expansion of the vertical gate by the provision of a generous radius at the junction with the casting. All these desirable features involve additional cutting off and dressing costs of course.

12.4.14 DIRECT GATING INTO THE MOULD CAVITY

If the casting engineer has successfully designed the running system to provide bottom-gating with minimal surface turbulence, then the casting will fill smoothly without the formation of film defects. However, the battle for a quality casting may not yet be won. Other defects can lie in wait for the unwary!

For the majority of castings, the gate connects directly into the mould cavity. I call this simply 'direct gating'. In most cases it is allowable, or tolerable, but it sometimes causes other problems because of the effect it has on the solidification pattern of the casting.

12.4.15 FLOW CHANNEL STRUCTURE

If the filling is slow, the melt may carve its own path through the solidifying casting, conveying hot metal up the flow channel, and allowing it to reach and spread out along the advancing liquid front. The channel and the portion of the mould in contact with it, become particularly hot, therefore taking significantly longer to cool. Thus, within an otherwise excellent casting with fine microstructure and low porosity, the channel region is coarse grained and with noticeable porosity (Figure 12.37). This structure may cause a casting being made to a strict quality specification to be scrapped. It is

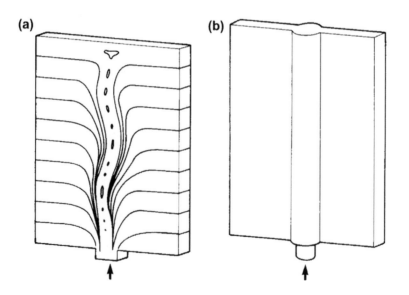

FIGURE 12.37

(a) Direct bottom-gated vertical plate and (b) the use of a boss to assist feeding after casting is filled.

possible that many so-called shrinkage problems (for which more or less fruitless attempts are made to provide a solution by extra feeders or other means) are actually residual flow channels that might be cured by changing ingate position or size, or raising fill rate. No research appears to have been carried out to guide us out of this difficulty.

The flow channel structure is a standard feature of castings that are filled slowly through a gate planted directly in their base but, amazingly, this serious limitation to structure control seems to have been widely overlooked. It is seen in a computer prediction of a thin wall steel casting in Figure 5.25, and seen through a glass window for cast iron by Xu and Mampaey (1994). Elliott and Mezoff (1947 report the phenomenon in Mg alloys castings. This widespread effect is discussed in more detail in Section 5.1.5.

If the flow channel is particularly well developed, it will almost certain remain as an elongated hot spot in the thin wall, and so will require to be fed. This will probably mean the provision of a tall feeder sat on the runner bar, and the increase of the areas of the runner and gate to ensure that the feeder has a feed path to deliver feed metal to the base of the flow channel after much of the remainder of the casting has frozen. Such feeding uphill is not recommended because it is not always reliable, but in the case of a well developed flow channel, it is the only chance to recover some solidarity of the casting.

Nevertheless, in general, the problem is reduced by filling faster (if that is possible without introducing other problems). However, even fast filling does not cure the other major problem of bottom gating, which is the adverse temperature gradient, with the coldest metal being at the top and the hottest at the bottom of the casting. Where feeders are placed at the top of the casting this thermal regime is clearly unfavourable for effective feeding. If bottom-gating into the mould cavity is used the unfavourable feeding regime simply has to be accepted, using enlarged, less efficient feeders.

12.4.16 INDIRECT GATING (INTO AN UP-RUNNER/RISER)

The problem with direct bottom-gating into the mould cavity arises because the hot metal that is required to fill the casting is gated directly into the casting and has to travel through the casting to reach all parts. There is an interesting indirect gating system that solves the major features of the bottom-gating and flow channel problem.

The solution is *not* to gate into the casting: a main flow path is created *outside* the casting. It is called a *riser* or *up-runner*. Metal is therefore diverted initially away from the casting, through the riser, only entering subsequently by displacement sideways from the riser as fresh supplies of hot metal arrive. The fresh supplies flood up into the top of the riser, ensuring that the riser remains hot, and that the hottest metal is delivered to the top of the mould cavity. The system is illustrated in Figure 12.38. The system has the special property that the riser and slot gate combination acts not only to fill but also to feed. (Risking a repeat of policy stated earlier as a reminder: the reader will notice that the use of the term 'riser' in this book is limited to this special form of feeder which also acts as an 'up-runner', in which the metal *rises* up the height of the casting. It is common in the United States to refer to conventional feeders placed on the tops of castings as risers. However, this terminology is avoided here; such reservoirs of metal are called feeders, not risers, following the simple logic of using a name that describes their action perfectly, and does not get confused with other bits of plumbing such as whistlers, up which metal also rises!)

The upper parts of the casting in Figure 12.38 will probably also require some feeding. This is easily achieved by planting a feeder on the top of the riser, as a kind of riser extension. This retains all the benefits of the system because its metal is hot, and hotter metal below in the riser will convect into the feeder.

The disadvantages of the riser and slot gate system are as follows:

1. The considerable cutoff and finishing problem because the gate often has to be sited on an exterior surface of the casting, and so requires much subsequent dressing to achieve an acceptable cosmetic finish.
2. There appears to be no method of predicting the width and thickness of the gate at the present time. Further research is required here. In the normal gate where it is required to freeze before the casting section to avoid the hot-spot problem in the junction, the thickness of the gate is held to half of the casting thickness or less. However, in this case the gate is really equivalent to a feeder neck, through which feed metal is required to flow until the casting has solidified. Whereas a thickness of double the casting thickness would then be predicted to avoid the junction hot spot under conditions of uniform starting temperature, the pre-heating effect of the gate because of the flow of metal

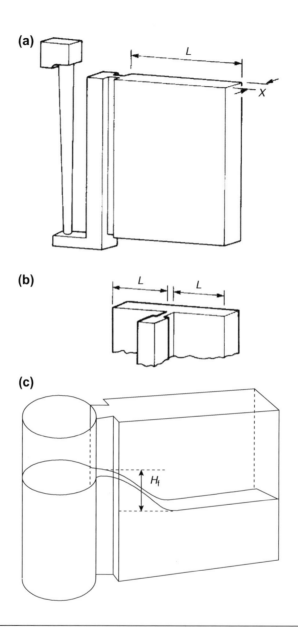

FIGURE 12.38

Riser and slot gate to both gate and feed a vertical plate from (a) its side, or (b) its centre. (c) The danger of causing a drop H_f too severe if the slot gate is too narrow.

through it might mean that a gate as narrow as half of the casting section may be good enough to continue feeding effectively. There are, unfortunately, no confirmatory data on this at the present time.

It is important to caution against the use of a gate which is too narrow for a completely different reason; if filling is reasonably fast, then the resistance to flow provided by a narrow slot gate will cause the riser to fill up to a high level before much metal has had a chance to fill the mould cavity. The dynamics of filling and surface tension, compounded by the presence of a strong oxide film, will together conspire to retain the liquid in the riser for as long as possible. The metal will therefore spill through the slot into the casting from an elevated level (the height of fall H_f shown in Figure 12.38(c)). Again, our no-fall rule is broken. It is desirable, therefore, to fill slowly, and/or to have a gate sufficiently wide to present minimal resistance to metal flow. In this way the system can work properly, with the liquid metal in the riser and casting rising substantially together. Metal will then enter the mould gently.

It is also important for the gate not to be thinner than the casting when the casting wall thickness reduces to approximately 4 mm. A gate 2 mm thick would hold back the metal because of the effect of surface tension and the surface film allowing the metal head to build up in the riser. When the metal eventually breaks through, the liquid will emerge as a jet, and fall and splash into the mould cavity. For casting sections of 4 mm or less the gate should probably be at least as thick as the wall.

In general, it seems reasonable to assume that conditions should be arranged so that the fall distance h in Figure 12.38(c) should be less than the height of the sessile drop. The fall will then be relatively harmless.

For thinner-section castings (for instance, less than 2 mm thickness) made under normal filling pressure, the feeding of thin-section castings can probably be neglected (as will be discussed in Section 6). Thus any hot-spot problems can also be disregarded, with the result that the ingate can be equal to the casting thickness. Surface tension controls the entry through the gate and the further progress of the metal through the mould cavity, reducing the problems of surface turbulence. Fill speed can therefore be increased.

A further important point of detail in Figure 12.38(a) should be noted. The runner turns upward on entry to the riser, directing the flow upwards. A *substantial* upward step is required (the step shown in the figure is probably not sufficient) to ensure this upward direction to the flow. If the provision of this step is neglected causing the base of the runner to be level with the gate and the base of the casting, metal rushes along the runner and travels unchecked directly into the mould cavity as in Figure 11.4(a). A flow path would then be set up so that the riser would receive no metal directly, only indirectly after it had circulated through the casting. The base of the casting would receive all the heated metal, and the riser would be cold. Such a flow regime clearly negates the reason for the provision of the system! Many such systems have failed through omitting this small but vital detail.

What rates are necessary to make the system work best? Again we find ourselves without firm data to give any guide. We can obtain some indication from the following considerations.

The first liquid metal to flow through the slot gate and along the base of the cavity travels as a stream. Being the first metal travelling over the cold surface of the mould, it is most at risk from freezing prematurely. Subsequent flow occurs over the top of this hot layer of metal and therefore does not lose so much heat from its under surface. Thus if we can ensure that conditions are right for the first metal to flow successfully, then all subsequent flow should be safe from early freezing. In the limiting condition where the tip of the first stream just solidifies on reaching the end of the plate, it will clearly have established the best possible temperature gradient for subsequent feeding by directional solidification back towards the riser. Subsequent layers overlying this initial metal will, of course, have slightly less beneficial temperature gradients because they will have cooled less during their journey. Nevertheless, this will be the best that we can do with a simple filling method; further improvements will have to await the application of programmed filling by pumped systems.

Focusing our attention, therefore, on the first metal into the mould, it is clear that the problem is simply a fluidity phenomenon. We shall assume that the height of the stream corresponds to the height h of the meniscus which can be supported by surface tension. If the distance to be run from the gate is L and the solidification time of the metal is t_f in that section thickness x, then in the limiting condition where the metal just freezes at the limit of flow (thus generating the maximum temperature gradient for subsequent feeding):

$$t_f = L/V \tag{12.2}$$

where V is the velocity of flow (ms^{-1}) of the metal stream. The corresponding rate of flow Q (kgs^{-1}) for metal of density ρ (kgm^{-3}) is easily shown to be:

$$Q = Vhx\rho$$
$$= Lhx\rho/t_f \tag{12.3}$$

At constant filling rate, the time t to fill a casting of height H is given by:

$$t = t_f(H/h) \tag{12.4}$$

Considering now the first length of melt to travel along the base of the mould, a 10 mm thick bar in Al-7Si alloy would be expected to freeze completely in about 40 s giving a flow life for the solidifying alloy of perhaps 20 s. The meniscus height h is approximately 12.5 mm, and so for a casting H = 100 mm high, eight times the meniscus height, the pouring time would be $8 \times 20 = 160$ s, or nearly 3 min. This is a surprisingly long time, and almost certain to be overestimated because the horizontal flows imagined for this exercise would probably result in some visible, but perhaps not especially important, horizontal laps.

Anyway, the conclusion is not likely to be particularly accurate, but does emphasise the important point that relatively thin cast sections do not necessarily require fast filling rates to avoid premature solidification. What is important is the steady, continuous advance of the meniscus.

Naturally, however, it is important not to press this conclusion too far, and the previously first-order approximation to the fill time probably represents a time that might be achievable in ideal circumstances: in fact, if the rate of filling is too slow, then the rate of advance of the liquid front will become unstable for other reasons:

1. Surface film problems may cause instability in the flow of some materials in the form of unzipping waves, as is explained in Chapter 3. Film-free systems will not suffer this problem, and vacuum casting may also assist.
2. Another instability that has been little researched is the flow of the metal in a pasty mode (Section 5.1.5) that shows the curious behaviour in which flow channels take a line of least resistance through the casting, abandoning the riser.

12.4.17 CENTRAL VERSUS EXTERNAL SYSTEMS

Most castings have to be run via an external running system as shown in Figures 12.21 and 12.23(a). Although this is satisfactory for the requirements of the running system, it is costly from the point of view of the space it occupies in the mould. This is especially noticeable in chemically bonded moulds, whose relatively high cost is, of course, directly related to their volume, and whose volume can be modified relatively easily because the moulds are not usually contained in moulding boxes, i.e. they are boxless. Naturally, in this situation, it would be far more desirable if the running system could somehow be incorporated inside the casting, so as to use no more sand than necessary.

This ideal might be achieved in some castings by the use of direct gating in conjunction with a filter as discussed later (Section 12.8.7).

For castings that have an open base, however, such as open frames, cylinders or rings, an excellent compact and effective solution is possible. It is illustrated for the case of cylinder and ring castings in Figures 12.22 and 12.23. The runners radiate outwards from the sprue exit and connect with vertical slot gates arranged as arcs around the base of the casting. Ruddle and Cibula (1957) describe a similar arrangement, but omit to show how it can be moulded (with all due respect to our elder statesmen of the foundry world, their suggested arrangement looks unmouldable!), and omit the upward gates. The vertical gates are an important feature for success, introducing useful friction into the system, and making for easy cutoff.

The vertical gates have a further important function: they allow some flexibility so that the contraction of the radial spider runners does not pull the sides of the casting inwards as they cool. The tendency for this system to distort the casting is a serious disadvantage. For this reason, the vertical gates need to be as long as possible to reduce distortion to a minimum. In addition, it is often useful to use only a single radial runner and gate.

Feeders can be sited on the top of the cylinder if required. Alternatively, if the casting is to be rolled through 180° after pouring, the feeding of the casting can take the form of a ring feeder at the base (later to become the top, of course). The ring can be filled tangentially, using now a single ingate because hot-spot creation in the feeder is of course not a problem. This is especially valuable because it eliminates any distortion from the inwards pull of the radial runners.

Experience with internal running has found it to be an effective and economical way to produce hollow shapes. It is also effective for the production of other common shapes such as gearboxes and clutch covers, where the sprue can be arranged to pass down through a rather small opening in one half of the casting and then be distributed via a spider of runners and gates on the open side. Care is needed however, because the long radial runners contract significantly, threatening distortion.

A final caution should be noted:

It has been observed that aluminium alloy castings of 300 mm or more internal diameter containing a central sprue exhibit a patternmaker's contraction considerably less than that which would have been expected for an external system. This seems almost certainly to be the result of the expansion of the internal core as a result of the extra heating from the internal running system. For a silica sand core this expansion can be between 1 and 1.5%, effectively negating the patternmaker's shrinkage allowance, which is normally between 1 and 1.3%. Thus the casting suffers an oversize bore. The use of a non-silica aggregate for the core eliminates this concern.

12.4.18 SEQUENTIAL GATING

When there are multiple impressions on a horizontal pattern plate, it is usually unwise to attempt to fill all the cavities at the same time. (This is contrary to the situation with a vertically parted mould, in which many filling systems specifically target the filling of all the cavities at once to reduce pouring time.)

The reasoning in the case of the horizontal mould is simple. The numerous cavities are filling at a comparatively slow rate and not necessarily in a smooth and progressive way. In fact, despite an otherwise good running system design, it is likely that filling will be severely irregular, with slopping and surging, because of the lack of constraint on the liquid and because of the additional tendency for the flow to be unstable at low flow rates in film-forming alloys. The result will be the non-filling of several the impressions and doubtful quality of the others.

Loper (1981) has provided a solution to this problem for multiple impressions on one plate as shown in Figure 12.39(a). He uses runner dams to retard the metal, allowing it time to build up a head of metal sufficient to fill the first set of impressions before overcoming the dam and proceeding to the next set of impressions, and so on.

The system has only been reported to have been used for grey iron castings in greensand moulds. It may give less satisfactory results for other metal-mould systems that are more susceptible to surface turbulence. However, the design of the overflow (the runner down the far side of each dam) could be designed as a miniature tapered down-runner to control the fall, and so reduce surface turbulence as far as possible. Probably, this has yet to be tested.

Another sequential-filling technique, 'horizontal' stack moulding, has also only so far been used with cast iron. This was invented in the 1970s by one of our great foundry characters from the United Kingdom, Fred Hoult, after his retirement at age 60. It is known in his honour as the 'H Process'. Figure 12.39(b) outlines his method. The progress of the metal across the top of those castings already filled keeps the feeders hot, and thus efficient. The length of the stack seems unlimited because the cold metal is repeatedly being taken from the front of the stream and diverted into castings. (The reader will note an interesting analogy with the up-runner and slot gate principle; one is horizontal and the other vertical, but both are designed to divert their metal into the mould progressively. The same effect is also used in the promotion of fluidity as described by Hiratsuka 1966.) Stacks of 20 or more moulds can easily be poured at one time. Pouring is continued until all the metal is used up, only the last casting being scrapped because of the incomplete pour, and the remaining unfilled moulds are usable as the first moulds in the next stack to be assembled.

The size of castings produced by the H Process is limited to parts weighing from a few grams to a few kilograms. Larger parts become unsuitable partly because of handling problems because the moulds are usually stacked vertically during assembly, then clamped with long threaded steel rods, and finally lowered to the floor to make a horizontal line. Larger parts are also unsuitable because of the fundamental limitation imposed by the increase of defects as a necessary consequence of the increased distance of fall of the liquid metal inside the mould, and possibly greater opportunity to splash in thicker sections.

The benefits of the horizontal stacking of moulds can be contrasted with the traditional stack moulding technique, in which the stack is vertical (Figure 12.39(c)). This is a way of casting many identical moulds at the same time. However, it is an awful technique, designed to make poor quality castings. The sprue tapers cannot be organised into a single tapering channel, so the fall of the melt is extremely turbulent. Uncontrollable entrainment defects are therefore to be expected. The technique is not recommended.

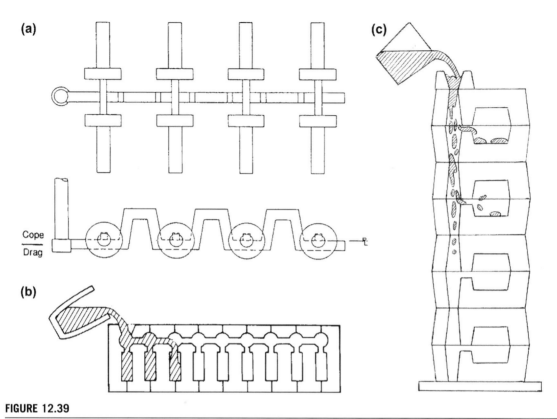

FIGURE 12.39

(a) Sequential filling for several impressions on a pattern plate *(after Loper, 1981)* and (b) sequential filling for horizontal stack moulded castings *(H Process)*. (c) Vertical stack moulding which has little to recommend it if good castings are required.

12.4.19 PRIMING TECHNIQUES

There have been several attempts over the years to introduce a two-stage filling process. The first stage consists of filling the sprue, after which a second stage of filling is started in which the runner and gates etc. are allowed to fill.

The stopping of the filling process after the filling of the sprue brings the melt in the sprue to a stop, ensuring the exclusion of air. After a delay of a few seconds, the melt is then allowed to start flowing once again. This second phase of filling has the full head H of metal in the sprue and pouring basin to drive it, but the column has to start to move from zero velocity. It reaches its 'equilibrium' velocity $(2g\,H)^{1/2}$ only after a period of acceleration. Thus the early phase of filling of the runner and gates starts from a zero rate, and has a gradually increasing velocity. The action is similar to our *'surge control'* techniques described earlier.

The benefits of the exclusion of air from the sprue, and the reduced velocity during the early part of stage 2, are benefits that have been recorded experimentally for semi-solid (actually partly solid) alloys and for different types of metal matrix composite. These materials are otherwise extremely difficult to cast without defects, almost certainly because their entrainment defects cannot float out but are trapped in suspension because of the high viscosity of the mixture. Weiss and Rose (1993) and Cox et al. (1994) developed a system in which the advance of the Al alloy metal matrix composite was arrested at the base of the sprue by a layer of ceramic paper supported on a ceramic foam filter (Figure 12.40(a)). When the sprue was filled, the paper was lifted from one corner by a rod, allowing the melt to flow through the filter and into the running system. These authors call their system *'interrupted pouring'*. However, the name *'two-stage pour'* is recommended as being more positive and less likely to be interpreted as a faulty pour as a result of an accident.

The two-stage pour was convincingly demonstrated as beneficial by Taghiabadi and colleagues (2003) for both partly solid and conventional aluminium casting alloys. These authors used Weibull statistics to confirm the reality of the benefits. They used a steel sheet to form a barrier, as a slide valve, in the runner. After the filling of the sprue, the sheet was withdrawn, allowing the mould to fill.

Wildermuth (1968) used a sheet metal slide gate coated with fireclay to close the ingate to his sand moulds (Figure 12.40(b)). This was successful to make significantly cleaner castings with fewer cope defects in both ductile iron and steel plate castings.

A second completely different embodiment of the two-stage filling concept is the *snorkel ladle*, sometimes known as the *eye-dropper ladle*. It is illustrated in Figure 12.40(c) and (d). The device is used mainly in the aluminium casting industry, but would with benefit extend to other casting industries. Instead of transferring metal from a furnace via a ladle or spoon of some kind, and pouring into a pouring basin connected to a sprue, the snorkel dips into the melt, and can be filled uphill through its open base nozzle simply by dipping sufficiently deeply. The stopper is then lowered to close the nozzle and the ladle is lifted clear from the melt. It is then transferred to the mould where it can deliver its contents into a conventional basin and sprue system, or, in the mode recommended here, lowered down through the

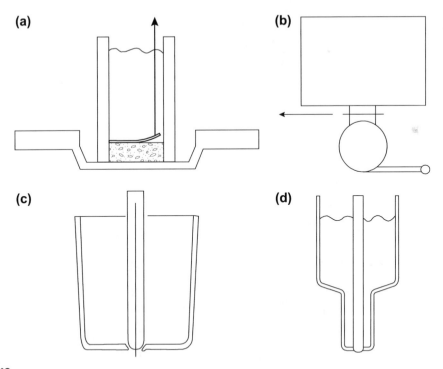

FIGURE 12.40

Two-stage filling techniques: (a) ceramic paper seal on top of ceramic foam filter, lifted by wire; (b) steel slide gate at entrance to runner; (c) stopper in the base of a ladle; (d) extended snorkel and stopper.

mould to reach and engage with the runner. Only then is its stopper raised and the melt delivered to the start of the running system with minimal surface turbulence. It is a variety of the contact pour technique in which the contact is made not at the top of the sprue but at its base. Effectively the nozzle extension on the ladle is the sprue. In this way the damaging fall of the melt down the sprue is completely avoided. The approach is capable of producing excellent products.

Two-stage filling in its various forms seems to offer real promise for many castings.

12.4.20 TANGENTIAL FILTER GATE

It was specifically designed by me early in 2013 to eliminate the troublesome bubbles which found their way through the filling system, probably as the residues of the maelstrom during priming and produced isolated defects in otherwise perfect castings. The concept is illustrated in Figure 12.41. Bubbles tend to flow past rather than building up against the surface of the filter where they are in danger of being forced through if the buildup becomes sufficient. Designs in which the runner has terminated at the filter do not work because bubbles build up at the end of the runner; the runner requires to extend beyond the filter, not encouraging bubbles to stop and accumulate adjacent to the filter. Figure 12.42 illustrates an un-wetted patch in a filter from a steel mould. The incomplete penetration of a filter is probably common because of the sensitivity of capillary repulsion to the precise size of pores, explaining how bubbles might easily pass through many filters. Although not yet properly researched, early signs show promise for the tangential filter in eliminating the ingress of bubbles into the casting. Naturally, the tangential filter performs its other duties including the reduction of velocity, and filtering out 'schoolboy howler' type pieces of rubbish which occasionally find their way into the casting via the gates. It is a simple concept, and may be found to work well enough. It would be expected to work particularly well with a vortex surge control flow-off similarly to the arrangement used by the Trident Gate described next. However, in the meantime, the problem of persuading bubbles to keep out of the casting was felt so demanding that a more sophisticated solution has enjoyed more development in the form of the Trident Gate.

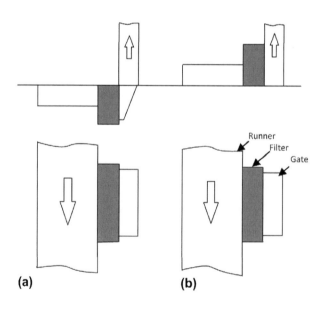

(a)   **(b)**

FIGURE 12.41

Tangential filter gate. Side and top views of gate moulded (a) below the joint and (b) above the joint.

FIGURE 12.42

Non-wetted patch in a filter from a steel casting, illustrating how bubbles may be sometimes encouraged to pass through filters.

Courtesy S Scholes.

12.4.21 TRIDENT GATE

The Trident Gate was developed in 2013 as a collaboration with Bob Puhakka as a result of lingering priming problems with the otherwise excellent Vortex Gate. The gate takes its name from one of its forms in which it possesses three upright spikes (two gates and one bubble trap) resembling Neptune's trident. Its more usual form has only one gate, but the name 'Bident' is unappealing, so 'Trident' tends to be used for both types.

The gate has two filters (Figure 12.43). The first filter is placed horizontally, forming the top of the runner, parallel with the flow. Clearly, bubbles now do not build up against its surface but in general simply flow by. The second is placed

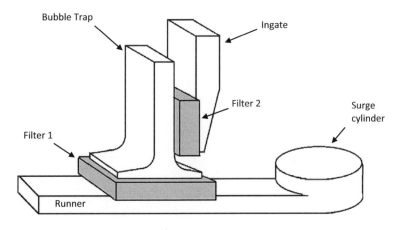

FIGURE 12.43

The so-called Trident Gate can achieve its three prongs on occasions if the Filter 2 is doubled up on the other side of the bubble trap to double the area of gate. The runner preferentially ends in a vortex surge cylinder to reduce jetting through the system.

vertically, so that any bubbles penetrating the first filter are delayed by the second, allowing them time to float up the face of the filter (even though 'dry' patches in filters will enable bubbles to penetrate more easily, filters will always present some difficulty to the transmission of bubbles) into the bubble trap volume provided previously. The melt progressing through this second filter is therefore has a good chance to be completely free from bubbles.

The Trident can be formed from a vertically split core, assembled together with its filters, and stored before use. The core is simply placed on the mould joint, above the runner which will be formed in the base part of the mould, and moulded in to the sand. The mould is then assembled after stripping off the tape which protected the lower face of the filter and the top gate exit. The Trident needs a well designed flow off to one or more vortex surge control cylinders assisting the gate to fill at a reasonable rate without spurting or turbulence.

When used in this way, the gate lives up to its promise of delivering excellent castings which appear completely free of visual or non-destructive testing faults. Furthermore, properties of castings appear to be excellent. Even so, at this time it is not clear whether the Trident is an over-complication; the tangential filter gate clearly deserves investigation.

12.5 SURGE CONTROL SYSTEMS

The flowing of metal past the gates and into some kind of dump has been widely used to eliminate the first cold metal, diverting it away, together with any initial contamination by sand or oxide. However, it is not the purpose of this section to discuss such devices used as dross and slag traps, but explore techniques for avoiding the violent surge of melt into the mould if the sprue, runner and gate link directly into the casting.

Usually, the first metal through the gate is a transient jet, the metal spurting through when the runner is suddenly filled. This is not a problem for castings of small height where the jet effect can be negligible, but becomes increasingly severe for those with a high head height. For a tall casting, the velocity at the end of the runner is high so the momentum of the melt, shocked to an instant stop, causes the metal to explode through the far gate and enter the cavity like a javelin. This damaging initial transient can occur despite the correct tapering of the runner because the taper is designed to distribute the flow evenly into the mould only after the achievement of steady state conditions.

If the metal were simply to bypass the ingate and continued to travel along the runner, the natural buildup of back pressure by friction along the length of the flow would slightly pressurise gates, and causing them to fill a little. The effect is seen in Figure 12.29. If this back-pressure could be increased in some way, this gentle filling might be controlled to advantage.

The design of a *flow-off* device, diverting the initial flow away from the casting, is capable of some sophistication, and promises to be a key ingredient, particularly for large, expensive one-off castings. It is a valuable technique for the reduction of the shock of the sudden filling of the runner and the impact of metal through the gates. This section introduces the concept of surge control systems.

A gate that channels the initial metal into a dump *below* the level of the runner is probably the least valuable form of this technique (Figure 12.44(a)). The downward-facing gate will continue to fill without generating significant back-pressure, the metal merely falling into the trap, until the instant the trap is completely filled. At that instant, the shock of filling is then likely to create, albeit at a short time later, the spurting action into the mould that it was designed to avoid. However, by this time, the gates and perhaps even the mould are likely to contain some liquid, so there is a chance that any deleterious jetting action will to some extent be suppressed. This flow-off dump has the benefit of working as a classical dross trap, of course.

A taper before the trap can be useful to prevent the back-wave from reversing debris out of the trap because there is only room for the inflow of metal (Figure 12.44(c)).

An improved form of the device is easily envisaged. A gate into a dump moulded *above* from the runner has a more positive action (Figure 12.44(d)). It provides a gradual reduction in flow rate along the runner because it generates a gradually increasing back-pressure as it fills, building up its head height. When placed at the end of the runner, the gate acts to reduce temporarily the speed into the (real) gates by providing additional gate area, and is valuable to reduce the unwanted final filling shock by some contribution to reducing speed. However, the narrow section, effectively a runner extension, seen in this image will fill so quickly that the desired gradual buildup of flow through the gate will be minimal. A further danger from something that resembles a simple runner extension is that such a device acts as a 'U' tube, so that

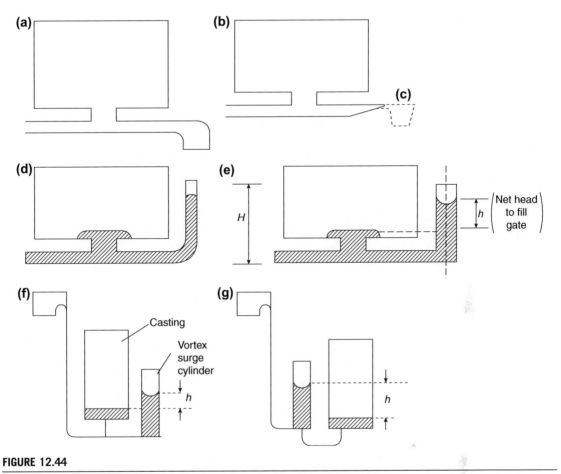

FIGURE 12.44

Bypass designs: (a) dross trap type not especially recommended; (b) without dross trap; (c) with non-return trap; (d) vertical runner extension for gravity deceleration (but care required to include damping of 'U-tube' oscillations); (e,f) surge control systems using a terminal vortex surge riser; (g) surge control system with in-line vortex and axial (central) outlet.

dynamic transient problems are to be expected, such as the melt overshoots to a high level in the tube, falls back too far, rebounds to a high level etc., oscillating wildly up and down. It is of little use to provide a steady back pressure.

Additional volume of the dump is an advantage to delay the buildup of pressure to fill the gate. The economically minded casting engineer might find that some castings could be made as 'free riders' in the mould at the end of such gates. The quality may not be high, especially because of the impregnation of the aggregate mould by the momentum of the metal. Even so, the part may be good enough for some purposes, and may help to boost earnings per mould.

A more sophisticated design incorporates all the desirable features of a fully developed surge control system. It consists of extending the runner into the base of an upright circular cylinder, which the runner enters tangentially (Figure 12.44(e) and (f)). The height and diameter of the cylinder are calculated to raise the back-pressure into the gates at a steady rate (avoiding the application of the full head from the filling system) for a sufficient time to ensure that the gates and the lower part of the casting is filled to a depth d sufficient to suppress a damaging jump (Figure 12.45(a)).

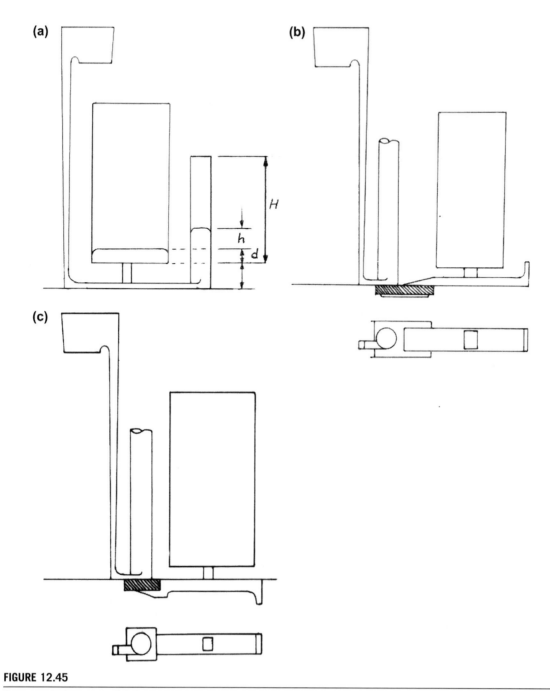

FIGURE 12.45

Surge control systems: (a) simplest system in which surge riser is placed 'second'; (b) primary position for surge riser, with runner in cope; (c) primary surge riser but runner in drag.

When the cylinder (a kind of vortex dump) is completely filled, only then does the full pressure of the sprue come in to operation to accelerate the filling of the mould cavity. The final filling of the dump may still occur with a 'bang' – the water hammer effect – announced by the shock wave of the impact as it flashes back along the runner at the speed of sound. However, this final filling shock will be considerably reduced from that produced by the metal impacting the end of a simple closed runner.

In practice, the levels in the riser and casting would be expected to rise together, the level in the riser leading that in the casting by the height difference h, which might be only a little more than 50 mm to overcome the friction through the gate and give the driving force for the rate of rise in the mould of 1 ms^{-1}.

Although the surge technique actually controls the speed of metal through the ingate, it is not called a speed control because its role is over soon after the start of the pour. The name 'surge control' emphasises its temporary nature.

An even more sophisticated variant that can be suggested is the incorporation of the surge control dump in line with the flow from the sprue (Figure 12.44(g)). The design of the dump as a vortex as before brings additional advantages: on arrival at the base of the sprue and turning into the runner at high speed, the speed is valuable to create a centrifugal action. This action is strongly organising to the melt, retaining the integrity of the front rather than the chaotic splashing that would have occurred in an impact into a rectangular space for instance. The rotary action also assists to centrifuge the entrained air, slag (and possibly some oxides) into the centre where they have opportunity to float if the cylinder is given sufficient height. The good quality melt is taken off from the centre of the base. The small fall down the exit of the surge cylinder is not especially harmful in this case because the rotational action assists the flow to progress with maximum friction down the walls of the exit channel. The system acts to take the first blast of high velocity metal, gradually increasing the height in the surge cylinder. In this way, a gradually increased head of metal is applied to the gates. Furthermore, of course, the metal reaching the gates should be free of air and other low density contaminants.

Options for rotary surge risers are seen in Figure 12.45; (a) is a secondary surge design, so-called because the surge riser comes second in line after the gate into the casting and (b) and (c) are primary surge systems, in which the exiting runner can emerge either in the cope or the drag as is convenient for the casting.

These surge control concepts promise to revolutionise the production of large steel castings, for which other good filling solutions are, in general, either not easy or not practical. The bypass and surge control devices represent valuable additions to the techniques of controlling not only the initial surge through the gates, but if their action is extended, as seems possible, they can make a valuable contribution to slowing velocities during the vulnerable early phase of filling.

The action of a bypass to double as a classical dross trap is described further in Section 12.7.

Finally, although the surge control concept can be valuable, it is probably an expensive option in terms of metal usage than the control of velocity using filters. However, the variety using no filter is almost certainly safer than the use of filters if the quality of the melt is in question. A blocked filter means a scrapped casting, whereas the secondary surge control will work despite any poor quality of the melt.

12.6 **VORTEX SYSTEMS**

The vortex has usually been regarded in foundries as a flow feature to be avoided at all costs. If the vortex truly swallowed air, and the air found its way into the casting, the vortex would certainly have to be avoided. However, in general, this seems to be not true.

The great value of the vortex is that it is a powerful organiser of the flow. Designers of water intakes for hydroelectric power stations are well aware of this benefit. Instead of the water being allowed to tumble haphazardly down the water intake from the reservoir, it is caused to spiral down the walls. At the base of the intake duct, the loss of rotational energy allows the duct to backfill to some extent (Figure 12.46). The central core of air terminates at the level surface of a comparatively tranquil pool, only gently circulating, near the base of the duct. Clearly, the spiralling central core of air terminates at the surface of the pool; it does not extend indefinitely through the system.

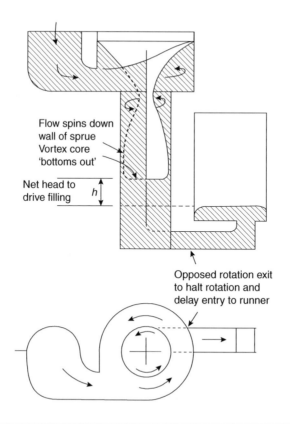

FIGURE 12.46

Vortex sprue.

After Isawa and Campbell (1994).

Several proposals to harness the benefits of vortices to running systems have originated in recent years from Birmingham, following the lead by Isawa (1994). They are potentially exciting departures from conventional approaches. Only initial results can be presented here. The systems merit much further investigation.

12.6.1 VORTEX SPRUE

The benefits of the vortex for the action of a sprue were first explored by Campbell and Isawa (1994), as illustrated in Figure 12.46. An aluminium alloy was poured off-axis, being diverted tangentially into a circular pouring basin. The melt spun around the outside of the basin, gradually filling up and so progressing towards the central sprue entrance. As it progressed inwards, gradually reducing its radius, its rotation speeded up, conserving its angular momentum like the spinning ice skater who closes in outstretched arms. Finally, the melt reached the lip of the sprue, where, at maximum spinning speed, it started its downward fall.

The rationale behind this thinking is that the initial fall during the filling of the sprue is controlled by the friction of a spiral descent down the wall of the sprue. Once the sprue starts to fill, the core of the vortex terminates near the base of the sprue; it does not in fact funnel air into the mould cavity. The hydrostatic pressure generated by this system, driving the flow into the runner and gates, arises only from the small depth of liquid at the base of the vortex, so that the net pressure head h driving the filling of the mould is small. Because the level of the pool at the base of the vortex and the level of metal in the mould cavity both rise together, the net head to drive the filling of the mould remains remarkably constant during the entire mould filling process. Thus the filling of the mould is necessarily gentle at all times.

Despite some early success with this system, it seems that the technique is probably not suitable for sprues of height greater than perhaps 200–300 mm, because the benefits of the spiral flow are lost progressively with increasing fall distance. More research may be needed to confirm the benefits and limitations of this design. For instance, the early work has been conducted on parallel cylindrical sprues because the taper has been thought to be not necessary as a result of the melt adopting its own 'taper' as it accelerates down the walls, becoming a thinner stream as it progresses. However, a taper may in any case be useful to favour the speeding up of the rate of spin, and so assisting to maintain the spin despite losses from friction against the walls. Also of course, the provision of a taper will assist the sprue to fill faster and increase yield. Much work remains to be done to define an optimum system.

12.6.2 **VORTEX WELL**

The provision of a cylindrical channel at the base of the sprue, entered tangentially by the melt, is a novel idea with considerable potential (Hsu et al., 2003). It gives a technique for dealing with the central issue of the high liquid velocity at the base of the sprue, and the problem of turning the right angle corner and successfully filling the runner. What is even better, it promises to solve all of these problems without significant surface turbulence. The vortex action also appears to take significant energy out of the metal, allowing the velocity along the runner to be modest, so that gates can be balanced and filled relatively easily.

A concept for a vortex well is seen in Figure 12.47(a). The vertical orientation is often convenient because the central outlet from the vortex can form the entrance to the runner, allowing the connection to many gates. Alternatively, the horizontal orientation (b) may be useful for delivery from a fast-delivering runner into a single vertical gate.

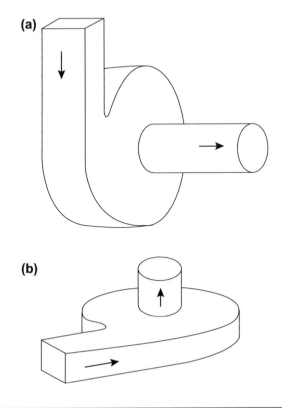

FIGURE 12.47

(a) Vortex well (with horizontal axis) and (b) vortex gate (with vertical axis).

Notice that the device works exactly opposite to the supposed action of a spinner designed to centrifuge buoyant inclusions from a melt. In the vortex well, the outlet to the rest of the filling system is the outlet that would normally be used to concentrate inclusions. Thus the device certainly does not operate to reduce the inclusion content. However, it should be highly effective in reducing the generation of inclusions by surface turbulence at the sprue base of poorly designed systems.

Once again, these are early days for this invention. Early trials on a steel casting of about 4 m in height have suggested that the vortex is extremely effective in absorbing the energy of the flow. In this respect, its action resembles that of a ceramic foam filter. To enable the device to be used in routine casting production, the energy absorbing behaviour would require to be quantified. There is no shortage of future tasks.

12.6.3 VORTEX RUNNER (THE OFFSET RUNNER)

The simple provision of an *offset sprue* causes the runner to fill tangentially, the melt spinning at high speed (Figures 12.17(e) and 12.48). The technique is especially suited to a vertically parted mould, where a rectangular cross-section sprue, moulded on only one side of the parting line, opens into a cylindrical runner moulded on the joint. The consequential highly organised filling of the runner is a definite improvement over many poor runner designs, as has been demonstrated by the Weibull statistics of strengths of castings produced by conventional in-line and offset (vortex) runners (Yang at al., 1998, 2000, 2003). The technique produces convincingly more reliable castings than conventional in-line sprues and runner systems.

Mbuya (2006) shows from mechanical test results analysed by Weibull technique that although the action of the vortex runner without a filter is better than the rectangular runner without a filter, it is less good than a rectangular runner with a filter. Such comparative studies would be helpful to decide whether other solutions such as the slot sprue/slot runner, or the vortex well, etc., might perform better. However, the vortex runner has the great benefit of

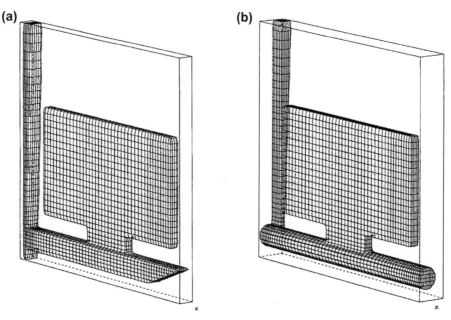

FIGURE 12.48

(a) A conventional runner, and (b) a vortex runner, useful on a vertically parted mould.

Courtesy X Yang and Flow 3-D.

simplicity (it is completely moulded, requiring no delay to the moulding cycle by the fitting of a filter) and low cost. It promises to be valuable for vertically parted moulds; it deals effectively with the problem of high velocities of flow in such moulds because of the great fall heights often encountered.

12.6.4 **VORTEX GATE**

When first used in about 2000, the vortex gate was hailed as a breakthrough for large Al alloy and steel castings. It was one of the few systems which was capable of taking a high velocity melt and reducing the velocity to gentle filling speeds without noticeable damage to the liquid metal.

Initially, the concept was merely that of the provision of a vertical cylinder in which the metal was introduced tangentially. The ratio of the area of the runner to the area of the top of the vortex cylinder gave the notional speed reduction achievable by the device. The simple cylinder was not particularly satisfactory: the melt arrived from the runner at high speed and immediately climbed high up the opposite wall of the cylinder during its first revolution of the cylinder. This was solved by extending the runner around the base of the cylinder. This 'runner extension' formed a 'step' which filled first around the base of the cylinder, constraining the melt to remain compactly circulating. When this lower flange on the cylinder was filled, the melt effectively overflowed the inside edge of the step, and expanded upwards in its circular motion around the cylinder. The action of the step caused the device to be called the 'vortex step gate' in its early days (Figure 12.49).

Later designs incorporated a filter at the top of the vortex cylinder simply held in place with nails or screws. This definitely aided the reduction of the unwanted residual rotation before entering the mould cavity, but also acted to hold back bubbles. The bubbles were the remnants of the first few seconds of priming flow. However, as a result of the centrifugal action, the bubbles were segregated into the centre of the cylinder from where they could not escape. If sufficient bubbles arrived, they would eventually grow to cover the base of the filter, and finally be forced through, creating a few isolated defects in the casting.

The vortex gate was capable of making castings of excellent quality, but was characterised by the appearance of a few isolated defects, mainly entrapped bubbles, among overall perfection. The fundamental limitation of this otherwise excellent fill system forced the re-think of the filling problem with a view to eliminating the possibility of bubbles accessing the mould cavity. The result was the tangential filter and the Trident Gate described earlier in Sections 12.4.20 and 12.4.21.

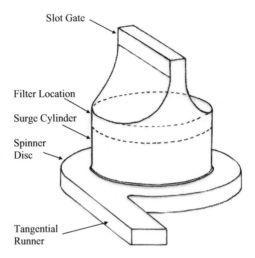

FIGURE 12.49

The vortex cylinder connected to a slot gate. It works better with a filter jammed or nailed into place.

12.7 INCLUSION CONTROL: FILTERS AND TRAPS

The term *'inclusion'* is a shorthand generally used for *'non-metallic inclusion'*. However, it is to be noted that such defects as tungsten droplets from a poor welding technique can appear in some recycled metals, especially Ti alloys; these, of course, constitute *'metallic inclusions'*. Furthermore, one of the most common defects in many castings is the bubble, entrained during pouring. This constitutes an *'air inclusion'* or *'gas inclusion'*.

That bubbles are trapped in the casting from the filling stage is remarkable in itself. Why did the bubble not simply rise to the surface, burst and disappear? This is a simple but important question. In most cases, the bubble will not have been retained by the growth of solid, because solid will, in general, not have time to form. The answer in practically all cases is that oxide films will also be present. In fact, the bubbles themselves are simply sections of the oxide films that have been entrained by the folding of the surface, but not folded perfectly face to face. The bubbles decorate the irregularly folded double films, as inflated islands trapped in the folds. Thus many bubbles, entangled in a jumble of bifilms, never succeed to reach the surface to escape. Even those that are sufficiently buoyant to power their way through the tangle may still not burst at the surface because it is likely to accumulate layers of oxide that bar its final escape.

This close association of bubbles and films (because they are both formed by the same turbulent entrainment process; they are both *entrainment defects*) is called by me *bubble damage*. We need to keep in mind that the bubble is the visible part of the total defect. The surrounding region of bifilms to which it is connected act as cracks, and can be much more extensive and often invisible. However, the presence of such films is the reason that cracks will often appear to start from porosity, despite the porosity have a nicely rounded shape that would not in itself appear to be a significant stress raiser.

Whereas inclusions are generally assumed to be particles having a compact shape, it is essential to keep in mind that the most damaging inclusions are the films (actually always double, unbonded films, remember, so that they act as cracks), and are common in many of our common casting alloys. Curiously, the majority of workers in this field have largely overlooked this simple fact. It is clear that techniques to remove particles will often not be effective for films and vice versa. The various methods to clean metals before casting have been reviewed in Section 1 as a fulfilment of Rule 1. The various methods to clean metals travelling through the filling systems of castings will be reviewed here.

12.7.1 DROSS TRAP (OR SLAG TRAP)

The dross trap is used in light-alloy and copper-based alloy casting. In ferrous castings, it is called a slag trap. For our purposes, we shall consider the devices as being one and the same.

Traditionally, it has made good sense to include a dross trap in the running system. In principle, a trap sited at the end of the runner should take the first metal through the runner and keep it away from the gates. This first metal is both cold, having given up much of its heat to the running system en route, and will have suffered damage by oxide or other films during those first moments before the sprue is properly filled.

In the past, designs have been along the lines of Figure 12.44(a). This type of trap was sized with a view to accommodating the total volume of metal through the system until the down-runner and horizontal runner were substantially filled. This was a praiseworthy aim. In practice, however, it was a regular joke among foundrymen that the best quality metal was concentrated in the dross trap and all the dross was in the casting! What had happened to lend more than an element of truth to this regrettable piece of folk law?

It seems that this rather chunky form of trap sets up a circulating eddy during filling. Dross arriving in the trap is therefore efficiently floated out again, only to be swept back along the runner on the back-wave, and finally through the gates and into the casting a few moments later! Ashton and Buhr (1974) have carried out work to show that runner extensions act poorly as traps for dirt. They observed that when the first metal reached the end of the runner extension it rose, and created a reflected wave which then travelled back along the top surface of the metal, carrying the slag or dirt back towards the ingates. Such observations have been repeated on iron and steel casting by Davies and Magny (1977) and on many different alloys in the author's laboratory using real-time radiography of moulds during casting. The effect has been confirmed in computer simulations. It seems, therefore, to be real and universal in castings of all types. We have to conclude that this design of dross trap cannot be recommended!

Figure 12.44(b) shows a simple wedge trap. It was thought that metal flowing into the narrowing section was trapped, with no rebound wave from the end wall, and no circulating eddy can form. However, video radiographic studies have shown that such traps can reflect a backward wave if the runner is sufficiently deep. Also, of course, the volume of melt that they can retain is very limited.

A useful design of dross trap appears to be a volume at the end of the runner that is provided with a narrow entrance (the extension shown in broken outline in Figure 12.44(b) and (c)) to suppress any outflow. It is a kind of wedge trap fitted with a more capacious end. In the case of persistent dross and slag problems, the trap can be extended, running around corners and into spare nooks and crannies of the mould. If the entrance section is substantially less than the height of a sessile drop, it will be filled by the entering liquid, but would be too narrow to allow a reflected wave to exit. It should therefore retain whatever material enters. In addition, depending on the narrowness of the tapered wedge entrance, to some extent the device should be capable of filling and pressurising the runner in a progressive manner akin to the action of a gate. This is a useful technique to reduce the initial transient momentum problems that cause gates to fill too quickly during the first few seconds. This potentially useful benefit has yet to be researched more thoroughly so as to provide useful guidelines for mould design.

The device can be envisaged to be useful in combination with other forms of bypass designs such as that shown in Figure 12.37.

12.7.2 SLAG POCKETS

For iron and steel castings, the term 'slag pocket' is widely used for a raised portion of the runner that is intended to collect slag. The large size of slag particles and their large density difference with the melt encourages such separation. However, such techniques are not the panacea that the casting engineer might wish for.

For instance, the use of traps of wedge-shaped design, Figure 12.50(a), is expected to be almost completely ineffective because the circulation pattern of flow would take out any material that happened to enter. On the other hand, a rectangular cavity has a secondary flow into which buoyant material can transfer if it has sufficient time and so remain trapped in the upper circulating eddy. This consideration again emphasises the need for relatively slow flow for its effectiveness. Also, of course, none of these traps can become effective until the runner and the trap become filled with metal. Thus many filling systems will have passed much if not all such unwanted material before the separation mechanisms have chance to come into operation. A further consideration that causes the author to hesitate to recommend such traps is that they locally remove the constraint on the flow of the metal, allowing surface turbulence. At this time, it is not clear whether traps cause more problems than they solve: all the slag in the trap might have been formed as a result of turbulence in the trap. The concern is that if it was a prolific slag generator it might have released slag into the casting.

Davis and Magny (1977) observed the filling or iron and steel castings by video radiography. They confirmed that most slag retention devices either do not work at all or work with only partial effectiveness. These authors made castings with different amounts of slag and tested the ability of slag pockets sited above runners to retain the slag. They found that rectangular pockets were tolerably effective only if the velocity of flow through the runner was below 0.4 m/s (interesting

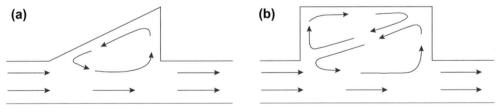

FIGURE 12.50

Various designs of slag pockets: (a) relatively ineffective self-emptying wedge; (b) rectangular trap stores buoyant phases in upper circulating flow. However, neither can be recommended because constraint over the flow is lost at these locations and surface turbulence encouraged.

that this is precisely the speed to avoid surface turbulence). All runners contain metal at much higher speeds: for metal which has fallen only 100 mm, the metal is already travelling four times too fast to permit slag pockets to work. It is necessary to conclude that slag pockets have been counter-productive; they should be avoided at all times.

12.7.3 SWIRL TRAP

The centripetal trap is an accurate name for this device, but rather a mouthful. It is also known as a whirl gate, or swirl gate, which is shorter, but inaccurate because the device is not really a gate at all. Choosing to combine the best of both names, we can call it a swirl trap. This is conveniently short, and accurately indicates its main purpose for trapping rubbish.

The idea behind the device is the use of the difference in density between the melt and the various unwanted materials which it may carry, either floating on its surface or in suspension in its interior. The spinning of the liquid creates a centrifugal action, throwing the heavy melt towards the outside where it escapes through the exit to continue its journey into the casting. Conversely, the lighter materials are thrown towards the centre, where they coagulate and float. The centripetal acceleration a_c is given by:

$$a_c = V^2/r \qquad (12.5)$$

where V is the local velocity of the melt and r is the radius at that point. For a swirl trap of 50 mm radius and sprue heights of 0.1 m and 1 m, corresponding to velocities of 1.5 and 4.5 m/s, respectively, we find that accelerations of 40 and 400 m/s^2, respectively, are experienced by the melt. Given that the gravitational acceleration g is 9.81 m/s^2, which we shall approximate to 10 m/s^2, these values illustrate that the separating forces within a swirl trap can be between 4 and 40 times that because of gravity. These are, of course, the so-called g forces experienced in centrifuges.

So much for the theory. What about the reality?

Foundrymen have used swirl traps extensively. This popularity is not easy to understand because, unfortunately, it cannot be the result of their effectiveness. In fact, the traps have worked so badly that Ruddle (1956) has recommended they should be abandoned on the grounds that their poor performance does not justify the additional complexity. One has to conclude that their extraordinarily wide use is a reflection of the fascination we all have with whirlpools, and an unshakeable belief, despite all evidence to the contrary, that the device should work.

Regrettably, the swirl trap is expected to be completely useless for film-forming alloys where inclusions in the form of films will be too sluggish to separate. Because some of the worst inclusions are films, the swirl trap is usually worse than useless, creating more films than it can remove. Worse still, in the case of alloys of aluminium and magnesium, their oxides are denser than the metal, and so, if centrifuged at all, will be centrifuged outwards, into the casting! Swirl traps are therefore of no use at all for light alloys (however, notice that the vortex sprue base, although not specifically designed to control inclusions, might have some residual useful effect for light alloys because the outlet is central). Finally, swirl traps seem to be difficult to design to ensure effective action. In the experience of the author, most do much more harm than good.

The inevitable conclusion is that swirl traps should be avoided.

The remainder of this section on swirl traps is for those who refuse to give up, or refuse to believe. It also serves as a mini-illustration of the real complexity of apparently simple foundry solutions. Such illustrations serve to keep us humble.

It is worthwhile to examine why the traditional swirl trap performs so disappointingly. On examination of the literature, the textbooks, and designs in actual use in foundries, three main faults stand out immediately:

1. The inlet and exit ducts from the swirl traps are almost always opposed, as shown in Figure 12.51(a). The rotation of the metal as a result of the tangential entry has, of course, to be brought to a stop and reversed in direction to make its exit from the trap. The disorganised flow never develops its intended rotation and cannot help to separate inclusions with any effectiveness.

   Where the inlet and exit ducts are arranged in the correct tangential sense (Figure 12.51(b)), then Trojan et al. (1966) have found that efficiency is improved in their model results using wood chips in water. Even so, efficiencies in trapping the chips varied between the wide limits of 50 and 100%.

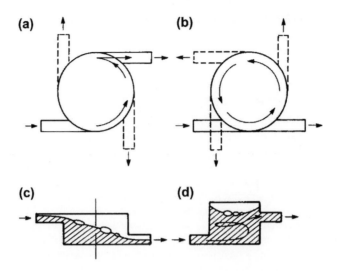

FIGURE 12.51

Swirl traps showing (a) incorrect opposed inlet and exit ducts; (b) correct tangential arrangements; (c) incorrect low exit; (d) correct high exit. Once again, I feel reluctant to recommend even a good design. They are all better avoided.

2. The inlet is nearly always arranged to be higher than the exit. This elementary fault gives two problems. First, any floating slag or dross on the first metal to arrive is immediately carried out of the trap before the trap is properly filled (Figure 12.51(c)). Second, as was realised many years previously (Johnson and Baker, 1948), the premature escape of metal hampers the setting up of a properly developed spinning action. Thus the trap is slow to develop its effect, perhaps never achieving its full speed in the short time available during the pour. This unsatisfactory situation is also seen in the work of Jirsa (1982), who describes a swirl trap for steel casting made from pre-formed refractory sections. In this design, the exit was again lower than the entrance, and filling of the trap was encouraged merely by making the exit smaller than the entrance. Much metal and slag almost certainly escaped before the trap could be filled and become fully operational. Because 90% efficiency was claimed for this design it seems probable that all of the remaining 10% which evaded the trap did so before the trap was filled (in other words, the trap was working at zero efficiency during this early stage). Jeancolas et al. (1971) report an 80% efficiency for their downhill swirl trap for bronze and steel casting, but admit that the trap does not work at all when only partly full.
3. In many designs of swirl trap, there is insufficient attention paid to providing accommodation for the trapped material. For instance, where the swirl trap has a closed top the separated material will collect against the centre of the ceiling of the trap. However, work with transparent models illustrates clearly how perturbations to the flow cause the inclusions, especially if small, to ebb and flow out of these areas back into the main flow into the casting (Jeancolas, 1971; Trojan, 1966). Also, of course, traps of such limited volume are in danger of becoming completely overwhelmed, becoming so full of slag or dross that the trap effectively overflows its collection of rubbish into the casting.

Where the trap has an open top, the parabolic form of the liquid surface assists the concentration of the floating material in the central 'well' as shown in Figure 12.51(d). The extra height for the separated materials to rise into is useful to keep the unwanted material well away from the exit, despite variations from time to time in flow rate. Some workers have opened out the top of the trap, extending it to the top of the cope, level with the pouring bush. This certainly provides ample opportunity for slag to float well clear, with no danger of the trap becoming overloaded with slag. However, the author does not recommend an open system of this kind, because of the instability which open-channel systems sometimes exhibit, causing surging and slopping between the various components comprising the 'U'-tube effect between the sprue, swirl trap and mould cavity.

It is clear that the optimum design for the swirl trap must include the features:

1. the entrance at the base of the trap,
2. the exit to be sited at a substantially higher level,
3. both entrance and exit to have similar tangential direction, and
4. an adequate height above the central axis to provide for the accumulation of separated debris.

In most situations, the only difference in height that can be conveniently provided in the mould will mean that the inlet will be moulded in the drag, and the exit in the cope. This is the most marginal difference in level and will be expected to be completely inadequate. At the high speeds at which the metal will enter the trap, the metal will surge over this small ledge with ease, taking inclusions directly into the casting, particularly if the inlet and outlet are in line as shown for one of the options in Figure 12.51(b). This simplest form of cope/drag parting line swirl trap cannot be expected to work.

The trap may be expected to work somewhat more effectively as the angle of the outlet progresses from 90 to 180°. (The 270° option would be more effective still, except that some reflection will show it to be unmouldable on this single joint line; the exit will overlay the entrance! Clearly, for the 270° option to be possible, the entrance and exits have to be moulded at different levels, necessitating a second joint line provided by a core or additional mould part.)

When using pre-formed refractory sections, or pre-formed baked sand cores, as is common for larger steel castings, the exit can with advantage be placed considerably higher than the entrance (Figure 12.51(d)).

These simple rules are designed to assist the trap to spin the metal up to full speed before the exit is reached, and before any floating or emulsified less-dense material has had a chance to escape.

For the separation of particulate slag inclusions from some irons and steels, *Castings* 1991 showed that a trap 100 mm diameter in the running system of moulds 0.1–1 m high would be expected to eliminate inclusions of 0.2 to 0.1 mm, respectively. The conclusion was that, when correctly designed, the swirl trap could be a useful device to divert unwanted buoyant particles away from ferrous castings.

We have to remember, however, that it is not expected to work for film type inclusions. Compared to particles, films would be expected to take between 10 and 100 times longer to separate under an equivalent field force. Thus most of the important inclusions in a large number of casting alloys will not be effectively trapped. Thus the alloys that most need the technique are least helped.

This damning conclusion applies to other field forces such as electromagnetic techniques that have recently been claimed to remove inclusions from melts. It is true that forces can be applied to non-conducting particles suspended in the liquid. However, whereas compact particles move relatively quickly, and can be separated in the short time available while the melt travels through the field. Films experience the same force, but move too slowly because of their high drag, and so are not removed.

In summary, we can conclude that apart from certain designs of bypass trap, other varieties are traps are not recommended. In general they almost certainly create more inclusions than they remove. Even the bypass variety of trap is probably of little use with a nicely designed, naturally pressurised filling system. Thus, finally, we reach the conclusion that few traps work, and none are needed. This is another traditional foundry issue that should be laid to rest.

Even so, we should make an exception for the vortex surge device required to get the Trident Gate to work well. That gate and surge bypass combination is worth making an exception for!

12.8 FILTERS

Over the years, there have been much work carried out to quantify the benefits of the use of filters. Nearly all of these have shown measurable, and sometimes important, gains in freedom from defects and improvements in mechanical properties. These studies are too numerous to list here, but include metals of all types, including Al alloys, irons and steels. The relatively few negative results can be traced to the use of unfavourable siting or geometry of the filter print. For positive and reliable results, these aspects of the use of filters cannot be overlooked. Special attention is devoted to them in what follows.

Filters take many forms: as simple strainers, woven glass or ceramic cloths and ceramic blocks of various types. Naturally, their effectiveness varies from application to application, as is discussed here.

12.8.1 STRAINERS

A sand or ceramic core may be moulded to provide a coarse array of holes, of a size and distribution resembling a domestic colander. A typical strainer core might be cylinder 30–50 mm diameter, 10–20 mm long, containing 10 or more holes, of diameter approximately 3–5 mm (Figure 12.52(a)). These devices have been traditionally used in an effort to prevent slag entering iron castings. The domestic colander used in the kitchen is usually used to strain aggregates such as peas. These represent solid spherical particles of the order of 5 mm diameter. Thus, when applied to most metal castings, the rather open design of strainers means that they can hardly be expected to perform any significant role as filters.

In fact, Webster (1966) has concluded that the strainer works by reducing the rate of flow of metal, assisting the upstream parts of the filling system to prime, and thereby allowing the slag to float. Slag can be given time to separate, and can be held against the top surface of the runner, or in special reservoirs placed above the strainers to collect the retained slag. Webster goes on to conclude that if the strainer only acts to reduce the rate of flow, then this can be carried out more simply and cheaply by the proper design of the running system.

This may not be the whole story. The strainer may be additionally useful to laminise the flow (i.e. cause the bulk flow to become more streamlined).

However, whatever benefits the strainer may have, its action to create jets downstream of the strainer, especially during priming, is definitely not helpful. The placing of a strainer in a geometry that will quickly fill the region at the back of the strainer would be a great advantage. Geometries to suppress jetting at the back of filters appear to be a key aspect of the siting of an effective filter. This concept is discussed later. The extruded or pressed ceramic filters with their arrays of parallel pores are, of course, equivalent to strainers with a finer pore size.

12.8.2 WOVEN CLOTH OR MESH

For light alloys, steel wire mesh or glass cloth (Figure 12.52(b)) is used to prevent the oxides from entering the casting. Cloth filter material has the great advantage of low cost. It can be bought in rolls up to 1 m wide and 100 m long, and is easily cut to size.

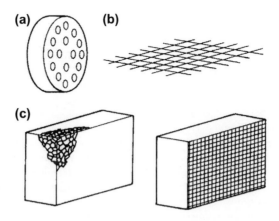

FIGURE 12.52

Various filters showing (a) a strainer core (hardly a filter at all); (b) woven cloth or mesh, forming a two-dimensional filter; (c) ceramic foam and extruded or pressed blocks, constituting three-dimensional filters.

The surprising effectiveness of these rather open meshes is the result of the most important inclusions being in the form of films, which appear to be intercepted by and wrap around the strands of the mesh. Openings in the mesh or weave are typically 1–2 mm which give good results. Many investigators have remarked on the curious fact that the cloth appears to be effective in filtering out inclusions down to a tenth of this size range. This mystery is immediately explained by the realisation that the filtered objects are films not particles, and on polished sections tangles of films can appear as isolated compact forms. Significantly, that such pore sizes are so effective is an independent confirmation of the minimum 1–2 mm size of the majority of films that cause problems in castings, particularly in light alloys.

The use of steel wire mesh is also useful to retain films. It has a great advantage over cloth because of its rigidity. A steel mesh can be placed in a running system with the full confidence that it will not deform and allow the melt to bypass the filter. The steel does not have time to go into solution during the filling of aluminium alloy castings, so that the material of the casting is in no danger of contamination. However, of course, the steel presents a problem of iron contamination during the recycling of the running system. Even the glass cloth can sometimes cause problems during the breakup of the mould, when fragments of glass fibre can be freed to find their way into the atmosphere of the foundry, and cause breathing problems for operators. Both materials therefore need care in use.

Some glass cloth filters are partially rigidised with a ceramic binder and some by impregnation with phenolic resin. (The outgassing of the resin can cause the evolution of large, centimetre-sized bubbles when contacted by the liquid metal. Provided the bubbles do not find their way into the casting the overall effect of the filters is definitely beneficial in aluminium alloys.) Both types soften at high temperature, permitting the cloth to stretch and deform.

A woven cloth based on a high silica fibre has been developed to avoid softening at high temperatures and might therefore be very suitable for use with light alloys. In fact at the present time its high temperature performance usually confines its use to copper-based alloys and cast irons. There are few data to report on the use of this material. However, it is expected that its use will be similar to that of the other meshes, so that the principles discussed here should still apply.

Despite the attraction of low cost, it has to be admitted that, in general, the glass cloth filters are not easy to use successfully.

For instance, as the cloth softens and stretches there is a strong possibility that the cloth will allow the metal to bypass the filter. It is essential to take this problem into account when deciding on the printing of the filter. Clearly, it is best if it can be firmly trapped on all of its edges. If it can be held on only three of its four edges, the vulnerability of the unsupported edge needs careful consideration. For instance, even though a cross-joint filter may be properly held on the edges that are available, the filter is sometimes defeated by the leading edge of the cloth bending out of straight, bowing like a sail, and thus allowing the liquid to jet past. All the filters shown in Figure 12.53 are at risk from this problem. It may be better to abandon cloth filtration if there is any danger of the melt jetting around a collapsing or ballooning filter.

When sited at the point where the flow crosses a joint, as in Figure 12.53(a) and (b), a greensand mould will probably hold the cloth successfully, the sand impresses itself into the weave, provided sufficient area of the cloth is trapped in the mould joint. In the case of a hard sand mould or metal die, the cloth requires a shallow print which must be deep enough to allow it room if the mould joint is not to be held apart. Also, of course, the print must not be too deep; otherwise, the cloth will not be held tight, and may be pulled out of position by the force of the liquid metal. Some slight crushing of a hard sand mould is desirable to hold the cloth as firmly as possible.

A rigidised cloth filter can be inserted across the flow by simply fitting it into the pattern in a pre-moulded slot across the runner and so moulding it integral with the mould (Figure 12.53(d)). However, this is only successful for relatively small castings. Where the runner area becomes large and the time and temperature becomes too high, the filter softens and bows in the force of the flow. Even if it is not entirely pulled out of position, it may be deformed to sag like a fence in a gale, so that metal is able to flow over the top. This is the reason for the design shown in Figure 12.53(e). The edge of the filter crosses the joint line, either to sit in a recess accurately provided on the other half of the mould, or if the upstand is limited to a millimetre or so, to be simply crushed against the other mould half. (The creation of some loose grains of sand is of little consequence in the running system, as has been shown by Davies and Magny (1977); loose material in the runner is never picked up by the metal and carried into the mould. The author can confirm this observation as

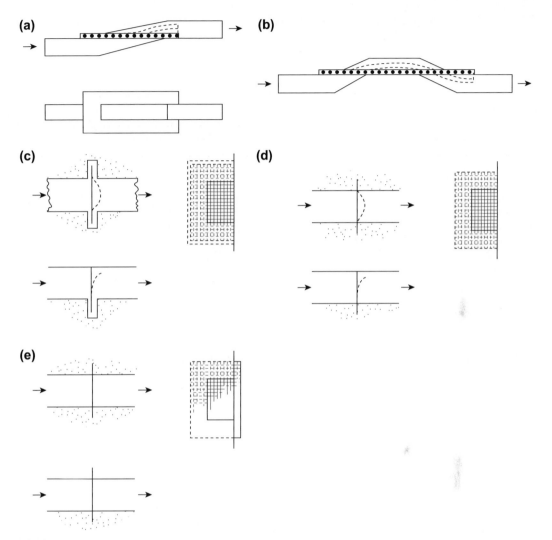

FIGURE 12.53

Siting of cloth filters (a) in the mould joint; (b) in a double crossing of the joint; (c) in a slot moulded across the runner; (d) in a slot cut in the runner pattern; (e) with an additional upstand across the joint plane to assist sealing.

particularly true for systems that are not too turbulent. The laminising action of the filter itself is probably additionally helpful.)

All the cloth filters used as shown in Figure 12.53 are defeatable because they are held only on three sides. The fourth side is the point of weakness. Failure of the filter by the liquid overshooting this unsupported edge can result in the creation of more oxide dross than the filter was intended to prevent! Increasing the trapped area of filter in the mould joint can significantly reduce these risks. It also helps to arrange for the selvage (the reinforced edge of the material) of the cloth to be uppermost to give the unsupported edge most strength; the ragged cut edge has little strength, letting the cloth bend easily, and allowing some, or perhaps all, of the flow to avoid the filter.

Geometries that combine bubble traps (or slag or any other low density phase) are shown in Figure 12.54 for in-line arrangements, and for those common occasions when the runner is required to be divided to go in opposite directions. With shallow runners of depth of a few millimetres, there is little practical difference in whether the metal goes up or down through the filter. Thus several permutations of these geometries can be envisaged. Much depends on the links to the gates and how the gates are to be placed on the casting. In general, however, I usually aim to have the exit from the filter into the runner below the joint; this fall distance can be less than the critical fall distance in the majority of slot shaped runners.

Cloth filters are entirely satisfactory where they can be held around all four sides. This is the case at the point where gates are taken vertically upwards from the top of the runner. This is a relatively unusual situation, where,

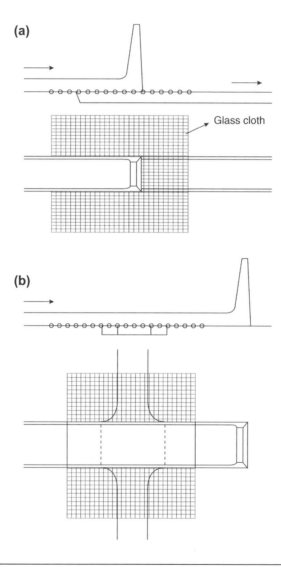

FIGURE 12.54

Uses of glass cloth filtration with a bubble trap (a) for an in line runner and (b) for transverse runners.

instead of a two-parted mould, a third mould part forms a base to the mould and allows the runner system to be located under the casting. Alternatively, a special core can be used to create an extra joint beneath the general level of the casting.

Another technique for holding the filter on all sides is the use of a 'window frame' of strong paper or cardboard that is bonded or stapled to the cloth. The frame is quickly dropped into its slot print in the mould, and gives a low cost rigid surround that survives sufficiently long to be effective. The outgassing of the paper may not be a problem because the gases have chance to escape via the mould. Even so, it would be reassuring to see published an independent X-ray video investigation to check this.

12.8.3 CERAMIC BLOCK FILTERS

Ceramic block filters of various types introduced in about 1980 have become popular and have demonstrated impressive effectiveness in many applications in running systems.

Unfortunately, much that has been written about the mechanisms by which they clean up the cast material appears to be irrelevant. This is because most speculation about the filtration mechanisms has considered only particulate inclusions. As has become quite clear over recent years, the most important and widespread inclusions are actually films. Thus the filtration mechanism at work is clearly quite different, and, in fact, easily understood.

In aluminium alloys, the action of a ceramic foam filter to stop films has, in general, not been recognised. This is probably because the films are so thin; a new film may be only 20 nm thick, making the doubled-over entrained bifilm still only about 40 nm. Thus the oxide bifilm, a thin ceramic, of thickness measured in nanometres wrapped around another piece of ceramic measured in micrometres (i.e. 1000 times thicker) cannot easily be detected under the microscope. This partly explains the curious experience of finding that a filter has cleaned up a casting, but on sectioning the filter to examine it under the microscope, not a single inclusion can be found.

The contributing effect, of course, is that the filter acts to improve the filling behaviour of the casting, so reducing the number of inclusions that are created in the mould during the filling process. This behaviour was confirmed by Din et al. (2003), who found only about 10% of the action of the filter was the result of filtration, but 90% was the result of improved flow.

A further widespread foundry experience is worth a comment. On occasions the quantity of inclusions has been so great that the filter has become blocked and the mould has not filled completely. Such experiences have caused some users to avoid the use of filters. However, in the experience of the author, such unfortunate events have resulted from the use of poor front ends of filling systems (poor basin and sprue designs) that create huge quantities of oxide films in the pouring process. The filter has therefore been overloaded, leading either to its apparently impressive performance, or its failure by blockage. The general advice given to users by filter manufacturers that filters will only pass limited quantities of metal is seen to be influenced by similar experience. The author has not found any limit to the volume of metal that can be put through a filter, provided the metal is sufficiently clean and the front end of the filling system is designed to perform well.

Thus, the secret of producing good castings using a filter is to team a good front end of the filling system together with the filter. (If the remainder of the filling system design is good, this will, of course, help additionally.) Little oxide is then entrained, so that the filter appears to do little filtration. However, it is then fully enabled to serve a valuable role as a flow rate control device. The beneficial action of the filter in this case is probably the result of several factors:

1. The reduction in velocity of the flow (provided an appropriately sized cross-section area channel is provided downstream of course). This is probably the single most important action of the filter. However, there are other important actions listed later.
2. Reduces the time for the back-filling of the sprue, thereby reducing entrainment defects from this source.
3. Smoothes fluctuations in flow.
4. Laminises flow, and thus aids fluidity a little.

5. The freezing of part of the melt inside the filter by the chilling action of the filter was predicted for Al alloys in ceramic foam filters by computer simulation and found by experiment by Gebelin and Jolly in 2002 and 2003. This phenomenon may be an advantage, because it may act to restrict flow, and so to reduce delivery from the filter in its early moments. The metal eventually re-melts as more hot metal continues to pass through the filter, allowing the flow to speed up to its full rate later during filling. (Interestingly, this advantage did not apply to pre-heated ceramic moulds where the pre-heat was sufficient to prevent any freezing in the filter.) This temporary freezing behaviour requires to be properly researched.

There are different types of ceramic block filter.

1. Foam filters made from open-cell plastic foams by impregnation with a ceramic slurry, squeezing out the excess slurry, and firing to burn out the plastic and develop strength in the ceramic. The foam structure consists of a skeleton of ceramic filaments and struts defining a network of interconnecting passageways.
2. Extruded forms that have long, straight, parallel holes. They are sometimes referred to as cellular filters; and
3. Pressed forms, again with long, straight but slightly tapered holes. The filters are made individually from a blank of mouldable clay by a simple pressing operation in a two-part steel die. As die-formed products they are the most accurately reproducible sizes of all the block filter types.
4. Sintered forms, sometimes known as bonded filters, in which crushed and graded ceramic particles are mixed with a ceramic binder and fired.

In all types the average pore size can be controlled in the range of 2–0.5 mm, although the sintered variety can achieve at least 2–0.05 mm. Insufficient research (other than that funded by the filter manufacturers!) has been carried out so far to be sure whether there are any significant differences in the performance between them.

An early result of Khan et al. (1987) found that the fatigue strength of ductile iron was improved by extruded cellular filters, but that the foam filters were unpredictable, with results varying from the best to the worst. Their mode of use of the filters was less than optimum, being blasted by metal in the entrance to the runner, and with no back protection for the melt. (We shall deal with these aspects later.) The result underlines the probable unrealised potential of both types, and reminds us that both would almost certainly benefit from the use of recent developments. In general, we have to conclude that the published comparisons made so far are, unfortunately, often not reliable.

For aluminium alloys, the results are less controversial because the filters are highly effective in removing films which have, of course, a powerful effect on mechanical properties. Mollard and Davidson (1978) are typical in their findings that the strength of Al-7Si-Mg alloy is improved by 50%, and elongation to failure is doubled. This kind of result is now common experience in the industry.

For some irons and steels, where a high proportion of the inclusions will be liquid, most filter materials are expected to be wetted by the inclusion so that collection efficiency will be high for those inclusions. Ali et al. (1985) found that for alumina inclusions in steel traversing an alumina filter, once an inclusion made contact with the filter it became an integral part of the filter. It effectively sintered into place; this action probably reflects the fact that neither the filter nor the inclusions were entirely solid at the temperature of operation because of the presence of impurities; they behaved as though they were 'sticky'. This behaviour is likely to characterise many types of inclusion at the temperature of liquid steel.

In contrast with this, Wieser and Dutta (1986) find that, whereas alumina inclusions in steel are retained by an alumina filter, even up to the point at which it will clog, deoxidation of steel with Mn and Si produces silica-containing products that are not retained by an extruded zirconia spinel filter. These authors also tested various locations of the filter, discovering that placing it in the pouring basin was of no use, because it was attacked by the slag and dissolved!

Block filters are more expensive than cloth filters. However, they are easier to use and more reliable. They retain sufficient rigidity to minimise any danger of distortion that might result in the bypassing of the filter. It is, however, important to secure a supply of filters that are manufactured within a close size tolerance, so that they will fit immediately into a print in the sand mould or into a location in a die, with minimal danger of leakage around the sides of the filter. Although all filter types have improved in dimensional control over recent years, the foam filter seems most difficult to control, the extruded is intermediate, whereas the pressed filter exhibits good accuracy and reproducibility as a result of it being made in a steel die; residual variation seems to be result of less good control of shrinkage on firing.

The combination of a bubble trap together with a filter is often useful. If a filter is placed alone in a runner, it will usually hold back the first bubbles that arrive. These will coalesce and, if sufficient bubbles are in the flow, the front of the filter becomes nearly covered with a massive bubble. At that point, the bubble is subjected to such a high pressure that it is forced through the filter and rattles up through the casting, creating damage. This is the reason for the placing of a bubble trap above the inlet side of a filter, so that bubbles can float clear of the face of the filter. In the bubble trap, they are seen in X-ray video studies of Al alloys to shrink and disappear into the sand mould. Examples are shown in Figures 12.54, 12.55 and 12.57.

12.8.4 LEAKAGE CONTROL

It is essential to control severe leakage past the filter. A 'rattling' fit with clearance of a millimetre or so is not a problem because the squeezing of the melt through such as small gap will ensure that it is both filtered and experiences sufficient drag to slow its flow. The edges of the filter print should continue to provide a final right angle turn for any relatively minor leakage around the filter.

The danger arises from a serious leak in which there is a straight line of sight opening which avoids the filter. The danger in this case is not from the bypass of unfiltered metal but the fact that the liquid metal is likely to jet through the gap, causing turbulence downstream of the filter.

For instance, Figure 12.55(e) shows a filter placement that at first sight appears simple and elegant. Unfortunately, it is a disaster. It crosses the mould joint, with the result that it has to be placed in the drag, sitting proud of the drag surface. The cope then has to be lowered so that the closing over the filter is carried out blind. It is not known whether the filter and or mould is crushed, or whether the filter is a loose fit, encouraging jetting of metal around the top of its print. The danger of attempting to site a filter across the mould joint is avoided in all other filter examples shown here. Only Figure 12.55(b) shows a filter that slightly protrudes above the joint. This is designed to create a gentle crush against the cope, ensuring a seal. Any loose sand generated by such crushing techniques has never, to the author's knowledge, posed any threat of sand inclusions in the casting (sand inclusions are never picked up from a runner as observed by Davis and Magny in 1977, but are generated by poor filling systems in which extreme turbulence is accompanied by copious entrainment of air. Sand inclusions are always cured by eliminating the turbulence and air entrainment in the filling system.).

All filter techniques described in this book use filters that are sat fully in the drag, so that they can be seen to fit well, giving some assurance that they will work as intended, or can be seen to fit badly, and corrective action taken in good time.

There are various techniques to reduce a leak bypassing a filter.

1. A seating of a compressible gasket of ceramic paper. This approach is useful when introducing filters into metal dies in which the filter is held by the closing of the two halves of the die. The variations in size of the filters, and the variability of the size and fit of the die parts with time and temperature, which would otherwise cause occasional cracking or crushing of the rather brittle filter, are accommodated safely by the gasket.

2. Moulding the filter directly into an aggregate (sand) mould. This is achieved simply by placing the filter on the pattern, and filling the mould box or core box with aggregate in the normal way. The filter is then perfectly held. In greensand systems or chemically bonded sands the mould material seems not to penetrate a ceramic foam filter more than the first pore depth. This is a smaller loss than would be suffered when using a normal geometrical print. However, the technique often requires other measures such as the moulding of the filter into a separate core, or the provision of a loose piece in the pattern to form the channel on the underside of the filter.

3. Use a filter printing geometry that is insensitive to leakage. Figure 12.57(b) and (c) are good examples where a leak past the filter is likely to be harmless. Any side leakage over the top of the filter from the sprue exit will be negligible because of the momentum of the flow, carrying most of the flow directly downwards through the filter. Contrast the situation with Figure 12.46 where an ill-fitting filter can easily allow liquid metal to force its way over the top of the filter and jet into the runner, introducing defects downstream of the filter. Figure 12.58 illustrates good examples of robust filter geometries.

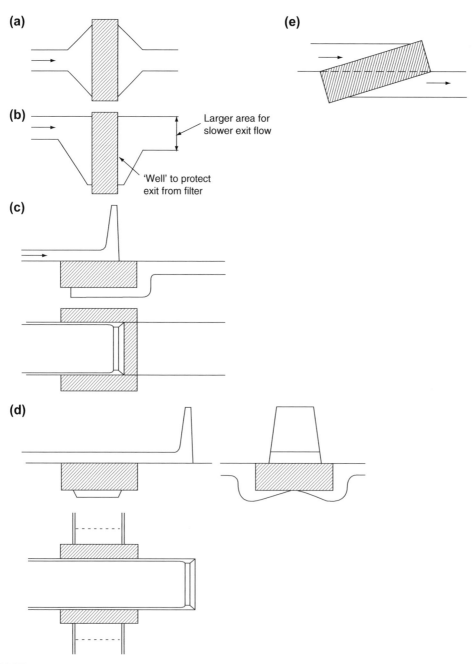

(a)

(b)

Larger area for
slower exit flow

'Well' to protect
exit from filter

(c)

(d)

(e)

FIGURE 12.55

Ceramic foam filter printing (a) conventional; (b) improved early back protection of filter by melt; (c) tangential in-line filtration in runner; (d) three views of tangential transverse runners with reduced fall and well volume; (e) A non-recommended cross-jointed filter (see text).

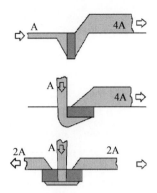

FIGURE 12.56

Examples of filters to ensure an expansion of four times from entrance to exit, taking advantage of the four times' reduction in velocity.

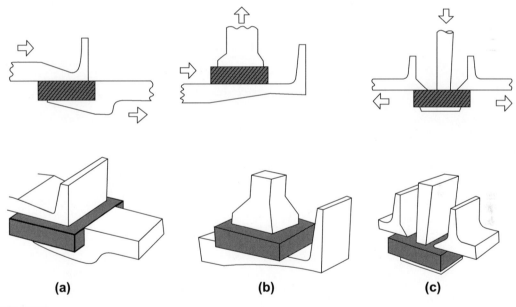

(a) **(b)** **(c)**

FIGURE 12.57

Filters with bubble traps: (a) in-line; (b) horizontal runner with filter beneath vertical gate; (c) recommended system for filter location at base of sprue. Several runners can be taken off in different directions.

There are other aspects of the siting of filters in running systems that are worth underlining.

1. Siting a filter so that some metal can flow over or adjacent to the filter for a second or two (into a slag trap for instance) before priming the filter is suggested to have the additional benefit that the pre-heat of the filter and the metal reduces the priming problem associated with the chilling of the metal by the filter (Wieser and Dutta, 1986). An example is seen in Figure 12.57(b). The chilling effect of the filter can be important for some thin wall castings of large area.

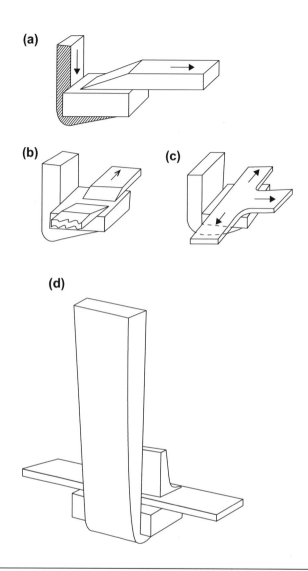

FIGURE 12.58

A possible connection of a sprue to a filter sited in the drag, with (a–c) showing one, two and three runners taken off in the cope; (d) shows the use of a bubble trap.

2. The area of the filter needs to be adequate. There is much evidence to support the fact that the larger the area (thereby giving a lower velocity of flow through the filter), the better the effectiveness of the filter. For instance, if the filter area is too small in relation to the velocity of flow then the filter will be unable to retain foreign matter: the force of the flow will strip away retained films like sheets from the washing line in a hurricane; particles and droplets will follow a similar fate. Having said this, it is important to bear in mind that filters in filling systems are all working well outside an optimum speed range to allow any effective filtration. Perhaps the main benefit from the increased filter area is the improved control of flow.

3. Many filter placements do not distribute the flow evenly over the whole of the filter surface. Thus a concentrated jet is unhelpful, being equivalent to reducing the active area of the filter. The tangential placement of a filter can also be poor in this respect because the flow naturally concentrates through the farthest portion of the filter and might even circulate backwards through the filter (Figure 12.56(a)). This inefficient flow regime is countered by tapering the tangential entrance and exit flow channels as illustrated in Figure 12.56(b), thereby directing the flow more uniformly through the filter. The provision of a bubble trap reduces the effectiveness of the taper but the presence of the trap is probably worth this sacrifice. (If the trap is not provided bubbles arriving from entrainment in the basin or sprue gather on the top surface of the filter. When they have accumulated to occupy almost the whole of the area of the filter the single large bubble is then forced through the filter, and travels on to create problems in the mould cavity. The trap is expected to be similarly useful for the diversion of slag from the filter face during the pouring of irons and steels.)

12.8.5 FILTERS IN RUNNING SYSTEMS

Mutharasan et al. (1981) find that the efficiency of removal of TiB_2 inclusions from liquid aluminium increased as the velocity through the filter fell from 10 mms^{-1} to 1 mms^{-1}. Later, the same authors found identical behaviour for the removal of up to 99% of alumina inclusions from liquid steel (Mutharasan et al., 1985). Ali (1985) confirms a strong effect of velocity, finding only at mms^{-1} velocities was a high level (96%) of filtration achieved in steel melts. However, all these authors seem unaware that their work is curiously inappropriate for castings; such creeping flow would never fill any normal mould. In the work by Wieser and Dutta (1986) on the filtration of alumina from liquid steel, somewhat higher velocities, in the range of 30–120 mms^{-1}, are implied despite the use of filter areas up to 10 times the runner area in an attempt to obtain sufficient slowing of the rate of flow. Even these flow velocities will not match most running systems. These facts underline the poverty of the data that currently exists in the understanding of the action of filters.

Wieser and Dutta go on to make the interesting point that working on the basis of providing a filter of sufficient size to deal with the initial high velocity in a bottom-gated casting, the subsequent fall in velocity as the casting fills and the effective head is reduced implies that the filter is oversize during the rest of the pour. However, this effect may be useful in countering the gradual blockage of the filter in steel containing a moderate amount of inclusions.

Although we have seen previously that filtration efficiency appears to be strongly dependent on inclusion and filter types, the casters of wrought Al alloys are aware of the necessity for the low velocities through filters, which is why their packed bed filters are so large, being nearly a cubic metre in volume. This snail pace flow rate to achieve good filtration performance by filters emphasises the fact that filters in running systems generally cannot possibly filter significantly because speeds are far too high. The research by Din and the author (2003) clearly showed that only about 10% of the cleanup of a casting was the result of filtration by the filter, whereas 90% of the cleanness benefit was the result of improved flow of the melt, avoiding the subsequent entrainment of oxide bifilms (see Section 12.8.7 on Direct Pour).

All the previous research emphasises the almost complete inability of filters to filter. In this account from now on, we shall abandon the notion that filters can filter and concentrate on what filters really do, and how we should use them to best effect.

One of the major concerns for the use of filters is the protection of the back of the filter. This condition may not be obvious at first sight, but is critical to good performance. If the melt is allowed to jet from the back of a filter, the downstream quality of the melt can be ruined. Loper and colleagues (1996) were among the first to draw attention to this problem. They described the need for what they called a 'hydraulic lock', a region behind the filter that flooded quickly to suppress jetting through the air. We need to return to this critical aspect of the use of filters in running systems.

For relatively small castings, and where the cost of the filter is a major issue, the best location when using a single filter is near the entrance to the runner, immediately following the sprue. The resistance to penetration of the pores of the filter by the action of surface tension is an additional benefit, delaying the priming of the filter until the sprue has at least partially filled. The frictional resistance to flow through the filter once it is operational provides a further contribution to the reduction in speed of the flow. This frictional resistance has been measured by Devaux (1987). He finds the head loss to be large for filters of area only one or two times the area of the runner. He concludes that whereas a filter area of twice the runner area is the minimum size that is acceptable for a thick-section casting, the filter area has to be increased to four times the runner area for thin-section castings. The pressure drop through filters is a key parameter that is not known with the accuracy that would be useful. Midea (2001) has attempted to quantify this resistance to flow but used only low flow velocities useful for only small castings. A slight improvement is made by Lo and Campbell (2000), who study flow up to 2.5 ms^{-1}. Even so, at this time the author regrets that it remains unclear how these measurements can be used in a design of a running system. A clear worked example would be useful for us all.

The filter positioned at the entrance to the runner also serves to arrest the initial splash of the first metal to arrive at the base of the sprue. At the beginning of the runner, the filter is ideally positioned to take out the films created before and during the pour.

For larger castings, particularly tonnage steel castings, the provision of several filters is useful; one filter per ingate. With dense melts such as steels, and where filters lead into vertical gates, there is much less danger of the priming of one filter and the filling of one ingate before the others, with the result that the melt preferentially fills the casting via the first ingate, whereas the other filters prevent the priming of other ingates. The metal therefore flows up through the first ingate, through the mould cavity, and pours down the empty, un-primed ingates to pour down on to the filters from the wrong side. There is increased danger of this occurring with light alloys, because a higher head, perhaps more than 100 mm in some cases, will be needed to prime the filter but the ingate height above the filter may be less than this, allowing the cavity to fill before pressure can be raised sufficiently high for the remaining filters to prime. The head to prime steel filters will be less than half of this value, reducing this danger for steel.

The clean liquid can be maintained relatively free from further contamination so long as no surface turbulence occurs from this point onwards. This condition can be fulfilled if the following occurs.

The melt proceeds at a sufficiently low velocity and/or is sufficiently constrained by geometry to prevent entrainment. This particularly includes the important provision to eliminate jetting from the rear face of the filter. This *essential* condition must be observed by only using those arrangements that ensure the melt covers and submerges the exit face of the filter as quickly as possible.

Every part of the subsequent journey for the liquid is either horizontal or uphill. The corollary of this condition is that the base of the sprue and the filter should always be at the lowest point of the running system and the casting. This excellent general rule is a key requirement.

In addition, the siting of a ceramic foam filter in a channel of a running system will reduce the flow rate as a result of two further factors

1. A reduction of the total cross-sectional area for flow by perhaps approximately 15% (because the porosity of the filter is roughly 85%).
2. Frictional loss. This is currently unknown, but might be at least a further 15%.Thus the volume flow rate Q may be reduced by a total of approximately 30%. This may be revealed by an extension of the fill time by 30%. (Insufficient work has been carried out on systems with and without filters to know whether this is accurate or not.)

A low velocity will be achieved after the filter if the cross sectional area of the runner downstream of the filter is increased in proportion to the reduction in speed provided by the filter.

In isolated observations of the velocity of flow through filters in generally non-pressurised flow conditions, the velocity appears to be reduced by approximately a factor of 5. If Q is reduced by a factor of 1.3, it follows that after the filter the area of the flow channel might be expanded by a factor of $5/1.3 = 3.85$.

If the filter had been placed in a filling system calculated to be naturally pressurised, it follows that the area downstream of the filter should remain (just) naturally pressurised after this expansion of 3.85. (Clearly a factor of 4 expansion is slightly too much, and some pressurisation might be lost. Although, fortunately, the danger downstream is much reduced as a result of the much reduced velocities, it seems prudent to recommend factors of expansion less than 4, even though a factor of 4 may be just workable in many situations.)

Although these areas are, I believe, essentially correct (although clearly some refinement of their accuracy might be needed if some good research on this subject were to be carried out) there are new dangers downstream. These include the following.

1. The melt may not initially fill the enlarged channel, but simply run along the bottom of the newly expanded runner, only reaching the correct velocity when the whole channel is finally filled.
2. The upper parts of the filter may jet metal, thus oxidising and contaminating the downstream melt, as a result of the channel being unfilled to the top, exposing the top of the filter for an extended time.

The previous description applies to the siting of a filter in the main runner, usually immediately after the sprue–runner junction.

If we now turn to the quite different situation in which filters are placed in each ingate, the situation seems quite different.

For instance, the behaviour of the filter in terms of the back-pressure it will generate is clearly a function of velocity V, and I suspect this function will in fact be V^2. Thus there is considerable change to the filling system as a whole when placing the filter in the high velocity region immediately following the sprue.

In contrast, in the ingates when using a naturally pressurised design, the velocities are significantly lower, so that back-pressure is greatly reduced. In any case, the gate area into the filter should have already been correctly sized, so no expansion on the output side would be required. The use of filters in gates is best employed in vertical gates because these ensure that the filter primes fully and is quickly protected by liquid on its exit face.

Like so much of gating design, these important details of the correct use of filters would greatly benefit from some systematic work evaluating the effects on fill times, sprue heights, velocities, flow rates and filter placement in runners or gates etc. Preferably, this would be carried out by computer simulation backed up by spot checks with real castings, and possibly even further backed up by more spot checks with video X-ray radiography or even water model systems.

Many of these concerns disappear when the Trident gate is used, together with the associated vortex surge cylinders to ensure a gentle priming phase.

12.8.6 TANGENTIAL PLACEMENT

Filters have been seen to be open to criticism because of their action in splitting up the flow, thereby, it was thought, probably introducing additional oxide into the melt.

There is some truth in this concern. A preliminary exploration of this problem was carried out by the author (Din and Campbell, 1994). Liquid Al alloy was recorded on optical video flowing through a ceramic foam filter in an open runner. The filter did appear to split the flow into separate jets; a tube of oxide forming around each jet. However, close observation indicated that the jets recombined about 10 mm downstream from the filter, so that air was excluded from the stream from that point onwards. The oxide tubes around the jets appeared to wave about in the eddies of the flow, remaining attached to the filter, like water weed attached to a grid across a flowing stream. The study was repeated and the observations confirmed by X-ray video radiography. The work was carried out at modest flow velocities in the region of 0.5–2.0 ms^{-1}. It is not certain, however, whether the oxides would continue to remain attached if speeds were much higher, or if the flow were to suffer major disruption from, for instance, the passage of bubbles through the filter.

What is certain is the damage that is done to the stream after the filter if the melt issuing from the filter is allowed to jet into the air. Loper and coworkers (1996) call this period during which this occurs *the spraying time*. This is so serious a problem that it is considered in some detail later. Loper describes the limited volume at the back of the filter as a *hydraulic lock*, the word *lock* appearing to be used in a similar sense to a lock on an inland waterway. In England, the length water impounded between two locks is called a 'pound'. Alternatively, an *air lock*, or a *vacuum lock*, is the impounded region of fluid (air or vacuum) between two gates or valves. Anyway, his meaning is clear. I discuss this problem in terms of the requirement to 'protect' the back of the filter with melt.

Unfortunately, most filters are placed transverse to the flow, simply straight across a runner (Figure 2.56(a)) and in locations where the pressure of the liquid is high (i.e. at the base of the sprue or entrance into the runner). In these circumstances, the melt shoots through a straight-through-hole type filter almost as though the filter was not present, indicating the such filters are not particularly effective when used in this way. When a foam filter suffers a similar direct impingement, penetration occurs by the melt seeking out the easiest flow paths through the various sizes of inter-connected channels and therefore emerges from the back of the filter at various random points. Jets of liquid project from these exit locations, and can be seen in video radiography. The jets impinge on the floor of the runner and on the shallow melt pool as it gradually builds up, causing severe local surface turbulence and so creating dross. If the runner behind the filter is long or has a large volume, the jetting or spraying behaviour can continue until the runner is full, creating large volumes of damaged metal.

Conversely, if the volume of the filter exit can be kept small and filled rapidly, the volume of damaged melt that can be formed is now reduced. Although this factor has been little researched, it is certain to be important in the design of a good placement for the filter.

Figure 12.56(a) shows a common but poor filter print design that would lead to extensive downstream damage – an extended spraying time in Loper's words. An improved geometry that assists the back of the filter to be covered with melt quickly is shown in Figure 12.55(b). Figure 12.55(c) is improved still further by placing the block filter tangentially to the direction of flow. The tangential mode has the advantage of the limitation of the exit volume under the filter, providing a geometrical form resembling a sump, or lowest point, so that the exit volume fills quickly. In this way, the opportunity for the melt to jet freely into air is greatly reduced so that the subsequent flow is protected. A further advantage of this geometry is the ability to site a bubble trap over a filter (Figure 12.55(c)), providing a method whereby the flow of metal and the flow of air bubbles can be divided into separate streams. The air bubbles in the trap are found to diffuse away gradually into sand moulds. For permanent moulds, the traps may need to be larger.

The volume adjacent to the filter exit is seen to be well controlled in the case of Figure 12.57(a) and (b), but Figure 12.58(a) could be a concern, because the single runner from the filter has to have sufficient area to retain the advantage of the flow reduction from the filter, and thus is rather thick. It may therefore not fill well, but may support spraying from the filter during most of the time to fill the complete length of the runner. Variant (b) is better with the two-way runner being only half the height. Variant (c) is probably excellent, with its three-way runner, each one third height, so that the region above the filter would be expected to fill completely in a dynamic situation, with the melt proceeding along the runners, filling the whole cross-section of each of the runners.

An additional benefit is that the straight-through-hole extruded or pressed filters seem to be effective when used tangentially in this way. A study of the effectiveness of tangential placement in the author's laboratory (Prodham et al., 1999) has shown that a straight-through-pore filter could achieve comparable reliability of mechanical properties as could be achieved by a relatively well-placed ceramic foam filter (Sirrell and Campbell, 1997).

Adams (2001) draws attention to the importance of the flow directed downwards through the filter. In this way, buoyant debris such as dross or slag can float clear. In contrast, with upward flow through the filter the buoyant debris collects on the intake face of the filter and progressively blocks the filter.

The tapering of both the tangential approach and the off-take from the filters further reduces the volume of melt, and distributes the flow through the filter more evenly, using the total volume of the filter to best advantage. A variety of filter placements, illustrating many of these key advantages are illustrated in Figures 12.57 and 12.58.

12.8.7 DIRECT POUR

Sandford (1988) showed that a variety of top pouring could be used in which a ceramic foam filter was used in conjunction with a ceramic fibre sleeve. The sleeve/filter combination was designed to be sited directly on the top of a mould to act as a pouring basin, eliminating any need for a conventional filling system. In addition, after filling, the system continued to work as a feeder. This simple and attractive system has much appeal.

Although at first sight, the technique seems to violate the condition for protection of the melt against jetting from the underside of the filter, jetting does not seem to be a problem in this case. Jetting is avoided almost certainly because the head pressure experienced by the filter is so low and contrasts with the usual situation where the filter experiences the full blast of flow emerging from the base of the much taller sprue.

Sandford's work illustrated that without the filter in place, direct pour of an aluminium alloy resulted in severe entrainment of oxides in the surface of a cast plate. The oxides were eliminated if a filter was interposed and if the fall after the filter was less than 50 mm. Even after a fall of 75 mm, after the filter relatively few oxides were entrained in the surface of the casting. The technique was further investigated in some detail (Din et al., 2003) with fascinating results illustrated in Figure 12.59. It seems that under conditions used by the authors in which the melt emerging from the filter fell into a runner bar and series of test bars, some surface turbulence was suffered and was assessed by measuring the scatter of tensile test results. The effect of the filter acting purely to filter the melt was seen to be present, but slight. The castings were found to be repeatable (although not necessarily free from defects) for fall distances after the filter of up to about 100 mm in agreement with Sandford. Above a fall of 200 mm, reproducibility was lost (Figure 12.59(a), (b) and (c)).

This interesting result explains the mix of success and failure experienced with the direct pour system. For modest fall heights of 100 mm or so, the filter acts to smooth the perturbations to flow, and so confers reproducibility on the casting. However, this may mean 100% good or 100% bad. The difference was seen by video radiography to be merely the chance flow of the metal, and the consequential chance location of defects.

The conclusion to this work was a surprise. It seems that direct pour should not necessarily be expected to work first time. If the technique were found to make a good casting, it should be used because the likelihood would be that all castings would then be good. However, if the first casting was bad, then all should be expected to be bad. The

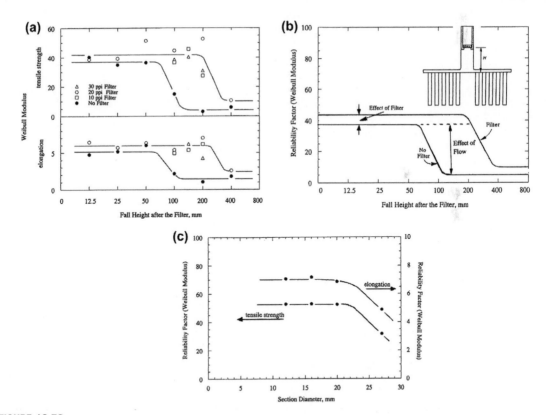

FIGURE 12.59

Direct pour filtration showing (a) the reduced reliability as the fall increases with and without a filter in place; (b) the interpretation of 'a'; (c) the reduced reliability as diameter of test bars is increased.

site of the filter and sleeve should then simply be changed to some other location to seek a different pattern of filling. This could mean a site only a few millimetres or centimetres away from the original site. The procedure could be repeated until a site was found that yielded a good casting. The likelihood is that all subsequent castings would be good.

However, the technique will clearly not be applicable to all casting types. For instance, it is difficult to see how the approach could reliably produce extensive relatively thin-walled products in film-forming alloys where surface tension is not quite in control of the spread of metal in the cavity. For such products, the advance of the liquid front is required to be steady, reproducible and controlled. Bottom-gating in such a case is the obvious solution. Also, the technique works less well in thicker section castings where the melt is less constrained after its fall from the underside of the filter. Figure 12.59(c) illustrates the fall in reliability of products as the diameter of the test bars increase from 12 mm to more than 20 mm. The larger sections would have offered reduced constraint to the falling metal, so suffering entrainment damage, and more scattered properties.

Even though the use and development of the direct pour technique will have to proceed with care, it is already achieving an important place in casting production. A successful application to a permanent moulded cylinder head casting is described by Datta and Sandford (1995). Success here appears to be the result of the limited, and therefore relatively safe, fall distance.

A development of the direct pour technique is described by Lerner and Aubrey (2000). For the direct pouring of ductile iron, they use a filter that is a loose fit in the ceramic sleeve. It is held in place by the force of impingement of the melt. When pouring is complete, the filter then floats to the top of the sleeve and can be lifted off and discarded, avoiding contamination of recycled metal.

12.8.8 SUNDRY ASPECTS

1. The dangers of using ceramic block filters in the direct impingement mode are illustrated by the work of Taylor and Baier (2003). The found that a ceramic foam filter placed transversely in the down-sprue worked better at the top of the sprue rather than at its base. This conclusion appears to be the result of the melt impact velocity on the filter causing jetting out of the back of the filter. Thus the high placement was favoured for the reasons outlined in the section above on the direct pour technique. This result is unfortunate, because if the filter exit volume had been limited to a few millilitres (a depth beneath the filter limited to 2–3 mm), the lower siting of the filter would probably have performed best.
2. It is essential for the filter to avoid the contamination of the melt or the melting equipment. Thus, for many years, there appeared to be a problem with Al-Si alloys that appeared to suffer from Ca contamination from an early formulation of the filter ceramic. If this problem actually existed, it has in any case now been resolved by modification of the chemistry of the filter material. In addition, modern filters for Al alloys are now designed to float, so that they can be skimmed from the top of the melting furnace when the running systems are recycled. This avoids the costly cutting out and separation of spent filters from recycled rigging to avoid them collecting in a mass the bottom of the melting furnace.

Interestingly, the steel gauzes used for Al alloys do not contaminate the alloy entering the casting. This is almost certainly the result of the alloy wrapping a protective alumina film over the wires of the mesh as the meniscus passes through. However, the steel will dissolve later if recycled via a melting furnace of course.

The recent introduction of carbon-based filters for steel does add a little carbon to the steel, but this seems negligible for most grades. Experience suggests that even ultra–low-carbon steels appear to avoid contamination from this source.

Xu and Mampaey (1997) report the additional benefit of a ceramic foam filter in an impressive 12-fold increase in the fluidity of grey iron poured at about 1400°C in sand moulds. They attribute this unlooked-for bonus to the effect of the filter in (1) laminising the flow and so reducing the apparent viscosity due to turbulence and (2) reducing the content of inclusions. One would imagine that films would be particularly important.

12.8.9 SUMMARY

So far as can be judged at this time, among the many requirements to achieve a clean casting, the key practical recommendations for the casting engineer can be summarised as follows.

1. Do not allow slag and dross to enter the filling system. This task is solved by eliminating the conical pouring basin and substituting an offset step basin or preferably changing to contact pouring.
2. Use a good filling system (well designed basin, or eliminated basin, and sprue) to avoid the creation of additional slag or dross that may block the filter.
3. Integrate the filter with a buoyant trap. The bubble trap should also work as a slag trap. The presence of the filter significantly aids the separation of a buoyant and a heavy phase. Where particularly dirty metals are in use, the trap will, of course, require the provision of sufficient volume and height on its upstream side to accommodate retained material, allowing slag and dross to float clear, and leaving the filter area to continue working unhindered. The volume requirement does, however, require being balanced by the shape of the trap, which needs a narrow aspect ratio; if it had a large open volume, the turbulence of the melt thrashing around in this open space could generate sufficient oxide to block the filter. At all times the melt requires to be constrained (i.e. subject to a gentle pressurisation).
4. Avoid the danger of the bypassing of the filter by poor printing. Mould-in the filter if possible.
5. Provide protection of the melt at the exit side of the filter, by rapid fill of this limited volume with liquid metal. A useful geometry to achieve this is the tangential placement of the filter, followed by a shallow well that can be filled quickly. For filters that fill uphill, the effective volume above should be as small as possible, so that an exit to a runner of thickness less than the sessile drop height is required, otherwise the whole runner may fill with jetted and damaged metal.

FILLING SYSTEM DESIGN PRACTICE

13

This section seeks to take all the information on filling system components and all the casting rules listed in earlier sections of the book and gather these together to see how we might achieve a complete, practical design of a filling system. The ability to design a quantitative solution, yielding precise dimensions of the filling channels at all points, is a key responsibility, perhaps *the* key responsibility, of the casting engineer.

When I am asked, 'What is the most important feature of the filling system design?' the reply is "Everything". As has been stated before, all gravity filling systems run with such high velocities that they are hypersensitive to any small error of mismatch, sizing or geometry. A small, innocent-looking expansion or ledge leads to a stream of entrainment defects. Every part of the system needs to be correct.

Naturally, computers are beginning to have some capability of optimising the design of filling systems (McDavid and Dantzig, 1989; Jolly et al., 2000). However, at this time, it is clear that most computer simulations of filling have to employ the liquid viscosity as an adjustable parameter, commonly using values for the metal viscosity that are many times the viscosity of the pure metal. The reason for this is almost certainly the presence of a dense suspension of bifilms which are probably the only features that can have any significant effect on the viscosity of melts. This adjustment of the viscosity is unsatisfactory in the sense that melts are starting to be prepared with very different bifilm contents, so that filling simulations by computer can be unexpectedly found to be unsatisfactory if the appropriate viscosity for the melt condition is not known.

The other significant limitation of computer simulations is that thermal and solute-driven convection is highly computationally intensive and is therefore generally neglected. Furthermore, of course, the damage that is introduced by oxides which in some cases appear to be capable of completely blocking the operation of both filters and feeders, is far from being simulated, and may ultimately be impossible. The list of potentially important factors required for such a model is daunting. These will include the rate of strengthening of the oxide and its fracture and tearing resistance, each of which are likely to be sensitive to alloy content, impurities, environment and age of the oxide etc. We are never likely to gather all the data that are needed or to have computers sufficiently powerful to take them all into account.

Anyway, until the time that the computer is tolerably proficient, it will be necessary for the casting engineer to undertake the duty of designing filling systems for castings. The complication of the procedure is not to be underestimated – if it were easy, the procedure would have been developed years ago. Many factors need to be taken into account. This short outline cannot cover all eventualities, but will present a systematic approach that should be generally useful.

In the meantime, on the shop floor, I always avoid a filling system which uses the conventional refractory tubing (colloquially known as 'pots' because they are made using the same fired clays as domestic pottery). I have never been happy with the results of using these pre-formed channels. They are never the right shape to distribute metal effectively and consequently never fill completely, resulting in energetic surface turbulence which inevitably leads to casting defects. This chapter advocates the use of completely sand-moulded filling channels. The extra cost involved because of the additional mould part to permit the running system to be moulded under the casting always pays off with a superior casting. In addition, of course, sand inclusions (the original but misguided reason for the use of pots) are never found provided the sand moulded system is correctly designed to be naturally pressurised at every point, as will be explained.

Complete Casting Handbook. http://dx.doi.org/10.1016/B978-0-444-63509-9.00013-3

13.1 BACKGROUND TO THE METHODING APPROACH

If a computer package is available to simulate the solidification of the casting, it is best to carry this out first. Most software packages are sufficiently accurate when confined to the simulation of solidification (it is the simulation of the flow of metal during filling, and other sophisticated simulations such as that of stress, strain and distortion that are more difficult, and the results often less accurate). A solidification simulation of the naked casting with the addition of no filling or feeding system will illustrate whether there are special problems with the casting itself. Figure 13.1 illustrates the formal logic of this approach. It lays out a powerful methodology that is strongly recommended.

If in fact there are no special problems, it is good news. Otherwise, if problems do appear for a long-running part, and if they can be eliminated by discussion with the designer of the component at this initial stage, this is usually the most valuable strategy. Such actions often include the shift of a parting line or the coring out of a heavy section or boss. The purpose of a modest one-time design change is to avoid, so far as possible, the ongoing expense for the life of the product of special actions such as the provision of chills or feeders or an extra core etc.

If, despite these efforts, a problem remains, the various options including additional chills, feeders or cores will require detailed study to limit, so far as possible, the cost penalties. The following section provides the background for the next steps of the procedure.

13.2 SELECTION OF A LAYOUT

First, it will be necessary to decide which way up the part is to be cast (this may be changed later in the light of many considerations, including problems of core assembly, desirable filling patterns, subsequent handling and de-gating issues etc.). If a two-part mould is to be used, the form of the casting should preferably be mainly in the cope, allowing gating at the lowest parts of the casting. This may prove so difficult that a third box part may be selected through which the running system could be sited under the casting. If some solution to the challenge of lowest point gating cannot easily be found, the risk of filling the casting at some slightly higher point may need to be assessed. Some filling damage might have to be accepted for some castings. Even so, it is unwelcome to have to make such decisions because the extent of any such damage is difficult to predict.

A heavy section of the casting needs special attention. This may most easily be achieved by orienting this part of the casting at the top and planting a feeder here. Alternatively, other considerations may dictate that the casting cannot be oriented this way, so arrangements may have to be made to provide chills and/or fins to this section if it has to be located in the drag.

When a general scheme is decided, including the approximate siting of gates and runners, the provision of feeders, if any, and the location of the sprue, a start can be made on the quantification of the system.

13.3 WEIGHT AND VOLUME ESTIMATES

The weight of the casting will be known or can be estimated. This is added to an estimate of the weight of the rigging (the filling and feeding system) to give an estimate of the total poured weight. Dividing this by the density of the liquid metal will give the total poured volume.

Unfortunately, of course, at this early stage, the weight of the rigging is clearly not accurately known because it has not yet been designed. However, an approximate estimate is nearly always good enough. Although a revised value can always be used in a subsequent iteration of the rigging design calculations to obtain an accurate value for the weight of the rigging, after some experience, an additional iteration will be found to be hardly ever necessary. If an iteration is carried out after the correct total weight of the rigging is known, only a change of a millimetre or two in many dimensions of the system will be found. These can usually be neglected.

Another question that constantly arises is 'Should the weight of metal in the pouring basin be included in the total poured weight?' The answer depends on the precise pouring technique.

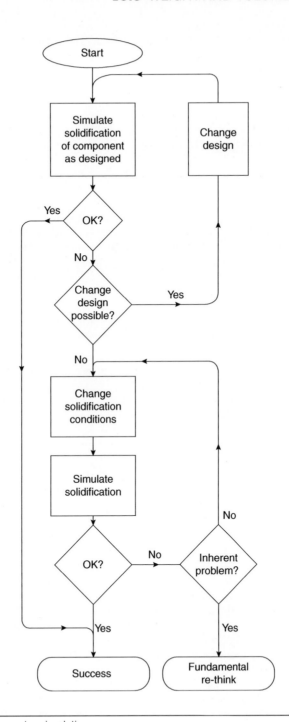

FIGURE 13.1

Methoding procedure for computer simulation.

1. A common approach is to time the pour from the start of pouring from the ladle. In this way, the melt is poured into the empty basin, gradually raising the level until the basin is sufficiently full. The stopper is raised to start the pour down the sprue, but pouring continues to ensure that the basin continuously topped up until the moment that the mould is full. The average fill rate is simply the total weight poured divided by the total time measured from the start of pouring of the basin.
2. Conversely, after the basin has been filled, there may be a short dwell time to allow the melt to settle. The filling of the casting should then be timed from the raising of the stopper. In those conditions in which the level of metal in the basin is required to be approximately equal to the maximum height of the casting + feeders, the basin must be maintained full during the entire pour. The weight of metal in the basin is no longer relevant to the calculation of the dimensions of the rest of the filling system. The average fill rate appropriate to the calculation of the filling system is the weight of the metal poured (minus the weight in the basin) divided by the time for the filling of the mould (neglecting the time to fill the basin).
3. If the metal in the basin is held back by a stopper, giving it time to settle, and if the basin is planted on the top of the mould, the casting is formed by eventually draining the melt from the basin. In this case, the filling time is clearly that time from raising of the stopper to the draining empty of the basin. The weight of metal in the basin is required to be included in the total weight poured.
4. An alternative filling system practice is to arrange for the basin to hold the complete charge (plus a small margin as explained later) for the running system and casting. After the basin is filled, a short delay of a few seconds is beneficial to allow as much detrainment of bubbles and oxides as possible. The stopper is then raised and the complete contents of the basin disappear into the entrance to the down-sprue. The time to empty the basin is the pour time. In this case, the average fill rate for the casting is the total weight of the casting plus rigging (minus any residual metal in the basin) divided by the fill time.

Those techniques, including the draining of the pouring basin, introduce an error in the design assumptions because the filling system is generally designed assuming a constant head pressure in the basin. Thus the draining of the basin will deliver a falling pressure at the end of the filling period. This is not usually serious because the filling system is not likely to depressurise sufficiently to introduce defects at this late stage, and the overall filling time is extended by only a second or so.

The draining does become a problem if the casting requires a relatively high fill rate during the last moments of the fill, or if it has narrow sections offering capillary repulsion to resist filling in top parts of the casting. These features would be good reasons to maintain a full basin and accept the loss of metal yield.

13.4 PRESSURISED VERSUS UNPRESSURISED

It was the famous researchers from the US Navy Research Laboratory, Johnson, Bishop and Pellini, who, in 1954 defined and classified filling systems as either pressurised or non-pressurised. This was a useful distinction.

In his book Castings (1991), the author recommended the achievement of velocity reduction by the progressive enlargement of the area of the flow channels at each stage, with the aim of progressively reducing the rate of flow. This is known as an *unpressurised* running system. The aim was to ensure that the gate was of a sufficient area to make a final reduction to the speed of the melt, so that it entered the mould at a speed no greater than its critical velocity. More recent research, however, with the benefit of video X-ray radiography and computer simulation, has demonstrated that this approach is wrong. The enlargement of the system, by, for instance, a factor of two to halve the velocity as the flow emerges from the exit of the sprue and enters the runner, usually fails to have any effect on the velocity, so that the runner is only half filled with this fast jet of metal. Worse still, the jet hits the far end of the runner, bouncing back to create a wave that rolls over the incoming fast jet; at the junction between these opposite streams bubbles and oxides are efficiently entrained. Thus, the unpressurised systems unfortunately behave poorly, degrading the metal, because much of the system runs only partly full. The other standard criticism (but incidentally of much less importance) was that unpressurised systems are heavy, thus reducing metallic yield, and thus costly.

In fact, video radiography reveals that at the abrupt increase in cross section at the base of the sprue on the entry to the runner, the entrainment of air occurs with dramatic effectiveness. This is because the melt jets along the base of the runner (not filling the additional area provided) and hits the end of the runner. The return flow creates an especially damaging situation: the rolling back wave which is formed (Figure 13.2(b)) can be long-lived, developing into a stable flow structure that resembles a *hydraulic jump*. The bubbles travel along the interface between the two opposing streams

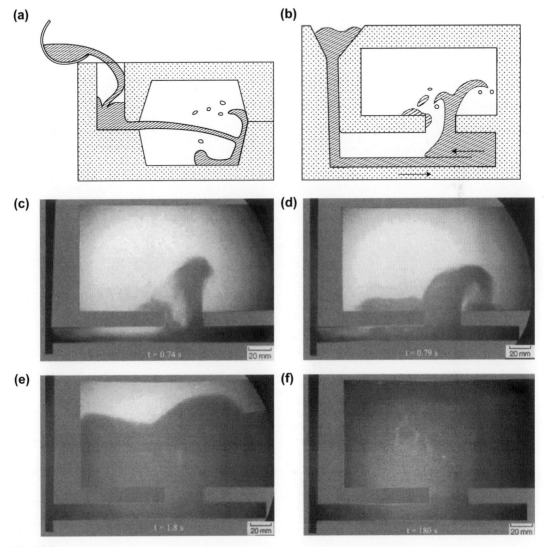

FIGURE 13.2

Mode of filling (a) a pressurised system, showing the jet into the mould cavity; (b) an unpressurised system, showing the fast underjet, and the rolling back wave in the oversized runner; (c–e) X-ray video frames of an Al alloy filling a mould 100 mm high 200 mm wide × 20 mm thick illustrating the unpressurised system; (f) the final casting showing subsurface bubbles and internal cracks.

(probably because of the presence of two non-wetting oxide films separating the two flowing streams) and progress to the ingate, usually collecting in a low-pressure zone on one side of the ingate, before proceeding to swim up through the metal in the mould cavity (Figure 13.2(e)). Naturally, these bubbles and oxides bequeath serious permanent damage to the casting (Figure 13.2(f)).

The cast iron foundryman had some justification therefore to champion his own favourite *pressurised* systems. For the benefit of the reader, the so-called *pressurised running system* is one in which the metal flow is *choked (i.e. limited by constriction)* at the gate; its rate of flow into the mould is to some extent controlled by the area of the gate, the last point in the running system (Figure 13.2(a)). This causes the running system to back-fill from this point and become pressurised with liquid, forcing the system to fill and exclude air. Thus, the system entrains fewer bubbles and oxides. However, it also forces the metal into the mould as a jet. Clearly, this system violates one of our principal rules because the metal is now entering the mould above its critical speed. The resulting splashing and other forms of surface turbulence inside the mould introduce their own spectrum of problems, different from those of the unpressurised system, but usually harming both the quality of the mould and the casting.

Thus, neither the unpressurised nor the pressurised traditional systems are seen to work satisfactorily. This is a regrettable appraisal of present casting technology.

Because for many years the pressurised systems were mainly used for cast iron, there were special reasons why the systems appeared to be adequate:

1. In the days of pouring grey iron into greensand moulds, the problems of surface turbulence were minimised by the tolerance of the metal-mould system. The oxidising environment in the greensand mould produced the liquid silicate film on the surface of the liquid iron. Thus, when this was turbulently entrained, it did not lead to a permanent defect (Castings, 2003). In fact, many good castings were produced by tipping the metal into the top of the mould, via an open feeder, using no running system at all! Nowadays, the use of certain core binders and mould additives cause solid graphitic surface films on the metal which reduce its tolerance to surface turbulence. The pressurised systems are therefore now producing defects when once they were working satisfactorily. This problem has become more acute as it has become increasingly common for irons to have alloy additions such as magnesium (to make ductile iron) and chromium (for many alloyed irons) which lead to strong oxide films on the liquid surface, leading to serious defects if entrained.

2. Over recent years the standards required of castings has risen to an extent that the traditional foundryman is shocked and dazed. Whereas the pressurised system was at one time satisfactory, it now needs to be reviewed. The achievement of quality is now being seen to be not by inspection, but by process control. Turbulence during filling introduces a factor that will never be predictable or controllable. This ultimately will be seen as unacceptable. Reproducibility of the casting process will be guaranteed only by systems that fill the mould cavity with laminar surface flow. At one time, this was achievable only with counter-gravity filling systems. Now, as we shall see, we can achieve some success with gravity systems, provided they are designed correctly.

The bold statement given by the author in Castings (1991) '*Unpressurised* systems are recommended; therefore, *pressurised* are not' is withdrawn. Recent research reveals that neither system is really satisfactory.

In summary, the unpressurised system had the praiseworthy aim to reduce the gate velocity to below the critical velocity. Unfortunately such systems usually ran only part-full, causing damage to the castings because of entrained air bubbles and oxides. The pressurised system probably benefited greatly from its ability to fill quickly and to run full, greatly reducing the damage from bubbles. However, the high velocity of the melt as it jetted into the mould created its own contribution to havoc. Further complications for both systems were the unsatisfactory conical basin, sprue designs and wells, all of which simply added to the existing fundamental problems.

The choke

Turning now to another hallowed dogma of running system design that requires to be laid to rest. This was the concept of the *choke*.

The choke was a local constriction designed to limit flow. In the non-pressurised system, the choke was generally at the base of the sprue, whereas the pressurised system was choked at the ingates into the mould. Unfortunately, a choke is an undesirable feature. Flow rates are usually already too high in most filling systems at the base of the sprue. Providing a constriction in the form of a choke speeds up the flow even more, causing it to emerge as a jet that does not expand to fill the downstream channel and so entrains air once again, with much consequential damage.

All the early unpressurised and pressurised systems were devised before the benefits of computer simulation and video X-ray radiography. They also pre-dated the development of the concepts of surface turbulence, critical velocity, critical fall height and bifilms. It is not surprising therefore that all these traditional approaches to the design of filling systems gave less than satisfactory results.

In the history of the development of filling systems, most of the early research was carried out in an effort to understand flow, but was of limited value because the emphasis was on steady-state flow through fully filled pipework, following the principles of hydraulics. This does, of course, sometimes occur late during the filling process. However, the real problems of filling are associated with the *priming* of the filling system, i.e. its behaviour before the filling system is filled. Thus these early studies give us relatively little useful background on which to base effective designs of filling systems for real castings.

A completely new approach is described in this book that attempts to address these issues. We shall abandon the concept of a localised choke. The whole of the length of the filling system should experience its walls in permanent contact and gently pressurised by the liquid metal. Thus, effectively, the whole length of the running system should be designed to act like a choke; a kind of *continuous choke* principle. A useful name is 'naturally pressurised' system, because the new design concept is based on designing the flow channels in the mould so as to follow the *natural form* that the flowing metal wishes to take, and at the same time develop a *natural backpressure* throughout the system as a result of the effect of frictional drag.

Gating ratios

It is necessary to divert briefly to examine a second red herring, the filling system as defined by its traditional gating ratios. When considering the relative areas of the key parts of the filling system, the sprue exit area is conventionally taken as 1, the runner cross section and the gate cross section areas are listed as comparisons. Thus 1:2:4 is a typical series indicating that the runner has twice the area of the sprue exit, and the gate has four times the sprue exit area. Another common gating ratio was 1:2:2. These ratios are frankly, unhelpful. These too should be laid to rest.

Other series of ratios that might have some helpful aspects are discussed later. However, I never find myself defining the filling system using ratios; the ratios are a natural outcome of the filling system design. They are an outcome, not an input.

If we imagine that after the metal emerges from the sprue and takes its right angle bend into the runner, the stream would be expected to lose some energy so that its velocity would fall by some amount, say 20% (this amount is not claimed to be accurate, but is taken here for clarity of illustration). This would allow us to expand the channel by this amount. We shall assume that the runner can remain parallel, retaining its area of 1.2 along its complete length (although, in reality, the gradual accumulation of the effect of friction along its length might add a further reduction in speed, say 10%, allowing in principle the area of the runner to be increased to 1.3 times the area of the sprue exit. However, for simplicity at this stage we shall neglect this effect). After turning through a further right-angle bend into the gate this turn might contribute a further velocity fall of 20%, giving a series of permissible area ratios of 1:1.2:1.4, although it will be noticed that the ingate velocity has only fallen by approximately 40% from that at the sprue exit.

If the assumed 20% expansion of area after each bend is not entirely allowed (for instance, if only 10% expansion were provided after each bend), the stream will experience a gentle pressurisation. If zero expansion of the runner was made after each bend, the melt would experience an even greater pressurisation. This pressurisation is in fact only a fairly modest pressure against the walls of the running system, but will be valuable to counter any effect of bubble formation and will act to support the walls of the running system against collapse (a special problem in large running systems for large castings). Thus, to be more sure of maintaining the system completely full, and slightly pressurised, a ratio of 1:1.1: 1.2 or perhaps even better, 1:1:1 might be used.

From the ratios, it is clear that the naturally pressurised system is partway between the pressurised and unpressurised systems. The value of ensuring that the whole of the filling system is at least gently pressurised was first appreciated by Jeancolas and his colleagues in 1962, but seems to have been neglected until now. This prophetic insight is a feature that should be part of all computer simulations of filling systems. If any part of the surface area of the filling system reveals an absence of pressure, the system requires revision to address this. A refinement would be to illustrate the pressurisation of the surface of the filling system, using contours of pressure. Any regions sufficiently low to present a risk of depressurisation in practice could indicate the need for action to address this. A robustly pressurised system is the target.

In the interests of maximising pressurisation to avoid air entrainment, I nearly always select the runner area equal to the sprue exit. This avoids any choice of losses at this junction which are in any case not precisely 20% and are probably not accurately known for the various different sprue–runner junctions that are used.

Moving on now to considerations of the area of the ingate, Table 13.1 indicates the ratios I normally use for a naturally pressurised system 1:1:n where the value 'n' can take almost any value (although normally between approximately 4 and 10), if care is taken to distribute the melt over the whole area of the gate. This is not always straightforward, but we shall devote space to a full consideration of the decision-making required for n.

However, at this point in our progress of examining the evolution of our running system design, the reader may already have noticed a major problem. The naturally pressurised system has no built-in mechanism for any significant reduction in velocity of the stream. Thus the high velocity at the base of the sprue is maintained (with only minor potential reduction due to frictional drag) into the mould. Thus the benefits of complete priming of the filling system to exclude air may be lost once again on entering the mould cavity.

This fundamental problem alerts us to the fact that the naturally pressurised approach requires completely separate mechanisms to reduce the velocity of the melt through the ingates. The options include the following.

1. The use of filters;
2. The provision of specially designed runner extension systems such as flow-offs;
3. A surge control system;
4. The use of a vertical fan gate at the end of the runner; and
5. The use of vortices to absorb energy whilst avoiding significant surface turbulence (such systems are probably not yet fully developed, but may represent future options).

We shall consider all these options in detail in due course, but the reader needs to be aware that, unfortunately, at this time, the use of naturally pressurised systems is in its infancy. In particular, the rules for such designs are not yet known for some features such as joining of round or square section sprues to rectangular runners. The rules for the incorporation of filters may even yet not be fully understood, and vortex systems in particular are definitely not yet understood, even though they appear to have much promise.

This creates a familiar problem for the foundry person: in the real world, the casting engineer has to take decisions on how to make things, whether or not the information is available at the time to help make the best choice. If he fails, it may not be his fault; it may be the fault of the absence of good rules defined by good research.

Insofar as the rules are presently understood for the majority of castings, they are set out later in this work, for good or for bad. I hope they assist the caster to achieve a good result. One day I hope we in the industry will all have the better answers that we need.

Table 13.1 Examples of Area Ratios (Sprue Exit Area:Runner Area:Gate Area)

Filling System	Examples of Area Ratios
Pressurised	1:0.8:0.6
	1:1:0.8
Unpressurised	1:2:4
	1:4:4
Naturally pressurised	1:1:n
Naturally pressurised with foam filter in gate	1:1:4
With speed reduction or bypass designs	1:1:10

In the meantime, computers are starting to be successful to simulate the flow of metal in filling systems. At the present time, such simulations are highly computationally intensive, and therefore slow and/or not particularly accurate. It is necessary to beware that some simulation packages are still highly inaccurate. However, time will improve the situation, to the great benefit of casting quality.

13.5 SELECTION OF A POURING TIME

A common concern is how can production rates be maintained high if metal velocities in the filling system need to be kept below the critical 0.5 ms^{-1}? Fortunately, this is not usually a problem because the time to fill a casting is dependent on the rate of mould filling measured as a volume per second, and can be fixed at a high level. At the same time, the velocity through the ingate, measured in metres per second, can be independently lowered simply by increasing the area of the gate (simple in principle, but sometimes harder in practice!). These considerations will become clearer as we proceed.

When faced with a new design of casting, the first question asked by the casting engineer is 'How fast should it be filled?' The selection of a pouring time is one of the most interesting moments in the design of a filling system for a new casting.

Sometimes, there is no choice. On a fast moulding line making 360 moulds per hour, there is only 10 s for each complete cycle, of which perhaps only 5 s may be the available time for pouring. (Although it is worth keeping in mind that even here, as a last resort, the pour time might be doubled if two pouring facilities were to be installed.)

When there *is* a choice, the pour time can often be changed between surprisingly large limits. Many castings are filled very much faster than necessary, and there are usually benefits to a reduction in this speed. One factor limiting the slowest speed is the rate of rise of the metal in some sections.

The surface of an Al casting becomes marked with striations because of the passage of transverse unzipping waves at a vertical rise velocity below 60 mms^{-1} (Evans et al., 1997). Although they can leave their witness on the surface of the oxidised surface of the casting, they are usually harmless to its internal structure. Even so, they may be unsightly on a cosmetically important casting, but if the alloy has an extra strong surface oxide, or is partially freezing because of a cool pour, the waves may lead to such severe surface horizontal laps that the casting becomes beyond repair. Types of geometries where this problem is most often seen are illustrated in Figure 13.3. A hollow cylinder cast on its side is a common casualty (Figure 13.3(a)) because of the sudden increase in area to be filled, reducing the rate of rise as the metal reaches the top of the core. The problem is also found on the upper surfaces of tilt castings if the rate of tilt is too slow and if, at larger angles, the area of the metal surface increases to further reduce the rate of rise as is seen in Figure 13.3(b).

Considerations that control the choice of rate of metal rise in steel castings indicate that these factors have yet to be properly researched (Forslund, 1954; Hess, 1974). In practice, a common rate of rise in a steel foundry making castings several metres tall and weighing several tonnes is 20 mms^{-1}. A further limit to the fill time of a steel casting is the possible collapse of the cope when subjected to radiant heat of the rising metal for too long. This problem is reduced by generous venting of the top of the mould via a top feeder as explained for Figure 4.7, and is further reduced by the practice of providing a white mould coat based on a material such as alumina or zircon, thus absorbing much less of the incident radiation.

Alternatively, another constraint on a choice of pour time is the consideration that it may be necessary to fill the mould before freezing starts in its thinnest section (or, more usefully, its smallest modulus). Thus an idea of the time available can be gained from the Figure 13.4. (Readers are encouraged to generate such diagrams for themselves for special casting conditions, using embedded thermocouples to determine the freezing times versus modulus relations, e.g. for cast iron in silicate-bonded sand moulds, or aluminium-based alloys in investment moulds at 600°C etc.)

Having made our choice of an approximate minimum fill time for the casting, it is instructive to consider whether this time could be doubled, or even doubled again. It is surprising how often this is a possibility. Whereas the experienced foundryman will hesitate to extend the pouring time of a familiar casting, his experience will be based usually on a poor filling system. Such systems generate problems such as slopping and surging and splashes ahead of the main body of

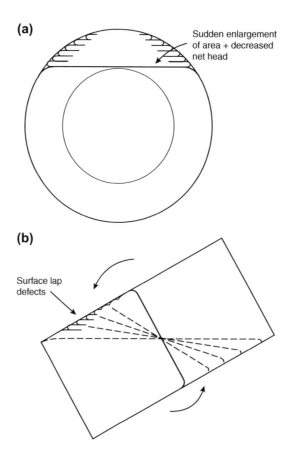

(a)

Sudden enlargement
of area + decreased
net head

(b)

Surface lap
defects

FIGURE 13.3

Common lap problems at reduced rates of rise, as a result of increased area of liquid surface in the mould (a) in a horizontal pipe or cylinder; (b) on the cope surface of a tilt casting.

melt. This generation of lap-type defects will give the appearance of having been filled too slowly; the erroneous diagnosis unfortunately leading to an erroneous corrective action that will make the problem worse.

In general, if there is a wide choice of time, for instance somewhere between 5 and 25 s, it is strongly recommended to opt for the maximum time, giving the minimum fill rate. This is because, compared with the faster fill rate, the selection of the slower rate reduces the cross section area of all parts of the filling system, in this case by a factor of 5. This is economically valuable, giving a great boost to yield. The filling channels shrink from appearing 'chunky' to appearing like needles (with the confident but erroneous predictions by all traditional onlookers that such systems will never fill!). In addition, there is the benefit that the slimmer filling system actually works better, improving the quality of the casting by giving less room for the metal to jump and splash. Random scrap from pouring defects is thereby reduced. These are important benefits.

On the arrival of a completely new design of casting, the choice of the time to fill the mould can sometimes be impressively arbitrary, with perhaps no one in the foundry having any clear idea on the time to use. Nevertheless, a value that seems reasonable can be tried and can always be modified, preferably extended, on a subsequent trial. Such modifications usually only require the filling system to be modified by a few millimetres.

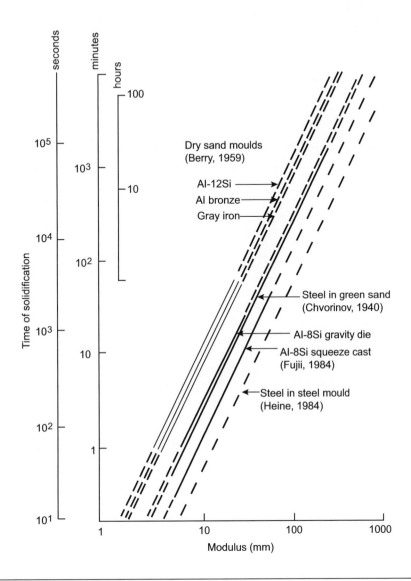

FIGURE 13.4

Freezing times for plates in different alloys and moulds.

An interesting benefit of lengthy filling times is the surface finish of the casting. The oxide on the surface of the melt has time to thicken, so that, if correctly bottom-gated, as the surface oxide slides around the rising meniscus to become the skin of the casting the liquid metal is mechanically protected from penetrating the mould. Also, of course, the development of metal pressure resulting from depth is also correspondingly delayed, further reducing penetration. The result is a beautiful shiny casting.

Another important fact is that, provided the pouring basin is kept filled to at least its designed level, the filling time is not allowed to vary by chance as in a hand-cut running system, and is not under the control of the pourer, but remains accurately under the control of the casting engineer.

13.6 THIN SECTIONS AND SLOW FILLING

It is perfectly normal to assume that thin sections will require to be filled as quickly as possible. However, we need to question this piece of conventional thinking.

It is certainly true that thin sections have to be filled before freezing if freezing would prevent the section filling. This much is obvious. However, the rest of the thinking is rather less obvious, but important for success. The factors that make slower filling usually more successful require to be properly understood.

1. If filling is extremely fast, it is likely to be extremely turbulent. Entry through the gate at speeds much above the critical velocity of 0.5 m/s will probably result in such surface turbulence that jetting and splashing occur. Droplets of melt that jump far above the rising metal and attach to the walls of the mould will oxidise and may freeze. Thus, when the bulk of the melt finally arrives the drops cannot be properly assimilated, forming an oxide lap or even a cold lap. An array of such lap features will lead the founder to think that the melt has been poured too cold, or too slowly, and may increase the temperature (unfortunately increasing the rate of oxidation) or increase the filling speed (unfortunately enhancing the turbulence). Either action will, of course, make the problem worse.

2. When filling at a speed at which the meniscus of the melt keeps itself together, that the melt travels together means that it 'keeps itself warm'. This cosy thought is not the whole story, however. The mechanism of slow filling of extensive thin sections ensures that the hot metal travels through the casting in a fairly straight line directly to the front, spreading laterally on arrival behind the advancing meniscus (Figure 12.32). This effect is extremely powerful in achieving the filling of thin sections. The effect has, perhaps, a minor downside, in that the flow of hot metal through the section to reach the front continues to maintain an open *flow channel*. This channel then freezes late and may lead to a coarser grain than is desirable (although such factors should not lead to any significant deterioration of properties providing the melt is clean). It may also lead to problems to obtain full soundness in this section, but once again, cleanness is a big help to keeping these problems at bay.

13.7 FILL RATE

Having selected a fill time, the *average fill rate* is, of course, simply the total poured weight divided by the total time, expressed in a convenient unit such as kgs^{-1}. This requires it to be converted to the average volume fill rate by dividing by the density of the liquid metal, giving a value in such units as m^3s^{-1}.

Even this value cannot be used directly because the filling system has to be increased in size to take the significantly higher rate of flow at the beginning of the pour. The *initial fill rate* is, of course, greater than the *average fill rate* because the high initial rate is not maintained. The metal slows as the mould fills, the fill rate finally falling to zero if the metal level in the mould finally reaches the same level as that in the pouring basin. To make allowance for this effect, it is convenient to assume that the *initial fill rate* is a factor of approximately 1.5 times higher than the *average fill rate*. This factor is actually precisely correct if the casting is a uniform plate with its top level with the pouring basin, as shown in Appendix 1. However, in general, the factor is reasonably accurate for most castings, being not particularly sensitive to geometry, as can be demonstrated by such exercises as checking the fill times of extreme examples such as a cone filled via its base compared to inverting it and filling via its tip.

A further factor needs to be taken into account; it is found by experience for many castings the frictional drag reduces flow rates in filling systems by approximately 30%, so that filling systems should be speeded up by the factor 1.3 to compensate. Combining the geometrical factor 1.5 with the friction factor 1.3

$$Design\ Factor = 1.5 \times 1.3 = 2.0$$

Thus a *design fill rate* is almost exactly twice the *average fill rate*. When we allow for density, we obtain the *design volume flow rate*, Q, preferably in units of m^3s^{-1}. This is the value to be used for defining the size of all of the remaining features of the filling system that we require to calculate. A word of caution is required at this point. The friction factor is dependent on the precise design of the filling system and so is expected to vary from one design to another. Special caution is required when employing vortex systems such as vortex gates; the dramatic speed reductions possible using these devices leads to design factors of up to 4.0. It would be valuable to carry out research to establish the design factors with greater certainty.

Incidentally, for a given volume flow rate, the mould will fill in the same time whether aluminium or iron is poured (Galileo would have known this). Thus, the system described later applies to all metals and alloys, perhaps to the surprise of many of us who have unwittingly accepted the dogma that each metal and alloy requires its own special system.

Later, when the first mould is poured, the filling should be timed with a stopwatch as a check of the running system design. The actual time should be within 10% of the predicted time. In fact, the agreement is often closer than this (Kotschi and Kleist, 1979), to the amazement of doubters of casting science!

After the first casting is produced, it may be clear that it needs a casting rate either slower (allowing some solidification during pouring) or faster (to avoid cold lap-type defects). These modifications to the rate can be easily and quickly carried out by minor adjustment (usually only millimetres of changes to dimensions are usually required) to the size of the filling channels. Again, it is useful to emphasise, provided the pouring basin is kept correctly filled, which frankly is easy, such changes remain under the control of the casting technologist (not the pourer).

13.8 POURING BASIN DESIGN

The most important issue to address when starting to design a pouring basin is 'Surely it would be better to eliminate the pouring basin?' The answer to this question is 'yes'. Therefore, if possible, the pouring basin should be abandoned and *contact pouring* adopted. It needs to be kept in mind that a good pouring basin design will only assist to eliminate bubbles and other less dense contaminants of the flow, but will never give sufficient time to allow the detrainment of bifilms. Unfortunately, it is clear when watching the filling of pouring basins that masses of bifilms are created in the pouring process. The basin should perhaps be considered an unnecessary evil. Contact pouring, if it can be implemented, avoids this major problem.

Anyway, in case the pouring basin cannot yet be abandoned, this section outlines the requirements of the best of the poor designs possible.

If the basin is designed with a response time of 1 s, it means that if pouring stops the basin will empty in 1 s, we have the volume per second emptying from the basin is Q, so that this equals the working volume of the basin filled to its minimum depth. For convenience, if the basin is a cube of side A, its volume is A^3. Thus in 1 s $Q = A^3$ or

$$A = Q^{1/3} \tag{13.1}$$

Naturally, a '2-s basin' is clearly double this volume. Now proceeding to choose its overall length and breadth dimensions, it is often convenient to adjust the shape by, for instance, halving the width and doubling the length. Taking this area and dividing into the known volume gives its depth. This gives the 'block' of liquid that the basin contains. It is important to remember now to double the height of the basin to give plenty of 'freeboard', so that the pourer can easily maintain the basin overfull, but never drop into the danger zone below the minimum height of liquid at which the filling system would become depressurised and thus be in danger of entrainment issues.

If a stopper is used and the basin is filled with the complete charge to fill the casting, the basin step becomes redundant because there is minimal transverse flow during the draining of the melt into the sprue. For this reason, the step can be eliminated with the benefit that the basin, with the help of a sloping floor, can completely drain empty. Care is needed for such a technique to ensure that there remains sufficient head pressure to fill the outlying regions near the top of the casting. The condition for draining the basin but not losing head height by the melt dropping down the sprue usually has knife-edge criticality. For this reason, a pyramidal or hopper shaped base to the basin allows nearly complete draining without significant loss of head condition to be less sensitively achieved.

13.9 SPRUE (DOWN-RUNNER) DESIGN

Now that an initial rate of pouring has been chosen, how can we achieve it accurately, limiting the rate of delivery of metal to precisely this chosen value? Theoretically it can be achieved by tailoring a funnel in the mould of exactly the right size to fit around a freely falling stream of metal, carrying just the right quantity of liquid metal (Figures 12.8 and 12.10). We call this our down-runner, or sprue.

The theoretical dimensions of the sprue can therefore be calculated as follows. If a stream of liquid is allowed to fall freely from a starting velocity of zero, then after falling a height h, it will have reached velocity v. (The height h always refers to the height from the melt surface in the pouring basin. This zero datum is one of the great benefits of the offset basin compared with the conical basin because the starting velocities can never be known with any accuracy when working with a conical basin.) Using basic conservation of energy; for a mass m falling with acceleration g, equating its loss of potential energy mgh with its gain in kinetic energy $mv^2/2$, mass m appears on both sides of the equation and can therefore be deleted, leaving us with

$$v = (2gh)^{1/2} \tag{13.2}$$

To obtain the sprue sizes, it is necessary to realise that the low velocity v_1 at the top of the sprue must be associated with a large cross-sectional area A_1. At the base of the sprue, the higher metal speed v_2 is associated with a smaller area A_2. If the falling stream is continuous it is clear that conservation of matter dictates that

$$Q = v_1A_1 = v_2A_2 = v_3A_3 \text{ etc.} \tag{13.3}$$

where subscripts 1, 2 and 3 can refer to any downstream location for the local values of the area of the stream and its velocity (for instance the area and velocity at the gates). Because the velocities are now known from the height that the melt has fallen (neglecting any losses from friction at this stage), and Q has been decided, each of the areas of the filling system can now be calculated.

In nearly all previous treatises on running systems, the important dimension of the sprue for controlling the precise rate of flow has been assumed to be the area of the exit. This part of the system has been assumed to act as 'the choke', regulating the rate of flow of metal throughout the whole running system. It is essential to revise this thinking. If the sprue is correctly designed to just touch the surface of the falling liquid at all points, the *whole sprue* is controlling (not just its exit). There is nothing special about the narrowest part at the sprue exit. We shall continue this concept so far as we can throughout the rest of the filling system, along the runner and through the gates; every section is providing some control. If we achieve the target of fitting the dimensions of the flow channels in the mould just to fit the natural shape of the flowing stream, it follows that no one part is exerting control. The whole system is all just as large as it needs to be; the channels of the filling system just touch the flowing stream at all points.

Even so, after such features as bends and filters and other complications, the energy losses are not known precisely. Thus, there is a sense in which the sprue (not just its exit, remember) is doing a good job of controlling, but beyond this point the precision of control may be lost to some extent after those features that introduce imponderables to the flow. (In the fullness of time, we hope to understand the features better. Even now, computers are starting to make useful inroads to this problem area.)

Thus, as long as the caster pours as fast as possible, attempting to fill the pouring basin as quickly as possible and keeping the basin full during the whole of the pour, then he or she will have no influence on the rate of filling inside the mould; the sprue (the whole sprue, remember) will control the rate at which metal fills the mould.

For most accurate results, it is best to calculate the sprue dimensions using the formulae given previously and using the alloy density to obtain the initial volume flow rate Q.

However, for many practical purposes, we can take a short cut. It is possible to construct a useful nomogram for Al (Figure 13.5) assuming a liquid density of 2500 kgm^{-3} and for the dense alloys based on Fe and Cu assuming a liquid density around 7500 kgm^{-3} (these figures are not especially accurate, but allow for a simple nomogram without the introduction of unacceptable errors). Thus, areas of sprues at the top and bottom can be read off and the sprue shape formed simply by joining these areas by a straight taper. Preferably, it is simple to read off areas of the sprue at other intermediate levels to provide a more accurately defined sprue having a curved taper. Recall that the heights are measured in every case from the level of metal in the pouring basin, regarding this as the zero datum.

The nomogram is easy to use. For instance if we wish to pour an aluminium alloy casting at an average rate of 1.0 kgs^{-1}, corresponding, of course, to a design rate of 2.0 kgs^{-1}. Figure 13.5 is used as follows. The 2.0 kgs^{-1} rate with a depth in the basin (the top level down to the level of the sprue entrance) of 100 mm, and a sprue length of 200 mm

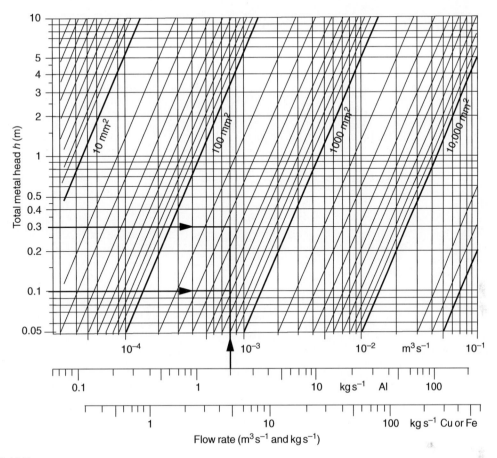

FIGURE 13.5

Nomogram giving approximate sprue areas (mm²) for light and dense metals as a function of flow rate and head height.

(total head height from the top level of the melt in the basin = 300 mm), then its entrance and exit areas can be read from the figure as approximately 580 and 330 mm², respectively. Remember that if the sprue shape is approximated to a straight taper, Section 12.2 advises to increase the area of the entrance by approximately 20% to compensate for errors, making the final straight taper sprue top close to 670 mm². However, a better approach is to form a nicely shaped hyperbolically tapered sprue which is easily defined from the nomogram, and is naturally accurate, avoiding any need for compensation for errors; it is then not necessary to add the 20% to the sprue entrance area.

As a check on the nomogram readouts for our aluminium alloy casting, we can now calculate the dimensions numerically using the equations given previously. At 1.0 kgs⁻¹ average fill rate, corresponding to a design rate 2.0 kgs⁻¹, assuming a liquid density of 2500 kgm⁻³, we obtain an initial volume flow rate $Q = 2.0/2500 = 0.8 \times 10^{-3}$ m³s⁻¹. We can calculate that the falls of 100 and 300 mm are seen to cause the melt to accelerate to a velocity of 1.41 and 2.45 ms⁻¹, giving areas of 566 and 327 mm², respectively. These values are in reasonable agreement with those taken directly from the nomogram.

The cross section of the filling system can, of course, be round or square, or even some other shape, provided the area is correct (we are neglecting the small corrections required as a result of increased drag as sections deviate further from a circle). However, in view of making the best junction to the runner, a slot sprue and slot runner are strongly recommended for most purposes because they can be linked together in a way to preserve the streamlined flow. (Multiple sprues might be useful to connect to several runners. Several such sprues would be expected to work better than one large sprue as a result of improved constraint of the metal during its fall as discussed in Section 12.2.)

If we were to choose a slot shape for the sprue, convenient sizes might be in the region of a 10×60 mm^2 entrance and a 6.5×50 mm^2 exit. (If two sprues were used as shown in Figure 12.13(a), these areas would, of course, be halved.)

Basin + sprue worked example

For a steel casting weighing 30 kg and intended to be poured in 10 s:

$$\text{Average pour rate} = 3 \text{ kg/s}$$
$$\text{Design pour rate} = 6 \text{ kg/s}$$
$$\text{Volume flow rate } Q = (6/7) \times 10^{-3} \text{ m}^3/\text{s (density of liquid } 7 \times 10^3 \text{ kg/m}^3)$$
$$= 0.86 \times 10^{-3} \text{ m}^3/\text{s}$$

Assuming a basin with a minimum working depth h_1 selected arbitrarily as 50 mm (but not forgetting that it would be safe to make this basin 100 mm deep, and aim to always fill above the 50 mm minimum)

$$\text{At sprue entrance velocity } V_1 = (2gh_1)^{1/2}$$
$$= (2 \times 9.81 \times 0.050)^{1/2}$$
$$= (2 \times 10 \times 0.050)^{1/2} \text{ approximately (good enough to within 1\%)}$$
$$= 1.0 \text{ m/s (compared with the accurate result of 0.99 m/s)}$$
$$\text{Because by definition we have } Q = V \times A$$

where V is the velocity and A the area of the liquid stream, so that:

$$\text{Sprue entrance area } A_1 = Q/V_1$$
$$= 0.86 \times 10^{-3}/1.0 \text{ m}^2$$
$$= 860 \text{ mm}^2$$
$$= 29 \text{ mm square or 33 mm diameter}$$

At sprue exit, if height of sprue is 200 mm, plus minimum height in basin 50 mm, total fall is a minimum of 250 mm.

$$\text{Velocity } V_2 = (2 \times 10 \times 0.250)^{1/2}$$
$$= 2.24 \text{ m/s}$$
$$\text{Sprue exit area } A_2 = Q/V_2 = 0.86 \times 10^{-3}/2.24 \text{ m}^2$$
$$= 0.384 \times 10^{-3} \text{ m}^2$$
$$= 384 \text{ mm}^2$$
$$= 400 \text{ mm}^2 \text{ with sufficient accuracy}$$
$$= 20 \text{ mm square}$$

The conversion factor to convert the side of a square to the diameter of a circle with the same area is given by the factor,

$$(4/\pi)^{1/2} = 1.128; \text{ hence, the equivalent sprue exit :}$$
$$= 20 \text{ mm} \times 1.128$$
$$= 22 \text{ mm diameter.}$$

Instead of either a square or round sprue, it will be easier to make a good connection with the runner using a rectangular sprue. Taking therefore the entrance area as 860 mm^2, we could use 10 deep × 86 wide, but this is probably inconveniently wide. Hence we finally chose approximately 15 × 60 mm (rounding up slightly) as an entrance. The exit area 384 mm^2, rounded up slightly to 400 mm^2, could then conveniently become 10 × 40 approximately, giving a useful taper that will allow the sprue pattern to be moulded easily, with a good draw angle on both faces of the rectangle.

If the decision were to go with the hyperbolic tapered sprue form, one of the few disadvantages is the reduction in draw angle as the sprue nears its base. For a tall sprue, the bottom sections become nearly parallel, making withdrawal from a mould a challenge because of the drag of the sand. If the drag problems result in clumps of sand pulled away from the sprue, such defects are likely to be important generators of defects as a result of the high velocities in tall sprues. To avoid these difficulties, the moulding of the hyperbolic sprue is best made as a longitudinally split, two-parted core which is lowered down through a large moulded hole in the cope and drag to meet the runner formed in the base part of the mould assembly. This option involves the problem of accurate alignment of sprue exit and runner entrance, which, if not perfect, unfortunately creates probably more problems than it solves.

13.10 **RUNNER DESIGN**

Taking a simple turn from the sprue exit into the runner, we retain the same area 400 mm^2 around the curve. The runner entrance will therefore have dimensions 40 mm wide × 10 mm deep. The inside radius of this turn should be at least approximately one or two times the thickness of the channel, thus we shall choose 10 mm, or if there is room on the pattern plate, 20 mm. Preferably the outside radius of the bend should be formed to give a parallel curve. If the outside curve cannot be moulded and has to be left as a right angle, then the inside curve radius would be better reduced back to 10–15 mm to avoid the otherwise large volume that would be created in the right-angle corner. Any unnecessary volume in the flow channels presents the risk of air bubble entrainment. The channels are required to be as small as possible, while still delivering the required flow. Their small area will constrain the flow, preferably giving it no chance to fold over on itself, or jump or splash. It should fill the channel immediately, progressing as a piston to push the air ahead.

The melt will be travelling close to 2.24 ms^{-1}. The problem remains, 'How to get the speed down from this value to at least 1.0 or preferably even 0.5 ms^{-1} at the gate *without* causing damage to the flow because of the speed reduction mechanism?' This is *the* central problem for the design of a good filling system. This central problem in the past either seems to have been overlooked or to have solutions proposed that do not work. At this point, we need to appreciate the possible solutions with some care.

We only have a limited number of strategies for speed reduction. At this stage of the development of the technology, the options appear to include the following.

1. Filtration
2. Several right-angle bends in succession (Jolly et al., 2000). This reduces speed but is in great danger of introducing turbulent entrainment of air and oxides.
3. A bypass runner design acting in a surge control mode, calculated to introduce the melt through the gate at the correct initial rate for a sufficient time to cover the gate with a depth of metal. (The rate through the gate increases later of course when the surge volume is full, but any jump or splash is now inhibited by the depth of metal in the mould above the gate.)
4. The neglect of the problem of speed reduction in the runner, and simply tapering the runner to balance flow through several gates, making the total area of gates sufficient to correspond to, say, 1 m/s.

Considering our first option: if the filter reduces the flow rate by a factor of 5, then in principle the runner exit from the filter could be increased in area by a factor of five. We shall choose a value of 3 to give a good margin of safety, helping to ensure that if the filter is placed near the runner entrance, the downstream part of the runner is properly filled and slightly pressurised. Its area would then be 3 × 400 = 1200 mm^2. The dimensions of the slot runner after the filter might then be 12 × 100 mm. This is an unsatisfactory size of runner. The large width would be in danger of collapse because of mould expansion for a high melting point material such as an iron or steel cast in a silica sand mould (although the runner would be expected to survive for an Al alloy). In addition, the depth at 12 mm is too deep because the height of the progressing liquid, now moving at a modest speed under 1.0 m/s, will be only about 8 mm. Thus, the runner will not initially fill the section and thus air cannot be pushed ahead in a kind of piston-like progression of the metal in the runner. This unsatisfactory size of runner forces us to abandon this approach.

It would perhaps be more convenient if the runner were divided into two runners of 10×60 mm, or better still, three runners of dimensions 8×50. More runners would help to constrain the melt even more, helping to avoid air and oxide entrainment by encouraging the once-through filling mode (i.e. no room for the damaging action of reflected back-waves). Much depends on the layout of the mould and the filling system. Two runners might be more conveniently filled from two separate sprues, possibly exiting from the opposite ends of a suitably modified single basin, complete with a central pouring well and vertical steps either side as illustrated in Figure 12.13.

An alternative arrangement can be envisaged. Instead of placing the filter close to the runner entrance, the runner is continued at its 10×40 size, and vertical gates rise from the runner, each containing a filter. The total area beyond the filters is required to be at least 1200 mm^2 to ensure the melt enters the mould cavity without jetting. Care needs to be taken with this solution to ensure that the large gate areas do not create hot spots at the junction with the casting. The gates may therefore require to be slimmed down to narrow slots having half the thickness of the local thickness of the casting (Chapter 5 section on Fins). The runner may require to be tapered depending on the particular geometry.

The second option, using a succession of right-angle bends, is only recommended if a good computer simulation package for flow in narrow filling systems is available to test the integrity of the flow (most simulation packages do not predict flows accurately; most cannot cope with thin sections; and most cannot cope with surface tension). If a proven software package to simulate flow is available, a reasonable solution can be found largely by trial and error along the lines of the development described by Jolly et al. (2000). This approach is not described further here.

The third option can be a good solution. Figure 12.37(a) illustrates the layout. The gradual filling of the surge riser will cause the metal in the gate to experience a gradually higher filling pressure. At the point at which the surge riser is filled, the pressure comes on to the gate from the full height of the metal in the pouring basin. At this instant, the casting should be filled to some depth at least 20–30 mm above the gate, so that any jetting into the mould when the full filling rate comes into effect will be to some extent suppressed. (The precise depth to suppress completely the formation of bifilms remains to be researched.)

As mentioned earlier, it is sensible to arrange the overflow, or surge riser, to be a cylinder and connected tangentially to the runner. In this way, by avoiding unnecessary turbulence and filling more progressively, the buildup of backpressure to fill the gates is smoothed, and, not unimportantly, a better quality of metal is preserved in the cast surge riser for future recycling within the foundry.

The careful sizing of surge risers to suppress the early jetting of melt through the gates is strongly recommended. To the author's knowledge, the technique has been relatively little used so far. More experience with the technique will almost certainly lead to greater sophistication in its use. Unfortunately, it is complicated to calculate the dimensions of the components of this system. I recommend therefore that the dimensions for the system are first guessed on the basis of experience, and then tested by computer simulation. One or two iterations will normally result in a closely optimum solution.

A recent development of this approach is the use of the *Trident Gate* (Section 12.4.20). The total exit areas from the gates remain unchanged from the values found in our calculation, but the two filters in each gate ensure tranquil, bubble-free filling. The tangential surge cylinders attached to the runner extensions, although essential for the effective operation of the Trident device, appear no longer sensitively critical in size as when used alone to control flow into the mould.

Our fourth option, to accept the high speed in the runner, and simply work out the areas of gates needed to transfer melt into the mould at around 1.0 m/s, actually works satisfactorily well in several cases. These circumstances include the following.

1. Where the ingates are taken vertically off the runner into the casting, the benefit of gravity to assist the gates to fill is useful, although the gates need to be reasonably tall to gain the full benefit of this effect. I try always to make vertical gates at least 50 mm tall, but prefer 100 mm if I can get it.
2. Those instances where it is known the ingate velocity is not under good control, and still dangerously high, but the geometry inside the mould cavity does not allow significant turbulence. Thus, if we can gate straight onto a core that

defines a narrow wall section, the metal can spread under control of the narrow section, being too constrained to be able to jump or splash. The spreading front increases its area and consequently slows as it fills the narrow wall. This works well.

13.11 **GATE DESIGN**

In general, it is essential that the liquid metal flows through the gates at a speed lower than the critical velocity so as to enter the mould cavity smoothly. If the rate of entry is too high, causing the metal to fountain or splash in the mould cavity the battle for quality has probably been lost. Unfortunately, this often happens even though the casting engineer has correctly calculated the area of gate required for a gentle filling action. This is because in gravity-filling systems, the high metal velocities along the runner are not helpful in allowing the gate to fill properly; if the gate is only half filled or less, the velocity through the remaining area will be correspondingly increased. It is sometimes not easy to arrange for the even filling of gates.

The gate should enter at the very base of the casting, if possible at right angles onto a thin section, as has been described earlier. Gating horizontally directly across a flat floor of a casting is to be avoided if possible – a thin jet of metal skating across a flat surface is a recipe for mould expansion defects of various sorts that will spoil the surface of the casting as illustrated in Figure 13.6. The casting will also be at risk from the formation of an oxide flow tube that may constitute a serious internal discontinuity in the casting. A further risk is the filling of features on the far side of the casting from the gate; this metal is relatively cold and is vulnerable to the formation of cold laps. If possible, it is usually far better to ingate vertically into one edge of the casting, the melt emerging via slots in the floor of the casting. Such ingating causes the metal to swell slowly from the gate, roll nicely over the floor and up the walls of the casting, while keeping the advancing front warm by flow channel behaviour, avoiding cold lap issues.

For our example casting, the velocity at the base of the sprue is 2.45 ms^{-1}. Thus to achieve 0.5 ms^{-1} through the gate(s), and if no friendly core is conveniently sited onto which to gate directly, we shall require an expansion of the area compared to that of the base of the sprue by an approximate factor of $A_2/A_3 = 5$. In terms of the gating ratio much loved by the traditionalists among us, we are using 1:1:5 for this casting. The use of this size of gate assumes that we are gaining no advantage from a bypass runner and surge rise design. If a good bypass design could be devised, an acceptable ratio might then become 1:1:1 effectively easing subsequent cutoff and reducing any possible problems of hot-spots or convection at this location.

Sometimes the bypass cannot be provided. Even if available, it may be useful to use both the bypass and the enlarged gate until such a time that our understanding of filling systems makes it clear that such belt-and-braces solutions are not required.

It is common for gates to be taken at right angles off a runner. Nafisi (2010) shows how the melt turns this right angle with difficulty, creating a vena contra and a vortex stirring pattern of flow in the gate. With gates at right angles to the runner, it is necessary to reduce this problem by providing a generous radius to the inside of the bend. For the future, more sinuous designs of runner/gate connections will become the norm, like branches of a tree, or perhaps more appropriately, branches of a river. Such solutions have to be developed in terms of useful designs that the foundry person can implement easily and quickly. With the numerical control machining of tooling, such options are looking attractive.

Where the gates form a T-junction with the casting, the maximum modulus of the gates should be half of that of the casting (if, as will be normal, no feeding is planned to be carried out via the gates). Thus, in general, the thickness of the gates needs to be less than half the thickness of the wall. This forces the shape of the gates to be usually of slot form. If made especially thin when casting alloys of good thermal conductivity, such as Al and Mg alloys and some Cu alloys, the gate can sometime be usefully employed to act as a cooling fin soon after the filling of the casting.

The other major consideration that must not be overlooked is the problem of the transverse, or lateral, velocity of the melt in the mould cavity as it spreads away from the ingate. This can easily exceed the critical velocity despite the velocity in the gate itself being correctly controlled. In this case a single gate may have to be divided to give multiple gates as described in Section 12.4.7.

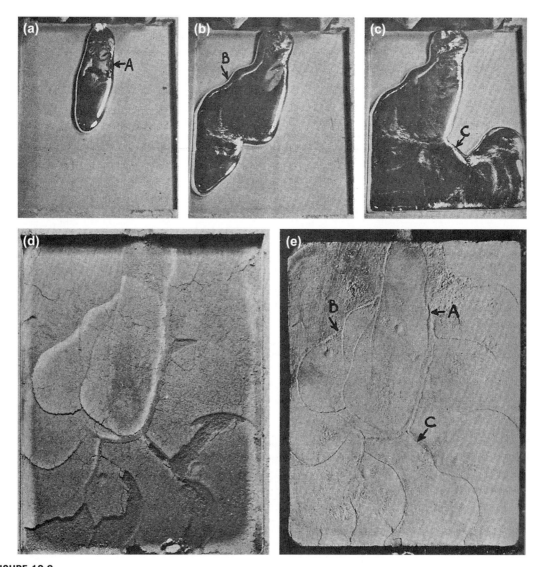

FIGURE 13.6

Relation between metal flow pattern over a horizontal surface and the development of rattails. (a–c) flow of metal; (d) the final mould surface; (e) the casting surface (image reversed to aid comparison) (Goad, 1959).

The area of the gate required to reduce the gate velocity to below the critical velocity, and the limitations of its thickness, sometimes dictates a length of slot significantly longer than the casting. In this situation, there little choice but to revise the design of the filling system, selecting a correspondingly longer fill time so that the gate can be shortened to fit the length available. Alternatively, a bypass runner design may be the solution. If a solution cannot be found, the conclusion has to be accepted that the casting as-designed cannot be made so as to enjoy reliable properties and performance. A serious discussion with the designer will probably be required.

PROCESSING (MELTING, MOULDING, CASTING, SOLIDIFYING)

MELTING

The economics of melting are key factor in foundry costing. For instance, if aluminium could be melted at 100% efficiency (i.e. no heat losses) to a temperature of 720°C, the energy required would be 320 kWh/t. For steel to 1600°C the requirement is nearly identical at 330 kWh/t. Naturally, real furnaces tend to fall a long way from this ideal, so that although some achieve in the region of 650 kWh/t, others are nearer double this. Furthermore, holding furnaces are essentially 100% *inefficient*, although for production and quality reasons their use may be valuable. In this account we shall not dwell much on the energy requirements and the economics, for it is the subject of many studies and publications and so is well understood.

In contrast, it is rare for the quality of melts to be considered. The quality can be significantly affected by the melting, holding and transfer techniques, and, of course, directly affects the quality and economics of production. This is the special subject of this section.

Steel founders have a wide choice of melting furnaces. Nowadays, only induction and arc type furnaces are generally employed in foundries. The older reverberatory roof designs such as the open hearth furnace are not generally used. Whatever the variety of furnace, from the point of view of quality of the liquid metal they produce, all varieties are probably capable of delivering a high quality product. The special reason for this is the large density difference between liquid steel and its oxides in suspension causing the oxides to float out relatively quickly.

It is regrettable that the oxides are usually efficiently re-introduced during the emptying of the furnace into the ladle, in which the melt performs a simulation of Niagara Falls. The saving feature of this trauma is that much, but probably not all, of the damage detrains within minutes, floating to the surface to join the slag layer before the melt is cast.

The situation is quite different for aluminium alloys in which the oxides (and possibly nitrides) are similar in density to the melt, and so are resistant to separation by flotation or sedimentation. Thus aluminium and its alloys present special difficulties if a relatively clean melt is desired. We shall give some space later to an examination of this problem.

The lightweight magnesium alloys represent the opposite extreme, in which the oxide is significantly denser than the melt, and so sinks rapidly. The problem in this case is to avoid disturbing this sediment if a good melt quality is required.

14.1 BATCH MELTING

14.1.1 LIQUID METAL DELIVERY

The delivery of liquid metal by tanker, often travelling many miles by road or rail from the smelter, is an obvious choice because, it saves cost to the smelting company of having to solidify it into ingots, and saves the cost to the foundry of re-melting the material in-house. The logic seems inescapable.

For deliveries of liquid iron there may not be a problem. The author is aware of only one such example of iron in the United Kingdom that travelled about 30 km by railway and appeared to be completely satisfactory.

For aluminium alloys, liquid delivery is more common. Even so, the author remains concerned that the cost savings may be more apparent than real. It is worth discussing the issues later.

For those foundries that produce one-off castings, or small batches, especially if the castings are in the tonnage size range, the arrival of a large quantity of metal at one time would be ideal. However, these are the very organisations that prefer to organise their own supplies of melt, keeping timing and quality issues under their own control. Also, of course, from a melt supplier's point of view, a smelter needs a regular, predictable delivery schedule, not huge quantities at long and irregular intervals.

In high volume operations, the author favours melt systems in which the melt level never changes outside closely controlled limits. The abrupt arrival therefore of a huge mass of metal into the system is problematic. If uncontrolled, it would cause large changes in metal level. Control measures involve costs in the form of large receiving furnaces that either tilt or are equipped with pumps or automatic ladling systems.

Typically, the melt is transferred from the tanker, empties down a chute and cascades into a large holding furnace. The furnace is a large capital item, and its life not made easy by the washing of melt up and down its walls. Also, although melt losses from such units are not generally made known, one would expect a loss by the conversion of several percentages of the melt to dross during this transfer. Up to half of the melting energy saving therefore disappears immediately in loss of metal. The depreciation, running and maintenance costs of the receiving furnace are a further loss.

For Al alloys, the loss that gives me most cause for concern, hidden and unsuspected, derives from the damage caused to the liquid metal by this traumatic transfer operation. The melt will be loaded with bifilms to the extent that it will be difficult to clean the melt sufficiently to make adequate castings. Down-stream processes such as rotary degassing and sink-and-float holding furnaces can probably only eliminate a percentage of inclusions. If, for instance, the processing is successful to remove 90% of solids, the remaining 10% may still be sufficient to degrade the castings to the point of unacceptability in terms of its ductility.

Thus, in summary, I find the supply of bulk liquid metals unattractive economically and technically.

14.1.2 REVERBERATORY FURNACES

Melting in a shallow pool can most conveniently be carried out by heating from above. The use of flames to achieve this is not helpful if the flame is directed to impinge directly on the metal. In this way, the melt is oxidised locally, and one would expect its hydrogen content to be increased. In addition, the transfer of heat in this way is not particularly efficient. It is far more effective to heat the roof of the furnace rather than the charge. This effect can be enhanced by ensuring that the flame is luminous (sometimes by adding carbonaceous additives to the fuel). In this way, the roof becomes incandescent.

This is how the furnace got its name: heat from a luminous flame radiated to the refractory roof of a furnace is reverberated (i.e. reflected or re-radiated) to the melt below. The Siemens open hearth furnace was an early example that made most of the high quality steel in the world for many years. Later types of furnace using electrical heating elements in the roof have taken up the name of electrical reverberatory furnaces ('electric reverbs' for short!). Such furnaces for Al holding can typically transfer up to 120 kW/m^2 to the melt.

The radiant roof principle has been useful for furnaces in which turbulence is at a minimum because the melt is heated from the top, leaving the bottom significantly cooler, causing a stratified temperature gradient that promotes stable conditions, reducing convected stirring to a minimum.

Such stratified stability can be counter-productive on occasions. The author has vivid memories of working a 1000 kg resistance-heated radiant roof furnace containing nominally Al-7Si-0.4Mg alloy. An Al-10Mg ingot was slipped into the bath on a Friday evening to raise the Mg level from its low value 0.3%Mg, where it had slowly fallen over the past week, back to 0.4%Mg. At the start of the next shift on Monday morning, the Mg content of the furnace was checked. Everyone was aghast to discover that it had not changed, but if anything, had fallen further. While we stood around baffled, a bright team member suggested that the ingot might have melted but its liquid metal might still be lying at the bottom of the furnace (its low temperature more than countering the fact that at 10%Mg the liquid might have been expected to float). This was found to be true. On stirring for a few minutes with a coated iron bar, the whole melt rose to its expected 0.4% Mg content.

For Al alloy holding furnaces, the stratification is eliminated by many users by several strategies; an immersed diffuser giving a train of bubbles will cause a local upwelling of the melt and thus a gradual turnover. Other furnaces use more aggressive systems such as centrifugal or electromagnetic pumps particularly if melting is required to occur in the furnace; melting in holding furnaces is otherwise extremely slow because the low thermal conductivity of liquid Al compared with solid Al. The author admits to all these uses being counter-productive to one of the main features of the Al alloy holding furnace; its utility in achieving some quiescence to allow the sink and float of inclusions, mainly bifilms of course, so as to achieve a good quality melt.

The rotary furnace ensures no such stratification of the charge. These are occasionally seen in aluminium foundries. In addition to ensuring the homogeneity of the melt, the design is reasonably thermally efficient because the heated roof is circulated under the melt where it can heat the charge most efficiently by direct contact. Such units can be useful for secondary smelters who wish to melt dross to recover the entrained aluminium alloy. The addition of a chloride flux based usually on common salt, sodium chloride, plus sometimes potassium chloride, significantly assists this process. For foundry use for Al alloys, however, the rotary furnace cannot be recommended because of the damaging amount of surface turbulence that it causes. The unit produces only rather poor quality metal.

14.1.3 CRUCIBLE MELTING (ELECTRIC RESISTANCE OR GAS HEATED)

In common with all bath and crucible type melting processes, crucible melting necessarily includes everything that is 'thrown into the pot'. This self-evident fact is often overlooked, but in fact conceals the important feature of this melting technique; the surface oxide on the *charge* necessarily forms part of the melt. These primary oxide skins can be the size of newspapers, as has been commented before. They represent a large content of thick oxides in suspension in the liquid metal. This is a good reason for not choosing a crucible furnace for metal supply in an aluminium foundry.

It is valuable to keep in mind that at least two populations of oxide bifilms are likely to be present in all melts: (1) the primary oxide skins from the charge and (2) the inherited population of bifilms contained in the charge materials. To these populations will be added those created during pouring and those created during surface turbulence inside the mould.

Panchanathan et al. (1965) found that when melting was carried out repeatedly on 99.5%Al progressively poorer mechanical properties were obtained. By the time the melt had been recycled eight times, the elongation values had fallen from approximately 30 to 20%. This is easily understood if the oxide content of the metal is progressively increased by repeated melting of castings.

The re-melting of aluminium alloy sand castings probably represents one of the worst cases because the oxide skin on a sand casting is particularly thick, having cooled from high temperature in the aggressively moist oxidising environment of the sand mould. When the skin is submerged in the melt, it can float about, substantially complete. One day, when re-melting scrapped cylinder heads by adding them in to one end of a combined melting and holding furnace, the author has seen the skin of a complete cylinder head fished out from the opposite end of the furnace, complete with all details such as combustion chambers and ports. (Someone joked that we should fill it with liquid metal and re-sell it.) It is easy to understand how the condition of some melts can be so bad as to resemble a slurry of old sacks. Unfortunately, this is not unusual.

The re-melting of aluminium alloy gravity die (permanent moulded) castings have an oxide skin that is much thinner as a result of the relatively inert environment provided by the metal mould, and seems to give less problems. Whether this is a real or imagined advantage in view of the damage that can be caused by any entrained oxide skin, irrespective of its thickness, is not clear at this time.

A further problem with crucible melting is that the debris in suspension usually continues to circulate for long periods because the heating of the crucible is via its walls, creating an upward convective flow adjacent to the walls and down in the centre of the melt. Although the rate of circulation is relatively gentle, the velocities will almost certainly exceed the Stokes settling velocity of the oxides. Thus oxides have little chance to separate by buoyancy, i.e. will be unable to sink or float.

A further aspect of crucible melting that has been little researched and remains only a suspicion at this time is the inexorable rise of iron contamination of Al alloys. Although iron tools are often blamed, many furnace operators take care to protect the implements from dissolution by a refractory wash coat. The suspicion is that at least some of the iron contamination arises from the reduction of iron impurities in the protective glazes or walls of some crucibles.

14.1.4 INDUCTION MELTING

Induction melting is another variety of crucible melting which suffers from the problem of all the surface oxide skins on the charge being included in the melt.

Although crucibles heated via their walls promote stirring of the melt, the induction furnace provides at least 10 times the velocity of stirring. Thus the original surface oxides incorporated into the melt are efficiently maintained in suspension; the strong stirring component ensures that none separate by buoyancy. Furthermore, there remains the suspicion that the stirring action may be sufficiently vigorous to shred oxides, or even manufacture new surface oxide and entrain it in some way.

The particularly violent stirring in induction furnaces of high power to weight ratio (more than approximately 10 kW/kg) is seen in crucibles in which the melt is partially levitated by the inductive forces. This action almost certainly introduces oxide by the continuous fluctuations in the area of the surface as the melt shudders and wobbles like an unsupported jelly. These conditions typify melting in water-cooled copper crucibles, the skull-melting process, so-called from the original hemispherical crucibles against which solidified a shell of solid during the melting process. On lifting this from the crucible after pouring, the residue resembled a skull cap. Thus skull melting would be expected to introduce masses of oxides (although this has never been proven).

14.1.5 AUTOMATIC BOTTOM POURING

Automatic bottom pouring is a technique used in the investment casting industry for the vacuum melting and casting of lost-wax moulds. It is a surprisingly simple process invented by Alec Allen in the United Kingdom in 1966. A hole in the base of a refractory melting crucible was covered by a cylindrical slug of metallic charge (often steels or Ni-base alloys) that was a close fit in the crucible. The slug was melted rapidly by induction. The rapidity of the melting meant that a low-cost disposable sand crucible was normally adequate as a melting crucible. The last part of the charge to melt was the portion over the hole. When this melted, the liquid metal poured through into the mould below. A variant of the process used a small coin-sized disc that fitted into a recess above the hole. The temperature of the pour could be controlled by selection of the thickness of the disc.

However, of course, a significant issue is the procurement of a pre-alloyed charge of exactly the size to fit the crucible and exactly the weight required by the casting. Moreover, of course, the Achilles' heel of this approach is simply the pour. Practically all processes that pour metal create defects. Regrettably, casting technology needs to get away from pouring.

The pushup base crucible process is a similar rapid 'just-in-time' melt production and casting technique that similarly uses pre-prepared 'logs' for melting, but achieves the avoidance of pouring by a commendably elegant technique. It is described with the counter-gravity casting processes in Section 16.33.

The automatic bottom pouring approach might be workable when linked with the concept of contact pouring. The worst effects of the pouring are then eliminated. A modified process might be capable of the production of excellent castings.

14.2 CONTINUOUS MELTING

14.2.1 TOWER (SHAFT) FURNACES

The tower or shaft furnace is one of the most efficient melting units because it is a true counter-current heat exchanger. Cold charge enters at the top of the tower and slowly descends, while spent gases from the melting zone pass up the tower, preheating the charge before its arrival in the melting zone, and cooling the gases before they emerge from the top of the shaft.

The cupola; a shaft furnace for iron

Iron foundries have had the benefit of such a furnace for many years in the form of the cupola. Because the melted iron passes over a deep bed of incandescent coke, it can be heated to a high temperature if necessary. The skilled furnace manager can manipulate and hold the metal temperature and composition with impressive precision.

Only recently has the coke-fired cupola started to give way to induction melting, not primarily because of inefficiency, but because of the twin problems of obtaining good quality coke and controlling the effluent into the environment from the top of the stack.

Some moves have been made towards the gas-fired cupola. Here the melting stack is supported not by coke, but by a bed of refractory spheres. Unfortunately, the spheres are eroded quickly by attempts to produce metal at a high temperature, greatly increasing costs. Thus, the gas-fired cupola has much in common with dry hearth furnaces, in which liquid metal is best produced at temperatures only just above the melting point of the metal or alloy. The gas-fired cupola is therefore said to be a good melter, but poor superheater of metal. A duplexing operation is an efficient solution in which an induction-heated holding furnace raises the temperature of the cupola iron to that required for casting.

The sink and float issues that are so important in the melting of aluminium alloys hardly applies in the case of the melting of irons and steels. This is simply because of the great density difference that the heavy metals enjoy with respect to entrained defects. Detrainment of harmful entrained bifilms in these metals is fast not only because of the larger density difference, but also because of the much higher temperatures, causing the entrained surface films to be molten or at least partially molten. Thus, the entrainment defects are compact and resistant to any unfurling action, being either spherical liquid drops or compact sticky masses that can float out quickly despite strong convective stirring. Those microscopic silica-based bifilms that are predicted to be present in liquid iron (Section 6.5) seem, fortunately, to remain in suspension for much longer periods, although the Monday morning melt quality from an induction furnace usually requires some attention.

14.2.2 DRY HEARTH FURNACES FOR NON-FERROUS METALS

For aluminium alloys, the vertical shaft furnace has become popular in recent years for good reason. The base of the tower is usually a gentle slope of refractory on which the charge in the tower is partially supported (other support comes from the sloping walls of the shaft). In addition to its melting efficiency from the excellent use of the heat available from the combustion gases, the unit can provide unusually clean metal. This is because the outer oxide skins on the charge do not enter the melt, but are left in place on the dry hearth; the pieces of charge collapse like punctured balloons as the melt escapes as a little stream via a hole in the skin, and flows as a cool liquid, practically at its melting point, out of the skin of the melting charge, down the hearth, travelling through an extended oxide tube and into the bulk of the melt. After its arrival in a holding volume, where some homogenisation of the melt can take place, it is also brought up to a useful casting temperature. In the meantime, the outer oxide skins pile up on the dry hearth and are raked off from time to time via a side door.

The concern that this type of furnace might result in a high hydrogen content in the melt from the volume of preheating 'wet' combustion gases is understandable, but seems unfounded. It is true that as the moisture content of the environment of an Al alloy melt increases, the hydrogen level of the melt increases because of the well known reaction

$$3H_2O + 2Al = 3H_2 + Al_2O_3$$

Thus, one would expect from the thermodynamics that higher moisture would necessarily lead to higher hydrogen levels in the melt. However, above a certain limiting concentration of moisture, the character of the oxide on the solid metal in the charge changes, becoming more resistant to further oxidation and the consequent absorption of hydrogen. Thus, somewhat counter-intuitively, the very high moisture content of the spent gases acts to resist the rapid takeup of hydrogen by favourably affecting the kinetics, despite the unfavorable thermodynamics (see Figure 1.3).

A variant of this type of melting furnace is the use of a reverberatory type of dry hearth unit in which the heat of the melting gases is recovered by the use of recuperative burners (i.e. burners that recoup, or recover, the heat). Such combustion units operate in pairs, with one burner firing at a time, the hot spent gases being used to preheat the recuperator system for the second burner. When the incoming gases for the first burner have cooled its recuperator sufficiently, the heating switches to the second burner and so on. The useful feature of this design is that similar melting efficiencies to the tower design are claimed, but roof height requirements are greatly reduced, and the loading of the furnace can be achieved at floor level by fork lift truck. Interestingly, however, in their critical study of a 130 ton reverberatory furnace, Hassan and Al-Kindi (2014) find the recuperative seriously inefficient for a variety of practical problems which seem mainly associated with the horizontal aspect of the furnace, requiring doors to be opened for its operation, and which therefore impairs its efficiency.

In contrast, the dry hearth tower melter enjoys fundamental design benefits which appear to make it the best option for the melting of Al alloys.

These benefits have extended to the melting of copper in an Asarco type furnace. Again, the design is a simple tower, gas heated at its base, and equipped with a dry hearth.

The key feature of all the dry hearth melters is that they necessarily produce liquid metal at only just above its melting point. This may be an advantage or disadvantage depending on circumstances. Because of the very low temperature of the liquid metal, the melt has a relatively low gas content. For many casting systems, therefore, a second furnace, or combined second chamber, will be required to raise the temperature of the liquid up to a convenient casting temperature.

A second feature of the dry hearth is that the system can provide liquid metal 'on tap' i.e. as a just-in-time delivery system. Thus, a large inventory of liquid metal is not required in the melt supply. It can be switched on or off or simply increased or decreased in rate simply by turning up or down the burners at the base of the stack.

Some caution is required when using the furnace in this mode for those aluminium alloys containing high levels of copper. In this case, there is a danger related to such changes in rate; the chemical analysis of the delivered metal changes somewhat. This happens because on slowing down the rate of delivery the eutectic continues to melt and flow for longer, causing the melt to become enriched in Cu-rich solute. Conversely, on speeding up the delivery rate the first metal to melt is now depleted in eutectic, and so is lower in solute.

For this reason, the furnaces are better used continuously, with only gradual changes in rate so far as possible if high Cu alloys are being used, or are necessarily connected to a holding furnace in which the melt can be homogenised, as described later. Also, naturally, the furnace is design to work at greatest efficiency when working near its maximum rate. Thus slow or zero rate melting will be costly if carried out to excess.

The additional advantage of a dry hearth furnace for aluminium alloys is that foundry returns that contain iron or steel cast-in inserts (such as the iron liners of cylinder blocks or valve seats in cylinder heads) can be recycled. The insets remain on the hearth and can, from time to time, be raked clear, together with all the dross of oxide skins from the charge materials. (A dross consists of oxides with entrapped liquid metal. Thus most dross contains between 50% and 80% metal, making the recovery of aluminium from dross economically valuable.)

The benefits of melting in a dry hearth furnace are, of course, eliminated at a stroke by the misguided enthusiasm of the operator, who, thinking he is keeping the furnace clean and tidy and that the heap of remaining oxide debris sitting on the hearth will all make good castings, shoves the heap down the slope and into the melt. Unfortunately, it is probably slightly less effort to push the dross downhill rather than rake it out of the furnace through the side door. The message is clear, but we are required to remind ourselves frequently: Good technology alone will not produce good castings. Good training and vigilant management remain essential. My old foundry manager used to warn me: 'You need at least six pairs of eyes to run a foundry'.

14.3 HOLDING, TRANSFER AND DISTRIBUTION

The function of a holder is generally well known. It is a reservoir intended to smooth rate changes between melt production and consumption and allows the smoothing or adjustment of both composition and temperature. However, it should not be overlooked that holding furnaces can have a significant effect on melt quality because they allow time for detrainment of defects by sedimentation or flotation.

However, the effectiveness of the holder in achieving these goals depends strongly on the method of heating, and the operating practice.

For Al alloys, a hot roof furnace, for instance a reverberatory (i.e. roof reflecting) furnace, is always recommended. Here the heat transfer from electrical elements or from a fuel flame is obtained by heating the roof, which then radiates its heat down to the melt surface. The formation of dross is minimised by the uniform distribution of heat over the whole surface of the melt. (This action contrasts with the practice of directing a flame directly onto the surface of the melt. The higher local concentration of heat from the flame impingement is suspected to lead to a higher rate of hydrogen pickup. In addition, the higher temperature and the constant mechanical agitation of the surface will increase dross formation.)

However, heating from above causes a stratification of the metal in the bath. The stratification ensures that any added metal does not mix, but simply finds its own level according to its density, and then travels in a straight line to the exit point, bypassing all the other metal layers above and below which remain stagnant in the furnace. Its composition is marginally smoothed by a small amount of turbulent mixing, and a minor contribution from diffusive interchange with layers above and below. However, clearly, the smoothing action is not achieved with the effectiveness that might be expected because most metal in the furnace remains unaware that any other metal has passed through the system.

This problem is not expected to arise with a more recent design of holder that uses immersion heaters. The immersed, closed-ended tubes consist of a ceramic such as silicon carbide (SiC) that exhibits good thermal conductivity and good resistance to attack by liquid aluminium. The tubes can be heated internally by gas or by electrical elements. In this way, heat is provided directly and at considerable depth in the melt, with the result that the melt against the walls of the heater tube generate a convective upwards flow that drives a circulation in the whole bath.

Other techniques to circulate the melt are common, by, for instance, the passage of bubbles of inert gas from a diffuser or by the action of a rotary degasser. Other relatively common techniques include the use of an immersed electromagnetic pump or an electromagnetic pump sited on the outer wall of the furnace, but inducing a motion in the melt through the wall.

Holding furnaces for cast irons are generally heated by gas or induction or both. If induction is used, the melt will certainly be well mixed.

For those holders that are stirred by some means, any changes of temperature or chemistry are, of course, smoothed with maximum effectiveness. However, in the case of the production of Al-alloy melts, such stirring action acts in opposition to one of the most valuable functions of the holding furnace, that of allowing inclusions to sink or float.

For Al alloys, the principal and most damaging inclusion is the bifilm. Because this has close to neutral buoyancy, its settling velocity is often much lower than a velocity imposed by stirring. Thus, if the melt is stirred, the defect can remain in circulation, never settling. In the view of the author, the effects of small changes to the composition of the melt and its gas content are generally insignificant compared with the effect of the presence of bifilms. Thus, our thinking regarding priorities in the operation of a holding furnace require to be re-thought. On balance, I am of the opinion that a holder should be left to itself in view that most melt treatments are probably counter-productive. An Al alloy melt benefits from quiet neglect.

The Cosworth Process was perhaps the first to acknowledge that for liquid aluminium alloys the oxide inclusions in a holding furnace could be encouraged to separate simply by a sink and float principle; the metal for casting being taken by a pump from a point at about midway depth where the best quality metal was to be expected. Degassing by argon bubbling was almost certainly counter-productive to melt quality by creating a circulation that retained bifilms in suspension.

The holding of melts in a closed pressurisable vessel for the low-pressure die casting of aluminium, or the dosing the liquid metal, are usually impaired by the initial turbulent pour of the melt to fill the furnace. The total pour height is often of the order of a metre or more. Not only are new oxides folded into the melt in this pouring action, but those oxides that have settled to the floor of the furnace from the last filling operation are stirred in to the melt once again.

A design by the author, patented by the company, Alotech, illustrates that simple equipment such as a holding furnace for liquid aluminium is capable of considerable sophistication, leading to the production of greatly improved processing and products (Grassi et al., 2012). It avoids convection and the disadvantages of stratification by making the bath shallow. Degassing occurs passively by the diffusion of hydrogen out of the shallow layer of metal into a counter-current flow of nitrogen, thus avoiding any disturbance of the melt (such as would occur with rotary degassing for instance). The sedimentation of bifilms is encouraged to the maximum extent by the elimination of stirring. It takes the concept of melt cleaning and degassing to an ultimate level that might represent a limit to what can be achieved. In addition, the technique is simple, low capital cost, low running cost, has no moving parts and is operator-free. At this time, the technique is being applied only to aluminium alloys.

Although the practice of quiescently holding liquid aluminium is well known in the industry to improve the properties of the melt, Raiszadeh and Griffiths (2011) have been the only researchers to confirm this in the laboratory. They found their Al-7Si-0.4Mg ally to increase its Weibull modulus with only a 20 min holding time, even though they still found oxides on fracture surfaces. The clear implication is that longer times would have produced even better results.

The other feature of this technique for obtaining ultra clean and degassed metal is the transfer of the metal into the mould without the re-introduction of defects, necessitating a connection to a casting station by pipework with no moving parts, and casting by counter-gravity or other variety of non-turbulent transfer into moulds.

14.3.1 HOLDER FAILURE MODES

Most holding furnaces which I see appear designed to maximise the dangers of failure. Cracks in the refractory lining of the furnace, filled with liquid metal, give a high conductivity path for leakage of heat from the furnace, with the result that if the leak reaches the outer steel shell, that portion of the shell becomes extremely hot, and may be dangerous to touch. If the heat input to the furnace is limited, such leaks sometimes give problems to maintain temperature in the furnace. For this reason, the common design of furnace with an unheated extension to form a pump well, or dipping area, are especially troublesome. The metal in the extension derives its heat by conduction and convection from the main heated body of the furnace. I personally no longer favour such unheated extensions because, on the development of a leak, they sometimes become so cool that despite maximum heat input to the main body of the furnace, threatening the refractories and the life of the furnace, casting from the extension still remains impossible. There is then no option but to shut down, drain the furnace and repair the lining. This usually takes a week or more.

A new lining dries quickly if numerous holes are provided all over and under the furnace body shell. Steam and sometimes water escape from the holes to aid the drying process, which can otherwise take weeks, with metal remaining uncastable because of high hydrogen content. All holding furnaces should be *specified* with drying holes all over the steel shell, at intervals no greater than 500 mm; otherwise, such essential drying vents are rarely provided by the furnace manufacturer.

It is best never to shut down a furnace. If a furnace is shut down, cooling and shrinking of the lining causes it to pull away from the crushable back-up insulation and create a gap between the hot face and backup insulation. The loss of mechanical support endangers the lining. Much of the lining shrinkage is caused by the contraction of the aluminium in cracks of the lining. Aluminium alloys have a linear contraction of about 1.3% from the melting temperature down to room temperature. For this reason, it is best to keep holding furnaces close to their operating temperature throughout their life. Over shutdown periods, the temperature might be reduced a little. It is safest to cool the furnace only to just below the solidus temperature so that no run-out disasters can occur while the furnace is unattended. The refractories therefore remain dry and ready for a quick start to production after the shutdown period.

14.3.2 TRANSFER AND DISTRIBUTION SYSTEMS

The tilting of large furnaces and the pouring of melts into ladles to be transported by overhead crane or forklift truck is damaging to the metal and dangerous to personnel. The melt losses on each pour are probably of the order of 1% each time, representing a significant hidden loss. If furnaces can remain static, and if melts can be piped around the foundry without moving parts, without damage to the melt, and never seen by operators, this is an ideal. This ideal standard is often not attained for a variety of reasons.

Siphons are widely used in Al reduction plants to transfer the newly won metal out of the reduction cells for further processing. Siphons have occasionally been used in foundries. However, although these devices might appear attractive at first sight, their design is generally poor, involving the displacement of the melt to a significant height by vacuum, only to allow it to cascade down a uniform tube into the awaiting vessel (Pechiney, 1977). Thus existing designs degrade the melt significantly.

For low melting point metals such as lead and zinc, it is possible to arrange distribution conveniently and successfully via heated pipes. The concept of metal on tap at various points around the foundry has also been applied to a more limited degree for liquid magnesium alloys.

For liquid aluminium alloys, the distribution by pipe is less easily achieved, but can be arranged via a horizontal U-shaped channel generally known as a launder. These long horizontal channels resemble Roman aqueducts, be many tens of metres in length and generally heated by an electrical resistance element in the roof lids of the channel.

Good insulation and tight-fitting lids ensures that the power consumption is only a kilowatt or so per metre. In addition, the melt is protected from the atmosphere to some degree, reducing oxidation. The launder can connect melting and holding furnaces and all the casting stations without the need for human interference. Thus at the point of delivery of the melt the gentle flow will have had chance to detrain bifilms, and so is expected to be in excellent condition and at the correct temperature.

Naturally, a launder system is only of use in foundries casting one or perhaps two alloys. Also, any risk of loss of power requires an emergency generator. However, where appropriate, a launder distribution has many advantages (Sieurin, 1974, 1975). In practice, it is a pity that the benefits of this excellent distribution system are often lost at the point of delivery by a waterfall into a mould or by a turbulence-inducing dipping bucket system. It is not necessary to organise the foundry so badly.

14.4 MELT TREATMENTS
14.4.1 DEGASSING

Gases dissolved in melts are disadvantageous because they precipitate in bifilms and cause them to unfurl. Having unfurled and at that stage starting to behave as a slightly open crack, the precipitation of yet more gas might further inflate the bifilm to create porosity. Each of these stages is accompanied by the progressive reduction in the mechanical properties of the cast product.

The degassing treatments for steels include deoxidation by various chemical additives such as Si, Mn and Al and possibly the use of a carbon boil to reduce hydrogen; the CO bubbles flushing the hydrogen from the melt. Alternatively, reduced pressure and the bubbling of inert gas through the melt in an argon/oxygen decarbonisation (AOD) vessel is relatively widely used in larger steel foundries casting special steels. These treatments are outlined in Chapter 6.6.

Degassing treatments for the various copper alloys are dealt with in Section 6.4.

The remainder of this section concentrates mainly on the various treatments for the light alloys Al- and Mg-based. For these metals there is no way to degas by chemical reaction. The only options are various physical techniques.

Passive degassing

Degassing passively is degassing by doing nothing. One just waits. This is a technique that is generally surprisingly successful. It is based on the fact that even if the atmosphere is practically saturated with water vapour, the hydrogen content of the melt in equilibrium with the atmosphere is not especially high and remains suitable for most products.

It does require that the melt is not surrounded by a blanket of wet refractories in, for instance, a crucible furnace that has been allowed to sit unused for a day or more. Because refractories can absorb up to 10 wt% water, the use of a furnace started up from cold is a serious risk. The outgassing of the refractories will provide a 100% water vapour atmosphere to surround the crucible and will almost certainly permeate the crucible to attack the melt. Keeping furnaces hot to avoid moisture pick up is to be recommended.

Even an induction furnace can continue to give a 100% water vapour environment to a melt for days or possibly weeks after the installation of a new lining. This is because the water is driven away from the melt, through the refractories and condenses on the water-cooled induction coils. It can remain here for a long time, constituting a reservoir to maintain the 100% water vapour atmosphere that will permeate the refractories of the furnace.

The reader should note that these sources of water vapour in the body of the furnace counter those degassing processes that attempt to remove gases by protecting the melt surface or by bubbling through the bulk, as are discussed later. Whether these surface or bulk degassing processes will be effective will depend on whether they exceed the rate of re-gassing via the surfaces in contact with the refractories that contain the 100% water vapour. Even if the degassing process is reasonably successful, as soon as degassing is stopped, re-gassing will continue unabated, raising the gas levels yet again (Chen et al., 1994).

Clearly, there is no substitute for a dry furnace. Perhaps induction furnace suppliers could think about heating their induction coils until water is eliminated. It would be a great help to founders.

Liquid argon shield

The dripping or pouring of a small stream of liquid argon onto the surface of a melt has been carried out for relatively small melts, particularly in investment foundries, for many years (Anderson et al., 1989). The liquefied gas is of course much lighter and perfectly immiscible with the liquid metal, but is heavier than air, and thus spreads over the surface of the melt. It boils off steadily, expanding by 600 times in its volume, thus carrying away volatile emissions from the environment of the surface. The technique has been applied with apparent success to melts of steels, Ni-base and Al-base alloys (Figure 14.1).

Although the liquefied gas is at cryogenic temperatures, its effect on the cooling of the melt is negligible because of its low thermal capacity, and the relatively poor rate of heat exchange with the melt. The liquefied gas gains heat only by direct contact with the melt; the radiant heat passing through the transparent liquid ineffectively.

The method is not recommended for induction melting involving vigorous stirring in view of the danger of entrapping the liquefied gas under the melt to cause ejection or explosion of the melt. There is also claimed to be occasional problem with the condensation of moisture from the atmosphere contaminating the addition (Zurecki and Best, 1996).

Gaseous argon shield

It is often found to be counter-productive to inject an inert gas onto the surface of a melt in the hope that this might somehow protect the melt and thus reduce its gas content. Usually, the gas content increases. The reason for this is that the injected gas entrains with it at least an equal volume of air, which is now driven across the melt surface, equilibrating with the melt. Thus the indiscriminate addition of inert gas introduces the environmental gas, usually air, undermining any benefit.

Zurecki and Best (1996) describe the blanketing of ferrous and non-ferrous melts with gaseous argon. Significantly, the argon is injected tangentially against the melt surface and is circulated and contained within a lightweight refractory cone (Figure 14.2). The injection of the gas parallel to the inner walls of the cone reduces the entrainment of air, and the swirling technique organises the gas so that the rate of gas usage is highly efficient. The heated and expanded gas finally escapes from the top of the cone. The authors describe how the technique improves metal yield and the recovery of alloy additions. It also reduces slag build up and crucible maintenance.

Liquid argon

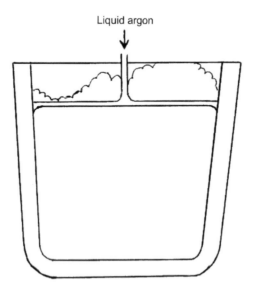

FIGURE 14.1

Protection of the surface of the melt by the addition of liquid argon.

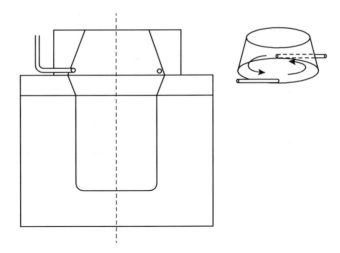

FIGURE 14.2

Protection of the surface of the melt by delivery of argon gas via a swirling action.

These techniques for degassing from the open surface of the melt rely, of course, on convection within the bulk to circulate the melt to within a diffusion distance of the surface.

Bulk degassing by bubbles

Traditionally, the steel industry has used the naturally occurring bubble technique of a 'carbon boil', in which the creation and floating out of carbon monoxide bubbles from the melt carries away unwanted gasses such as hydrogen (passively by simple flushing action) and oxygen (actively by chemical reaction with carbon). The nitrogen in the melt may go up or down depending on its starting value and the nitrogen content of the environment above the melt, because it will tend to equilibrate with the environment.

Hydrogen is a big issue for steel foundries making very large castings such as backup rolls for rolling mills. These castings are too large, having diffusion distances too great for hydrogen which could therefore not diffuse out and escape during subsequent heat treatments. The castings were therefore in danger of suffering failure by hydrogen cracking. To reduce the hydrogen level before casting, a typical practice would be to take a sample at melt-out to check the hydrogen content. If this was above 4 ppm, then a 0.4% carbon boil had to be induced, before taking off the oxidising slag and decarburising under a reducing slag. This had to be carried out as quickly as possible because the hydrogen would start to climb, threatening to return the analysis to its starting value while the furnace lining continued to deteriorate. The furnace hands would explain that these problems always seemed to occur on a Friday when everybody wanted an early finish.

The more recent oxygen steelmaking processes in which oxygen is injected into the melt certainly reduce hydrogen and nitrogen, but, of course, raise the oxygen in the melt. The oxygen requires to be subsequently reduced either chemically by reaction with C or other deoxidisers such as Si, Mn or Al etc., or use even more modern techniques such as AOD (argon-oxygen decarburisation) in special vessels. We shall not dwell further on these sophistications because these specialised techniques, claimed to be well understood and well developed for steel, are almost unused elsewhere in the casting industry. Even so, there is some recent evidence that the simple use of non-entraining filling system designs for castings might eliminate the need for these complex and expensive treatments (Puhakka, 2011). If proven to be true this would be a revolutionary step forward for steel casters.

By comparison, the approaches to the degassing of aluminium alloys, containing only hydrogen, has for many years been primitive. Only recently have more effective techniques been introduced, and even these involve some uncertainty and appear to introduce some counter-productive effects.

For instance, until approximately 1980, aluminium was commonly degassed using immersed tablets of hexachloroethane (C_2Cl_6) which thermally degrade in the melt to release large bubbles of chlorine and carbon (the latter as smoke). Alternatively, a primitive tube lance was used to introduce a gas such as nitrogen or argon. Again large bubbles of the gas were formed. These techniques involving the generation of large bubbles were so inefficient that little dissolved gas could be removed, but the creation of large areas of fresh melt to the atmosphere each time a bubble burst at the surface of the melt provided an excellent opportunity for the melt to equilibrate with the environment. Thus on a dry day the degassing effect might be acceptable. On a damp day, or when the flue gasses from the gas-fired furnace were suffering poor extraction, the melt could gain hydrogen faster than it could lose it. This poor rate of degassing, combined with the high rate of re-gassing from interaction with the environment, led to variable and unsatisfactory results. Further complication arose from the significant entrainment of new oxides and chlorides from the melt surface, as a result of the major disturbance from the bursting of large bubbles of escaping gas. The creation of oxides was probably enhanced if the tablets were slightly damp. Overall, it is likely that most melts were significantly damaged by the use of hexachloroethane tablets. This primitive technique cannot be recommended.

The introduction of inert gas into a melt using a lance has mixed results. The technique is inefficient because of the large bubble size, giving a relatively small total area for the exchange of gases. In addition, the bursting of large bubbles at the melt surface creates opportunity for enhanced equilibration with the environment, plus the creation of fresh oxide bifilms as the bubble opens to the atmosphere.

The situation is greatly improved by the provision of a porous refractory diffuser on the end of the lance, or the setting of a porous brick diffuser in the base of the furnace or ladle. The much finer bubbles are significantly more effective in eliminating gas because of their significantly larger total surface area, and the smaller distance between bubbles reduces the diffusion distance for hydrogen to escape.

Even so, a considerable part of the benefit of degassing is not merely the degassing action. The action of the bubbles to eliminate oxide bifilms has in general been overlooked. It is assumed that the millions of tiny bubbles attach to oxides in the melt and float them to the top, where they can be skimmed off. Thus the melt is cleared, at least partially, of oxide bifilms. This will almost certainly result in a greater benefit to the castings than the mere reduction of gas content.

The size of bubbles in relation to the efficiency of removal of films is probably critical. For instance, large bubbles will displace large volumes of melt during their rise to the surface, their laminar streamlines displacing bifilms sideways, so that the bifilm and bubble never make contact (Figure 14.3(b)). Haugland and Engh (1977) use a water model to observe particles sliding around a bubble surface in the boundary layer of the bubble, but still avoiding contact.

On the other hand, small bubbles will displace relatively little liquid, and so be able to impact on relatively large bifilms in their path. Thus large bifilms will be contacted, collecting smaller bubbles beneath them and will be buoyed up to the surface. In this way, the large oxide skins from the charge materials, measuring fractions of a metre across, are easily and quickly floated out. This is a major benefit that castings did not enjoy until the arrival of effective bubble techniques.

It follows that the smallest oxide that can be removed by a bubble is approximately the same size as the bubble, thus there is definitely an advantage to move to smaller bubbles to scavenge the melt clean down to smaller oxide sizes. Because the smallest bubbles in melts are usually measured in millimetres as seen from Figure 10.12, it follows that oxides smaller than this will be expected to be unaffected by this flotation process.

The reader will by now appreciate that although these techniques are universally known as 'degassing techniques' because their use results in reduced porosity in the casting, a powerful but largely unappreciated benefit is the removal of significant quantities of oxide bifilms. This benefit was explicitly appreciated by the researchers Laslaz and Laty as long ago as 1991. Curiously, however, this important fact remains stubbornly and consistently overlooked, to the great detriment of explanations of how bubble 'degassing' works to improve the soundness of castings.

Rotary degassing of Al alloys

Rotary degassing appeared to come to the rescue of the Al casting industry. The use of a rapidly rotating rotor to chop jets of inert gas into fine bubbles that rose as clouds through the melt raised the rate of degassing in comparison with the normal rates of re-gassing. Thus, effective degassing could be reliably achieved in times that were acceptable in a production environment.

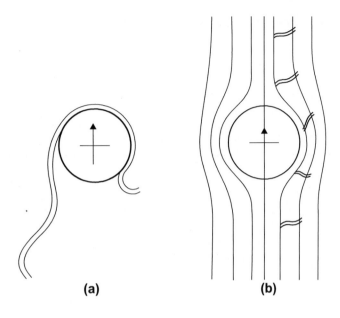

(a)   **(b)**

FIGURE 14.3

For given size of bubble, (a) larger sized bifilms are easily floated out, whereas (b) small bifilms follow the streamlines of the rising bubble and so do not contact the bubble.

The use of an 'inert' gas is a convenient untruth. Even sources of high purity inert gas contain sufficient oxygen as an impurity to create an oxide film on the inside of the surface of the bubbles. Thus millions of very thin bifilms must be created during the degassing process. For a new rotor, or after a weekend, the rotor and its shaft will have absorbed considerable quantities of water vapour (most refractories can commonly absorb water up to 10% of their weight). Furthermore, the tubing conveying the gas from the gas bottle or other source is usually not clean, comprising long lengths of rubber or plastic tubing. These materials absorb and therefore leak large quantities of volatiles into the degassing line. Thus, by the time the gas arrives in the melt, it is usually not particularly inert. Also of course the bursting of bubbles at the surface of the melt is certain to introduce some oxides and possibly nitrides. At this time no one knows whether the inclusions created as by-products of degassing are negligible or whether they seriously reduce properties. What is known is that despite expected counter-productive effects such as the generation of millions of smaller oxides, aluminium alloys are usually improved by rotary degassing. Whether the counter-productive effects can be reduced to secure further improvements appears never to have been investigated. Another doctoral research topic for someone.

Rotary degassing was introduced as an attempt to reduce bubble size and therefore improve degassing effectiveness. This seems to have been achieved, but is but one benefit to be set against several disadvantages and uncertainties. Despite a great amount of research into rotary degassing many imponderables remain, especially concerning the oxide population of the melt, which most researchers overlook in their preoccupation with the gas content.

For instance, it seems that before degassing the melt contains mainly spinel bifilms, large but few in number, suggestive that these represent the oxide skins from the charge. After rotary degassing, the melt contains masses of fine, pure alumina bifilms (Enright, 2001). The mass of fine alumina bifilms created during the degassing process seems likely to be the reason why inclusion monitoring techniques such as pressurised filtration are overwhelmed and incapable of giving sensible answers for some minutes after treatment.

The alumina bifilms may have been created either on the inner surfaces of bubbles from the oxygen and moisture impurities in the (so-called) inert gas (for instance, among other prolific sources, there will be much contamination from the inner channels of the rotary degasser). Alternatively, the pure alumina might originate from the waft of fresh air that will enter the bubble as it bursts at the melt surface. These questions remain unanswered at this time.

Also unanswered is the question of possible re-entrainment of oxides after having been carried to the surface. It is easy to imagine that large oxide skins will retain sufficient trapped bubbles to remain buoyant, allowing them to join the layer of dross as relatively permanent contributors, and thus be scraped off. However, it seems also possible to envisage that small oxides buoyed up by a bubble might easily re-entrain once their bubble buoyancy aid has burst.

Pouly and Wuilloud (1997) draw attention to the fact that some rotor designs work counter-productively, increasing the number and volume of inclusions as measured by the porous disc filtration analysis technique. Dispinar et al. (2010) carried out some excellent experiments in which the number and size of oxide bifilms was counted in addition to measurements of the hydrogen content of an Al alloy. He first upgassed a melt by adding water vapour to the degasser while monitoring hydrogen and oxides. Both increased. He then degassed with pure inert gas, finding the hydrogen decreased, but the oxide content continued to increase. He attributed this result to the vortex formed around the central shaft of the degasser which was probably entraining air into the melt. However, further work might reveal that such behaviour may be usual whether or not a vortex is formed. Water models of rotary degassers clearly show that even the baffle board introduced to arrest the rotation of the melt to avoid vortex formation can cause severe air entrainment in its own wake; it appears to be capable of creating more problems than it solves.

Pouly and Wuilloud also report that the use of chlorine (nowadays avoided for environmental reasons) was not always helpful. Martin et al. (1992) find that if nitrogen alone is used, AlN is formed creating a 'wet' dross (i.e. a dross high in liquid aluminium alloy). It seems likely that the AlN is in the form of aluminium nitride bifilms; otherwise, the retention of entrapped metal is not easily explained.

Other chemical interactions are noted by Khorasani (1995) who found 10% of the Sr in solution in the melt was lost within 3 min, indicating the huge rate of reaction either with the nitrogen used for degassing, or with oxygen from contamination of the gas or from bursting of bubbles at the surface.

There is no doubt that the addition of fluxes, mainly chlorides and fluorides, together with the bubble action, greatly appears to clean the melt. This may be the result of the wetting of oxides, causing them to adhere and be assimilated by the liquid flux at the melt surface, reducing re-entrainment. The result of this research, carried out by the world's large aluminium suppliers, is viewed to be too valuable to be allowed to be published for the benefit of society. This antisocial behaviour is practised even though it is well known that technical leads in industry can in practice only be held without patent protection for only a few years, after which competitive advantage is lost because all competitors have become aware of the technology.

After treatment, after the rotor has been raised out of the melt and the surface skimmed, a test of the efficiency of the cleaning action is simply to look at the surface of the melt. If the cleaning action is not complete, during the next few minutes small particles will arrive at the surface, under the surface oxide film, which is seen to pop upwards, announcing each new arrival. The 'degassing' treatment can be repeated until no further arrival of debris can be seen. The melt surface then retains its pristine mirror smoothness. The melt can then be pronounced as clean as the treatment can achieve. This probably means the elimination, or near elimination, of the primary spinel skins, but its replacement by millions of fine alumina bifilms.

That pressure die castings benefit from reduced leakage problems if the melt has been treated by rotary degassing (Hairy et al., 2003) is clear evidence for the action of rotary degassing to remove large oxides, probably primary skins from charge materials. (Leakage defects can have little to do with degassing because hydrogen bubbles cannot be envisaged to affect leakage problems, whereas large oxide bifilms can provide excellent leak paths.)

An experience by the author illustrates some of the misconceptions surrounding the dual role of hydrogen and oxide bifilms in aluminium melts. An operator used his rotary degasser for 5 min to degas 200 kg of Al alloy. The melt was tested with a reduced pressure test (RPT; see later) sample that was found to contain no bubbles. The melt was therefore deemed to be degassed. The melt was transferred in its ladle via fork lift truck to a low-pressure die casting furnace, into which it was poured. The melt in the low-pressure furnace was then tested again by RPT, and the sample found to contain many bubbles. The operator was baffled. He could not understand how so much gas could have re-entered the melt in only the few minutes required for the transfer.

The truth is more complicated, and unfortunately more uncertain. It may have been that the melt was insufficiently degassed with only 5 min of treatment. In fact, with a damp rotor, the gas level is likely to rise initially (getting worse before it gets better!). The RPT showed no bubbles *not* because the hydrogen was low, but because the short treatment had clearly been sufficiently successful to remove a large proportion of the bifilms that were the nuclei for the bubbles. This metal, still retaining a high level of hydrogen, was then poured from a considerable height, reintroducing copious quantities of oxide bifilms that act as excellent nuclei, so that the RPT could now reveal the high hydrogen content.

A second scenario is a possibility. If the melt was sampled for the RPT test immediately after the rotary degassing action, the hydrogen may not have been reduced significantly, as in the previous explanation. In addition, the remaining bifilms would, at that early stage, still have been uniformly dispersed throughout the melt. Thus, in addition to the hydrogen content, the bifilm content may also have remained high. Thus, a single sample from near the surface would have shown the presence of few, if any bifilms, and thus few, if any pore indications on the RPT sample. During transfer to the furnace, the melt would have some time for the bifilm population to separate out, with some sinking and some floating. Thus a second RPT a few minutes later would have shown a massive content of bifilms that had segregated close to the top surface. These would have been transferred to the furnace, adding to those created during the pour and those stirred up once again from the bottom of the furnace. After a delay of a few more minutes, a further RPT test taken from the top of the melt would sample the newly enhanced bifilm population that had segregated to the surface of the melt, creating numerous pore indications on the sample and giving the impression of a higher gas content to anyone who was unaware of the possible changes in bifilm population.

Hu et al. (2008) are among the few to study the rotary degassing of Mg alloy, deriving a model for the rate of degassing as a function of parameters such as the rate of rotation of the rotor etc. They point out that although Mg has a high solubility for hydrogen, the small difference in solubility in the solid (its partition coefficient 0.41 contrasts with that for Al 0.05) causes it to be widely reported that there is little gas problem with Mg. They do report a problem in Mg alloys for the attainment of good ductility, although, as usual, these authors overlook the important distinction whether this is as a result of hydrogen gas or oxides.

I have to confess to feeling diffident about this section concerning rotary degassing. The technique is widely used in the Al casting industry and so much has been written, but so little is known. I personally am reluctant to use rotary degassing while its action is so obscure, and an optimum procedure for reduction of gas and oxides does not yet appear to have been defined. After so many years of research and development, this situation is a disappointment. I look forward to it being remedied soon.

Vacuum degassing

Treatments to reduce the gas content of melts include vacuum degassing. Such a technique has only been widely adopted by the primary steel and Ni-base alloy industry. If a melt were simply to be placed under vacuum, the rate of degassing would be low because the dissolved gas would have to diffuse to the free surface to escape. The slow convection of the melt will gradually bring most of the volume near to the top surface, given time. The process is greatly speeded by the more powerful stirring action of induction melting. Additionally, the introduction of the flushing action from millions of small bubbles of inert gas via a porous plug stirs the melt rapidly to bring bulk liquid up to the surface, and the bubbles create additional degassing surface. The process is then accomplished rapidly.

The steel casting industry has in general dropped vacuum degassing and has taken up the more flexible AOD process mentioned previously. The process is described in more detail in Section 6.6.

14.4.2 DETRAINMENT (CLEANING)

Natural flotation and sedimentation

The very light metal magnesium is lighter than its oxide, MgO. Thus, the oxide readily sinks, so that clean magnesium alloys are relatively easily achieved within a few minutes, provided the melt is carefully neglected and enthusiastic stirrers of sediments can be kept away.

The dense metal based on zinc, copper and iron all have oxides that are lighter and therefore float to the surface of the melt. The higher melting point irons and steels achieve this especially effectively because their oxides are sometime liquid and so float out as especially compact spherical droplets, or partially liquid, so that compact bifilms resemble sticky lumps that cannot unfurl, thus remaining compact and achieving high flotation velocities, clearing the melt rapidly after a pouring or stirring action.

Thus, steel melts in the furnace are probably of good quality, but are disastrously damaged during the pour into the ladle. The melt then recovers to some extent during the transfer of the ladle to the casting station, and during the usual subsequent wait to allow its temperature to fall to the casting temperature. Unfortunately, defects are again re-introduced during the final casting operation. Although many continuous casting operations do take precautions to avoid a final re-contamination with oxides, it seems probable that in practice even greater care is really required. Clearly, we achieve cleaner steels than we deserve from our primitive bucket technology. For the future, really excellent steels should be attainable, cultivating the good natural behaviour of this useful engineering material.

In the meantime, only the aluminium alloys remain reluctant to clean their melts by flotation or sedimentation. Although the oxide is slightly denser than the liquid, its structure as a bifilm containing some entrained air, and probably gathering some hydrogen from the melt, causes the oxides to be often of nearly neutral buoyancy. Thus, Al alloys are renowned for retaining their population of oxide bifilms for hours or days. The retention is enhanced because of natural convection that characterises all furnace designs, and can sometimes be further enhanced by such treatments as bubble degassing which cause effective stirring and over-turning of the whole melt.

Only the Alotech holding furnace for Al alloys is designed to eliminate convection (by a design protected by patent pending at the time of writing) and thus provide conditions for the natural and passive sedimentation of oxides. With the use of a counter-current of dry gas, it also ensures the passive degassing of the melt. The use of passive systems for melt quality enhancement has been so far overlooked, but has to be one of the most attractive ways forward (Grassi et al., 2012).

Aided flotation

Flotation is a process widely used in the mineral processing industry to float out specific components of a slurry type mixture. Such a process clearly operates during bubble degassing of metal, although the researches so far have concentrated on the removal of gas rather than the floating out of oxides. Other important features that have been neglected in all this work include the rate of re-gassing from the environment or from a damp rotor.

As has been described previously, it is known that during the course of degassing of an Al melt with nitrogen, that the first oxides that are removed from the melt are mainly spinels, probably the oxide skins originating on the charge materials. After some minutes, the spinels are replaced by millions of fine, pure alumina films. Thus the bifilm population is completely changed by rotary degassing.

In terms of the general quality of the castings, mechanical properties would be improved because the old thick oxides would constitute massive defects. Their replacement by the smaller alumina bifilms, despite their enormously greater numbers, these would be expected to unfurl less easily, and confer better properties because of the smaller maximum defect size. Nevertheless, it need hardly be emphasised, that the properties are likely to be even better without this huge population of small cracks.

Aided sedimentation

The sedimentation of oxide bifilms in light alloy melts by the formation of heavy precipitates on them is the only example known to the author of the cleaning of melts by aided sedimentation. This phenomenon is so important that we shall dwell on it at length while noting that other examples may remain to be discovered in other alloy systems.

The sedimentation of oxides (actually bifilms, of course) in an Al alloy by the formation of titanium-rich precipitates on them, causing them to drop like stones to the bottom of the melt was observed by Mountford and Calvert in 1959. They looked into the melt by means of ultrasonic reflections, noting that on cooling the melt to cause Ti compounds to supersaturate and precipitate from solution, the liquid miraculously cleared of a fog of suspended particles, transferring to the bottom of the melt to form a layer of sediment rich in oxides and Ti compounds. When the sediment was disturbed before pouring, the fog returned, and polished sections of the castings exhibited tangles of oxides containing large Ti intermetallics and associated porosity.

While studying the 'fade' of grain refiners, Schaffer et al. (2004) observe the strange settling behaviour of grain refining particles in Al alloys, surprisingly completing in only a few minutes, but offer no explanation. Chu (2002) finds the TiB_2 particles are associated with spinel oxides and Srimanosaowapak and O'Reilly (2005) confirm that $TiAl_3$ particles assist to sediment oxides during grain refinement. This effect has been used for many years (Cook et al., 1997) to sediment impurities from electrical conductivity grade aluminium; Cook uses 3:1 TiB additions (instead of the more usual 5:1 TiB ratio) which precipitates as a boride-rich sludge, leaving the metal with high electrical conductivity. Vanadium borides were rather less effective than titanium borides, giving settling times of 5–6 h. These long times may have been the result of rather more vigorous convection stirring in their experiment. Wuilloud (1994) describes how Alusuisse have cleaned melts in only 50 min by the addition of grain refiner. Without grain refiner, 80% of the 'impurities' still remained in suspension after this time.

Interestingly Wuilloud reports that the addition of the grain refiner to the launder to the direct chill (DC) casters resulted in a reduction in quality of the cast metal. Nadel et al. (2007) appear unaware of the Alusuisse technique, assuming that grain refinement would reduce the cracking observed in DC ingots. They found, in agreement with Wuilloud, that grain refinement in the furnace did grain refine the ingot and eliminate cracking. However, grain refinement in the rapidly moving stream of metal in the launder, transferring the melt to the DC caster, did, once again, grain refine, but the ingot was cracked. Clearly, when adding to the launder, the bifilms had no time to settle. Thus the elimination of cracking was the result of elimination of bifilms rather than the result of the finer grain size (Campbell, 2008).

Following treatment by rotary degassing, Schaffer and Dahle (2009) find the loss of grain refinement was due to the sedimentation of TiB_2 particles, presumably on oxides. This contrasts with the findings of Khorasani (1995) who found evidence of flotation of TiB_2 in the dross layer. This somewhat messy situation might be explained by the sedimentation of large TiB_2 precipitates on smaller oxides, but flotation of smaller TiB_2 particles on larger oxides. Alternatively, the situation may be more complicated by some precipitation on spinels and some on the pure alumina particles. The presence of changing levels of hydrogen in solution will complicate matters further by inflating bifilms and changing their buoyancy by different degrees.

Only a limited amount of work appears to have been carried out on Mg alloys to study the cleaning action of aided sedimentation. As long ago as 1969, Fruehling and Hanawalt studied the fracture surfaces of 12 mm diameter test bars, in a demonstration that a protective atmosphere could be at least as effective as a flux in achieving a clean melt. They showed that oxide skins would separate from Mg alloys, achieving metals with completely clean fractures, in only a matter of minutes. Simensen (1981) aided gravity by centrifuging both Al and Mg alloy melts, precipitating out TiB_2 and VB_2 in Al alloys and Al_4C_3, Al_3Zr and $(Fe,Mn)_3Si$ in Mg alloys. Exceptions were low-density agglomerates of oxides and chlorides which segregated to the rotation axis. The technique, although scientifically interesting, can hardly be judged a useful foundry tool.

Filtration

Filtration is perhaps the most obvious way to remove suspended solids from liquid metals. However, it is not without its problems, as we shall see.

The action of a filter in the supply of liquid metal in the launder of a melt distribution system in a foundry or cast house (i.e. a foundry for continuous casting of long products) is rather different from its action in the running system of a shaped casting.

The filters used in the running systems of castings act for only the few seconds or minutes that the casting is being poured. The velocities through them are high, usually several metres per second, compared with the more usual 0.1 m/s rates in cast house launders. The filtration effect in running systems is further not helped by the concentration of flow through an area often a factor of up to 100 smaller than that used in launder systems. The total volume per second per unit area rates at which casting filters are used are therefore higher by a factor of at least 1000 times. It is hardly surprising therefore that any filtration action in the filling systems of castings is therefore practically zero. However, the effect on the flow is profound. The use of filters in running systems is dealt with in detail later in the book. We concentrate in this section on filters in launder systems, which, in this case, do appear to achieve some filtration.

The treatment of tonnage quantities of metal by filtration has been developed only for the aluminium casting industry. The central problem for most workers in this field is to understand how the filters work because the filters commonly have pores sizes of around 1 mm, whereas, puzzlingly, they seem to be effective to remove a high percentage of inclusions of only 0.1 mm diameter. Most researchers write at length, listing the mechanisms that might be successful to explain the trapping of such small solid particles. Unfortunately, these conjectures are probably not helpful because without exception all the conjectures to the present day relate to the effect that filters may have on arresting *particles*. These conjectures are not repeated here. We have no interest in stopping particles. For engineering castings, particles generally have little or no influence on properties.

The fact is that the important solids being filtered in aluminium alloys are not particles resembling small solid spheres, as has generally been assumed. The important inclusions are *films* (actually always *double films* that we have called *bifilms*). Once this is appreciated, the filtration mechanism becomes much easier to understand. The films are often of size 1–10 mm, and so are, in principle, easily trapped by pores of 1 mm diameter. Such bifilms are not easily seen in their entirety in an optical microscope, the visible fragments that happen to intersect the polished surface appearing to be much smaller, explaining why filters *appear* to arrest particles smaller than their pore diameters.

We need to take care. This explanation, whilst probably having some truth, may oversimplify the real situation. In Section 1, the life of the bifilm was described as starting as the folding in of a planar crack-like defect as a result of surface turbulence. However, internal turbulence wrapped the defect into a compact form, reducing its size by a factor of 10 or so. In this form, it could pass through a filter; finally opening once again in the casting as the liquid metal finally comes to rest, and bifilm-opening (i.e. unfurling) processes start to come into action.

In more detail now, the trapping of compact forms of bifilms is explained by their irregular and changing form. During the compacted stage of their life, they will be constantly in a state of flux, ravelling and unravelling as they tumble along in the severely turbulent flow in the channels of the running system. Two-dimensional images of such defects seen on polished sections always show loose trailing fragments. Thus, in their progress through the filter, one such end could become attached, possibly wrapping over a web or wall of filter material. The rest of the defect would then roll out, unravelling in the flow, and be flattened against the internal surfaces of the filter, where it would remain fixed in the tranquil boundary layer. The thin oxide ceramic film would be essentially invisible against the coarse thickness of the body of the ceramic filter, explaining why polished sections of many filters appear free from inclusions after they appear to have completed a useful job of cleaning up a casting. The other reason why the filter often looks free from inclusions is that the major task of the filter is not filtration but the slowing of the velocity of flow, reducing the damage that the melt can do to itself from entering the mould cavity at too high a speed, so creating damaging oxide bifilms in the casting.

Packed beds

In practice, in DC continuous casting plants, filters have been used for many years, sited in the launder system between the melting furnaces and casting units. Commonly, the filtration system is a large and expensive installation that may approach a cubic metre in volume. The filter system typically comprises at least one crucible furnace that is in two parts. One half is filled with refractory material such as alumina balls, or tabular alumina. The melt flows down half of the crucible, through a connecting port, and up through the deep packed bed of refractory particles. Such systems have been shown to be effective in greatly reducing the inclusion count (the number of inclusions per unit area). However, it is known that the accidental disturbance of such filter beds releases large quantities of inclusions into the melt stream. This has also been reported when enthusiastic operators see the rising upstream level of the melt indicating the filter is becoming blocked. They consequently stir the bed with iron rods to ease the flow of metal. It is not easy to imagine actions that could be more counter-productive.

Some work has been carried out on filtering liquid aluminium through packed beds of alumina (Mutharasan et al., 1981) and through bauxite, alumino-silicates, magnesia, chrome-magnesite, limestone, silicon carbide, carbon and steel wool (Hedjazi et al., 1975). This latter piece of work demonstrated that all of the materials were effective in reducing macro-inclusions. This is perhaps to be expected as a simple sieve effect. However, only the alumino-silicates were really effective in removing any micro-inclusions and films, whereas the carbon and chrome-magnesite removed only a small percentage of films and appeared to actually increase the number of micro-inclusions. The authors suggest that the wettability of the inclusions and the filter material is essential for effective filtration.

An interesting application of a chemically active packed bed is that by Geskin et al. (1986), in which liquid copper is passed through charcoal to provide oxygen-free copper castings. It is certain, however, that the charcoal will have been difficult to dry thoroughly, so that the final casting may be somewhat high in hydrogen. Because of the low oxide bifilm content, and possibly the low oxygen content in solution, it is likely that hydrogen pores will not be nucleated (Section 6.4). The hydrogen is expected therefore to stay in solution and remain harmless.

Alternative varieties of filters

Other large and expensive filtration systems include the use of a pack of porous refractory tubes, sealed in a large heated box, through which the aluminium is forced. The pores in this case are between 0.25 and 0.05 mm, with the result that the filter takes a high head of metal to prime it. However, the technique is not subject to failure because of disturbance, and guarantees a high quality of liquid metal.

Other smaller and somewhat cheaper systems that have been used include a box housing a large area (usually at least 300×300 mm) ceramic foam filter sited permanently below the surface of the melt. The velocities through the filter are usually low, encouraged by the large area of the filter. The filter is only brought out into the air to be changed when the metal level on either side of the filter box shows a large difference, indicating that the filter is becoming blocked.

The efficiency of many filtration devices can be understood when it is assumed that the important filtration action is the removal of films (not particles). Thus glass cloth is widely used to good effect in many different forms in the Al casting industry. A woven silica fibre cloth is similarly effective for cast iron.

Practical aspects

It is typical of most filtration systems that the high quality of metal that they produce (often at considerable expense) is destroyed by thoughtless handling of the melt down-stream of the filter.

Even the filter itself can give difficulties in this way. Many have wondered whether the filter itself causes oxides because the flow necessarily emerges in a divided state and therefore must create double films in the many confluence events. Video observations by the author on a stream of aluminium alloy emerging from a ceramic foam filter with a pore diameter close to 1 mm have helped to clarify the situation. It seems true that the flow emerges divided as separate jets for even modest velocities around 500 mms^{-1}. However, within a few millimetres (apparently depending on the flow rate of the metal), the separate jets merge. Thus the oxide tubes formed around the jets appear to be up to 10 mm long, but they remained attached to the filter. The oxide tubes did not extend further because, of course, after the streams merged oxygen was necessarily excluded. The mini forest of 10 mm long tubes was seen to wave about in the flow like underwater grass. It is possible that more rapid flows might cause the grass to detach as a result of its greater length and the higher speed of the metal. This seems possible in running systems of tall castings where the velocities can be extremely high. Further work will be necessary to clarify this.

In the meantime, the problem is avoidable by ensuring that the exit face of the filter quickly becomes covered, effectively protected by melt quickly after the filter becomes primed. This is an important aspect of the designing of effective filters into the running systems of shaped castings, as is discussed in Section 12.8.

14.4.3 ADDITIONS

Additions to melts are made for a variety of reasons. These can include additives for chemical degassing (as the addition of Al to steel to fix oxygen and nitrogen) or grain refinement (as in the addition of titanium and/or boron or carbon to Al alloys). Sometimes it seems certain that the poor quality of such additions (perhaps melted poorly and cast turbulently, and so containing a high level of oxides) can contaminate the melt directly. It is noteworthy that many such additions are made at a late stage, after all refining actions have been completed, with the result that the primary oxide skins of such additions contribute significant defects which have automatically avoided the refining stages.

Indirectly, however, the person in charge of making the addition will normally be under instructions to stir the melt to ensure the dissolution of the addition and its distribution throughout the melt. Such stirring actions can disturb the sediment at the bottoms of melts, efficiently re-introducing and re-distributing those inclusions that had spent much time in settling out. The author has vivid memories of wrecking the quality of early Cosworth melts in this way: the addition of grain refiners gave wonderfully grain refined castings, and should have improved feeding and therefore the soundness of the castings. In fact, all the castings from those melts were scrapped because of a rash of severe microporosity, initiated on the stirred-up oxides that constituted the sediment in the holding furnace.

Additions of Sr to aluminium melts have often been accused of also adding hydrogen because of the porosity that has often been noted to follow such additions. Here again, it is unreasonable to suppose that sufficient hydrogen could be introduced by such a small addition. Section 6.3.5 describes how the action of Sr is particularly complicated. However, all the detailed mechanisms described in Section 6.3.5 explaining the enhanced porosity will be certainly made worse by stirred sediment.

14.4.4 POURING

Most foundries handle their metal from one location in the foundry to another by ladle. The metal is, of course, transferred into the ladle by pouring, and out of the ladle by pouring, into a further ladle or into the mould by pouring. At every pouring operation, it is likely that large areas of oxide film will be entrained in the melt. Furthermore, because pour heights are usually not controlled, the amount of oxide introduced in this way will vary from one occasion to the next (Prakash et al., 2009).

Not all pouring necessarily results in damage to melts. It is known that pours from heights less than the height of the sessile drop cannot entrain the surface oxide. However, such heights are very low, being 16 mm for Mg, 13 mm for Al and only 8 mm for dense metals such as copper-base, and iron and steel alloys.

However, this theoretical limit, whilst absolutely safe, may be exceeded for some metals with minimum risk. As long ago as 1928, Beck described how liquid magnesium could be transferred from a ladle into a mould by arranging the pouring lip of the ladle to be as close as possible to the pouring cup of the mould, and be relatively fixed in position. In this way the semi-rigid oxide tube that formed automatically around the stream of pouring metal remained unbroken, and so protected the falling stream.

Early observations by Turner (1965) noted that air was taken in to the melt, reappearing as bubble on the surface when the pouring height exceeded about 90 mm. This was closely confirmed by Din et al. (2003) on Al 7Si 0.4Mg alloy, demonstrating that in practice the damage caused by falls up to 100 mm appears controlled and reproducible, although this good behaviour was probably only achieved provided the melt had first gone through a ceramic foam filter. Above 200 mm, Din and colleagues found that random damage was certain whether a filter had been in place or not. At these high energies of the plunging jet, bubbles are entrained, with the consequence that bubble trails add to the total damage in terms of area of bifilms. In summary, it seems that fall heights of perhaps 10–20 mm are completely safe, whereas up to 100 mm may be successful but are a risk, but somewhere in the region of 100–200 mm severe harm can be expected.

In steel foundries using bottom-pour ladles, the velocities of the pour are always in excess of 4 ms^{-1}, introducing the danger of massive air entrainment and oxide generation (Figure 12.1). This is the major challenge facing steel casters and currently causes extensive damage to the casting, requiring costly upgrading procedures. Similar damage is caused during ingot casting, leading at this time to the near-universal production of poor quality ingot steels. It is hoped that this inexcusable situation will be improved soon.

For this reason, the inert gas shroud has been adopted in some steel foundries where it takes the form of a protective shield, rather like a collar, around the metal stream issuing from a bottom-poured ladle. It encloses an inert gas environment, usually argon (Figure 14.4). With its use, Wells and Oleksa (1988) report a two-thirds reduction in oxides in steel castings and reduced cleaning of the ladle nozzle. The Steel Foundry Society of America (SFSA, 2000) reports further developments in which macro-inclusions were reduced by 90% by use of a shroud.

It is difficult to believe that a user would think that the short distance between the ladle and the conical basin was influential in any substantial reduction of the oxidation of the melt. Usually, the time involved in this short journey will be probably only a few hundred milliseconds. It is not easy therefore to escape the conclusion that users were in fact tacitly acknowledging the air pump action of the conical basin. The shroud therefore encourages argon to be sucked into the cone instead of air, assuming that the rate of delivery of argon is sufficient (because venturi pumps usually transfer roughly equal volumes of pumping fluid and entrained fluid).

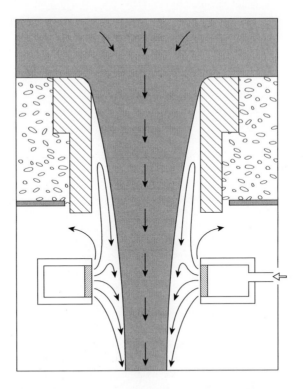

FIGURE 14.4

The use of a protective gas shroud (usually argon) around the metal stream from a bottom-pour ladle, illustrating the entrainment of protective gas into the annular gap between the stream and a parallel bore nozzle, thus reducing oxidation in the nozzle.

The beneficial action of an argon shroud is that the reactive gas is simply replaced by an inactive gas. Thus although volumes of bubbles will continue to be entrained with the flow, they at least do not react to produce oxides or nitrides.

In fact, of course, the shroud will never be completely protective for various reasons: the gas itself will be contaminated with oxygen, water vapour and other gases and volatiles in the plumbing system that delivers the gas. More important still, the seal of the shroud around the stream cannot be made proof against leakage of air; and finally the outgassing from the mould, especially in the case of an aggregate (sand) mould, will normally totally overwhelm any attempt to provide a protective environment. Monroe and Blair (1995) draw attention to the mechanical sealing problems.

Even so, when used appropriately, the shroud is useful. It greatly reduces re-oxidation problems of steels during casting as demonstrated by research carried out by the Steel Foundry Society of America. The result emphasises the damage done by the emulsion of steel and air bubbles that characterises the average poorly designed casting system.

The shroud has been taken to an extreme form as a silica tube, more than a metre long, mounted directly to the underside of a bottom-pour ladle (Harrison Steel, USA, 1999). The tube acts as a re-usable sprue and is inserted through the cope and lowered carefully through the mould so that its exit reaches the lowest point of the filling system, usually at the drag. The stopper is then opened. If the seal between the ladle and tube is good, the filling rate of the mould is high. (If leakage of air occurs at the seal the rate of mould filling is significantly reduced, implying the strong pumping action of the falling stream to create a vacuum in the upper part of the tube, drawing in air if it can, and thus diluting the falling stream with air.) Several castings in succession can be poured from one tube. However, after the tube cools the silica cracks into fragments and must be replaced. Although this solution to the protection of the metal stream from oxidation is to be admired for its ingenuity, it does appear to the author to be awkward in use. In addition, the leakage problem is always an attendant danger.

In Figure 14.4 the use of a nozzle with an internally straight bore is seen to be a problem. An annular gap necessarily forms in which air is rapidly drawn down by frictional contact with the falling jet; consequently, air is drawn up to replace it. This extremely rapid circulation of air in this confined region explains the buildup of oxides leading the eventual constriction of the nozzle. The tapered nozzle avoids this problem by ensuring contact with the melt as it accelerates down the nozzle; there is simply no room for the air. Effectively, the tapered nozzle is a short sprue and benefits from pressurisation of the melt against its walls to keep any gap closed and keep air out. (It seems paradoxical that one would expect the tapered nozzle to be easier and less costly to produce than the straight bore nozzle: we clearly go to great lengths to make life difficult for ourselves.) It seems to me that the alternative theories of nozzle blockage resulting from inclusions in the steel (Ni et al., 2014) is less likely, or at the most, a marginal contribution to blockage.

Moving on to the pouring issues in most foundries, the lower the pour heights, the less damage is suffered by the melt, and because less metal is oxidised, melt losses are directly reduced. Sometimes big savings in melt losses can be achieved in the foundry by simply reducing pour heights. Ultimately, of course, it is best to avoid pouring altogether. In this way, losses are reduced to a minimum and the melt is maintained free from damage.

Until recently, such concepts have been regarded as pipe dreams. However, the development of the Cosworth Process has demonstrated that it is possible for aluminium alloy castings to be made without the melt suffering any pouring action at any point of the process. Once melted, the liquid metal travels along horizontal heated channels, retaining its constant level through the holding furnace, and finally to the pump, where it is pressurised to fill the mould in a counter-gravity mode. Such technology might also be applied to magnesium alloys.

The potential for extension of this technology to other alloy systems such as copper-based or ferrous alloys is less clear. This is because many of these other alloy systems either do not suffer the same problems from bifilms, or do not have the production requirements of some of the high volume aluminium foundries. Thus, in normal circumstances, many irons and steels are relatively free from bifilms because of the large density difference between the inclusions and the parent melt, encouraging rapid flotation. Alternatively, many copper-based and steel foundries are more like jobbing shops, where the volume requirement may make the justification of a counter-gravity system more difficult. The technical challenges are also non-trivial; high technology pumps, if they were available, would have problems to survive the oxidation and thermal shock problems of a stop-go production requirement, although a low-pressure system delivering melt directly from the melting furnace into the base of moulds seems a relatively straightforward and attractive option. The reader is recommended to Section 16.3 for more details.

As a well known instance, the counter-gravity Griffin Process has been impressively successful for the volume production of steel wheels for rail rolling stock. The process produces wheels that require no machining (apart from the centre hub); the outer cast rim runs directly on the steel rail. The products out-perform forged steel wheels in terms of reliability in service, earning the process 80% of the freight rail wheel market in the United States. What a demonstration of the soundness of the counter-gravity concept! It contrasts dramatically with our current steel castings produced world-wide by gravity pouring, in which defects and expensive upgrading of the casting are the norm.

14.5 CAST MATERIAL
14.5.1 LIQUID METAL

It has become fashionable to criticise the traditional approaches to the production of castings by the filling of a mould with liquid metal. The liquid casting approach probably deserves such criticism, earned by its abject failure on occasions to deliver castings of respectable quality. This has been a regrettable and lengthy 'dark ages' from which it is hoped the industry is gradually emerging. It will be long and hard for the casting community to reverse its unflattering image. A new professionalism will only be recognised by results in the form of on-time deliveries of good castings.

In the meantime, it is understandable that the casting of liquid metal has been thought to be fundamentally problematic (in contrast to the central message of this book). Thus, huge expenditure and effort has been carried out on the casting of various partially solid alloys and liquid/solid mixtures (misleadingly named semi-solid mixtures because they are usually not 50% solid mixes), for which there are undoubted benefits.

Even so, the author remains convinced that despite all the well-known difficulties with the casting of liquid metals, the difficulties are worth living with, and ultimately, the benefits to properties hard to beat. For this reason, this book remains strongly committed to the production of castings by the casting of liquid metal, convinced that such a production route will provide the major benefits to the majority of customers for the foreseeable future.

This is not to deny the special features and advantages enjoyed by some non-liquids, or perhaps we should say, partially solid routes to casting manufacture.

14.5.2 PARTIALLY SOLID MIXTURES

Partially solid mixtures have been widely known as '*semi-solid*' mixtures. I have to say I try avoid such inaccurate names. Imprecision in our words can become wooliness in our thinking. I agree the accurate name does not alliterate to help it off the lazy tongue, but it is correct, and in my view therefore, worth the effort.

The advantages of casting partially solid mixtures are well known, and include their use in high-pressure die casting, in which the slurry seems more resistant to disintegration, and thus avoids the worst excesses of entrainment usually suffered by the high pressure die casting technique when applied to liquid metals. Furthermore, because a significant fraction of the heat content has been removed from the mixture before casting, the metal dies run cooler. They therefore suffer reduced crazing due to thermal shock or thermal fatigue, and so last longer. Freezing is also faster, thus increasing productivity. Although reduced shrinkage porosity is also cited as an advantage because, once again, a significant fraction of the cast material has already solidified, this is almost certainly more apparent than real; the so-called shrinkage porosity in pressure die castings was mainly misdiagnosed entrainment problems including air and oxides. However, of course, there has to be some small component of reduced shrinkage problems, but in such thin-section castings this would be expected to be practically negligible. Anyway, the undoubted overall improved soundness confers improved and more reliable mechanical properties, and improved leak tightness.

The disadvantages of the processing of partially solid mixtures relates to the increased cost of processing of the mixtures, and for some approaches, the increased control necessary to form a correct mix (some approaches have automatic, intrinsic, built-in control, such as the approach by Rheometal (Granath et al., 2008), using not temperature control, but a method of mixtures based on enthalpy equilibration, melting a known amount of solid in a known amount of liquid. The oxide skin from the melted solid addition appears not to have raised any problems so far as is known to date, but probably requires further investigation.).

During the mixing processes in which the dendrites are grown and fragmented, some but not all of the processes tend to introduce air and oxides into the mixture. The result is, of course, a mix that now contains a dispersion of oxide bifilms that will limit the properties achievable from the final casting. The beneficial aspect of the defect dispersion is that it is probably rather uniform in its extent and in the size of the individual defects. Thus the properties may not be optimum, but the product will probably be uniformly reliable or unreliable.

Rheocasting

Rheocasting is technique for producing a partially solid alloy but flowable alloy by careful cooling of the liquid alloy into the liquid + solid range while mechanically stirring to ensure that dendrites are broken up, forming compact, near-spherical forms. Such an approach was discovered by Spenser Mehrabian and Flemings at MIT in 1972. The process continues to find niches in the industrial and commercial world in the second millennium.

Thixocasting

Thixocasting was a variant of the rheocasting approach in which the partially solid mixture was precast in the form of semi-continuously cast 'logs'. The logs required to be sawn to correct length, reheated to the casting temperature, and finally cast. This approach has always suffered from the cost of the separate melting and rheocasting process steps. Furthermore, the foundry returns (runners, feeders and scrap castings) present a problem for recycling. The just-in-time processes for the production of partially solid mixes avoid both of these issues.

Just-in-time slurry production

New concepts require agonisingly long development periods before they become accepted by users. Companies such as Alcan Canada (Doutre et al., 2000; Lashkari and Ghomashchi, 2007), Ube Japan (Lee and Kim, 2007), Idra Italy (Yurko et al., 2003) and Rheometal, Sweden (Granath et al., 2008), among others (Midson, 2008), have continued to develop novel just-in-time slurry production techniques to counter one of the challenges of the process, the production of controlled qualities of slurry at commercially attractive production rates, and commercially attractive costs.

Strain-induced melt activation

Strain-induced melt activation produces a thixotropic condition by first deforming the solid alloy followed by a heat treatment into the mushy zone. The approach avoids melting and the entrainment of air and oxides suffered by competitive techniques for slurry production. Chen et al. (2010) describe its use for magnesium alloy AZ91, using equal channel angular pressing to produce a high but uniform strained condition, and thereby achieve impressive results. Even so, it seems likely its potential for commercial exploitation will be limited by its challenging economics, in common with other semi-solid routes.

Mg alloy slurry production from granules

The work by Fan et al. (2007), Czerwinski (2008) and others has demonstrated the feasibility of producing rheocastings in Mg alloys by the processing of alloy granules through a twin screw injection moulding machine as part of a pressure die casting process. The processing of Mg granules via a route more normally associated with the processing of plastics is possible partly because of their low strength and hardness, and partly because Mg reacts with the steel screws to only a limited extent, compared for instance, to Al alloys which dissolve iron readily. Once again, the uniform dispersion and shredding of the oxide skins from the original granules is seen to deliver a product that has uniformly reliable, if not optimum, properties. In fact, the precipitation of the so-called β-$Mg_{17}Al_{12}$ phase at grain boundaries in thixotropic Mg alloys containing over 7%Al has been cited (D'Errico, 2008) as limiting ductility. It seems almost certain that this phase is precipitating on bifilms generated by the mixing process, and pushed ahead of the growing grains, effectively pre-cracking the grain boundaries.

Metal matrix composites

Solid/liquid mixtures can be made by the addition of particles of solid into the melt. The exploration of such mixtures was pioneered by Pradeep Rohatgi, who over the years since his time in 1965 in the International Nickel Company Laboratories, New York, has explored mixtures of practically everything with everything (see his Rohatgi Symposium reported by Gupta, 2006). Most matrices studied so far are those of Al alloys, but other work includes Zn, Mg and, later, superalloys (Rohatgi, 1990). The particle additions include graphite, SiC, Al_2O_3 and low cost additives such as fly ash, a waste product from power plants. Thousands of tonnes of such metal matrix composites (MMCs) are now processed for use in the transport industries.

The process most used to make mixtures is the so-called vortex method in which the solid additions are fed into a central vortex made in a rapidly stirred liquid metal. Some work has also been carried out using particles in suspension in a carrier gas, introduced under the melt surface from a lance.

All processes that introduce particles from the outside into the interior of the melt suffer the problem of having to penetrate the oxide skin on the melt. This means that particles usually enter the matrix as clumps, each clump being enclosed by a wrapping of the oxide film necessarily entrained from the surface (Figure 2.3). The problem with such clumps is that the reservoir of air they contain continues to thicken and strengthen their surrounding oxide envelope. The vigorous stirring required to break these large agglomerates down into smaller agglomerates automatically introduces yet more oxides into the melt via the central vortex. Tian et al. (1999) study clusters of alumina particles in liquid Al, finding clusters from one particle of 10 μm in diameter to clusters of particles 100 μm in diameter. In each case, the clusters are enclosed by amorphous alumina films.

The MMC studies by Emamy et al. (2009) also illustrate particle clusters inside oxide 'paper bags'. There is evidence from this work that the loss of fluidity with higher additions of particles is not primarily the result of the higher percentage of particles but the result of increased oxide bifilm content from the vortex stirring process. The huge increase in bifilm population in such vortex-produced MMCs is also seen to degrade ductility, even though, paradoxically, there may be some slight increase in strength.

Badia (1971) explores the production of MMCs in Al and Zn matrices by additions of a variety of solids including SiC, Al_2O_3, SiO_2 and graphite. He found that coating of the particles with Ni was essential to achieve some kind of wetting so that the additions would remain in suspension; otherwise, the additions appeared to surface once again, rejected from the melt as a dry powder. The action of the Ni is not completely clear and may depend not only on its metallic nature, but also on its powerful exothermic reaction with Al to form nickel aluminides. Any benefit to the MMC of the introduction of a lightweight phase is lost however, because the 5 μm Ni coat on particles 5 μm in diameter corresponds to an addition containing 50% Ni. Badia used the vortex addition technique and subsurface injection from a lance, but preferred the vortex approach because continued stirring with the rotor could maintain the addition in suspension.

Practically all of the previous studies have focussed on the use of particle additions to a melt. Occasional exploratory studies have included a variety of fibres (Rohatgi, 1990) and SiC whiskers (Das and Chatterjee, 1981) for which surprising gains in strength have been recorded for levels of only 0.5 volume percent addition.

In contrast to these laboratory studies, in which the introduction of particles was attempted with relatively inefficient protective atmospheres, Alcan instigated a major production initiative in which they produced an Al alloy – SiC MMC in tonnage quantities in a high vacuum environment (Hammond, 1989) – although at the time the production conditions were a commercial secret. The uniform dispersion of the particulate phase in Duralcan, as it came to be known, indicated its relative freedom from oxide films, although work by Emamy and the author did reveal some residual bonding problems, even though these were much less than those commonly seen elsewhere. Later work (Hoover, 1991) confirmed the excellent specific stiffness, good tensile properties including fatigue, and good retention of properties at elevated temperature.

In situ metal matrix composites

The formation of MMCs by the precipitation of the strong, hard particles in situ in the matrix by a metallurgical reaction during solidification is an attractive technique. In this case, the strengthening particles suffer no poor cohesion with the matrix because of an intervening oxide film and its associated thin layer of air; the particles are bonded to the matrix with atomic perfection, having grown atom by atom from the liquid.

The difference in the bonding of an in situ–generated MMC based on TiB_2 particles in an Al alloy matrix, and Duralcan (a good quality extrinsically generated MMC based on SiC particles) was shown clearly in work by Emamy and Campbell (1995). A non-fed casting, designed to create a reduced internal pressure, possibly a slightly negative pressure (corresponding to a hydrostatic tensile stress) was seen to cause significant dispersed microporosity in Duralcan, but very little in the in situ MMC. The difference between the relatively poor bond and the perfect bond was clear.

Hadian et al. (2009) studies the system $Al-Mg_2Si$ and the refinement of its Mg_2Si particles by more rapid freezing and by the addition of Li. Although the Mg_2Si particles are usually regarded as brittle, this study is notable for its image of an Mg_2Si particle cracked (one assumes by the presence of a central bifilm on which it formed) but plastically peeled open, revealing the particle to be impressively ductile (Figure 6.22). This is expected to be a universal feature of in situ MMCs; all the hard, strong particles will probably deposit on a bifilm, and thus all will be expected to possess an in situ crack, despite the intrinsic strength and crack resistance that they would be assumed to possess.

The common Al-Si system similarly behaves as an in situ MMC in which the silicon particles act as the hard, strengthening phase in the ductile Al matrix. Once again, the presence of central cracks in the Si is attributed to the formation of the Si particles on bifilms in suspension in the melt. It is known that clean melts cannot form Si particles in this way, with the result that the Si is forced to precipitate at a lower temperature and a different form, as a finely spaced eutectic which we call a 'modified' Al-Si eutectic. The addition of such elements as Sr and Na similarly act to inhibit precipitation on the oxide bifilms, thus encouraging the modified structure (Campbell, 2009). When the Si phase forms in this way, it can now no longer be regarded as separate particles, even though this is its appearance on a polished micrographic section. The Si has the form of a continuous branching growth known as a coral form. This form of the Al-Si MMC can be particularly strong and tough if bifilms are somehow removed by careful processing because the development of the MMC structure now no longer depends on their presence.

14.6 RE-MELTING PROCESSES

In the quest for really clean steels, so-called secondary re-melting processes have been developed in which steel is prepared in the form of an electrode which is electrically re-melted drop by drop and progressively re-solidified into a new ingot. Although many forms of melting have been employed from time to time, including plasma and electron beam, the two main processes used world-wide are vacuum arc re-melting and electro-slag re-melting processes, sometimes called vacuum arc refining and electro-slag refining.

14.6.1 ELECTRO-SLAG RE-MELTING

The heat source in this process is the resistance heating of the layer of liquid slag between the electrode and the ingot (Figure 14.5). The hot slag remelts the tip of the electrode drop by drop so as to build up an ingot.

A thin layer of slag freezes against the water-cooled wall of the mould, creating a protective layer, so the liquid metal does not directly contact the mould (Figure 14.5). Thus the melt is not strongly chilled by the cold wall, and cold laps are thereby avoided. The liquid oxide environment (the slag layer) of the liquid metal ensures the absence of solid oxide films on the surface of the liquid metal, so that oxide laps are also avoided. As a result of the avoidance of both varieties of laps, electro-slag re-melting (ESR) material has a fundamentally crack-free nature. This is clearly demonstrated by an extreme case provided by one of the more difficult-to-work materials such as the Ni-base alloy, Waspalloy. Vacuum arc re-melting (VAR) ingots of Waspalloy are hard to forge without cracking despite having their surface deeply peeled, compared with ESR ingots which forge like butter despite having none of their surface removed.

Although it was generally assumed that the melt was cleaned from inclusions by dissolution of inclusions in the slag layer as droplets of the melt passed through the slag, a study of a cross section of an electrode taken out of the slag and sectioned down its centreline clearly shows that all of the inclusions go into solution at the very high temperatures before melting occurs. The subsequent relatively clean structure of the ingot, displaying only extremely fine dispersed inclusions if any, is the result of the relatively rapid solidification encouraged by the water cooling of the mould and its

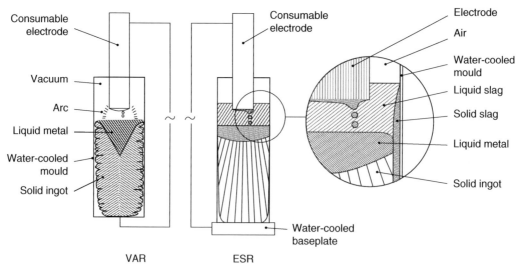

FIGURE 14.5

Comparison of the two major ferrous alloy refining systems.

baseplate. However, of course, some chemical interaction occurs with the slag layer, so that, for instance, some sulphur can be absorbed from the metal, but in general these seem to be less important actions.

However, I am not aware of any attempt ever to upgrade the quality of ESR electrodes which might similarly enhance the quality of the final ESR ingot. So far as can be ascertained, all trials to evaluate ESR have been carried out with poor electrode quality, usually produced by the top-pouring of an open-top ingot in air, consequently hampering the performance of the ESR process. The bifilm and major lap defects on the electrodes can be seen with the unaided eye from a distance of at least 100 m. Perhaps ESR has suffered these appalling electrodes because it was generally assumed that it had the capability of dealing with any quality of electrode material. It is true it is a good process, but like all processes, it has its limits. As a starting electrode, an ingot top-poured in air is beyond reasonable limits.

Thus it is unfortunate that current ESR ingots almost certainly contain oxide bifilms as a result of the choice of an extremely poor quality starting material. Effectively, although the process has impressive potential, it is currently usually overwhelmed by the overload of defects from a poor electrode. I have seen an ESR ingot containing fragments of unmelted electrode material the size of my fist. Such fragments have detached from the electrode as a result of the massive oxide film defects created by the poor electrode casting technique. It would not have been difficult or costly to cast the electrode well, in air, resulting in practically zero defects from the final ESR ingot. It is with great regret that I have to report I have never seen an ESR electrode formed from a casting process which I could approve as having been cast well.

Naturally, it is necessary to address the additional reasons usually levelled against ESR by devotees of VAR.

1. The danger of increased hydrogen. The concern regarding hydrogen content is probably an illusion for most products which are sufficiently small to allow the gas to diffuse away. Larger ESR diameter ingots may retain their hydrogen which may or may not be a concern, but the problem would be avoidable by relatively easy containment and protection of the melting environment using an argon shroud around the open top of the mould.
2. VAR enjoys a much higher temperature gradient during solidification, allowing the production of larger diameter ingot without channel defects (freckles). Once again, this is relatively easily addressed in ESR by the move from full size steel moulds to the proven use of short, water-cooled collar moulds made from steel, somewhat analogous to the water-cooled copper moulds used for continuous casting. The ingot emerging from the base of such moulds can then be cooled by the direct impingement of water jets, providing a higher rate of heat transfer than can be achieved via the thermal barriers of the VAR mould, thus exceeding VAR's ability to keep channel defects at bay.
3. A related rate of cooling problem for the conventional full size ESR mould is the relatively low rate of solidification compared with VAR, resulting in a larger dendrite arm spacing, which in turn significantly increases solution and homogenisation treatment times. As for (2) previously, the move to a collar mould is a complete solution.

14.6.2 VACUUM ARC RE-MELTING

Special steels requiring high cleanness such as bearing steels have been vacuum induction melted (VIM)-VAR processed since the 1970s to eliminate the oxide inclusions thought to promote fatigue.

The VAR process consists of re-melting the electrode in a vacuum, striking an arc between the electrode and the water-cooled base of the mould. The arc provides the heat source, melting the electrode drop by drop to build up the new ingot (Figure 14.5). All this takes place in a water-cooled evacuated mould.

However, the production of the electrode is usually somewhat better carried out than that for the average ESR ingot. VAR is routinely carried out with some pre-refining process for the production of a reasonable quality of starting material. The electrode is usually melted and cast in a VIM process. However, it is regrettable that the casting process is typically a conventional pour, allowing the melt to fall through a height of between 2 and 3 m into the open top of a vertically standing ingot mould. The pouring process creates masses of bifilm defects as a result of its vacuum environment being insufficiently good to avoid oxide formation during the pour (it is appropriate to view the 'vacuum' as merely dilute air). The result is an electrode which is marginally better in terms of defect concentration compared with the dreadful air top-poured electrode described earlier. Thus, the provision of this vacuum top-poured ingot as an electrode for the starting point of the subsequent VAR melting and solidification process naturally causes the VAR process to suffer an unnecessarily disadvantaged starting point. For this reason, although VAR steels are reasonably good, they are less good than they should be.

Furthermore, because the vacuum of the VAR process is also rather poor, the VAR ingot tends to be formed of layers of metal pancakes separated by laps which are in fact bifilm cracks. The pancakes arise as a result of the slow upward filling, fed by the dribble of metal, feeding the spread of successive layers which reach the crucible wall and freeze. Each layer has time to grow a thick oxide layer before the next layer rolls over it, laying down its relatively young, thin oxide on top of the old thick oxide to create a horizontal bifilm, separating each pancake layer from its neighbours. The horizontal bifilms occupy large fractions of the area of the ingot, causing serious cracking problems during the forging of many highly alloyed Ni-base alloys despite significant thicknesses of the ingot surface being 'peeled' off before forging.

As is to be expected, there is a mass of evidence for the tolerably good properties of VAR material. The fair performance of such materials is not in dispute (even though of course the multiple re-melting and peeling operations, together with cracking failures during working, are extremely inefficient and therefore costly in terms of energy and materials).

Despite these problems, it is easy to understand the superficial attraction of the VAR process: the 'vacuum' connotation suggests a perception of cleanness. The relatively good performance of VAR has been aided by its common provision of a VIM electrode. This benefit has been rare for ESR which has clouded comparisons of the two re-melting systems in favour of VAR. As we have seen previously, the other perceived advantages of VAR such as low hydrogen, fast cooling and ability to reduce channel segregates in large diameter ingots are all easily addressed by ESR. It follows that given a level playing field in which VAR and ESR would be provided with equal quality starting material for their electrodes, the quality of ESR would be expected to be significantly superior. The usually standard choice of VAR for most aerospace and other critical applications is therefore seen to have been a costly historical mistake, costing the aerospace and defence industries dearly over past decades. It would not be too late to make a change.

VAR has enjoyed an unearned reputation as a refining or secondary melting process, whereas ESR could in general provide a superior product. As a result, my personal view is that VAR should in general be abandoned as an immensely costly but fundamentally unsatisfactory processing route.

MOULDING

Figure 10.2 illustrates the complexity of the casting industry by the wide variety of its moulds. The use of permanent metal moulds brings the advantages of rapid freezing, giving castings with good mechanical properties, but also the penalty of having to wait for the casting to freeze before opening the mould and the ejection of the casting. For moulds manufactured and destroyed each time from aggregates, the cooling is slower usually giving poorer properties to the casting, but there is no waiting time for solidification. Productivity is now governed by the speed at which moulds can be made. This can be impressively high for highly productive green sand plants; a mould every 10 s is possible. Furthermore, it is common for the mould to contain more than one impression.

This broad brush description requires much, very much, qualification as is attempted in this chapter.

15.1 INERT MOULDS AND CORES

Very few moulds and cores are really inert toward the material being cast into them. However, some are very nearly so, especially at lower temperatures.

Metal moulds have the great advantage that they are relatively inert toward the liquid metal. They are usually used warm and dry, so their intrinsic hydrogen content is low, in sharp comparison to the various aggregate moulds that contain sufficient moisture and other volatile and reactive components to make copious and generous hydrogen contribution to the cast metal.

The usefulness of a relatively inert mould is emphasised by the work of Stolarczyk (1960), who measured approximately 0.5% porosity in gunmetal cast into steel-lined moulds, compared with 4.5% porosity for identical test bars cast in greensand moulds.

For light alloys and lower-temperature casting materials, investment moulds are largely inert. Interestingly, dry sand moulds (i.e. greensand moulds that have been dried in an oven) have been found to be largely inert, as shown by Locke and Ashbrook (1950).

Carbon-based and graphite dies have been found useful for zinc alloys. However, their lives are short for the casting of aluminium alloys because of the degradation of the carbon by oxidation. All the more impressive therefore is the use of graphite moulds for steel, used for the casting of millions of railroad wheels by the Griffin Process. Carbon-based moulds are used for the casting of titanium alloys in vacuum. Oxidation of the mould is reduced by the vacuum environment, but the contamination of the surface of the titanium casting with carbon is severe, promoting the formation of an outer layer of the alloy where the titanium alloy alpha-phase is stabilised. This surface layer is known in titanium castings as the *alpha-case*. It usually has to be removed by machining or chemical dissolution.

15.1.1 PERMANENT METAL MOULDS (DIES)

The use of a mould that is permanent has always been attractive to founders ever since the use of hollowed out pieces of stone used to cast multiple copies of bronze axes and arrow heads.

The concept remains popular because usually the customer will pay for it, and once paid for, the mould then costs little more in service, provided no catastrophe overtakes it. Unfortunately, it has to be noted that catastrophic failure, often as a result of thermal fatigue aided by internal stress from its heat treatment, is not unknown.

Complete Casting Handbook. http://dx.doi.org/10.1016/B978-0-444-63509-9.00015-7

The practice of using permanent moulds is, with some significant exceptions that will be listed later, usually confined to low melting point alloys; aluminium alloys being the most common. Other cast materials include alloys of magnesium, zinc and lead. Only rarely are permanent moulds used for the higher temperature alloys such as brasses and bronzes and cast iron.

The major metallurgical benefit of a metal mould, compared with an aggregate mould, is the relatively rapid freezing, conferring higher mechanical properties. Moreover, the liquid metal and mould are practically inert toward each other, so that there are no significant chemical reactions. This is an additional and significant metallurgical benefit of metal moulds that is often overlooked.

The major challenge provided by permanent moulds is that the mould has to be designed to disassemble to release the casting. This obvious point is not to be underestimated because for many castings this presents major limitations to both the casting and the filling and feeding systems. Sand castings have no such problem because the mould is destroyed to release the casting, thus allowing geometries of almost any complexity or sophistication.

Gravity and Low-pressure die-casting dies (Permanent Moulds)

For dies filled simply by pouring under gravity, they are not surprisingly called *gravity dies* (known in the United States, also not surprisingly, as *permanent moulds*). The service conditions for a gravity die are not too severe because the die is to some extent protected from direct contact with the melt by the application of a die coat.

Low-pressure die casting dies are extremely similar; being made from iron or steel, depending on the service requirements, and once again enjoy the protection of a ceramic die coating applied as a wash by painting or spraying.

One of the most notable high temperature developments of low-pressure die casting is the famous Griffin Process for steel railway rolling stock. In this case, the dies are machined from graphite. Although a successful process has been built on this process, the machining of graphite is always a somewhat messy and dirty occupation. There seems little doubt that the process might have been operated with greensand with practically no loss of accuracy and no loss of performance of the steel casting. It would be interesting to see a comparative evaluation of the two types of mould for this high volume product.

High-Pressure die-casting dies

High-pressure die casting (HPDC) is known in the United States, confusingly, as die-casting.

Dies for pressure die casting are hardly inert, partly because of the gradual dissolution of the unprotected steel die, but mainly because of the overwhelming effect of the evaporation of the die-dressing material. This may be an oil- or water-based suspension of graphite sprayed onto the surface of the die that is designed to cool and lubricate the die between shots. The gases found in pores in pressure die castings have been found to be mainly products of decomposition of the die lubrication, and the volume of gases found trapped in the casting has been found to correspond on occasions to nearly the volume of the die cavity.

A little-known problem is the boiling of residual coolant trapped inside joints of the die. Thus as liquid metal is introduced into the die, the coolant, especially if water-based, will boil. If there is no route for the vapour to escape via the back of the die, vapour may be forced into the liquid metal as bubbles. If this happens, it is likely that at least some of these bubbles will be permanently trapped as blow holes (note the *correct* and rather rare use of the term 'blow hole' in this instance) in the casting. This problem is common in pressure die and squeeze casting processes.

The recent development in pressure die casting to separate the functions of (1) cooling of the die and (2) its lubrication, is seen as a positive step toward solving this problem. The approach is to use more effective cooling by built-in cooling channels, whereas lubrication is achieved by the application of minute additions of waxes or other materials to the shot sleeve.

When used for its common purpose of casting aluminium or magnesium alloys the dies are necessarily made from good quality hot work tool steel, H13. In this case grey iron would not have the adequate surface integrity because of the presence of graphite flakes in the material, and would not have the strength or fatigue resistance. No protective coating is applied to the surfaces of the die. The result is an excellent surface finish. The danger of cold laps is reduced by extremely rapid filling, and by high pressure that is subsequently applied to ensure faithful reproduction of the profile of the mould. Even so, of course, the high speed of filling causes other serious problems that we shall consider later.

Like all dies, however, it is regrettable that they are not really permanent. An aluminium casting weighing up to 1 kg might be produced tolerably well for up to 80,000 or 100,000 shots before the die will need to be refurbished as a result of heat checking (cracks from thermal fatigue of the die surface that produce a network of unsightly raised fins on the cast surface).

Magnesium alloy HPDCs are much kinder to dies, giving perhaps five times or more life. This is partly the result of the lower heat content of Mg alloys and partly because of the low solubility of iron in liquid magnesium that reduces its action to dissolve the die surface. In contrast, liquid Al has a relatively high solubility of iron, resulting in a phenomenon known as 'soldering' of the casting to the die. This action is particularly destructive to dies. It is countered to some extent by the addition of high levels of iron to HPDC alloys, thus pre-saturating the liquid metal with iron and so greatly reducing the tendency for the melt to dissolve the die.

Attempts have been made to cast molten stainless steels by HPDC. In this case the only die material capable of withstanding the rigours of this process was an alloy of molybdenum. Even these 'moly' dies suffered early degradation by thermal fatigue cracking, threatening the commercial viability of the process. So far as the author is aware, partially solid (inaccurately called 'semisolid') stainless steels have not been extensively trialed. These cooler mixtures that have already given up practically half of their latent and specific heat during freezing would greatly benefit die life. The commercial viability of the somewhat more expensive mixture would remain a potential show-stopper of course.

Because of the very high pressures involved in HPDC internal cavities are usually limited to those formable by straight withdrawable steel cores. If the core cannot be withdrawn, then the usual alternatives such as sand cores are totally unsuitable because they are penetrated by the liquid metal. One of the very few options to make undrawable cavities in HPDCs is salt cores, described later.

15.1.2 SALT CORES

Salt cores are a class of components that once again can be extremely dry and can be used hot. Provided the salts are mainly chlorides and fluorides, these highly stable compounds have very little reaction with most molten metals.

They are useful to provide the undercuts and other difficult-to-mould features permanent mould and in HPDC of Al alloy parts, where simple two-part dies are preferred for robustness and mechanical simplicity. The core is subsequently dissolved out by immersion in water. Salt cores have been used for many years worldwide for the production of oil galleries in Al alloy gravity cast pistons. There is still interest in Japan on improving these core materials still further (Yaokawa, 2007).

Loper and colleagues (1985) review how salt has been used in various ways to make cores in Al alloy and grey iron castings. They describe many different mixtures of salts, including NaCl, KCl and $CaCl_2$. The only problem they reported was the loss of some mechanical properties of the cores during periods of high humidity. The cores were dissolved out by simple immersion in cold water, but faster in hot water, and faster still by pressurised water jet.

1. Solid salt cores have been made by the casting of molten salt using a gravity die (permanent mould) process.
2. Solid salt cores can be made by pressure die casting. Their successful use for internal combustion engines includes such achievements as closed deck cylinder blocks, as demonstrated by the Mercury Marine company in the United States since 1976. The heat content of the salt is sufficiently low that the dies have never been observed to suffer from any heat checking (cracking) problem. The cores are used almost immediately, being ejected from the first machine and loaded into a second pressure die casting machine standing nearby to make an Al casting. The cores may be subjected to a temperature equilibration treatment in an oven, the high coefficient of thermal expansion of the salt allowing an adjustment of size of the core if necessary.

One of the major advantages of the salt core technique in HPDC is its flexibility. It allows expensive tooling to be adapted, simply changing the salt core dies to revise the inner features of the casting, for a fraction of the cost of complete new tooling. With changes to cylinder bore diameter, or iron sleeves, etc. a block can be refurbished quickly and at minimal expense, compared with conventional HPDC tooling. Also, of course, the cores can provide much more complex internal detail. Major complex parts for small engines such as closed deck cylinder blocks could not be made without the use of such cores.

The cores are characterised by excellent surface finish and accuracy. Core breakage problems appear to be practically unknown. Furthermore, the salt casting is optimised for internal porosity simply by holding it against a light because it is translucent. No costly X-ray radiography is required.

Although to make an HPDC it is first necessary to make 'a second HPDC' tool for the core, the salt casting costs only about one quarter or one fifth of the cost of the Al die casting so the economics are not discouraging. The rate of production of cores matches the rate of production of Al castings, and the rate of washing out is twice this rate.

3. Salt cores formed of aggregates bonded with salt have sometimes been used. The aggregates have been alumina, iron oxide or silica sand and up to 40% by volume of salt has been used.
4. Other varieties of salt core have been formed from mixture of grains of an inert aggregate with grains of salt both bonded with a resin binder. The best solution rates were achieved using a similar size of sand and aggregate grains because this maximised the permeability of the core. The production technique has the benefit of being a cold process using normal core production techniques.

These versatile moulding materials probably deserve wider use throughout the industry.

15.1.3 CERAMIC MOULDS AND CORES

The other category of inert moulds are the ceramic moulds. Thus lost wax moulds are fired at high temperature (usually 1000°C or above) and thus are especially dry. In addition, of course, ceramic moulds can be used at high temperature, so there is a greatly reduced chance for reactions with volatile mould components.

Ceramic moulds are such a specialised technology that only an outline can be given here. A tremendous and unique advantage is the freedom to select an appropriate mould temperature for casting. Ceramic moulds are usually fired to a relatively high temperature before casting. Thus they are generally free from outgassing and reactivity problems with melts. Ceramic moulds for light metals are often pre-heated to a temperature in excess of the freezing point of the melts so that there are no problems of fluidity – if a sufficiently long fluidity test could be devised the melt would run for ever! For Ni- and Co-based superalloys for single crystal turbine blades, the moulds are similarly held above the freezing point of the melts during filling and only slowly withdrawn into a cooling zone to grow the crystal under conditions of controlled speed and temperature gradient.

Investment Shell (Lost Wax) Moulds

As a monarch might be *invested* in the *vestments* of royalty, the verb *to invest* means *to coat, cloak or enrobe* and thus nicely enshrines the concept of the making of an *investment coating* around a wax pattern which is *invested* (i.e. *coated* by dipping) in successive layers of a slurry of finely powdered mineral, most commonly zircon or aluminium silicate (mullite), and binder (typically a very fine dispersion (a sol) of silicic acid). A coarser mineral (stucco) is applied to each layer of slurry before drying, so as to gradually build up a shell. The investing process is the special feature that confers the admirable surface finish and fine detail that investment castings enjoy. After drying, the shell is steam heated, typically in an autoclave to melt out the wax, giving the process its other well known name, the *lost wax* process. Having lost its wax, the shell is then fired to (1) burn out any residual carbon remaining from traces of wax and (2) develop the full strength of the shell. Strength development during firing comes about as a result of the silicic acid losing its water and being converted to a ceramic substance (cristobalite). Sometimes the shells are cooled to room temperature once again and checked for cracks which are then repaired. Otherwise, the shells are taken directly at the firing temperature and filled with metal.

When the casting has cooled to room temperature the shell has to be removed. This can present a problem. Many shells are stronger than the light alloy castings they contain, so that mechanically breaking off the shell can easily damage the casting. Analogously, the Ni-based single crystals are vulnerable, because any impact damage can cause a recrystallisation event during the subsequent heat treatment, thereby scrapping the blade.

Other disadvantages with the lost wax process include high unit cost, high energy consumption and that the spent shells are not reclaimed, although work is in progress to redress this issue. It is also difficult and expensive both to make cores for investment casting and to remove them afterwards. One common approach is to prepare a core by injecting a hot dispersion of slurry and wax into a core box cavity, cooling the resultant product and then firing it. The core can be removed from the casting mechanically or by washing with caustic soda acid, techniques which are only feasible for ferrous metal castings.

Yet more potential problems should be highlighted: accuracy starts to be a significant problem for larger investment castings. This mainly results from distortions that occur in the wax before investment. For this reason, the provision of costly jigs and fixtures to straighten wax patterns is sometimes essential to guarantee accuracy.

Graphite-based investment shells have been used, particularly for very reactive metals such as titanium and zirconium (Howe, 1965).

Several recent developments are promising to revolutionise the lost wax process. These include new low-cost aggregate systems, new binder systems, techniques enabling shells and cores to be made far more quickly and cheaply and shells and cores that can be removed by washing with water after casting. This may lead to a growth in the popularity of lost wax investment casting, particularly for light metals.

Investment 2-Part Block Moulds (Shaw Process)

The investment block moulds, originally called the Shaw Process, originated in England by two British scientists, Clifford and Noel Shaw, in 1938, the year of my birth.

Although the process has generally fallen into disuse because of the apparent economy of simply making a lightweight shell, it has interesting strengths and advantages. For instance, a key difference between the shell and the block processes is that the block mould is made up as a cope and drag by the use of normal patternwork (not a wax pattern).

Ceramic slurry is prepared as a colloidal suspension of silica in alcohol, to which is added various ceramic fillers to make a smooth cream. The slurry is poured (invested) over a pattern, filling up to the top of a surrounding frame. After the mould is nearly set, but retains some flexibility as a gel, it is stripped from the pattern and placed on a board to dry and develop its green strength. At this stage the alcohol is flamed off, causing the surface of the mould to develop its characteristic micro-crazed structure, conferring essential permeability and thermal shock resistance to the mould. In this way, a drag half of a mould can be made. The cope half is made similarly. The two halves are then assembled, fired and cast. The cope and drag technique for mould assembly allows cores to be placed. Errors from the distortion of the wax pattern are also avoided. Lubalin and Christensen (1960) give a good description of the process and the wide range of castings that can be made.

One can envisage a more modern variant of this process in which the mould halves are injection moulded and thus produced rapidly for volume production with a fully robotised process. This concept seems ripe for re-examination and fuller exploitation.

Plaster Investment Block Moulds

Plaster of Paris ($CaSO_4 \cdot \frac{1}{2}H_2O$) is used as a mould material to make block moulds by a variant of the lost wax investment process. Plaster formulations contain 15–25% plaster, the remainder being inert fillers such as quartz and cristobalite and often Portland cement.

After energetic mixing of the powder mixture with water to make a paste, the mixture can be subjected to a vacuum and mechanical vibration to encourage the detrainment of bubbles, after which the mixture is poured around the wax pattern. The plaster takes up water of crystallisation, converting to gypsum ($CaSO_4 \cdot 2H_2O$) and turns solid. Controlled slow drying follows, then de-waxing, and final firing at about 700°C (any higher temperature would cause the gypsum to lose its water of crystallisation and break down to a powder once again). In this way a one-piece block mould is made by a variant of the lost wax process.

The combination of this process with the application of a reduced pressure applied to the base of the permeable moulds (the steel box is open at its base) is the principle of the vacuum-assisted casting, allowing the pouring in air of extremely thin-walled castings only a fraction of a millimetre in wall thickness. The natural high fluidity noted in this process arises because of the high temperature of the mould (above the freezing point of the alloy, so that 'fluidity' in terms of flow distance is effectively infinite) and its extremely low thermal diffusivity. The process is ideal for aluminium and magnesium alloys, but can also produce some copper-based products if the castings are not too heavy.

The castings can be de-moulded by placing the moulds in water, causing the mould material to spall, pieces jumping off in a series of energetic and noisy explosions. The final traces of wet mould are brushed or washed away.

15.1.4 MAGNETIC MOULDING

The concept of holding a mould of fine steel shot together as a rigid mould by the application of a magnetic field has an alluring attraction. This apparently attractive process was invented in 1971 (Wittmoser 1971, 1972 and 1975), but never seems to have come into general use. Steel shot is 'frozen' into position around the pattern by the application of a magnetic field. After the casting is poured and solidified, the magnetic field is switched off, allowing the mould to disintegrate and the casting to be removed.

The reality, however, may be something else. Since Wittmoser first publicised his work in Germany relating to magnetic moulding in 1975, little has been done to extend and confirm this early sign of promise; this early failure to develop the process has never been explained. Wittmoser poured cast iron into lost foam moulds surrounded by steel shot in an Al alloy moulding box, the shot being held in place by a powerful magnetic field. This early work was almost certainly undermined to some extent by the mould material exceeding the Curie temperature of mild steel at approximately 900°C. Above this temperature, the ferromagnetic property is lost, so that parts of the mould would have become effectively non-magnetic. This may have explained some of the difficulties with the original experiments. Later workers (Geffroy et al., 2006) trialed Al alloys. At this lower casting temperature, the process might have been expected to have achieved more success. However, all these early trials continued to use a lost foam pattern as a starting point. This starting point may have been necessary to support the steel shot, but the process appears never to have been tested without this support, so remains unclear. If the magnetically held mould did not require the foam pattern for support it would have been valuable to know.

A further important disadvantage, apparently never mentioned in print, would be the cost of maintaining the powerful magnetic field that would have been required. In times gone by, this would only have been possible with the use of powerful electromagnets, thus consuming significant amounts of power for all the time the mould is made, assembled and cast. A hold-up on the casting line would have been particularly expensive. Nowadays a sufficiently powerful field may be available with modern permanent magnets, such as the neodymium-iron-boron types. As with all permanent magnets, such devices can be effectively switched on or off by the use of keepers. This possible option appears never to have been tested. Even if viable, the extensive use of such magnets in a foundry would probably represent an eye-watering capital outlay.

Although claims have been made that the technology combines the benefits of sand moulding with the rapidity of freezing of metal mould, Geffroy finds this to be largely untrue. The solidification rate is much nearer to that of sand moulds. This follows from the reduced contact between steel particles and the lower overall bulk density of the packing of the steel.

This process might be significantly more attractive if freed from the lost foam approach. The ability to make conventional 'lost air' magnetically moulded castings in copes and drags may yet hold promise. However, finally, we have to conclude that we all remain in the dark about the viability and economic value of this technology.

15.2 AGGREGATE MOULDING MATERIALS

Ken Harris (2008) has passed on to me a useful little formula from Dr Robert Sparks (2000). It is approximately

$$\text{AFS Number} \times \text{micrometers} = 15,000$$

Thus the widely quoted American Foundry Society fineness number for aggregates, particularly sands, can be approximately converted quickly into average grain diameters. Thus, we obtain some rough equivalents that are useful to commit to memory

$$\text{AFS } 50 = 300 \ \mu\text{m}$$

$$\text{AFS } 75 = 200 \ \mu\text{m}$$

$$\text{AFS } 100 = 150 \ \mu\text{m}$$

$$\text{AFS } 150 = 100 \ \mu\text{m}$$

$$\text{AFS } 200 = 75 \ \mu\text{m}$$

When the molten metal enters an aggregate mould, the mould can react violently. Frenzied activity crowds into this brief moment of the birth of the casting: buckling, outgassing, pressurisation, cracking, explosions, disintegration and chemical attack. After all these possibilities, the survival of a saleable casting should perhaps be a matter for surprise. In this section, we examine the options for the casting engineer to ensure that the moulding and casting processes are appropriate, economic and under control.

Non-permanent moulds made from particulate materials that are designed to be broken up to release the casting after it has solidified are commonly known as sand moulds. However, there are now so many options for aggregates that are no longer really 'sands' that it hardly seems appropriate to continue to call them sand moulds. We should be calling them aggregated moulds (but the reader will notice occasional lapses by the author).

The many aggregates that can serve as adequate moulding sands have been reviewed many times. The reader who would like more detail is referred to reviews by Middleton (1964), Garner (1977), Rassenfoss (1977) and Harris (2005).

The nature of aggregate moulds is nicely illustrated by the observation that patterns transferred from old jolt-squeeze machines to new high pressure moulding machines produced castings that were undersize (simply because the mould were harder. The earlier softer moulds yielded under the weight of the melt, or the pressure generated by the graphite precipitation when casting iron, so becoming oversize).

15.2.1 SILICA SAND

Naturally, the most common aggregate still in use is silica sand because of its worldwide availability, appropriate particle size and distribution and high melting point. Even so, the situation is undergoing rapid change as a result of the danger of breathing in silica dust leading to the danger of silicosis, an incurable and sometimes deadly affliction. At the time of writing, silica sand in foundries is already banned in New South Wales, Australia.

In addition, the alpha quartz to beta quartz transition at around 530°C is accompanied by a total volume expansion of about 2.5% leading to length changes in the mould of around 1.5%. This is a serious volume change and gives the founder many problems of mould failure and loss of control over casting accuracy. For instance, in a mould 600 mm long with a delicate central core, when surrounded by liquid aluminium at 700°C, the core would easily exceed 530°C, but the mould would only achieve less than half of this increase, making a differential expansion of the order of 1.0%. Thus the core would expand relative to the mould by $600 \times 1/100 = 6$ mm. For a modern lightweight casting with walls at each end of the casting only 4 mm thick, the walls might be only 1 mm thick if the expansion were symmetrical. If not, the wall thickness could easily be 2 mm at one end and zero mm at the other. It is understandable that customers do not warm to the offer of windows in their castings in exchange for walls.

The problem of the large and non-linear expansion of silica sand is even worse in ferrous castings, particularly where castings are made with new sand. However, repeated reuse (re-cycling) of silica sand can reduce the problem because the proportion of more stable forms of silica such as cristobalite tends to increase each time the sand is heated to higher temperatures. Light-alloy foundries do not generally heat enough of the sand sufficiently to gain this benefit, although light metal foundries that use thermal sand reclamation methods may do so.

The thermal shock responsible for the change of crystalline form of silica leads to increased fracture of the sand grains. This will cause a good quality, round-grained silica sand to become steadily more angular in shape, thereby increasing binder demand for the same sieve analysis, impairing surface finish of the castings, and wearing out tooling and plant at an enhanced rate (as witnessed by the huge accumulation of iron filings on the magnetic separator in the sand re-cycling system).

There are other natural sands that do not have these problems. These include zircon (zirconium silicate, a white, yellow or orange sand mainly produced in Florida and Australia, and rather expensive because of its rarity). Another is chromite, usually available as a black, crushed mineral. The other common crushed moulding aggregate is olivine (that derives its name from its olive colour when new, although turns brown after use for casting) and widely used in steel casting both to reduce expansion and metal/sand interactions. These three products contain little or no free silica and have relatively low and linear thermal expansion coefficients.

Finally, it is of interest that some silica sands, such as those produced in Sweden, contain quantities of feldspars (up to 25–30%) that help to reduce the deleterious effects of silica expansion. These feldspars work in two ways: (1) by virtue of their lower, more linear expansion and (2) because some melt at lower temperatures than silica, thus counteracting the general expansion of the mould or core from which they are made. These facts do however argue against these sands being used for high temperature work, e.g. large steel castings.

Some suppliers offer additives that can be mixed in various proportions with silica sand to reduce the effects of the expansion caused by the change in crystalline structure. Common to many of these is a content of iron oxide (Naro, 2000). These would appear to function in a similar way to the naturally occurring feldspar/silica sands described previously; however, they may have a role to play where such sands are not locally available. A novel approach showing much promise uses additions of sodium bicarbonate to urethane-bonded sand. The approach appears to work by generating a foam structure in the bond material which allows the core sufficient plasticity to avoid cracking (Schrey, 2007).

15.2.2 CHROMITE SAND

Chromite is a widely used aggregate valued for its relatively low thermal expansion and good chilling power, particularly for steel castings.

However, the general problem with chromite is only found in steel casting operations where it is commonly employed as a facing sand for silica moulds. Here the surface of the casting can become coated in so-called 'chromite glaze', the result of chromite being fluxed by silica or clay, or simply by the decomposition of chromite as discussed later. This is difficult to remove and results in a poor finish to the casting (Petro and Flinn, 1978).

The chromite glaze problem has been assumed to be the formation of fayalite (ferrous orthosilicate Fe_2SiO_4 with a melting point of 1205°C) as the reaction product between silica and chromite. Fayalite is not magnetic and so cannot be separated from the sand magnetically. During the re-cycling of the sand, this low melting point constituent therefore builds up, lowering the refractoriness of the moulding aggregate. It seems likely that another ferrous silicate, grunerite ($FeSiO_3$), also non-magnetic, with a melting point of about 1150°C may also contribute to the impairment of the foundry sand for ferrous castings. These issues have yet to be properly researched and clarified.

An additional problem with chromite is that it is not particularly stable at steel casting temperatures. In oxidising conditions, it slowly oxidises to Fe_2O_3 and Cr_2O_3 but in the presence of carbon from organic binders its iron oxide constituent can reduce to liquid iron. Drops of liquid iron form microscopic beads on the surfaces of the sand grains (Scheffer, 1975; Biel et al., 1980) causing the sand to become highly magnetic (as quickly tested by the reader with a hand-held magnet). The reducing conditions are of course avoided by the use of carbon-free inorganic binders such as silicates. Whether oxidised or reduced, the weakened grains subsequently crumble to fragments.

Most chromite sand is derived from crushed ore, and so as well as being therefore rather angular, dust is created during conveyance which in turn increases the risk of reaction with silica sand. In general, it is estimated that 20–30% of chromite sand is lost each time that the sand is reused.

15.2.3 OLIVINE SAND

Olivine sand is sometimes used as an alternative to silica, primarily by foundries that cast high-alloy steels. Olivine possesses several advantages compared with silica, the most important being a lower and more uniform thermal expansion and inertness toward certain difficult alloys, notably manganese steels. Because olivine does not react with chromite, unlike silica, it can be tolerated at far higher levels in reclaimed chromite sand, and vice versa. Furthermore olivine with low loss on ignition (LOI), having a low serpentine content, is not regarded as being hazardous to health. Olivine sand does not react with alkaline binders. Some foundries use a 'dead-burnt' serpentine/olivine which is brown in colour and behaves similarly to ordinary olivine.

In spite of these valuable advantages, olivine is not widely used for two main reasons. The most important is that it is incompatible with the common binder system, furan no-bake (FNB), and requires specially formulated phenol-urethane no-bake (PUNB) binders. This means that olivine is used almost solely with alkaline binders and this leads to low sand reclamation rates with conventional equipment. A secondary disadvantage suffered by olivine sand is that it is, like chromite, made from crushed rock and thus possesses an angular grain shape that causes dusting during conveying and mixing. Finally, the refractoriness of olivine is related to its iron content which for steel casting should not be greater than 8% and for certain goods e.g. large carbon steel castings, maximum 6%.

15.2.4 ZIRCON SAND

Zircon sand is a light colour and generally in the form of fine rounded grains. Its extremely low and linear coefficient of expansion allows it to make extremely accurate cores and even whole moulds. Its high bulk density almost exactly matches the density of liquid aluminium, so that cores have close to neutral buoyancy which makes zircon the ideal aggregate for delicate and poorly supported cores in Al alloy castings.

Zircon has a fairly high thermal diffusivity compared to silica, giving it greater chilling power and faster freezing of castings. However, this effect is not great, and in any case, greater chilling power is a two-edged sword; although conferring slightly improved properties for the casting, the achievement of thin-walled sections is made more difficult.

The high density of zircon (nearly double that of silica sand) leads to significantly greater mechanical handling challenges in the foundry. Robots for conventional sand moulds will usually find zircon moulds too heavy to lift and sand silos will require strengthening. Vibrating equipment and pneumatic conveying systems require twice the energy.

The rate of wear of patternwork and tooling is a serious consideration when using denser moulding materials. Zircon in particular is found to be especially abrasive toward tooling as a result of its hardness and its high density thus requiring high blow pressures on core blowing machines to fill core boxes. As the percentage of broken zircon grains increases with cycling around the sand system, the increasing number of sharp edges greatly increases the rate at which it can inflict wear damage on the core boxes and other expensive tooling and plant. Although founders that use zircon complain about its high cost, the real hidden cost is the rate of destruction of the tooling.

15.2.5 OTHER MINERALS

There seems no shortage of minerals that would make good moulding materials (Ramrattan et al., 1996). Of growing interest is the possibility of using quite new minerals for moulding that could be available as fines, the unwanted by-products of crushed aggregates for other purposes such as road making, and thus in plentiful supply and at low cost. One such material is Norite. Another might be anorthosite. When it is realised that there are probably 10–20 local quarries for every foundry, supplies of such materials seem virtually unlimited (Harris, 2005). In addition, the relative closeness could mean considerable savings for some casting operations that are hundreds of kilometres from a source of the more traditional sands, and for whom the transport cost of the sand is the major part of the mould costs.

From time to time investigators have reported the benefits of moulding in various kinds of bonded metal shot. For instance, Al shot has been used for the casting of Al alloys, and steel shot is commonly used for many alloys. The rusting corrosion of these metals is a source of trouble that prevents the easy re-cycling of the material. Also, of course, for steel shot the moulding material is heavy and requires strengthened storage silos, and expensive, heavy duty cranes and mould handling equipment.

There has been a long-standing interest in moulding with hollow ceramic spheres that originate as a component of flue-gas dust from fuel-fired power stations. The particles are occasionally known as cenospheres. They are extraordinarily light and insulating and have a melting point in the region of 1400°C. Because of their extremely low thermal diffusivity metals tend to run twice or three times as far in such moulds. They have wide uses for the casting of many metals up to the melting point of cast irons (Showman 2005, 2007). Their suitability for re-cycling at a reasonable rate is questionable however, because the spheres are relatively easily broken during mould compaction or possibly by the thermal shock of casting.

15.2.6 CARBON

Carbon has been used for moulds in a variety of ways. Investment moulds have been mentioned previously. However, its use as an aggregate represents a major application.

The aggregate particles can be round, fine grains that can be bonded with clay and water mixes (Saia and Edelman, 1968) or enhanced with syrups (Kihlstadius, 1988) to make a 'greensand'. Alternatively, they can be bonded simply with a resin binder (Clausen, 1992). The light weight of graphite does not provide it with much thermal capacity, but its excellent thermal conductivity can be useful. Its stability of dimensions is the result of its extremely low coefficient of thermal expansion, and its chemical stability has found good use in the casting of the extremely reactive Mg-Li alloys (Saia and Edelman, 1968), copper-based alloys (Clausen, 1992), titanium alloys (Pulkonik, 1967) and uranium (Edelman and Saia, 1968). Practically, of course, in the foundry environment, it is welcome as a result of its light weight, greatly easing the handling of moulds and cores.

For ferrous alloys, most 'carbon sand' has an uncomfortably high sulphur content, sometimes as high as 7 wt% (Gentry, 1966). This can contaminate the surface layers of steel castings, and lead to loss of nodularity in the surface layers of nodular cast irons. The sulphur content can be reduced during the manufacture of the sand, boiling off the sulphur at treatment temperatures in the region of 1000°C.

Another major use of carbon is in permanent moulds, machined from solid blocks of graphite. Mihaichuk (1986) describes the successful use for the casting of ZA (zinc-aluminium) alloys. The process is also highly successful for the casting of steel rail road wheels as seen in the Griffin Process.

The author has only one experience of attempting to introduce the large-scale use of carbon in the foundry. This was a trial using carbon fibres for reinforcing delicate water jacket cores for high-performance cylinder heads. The fibres flew about in the atmosphere, were trodden and crushed underfoot and reduced to powder at several other stages of manufacture. After a week or so, the powder eventually found its way into our numerous electric motors and other equipment around the plant and shorted these out, bringing the plant to a stop. The trial was immediately discontinued and never repeated.

15.2.7 SYNTHETIC AGGREGATES

Recently, some synthetic aggregates have become available which appear particularly attractive; they are formed from such stable inert refractories as alumina, mullite or bauxite and are formed into spherical grains which have excellent flowability and packing density. Because of their intrinsic hardness and wear resistance, they do not break down or wear so as to create dust. Their spherical form bonds efficiently, requiring reduced levels of mould binder. If the foundry can discipline itself to reduce sand losses to nearly zero, these attractive aggregates can be afforded as one-off capital items, otherwise aggregate costs are sufficiently high that even a small percentage loss could prove uneconomic.

It may be necessary to select the variety of synthetic aggregate with some care. One example is formed from a molten oxide mineral, atomised by a jet of air. This production route gives a solid spherical grain which will be expected to perform well as a casting sand.

At least one other synthetic aggregate is formed by a solid-state route: aggregating powders are rolled around on an oscillating tray with a temporary sticky organic binder. The aggregated spheres are heated to burn off the temporary binder and are sintered. The sintering is only sufficient to produce rather porous spherical particles. Some early experience with the casting of aluminium has indicated that the mould binder gets into the pores of the spherical particles, and on contact with the hot metal boils and explodes out of the pores, peppering the surface of the casting with porosity in the form of micro-blows.

15.3 BINDERS

Aggregate moulds such as sand moulds require the individual particles to be held together with some kind of binder. There are very many binders from which to choose but the broad categories are listed later.

15.3.1 GREENSAND (CLAY + WATER)

Greensand is perhaps the most widely used moulding medium, consisting of the aggregate, bonded with a mixture of mainly clay and water.

Greensand can be almost any colour including brown, red or black, but I have never known it to be green. It should be written as one word, 'greensand', not as 'green sand'. Its 'greenness' refers to its weak, plastic condition in contrast to that of dried moulds which are extremely hard and rigid. The greenness concept is common to other industries such as ceramics where a ceramic component has only 'green strength' until it is fired to melt the glaze bond, after which it becomes as hard and brittle as glass. The origin of the concept of greenness almost certainly lies in forestry and timberwork, where the wood is green and flexible when full of sap, unseasoned, before drying. In the 1850s in Sheffield, hand files were made of steel carburised and hardened to a limited depth, retaining a soft centre known as 'the sap'.

Greensand always has been and remains one of the most important and common binder systems. It is low cost, environmentally friendly (because there can be minimal toxic additions) and re-cyclable. However, most important of all, it is extremely fast. Compared with chemically bonded moulds which can be produced at perhaps 60–100 per hour, modern greensand plants can produce 600 moulds per hour. These awesome rate follows from simple fundamentals: once all the grains are evenly coated with the clay/water binder, the bond is formed simply by forcing the grains together. Thus the bond is formed instantly simply by pressure. No other process can compete for speed of moulding. Busby (1996) gives a good review of greensand versus other moulding processes.

The disadvantages are that the mould is somewhat plastic (i.e. remains rather 'green') so that if the aggregate is not particularly well-consolidated, the mould can distort, producing an inaccurate casting. For instance, the mould can sag under the weight of heavy cores, or the weight of the casting, or soft spots can yield under the pressure of liquid metal leading to local swells on the casting (the pressure can be high from the depth of metal in a tall casting, and from the expansion of graphite in cast irons).

These disadvantages are now countered rather successfully by modern moulding plants that in addition to their high speed, give a mould that is impressively well-consolidated. Whereas a mould compacted by hand could show a complete hand print (including the finger prints on all fingers), a really strong mould made on a modern automatic greensand moulding plant is so hard that only a rather limited thumb print is all that the average person can easily make.

The limit to the hardness of greensand moulds is set by the pressure at which the whole mould and its steel moulding box will distort elastically, causing the mould to spring into a new shape on being stripped from the pattern. This limit seems to be about 1.0 MPa (150 psi). It creates problems for the founder because such elastically distorted mould halves do not close properly. The faces of the mould bow elastically into a convex form, causing the two mould halves to impinge in their central areas, producing features such as sharp edges in the centre of the mould face to break off. This effect is generally controlled by the relief of the tooling to give the region adjacent to such edges sufficient clearance to avoid direct contact, but not sufficient for the metal to penetrate to cause expensive flash. Such clearance is generally in the region of 0.1 mm and 5–10 mm wide.

Cores may absorb water from the greensand, particularly if a delay happens between coring up and casting, as when a stoppage on the line occurs. Similarly, condensation may occur on metal chills. Condensation occurs even in rather dry resin–bonded sand moulds if left closed overnight. This is because the moulds cool during the night, but during their re-heating in the morning vapours are driven inwards by the inward flow of heat, thus condensing the volatiles onto the interior metal chills, still cold from their overnight cool. If cored-up moulds are required to be stored overnight they should be left open. Condensation problems are then avoided.

There is a strong move nowadays to reducing or eliminating the organic additions to greensands (Grehorst and Crepaz, 2005). Organics heat up and pyrolyse or burn, often giving off unhealthy emissions. Thus the motivation is to return to inorganic mixes based on simpler clay/water binders. This subject is too extensive to be reviewed in this work. The interested reader is referred to the important series of researches published over several years by Heine and Green (1989–1992).

Nearly all the important engineering metals continue to be cast in greensand, including Al, cast iron and steel.

For a small number of hand-moulded castings, a special form of greensand is made up using oil instead of water. Oil-bonded greensand is a traditional moulding material still popular with model-makers as a result of its plasticity, conferring easy moulding and yielding a good surface finish of the castings. The oil/clay/sand mix has such good flowability that it is appropriate for small parts, but, as a result of it being generally 'greener' than water-based greensands, it is not suitable for castings weighing more than about 1 kg (Butler and Lund, 2003).

15.3.2 DRY SAND

Dried greensand cores were one of the traditional solutions to the unsuitability of greensand as a core material. The drying process increased the strength enormously and reduced the outgassing from the relatively high water content (often in the region of 2–4 wt%) of greensand. The cores had to be supported in their green state on cradles called 'driers' before being transferred to drying ovens. The process was akin to that used for linseed oil-bonded cores. Naturally, the drying process was lengthy and energy intensive. Furthermore, the hardening 'out of the box' meant that core distortion was a common problem despite the hundreds or thousands of 'driers' that were in use at any one time.

Dry sand cores and oil baked cores have now been practically entirely superseded by the various chemical binder systems. These later processes harden cores in the core box, thus enjoying the benefit of a huge increase in accuracy. The appellation 'no-bake' applied to many of the modern binder systems is a reminder of the slower, less convenient binders dating from earlier days, based on oils that required baking in a core oven for many hours.

15.3.3 CHEMICAL BINDERS

There are now so many chemical binder systems that the subject it is not possible to review the field here. For those wanting to review the first efforts to use a variety of vegetable and mineral oils and resins, the review by Middleton (1965) is interesting. Later reviews of more modern binder systems are provided by Webster (1980) and by Archibald and Smith (1988), but the reader needs to be aware that binder development is proceeding apace; it is now difficult to keep abreast of these advances.

A perfect binder would have low viscosity so that the aggregate mix had good fluidity, but in its cured form was strong, so that little addition was needed, retaining the void content of the aggregate to maximise permeability. It should also react minimally with the melt. However, of course, the chemical interactions increase in number and severity with increasing temperature, so that whereas aluminium may not suffer too much, irons and steels can be significantly damaged by interactions with the mould. The reactions occur with both the mould surface and with the atmosphere formed in the mould during filling with the hot metal. Working our way up the temperature spectrum of casting alloys we have the following.

1. Low-melting-point lead and zinc alloy liquids generally cast at temperatures up to 500°C are too cool to cause significant reactions.
2. Magnesium and aluminium have casting temperatures commonly up to 750°C. They react with water vapour and various organics to produce the solid oxide skin and free hydrogen that can diffuse into the melt. These reactions continue for some time after solidification and during cooling. This source of hydrogen is likely to be important in growing pores that are located just under the casting surface. We shall return to the subject of the growth of subsurface porosity later. Despite this reactivity at the surface of the liquid metal, it is worth noting that the temperature at which light alloys are cast does not lead to extensive breakdown of the chemical constituents of the mould, as is clear from Figures 4.1 and 4.4. Thus light alloys have the widest free choice of binder systems.
3. Copper-based melts up to 1300°C take part in several important reactions as will be described later.

4. Irons and steels (including Ni-and Co-base alloys) in the range 1400–1600°C are especially reactive in many ways. Aggregates and binders start to be a problem, greatly constraining the choice of materials for the steel caster. We shall devote some space to adequate solutions.
5. Titanium and zirconium in the range 1600–1700°C are so reactive they are problematic to cast into moulds of any type. Reactions with most moulding materials cause the formation of the troublesome alpha-case on titanium alloys, in which the alpha-titanium phase is stabilised by interstitial elements (oxygen and/or carbon) absorbed from the breakdown of the mould. The alpha-case usually has to be removed by machining or chemical dissolution.

There are many different types of chemical binders. They divide into two main types: organic and inorganic. The molecular chemistry for keen chemists is described in detail in the work edited by Webster (1980). We can only give an outline here, concentrating on their practical application in foundries.

15.3.3.1 Furans

Sulphonic acid cured furan no-bake (FNB) binders are based upon furfuryl alcohol (2-furylmethanol) and are very widely used in ferrous casting because they provide excellent mould and core strength, cure rapidly and allow the sand with which they are used to be reclaimed at fairly high yields, generally 75–80%, where due allowance is made for the need to keep total sulphur content below 0.1%. In passing, it may be noted that furan foundries generally exceed this sulphur level to minimise sand consumption and accept poorer surface properties of the casting! Foundries that wish to reduce the sulphur content of a conventionally reclaimed furan bonded sand do have several alternatives to using a sulphonic acid, and both phosphoric and lactic acid are sometimes used for this purpose. These are, however, weaker acids and will cure furan resins more slowly than the sulphonic acids, and particularly so for lactic acid. The use of phosphoric acid may necessitate a lower reclaim rate.

Furan binders are more or less anhydrous and commonly used at a rate of about 1% on (silica) sand. Experience shows that properly reclaimed furan-bonded sand that contains less than 0.1% total sulphur can be used with phenolic urethane no-bake (PUNB) binders.

The use of furan binders is associated however with several casting defects, the most serious being sulphur damage, which leads to poor surface finish in some steels, and destruction of near-surface graphite spheroids in ductile irons. The presence of urea-based polymers in some types of furan binders may also lead to nitrogen damage. As mentioned elsewhere, furan binders appear also to reduce chromite sand, forming iron metal.

Finally, and giving rise to most concern, furan binders have been found to be carcinogenic and decompose at casting temperatures with the emission of toxic gases; they contribute significantly to air pollution and the general blackening of the foundry environment, particularly the windows, thus further contributing to our dungeon-like furan foundries.

15.3.3.2 Alkaline phenolics

Alkaline phenolic (A-P) also called phenolic ester binders are fairly widely used in non-ferrous and steel casting, particularly where the use of furans can lead to difficulties, e.g. where alloys are cast that are sensitive to nitrogen or sulphur compounds and where hot tearing is prevalent. A-P binders contain about 50% water and are typically used at an addition rate of about 1.3% on silica sand. A-P binders are typically cured by the addition of an ester, such as di- or tristearin, by injecting methyl formate gas.

A-P–bonded sand is more difficult to reclaim conventionally than furan-bonded sand; reclaim rates being typically 70–75%, leading to higher volumes of waste per tonne cast. Reclaim rates are somewhat lower with silica sand than with olivine because the alkali in the binder reacts with the silica to form a surface film of sodium silicate that is highly resistant to attrition during conventional processing. Alkaline residues in the sand make it difficult to use reclaimed A-P bonded sand with PUNB or FNB binders. Like furans, alkaline phenolic binders may also cause deterioration of chromite sand and both the binders and their pyrolysis products are considered to be as toxic.

15.3.3.3 Phenol-urethane (cold box)

At the time of writing this binder system is probably the most widely used in the non-ferrous casting industry.

It can be used in an 'air hardening' form (the curing process has in fact nothing to do air) in which its excellent bench life, followed by a sudden and rapid curing is convenient for moulders, and the timing is widely adjustable by varying the catalyst addition at the mixer.

It is also important as a core binder in which the resin is hardened by an amine gas. The gassing time is usually a few seconds, but is required to be followed by an air purge that can last several times longer. The whole cycle for the production of an average core weighing a kilogramme or so would be in the range of 30–60 s. This rapidity of production, together with its high strength, requiring low resin additions, has promoted this system to number one choice at this time. The reek of amine gas is tenacious however, and can be unpopular on arrival home at the end of the day. The amines in the exhausted gases need to be removed by washing with dilute acid, which adds costs.

15.3.3.4 Sodium silicates (waterglass)

The inorganic types are typified by one of the earliest of the chemical binders, sodium silicate, or waterglass. Several good reviews are available (Srinagesh, 1979; Harris, 2004). This substance can be 'cured' (better: 'set') by passing carbon dioxide gas through the mould to precipitate silica. For this reason the process is widely referred to as the CO_2 *process*. However, many foundries prefer to use esters such as di- and tristearin to accomplish this. Ester curing is easier to control and more uniform, although slower and more expensive than CO_2. That sodium silicate binders are not thermally degraded and may indeed melt during casting can cause moulds and cores made with these binders to be difficult to break down after casting, so it is often difficult to remove the casting from the mould and even more difficult to remove the core from out of the casting. Many Al-alloy castings have been scrapped as a result of internal cores that are rock hard, and impossible to dislodge.

These properties also make silicate bonded sand difficult to reclaim at high yield by conventional mechanical methods, whilst thermal reclamation is more or less useless. A typical reclamation yield for a standard shake-out + screening operation may be as low as 50%.

It has been such historical difficulties that forced the ferrous casters to move away from silicates to organic binders. The main reason for the move appears to be the difficulties associated with core removal and sand reclamation, particularly with the older silicate systems that required additions of 4–6% of binder.

Silicate binders are inorganic bulk products, not based on oil, and are thus much cheaper both per tonne and per tonne of metal cast than the other binders considered here. The newer systems provide in most cases more than adequate strength and are typically used at the rate of 2.5–3%.

Silicate binders possess several technical advantages, including greater resistance to hot tearing of the casting, almost complete lack of emissions, a considerable chill effect that can make the use of chromite facing sand unnecessary for steel casting and yield castings having an excellent surface finish. Silicate binders can also be formulated to cure as rapidly as furans, although the silicate process is better described as 'setting' rather than 'curing'.

However, even with the modern silicate systems, silicate-bonded sand is still difficult to reclaim. Furthermore, the presence of alkaline binder residues means that conventionally reclaimed silicate-bonded sand cannot be used with FNB or PUNB binders. The limited bond strength of silicate binders may restrict the size of boxless moulds that can be lifted and moved.

More recently, however, there have been considerable advances in the use of sodium silicate, allowing it to become the binder of choice for both light alloy and steel castings. The use of hot air to dry the binder has the effect of providing a uniform set of the binder without limiting its bench life. It has the great benefit of environmental friendliness, low cost and low gas evolution on casting. Heating of the cores to about 300°C removes the remaining water of crystallisation, making the cores virtually inert, free from outgassing problems. This is a massive benefit, allowing the use of core designs impossible in other processes. In addition, for Al-alloy casting, the binder can be made soluble in water, allowing the moulds and cores to be removed easily after casting.

Other recent developments in inorganic binder systems include the use of mixtures of silicates with phosphates and even borates. The sodium polyphosphate glass binder deserves special mention, simply requiring warm air to remove excess water to cure the bond. This process is attractive for difficult-to-vent cores because of its extremely low gas evolution on casting (Armbruster and Dodd, 1993). These recent inorganic materials promise a revolution in casting techniques.

Concerns about the costs and environmental aspects of organic binders and thermal reclamation have led some of our major light-metal foundries to devote a growing amount of attention to silicate binders and advanced mechanical reclaim systems. Inorganic silicate binders are commonly used in ferrous and particularly steel casting because of their low cost and excellent casting properties. However, the major growth area for these binders is light-metal casting. The benefits will be enhanced if foundries were to use non-silica sands. Because many such sands have a rather angular grain shape, it may be a good idea to use a grain-rounding attrition process, to minimise binder consumption and improve the surface finish of cast parts.

Cores made with silicate binders for ferrous casting can, if necessary, be removed in the dry state by vibration, although this might damage light metal parts, in which case a water-soluble core system might be preferred. New technology is becoming available that promises to provide rapid manufacture of cores with minimal gas development during casting, minimal environmental impact and that can be removed from the finished casting by washing out with water.

The recent re-introduction of the inorganic sodium silicate binders has been a success story. However, not all inorganic salts are suitable as binders. For instance, many have large quantities of water of crystallisation that is released when heated by contact with the liquid metal, causing blow defects in the castings. An example is $MgSO_4$ which contain five molecules of water for each molecule of magnesium sulphate. Although common salt, NaCl, has no combined water, it has not so far been found to have useful binding properties.

15.3.3.5 Hot box and warm box processes

The hot box and warm box processes use a liquid thermosetting resin binder. The mix is blown into a heated steel core box and allowed to cure for 10–30 s before ejection. Some post-curing may be needed outside the box, aided by the exothermic heat provided by the chemical reaction. The process has enjoyed wide popularity for the production of water jacket cores for such castings as automotive cylinder heads and blocks as a result of the strength of the cores and their relatively low volatile content, reducing blow problems.

15.3.3.6 Croning shell process

Often known simply as the *shell process*, it was patented in Germany by Johannes Croning in 1943 for making moulds or cores. The binder was based on a thermosetting resin. Mariotto (1994) gives a good review of the development of the process. The sand grains are pre-coated with the resin, which can then be handled as dry, free-flowing sand. The sand is put into contact with a heated metal pattern, against which the resin melts and sets after about 30 s. The remainder of the unheated sand further away from the pattern can be poured away to leave the shell to complete its hardening process, after which it can be ejected from the hot pattern. Mould halves can be made in this way, or hollow cores can be made if there is a print sufficiently large that can be used to pour out the un-reacted sand. If there is no large core print, the core can be produced solid, at additional cost of material, and possibly introducing the possibility of core blows if the core cannot be vented in some other way. Taft (1968) describes how the process was used with great success in his family foundry using 100% zircon sand, and specialising in hydraulic castings in grey iron. These casting benefited by improved dimensional accuracy, soundness and more easily cleared cored passageways, eliminating the risk of the contamination of hydraulic circuits.

When moulds are made by putting together two mould halves, this shell mould can be cast with its parting line oriented horizontally or vertically. For vertical orientation, a good filling system can be provided that runs around the parting line; even a good pouring basin design can also be incorporated. The problem with vertical orientation is that the mould halves require to be supported by backing up with steel shot or other technique. If shot is used, it is usually consolidated by vibration although too much vibration can crush the mould. The mould backup requires some engineering and associated cost, but has the potential for making excellent castings.

If the two shell halves are put together to make a horizontal mould, the filling system now becomes problematic to the point that I normally recommend owners of such lines to shut them down. The sprue is now necessarily wrongly tapered and no pouring basin of any respectable design can be incorporated. Thus the castings are necessarily at significant risk from entrainment defects. Such defects are therefore common in horizontal shell moulded castings. Some reduction of defects can be made by such means as the provision of filters at the base of the sprue or at the entrance to the runner, so as to assist the sprue to back-fill as effectively as possible. Overall, however, despite its relatively wide use and evident surface finish advantages, the horizontally-cast shell process is not a casting system to be recommended.

15.3.4 EFFSET PROCESS (ICE BINDER)

The Effset process was invented by Fred Hoult in the 1970s, together with the H-Process. Impressively, this double achievement was made after his retirement at the age of 60 (I am encouraged that there may be hope for me yet). Silica sand mixed with a little clay (2–5%) and selected amounts of water allowed moulds to be made by consolidating in a core box. The moulds had sufficient green strength to be stripped from the box and laid out on a specially profiled support like a cradle. The mould on its support was transferred into a cryogenic chamber where liquid nitrogen was rained on to the moulds. Only sufficient nitrogen was applied to freeze the surface of the moulds, which were then extracted from the cryogenic chamber, assembled (usually in a stack) and cast. I have happy memories of a conference in which Fred was projecting an image of his process onto the screen. The image was of a stack of moulds that had just been filled with liquid iron, revealing hoar frost still over parts of the mould, and gentle clouds of white steam rising from disintegrating moulds. In his characterful English accent, Fred announced with tremendous enthusiasm, to the entertainment of all, "You can *see* it doesn't smell!"

The process was used in Fred's Rotherham foundry for many years, giving excellent surface finish and showing no problems of chill (i.e. white iron) in his iron castings (Hoult, 1979). Omura and Tada (2012) find that the fluidity of aluminium alloys in frozen moulds is higher than that in greensand. However, the Effset process has never been taken up in volume production. It seems the high energy cost involved in the use of liquid nitrogen inhibits its use. An effort at re-examination of ice as a binder was provided by Abbas and colleagues (1993). Even so, as energy costs and cryogenic gas availability changes the environmental benefits of this process suggest it should be kept under review.

15.3.5 LOAM

Loam is a kind of mud. It is a mix of sand, clay and water, plus on occasions time-honoured fibre strengthening additions such as manure from horses fed with a high diet of straw (horses chew the straw, chopping the fibres into exactly the right lengths, whereas cow manure is of no use because the fibre content, fibre length and strength based on grass is not helpful). The sticky and sloppy mix is usually applied by hand to build up a mould or core, and is particularly useful for the forming of moulds for bells which are formed to final shape by a *sweep*, a kind of *strickling* action made by a shaped blade rotated around a central axis. Additionally, the loam surface is conducive to finishing by *sleeking* with a wet brush. At that stage, letters and ornamentation can be carefully impressed into the surface for the insignia and decoration of the casting. The shape is then transferred to an oven where it is subjected to a careful and thorough drying treatment taking typically several days. This is not a process designed for high production.

An interesting historical account of John Wilkinson, and English iron master from the seventeenth century, describes how his iron cannons out-performed all other cannons. His were cast solid and the bore machined afterwards, compared with the normal manufacturing route in which cannons were cast with a loam core to form the bore. To the solidification scientist, this may have been surprising because the long freezing time of the solid casting should have yielded a coarser and weaker structure. However, this expectation overlooks the casting problems associated with loam cores. The core would have probably have been formed around an iron rod for strength, then wrapped with rope to provide some permeability and allow gases to escape and finally sleeked with loam and dried. The drying process could never be complete, because the clay itself would contain a significant volume of water in its crystal structure overwhelming the ability of the rope to convey away the volume of gases. Thus the core would be expected to outgas violently, causing the melt to boil, probably fountaining out of the mould, but leaving behind a mass of entrained bifilm and bubble defects that would have greatly weakened the casting.

The long and slender aspect ratio of the cannon core meant that its narrow rope vent was hopelessly inadequate. The bell survives because its loam core is usually made hollow, on a perforated steel former of large area, and outgases easily via its large base.

15.3.6 CEMENT BINDERS

The mixture of sand, cement and water forms an extremely hard mould capable of withstanding the loads imposed by very large castings, particularly ships' propellers. It was popular in the days when one of the only alternative mould materials was greensand which was far too weak for such products (Menzel, 1937). Later work summarised much earlier studies and investigated improved mixes and enhanced breakdown after casting (Wallace and Hrusovsky, 1979). The process is still in use for large heavy castings such as propellers in Ni/Al bronze and 40 tonne ductile iron gears.

15.3.7 FLUID (CASTABLE) SAND

Castable sand was invented in Russia in the 1960s by Liass and Borsuk (Liass, 1968). A small amount of wetting agent added to the sand caused a foaming action during mixing. The special feature of the process was the low liquid content, the fluid properties of the mix being achieved by suspending the sand grains in an air/liquid foam, reducing the friction between grains and allowing the sand mix to flow rather like water. As such, it could be poured into a moulding box, where the kinetic energy of the fall would provide the necessary compaction. The mix had a fluid life of around 1 min, and would subsequently harden by a chemical reaction. The process was analogous to the forming of lightweight concretes, and similar to the production of foamed plaster moulds.

Setting mixtures such as sodium silicate and dicalcium silicate were originally used, later superseded by various cement mixes. The process attracted world attention in the early days because of its contribution toward a cleaner foundry environment, reduced labour requirement and a higher productivity. Mixing a 250 kg batch would take around 4 min, and could be discharged into a mould or core box, building up large moulds or cores in layers (with nearly undetectable joints between the layers) or could of course be discharged into a rapid succession of smaller boxes. The application of a wax parting agent to the pattern-work was necessary (Hassell, 1980).

But the use of foam caused the moulding material to have relatively poor resistance to metal penetration resulting in poor surface finish to the castings (Brown, 1970; Nicholas, 1972). Later developments using no wetting agent improved on this problem (Tsuruya, 1974). Even so, the process seems only ever been applied to simple-shaped castings. These include ingot moulds, slag ladles and machine bases, but has on occasions been extended to the production of medium to heavy cores for large valve castings using a furane resin binder (Hassell, 1980).

Despite the enthusiasm that accompanied its launch, fluid sand is probably no longer used today.

15.4 OTHER AGGREGATE MOULD PROCESSES
15.4.1 PRECISION CORE ASSEMBLY

The use of hard cores hardened 'in the box' has led to the possibilities of highly accurate cores and moulds that can be assembled with extreme precision, often measure in micrometres. Thus the name 'precision core assembly' arose to describe this new standard of achievement. The castings from this process were some of the first to team up successfully with robotised pickup and machining systems, although this achievement is now shared by many of the high-pressure greensand lines, and, of course, by-products of the various forms of HPDC.

15.4.2 MACHINED-TO-FORM

This process is not a moulding process at all. A rigid block of chemically bonded sand is simply machined by programmed robot or machining centre. In this way a mould can be shaped with tolerable accuracy and with tolerable surface finish. It is a technique for rapid prototyping, in which the digital model sitting in the computer memory is used to produce a mould directly, eliminating the lengthy intermediate stage of the production of dies or patterns. The process can of course also be used for the production of a limited series of parts. Critics of the process complain that the robot or machining centre wears out rapidly (despite the best protective measures) as a result of the creation of fine abrasive ceramic dust that the cost of replacement of the machining facilities should form a major component of the cost of the mould, but is frequently overlooked. Advocates of the process say this problem is now solved.

15.4.3 UNBONDED AGGREGATE MOULDS

Both the lost foam process and the V Process use unbonded aggregates. These moulding processes are integrally defined together with the casting process. As a result, they are dealt with in Chapter 16, *Casting*.

15.5 RUBBER MOULDS

It is not easy to describe rubber as an engineering material for moulds. Even so, for a relatively few zinc alloy castings it can be excellent. A liquid silicone rubber is poured around the pattern, contained within a frame and compressed between platens. After curing, a knife is used to slice into the rubber, opening up a cut sufficiently wide that the rubber can be peeled back to release the pattern. The mould is closed and replaced between the platens, constraining the mould to retain its overall shape. The low melting point alloy can be poured directly into the rubber, with surprisingly faithful accuracy for the casting. Usually, the process seems confined to castings of up to 100 g, although reports of castings up to 1 kg are not unknown.

As an alternative to using cast rubber form as a mould, many founders instead use the cast rubber shape as a pattern, and produce from it a plaster mould. Such moulds are used, for instance, to cast millions of Al alloy turbocharger impellers per year because for these small 100 g castings the rubber pattern will flex, allowing it to be withdrawn vertically from the cured plaster (the complex helical shape of some blades cannot be perfectly withdrawn even if the pattern were twisted on withdrawal) creating accurate castings with excellent surface finish.

15.6 RECLAMATION AND RE-CYCLING OF AGGREGATES

The greensand system, based essentially on a clay/water bond, has enjoyed the attention of enormous development, with the result that whole books are devoted to this important technology. We shall not attempt to reproduce this extensive fund of knowledge.

The reclamation and re-cycling of greensand has always been a major factor in the success of this moulding system. The burning out of a percentage of the clay in ferrous foundries has been dealt with by the dumping of a similar percentage and top up with new sand and binder. New sand introduced via cores, if any, can provide all or part of the top up. Even so, despite its enormous importance, it has to be admitted that the re-cycling of greensand has the problems of being both capital intensive, requiring large plant and equipment for major foundries, and energy intensive, as a result of the energy of crushing, sieving, and re-mixing (the word 'mulling' is often used, the process being carried out in a 'muller'. The word 'mull' being the German word for 'mill' as used by millers for milling grain to make flour).

One of the dangers of the reclamation of greensand has been the addition of uncontrolled water makeup; if this has contained chlorides, as is common in drinking water, the gradual buildup of chlorides in the re-cycled sand seriously reduced the strength of the bond. Boenisch (1988) was the first to point out this problem which was quickly confirmed in independent tests by Hold and coworkers (1989).

Turning now to those foundries using chemical binders, the re-cycling of chemically bonded sands has provided new challenges to the industry. In this section we shall confine ourselves to the complexities of foundries with a variety of chemical bonding processes which in general have so far not been treated in foundry textbooks.

A conventional sand reclamation line will typically consist of a shake-out unit coupled with a vibrating sieve that functions as a low energy lump crusher. Some foundries employ a mill to break down small lumps to individual grains and remove binder residues. The sieved sand can then passed through a sand cooler, often a fluid bed that also functions as a deduster.

15.6.1 AGGREGATE RECLAMATION IN AN AL FOUNDRY

The standard reclamation system works satisfactorily with sand bonded with organic binders as long as the binder is fully thermally degraded, which is normally the case for steel cast at low sand/metal ratios. However, in an Al alloy foundry, resin binder residues are only partly degraded and alkaline binder residues are fused onto the surface of the sand grains. Conventional reclamation imparts relatively little attrition energy and cannot efficiently remove these residues.

Most sand reclamation equipment is designed to target a low LOI. The LOI of the sand consists of four components: (1) char; (2) partially degraded binder films; (3) un-degraded binder films and (4) carbon as a thin film around the sand. Attrition processes remove components (1) and (2) efficiently, but is less effective with (3). Component (4) is not removed and is in fact a desirable component in the sand! Thermal reclaim removes, of course, all four of these components, and could therefore be regarded as overkill.

Grain attrition processes appear to work well enough for binders that are brittle after casting, such as furans. These binders spall off the grains as shell-like fragments forming a light dust, allowing much of the binder to be removed via the dust extraction system. However, attrition can be difficult to use with binders that remain in an elastic or plastic state after use because such binders can become soft during grain compaction and redistribute themselves, smearing onto other sand grains.

The rather low casting temperatures in an Al foundry mean that the residual binder adhering to the PU-bonded sand grains risks still being non-brittle. This means that it cannot easily be removed by the attrition forces that are encountered during the lump reduction and screening processes (so-called primary attrition) widely used. For this reason, many light metal foundries that use PU binders have chosen to reclaim their sand thermally.

Embrittlement of phenol-urethane (PU) binders starts at about 200°C but proceeds rapidly at higher than 350°C. The reclaim rate for sand processed by secondary attrition will therefore, for a given LOI (or more accurately that large proportion of the LOI that derives from un-degraded binder), depend upon the sand's thermal history, which in turn is a function of sand-to-metal ratio. Because the heat capacities of aluminium and silica are roughly similar on a kilogram-for-kilogram basis, this suggests that PU binder would be sufficiently embrittled at sand-to-metal ratios of up to about 4 to be removed, allowing the achievement of rates of sand reclamation characteristic of thermal processing.

After the reduction of the sand to grain size, secondary attrition ('scrubbing') can expose the sand particles to higher energy processing. Grain impact techniques throw particles against either (1) a fixed target such as a steel plate or (2) against other sand particles (particle-on-particle, [POP], attrition). Consider these systems in order.

Impact attrition systems accelerate the sand particles to high speed, using compressed air or a rotating disc. The grains are impacted onto a stationary steel target to dislodge the binder. Sand processed by air attrition will almost always suffer a disproportionate loss of the largest particles. Naturally, the system is self-destructing, particularly if high air speeds are employed to remove recalcitrant binder residues, and it is not uncommon for parts to be replaced every 1–2 months. Use of a fluid bed deduster additionally results in loss of useful sand from the bed because of fluctuating conditions. A further limitation is the size of the individual units, typically 0.5–2 tonnes per hour, so that several must be mounted in parallel if a higher capacity is needed.

When a grain is subjected to impact, the collision force will concentrate at the binder–grain interface thereby detaching the surface layer in accordance with its brittleness. However, if the collision is too energetic, it may cause the sand grains themselves to fracture. Poor quality silica sand will have grains that contain defects and thus more likely to break, each sand variety having its own effective fracture threshold. The more energetic the collision, the more sand grains will be broken and thus the greater the loss of sand as fines. A good attrition system requires an impact velocity that can be adjusted according to the individual sand/binder system. A single collision between a sand grain and an immovable plate will, if it is to remove the binder residue effectively, require that the grain has a higher energy than if it were to experience multiple collisions at close to the fracture threshold.

POP attrition conditions provide the desirable multiple collisions between grains. Furthermore, these minor impacts will be mainly glancing collisions which absorb less energy than when one of the 'targets' is fixed. In addition, glancing collisions cause particles to spin rapidly, so that subsequent collisions become a mass of buzzing grains. The frictional energy dissipated by spinning and chattering against their neighbours is usefully employed to erode away fragments of binder and round off corners of grains.

Attrition by particle-on-particle collision can be nicely controlled, giving a scrubbing action to the grains. Binder residues are stripped from the sand as small flakes, up to 0.2 mm in size, as well as dust. The process will yield reclaim rates of around 90% or more with all chemically bonded sand systems. The quality of sand processed in the equipment is steadily improved, becoming steadily more rounded and will generally be better than new sand, especially for crushed aggregates.

In some foundries, a useful compromise is reached by reclaiming all sand by POP but using a small thermal unit to provide sufficient sand for cores only. Sand used in moulds will then consist of perhaps 12% new sand (from approximately 7% reclaim loss + 5% general losses in the foundry) and 88% mechanically reclaimed sand. If we assume that 10% of the total sand in circulation is core sand, the mould sand will consist of 12% new sand, 10% sand with low LOI from cores and 78% reclaimed mould sand. Provided the foundry runs at sand-to-metal ratios lower than about 8 this should reduce LOI sufficiently to allow moulds to be made without problems.

Attrition requires to be followed by a classification process. In many foundries, this is merely a fluid bed acting as a deduster, but the action of the fluid bed is sensitive to variations in both feed rate and particle size distribution. Both these factors need to be carefully controlled if the system is to function optimally. Most foundries that use fluid bed equipment tend to set the air speed for average or even worst case conditions, leading to excessive loss of useful sand if the feed should contain more fines or is processed at higher rates.

In contrast with a fluid bed, a cross-flow air classifier is a true classifier, using laminar air flow for separating useful sand from fines, flakes and oversized particles. The design is simple and relatively low cost, but has been demonstrated to be robust and tolerates far wider variations in feed rate and particle size distribution than a fluid bed deduster. Sand and fines fall into categories which allow the sand distribution to be controlled as desired.

At the time of writing, there are clear moves toward (1) mechanical as opposed to thermal reclaim and (2) inorganic as opposed to organic binders. There are several good reasons for preferring mechanical reclaim, with perhaps the most important being the following.

1. Mechanical systems are more flexible as regards feed and can be used to reclaim sand bonded with organic, alkaline phenolic and inorganic binders.
2. When used with sand bonded with organic binders, POP mechanical systems will provide yields of high quality sand similar to that provided by thermal reclaim, i.e. reclaimed sand that can replace new sand.
3. Thermal systems are ineffectual for reclaiming inorganic bonded sand and do not offer particularly good yields.
4. POP mechanical systems improve the sand's grain shape and particle size distribution, allowing binder consumption to be reduced and improving the surface finish of cast parts; this is particularly valuable for foundries located in areas having poor quality local sand.
5. Mechanical systems are far cheaper to install and operate than thermal systems.

15.6.2 AGGREGATE RECLAMATION IN A DUCTILE IRON FOUNDRY

As an example of the complexities of some real foundries, it is worth considering the problems of reclamation of core sand in a ductile iron foundry using greensand, and furan-bonded sand for moulds and furan- and PU-bonded sands for cores.

This mix of bonding processes arises from fundamental considerations: a greensand line cannot, of course, use cores made in greensand because the material is relatively weak, and contains up to 10 volume % of binder in the form of moist clay. (Most of the voids between the sand grains are filled with clay, making the mix nearly impermeable, and its high water content therefore leading to inevitable outgassing problems from cores.)

Thus cores for greensand lines are typically chemically bonded. The high strength of the chemical binders means that only a relatively small percentage addition of binder is required (often approximately 1 wt%, corresponding to approximately 3 volume %), which, together with a larger and closer grading of sand particle sizes, gives tolerable permeability, allowing outgassing to occur more easily via the core prints.

Alongside the greensand system, a POP type secondary reclaim system together with an efficient cross-flow air classifier can reclaim up to 94% of organically bonded mould and core sand from iron casting. The reclaimed sand can have a low LOI and, more importantly, a low total sulphur content ($\leq 0.12\%$). Once loose char, dust and fines are removed from the sand, total sulphur is also reduced, and constitutes the prime determining factor for making good ductile iron castings with this binder system and for using reclaimed furan-bonded sand with PU binders.

Experience shows that a foundry using such a reclaim system to make sand for PU-bonded cores (and moulds) from reclaimed (furan-bonded) sand experiences the following benefits:

1. Lower scrap rate because of lower thermal expansion of reclaimed sand compared with new sand.
2. Lower binder consumption than with standard reclaimed sand because the POP system will provide sand with better grain shape and optimised particle size distribution (giving higher bulk density but with better gas permeability).
3. Minimised sand consumption and dumpsite costs. Of course, the energy consumption, which for a small plant is about 3.5 kWh per tonne processed, is far lower than a thermal unit (approximately 200 kWh per tonne).
4. A better working environment

A suggested optimisation of the foundry reclaim system might be as follows.

1. All chemically bonded sand to be reduced to grain size through primary attrition, then mixed with the required amount of new makeup sand to replace losses, and finally put through secondary attrition (preferably POP type).
2. This sand (so-called 'unit' sand) is then used for *all* cores, i.e. those used in both greensand and chemically bonded sand moulds. Reclaimed sand has a lower thermal expansion than new sand and will have a grain shape and particle size distribution that is as good as or better than new sand. Furthermore, the addition of new sand to this unit sand will reduce LOI to well below any level that might cause problems for ductile iron chemistry.
3. New sand is used for chemically bonded moulds, although of course reclaimed sand could be used here, too.

An example known to the author was a furan/PU foundry that ran a POP attrition plant, reclaiming 94% of the furan sand and using this to make its PU-bonded cores without any problems arising (something that it had been unable to do before the POP unit). Although a maximum 1% LOI for reclaimed furan sand was traditionally thought to be the limit when using PU binders, the LOI was found to fall from about 3.5% to only 2.6% (i.e. about 1/3), but, somewhat surprisingly, was observed to cause no problems. On closer examination, it was found that the total sulphur content had fallen from more than 0.3–0.12% (i.e. 60% reduction), almost certainly accounting for the improved behaviour.

LOI on its own can, of course, be misleading; for example the presence of the carbon film around the sand grains (which shows up as LOI) can be beneficial, effectively insulating the grain from chemical attack by the metal and its vapours.

15.6.3 AGGREGATE RECLAMATION WITH SOLUBLE INORGANIC BINDERS

The use of inorganic binders that are soluble in water has great attraction in the foundry. The chemicals are without smell, clean and mostly free from any toxicity problems.

The most widely successful of the inorganic binder systems have been based on sodium silicate. The curing action has traditionally been effected by carbon dioxide gas. The 'CO_2' process, as it has become known, has a long history. It can deliver hard moulds tolerably quickly. One of its major drawbacks is the tendency of the silicate binder to melt at casting temperatures, fritting like a glass to make cores extremely problematic to remove.

Alternatively, for light-alloy casting, the mould can be cured simply by drying. This approach has the great advantage that the binder remains soluble in water. Thus after casting, the mould can be ablated (i.e. eroded off by water jets) off the casting without damage to the casting. This is especially valuable in an Al alloy foundry in which the castings are relatively weak and soft, and so easily damaged, when breaking out the moulds. The ablation casting process, patented by Alotech in about 2006, takes advantage of the solubility, removing the mould while the casting is still at least partially molten, hence greatly enhancing the rate of solidification.

By removing the mould by jets of water, the atmosphere of the foundry is freed from dust, eliminating the huge and expensive dust extraction systems. The elimination of fume means that operators enjoy clean conditions, in contrast to most resin-bonded binder systems in which the pungent fumes unpleasantly cling to skin and permeate fabrics.

Sodium silicate does contain some combined water, but this does not seem to create significant problems for Al castings. Measured gas evolution from silicate-bonded cores has been found to be low (Wolff and Steinhauser, 2004). For steel casting, the release of water vapour at the higher temperature seems a positive advantage because its evaporation conveys heat away from the casting, causing silica sand to achieve the chilling power of chromite sand. Several steel foundries have abandoned the use of chromite as a facing sand, adopting instead a simpler and less costly one-piece mould bonded with a sodium silicate binder.

The reasons which traditionally have inhibited the widespread adoption of silicate binders in foundries include the following.

1. Shake-out difficulty for cores. This problem has been addressed by (a) organic additions to the silicate and (b) in quite separate developments avoiding carbon dioxide, using drying in air.
2. Long cycle times for curing by drying have meant that a core weighing a kilogram or so may require several minutes to be dried sufficiently to be useable. This contrasts with PU binder systems, currently the norm in the industry, that require perhaps only 5 s cure by an amine vapour, followed by 30 s or so of purge air to remove the vapour, giving a total cycle time in the core machine of approximately 40 s. However, developments are taking place to reduce the cycle time for the drying of silicates by using higher temperature and faster air flows, together with reduced sand thickness using shell techniques.
3. Large energy requirement of curing by evaporation of water. It is true that the energy required to evaporate water is unusually high (2.26 MJ/kg). However, most of the water from re-cycled sand can be eliminated by simple draining, or possibly centrifuging which is not energy intensive, leaving only a percent or two to be removed by evaporation. Each percentage of moisture per tonne of sand requires approximately 6.4 kWh to convert it to vapour, so that a maximum of about 12 kWh/tonne would be needed for 2% of residual moisture. (This compares with the energy requirement of at least 200 kWh/tonne for thermal reclamation at 700°C.) Furthermore, because the binder to be added to the reclaimed sand is water-based, not all the residual moisture need be removed because the added binder can be concentrated, achieving the correct mix when diluted with the residual moisture already present in the returning sand.

Thus it can be seen that inorganic binders appear to be addressing the issues that originally prevented their use. It is expected that they will become progressively more important in the casting industry during the next few years.

15.6.4 FACING AND BACKING SANDS

Occasional foundries using chemical binders have simplified their operations to the point that only a single sand, sometimes called a 'unit sand', is used for both moulds and cores, together with a single binder system. This unusual achievement has clear benefits to the economics and management of the foundry.

However, several ferrous foundries, particularly those making larger castings in relatively low numbers, will often use a facing sand and a backing sand. Chromite and zircon sands have been commonly used to face moulds and cores for two reasons: (1) they act as a chill and help to stop penetration by molten metal and (2) they act as a barrier to stop silica backing sand coming into contact with the molten metal. This latter point is especially useful when casting high-manganese steel because silica reacts with manganese steel to form a low melting point manganese silicate, causing a severe burn-on problem. In contrast, olivine is perhaps the most resistant of all aggregates to attack by manganese steel.

When chromite is used as a facing sand in silica sand moulds, or in cores, silica and chromite unfortunately become mixed during the re-cycling of the sand. Although fluid bed separation has been reported (Perbet, 1988), the more usual magnetic separation represents a significant investment in the re-cycling system. Chromite is slightly magnetic so that 50–75% can be separated out as a roughly 80–85% rich fraction (Sontz, 1972). Normally two magnets are required: the first to remove unwanted iron and steel (it is always amazing how much iron and steel gets into sand re-cycling plants) and the second a more powerful (and more expensive) rare-earth magnet that is required to remove the magnetically weaker chromite. If the chromite and silica are not effectively separated during re-cycling, the sand becomes practically unusable for steel casting, as discussed in the section on chromite.

Steel foundries have used zircon for facing because they wish to employ a silica sand backing. They have shied away from chromite to avoid the risk of (what is assumed to be) fayalite formation and the hassle, cost and uncertainty of magnetic separation. Zircon cannot be reclaimed separately, and thus is lost into the backing sand after re-cycling. The rising cost of zircon as world supplies diminish is now closing this option.

The re-cycling of the silica backing sand can be carried out at or better than 90% efficiency (particularly for alkali phenolic bonded sand) with modern reclaim systems that can simultaneously improve the grain shape by a progressive rounding action during each cycle. However, the effects from any buildup of low melting point phases when using chromite with improved reclaim systems do not appear to be known at this time.

An interesting potential solution for steel foundries is to continue with chromite for facing but replace their silica backing sand with olivine, operating a single stage magnetic separation. This completely avoids the risk of iron silicate formation and will allow foundries to increase their re-cycle yields of both chromite (no longer lost as an iron silicate) and olivine (because of lower binder residues). In summary, a facing sand of typically 80% chromite will not be degraded by reaction with its 20% content of olivine because this mixture is effectively inert (in contrast to an 80/20 chromite/silica mixture). Also, of course, the chromite content of the olivine backing sand is also not damaging.

Silica-zircon foundries could change to olivine-chromite and need only invest in a simple magnetic separation stage. These are relatively modest investments that would be repaid rapidly. The more expensive olivine is envisaged to be more than offset by the higher rates of reclamation that are offered by POP re-cycling systems.

Finally, we should note that the benefits of the unit sand concept are enjoyed by some steel foundries who have found a facing sand to be unnecessary. They have discovered that by using a silicate-based binder they are able to avoid using the chromite facing sand. The reason for this seems to be the significant chill effect provided by the silicate binder: its free water together with its water of crystallisation is driven off and the superheated steam carries with it a huge transfer of latent heat into the depths of the mould, as discussed in Chapter 4 on evaporation and condensation zones.

CASTING

The casting process is the point in manufacture when most of the defects are introduced into the cast part. It is true that the melt might be the wrong composition, and the mould might be the wrong shape or fall apart, but these misfortunes are unusual, and they are usually modest compared with the damage that is inflicted by most filling system designs. Anyway, we shall hope that the earlier chapters in this book have been observed so far as possible, such that these problems are minimised.

In the description of rule 2, the problems of pouring by gravity were emphasised, leading to solutions for gravity pouring which could only be described as damage limitation exercises. This contrasted with the process that minimises the effect of gravity, tilt casting and the elimination of gravity in 'level transfer' and 'counter-gravity' filling systems. Only tilt, level transfer and counter-gravity can produce ideal transfer of metal into the mould, thus manufacturing the best quality of casting. These points are so fundamental that they will arise time after time as we examine the various casting systems. No apology is offered for repetition.

16.1 GRAVITY CASTING

Today's technology for the production of most castings involves gravity pouring. If this were not bad enough in itself, the design of the filling systems usually make this bad situation much worse. This must be kept in mind. Only relatively rarely are castings produced with filling system designs using the principles set out in this book. Naturally, we all hope this poor situation will improve.

16.1.1 GRAVITY POURING OF OPEN MOULDS

Most castings require a mould to be formed in two parts: the bottom part (*the drag*) forms the base of the casting and the top half (*the cope*) forms the top of the casting. I remember these names because the drag can sometimes be moved by being *dragged* along the floor. The cope is reminiscent of coping stone, the stone topping the wall, or from the bishop's cope, the name for his cloak.

Some castings require no shaping of the top surface. In this case, only a drag is required. The absence of a cope means that the mould cavity is open, so that metal can be poured directly in. The foundryman can therefore watch the flow of metal and actively direct it around the mould using his skill during pouring to encourage the flow to fill all parts (Figure 11.1).

Such open top moulds represent a successful and economical technique for the production of aluminium or bronze wall plaques and plates in cast iron, which do not require a well-formed back surface. The first great engineering structure, the Iron Bridge built across the River Severn by the great English ironmaster Abraham Darby in 1779, had all its main spars cast in this way. This spectacular feat is not to be underestimated, with its main structural members more than 23 m long cast in open top sand moulds; it heralded the dawn of the modern concept of the structural engineering casting and still stands to this day.

Other viscous and poorly fluid materials are cast similarly, such as hydraulic cements, concretes and organic resins and resin/aggregate mixtures to form 'resin concretes'. Molten ceramics such as liquid basalt are poured in the same way, as witnessed by the cast basalt curb stones outside the house where I once lived, that have lined the edge of the road for more than 100 years and whose maker's name is still as sharply defined as the day it was cast.

Complete Casting Handbook. http://dx.doi.org/10.1016/B978-0-444-63509-9.00016-9

The remainder of this section concentrates on the complex problem of designing filling systems for castings in which all the surfaces are moulded i.e. the mould is closed. In all such circumstances, a *bottom-gated* system is adopted (i.e. the melt enters the mould cavity from one or more gates located at the lowest point, or if more than one low point, at each lowest point).

16.1.2 GRAVITY POURING OF CLOSED MOULDS

The series of funnels, pipes and channels to guide the metal from the ladle into the mould constitutes our liquid metal 'plumbing', and is known as the *filling system* or *running system*. Its design is crucial; so crucial, that the chapters on the design of the filling system are by far the most important sections of this book.

In addition to them being categorised as pressurised, naturally pressurised and nonpressurised, filling systems are commonly also can be categorised as top gated, gated at the mould joint, or bottom-gated. It will by now be clear that in this book the only pressurisation recommended is natural pressurisation, and the only gate location recommended is bottom-gated.

However, the reader needs to keep in mind that the elimination of a running system by simply pouring into the top of the mould (down an open feeder, for instance) may be a reasonable solution for the filling of a closed mould in some cases. Although apparently counter to much of the teaching in this book, there is no doubt that a top-poured option has often been demonstrated to be preferable to some poorly designed running systems, especially poorly designed bottom-gated systems. There are fundamental reasons for this that are worth examining right away.

In top pouring via, for instance a feeder, the plunge of a jet into a liquid is accompanied by relatively low shear forces in the liquid, since the liquid surrounding the jet will move with the jet, reducing the shearing action. Thus, although damage is always done by top pouring, in some circumstances it may not be too bad, and may be preferable to a costly, difficult, or poor bottom-gated system.

In poor filling system designs, velocities in the channels can be significantly higher than the free-fall velocities. What is worse, the walls of the channels are stationary and so maximise the shearing action, encouraging surface turbulence and the consequential damage from the shredding and entraining of bubbles and bifilms.

Ultimately, however, a bottom-gated system, if designed well, has the greatest potential for success.

It is all the more disappointing therefore that many of our horizontally parted moulds produced on automatic moulding systems practically dictate the use of a poor filling system, resulting in the high-rate production of poor castings to the extent that Al-alloy casting cannot generally be produced on automatic greensand moulding lines. These current designs of automated moulding plants seem perversely designed to produce scrap rather than produce castings. Iron casting is fortunately more tolerant of such abuse of casting principles. The manufacturers of such systems will often deny this assertion, pointing out that a conical basin is not necessarily the only option, and that a three-dimensional milling machine could cut an offset step basin while the cope was upside down. Similarly, a tapered sprue could be cut or moulded at the same time. However, the author has never yet seen such solutions put into practice. They would be a welcome development.

Shrouded Pouring

The problem of the voluminous entrainment of air by the conical pouring basin has often been tackled by the provision of a shroud. This can be a steel or refractory cylinder that surrounds the outlet from a bottom-poured ladle, and nests in the funnel-shaped pouring basin. By this means, the inward flow of air into the conical funnel is reduced. However, even if the shroud were 90% efficient, which would be a surprisingly high efficiency, the remaining 10% of entrained air would still be a serious problem.

Some shrouds have an inert gas piped into them in an additional effort to protect the metal stream. This is clearly helpful, but once again inefficient. Plenty of air will still find its way into the metal stream. In any case, the bursting of bubbles of inert gas will still create some turbulence and splashing in the metal rising in the mould cavity, with consequent creation of defects.

In conclusion, shrouds reduce entrainment problems, but are inefficient. I avoid such unsatisfactory solutions.

Contact Pouring

The sitting of the bottom-pour ladle directly on its mould, so as to align the ladle nozzle and the entrance to the down-sprue of the casting, then raising the stopper to allow the ladle to drain, filling the casting, is the ultimate technique to eliminate the entrainment of air. It is simple, uses no costly inert gas and is 100% efficient. For shaped casting production, the technique has the added advantage that a pouring basin is eliminated. These are all really powerful advantages.

I always thought the contact pour technique would be useful and important, but I have in the past underestimated its importance. I now have come to realise its importance. It is an excellent technique which would, at almost no cost, revolutionise most foundries at a stroke. I now recommend it strongly for most foundries. It represents a major step towards excellent quality of castings, as a kind of low cost halfway development on the road to fundamental and revolutionary changes such as counter-gravity filling of moulds.

If only one casting is to be made from a ladle, effectively emptying the ladle, it is straightforward to calculate a suitable filling system using the head height in the ladle as the initial head height for driving the filling system. However, if several castings are to be filled from one ladle, the reducing head height in the ladle may cause the first casting to fill rather too fast and the last casting to fill too slowly. This must be checked by the casting engineer. Usually, however, the system is particularly tolerant, allowing five or six pours without problem because the changing height effect is reduced by the square root relation between height and velocity (Eqn (13.2)).

The contact between the nozzle in the base of the ladle and the entrance to the sprue is easily sealed with a layer of ceramic fibre blanket. However, the alignment of the nozzle with the sprue entrance can be a minor technical challenge. When pouring steel the alignment is easily seen from the pool of light radiated down from the white-hot stopper, shining on the white ceramic fibre surround of the sprue. For the lower melting point metals line of sight can be good enough but slow, whereas an engineering solution is usually best: laser guidance with feedback control to the crane should not be a problem in these modern times.

The reports of the benefits in the literature of contact pouring are universally positive. Examples include Jeancolas and coworkers (1962) and Schilling and Riethmann (1992). The author also has used the technique to excellent effect for a 50 tonne steel casting that proved to be free from defects as a result of zero air entrainment in the filling system (Kang et al., 2005).

Automatic Bottom Pouring

The automatic bottom pouring technique was invented by Alec Allan in the United Kingdom in 1963 (Allan, 1963, 1968). Although its name refers specifically to the use of a crucible with a hole in its base, through which the melt is poured when it is finally molten, the technique as practised to this day tends to employ only low cost, one-shot consumable crucibles, often made from simple bonded silica sand or from lightweight ceramic fibre. A specially shaped slug of metal which is a close fit in the crucible is melted extremely rapidly, usually within approximately 60 s. The base of the charge melts last, so the melt pours automatically, precisely when the last metal is melted. Sometimes the process is used with a penny-shaped metal disc in the exit nozzle of the crucible, so that the superheat at which the melt is poured can be controlled by the thickness of the disc.

The mould is placed, of course, under the crucible. This simple in-line arrangement involves no moving parts, and thus can provide extremely low-cost melting and casting systems, especially for vacuum melting and casting. The cylindrical vacuum chamber can be a fibreglass-resin tube, allowing the induction coil to be placed outside, once again simplifying the design by avoiding electrical connections through the wall of the chamber. The rather poor vacuum held by the fibreglass tube seems to be no problem for the melting and casting of most alloys. The process continues to be used, being highly productive for the casting of Ni-base superalloys for turbocharger wheels, and iron and steel castings, but Allan also describes its use for magnesium.

Although impressively simple, it is a pity that it is a *pouring* process, consequently demonstrating a constantly varying proportion of scrap from turbulently filled moulds. We really need to get away from *pouring* metals.

Having said this, there may be a way forward for automatic bottom pouring if the melting crucible could be placed in contact with the mould to achieve *contact pouring*. If this process could be successfully developed, it would be immensely attractive. An opportunity awaits someone.

Direct Pour

This name was chosen by the Foseco Company to describe the use of a ceramic fibre sleeve, fitted with a ceramic foam filter in its base, place directly onto and in contact with the casting as a substitute for a combined filling and feeding system. As such, of course, it could be convenient and highly economical (Sandford, 1993). However, despite several notable successes, the system has not always yielded good results.

This technique has been researched. The results are summarised in Figure 12.55, showing that the technique is mainly effective for relatively small falls after the filter in stabilising and making reproducible the casting conditions (the filtering action is insignificant). Because the stable regime promoted by the system may not yield a good casting, this means that if the first trial location for the direct pour sleeve on the casting is not good, all castings are likely to be unacceptable. Conversely, if the first trial location does yield a good result, all results are likely to be good. In the case of the poor result, repeated different locations should be tried if possible. If one is found to work it is probable that all subsequent castings will be good.

Gravity dies (Permanent Moulds)

For dies filled simply by pouring under gravity, in the United Kingdom they are not surprisingly called *gravity dies* (known in the United States as *permanent moulds*).

That the mould is hot and extremely dry means that the liquid aluminium and mould are practically inert towards each other. This is a significant benefit of hot dry metal moulds that is often overlooked. For this reason, subsurface porosity as a result of high hydrogen levels is rarely seen in casting from metal moulds. The use of permanent moulds is usually confined to low melting point metals (e.g. Zn, Pb) and the light alloys (Mg and Al). Only relatively rarely are permanent moulds used for the higher temperature alloys such as brasses and bronzes, and cast iron, although even for these metals, metal moulds can be valuable and successful.

The benefit of high temperature of the mould introduces a challenge for the control of the temperature, of course. Thus the die temperature falls when the casting cycle is interrupted by the fairly common occurrence of metal becoming trapped in the die, or coating repair, or sometimes even the arrival of a new batch of cores that have to be signed for and wheeled into place.

Cast iron or steel moulds used in gravity die casting (permanent moulding) or low-pressure die casting (low-pressure permanent moulding) of aluminium are coated with an oxide wash of rather variable thickness, in the range 0.5–2 mm thick, infringing on the accuracy of the casting. Its purpose is to reduce the thermal shock to the die, and by thus reducing the rate of heat transfer, allow the die to fill without premature freezing. Without the die coat it would be difficult if not impossible to fill the die without the formation of cold laps.

Because the die coat reduces the severity of the thermal loading, the die material can sometimes be grey cast iron. This is welcome because of the ease of obtaining a block of starting material, suitably shaped with a contoured back if necessary, and because of its excellent and easy machining. It has to be admitted that grey iron has limited strength and limited fatigue resistance. Thermal fatigue usually sets in after thousands of casts. This limit to die life can be an important threat to surface finish as the die ages, developing multiple small cracks, often called checks, or occasional large cracks resulting in sudden catastrophic failure. Such failures have disastrous effects on production because dies take time to replace. Such failure is commonly associated with heavy sections of the casting, such as a heavy boss. The iron or steel die in this region suffers from repeated transformation to austenite and back again. The large volume change accompanying this reaction corresponds to a massive plastic strain of several percent, so that steel, and more particularly cast iron, suffers thermal fatigue. The severe strains of thermal fatigue often lead to failure after relatively few cycles.

For dies that are subject to the rigours of volume production and where total reliability is required, the dies are machined from steel. Usually a special grade of hot work die steel, commonly an H13 grade containing 4% Cr is chosen, that is preferably characterised by an especially fine grain size.

For the production of simple two-parted castings, gravity dies are usually constructed with a vertical split line. The moving die half (if only one half moves) is usually arranged to retain the casting. Thus the action of opening the die causes the casting to impact on the ejector pins and so release the casting from the moving side. The casting may be caught on a tray and swung out from between the die halves. Alternatively, it may be simply picked up from where it has fallen. I only once saw a die arranged wrongly by mistake, with the casting retained in the fixed die half. It was an embarrassment to the die designer and a constant source of problems for the die operator until the die could be turned around.

Complex dies such as for automotive cylinder heads are usually provided with vacuum connections to the backs of core prints so as to suck gases out of difficult-to-vent cores such as water jacket cores. The vents usually block by the condensation of tars and other volatiles after 15–25 castings, depending on the casting. Sometimes knockout vents are employed to allow the vents to be cleaned out by drilling. The stopping of production to clean out the vents is a non-trivial hindrance to production rates using gravity dies. The campaign life of a die is usually dictated by the cleaning-out of vents rather than the maintenance and repair of the die coat. If vents are not maintained fully operational, blows from cores increase to become a major source of scrap.

More recently there is hope that difficult-to-vent cores might become producible with the new generation of silicate binders that do not create condensable outgassing products.

Binders that do not create fume on casting are very much to be desired. Fume in the immediate environment of operators is a problem in gravity die operations. The opening of a die containing cores is usually accompanied by a cloud of fume, most of which avoids the fume extraction hood because of side draughts. Sand castings tend to avoid the worst of such fume emissions because the casting and cores remain in the mould for longer and are cooler by the time the mould is opened. In any case, the sand mould is opened in a completely closed environment so that operators can work in a cleaner atmosphere.

The provision of a small steel mesh filter, with an approximately 2 mm mesh opening, is often usefully incorporated into the runner. This can be extracted from the runner while the metal is still hot, so is not recycled with the foundry returns, thus avoiding contaminating the metal with iron.

Turning to the possibility of the casting of cast irons in cast iron dies, the subject is studied at some length by Jones et al. (1974). These authors used grey iron dies coated only with acetylene soot (although the author is aware of other foundries that also coat with an oxide wash in addition to the application of fresh layer of soot after the ejection of each casting). They found good results for grey iron castings at slightly hypereutectic composition, well inoculated, poured close to 1350°C and with mould temperatures up to 300°C. It was also important that the moulds were of good quality, avoiding flash. In this way, carbide formation in the iron could be avoided. They claimed dimensional reproducibility, surface finish and strength properties were consistently superior to comparable aggregate-moulded castings.

These authors went on to check ductile irons. The greater tendency for these irons to solidify with carbides required a higher mould temperature up to 450°C. Even so, conditions for achieving ductile irons completely free from carbide were not found, but the heat treatment to eliminate the carbides was relatively short because of the fine structure.

The production of malleable irons, cast originally of course as white irons, was not considered attractive mainly because of the problem of achieving sound castings without the use of large feeders.

Carbon-based and graphite dies have been found useful for zinc alloys. However, for aluminium alloys, the lives of carbon moulds are short because of the degradation of the carbon by oxidation. (All the more impressive therefore is the use of graphite moulds for steel, used for the casting of millions of railroad wheels by the Griffin Process, described in the next section.) Graphite is often used as a moulding material for titanium, but any benefits of inertness are lost by the reaction to form the so-called alpha case. The details are described in Chapter 6 under the various cast metals.

16.1.3 TWO-STAGE FILLING (PRIMING TECHNIQUES)

There have been several attempts over the years to reduce some of the problems of gravity filling by the introduction of a two-stage filling process. The first stage consists of filling the sprue, after which a second stage of filling is started by the opening of a valve by which the runner and gates etc. are allowed to fill.

After the filling of the sprue, a short dwell for a few seconds with the metal at rest allows bubbles to separate, excluding air, but also allowing the oxide damage created during the first part of the pour to separate, probably becoming glued to the walls of the running system. After this few seconds of quiescence, the melt is allowed to start flowing once again. This second phase of filling has the full head H of metal in the sprue and pouring basin to drive it, but the column has to start to move from zero velocity. It reaches its steady state velocity $(2gH)^{1/2}$ only after a period of acceleration. Thus, the early phase of filling of the runner and gates starts from a zero rate and has a gradually increasing velocity, although at the relatively fast rate of gravitational acceleration this is probably only a minimal benefit.

The benefits of the exclusion of air from the sprue, and the reduced velocity during the early part of stage 2, are benefits that have been recorded experimentally for semi-solid (actually partly solid) alloys. These materials are otherwise extremely difficult to cast without defects, almost certainly because the defects normally entrained during the pour cannot float out of the casting but are trapped in suspension because of the high viscosity of the mixture.

Workers from Alcan (described by Weiss and Rose, 1993 and Cox et al., 1994) developed a system in which the advance of the melt was arrested at the base of the sprue by a layer of ceramic paper supported on a ceramic foam filter (Figure 12.45(a)). When the sprue was filled the ceramic paper was lifted from one corner by a rod, allowing the melt to flow through the filter and into the running system. These authors call their system 'interrupted pouring'. However, the name 'two-stage pour' is recommended as being more positive, and less likely to be interpreted as a faulty pour as a result of an accident.

The two-stage pour has been convincingly demonstrated as beneficial by several investigators. For instance, as early as 1968, Wildermuth describes its use with cast iron and steel using a sheet metal slide valve coated with fireclay, finding that the technique gave significantly cleaner castings. Taghiabadi and colleagues (2003) used the technique for both partly solid and conventional aluminium casting alloys. These authors used Weibull statistics to confirm the reality of the benefits. They used a steel sheet to form a barrier, as a slide valve, in the runner (Figure 12.45(b)). After the filling of the sprue, the sheet was slid aside, opening the runner and allowing the mould to fill.

A fascinating major benefit of the two-stage filling of the down-sprue is that no matter how dreadful the design of the down-sprue, once primed during the first stage of filling, it tends to remain full, operating perfectly with zero entrainment of air. This can easily be demonstrated by a water model.

A second completely different incarnation of the two-stage filling concept is the *bottom-pour ladle*. It is illustrated in Figures 12.3 and 12.45(c). The device is used mainly in the aluminium casting industry, but has been extended with benefit to other casting industries. Instead of transferring metal from a furnace via a ladle or spoon of some kind and pouring into a pouring basin connected to a sprue, the ladle dips into the melt and can be filled uphill via the bottom nozzle. The ladle nozzle is then closed by a stopper and the ladle transferred to the mould. The stopper is raised and can deliver the contents of the ladle into a conventional pouring basin. Preferably, however, it can be lowered onto and sat on the mould, aligning precisely with the entrance to the sprue, and thus delivering the melt directly in the sprue by the *contact pour* technique. This technique is strongly recommended in place of pouring into a pouring basin.

Alternatively again, if the ladle is equipped with an extension (Figure 12.45(d)), converting it to a *snorkel ladle*, the nose of the snorkel can be lowered down through the mould to reach and engage with the runner. Only then is its stopper raised and the melt delivered to the start of the running system with minimal surface turbulence. The approach is capable of producing excellent products. Two-stage filling in its various forms seems to offer real promise for many castings.

16.1.4 VERTICAL STACK MOULDING

The stacking of identical aggregate moulds and pouring into the top has been used for several products, for instance for cast iron piston rings. Although cast iron has a relatively low susceptibility to entrainment problems compared with many other metals, such mistreatment ensures that even this metal does not survive unscathed. Bubbles and oxides are necessarily entrained during the tumbling of the melt down the necessarily non-tapered irregular sprue (Figure 12.49(c)), and these defects find their way randomly into castings, requiring much subsequent testing and sorting. This awful process cannot be recommended. It throws all the principles of good filling design to the wind. Both the castings and the profitability of the foundry suffer lamentably.

16.1.5 HORIZONTAL STACK MOULDING (H PROCESS)

Foundries that are set up based on the H process (another of Fred Hoult's inventions; Hoult, 1979) have a pleasing simplicity. The moulding section of the foundry consists of relatively small core machines making the moulds by blowing into a corebox that forms both the front and back surfaces of the casting (Figure 12.49(b)). The use of identical cores forming both the front and back surfaces of the mould in this way effectively doubles the rate of production of moulds.

The moulds are stacked vertically in a frame, and then clamped, and the frame is turned horizontal and transferred to the casting area. After casting and cooling, the sand can be mainly recycled, and the castings separated from the running system by simply breaking them off.

The overflow of metal from one mould to the next necessarily involves a fall inside the mould cavity. This fundamental problem leads to the entrainment of severe defects if the fall is much over 200 mm, explaining the fact that H Process castings are normally limited in height to about 200 mm and may additionally explain why the process appears to be also limited to cast iron which has least sensitivity to entrainment problems. The one good feature of the fall of metal in the mould is that the fall distance is also minimised (as opposed to castings poured from above the mould via a conical basin, where velocities can become very high) so that the entrainment damage is limited. The production of Al alloy castings by this process would be expected to be problematic.

A computer study by Xiao (1998) shows that if the mould stack is perfectly horizontal, the melt dribbles into more distant cavities, possibly spoiling these castings. If the mould stack is tilted by about 7°, causing the melt to progress upwards from one cavity to the next, this problem is reduced. Furthermore, if the runner linking the cavities has a flat base, as opposed to being of a circular section, that also helps.

16.1.6 POSTSCRIPT TO GRAVITY FILLING

Later we consider an appraisal of counter-gravity filling of moulds, but it is salutary to consider why anyone would go to this trouble when gravity filling appears to be so easy.

There are some advantages to the use of gravity to action the filling of moulds by simply pouring metal. It is simple, low cost and completely reliable because gravity has never been known to suffer a power failure. It is with regret, however, that the advantages finish here, and the disadvantages start. Furthermore, the disadvantages are serious.

Nearly all the problems of gravity pouring arise as a result of the velocity of the fall. After a trivial fall distance corresponding to the few millimetres of the critical fall height, gravity has accelerated the melt to its critical velocity. Beyond this point, the continued acceleration of gravity results in the metal going too fast and having too much energy. Thus the danger of entrainment defects grows alarmingly with the increasing distance of the fall. Because the critical fall distance is so small, nearly all actual falls exceed this limit. In other words, the energy content of the melt, when allowed to fall even only relatively small distances under gravity, are nearly always sufficiently high to cause the breakup of the liquid surface. (It is of little comfort at this time to know that foundries on the moon would fare a little better.)

These high speeds result in gravity filling systems being hypersensitive to small errors. For instance, a tiny mismatch of a millimetre or so in the sprue can easily lead to the creation of volumes of entrainment defects.

A second fundamental drawback of gravity filling is the fact that at the start of pouring, at the time the melt is first entering the ingates, the narrowest part of the mould cross-section where volume flow rate should be slowest, unfortunately, the speed of flow by gravity is highest. Conversely, at a late stage of filling, when the melt is at its coldest and approaching the top of the mould cavity the speed of filling is slowest, endangering the casting because of cold laps and misruns. Thus filling by gravity gives a completely inappropriate filling profile.

Thus, to some extent, there are always problems to be expected with castings poured by gravity. The long section on filling system design in this book is all about reducing this damage as far as possible. It is a tribute to the dogged determination of the casting fraternity that gravity pouring, despite its severe shortcomings, has achieved the level of success that it currently enjoys.

Even so, there is no shortage of viable alternative processes that do not suffer these disadvantages. These are described later and are recommended. The good-natured behaviour of several of these processes, including some tilt processes and counter-gravity systems that use low filling speeds, errors of mismatch of flow channels and sharp, un-radiused corners etc. all become irrelevant.

16.2 HORIZONTAL TRANSFER CASTING

The quest to avoid the gravity pouring of liquid metals has led to systems employing horizontal transfer and counter-gravity transfer. These solutions to avoid pouring are clearly seen to be key developments; both seem capable of giving competitive casting processes that offer products of unexcelled quality. The major approaches to the first approach, horizontal transfer, are described below.

16.2.1 LEVEL POUR (SIDE POUR)

The *level pour* technique was invented by Erik Laid (1978). At that time, this clever technique delivered castings of unexcelled quality. It seems a pity that the process is not more generally used. This has partly occurred as the result of the process remaining commercially confidential for much of its history, so that relatively little has been published concerning the operational details that might assist a new user to achieve success. Also, the technique is limited to the type of castings, being applied easily only to plate, box or cylinder type castings where a long slot ingate can be provided up the complete height of the casting. In addition, of course, a fairly complex casting station over a deep floor pit is required.

The arrangement to achieve the so-called level filling of the mould is shown in Figure 16.1. An insulated pouring basin connects to a horizontal insulated trough that surrounds three of the four sides of the mould (a distribution system reminiscent of a Roman aqueduct). The melt enters the mould cavity via slot gates that extend vertically from the drag to the cope. Either side of each slot gate are guide plates that contain the melt between sliding seals as it flows out of the (stationary) trough and into the (descending) mould.

Casting starts with the mould sitting on the fully raised mould platform, so that the trough provides its first metal at the lowest level of the drag. The mould platform is then slowly lowered whilst pouring continues. The rate of withdrawal of the mould is such that the metal in the slot gate has time to solidify against a water-cooled chill positioned underneath the melt entrance. Thus the melt in the open slot is frozen before it emerges from out of contact with the chill, sliding out into the open air.

In one of the rare descriptions of the use of the process by Bossing (1982), the large area of melt contained in the pouring basin and distribution trough is claimed to smooth the rate of flow from the point of pour to delivery into the mould, despite variations in the rate of pour into the launder system from the ladle. Also in this description is an additional complicated distribution system inside the mould, in which tiers of runners are provided to minimise feeding

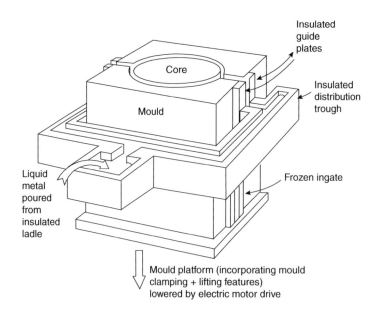

FIGURE 16.1

Level pour technique.

distances and maximise temperature gradients. In general, such sophistication would not be expected to be necessary for most products.

16.2.2 CONTROLLED TILT CASTING

It seems that foundrymen have been fascinated by the intuition that tilt casting might be a solution to the obvious problems of gravity pouring. The result has been that the patent literature is littered with re-inventions of the process decade after decade. Even now, the industry has many completely different varieties of tilt process, so the reader needs to be cautious about exactly what tilt process is being referred to.

Furthermore, the deceptive simplicity of the process conceals some fundamental pitfalls for the unwary. The piles of scrap seen from time to time in tilt-pour foundries are silent testimony to these hidden dangers. Generally, however, the dangers can be avoided, as will be discussed in this section.

Durville Process Casting

The most common form of tilt casting is a process with the unique feature that, in principle, liquid metal can be transferred into a mould by simple mechanical means under the action of gravity, but without surface turbulence. It therefore has the potential to produce very high quality castings. This was understood by the Frenchman, Pierre Gaston Durville (1874–1959), and applied by him for the casting of aluminium bronze in an effort to reduce surface defects in French coinage.

The various stages of liquid metal transfer in Durville's process are schematically illustrated in Figure 16.2(a). A melting crucible and a mould are fixed opposite to each other on a rotatable platform. A short channel section connects the two. The crucible is charged with metal, which is melted and skimmed clear of dross, and the platform is rotated. Some operators of this process have melted elsewhere and transferred the melt to the Durville unit, pouring the melt into the crucible, thus introducing damage. In the process as originally conceived by Durville, the metal is melted in the crucible fixed in the tilt machine. No pouring under gravity takes place at all. Also, because he was casting large ingots in open-ended moulds for subsequent working, he was able to look into the crucible and into the mould, observing the transfer of the melt as the rotation of the mould progressed. In this way, he could ensure that the rate of rotation was correct, carefully adjusting it all the time, to avoid any disturbance of the surface of the liquid. During the whole process of the transfer, careful control ensured that the melt progressed by 'rolling' in its skin of oxide, like inside a rubber sack, avoiding any folding of its skin by disturbances such as waves. The most sensitive part of the transfer was at the tilt angle close to the horizontal. In this condition the melt front progresses by expanding its skin of oxide, whilst its top surface at all times remains horizontal and tranquil. At this critical stage of metal flow, the rate of rotation must be a minimum. If this stage is not kept under good control, the metal surges into the mould as a wave, and splashes upwards against the rear face of the mould, and is damaged during its fall by entraining oxides.

Clearly Durville understood what he was doing. He avoided any pouring action to fill the crucible or ladle, and he controlled the rate of rotation of the assembly to ensure horizontal transfer. Few that claim to follow his process have understood these critical aspects.

The running system for tilt castings, if any, does not necessary follow the design rules for gravity casting because gravity is only marginally influential in this process. The rate of filling of the casting is under the control of the rate of tilt, not necessarily the channels of the filling system. Furthermore, of course, after the mould is filled the filling channels can be used as feeding channels, and thus be sized appropriately. All this is quite different to the design procedure for a gravity-filled mould.

In the United States, Stahl (1961) popularised the concept of 'tilt pouring' for aluminium alloys into shaped permanent mould castings. The gating designs and the advantages of tilt pouring over gravity top pouring have been reviewed and summarised in several papers from this source (Stahl, 1963, 1986, 1989). These benefits include the following.

1. The control over the rate of filling helps to control flash. This is because the rate of increase of pressure due to the head of metal is rather slow, starting from zero, in contrast to normal gravity-poured castings, where there are high

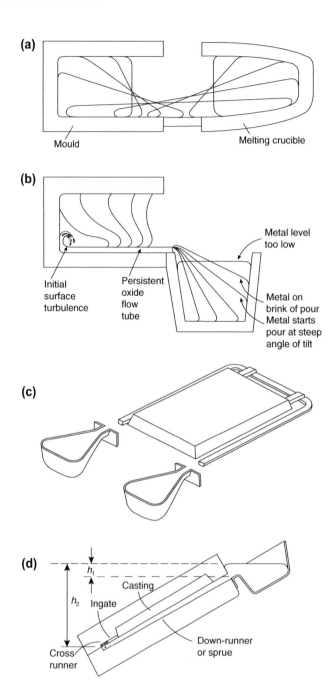

FIGURE 16.2

Tilt casting process (a) Durville; (b) semi-Durville; (c) twin-poured tilting die *(adapted from Nyamekye, 1994.)* and (d) outline of tilt running system at the critical moment that metal reaches the far end of the 'sprue' after the melt has effectively fallen height h_2, unfortunately encouraging damage by surface turbulence.

dynamic pressures and a high rate of attaining the full hydrostatic pressure. The reduced flash is important for castings such as gratings and grills that are otherwise difficult to dress. (This advantage is, of course, available to other gravity-casting techniques, provided a gentle fill is employed; it is less easy to achieve in any of the injection or low-pressure techniques, where the driving force is high, and generally under poorer control.)

2. Automated casting can be arranged relatively easily, with the benefit of consistent results.
3. Two cups can be filled consecutively by one operator, or, alternatively, large cups can be filled by successive charges from a ladle that can be handled by one person. Thus very large castings can be easily produced by one caster. This seems to be a unique benefit for tilt casting, making for considerable economies compared with normal gravity-poured moulds, where several casters may be required at one time, pouring simultaneously, or in an unbroken succession, possibly into several down-sprues.
4. The setting of cores has the flexibility of being carried out in either the horizontal or vertical attitude, or even in some intermediate position, depending on the requirements of the cores.
5. For the ejection of the casting the operator similarly has the option of carrying this out vertically or horizontally.

A useful 'bottom-gated' tilt arrangement is shown in Figure 16.2(c). Here the sprue is in the drag, and the remainder of the running and gating system, and the mould cavity, is in the cope. Care needs to be taken with a tilt die to ensure that the remaining pockets of air in the die can vent freely to atmosphere. Also, the die side that retains the casting has to contain the ejectors if they are needed.

Dies made for the Stahl casting method feature a tapered sprue carrying metal down to a gate near its base. A typical arrangement is shown in Figure 16.2(c) and (d). Effectively, the running systems are of the bottom-gated type, which give less good temperature gradients, or side-gated, which is some improvement. Figure 16.3 shows a die with side gates, but arranged so as to allow the melt to fall inside the mould cavity. The transfer of the mould cavity into the top half of the die would have eliminated this relatively small defect (but whose consequences are not easily predicted, and might therefore be non-trivial).

Occasionally, the metal will be poured directly into the mould cavity, eliminating runners entirely and giving the best temperature gradient but poorest filling. It is significant that Stahl reports the disadvantage of the appearance of the flow path down the face of the casting in this case, outlined by streaks of oxide. Such features are a concern because they indicate the presence of flow tubes, thus possibly constituting serious defects. This is a symptom of poor downhill filling, not employing the controlled horizontal fill approach to be described later.

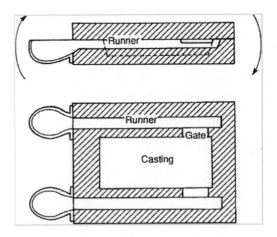

FIGURE 16.3

Tilt pour system that would have benefited from the mould cavity mainly in the upper half of the die, avoiding the fall of metal from the gate. Vents would then have been necessary.

In an effort to understand the process in some depth, Nguyen (1986) simulated tilt casting using a water model of liquid metal flow, and King and Hong (1995) carried out some of the first computer simulations of the tilt casting process. They found that a combination of gravity, centrifugal and Coriolis forces govern tilt-driven flow. However, for the slow rates of rotation such as are used in most tilt casting operations, centrifugal and Coriolis effects contribute less than 10% of the effects resulting from gravitational forces, and could therefore normally be neglected. The angular velocity of the rotating mould also made some contribution to the linear velocity of the liquid front, but this again was usually negligible because the axis of rotation was often not far from the centre of the mould.

However, despite these studies, and despite its evident potential, the process has continued to be perfectly capable of producing copious volumes of scrap castings.

The first detailed study of tilt casting using the recently introduced concepts of critical velocity and surface turbulence was carried out in the author's laboratory by Mi (2000). In addition to the benefits of working within the new conceptual framework, he had available powerful experimental techniques. He used a computer controlled, programmable casting wheel onto which sand moulds could be fixed to produce castings in an Al-4.5%Cu alloy. The flow of the metal during the filling of the mould was recorded using video X-ray radiography, and the consequential reliability of the castings was checked by Weibull statistics.

Armed with these techniques, Mi found that at the slow rotation speeds used in his work the mechanical effect of surface tension and/or surface films on the liquid meniscus could not be neglected. For all starting conditions, the flow at low tilt speeds is significantly affected by surface tension (most probably aided by the effect of a strong oxide film). Thus below a speed of rotation of approximately 7 degree per second the speed of the melt arriving at the end of the runner is reduced in a rather erratic way. Gravity only takes control after tilting through a sufficiently large angle.

As with all casting processes, if carried out too slowly, premature freezing will lead to misrun castings. One interesting case was found in which the melt was transferred so slowly into the runner that frozen metal in the mouth of the runner acted as an obstructing 'ski jump' to the remaining flow, significantly impairing the casting. At higher speeds, however, although ski jumps could be avoided, the considerable danger of surface turbulence increased.

The radiographic recordings revealed that the molten metal could exhibit tranquil or chaotic flow into the mould during tilt casting, depending on (1) the angle of tilt of the mould at the start of casting and (2) the tilting speed. The quality of the castings (assessed by the scatter in mechanical properties) could be linked directly to the quality of the flow into the mould.

We can follow the progress of the melt during the tilt casting process. Initially, the pouring basin at the mouth of the runner is filled. Only then is the tilting of the mould activated. Three starting positions were investigated:

1. If the mould starts from some position in which it is already tilted downward, once the metal enters the sprue it is immediately unstable, and runs downhill. The melt accelerates under gravity, hitting the far end of the runner at a speed sufficient to cause splashing. The splash action entrains the melt surface. Castings of poor reliability are the result.

2. If the mould starts from a horizontal position, the metal in the basin is not usually filled to the brim and therefore does not start to overflow the brim of the basin and enter the runner until a significant tilt angle has been reached. At this stage, the vertical fall distance between the start and the far end of the runner is likely to be greater than the critical fall distance. Thus, although slightly better castings can be made, the danger of poor reliability remains. This unsatisfactory mode of transfer typifies many tilt casting arrangements as seen in Figure 16.2(d) and particularly in the so-called semi-Durville type process shown in Figure 16.2(b). The oxide flow tube from the downhill flow of the stream, plus the entrainment from the splash at the end of the runner can be serious defects. Durville would not have liked to see his name associated with this awful process.

3. If, however, the mould is initially tilted slightly uphill during the filling of the basin, there is a chance that by the time the change of angle becomes sufficient to start the overflow of melt from the basin, the angle of the runner is still somewhat above the horizontal (Figure 16.4). The nature of the liquid metal transfer is now quite different. At the start of filling the runner, the meniscus is effectively climbing a slight upward slope. Thus its progress is totally stable, its forward motion being controlled by additional tilt. If the mould is not tilted further, the melt will not advance. By extremely careful control of the rate of tilt, it is possible in principle to cause the melt to arrive at the

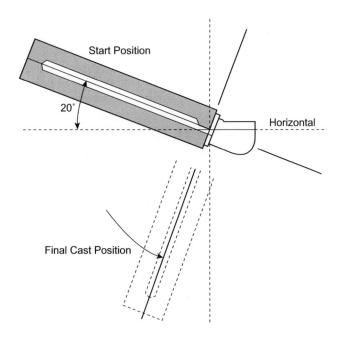

FIGURE 16.4

A tilt pour die starting at a 20° positive tilt, designed to encourage the 'runner' to fill uphill (this is a convenient optical illusion), ensuring that the melt reaches the far end of the runner at a controlled speed.

base of the runner at zero velocity if required. (Such drastic reductions in speed would, of course, more than likely be counter-productive, involving too great a loss of heat, and are therefore not recommended.). Even at quite high tilting speeds of 30 degrees per second as used by Mi in his experimental mould, the velocity of the melt at the end of the runner did not exceed the critical value 0.5 ms^{-1}, and thus produced sound and repeatable castings.

The unique feature of the transfer when started above the horizontal in this way is that the surface of the liquid metal is close to *horizontal* at all times during the transfer process. Thus in contrast to all other types of gravity pouring, this condition of tilt casting does not involve pouring (i.e. a free *vertical* fall) at all. It is a *horizontal* transfer process. It will be seen that in the critical region of tilt near to the horizontal, the liquid transfer occurs essentially horizontally. Durville would have approved.

Thus the optimum operational mode for tilt casting is the condition of horizontal transfer. Horizontal transfer requires the correct choice of starting angle above the horizontal, and the correct tilting speed as found by Cox and Harding (2007). These authors noted that a really accurately controlled tilt speed increased the two-parameter Weibull modulus (the reliability) of Al alloy castings from 2 to 55, an enormous increase. Conversely, a poor choice of rotation parameters created significant surface turbulence.

An operational map can be constructed (Figure 16.5), revealing for the first time an operational window for the production of reliable castings. It is recognised that the conditions defined by the window are to some extent dependent on the geometry of the mould that is chosen. However, the mould in Mi's experiments was designed to be close to the size and shape of many industrial castings, particularly those for automotive applications. Thus although the numerical conclusions would require some adaptation for other geometries, the principles are of general significance and are clear: there are conditions, possibly narrowly restricted, but in which horizontal transfer of the melt is possible, and gives excellent castings.

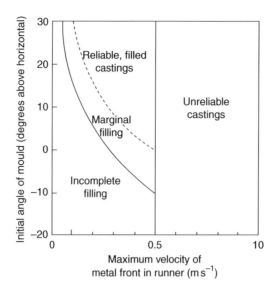

FIGURE 16.5

Map of variables for tilt pouring, showing the operational window for good castings (Mi et al., 2002).

The problem of horizontal transfer is that it is slow, sometimes resulting in the freezing of the ski jump at the entrance to the runner, or even the non-filling of the mould. This can usually be solved by increasing the rate of tilt *after* the runner is primed. This is the reason for the extended threshold, increasing the window of possible filling conditions from the dotted curve to the full line curve on the process map (Figure 16.5). A constant tilt rate (as is common for most tilt machines at this time) cannot achieve this useful extension of the filling conditions to achieve good castings. Programmable tilt rates are required to achieve this solution.

A final danger should be mentioned. At certain critical rates of rise of the melt against an inclined surface, the development of the transverse travelling waves seems to occur to give lap problems on the cope surface of castings (Figure 13.3). In principle, such problems could be included as an additional threshold to be avoided on the operational window map (Figure 16.5). Fortunately, this does not seem to be a common fault. Thus in the meantime, the laps can probably be avoided by increasing the rate of tilt during this part of the filling of the mould. Once again, the benefits of a programmable tilt rate are clear.

In summary, the conclusions for tilt casting are as follows.

1. If tilt casting is initiated from a tilt orientation at or below the horizontal, during the priming of the runner the liquid metal accelerates downhill at a rate out of the control of the operator. The metal runs as a narrow jet, forming a persistent oxide flow tube. In addition, the velocity of the liquid at the far end of the runner is almost certain to exceed the critical condition for surface turbulence. Once the mould is initially inclined by more than 10° below the horizontal at the initiation of flow, Mi found that it was no longer possible to produce reliable castings by the tilt casting process.
2. Tilt casting operations benefit from using a sufficiently positive starting angle that the melt advances into an upward sloping runner. In this way, its advance is stable and controlled. This mode of filling is characterised by horizontal liquid metal transfer, promoting a mould filling condition free from surface turbulence.
3. Tilt filling is preferably slow at the early stages of filling to avoid the high velocities at the far end of the running system. However, after the running system is primed, speeding up the rate of rotation of the mould greatly helps to prevent any consequential non-filling of the castings.

16.2.3 ROLL-OVER AS A CASTING PROCESS

There are casting processes in which the mould is planted upside-down over the mouth of a small melting unit, usually a small induction furnace, and the whole assembly is rotated swiftly through 180°. This technique is widely used in investment casting. However, it is not a process that appears to enjoy much control. It is in reality a kind of dump process for liquid metal. The procedure might not be so bad with a carefully designed filling system in the mould, but this is rarely provided at this time; the melt simply tumbles into the mould. There is little to commend this approach.

The Invocast method by Butler (1980) provides an interesting development which has similarities to the Stahl tilt-pouring method, in that the first action is to fill a pouring cup, and the next is the rotation of the die, which causes the die cavity to fill. It is shown in Figure 16.9. However, there is an important improvement in that the metal does not necessarily run downhill (this will vary somewhat from casting to casting) but fills the cavity progressively as rotation proceeds. Finally, the feeder is sited on top where it can feed most effectively. The preheating of the running system adds to the effectiveness of its action when it reverts to become the feeding system.

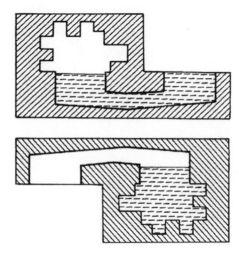

FIGURE 16.6

Inversion casting of a gravity die, showing the start and finish stages.

After Butler (1980).

Rotocast Process

The rotocast process, first used for the production of Al alloy cylinder heads by the Mandl and Berger Company, Austria, is another casting process executed by a roll-over transfer. It is described by Grunenberg et al. (1999). A cylindrical ladle is filled with metal (by pouring of course, so some defects will be added by this initial action) and is then offered up to a permanent mould (Figure 16.6). The mould is then rotated over about its longitudinal axis, transferring the melt from the ladle into the mould. Because the ladle and mould are a pressure-tight system, pressure can be applied during the freezing of the casting. This is said to reduce the feeder size in the sand top core, even though pressure timing and magnitude have to be carefully controlled to avoid metal penetration of the sand cores.

In this same report, these authors go on to describe a second (un-named) process which they seem to consider at the time to be an improvement on the simple rotocast process. The system employs a permanent mould with a top sand core to contain the feeders. This is a variety of the semi-Durville process (Figure 16.7) and shares its same problems; the serious limitations of this incomplete application of Durville's technique are clearly seen: the degree of filling of the ladle, varying from time to time, dictates a varying angle at which the pour can start, resulting in a variable initial

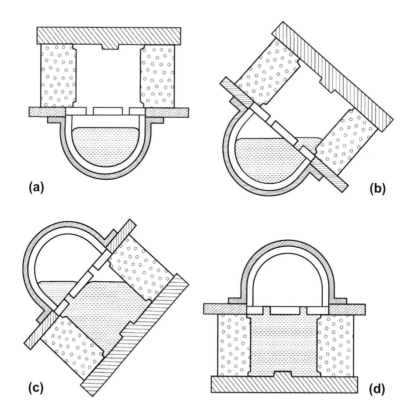

FIGURE 16.7

Rotocast process used for automotive cylinder heads and blocks. (a) start position; (b) metal entering mould; (c) mould almost filled; (d) roll-over casting cycle complete. The roll-over action takes about 10 s.

After Schneider (2006).

downhill flow of the melt, often at a steep angle, achieving high and damaging velocities. A further feature is the use of pressurisation after filling in an effort to reduce feeder size. This need for this complication reveals the poor metal quality, requiring bifilms to be kept closed during freezing. It would surely be better to opt for a technique to improve metal quality and improve filling. I am not aware of this process still being used. Perhaps the users finally had second thoughts about the reality of its capabilities.

16.2.4 ROLL-OVER AFTER CASTING (SOMETIMES CALLED INVERSION CASTING)

Inversion casting is not tilt casting. In fact, it might better be described as a solidification control process rather than a casting process because the pouring action is carried out as a normal gravity fill. It is only after the mould is filled that the whole assembly is rotated through 180° to encourage solidification under improved temperature gradients, and so aid the feeding of the casting.

In 1935, Batty was probably the first to describe simple inversion casting. He gives the example of a heavy-walled cylinder casting that was cast standing upright as shown in Figure 16.8. The gating system was a classical bottom-run method for optimum filling. This would normally result, of course, in a temperature regime in which the highest temperature was at the base of the casting, and any top feeders would be too cold to be effective. He solves this problem by

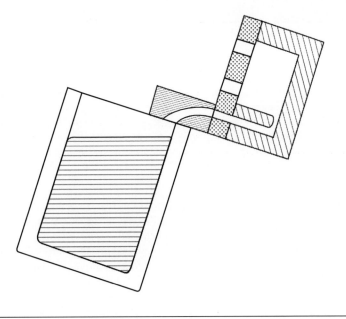

FIGURE 16.8

A tilt-casting process in which the ladle connects directly to a permanent mould with a sand top core containing the feeders (Grunenberg, Escherle and Sturm, 1999). The danger of some downhill flow inside the mould can be seen.

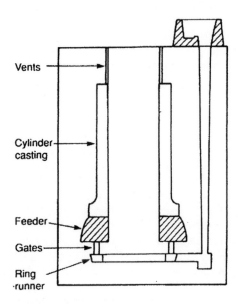

FIGURE 16.9

Cylinder casting-arranged as an inversion casting.

Adapted from Batty (1935), but core and mould assembly details omitted for clarity.

simply turning the whole mould over immediately after casting. The ring runner is then the feeder (and may need to be somewhat larger than normal to fulfil this additional role) and acts effectively because it is preheated, and because the temperature gradients in the casting are all favourably directional towards the feeder.

Let us review this worthy effort by Batty. We now know that the well at the base of the sprue would have introduced significant entrainment defects; thus, this should be eliminated. Personally, I would have eliminated all this fussy runner and ingate complication at the base of the casting, simply turning the sprue and entering the casting tangentially, thus spinning the metal up inside the casting. Finally, Batty's favourable temperature gradients for feeding after roll-over would only be obtained if especially slow pouring was used, thus threatening the formation of cold laps or unfilled castings. In this case, because of the relatively heavy wall section of this casting I would not have employed a roll-over. I would have allowed the casting to have inverted its temperature gradient by itself, simply by the natural action of convection. The feeder then should have been located at the top of the stationary (non-roll-over) casting.

Cosworth Process

After the counter-gravity filling of the mould, the Cosworth Process uses the inversion concept whilst keeping the liquid metal under pressure from the pump during the roll-over. In addition to the feeding of the casting being improved by the filling system becoming the feeding system, the mould, now filled with liquid metal, can be immediately disconnected from the pump and moved on to a cooling conveyer. The casting station is then released for the production of further castings (previously, the casting was fed during freezing by keeping it on the casting station for up to 5 min). The technique speeds up productivity by more than a factor of 5 (Smith, 1986). Green (2005) draws attention to the importance of maintaining the pressure on the melt during the roll-over because otherwise the metallostatic pressure because of depth alone (particularly when, during rotation, the mould is sideways and so generating a much reduced head pressure) may be insufficient to hold expanding gases in cores, therefore leading to the possibility of core blows. Processes that employ roll-over but which are not capable of maintaining the pressurisation of the liquid are therefore at risk.

Core Package System (CPS)

The so-called *core-package system* with its inappropriately widely encompassing name, is in fact a rather specialised form of precision sand core assembly process involving the filling of the sprue by contact pouring by offering up the mould assembly to the underside of a launder (Schneider, 2006). The stopper sealing the nozzle in the bottom of the launder is raised and the mould is filled. This works well. After the mould is filled, it is lowered to disengage with the launder, and a small arm automatically closes the sprue with a cap to prevent it emptying during the roll-over. The mould is then rolled over so that the reservoir under the casting is brought to the top to act as a feeder. However, during filling, the emptying of the high velocity metal into feeder volume creates major problems because the constraint over the liquid is lost, allowing the melt to jet and ricochet throughout the volume, entraining major defects (Figure 16.10). This otherwise good casting process contrasts of course with the true counter-gravity filling of the feeder volume which is gentle and controlled and has the potential to avoid the creation of any defects.

Figure 16.10 does illustrate the other important aspect of the roll-over processes: that chills can be used to good effect to enhance temperature gradients and so enhance feeding. As the figure illustrates, the use of chills dictates which way up the casting is filled; for cylinder heads, the chill is desired on the combustion face, and specifically between exhaust valve seats in four-valve per cylinder engines. For cylinder blocks, the fire face is relatively lightly loaded compared with the main bearings, whose fatigue performance greatly benefits from being chilled.

The roll-over principal, when used appropriately, either as a casting process or post-casting process, seems to be a powerful solution to some of the problems of filling and/or feeding castings. It strongly deserves to be more widely used.

16.3 COUNTER-GRAVITY

My toes curl when I hear the counter-gravity process called '*counter-gravity pouring*'. Pouring it is not. Furthermore, such inaccuracy undermines the thinking that is fundamental to the whole concept of the avoidance of pouring. The word pouring must be eliminated from the vocabulary of those rational souls employing the counter-gravity process. In its place, we need the appropriate and heart warming phrases 'counter-gravity filling' or 'counter-gravity casting'.

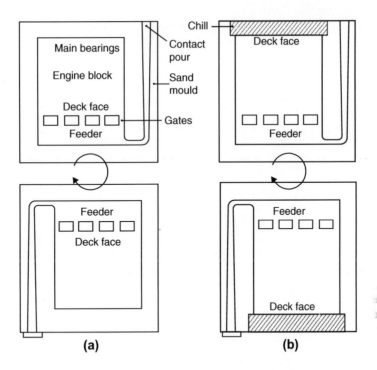

FIGURE 16.10

The so-called core package system (CPS) for engine blocks uses good filling via contact pour, and roll-over for feeding, but the high velocity flow into the volume of the feeder can cause major problems. The feeder can act via the ingates (a) alone; or (b) with a chill to aid feeding.

After Schneider (2006).

Over the past 100 years and more, the fundamental problems of gravity filling have prompted casting engineers to dream up and develop counter-gravity systems.

Numerous systems have arisen. The most common is *low-pressure casting*, in which air or an inert gas is used to pressurise an enclosed furnace, forcing the melt up a riser tube and into the casting (Figure 16.11). Other systems use a partial vacuum to draw up the metal. Yet others use various forms of pumps, including direct displacement by a piston, by gas pressure (pneumatic pumps), and by various types of electromagnetic (EM) action.

The fundamental action at the heart of the counter-gravity concept is that the liquid meniscus can be made to rise at a rate at which its oxide is pinned against the mould walls, the oxide becoming the skin of the casting (Figure 2.2(a)). The oxide on the surface of the advancing meniscus never becomes entrained into the bulk of the melt, creating defects in the casting. Only perfectly executed horizontal filling can match this extraordinary perfection of filling control. Counter-gravity is among the few (the very few) filling processes that can produce uniquely perfect castings.

Clearly, with a good counter-gravity system, it is possible to envisage the filling of the mould at velocities that never exceed the critical velocity, so that the air in the mould is pushed ahead of the metal, and no surface entrainment occurs. The filling can start gently through the ingates, speed up during the filling of the main part of the mould cavity, and finally slow down and stop as the mould is filled. The final deceleration is useful to avoid any final impact at the instant the mould is filled. If not controlled in this way, the transient pressure pulse resulting from the sudden loss of momentum of the melt can cause the liquid to penetrate any sand cores, and can open mould joints to produce flash, and generally impair surface finish.

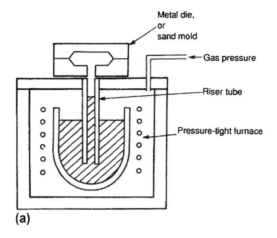

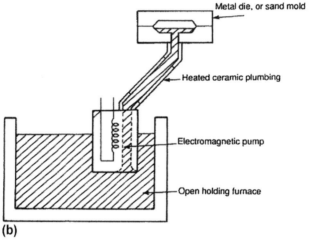

FIGURE 16.11

Counter-gravity castings by (a) conventional low-pressure casting machined using a sealed pressure vessel; (b) electromagnetic pump in an open furnace.

When using a good counter-gravity system, good filling conditions are not difficult to achieve. In fact, in comparison with gravity pouring, in which it is sometimes difficult to achieve a good casting, counter-gravity is such a robust technique that it is often difficult to make a bad casting. This fundamental difference between gravity and counter-gravity filling are not widely appreciated. In general, those who are familiar with gravity filling but have finally accepted a change to counter-gravity are suddenly amazed by the powerful benefits.

A counter-gravity-filled investment casting method that crams a 100 or more steel castings on one wax assembly makes millions of automotive rocker arm castings per year. These show a reject rate in the fractions of a part per million parts, illustrating the astonishing reliability that can be achieved by counter-gravity. Interestingly, however, this mode of casting was devised by Chandley (1976) not initially for its high reliability but because of its low costs. Further developments of this process by the Hitchener Company into aero engine turbine components among other markets have been impressive. They are reviewed by Shendye and Gilles in 2009. Griffiths (2007) confirms that counter-gravity filling of a low alloy steel produced castings with a higher Weibull modulus (higher reliability) than even well-designed gravity filling systems. A vertical stack of aggregate moulds filled with nodular or highly alloyed cast irons produced high volumes of castings with reliable 2.8 mm thickness walls (Purdom, 1992). Brasses are cast into permanent moulds to produce dense, leak-free domestic plumbing fittings required to take a high polish and faultless chromium plate (Lansdow, 1997).

That is not to say that the counter-gravity technique is not sometimes used badly. A common poor practice is a failure to keep the metal velocity under proper control. Entering the mould too quickly, even with a counter-gravity system, can make impressively bad castings. However, in principle, the technique *can* be controlled, in contrast to gravity pouring where, in principle, control is often difficult or impossible.

A concern often expressed about counter-gravity is that the adoption of filling speeds below the critical speed of approximately 0.5 ms^{-1} will slow the production rate. Such fears are groundless. For instance if the casting is 0.5 m tall (a tall casting) it can, in principle, be filled in 1 s. This would be a challenge!

In fact, the unfounded fear of the use of low velocities of the melt leading to a sacrifice of production rate follows from the confusion of (1) flow velocity (usually measured in ms^{-1}) and (2) melt volume flow rate (usually measured in m^{3}s^{-1}). For instance, the filling time can be kept short by retaining the slow filling velocity but increasing the volume flow rate simply by increasing the areas of the flow channels. Worked examples to emphasise and clarify this point are given in Chapter 13 dealing with the calculation of the filling system.

The other problems relate to the remainder of the melting and melt handling systems in the foundry, that are often poor, involving multiple pouring operations from melt furnaces to ladles and then into the counter-gravity holding vessel. A widespread re-charging technique for a low-pressure casting unit is illustrated in Figure 16.12; much of the entrainment damage suffered in such processes usually cannot be blamed on the counter-gravity system itself.

The lesson is that only limited success can be expected from a foundry that has added a counter-gravity system on to the end of a badly designed melting and melt handling system which generates defect-laden metal. There is no substitute for an integrated approach to the whole production system. Some of the very few systems to achieve this so far have been

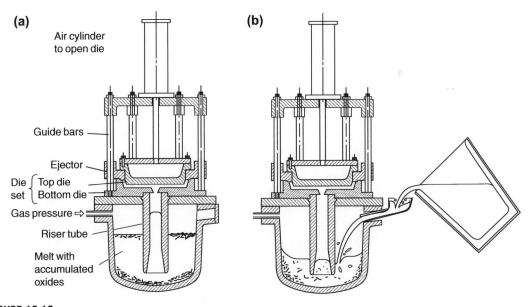

FIGURE 16.12

Low-pressure casting unit showing (a) sink and float of oxide; and (b) the conventional poor filling technique that re-entrains debris.

the processes that the author has assisted to develop; the Cosworth Process and, perhaps at some future date, in the Alotech ablation processes. In these processes, when properly implemented, the liquid metal is allowed to settle to eliminate a proportion of its defects in suspension, and subsequently is never poured, never flows downhill, and is finally transferred into the mould without surface turbulence. It is not difficult to arrange these highly beneficial features of a melt system.

Finally, the concept of an integrated approach necessarily involves dealing with convection during the solidification of the casting. The problem is highlighted by the author as casting rule 7. This serious problem is usually completely overlooked. It has been the death of many otherwise good counter-gravity systems, but is specifically addressed in the Cosworth Process by roll-over after casting.

16.3.1 LOW-PRESSURE CASTING

Low-pressure die casting

The low-pressure die casting (low-pressure permanent mould) process is widely used for the casting of automotive parts such as wheels and cylinder heads which require good integrity and, for wheels, good integrity and good cosmetic appearance when finely machined or polished.

Its name arises from its relatively low pressure (perhaps 0.3–0.5 bar) required to lift the metal into the mould (note that 1 bar will raise an Al melt approximately 4.5 m) followed sometimes by the application of a higher pressure stage in an effort to reduce any porosity in the casting. During this stage, the pressure might be raised to 1 bar. Even these highest pressures are typically only about 1% of the pressures used in high pressure die casting (HPDC).

The process enjoys several advantages, including ease of automation, so that despite its cycle time being longer than that of gravity permanent mould systems, one man can usually control several machines. In addition, in common with most counter-gravity systems, it has metallic yields in the range 80–95%, compared with 50–75% for gravity-poured systems that require oversized feeders as a result of the damage that their filling systems introduce. In common with gravity pouring, sand cores are not a problem provided the applied pressure is under good control (otherwise, with overpressure being sometimes applied to prevent bifilm opening to create microporosity, sand cores can become impregnated with metal. Clearly, if the melt is of good quality, overpressure to suppress porosity is not required).

Low-pressure die casting dies can be made from cast iron or steel, depending on the service requirements, and enjoy the protection of a ceramic die coating applied by spray.

In most low-pressure permanent mould casting the die is split horizontally, with the casting retained in the upper half (ejectors cannot usually be placed in the lower half because of the presence of the furnace under the die). As the upper half is raised after the freezing of the casting, the ejectors impinge against the stripping plate to eject the casting. In this case, the casting is usually caught on a tray swung under the upper die as it is raised. The tray and casting are then swung clear, presenting the casting outside the machine for onward processing.

The dies are often water-cooled by internally drilled water channels. Cooled dies are expensive and complicated, but hold the record for productivity of certain types of casting such as automotive wheels. An uncooled die for a wheel might have a cycle time of 5 min, whereas a cooled die can be less than 2 min. For larger castings, such as some cylinder blocks, a cycle time of 15 min may be usual.

One of the most notable high-temperature developments of low pressure die casting is the famous Griffin Process for steel railway rolling stock (Figure 16.13), described by Hursen (1955). It makes an interesting comparison to the use of the process for Al alloys. The insulated ladle in the pressure vessel can hold up to 7 tonnes of liquid carbon steel. The large density difference between the steel and buoyant defects such as bubbles, bubble trails and other entrained oxides encourages such materials to float out relatively quickly, so that the topping up of the furnace does not necessarily introduce permanent damage; after the metal has been poured into the pressurised vessel, the entrained defects quickly float away from the bottom of the melt where the intake to the riser tube is located. In addition, little or no sediment is to be expected. Thus a good quality of steel has had chance to develop at this location before the filling of the next casting. In this way, a

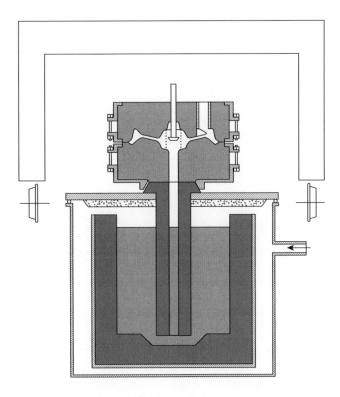

FIGURE 16.13

The counter-gravity casting of steel railroad wheels by the Griffin Process. The mould transfer car (transporting frame shown in outline) carries the mould along the production track, and houses all the gear for lowering, raising and clamping down of the mould and operating the stopper.

After Hursen (1955).

high-integrity, safety-critical product can be routinely produced. Even so, one can imagine that the deoxidation practice, leaving different amounts of Si, Mn and Al, plus others such as Ca, could influence the flotation time significantly.

In this case, the dies are machined from graphite. Although a successful process has been built on this process, the machining of graphite is always a somewhat messy and dirty occupation. There seems little doubt that the process might have been operated with greensand with practically no loss of accuracy and no loss of performance of the steel casting. It might be claimed that greensand may be only slightly less messy and inconvenient, although it would be interesting to see a comparative evaluation of the two types of mould for this high volume product.

In practice, the process is impressively automated and productive. Moulds arrive automatically on a transfer car rolling on a track, and are indexed automatically into place on the casting unit. The filling of the mould with steel at 1620°C is registered by observing the top of one of the three or four sand-lined feeders. When the metal arrives at the feeder tops, the filling is stopped and the timing of other operations is started (the somewhat variable filling time arises because of the falling level of the melt in the holding furnace). After filling, no waiting is required for solidification because the graphite stopper is lowered to seal off the mould (Figure 16.13) and the mould is automatically indexed to the cooling line. The ladle can cast about 20 wheels before being returned to the melting furnace where its remaining metal is returned and the holder can be topped up. The centre of the casting containing the stopper (shown as dotted lines in Figure 16.13) is removed by flame cutting. Only this central bore is machined. The remainder of the casting, including the tread of the wheel that runs directly on the rail, remains cast to size. The cast wheels at least equal if not exceed the reliability of forged and rolled steel wheels. I find all this a deeply impressive achievement.

Unfortunately, in spite of the excellent potential of the low pressure system, the process has some fundamental problems that can produce significant amounts of scrap. The problems arise from several sources.

It is necessary to take stock of the many problems associated with low-pressure casting:

1. The re-filling of the pressurised furnace with metal (Figure 16.12(b)) is usually a major source of entrainment defects that causes all the usual problems of fine scattered porosity often suffered by the process.

   There is one variety of low-pressure casting unit that goes some way towards solving the problem of the massive damage introduced by the conventional topping up operation. It has a separate crucible and furnace assembly that can be detached from the casting machine. This is usually achieved by the pressurised furnace assembly containing the crucible being indexed out from under the machine. Up to three such assemblies are required per casting machine so that one can be melting, another treating the melt by, for instance, rotary degassing and passive holding and the third is in place in the machine making castings. This system is clearly a significant improvement on re-filling of the pressurised vessel by gravity pouring through a port in its side.

2. In the past, many low-pressure die casting machines have been so poorly controlled on flow rate, that the speed of entry into the die could greatly exceed the critical velocity, thus negating one of the most important potential benefits of the low-pressure system. Oxide bifilms and possibly bubbles are generated inside the mould, having little or no time to separate from the casting. Fortunately, most modern machines have significantly improved control so this is no longer a problem.

   I recall an instance in the Cosworth foundry when the melt was pumped too quickly into the mould by mistake. The result was disaster, with the X-ray radiograph displaying hot tears, gas bubbles, apparent shrinkage porosity and sand inclusions. This contrasted with normally defect-free castings when the ingate velocity was correctly controlled below the critical velocity of 0.5 m/s.

3. The melt is usually allowed to fall down the riser tube after the solidification of each casting. This fall leaves a skin of oxide on the inside of the riser tube, which can detach and spoil the subsequent casting, although a filter of some kind at the mouth of the mould is usually in place to help to reduce this problem. Pechiney, France, was probably the first in an attempt to address this issue by providing an inert gas environment for the top of the riser tube (Charbonnier, 1983). However, the problem is best solved by keeping the riser tube filled to within a few millimetres of the top, with the melt ready to enter the next casting. Unfortunately, there is significant resistance within the industry to keep pressurised furnaces pressurised to achieve this mode of operation, said to be a concern of safety. However, processes using EM pumps achieve this maintenance of the level without difficulty and thereby avoid unnecessary oxide generation.

4. The second serious feature of the sudden release of pressure is fall of the melt in the rise tube creating the consequential 'whoosh' of melt issuing from its base, efficiently stirring up all those oxides that have settled to the floor of the furnace since the last casting. Thus defects in suspension in the melt are never allowed to settle quiescently on the floor of the furnace; inclusions are re-stirred into suspension between every casting. A controlled, gradual release of pressure would easily solve this problem.

5. The third serious problem arising from the sudden release of pressure is little known. It is further compounded by the application of a higher pressure than is needed to fill the mould in an attempt to induce higher soundness. These actions can be highly counter-productive if the melt is contained in a standard refractory lining in the pressurised furnace. This is because the lining becomes impregnated with the pressurising gas; if the pressure is very high, the volume of impregnated gases is substantial. On the release of pressure after solidification of the casting, the refractory lining releases its gases which bubble up through the melt, creating bubble trails. If air is used for pressurisation the melt is filled with bubble trails of oxide. Nitrogen pressurisation results in nitrides and is known to bring the melt into an uncastable condition after only a few pressurisations. Even argon pressurisation will create some problems as a result of the contamination of oxygen and nitrogen residues in the lining.

   In explanation, although the pressurising gas cannot penetrate the lining covered by the liquid metal, it penetrates the lining from its back, and has several minutes to saturate the lining throughout its thickness at the high pressure.

On rapid release of the pressure, the gas naturally escapes rapidly in whatever way it can, therefore from both front and back, creating bubbles from the front.

The use of minimal pressures, merely for filling rather than enhancing soundness, reduces this serious problem. The use of hydrostatically pressed crucible for containment of the melt rather than a rammed refractory lining is a further effective solution.

6. New riser tubes offering longer life might now create conditions in which the sediment in the furnace now has time to build up to levels for which entrainment of the sediment in the casting becomes unavoidable.

7. Convection of melt up the riser tube and into the mould, delaying the freezing of the casting, can be a serious contributor to poor productivity and poor solidification structure in the casting. I recall an operator who related to me that he timed the freezing of his cylinder head casting, then degassed the melt with an inert gas. On re-checking the freezing time, it was found to have increased. He concluded the reduced hydrogen affected the freezing behaviour. He degassed again and the freezing increased again. This happened three or four times (I forget exactly how many). Clearly, the variations in hydrogen content could not explain these results. I concluded that his original melt convected very little because of its high viscosity resulting from its content of oxide bifilms in suspension. This slurry would move only sluggishly in the rather narrow riser tube. After 'degassing', which would have removed a significant proportion of oxides, the higher fluidity of the melt, as a 'thinned' slurry, would have led to increased thermal convection in the riser tube, conveying heat from the melt in the furnace, up the riser tube and into the casting, delaying its solidification. The progressive cleaning of the melt and progressively worsening productivity illustrated the influence exerted by two powerful factors normally overlooked in casting operations: oxide bifilms and thermal convection.

Riser tubes that deliver the melt directly into the mould cavity (rather than turning some right-angle corners in a distribution system) are particularly at risk from the convection problem. The use of a filter at the top of the riser tube is a help, but usually does not completely cure the problem. Muller and Feikus (1996) in an excellent article describe the use of a new design of spreader pin sited at the top exit of the riser tube. This device helped to remove some heat from the casting at the top of the riser tube, assisting to control the convective flow of hot metal into the casting at this point, and thus increasing casting quality and productivity.

As a result of all these problems, many of which are not addressed as suggested previously, the performance of many low-pressure casting operations is often disappointing. For serious operators, additional suggestions by Muller and Feikus (1996) are recommended reading.

Medium-Pressure Die-Casting Process

A potentially interesting variant of the low-pressure process has been described as the medium-pressure casting process by Eigenfeld (1988). Essentially the mould is filled exactly as the low-pressure process, but after the mould is filled and the casting partly solidified, the casting is pressurised by one or more small pistons, well known in HPDC as squeeze pins, that collapses the casting surface locally. The internal pressurisation of the melt in this way will reduce the 'air gap' around the casting and thus increase the freezing rate. The process is claimed to give better properties than conventional LP die casting for this reason.

Counter-pressure Casting process

A complication of the usual low-pressure die casting process has been introduced as the *counter-pressure casting process*, in which a pressurised chamber is lowered over the die to apply pressure counter to the pressure in the furnace chamber, so that only the differential pressure raises the melt up the riser tube (Balewsky and Dimov, 1965). The counter-pressure, usually between 3 and 10 bar, acts on the liquid during mould filling and solidification to help to suppress the formation of porosity. The lowering of the counter-pressure chamber and its pressurisation with gas would be expected to slow the production cycle, but Wurker and Zeuner (2004, 2006) claim otherwise. The development is clearly devised as a logical approach to reduce the problems arising from current poor melt practice. However, if the cast metal had a reasonable quality, free from oxide bifilms, no porosity could be generated, so that no counter pressure would be needed.

Vacuum riserless casting and/or pressure riserless casting

Vacuum riserless casting and/or pressure riserless casting is a process using permanent water cooled metal moulds filled by vacuum and/or pressure, so that a riser (i.e. feeder) is not required (Spada, 2004).

This author has to admit to being biased; he cannot get to like this process. It seems to him to be unnecessarily and unbelievably complicated. The permanent mould, possibly sprayed with release agent between every cast, is filled partly or entirely by the application of a vacuum to the mould, drawing metal up from the furnace sited below the mould. Any leakage of the riser tube(s) will introduce air bubbles into the casting. The melt may rise up more than one riser tube, introducing the danger of convection between the tubes (melt rising up one and descending down another, the heat thus transferred from the furnace delaying the solidification of the casting, and impairing structure), although, fortunately, the strong water cooling of the metal mould will reduce this danger. Having filled the mould, additional pressure can now be applied from below to help to suppress pore formation and increase cooling rate by pressurising the casting against the mould. When releasing the pressure, the residual melt in the riser tubes falls back into the furnace, potentially increasing oxide, but certainly re-stirring settled oxides, maintaining them in suspension. Fortunately, the process can be used with pre-treated melts with exchangeable furnaces, or better still, with furnaces kept full from a heated launder distribution system (Spada, 2004, 2005). It is hoped the units use hydrostatically pressed crucibles to reduce the otherwise major impairment from bubble damage to the melt.

The control problems are daunting, involving vacuum cycles, pressure cycles, and complex water cooling, in addition to the normal cycles involving opening and closing the die etc. The capital and maintenance costs must also be daunting. The castings can be excellent of course, but the workforce has to be unusually diligent and well above average capability to succeed with this system. They have my best wishes.

T-Mag Process

A recent interpretation of the tilt casting process for Mg alloy castings named *T-Mag* (presumably T for tilt and Mag for magnesium) has been demonstrated by CSIRO in Australia (Nguyen et al., 2006) using a permanent mould connected to the melting furnace via a heated transfer tube (Figure 16.14). When tilting the assembly, the melt enters the die via a bottom gate, thus having the potential to fill the cavity smoothly. It is therefore better described as a counter-gravity process. The casting solidifies under the hydrostatic pressure because of the height difference ΔH. After solidification, the assembly is rotated back to its vertical position, lowering the melt gently in the riser tube (in contrast to most low-pressure units), the die is opened, and the casting ejected. The casting has the short 'carrot' ingate typical of low-pressure castings, and thus enjoys a good metallic yield in addition to its reasonably high quality. At this time, the direct addition of ingots or foundry returns to the melt introduces the oxide skins of these charge materials, which are almost certainly troublesome with respect to the attainment of the highest and most reproducible properties (Wang, Lett et al., 2011). Thus although at this time performing better than most Mg casting processes, it seems the process is in its early days of development and it is to be hoped that the process will receive further development to achieve its full potential.

16.3.2 LIQUID METAL PUMPS

A pump can be an excellent way to transfer metal into a mould. There are many varieties of pump, but as with all pumps, they are characterised by their particular characteristic curves. These are graphs of the relation between head height and volume flow rate as shown in Figure 16.15. The characteristic curves require to be treated with some caution; they show the *potential* of the pump. When programming the pump to fill some odd-shaped volume, such as a casting, the setting of the pump power to achieve a certain metal height at a certain time will not, in general, deliver metal at that height at that time. This is simply because it takes *time* for the melt to be accelerated by the pump and *time* for the mould cavity to fill the part of the casting volume up to that level. This wide misunderstanding of the necessary lag between the programmed and achieved conditions is easily dealt with by the computer, which can apply Newton's laws of motion to the metal without difficulty or complaint, and get the programming exactly right first time. Alternatively, the problem is dealt with, within the limits of the power of the pump, by feed-back control as described in Section 16.3.5.

We shall consider the main pump options for use in foundries next.

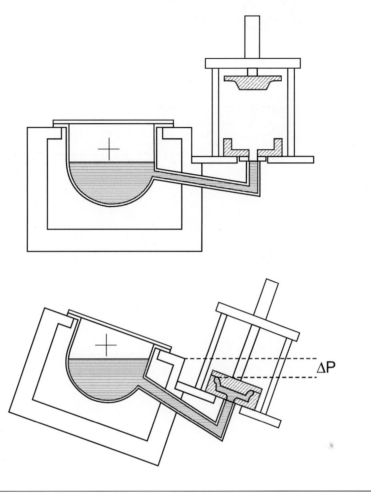

FIGURE 16.14

The 'counter-gravity' casting of Mg alloys using the tilting furnace of the T-Mag technique. The figure shows a permanent mould, but would be expected to work perfectly well with a sand mould, particularly if convection were controlled. Solidification occurs under pressure ΔP. The unit reverts to its upright mode to open the die and eject the casting.

Centrifugal pumps

A brief history of centrifugal pumps is given by Sweeney (1964). He describes the first pumps for molten aluminium in Cleveland, Ohio, in 1945.

Centrifugal pumps usually have carbon based rotors, and can have high volumetric capacity, although the characteristic curves shown in Figure 16.15(a) show the output of a rather small pump with a similar characteristic curves to the EM pump PG450 used for making castings (Figure 16.15(c)). Much larger centrifugal pumps are available for transferring tonnage quantities of melts between furnaces because they can transfer Al at up to 100 kg/s and up to 7 m high. Centrifugal pumps are currently being evaluated for use for filling moulds in Cosworth type operations.

Electromagnetic pumps

The great benefit of EM pumps is that they contain no moving parts and their life in a melt can usually be measured in months. Their relatively high initial costs are usually offset by good reliability and relatively modest maintenance and

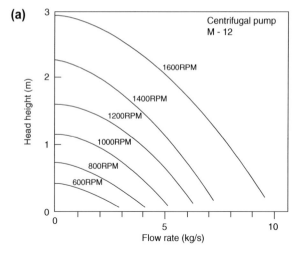

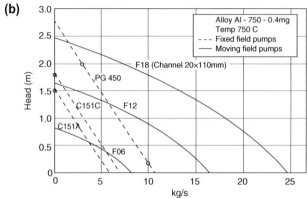

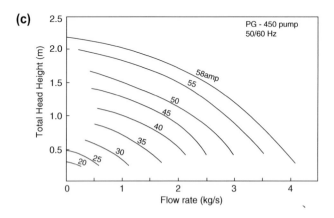

FIGURE 16.15

Characteristic curves for the performance of (a) centrifugal pump; (b) two different varieties of electromagnetic pump at full power; (c) an electromagnetic pump at various power levels.

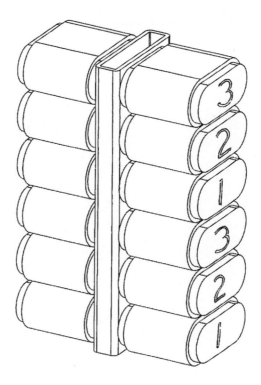

FIGURE 16.16

The moving field electromagnetic pump with simple ceramic tube delivery system.

running costs. Their use in the relatively non-aggressive environments for low melting point metals tin and lead is easily understood. Use for liquid Mg has been more problematical because only ferritic stainless type steels can withstand immersion in Mg and the parts out of the melt have to withstand oxidation at these higher temperatures for long periods. The ferromagnetism of the alloys does hamper the transfer of magnetic fields to act as driver to the melt. Thus there are plenty of fundamental problems for Mg.

There are practical engineering solutions for liquid aluminium alloys making the use of EM pumps for Al now relatively well established. There are two main varieties of EM pumps; both rely on the principles of induction.

a. The three-phase moving field pump. This simple design has at its heart a straight ceramic tube containing the melt. The tube is surrounded by a series of electromagnets, each magnetic coil connected in turn to a single phase of the three-phase electricity supply, with the phases connected to the magnets in the order 1, 2, 3, 1, 2, 3 etc. along the length of the tube. As the phases cycle, so the magnetic field progresses along the length of the tube dragging the liquid metal with it (Figure 16.16).

In common with all pumps, the performance of EM pumps is described quantitatively by a characteristic curve giving its rate of delivery at various heights. Thus its maximum rate is delivered at zero height, but its rate of delivery falls to zero at its maximum delivery height. At this height, the melt is working hard to just support the column of liquid metal in a kind of stalled mode (Figure 16.15(c)). An increase in electrical current to the coils will increase performance, moving the characteristic curve out to higher rates and greater heights. Thus a family of curves can normally be generated for a series of different currents. This particular set of curves was generated by placing pumps in series. The curves F16, F12 and F18 correspond to one, two or three pump modules assembled together, end to end. Clearly, the volume capacity of the moving field pumps is vastly greater than is required for casting

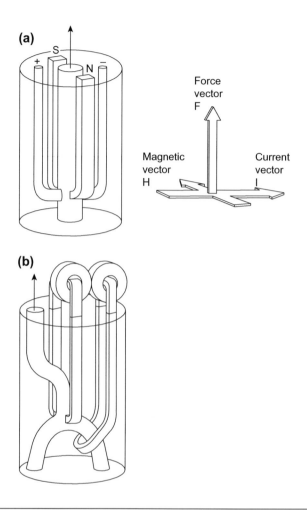

FIGURE 16.17

The fixed field pump built from castable refractory blocks designed for part-way immersion in the melt for periods of months: (a) the principal of employing three vectors at right angles; (b) a simplified model of a real pump.

production, being mainly intended for transfer of melts between furnaces. However, the pumps are easily downgraded by the use of a smaller cross-section central tube. The overheating that might occur during their use in a 'stalled' condition, retaining power input to continue holding metal in a mould that is already full, but taking time to solidify, is relatively easily overcome by additional cooling (otherwise the melt, now stationary, in the moving field becomes highly overheated and might melt parts of the pump).

b. The AC (alternating current) single-phase stationary field pump (Figure 16.17). This pump is a clever but complex design, involving more sophisticated and costly maintenance. It relies on the inductive principle enshrined in Faraday's left hand motor rule. (The mutual relation between the three vectors at right angles is given by the thumb and principal two fingers on the left hand held out at right angles, denoting 'first finger = field; second finger = current; thumb = motion.) Thus, the interaction of the magnetic field vector at right angles to the electrical current vector produces a force vector in the liquid at right angles to both (Figure 16.17).

To understand the AC design, it is perhaps helpful to first consider what an equivalent direct current (DC) design might look like. In Figure 16.17(a), the DC pump has a permanent magnet to give a field vector across the working volume of the liquid metal. Electrodes placed either side are connected to a continuous direct current to give the current vector at right angles to the magnetic field. The liquid in the working volume now experiences a body force, moving it upwards. In principle, as the current is increased the flow would similarly respond. Unfortunately, such a simple incarnation of the EM pump would work excellently for about half an hour if we were lucky, after which the electrodes would be destroyed, having dissolved away in the liquid Al.

The AC pump design avoids any contact by a clever design. It similarly uses an EM field across the working volume, conducted down to this level via soft magnetic armatures, and at the same time induces a current in a loop of liquid metal (once again conducting a field down to the loop via soft magnetic links) that connects outside of the pump via the bulk melt, this providing the electrical current at right angles to the magnetic field. The current in the liquid metal loop is effectively a single turn transformer, linked magnetically to the multi-turn coil above. The motion of the melt is at right angles to both current and field vectors and is taken up a vertical channel, taking its supply from the arms of the loop which connect to the melt. Because the magnetic and current vectors are in phase, they both reverse at the same time, fortunately (as you can demonstrate to yourself with the left hand rule) maintaining the force in the same direction. (Physicists will know that multiplying two negative vectors gives a positive vector.) The pulsating force is smoothed by the mass of metal that is moved, although the 60 Hz electrical supply in the United States gives noticeably better performance than the 50 Hz available in the United Kingdom. Typical characteristic curves for three of these pumps working at maximum output are also shown in Figure 16.15(b). The operation of one of the pumps at variable current is shown in Figure 16.15(c). These pumps are ideally suited in the pressure and flow rate range for making automotive and aerospace castings, and have therefore been in use for variants of the Cosworth Process since about 1980.

Given reasonably clean metal, the pumps continue to work in the melt for at least 6 months (one pump the author used lasted 15 months!). However, attempts to introduce a chlorine gas mixture in the rotary degassing stage before the pump caused pumps to block with a mixture of liquid chlorides and solid oxides within a few days (Sokolwski et al., 2003).

Cosworth Process

In 1978, the Cosworth Process was the first foundry to be set up to deliver good-quality, bifilm-free metal into a mould. As a result, it was the first to use an EM pump for a production plant. It was funded by Cosworth Engineering to provide the cylinder heads and blocks for their highly successful Formula 1 racing engines, although later diversified its achievements to aerospace castings. Initially the process used zircon for its precision sand core assembly process, although silica was later successfully by licensees. The first form of the process shown in Figure 16.18(a) ran into severe difficulties because of convection problems that prevented the freezing of the castings and nearly brought the company to a stop. The later variant of the process shown in Figure 16.18(b) used a roll-over technique after filling the mould. This cured the convection problems and simultaneously increased production from one casting in a mould every 5 min to two castings per mould every 45 s. The roll-over was the breakthrough that projected the process into a world leading position for making V-shaped cylinder blocks. The process continues to be used in the United Kingdom and in Windsor, Ontario, Canada, originally by Ford, now taken over by Nemak. GM and other major producers have also adopted the process. The process is also capable of making excellent cylinder head castings and aerospace products.

A recent history of the process by me (Quality Castings, 2015) revealed the rather shocking realisation that no one has yet set up the process correctly so as to achieve its full potential. In general, the metal melting and preparation has been poor, contributing to less than optimum properties of the castings. To ensure freedom from bifilms the melting should be carried out in a dry hearth tower furnace, all pouring transfers eliminated, and cleanness enhanced by either good filters or sedimentation in a properly designed holding furnace. The several attempts to set up the process without a proper understanding of these simple principles have led to failure.

The Cosworth Process was the inspiration for many subsequent plants. Rover (UK) used the process in two plants, one using the moving field pump and the other using the stationary field pump. They named this the *low-pressure sand (LPS)* system. The unfortunate lack of roll-over for these plants meant that casting soundness was never under proper control, with the result that 100% impregnation of castings was necessary. Even so, the process was successful to produce the challenging castings required for the K-series engine, a lightweight thin-walled design that was years ahead of its time, and for which no other casting process was at that time capable.

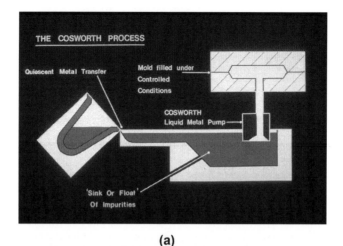

(a)

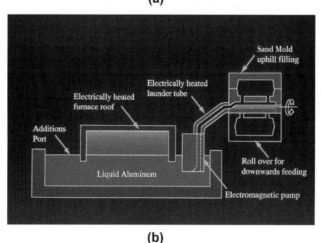

(b)

FIGURE 16.18

(a) Cosworth Mk I Process which was almost unworkable as a result of convection problems; (b) Cosworth Mk II Process solving convection and productivity by roll-over after mould filling.

Pneumatic Pumps

At the start of the Cosworth development, a simple pneumatic pump (Figure 16.19(a)) was used as a backup in case the newly arrived EM pump did not perform. In the event, the EM pump worked perfectly and the pneumatic pump was neglected. In retrospect this neglect was probably an error. The pneumatic pump made castings of equivalent excellence. Later, an even simpler and higher capacity pump avoiding the complication and volume reduction of the inner vessel was developed (Figure 16.19(b)). Note that the double stopper allows the melt to be retained at the top of the riser tube between castings; this is essential for maintaining the riser tube free from oxide. Another major advantage with the pneumatic concept is the absence of turbulence, in contrast with both centrifugal and EM pumps. An inert gas is required to pressurise the pump to avoid overmuch oxidation of the interior. However, at intervals, as necessary, when the refractories have reached the end of their life, these relatively low cost components immersed in the liquid metal can be discarded and replaced. The relatively complex top plate with its mechanical actions, bellows and sliding seals is a

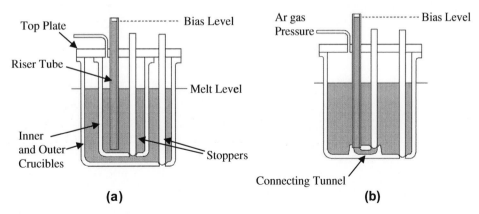

FIGURE 16.19

Pneumatic pumps by the author: (a) pump I: the early pump for low melting point metals including aluminium. (b) Pump II is simpler and less costly development, delivering a much increased volume. The great advantage of pneumatic pumps is the significant freedom from high speed turbulence.

permanent feature. Modest leakages from the pump are to be expected but are relatively unimportant because the gas flow is controlled by pressure rather than flow rate. The extreme simplicity and low cost of the pneumatic pump should help to promote a wide uptake.

16.3.3 DIRECT VERTICAL INJECTION

Direct-acting piston displacement pump

Direct-acting piston pumps for lead and zinc appear to be successful. However, their attempted use for liquid Al alloys has so far not resulted in success. Although such wear-resistant and Al-resistant materials as SiC can be finely and accurately ground to make cylinders and pistons, the damage caused by particles of alumina ensures that the pumps quickly wear and fail. Sweeney (1964) describes how attempts to produce cylinder and piston pumps were rapidly abandoned because of oxide problems. Valves such as the ball check type similarly proved unreliable because of oxides. Such failure is hardly surprising. Films of alumina of thickness measured in nanometres will probably always be present in alloys of liquid Al, and will easily find their way into the most accurately fitting parts, leading inevitably to scoring and wear.

Lost Crucible Technique

A rather different type of direct-acting displacement approach started development under the name '*Rimlock*' (rapid induction melting lost crucible process) as described by Bird and Savage (1996). Commercial changes caused the same project to be continued under a new name, '*Crimson*' (Jolly, 2008). This curious acronym represents 'constrained rapid induction melted single shot net shape casting process'. Despite its name, the process appears to have good potential, representing a significant departure from conventional foundry thinking. In its original form it uses a one-shot ceramic fibre crucible, into which is placed a pre-cast cylindrical slug of metal as a charge (Figure 16.20). The charge material is a sawn off log from continuously cast billet, and thus is of reasonable quality. The induction melting is then carried out at high speed, taking perhaps 60 s, the idea being to subject the melt to minimum time for hydrogen pick-up, or oxidation of alloying elements. When up to casting temperature a piston pushes up the base of the fibre crucible, pushing out the base, and continues to rise, pushing the melt upwards and into the mould cavity sited on the top of

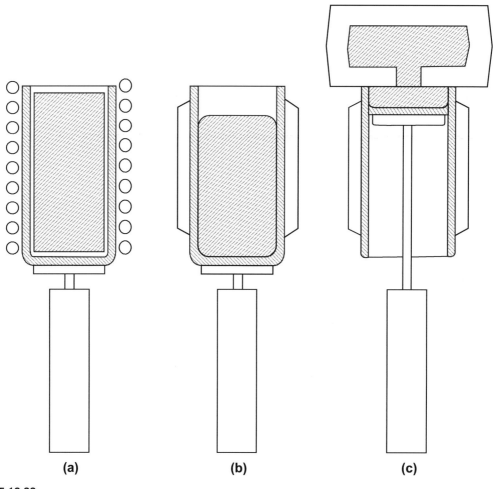

(a)   (b)   (c)

FIGURE 16.20

Counter-gravity with a sacrificial fibre crucible showing (a) rapid melting; (b) clamping to support crucible; (c) casting by pushing up the base of the crucible. After the mould is filled the whole assembly can be rotated through 180° to use the residual melt in the crucible as a feeder.

the crucible. When the mould is full the whole crucible, piston and mould assembly can be rotated through 180°. The temperature gradient in the mould is now favourable for feeding, which can be maintained under pressure from the piston if necessary. More recently the process has been developed for use with a re-useable refractory crucible and moveable base.

This is a process enjoying numerous advantages that clearly demands greater attention. One such advantage is the very directness of injection, so that feedback control from sensing the melt level in the mould is unnecessary. The piston position gives it exactly. Thus a piston actuator driven electrically (by rack and pinion or by screw etc. rather than indirectly by air or hydraulics) can be linked directly to computer control. A modern development is targeting the use of re-useable ceramic crucibles with a sliding base.

The technique seems ideal for light metals, particularly magnesium alloys, because the inventory of liquid metal in the foundry is minimised, being a just-in-time melt preparation procedure. This would be an important and welcome safety feature in Mg casting facilities.

16.3.4 PROGRAMMABLE CONTROL

The varying cross-sectional areas of the metal as it rises in the mould pose a problem if the fill rate through the bottom gate is fixed (as is approximately true for many counter-gravity filling systems that lack any sophistication of programmable control). Naturally, the melt may become too slow if the area of the mould increases greatly, leading to a danger of cold laps or oxide laps. Alternatively, if the local velocity is increased above the critical velocity through a narrow part of the mould, the metal may jet, causing entrainment defects.

Counter-gravity filling is unique in having the potential to address this difficulty. In principle, the melt can be speeded up or slowed down as required at each stage of filling. Even so, such programming of the fill rate is not easily achieved. In most moulds there is no way to determine where the melt level is at any time during filling. Thus, if the pre-programmed filling sequence (called here the filling profile) gets out of step, its phases occurring either early or late, the filling can become worse than that offered by a constant rate system. The mis-timing problems can easily arise from splashes that happen to start timers early, or from blockage in the pump or melt delivery system causing the time of arrival of the melt to be late.

16.3.5 FEEDBACK CONTROL

The only sure way to avoid the difficulties of the driving signal and actual metal level getting out of synchronisation is to provide feedback control from knowledge of the height of the metal. This involves a system to monitor the height of metal in the mould, feeding this signal back to the delivery system, and forcing the system to adhere to a pre-programmed fill pattern. Good feedback control solves many of the filling problems associated with casting production.

Although Lin and Shih (2003) describe a closed loop control system for a low-pressure casting machine, they monitor the pressure response in the low-pressure furnace chamber, using this as feedback to control the level of their pressurisation profile. This is a useful approach for monitoring and compensating for the changes of level in the holding furnace, which is otherwise difficult to monitor accurately; clearly, as castings are produced and the level falls, increasing pressure will be required to fill the castings. Although improvements to their castings were reported, the *rate* of rise of the melt is still unknown and uncontrolled; it will be variable with (1) different castings, (2) different dies with (3) different amounts of leakage of pressurised air from the furnace leading to different back pressure during filling, and (4) possible blockages to the flow path, especially if a filter is used in the riser tube.

A related system for the monitoring of height is the sensing of the pressure of the melt in the melt delivery system. This has been attempted by the provision of a pocket of a few cubic centimetres of inert gas above the melt contained in the permanent plumbing of the liquid metal delivery system close to the pump, and connected to a pressure transducer via a capillary. The system appears to be no longer used, as a result of practical difficulties associated with the blockage of the capillary.

A non-contact system used by the Cosworth sand casting process senses the change in electrical capacitance between the melt and the clamp plate holding down the mould when the two are connected as a parallel plate condenser. This system has been used successfully for many years. However, it is not necessarily recommended because capacitance is powerfully affected by moisture. The Cosworth system with relatively dry moulds countered this problem by zeroing the signal at the instant the melt entered the mould. This was set as the start point for the remainder of the fill profile.

Probably, the use of inductance to monitor the rising of the melt, using an inductive loop above the mould, might have given a more reliable signal because inductance is not affected by moisture. To the author's knowledge, this has never been tested.

The author has demonstrated that archaeological ground-penetrating radar will penetrate aggregate moulds and will deliver a signal indicating the progress of the metal as it rises in the mould. However, a higher frequency system might be beneficial to increase accuracy; an accuracy of defining the level of the melt to at least ± 10 mm is really required.

However, it is a pity that feedback control is little used at this time. The lack of proper control in counter-gravity leads to unsatisfactory modes of filling that explains many of the problems with this otherwise excellent technique.

16.3.6 FAILURE MODES OF LOW-PRESSURE CASTING

Most counter-gravity systems are quite safe, particularly the direct piston action filling, and filling by EM pumps.

I am often asked the question about the danger of leakages from the mould when using an EM pump. I always recall the experience of walking past the casting machine and seeing a gentle flow of liquid aluminium emerging from the base of the mould and spreading over the casting table. With some concern, I pointed this out to Trevor, our casting man, who was sitting reading the newspaper at the time. He carefully folded his newspaper, got out from his chair, walked to a distant stack of ingots, brought one back and placed it against the mould to freeze the stream. He then sat down and continued reading his newspaper.

The purpose of this story is to show how relaxed and safe counter-gravity casting can be. The pressures in the mould are only those that would have been experienced with conventional gravity casting – otherwise, in a sand mould penetration of the mould by metal would occur.

Even in permanent mould low-pressure casting with pressure intensification, pressures are only approximately 1 bar. Pressures in compressed air systems used extensively around the foundry are usually 7 times this value.

Even the most significant danger with a low-pressure system for casting Al or Mg is rather tame; it occurs if the melt level in the furnace becomes too low. A bubble escaping up the riser tube expands as it rises, and can eject metal from the top. However, any danger is avoided because the furnace is only pressurised when the mould is in place, thus the ejected metal, plus air, is easily contained by the strong metal mould. The only item that suffers from this event is the casting.

With dense alloys based on Cu or Fe the situation can be different. As a result of the greater density of these melts, the pressures required are nearly three times those required for light alloys. This is still not a problem for those processes that use a relatively hard pneumatic system to drive the casting process. Such processes are akin to that used by the Griffin Process for the low-pressure casting of steel railroad wheels. In this process, the metal is contained in a ladle which is a relatively close fit in its pressurised container. Thus the process is under relatively good control because relatively little air is required to change the pressure in the container.

It is a mistake to place the steel melting furnace in a large pit, pressurising the pit to force the melt up the riser tube. In this situation at least 10 times the volume of air is required to pressurise the furnace. This introduces significant delay in the response of the casting process, making the control 'spongy' and uncertain. Furthermore, this uncertainty in the control of pressure is worsened because cold air introduced to deliver a certain amount of pressure is subjected to significant heating and further uncontrolled expansion in the vicinity of the furnace and the liquid metal. The heating of the gas might increase its pressure by a further factor of two or more. Thus the energy contained in the pressurised gas will be between 30 and 60 times that required for a light alloy casting. If the furnace level now becomes too low, so that air escapes up the riser tube, the expansion of this highly compressed air will eject metal and destroy a sand mould, taking the foundry roof with it. At least one such event has happened in a UK development facility.

This danger is avoided for steel castings by reducing the volume of pressurised air. The building of special induction furnaces with pressurised steel shells is expected to be extremely safe. Furthermore, the control of the melt is improved, being a hard pneumatic system giving more immediate casting process response. Such pressurised furnaces would revolutionise steel foundries. No ladles would be needed, and few would ever see any liquid metal. They might read their newspapers while the casting was made.

16.4 CENTRIFUGAL CASTING

Centrifuge Casting

Casting on a rotating centrifuge table is commonly used for Ti alloys and Nb alloys. The melting furnace and rotating table are enclosed in a vacuum chamber. Moulds to make shaped castings are placed around the periphery of the centrifuge and are connected by a spider of radial runners to a central sprue. This is perhaps usefully called centrifuge casting.

Centrifuge casting is used for the casting of high melting point metals melted in a water cooled copper crucible (the 'cold crucible' technique) because, it is believed, the superheat is too low for simple gravity casting, so the mould is required to be filled as quickly as possible. This dubious logic has led to the universal occurrence of entrainment defects in all titanium castings, requiring them all to be subjected to an expensive HIPping (hot isostatic pressing) process to close up voids. Fortunately for Ti and its alloys, oxygen is soluble in Ti with the result that the HIP process not only closes defects, but encourages them to go into solution. We can be grateful that poor Ti casting technology can nevertheless result in a good (if expensive) product.

Guo (2001) studied the filling of automotive exhaust valves with TiAl alloy by this technique, showing that the radial runner only fills along one side with a thin jet, leading to severe turbulence. Other researchers (Changyun, 2006 and Jakumeit, 2007) also show that the direct siting of the radial runner (equivalent to top gating) onto the casting results in very poor filling. Somewhat improved filling is obtained by the equivalent of bottom gating into the mould cavity, in which the runner bypasses the mould and turns back, entering the mould from its outermost side. Even in this case, the results were still impressively turbulent because the velocities are so high. Once again, as is common for castings of all types, the filling systems are greatly oversized, and thus have only a small proportion of their area conveying any liquid metal, allowing a disappointing degree of turbulence from unconstrained flow of the melt. The design of a radial runner of a size to completely fill with liquid along its length, in which the melt is subject to increasing acceleration as it progresses outwards, represents an interesting challenge. The channels can be predicted to be of extreme thinness, and experience extreme drag forces from the walls, whose roughness will be of a similar scale to the dimensions of the channel. Thus there are plenty of unknowns to be explored. Perhaps the answer is that centrifuge casting cannot be optimised. Perhaps it should be abandoned. Alternatively, perhaps we shall have to live with centrifuge casting together with its defects and the expense of HIPping.

Jewellery is similarly produced by centrifuge casting to overcome the capillary repulsion experienced by melts when attempting to fill such narrow channels and fine detail. This process is better engineered because the melt, being at the end of the long rotating arm, travels along only a short radial runner into the mould. It seems, with good reason, both casters and wearers are content with the process as it is.

True Centrifugal Casting

What I call 'true' centrifugal casting uses the centrifugal action to hold the melt against the wall of a rotating mould. For interested readers, there is much to be learned from the practical and entertaining account given in an article by one of the leading UK companies (Gibson, 2010) which contrasts with the practical but sober descriptions by the Americans Zuehlke (1943) and Janco (1988).

Rings, hollow cylinders, tubes, pipes and large motor housings are often cast using the centrifugal casting process for several reasons. There is no central core required and the wall thickness can therefore be controlled simply by controlling the volume of metal poured. Neither running system nor feeders are required so the metallic yield can be close to 100%. Rollers or bushings have high concentricity, straightness and uniform wall thickness. Rolls are often cast using a double pour, in which the outer working surface of the roll is composed of a hard wearing alloy, or corrosion resistant alloy, whereas the interior backing alloy is a lower cost cast iron or steel. Centrifugally cast rolls are used in the paper, printing, plastics, foil and textile industries.

The centrifugal action exerts an acceleration v^2/r on the metal whilst it solidifies, where v is the linear velocity of the metal and r is the radial distance from the axis of rotation. This acceleration can be 100 or more times the acceleration from gravity g. Thus contact with the wall of the rotating die is good, and the strong temperature gradient which results assists the feeding under the enhanced g force. The inward advance of the solidification front can push remaining defects (particularly oxide films that may not have centrifuged to the centre as a result of their high drag and consequential low Stokes velocity) concentrating them in the inner bore, which is often cleaned up by machining.

However, the pouring of metal into spinning moulds usually involves considerable surface turbulence as the metal surges into the mould and is then whisked up to speed. In horizontally spun castings the melt may slip for some time, delaying its acceleration up to the mould speed, and so will rain from the upper parts of the mould. If this problem continues until the casting is nearly solid laps and oxide defects are created. Grey cast iron pipes are relatively free from any substantial problems which result from this trauma. This is less true for ductile iron pipes because of the presence of the magnesium-rich strong oxide films. It becomes a major problem for higher-temperature materials such as stainless steels, where deep laps are sometimes seen. These form spiral patterns on the outside of spun-cast tubes. The problem is worst in relatively thin-walled tubes which solidify quickly. In thicker-walled tubes the problem seems to disappear, probably as a result of the extra time available for the re-solution of some of the less stable constituents of the surface film, or possibly the washing away of the film from the surface. If the film is mechanically removed in this way, then it will be expected to be centrifuged to the inner surface of the casting to join and be assimilated with the other low-density impurities such as slags, inclusions and bubbles, which is one of the well-known valuable features of the spin-casting process.

The surface turbulence generated during the pouring of centrifugal castings of all types seems to be a general major fault of centrifugal casting, and seems to be generally ignored. This is surely a mistake. The process spends all of its efforts on centrifuging out those defects that should never have been put in. Worse still, not all of the defects introduced by the pour are eliminated by the centrifugal action. The melt delivery spout might be modified to greatly reduce these problems. Some actions expected to reduce turbulence are as follows.

1. The melt delivery spout should be shaped to deliver a ribbon of melt parallel to the rotating surface and as close as possible to this surface, reducing unwanted excess fall distance of the stream.
2. The spout should guide the flow directly onto the walls in the direction of rotation.
3. The velocity of delivery of the metal (see Eqn (13.2)) should match the speed of travel of the wall. Wall speeds are typically in the range 2–5 m/s, and so are easily matched by velocities from a good filling system.
4. The conical basin delivery into the filling system must be abandoned, together with other aspects of the launder and spout design, so as to fulfil the requirements of a good naturally pressurised filling system. In this way, only metal will be delivered via the filling system, not a mix of metal and air. However, even natural pressurisation might have to be abandoned because at high delivery speeds the delivery system may require to be highly pressurised, reducing the final orifice area, to speed up the flow (of air-free metal) to match the linear wall speed.

When fully up to speed, the liquid in a vertical axis mould takes up a paraboloidal shape, with the result that the wall at the base of the cylinder is thicker than that at the top. Figure 16.21 shows experimentally measured profiles for different speeds by Donoho in 1944, and Figure 16.22(c) shows the effect as more commonly seen. We can calculate the extra machining allowance, b, which is needed to produce a parallel bore. The attempted solution to this problem in *Castings* 1st edition (1991) resulted in equations of the correct form, but whose constant term was regrettably in error. I am happy to confirm that the version here is correct, and further corroborated by comparison with the valuable experimental data discovered in the work recorded by Donoho.

When the liquid surface is in balance, it experiences the simultaneous accelerations of gravity g downwards, and the centrifugal acceleration v^2/r inwards towards the rotation axis, where v is the linear speed of the surface at that point, and r is the radial distance from the axis. The liquid takes up a parabolic form. The equation of the curve given by Donoho is

$$h = 0.0000142N^2r^2$$

where N is in rotations per minute (rpm) and R is in inches. This equation is translated into SI units (and, incidentally, thereby takes up the pleasing simplicity of equations and parameters that can be usually enjoyed in the SI system)

$$h = 2N^2r^2 \tag{16.1}$$

where h is the distance up the vertical axis in metres and N is the rate of rotation in revolutions per second. We can integrate to obtain the relation between h and r as:

$$h_t - h_b = 2N^2\left(r_t^2 - r_b^2\right)$$

Furthermore, $h_t - h_b = H$, the height of the cylindrical casting, and $\left(r_t^2 - r_b^2\right)$ can be rewritten in terms of its factors $(r_t + r_b)(r_t - r_b)$. This in turn can be written to a close approximation if the machining thickness $b << r$

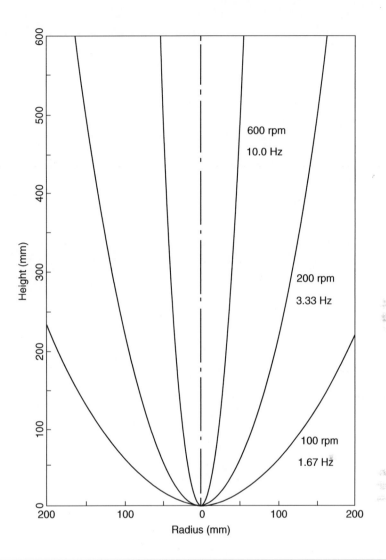

FIGURE 16.21

Profiles of spinning melts found experimentally by Donoho (1944).

$$(r_t + r_b) = 2R = D$$

where R is accurately the average radius, but is closely similar in size to the radius at the top of the cylinder, where, of course, 2R is the diameter D. Furthermore

$$(r_t - r_b) = b$$

in terms of the extra machining b which will be required to make the bore parallel. Approximating again if $b \ll D$ we obtain an explicit relation for b

$$b = (1/2N^2)(H/D) \tag{16.2}$$

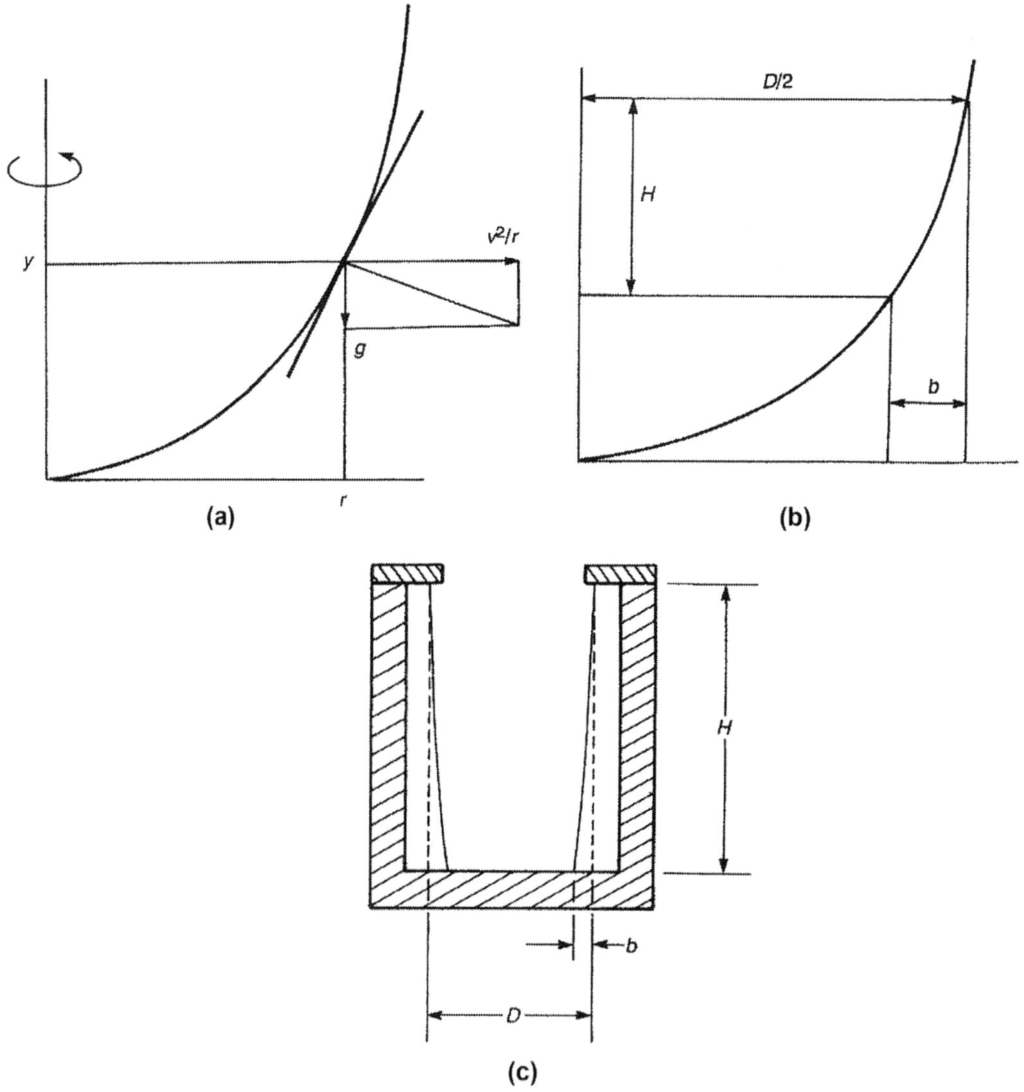

FIGURE 16.22

Vertical axis spinning: (a, b) generation of parabola; (c) consequential additional machining allowance required on the internal bore of a spun-cast tube.

Thus it is clear that when calculating the extra machining b required for a vertically spun tube, the only important parameters are N (which is, of course, intuitively obvious!) and the ratio H/D, which greatly simplifies predictions. The approximate Eqn (16.2) is the basis of Figure 16.23. The figure illustrates the regime at the top of the figure where machining allowance is less than 1 mm, so that extra machining allowance is practically negligible, and the products have virtually parallel bores. For the regime below the 10 mm line, the extra machining for those alloys which are difficult to machine may become so severe that other routes such as horizontal spinning or conventional static casting using an internal core may have to be considered.

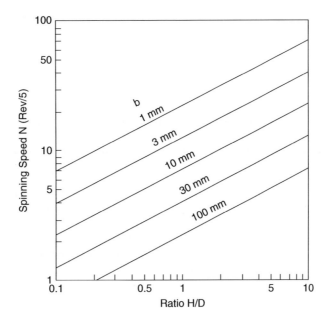

FIGURE 16.23

Machining allowance b from Eqn (16.2) in terms of speed and height/diameter ratio.

The centrifuging of the drosses caused by pouring into the bore of the part represent loss of metal and increased difficulties of machining, increasing the amount of metal suggested by the calculation of the machining allowance *b*. The provision of better pouring systems promises better products at lower cost.

16.5 PRESSURE-ASSISTED CASTING

The application of pressure during the solidification of a casting has, in line with natural expectations, generally been beneficial to soundness and mechanical properties.

In terms of the classical theory of the formation of pores, whether of gas or shrinkage origin, the application of pressure suppresses porosity as described in Chapter 7. In practical detail, it seems that the main reason for the suppression of porosity is probably the effect of pressure to keep bifilms firmly closed, thus negating the action of the various mechanisms that would drive them to open. The maintenance of the population of bifilms in their compact form maintains the reasonably high soundness and properties of the casting. This is the action of hot isostatic pressing (Section 19.3, HIPping).

A second indirect but additive effect may be the action of pressure to maintain better contact between the casting and the wall of the mould, so facilitating heat transfer. Once again, this is an effect that results in the bifilms being frozen in their compact form, before they have much chance to unfurl.

These benefits of the application of pressure have been widely explored and widely reported, as seen for instance in the work of Berry and colleagues (1999 and 2005). These authors survey the long history of the use of pressure back to 1922, although they note that Whitworth patented the process in the 1850s. In particular, they list the benefits of pressure applied to sand castings of steel and aluminium by (1) externally pressurising the casting by solidification in autoclaves or (2) internally pressurising the casting by pressurising the metal in the feeder. Both techniques work, but pressurisation of the feeder would be expected to be especially effective because of its additional actions (a) to establish a pressure gradient to aid feeding and (b) pressurising the casting internally to suppress the formation of any surface-initiated porosity. The benefit of avoiding surface-initiated porosity is not enjoyed by most other general applications of pressure that act both on the outside and inside of the casting.

However, the application of pressure to a feeder noted in earlier work by Berry and Watmough (1961) required caution because an application too early or too severe would result in swelling of the casting and penetration of the sand mould by metal (the pressurisation in an autoclave would not give this problem). A delayed application of pressure would allow the buildup of a strong oxide film, and possibly some surface solidification, resisting the penetration of the liquid metal into the mould. Interestingly, only modest pressures of up to 1 atm were found to be effective in greatly reducing porosity and improving properties.

The use of pressure to suppress pore formation is common to the counter-pressure casting system in which an overpressure of up to 6 bar is applied during the counter-gravity filling. Filling occurs by a relatively small differential pressure (Wurker and Zeuner, 2004).

Experiments are regularly reported in which the application of pressure during solidification is found to benefit castings, but the experiments are often fundamentally flawed. For instance, Mufti et al. (1995) describe solidification of Al alloy casting in a chamber pressurised to 20 bar (2 MPa) that is found to suppress porosity. However, unfortunately, the pouring of the melt from a height of about a metre creates the very problems that the pressure freezing has subsequently to attempt to cure. It is probable that careful mould filling without the application of any pressure would have resulted in sounder castings.

The application of pressure in the case of HPDC is a similar matter; the vicious turbulence of the filling cycle generates massive defects in the form of bifilms and bubbles. In this case the action of pressure is the attempt to reduce the size and deleterious action of these defects. Squeeze casting and counter-pressure casting processes are not so disadvantaged; their tolerable or even good filling retains whatever quality of the melt has already been achieved, so that the pressure is able to act in a fully positive way. These processes are discussed next.

16.5.1 HIGH PRESSURE DIE CASTING

HPDC is known in the United States, confusingly, as die-casting. The reader is recommended to the excellent book by Arthur Street (2nd edition, 1986) for a mass of historical and technical details. In this short account, we can only highlight the fundamental issues.

As readers will be well aware, HPDC is characterised by excellent surface finish and accuracy (at least when the die is new). Although the tooling is expensive, the productivity is high and part price is thereby modest. As such the process is highly popular, so that more than 50% of all Al alloy castings are produced by HPDC.

When used for its common purpose of casting aluminium or magnesium alloys the dies are necessarily made from good quality hot work tool steel, H13. (In this case, grey iron would not have the adequate surface integrity because of the presence of graphite flakes in the material, and would not have the strength or fatigue resistance.) No protective coating is applied to the surfaces of the die. The result is an excellent surface finish. The danger of cold laps is reduced by extremely rapid filling, and by high pressure that is subsequently applied to ensure faithful reproduction of the profile of the mould. Even so, of course, the high speed of filling causes other serious problems that we shall consider below.

Like all dies, however, it is regrettable that they are not really permanent. An aluminium casting weighing up to 1 kg might be produced tolerably well for up to 80,000 or 100,000 shots before the die will require to be refurbished as a result of heat checking (cracks from thermal fatigue of the die surface that produce a network of un-sightly raised fins, which can be razor-sharp, on the surface of the casting).

Magnesium alloy HPDC is much kinder to dies, giving perhaps five times or more life. This is partly the result of the lower heat content of Mg alloys, and partly because of the low solubility of iron in liquid magnesium that reduces its action to dissolve the die surface. In contrast, liquid Al has a relatively high solubility of iron, resulting in a phenomenon known as 'soldering' of the casting to the die. This action is particularly destructive to dies not only destroying the die surface by dissolution, but the enhanced heat transfer that results from this intimate contact can then take further toll. The soldering problem is countered to some extent by the addition of high levels of iron to HPDC alloys, thus pre-saturating the liquid metal with iron and so greatly reducing the tendency for the melt to dissolve the die.

Attempts have been made to cast molten stainless steels by HPDC. In this case, the only die material capable of withstanding the rigours of this process was a molybdenum-based alloy. Even these 'moly' dies suffered early degradation by thermal fatigue cracking, threatening the commercial viability of the process. So far as the author is aware, partially solid (inaccurately called 'semi solid') stainless steels have not been extensively trialed. These cooler mixtures have already given up practically half of their latent and specific heat during freezing and would therefore greatly benefit die life. The commercial viability of the somewhat more expensive mixture would remain a potential show-stopper of course.

Turning now to the process itself, there are two main varieties of HPDC: hot chamber and cold chamber. The hot chamber machines are normally used to cast zinc and magnesium alloys, whereas the cold chamber machines mostly make aluminium alloy castings. There are numerous texts describing the technical details of the machinery involved. In fact, the dominance of the machinery aspects of the work dictate that most operators of HPDC do not consider themselves foundry people at all; they consider themselves engineers. The rift between the attitudes of the personnel involved in these two casting industries has been unhelpful in the past. After all, what do HPDC engineers need to know about sands, binders, insulated feeders and the like? However, there are signs of improved relations with developments such as indirect squeeze casting etc., proving the benefits of cross-fertilisation of ideas between the two technologies. Long may this continue and prosper.

HPDC is an injection casting technique. The usual design of cold-chamber machines has a horizontal-stroke injection (Figure 16.24) which is programmed to pressurise the melt as shown in Figure 16.25. However, it is regrettable that optimally controlled filling has proved elusive in this particular mode of injection. In fact, in the past, most HPDC machines have used injection strokes that cause much damage to the integrity of the finished casting. Typically the metal stream jets or sprays to the far end of the die cavity, and then ricochets and splashes backwards down the walls, sealing off routes for the venting of gases from the die cavity, and so entrapping the residual air inside the casting. Not only the air in the cavity is entrained, but so also are the vapours boiling off from the die lubricant and/or coolant.

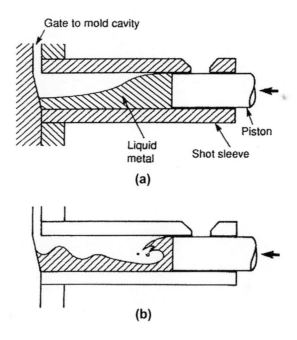

FIGURE 16.24

Injection of metal in a horizontal shot sleeve of a cold chamber die castings machine comparing (a) controlled and (b) uncontrolled first stages of injection.

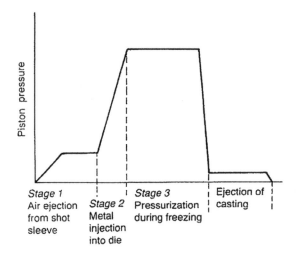

FIGURE 16.25

Typical injection stages during pressure die casting.

FIGURE 16.26

A pressure die casting in Zn-27Al alloy showing a polished section through a heavy section.

The result is a casting which exhibits the typical 'aero chocolate' structure – a tolerably good skin but full of air bubbles just beneath the surface (Figure 16.26). So much air is entrained in most pressure die castings that Hangai and coworkers (2010) remelt scrap castings as an optimum charge material to make Al alloy foam. In detail, it seems that the average high pressure casting probably has two populations of bifilm defects. These will be (1) oxide flow tubes that have formed around the incoming jets, separating the casting into longitudinal isolated zones as is clear in Figure 2.31; and (2) compact convoluted bifilms resulting from less organised turbulence during filling. That the flow tubes are likely to be often aligned parallel to stresses explains the good performance of some castings. However, the unpredictability of turbulence and consequential bifilms would normally make the application of HPDC to safety-critical parts an unacceptable risk.

Koster and Goehring (1941) have demonstrated the filling action by injecting Wood's metal into transparent dies. They confirm that the spread of the liquid metal around the cavity causes the vents to be sealed first, so that the high pressures are merely required to compress the entrapped gases. The castings have a structure which is full of discontinuities which result from the surrounding of the initial spray of oxidised and already frozen droplets by liquid metal which arrives and freezes subsequently, as can be discerned from interesting micrographs by Bonsack (1962).

For these reasons, pressure die castings have to be treated differently to most other types of castings. For instance:

1. Pressure die castings cannot be subjected to a normal solution heat treated because the reduced strength and creep resistance of the metal at high temperature allows the entrapped gases to expand, causing blisters on the surface or even gross distortion of the casting. The use of pressure die castings for structural parts is therefore restricted because not only are the castings limited in strength by high internal porosity and planes of additional weakness caused by bifilms, but also the presence of these defects in turn prevents any subsequent strength benefit from heat treatment. (Recent work to counter this problem is described later.)
2. Pressure die castings should not be machined at all if possible. Even light machining cuts are likely to penetrate the relatively sound skin (which is actually not particularly sound), encountering the unsatisfactory structure beneath. Deep cuts or drilled holes should be expected, therefore, to connect with the internal network of porosity and so cause leakage of fluids via distant points on the casting.
3. Subsequent chemical treatments in aqueous media, such as electroplating or anodising, are made less easy because of penetration of the aggressive liquids into pores or laps through the relatively sound skin, and the subsequent unsightly spots of corrosion which are caused by the liquid slowly leaking out over a long period.
4. Pressure die castings cannot be welded; in any such attempt the weld pool excels itself in an energetic imitation of Mount Etna!

In view of this apparently damning list of shortcomings, it is nothing short of amazing that pressure die casting has achieved such an important place in manufacturing. In the Western world, aluminium alloy pressure die castings make up more than half of the tonnage of all aluminium alloy castings. The great success of pressure die castings derives from the advantages of accuracy, surface finish, and the ability to reproduce detail, all at low cost and reasonable production rates. It is not surprising, therefore, that much effort has been expended on attempts to overcome the important disadvantages of the process.

A first attempt to improve on the filling of the die came with the adoption of three-stage injection (Figure 16.25). The stages are:

1. A controlled acceleration of the piston to cause a wave of liquid metal in the shot sleeve to expel the air ahead (not trap a bubble of air which was then injected at a part-full stage) as shown in Figure 16.24.
2. A more rapid fill of the die.
3. A high pressure consolidation.

This three-stage technique, whilst being a great step forward in avoiding air injection during the first stage, was still far from perfect during the second stage; the high entrapment of air required high consolidation pressures during the final stage.

Recently, much work has been carried out in an attempt to optimise gating designs and fill rates to promote the progressive filling of the die, to expel air ahead of the metal. In addition entrapment of vapour from coolant sprayed onto the die can be reduced by providing the die with internal cooling passages. Furthermore, there is some move away from water-borne graphite sprayed lubricant towards less volatile substances such as waxes. Substantial progress has also been made using computer simulation to ensure that the pattern of mould filling avoids entrapping major volumes of entrapped gases.

A novel design of runner for HPDC uses a significantly smaller cross-sectional area and has been reported to improve casting soundness (Gunasegaram, 2007). It seems to me to be likely to be the result of improved priming and exclusion of air from the shot sleeve, thus sparing the casting from massive defects that subsequent filling and pressurisation attempts to heal. The action of improvement is unlikely to be any reproducible mechanism within the mould cavity itself because conditions are so dramatically turbulent; one can envisage that the narrower runner will occasionally but not always benefit the filling of the mould, even though there may be additional beneficial effects from a refined structure, such structural benefits to strength and performance are likely to be small (see Section 9.4).

Fundamentally, the Weber numbers are so far from optimum (see Chapter 2) that it is difficult to understand how a really satisfactory filling solution can be achieved. The production of high-quality castings by the use of horizontal shot sleeves for the injection of metal into dies remains a largely unsolved quest. Nevertheless, the huge market tells us that these castings are adequate, if not more than adequate, for the purposes of most customers. Having said this, significant efforts continue in an attempt to improve HPDC further as noted next.

Squeeze Pins
Pressure die castings can be made more sound in local areas by the use of squeeze pins. Such techniques are in the category of keeping the customer happy by appearing to provide a sounder casting. It is doubtful if strength is improved, but leak tightness would, of course, benefit. When investigating the technique for Al-Si-Cu alloys, Wan and colleagues (2002) found that the Cu was squeezed out from the highly pressurised region, reducing in concentration from 2.6% to 1.6% Cu, and was segregated to the surface regions around the pin.

'Pore free' Process
The filling of the die with oxygen before the injection of the liquid metal greatly reduces the entrained porosity as a result of the entrapped oxygen reacting with the metal to produce a compact solid oxide. The technique has been mainly used for Al casting, but is also reported for Zn and Pb alloys. However, of course, the oxygen is an additional cost, and productivity is reduced because of the additional time taken to flush with oxygen. Furthermore, bifilms might be expected to be made worse, impairing both mechanical properties and machining although I could find no report of this. The process was useful in the early days of its invention before computer simulation and vacuum techniques, but is much less used today. A good summary of the process is given by Arthur Street (1986).

Vacuum Processes
One of the many systems for evacuating the shot sleeve and die is shown in Figure 16.27. The evacuation of the die leads to the melt being drawn up from the holding furnace and filling the shot sleeve. In itself, the counter-gravity filling of the shot sleeve probably significantly reduces defects in the casting. The movement of the piston cuts off the supply of metal from the holder and fills the cavity in the usual way. Naturally, porosity is significantly reduced. Some users claim that the castings are sufficiently sound to be heat treated and welded, although some users have been less than satisfied. Progress probably continues on this issue.

Heat treatment of HPDC
Because rate of diffusion is the process controlling the dendrite arm spacing, refining the spacing with faster cooling, and the rate of diffusion controls the length of heat treatment times, with finer dendrite arm spacing speeding the rate of homogenisation, the rates compensate to the degree that the faster cooling means faster heat treatment.

Because of the fast solidification times for HPDC, the heat treatment homogenisation time can be correspondingly reduced; no longer measured in hours but in minutes. Very short homogenisation times does not permit blister formation to become too advanced, and may therefore remain acceptable. The ageing treatment is at a much lower temperature at which no danger of blistering is experienced, so that a useful T6 strengthening heat treatment can be carried out in full which can double as-cast strength and toughness (Lumley, 2013).

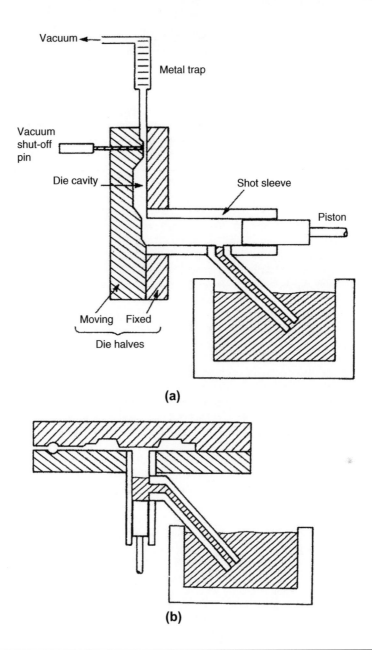

(a)

(b)

FIGURE 16.27

Vacuum delivery systems to the shot sleeves of pressure die casting machines (a) horizontal cold chamber;
(b) vertical injection type machine.

Naturally, the use of vacuum during the filling of the die assists significantly by reducing the internal gas content of the casting, permitting longer treatments without danger of un-sightly blisters or distortion.

16.5.2 SQUEEZE CASTING

There are two varieties of squeeze casting. Both subject the solidifying casting to a high pressure partly to suppress the formation of porosity, and partly to achieve good thermal contact between the casting and the die to achieve rapid solidification in the quest for good properties and good productivity.

Indirect Squeeze Casting

The indirect squeeze casting process uses a HPDC machine, but oriented with a vertical shot tube so that the die can be filled via a bottom gate by a kind of counter-gravity process. The machine, in common with conventional HPDC machines, is complex and capital intensive. Installation costs and general difficulty of maintenance are non-trivial matters because its height usually must be set in a deep pit. (The machine contrasts sharply with the simplicity of the forge required for direct squeeze casting discussed later.)

Even so, the indirect process enjoys the important benefit of a counter-gravity filling technique that solves at a stroke many of the problems of the direct process. The simplicity of the filling process makes for easy and precise control. The melt is transferred upwards into the closed die by the vertical displacement of the piston in the shot sleeve. Pressure during freezing is applied directly by the piston. Generation of oxide defects during filling are avoided by keeping the ingate velocity lower than the critical 0.5 m/s. This critical velocity has been confirmed by many investigators (for instance, Xue and Thorpe, 1995, and Itamura, 2002). This is aided by increasing the area of the gate to such a size that the gate must be cut off by sawing rather than simply fracturing or clipping (shearing) as is usual for conventional HPDC. The size of the gate means that metallic yield is often only about 50% for this process.

The counter-gravity filling ensures that the oxide on the liquid meniscus is laid out against the surface of the die as the melt rises, thus protecting the die from direct contact with the metal. For this reason, no welding or 'soldering' occurs, so that low-Fe high-performance alloys can be cast (in contrast to HPDC). In addition, the absence of the momentum impact at the end of the filling stroke gives better dimensional control; no die 'bounce', no pushing back of pins, no flash, allowing greater tolerance for tool fits, encouraging the use of moving parts in the die including squeeze pins. Even so, naturally, such tooling is expensive.

Modest problems reported from time to time include criticism that (1) the narrowest walls are limited to perhaps 4 mm thickness and (2) the machines are slower than conventional HPDC machines. Although some of the slowness is clearly attributable to the wait required for the heavier sections to freeze and cannot therefore be claimed as a justified criticism; any permanent mould process would have suffered a similar disadvantage. Overall, these seem to me to be reasonable penalties if the customer's priorities are the reliability of the casting.

However, it has to be reported that the process has suffered some lack of reliability of its castings as a result of the entrainment of oxide bifilms during the pouring of the melt into the shot sleeve. This fall of only 50–100 mm entrains bubbles (and unseen bifilms of course) which appear on the cope surface, and were originally tolerated by using the drag as the best surface, or providing more machining allowance to the cope. These unsatisfactory solutions have finally been bravely tackled by the counter-gravity filling of the shot sleeve itself via a side port as illustrated schematically in Figure 16.27(b) (Okada et al., 1982). The connecting up of furnaces and casting machines in this way is known to be difficult, and in practice it is not known how successful this technique has been in the years subsequent to its introduction.

Direct Squeeze Casting

When shaping a solid piece of metal by closed die forging, the die is initially open. The work piece is placed in the lower die half, and the top die is then brought down to engage with the work piece. The application of pressure between the slowly closing die halves causes the solid to flow plastically within the constraints of the die, being displaced to fill the outer sections of the die cavity.

There are several casting processes that have much in common with this shaping technique. Their common features are the pouring of the liquid metal into the bottom half of an open die or mould, and the subsequent closing of the die or mould so as to displace the liquid into the extremities of the cavity (Figure 16.28).

This process for the 'forging' of liquid metal has a long and complicated history. It was first suggested by Chernov in 1878 but appears to have been first used before 1930 under the name of Cothias Process in the United Kingdom (Chambers, 1980), although Welter in Germany was studying the effect of pressure on solidifying metals in 1931. However, it was in Russia during the 1960 and 1970s under the name of extrusion casting (Plyatskii, 1965) that the bulk of the early work to develop the process the process was carried out.

In general, no running system is required. Furthermore, because the liquid displacements are rather limited, no great flow distances are involved, so that fluidity, which is normally such an advantage for normal casting alloys, is no longer required. For this reason, squeeze casting, as it has become known, has the unique advantage over all other casting processes of being not limited to casting alloys. In fact, it can operate very satisfactorily with wrought alloy compositions and so benefit from the considerably higher strengths that are attainable with these materials. The users of the process sometimes emphasise this fundamental difference by avoiding the description '*squeeze casting*' and instead calling the process '*squeeze forming*'.

Additional benefits to the metallurgical structure of squeeze-cast material result from the high pressure that is applied during the freezing of the casting. The rate of cooling of gravity and other die castings is normally slowed by the presence of an air gap which forms between the casting and the die as the casting cools and contracts away from the die. This does not occur to such an extent in squeeze casting and the consequent improved cooling rate results in a significantly finer cast structure. Figure 5.8 shows how the freezing rate is twice that of an equivalent gravity die casting, and nearly 10 times faster than conventional sand castings. The application of the high pressure during solidification also tends to suppress the formation of porosity. Squeeze castings are probably more sound than most types of casting.

As always, however, freedom from defects is not infallibly guaranteed. For instance, lap defects are easily formed if the displacement velocity is not correct, especially if a waterfall effect occurs during the displacement of the liquid during die closure. Hong (2000) investigates some of these problems. Even porosity is possible in a heavy section, especially if it is surrounded by thin sections that solidify first, holding the die halves apart, and thus prevent the application of the full die closure pressure to the heavy section during its later solidification.

Interestingly, Herrera and the author (1997), when studying the mechanical properties of direct squeeze forming of copper alloys compared with forgings, the squeeze castings were highly variable and unreliable. The poor results were thought to reflect the inability of the process to close up serious oxide bifilms defects introduced by the poor filling of the die. This result would not be expected to be confined to copper alloys, but would be expected to be a general result applying to squeeze cast alloys of all metals; it reflects the problems introduced by poor filling of the die.

Furthermore, the very high temperature gradient generated by the pressurisation of the casting against the face of the die leads to dendritic segregation (inverse segregation) as described in Section 5.3.3. Even worse, the eutectic can be squeezed out of the alloy, appearing as an un-sightly exudate on the surface of the casting. Although this effect is accentuated with strongly segregating solutes, such as the element Cu in Al alloys, the effect can be so pronounced when using high squeeze pressures that it even occurs in alloys that usually show no segregation problems such as the normally well-behaved Al-7Si-0.4Mg alloy (Britnell and Neailey, 2003).

Squeeze casting, in common with other die processes, has generally been limited to relatively low melting point materials such as zinc-, magnesium- and aluminium-based alloys. Although gravity die casting has been extended to cast iron to a limited degree, such an extension is difficult to envisage for squeeze casting, where the enhanced rate of heat transfer would almost certainly result in rapid deterioration of the die.

The direct squeeze process uses a forging press to make castings. A forge is a simple machine costing a fraction of the capital investment of the indirect squeeze forming machine. Even so, the advantages probably finish at that point. This is because, in general, despite the enormous potential of the process as described previously, its performance in a production environment is often deeply disappointing. It seems that scrap arises from several causes that are not easily overcome. These include the following.

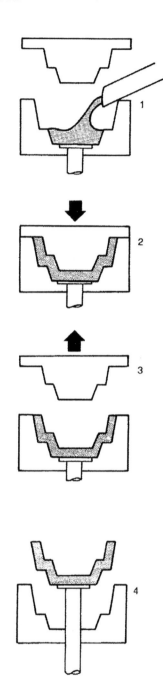

FIGURE 16.28

Squeeze casting.

1. poor melt control before casting (clearly this is not the fault of the casting process!);
2. poor transfer into the die, creating oxide dross in the casting (a bottom-stoppered transfer ladle is recommended to deliver melt into the bottom of the die, preferably into a pocket in the base of the die so that the melt can be constrained during the initial spurt on transfer, and general turbulence in the die can be suppressed);
3. boiling of mould dressing/lubricant in die joints (recommended to provide generous vents out of the back of the die, or ensure nearby cooling water channels to prevent boiling);
4. relatively thick walled castings (compared with HPDC) because of the time required to close the die;
5. dosing weight requires reasonable control to ensure control of casting thickness.

The great simplicity of the process cries out for it to be properly developed to achieve its full potential.

Press Casting

The forging technique has been extended to sand castings, but not for the purpose of applying high pressure during freezing, which would, of course, only lead to metal penetration into the sand, but simply to move the two halves of the mould together immediately after casting so that a thinner section can be cast.

In this way, cast iron gutters were cast in the United Kingdom for more than 50 years (Chadwick and Yue, 1989). For more sophisticated Al alloy products, Miller (1967) describes a Boeing development that demonstrated how aircraft castings of large surface area could be cast at 0.5 mm thickness. The method is, of course, limited to configurations where the thin-wall portions of the casting lie in nearly horizontal planes that do not overlap. High pouring temperatures were required, and excess melt was expelled into feeders and reversed into the filling system during closure of the mould. This movement of high temperature metal was claimed to remelt sufficient solidified skin to permit the full closure of the mould to obtain success in achieving the thinnest sections.

Terashima et al. (2008) outline a further development of this concept using greensand moulds. The open mould receives the melt which is poured into the drag, and the cope is then lowered into place. The process specifically avoids pressurising the melt to any significant extent to reduce sand penetrated by the metal and the impairment of surface finish (Tasaki et al., 2008).

The forging approach to the filling of a sand casting does appear to be an interesting process because fluidity problems are reduced as a result of the only limited displacement of the melt into outer regions of the mould at extremely gentle speeds. This clearly contrasts with conventional filling systems in which flow has to take place through long filling channels and then enter the mould cavity at relatively high speed via limited locations. The complete absence of a filling system (and a feeding system if the wall thickness is relatively small and uniform) is another powerful advantage. The press casting technique promises great potential.

Cast Pre-forms for Forging

There are possibly good reasons for the casting of a material such as ductile cast iron and subjecting this material to hot rolling or forging as described by Neumeier and colleagues (1976). Such a product would have properties that would not be easily achievable by any other process route.

However, I have come across several applications of Al alloy castings as part of development projects loudly heralded as being 'the way forward', in which the casting were used as pre-forms for a subsequent forming process in a closed die forging operation. Such developments have never succeeded. The reason is that the casting defects are typically oxide bifilms introduced by the poor casting technique. These defects are not repairable by forging, with the result that the final parts inherit the variable and often poor mechanical properties of the casting.

The cast + forge production route is in any case fundamentally illogical. If the casting had been made without defects, it would have been reliably strong, making a forging step unnecessary because the properties could not have been further improved (the work hardening involved in such simple forging steps is relatively trivial).

16.6 **LOST WAX AND OTHER CERAMIC MOULD CASTING PROCESSES**

The 'lost wax process' was assumed by the lady translator to refer, tragically, to the lost wax-process. It was only much later she realised it should be referred to as the lost-wax process.

Moving on from quaint semantic issues to more serious matters, in the moulding section, it was mentioned that the time of writing, the investment casting industry and its customers suffer from the use of probably the worst mould filling systems anywhere in the casting industry, seemingly devised to deliver the most defective castings possible.

This regrettable fact arises from the conventional way a wax tree is assembled, usually with a central conical pouring basin. As explained earlier in this work, the conical basin is one of the worst possible features. Moreover, the filling channels are nearly always vastly oversize because they have been built to maintain the strength of the wax assembly rather than designed for control of flow during filling. (We shall see how it is not so difficult to devise designs that are mechanically strong and combine good filling designs.)

The cases of poor filling are maximised for vacuum casting, where the melting furnace is typically a metre or more above the mould, creating masses of entrained oxide bifilms as a result of the severe surface turbulence when falling from this height. (This point is covered in Section 16.10 'Vacuum Melting and Casting'.)

Furthermore, convection is common in investment shell castings, creating heterogeneous, irregular structures, containing both fine and coarse regions in adjacent regions of the same casting. Those regions with coarse structures being expected to exhibit poorer properties of course, but in any case often failing customer specifications for grain size, even though the grain size may have little effect on properties.

The one good feature of current filling system designs, that happens by chance to prevent lost wax castings from being worse than they are, is the use of small-sized ingates. Ingates of small area do not permit large bifilms to enter, especially those strong oxide films formed by certain stainless steels as found in the work by Cox et al. (2000). The effect is analogous to the blocking of flow into flash by the bridging of oxide films at the entrance to the thinner section. The effect explains why melts of Al alloys recently grain refined with Ti + B additions to sediment bifilms form a clean melt which suffers significantly increased metal flash on the castings, and increased dressing costs.

In summary, the beautiful surface finish and fine detail of lost wax castings presents an allure in common with HPDCs, in that the surface appearance often covers a poor internal structure with properties much less good than could be achieved. It might be unjust to call these products 'whited sepulchres', but what can be said is that most lost wax castings have some considerable way to go to become properly reliable.

Figure 16.29(a) and (b) illustrate a type of casting cluster which is commonly used, but whose central post is no longer the sprue. The sprue has become a slim piece of sheet wax, cut with a knife to its calculated tapered profile. The vents off the top of the moulds do not connect with the pouring basin so that melt cannot spill down the vents and into the mould, nor can a convective loop be set up to delay the freezing of the casting. Thus these connections have to be plugged at some point to prevent the flow of metal (an offshoot will allow the escape of air if necessary). Figure 16.29(a) shows the concepts of minimum depths in the pouring basin that require to be met to ensure the pressurisation of the filling system; Figure 16.29(b) shows an improved system in which the accessible entrance to the sprue can be sealed with a stopper until the basin is filled and some seconds of delay are imposed before allowing the metal to fill the casting. If this basin can be sized to hold the entire volume of melt required to fill the casting, so much the better.

With some notable exceptions, including probably vacuum investment casting which would require a quite different design of vacuum furnace, these advances are often neither difficult nor costly to make, and promise a better future for both producers and users.

Shaw Process (Two-Part Block Mould)

Shaw process moulds are made by pouring a ceramic slurry into fairly conventional core boxes or patternwork. Thus a cope and drag can be made. The process is described in Section 15.1.3.

The cope and drag technique for mould assembly has advantages in that cores, filters and chills are easily placed. Conventional filling system designs can usually be implemented, for instance the placement of a runner around a horizontal joint line. That the mould can be placed horizontally usually means that the velocities in the filling system are relatively low, especially from lip-poured hand-held crucibles, so that relatively little damage is introduced compared to vacuum-cast investment moulds where the fall heights of the melt are a disaster. The development of the

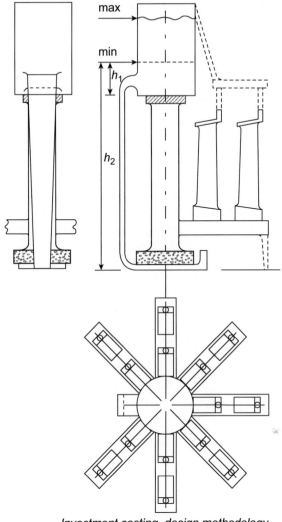

Investment casting, design methodology.

(a)

FIGURE 16.29

Investment mould using (a) an offset basin, natural pressurisation and bottom gating; and (b) an improved design using a stopper, with a basin preferably sufficiently large to make the whole casting without top-up of the basin.

process as a relatively thin shell rather than a block mould has improved the economics (Ball, 1991, 1998). There is no record that the process has ever used a good design of pouring basin, but with very small castings, particularly if poured rather slowly, require such narrow sprues that a conical basin may be acceptable, because surface tension will assist to keep air out of the sprue. Finally, the casting is relatively easily extracted by separating the mould halves after solidification.

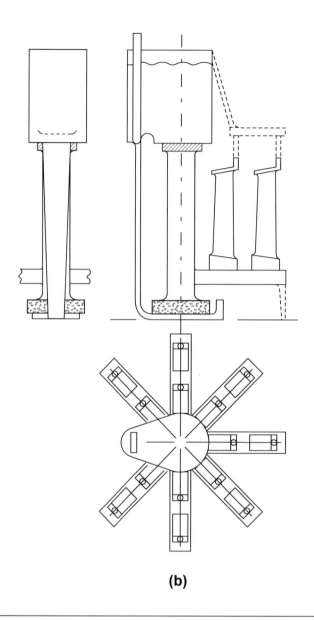

(b)

FIGURE 16.29 Cont'd

Plaster Investment (Integral Block Moulds)

The combination of this process with the application of a reduced pressure applied to the base of the permeable moulds (the steel box containing the plaster mould is open at its base) is the principle of the *vacuum-assisted casting process*, allowing the pouring in air of extremely thin-walled castings only a fraction of a millimetre in wall thickness.

From the point of view of achieving high properties from castings made by vacuum assistance in plaster moulds there is a concern that the extremely thin walls fill by a curiously deleterious process I have called micro-jetting. This was first observed by Evans (1997) in plaster moulds and seems to arise because of the small dimensions of the advancing meniscus, having an area of oxide that is beginning to be too small to contain serious defects. Thus this relatively strong oxide film resists the advance of the melt, building up a backpressure. When the back-pressure builds up sufficiently, the film bursts, allowing a jet of melt to shoot forward. This happens repeatedly, with the melt not advancing in the conventional way by the smooth progression of a smoothly curved meniscus, but by a tangled mass of jets, all surrounded by their tube of oxide. The creation of masses of oxides in this way was thought to be the reason for the low reliability (in terms of Weibull modulus) of cast test pieces in Evans's work.

Clearly, more research is required on the phenomenon of micro-jetting. It is not certain whether this is a fundamental limitation to the attainment of good properties in casting sections thinner than a few millimetres. Also, it is not clear at this time whether the phenomenon is limited to plaster moulds or is a feature common to thin sections of any mould. If true, this would be a serious threat to the attainment of high reliabilities in such parts as turbine blades. Although these parts currently have good properties, it may be that even higher properties might be attainable if micro-jetting could be avoided.

16.7 LOST FOAM CASTING

The lost foam technique is a clever and intriguing process for the production of castings, first announced by Wittmoser in 1968. However, whereas precision sand (otherwise known as core assembly) casting is evidently straightforward, the apparent simplicity of the lost foam casting process is misleading. Furthermore, lost foam casting is not without its own special (and serious) problems.

The starting point is the production of polystyrene beads, which are blown into tooling (the patternwork) and expanded in situ by steam to fill the tooling. Heat is then extracted, cooling the expanded foam and causing it to gain strength so that it can be ejected from the tooling. Naturally the tooling has to have high thermal conductivity to maintain a good rate of production when oscillating between the hot and cold parts of its cycle. It is therefore usually made of aluminium alloy, and is sculpted on its reverse side to reduce its thickness as far as possible. The foam pattern is a replica of the finished casting. By gluing together separately produced foam components, complex geometries can be assembled which can be difficult to reproduce by conventional casting techniques. The assembly is then attached to its filling system and dipped into (or poured over by) a refractory wash.

After coating (investing) the foam pattern with the ceramic slurry, the coat takes time to dry. When dry the assembly is ready to cast. Clearly the thin invested shell around the foam will not support the weight of the metal pouring into the foam, so it is backed up with an aggregate. Thus the purpose of the aggregate in this case is mainly to support the ceramic shell and to conduct heat away from the casting.

Early problems of the lost foam production of cylinder heads and other castings requiring some degree of precision were found to be associated with silica sand used for backing, because the alpha/beta quartz transition at around 570°C created unacceptable distortions. Manufacturers therefore changed to artificial ceramic beads based on alumina or mullite ceramics that have a much reduced and uniform coefficient of expansion. This move has been rewarded by an immediate improvement in casting accuracy and reproducibility.

A further advantage of the ceramic beads is that their spherical shape allows them to flow easily, filling all parts of the pattern, sometimes required to be uphill into blind holes, usually with some encouragement from carefully controlled vibration of the mould. The upper surface of the pouring cup attached to the filling system is arranged to remain clear above the aggregate. The cup is then filled with liquid metal, which proceeds to vapourise the foam, progressively replacing it as it advances, thus forming the casting. The final benefit of using an unbonded aggregate is seen after the castings have solidified, when the casting is removed from its mould simply by releasing the bottom doors on the box, or by turning the mould box over and tipping everything out. The casting is caught on a grid, and the aggregate pours through the grid to start its recycling process.

A nice article by Walling and Dantzig (1994) suggests that in the cast of the casting of Al alloys, the foam melts and collapses but very little breakdown of the polystyrene occurs. The rate of pattern elimination at 60–80 mm/s (not the design of the filling system) controls the rate of advance of the liquid metal. The more rapid melting and collapsing of the foam during the casting of iron, plus much greater breakdown of the polystyrene to produce volumes of styrene vapour lead to significantly greater cooling of the iron, but allows its filling rate to be controlled by the filling system.

The volumes of gaseous products evolved from the foam must contain a significant proportion of air because the foam itself contains a high percentage of air, which in turn will expand further when heated by the melt. This large volume of gases must be vented through the ceramic coating, which is why the permeability of the coating appears to be of great importance for the control of the process.

With respect to the competition between castings produced by lost foam processes and other conventional 'lost air' processes there are several issues. These are not easily compared because all processes are operated rather differently by different manufacturers, with different degrees of competence and with different overhead costs, giving different apparently final costs. However, even allowing for this, there are several differences which are fundamental; these are discussed next.

Tooling for lost foam process used to be regarded as complicated as a result of the back of the tool having to be sculpted to reduce overall thickness so that heat could be transferred from the foam at the maximum rate to maintain productivity of the foam patterns. The wide use of computer-aided design and machining has reduced this problem. The tooling is typically machined from aluminium alloy, and so is relatively light in weight. It also has a long life because the polystyrene beads exact almost no wear. Even so, in appearance, the tooling is a mass of tubing for steam and water supplies, in addition to pullbacks and other motions now beautifully engineered with various varieties of linear actuators using roller bearing sleeves and similar bearings as a result of its life in a light engineering, rather than a foundry environment.

Conversely, tooling for nearly every other casting process does suffer wear. Most other tooling suffers either sand or hot metal or both, in addition to foundry personnel! The conventional practice is to introduce hardened steel wear plates into core box tooling which can be replaced from time to time as necessary. However, as designs of core blowing machines improve the fluidisation of sand before blowing, the problem is gradually being reduced. Ultimately, although wear remains a potential threat to accuracy an adequate standard of casting can be achieved with the application of statistical process control techniques to monitor and predict the gradual changes in dimensions of the casting.

The lost foam process is renowned for the accuracy with which it can maintain wall thickness. Also, for small castings, particularly bracketry, the ability to cast bolt holes and to work with zero draft angles so that machining of some faces can be avoided altogether is a further powerful advantage.

However, for larger castings, the slow drift in dimensions of the foam from loss of pentane during the ageing of the beads after the pre-expansion is a constant threat to overall accuracy. Assembly of foam patterns can also introduce errors. Small scale errors include the glue bead which is particularly unwelcome to the designers of such features as aerodynamic ports in cylinder heads. An additional well-known problem contributing to large scale dimensional control and distortion is the problem of maintaining the shape and size of the flimsy polystyrene pattern during the time it is being covered with about half a tonne of sand. It is practically inconceivable that the application of the backing aggregate, no matter how well controlled, can be accomplished without some distortion.

For certain applications, the absence of a parting line in the mould can be a great advantage (the parting line on the polystyrene pattern still exists of course, but this can be well controlled and so is not usually a problem).

Turning to the issue of surface finish, although the industry has made great strides in improving finish of lost foam castings over recent years, there remains from time to time on the patterns an area of poorly consolidated beads, giving a strong 'orange peel' effect on patches of the casting. These impressed patterns in the surface have been the subject of widespread discussion for many years. There has been talk of the roots of such impressions being sites for fatigue initiation but I am not aware of this ever having been demonstrated.

The real problem with the re-entrant pockets in the surface is their content of refractory wash and possibly entrapped sand grains. If these are released into water and oil ways of castings during service, this is clearly a serious matter. This problem is especially acute at the junction of glued patterns, where a local absence of glue will allow the ingress of the refractory slurry, resulting in a casting defect of some kind, depending whether the fin of refractory remains in place to penetrate the wall, or whether it is washed off into the liquid metal to appear elsewhere as an inclusion.

It is necessary to bear in mind that conventional core assembly also suffers cores with occasional poorly compacted sand, resulting in localised areas which may exhibit metal penetration and retained grains of sand in the surface. There seems to be relatively little to choose between the processes on these grounds.

The minimum wall thickness of lost foam castings is typically about 3.5 mm, being dictated by the ability to blow and consolidate the polystyrene beads; a minimum section is said to be about three bead diameters. However, the wall thickness is also dictated to some extent by the fluidity of the metal to fill the section. This appears to be reduced by the evaporation of the polystyrene which is endothermic (requires heat) and thus chills the liquid metal. The permeability of the pattern coating also influences the resistance offered by the pattern to the advancing liquid.

Worse still, the resistance offered by the foam to the advance of the metal will result in the metal taking different routes into the mould on different occasions depending on the precise density of different parts of the pattern. This may result in the serious situation that entrapment of the foam may occur, enclosing foam decomposition products in the casting. Such defects are not uncommon in lost foam products. Even without wholesale entrapment of decomposition products, the advancing surface of the liquid metal seems to have a considerable thickness of a film, of somewhat uncertain composition, but probably mostly oxide which, on meeting other liquid fronts, in a confluence of fronts results in a serious lap defect known generally within the industry as a 'fold'. It is effectively a highly visible and serious bifilm defect.

The appearance of dispersed gas porosity, entrained decomposition products and film/lap defects is to be expected in lost foam castings. These are serious defects which exceed those to be found in most competitive casting processes. The development of foams which give less undesirable reaction products is reported from time to time. Nevertheless, the majority of the industry appears to be still working with polystyrene as a pattern material which is known to give these serious problems.

Furthermore, the following of the casting process with pressurisation of the casting in a pressure vessel immediately after casting is a further treatment that improves the reliability of lost foam products, although coming at the expense of additional processing complexity and cost (Garat, 1987, 1991). Although this technique clearly improves the castings a little its effectiveness is necessarily limited, and cannot achieve the properties achieved by well-made 'lost air' castings.

From observations of the filling of lost foam castings by X-ray radiography, it is clear that a high proportion of damage to the melt comes from the initial fall down the foam-filled sprue. This is particularly chaotic, with liquid metal fighting its way downwards in fragmenting and turbulent masses against the turbulent pockets of liquid styrene and vapour powering and meandering upwards. If the sprue is narrowed, as in conventional 'lost air' casting to reduce turbulence, the foam extracts so much heat from the advancing front that the melt freezes, never reaching the base of the sprue. This emphasises the nature of fluidity in lost foam moulds: the front advances by fresh melt arriving at the front by its momentum as illustrated in Figures 3.24 and 5.24(b). This important mode of flow requires plenty of space, with generous-sized filling systems, so that constrained systems such as the naturally pressurised filling designs recommended elsewhere in this book do not work for lost foam castings.

There are numerous studies to illustrate that all attempts to fill lost foam moulds by any kind of gravity pouring technique introduce masses of defects into castings. For Al alloy castings, Katashima (1989) uses a silica window in the mould, revealing horrifyingly messy filling conditions in which islands of foam fragments are surrounded by metal. Bennett (2000) and Tschopp (2000) illustrate micrographs and fractographs of clear fold type defects (i.e. bifilms) and blisters on the surfaces of the castings (again, double oxide defects inflated probably by outgassing vapours). Similarly, Carlson (1989) and Shivkumar (1989) report oxide films and black films, with diameters sometimes measured in centimetres, on fracture surfaces. For iron castings Gallois (1987) identify lustrous carbon films. Metallurgically, therefore, lost foam castings poured by gravity usually contain large numbers of serious defects. The author has to admit he has not earned himself popularity for drawing attention to this relatively lamentable performance over many years (Campbell, 1991).

Counter-gravity filling of lost foam moulds immediately suggests itself as a possible remedy, as actually confirmed by X-ray radiographic studies. However, even this route is not without its challenges because Ainsworth and Griffith (2006) show that a rate of advance of the front of only 5 mm/s in an Al-10Si alloy is the maximum velocity that can be tolerated before the front advances irregularly and entrains foam and its degradation products. Moreover, this velocity was found to be too low to allow the complete filling of their test mould. Furthermore, Weibull analyses of the mechanical properties indicated that even at such low filling speeds the reliability of the castings did not reach those of normal 'lost air' castings.

Fan and Ji (2005) used counter-gravity filling of Mg alloy AZ91E, but found rather indifferent properties, which they considered a good result because the properties were as good as their indifferent 'lost air' castings.

Clearly, the production of lost foam castings cannot yet achieve the quality that reasonable quality conventional 'lost air' castings can achieve, even when produced at glacially slow filling rates that threaten the filling of the mould, or even when subjected to a variety of post-casting HIPping process. Clearly, the process is not yet fully out of serious problems. Perhaps further development to solve its problems may be successful.

In the meantime, admittedly, not all castings require perfection, and quality needs to be appropriate to the product. With all its faults, lost foam remains highly appropriate for such castings as motor housings equipped with closely spaced cooling fins where dressing of flash (extensively required for instance at the parting line of greensand moulded castings) is not easily accomplished between each fin, and does not exist for lost foam products. Other products having long, complex internal passageways may lend themselves to lost foam, whereas to provide such passages by the use of cores might be practically impossible.

Even so, the launch of this process onto an unsuspecting foundry industry before its full development into a reliable process is a charge often heard. The charge can be countered, however, because, clearly, the buyer should have carried out his due diligence, and this would not have been difficult. Sadly, many lost foam operations have struggled and finally shut down their operations as a result of these failures. To close a disappointing chapter in the history of the casting industry, it would be good to think that lost foam might one day emerge to become the reliable process we all would like it to be.

As a postscript to this less than flattering account, based perhaps on my expectations and aspirations that might be set rather high, if not too high, the reader might wish to see the excellent account given by Donahue and Anderson (2008) which is much less pessimistic and does not regard lost foam defects so negatively. The company they work for has a successful lost foam production line which incorporates pressurisation immediately after casting.

Replicast

Replicast is a process in which the invested coating of the foam is built up rather more thickly than for conventional lost foam casting, so that although it remains fragile it can be self-supporting. The foam pattern with its thickened ceramic shell is then fired to burn out the foam. The shell is immersed in an unbonded aggregate to support the shell during casting, and the shell can then be filled with liquid metal (Ashton, 1991). Naturally, the process has many of the benefits of lost foam – plus many of its disadvantages. The major benefit is the elimination of foam, eliminating the major defects that characterise lost foam castings. As a kind of investment casting process, the process has traditionally suffered from the problem typical of most investment casting processes; the provision of a poor filling system. This is, of course, probably not endemic to the process but appears not to have been explored at the time of writing. Developments here are awaited with interest.

16.8 VACUUM MOULDING (V PROCESS)

The vacuum moulding process, or V process, is an interesting hard sand moulding process which can maintain the bonding and rigidity of the mould without requiring control over materials of imprecise chemistry such as resins. It was invented in Japan in 1972 (Akita, 1972) and immediately attracted worldwide interest.

In summary, it uses a dry, unbonded, free-flowing aggregate, usually silica sand. The mould is made in a special moulding box, sealed on top and bottom surface by a plastic film, and the dry, unbonded, free-flowing sand is consolidated by the application of a vacuum (actually a partial vacuum in the region of 0.5 atm). It is significantly different from other moulding processes, and thus possesses several unique advantages plus, of course, some unique disadvantages.

In practice, a plastic film is heated by overhead radiant heaters to soften the film, making it more stretchable. It is then drawn down onto the pattern by the application of a vacuum, drawing air through vents planted in the surface of the pattern. The vents connect through to a hollow base plate through which the vacuum is drawn. In particular, vents on the pattern need to be sited at the bottoms of recesses, or in sharp corners to ensure that the film sucks completely down, faithfully reproducing the contours of the pattern.

The film is subsequently coated by spraying with a ceramic coat. In the move from alcohol- to water-based coatings, the drying time has become a significant disadvantage. When the coating is dry, a moulding box is lowered onto the pattern, and dry sand is poured into the box, being consolidated against the plastic film by vibration. A second plastic film is applied, covering and sealing the top of the mould, and a vacuum is drawn through the walls of the moulding box. The mould then becomes impressively hard. So long as the vacuum is connected to the box the mould can be handled like any other mould. A major disadvantage of some interpretations of the process is the flexible connection to the mould. The floor of the foundry can resemble a snake pit of trip hazards. Fortunately, some good engineering designs for automated plants avoid this problem. Others have valves on the mould that can be turned off to allow the mould to be disconnected from the vacuum line, but this introduces the risk that small leaks in the film or elsewhere will cause the mould to collapse as the vacuum is gradually lost. If this happens during the handling of the mould, the event can be a hazard.

The mould enjoys the benefits of other hard sand processes such as the core assembly process, in claiming to produce among the most dimensionally accurate castings.

A remarkable and useful feature of the V process is that because of the presence of the plastic film which confers a glossy, slippery, almost sensuous smoothness, to the mould surface, moulds can be stripped from patterns with almost zero force without the aid of any taper or draft. Even slight negative draft of a few degrees can be tolerated because the mould can spring back elastically if the deformation is not too severe. The positioning of cores is also delightful because the core slides into place into zero-clearance prints and is gripped by the mould in a unique way. The puncturing of the plastic film at the base of the core print before the insertion of the core ensures that the core can vent; otherwise, core is sealed from the outside world so that core blows could become a serious possibility. (Small holes in the plastic films appear to be harmless to the process while the vacuum continues to be drawn.)

There seems to be little problem with the silica sand aggregate, but although the *moulds* are of excellent accuracy, the *casting* can suffer distortion as a result of the heating up of the sand and its expansion because of the alpha/beta quartz transition. I am not aware of anyone using a non-silica sand for this process, even though the process would be ideally suited to an improved aggregate.

The use of an unbonded sand eliminates, of course, the cost of a binder which is particularly valuable for large castings requiring only a modest rate of production. For this reason, the process has achieved good success for the production of iron castings for piano frames, counter-weights for fork lift trucks and slag buckets for steelworks and the like.

During the filling of the mould with cast iron, the hot metal radiates heat ahead to vapourise the film before its arrival, the vapour recondensing in the nearby cool surface grains of the aggregate to create a temporary binder to that localised region, and thus assisting to avoid the collapse of the surface during the few seconds that the support of the film + vacuum combination is lost. Any film that survives evaporation is submerged by the iron usually without problem as a result of the high density and high surface tension of iron, although Schmied (1988) draws attention to occasional blow holes and spatter from the vaporisation of the film.

These problems are significantly more acute for Al alloys. Negligible heat is radiated ahead by the advancing Al front, so that the plastic film is always over-run by the liquid metal. The film boils and the vapours easily penetrate the metal as oxidising bubbles, forming bubble trails and laminations right through the cast section. The consequential oxide films and bubble trail defects are often misinterpreted as entrained plastic film.

The spray coating of the mould is a significant disadvantage of the process. It seems to be generally required to prevent collapse of the mould during the pouring of the liquid metal. Because the coating is applied to the plastic film on its sand side, away from the pattern, the coating does not affect accuracy. It acts to maintain the surface finish of the casting, reducing metal penetration by the suction of the vacuum. It also supports the maintenance of the vacuum, and hence the integrity of the mould, during the early stages of the filling of the mould. A coating is a significant disadvantage of course, because it costs money, requires drying time thus requires floor space and affects production rate; all of these resources are usually at a premium.

The 15–20% slower cooling rate of evacuated moulds is an advantage for the running of thin sections such as bath tubs (Clark, 1989), but a disadvantage for those products requiring high mechanical properties.

Whereas the moulding of bath tubs does not present undue challenges, there are a large number of cases in which the process will not mould for what appears to be a trivial but enormously frustrating reason. If two upstanding features on the pattern stand nearer together than the depth between them, then the plastic film bridging the upstanding features has difficulty to stretch to reach the bottom of this hollow, the elongation being required is 200% (i.e. stretching to a distance of three times its original length). The film will usually tear, and the mould is spoiled. It is surprising how often this troublesome issue of 1:1 spacing-to-depth ratio occurs. Apparently innocent shapes on the pattern can bring the process to a stop. The problem can be overcome by human intervention, applying a 'patch' of plastic film to seal the local leak.

Another concern about the process generally overlooked is the presence of the partial vacuum inside the mould cavity itself. In principle, the mould cavity should not be at a reduced pressure. In fact, if it has feeders open at the top, it will originally be at atmospheric pressure. However, as a result of small leaks in the film, parts of the mould will have air sucked out. Thus as the melt reaches these parts, and if the filling metal cuts off connections to the atmosphere, the melt will accelerate, possibly becoming locally turbulent. Thus the advantages of a nicely designed filling system are potentially lost, the control over the flow being lost in randomly different regions of the mould on different occasions, leading to inconsistent quality of the products.

The vacuum pumps work in an environment full of abrasive silica dust, so that pumps are usually of the water-ring seal variety, avoiding rubbing sealing surfaces. Even so, maintenance of the pumps seems to be a significant process cost. The energy to drive the pumps is another major cost. Other practical costs are listed in the frank article by Enderle (1979).

After pouring and solidification, the casting is released from the mould simply by turning off the vacuum and releasing air into the mould. The mould collapses onto a grid, freeing the casting from its mould with minimum effort.

The sand falls through the grid for recycling. Only 1–2% sand losses are reported (Engels and Schneider, 1986), although this figure seems low for iron castings in view of the dust nuisance reported at shake-out. The dust may result from sand fracture at the higher temperature of iron and steel castings, plus the reversion of the coating to powder when dry, all carried aloft by the strong convection of the air around hot castings.

While the dust ascends skywards, accompanied by a small mushroom cloud of unpleasant smoke and fume from the incomplete burning and pyrolysis of the film, the residual plastic film from the outer areas of the mould remains in a heap on the grid. The film is partly blackened and sticky, with adhering aggregate and remains of broken cores as a tangled, dusty and gritty mess requiring disposal. The volumes of spent plastic for disposal are impressive. It is a messy conclusion to an otherwise elegant and relatively clean process.

16.9 VACUUM-ASSISTED CASTING

Moulds poured in air may be assisted to fill by the application of vacuum to some part of the mould.

A common use is in plaster block moulds. These are formed in steel boxes with open tops and open bases. The open top exposes the sprue entrance for the gravity pour, plus any vents or feeders. The open base is placed on a manifold connected to a vacuum line, and roughly sealed to prevent excessive leakage of air into the vacuum line. Air is drawn through the permeable mould, so that after pouring, the metal is assisted in its flow, effectively being pulled through the mould into all the finest features and corners Figure 16.30. The technique achieves the filling of extremely narrow sections in the region of 0.5 mm.

In a very different application, a sand mould filled by a bottom-gated gravity system may have problems filling completely; the net head for filling decreasing as the melt level builds up in the mould cavity. Some bosses or ears at the top of the casting may have problems to fill. The application of a gentle suction from a domestic vacuum cleaner, applied locally to the top of the sand mould in the region of the unfilled features will greatly assist the filling. This technique has the benefit of not affecting the filling of the main portion of the mould, which is probably filling satisfactorily. The effect is only felt by the metal as it nears the source of suction, so that the ever-decreasing volume above the melt experiences a concentration of the suction effect, accelerating the melt into these remaining volumes.

The great power of a modest amount of suction to transform the filling of moulds is easily explained. For instance, a perfect vacuum of 1.0 atm, as every reader should know, would be capable of sucking up 760 mm height of mercury (whose density is 13,500 kg/m^3). For liquid aluminium of density, only 2400 kg/m^3 this level of suction could raise the level of aluminium by the huge height 4.2 m. Thus, to make a few centimetres difference to the height of an Al casting, we need a maximum of only 0.1 atm (1.5 psi); even this tiny pressure difference can raise the liquid height by an impressive 420 mm.

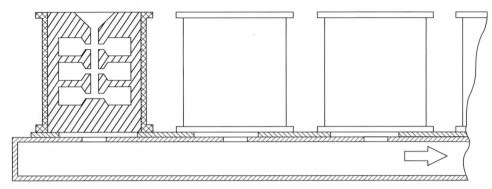

FIGURE 16.30

Vacuum-assisted casting, showing plaster moulds sitting on a vacuum manifold lined up for pouring.

For the ferrous metals and copper-based alloys whose densities are in the region of 7000–8000 kg/m^3 an atmosphere corresponds to a height of approximately 1.3 m.

Vacuum-assisted casting is often misleadingly called vacuum casting, implying the assumed benefits of low gas content and an expensive production process. Buyer beware.

16.10 VACUUM MELTING AND CASTING

In the shaped casting industry, although some melting and casting of Ti and Nb alloys is carried out in vacuum chambers, often under an inert gas atmosphere, the majority of the melting and casting under vacuum (or inert atmosphere) is carried out by the Ni- and Co-based superalloy casters. This and following sections deal mainly with this industry.

When melting under vacuum, the melt attempts to equilibrate with the vacuum environment, thus losing its volatile content, such as volatile alloys and gases. The volatile metals re-condense as dust over the internal surfaces of the vacuum chamber, occasionally flaring off as a flame gently licks around the interior of the furnace, burning the dust to oxide when the vacuum is released and the door is opened to let in air. Vacuum chambers therefore become very dirty environments, coated in black dust. From time to time if burning does not occur naturally the dust may require to be intentionally set alight to become a relatively inert oxide, and then be carefully cleaned off with a suction cleaner.

The loss of alloying elements sometimes has to be compensated by alloy additions, although some losses (such as heavy element contaminants including Pb, As etc.) are of course usually beneficial.

The fundamental problem with the 'vacuum' is that it is far from perfect. It seems best to view it as 'dilute air'. The residual oxygen and nitrogen in the dilute air is effectively 'gettered' by the active elements Al and Cr which are universally present in high temperature alloys based on Ni and Co. The pouring of the melt in the dilute air entrains bifilms into the melt in generous amounts. The entrained bifilms are rich in very stable oxides, alumina, chromia and titania. The evidence for this behaviour is compelling. It is set out in Section 6.72.

The foundry generally melts pre-alloyed metal purchased from an alloy manufacturer. The alloy producer mixes pure liquid nickel with the alloying elements, melts it in a large induction furnace to achieve a homogenous melt, then casts it into convenient 'sticks' or 'logs' of various diameters for remelting and casting by foundries. Unfortunately, massive damage is done to the alloy by the way the alloy is cast into sticks. This is carried out by pouring from the tilting furnace into a series of launders, the melt falling from launder to launder, and finally falling into tall steel tubes. These pouring events, involving a fall totalling several metres, entrain quantities of oxide films, plus probably nitride films. Thus the alloy arrives at the foundry already in a poor condition before being remelted by the foundry and poured for the final time to make shaped castings.

Even if the Ni-base alloy producer were to improve his casting process, the casters of the shaped casting are, at this time, unfortunately likely to undo any good that the alloy caster could provide. This disappointing situation

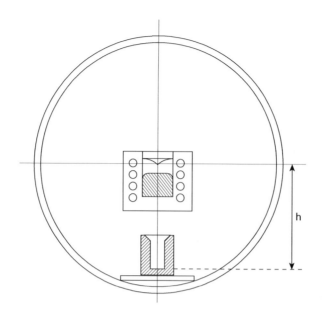

FIGURE 16.31

The unfavourable pouring geometry of a typical vacuum melting and casting furnace.

follows because of the geometry of conventional vacuum melting and casting furnaces used in foundries because the melting crucible is typically a metre or more above the mould. Figure 16.31 shows the problem in the form of the fall distance h. Metal entry velocities into the filling system are therefore often in the region of 4–5 m/s. The entrance into the filling system is always a conical funnel, concentrating and accelerating the melt, together with its entrained vacuum, even further, so that damage from turbulence is practically guaranteed. This is the regrettable situation for the current manufacture of nearly all of the world's Ni- and Co-base alloy turbine blades for aircraft engines. For single crystal turbine blades, the melt suffers a similar traumatic filling mode, but moulds are held above the freezing point of the melts during filling, and only slowly withdrawn into a cooling zone to grow the crystal under conditions of controlled speed and temperature gradient. The holding of the melt for this relatively long period probably helps to float out a large proportion (but clearly not all) of the oxide bifilms entrained during the pour. As well, the directional solidification will push oxides ahead rather than entraining them in the solid. Thus there are good reasons why the single crystals blades appear to be more free from bifilms failures than polycrystalline blades. The lack of bifilms is the real reason for the good properties of the single crystal (not that it lacks grain boundaries).

There is an interesting situation with the Ni-base alloys containing hafnium (Hf). It seems that when these alloys are cast, a curious unexplained glassy surface defect on the blades that appears to run down the casting surface like a congealed river. If a filter is placed in the filling system, the defect does not occur. It seems likely that this represents an attack of the ceramic mould by HfO surface films generated by turbulence. The HfO will most likely form a low melting point oxide mixture with the components of the mould, often mainly Al_2O_3. If the filter is present, any HfO already present will be reduced by filtering out, and the reduced turbulence after the filter will not generate more HfO.

Clearly, this is an industry that is very much in need of improved melt handling technology and would greatly benefit from a transition to better mould filling techniques. These are already well proven and demonstrated for aerospace high temperature alloys in the case of gravity systems that reduce defects by approximately a factor of 10 (Li, 2004) and better still, counter-gravity systems (Shendye and Gilles, 2009).

CONTROLLED SOLIDIFICATION TECHNIQUES

17.1 CONVENTIONAL SHAPED CASTINGS

Most castings will solidify from the outside and will grow grains that will finally merge and consolidate in the centre of the casting. Central regions that form hot spots not in communication with a source of feed metal will cause some local shrinkage that may take the form of internal porosity if the porosity can be initiated, meaning in practice that bifilms will be required to be present. Alternatively, in the absence of bifilms (i.e. very clean metal), local shrinkage may create external porosity in the shape of surface sinks.

Although many foundry texts recommend the provision of a feeder to counter this problem, my personal approach is to avoid the provision of a feeder if at all possible. I prefer to use some kind of chill. This may take the form of a substantial block of cold metal forming part of the mould, or might even be an internal chill around which the melt freezes. More usually, however, I opt for the provision of cooling fins.

Cooling fins work excellently for high-conductivity alloys such as Al and Mg alloys and some Cu-based alloys (although not particularly well for aluminium bronzes). Unfortunately, the concept works so poorly for steels that fins cannot be recommended. For grey cast irons, the situation is somewhat intermediate, with fins often working sufficiently well to strike back white iron locally into the casting; such an effect has been used historically for the tips of cams on camshafts for automotive engines, although there seems to have been a move away from this technique to the placing of shaped chills in the interests of greater reproducibility and control. More recently, there is interest in the production of wear faces on cams not by solidification control but by subsequent heat treatment of the casting by local application of rapid induction heating.

In the special case of conventionally poured and solidified polycrystalline turbine blades, the fundamental problem in the casting of these alloys in a conventional vacuum induction furnace is seen in Figure 16.30. The height 'h' of the fall of metal into the mould, followed by freezing from all directions, means that bifilms created by the fall have little chance to escape, but become trapped in the casting. As a result of them being pushed by advancing dendrites, they naturally finish up in grain boundaries. As a result, during creep or other tensile failure modes, the castings fail by 'cavitation' at the boundaries, the growing cavities gradually joining to lead to complete rupture. This explains the relatively poor properties of polycrystalline equiaxed turbine blades compared with other grain structures (Figures 17.1 and 17.2) as described in the following sections.

17.2 DIRECTIONAL SOLIDIFICATION

The unidirectional solidification of castings has a long history, promoted initially by the desire to obtain high soundness; the high temperature gradient shortening the pasty zone so that feeding could occur to the roots of the dendrites with greater efficiency. Directional solidification (DS) is mostly used to imply unidirectional solidification. It has been used for permanent magnet manufacture, but its main use has been for turbine blades for jet engines.

In practice, DS has nearly always been carried out vertically in an upward direction. It was not suspected that this particular mode of freezing would yield an additional important effect; a significant reduction in the bifilm population

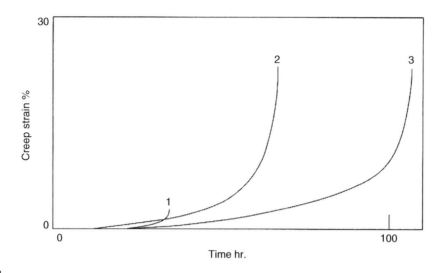

FIGURE 17.1

Creep behaviour of Ni-base alloy Mar M200 at 980°C and 207 MPa cast as (1) conventional equiaxed structure; (2) directional solidification (DS) columnar structure; (3) single crystal.

After Versnyder and Shank (1970).

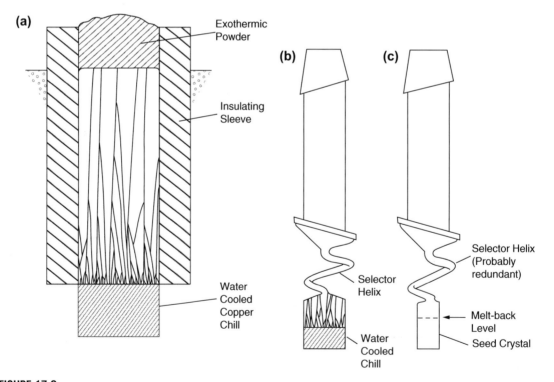

FIGURE 17.2

(a) Directional vertical solidification of a steel casting *(after Polich and Flemings, 1965)*; (b) single crystal by selection from a chilled DS base; (c) single crystal grown from a seed and therefore accurately oriented.

This occurs partly (1) by the vertically directed freezing allowing bifilms to float upward, away from the freezing front; and (2) because of the unusually long time available for this flotation because of the relatively slow rate of advance of the front during DS; and (3) any residual bifilms will have the best chance to be pushed ahead of the advancing solidification front, thereby keeping the solid relatively free from serious defects.

Work at MIT first published in 1965 for steel castings, by various combinations of the authors Nereo, Polich and Flemings, quickly showed benefits to the strength, toughness and fatigue resistance revealed by directional solidification. Later work (Flemings and Mehrabian, 1970; Hurtuk and Tzavaras, 1975) has further confirmed the beneficial effects of DS for steel castings. The melting and pouring in this work was all carried out in air into relatively simple moulds (Figure 17.2(a)) approximately 100 mm diameter and 230 mm tall so that significant oxidation and bifilm contamination of the melt would be expected, particularly during the final pour into the moulds.

Significantly, the benefits to properties were a strong function of dendrite arm spacing (DAS) in non-directionally solidified castings but *not* for DS castings. This result is difficult to explain unless one assumes that bifilms are trapped amongst grains during equiaxed freezing but not in DS material. Thus for the relatively clean DS material, the properties are not affected by DAS as seen in Figure 9.26 confirming that mechanical properties are a function of bifilms; the *apparent* strong effect of DAS in most castings and many alloys is the result of presence of unfurling bifilms.

This work on steels coincided with a technically analogous, but, as it turned out, far more important and exciting breakthrough in Ni-base superalloys. Versnyder and Shank (1970) review the early development of DS and single crystals for blades and vanes for use in aircraft turbine engines. Their world-famous summary of their creep behaviour is seen in Figure 17.1. These early studies also indicated that DS polycrystal castings possessed up to 10 or 100 times greater fatigue life compared with equiaxed polycrystal castings (Leverant and Gell, 1969). Despite this spectacular advance, Zhou and Volek (2007) draw attention to the fact that performance, particularly in creep, remained limited by crack formation along both the longitudinal and transverse grain boundaries of the DS castings, although the failure of transverse boundaries is most usually cited as the driving force for the development of single crystals, in which, of course, all grain boundaries were eliminated. Because, of course, grain boundaries are enormously strong, it seems most likely that these grain boundary failures were not the result of failures of the grain boundaries themselves but the result of bifilms segregated to the boundaries by dendrite pushing. This seems especially likely in view of the extreme turbulence associated with the manufacture of these vacuum cast blades, so that the presence of bifilms is guaranteed. They are such serious defects that it is reasonable to suppose they will have a major effect on properties.

If, as seems reasonable and logical, it is accepted that grain boundaries are strong, the Versnyder and Shank result shown in Figure 17.1 cannot be explained by conventional metallurgy. The DS and single crystal technology for turbine blades in jet engines is, perhaps, the best evidence yet for the profound effect of bifilms in castings.

It is sobering to realise that the poor properties shown by the equiaxed casting in Figure 17.1 are typical of the bifilm-crammed structures of conventional castings which we produced in our aluminium, steel and titanium foundries every day; in fact, of course, in practically all our foundries of all types. This is where metallurgy is today; we are currently producing poor metallurgical products which have enormous potential for improved properties as we achieve progressively better ways to bring bifilms under control for the first time.

17.3 SINGLE CRYSTAL SOLIDIFICATION

The benefits of DS are brought to complete fruition in single crystal growth.

By the provision of a crystal selector such as a helical channel (known as the 'pig tail'), a single grain orientation can be selected close to the <001> favoured dendrite growth direction. Grains growing in other directions impinge on the walls of the helix and are eliminated (Figure 17.2(b)). Alternatively, the provision of a seed crystal allows any direction of growth to be selected, effectively making the pig tail redundant (even though to this date the pig tail seems to have survived as in Figure 17.2(c) for no good reason that anyone can think of).

After the melting and pouring of the metal into the mould, the mould is withdrawn slowly from a hot zone at a temperature above the melting point of the alloy into a cold zone. The two zones are separated by a baffle designed to fit around the mould as closely as possible to maintain the temperature gradient as high as possible. Withdrawal for an average turbine blade might take up to 2 h.

By the production of a single crystal, grain boundaries (thought to be the weak features of the structure) could be eliminated. In addition, the alloy could be developed for maximum properties such as strength and oxidation resistance because the absence of grain boundaries meant that complications of the control of grain boundary precipitates were avoided. Thus turbine blades and vanes for aircraft engines took a further leap forward, beyond the attainments of the DS castings, and are now used for the most demanding locations in turbines, withstanding extremes of stress and temperature, justifying their name 'superalloys'.

It is a pity that at this time the casting of these excellent materials has to suffer casting conditions that risk the introduction of defects. The risk is significant because of the huge fall in properties that is clearly possible as illustrated in Figure 17.1 (interpreting the figure as illustrating the effects of bifilm rather than the effects of grain boundaries). The latest vacuum furnace designs have reduced the height h (Figure 16.30) of the melting crucible above the mould, but it remains inescapable that the fall will spell problems for good control of both structure and properties.

The target to achieve a single crystal is often undermined by the growth of various kinds of secondary or stray crystals. Carney and Beech (1997) found oxides at the root of most such misaligned crystals, as might be expected if bifilms were present, simply because crystals would not, in general, be expected to grow through bifilms. Although the directional growth of single crystals would be expected to push many bifilms ahead, thus clearing the casting of many if not most defects, some bifilms that stretch into or across the casting section will be expected to be attached to the mould walls, and thus be immovable. Such barriers to growth will lead to the undercooling of the melt on the far side of the barrier as the mould continues to be withdrawn into the cool zone of the furnace. Eventually, the undercooling will become sufficient to nucleate a new grain, having no necessary orientation relation to the original growth direction. This stray crystal would probably cause the casting to be rejected.

It is important to note that the bifilm barriers may not only be oxides. Ford and Wallbank (1998) found that all castings with additions of nitrogen to the melt formed stray crystals, suggesting the existence of nitride-based bifilms.

In passing, we should note that the seed crystal is covered with an oxide film which appears to be no problem to the orientation of the seeded growth. This is probably as a result of the oxide on the seed being an old, thick oxide and/or nitride that grew in the solid state. It would be expected to be highly fractured and porous, unlike most bifilms which would be only micrometres or nanometres thick, and probably, like most thin films in their early phase of growth, rather perfect.

Other factors emerge to erode the perfection of growth of the single crystal.

1. Although the advance of the front drives the solidification of the dendrites in a vertical direction, these long stalks of solid grow relatively true to their <001> direction. However, when they arrive at a point in the casting where they have to grow sideways, the cantilevered weight of the dendrite (denser than the liquid by a few percent of course) can be supported for a limited distance. Beyond this, the bending moment at the root of the dendrite causes the sideways arm to hinge at the root, until it falls, its fall arrested by the mould or other dendrites. In this way, low-angle boundaries are created (Newell et al., 2009). The situation is analogous to the formation of the branched columnar zone in steel ingots, despite the wall-like secondary arms creating box-girder type morphology of low-carbon steel dendrites which helps to resist bending. This fact suggests that different Ni-base alloys may have dendrite morphologies which will resist this problem to different degrees.

2. The microsegregation between dendrite arms accumulates near the roots of the advancing array. If this is denser than the matrix alloy, this liquid tends to flow to the edges of the mould, creating a convex front. If the segregated liquid is lighter than the bulk alloy, the situation is unstable, and plumes of segregated liquid can ascend through the mesh, dissolving the mesh as it goes, forming a channel segregate. The dendrite arms that are melted off tumble into new orientations in the channel, giving the name 'freckle defect' when studied in the etched condition. Ever since McDonald and Hunt uncovered the mechanism for the formation of the channel segregate in 1969, there has been a huge amount of work on this subject, particularly in relation to superalloys (for instance Purvis et al., 1994) and more generally reviewed by Beckermann (2002). A key parameter now included so far as possible in new alloy

designs is the balancing of elements in an attempt to make the interdendritic liquid as near neutral buoyancy as possible. A second powerfully controlling factor is the spacing of primary dendrite stems; fine spacing provides useful viscous drag to help dampen interdendritic flow.

Although the Ni-base superalloys have excellent resistance to oxidation at high temperature, their resistance must be enhanced further for the highest temperature applications. This is achieved by various kinds of coatings, often by a kind of aluminising process, in which aluminium is diffused into the surface of the casting and then subjected to oxidising conditions to generate a strong protective layer of alumina. The beauty of such a coating is that it is mainly metallic, having some ductility and thus resistant to damage such as spalling experienced by brittle or poorly adherent coatings. Also, because of its reserve of aluminium in depth, it is self-healing if damaged.

We need to note, however, that in their studies of the protection of Rene 80 with an aluminide coating, Rahmani and Nategh (2010) observed that cracks in the casting associated with grain boundaries and carbides (features strongly suggesting the presence of bifilms) unfortunately overwhelmed the effects of the coating because the cracks from the matrix extended through the coating. Once again, it seems that bifilms are likely to be responsible for this major potential failure mode of an otherwise excellent technology, making improved casting technology an urgent issue for superalloy castings.

If it is true that single crystal blades derive their extraordinary properties mainly by the adventitious reduction of bifilms, it follows that there are almost certainly much cheaper ways to obtain bifilm-free alloys by sensible melting and casting techniques. Thus polycrystalline blades, produced quickly and cheaply, could in principle equal the creep properties of single crystals, although they would not have the unique directional properties of single crystals which are exploited in some blade designs to improve stiffness and vibrational response of course.

Alternatively, of course, an improved melting and casting approach could bring further benefits to single crystals, eliminate the current risks to integrity, providing manufacturing economies and possibly further improving properties to give extended times between service inspections of engines.

17.4 **RAPID SOLIDIFICATION CASTING**

The use of the term rapid solidification has traditionally been attributed to such processes as splat cooling of perhaps a gram of material between cold anvils, or the production of ribbon off a rapidly rotating wheel. These process give cooling rates in the region of millions of degrees per second, but the product is hardly a shaped casting, but rather more a heap of bits or ribbons requiring some other process for consolidation.

Here we shall content ourselves with more limited rates of cooling, at rates in the region of perhaps a hundred degrees per second. These more modest rates are still capable of delivering excellent if not astonishing properties, but at the same time, of course, can directly deliver engineering products such as shaped castings.

Sophia and Hero processes

During freezing of a conventional shaped casting, the casting contracts away from the mould, and the mould expands, opening the so-called 'air gap' between the casting and the mould (possibly containing a variety of gases except air). The gap is a major barrier to the rate of escape of heat from the casting. Thus the air gap, above nearly all other resistances to cooling, controls the rate of cooling. The metallurgist will be concerned about the fineness of the microstructure, particularly the DAS, but as we have seen the mechanical properties of castings seem mainly dependent on the unfurling of bifilms. Thus, an absence of bifilms, if we could achieve this, would make rapid cooling unnecessary to achieve good properties, but it is important not to overlook that the finer dendrite spacing will be a great benefit to the reduction in heat treatment times.

Some Al alloy investment castings are subjected to a more rapid solidification by being immersed in a coolant immediately after the mould is filled with liquid metal. The Sophia Process is known to involve quenching the mould and its liquid metal contents into a soluble oil. At the same time, a pressure is applied to close porosity so far as possible. These actions significantly increase the properties of casting, and make any further enhancement by, for instance, hot isostatic pressing, to be unnecessary.

The Hero Process used by Tital in Germany is thought to be similar. However, the details of the process remain closely guarded despite the claimed original patent now likely to be nearly or actually expired.

The quenching of the complete investment mould and its liquid metal into a cooling liquid certainly raises properties, but is clearly a messy process, especially if the quenchant is an oil. Furthermore, of course, the efficiency of heat transfer is also limited by the presence of the investment shell that remains in place during freezing. These issues are largely solved by the following sand casting process, even though, of course, an aggregate mould process cannot match the surface finish enjoyed by the investment casting.

Ablation casting

I first came across the word 'ablation' in relation to a NASA shuttle space vehicle re-entering the Earth's atmosphere. To avoid the vehicle heating up and being destroyed, its leading surface was covered with ceramic tiles that were designed to heat up and vaporise, the vapour carrying away and leaving behind the frictional heat generated by the atmosphere so that the space craft itself did not heat up. Thus the term ablation denotes an eroding away and carrying away process.

Ablation casting is a new approach to the making of an aggregate moulded casting in which the fundamental limitation to heat flow presented by the air gap is removed by the application of coolant direct onto the surface of the solidifying casting. As a necessary prior step the mould is removed to allow direct access of coolant. This is achieved by the application of a solute directed onto the mould, but with the mould aggregate bonded with a binder soluble in the coolant. The mould is thereby dissolved away and removed in the flow of the coolant. Thus, unlike Sophia or Hero, in ablation casting, the mould does not obstruct cooling.

In practice, therefore, the liquid metal is poured into an aggregate mould bonded with a water-soluble binder. While the metal is still molten, the mould is ablated away by water jets (Figure 17.3(a)). Because the mould is progressively removed, disappearing steadily as ablation advances along its length, the water is enabled to contact the metal casting directly. A high temperature gradient is progressed through the casting, assisting to eliminate porosity especially in thick sections. The outcome is solidification of a shaped casting under (1) an unprecedented temperature gradient and (2) high solidification speeds. A unique microstructure characterised by extreme soundness and fineness of the last phases to solidify is thereby produced. Porosity can be controlled to be extremely low or effectively zero even in such features as isolated bosses. This is because freezing conditions are now no longer under the control of the mould, but under independent control from ablative cooling which is managed and directed by the casting engineer.

The solidification can be directed to a final corner of the casting where a feeder is located, possibly planted off the casting on a short extension, as an extended feeder neck. Because the freezing front arrives at this point fairly quickly, the feeder can be arranged to drain almost completely, its final form being only a metal skin like an empty paper cup. Feeding can therefore be extremely efficient, so that metal yields can be high.

At the completion of ablation, the mould has been completely removed and the casting sits on the ablation station clean and cold, ready to be picked up immediately for further processing (Figure 17.3(b)).

The mechanical properties achieved by the process are equal to or surpass the best competitive processes such as squeeze casting. This is perhaps not so surprising because the pouring conditions for most squeeze products are poor, introducing defects. In addition, cooling conditions for most squeeze castings are heterogeneous, with only parts of the casting feeling the compressive force holding the casting and mould in contact. Ablation casting can use any form of filling, with the current gravity filling processes naturally suffering defects. Nevertheless, despite the general use of gravity filling during this early stage of development of the process, rapid solidification freezes in bifilms in their compact state so that properties remain good. Also, it is worth noting that the company, Alotech Limited, generally applies filling systems that conform to the principles of this book, which has to be a further significant advantage over most competition at this time. Even so, the properties are surprisingly good, leading one to suspect some additional factor. It seems likely that the unusually powerful temperature gradient of the process may be efficient in assisting the freezing front to push bifilms ahead, effectively cleaning large areas of the casting. When the process is further developed, perhaps introducing a filling process such as counter-gravity, properties from ablation should give yet further improvement, even though, of course, competitive processes should also improve if provided with such advantageous filling systems.

FIGURE 17.3

Ablation casting of an Al alloy; (a) eroding the mould away and cooling the casting; (b) the final casting, cold and clean.

Additional advantages for ablated castings appear to be their highly competitive cost. This arises mainly because the silicate-based binder is much less costly than competitive resin-based binders, but also benefits from the relatively low cost of capital equipment and modest cost of tooling. Further benefits arise as a result of the water-based technology. The process produces no smell, no fume and no dust (the recycled aggregate is mostly either wet or damp), saving the installation of costly fume and dust extraction equipment.

The optical microstructure of an Al-7Si-0.4Mg test casting that was allowed to solidify without ablation is shown in Figure 6.9(a), and is typical of a non-modified Al-Si eutectic. Typical ablation-cooled structures are shown in Figures 17.4 and 17.5 illustrating the extreme fineness of the eutectic spacing that is difficult to resolve even at 1000 magnification in the optical microscope (but only limited regions of the dendrite spacing have been refined).

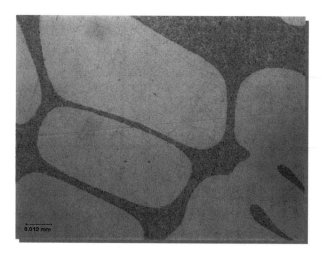

FIGURE 17.4

Structure of an Al-7Si-0.4Mg alloy achieved by ablation, without added modifiers such as Sr or Na. In the optical microscope, the dendrites are seen to be coarse, but the Al-Si eutectic is hardly resolvable even at 1000×.

The Si inter-particle spacing is under 1 μm in this structure. Note that no Na or Sr has been added to achieve this apparently well-modified structure.

Figure 17.5 illustrates the extent of freezing that has taken place before the ablation cooling has had time to reach these regions. Thus the secondary DAS is in general rather coarse, and typical of a sand casting of that section thickness. In some instances, particularly in the centres of sections, it is at first sight curiously perverse that the DAS is fine because fine structures as a result of rapid cooling are normally limited to the outer skin of the casting. The fine central regions are of course the result of ablation arriving in time to limit dendrite arm coarsening in these regions. Furthermore, such regions would often contain porosity, thus weakening the structure, whereas ablation tends to yield sound and fine grained interiors that would be expected to have enhanced properties.

It is extremely significant that the improved properties are clearly not directly related to DAS, which in general is not improved. Improved properties appear to be a result only from improved control of the solidification of the eutectic. The current preoccupation focussing on the achievement of a fine DAS is therefore seen in general to be a misunderstanding (although of course it is correct that DAS indicates the freezing time of the dendrites).

Alloys that are traditionally difficult to cast because of hot tearing problems such as the Al-4.5Cu series alloys prove to be relatively straightforward with ablation, although the rapid and directional solidification must be carefully controlled to avoid dendritic segregation problems (inverse segregation). Furthermore, the wrought alloys including the 6000 and 7000 series have been demonstrated to be capable of being made into shaped castings. Much of this success with otherwise difficult casting alloys arises because the casting does not build up long-range stresses during casting: the casting ahead of the freezing front remains molten and can therefore effortlessly accommodate strain without causing stress.

Perhaps surprisingly, ablation has produced excellent Mg alloy castings which have displayed properties well in excess of those of any other competing Mg casting process at this time.

At the time of writing, the process is protected by several patents, and is, in common with many new processes, probably somewhat more difficult than it looks, even though its difficulties are clearly worth overcoming. It is still undergoing rapid development, and at this time is just starting its first series production castings.

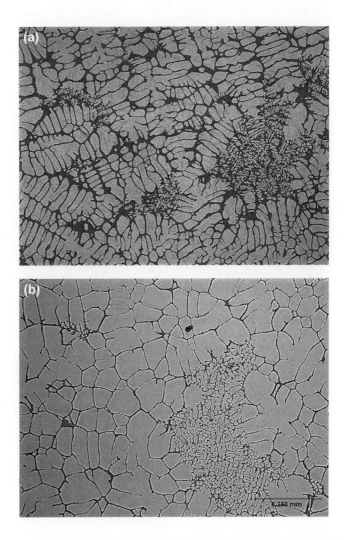

FIGURE 17.5

The dual dendrite structure often observed in ablation-cooled castings of Al alloys as a result of natural cooling finally caught up by ablation cooling; (a) in an Al-7Si-0.4Mg alloy containing significant eutectic phase; (b) in an Al-4.5Cu solid solution alloy exhibiting no significant eutectic phase.

For the future of ablation, a combination of the moulding and solidification technology with good filling, possibly via a counter-gravity process, promises to be a powerful process for both Mg- and Al-based castings. It is not known at this time whether the process can be extended to encompass higher melting point copper and ferrous alloys. As we continue to note in this book, there remains no shortage of future challenges.

DIMENSIONAL ACCURACY

18

The casting which we make is, of course, never quite perfect in terms of size and shape. To allow for this, tolerances are quoted on engineering drawings. So long as the casting is within tolerance, it will be acceptable.

Some reasons for the casting being out of tolerance include elementary mistakes like the patternmaker planting the boss in the wrong place. This leads to an obvious systematic error in the casting, and is easily recognised and dealt with by correcting the pattern. It is an example of those errors which can be put right after the first sample batch of castings is made and checked. (Even in this simple case, care may be needed if great accuracy is required, as is explained a little later.)

Another common systematic error in castings is the wrong choice of patternmaker's contraction allowance. The contraction of the casting during cooling in the mould is often of the order of 1–2%. However, it depends on several factors, particularly strongly on the strength of the mould and the cores. For instance, in an extreme case, a perfectly rigid mould will fix the casting size; in such a situation, the casting simply would have to stretch during cooling because it would be prevented from taking its natural course of contracting. Such a situation is common for thin wall castings made in steel moulds, such as high pressure die castings (HPDCs). However, in other casting processes in which the mould can yield to some extent, there is often uncertainty. The choice of contraction allowance before the making of the first casting sometimes has to be decided in the absence of previous similar castings that would have provided a guide. Thus, the chosen value for contraction is often not exactly right. This point is taken up at length in a later section on 'net shape', with recommendations on how to live with the problem.

Other errors are less easily dealt with. These are random errors. No two castings are precisely alike. The same is true for any product, including precision-machined parts. The International Standards Organisation (ISO) Standard (1984) for casting tolerances indicates that, although different casting processes have different capabilities for precision, in general the inaccuracies of castings grow with increasing casting size, and the standard therefore specifies increasing linear tolerances as linear dimensions increase. (Nevertheless, it is worth pointing out that the corresponding percentage tolerance actually falls as casting size increases.) Although other work on the tolerance of castings suggests that the ISO standard has considerable potential for further improvement (Reddy et al., 1988), it is the only European standard at this time. This also has to be lived with.

Because of the effects of random errors being superimposed on systematic errors, where great precision is required it is of course risky to move the boss into an apparently correct location simply after the production of the first trial casting. Figure 18.1 illustrates that the random scatter in positions might mean that the boss appeared to be in the correct place first time if the casting happened to be casting 1 in Figure 18.1, or might have been more than twice as far out of place, compared with its average position, if the first casting had been number 2 in Figure 18.1. A sample of at least two or three castings is really needed, and preferably 10 or 100. The mean boss location and its standard deviation from the mean position can then be known accurately and the appropriate actions taken.

At the present time, it will be of little surprise to note that such exemplary action is not common in the industry. This is because companies are not generally equipped with a sufficient number of fast, automated three-coordinate measuring machines. As such standards of measurement become more common, so the attainments in terms of accuracy of castings will increase. However, it is a pleasure to note that advances in metrology are occurring fast and furiously. In particular, the use of lasers and computers has revolutionised measurement of the exterior features of castings. For those fortunate enough to be able to afford it, a similar revolution has happened in X-ray radiographic tomography, so that

Complete Casting Handbook. http://dx.doi.org/10.1016/B978-0-444-63509-9.00018-2

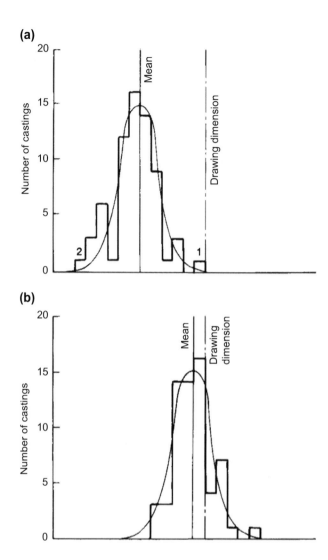

FIGURE 18.1

Statistical distribution of casting dimensions (a) before and (b) after pattern development.

Based on Osborn (1979).

three-dimensional views of the interiors of castings can now be measured with confidence to ensure that hidden wall thicknesses are correct, and that interior cores have not moved or distorted during casting. A further benefit is to ascertain that all core sand has been removed even from inaccessible pockets inside the casting.

Turning now to the accuracy of castings themselves: one of the fundamental requirements for an accurate casting is an accurate rigid pattern and an accurate rigid mould. This may seem self-evident, but it is comparatively rarely achieved.

Figure 10.2 shows several process routes for the manufacture of castings, but in Figure 18.2 they are assessed in terms of their potential for accuracy, and this potential quantified for one particular casting in Figure 18.3. In principle, only the

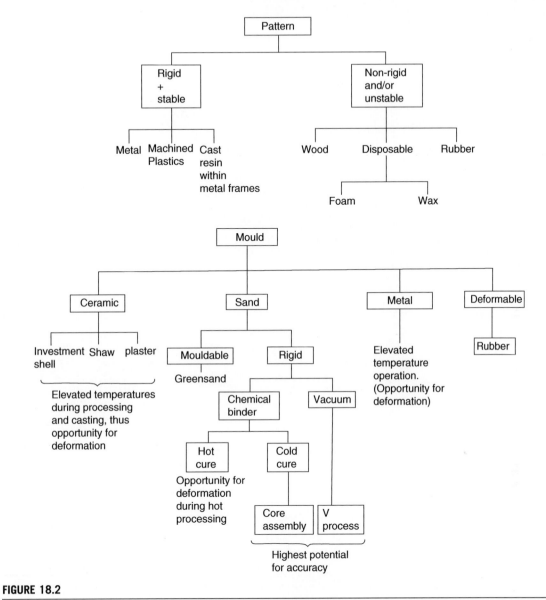

FIGURE 18.2

Patterns and moulds categorised according to their potential for accuracy.

rigid patterns in combination with the hard aggregate mould routes (core assembly and V process) meet all the requirements of accuracy, rigidity, room temperature formation and curing against the patternwork, and room temperature assembly of moulds and cores.

This is not to say that other processes are inappropriate. Some will be adequate in accuracy and have the benefit of low cost. Others will be available locally to the customer, and the foundry may have the capacity to accept the order at that time. There are many factors that influence the choice of an appropriate manufacturing route and choice of manufacturer.

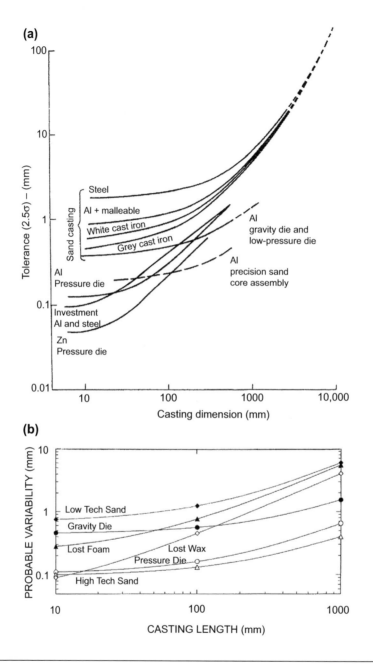

FIGURE 18.3

The average tolerance (taken as 2.5σ) exhibited by various casting processes (sand, and gravity and low-pressure die from IEE TS7I; investment from BICTA 1990; pressure die from ZDA 1990; sand core assembly data are tentative, from early production experience). Relation of casting dimensions with (a) potential variability of a hollow 5 mm wall thickness casting of overall size 250 × 200 × 200 mm, and (b) probable variability of a similar but solid casting (Campbell, 2000b).

The ISO standard for casting accuracy gives the general trend of increase in the size of random errors as casting size increases. However, the casting designer and engineer require much more detailed knowledge of the sources of individual contributions to the final total error. The remainder of this chapter is an examination of these contributions. The reduction of these errors allows the production of castings which can considerably exceed the minimum accuracy requirements specified in the ISO.

The use of computers has of course greatly aided the quest for more accurate and reproducible tooling production and replication. Computer-aided design/computer-aided manufacture has itself benefited from the rationalisation and standardisation of data transfer in the form of Standard Triangulation Language, International Graphical Exchange Standard, and Standard for Exchange of Product Information. The world-wide adoption of such standards has allowed tool sourcing to become a world-wide activity.

18.1 **THE CONCEPT OF NET SHAPE**

A 'net shape' product is one whose shape requires no further modification and thus is ready for use. The concept is one which is usually applied only to parts which are required to assemble and fit closely with other parts.

Most soft plastic toys are made net shape, but the shape is often not critical and the concept has no relevance. However, it is critical in the case of components such as Lego, where each injection moulded plastic part is required to fit accurately into other parts without any further processing; clearly, in this case, no machining can be contemplated on grounds of cost. The component must be net shape.

For many metal components, with perhaps the exception of some powder forming routes, the part can rarely be finished exactly to the required final tolerance in a single forming operation. Thus, in general, a forming operation such as forging or casting is carried out to produce a 'near-net shape' product, which is subsequently brought into the required tolerance by a finishing operation of some kind.

If we try to evaluate the various casting processes for their capability in the production of net or near-net shape products, the problem resolves to 'what accuracy can the various casting processes achieve?'

Answers to this question are tasks not easily limited to a manageable scale. For instance, accuracy is required in terms of straightness, flatness, concentricity, etc. Here we shall consider only the ability to reproduce a length.

Sources of error come in various forms: (1) random errors as a result of variations in processing and (2) systematic errors of many types, including the boss in the wrong place, and more general problems that apply uniformly to parts or to the whole of the product. For instance, these sources include such important features as the mismatch between mould halves. Also included in this category are features such as the thickness of a mould coating when applied uniformly (well or badly) to the whole or part of a mould by spraying or dipping. Systematic errors occur resulting from expansion or contraction during processing, and which are thus a function of the size of the component.

There is also the fundamental difficulty associated with the fact that an intrinsically poor processing route could be carried out with extreme care by one manufacturer, giving a satisfactory product in spite of the technology. This contrasts with the situation in which an intrinsically reproducible process is carried out with ineptitude or even crass irresponsibility by another, therefore giving an unacceptable product. Thus, although in this study the fundamental capability of the process will be emphasised, the ability of the manufacturer to overcome deficiencies in the process by intelligent and diligent effort must not be underestimated. Even so, the view is taken here that processes which are intrinsically reproducible are to be favoured over those which are not. Also, some attempt will be made to place limits on the achievements of both good and bad practice to assess the extent of the tolerance problem.

The only previous comparative study of the accuracy of the various casting processes has been carried out by the author (2000). However, because of the paucity of data on the accuracy of castings produced by different processes (notwithstanding the existence of the ISO, 1984 standard and one notable experimental attempt by the Institute of British Foundryment – IBF, 1979), he found himself driven to making estimates of accuracy of the different processes based merely on his own experience. This, clearly, is hardly satisfactory, but is judged to be better than nothing at this stage, and therefore comes with no apologies. This section considering the problems of net shape is based on the original article, and is offered as a preliminary study, outlining the concepts involved, and providing a framework in which the mechanics of the problem can be understood, and into which better data can be fitted as it becomes available.

18.1.1 EFFECT OF THE CASTING PROCESS

Inaccuracy can be introduced into the product at several stages of the casting process which typically involves the sequential production of a series of shapes, one formed intimately against the other, thus producing a series of positive and negative forms which finally give the desired positive shape. As the number of steps increases, there is an increasing chance to introduce error. The main processes are considered in order of increasing complexity, and by implication, vulnerability to error, later.

Exterior shapes

Die casting is the most direct of the casting processes. Here, a metal mould (a near-net-negative of the required shape) is filled with liquid metal which is allowed to solidify to give the final desired positive form. Pressure die casting involves the injection of the metal at high speed and high pressure into the die cavity, whereas low pressure die casting and gravity die casting use similar cast iron or steel dies which are covered with a protective refractory coating, and which are filled relatively quiescently.

Sand casting involves the additional step of having a positive pattern which is used to produce a negative sand mould to produce the positive cast form. Two main types of sand mould are used: (1) greensand and (2) chemically bonded sand. Although sand has a poor image, and is in fact often used badly and under poor control, it has considerable potential for reproducibility. It often does not require a protective mould coat and thus retains the accuracy of the pattern. Because the pattern never contacts the liquid metal it can be long-lived if it suffers little wear, retaining its original accuracy for up to 10 times longer than die casting processes.

Lost wax (i.e. investment) casting involves a further additional step, because a negative die is required, usually accurately machined from an aluminium alloy. This die to form the wax pattern is long-lived and retains its accuracy well. The die is filled with liquid wax to make a positive pattern. A ceramic shell mould is formed around this, and the wax melted out, to produce a negative into which metal is finally poured to create the final positive shape. Significant scope for variability exists in (1) the expansion and/or distortion of the wax pattern, (2) the stresses of the dewaxing operation, (3) the firing of the shell where sintering and shrinkage of the shell occurs, (4) the expansion and phase changes in the shell which occur on casting (particularly when pouring high temperature alloys and steels) and (5) the variable restraint of the mould on cooling as a result of the prior variations in chemistry and sintering of the ceramic shell.

Lost foam casting is some ways similar to lost wax, but where the disposable positive pattern is formed from polystyrene expanded inside a machined aluminium die. Again, the die is accurate and the least subject to errors of wear compared with all the casting processes. The foam pattern may be subject up to 0.8% shrinkage (depending on the type of foam (Brown, 1992)) and may require being assembled and glued together from individual foam parts. The whole is then coated with a ceramic slurry which is allowed to dry. The assembly is finally placed in a box, loose dry sand is vibrated into place around the pattern to support it during casting and extract the heat during freezing. This action of pouring and vibrating sand into place in this way usually introduces measurable distortion of the flimsy polystyrene pattern. The metal is poured into the foam, displacing it to form the final positive shape. The mould may, of course, distort during the casting and cooling process.

Interior shapes

Hollow parts of castings which can be formed by simple withdrawable shapes can be introduced into die castings. In this case, the core is a permanent feature of the die tooling and is usually made from steel, and is sometimes water cooled. Pressure die cast engine blocks are among the most ambitious aluminium alloy castings made by this process.

More complex interior shapes cannot be withdrawn and thus require to be formed by disposable cores. The most common type is the resin-bonded sand core, in which the resin is designed to break down after casting, allowing the sand to flow out of the cored cavity. Alternatively, in a few instances, cores can be dissolved out (such as salt core by water, or silica core by hydrofluoric acid in the case of Ni-based turbine blades).

The core is supported inside the mould cavity by its prints. These are (positive) core extensions which locate in (negative) print locations in the mould or die. For sand casting, the location of the core can be precise because assembly is all at room temperature, and because the core-to-mould interface is usually kept clear of uncertainties such as a variable thickness of core coating. For gravity die casting, the sand core is often poorly located because the thermal distortion of the die causes the print negatives to be out of location. In addition, the die has a variable thickness of die coat which has been sprayed on, some of which finds its way onto the print locations, causing further deterioration of their precision. Worse still, especially if core loading of the die is lengthy or is delayed, the heat of the die can sometimes soften, or even cause to break down the resin binder in the sand core, causing the core prints to disintegrate and the core to sag.

Once surrounded by liquid metal, the core will be subject to buoyancy forces which will tend to make it float. This is not so bad in liquid magnesium where the density difference with silica sand is near zero. Similar neutral buoyancy applies for the aluminium/zircon sand system. Buoyancy becomes increasingly problematic for the common systems such as silica sand together with liquid aluminium or iron or other dense metals such as copper and steel, the latter systems more particularly so because of the higher casting temperatures involved. Flotation forces have to be withstood by adequate mechanical support of the core, usually by good print design and location, or, often less desirably, by internal steel reinforcement or metallic supports (chaplets) which are cast into the product as permanent features. To ensure good fusion between the insert and the casting is not always easy. Sometimes such features are subsequently machined out.

18.1.2 EFFECT OF EXPANSION AND CONTRACTION

In greensand moulding systems the sand is bonded automatically by the compaction of the sand. However, at compaction pressures above about 1 MPa (10 bar) the sand mould deforms elastically, and, on withdrawing the pattern, is subject to 'spring back'. This general distortion of the mould leads to numerous difficulties relating to core assembly and mould closure. This, to the author's knowledge, is the only common moulding system exhibiting distortion resulting from stress. Most distortions in casting processes arise because of thermal expansion as is discussed later.

When the metal first enters the mould, the mould, and more particularly, the cores, heat up and expand. Thus the mould cavity enlarges and will probably suffer some distortion. This is reasonably reproducible in sand moulds because the pouring temperature is normally under good control (usually ±10°C or better) and the moulding sand is always close to room temperature. Thus the starting conditions are usually reproducible. This is less true for the various processes using metal moulds, where the die temperature of perhaps 300–400°C is often poorly controlled, varying by as much as ±100°C or more.

As the casting cools, it contracts, with the result that some parts of the casting are subject to tensile extension or compression because of geometrical constraint of the mould. The constraints exerted by the disposable sand mould are perhaps less severe than those of the metal die, but in general the casting stays in the sand mould longer and the constraint therefore operates for longer.

Although an attempt will be made to estimate these expansions and contractions, the final result corresponds to the patternmaker's contraction allowance. This is the value based on hundreds of years of experience by patternmakers and toolmakers, and is the allowance they have to provide, making the pattern a little larger than the final casting. Unfortunately, the factor is sensitive to geometry of the casting, and mistakes are often made in the correct choice of the allowance, which can vary between 0% and 1.3% for aluminium castings and up to 2.4% for steel castings. Some recent work in the author's laboratory have highlighted the uncertainties of this factor for cast aluminium and cast irons (Nyichomba and Campbell, 1998). Estimates of this allowance are made in Section 18.5.

In the pressure die casting process, the casting is normally thin-walled and thus rather weak. As it cools in the rigid steel die, contracting onto projections of the geometry, it is therefore forced to stretch plastically. The size of the casting is thus mainly controlled by the time to ejection; larger castings are produced if ejected later.

The hot casting is cooled to room temperature in a variety of ways. This may occur either as a series of individual castings on a cooling conveyor, or as a heap in a bin, or by an immediate quench into water. These post casting operations are likely to create different patterns of distortion. Likewise, heat treatment, even natural ageing, affects many products, particularly heat treatable aluminium alloys. The latter grow slightly (of the order of 0.05%, i.e. 0.5 mm per metre) as hardening precipitates form.

18.1.3 EFFECT OF CAST ALLOY

Zinc castings are most commonly supplied as pressure die castings. As we have seen earlier, the thermal distortions of both die and casting are the least of any of the casting processes because (1) the casting temperatures are low, creating low expansion problems for the die and minor contraction problems for the casting; (2) the dies are accurately machined from steel; (3) the dies use no protective die coating and (4) are supported in a close fitting and rigid steel bolster. Thus zinc pressure die castings are intrinsically capable of meeting many net shape requirements. This is a clear-cut example of near-automatic achievement of net shape by zinc pressure die castings.

The light alloys based on magnesium and aluminium can be cast in moulds of all types. For HPDC, the dies are steel and no protective mould coat is used, so maximising accuracy. For low pressure and gravity die castings, the die is cast iron or steel but is protected by a ceramic die coat. The thickness of the coating is not easily controlled, thus limiting accuracy, plus in addition the dies themselves are rather rough and inaccurate compared to HPDC dies. The dies are also often subject to distortion partly because the die is free-standing (i.e. not supported by a surrounding steel bolster as in the case of HPDC). For sand castings, the moulds can be used without a protective coating, and thus retain their dimensions.

Cast iron is most commonly poured into greensand moulds with no protective coating. Chemically bonded sand moulds and cores on the other hand usually require a refractory wash coat to obtain an acceptable surface finish. Although cast iron dies are possible for cast iron, their use in Western Europe is limited to specialised casting operations, where, again, a die coat is required. At this time, iron dies for iron casting is more widely used in Eastern Europe.

Steel is cast into aggregate moulds. For modest sizes of products, greensand is widely used without a mould coat. Increasingly often, however, steel is being cast into chemically bonded sand moulds with a protective mould coat to achieve an acceptable finish. The coating ranges in thickness from 0.2 to 0.6 mm, being usually 0.3 ± 0.1 mm. The high temperature of steel casting leads to considerable interaction of the surface of the casting with its environment, leading to sand burn-on and oxidation, although much of these problems would be solved immediately by the application of a good mould filling system. Many large steel castings lose up to 0.8 mm or more of their surface during heat treatment because of the flaking off of oxide from the outside surface of the casting, and approximately 0.4 mm from interior surfaces. Because of these interactions, plus the application of mould coatings to prevent them, steel castings are normally the furthest removed from the net shape concept. On top of this, the rough-and-ready patterns often seen in the cast steel industry clearly do not help the attainment of accuracy. For steel components weighing up to a few kilograms, the favoured route for near net shape is therefore lost wax casting in vacuum.

18.2 MOULD DESIGN
18.2.1 GENERAL ISSUES

The problem for the casting engineer is to achieve a successful design of the mould. This problem is not to be underestimated because it requires the simultaneous solving of a list of issues including.

1. The design of the mould and core assembly can be a problem in itself. It is not uncommon to find that it is impossible to assemble the cores because of some shape feature of neighbouring cores has been overlooked. It is all too easy to stumble into such pitfalls in a complex core assembly. When the first set of cores are made from the new patternwork, with its shining new varnish and paintwork, the discovery of such 'passing problems', where one core will not pass another and so fit into the assembly, are greeted with embarrassment and dismay.

The other common problem for the casting engineer and toolmaker is the design of the assembly so that cores fit, in logical order, only into the drag if possible (Figure 18.4). Cores in the cope are not usually an option for horizontally parted greensand moulds, because, if the sand strength is not high, they are in danger of falling out of their prints when the cope is turned over and closed onto the drag before casting. Gluing cores into the cope is possible in the case of strong chemically bonded sand moulds. However, gluing takes time and is therefore costly and introduces the danger that any excess glue that exudes out of the join may cause a blow hole defect in the casting if it contacts the metal. In addition, glue applied to a core print may prevent the core from venting, leading to a blow defect from the core. The use of glues should therefore be avoided if at all possible.

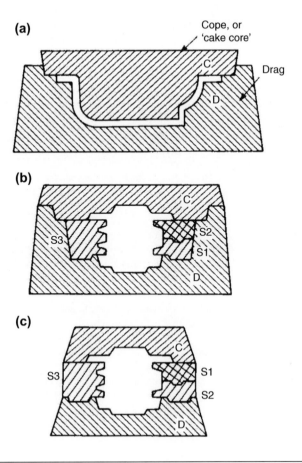

FIGURE 18.4

(a) A simple cake core and drag assembly; (b) a cope and drag with side cores, all located in the drag; (c) an apparently lower-cost alternative to (b), but resulting in loss of dimensional control.

It is common for complex core assemblies to be assembled at a separate station sited off the mould assembly line. Core assembly can then be accurate because the assembly is built up in a jig. The cores are designed to be lifted by the jig, transferring from the assembly station and lowered into the mould as a complete package. Castings that require lengthy core assembly times are not thereby allowed to slow the cycle time of the moulding line.

2. The filling system. The provision of a good filling system and its integration with the rest of the mould and core system is sometimes not easy, and in some cases the additional trouble or expense to provide a good filling system is by-passed. (The minefield of poor castings and high scrap rates is always entered for apparently good reasons.) The filling system design forms a major part of this book. It is mandatory reading. Its rules are strongly recommended to be followed in all cases.

3. The feeding system. Naturally, following the first rule for feeding, it is clearly best if feeders can be completely avoided. I attempt to avoid feeders by the use of chills, fins or pins. However, if feeders are necessary, they are required to be placed high on the casting to feed downwards under gravity. This may force the whole casting to be turned over to orient its heavy sections at the top. Another one of the key issues is to place the feeders so that they are easy to cut off or machine away subsequently.

4. The avoidance of infringement of any of the 10 rules. For instance, convection considerations might force the issue of rotating the mould through 180° after filling. This action usually confers other benefits and makes the integration of the filling and feeding systems powerfully effective and economic. It is a strategy to be recommended, although it is essential to observe the precautions listed in the section on rollover systems.

However, sometimes the solution to all these issues is not straightforward. For instance, much time may be spent attempting to solve the issues with the casting oriented in one direction, only to realise that such an orientation involves insoluble problems. The casting is then turned upside down and the exercise is repeated in the hope of a better outcome. Such experiences are the day-to-day routine of the casting engineer.

Furthermore, the complexity of the issues is not easily solved at this time by computer. There have been many such attempts, but it is fair to say at this stage such efforts have not been developed to such a degree that the professional casting expert is offered any significant help. However, of course, we can look forward to the day when the computer is sufficiently capable to provide useful solutions.

18.2.2 ASSEMBLY METHODS

The design of the mould assembly often simply involves two parts: a cope and a drag, such maximum simplicity makes for maximum accuracy. But for more complex castings the mould assembly can be complicated, requiring many parts, and requiring much discussion between the pattern shop, foundry, and casting designer to find an appropriate solution. Accuracy can now become illusive and troublesome.

The simplest form of construction of moulds, which was popular with many foundrymen, consisted of a drag with a cope in the form of a 'cake core' as shown in Figure 18.4(a). This simple construct did not allow for the placement of any useful bottom-gated running system, and the top pouring which had to be accepted as a consequence was generally acceptable for grey iron castings in greensand moulds. It did not give good results for the majority of casting alloys. The arrangement would, however, have been excellent for a counter-gravity filling technique.

As a general rule, it is useful to ensure that even in the most complicated of mould assemblies, the design of the assembly consists principally of a drag and a cope, and that all parts are interrelated via a single mould part, which will normally be the drag (Figure 18.4(b)). In this case, the assembly of cores will be in the drag, with each core located separately, and the final operation will simply consist of closing with the cope (sometimes called 'a cake core'). The cope also should have features which contact and locate directly on the drag as the key mould part.

Such simple rules are easily forgotten. Figure 18.4(c) shows how it would have been easy to save some sand by abandoning the deep drag construction, and having a core assembly which consists simply of a random heap of assorted shapes: one layer of cores tottering upon another, with finally the cope perched precariously on top. The overall accuracy of the casting now suffers from the accumulation of errors introduced by the intermediate side cores S1, S2 and S3, providing a poor and variable match between the top and bottom features of the casting.

Figure 18.4(b) shows a simple type of cope-drag arrangement with side cores, all located in the drag, apart from side core S2 which rests on S1. It was judged that the small accumulation of errors in the positioning of S2 would be acceptable in this case. If the positioning of S2 had been critical, it could still have been located in the drag by stepping the contour of the drag appropriately.

The dimensional problems which arise in setting cores are examined by Skarbinski (1971). In general, it needs to be said about the accurate printing of cores that the core print should be designed assuming that the core will be produced with errors in its size and shape, and will have to fit into a mould which will also have suffered some distortion during its manufacture. The print also sometimes has to restrict the movement of the core during mould filling because of the build-up of pressure against its front face, and because of buoyancy tending to make it lift. It may also have to allow the relative movement of the core in its print to permit the thermal expansion of the core. All this is a seemingly impossible task to achieve accurately. However, it is usually solvable within practical limits by applying the following simple rules:

1. The print requires *tolerance* where it needs to fit (i.e. must not be made size-for-size; that would have produced an interference fit. The only exception to this is the heights of cores where cores are stacked one on top of the other because in this case the accumulation of errors must be kept to a minimum).

2. The print requires *clearance* where it is not required to fit, and where expansion clearance is required.

3. The number of cores should be reduced to a *minimum*, moulding as much as possible directly in the cope and drag.

Rules often appear pedantic or even pedestrian when they are spelled out! However, the application of the rules involves much work which, unfortunately, is often neglected during the urgency involved in the design and manufacture of patternwork. Although some prints are easily designed, others require much thought, and compromises have to be carefully assessed. Every print requires individual detailed design. It is attention to details such as these that makes the difference between the inadequate and the excellent casting.

Nevertheless, these problems are eliminated if the use of the core can be avoided altogether as suggested by following the useful rule 3. Not only does the application of this principle reduce dimensional errors, but also the addition of each core involves considerable extra tooling cost, and an additional cost in the production of the casting, sometimes approaching the cost of the production of a cope or drag; it is effectively the third piece of sand to be added to the two original mould halves; thus, the costs at this stage may increase by 50% for the addition of the first core. At other times, a small core can save money by avoiding extra complexity of the tooling. Each case needs separate evaluation.

The perhaps unexpectedly high addition to total casting costs resulting from the use of cores arises from the accumulation of several minor operations, most of which are usually overlooked. For instance, the core needs to be scheduled, made, perhaps on a capital-intensive core-making machine, stored, de-flashed, retrieved from storage, transported to the moulding line, and then correctly assembled into the mould. Errors arise as a result of the incorrect core being made or transferred, or sufficient are broken in storage or transit to cause the whole process to be repeated, or its assembly into the mould gets forgotten at the last moment! Cores are therefore almost certainly more expensive than most foundry accounting systems are aware of. (The costs of chills, and of scrapped castings, are similarly illusive and therefore generally underestimated or completely overlooked.)

A further use of cores, in addition to their obvious purpose in providing detail which cannot be moulded directly, is that the running system can often be integrated behind and underneath them, the main runner and gates being located beneath side or end core(s). This is a valuable facility offered by the use of a core and should not be overlooked. In several castings the addition of a core may be for the sole purpose of providing a good running system. Such a core is often money well spent! It could make the difference between a crippling scrap rate and trauma-free production.

Where a complex array of internal cores has to be loaded into a drag, as in the case of the assembly of an automotive crankcase for instance, it is common to arrange for the cores to be preassembled in an assembly jig, and for the complete core package to be lowered into the drag. This saves time on the assembly line.

The problem with the automation of core-assembly systems is finding the core again with sufficient accuracy so that it can be picked up once again after it has been put down on, for instance, a conveyor or a storage rack. This is a difficult job for a robot because extreme accuracy is required, and the cores are often of extreme delicacy. Although this problem is being increasingly well addressed by vision systems and improved robots, clearly, one method of solving this problem is never to put the cores down in the first instance. This simple solution is powerful.

Schilling (1987) has succeeded in developing this concept with a unique system of making and assembling cores in which the cores are not released from one half of the opened core-box until the other half of the core has already been located in the core-assembly package. In this way, the cores are assembled completely automatically and with unbelievable precision. Cores are located to better than 0.03 mm, allowing them to be assembled with clearances which are so small that the cores could not be assembled by hand. In fact, the cores are sprung into place with interference fits. The rigorous application of this technique means that castings need to be designed for the process because the assembly of each core is by vertical placement over the previous core. For instance, any threading of cores in through holes in the sides of other cores, such as often occurs with port cores through the water jacket core of a cylinder head casting, is not possible. This disadvantage will limit the technique to partial application, loading some but not all cores of a cylinder head, for instance. Even this would be an important advance.

The fundamental objection to this super-accuracy technique is the requirement for the plant to run with sufficient reliability. Because the cores are never released, but always under the control of a core-box, there can be no buffer stock of cores between operations. A typical core assembly system involving this principle might involve at least 30–40 machines including core blowers and robots etc. Even if the machines were of excellent reliability, each having an up-time of 95–98%, the multiplication of all those reliabilities results in an overall reliability for the whole operation of usually less than 50%. This is a serious disadvantage that requires an order of magnitude increase in maintenance effort and skill to reduce this drawback. This substantial challenge makes this system difficult to recommend.

A final note in this section relates to cope-to-drag location. This is, of course, of primary importance. Failure to achieve good location results in a mismatch defect. For the case of precision core packages, the sand mould is not contained in a box, and thus is located directly with sand-to-sand locations. Because this is defined from the pattern-work, the location relates perfectly to the casting details, and mismatch is therefore not possible.

In foundries using moulds contained in moulding boxes, however, mismatch is unfortunately all too common and is usually the result of the use of worn pins and bushes which are used to locate the boxes. Southam (1987) analyses the effect of the errors involved in the pin and bush location system. These are numerous and serious. The pin to bush clearance is typically 0.25 mm, and given an apparently acceptable additional wear of 0.35 mm, he finds that the total possible mismatch between cope and drag moulds is as much as 1.5 mm. (Mismatch is a lateral location error, and not to be confused with the vertical precision with which cope and drag meet which is normally of the order of ±0.1 mm.)

He proposes, therefore, a completely different system, in which mould closure is carried out in a special station, where the cope and drag boxes are simply guided by wear blocks fixed to the outside edges of the box. These slide against two guides on the long side of the box, and one guide against the narrow side of the box during moulding and closing operations. The boxes are held against the guides by light pressure from springs or pneumatic cylinders. The system appears deceptively simple, but actually requires a certain amount of good engineering to ensure that it operates correctly on mould closure, as Southam describes. Although Southam calls his method the three-point registration system, it is in reality a classical six-point location system because he uses a further three points to locate the drag in a parallel plane to the cope during closure (the six-point location system is discussed in relation of casting datums as good casting rule 10).

The ability to locate cope to drag with negligible error has several benefits, which Southam lists. The maintenance and replacement of worn pins and bushings is a foundry chore and expense which is eliminated. Instead, only three guides on the closer and three each on the cope and drag pattern need to be checked, and the effect of wear of these parts on mismatch is minimal because the resultant displacement is largely self-compensating. In addition, the foundry will be capable of producing castings with thinner walls, reduced dressing, reduced machining, and improved appearance.

18.3 MOULD ACCURACY
18.3.1 AGGREGATE MOULDS

Buyers of castings have been traditionally prejudiced against sand castings as opposed to the various varieties of metal moulded castings because they have long memories. They recall, or perhaps imagine they recall, the time when most sand castings were made individually by a bench moulder or floor moulder, equipped with a wheelbarrow, watering can and shovel to produce his greensand mix, together with an ancient and battered wood pattern that had seen better days. Working practices encouraged additional variability, for example by rapping or vibrating the pattern to ease its withdrawal from the mould. Lifting off by hand or crane produced similar inaccuracies, as did errors in draft angle from hand-made patterns. All these variations would usually result in an oversize mould. The addition of mould and core washes, and variations in the density of hand ramming of the moulds, would add further variation.

A typical problem in hand-moulded castings was the variability of core location. This arose because a core laid in a print on the main cope/drag joint would usually be a poor fit. If it sat proud of the surface, then on the closing of the mould the core will be forced to settle, by deforming either the cope or the drag. Depending on which happens to have been more densely rammed that day, the core will be accommodated more in the drag or the cope, so that detail on the casting will vary up or down from day to day, and from casting to casting.

Originally, most sand moulds were made using greensand. This is a mixture of sand, clay and water, often in the form in which it was dug out of the local hillside. As such, it would be termed a *natural greensand*. Such natural products were a pleasure to work with, but its variable content of clay, and variability in the type of clay, together with variations in water content, necessarily resulted in rather variable castings.

Later, more controlled greensand mixtures were created by the use of clean, washed silica sands that were mixed with selected clays and other additives to form *synthetic greensands*. These dough-like mixtures were rammed around the pattern to form the mould. The synthetic sands could give rather harder and more reproducible moulds. Additionally, developments away from hand compaction to small-scale automation, such as jolt-squeeze compaction, have slowly given way to massive, fully automated plants with sophisticated high rate compacting mechanisms that produce very high-density moulds of impressive rigidity. Such moulding methods produce the highest accuracy yet achieved by the greensand route.

Studies like those of Bates and Wallace (1966) show how the greensand mould deforms less during casting as the mould density increases, whereas those of Rao and Roshan (1988) confirm that the final castings improve in accuracy as the compaction is improved.

Naturally, the use of a well-controlled synthetic greensand, with a metal pattern, and the mould reproduced by an automated moulding process yielding uniformly hard moulds, was a substantial step forward. Such systems continue to be widely used today for large-volume production with the highest rates of production and highly competitive economics. Despite these benefits, it may be no surprise that the high-pressure moulding process has brought its own crop of new problems in the quest for accuracy. These include distortion of tooling if not adequately constructed, and of spring-back of the mould because of the elastic behaviour of the highly compacted and highly stressed sand mould.

One of the consequences of requiring an accurate pattern and accurate mould is the requirement for *room temperature operation*. Clearly, for all those pattern and moulding processes that have to operate at high temperature raise the danger of distortion of the tooling, and poor temperature control leading to dimensional variation.

A further step in the production of accurate, very hard moulds, has been developed in recent years. This involves the use of chemically bonded sand. The ability to cure the binder at room temperature has had a further important benefit; the tooling remains accurate, suffering neither expansion nor distortion by significant changes in temperature. In fact, providing the tooling is made from metal and/or resin, the tooling never deviates by more than a few degrees from the temperature at which the toolmaker made it. These, together with the fact that the mould or core can cure while in contact with the tooling, are the two vital steps in the construction of accurate moulds. They have revolutionised the concept of the production of accurate castings.

One of the major advantages for accuracy enjoyed by the aggregate mould is not only its reproducible temperature during its formation, but also its reproducible temperature at casting. This is always reasonably close to room temperature. A temperature variation of 20°C will produce a change in length of only approximately 0.02%, or 0.1 mm in 500 mm. This is a useful stable base from which to start, and contrasts with the problems suffered by metal dies that are not easily controlled to this precision. In fact, a variation of die temperature of 100–200°C would not be uncommon, with a consequent change of size of up to 0.2%, or 1 mm in 500 mm. This represents a serious error in a casting that is required to maintain a wall thickness of 2.0 mm. Figure 18.2 is a brief summary of the competing processes.

Nowadays, the hard sand core assembly technique is perhaps the most accurate of all the casting processes, especially for castings exceeding 100 mm in size (Figure 18.3).

Such statements are not easily justified, however, because individual foundries differ widely in their capability, often simply because of attention to detail. Also, individual castings suit some processes and not others. Moreover, as a result of the new standards set by precision sand processes the competitive processes have improved markedly. Thus although it is true to say that sand casting has come a long way in recent years, so have all the others!

Thus, although our castings buyer no longer has any justification for rejecting aggregate moulded castings, he might still be justified to select a permanent mould of some kind. The phrase 'metal die' continues to have an *aura* of engineering excellence, and 'pressure die casting' has the *impression* of having an advantage of consolidation by pressure. Producers of sand castings clearly need to counter these beguiling marketing ploys, and market their own solid and reliable new achievements.

Core making has gone through a similar revolution. The two main original processes for making cores employed dry sand (i.e. dried greensand) and linseed oil–bonded sands. Both processes had in common the transfer of the core from the core box in its green, or soft, state directly into a specially shaped cradle or carrier, often called, rather confusingly, a 'drier'. The core at this stage was so weak that it would deform at the touch of a finger. It was then hardened by drying in an oven for an hour or more. Naturally, the opportunities for loss of accuracy were plentiful. Neither process is much used today, to the relief of most casting engineers.

A great improvement in accuracy came with the development of chemical binders that would allow the core to be hardened in the core box. In this way, cores of very precise form could be guaranteed and are nowadays commonplace in the industry.

The original chemical-binder process used sodium silicate that was cured by carbon dioxide gas, often known as 'the CO_2 process' for short. This process was useful for moulds, but never popular for cores for the casting of Al and iron alloys because the heat of casting fused the silicate bond to the sand grains, producing a glassy bond of such strength that the core could not be subsequently removed from the casting. Its strength resembled that of concrete, and for many aluminium alloy castings the heat treated core exceeded the strength of the casting. Attempts to remove the core therefore often led to the destruction of the casting!

The Croning shell process, often known for short simply as the shell process, was another step forward. Here the binder was a phenolic resin, pre-coated onto the sand grains and solidified as a dry coating, thus allowing the sand to be blown or dumped in a dry, free-flowing state onto the pattern. The pattern had to be heated to melt and cure the resin in the thin shell of sand in contact with the pattern. Because of the use of heat for curing there was the danger of distortion of the tooling, and the biscuit-like core was sufficiently thin that it too was subject to distortion if not carefully handled and stored. For aluminium alloy castings, the heat input from modern thin-walled castings gives problems because of the incomplete breakdown of the resin after casting, with the result that de-coring is returning as a problem, although developments continue by suppliers of shell sand in an attempt to overcome this.

The hot box process again involves the curing of the core inside the box, and in general gives much improved de-coring performance. However, the problems of heating and controlling the temperature of core boxes, and the distortion of tooling, and the rate of diffusion of heat into sand, all remain to limit the size of core which it is feasible to produce by this technique. It remains a useful and economical process for cores where such limits are acceptable.

The various cold (or more accurately, room temperature) curing processes are not subject to these limitations. Large products can be produced so that not only internal cores for castings, but also whole moulds, can be manufactured by these processes. This has allowed the opportunity for the creation of the 'core-assembly' technique for the production of a mould. Because the tooling and the mould are all at room temperature, no thermal expansion or distortions arise to impair accuracy. For this reason the core-assembly technique is probably the most accurate moulding process in use today and can produce the most accurate aggregate mould castings. The process rivals the accuracy (if not the surface finish) of other near-net shape processes such as investment casting and HPDC (Figure 18.3).

The control problem faced by users of chemically bonded sands is that of controlling the strength of the cured sand. Continuous mixers are notoriously difficult to control; the quantities of sand and resin that are processed through the mixer change erratically, so that the quality of mixed sand is often not reproducible from time to time during the day. This will, of course, alter the strength of the moulds and cores, resulting in changes to the size and shape of the casting. Some progress is being made in using continuous monitoring and feedback control on the mixers. However, the problem is resolved to a certain extent by the use of batch mixers, where ingredients can be weighed into a mixing bowl. This still leaves the problem that the chemistry of the components of the binder themselves needs to be controlled within close limits. This is an area of continuing concern because organic resins are not easily characterised nor the chemical processing by the resin supplier controlled to the degree a founder might wish. The problem may already be limiting the accuracy of castings produced by this process route.

One interesting hard sand moulding process not requiring control over materials of imprecise chemistry such as binder resins, is the vacuum film moulding process, or V process, in which the mould is sealed on all sides by a plastic film, and the sand is consolidated by the application of a vacuum. An investigation by Grote (1982) into the reproducibility of greensand, cold set and V process, confirmed that cold set (a room temperature cured resin binder) and V process had similar good reproducibility, and both were better than greensand, although his version of greensand was probably less than an optimum system!

A final problem remaining for the sand mould is that of retaining its accuracy when subjected to the shock of being filled with molten metal. This problem includes phenomena such as the thermal distortion of the mould, its deformation under the weight of metal pressing against its walls, and the tendency for cores to float. All these matters are now usually controllable within close limits, particularly for a wide variety of non-silica aggregates and binders. For silica sand, however, we cannot avoid the 1.5% linear expansion at 570°C as it experiences its alpha to beta quartz transition. This is a major factor affecting the accuracy and reproducibility of castings, as illustrated, for instance, in the work by Jennings et al. (2001) in which he found that castings made in zircon sand which suffers no such phase change enjoyed a standard deviation 10 times less.

18.3.2 CERAMIC MOULDS

Ceramic moulds as used in investment casting have enjoyed the reputation of being the most accurate of all the various mould types available. This is a curious perception which may be the result of the process being for many years limited to the production of rather small castings where dimensional problems were naturally too small to be of concern. It probably also relates to the wonderful appearance of investment castings, with such excellent surface finish and sharp, fine detail, giving the comforting impression of accuracy.

However, when ceramic moulds are produced in sizes similar to those of other casting processes, such as precision sand core assembly processes, the dimensional problems of investment castings are seen to be non-trivial. This is hardly surprising in view of the multiple operations involved in the production of a ceramic mould, each of which can introduce errors. These include the injection of the wax into the aluminium die; the temperature of both the wax and the die affect the size and distortion of the wax pattern. The temperatures of the slurry tank and the drying room will similarly contribute an effect during the building up of the shell. The firing of the ceramic involves a certain amount of contraction of the ceramic as the bond is created. The temperatures of the mould and the metal at the time of casting are other key influences. During the time that the molten metal is in the mould, the mould suffers creep as it sinters and softens, often leading to swelling or other distortions of the casting.

Beeley (1972) shows how the expansion characteristics of different ceramic shell materials are strongly affected by how much silica is contained. He also quotes Carter, who showed that a Co–Cr alloy in an investment mould shrinks by 1.6% when the mould is at room temperature, but less than 0.3% when heated to 1000°C. Clegg and Das (1987) evaluate similar problems with Shaw process moulds, finding that the linear dimensions of the mould are greatly affected by the time and temperature used for firing, by the composition of the ceramic, and by the size of casting which is produced.

To summarise the dimensional problems of ceramic moulds, it seems a tribute to the dedication and perseverance of the industry, rather than to the intrinsic merits of the process, that all of these problems are held within tolerance sufficiently well that an acceptable product can usually be made.

18.3.3 METAL MOULDS (DIES)

Castings produced in zinc-based alloys by the pressure die casting process represent a level of accuracy which is hard to beat (Figure 18.3), representing one of the casting industry's most accurate products. This is because of the low temperatures involved. The metal die is accurate, and distorts very little, and the casting itself contracts little during its limited cooling to room temperature. (By extension of this reasoning, plastic injection mouldings can be similarly highly precise.)

For higher melting-point materials, however, the problems mount rapidly. Table 18.1 shows how metal dies suffer increasing problems with thermal shock and fatigue as the temperature of the casting alloy increases. The difficulties are such that although Mo dies have been tried for a stainless steel, the results were not satisfactory and so far as is known no one now uses the process. Even for lower temperature materials such as Al alloys, the degradation of the surface of the die by the development first of fine cracks, leading to general crazing, and finally resulting in major disintegration of parts of the surface, leads to inaccurate and roughened products long before the die becomes so bad that the decision to abandon or rework it becomes unavoidable. These problems are common in long-running aluminium alloy castings made by HPDC and gravity die (permanent mould) casting. Total failure of the die by catastrophic cracking is also sometimes suffered.

Magnesium presents less of a problem when cast into metal moulds because the thermal capacity of magnesium is significantly lower than that of aluminium, resulting in reduced thermal shock to the die. In addition magnesium

Table 18.1 Life of Metal Dies

| Alloy Base | Die Life | | | | | |
	Rubber	Graphite	Gravity (Permanent Mould)	Low Pressure	High Pressure	Squeeze
Zn	Good	Good	Excellent	Excellent	Excellent	Excellent
Mg	–	–	Excellent	Excellent	Excellent	Excellent
Al	Limited	Limited	Satisfactory	Satisfactory	Tolerable	Good
Cu	–	–	Tolerable	Tolerable	Rare	Tolerable
Cast iron	–	–	Tolerable	Rare	–	–
Stainless steel	–	–	–	–	Mo-based die	–
Steel	–	Griffin only	–	Griffin only	–	–

dissolves iron less readily than aluminium, so that 'soldering' to the die, a localised welding effect, with the consequent gradual removal of die surface, is reduced. The attempt to solve the soldering problem in aluminium alloys by increasing the iron content of the alloys to near-saturation levels is only partially successful because some sticking of the casting together with erosion of the die remains common.

The standard of construction of dies for HPDC is necessarily extremely high. This is a consequence of the requirement for the die to fit into, and be operated within, the massive (and expensive) machine tool that is the HPDC machine. The accuracy of fit between the various parts of the die also needs to be excellent to prevent the penetration of liquid metal at the high pressures used for injection. Between each shot, the die is sprayed with a coolant and lubricant and/or parting agent. The parting agent gives minimal build-up problems on the die, contributing significantly to the ability of HPDC to maintain the accuracy of its products (unlike, for instance, gravity die casting, where a rather thick die coat is used).

In fact, it seems that for HPDC, one of the most significant variables affecting the size of castings is the temperature at which the castings are ejected from the die. This is because if ejected early, the casting is hotter and thus cools and contracts over a greater temperature range without the constraint of the die, thus finishing smaller. A casting ejected late cools more in the die, so its contraction is restrained by the rigidity of the die; the casting is stretched plastically, as by medieval torture on a rack, thus finishing larger.

In low-pressure and gravity die casting, the standard of die construction is considerably lower. This is because the die does not need to be fitted into a precise machine, but is usually serviced by simple bolt-on actions such as rack and pinion, or simple hydraulic cylinders which actuate opening and closing of the die. The gravity die casting industry is therefore less capital-intensive, and grows more easily in a piecemeal manner. It is more labour-intensive than the HPDC industry. The two processes are not often mixed in the same foundry.

The lower rates of filling of the die under gravity require the die to be coated with an insulating ceramic layer known as a die coat. This is usually sprayed on at the start of a shift and cleaned off by grit blasting before renewal. The coating may become damaged during the production of a run of castings and might require running repairs by the operator. The die coat is therefore a major variable in the size of gravity and low-pressure die castings. Its thickness of application will vary from day to day because the spray techniques for its application are not easily controlled. Also, parts of the coating on faces with minimal relief or draw will wear during the production run, giving slowly increasing sizes of castings.

The die itself is subject to wear in many ways. In particular, the working face will be slowly eroded by repeated cleaning and recoating, in addition to any thermal damage it may be suffering. The moving parts of the die include slides and other mating components which are subject to general sliding and abrasive wear, particularly if sand cores are used in the mould.

Table 18.1 indicates some of the different materials used for dies. For HPDC dies, the construction material is usually a special hot-work tool steel containing chromium for greater strength and hardness, oxidation resistance and good response to nitriding. It is usually heat treated to achieve its optimum strength and toughness, and finally nitrided to give an extremely hard-wearing surface. Grey iron would not be adequate for pressure dies because it lacks strength and fatigue resistance. Furthermore, its graphite flakes would tend to open up by oxidation, degrading the integrity of surface which would be needed against the penetrating action of liquid metals at high pressure.

Grey iron is, however, widely used for gravity and low-pressure dies. This is because the filling of moulds under gravity or low pressure introduces only gentle pressurisation of the surface by the melt, and the surface is protected by the ceramic die coat. Also, the iron is highly castable, allowing the rough die parts to be produced quickly and usually without problems of porosity or other unsoundness; the castings are easily machined to final size and shape; the material has good thermal conductivity and fair resistance to thermal shock. Weight is easily reduced by sculpting the back of the die to give a mould of even thickness.

Graphite is occasionally chosen as a die material. It is easily machined and is not wetted or attacked by most metals (magnesium would be a spectacular exception!). For zinc alloys, therefore, it has been used with success, although care is required in longer runs to avoid damaging the die by mishandling because its strength is low. The material is not useful for aluminium alloys, because the graphite oxidises increasingly rapidly at temperatures above about 350°C. After two or three casts of aluminium alloy, the die is unusable. Treatment of the surface by a protective silicate wash extends the die life a little.

This experience makes the use of graphite as a die material in the Griffin process all the more impressive. Here, railway wheels are cast in steel displaced into the die by low pressure. The wheels are produced closely to net shape, so that they require little machining (Kotzin, 1981).

18.4 **TOOLING ACCURACY**

Tooling is taken to include the pattern and its core boxes, or the die, and any measuring or checking jigs and gauges.

The wear of patterns used in sand casting causes the casting to become undersize, whereas the wear of dies used in die casting causes the casting to become gradually oversize. An effect opposite to wear happens as a result of build-up problems on the tooling. Patterns for sand casting are subject to the deposition of small amounts of sand and binder, and the gradual accumulation of release agents. Dies may accumulate layers of die coat.

Distortion is another problem. Wood is a useful and pleasant material for pattern construction. It is easily worked, light to handle and easily and quickly repaired or modified as necessary. Even so, it is not a contender for accurate work because of its tendency to warp and shrink. A good patternmaker will attempt to reduce such movement to a minimum by the careful use of ply, taking care to join layers to cross grain where possible, and affixing strengthening battens. The use of various stabilised woods and simulated wood-like plastic materials has also helped considerably (Barrett, 1967). Nevertheless, the ultimate stability in tooling is only achieved with the use of metal or cast resin in metal frames.

Cast-resin patterns which are backed with wood frames are less reliable; the warpage of the wood distorts the internal resin shape, usually within a month or so. After a year, the tooling is seriously inaccurate, so that cores will not assemble properly. In contrast, resin patterns cast into aluminium alloy frames for strength and rigidity, are usually extremely reliable. However, some resin systems such as polyurethanes tend to suffer from the absorption of solvents from the chemical binders in the sand, and so can suffer swelling and degradation (Gouwens, 1967).

The working temperature of tooling affects the casting size directly; a warm pattern will give a slightly larger casting. If we consider an epoxy resin core-box cast into an aluminium alloy frame, the box will largely take its size from the temperature of the metal frame (i.e. not the internal lining of epoxy resin) which has a coefficient of expansion of $20 \times 10^{-6} \, K^{-1}$. If the temperature at the start of the Monday morning shift is 10°C, and if the returning sand creeps up to 30°C by the end of the morning, then for a 500 mm long casting the 20°C temperature rise will cause the castings to grow by $20 \times 20 \times 10^{-6} \times 500 = 0.2$ mm. This is not large in itself, but when it is added to other random variables the uncertainty in the final casting length becomes increasingly out of control.

Anderson (1987) has emphasised the important requirement that for the most accurate work the pattern or die should be used as an adaptive control element in production. Thus it needs to be built in such a way that it can be modified to produce the required size and shape of the casting. The use of patterns split transversely across their major length is common. The prior insertion of a spacer in this split allows the spacer to be removed and replaced by shims thinner or thicker as necessary. Such simple techniques involve only modest extra expense during the construction of the pattern but are a reassurance against the possibility of major expensive rebuilds later.

18.5 CASTING ACCURACY

For a good general review of accuracy, precision and tolerance concepts in casting manufacture the interested reader is recommended to the paper by Anderson (1987). We shall rely on some of his key points in this section.

18.5.1 UNIFORM CONTRACTION

This section is an examination of the various effects that follow from the strains and consequent stresses in the casting/mould system as a result of the linear contraction of the solid casting as it cools. Uniform contraction and distortion are considered separately. A distorted casting can usually be straightened, whereas a casting that is uniformly oversize or undersize is scrap.

Following Figure 7.1, when the contraction in the liquid state and most of the contraction on freezing will have been completed, the casting has cooled sufficiently to develop some coherence as a solid. Further cooling in the solid state will cause the casting to contract as a whole.

The point at which the casting develops its solid-like character is different for long- and short-freezing-range alloys. For short-freezing-range material the point is reached when the casting has developed a solid skin. For the case of long-freezing-range alloys, the point is marked by the development of a coherent skeleton of solid dendrites.

At this stage of freezing because the casting fits the mould rather well, having been poured in as a liquid (and therefore at that earlier stage fitting perfectly!), as it cools further and contracts, something has to give.

It is not the case that either the casting or the mould will yield. Both yield. Following Newton, the action of the casting on the mould causes an equal and opposite action of the mould on the casting. The degree of yielding of the casting and mould depends on the relative strengths of each. Naturally, this varies greatly from one casting/mould system to another.

The contraction of the casting from its freezing temperature to room temperature can cause the patternmaker sleepless nights. This is because the pattern must be made oversize by an amount known as the contraction allowance (or patternmaker's shrinkage allowance), so that the casting will finally finish at the correct size at room temperature. However, the patternmaker often does not know exactly what allowance to use when he starts to construct the pattern.

This was not so important in the 'bad old days' (perhaps 'the good old days'?) when castings were regarded only as 'rough castings' having plenty of machining allowance. However, now that greater accuracy is being sought in the quest for a 'near-net shape' product, the problem has become serious; the patternmaker risks finding that the wrong allowance was chosen only after the first casting is made! This is the emergency scenario when the tooling has to be modified or remade, but the tooling budget has already been spent, and the deadline for delivery is about to be passed. The contents of this small chapter are recommended to the reader as the only procedure known to the author to avoid this disaster.

The Imperial System of measurements gave us a vast legacy of choice of presentation of patternmaker's shrinkage allowance data. For instance, for simple and heavy aluminium alloy castings many are made using an addition of '5/32 inch per foot' to all the linear sizes. This corresponds to the widely used contraction allowance of '1 in 77'. The author has given up these various units and ratios in favour of a simple percentage, in this case 1.30%.

For other aluminium alloy castings with larger internal cores, such as cylinder blocks, the allowance is only 1/8 inch per foot, or 1 in 96, or, as recommended here, 1.04%.

Other aluminium alloy castings such as sumps (oil pans) and thin-walled pipes contract even less. A contraction of 0.60% is common.

Whereas the patternmaker would originally have chosen a special wooden rule whose scale was expanded by the correct amount so that he could read the dimensions directly, without conversion problems, this clearly limited him to specific contraction values. Nowadays, the greater accuracy requires that intermediate values must be chosen, such as

1.15, or 1.20% etc. for different castings. These are now easy to apply with the use of electronic and digital measuring instruments, which can be programmed for any value of contraction. In fact, a virtual solid model developed in the computer should be capable of allowing for three different contractions along each of the three perpendicular axes.

The different contractions are the result of different degrees of constraint by the mould during cooling. For instance, in the case of zero constraint, a casting such as a straight bar will contract freely to its maximum extent. We can therefore calculate this rather easily, assuming an average linear contraction. For instance for Al-Si alloys the coefficient of thermal expansion is close to 20.5×10^{-6} C^{-1} and the total cooling from 660 to 25°C. From this, we can predict the contraction as $20.5 \times 10^{-6} \times 635 = 0.0130$ or 1.30%, in exact agreement with practice.

Turning now to the case of high mould constraint, it is possible to envisage an ideal case in which a large box casting with thin walls was cast around a large, rigid sand core. If the wall thickness of the casting is imagined to be vanishingly thin, like a sheet of paper, then its strength will be negligible and the core not compressed at all. Thus the casting will not be allowed to contract; its paper-thin walls will be forced to stretch. We can therefore envisage in principle the case of infinite constraint in which the casting contraction is zero.

In practice, of course, the real world is filled with casting/mould combinations that lie intermediate between the case of zero and infinite constraint; i.e. partway between 0 and 1.30% contraction in the case of aluminium alloy castings.

How can we obtain an estimate of the degree of constraint, so as to be able to predict the contraction allowance exactly? This is the patternmaker's problem. It is a difficult question, to which there is no accurate answer at this time. However, we can obtain a useful estimate by the following procedure which, fortunately, is good enough for many purposes.

In the case of the straight parallel-sided bar casting made in a sand mould, the casting suffers no constraint. We can define this as being a fully dense metal casting, in the case of aluminium having a density of about 2700 kgm^{-3}. This casting will contract the maximum amount, which for aluminium contracting from its melting point is 1.30%. In contrast, our thin-walled box casting has maximum constraint, contains maximum sand, and has (when cast and finally emptied of sand) a density of practically 0 kgm^{-3}. This simple theory gives us the two extreme points on our calibration curve given in Figure 18.5.

Intermediate points are found from measurements on actual castings, taking the volume of the casting divided by the overall volume occupied by the envelope of the casting. The envelope is the shape given by a tight-fitting rubber balloon stretched over the casting. This gives a measure of the amount of restraining sand it contains, compared with the amount of metal in the casting.

When this is carried out accurately, it is found that different varieties of casting are found to lie on a family of approximately parallel curves, all starting and finishing at our theoretical points as shown in the figure. Thus the procedure is not absolute, it does not yield a single universal curve. Nevertheless, it is a helpful guide in the absence of any better alternative at the present time.

In the case of steel castings, the famous result shown in Figure 18.6 can be explained for the first time. Following the procedure that was outlined for aluminium: for the straight bar, the average thermal contraction of steel is around 16×10^{-6} C^{-1} and the cooling range from freezing point to room temperature is close to 1500°C. Thus the contraction is $16 \times 10^{-6} \times 1500$, which is 2.4%, in agreement with the measured value. We can plot this at the full density of steel of approximately 7850 kgm^{-3} to define our theoretical point, coincident with our measured point, to define the zero constraint condition. The other theoretical point is, of course, the origin (zero contraction at zero envelope density) as before. Working out the area of the sand mould envelopes of the dumbbell and H shapes in Figure 18.6, and dividing by the area of the casting, allows us to plot the two remaining points, giving the nearly linear relation in Figure 18.7.

The Al alloy contraction result from Figure 18.5 is also shown on Figure 18.7 for comparison. Until better methods become available, it seems reasonable to suppose that each foundry will have to determine for itself an equivalent of Figures 18.5 and 18.7 for each of its processes. For instance, it is well known that the values of contraction allowance for greensand are dependent on the hardness of ramming. Similarly, the percentage of binder in chemically bonded sands significantly affects the contraction of the casting. A standard trick to reduce the constraint provided by a central core is to reduce its binder level or to make it hollow.

These relations for sand moulds and sand cores are not expected to apply accurately to metal dies. Here the casting is subject to high mould constraint up to the time of ejection. Clearly, the casting contracts freely only after this instant.

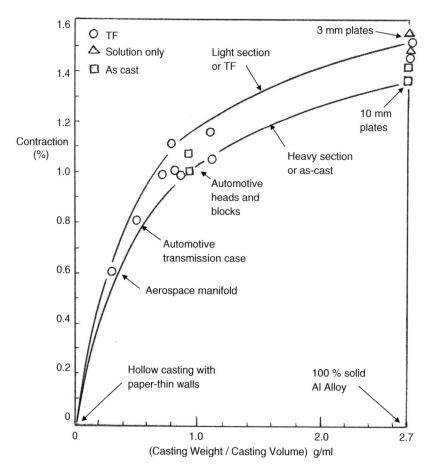

FIGURE 18.5

An experimental result from an automotive and aerospace foundry showing how some castings hardly shrink in size at all when cast, whereas other shrink almost the full theoretical amount of the solid metal. The resistance to shrinkage provided by core and mould geometry accounts for the difference.

For Al-Si alloys cast in gravity and low-pressure dies, the contraction varies between 0.75% for low-silicon alloys, and 0.5% for eutectic (approximately Al-11Si) alloys (Street, 1986), although much of the industry seems to work generally at 0.6%. These low values reflect the high resistance of the die to the contraction of the casting. However, much lower contractions, effectively zero, are occasionally found for thin boxes and window-frame–type castings.

For pressure die casting in magnesium, the contraction allowance is 0.7%, whereas for aluminium alloys it is close to 0.5%. The value is at the lower end of the range for gravity and low-pressure dies, indicating the even greater constraint in high-pressure die design.

These figures for the contraction allowance of die castings are the result of the prior expansion of the die from room temperature to its working temperature, and the subsequent contraction of the casting after its ejection out of the die (we shall assume that its contraction whilst in the die is negligible). We can estimate this quantitatively, taking the die working temperature as roughly 350°C on average (the hot face will be nearer 450°C, but the interior of the die may

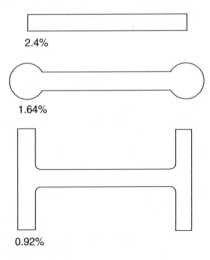

2.4%

1.64%

0.92%

FIGURE 18.6

Contraction of steel greensand castings, showing widely different contractions.

(Steel Castings Handbook, 1970)

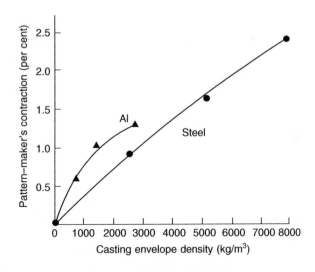

FIGURE 18.7

Contraction allowance for aluminium and steel castings as a function of mould constraint.

be water-cooled), ambient temperature as 25°C, and the temperature of the casting at ejection approximately 500°C, we have:

$$\text{Total casting contraction} = \text{Die Expansion} - \text{Casting Contraction after ejection}$$

$$= (350 - 25) \times 11.7 \times 10^{-6} - (500 - 25) \times 20.5 \times 10^{-6}$$

$$= 0.60 \text{ per cent}$$

Using these rough assumptions and simple logic, the answer is seen to be precisely correct. Furthermore, it is clear, as confirmed by experience, that the size of pressure die castings is controlled by the time of ejection of the casting from the die.

For lost-wax castings, the problem is compounded by having to take account of the expansion of the aluminium die into which the hot wax is injected, the contraction of the wax pattern, the expansion of the ceramic shell, and the final contraction of the casting itself. This complicated equation is a major source of uncertainty in the accuracy of what is known as 'precision casting'. Regrettably, it is the reason why many so-called precision castings suffer dimensional out-of-tolerance problems.

It is important to note that the pattern contraction allowance can often be different in different directions, because of different geometrical constraints offered by the casting. Thus each of the x, y and z axes may require a different value.

It is also important to remember that the casting contraction can be greatly affected by the precipitation of gas during solidification. Girshovich et al. (1963) draw attention to this problem in aluminium-, copper- and ferrous-based alloys. The author has sobering and unforgettable experience of a major pickup of hydrogen gas in a 1000 kg holding furnace after the addition of Sr to an Al-7Si-0.4Mg alloy. Over the next 3 days, the castings suffered up to 3 volume% porosity, therefore growing linearly by 1% (the 500 mm long castings growing by 5 mm!). The castings were all outside their machining allowance and were consequently scrapped. Strontium addition was immediately discontinued at that time! The subsequent introduction of better degassing techniques has allowed the question of Sr addition to be revisited. In a related experience with steels, Schurmann (1965) describes how rimming steel ingots that failed to develop any significant rimming action tended to grow a solid crust over the ingot top. The internal pressure in the ingot could not then be relieved by the escape of gas, so the ingot swelled.

Finally, the reader would be forgiven for assuming that the size of the casting was fixed when at last it reached room temperature. However, this is rarely true. For instance, Figure 18.8 shows how the common zinc pressure die casting alloys continue to shrink in size for the first 6 months or so. Alloy A is then fairly stable for the next few decades. Alloy B, on the other hand, starts to reverse its shrinkage after about the first year. At 100°C these changes can be accelerated by about a factor of 250, effectively compressing years into days, as the time axis shows.

The zinc die casting alloy ZA27 (Zn-27%Al) shrinks only about one-tenth of the amount of the lower-aluminium zinc alloys (8 and 12%Al), but its expansion is greater, as seen in Figure 18.9. (Incidentally, the time axis in Figures 18.8 and 18.9 is calculated assuming the useful relation for reaction rates of a factor of two increase in reaction rate for every 10°C rise in temperature.)

Aluminium alloy castings also show size changes. For instance, Al-7Si-0.5Mg alloy contracts by 0.1–0.2% after solution treatment and quenching as alloying elements are taken into solution. The castings grow by 0.05–0.15% during ageing as the alloying elements precipitate once again (Hunsicker, 1980). Gloria et al. (2000) find that Al-8Si-3.3Cu-0.2Mg expands during solution treatment as the $CuAl_2$ phase dissolves.

Figure 18.10 shows the Al-17Si alloy used for wear-resistant applications exhibiting considerable growth at temperatures high enough to allow silicon to precipitate. This growth in service can be reduced by a preage at a minimum of 230°C for 8 h (but a treatment of 260°C for 1 h would be closely equivalent, using our 10°C rule equivalent to doubling the reaction speed).

Even higher percentage growths are exhibited by white cast irons. At high temperatures, greater than approximately 900°C, the breakdown of the metastable cementite to stable graphite takes several days. During this period, the linear dimensions of the casting will grow by up to 1.6% (Johnson and Nohr, 1970). Grey irons will also grow at temperatures down to 350°C, and growth can be catastrophic if the iron is cycled repeatedly through the ferrite/austenite phase change. Walton (1971) observed a growth of 3.5% in only 500 h in a grey iron subjected to cyclic heating to 800°C. Growth can be further enhanced by the internal oxidation of the material. Angus (1976) gives more details of the growth of cast iron at elevated temperatures, and the means by which it can be controlled by control of structure and chemical composition of the iron.

These are just a few examples of the growth and/or shrinkage of castings that can occur in the solid state because of microstructural changes occurring within the alloy. The casting engineer needs to be on guard against such problems.

The changes cause the patternmaker problems when attempting to decide what contraction allowance to use when constructing the pattern. His decision may be right or wrong, depending on whether the foundry check the dimensions of the casting before or (more correctly) after heat treatment, and will depend on the service conditions.

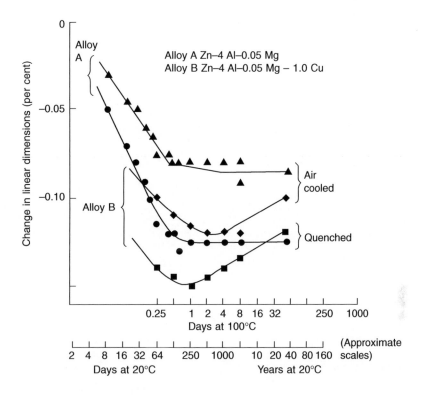

FIGURE 18.8

Zinc pressure die casting alloys (Zn-4Al) showing accelerated ageing at 100°C, or slow shrinkage followed by expansion in alloy B taking the place of decades at 20°C.

Data from Street (1977).

If the allowance was chosen wrongly, giving an undersized casting in grey cast iron, then a heat treatment may save the day by growing the casting by up to 1% or so. However, such good fortune is rare. The choice of contraction allowance before making the casting will remain a difficult and risky decision, and will become more difficult as demands on casting accuracy increase.

18.5.2 NONUNIFORM CONTRACTION (DISTORTION)

If the casting was cooled at a uniform rate and with a uniform constraint acting at all points over its surface, then it would reach room temperature perfectly in proportion, perhaps a little large, or a little small, but not distorted.

In practice, of course, this utopia is never realised. Usually the casting is somewhat large, or somewhat small, and is not as accurate a shape as a discerning customer would prefer. Occasionally, it may be very seriously distorted. We shall examine the reasons for these factors and see to what extent they can be controlled.

Mould constraint

Again, wishing ourselves into utopia, we can envisage that if the constraint by the mould were either zero or infinite, in both cases the casting would be of predictable size and correct shape.

In reality, of course, different parts of the casting experience different degrees of constraint by the mould. One of the most common examples of this problem is a simple five-sided box with its sixth side remaining open, as shown in

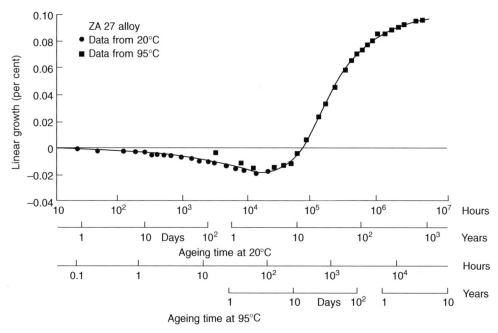

FIGURE 18.9

Zn-27Al alloy dimensional changes with time.

Data from Fakes and Wall (1982).

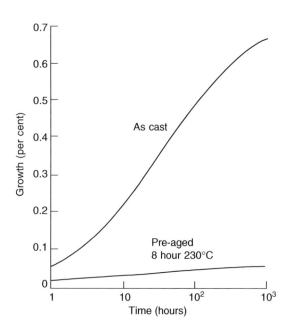

FIGURE 18.10

Permanent growth of A390 (Al-17Si alloy) at 165°C with time (Jorstad, 1971).

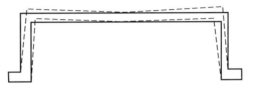

FIGURE 18.11

Distortion of an open-sided box casting during cooling as a result of uneven mould constraint.

Figure 18.11. The closed face wishes to contract as a straight length, whereas the vertical sides have maximum constraint. The result is a compromise, with the straight side shortening with an effective contraction allowance of perhaps 1.2% in the case of an aluminium alloy casting. The vertical sides will be restrained from pulling inwards and so have an effective contraction allowance of perhaps only 0.9%, or perhaps as low as zero if the walls are very thin. During cooling the casting therefore develops a bowed shape with non-parallel walls. For a hard, chemically bonded sand mould the camber will be approximately 1 mm in the centre of the long side of an open box 500 × 100 × 100 mm with walls 4 mm thick.

The box casting may be cast somewhat straighter by several techniques well known to the foundry technologist.

1. The centre core can be made weaker, reducing its constraint on the contraction of the casting. This is achieved either by reducing the percentage addition of binder, or, usually more conveniently, by hollowing out the centre core. The thin shell of sand thereby becomes hotter, giving greater breakdown of the binder in the case of an organic binder, and so allowing the core to collapse earlier.
2. Tie bars can be connected across the open side of the box, thereby holding the walls in place and balancing the effect of the contraction of the closed side. The tie bar need not be a separate device. It can be the running system of the casting, carefully sized so as to carry out its two jobs effectively.

This raises the important issue of the influence of the running and feeding system. Unfortunately, these appendages to the casting cannot be neglected. They can be used positively to resist casting distortion as previously. Alternatively, they can cause distortion and even tensile failure as shown in the simple case in Figure 18.12(a). In addition, if this casting is filled at the flange end, leaving the plate free to contract along its length as shown in Figure 18.12(b), the problem is solved. However, note the important point that whether or not casting 'a' has suffered any tensile failure, it will measure somewhat longer than casting 'b'. Thus different pattern contraction allowances are appropriate for these two different constraint modes.

In more complex castings, the effect of geometry can be hard to predict and harder to rectify if the casting is particularly badly out of shape. Especially for large, thin-walled castings requiring close dimensional tolerance it may be wise to include for a straightening jig in the tooling price. This will be an expensive piece of tooling, usually resisted by the customer.

Casting constraint

Even if the casting were subjected to no constraint at all from the mould, it would certainly suffer internally generated constraints as a result of uneven cooling. The famous example is the mixed-section casting shown in Figure 18.13(a). If a failure occurs it always happens in the thicker section. This may at first sight be surprising. The explanation of this behaviour requires careful reasoning, as follows.

First, the thin section solidifies and cools. Its contraction along its length is easily accommodated by the heavier section, which simply contracts under the compressive load because it is hot, and therefore plastic, if not actually still molten. Later, however, when the thin section has practically finished contracting, the heavier section starts to contract. It is now unable to squash the thin section significantly, which has by now become rigid and strong. The result is the possible bending of the thin section as tension builds up in the thick section. Under its tensile stress, the thick section might therefore stretch plastically, or hot tear, or cold crack.

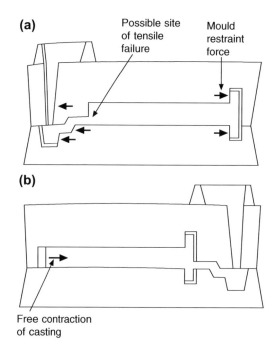

FIGURE 18.12

(a) Effect of the filling and feeding systems imposing constraint on the contraction of a casting. (b) Applying the filling system to the opposite end of the casting eliminates the problem, permitting the casting to contract freely.

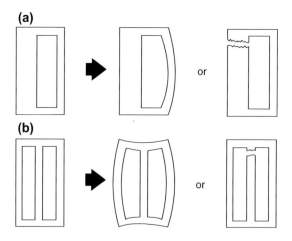

FIGURE 18.13

(a) Thick/thin section casting showing tensile stress in the thick section; (b) even-walled casting showing internal tensile stress.

The example shown in Figure 18.13(b) is another common failure mode. The internal walls of a casting remain hot for longest even though the casting may have been designed with even wall sections. This is, of course, simply the result of the internal sections being surrounded by other hot sections. The reasoning is therefore the same as that for the thick-/thin-section casting previously. The outer walls become cool and rigid, and the internal walls of the casting suffer tension at a later stage of cooling. This tension may be retained as a residual stress in the finished casting, or may be sufficiently high to cause catastrophic failure by tearing or cracking.

The same reasoning applies to the case of a single-component heavy-section casting such as a solid ingot, billet or slab, and especially when these are cast in steel, because of its poor thermal conductivity. The inner parts of the casting solidify and contract last, putting the internal parts of the casting into tension (notice it is always the inside of the casting that suffers the tension; the outside being in compression). Because of the low yield point of the hot metal, extensive plastic yielding occurs at high temperatures. However, as the temperature falls, the stress cannot be relieved by plastic flow, so that increasing amounts of stress are built up and retained.

An example shown in Figure 18.14 shows the kind of distortion to be expected from a box section casting with uneven walls. The late contraction of the thicker walls collapses the box asymmetrically (the casting is at risk from tensile failure in the thicker walls, but we shall assume that neither tearing nor cracking occurs in this case). There is clearly some strong additional effect from mould constraint. If the central core were less rigid, then the casting would contract more evenly, remaining more square.

There is an important kind of distortion seen in plate shaped castings which have heavy ribs adjoining the edges of the plate, or whose faces are reinforced by heavy-section ribs. It is often seen in thin-section boxes that have reinforcing ribs around the edges of the box faces. The general argument is the same as before: the thin, flat faces cool first, and the subsequent contraction of the heavier ribs causes the face to buckle, springing inwards or outwards. This is known as 'oil-can distortion'; an apt name which describes the exasperating nature of this defect, as any attempts to straighten the face cause it to buckle in the opposite direction, taking up its new reversed curvature. It can be flipped backwards and forwards indefinitely, but not straightened permanently. Once a casting exhibits oil-can distortion, it is practically

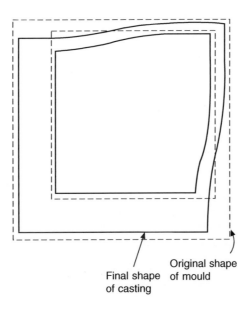

Final shape of casting   Original shape of mould

FIGURE 18.14

Distortion in an uneven box section casting because of combined casting and core constraint.

impossible to cure. The effect is more often seen after quenching from heat treatment, where, of course, the rate of cooling is greater than in the mould, and where the casting does not have the benefit of the support of the mould.

Oil-can distortion may be preventable by careful design of ribs to ensure that their geometrical modulus (i.e. their cooling rate) is similar to, or less than, that of the thinner flat face. Alternatively, the cooling rate from the quench needs to be equalised better, possibly by the use of polymer quenchants and/or the masking of the more rapidly cooling areas.

In ductile iron castings that exhibit expansion on solidification resulting from graphite precipitation, the expansion can be used to good effect to reduce or eliminate the necessity for feeders, particularly if the casting cools uniformly. Tafazzoli and Kondic (1977) draw attention to the problem created if the cooling is not uniform. The freezing of sections that freeze first leads to mould dilation in those regions that solidify last. Although these authors attribute this behaviour to the mismatch between the timing of the graphite expansion and the austenite contraction, it seems more likely to be the result of the pressure within the casting being less easily withstood by those portions of the mould that contain the heavier sections of the casting. This follows from the effect of casting modulus on mould dilation; the lighter sections are cooler and stronger, and the thicker sections are hotter and thus more plastic. Any internal pressure will therefore transfer material from those sections able to withstand the pressure to those that cannot. The thinner sections will retain their size while the thicker sections will swell. Tafazzoli and Kondic recommend the use of chills or other devices to encourage uniformity.

In a classic series of articles, Longden (1931–1932, 1939–1940, 1947–1948, 1948) published the results of measurements that he carried out on grey iron lathe beds and other machine beds. The curvature of a casting up to 10 m long could result in a maximum out-of-line deviation (camber) of 50 mm or more. Longden summarised his findings in a nomogram that allowed him to make a prediction of the camber to be expected on any new casting. The reverse camber was then constructed into the mould to give a straight casting. Although Longden's nomogram is probably somewhat specific to his type of machine tool bases, and therefore not useful as a general predictive tool for other castings, it is presented in Figure 18.15 as an example of what can be achieved in the prediction of casting distortion. It is to be expected that castings of other types may exhibit a similar relationship.

Pursuing this nice but very specific example, it is possible to show that Longden's graphical summary can be converted into a quantitative form, where length L and depth D are in metres and wall section thickness w and camber c are in millimetres. Simplifying Longden's nomogram within the limits of accuracy of the original data and making the further approximation that the slight curved lines on the left-hand side are straight lines through the value $L = 1.5$ m, then with perhaps about 10% accuracy for wall thickness w from 10 to 40 mm the camber is given by:

$$c = (L - 1.5)(7.62 - 1.073w) - 660D + 310$$

and for wall thickness w from 40 to 70 mm:

$$c = (L - 1.5)(144 - 2.03w)(1 - D)$$

We now move on to a further type of internal constraint that appears to be universal in castings of all sizes and shapes, and which is rarely recognised, but was investigated by Weiner and Boley (1963) in a theoretical study of a simple slab casting. They assumed elastic–plastic behaviour of the solid, and that the yield point of the solid was zero at the melting point (not quite true, but a reasonable working approximation) and increased as the casting cooled. They found that plastic flow of the solid occurs at the very beginning of solidification. The stress history of a given particle was found to be as follows. On freezing, the particle is subject to tension, and because the yield stress is initially zero, its behaviour is at first plastic. As it cools, the tensile stress on it increases and remains equal to the yield stress corresponding to its temperature until such time as the rate of increase of stress upon it is less than the rate of increase of its yield stress. It then starts to behave elastically. Soon after, unloading begins, the stress on the particle decreases rapidly, becoming compressive, and finally reaches the yield stress in the opposite direction. Its behaviour remains plastic thereafter. Weiner and Boley's analytical predictions have been accurately confirmed in a later numerical study by Thomas and Parkman (1998).

To sum up their findings, in a solidifying material there will be various deformation regimes. These are (1) a plastic zone in tension at the solidification front because the strength of the solid is low; (2) a central region where the stresses are in the elastic range; and (3) a zone at the surface of the casting where there is plastic flow in compression. The overall scheme is illustrated in Figure 18.16.

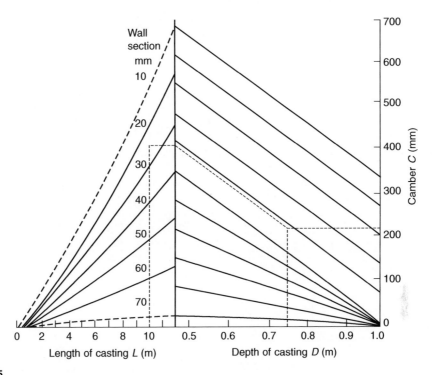

FIGURE 18.15

Camber nomogram based on data from Longden (1984) for machine tool bed castings in grey iron. A 10 m long casting with 25 mm thick side walls 750 mm depth will show approximately 210 mm camber (deviation from straight).

The propagation of the tensile plastic region, the central elastic zone, and the compressive plastic zone are reminiscent of the propagation of the various transformation zones through the sand mould. These waves of strain and stress spread through the newly solidified casting, remaining parallel to the solidification front and remaining at the same relative distances, as illustrated in Figure 18.16.

If the yield stress were not assumed to be zero at the freezing point, but were to be given some small finite value, then the analysis would be expected to be modified only very slightly, with a narrow elastic zone appearing at the solidification front on the right-hand side of Figure 18.16.

The analysis will be fundamentally modified for materials that undergo certain phase changes during cooling. If the crystallographic rearrangement involves a large enough shear strain, or change of volume, as is common in steels cooling through the γ to α transition for instance, then the material will be locally strained well above its yield point, adding an additional plastic front which will propagate through the material. The phenomenon is known as *transformation induced plasticity*. This additional opportunity for the plastic relief of stress will fundamentally alter the distribution of stress as predicted in Figure 18.16. However, the prediction is expected to be reasonably accurate for many other metals such as zinc-, aluminium-, magnesium-, copper- and nickel-based alloys, and for those steels that remain single phase from solidification to room temperature.

The high internal tensions predicted by this analysis will be independent of, and will be superimposed on, stresses that arise as a result of other mould and/or casting constraints as we have discussed previously. It is not surprising, therefore, to note that on occasions castings fail whilst cooling after solidification.

Drezet et al. (2000) showed that the elastic–plastic model would not be fundamentally altered if creep flow behaviour were assumed instead of the elastic–plastic flow behaviour with yield stress a function of temperature. They found that

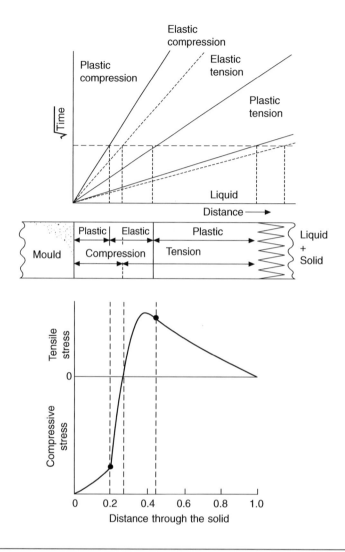

FIGURE 18.16

Elastic/plastic regimes in a simple slab casting.

After Weiner and Boley (1963).

such simulations were insensitive to the rheological model employed, but that the deformation was mainly a simple function of the thermal contraction and the conditions for continuity.

Richmond and Tien (1971) and Tien and Richmond (1982) criticise the model by Weiner and Boley on the grounds that they do not take account of the friction at the casting–mould interface. When Richmond and Tien include this they find that the casting–mould interface is no longer in compression but in tension. This is almost certainly true for large castings in metal moulds such as steel ingots in cast iron ingot moulds, where the pressure between the mould and casting is high, the friction is high, and the mould is rigid. These authors explain the occurrence of surface cracks in steel ingots in this way. However, Weiner and Boley are likely to more nearly correct for smaller castings in sand moulds. Here the interfacial pressure will be less, and the surface more accommodating, and the air gap ensuring that the casting and mould are not in contact in some places. All these factors will reduce the restraint resulting from friction. Thus, their analysis remains probably the most appropriate for medium-sized shaped castings.

18.5.3 PROCESS COMPARISON

It is not easy making a comparison between the various casting processes to ascertain their comparative capabilities in terms of accuracy, or possibly, reproducibility. There are many separate factors that influence the capabilities of different processes, to the extent that it may seem rather surprising that a casting achieves any approach to accuracy at all. However, it is a pleasure to note that the majority of casting processes do rather well, and some excellently. That is not to exclude the certainty that most could probably do better!

There have been several studies attempting to quantify what tolerances are achieved in practice. Useful researches were carried out in Sweden by Villner (1969, 1974) and in the United Kingdom by the Institute of British Foundrymen (IBF Technical Subcommittee T571, 1969, 1971, 1976, 1979). These workers took production castings from several foundries, covering a wide variety of different metals and different processes. Linear measurements taken from a large number of castings were subjected to multiple regression analysis, a statistical technique used to assess quantitatively the effects of a large number of factors simultaneously. The technique identified the following factors as important.

1. The total area of cores projected in the plane of the mould joint.
2. The wall thickness, indicating that thinner walled castings were in general more accurate.
3. The mould joint was generally closed accurately to within approximately 0.1 mm showing that the main mould joint was usually one of the most accurate of the mating interfaces within the mould assembly, contributing only a very small error to any final dimension across this joint. (Note that lateral error, known as mismatch across the joint, was a separate problem. Mismatch could often be more than 10 times the closing error, and could easily scrap castings.)
4. The great value of keeping critical dimensions or reference points all within one half of the mould, or totally within one core was demonstrated. This was a large factor reducing variability.

Apart from the three continuous variables, 'drawing dimension', 'projected area of core' and 'general wall thickness', there were additional factors contributing to variation such a metal type, pattern and mould distortions, temperature variations, coating thickness etc.

Regardless of these numerous effects, the main result was that the standard deviation, sigma (σ), increased with size of the dimension. Although the regression equation derived by these investigators indicated that σ increases linearly with the dimension, this was only, of course, the result of the regression analysis itself being carried out assuming linearity, although the assumption of linearity was probably not too bad as a working approximation at this stage.

Taking therefore an example of a hollow casting 5 mm wall thickness, of overall size approximately $250 \times 200 \times 200$ mm, containing a large body core, Figure 18.3(a) shows the result for 2.5σ, where σ is the expected standard deviation (2.5σ encloses 99% of all expected results). A second study by the author in 2000 considered only probable variation in the length of a simple solid casting using σ as a measure. The results of this rather different casting are shown in Figure 18.3(b). The similarities between the two plots are probably significant because not only length is seen to dominate.

In fact, from Figure 18.3(b), is clear that the variability of processes such as gravity die casting are dominated by factors which are not functions of length and probably include such variable factors as die coating. This contrasts with processes such as lost foam, and more particularly, lost wax, whose variability increases with casting size, reflecting the importance of thermal expansion problems in these processes. Thus lost wax can retain its common name '*precision casting process*' only for very small castings. For large castings, the accuracy of the process becomes no better than low-technology sand casting (although it retains its excellent surface finish and definition of course).

The processes which shine out as having intrinsically repeatable dimensions are pressure die casting and high-technology sand casting. There are good reasons for this. Pressure die casting is a simple, direct process, operated in a rigidly supported steel die and uses no die coat. Sand casting involves more steps, but all the steps are carried out at room temperature so that no significant expansion errors accumulate.

The position of gravity die being systematically below that of the variability of low-technology sand casting was one of the major reasons for the historical choice of 'die casting' as opposed to sand casting for many automotive applications. Now, interestingly, recent advances in sand moulding, both in greensand and chemically bonded sands, have overtaken the accuracy possible in gravity die. Even so, the poor image of sand casting lives on in the engineering profession, whereas 'die casting' has the image of cleanness and precision! It is clear that a re-education of the engineering profession, along with ourselves as founders, will take a generation or more.

From Figure 18.3(a) and (b) the increase in accuracy for a 500 mm long aluminium alloy casting when changing from gravity die to a precision sand process is a factor of between 2 and 4. This is roughly in line with measurements carried out on production runs of cylinder heads by Ford of America who found on average that an accurate sand process yielded castings more than twice as accurate as their standard supplies of gravity die castings. This was a critical finding which led to the adoption of aggregate moulds using the Cosworth Process in North America for automotive cylinder blocks instead of their previous reliance on permanent mould castings.

In conclusion, pressure die casting and high-technology sand casting are the processes with the greatest capability for reproducibility of dimensions. The lost wax process is accurate for small components, but as the casting size increases becomes rapidly poorer, eventually becoming as bad as low-technology sand casting at large sizes. Gravity die casting and lost foam casting show intermediate performance.

A word of warning about Figure 18.3(a) and (b) is probably necessary. Such comparative diagrams are intended to give a general overview of the capability of the various processes, and to this extent they are useful and fair. In particular instances, though, certain foundries may achieve much better results than the norm and, regrettably, some much worse! Also, some castings of regular shape are more easily kept to close tolerance, whereas flimsy or complex shapes may prove very difficult. Thus figures such as Figure 18.3 require to be treated with some caution.

Finally, we need to remind ourselves that cast products are not bought simply for their dimensional reproducibility. Surface finish, internal integrity and many other factors, not the least of which is cost, are important. This section has been aimed at clarifying and quantifying so far as possible only the ability of a process to achieve a level of dimensional control.

18.5.4 GENERAL SUMMARY

The main factors which control the accuracy of the final casting are briefly listed as a summary.

1. Pattern (or tooling) inaccuracy.
2. Mould inaccuracy.
3. Mould expansion and/or contraction because of temperature or pressure.
4. Casting expansion because of precipitation of less dense phases such as graphite or gases.
5. Casting contraction on freezing (solidification shrinkage) causing local sinks.
6. Casting contraction on cooling leading to (a) different overall casting size, depending on the constraint by the mould, and (b) distortion if unevenly constrained or unevenly cooled.
7. Casting overall change of size on heat treatment or on slow ageing at room temperature.
8. Casting distortion if unevenly cooled by an inappropriate quenchant or too rapid quench from heat-treatment temperature.
9. Casting distortion caused by shot blasting. The compressive stresses introduced into the surface by a peening effect can lead to the distortion of the casting as outlined in Chapter 19, *Postcasting Processing*.

18.6 METROLOGY

The technology of metrology has advanced so far over recent years, and continues to advance, such that I seriously thought that this section should be abandoned. However, not all of us have the benefits of modern, computer-controlled, remote-scanning laser techniques that operate as a three-dimensional coordinate measuring machines. Even for those that do have the luxury of remote optical scanning for dimensional inspection, Minetola et al. (2012) explain the concerns of potential pitfalls. Thus, in this transitional period, while many of us struggle with our traditional systems, and possibly struggle with the new technology, there is perhaps room for this section for a year or two more. Those with laser systems or X-ray radiography tomography are permitted to skip this section.

Even if it were possible to produce an absolutely accurate casting, it would not be possible to prove it! This apparently curious statement is the consequence of errors which occur during measurement. Inexact measuring of the casting will cause the random deviations in the measurements, causing dimensions of the casting to appear too large or too small. Svensson and Villner (1974) point out this problem and work out the influence of measuring accuracy on the apparent dimensional accuracy of the casting.

It is clear that even if the casting has dimensions which are perfectly correct; even careful measurement will introduce a certain amount of apparent error, and careless measurement will, of course, introduce even more. The more recent introduction of large-size three-dimensional coordinate measuring machines has significantly reduced these errors, which have been such a traditional problem within the industry.

Even so, problems will remain. For instance, the Swedish workers point out that for small dimensions, and where high accuracy is required, the surface roughness will influence the apparent accuracy of the casting. Thus a change in the surface finish from 75 to 200 μm will give an increase of one tolerance grade in the ISO system.

The surface finish influences the measurement and location processes in other ways. For instance, the modern touch probes, which locate dimensions on the casting with the most delicate of contact pressure, effectively only measure to high spots, thus biasing the measurements in one direction: exterior dimensions on the casting are measured oversize, and cored holes appear undersize.

Results from mark-out equipment using a mark-out table and a mechanically scribed line tend to give more averaged results because minor surface irregularities are cut through.

Similarly, when castings are clamped on to their location points, the small area of the contact points, typically a 5 mm diameter and the high loads which can be exerted by the clamps, ensure that the locating jig point actually indents the surface of the casting by up to 0.25 mm for some aluminium alloy sand castings. Harder materials such as cast irons will, of course, indent less. All surface irregularities are effectively locally smoothed and averaged in this operation. The indentation effect sets an upper limit to the accuracy and repeatability with which castings can be picked up for measurement or machining.

A traditional method of checking the profile of a casting is by the use of template gauges. These are typically sheets of metal which have been cut to the correct contour. On applying them to the casting, the contour on the casting can be seen to be correct or not, depending on the clearance which can be seen between the two. This is an analogue technique which can no longer be recommended in these modern times. The gauges are expensive to make. They are also subject to wear and thus need to be checked regularly and occasionally replaced. However, what is much more serious, they are difficult to use in any effective way. This is because in practice the contours never match exactly. The problem for the user then is how inaccurate can the contour of the casting be allowed to become before remedial action must be taken?

The use of 'go/no go' gauges removes the matter of judgement. However, the gauges are again subject to wear, and thus require the cost and complexity of a calibration system. More fundamentally, their use is similarly not helpful in terms of providing useful data to assist process control.

All these difficulties can be removed by the use of a much simpler technique: the use of simple goalpost fixtures which straddle the casting and which are equipped with one or more spring-contact probes, such as dial gauges. The readings from the gauges are read and recorded. The operation becomes even simpler with the use of digital read-out devices (Figure 18.17). Linear transducers are easily fitted and operated, and give an immediate numerical signal of the degree of inaccuracy. Laser techniques give an even faster and contactless benefit involving no wear.

The goalpost would be calibrated and stored on a standard casting and thereby always be seen to be in calibration by being set to zero in this position. (For calibration away from the zero, other readings can be obtained by the insertion of slip gauges under the probe.)

The use of digital electronic read-out in this way allows its incorporation into data-logging and quality-monitoring systems, such as statistical process control. By watching the trends on a daily or weekly basis, the gradual drifts in casting dimensions can be used to predict, for instance, that tooling wear will reach a level which will require the tooling to be replaced in 3 months' time. Such prior warning allows the appropriate action to be planned well in advance.

Even the use of goalpost systems is outdated by modern digital laser-reading systems, in which the contour is held as a virtual contour inside the computer. Even complex free-form surfaces typical of aerofoils can now be modelled remotely on a screen; the products scanned and checked for conformity within seconds; and the results immediately forwarded electronically to the customer in any part of the world.

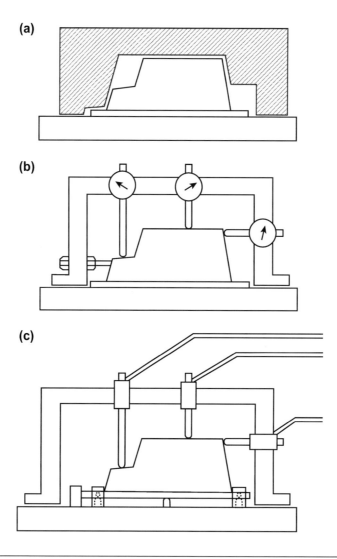

FIGURE 18.17

Checking techniques for size and shape of castings; (a) template; (b) analogue quantification using dial gauges; (c) digital quantification using linear displacement transducers with the casting on a six-point jig. Modern equivalents are laser non-contacting techniques.

Very large castings benefit particularly because the handling of mechanical gauges several meters long, while attempting to measure to fractions of millimeters become difficult if not impossible. The use of a remote laser technique is quick, easy for the operator, and amazingly accurate. Parts of up to 4 m in size can be measured to within 0.03 mm (Tigges, 2010).

Digital techniques have introduced a new level of capability in terms of the accurate measurement of castings. This has raised the stakes for the founder, who now is challenged to improve casting accuracy yet further.

POST-CASTING PROCESSING

The cutting off of gates and feeders is a chore, significantly aided by not having feeders, or even running systems, if possible.

Zinc castings are special because they can be cleaned from flash by cryogenic tumbling as a result of embrittlement of the alloy by cooling to $-196°C$ in liquid nitrogen, or possibly using a thermal deburring operation in which the castings are momentarily heated to $3000°C$ in a sealed chamber in which a mix of natural gas and oxygen are ignited to burn off flash and melt corners (Birch, 2000a).

High pressure die castings (HPDCs) of Zn, Mg and Al alloys can be clipped as a result of (1) their high accuracy and (2) their cleanness. The clipping of sand castings might be expected to result in a very blunt clipping tool, bringing the press to a full stop. However, these disadvantages cannot be too great because the process is currently being promoted for aluminium and iron sand castings (SERF, 2002; Ricken et al., 2009). The process is highly productive, taking seconds (not minutes by grinding for instance). Clean clippings are efficiently remelted, again contrasting with grindings. Operator problems such as white finger are also avoided.

Other alloy systems are much less easy. Feeder heads on steel castings can sometimes be removed by a powerful hammer blow, or with hydraulic wedges, although I suspect these fracturing processes are successful because of the density of bifilms, created by the poor filling technique, concentrating high up in the casting, particularly in the region of the feeder neck. The impact or wedge removal technique would probably not be possible with good quality well-made castings. Alternatively, large steel feeder heads are most often removed by flame or arc cutting. The troublesome cracks generated by the process are avoided if the filling system is improved so as to avoid oxide entrainment (founders are amazed and delighted that the cracks no longer appear after a good naturally pressurised filling system is applied to the casting!).

The removal of cores can be a challenge. Mechanical shock and vibration systems are widely used, as are high pressure water systems. Donahue (2006) describes a novel high velocity jet of liquid nitrogen, of similar density to water, as being highly effective for the removal of residual shell and core material from ceramic investment moulded castings.

Having completed de-moulding, de-gating, de-flashing and removing feeders, there is often still much to do to get the casting into the condition the customer has ordered (working in a foundry is a challenge).

This final chapter deals with these final issues. When these are complete, it is always a pleasure to see the castings going out of the door. (However, it is worth bearing in mind that finally getting paid for them is more certain if the castings are good.)

19.1 SURFACE CLEANING

A common treatment for the cleaning of castings of all types was *sand blasting*, in which particles of sand are entrained in a powerful blast of air directed at the casting. The process is now almost universally unused as a result of the concern about the generation of harmful silica dust.

Treatments nowadays include *shot blasting*, in which steel or iron shot is used, centrifuged to high speed by spinning wheels in so-called airless blasting. Stainless steel shot is, naturally, significantly more expensive than conventional carbon steel shot, but has the advantage for aluminium castings that they remain bright and free from rust stains.

Complete Casting Handbook. http://dx.doi.org/10.1016/B978-0-444-63509-9.00019-4

Other treatments include *grit blasting* in which the grit is commonly a silica-free hard mineral such as alumina. The many other blast media include glass beads, plastic particles, walnut shells—almost anything seems to be used.

The particular benefit of blasting with a heavy shot such as steel is that the surface of the casting is deformed by the impact of each particle, generating a localised compressive stress at the site of impact. This is a peening process. Fathallah et al. (2000) finds 100% coverage of the surface is optimum to improve the fatigue resistance of a steel, whereas at 1000% coverage, the overlaps and scaling of the surface damage reduces the fatigue resistance once more.

Many other studies have confirmed that the compressive stresses in the surface of shot-blasted or shot-peened castings enjoy an improved fatigue performance as a result of the inhibition of the growth of surface cracks when subject to compressive stress. Improvements to alloy Al-7Si-0.5Mg were shown by Pitcher and Forsyth (1982), Ji et al. (2002) showed the benefits for ductile iron; Naro and Wallace (1967) improved steel castings. These experimental demonstrations have been supported by some excellent theoretical models by Fathallah et al. (1998) and Evans (2002).

Although the improvement of fatigue resistance following shot peening is perhaps to be expected, the effect of grit blasting is less easy to predict, because the fine notch effect from the indentations of the grit particles would impair, whereas the induced compressive stresses would enhance fatigue life. From tests on aluminium alloys, Myllymaki (1987) finds that these opposing effects are in fact tolerably balanced, so that grit blasting has little net effect on fatigue behaviour.

Kasch and Mikelonis (1969) noticed that the compressive stresses introduced into the surface by the peening effect caused significant movement of the casting on machining. Mroz and Goodrich (2006) found for both grey and ductile irons the depth of cold work and the residual stress did not change significantly after the first shot blast cycle.

However, the distortion effect is used to good effect in the sheet metal industry to induce the controlled forming of curved surfaces; aircraft wing panels are formed from flat sheets in this way; the flat product gradually curls up, becoming convex towards the direction of the impingement of the shot. The distortion effect of the peening treatment is controlled by the use of Almen strips. These are small strips placed alongside the article to be peened, and the curvature of the strips (measured as the arc height) is measured as a function of processing time (Cao et al., 1995). It seems that the use of controlled peening to adjust or vary the shapes of castings seems, so far as I am aware, not used at this time, but might be a useful technique to generate subtle curves on castings that might be difficult to mould.

Sand retention on as-cast surfaces of sand castings can be a problem. Figure 19.1 shows such a surface for an Al alloy cast in a zircon sand mould. A 10 kg cylinder head casting might have 0.5–1 g of sand retained on its surface. This compares with values of up to 0.05 g for a permanently moulded cylinder head which would have relatively few sand cores.

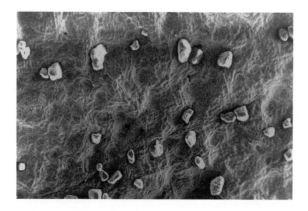

FIGURE 19.1

Surface of an Al alloy sand casting, showing adhering sand.

After a blasting treatment, the surface of the casting will have achieved some degree of uniformity of appearance and texture. For sand castings, however, not all of the sand particles are removed; a significant percentage of adhering particles are pounded in to the surface, thus becoming even more attached by the plastic flow of metal around and over them. This residual adhering sand is not important and not noticed for most applications. It might even be slightly beneficial for applications requiring paint adherence. However, it does affect the rate of wear of machine cutting tools if extensive machining of the surface is required.

The elimination of such sand from Al alloy sand castings can be achieved by a caustic etch. This seems to release most of the particles and improve machinability. However, of course, such chemical treatments are to be avoided if at all possible for environmental reasons.

When grit-blasting Al alloy castings, whether sand castings or permanent mould castings, residual grit is always found to be embedded in the surface. Simply picking up the casting and dropping it on to a cast iron surface table will reveal a ghostly outline of the casting on the table delineated by the grit that has been shaken out by the impact of the fall.

Little research appears to have been carried out on the prevention or reduction of sand retention on castings. We may speculate that, as suggested in Section 9.12 on surface finish, any process that thickens the oxide on the advancing melt would be beneficial. Thus slower filling will allow the oxide to grow thicker, and the surface will experience a slower build-up of pressure, delaying mould penetration while the surface oxide thickens further. Chemical additions to the alloy may also help, as suggested for Sr addition to Al-Si alloys. Finally, of course, the final overpressure applied by counter-gravity filling systems will also require to be reduced so far as possible.

19.2 HEAT TREATMENT

The heat treatment of castings is one of the most expensive of all the post-casting operations. Furthermore, high temperature heat treatments are, of course, the most expensive of these. The high temperature treatments include (1) stress relieving followed by a slow cool, and (2) solution or homogenisation treatments followed by a quench. In addition to the high costs, the treatments requiring quench are significant dangers to the integrity and accuracy of the casting as we have discussed. There are both considerable economic and technical incentives to avoid these processes if at all possible.

The heat treatment of ferrous castings is a huge subject that cannot be covered here. In relation to castings, it is worth mentioning that alloy steel castings require to be homogenised at a temperature of at least 1300°C (Flemings, 1974). These are challenging temperatures, requiring highly specialised furnaces with vacuum or protective atmosphere, and specialised quenching facilities.

Similarly, for light alloys, the solution treatment is also expensive. It is performed at temperatures as near as possible to the melting temperature of the alloy, generally in the region of 500–550°C followed by a quench to retain solutes in solution.

The high temperature is a risk from the possibly incipient melting (or worse, wholesale melting). The quench is risky from the point of view of introducing several problems including residual stress and distortion. We shall devote some space to considering the important questions as to how these risks can be reduced or avoided.

19.2.1 HOMOGENISATION AND SOLUTION TREATMENTS

For a single-phase alloy, Flemings (1974) describes a simple and elegant model of the microsegregation, or coring, present in the dendrites, and how it can be reduced by a high-temperature heat treatment termed homogenisation. He defines a useful parameter that he calls the index of residual microsegregation, δ, as

$$\delta = \frac{C_M - C_m}{C_M^O - C_m^O} \tag{19.1}$$

where C_M = maximum solute concentration of element (in interdendritic spaces) at time t, C_m = minimum solute concentration of element (in centre of dendrite arms) at time t, C_M^O = maximum initial concentration of element, and C_m^O = minimum initial concentration of element.

The parameter δ is precisely unity before any homogenisation treatment. If homogenisation could be carried out to perfection, then δ would become precisely zero. After any real homogenisation treatment, δ would have some intermediate value that depends on the dimensionless group of variables Dt/l^2. Here, D is the coefficient of diffusion of the homogenising element, t is the time spent at the homogenising temperature, and l is the diffusion distance, of the order of the dendrite arm spacing (DAS). Assuming a sinusoidal distribution of the concentration of the element across the dendrite, Flemings finds the solution, approximately, as:

$$\delta = exp\left(-\pi^2 Dt/l_o^2\right) \tag{19.2}$$

where $l_o = (DAS)/2$. Equation (19.2) is useful for the approximate prediction of times and temperatures required to homogenise a given cast structure. Flemings shows that, for a low-alloy steel, carbon is always homogenised by the time that the steel is heated to about 900°C for all normal values of DAS because of its high value for D (see Figure 1.4(c)). However, for the substitutional elements manganese and nickel, little homogenisation occurs below 1100°C, and for homogenisation to be about 95% complete ($\delta = 0.05$) requires 1 h at 1350°C if DAS = 50 μm. The fine DAS value is obtained by ensuring that such critical parts of the casting are within 6 mm of a chill. If no chill is used and the DAS = 200 μm or more, then practically no homogenisation is achieved ($\delta = 0.95$) at this time and temperature.

Flemings emphasises that the normal so-called homogenisation treatments for steels based on temperatures of 1100°C achieve only the homogenisation of carbon. The more recent use of vacuum heat-treatment furnaces capable of 1350°C and higher has produced very large improvements in the mechanical properties of cast steels.

The term *homogenisation treatment* is reserved for treatments designed to smooth out concentration gradients within a single-phase alloy.

The term *solution treatment* applies to those treatments designed to take into solution (i.e. dissolve) one or more second phases. These are also discussed by Flemings (1974). His presentation is summarised next.

Flemings considers a binary alloy containing a non-equilibrium eutectic. The dendrites are again assumed to be cored, having a sinusoidal distribution of solute as before, but containing interdendritic plates of divorced eutectic; for instance, in the case of the Al-4.5Cu alloy, a single plate of $CuAl_2$ phase separates the cored aluminium-rich dendrites. Dissolution of the interdendritic second phase is assumed to be limited by diffusion in the α-phase. If f and f_o are the volume fractions of eutectic at times t and t_0 respectively, then the answer is similar to that seen in Eqn 9.10, approximately:

$$f/f_0 = exp\left(-2.5Dt/l_o^2\right) \tag{19.3}$$

If the DAS is 100 μm, and if impurity levels in the alloys are kept low, so that solution temperatures within 10 or 20°C of the melting point can be employed without danger of melting the alloy, then a 10 h solution treatment is needed to dissolve all the second phase. This large DAS is easily reduced to below 30 μm by chilling the casting, so that times shorted than 1 h are easily attained.

Experimental tests of the theory show good agreement, particularly at short times (Figure 19.2). At long times, the dissolution of the last traces of segregate require more time than the simple theory predicts because more segregate exists between primary arms than between secondary arms, and so the last remnant of solute must diffuse over larger distances than simply the secondary DAS.

19.2.2 HEAT TREATMENT REDUCTION AND/OR ELIMINATION

There is strong motivation to abandon or significantly modify the high temperature solution treatment for Al alloys. This has taken three different options:

1. The total avoidance of any artificial heat treatment, using those alloys that exhibit natural ageing at ambient temperatures such the 7xx alloys based on the Al-Zn-Mg series (Sigworth et al., 2006). At room temperatures, the development of properties is particularly slow, taking days or months, but only an hour or less at economic ageing temperatures in the region of 100°C.

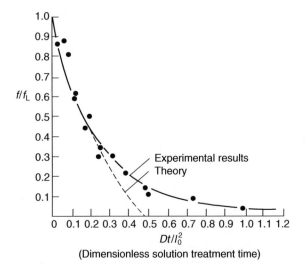

FIGURE 19.2

Rate of solution of eutectic in Al-4.5Cu alloy as a function of (dimensionless) time.

Data from Singh et al. (1970).

2. Avoidance of solution heat treatment and quenching, but using heat treatments that involve only ageing at temperatures in the region of only 150–250°C. These treatments often take several hours, although I personally aim to get the treatment done within an hour at most by opting for higher temperatures.

3. Using only minimal solution treatment times and temperatures, followed by quenching and minimal ageing. For instance, a typical solution heat treatment for the common structural engineering alloy Al-7Si-0.4Mg might be 6–12 h at 530°C followed by quenching and finally ageing for up to 12 h at 160°C. Personally, I always aim for a treatment not longer than 1 h on the grounds that, if cast well, the alloy will display more than adequate properties to pass all specifications. The various ways to reduce these excessive treatment times (optimised by metallurgists in an age of plentiful and low cost energy, and now completely inappropriate) are discussed later. In the meantime, as an instance, the treatment described previously would give nearly equivalent results at 1 h at 540 or 545°C if possible, and 1 h at 200°C.

The use of the rule 'a 10°C rise doubles reaction rates' is illustrated in Figures 18.8 and 18.9. The curves of strength versus time all collapse onto a single curve if allowance is made for the 10°C rule. It is strongly recommended that all heat treatment results are presented in this way. Occasionally, the curves will not precisely superimpose, but this effect is easily distinguished by a fanning outwards of the curves, or the generation of a set of curves that are clearly part of a family, and so largely parallel or uniformly diverging. As an example the treatment of an Al alloy at 180°C for 4 h is more or less exactly equivalent to a treatment at 190°C for 2 h, or 200°C for 1 h. These are all effectively *equivalent* treatments. One would expect an identical microstructure and identical properties from all of these treatments because they are *equivalent*. Although, of course, there is a limit to how far this translation of the treatment can be pushed to higher temperatures and shorter times, for most practical purposes on the shop floor, it is often surprisingly effective over many iterations.

From time to time, various stepped solution treatments are proposed. From the point of view of achieving the very best properties, such treatments are probably necessary. For instance, Gauthier and Samuel describe solutionising at 515°C for 12 h followed by 540°C for a further 12 h which gives excellent strength and ductility for the common automotive alloy 319.2 (Al-6Si-3.5Cu). Their optimum ageing treatment is 155°C for 12 h. However, such a daunting high-cost treatment for such a low-cost alloy can hardly be recommended. For instance, the ageing treatment can be seen to be approximately equivalent to 1 h at 200°C. By contrast, Rometsch et al. (2001) study the solution treatment of 356 alloy (Al-7Si-0.4Mg), finding with a DAS of 40 μm that the alloy can be solution treated at 540°C in a time somewhere between 4 and 15 min.

For those alloys produced by HPDC that would blister and distort with even my kind of reduced treatment, Mehta (2008) reports that treatments as short as 10–15 min at temperatures of only 430–490°C, followed by quenching into water, result in useful properties when combined with ageing treatment for T4 condition (22°C), T6 (150°C) and T7 (200°C). Lumley et al. (2007) confirm the top temperature of 290°C for 15 min for 380 alloy to avoid blisters, although the 20 h age at 150°C should be substituted by 15 min at 210°C to create a useful energy saving assuming the 10°C rise equivalence to doubling rate.

Thus the conventional wisdom that HPDC material cannot be heat treated would be normally true for conventional HPDC and conventional solution heat treatments, but it seems it may be possible to develop a useful degree of improvement as outlined previously. The reduced temperatures and times are possible because of the extremely fine DAS. In addition, improvements in HPDC involving evacuation of the die and careful control of the shot stroke all contribute to increasing the soundness and reduced bifilm content of HPDC material, further allowing the benefit of some kind of solution treatment if necessary.

Another related aside might usefully be introduced here. It is unhelpful to see heat treatment results plotted on a linear time scale because the short times are necessarily too compressed to be properly discernible. I find a logarithmic plot based on the factor of 2 extremely convenient for the plotting of heat treatment results. Thus the time axis has equal divisions labelled 1, 2, 4, 8, 15 and 30 s; 1, 2, 4, 8, 15 and 30 min; and 1, 2, 4, 8, 16, 32 h etc. The small error introduced by using 15 instead of 16 in this series can easily shown to be quite negligible in relation to the other inevitable experimental inaccuracies that will be naturally present in the plots. Another useful approximate log base two scale I sometimes use is 1, 2, 4, 8, 16, 32, 64, 125, 500, 1000, 2000, 4000 etc. Figure 18.8 shows an example of this kind of plot, combined with the 10°C effect of the doubling of reaction rates, allowing all of the different temperature results to be collapsed, with acceptable accuracy, onto a single curve. This technique for recording all kinds of diffusion-rate-limited reactions is warmly recommended.

19.2.3 BLISTER FORMATION

A further problem can occur for those alloys cast turbulently, such as by HPDC. The hydrogen absorbed from the atmosphere of the heat treatment furnace can diffuse into the part, inflating internal bifilms. Thus blisters can occur (a localised distortion) if the bifilm is close to the surface, whereas if the inflated bifilms are deeper inside the casting distortion of the whole casting can occur. Figure 19.3 shows blisters developed on a rolled product produced from a defective cast ingot. Blisters are common on plaster moulded castings of Al alloys produced from relatively dirty metal and are even seen on the surface of premium quality forgings produced from less-than-well-cast products.

In the past, of course, the essential role of the bifilm has not been appreciated because of all attention being focussed on the role of hydrogen. This is understandable because the introduction of a little $NaBF_4$ in the furnace atmosphere completely eliminates the problem (Dorward, 2001). This illustrates that some kind of absorption mechanism is taking place in which the absorption of hydrogen at the casting surface is inhibited by some kind of passivating action of the fluoride. In similar work, Zurecki (1996) finds that low levels of sulphur hexafluoride gas suppress blister formation during the heat treatment of Al-Mg alloys. Other halogen gases (those containing Cl and/or F) would also be expected to be effective to different extents. These solutions to blister problems are unwelcome because of the toxicity of the chlorides and fluorides.

As a personal experience of these problems, Al alloy castings produced by counter-gravity casting never displayed blisters from heat treatment even in atmospheres containing high levels of water vapour (the work basket was unfortunately constructed from square tubular steel, with the result that the tubes filled with water during the quench, but boiled off when returned into the furnace, giving a nearly pure steam atmosphere). However, gravity poured products treated in the same furnace, even after heavy forging, would, on occasion, produce crops of unsightly blisters, causing the parts to be scrapped.

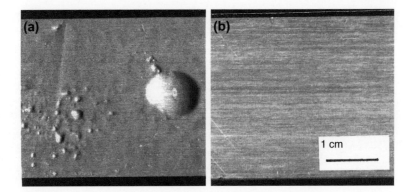

FIGURE 19.3

Surfaces of stretched sheet of 7475 (Al-Zn-Mg-Cu) alloy heated to 515°C (a) in air atmosphere; (b) with NaBF$_4$ in furnace atmosphere (Dorward, 2001).

Blister defects in rolled products are likely to be the consequence of the decoherence of near-surface films, often aided by the precipitation of hydrogen into such ready-made fissures. Celik and Bennett (1979) have shown that inclusions are implicated, and filtering of the melt before casting is an effective way to control the blistering behaviour of the rolled product. This observation is further confirmation that rolling is not especially effective in bonding or welding such defects to render them harmless.

19.2.4 INCIPIENT MELTING

The use of solution treatment temperatures as high as possible, taking advantage of the general rule that an increase of 10°C will double the rate of reaction, is of course limited to the temperature at which the alloy will melt. The sight of the castings flowing out from under the door of the heat treatment furnace is to be avoided if at all possible.

However, well before the castings become too hot to melt entirely, they can suffer from overheating; a phenomenon known as 'incipient melting'. This phenomenon was a puzzle in the sense that on heating above some low melting point constituents of the alloy, that might constitute only 1–2 volume % of the alloy, these constituents would clearly melt. However, on subsequent cooling, they should re-solidify so as to result in no net effect. However, this was untrue for most alloys and most castings. Their mechanical properties, particularly ductility, would be significantly impaired by overheating.

An exception known to me was the early Cosworth castings, which were exceptionally clean and did not suffer loss of properties if heated above the melting point of some phases, but were in fact observed to continue to improve in properties. Researchers have occasionally reported (for instance, Zhang et al., 1998) that temperatures up to 560°C have been perfectly acceptable for Al-7Si-0.4Mg alloy in contrast to most of the industry that works at a maximum temperature of 540° or perhaps 545°C, in view of the known melting points of Fe-rich phases in the region of 550–555°C. This gave a clue that the presence of bifilms, as an uncontrolled influence, may be a factor contributing to the permanent degradation of properties when overheated. In confirmation of this suspicion, the author has proposed a mechanism that appears to fit the facts (Campbell, 2009a). This is outlined next.

On heating over the melting point of some of the low melting point grain boundary constituents, these phases will melt, and so expand in volume. This expansion is not easily contained because both the newly arrived liquid phase and the surrounding solid are both effectively incompressible, resulting in the plastic deformation of the surrounding solid matrix. These millions of tiny expansions will result in a tiny growth of the whole casting. The resulting high compression stress around the melting phases will ensure that any bifilm on which the phases formed will remain tightly shut. However, on subsequent cooling and solidification, the previously melted phase will now occupy less

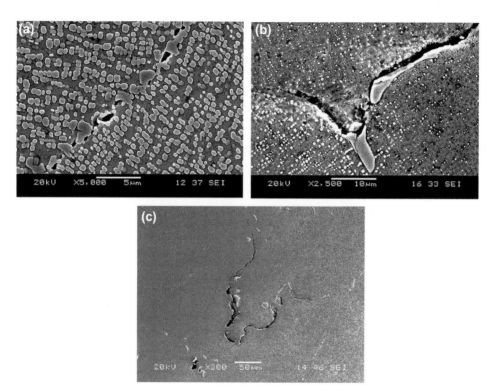

FIGURE 19.4

Grain boundary melting phenomena in the HAZ of welded Ni-base alloy 738LC showing (a) cracked and decohered grain boundary carbides; (b) asymmetric features of the crack and melted regions suggestive of an asymmetric bifilm; (c) a convoluted crack reminiscent of an opened bifilm (Sidhu et al., 2005).

volume as a result of natural solidification shrinkage. The overall plastic growth of the casting is not reversed because the nearby bifilm accommodates this reversed stress by simply opening (because there is no bonding between its internal surfaces). This is vastly easier than the problem of reversing all the plastic flow. Thus, the alloy now suffers from grain boundary porosity in the form of expanded bifilms.

The presence of bifilms at the sites of incipient melting is strongly indicated by Ni-base alloys, and is common in heat affected zones (HAZ) of welds that experience temperatures not only in the incipient melting range but naturally up into the complete melting region. Figure 19.4 shows instances of incipient melting in the HAZ of welded 738LC Ni-base superalloy (Sidhu et al., 2005). In the case of Figure 19.4(a), the bifilm is thin and symmetrical, suggesting it is a rapidly entrained alumina type; the $M_{23}C_6$ particle has cracked through its centre, and grain boundary γ' particles also contain traces of cracks. In contrast, Figure 19.4(b) shows MC carbides on only one side of the crack, suggesting the originating bifilm to be asymmetrical, perhaps an alumina plus a spinel? Figure 19.4(c) shows a typical convoluted bifilm, opened by the melting of the grain boundary phase in the heat affected zone of the weld.

19.2.5 FLUID BEDS

Over recent years, there has been significant interest in the use of fluidised beds for heat treatment, particularly for Al alloys (Chaudhury and Apelian, 2006). The advantages include extreme uniformity of temperature and good heat transfer leading to rapid heat up of parts. In contrast, a muffle furnace with circulating air often has a problem to achieve the 2.5°C uniformity required for demanding work, in which all nine checked points—the eight corners and the centre—of the furnace are required to be within the 2.5°C tolerance.

19.2.6 QUENCHING

When used for quenching, the rate of cooling in a fluid bed is only about half that of a water quench. A slower quench is almost certainly a benefit for some castings, where the water quench is usually too fast, creating severe residual stress problems. In fact, the relevant question is, 'is the fluid bed quench slow enough?' In general, as we have seen in Section 10.9, the loss of properties as a result of a slower quench is relatively minor compared to the overall benefit, which can be extremely valuable. Ragab et al. (2013) describe its use for quenching of Al-7Si-0.4Mg alloy.

Ramesh and Prabhu (2013) propose the use of a cylindrical probe which can be used to quantify a quench action. They define a dimensionless cooling parameter

$$D^2(dT/dt)/\alpha\Delta T$$

where D is the probe diameter, dT/dt is the cooling rate, α is the probe diffusivity, and ΔT is the initial temperature difference between the probe and the quench medium.

For those who want to know the response of properties to differing quench rate, the technique described by Dolan et al. (2005) is recommended. These researchers used a variant of the Jominy end quench technique, usually reserved for the investigation of steel heat treatments, in which water is jetted on to the end of a pre-heated 25 mm diameter bar of the alloy to be tested. Thermocouples are inserted in holes drilled at intervals of 2, 38 and 78 mm long, the length of the bar to ascertain the spectrum of cooling rates with increasing distance. These are correlated with hardness measurements at 2 mm intervals along flats machined on both sides of the bar. In addition, of course, the structural changes induced by different quench rates can also be studied.

19.3 HOT ISOSTATIC PRESSING

In his history of hot isostatic pressing (HIPping), Mashl (2008) describes how pressing from all directions on a component, using a gas at high pressure, was found to densify ceramics as early as the 1950s. In 1970, it was first used on Al alloy castings, and within 5 years had started to become widely used for light alloy aerospace castings. In a significant historical paper, Coble and Flemings (1971) confirmed that pores, if sufficiently fine, and in a fine-grained matrix, would gradually disappear given a few tens of hours and a temperature high enough for a 'sintering reaction' to occur. They found that the application of modest pressure, about 20 atm (2 MPa), greatly assisted the process. The development of HIPping was the outcome. Much higher pressures were employed, usually nearer 1000 atm (100 MPa), and temperatures as near the melting point as was practical. The optimum conditions for HIP have been defined in an elegant study by Arzt et al. (1983) in which they characterise tool steel, superalloys, alumina and ice.

The significant improvement in mechanical properties, particularly average fatigue life, reported for certain alloys after HIPping is probably at least in part due to the contribution towards the de-activation of entrained double oxide film defects (bifilms) as fatigue crack initiators. We shall evaluate the evidence here for HIPping as a solid-state process for the closing of pores and bifilms.

When a cast Al-7Si-0.3Mg alloy with oxide film defects, like that shown in Figure 19.5(a), is subjected to HIPping treatment, the applied pressure at temperature close to its melting temperature induces a substantial plastic deformation in the casting causing the defects to collapse and their surfaces to be forced into contact (Figure 19.5(b)). This plastic collapse phase occurs almost immediately, probably requiring only a few seconds or a few minutes. The slightly extended time may be required for the residual air to be reacted to form oxides and nitrides. At this stage, the pores and cracks are closed but, of course, not necessarily bonded. Thus only a modest, if any, benefit from HIPping may be expected at this early stage. Also noticeable in Figure 19.5(b) are residual pores among the collapsed oxide bifilms that appear to have survived the complete HIP process. It seems likely they contain the insoluble 1% argon constituent from the air. Such pores are never expected to close, but will contain the argon at the huge pressure experienced during the HIP process and

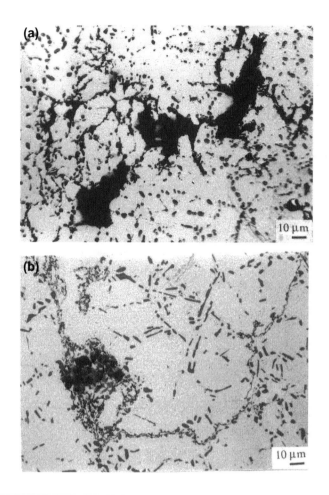

FIGURE 19.5

Optical micrograph of an Al-7Si-0.4Mg alloy showing (a) a network of oxide films and associated pores in the as-cast condition; (b) the network of films and collapsed pore in the HIPped condition (Nyahumwa et al., 2000).

will pressurise most of their associated bifilm, and thus will expand again if the temperature is raised sufficiently. Such pores might be expected to re-grow to some extent, triggering failure during such processes as creep or super plastic forming.

For those Al alloys containing no Mg for which the oxide is the very stable ceramic, alumina, Al_2O_3, there is little or no benefit to be expected from extending the Hip cycle. Similarly, for those alloys containing more than approximately 1%Mg, once again the oxide is extremely stable, being pure magnesia, MgO. For such alloys, no bonding is expected to develop between the extremely stable ceramic surfaces dividing the central interface of the bifilm.

The situation is interestingly different for many aluminium alloys of intermediate Mg content in the range of approximately 0.05–0.5%, as for instance typified by Al-7Si-0.4Mg alloy. For such alloys, an alumina (Al_2O_3) film is first formed during the rapid entrainment process. However, after an incubation period that is a function of time and temperature, the entrained film transforms to magnesium aluminate spinel, $MgAl_2O_4$, (Aryafar et al., 2010). This involves a volume change and atomic re-arrangement of the crystal structure that would be expected to encourage diffusion bonding across any oxide–oxide interface that happened to be in contact. Analysis of bifilms that have acted as fatigue crack initiation sites have confirmed their conversion to a spinel and confirms that fatigue properties are improved by HIPping (Nyahumwa et al., 1998, 2000). A further positive finding from this work was that compared to filtered castings, the unfiltered but HIPped castings exhibited higher fatigue performance, despite larger maximum defect sizes, implying some degree of bonding across the crack. The application of HIP to castings shown in Figure 9.34(a) resulted in fatigue test samples that did not fail; the run-outs at the stress 150 MPa reached nearly 10^8 cycles (not shown). At the higher stress of 240 MPa, improved fatigue lives were still recorded (Figure 9.34(b)) although it is of interest that in all failures, with the possible exception of the most resistant specimen, the fatigue failures still occurred from oxides, being probably healed or partly healed bifilms.

The liquid-state healing mechanism as described in Section 2.4, and the HIP solid-state healing mechanism are suggested to be analogous. However, there are interesting differences. In the liquid state, where bifilm are floating freely in the melt, the healing mechanism operates at high temperature (i.e. approximately 700°C for an aluminium alloy) and with only the moderate external applied pressure (<0.1 MPa) because of depth in the liquid. When the casting has solidified, the solid-state healing process due to HIP operates at lower temperature but very much higher pressure (i.e. 100 MPa at 500°C for Al-7Si-Mg alloy). In both the liquid and solid conditions, pressure inside the bifilm will be expected to fall as the oxidation reaction proceeds. When all the oxygen is consumed, the nitrogen is subsequently consumed to form nitrides (Raiszadeh and Griffiths, 2008).

Whether bifilms are actually 'healed' (i.e. effectively welded) clearly relates to the chemistry of the alloy and its films. Some bifilms appear to heal as we have seen previously, whereas others are resistant. These important factors are not well researched at this time. Thus, HIPping has limitations that do not appear to be widely known or understood. However, evidence for possible mechanisms is discussed below.

In contrast to this beneficial action of HIPping found by Nyahumwa and others as described previously, Wakefield and Sharp (1992) observed HIPping to have no beneficial effect on the fatigue properties of Al-10 Mg alloy castings, despite the closure of pores and cracks. The inference is that the bifilms in this alloy proved impossible to de-activate, resisting effective welding. To restate and expand on the phenomenon, this is attributed to the magnesia (MgO) film formed during oxidation of Al-10 Mg and entrained during casting. The magnesia film is (1) thicker and (2) has a stable structure that does not transform during HIPping (Nyahumwa et al., 1998, 2000). This lack of any substantial atomic movement explains the inert nature of the highly stable MgO compound. (Similarly, again, as discussed previously, it would be expected that Al alloys with very low Mg contents, that would be expected to contain entrained alumina films would be similarly resistant to HIPping because of the great stability of alumina in the absence of sufficient Mg to convert it to a spinel structure.)

A further sobering example can be quoted from failed attempts to hip Ni-base superalloys. During the early development of the Pegasus engine for the Harrier Jump Jet, 25 polycrystalline Ni-based alloy turbine blades that had previously been scrapped because of their content of porosity were subjected to HIPping, and were fitted to a test engine alongside sound blades to evaluate whether HIPping might be a satisfactory reclamation technique for blades. The HIPped blades failed within a few hours, damaging the engine and forcing a rapid shut down of the test. The failures had occurred by creep cavitation at the grain boundaries of recrystallised regions in the centre of the castings. Almost certainly the original porosity would be caused by bifilms, probably of aluminium and/or chromium oxides entrained by the severe turbulence that is usual during the vacuum casting process. (As we have observed many times, the vacuum is known to contain plenty of residual air to ensure the creation of surface films.) The great stability of the films, formed at the high casting temperature, would ensure that they were resistant to any re-bonding action. The recrystallisation would have happened because of the large plastic strains that were a necessary feature of the collapse of the porosity. However, the subsequent grain growth would expand grains up to local barriers such as bifilms. Thus, the bifilms, effectively unbonded, and so acting as efficient cracks, were automatically located at the grain boundaries from where the failures were seen to occur.

The closure of internal cavities usually causes negligible changes to the overall dimensions of the casting if the pores are small and/or deep seated. For large or near-surface pores, however, the collapse of the surface of the casting in the form of a localised sink may scrap the casting if the depression exceeds the machining allowance. In a severe case, the surface may puncture, opening up the internal cavity to the surface (Zeitler and Scharfenberger, 1984).

Naturally, HIPping cannot work if the pores are already connected to the outer surface of the casting. Such pores will never heal. Unfortunately, this is all too common. The existence of the various forms of surface-connected porosity is well known to those who HIP castings. In particular, pressure die castings are woefully resistant to the benefits of HIPping because of their many surface-connected bifilms (traversing the so-called 'dense' outer layers of the casting). However, many gravity filled and even counter-gravity filled castings exhibit surface-connected pores as a result of bifilms intersecting the surface.

In addition, although an excellent start, even good filling of the castings will not guarantee a satisfactory response to HIP treatment if the alloy has a long freezing range and is poorly fed. As we have seen in Section 7.1.5, poor feeding in a long freezing range alloy can easily lead to conditions in which residual liquid is sucked away from the surface of the casting, subsequently expanding as shrinkage porosity inside the casting, to create surface-connected porosity.

For reliable HIP response, the surface of the casting must be sound.

Finally, therefore, although the mechanical properties of castings usually exhibit an improvement, in the sense that their *average* properties are raised, the Weibull modulus most often *falls*. This is a direct result of the closure and welding of some defects, improving the properties of some castings, but leaving some castings unaffected, for the various reasons we have seen. Thus the *scatter* of properties is increased. Regrettably, this is one of the greatest disadvantages of HIPping, often overlooked.

19.4 MACHINING

There are numerous text books on machining, so we shall limit our attention to only a few of those aspects of relevance to castings.

An early experience of mine taught me that casting defects can have a profound effect on the difficulties experienced by the machinist. A Ni-base superalloy turbine blade for a land turbine was top poured through the root of the blade, creating oxide flow tube defects that opened into vertical cracks when grinding the 'fir tree' roots. The founder asserted the problem was the result of severe grinding, the heat of grinding producing grinding cracks. However, in fact, even the lightest grinding had difficulty to eliminate the problem.

Turning to the wide field of cast irons, Teague and Richards (2010) review the information on the age strengthening of grey irons, finding that ageing for 5 days at room temperature greatly improves tool life. Apparently, the effect is the result of the precipitation of nitrides. They report an investigation that attempted to reduce this time by ageing at 271°C for 75 min which yielded no clear result. It seems to me this attempted treatment was way over the top. Taking the 5 days at room temperature as 120 h at 20°C, and using the useful approximation that a 10°C rise in temperature will double the rate of reaction, this gives equivalent treatments at 60 h for 30°C, 30 h for 40°C etc.; finally indicating 1 h at 90°C. This extrapolation of times and temperatures makes no claim to be more than an approximation, but I have often found it usually to be impressively accurate. It seems more modest times and temperatures have not yet been tested.

Machining of grey cast iron can pose additional difficulties (Hoff and Anderson, 1968) because the skin typically contains the following.

1. No graphite but only fine pearlite (equivalent to a hard steel)
2. Oxides
3. Sand grains

For this reason, machinists are usually keen to ensure that the first machining cut is taken well under the casting skin. This is not always possible, however, and the effects of decarburisation and inclusions are noted below.

Rickards (1975) has researched the problems of poor machinability as a result of the formation of graphite-free surface layers on grey iron castings made in greensand moulds. It seems the problem is affected by section size and the carbon content of the sand. For instance for a 300 mm section thickness, the graphite-free layer can reach 1 mm. He found 5% coal additions to the sand to roughly halve the problem, and additions up to 10% would control the problem in thin-walled castings, but could cause a carburised layer in castings with heavy sections up to 100 mm thick.

The author recalls those working in a machine shop drilling ductile iron crankshafts who could tell the difference between those castings made with good and bad filling system designs simply by listening as they walked among the machines. Those that were cast with significant surface turbulence contained inclusions rich in magnesium oxide, creating the characteristic high pitched singing noise that the machinists knew signalled a poor finish and short tool life.

The effect of oxides in castings is well known to machinists. The alumina inclusions in light alloys and mixed oxides in iron and steel castings are all known to cause defects on the machined surface by dragging out, leaving unsightly grooves. Worse still, the cutting edge of the tool is often chipped or blunted by encounters with such hard particles, that are in general vastly harder than the cast alloy, and even harder, in many cases, than the tool tip.

For sand castings of all types, but particularly the softer light-alloy castings, and particularly when turbulently filled, the presence of residual sand on the surface of the casting is often a problem. For iron castings, it is widely appreciated that the first cut should be sufficiently deep to get below the glassy surface inclusions.

In light alloys cast in sand, the complete removal of all sand is practically impossible short of dissolving away part of the surface in an acid or alkaline bath. Normally, some residual surface grains are broken and hammered into the surface by the so-called cleaning processes such as shot blasting. Grit blasting can be even worse because particles of grit (often hard, angular, abrasive grains of alumina) remain embedded in the surface.

Because of this residual surface aggregate problem, the slightly impaired machinability of sand castings of all types, ferrous and non-ferrous, compared with die (permanent mould) castings is one of the very few potential disadvantages of sand castings. This disadvantage seems little known and publicised, but needs to be acknowledged and accepted by all the parties in the casting supply chain to save future heartache and cancelled orders.

It is worth emphasising once again benefits of gentle uphill filling of mould cavities. The beautiful glossy shine that iron castings exhibit when filled by non-turbulent bottom-gated systems illustrates the protection provided by the lustrous carbon film acting as a mechanical barrier between the liquid metal and the mould. Because the liquid metal now no longer comes into contact with the sand mould, the surface of the casting should be completely free from embedded sand grains, and should exhibit excellent machinability. Similar benefits are achieved for most casting alloys by the insulating action of the oxide film when carefully uphill filling moulds. At this time, I am not sure that this prediction has ever been tested.

For Al alloys, the machining of high Si alloys (in the range approximately 10–20% Si) is well known to require diamond tipped tools. However, what is less well known is that even the lower Si alloys present some degree of tool wear. Armstrong and Martin (1974) draw attention to the excellent machinability of the 7%Zn alloy (771.0) that requires only a low temperature age (180°C for about 4 h) followed by an air cool (not a quench), resulting in a highly stable casting with low internal stress that maintained its highly accurate dimensions after machining, and enjoyed tool lives up to 20 times longer than the normal Si-containing alloys, plus machining speeds up to five times faster.

Dasch et al. (2009) describes investigations into the dry machining of Al alloys which would have significant environmental benefits compared with current wet machining methods, but has been prevented by the build-up of hot Al on tool tips. He finds only 0.15%Sn addition to an Al alloy achieved a 1000 fold increase in tool life, in addition to the ability to machine dry. Higher cutting speeds enhanced the effect, such that faster drilling worked better than low speed drilling.

For the future, the advent of other machining techniques will become increasingly important. These include machining or cutting by fluid bed, water jet, flame, plasma, oxygen, laser, spark and abrasive linishing. The smoothing of complex internal passageways in castings is sometime accomplished by forcing through abrasive mixtures of SiC in silicone putty. Even so, our current cutting technology based mainly on single point or multipoint cutting tools is likely to be with us for a long time yet.

As an interesting and potentially important aside, designers of light alloy castings often despair at the inability of threaded holes in castings to avoid stripping out of the threads when subjected to large loads. Thus relatively expensive heli coil inserts are specified. These strong steel inserts spread the pull-out load and significantly strengthen the fastener. However, these are costly and inefficient solutions. The author has experience of M8 self-tapping bolts (thread-rolling type rather than the thread cutting type) in an Al alloy casting that enjoyed no cost to machine the thread, and the better thread engagement and work hardening of the surrounding matrix resulted in 8 times greater pull-out load. The benefits of thread-forming fasteners in net-shape cast holes of light alloys has been further researched by Paxton et al. (2006). The work hardened region around the fastener acts as its own integral helicoil, spreading the pull-out load. Furthermore, the avoidance of a second metal minimises the potential for corrosion problems.

Although cast holes can be envisaged for HPDC, small diameter cast holes are not attractive for aggregate moulded castings because the small sand projections on cores and mould would be vulnerable to being slightly ill-formed or even broken off. Thus, for sand castings, it would be preferable to machine the tapping sized holes in the casting. This seems a relatively small penalty, still enjoying the benefit of avoiding tapping, and still achieving a greatly superior engineering feature which may be particularly important for certain critical castings.

19.5 PAINTING

For those painting processes where the paint is cured by heat, such as powder coating, the curing cycle is usually integrated with the heat treatment requirements for the casting, and thus constitutes the final heat treatment stage (the ageing) of the product. Aluminium alloy wheels for cars are usually heat treated and painted in this combined process.

However, problems arise if the casting exhibits any surface-connected porosity. This can cause a defect in the smooth surface of the paint known as a paint crater.

Before painting, the surface of the casting is usually machined to a bright finish and may be further mechanically polished and subsequently subjected to cleaning with a solvent. The paint powder is usually applied electrostatically to achieve a tolerably uniform deposit on a complex shape. During the heating of the casting for the baking cycle, the powder melts and slowly cures, while air expands from surface-connected cavities in the casting and bubbles through the melting paint. At a later stage, when the paint becomes viscous, the collapsing walls of the last blister tend to remain 'frozen' in place, becoming the crater wall (Figure 19.6(a)). Micks and Zabek (1973) investigated paint craters originally assuming them to be faults in the paint application process. However, they found that the craters were invariably linked to surface-connected porosity. The central hole in the crater was always connected to an internal cavity which was found to remain free from paint (Figure 19.6(b)).

Unfortunately, such craters are a serious source of scrap for aluminium alloy road wheels for cars, and kitchen utensils because the cosmetic requirement for such castings from the car owner and the cook, is simply perfection.

For Mg alloy castings, particularly those for aerospace applications, the special surface coatings to protect the casting from corrosion are a significant contribution to the cost of the product, and, because Mg alloys are so light, even make a contribution to its weight. The maintenance of the coating, often requiring replacement after weld repair or other disturbance represent further discouraging costs to the selection of these otherwise attractive alloys.

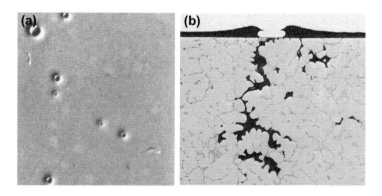

FIGURE 19.6

(a) Paint craters of varying sizes on the surface of an Al-5Si-0.5Mg alloy casting, acrylic powder coated and baked at 230°C (b) Optical micrograph of a section through a crater, showing its connection into surface-connected porosity (Micks and Zabek, 1973).

19.6 **PLASTIC WORKING (FORGING, ROLLING, EXTRUSION)**

The effects of bifilm populations in steels, and more particularly Ni-base alloys, can be so profoundly damaging that no plastic working can be applied; the cast ingot falls into pieces on the first stroke of the forge. Fortunately, most cast metals are not so bad, and plastic working can be successfully carried out as is discussed below.

It seems curiously perverse that most metallurgical texts continue to foster the erroneous assumption that working eliminates casting defects. Usually, the strains involved in most working operations are too low to effect any significant welding of faults. Bifilms in general are merely pushed around, if anything, growing worse before (if ever) getting better. The growth of casting defects by plastic working is well known to those who work in the forging industry. There are good reasons for this behaviour.

During the working of cast material, for instance by rolling, it is to be expected that the defects will be elongated in the direction of working. The elongation of the defect necessarily rotates it, aligning the defect along the rolling direction. This is clearly seen in the lengthening of graphite nodules during the rolling and forging of ductile iron (Neumeier et al., 1976).

In addition, of course, elongation of the work piece increases the area of its defect. The newly extended surfaces in the bifilm would necessarily be oxidised by the remnant of air entrained in the defect. The entrapped air would be contained partly amid the microscopic pores between the crystals of the oxide, and partly in reservoirs formed by folds and in expanded regions constituted by those bubbles entrained as part of the bifilm. During rolling, the continued oxidation of the expanding surfaces would hinder the welding of the interfaces that were being newly created, and by this means assisting the defect to grow as a crack, possibly to many times its initial size. The great effectiveness of the creation of this newly oxidised extension to the crack would be a consequence of the small oxygen requirement; the oxide would grow to a thickness of only nanometres at the hot working temperature (which is, naturally, much lower than the temperature at which the film was first formed on the melt). In other words, the entrained residual oxygen is highly efficient at these temperatures to continue the oxidation of new area as it is formed.

Miyagi et al. (1985) used ultrasonics to observe this increased area of cavities in 5083 alloy (Al-Mg type) during the early stages of hot rolling. Micro-cavities near the surface of the rolled plate seemed to close early, but those nearer the centre were long-lived. Micro-examination revealed that they were smooth sided, and appeared to have opened and expanded along grain boundaries. (The association with grain boundaries is, of course, a feature to be expected of bifilms.) They were seen to reduce in size only after reductions of more than 50%.

Later, if the extension were sufficiently large to consume the remaining air, further extension would result in the welding up of new extensions to the crack. It seems likely that normal forging and rolling do not reach this stage for most alloys, and so do little to heal defects, simply extending and re-aligning them as described previously. Processes such as extrusion and multi-pass rolling may sometimes account for a more complete elimination of the casting defects because of the much greater strains that can be involved. Even so, it seems that many defects remain, as indicated, for instance, by the evidence from corrosion (Section 9.10).

Work by Harper (1966) on the hot rolling of wire-bar copper showed that rounded bubbles of gas are collapsed asymmetrically, becoming pinched off, and forming what appear to be trails of minute beads on a string, the beads being microscopic bubbles only 30–40 nm diameter. It is hard to avoid the conclusion that the bubbles are the residual insoluble argon sitting in the collapsed bifilm, or even the collapsed bubble trail, resembling a string on the two-dimensional polished section.

Many bifilms will be already linked to the surface. Alternatively, during working operations such as rolling, they can become linked to the air by local plastic failure of ligaments of sound material separating the defect from the outside world. Thus, a corroding environment is expected to penetrate most near-surface entrainment defect cracks in cast, forged, rolled, and even extruded materials as is seen in Section 9.10.

The huge amount of work on the behaviour of inclusions in steels during working cannot be covered in this short account. However, the early work by Wojcik et al. (1967) was notable because they were able to show that their steels, destined for piercing to make tubes, contained a distribution of inclusions that they found to be log normal. This finding allowed them to extrapolate their inclusion counts from relatively few microsections to assess the probability of the presence in the billets of a small number of inclusions sufficiently large to cause wall defects in the tubes.

Charles and Uchiyama (1969) and Wardle and Billington (1983) studied the plasticity of inclusions in steels as a function of composition and temperature, showing the regimes in which the inclusions would either flow, thereby elongating, or fracture, so forming chains of fragments. Leduc et al. (1980) checked the density of cast low carbon steels during hot rolling, but did not find any evidence for a fall in density during the early stages of rolling similar to that found for the non-ferrous materials. They found a progressive increase in density, becoming fully dense at about 75% reduction. Examination of the microstructure indicated that all the pores were eliminated at that stage. This is probably understandable in view of the low residual levels of Al in the steels (0.02–0.04%) and absence of other strong oxide forming elements. Any surface oxide would therefore have been expected to be a liquid iron-manganese silicate, and so any bifilms, if present, would have been easily welded shut.

19.7 IMPREGNATION

The impregnation process is a sealing technique, designed to seal porosity and eliminate leakage problems in castings.
Clearly, good casting processes do not need impregnation.
However, many do.
For instance HPDCs are highly prone to leakage problems because of their high content of bifilms and inflated bubble trails. Similarly, low pressure castings that have not had the benefit of the roll-over action following the filling of the mould are sometimes prone to an interconnected type of shrinkage porosity as a result of convection problems.
Although some casting operations use impregnation only to seal those castings that are shown to leak, others do not carry out an initial sorting test, but simply apply impregnation to all castings. Although, like heat treatment, impregnation is often carried out by specialist offsite operators at a price usually based on the weight of the casting, it is sometimes installed as an integral part of the foundry operation.
The impregnation process involves the placing of the casting in a vessel from which the air is evacuated. The casting is then lowered into the sealing liquid. When fully immersed, air is re-introduced into the vessel to restore atmospheric pressure. In this way, the liquid is forced into the evacuated pores in the casting. The casting is then raised out of the liquid and is allowed to drain. Excess sealant is then washed off and the sealant is cured.
There are two main curing processes based on two different sealing systems. These are (1) sodium silicate cured with the addition of a catalyst to the liquid and (2) a thermosetting resin hardened by a subsequent low temperature heat treatment.
A recent development is the streamlining of the process by the use of special sealants, allowing the process to be carried out in a single vessel with only a short treatment time (Young, 2002).
In a small percentage of cases, leakage of the casting is not cured by the first impregnation. Some casters and their buyers allow two or even three such attempts. Usually, if the casting still leaks after repeated attempts to seal it, it is finally scrapped.

19.8 NON-DESTRUCTIVE TESTING

Although destructive testing is the most discriminating and reliable way to determine the suitability of a casting for service, the destroyed casting is somewhat unfit for service. Clearly, it is essential to have reliable tests that do not involve destroying the product. Naturally, this criterion puts severe limitations on the appropriateness of the various approaches to the solution to this problem.

19.8.1 X-RAY RADIOGRAPHY

Radiography is a powerful technique for checking the presence of major defects such as open cracks, porosity or foreign (entrained) inclusions. The traditional radiographic limit to detection has been approximately 2% of the section thickness, although modern techniques are providing ever greater sensitivity of detection. If 1% could be achieved this would be 100 μm in a 10 mm section. This is a substantial defect, especially if it is a crack. In general, however, a crack can usually only be detected if it is aligned with the viewing direction. Even then, if the crack is tightly closed, the crack can remain undetected. Thus it has to be concluded that radiography cannot offer reliability for the detection of cracks.

It is well to remember that the features noted on radiographs generally appear to be shrinkage porosity are nearly always *not* shrinkage, but clusters of bifilms together with their entrained internal air. Rounded pores generally identified as 'gas' are also most often entrained *air bubbles*. The reader needs to be aware that at the time of writing practically all radiographic inspections are diagnosed incorrectly. The accurate diagnosis of radiographs cannot be expected for many years.

The powerful new technique of three-dimensional radiographic examination by tomography can be attractive if the cost and the significant time required are available. Although the technique is clearly capable of identifying serious defects, the additional capability to assess inaccessible wall thickness deep inside complex castings or the presence of residual core material in an inaccessible passageway, are major potential advantages.

19.8.2 DYE PENETRANT INSPECTION

Dye penetrant inspection is a sensitive technique for surface-connected defects such as porosity and cracks. In this process, the casting is bathed or immersed in a dye, a liquid designed to be highly wetting so as to penetrate small holes and fissures rapidly. The castings is withdrawn from the die, washed and coated with a 'developer' in the form of a fine white powder that can absorb the dye by capillary attraction as a blotting paper, sucking the die out of its crevice and into the developer layer where it can be seen. The most sensitive dyes fluoresce in ultraviolet light, and are therefore viewed in a darkened area.

The dye penetrant inspection approach is of course limited to the detection of surface-connected defects; any internal defects, which may be major faults, remain undetectable.

19.8.3 LEAK TESTING

Section 9.11 describes how the leakage of castings is almost never from 'porosity' resulting from shrinkage or gas. Section 7.2.2 illustrates how leaks can result from core blows that have effectively punched holes through the walls of the casting, creating serious leaks. In general, however, leaks appear to be mainly the result of the presence of bifilms. Figure 19.7 shows an approximately 20 mm long bifilm in an Al alloy casting which clearly had no problem to bridge the 5 mm wall thickness even though the path was long and meandering.

The finding of such leaks by leak testing is a costly and time-consuming activity which those counter-gravity casting operations that do not make leaky castings have the good fortune to avoid. Interestingly, the leak shown in Figure 19.7 was discovered not by leak testing, but by a resonance test which is far quicker and easier, besides, of course, also discovering any other seriously deleterious features from which the casting may be suffering. This indicates an interesting future direction for the general testing of castings.

However, for those casters that need to conduct a leak test the many techniques are listed in some detail in Section 9.11. By far the commonest system is the time-honoured bubble test by pressurisation of the casting with air while immersing the casting in water. However, the test is easier described than done. Leakages from seals as a result of imperfect dressing of prints can give plenty of additional bubbles that boil to the surface of the water, obscuring any observation of the relatively tiny leaks that the test was designed to find.

Another aspect of the test which gives cause for some concern is the procedure that sets aside any failed casting, for instance an aluminium cylinder head, with a view to repeat the test on the following day. Often, the casting will pass the test on the following day, almost certainly as a result of the leak path being sealed by corrosion products. The rapidity of the corrosive action is likely to result from the association of intermetallic particles that have precipitated on the bifilm, creating efficient corrosion couples. Having reflected for some years on the honesty of this procedure, I find I have never been aware of any failures of these castings in service. Thus, it seems a proven technique for the rescue of expensive and valuable products. Perhaps the procedure would be less acceptable for products that had to withstand high pressures.

Overall, however, the shop floor experience that good quality melts transferred into moulds by appropriate counter-gravity casting operations do not suffer leakage problems is powerful evidence that the leaks are mainly the result of

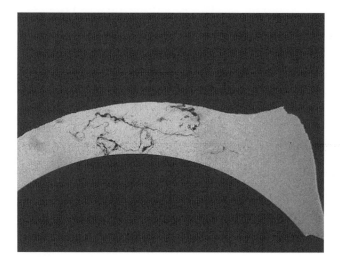

FIGURE 19.7

A leak path formed by an oxide bifilm, causing the master brake cylinder to be rejected.

Courtesy Magnaflux Quasar.

bifilms. The clear message is that these features are avoidable using good filling technology as described in this book, and leak testing could, with relief, be yet another foundry chore consigned to the rubbish bin of history.

19.8.4 RESONANT FREQUENCY TESTING

Parts have been tested by checking their resonant frequency for many years. From time to time, various forms of the 'ding it and listen' method, sometimes enhanced with electronic equipment, have been tried but have not in general been demonstrated to be sufficiently discriminating for testing the serviceability of castings. The normal variations within castings often changed the 'ringing frequency' more than unacceptable changes in material or structural integrity.

It is worth devoting some space to one particular development of the resonance test that has become available over recent years to illustrate the degree of sophistication that has been necessary to evolve a useful and powerfully discriminating test. The technique is outlined next.

A casting is subjected to a swept spectrum of vibrational frequencies and its response is recorded as a frequency spectrum. Among the thousands of resonance peaks observed in the spectrum, about 50 prominent and recognisable peaks are selected as potentially useful for further analysis. These 50 frequencies are recorded for each casting in a sample 'teaching' set of 100 or more castings, some of which are normal, acceptable castings and some of which have been rejected for various reasons. These samples provide the resonant frequency information to 'teach' the system to accept changes in frequencies that are the result of normal process changes (such as slight changes in size or shape, chemical composition etc.) which do not affect its serviceability. The software selects and analyses about six resonant frequencies to incorporate the possible variations into one comprehensive 'pattern' that can encompass all the acceptable variations.

The presence of the unacceptable castings in the sample set is to ascertain whether the frequency pattern that accepts all the good parts is in fact capable of rejecting all of the castings in which there is a known defect. If defects are present, the frequency pattern of the unacceptable casting will change in ways that are different from those for normal castings. The amount of change is usually a measure of the seriousness of the defect. This concept relies on the quality assurance teaching phase defining as 'acceptable' those parts that are structurally acceptable, as measured by metallurgical, destructive and non-destructive testing. An iterative 'teaching' process is undertaken, so that suitable frequency patterns

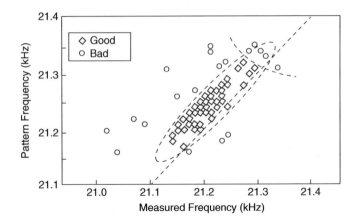

FIGURE 19.8

A representation of a sophisticated resonant frequency sorting technique, here illustrating a two-dimensional elliptical and a curved boundary, separating good parts from bad. In practice, multi-dimensional criteria are used.

Data courtesy Magnaflux Quasar.

are progressively identified and confirmed by separate testing as useful indicators. Because of the number of castings required for this setting-up phase, the technique is limited to long series production parts, but achieves zero or near-zero false acceptance errors in millions of components. Typical production testing time is a few seconds per casting in a fully automated, computer controlled sorting operation.

A typical distribution chart of testing for leaking master brake cylinders is shown in Figure 19.8. 'Perfect' castings would be expected to fall on the 45° line for which observed and predicted vibration frequency pattern are identical. However, a spread of vibration pattern (1) within the region defined by the ellipse and (2) to the left of a second criterion is found to deliver acceptable products. Frequencies outside these two boundaries, 'the acceptance window', were always found to have some serious fault, although many had been accepted by X-ray and liquid dye penetrant inspection. Figure 19.8 shows only two discriminating frequency boundaries as a simplifying example, whereas in fact the technique always uses a pattern combining about six resonances analysed by sophisticated software. In view of the ability of the technique to allow for the harmless changes caused by slight process variations, it has been called 'process-compensated resonance testing'.

Current developments include the exploration of computer models of the component under test, to ascertain whether the 'teaching' stage, and the exploration of the effects of different kinds and sizes of defects and other changes, may be carried out in the computer.

Castings with structural weakness (previously accepted because the major defects were effectively invisible bifilms that gave little or no classical non-destructive testing indications) are rejected. Conversely, castings that would have been rejected for negative non-destructive testing indications are accepted if they are assessed to be structurally acceptable.

A further interesting capability of the technique, known as resonant ultrasound spectrometry, is the accurate measurement of the elastic (Young's) modulus, as well as other elastic properties such as shear modulus and Poisson's ratio. The elastic moduli are reduced if there is a large population of small bifilms present in the casting. The casting can be rejected if the elastic moduli fall sufficiently to indicate a serious concentration of bifilms that would, for instance, impair the ductility and toughness of the alloy. The ability of the technique to assess ductility and toughness without resort to the expense and delay involved with a destructive tensile test is a further impressive advantage.

The concept provides a quality assurance method that predicts the structural performance of the casting, reduces false reject scrap, and reduces testing costs. This new, fast and low-cost technique appears to out-perform our existing non-destructive testing techniques, and promises to be powerful for future high production and safety-critical products.

Appendix I

THE 1.5 FACTOR

Experimental results for side gated 99.8 Al plate castings plotted in Figure A1 show that casting time t_c may be estimated for the plates and other castings from an equation:

$$t_c = 1.5 \times \text{ Casting volume/Initial casting rate} \tag{1}$$

This is equivalent to

$$\text{Initial fill rate/Average fill rate} = 1.5 \tag{2}$$

These experimental results give support to the value of 1.5 chosen by previous authors, particularly those of the British Non-Ferrous Research Association (now no longer with us) researching for the UK Admiralty (Ship Department 1975).

Exploring the 1.5 factor further by a theoretical approach is not quite so straightforward, but an attempt is outlined here.

Considering Figure A2, the velocity at the base of the sprue is given by

$$V_2 = (2gH)^{0.5}$$

If the areas of the base of the sprue is A_2 and the mould cavity is of uniform area A_C the initial velocity of rise in the mould will be given by

$$V_i = (A_2/A_C) \cdot (2gH)^{0.5}$$

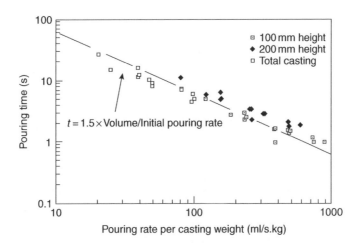

FIGURE A1

Experimental demonstration of the relation between initial and average filling rates.

Data from Runyoro and Campbell (1992).

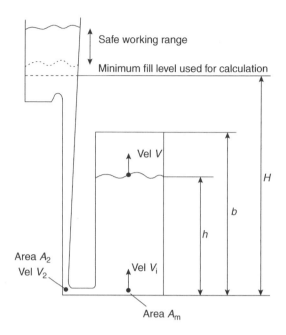

FIGURE A2

Schematic view of the filling of a uniform section casting.

Similarly, at some later instant, when the melt has reached height h, the net head driving the filling is now reduced to (H−h) so that the rate of rise is now

$$V = (A_2/A_C) \cdot (2g(H - h))^{0.5}$$

Substituting dh/dt for the rate of rise V, rearranging and integrating between the limits of time $t = 0$ at $h = 0$, and $t = t_c$ at $h = b$, we find the casting time (the time to fill the mould) t_c is given by

$$t_c = (A_C/A_2) \cdot (2/g)^{0.5} \left[H^{0.5} - (H - b)^{0.5} \right]$$

Now writing simple definitions of the initial rate of castings Q_i and the average rate of casting Q_{av} in such units as volume of liquid per second, defined by the appropriate velocity times the area, we have

$$Q_i = A_2 \cdot (2gH)^{0.5}$$
$$Q_{av} = A_C \cdot b/t_c$$

It follows that

$$Q_i/Q_{av} = 2H^{0.5} \left[H^{0.5} - (H - b)^{0.5} \right]/b$$

This solution to the filling problem is interesting. There are various combinations of H and b that can fulfil the conditions defined by the equation. For instance, if $H = b$, then $Q_i/Q_{av} = 2$, which is actually an obvious result, meaning simply that the average is half of the start and finishing rates.

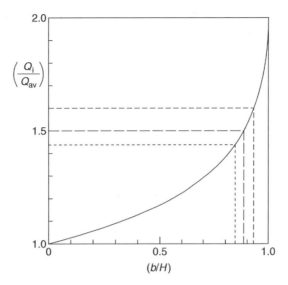

FIGURE A3

The relation between the initial and average fill rates for a uniform section casting as a function of the relative heights of the casting and the pouring basin.

On the other hand, $Q_i/Q_{av} = 1.5$ only when $b = 0.89H$. This represents an intriguing result, indicating that for most castings the top of the pouring basin is on average only about 10% higher than the height of the casting. Thus it seems the factor 1.5 is quite fortuitous, and results simply from the geometry we happen to select for most of the castings we make. If, in general, we were to raise (or lower) our pouring basins in relation to the tops of our castings, the factor would have to be revised.

However, all is not as bad as it seems. Figure A3 shows that the factor 1.5 does not change rapidly with changes in relative height of basin, varying over reasonable changes in basin height of b/H from 85% to 95% from roughly 1.45 to 1.60. These changes are of the same order as errors arising from other factors such as frictional losses etc. and so can be neglected for most practical purposes.

Appendix II

THE BERNOULLI EQUATION

Daniel Bernoulli represents the revered name in flow. He published his equation in 1738 in one of the first books on fluid flow. This magnificent result is the one used for all descriptions of flow in pipes and channels. Whole books are devoted to its application.

There are, of course, excellent examples of the power of Bernoulli's equation. Sutton (2002) has made good use of the equation to describe the pressures along a long runner, explaining the early partial filling of gates at different positions along the runner, and thus resulting in part-filled castings. He used the equation in its simplest form, derived from a statement of conservation of energy along a flow tube as illustrated schematically in Figure A4:

$$p_1/\rho g + v_1^2/2g + z_1 = p_2/\rho g + v_2^2/2g + z_2 = \text{constant}$$

Where all the component terms of this equation have units of length, conveniently metres. For this reason, each term can be regarded as a 'head'. Thus $p/\rho g$ = pressure head, $v^2/2g$ = kinetic or velocity head, and z = potential or elevation head.

In application to the running system used by Sutton (Figure A5) at location 1, the height above the centreline of the runner is 0.5 m, the kinetic head at this point is zero because the melt has zero downwards velocity, and the elevation head z is considered zero because the runner is horizontal. At point 2, the pressure head requires to be known because this is the pressure raising the melt level in the vertical ingate. The elevation is zero once again, and the velocity head is close to 0.25 m, easily deduced from the total fall height and allowing for a small loss factor of 0.70 (probably overestimated because I think

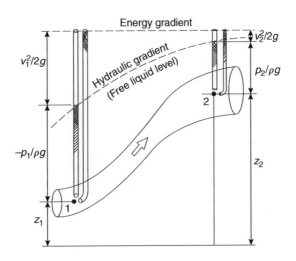

FIGURE A4

A pictorial representation of the factors in the Bernoulli equation.

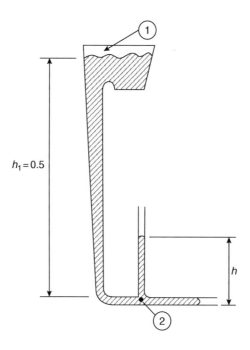

FIGURE A5

An example of the use of the Bernoulli equation by Sutton (2002) to calculate the rise of metal in a vertical gate.

this should be more like 0.80 or even 0.85) as a result of the turn at the base of the sprue. Thus the Bernoulli equation becomes

$$0.5 + 0 + 0 = p_2/\rho g + 0.25 + 0$$

Thus

$$p_2/\rho g = 0.25 \text{ metre}$$

It is not necessary to find p_2 alone; the whole term is the height distance. Thus, this answer would be the same for aluminium or iron.

Sutton found that because of this kinetic head, ingates were filling before the runner was fully filled. The first impression in his multi-impression mould was only about 200 mm above the runner so that metal entered the mould cavity under only about 50-mm net head. The result was a premature dribble into the cavity that quickly froze. The arrival of melt at the intended full flow rate a few seconds later was too late to remelt and thus assimilate the frozen droplets. An apparently mis-run casting was the result.

In general, however, the application of the Bernoulli equation to filling systems is not quite so straightforward as has sometimes been assumed. There are various reasons for this.

1. In general, Bernoulli's equation relates to steady state flow. However, of course, in filling systems, most of the interest necessarily lies in the priming of the flow channels. In this situation, the surface tension of the advancing meniscus can be important, as enshrined in the Weber number. If the priming is not carried out well, the casting is likely to suffer severely.

2. The surface tension of liquid metals is more than 10 times higher than that of water, and even higher still compared with most organic liquids. Thus pressures resulting from surface tension have been neglected and are neglectable for such common room temperature liquids on which most flow research has been conducted. The additional pressure generated because of the curvature of the meniscus at the flow front, and the curvature at the sides of a flow stream affect the behaviour of metals in many examples involved in the filling of moulds. For instance, at the critical velocity that is targeted in mould filling, the effects of surface tension and flow forces are equal. At velocities lower than this, surface tension dominates.

3. The presence of the oxide (or other thin, solid film) on the surface of an advancing liquid is a further complication, and is not easily allowed for. The flow adopts a stick-slip motion as the film breaks and re-forms. The advance of the unzipping wave is a classic instance that could not be predicted by a purely liquid model such as that described by Bernoulli.

4. The frictional losses during flow, which can be explicitly cited in Bernoulli's equation, are known to be important. However, in general, although they are assumed to be known, they have been little researched in the case of the flow of liquid metals. Furthermore, it is unfortunate that most of the research to date in this field has used such poor designs of filling systems that the existing figures are almost certainly misleading. The losses need to be confirmed by new, careful, accurate studies, supplemented by accurate computer simulation together with video X-ray radiography of real flows.

5. The presence of oxide films floating about in suspension is another uncertainty that can cause problems. The density of such defects can easily reach levels at which the effective viscosity of the mixture can be very much increased (although it is to be noted that viscosity does not appear explicitly in the Bernoulli equation). The suppression of convection in such contaminated liquids is common. Flow out of thick sections and into very thin sections can be prevented completely by blockage of the entrance into the thin section.

From this list, it is clear that the application of Bernoulli is more accurate for thicker section flows where surface effects and internal defects in the liquid are less dominant. As filling systems are progressively slimmed, and casting sections are thinned, Bernoulli's equation has to be used with greater caution.

As a result of the problems of the application of Bernoulli to the priming of the filling system, it has been relatively little used in this book because the concentration of effort has focussed on the control of the priming of the system. The subsequent flow of the system when completely filled, as nicely described by Bernoulli, is, with the greatest respect to the great man and his magnificent equation, much less important.

Appendix III

The recording of the lists of choices when drafting up a new methoding design for a casting must be carefully recorded. If there is found to be any problem with the first design, a second iteration can be worked out in the light of what appears to be needed (for instance, the casting may be required to fill a little faster). Table A1 can be used as a hard copy, allowing up to four iterations for a new filling system design. Alternatively, of course, the table can form the basis of a computer spread sheet, so that iterations can be performed rapidly and recorded digitally.

Table A1 Running System Record						
Casting Name			**Design 1**	**Design 2**	**Design 3**	**Design 4**
Part Number			**Signature**	**Signature**	**Signature**	**Signature**
Customer			**Date**	**Date**	**Date**	**Date**
Alloy						
Casting weight	M_C	kg				
Rigging weight	M_R	kg				
Total pour weight	$M_C + M_R = M$	kg				
Fill time selected	t	s				
Average flow rate	M/t	kg/s				
Design flow rate	2M/t	kg/s				
Volume flow rate	$Q = 2M/\rho t$	m^3/s				
Height in basin	h	m				
Working basin depth	h to 2h	m				
Velocity into sprue	$V_1 = (2gh)^{1/2}$	m/s				
Area sprue top	$A_1 = Q/V_1$	m^2				
Velocity exit sprue	$V_2 = (2gH)^{1/2}$	m/s				
Area sprue exit	$A_2 = Q/V_2$	m^2				
Radius to runners	$R = A_2^{1/2}$	m				
Number of runners	n					
Area of each runner	A_2/n	m^2				
Select critical velocity	$V_C = 0.5-1.0$	m/s				
Total gate area	$A3 = A_2 \cdot V_2/V_C$	m^2				
Basin volume (1 s)	Q	m^3				

Appendix IV

RATE OF DELIVERY OF STEEL FROM A BOTTOM-POUR LADLE

The following is an example of how the nomogram is used.

A ladle contains 5000 kg of steel, from which we wish to pour a casting of total weight 1250 kg. Thus, in Figure A6, we follow the arrows from the start point to junction A. From here, a horizontal line connects to the next figure, where we select a pouring nozzle for the ladle of 60-mm diameter. At this junction B, we drop a vertical line down to intersect with the line denoting that our ladle is about 1.5 m internal diameter. From

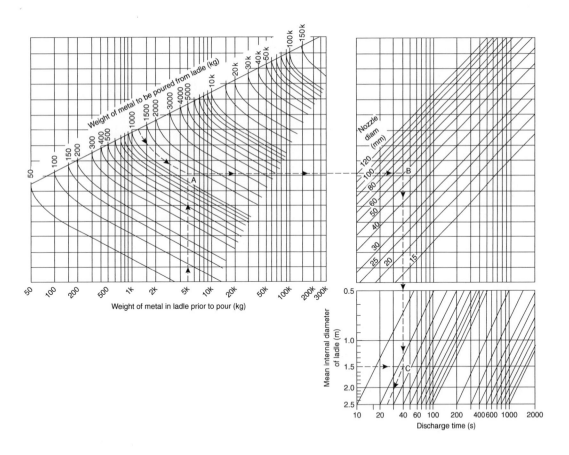

FIGURE A6

Rate of delivery of metal from a bottom-pour ladle.

this junction C, we continue with a parallel line to the family of sloping lines, to find that our casting will pour in approximately 23 s.

Interestingly, the reader can check that the next 1250 kg casting in line (now starting with a ladle of 5000 − 1250 = 3750 kg) will be found to pour in about 29 s, and the next in 34 s, and the next in 77 s, as the ladle progressively empties.

References

Abbas, M., Ray, N.K., Deb, P., Mallick, J.P., 1993. Indian Foundry J. 39 (9), 12–16.

Adams, C.M., 1958. Thermal considerations in freezing (Chapter). In: Liquid Metals and Solidification. ASM, Cleveland, Ohio, pp. 187–217.

Adams, A., 2001. Mod. Cast. 91 (3), 34–36.

Adams, C.M., Taylor, H.F., 1953. Am. Found. Soc. Trans. 61, 686–696.

Adefuye Segun, 1997. The Fluidity of Al-Si Alloys (Ph.D. thesis). University of Birmingham, UK (Supervisor N R Green).

Afseth, A., Nordlien, J.H., Scamans, G.M., Nisancioglu, K., 2000. In: ASST 2000 (2nd Internat. Symp. Al Surface Science Technology) UMIST Manchester, UK. Alcan + Dera, pp. 53–58.

Agema, K.S., Fray, D.J., 1988. Agema Ph.D. thesis, Dept Materials, Cambridge, UK.

Aguilar-Santillan, J., 2009. Metall. Mater. Trans. B 40B (3), 376–387.

Ahamed, A.K.M.A., Kato, H., 2008. Int. J. Cast Met. Res. 21 (1–4), 162–167.

Ahn, J.-H., Berghezan, 1991. Mater. Sci. Technol. 7 (7), 643–648.

Ainsworth, M.J., Griffiths, W.D., 2006. TAFS 114, 965–977.

Akita, K.K., UK Patent 1397821 filed 10 April 1972.

Alexopoulos, N.D., Tiryakioglu, M., 2009. Mater. Sci. Eng. A. http://dx.doi.org/10.1016/j.msea.2008.12.026.

Ali, S., Mutharasan, R., Apelian, D., 1985. Met. Trans. 16B, 725–742.

Aliravci, C.A., Gruzleski, J.E., Dimayuga, F.C., 1992. TAFS 100, 353–362.

Allen, D.I., Hunt, J.D., 1979. In: Solidification and Casting of Metals. Sheffield Conference, 1977, Metals. Soc., pp. 39–43.

Allen, A.G., Howard, J.C., Howard, J.F., Neenan, P.A. UK Patent Specification 1013851. Application date 31 January 1963.

Allen, N.P., 1932. J. Inst. Met. 49, 317–346.

Allen, N.P., 1933. J. Inst. Met. 51, 233–308.

Allen, N.P., 1934. J. Inst. Met. 52, 192–220.

Allen, A.G., 25 January 1968. Foundry Trade J. 159–161.

Almarez, G.M.D., Aburto, A.D., Gomez, E.C., January 2014. Metall. Mater. Trans. A 45A, 280–286.

Alonso, G., Stefanescu, D.M., Suarez, R., Loizaga, A., Zarrabeitia, G., 2014. Int. J. Cast Met. Res. 27 (2), 87–100.

Alsem, W.H.M., van Wiggen, P.C., Vader, M., 1992. Light metals. In: Cutshall, E.R. (Ed.), The Minerals. Metals and Materials Soc., USA, pp. 821–829.

Altstetter, J.D., Nowicki, R.M., 1982. TAFS 90, 959–970.

American Foundrymen's Society, 1987. Green Sand Additives. AFS, Detroit, USA.

Anderson, J.V., Karsay, S.I., 1985. Br. Found. 492–498.

Anderson, S.H., Foss, J.W., Nagan, R.M., Jhala, B.S., 1989. TAFS 97, 709–722.

Anderson, G.R., 1985. In: Patt-tech 85 Conference. Oxford, UK.

Anderson, G.R., 1987. TAFS 95, 203–210.

Andresen, P.L., Chou, P.H., Morra, M.M., Nelson, J.L., Rebak, R.B., 2009. Metall. Mater. Trans. A 40A (12), 2824–2836.

Andrew, J.H., Percival, R.T., Bottomley, G.T.C., 1936. Iron Steel Spec. Rep. 15, 43–64.

Angelov, G., June 1969. Russ. Cast. Prod. 282–284.

Angus, H.T., 1976. Cast Iron. Butterworths, London.

Anson, J.P., Gruzleski, J.E., 1999. Trans. Am. Found. Soc. 107, 135–142.

Anson, J.P., Stucky, M., Gruzleski, J.E., 2000. Trans. Am. Found. Soc. 108, 419–426.

Antes, H.W., Norton, J.T., Edelman, R.E., 1958. TAFS 66, 135–142.

Appa Rao, G., Srinivas, M., Sarma, D.S., 2004. Mater. Sci. Technol. 20, 1161–1170.

Archibald, J.J., Smith, R.L., 1988. "Casting" from Metals Handbook, vol. 15. ASM, 214–221.

Armbruster, D.R., Dodd, S.F., 1993. TAFS 101, 853–856.

Armstrong, G.L., Martin, W., 1974. TAFS 82, 253–255.

Arnold, F.L., Prestley, J.S., 1961. Trans. Am. Found. Soc. 69, 129–136.

Arnold, F.L., Jorstad, J.L., Stein, G.E., 1963. Curr. Eng. Pract. 6, 10–15.

Aryafar, M., Raiszadeh, R., Shalbafzadeh, A., 2010b. J. Mater. Sci. 45, 3041–3051.

Arzt, E., Ashby, M.F., Easterling, K.E., February 1983. Metall. Trans. A 14, 211–221.

Asbjornsonn, 2001. Ph.D. thesis, Department of Materials, University of Nottingham.

Ashton, M.C., Buhr, R.C., 1974. Phys. Met. Div. Internal Report PM-1-74-22. Canada Dept Energy Mines and Resources.

Ashton, M.C., 1990. In: SCRATA 34th Conference. Sutton Coldfield, Paper 2. Steel Castings Research and Trade Assoc, Sheffield, UK.

Ashton, M.C., January 1991. Met. Mater. 12–17.

Askeland, D., Holt, M.L., 1975. TAFS 83, 99–106.

Atwood, R.C., Lee, P.D., 2000. In: Sahm, P., Hansen, P. (Eds.), Modeling of Casting Welding and Advanced Solidification Processing Conf. Aachen.

Avey, M.A., Jensen, K.H., Weiss, D.J., 1989. TAFS 97, 207–212.

Aylen, P., Foulard, J., Galey, J., 1965. TAFS 73, 311–316.

Bachelot, F. 1997. M.Phil. thesis, University of Birmingham, UK.

Bachmann, P.K., Messier, R., 1984. Chem. Eng. News 67 (20), 24–39.

Backerud, L., Chai, G., Tamminen, J., 1990. Solidification Characteristics of Aluminum Alloys, Volume 2, Foundry Alloys. AFS/Skanaluminium (printed in USA).

Badia, F.A., Rohatgi, P., 1969. TAFS 77, 402–406.

Badia, F.A., 1971. TAFS 79, 347–350.

Bahreinian, F., Boutorabi, S.M.A., Campbell, J., 2005. In: Tiryakioglu, M., Crepeau, P.N. (Eds.), Shape Casting: The John Campbell Symposium. TMS, pp. 463–472.

Bailey, A., Davenport, A.J., April 2002. Final Year Project. University of Birmingham, Dept. Metall, UK.

Bak, C., Degois, M., Schissler, J.M., 1980. TAFS 88, 301–312.

Baker, W.A., 1945. J. Inst. Met. 71, 165–204.

Baker, W.F., 1986. TAFS 94, 215–218.

Bakhtiarani, F.N., Raiszadeh, R., 2011. Met. Mat. Trans. B 42 (2), 331–340.

Balewsky, A.T., Dimov, T., 1965. Br. Found. 78 (7), 280–283.

Baliktay, S., Nickel, E.G., 1988. In: Seventh World Conf. Invest. Casting, Munich paper 10.

Ball, R., May 1998. Foundryman 157–160, 175.

Ball, R., Hardcastle, P., January 1991. Foundry Trade J. 11 (25), 52–53.

Barbe, L., Bultink, I., Duprez, L., De Cooman, B.C., 2002. Mater. Sci. Technol. 18, 664–672.

Barkhudarov, M.R., Hirt, C.W., January–February 1999. In: Die Casting Engineer, pp. 44–47.

Barlow, G., 1970. Ph.D. thesis, University of Leeds.

Barrett, L.G., 1967. TAFS 75, 326–329.

Barton, R., 28 February 1985. Foundry Trade J. 117–126.

Basdogan, M.F., Bennett, G.H.J., Kondic, V., 1983. Solidification technology in the foundry and cast house. In: Warwick Univ. Conf. 1980. Metals Soc. Publication, pp. 240–247.

Bastien, P., Armbruster, J.C., Azou, P., 1962. In: 29 Internat. Found. Congress Detroit, pp. 400–409.

Bates, C.E., Monroe, R.W., 1981. TAFS 89, 671–686.

Bates, C.E., Scott, W.D., 1977. TAFS 85, 209–226.

Bates, C., Wallace, J.F., 1966. TAFS 74, 174–185.

Batty, O., 1935. TAFS 43, 75–106.

Bean, X., Marsh, L.E., 1969. Metal. Prog. 95, 131–134.

Beck, A., Schmidt, W., Schreiber, O., 1928. US Patent 1788185.

Beckermann, C., 2002. Int. Mater. Rev. 47 (5), 243–261.

Beech, J., April 1974. The Metallurgist and Materials Technolo-gist, pp. 129–232.

Beeley, P.R., Smart, R.F., 2009. Investment Casting. Maney Materials Science.

Beeley, P.R., 1972. Foundry Technology. Butterworths, London.

Beeley, P.R., 2001. Foundry Technology, second ed. Butterworth-Heinemann, London.

Benaily, N., 1998. Inoculation of Flake Graphite Iron (M.Phil. thesis), Univ. Birmingham, UK.

Bennett, S., Moody, T., Vrieze, A., Jackson, M., Askeland, D.R., Ransay, C.W., 2000. TAFS 108, 795–803.

Benson, L.E., June 1938a. Foundry Trade J. 527–528.

Benson, L.E., June 1938b. Foundry Trade J. 543–544.

Benson, L.E., 1946. J. Inst. Met. 72, 501–510.

Beretta, S., Murakami, Y., 2001. Metall. Mater. Trans. B 32B, 517–523.

Berger, R., 1932. Fonderie Belge 17.

Berry, J.T., Taylor, R.P., 1999. TAFS 107, 203–206.

Berry, J.T., Watmough, T., 1961. TAFS 69, 11–22.

Berry, J.T., Kondic, V., Martin, G., 1959. TAFS 67, 449–476.

Berry, J.T., Watmough, T., 1961. TAFS 69, 11–22.

Berry, J.T., Luck, R., Taylor, R.P., 2005. In: Tiryakioglu, M., Crepeau, P.N. (Eds.), Shape Casting: The John Campbell Symposium. TMS, pp. 113–122.

Berthelot, M., 1850. Ann. Chim. 30, 232.

Bertolino, M.F., Wallace, J.F., 1968. TAFS 76, 589–628.

Betts, B.P., Kondic, V., 1961. Br. Found. 54, 1–4.

Bex, T., November 1991. Mod. Cast. 56.

Bhaumik, S., Moles, V., Gottstein, G., Heering, C., Hirt, G., 2010. Adv. Eng. Mater. 12 (3), 127–130.

Biel, J., Smalinskas, K., Petro, A., Flinn, R.A., 1980. TAFS 88, 683–694.

Bindernagel, I., Kolorz, A., Orths, K., 1975. TAFS 83, 557–560.

Bindernagel, I., Kolorz, A., Orths, K., 1976. AFS Int. J. Cast Met. 1 (4), 42–45.

Birch, J., March 2000a. Diecast. World 18–19.

Birch, J., September 2000b. Diecast. World 174 (3570), 28.

Birch, J., March 2000c. Diecast. World 174, 35.

Bird, P., Savage, W., 1996. TAFS 104, 321–324.

Bird, P.G., 1 March 1989. Foseco Technical Service Report MMP1.89. Examination of the Factors Controlling the Flow Rate of Aluminium through DYPUR Units.

Birol, Y., 2012. Mater. Sci. Technol. 28 (8), 924–927.

Bishop, H.F., Ackerlind, C.G., Pellini, W.S., 1952. TAFS 60, 818–833.

Bishop, H.F., Myskowski, E.T., Pellini, W.S., 1955. TAFS 63, 271–281.

Bisuola, V.B., Martorano, M.A., 2008. Metall. Mater. Trans. A 39A (12), 2885–2895.

Biswas, P.K., Rohatgi, P.K., Dwarakadasa, E.S., 1985. Br. Found. 78, 511–516.

Biswas, P.K., Pillai, R.M., Rohatgi, P.K., Dwarakadasa, E.S., 1994. Cast. Met. 7 (2), 65–83.

Boenisch, D., Patterson, W., 1966. TAFS 74, 470–484.

Boenisch, D., 1967. TAFS 75, 33–37.

Boenisch, D., March 1988. In: Conference "Ensuring Quality Castings". BCIRA, Univ. of Warwick, pp. 29–31, paper 18.

Bonsak, W., 1962. TAFS 70, 374–382.

Boom, R., Dankert, O., Veen van, A., Kamperman, A.A., October 2000. Metall. Mater. Trans. B 31B (5), 913–919.

Bossing, E., 1982. TAFS 90, 33–38.

Bounds, S.M., Moran, G.J., Pericleous, K.A., Cross, M., 1998. In: Thomas, B.G., Beckerman, C. (Eds.), Modelling of Casting, Welding and Solidification Processing Conf. VIII, San Diego. TMS, pp. 857–864.

Boutorabi, S.M.A., Din, T., Campbell, J., 1992. Univ. Bham, unpublished work.

Bower, T.F., Brody, H.D., Flemings, M.C., 1966. Trans. AIME 236, 624.

Bracale, G., 1962. TAFS 70, 228–252.

Bradley, F.J., Hooper, J.A., Kannan, S., Balakrishna, J.V., Heinemann, S., 1992. TAFS 100, 917–923.

Bramfitt, B.L., 1970. Met. Trans. 1, 1987–1995.

Brandes, E.A., Brook, G.B. (Eds.), 1992. Smithells Metals Reference Book, seventh ed. Butterworths.

Brauer, H.E., Pierce, W.M., 1923. Trans. Am. Inst. Min. Metall. 68, 796–832.

Bridge, M.R., Stephenson, M.P., Beech, J., 1982. Met. Technol. 9, 429–433.

Bridges, D., 1999. Wheels and Axles. I. Mech. E. Seminar, London.

Briggs, C.W., Gezelius, R.A., 1934. TAFA 42, 449–476.

Briggs, L.J., 1950. J. Appl. Phys. 21, 721–722.

Brimacombe, J.K., Sorimachi, K., 1977. Met. Trans. 8B, 489–505.

Britney, D.J., Neailey, K., 2003. J. Mater. Process. Technol. 138, 306–310.

Bromfield, G., July 1991. Foundryman 261–265.

Brondyke, K.J., Hess, P.D., 1964. Trans. Met. Soc. AIME 230 (7), 1542–1546.

Brookes, B.E., Berckermann, C., Richards, V.L., 2007. Int. J Cast Met. Res. 20 (4), 177–190.

Brown, N., Rastall, D., 1986. European Patent Application No 87111549.9 filed 10 August 1987. p. 9.

Brown, J.R., 1970. Br. Found. 63, 273–279.

Brown, J.R., 1992. Met. Mater. 8, 550–555.

Bryant, M.D., Moore, A., 1971. Br. Found. 64, 215–229, 306–307.

Buhrig-Polackzek, A., Santos de, A., 2008. In: Metals Handbook, Volume 15, Casting. ASM, Ohio, USA, pp. 317–329.

Burchell, V.H., 1969. Br. Found. 62, 138–146.

Burton, B., Greenwood, G.W., 1970. Met. Sci. J. 4, 215–218.

Busby, A.D., 1996. TAFS 104, 957–968.

Butakov, D.K., Mel'nikov, L.M., Rudakov, I.P., Maslova Yu, N., 1968. Lit. Proizv. 4, 33–35.

Butler, T., Lund, J.N., February 2003. Mod. Cast. 40–42.

Butler, C.J., 1980. UK Patent GB 2020714.

Byczynski, G.E., Cusinato, D.A., 4–5 April 2001. In: First International Conf. Filling and Feeding of Castings. University of Birmingham, UK and Int. J. Cast Met. Res. 2002, 14 (5), 315–324.

Caceres, C.H., 2000. TAFS 108, 709–712.

Caceres, C.H., 2004. In: Nie, J.F., et al. (Eds.), Proc. 9th Internat. Conf. Al Alloys. Inst Materials Engineering Australia Ltd, pp. 1216–1221.

Caine, J.B., Toepke, R.E., 1966. TAFS 74, 19–22.

Caine, J.B., Toepke, R.F., 1967. TAFS 75, 10–16.

Campbell, J., Bannister, J.W., 1975. Met. Technol. 2 (9), 409–415.

Campbell, J., Caton, P.D., 1977. In: Institute of Metals Conference on Solidification, Sheffield, UK, pp. 208–217.

Campbell, J., Clyne, T.W., 1991. Cast. Met. 3 (4), 224–226.

Campbell, J., Isawa, T., 1994. UK Patent GB 2284168 B (Filed 4 February 1994).

Campbell, J., Naro, R.L., 2010. Lustrous carbon in grey iron. TAFS 114, 6 paper 10–36.

Campbell, J., Olliff, I.D., June 1971. AES Cast Met. Res. J. 55–61.

Campbell, J., Tiryakioglu, M., 2010. Effect of Sr on Porosity. Met. Mat. Trans. 26 (3), 262–268.

Campbell, H.L., 1950. Foundry 78, 86, 87, 210, 212, 213.

Campbell, J., 1967. Trans. Met. Soc. AIME 239, 138–142.

Campbell, J., 1968a. The Solidification of Metals. ISI Publication 110, pp. 19–26.

Campbell, J., 1968b. Trans. Met. Soc. AIME 242, 264–268.

Campbell, J., 1968c. Trans. Met. Soc. AIME 242, 268–271.

Campbell, J., 1968d. Trans. Met. Soc. AIME 242, 1464–1465.

Campbell, J., 1969a. Feeding Mechanisms in Castings. Cast. Met. Res. J. 5 (1), 1–8.

Campbell, J., 1969b. The non-equilibrium freezing range and its relation to hydrostatic tension and pore formation in solidifying binary alloys. Trans. AIME 245, 2325–2334.

Campbell, J., 1971. Metallography 4, 269–278.

Campbell, J., 1980. Solidification technology in the foundry and casthouse. In: Warwick Conference, Metals Soc. Publication 1981, vol. 273, pp. 61–64.

Campbell, J., 1981. Int. Met. Rev. 26 (2), 71–108.

Campbell, J., 1988. Mater. Sci. Technol. 4, 194–204.

Campbell, J., 1991a. Cast. Met. 4 (2), 101–102.

Campbell, J., 1991b. Castings. Published by Butterworth Heinemann (now Elsevier).

Campbell, J., September 1991c. Metals and Materials, p. 575.

Campbell, J., 1994. Cast. Met. 7 (4), 227–237.

Campbell, J., 2000a. Ingenia 1 (4), 35–39.

Campbell, J., 2000b. The concept of net shape for castings. Mater. Des. 21, 373–380.

Campbell, J., 2003. Castings. Elsevier, Oxford, UK pp (a) 178–181; (b) 161–162; (c) 158–160.

Campbell, J., 2006a. Mater. Sci Technol. 22 (2), 127–145 and (8), 999–1008.

Campbell, J., 2006b. Modeling of Entrainment Defects during Casting. In: San Antonio, USA. TMS Annual Congress.

Campbell, J., 2007. AFS Int. J. Metalcast. 1 (1), 7–20.

Campbell, J., 2008. Mater. Sci. Technol. 24 (7), 875–881.

Campbell, J., 2009a. Incipient grain boundary melting. Mater. Sci. Technol. 25 (1), 125–126.

Campbell, J., 2009b. A hypothesis for graphite formation in cast irons. Metall. Mater. Trans. B 40B (6), 786–801.

Campbell, J., 2009c. Stress corrosion cracking of Mg alloys. Metall. Mater. Trans. A 40A (7), 1510–1511.

Campbell, J., May 2009d. Discussion of "Effect of Sr and P on Eutectic Al-Si Nucleation and Formation of beta-Al$_5$FeSi in Hypoeutectic Al-Si Foundry Alloys". Metall. Mater. Trans. A 40A, 1009–1010.

Campbell, J., 2010a. Stress corrosion cracking of stainless…. Metall. Mater. Trans. A 41A (5), 1101.

Campbell, J., 2011a. In: Tiryakioglu, M., Crepeau, P., Campbell, J. (Eds.), The Origin of Griffiths Cracks. Shape Casting Symposium TMS Annual Congress, San Diego, CA.

Campbell, J., 2011b. The origin of Griffith cracks. Metall. Mater. Trans. B 42B, 1091–1097.

Campbell, J., September 2014. Metall. Mater. Trans. A 45A, 4193.

Campbell, J., 2015. Quality Castings: A Personal History of the Cosworth Casting Process. AFS.

Cao, X., Campbell, J., 2000. Am. Found. Soc. Trans. 108, 391–400.

Cao, X., Campbell, J., July 2003. Metall. Mater. Trans. A 34A, 1409–1420.

Cao, W., Fathallah, R., Castex, L., 1995. Mater. Sci. Technol. 11 (9), 967–973.

Cao, P., Qian, M., StJohn, D.H., Frost, T.M., 2004. Mater. Sci. Technol. 20 (5), 585–592.

Cao, X., 2001. Ph.D. thesis. Department of Metallurgy, University of Birmingham, UK.

Cao, X., 2009. Personal communication.

Capello, G.P., Carosso, M., May 1989. AGARD Report, No. 762.

Cappy, M., Draper, A., Scholl, G.W., 1974. TAFS 82, 355–360.

Carlberg, T., Fredriksson, H., 1979. In: Solidification and Casting of Metals Conf., Univ. Sheffield, UK, 1977. The Metals Society, pp. 115–124.

Carlson, C., Beckermann, C., 2009. Metall. Mater. Trans. A 40A, 163–175 and 3054–3055.

Carlson, K.D., Beckermann, C., 2008. Proc 62nd SFSA Conf. Paper No 5.6, Steel Founders' Soc of America.

Carlson, B.E., Pehlke, 1989. TAFS 97, 903–914.

Carlson, G.A., 1975. J. Appl. Phys. 46 (9), 4069–4070.

Carne, C.A., Beech, J., 1997. In: Beech, J., Jones, H. (Eds.), Solidification Processing. Sheffield University, UK, pp. 33–36.

Carte, A.E., 1960. Proc. Phys. Soc. 77, 757–769.

Carter, S.F., Evans, W.J., Harkness, 1. C., Wallace, J.F., 1979. TAFS 87, 245–268.

Celik, M.C., Bennett, G.H.J., April 1979. Met. Technol. 138–144.

Chadwick, H., Campbell, J., 1997. University of Birmingham, unpublished research.

Chadwick, G.A., Yue, T.M., January 1989. Met. Mater. 6–12.

Chadwick, G.A., Yue, T.M., 1991. Hi-Tech Metals R&D Ltd, Southampton, personal communication. (See Yue, T.M.).

Chadwick, P., 1963. Int. J. Mech. Sci. 5, 165–182.

Chadwick, H., 1991. Cast. Met. 4 (1), 43–49.

Chalmers, 1953. Quoted by Flemings, M.C., 1974

Chakrabarti, I., Campbell, J., 2000. University of Birmingham, unpublished research.

Chamberlain, B., Zabek, V.J., 1973. TAFS 81, 322–327.

Chambers, L.W., 1980 (April 24th). Foundry Trade J. 802.

Chandley, G.D., 1976. TAFS 84, 37–42.

Chandley, G.D., 1983. TAFS 91, 199–204.

Chandley, G.D., 1989. Cast. Met. 2 (1), 2–10.

Chandley, D., 2000. Metall. Sci. Technol. 18 (1), 8–11.

Chandravanshi, V., Sarkar, R., Kamat, S.V., Nandy, T.K., January 2013. Metall. Mater. Trans. A 44A, 201–211.

Chang, J.-K.(B.), Taleff, E.M., Krajewski, P.E., 2009. Metall. Mater. Trans. A 40A (13), 3128–3137.

Changyun, L., Shiping, W., Jingjie, G., Hengzhi, F., 2006. Int. J. Cast Met. Res. 19 (4), 237–240.

Charbonnier, J., Perrier, J.J., January 1983. Giesserei 70 (2), 50–55.

Charles, J.A., Uchiyama, I., July 1969. J. Iron Steel Inst. 207, 979–983.

Chaudhury, S.K., Apelian, D., 2006. Int. J. Cast Met. Res. 19 (6), 361–369.

Chechulin, V.A., 1965. In: Gulyaev, B.B. (Ed.), Gases in Cast Metals. Consultants Bureau Translation, pp. 214–218.

Chegini, S., Raiszadeh, R., 2014. Int. J. Cast Met. Res. 27 (6), 349–356.

Chen, X.G., Engler, S., 1994. TAFS 102, 673–682.

Chen, X.-G., Fortier, M., 2010. J. Mater. Process. Technol. 210, 1780–1786.

Chen, Q.Z., Knowles, D.M., 2003. Mater. Sci. Technol. 19 (4), 447–455.

Chen, C.O., Ramberg, F., Evensen, J.D., 1984. Metal. Sci. 18, 1–5.

Chen, G.J., Liu, S.H., Ren, B.L.T., 1989. TAFS 97, 335–338.

Chen, X.G., Klinkenberg, F.J., Ellerbrok, R., Engler, S., 1994. TAFS 102, 191–197.

Chen, T.J., Ma, Y., Li, Y.D., Lu, G.X., Hao, Y., 2010. Mater. Sci. Technol. 26 (10), 1197–1206.

Chen, X.-C., Shi, C.-B., Guo, H.-J., Wang, F., Ren, H., Feng, D., December 2012. Metall. Mater. Trans. B 43B, 1593–1607.

Chen, Z., Mo, Y., Nie, Z., August 2013. Metall. Mater. Trans. 44A, 3910–3920.

Cheng, X., Yuan, C., Blackburn, S., Withey, P.A., 2014. Mater. Sci. Technol. 30 (14), 1758–1764.

Chernov, D.K., December 1878. Reports of the Imperial Russian Metallurgical Society (see Russkoe Metalurgicheskoi Obshchestro 1, 1915).

Chiesa, F., 1990. TAFS 98, 193–200.

Chisamera, M., Riposan, I., Barstow, M., 1996. TAFS 104, 581–588.

Cho, J.-I., Loper, C.R., 2000. Trans. Am. Found. Soc. 108, 359–367.

Cho, Y.H., Lee, H.-C., Oh, K.H., Dahle, A.K., 2008. Metall. Mater. Trans. 39A (10), 2435–2448.

Choudhury, A., Blum, M., Scholz, H., Jarczyk, G., Busse, P., February 1999. In: Proc. 1999 Internat. Symp. Liquid Metal Processing and Casting; Santa Fe, NM, pp. 244–255.

Chu, M.G., 2002. Light Met. 899–907.

Chung, Y., Cramb, A.W., 2000. Metall. Mater. Trans. B 31B, 957–971.

Church, N., Wieser, P., Wallace, J.F., 1966. TAFS 74, 113–128. Also in Br. Found. 1966 59, 349–363.

Churches, D.M., Rundman, K.B., 1995. TAFS 103, 587–594.

Chvorinov, N., 1940. Giesserei 27, 177–186, 201–208, 222–225.

Chvorinov, N., 1940. Giesserei 10, 177–186, 201, 222 and 27 (31 May) 201–208.

Cibula, A., 1955. Proc. IBF 45, A73–A90. Also in Foundry Trade J. 1955 98, 713–726.

Clark, J.C., April 1989. BCIRA Report Number 1769, 181–193.

Clausen, C., August 1992. Mod. Cast. 39.

Claxton, K.T., 1967. The Influence of radiation on the inception of boiling in liquid sodium. In: UK Atomic Energy Authority Research Group Report AERE-R5308. Also in Proc. Internat. Conf. on the Safety of Fast Reactors (CAE). Aix-en-Provence, September 1967, p. II-B-8-1.

Claxton, K.T., 1969. Private communication. UKAEA, Harwell, UK.

Clegg, A.L., Das, A.A., 1987. Br. Found. 80, 137–144.

Clyne, T.W., Davies, G.J., 1975. Br. Found. 68, 238–244.

Clyne, I.W., Davies, G.J., 1979. In: Solidification and Casting of Metals. Metals Soc. Conference, Sheffield, 1977, pp. 275–278.

Clyne, T.W., Davies, G.J., 1981. Br. Found. 74, 65–73.

Clyne, T.W., Kurz, W., 1981. Met. Trans. 12A, 965–971.

Clyne, T.W., Wolf, M., Kurz, W., 1982. Met. Trans. 13B, 259–266.

Clyne, T.W., 1977. Ph.D. thesis, University of Cambridge.

Coble, R.L., Flemings, M.C., 2 February 1971. Met. Trans. 409–415.

Cochran, C.N., Belitskus, D.L., Kinosz, D.L., 1977. Metall. Trans. 8B, 323–332.

Cole, G.S., Cisse, J., Kerr, H.W., Bolling, G.F., 1972. TAFS 80, 211–218.

Cole, G.S., 1972. TAFS 80, 335–348.

Colwell, D.L., 1963. TAFS 71, 172–176.

Cook, R., Kearns, M.A., Cooper, P.S., 1997. In: Huglen, R. (Ed.), Light Metals. TMS, pp. 809–814.

Cotton, J.D., Clark, L.P., Phelps, H.R., June 2006. J. Met. 13–16.

Cottrell, A.H., 1964. The Mechanical Properties of Matter. Wiley, p. 82.

Couture, A., Edwards, J.O., 1966. TAFS 74, 709–721, 792–793.

Couture, A., Edwards, J.O., 1967. AFS Cast Met. Res. J. 3 (2), 57–69.

Couture, A., Edwards, J.O., 1973. TAFS 81, 453–461.

Cowen, C.J., Boehlert, C.J., 2007. Metall. Mater. Trans. A 38A, 26–34.

Cox, M., Harding, R.A., 2007. Mater. Sci. Technol. 23 (2), 214–224.

Cox, M., Wickins, M., Kuang, J.P., Harding, R.A., Campbell, J., 2000. Mater. Sci. Technol. 16, 1445–1452.

Cox, M., Harding, R.A., Green, N.R., Scholl, G.W., 2007. Mater. Sci. Technol. 23 (9), 1075–1084.

Creese, R.C., Sarfaraz, A., 1987. TAFS 95, 689–692.

Creese, R.C., Sarfaraz, A., 1988. TAFS 96, 705–714.

Creese, R.C., Xia, Y., 1991. TAFS 99, 717–727.

Crossley, F.A., Mondolfo, L.F., 1951. J. Met. 3, 1143–1154.

Cunliffe, E.L., 1996. The minimal gating of aluminium alloy castings (Ph.D. thesis), Metals and Materials Department, University of Birmingham, UK.

Cunningham, M., 1988. Stahl Speciality Co, Kingsville, MO, USA. Private communication.

Cupini, N.L., Prates de Campos Filho, M., 1977. In: Sheffield Conf. "Solidification and Casting of Metals" Metals Soc. 1979, pp. 193–197.

Cupini, N.L., de Galiza, J.A., Robert, M.H., Pontes, P.S., 1980. Solidification technology in the foundry and cast house. In: Metals Soc. Conf, pp. 65–69.

Czerwinski, F., November 2008. J. Met. 82–86.

D'Errico, F., Rivolta, B., Gerosa, R., Perricone, G., November 2008. J. Met. 70–75.

Dai, X., Yang, X., Campbell, J., Wood, J., 2003. Mater. Sci. Eng. A354 (1–2), 315–325.

Dantzig, J.A., Rappaz, M., 2009. Solidification. EPFL Press, Lausanne, Switzerland, pp. 486–490.

Das, A.A., Chatterjee, S., 1981. Metall. Mater. Technol. 13 (3), 137–142.

Das, C.R., Albert, S.K., Bhaduri, A.K., Murty, B.S., May 2013. Metall. Mater. Trans. A 44A, 2171–2186.

Dasch, J.M., Ang, C.C., Wong, C.A., Waldo, R.A., 2009. J. Mater. Process. Technol. 209, 4638–4644.

Dasgupta, S., Parmenter, L., Apelian, D., 1998. In: AFS 5th Internat. Conf. Molten Aluminum Processing, pp. 285–300.

Datta, N., Sandford, P., 1995. In: 3rd AFS Internat. Permanent Mold Casting of Aluminum Conference Paper 3, p. 19.

Davidson, R.M.R., 1990. Mater. Perform. 29 (1), 57–62.

Davies, G.J., Shin, Y.K., 1980. Paper 78. In: Solidification Technology in the Foundry and Casthouse. Metals Soc. Conference, Warwick, pp. 517–523 (Published Metal Soc. 1983).

Davies, I.G., Dennis, J.M., Hellawell, A., 1970. Metall. Trans. 1, 275–280.

Davies, V. de L., 1963. J. Inst. Met. 92, 127.

Davies, V. de L., 1964–1965. J. Inst. Met. 93, 10.

Davies, V. de L., 1970. Br. Found. 63, 93–101.

Davis, K.G., Magny, J.-G., 1977. TAFS 85, 227–236.

Davis, K.G., Internat, A.F.S., March 1977. Cast Met. J. 23–27.

Dawson, J.V., Kilshaw, J.A., Morgan, A.D., 1965. TAFS 73, 224–240.

Dawson, J.V., 1962. BCIRA J. 10 (4), 433–437.

Daybell, E., 1953. Proc. Inst. Br. Found. 46, B46–B54.

De Sy, A., 1967. TAFS 75, 161–172.

Delamore, G.W., Smith, R.W., 1971. Met. Trans. 2, 1733–1743.

Delamore, G.W., Smith, R.W., Mackay, W.B.F., 1971. TAFS 79, 560–564.

Denisov, V.A., Manakin, A.M., 1965. Russ. Cast. Prod. 217–219.

Dennis, K., Drew, R.A.L., Gruzleski, J.E., 2000. Alum. Trans. 3 (1), 31–39.

Devaux, H., 1987. In: Moreau, R.J. (Ed.), Measurement and Control in Liquid Metals Processing. Nijhoff, The Netherlands, pp. 107–115.

DeYoung, D.H., Dunlay, M.J., 2002. US Patent 6334978.

Dickhaus, C.H., Ohm, L., Engler, S., 1993. TAFS 101, 677–684.

Dietert, H.W., Fairfield, H.H., Brewster, F.S., 1948. TAFS 56, 528–535.

Dietert, H.W., Doelman, R.L., Bennett, R.W., 1970. TAFS 78, 145–156.

Dimayuga, F.C., Handiak, N., Gruzleski, I.E., 1988. TAFS 96, 83–88.

Dimmick, T., 2001. Mod. Cast. 91 (3), 31–33.

Din, T., Campbell, J., 1994. University of Birmingham, UK, unpublished work.

Din, T., Rashid, A.K.M.B., Campbell, J., 1996. Mater. Sci. Technol. 12 (3), 269–273.

Din, T., Kendrick, R., Campbell, J., 2003. TAFS 111, 91–100 (paper 03-017).

Dinayuga, F.C., Handiak, N., Gruzleski, J.E., 1988. TAFS 96, 83–99.

Dion, J.-L., Couture, A., Edwards, J.O., 1978. TAFS 86, 309–314 and AFS Int. Cast Met. J. 1979 4 (2), 7–13.

Dion, J.-L., Fasoyinu, F.A., Cousineau, D., Bibby, C., Sahoo, M., 1995. TAFS 103, 367–377.

Dionne, P., Dickson, J.I., Bailon, J.P., 1984. TAFS 92, 693–701.

Disa Industries, July 2002. Foundry Trade J. 27.

Dispinar, D., Akhtar, S., Nordmark, A., Di Sabatino, M., Arnberg, L., 2010. Mater. Sci. Eng. A 527 (16–17), 3719–3725.

Dispinar, D., Campbell, J., 2004. Int. J. Cast Met. Res. 17 (5), 280–286 and 278–294.

Dispinar, D., Campbell, J., April 2005. University of Birmingham, UK, unpublished work.

Dispinar, D., Campbell, J., 2006. Int. J. Cast Met. Res. 19 (1), 5–17.

Dispinar, D., Campbell, J., 2007. J. Mater. Sci. 42, 10296–10298.

Dispinar, D., Campbell, J., 2011. Mater. Sci. Eng. A 528 (10), 3860–3865.

DiSylvestro, G., Faist, C.A., 1977. TAFS 85, 627–642.

Divandari, M., Campbell, J., 11–13 October 1999. In: AFS 1st Internat. Conf. on Gating Filling and Feeding of Aluminum Castings, pp. 49–63.

Divandari, M., Campbell, J., 2000. Alum. Trans. 2 (2), 233–238.

Divandari, M., 1998. University of Birmingham, UK, unpublished work.

Dixon, B., 1988. In: Solidification Processing Conference 1987. Inst. Metals, pp. 381–383.

Dodd, R.A., Pollard, W.A., Meier, J.W., 1957. TAFS 65, 100–117.

Dodd, R.A., 1950. Ph.D. thesis, Dept. Industrial Metallurgy, Univ. Birmingham, UK.

Dodd, R.A., 1955a. Hot Tearing of Casting: A Review of the Literature. Dept. Mines Ontario, Canada. Research Report PM184. Also in Foundry Trade J. 1956, 101, 321–331.

Dodd, R.A., 1955b. Hot Tearing of Binary Mg-Al and Mg-Zn Alloys. Dept Mines Ontario, Canada. Research Report PM191.

Dolan, G.P., Flynn, R.J., Tanner, D.A., Robinson, J.S., 2005. Mater. Sci. Technol. 21 (6), 687–692.

Donahue, R., Anderson, K., 2008. ASM Handbook "Casting", vol. 15, pp. 640–645.

Donahue, J., June 2006. INCAST 11.

Dong, S., Niyama, E., Anzai, K., 1995. ISIJ Int. 35, 730–736.

Donoho, C.K., 1944. Trans. Am. Found. Assoc. 52, 313–332.

Doremus, G.B., Loper, C.R., 1970. TAFS 78, 338–342.

Dorward, R.C., 2001. Oxid. Met. 55 (1/2), 69–74.

Double, D.D., Hellawell, A., 1974. Acta Metall. 22, 481–487.

Doutre, D.A., Hay, G., Wales, P., European Patent 20000951137 filed 26 July 2000.

Doutre, D., 1998. In: Wilkinson, D.S., Poole, W.J., Alpes, A. (Eds.), Advances in Industrial Materials. The Metallurgical Soc. of CIM.

Draper, A.L., Gaindhar, J.L., 1975. TAFS 83, 593–616.

Draper, A.B., 1976. TAFS 84, 749–764.

Drezet, J.M., Commet, B., Fjaer, H.G., Magnin, B., 2000. In: Sahm, P.R., Hansen, P.N., Conley, J.G. (Eds.), Modeling of Casting, Welding and Advanced Solidification Processes IX, pp. 33–40.

Drouzy, M., Mascre, C., 1969. Metall. Rev. 14, 25–46.

Drouzy, M., Jacob, S., Richard, M., 1980. Int. Cast Met. J. 5 (2), 43–45.

Druschitz, A.P., Chaput, W.W., 1993. TAFS 101, 447–458.

Durrans, I., 1981. Thesis. University of Oxford.

Durville, P.H.G., 1913. Br. Patent 23719.

Eastwood, L.W., 1951. In: AFS Symposium on Principles of Gating, pp. 25–30.

Edelman, R.E., Saia, A., 1968. TAFS 76, 222–224.

Edelson, B.J., Baldwin, W.M., 1962. Trans. ASM 55, 230.

Eigenfeld, K., 1988. 8th Europe Kolloquium der ne. Metallgiesserei im CAEF Paris.

Einstein, A., 1906. Ann. Phys. 19, 289 and 1911, 34, 591.

Elliott, H.E., Mezoff, J.G., 1947. TAFS 55, 241–253.

Elliott, H.E., Mezoff, J.G., 1948. TAFS 56, 223–245, 279–285.

Ellison, W., Wechselblatt, P.M., 1966. TAFS 74, 350–356.

El-Mahallawi, S., Beeley, P.R., 1965. Br. Found. 58, 241–248.

El-Sayed, M.A., Salem, H.A.G., Kandeil, A.Y., Griffiths, W.D., August 2014. Metall. Mater. Trans. B 45B, 1398–1406.

Emadi, D., Whiting, M., Djurdjevic, M., Kierkus, W.T., Sokolowski, J., 2004. MJoM 10, 91–106.

Emami, S., Sohn, H.Y., Kim, H.G., August 2014. Metall. Mater. B 45B, 1370–1379.

Emamy, G.M., Campbell, J., 1995a. Int. J. Cast Met. Res. 8, 13–20.

Emamy, G.M., Campbell, J., 1995b. Int. J. Cast Met. Res. 8, 115–122.

Emamy, G.M., Campbell, J., 1997. Trans. AFS 105, 655–663.

Emamy, G.M., Taghiabadi, R., Mahmudi, M., Campbell, J., 2002. In: Statistical Study of Tensile Properties of A356 Aluminum Alloy, Using a New Casting Design. ASM 2nd Internat. Al Casting Technology Symposium, 7–10 October, 2002, Columbus, Ohio.

Emamy, M., Abbasi, R., Kaboli, S., Campbell, J., 2009. Int. J. Cast Met. Res. 22 (6), 430–437.

Enderle, R.J., 1979. Trans. Am. Found. Soc. 87, 59–64.

Engels, G., Schneider, G., 1986. Cast. Plant Technol. (2), 12–20.

Enright, P., Hughes, I.R., November 1996. Foundryman 390–395.

Enright, N., Lu, S.Z., Hellawell, A., Pilling, J., 2000. TAFS 108, 157–162.

Enright, P., 2001. Private communication.

Evans, J., Runyoro, J., Campbell, J., 1997a. In: Beech, J., Jones, H. (Eds.), Solidification Processing. Dept. Engineering Materials, University of Sheffield, pp. 74–78.

Evans, J., Runyoro, J., Campbell, J., 1997b. In: SP97 4th Decennial International Conf. on Solidification Processing, Sheffield, pp. 74–78.

Evans, M.-H., Richardson, A.D., Wang, L., Wood, R.J.K., 2013. Effect of hydrogen on butterfly and white etching crack (WEC) formation under rolling contact fatigue (RCF). Wear. http://dx.doi.org/10.1016/j.wear.2013.03.008i.

Evans, E.P., 1951. BCIRA J. 4 (319), 86–139.

Evans, R.W., August 2002. Mater. Sci. Technol. 18, 831–839.

Fan, Z.-T., Ji, S., 2005. Mater. Sci. Technol. 21 (6), 727–734.

Fan, Z., Ji, S., Fang, X., Liu, G., Patel, J., Das, A., 2007. In: Crepeau, P.N., Tiryakioglu, M., Campbell, J. (Eds.), Shape Casting: The 2nd Internat. Symp. TMS, pp. 299–306.

Fang, O.T., Granger, D.A., 1989. TAFS 97, 989–1000.

Fasoyinu, F.A., Sadayappan, M., Cousineau, D., Sahoo, M., 1998a. TAFS 106, 721–734.

Fasoyinu, F.A., Sadayappan, M., Cousineau, D., Zavadil, R., Sahoo, M., 1998b. TAFS 106, 327–337.

Fathallah, R., Inglebert, G., Castex, L., 1998. Mater. Sci. Technol. 14, 631–639.

Fathallah, R., Sidham, H., Braham, C., Castex, L., August 2000. Mater. Sci. Technol. 19, 1050–1056.

Faubert, G.P., Moore, D.J., Rundman, K.B., 1990. TAFS 98, 831–845.

Feest, E.A., McHugh, G., Morton, D.O., Welch, L.S., Cook, I.A., 1983. In: Solidification Technology in the Foundry and Cast House. Warwick Univ. Conf. 1980. Metals Soc. Publication, pp. 232–239.

Felberbaum, M., Dahle, A., 2011. In: Cast House Symposium. TMS Annual Congress, San Diego, CA.

Feliu, S., Flemings, M.C., Taylor, H.F., 1960. Br. Found. 53, 413–425.

Feliu, S., Luis, L., Siguin, D., Alvarez, J., 1962. TAFS 70, 838–844 (1964) 71, 145–157; (1964) 72, 129–137.

Feliu, S., 1962. TAFS 70, 838–844.

Feliu, S., 1964. TAFS 72, 129–137.

Fenn, D., Harding, R.A., 2002. University of Birmingham, UK, unpublished work.

Fernandez, E., Musalem, P., Fava, J., Flinn, R.A., 1970. TAFS 78, 308–312.

Feurer, U., 1976. Giesserei 28, 75–80 (English translation by Alusuisse).

Fischer, R.B., 1988. TAFS 96, 925–944.

Fisher, J.C., 1948. J. Appl. Phys. 19, 1062–1067.

Flemings, M.C., Mehrabian, R., 1970. TAFS 78, 388–394.

Flemings, M.C., Nereo, G.E., 1967. TMS-AIME 239, 1449–1461.

Flemings, M.C., Conrad, H.F., Taylor, H.F., 1959. TAFS 67, 496–507.

Flemings, M.C., Niiyama, E., Taylor, H.F., 1961. TAFS 69, 625–635.

Flemings, M.C., Mollard, F.R., Niyama, E.F., Taylor, H.F., 1962. TAFS 70, 1029–1039.

Flemings, M.C., Poirier, D.R., Barone, R.V., Brody, H.D., April 1970. J. Iron Steel Inst. 371–381.

Flemings, M.C., Mollard, F.R., Niyama, E.F., 1987. TAFS 95, 647–652.

Flemings, M.C., Kattamis, T.Z., Bardes, B.P., 1991. TAFS 99, 501–506.

Flemings, M.C., 1963. In: 30th Internat. Found. Cong. Prague, pp. 61–81 and Br. Found. (1964) 57, 312–325.

Flemings, M.C., 1974. Solidification Processing. McGraw-Hill, USA.

Fletcher, A.J., Griffiths, W.D., 1995. Mater. Sci. Technol. 11 (3), 322–326.

Fletcher, A.J., 1989. Thermal Stress and Strain Generation in Heat Treatment. Elsevier.

Flinn, R.A., Van Vlack, L.H., Colligan, G.A., 1986. TAFS 94, 29–46.

Flood, S.C., Hunt, J.D., 1981. Met. Sci. 15, 287–294.

Fomin, V.V., Stekol'nikov, G.A., Omel'chenko, V.S., 1965. Russ. Cast. Prod. 229–231.

Fonda, R.W., Lauridsen, E.M., Ludwig, W., Tafforeau, P., Spanos, G., 2007. Metall. Mater. Trans. A 38A (11), 2721–2726.

Ford, D.A., Wallbank, J., 1998. Int. J. Cast Met. Res. 11, 23–35.

Forest, B., Berovici, S., 1980. In: Solidification Technology in the Foundry and Casthouse. Metals Soc. Conf. Warwick, Paper 93, pp. 1–12.

Forgac, J.M., Angus, J.C., June 1981. Metall. Trans. B 12B, 413–416.

Forgac, J.M., Schur, T.P., Angus, J.C., 1979. J. Appl. Mech. 46, 83–89.

Forslund, S.H.C., 1954. In: 21st Internat. Foundry Congress, Florence, Paper 15.

Forsyth, P.J.E., 1995. Mater. Sci. Technol. 11 (3), 1025–1033.

Forsyth, P.J.E., 1999. Mater. Sci. Technol. 15 (3), 301–308.

Forward, G., Elliott, J.F., 1967. J. Met. 19, 54–59.

Fox, S., Campbell, J., 2000. Scr. Mater. 43 (10), 881–886.

Fox, S., Campbell, J., 2002. Int. J. Cast Met. Res., 14 (6), 335–340.

Franklin, A.G., Rule, G., Widdowson, R., September 1969. J. Iron Steel Inst. 1208–1218.

Fras, E., Gorny, M., Lopez, H.F., 2007a. TAFS 115, 1–17.

Fras, E., Wiencek, K., Gorny, M., Lopez, H.F., 2007b. Trans. Met. Mater. A 38A (2), 385–395.

Fras, E., Gorny, M., Lopez, H.F., 2012. Metall. Mater. Trans. 43A (11), 4204–4218.

Frawley, I.J., Moore, W.F., Kiesler, A.L., 1974. TAFS 82, 561–570.

Fredriksson, H., Lehtinen, B., 1979. In: Solidification and Casting of Metals. Metals Soc. Conf. Metals Soc, Sheffield, pp. 260–267.

Fredriksson, H., Haddad-Sabzevar, M., Hansson, K., Kron, J., 2005. Mater. Sci. Technol. 21 (5), 521–529.

Fredriksson, H., 1984. Mat. Sci. Eng. 65, 137–144.

Fredriksson, H., 1996. In: Ohnaka, I., Stefanescu, D.M. (Eds.), Solidification Science and Processing. The Minerals, Metals and Materials Society.

French, A.R., 1957. Inst. Met. Monogr. Rep. 22, 60–84.

Freti, S., Bornand, J.D., Buxmann, K., 12–16 June 1982. Light Metal Age.

Friebel, V.R., Roe, W.P., 1963. TAFS 71, 388–393.

Frommeyer, G., Derder, C., Jimenez, J.A., 2002. Mater. Sci. Technol. 18 (9), 981–986.

Fruehling, J.W., Hanawalt, J.D., 1969. TAFS 77, 159–164.

Fruehling, J.W., Hanawalt, J.D., 1969. TAFS 77, 159–590.

Fuji, N., Fuji, M., Morimoto, S., Okada, S., 1984. J. Jpn. Inst. Light Met. 34 (8), 446–453 (Met. Abstr. 51-1657).

Fuji, M., Fuji, N., Morimoto, S., Okada, S., 1986. J. Jpn. Inst. Light Met. 36 (6), 353–360 (Transl. NF 197).

Fuoco, R., Correa, E.R., Correa, A.V.O., 1995. TAFS 103, 379–387.

Fuoco, R., Correa, E.R., Andrade Bastos, M.de, 1998. TAFS 106, 401–409.

Gadd, M.A., Bennett, G.J., 1984. In: Fredriksson, H., Hillert, M. (Eds.), The Physical Metallurgy of Cast Iron. North-Holland, p. 99.

Gagne, M., Goller, R., 1983. TAFS 91, 37–46.

Gagne, M., Paquin, M.-P., Cabanne, P.-M., 2008. World Foundry Congress, 68th, pp. 101–106.

Gall, K., Horstemeyer, M.F., van Schilfgaarde, M., Baskes, M.I., 2000. J. Mech. Phys. Solids 48, 2183–2218.

Gallois, B., Behi, M., Panchal, J.M., 1987. TAFS 95, 579–590.

Gammelsaeter, R., Beck, K., Johansen, S.T., 1997. In: Light Metals (Conference), pp. 1007–1011.

Garat, M., Guy, S., Thomas, J., 1991. Foundryman 84 (1), 29–34.

Garat, M., European Patent Application 0274964 Filed 16 November 1987 (in French).

Garbellini, O., Palacio, H., Biloni, H., 1990. Cast. Met. 3 (2), 82–90.

Garcia-Garcia, O., Sanches-Araiza, M., Castro-Roman, M., Escobedo, B.J.C., 2007. In: Crepeau, P.N., Tiryakioglu, M., Campbell, J. (Eds.), Shape Casting: 2nd International Symposium. TMS, pp. 109–116.

Garnar, T.E., 1997. TAFS 85, 399–416.

Gauthier, J., Samuel, F.H., 1995a. TAFS 103, 849–857.

Gauthier, J., Louchez, P.R., Samuel, F.H., 1995b. Cast. Met. 8 (2), 91–106.

Gebelin, J.-C., Griffiths, W.D., 2007. In: Jones, H. (Ed.), Proc 5th Decennial Internat. Conf. on Solidification Processing. (SP07) Sheffield, UK, pp. 512–516.

Gebelin, J.-C., Jolly, M.R., 2002. TAFS 110, 109–119.

Gebelin, J.-C., Jolly, M.R., 2003. In: Modeling of Casting, Welding and Solidification Processing Conference X, Destin USA.

Gebelin, J-C., Jolly, M.R., Jones, S., December. INCAST 13 (11), 22–27.

Gebelin, J.-C., Jolly, M.R., Hsu, F.-Y., 2005. In: Tiryakioglu, M., Crepeau, P.N. (Eds.), Shape Casting: The John Campbell Symposium. TMS, pp. 355–364.

Gebelin, J.-C., Jolly, M.R., Hsu, F.-Y., 2006. Int. J. Cast Met. Res. 19 (1), 18–25.

Geffroy, P.-M., Lakehal, M., Goni, J., Beaugnon, E., Heintz, J.-M., Silvain, J.-F., 2006a. Metall. Mater. Trans. A 37A, 441–447.

Geffroy, P.-M., Pena, X., Lakehal, M., Goni, J., Egizabal, P., Silvain, J.-F., February 2006b. Fonderie Fondeur d'aujord'hui 252, 8–19.

Gelperin, N.B., 1946. TAFS 54, 724–726.

Genders, R., Bailey, G.L., 1934. The Casting of Brass Ingots. The British Non-Ferrous Metals Research Association.

Gentry, E.G., 1966. TAFS 74, 142–149.

Gernez, M., 1867. Philos. Mag. 33 (4), 479.

Geskin, F.S., Ling, E., Weinstein, M.I., 1986. TAFS 94, 155–158.

Ghomashchi, M.R., Chadwick, G.A., 1986. Met. Mater. 477–482.

Ghomashchi, M.R., 1995. J. Mater. Process. Technol. 52, 193–206.

Ghosh, S., Mott, W.J., 1964. Trans. Am. Found. Soc. 72, 721–732.

Giese, S.R., Stefanescu, D.M., Barlow, J., Piwonka, T.S., 1996. Part II, TAFS 104, 1249–1257.

Girshovich, N.G., Lebedev, K.P., Nakhendzi Yu, A., April 1963. Russ. Cast. Prod. 174–178.

Giuranno, D., Ricci, E., Arato, E., Costa, P., 2006. Acta Mater. 54, 2625–2630.

Glatz, J., December 1996. Materials Evaluation, 1352–1362 and INCAST June 1997, 1–7.

Glenister, S.M.D., Elliott, R., 1981. Met. Sci. 15 (4), 181–184.

Gloria, D., Hernandez, F., Valtierra, S., Cisneros, M.A., October 2000. In: 20th ASM Heat Treating Soc. Conf. Proc, St Louis, MO, pp. 674–679.

Goad, P.W., 1959. TAFS 67, 436–448.

Godding, R.G., 1962. BCIRA J. 10 (3), 292–297.

Godlewski, L.A., Zindel, J.W., 2001. TAFS 109, 315–325.

Godlewski, L.A., Su, X., Pollock Tresa, M., Allison, J.E., 2013. Metall. Mater. Trans. A 44A, 4809–4818.

Goklu, S.M., Lange, K.W., April 1986. In: Proc. Conf. Process Technology, vol. 6. Iron and Steel Society, Washington, USA, pp. 1135–1146.

Gonya, H.J., Ekey, D.C., 1951. TAFS 59, 253–260.

Goodwin, F.E., 2008. ASM Handbook. In: Casting, vol. 15, pp. 1095–1099.

Goodwin, F.E., 2009. Cast Met. Diecast. Times 11 (4), 14.

Goria, C.A., Serramoglia, G., Caironi, G., Tosi, G., 1986. TAFS 94, 589–600 (These authors quote Kobzar, A.I., Ivanyuk, E.G., 1975. Russ. Cast. Prod. 7, 302–330).

Gorshov, A.A., 1964. Russ. Cast. Prod. 338–340.

Gould, G.C., Form, G.W., Wallace, J.F., 1960. TAFS 68, 258–267.

Gouwens, P.R., 1967. TAFS 75, 401–407.

Graham, A.L., Mizzi, B.A., Pedicini, L.J., 1987. TAFS 95, 343–350.

Granath, O., Wessen, M., Cao, H., 2008. Int. J. Cast Met. Res. 21 (5), 349–356.

Grandfield, J.F., Nguyen, T.T., Rohan, P., Nguyen, V., 2007. In: Jones, H. (Ed.), Proc. 5th Decennial Internat. Conf. on Solidification Processing. (SP07). Sheffield, UK, pp. 507–511.

Grassi, J.R., Campbell, J., Shaw, C.W., 6 November 2012. Integrated Quiescent Processing of Melts. US Patent 8303890.

Graverend, J-B.Le, Cormier, J., Kruch, S., Gallerneau, F., Mendez, J., 2012. Metall. Mater. Trans. A 43A (11), 3988–3997.

Gray, R.J., Perkins, A., Walker, B., 1977. In: Sheffield Conference "Solidification & Casting of Metals", pp. 300–305 (Published by Metals Soc. 1979 Book 192).

Greaves, R.H., 1936. ISI Spec. Rep. 15, 26–42.

Grebe, W., Grimm, G.P., 1967. Aluminium 43, 673–683.

Green, N.R., Campbell, J., 1994. Trans. AFS 102, 341–347.

Green, R.A., Heine, 1990. TAFS 98, 495–503.

Green, N.R., Tomkinson, A.M., Wright, T.C., Evans, J.P., Fuchs, U., Tschegg, S., 2005. In: Tiryakioglu, M., Crepeau, P.N. (Eds.), Shape Casting: The John Campbell Symposium. TMS.

Green, R.A., 1990. TAFS 98, 947–952.

Green, N.R., 1995. University of Birmingham, UK, unpublished work.

Greer, A.L., Bunn, A.M., Tronche, A., Evans, P.V., Bristow, D.J., 2000. Acta Mater. 48 (11), 2765–3026.

Grefhorst, C., Crepaz, R., 2005. Casting Plant and Technology International 1, pp. 28–35.

Griffiths, W.D., Lai, N.-W., 2007. Metall. Mater. Trans. 38A, 190–196.

Griffiths, W.D., Cox, M., Campbell, J., Scholl, G., 2007. Mater. Sci. Technol. 23 (2), 137–144.

Grill, A., Brimacombe, J.K., 1976. Ironmak. Steelmak. 3 (2), 76–86.

Grote, R.E., 1982. TAFS 90, 93–102.

Groteke, D.E., 1985. TAFS 93, 953–960.

Groteke, D.E., 2008. Personal communication to JC.

Grube, K.R., Kura, J.G., 1955. TAFS 63, 35–48.

Grunengerg, N., Escherle, E., Sturm, J.C., 1999. TAFS 107, 153–159.

Grupke, C.C., Hunter, L.J., Leonard, C., Nath, R.H., Weaver, G.J., 2011. Quality assurance with process compensated resonant testing. TAFS paper 11-036.

Gruzleski, J., Handiak, N., Campbell, H., Closset, B., 1986. TAFS 94, 147–154.

Gruznykh, I.V., Nekhendzi, Yu A., 1961. Russ. Cast. Prod. 6, 243–245.

Gunasegaram, D., Givord, M., O'Donnell, R., Finnin, B., December 2007. Foundry Trade J. 362–365.

Guo, J., Sheng, W., Su, Y., Ding, H., Jia, J., January 2001. Int. J. Cast Met. Res.

Gupta, N., Satyanarayana, K.G., 2006. The solidification processing of metal-matrix composites: the Rohatgi symposium. JOM: J. Miner. Met. Mater. Soc. ISSN: 1047–4838 58 (11), 92–94.

Gurland, J., 1966. Trans. AIME 236, 642–646.

Gurland, J., 1979. Application of the hall-petch relation to particle strengthening in spheroidized steels and aluminum-silicon alloys. In: Proceedings of the 2nd Conf. on Strength of Metals and Alloys, pp. 621–625.

Guthrie, R.I.L., 1989. Engineering in Process Metallurgy. Clarendon Press, Oxford, UK.

Guven, Y.F., Hunt, J.D., 1988. Cast. Met. 1 (2), 104–111.

Ha, M., Kim, W.-S., Moon, H.-K., Lee, B.-J., Lee, S., 2008. Metall. Mater. Trans. A 39 (5), 1087–1098.

HabibollahZadeh, A., Campbell, J., 2003. Paper 03-022 Trans. Am. Found. Soc. 10.

Habibollahzadeh, A., Campbell, J., 2004. Int. J. Cast Met. Res. 17 (3), 1–8.

Hadian, R., Emamy, M., Campbell, J., 2009. The modification of cast Al-Mg$_2$Si metal matrix composite by Li. Metall. Mater. Trans. B 40 (6), 822–832.

Haginoya, I., 1976. J. Jpn. Inst. Light Met. 26 (3), 131–138 and Ibid 1981 31, 769–774.

Hairy, P., Longa, Y., Laguerre, C., et al., June/July 2003. Fonderie Fondeur d'aujourd'hui (226), 20–30. ALFED catalogue C/27918.

Hall, F.R., Shippen, J., 1994. Eng. Failure Anal. 1 (3), 215–229.

Hallam, C.P., Griffiths, W.D., Butler, N.D., 2000. In: Sahm, P.R., Hansen, P.N., Conley, J.G. (Eds.), Modeling of Casting, Welding and Advanced Solidification Processes IX.

Halvaee, A., Campbell, J., 1997. TAFS 105, 35–46.

Hammitt, F.G., 1974. Proc. 1973 Symp. Copenhagen. In: Bjorno, L. (Ed.), Finite-Amplitude Wave Effects in Liquids. IPC Science and Technology Press, pp. 258–262. Paper 3.10.

Hammond, D.E., August 1989. Mod. Cast. 29–33.

Hangai, Y., Kato, H., Utsunomiya, T., Kitahara, S., 2010. Metall. Mater. Trans. A 41A, 1883–1886.

Hansen, P.N., Rasmussen, N.W., 1994. In: BCIRA International Conference. University of Warwick, UK.

Hansen, P.N., Sahm, P.R., 1988. In: Giamei, A.F., Abbaschian, G.J. (Eds.), Modelling of Casting and Welding Processes IV. The Mineral, Metals and Materials Society.

Hansen, P.N., Sahm, P.R., 1988. In: Giamei, A.F., Abbaschian, G.J. (Eds.), Modelling of Casting, Welding and Advanced Solidification Processes IV. The Mineral, Metals and Materials Society.

Hansen, P.N., 1975. Ph.D. thesis. Part 2, Technical University of Denmark, Copenhagen (Lists 283 publications on hot tearing.).

Harding, R.A., Campbell, J., Saunders, N.J., 1997. Inoculation of ductile iron. In: Beech, J., Jones, H. (Eds.), Solidification Processing 97 Sheffield Conference, 7–10 July, 1997.

Harding, R.A., June 2006. In: 67th World Foundry Congress, Harrogate, UK. Paper 17.

Harinath, U., Narayana, K.L., Roshan, H.M., 1979. TAFS 87, 231–236.

Harper, S., 1966. J. Inst. Met. 94, 70–72.

Harris, K.P., September 2004. Foundry Pract. (Foseco) 242, 01–07.

Harris, K.P., 2005a. In: Tiryakioglu, M., Creapeau, P.N. (Eds.), Shape Casting: The JC Symposium. TMS Annual Congress, pp. 433–442.

Harris, K.P., July 2005b. Mod. Cast. 36–38.

Harrison Steel 1999. Personal communication.

Harsem, O., Hartvig, T., Wintermark, H., 1968. In: 35th Internat. Found. Cong. Kyoto, Paper 15.

Hart, R.G., Berke, N.S., Townsend, H.E., 1984. Met. Trans. 15B, 393–395.

Hartmann, D., Stets, W., 2006. TAFS 114, 1055–1058.

Hartung, C., Ecob, C., Wilkinson, D., 2008. Cast. Plant Technol. 2, 18–21.

Hashemi, H.R., Ashoori, H., Davami, P., 2001. Mater. Sci. Technol. 17, 639–644.

Hassan, M.I., Al-Kindi, R., 2014. J. Met. 66 (9), 1603–1611.

Hassell, H.J., 1980. Br. Found. 73 (4), 95.

Haugland, E., Engh, T.A., 1997. In: Light Metals (TMS Conference), pp. 997–1005.

Hayes, K.D., Barlow, J.O., Stefanescu, D.M., Piwonka, T.S., 1998. TAFS 106, 769–776.

Hayes, J.S., Keyte, R., Prangnell, P.B., 2000. Mater. Sci. Technol. 16, 1259–1263.

Hebel, T.E., 1989. Heat Treating, Fairchild Business Publication, USA.

Hedjazi, D., Bennett, G.H.J., Kondic, V., 1975. Br. Found. 68, 305–309.

Hedjazi, Dj., Bennett, G.H.J., Kondic, V., December 1976. Met. Technol. 537–541.

Heine, R.W., Green, R.A., 1989. TAFS 97, 157–164.

Heine, R.W., Green, R.A., 1992. TAFS 100, 499–508.

Heine, H.J., Heine, R.W., 1968. TAFS 76, 470–484.

Heine, R.W., Loper, C.R., 1966a. TAFS 74, 274–280.

Heine, R.W., Loper, C.R., 1966b. TAFS 74, 421–428 and Br. Found. 1967 60, 347–353.

Heine, R.W., Rosenthal, P.C., 1955. Principles of Metal Casting. McGraw-Hill Book Co Inc + AFS, NY, USA.

Heine, R.W., Uicker, J.J., 1983. TAFS 91, 127–136.

Heine, R.W., Uicker, J.J., Gantenbein, D., 1984. TAFS 92, 135–150.

Heine, R.W., Green, R.A., Shih, T.S., 1990. TAFS 98, 245–252.

Heine, R.W., 1951. TAFS 59, 121–138.

Heine, R.W., 1968. TAFS 76, 463–469.

Heine, R.W., 1982. TAFS 90, 147–158.

Hendricks, M.J., Wang, P., Bijvoet, M., 2005. Br. Invest. Casters Assoc. Bull. 47, 4–5.

Henschel, C., Heine, R.W., Schumacher, J.S., 1966. TAFS 74, 357–364.

Hernandes-Reyes, B., 1989. TAFS 97, 529–538.

Herrara, A., Campbell, J., 1997. TAFS 105, 5–11.

Herrera, A., Kondic, V., 1979. In: Solidification and Casting of Metals. Conf. Sheffield, 1977. Metals Soc, London, pp. 460–465.

Hess, K., March 1974. AFS Cast Met. Res. J. 6–14.

Hetke, A., Gundlach, R.B., 1994. TAFS 102, 367–380.

Heusler, L., Feikus, F.J., Otte, M.O., 2001. TAFS 109, 215–223.

Heyn, E., 1914. J. Inst. Met. 12, 3.

Hill, H.N., Barker, R.S., Willey, L.A., 1960. Trans. Am. Soc. Met. 52, 657–671.

Hillert, M., Lindblom, Y., 1954. The growth of nodular graphite. J. Iron Steel Inst. 148, 388–390.

Hillert, M., Steinhauser, H., 1960. Jernkontorets Ann. 144, 520–522.

Hillert, M., 1968. In: Merchant, H.D. (Ed.), Recent Research on Cast Iron. Gordon and Breach, NY, pp. 101–127.

Hillis, J.E., November 2002. In: Internat. Conf. on SF6 and the Environment.

Hiratsuka, S., Niyama, E., Anzai, K., Hori, H., Kowata, T., 1966. In: 4th Asian Foundry Congress, pp. 525–531.

Hiratsuka, S., Niyama, E., Funakubo, T., Anzai, K., 1994. Trans. Jpn. Found. Soc. 13 (11), 18–24.

Hiratsuka, S., Niyama, E., Horie, H., Kowata, T., Anzai, K., Nakamura, M., 1998. Int. J. Cast Met. Res. 10 (4), 201–205.

Hirt, C.W., 2003. www.flow3d.com.

Ho, K., Pehlke, R.D., 1984. TAFS 92, 587–598.

Ho, P.S., Kwok, T., Nguyen, T., Nitta, C., Yip, S., 1985. Scr. Metall. 19 (8), 993–998.

Hoar, T.P., Atterton, D.V., 1950. J. Iron Steel Inst. 166, 1–7.

Hoar, T.P., Atterton, D.V., Houseman, D.H., 1953. J. Iron Steel Inst. 175, 19–29.

Hoar, T.P., Atterton, D.V., Houseman, D.H., 1956. Metallurgia 53, 21–25.

Hochgraf, F.G., 1976. Metallography 9, 167–176.

Hodaj, F., Durand, F., 1997. Acta Mater. 45, 2121.

Hodjat, Y., Mobley, C.E., 1984. TAFS 92, 319–321.

Hoff, O., Andersen, P., 1968. In: 35th Internat. Found. Cong., Kyoto, Paper 8.

Hoffman, E., Wolf, G., 2001. Giessereiforschung 53 (4), 131–151.

Hofmann, R., Wittmoser, A., 16 November 1971, US Patents 3619866 and 3620286

Hofmann, F., 1962. TAFS 70, 1–12.

Hofmann, F., 1966. AFS Cast Met. Res. J. 2 (4), 153–165.

Hoffmann, J., 2001. Foundry Trade J. 175 (3578), 32–34.

Holt, G.S., Mitchell, C.J., Simmons, R.E., 1989. BCIRA Report 1779 and BCIRA J. July 1989, 291–296.

Holtzer, M., March 1990. Foundryman 135–144.

Hong, C.P., Shen, H.F., Lee, S.M., 2000. Metall. Mater. Trans. B 31B, 297–305.

Hoover, W.R., 1991. In: Hansen, N., et al. (Eds.), Proc 12th Riso Internat. Symp. Materials Science: Metal Matrix Composites – Processing, Microstructure and Properties, pp. 387–392.

Hoult, F.H., 1979. Trans. AFS 87, 237–240 and 241–244.

Hoult, F.H., 1979a. TAFS 87, 237–240.

Hoult, F.H., 1979b. TAFS 87, 241–244.

Howe Sound Co – Superalloy Group USA, 1965. British Patent 1125124.

Hsu, W., Jolly, M.R., Campbell, J., 2006. Int. J. Cast Met. Res. 19 (1), 38–44.

Hsu, F.-Y., Jolly, M.R., Campbell, J., 2005. In: Tiryakioglu, M., Crepeau, P.N. (Eds.), Shape Casting: The John Campbell Symposium. TMS, San Francisco, CA, pp. 73–82.

Hu, D., Loretto, M.H., 2000. University of Birmingham, UK, personal communication.

Hu, Z.C., Zhang, E.L., Zeng, S.Y., 2008. Mater. Sci. Technol. 24 (11), 1304–1308.

Hua, C.H., Parlee, N.A.D., 1982. Met. Trans. 13B, 357–367.

Huang, J., Conley, J.G., 1998. TAFS 106, 265–270.

Huang, H., Lodhia, A.V., Berry, J.T., 1990. TAFS 98, 547–552.

Huang, L.W., Shu, W.J., Shih, T.S., 2000. TAFS 108, 547–560.

Huber, G., Brechet, Y., Pardoen, T., 2005. Acta Mat. 53, 2739–2749.

Hudak, D., Tiryakioğlu, M., 2009. On estimating percentiles of the Weibull distribution by the linear regression method. J. Mater. Sci. 44, 1959–1964.

Hughes, I.C.H., 1988. Metals Handbook. In: Casting, vol. 15. ASM, Ohio, USA, 647–666.

Hull, D.R., 1950. Casting of Brass and Bronze. ASM.

Hultgren, A., Phragmen, G., 1939. Trans. AIME 135, 133–244.

Hummer, R., 1988. Cast. Met. 1 (2), 62–68.

Hunsicker, H.Y., 1980. Met. Trans. 11A, 759–773.

Hunt, J.D., Thomas, R.W., 1997. In: Beech, J., Jones, H. (Eds.), Proc. 4th Decennial Internat. Conf. on Solidification Processing (SP97). University of Sheffield, UK, pp. 350–353.

Hunt, J.D., 1980. University of Oxford, personal communication.

Hurtuk, D.J., Tzavaras, A.A., 1975. TAFS 83, 423–428.

Hurum, F., 1952. TAFS 60, 834–848.

Hurum, F., 1965. TAFS 73, 53–64.

Hutchins, M., 2007. Foundry Trade J. 181 (3645), 195.

Hutchinson, H.P., Sutherland, D.S., 1965. Nature 206, 1036–1037.

IBF Technical Subcommittee TS 17, 1948. In: Symp. Internal Stresses in Metals and Alloys, London 1947. The Inst of Metals, pp. 179–188.

IBF Technical Subcommittee TS 18, 1949. Proc. IBF 42, A61–A77.

IBF Technical Subcommittee TS 32, 1952. Br. Found. 45, A48–A56. Foundry Trade J. 93, 471–477.

IBF Technical Subcommittee TS 32, 1956. Foundry Trade J. 101, 19–27.

IBF Technical Subcommittee TS 32, 1960. Br. Found. 53, 10–13 (but original work reported in Br. Found. 1952, 45, A48–A56).

IBF Technical Subcommittee TS 35, 1960. Br. Found. 53, 15–20.

IBF Technical Subcommittee TS 61, 1964. Br. Found. 57, 75–89, 504–508.

IBF Technical Subcommittee TS 71, 1969. Br. Found. 62, 179–196.

IBF Technical Subcommittee TS 71, 1971. Br. Found. 64, 364–379.

IBF Technical Subcommittee TS 71, 1976. Br. Found. 69, 53–60.

IBF Technical Subcommittee TS 71, 1979. Br. Found. 72, 46–52.

Iida, T., Guthrie, R.I.L., 1988. The Physical Properties of Liquid Metals. Clarendon Press, Oxford, p. 14.

Impey, S., Stephenson, D.J., Nicholls, J.R., 1993. In: Microscopy of Oxidation 2. Cambridge Conference, pp. 323–337.

International Magnesium Association, 2006. Alternatives to SF6 for Magnesium Melt Protection. EPA-430-R-06-007.

Ionescu, V., 2002. Mod. Cast. 92 (5), 21–23.

Isaac, J., Reddy, G.P., Sharman, G.K., 1985. TAFS 93, 29–34.

Isawa, T., Campbell, J., November 1994. Trans. Jpn. Found. Soc. 13, 38–49.

Isawa, T., 1993. The Control of the Initial Fall of Liquid Metal in Gravity Filled Casting systems (Ph.D. thesis). University of Birmingham Department of Metallurgy and Materials Science.

ISO Standard 8062, 1984. Castings – System of Dimensional Tolerances.

Isobe, T., Kubota, M., Kitaoka 5., 1978. J. Jpn. Found. Soc. 50 (11), 671–676.

Itamura, M., Yamamoto, N., Niyama, E., Anzai, K., 1995. In: Lee, Z.H., Hong, C.P., Kim, M.H. (Eds.), Proc 3rd Asian Foundry Congress, pp. 371–378.

Itamura, M., Murakami, K., Harada, T., Tanaka, M., Yamamoto, N., 2002. Int. J. Cast Met. Res. 15 (3), 167–172.

Itofugi, H., Uchikawa, H., 1990. TAFS 98, 429–448.

Itofugi, H., 1996. TAFS 104, 79–87.

Iyengar, R.K., Philbrook, W.O., 1972. Met. Trans. 3, 1823–1830.

Jackson, W.J., Wright, J.C., September 1977. Met. Technol. 425–433.

Jackson, K.A., Hunt, J.D., Uhlmann, D.R., Seward, T.P., 1966. Trans. AIME 236, 149–158.

Jackson, R.S., 1956. Foundry Trade J. 100, 487–493.

Jackson, W.J., April 1972. Iron Steel 163–172.

Jacob, S., Drouzy, M., 1974. In: Internat. Foundry Congress, 41 Liege, Belgium. Paper 6.

Jacobi, H., 1976. Arch. Eisenhuttenwes. 47, 441–446.

Jacobs, M.H., Law, T.J., Melford, D.A., Stowell, M.J., 1974. Met. Technol. 1 (11), 490–500.

Jakumeit, J., Laqua, R., Hecht, U., Goodheart, K., Peric, M., 2007. In: Jones, H. (Ed.), SP07 Proc. 5th Decennial Internat. Conf. Solidification Processing, pp. 292–296.

Jaquet, J.C., 1988. In: 8th Colloque Europeen de la Fonderie des Metaux Non Ferreux du CAEF.

Jaradeh, M.M., Carlberg, T., 2011. Metall. Mater. Trans. B 42B, 121–132.

Jay, R., Cibula, A., 1956. Proc. Inst. Br. Found. 49, A126–A140.

Jayatilaka, A. de S., Trustrum, K., 1977. J. Mater. Sci. 12, 1426.

Jeancolas, M., Devaux, H., 1969. Fonderie 285, 487–499.

Jeancolas, M., Cohen de Lara, G., Hanf, H., 1962. TAFS 70, 503–512.

Jeancolas, M., Devaux, H., Graham, G., 1971. Br. Found. 64, 141–154.

Jelm, C.R., Herres, S.A., 1946. Trans. Am. Found. Assoc. 54, 241–251.

Jennings, J.M., Griffin, J.A., Bates, C.E., 2001. TAFS 109, 177–186.

Ji, S., Roberts, K., Fan, Z., February 2002. Mater. Sci. Technol. 18, 193–197.

Jian, X., Xu, C., Meek, T., Han, Q., 2005. TAFS 113, 131–138.

Jiang, H., Bowen, P., Kntt, J.F., 1999. J. Mater. Sci. 34, 719–725.

Jianzhong, L., 1989. TAFS 97, 31–34.

Jirsa, J., October 1982. Foundry Trade J. 7, 520–527.

Jiyang, Z., Schmidt, W., Engler, S., 1990. TAFS 98, 783–786.

Jo, C.-Y., Joo, D.-W., Kim, I.-B., 2001. Mater. Sci. Technol. 17, 1191–1196.

Johnson, W.H., Baker, W.O., 1948. TAFS 56, 389–397.

Johnson, S.B., Loper, C.R., 1969. TAFS 77, 360–367.

Johnson, A.S., Nohr, C., 1970. TAFS 78, 194–207.

Johnson, R.A., Orlov, A.N., 1986. Physics of Radiation Effects in Crystals. Elsevier, North Holland.

Johnson, W.C., Smart, H.B., 1979. In: "Solidification and Casting of Metals" Sheffield Conf. 1977. Metals Soc. Publication 192, pp. 125–130.

Johnson, W.H., Bishop, H.F., Pellini, W.S., 1954. Foundry 102–107 and 271–272.

Johnson, T.V., Kind, H.C., Wallace, J.F., Nieh, C.V., Kim, H.J., 1989. TAFS 97, 879–886.

Jolly, M.J., Lo, H.S.H., Turan, M., Campbell, J., Yang, X., 2000. In: Sahm, P.R., Hansen, P.N., Conley, J.G. (Eds.), Modeling of Casting, Welding and Advanced Solidification Processing IX, Aachen.

Jolly, M.J., 2008. Engineering School, University of Birmingham, UK, personal communication.

Jones, D.R., Grim, R.E., 1959. TAFS 67, 397–400.

Jones, C.A., Fisher, J.C., Bates, C.E., 1974. TAFS 82, 547–559.

Jones, S.G., April 1948. Am. Found. 139.

Jones, S., June 2005. Foundry Trade J. 179, 156–157.

Jones, S., October 2006. Foundry Trade J. 180 (3638), 267–270.

Jones, S., April 2006. INCAST 18–21.

Joo, S.-H., Jung, J., Chun, M.S., Moon, C.H., Lee, S., Kim, H.S., August 2014. Metall. Mater. Trans. A 45A, 4002–4011.

Jorstad, J.L., 1971. TAFS 79, 85–90.

Jorstad, J.L., 1996. TAFS 104, 669–671.

Kahl, W., Fromm, E., 1984. Aluminium 60 (9), E581–E586.

Kahl, W., Fromm, E., 1985. Met. Trans. B 16B (3), 47–51.

Kahn, P.R., Su, W.M., Kim, H.S., Kang, J.W., Wallace, J.F., 1987. TAFS 95, 105–116.

Kaiser, W.D., Groenveld, T.P., 1975. In: 8th Internat. Die Casting Congress, Detroit, Michigan, USA, pp. 1–9. Paper Number G-T75-084.

Kaiser, G., 1966. Ber. Bunsenges. Phys. Chem. 70 (6), 635–639.

Kallbom, R., Hamberg, K., Wessen, M., Bjorkegren, L.-E., 2005. Mater. Sci. Eng. A A413–A414, 346–351.

Kallbom, R., Hamberg, K., Bjorkegren, L.-E., 2006. In: World Foundry Congress, Harrogate, UK, paper 184/1-10.

Kang, X., Li, D.-Z., Xia, L., Campbell, J., Li, Y.Y., 2005. In: Tiryakioglu, M., Crepeau, P.N. (Eds.), Shape Casting: The John Campbell Symposium. TMS, pp. 377–384.

Karsay, S.I., 1971. Ductile Iron II; Engineering Design Properties Applications. Quebec Iron & Titanium (QIT) Corporation, Canada.

Karsay, S.I., 1980. Ductile Iron: The State of the Art 1980. QIT-Fer et Titane Inc, Canada.

Karsay, S.I., 1985. Ductile Iron Production Practices. AFS.

Karsay, S.I., 1992. Ductile Iron Production Practices; the State of the Art 1992. QIT Fer et Titane Inc., Canada and "Ductile Iron; the essentials of gating and risering system design" revised 2000 published by Rio Tinto Iron & Titanium Inc.

Kasch, F.E., Mikelonis, P.J., 1969. TAFS 77, 77–89.

Kaspersma, J.H., Shay, R.H., 1982. Met. Trans. 13B, 267–273.

Katashima, S., Tashima, S., Yang, R.-S., 1989. TAFS 97, 545–552.

Katgerman, L., 1982. J. Met. 34 (2), 46–49.

Kato, E., Metall. Mater. Trans. A 30A, 2449–2453.

Kay, J.M., Nedderman, R.M., 1974. An Introduction to Fluid Mechanics and Heat Transfer, third ed. CUP. pp. 115–119.

Kearney, A.L., Raffin, J., 1987. Hot Tear Control Handbook for Aluminium Foundrymen and Casting Designers. American Foundrymen's Soc., Des Plaines Illinois, USA.

Khalili, A., Kromp, 1991. J. Mater. Sci. 26, 6741–6752.

Khan, P.R., Su, W.M., Kim, H.S., Kang, J.W., Wallace, J.F., 1987. TAFS 95, 105–116.

Khorasani, A.N., 1995. TAFS 103, 515–519 and Mod. Cast. 1996 86, 36–38.

Kiessling, R., 1987. Non-metallic Inclusions in Steel. The Metals Society.

Kihlstadius, D., 1988. ASM Metals Handbook. In: Casting, vol. 15, 273–274.

Kilshaw, J.A., 1963. BCIRA J. 11, 767.

Kilshaw, J.A., 1964. BCIRA J. 12, 14.

Kim, S.B., Hong, C.P., 1995. In: Modeling of Casting, Welding and Advanced Solidification Processes VII. TMS, pp. 155–162.

Kim, M.H., Loper, C.R., Kang, C.S., 1985. TAFS 93, 463–474.

Kim, M.H., Moon, J.T., Kang, C.S., Loper, C.R., 1993. TAFS 101, 991–998.

Kim, D., Han, K., Lee, B., Han, I., Park, J.H., Lee, C., 2014. Metall. Mater. Trans. A 45A, 2046–2054.

Kirner, J.F., Anewalt, M.R., Karwacki, E.J., Cabrera, A.L., 1988. Met. Trans. 19A, 3045–3055.

Kita, K., 1979. AFS Int. Cast Met. J. 4 (4), 35–40.

Kitaoka, S., 2001. In: Conference: Light Metals (Metaux Legers): International Symposium on Light Metals as Held at the 40th Annual Conference of Metallurgists of CIM (COM 2001), Toronto, Ontario, Canada, 26–29 August 2001. Canadian Institute of Mining, Metallurgy and Petroleum, pp. 13–24. Xerox Tower Suite 1210 3400 de Maisonneuve Blvd. W, Montreal, PQ, Quebec H3Z 3B8, Canada, 2001.

Klemp, T., 1989. TAFS 97, 1009–1024.

Knott, J.F., Elliott, D., 1979. Worked Examples in Fracture Mechanics. Inst. Metals Monograph 4, Inst. Metals, London.

Kokai, K., 28 March 1985. Japan patent application by Toyota 60 54244.

Kolorz, A., Lohborg, K., 1963. In: 30th Internat. Found. Congress, pp. 225–246.

Kolsgaard, A., Brusethaug, S., 1994. Mater. Sci. Technol. 10 (6), 545–551.

Kondic, V., 1959. Foundry 87, 79–83.

Kono, R., Miura, T., 1975. Br. Found. 69, 70–78.

Koster, W., Goebring, K., 1941. Giesserei 28 (26), 521.

Kotschi, T.P., Kleist, O.F., 1979. AFS Int. Cast. Met. J. 4 (3), 29–38.

Kotschi, R.M., Loper, C.R., 1974. TAFS 82, 535–542.

Kotschi, R.M., Loper, C.R., 1975. TAFS 83, 173–184.

Kotsyubinskii, O. Yu, Gerchikov, A.M., Uteshev, R.A., Novikov, M.I., 1961. Russ. Cast. Prod. 8, 365–368.

Kotsyubinskii, O Yu, Oberman, Ya I., Gerchikov, A.M., 1962. Russ. Cast. Prod. 4, 190–191.

Kotsyubinskii, O. Yu, Oberman, Ya I., Gini, E.Gh, 1968. Russ. Cast. Prod. 4, 171–172.

Kotsyubinskii, O. Yu, 1961–1962. Russ. Cast. Prod. 269–272.

Kotsyubinskii, O. Yu, 1963. In: 30th Internat. Found. Cong, pp. 475–487.

Kotzin, E.L., 1981. Metalcasting and Molding Processes. American Foundrymen's Soc, Des Plaines, Illinois, USA.

Kraft, G., 1978. Metall. ISSN: 0075-2819 32 (6), 560–562.

Krinitsky, A.I., 1953. TAFS 61, 399–410.

Kron, J., Bellet, M., Ludwig, A., Pustal, B., Wendt, J., Fredriksson, H., 2004. Int. J. Cast Met. Res. 17 (5), 295–310.

Kruse, B.L., Richards, V.L., Jackson, P.D., 2006. TAFS 114, 783–795.

Kubo, K., Pehlke, R.D., 1985. Met. Trans. 16B, 359–366.

Kubo, K., Pehlke, R.D., 1986. Met. Trans. 17B, 903–911.

Kujanpaa, V.P., Moisio, T.J.I., 1980. In: Solidification Technology in the Foundry and Cast House, Warwick Conf, pp. 372–375 (Metals Soc. 1983).

Kulas, M.-A., Green, W.P., Taleff, E.M., Krajewski, P.E., McNelly, T.R., 2006. Metall. Mater. Trans. A 37A, 645–655.

Kunes, J., Chaloupka, L., Trkovsky, V., Schneller, J., Zuzanak, A., 1990. TAFS 98, 559–563.

Kuyucak, S., 2002. TAFS, 110.

Kuyucak, S., February 2008. In: 68th World Foundry Congress, pp. 483–487.

Lagowski, B., Meier, J.W., 1964. TAFS 72, 561–574.

Lagowski, B., 1967. TAFS 75, 229–256.

Lagowski, B., 1979. TAFS 87, 387–390.

Lai, N.-W., Griffiths, W.D., Campbell, J., 2003. In: Stefanescu, D.M., Warren, J.A., Jolly, M.R., Krane, M.J.M. (Eds.), Modeling of Casting, Welding and Solidification Processes-X, pp. 415–422.

Laid, E., 16 February 1978. US patent application number 878309.

Lalpoor, M., Eskin, D.G., Katgerman, L., 2009. Metall. Mater. Trans. A 40A (13), 3304–3313.

Lalpoor, M., Eskin, D.G., Katgerman, L., 2010. Metall. Mater. Trans. A 41A (9), 2425–2434.

Lane, A.M., Stefanescu, D.M., Piwonka, T.S., Pattabhi, R., October 1969. Mod. Cast. 54–55.

Lang, G., 1972. Aluminium 48 (10), 664–672.

Lansdow, P., 1997. Personal communication regarding casting plant from KWC, Unterkulm, Switzerland, seen at Maynal Castings limited, Wolverhampton, UK.

Larranaga, P., Asenjo, I., Sertucha, J., Suarez, R., Ferrer, I., Lacaze, J., 2009. Trans. Met. Mater. A 40A, 654–661.

Lashkari, O., Ghomashchi, R., April 2007. Mater. Sci. Eng. A 454–455, 30–36.

Laslaz, G., Laty, P., 1991. TAFS 99, 83–90.

Latimer, K.G., Read, P.J., 1976. Br. Found. 69, 44–52.

LaVelle, D.L., 1962. TAFS 70, 641–647.

Lawrence, M., February 1990. Mod. Cast. 51–53.

Ledebur, A., 1882. Stahl Eisen 2, 591.

Leduc, L., Nadarajah, T., Sellars, C.M., July 1980. Met. Technol. 269–273.

Lee, J.-K., Kim, S.K., March 2007. Mater. Sci. Eng. A 449–451, 680–683.

Lee, R.S., 1987. TAFS 95, 875–882.

Lees, D.C.G., 1946. J. Inst. Met. 72, 343–364.

Lerner, Y., Aubrey, L.S., 2000. TAFS 108, 219–226.

Lerner, S.Y., Kouznetsov, V.E., May 2004. Mod. Cast. 37–41.

Leth-Olsen, H., Nisancioglu, K., 1998. Corros. Sci. 40, 1179–1194 and 1194–1214.

Lett, R.L., Felicelli, S.D., Berry, J.T., Cuesta, R., Losua, D., 2009. TAFS 8, paper 09-043.

Levelink, H.G., van den Berg, H., 1962. TAFS 70, 152–163.

Levelink, H.G., van den Berg, H., 1968. TAFS 76, 241–251.

Levelink, H.O., van den Berg, H., 1971. TAFS 79, 421–432.

Levelink, H.O., Julien, F.P.M.A., June 1973. AFS Cast Met. Res. J. 56–63.

Levelink, H.O., 1972. TAFS 80, 359–368.

Leverant, G.R., Gell, M., 1969. TAIME 245, 1157–1172.

Levinson, D.W., Murphy, A.H., Rostoker, W., 1955. TAFS 63, 683–686.

Lewis, G.M., 1961. Proc. Phys. Soc. (London) 71, 133.

Lewis, R., Ransing, R., 1998. Metall. Mater. Trans. B 29B (2), 437–448.

Li, Y.X., Liu, B.C., Loper, C.R., 1990. TAFS 98, 483–488.

Li, Y., Jolly, M.R., Campbell, J., 1998. In: Thomas, B.G., Beckermann, C. (Eds.), Modeling of Casting, Welding and Advanced Solidification Processes VIII. The Minerals, Metals and Materials Soc., pp. 1241–1253.

Li, D.Z., Campbell, J., Li, Y.Y., 2004. J. Process. Technol. 148 (3), 310–316.

Li, J., Wang, B., Li, J., July 2014. ICASP 4: The 4th Internat. Conf. on Advances in Solidification Processes, Windsor, UK, pp. 8–11.

Li, Q.L., Xia, T.D., Lan, Y.F., Li, P.F., 2014. Mater. Sci. Technol. 30 (7), 835–841.

Liass, A.M., Borsuk, 18 December 1962. French Patent 1342529.

Liass, A.M., 1968. Foundry Trade J. 124, 3–10.

Liddiard, E.A.G., Baker, W.A., 1945. TAFS 53, 54–65.

Lin, H.J., Hwang, W.-S., 1988. TAFS 96, 447–458.

Lin, D.-Y., Shih, M.C., 2003. Int. J. Cast Met. Res. 16 (6), 537–540.

Lin, J., Sharif, M.A.R., Hill, J.L., 1999. Alum. Trans. 1 (1), 72–78.

Lind, M., Holappa, L., 2010. Metall. Mater. Trans. B 41B, 359–366.

Ling, Y., Mampaey, F., Degrieck, J., Wettinck, E., 2000. In: Sahm, P.R., Hansen, P.N., Conley, J.G. (Eds.), Modeling of Casting, Welding and Advanced Solidification Processes IX, pp. 357–364.

Liu, S., Loper, C.R., 1990. TAFS 98, 385–394.

Liu, P.C., Loper, C.R., Kimura, T., Park, H.K., 1980. TAFS 88, 97–118.

Liu, P.C., Li, C.L., Wu, D.H., Loper, C.R., 1983. TAFS 91, 119–126.

Liu, S.L., Loper, C.R., Witter, T.H., 1992. TAFS 100, 899–906.

Liu, F., Zhao, D.W., Yang, G.C., 2001. Metall. Mater. Trans. B 32B, 449–460.

Liu, L., Samuel, A.M., Samuel, F.H., Dowty, H.W., Valtierra, S., 2002. The role of Sr oxide on porosity. Trans. AFS 110, 449–462.

Liu, L., Samuel, A.M., Samuel, F.H., Doty, H.W., Valtierra, S., 2003a. The role of Fe in Sr-modified 319 and 356. J. Mater. Sci. 38, 1255–1267.

Liu, L., Samuel, A.M., Samuel, F.H., Doty, H.W., Valtierra, S., 2003b. Int. J. Cast Met. Res. 16 (4), 397–408.

Liu, Z., Wang, F., Qiu, D., Taylor, J.A., Zhang, M., September 2013. Metall. Mater. Trans. A 44A, 4025–4030.

Livingston, H.K., Swingley, C.S., 1971. Surf. Sci. 24, 625–634.

Llewelyn, G., Ball, J.I., 1962. GB Patent Application, Complete Published 1965, 987060.

Lo, H., Campbell, J., 2000. In: Sahm, P.R., Hansen, P.N., Conley, J.G. (Eds.), Modeling of Casting, Welding and Advanced Solidification Processes IX, pp. 373–380.

Locke, C., Ashbrook, R.L., 1950. TAFS 58, 584–594.

Locke, C., Ashbrook, R.L., 1972. TAFS 80, 91–104.

Locke, C., Berger, M.J., 1951. Steel Founders Soc. America Research Report 25 "The Flow of Steel in Sand Molds Part II".

Longden, E., 1931–32. Proc. IBF 25, 95–145.

Longden, E., 1939–40. Proc. IBF 33, 77–107.

Longden, E., 1947–48. Proc. IBF 41, A152–A165.

Longden, E., 1948. TAFS 56, 36–56.

Loper, C.R., Fang, K., 2008. TAFS 8, paper 08-065 (05).

Loper, C.R., Heine, R.W., 1961. TAFS 69, 583–600.

Loper, C.R., Heine, R.W., 1968. TAFS 76, 547–554.

Loper, C.R., Kotschi, R.M., 1974. TAFS 82, 279–284.

Loper, C.R., LeMahieu, D.L., 1971. TAFS 79, 483–492.

Loper, C.R., Miskinis, 1985. TAFS 93, 545–560.

Loper, C.R., Newby, M.R., 1994. TAFS 102, 897–901.

Loper, C.R., Saig, A.G., 1976. TAFS 84, 765–768.

Loper, C.R., Javaid, A., Hitchings, J.R., 1996. TAFS 104, 57–65.

Loper, C.R., 1992. TAFS 100, 533–538.

Loper, C.R., 1999. TAFS 107, 523–528.

Low, J.R., 1969. Trans. AIME 245, 2481–2494.

Lu, Y.H., Chen, Z.R., Zhu, X.F., Shoji, T., 2014. Mater. Sci. Technol. 30 (15), 1944–1950.

Lubulin, I., Christensen, R.J., 1960. TAFS 68, 539–550.

Ludwig, T.H., Schaffer, P.L., Arnberg, L., December 2013. Metall. Mater. Trans. 44A, 5796–5805.

Lukens, M.C., Hou, T.X., Pehlke, R.D., 1990. TAFS 98, 63–70.

Lukens, M.C., Hou, T.X., Purvis, A.L., Pehlke, R.D., 1991. TAFS 99, 445–449.

Lumley, R.N., Sercombe, T.B., Schaffer, G.B., 1999. Metall. Mater. Trans. 30A, 457–468.

Lumley, R.N., O'Donnell, R.G., Gunasegaram, D.R., Givord, M., 2007. Metall. Mater. Trans. A 38A, 2565–2574.

Lumley, R., May 2013. Foundry Trade J. 106–107.

Lynch, R.F., Olley, R.P., Gallagher, P.C.J., March 1977. AFS Int. Cast Met. J. 61–86.

Ma, Z., Samuel, A.M., Samuel, F.H., Doty, H.W., Valtierra, S., 2008. Mater. Sci. Eng. A 490, 36–51.

Mae, Y., Sakonooka, A., 1985. Met. Soc. AIME, TMS, Paper 0197 7 (US Patent 4808243, 1987).

Maeda, Y., Nomura, H., Otsuka, Y., Tomishige, H., Mori, Y., 2002. Int. J. Cast Met. Res. 15, 441–444.

Maidment, L.J., Walter, S., Raul, G., 1984. In: 8th Heat Treating Conf., Detroit.

Majumdar, I., Raychaudhuri, B.C., 1981. Int. J. Heat Mass Transf. 24 (7), 1089–1095.

Malzer, G., Hayes, R.W., Mack, T., Eggler, G., 2007. Metall. Mater. Trans. A 38A (2), 314–327.

Mampaey, F., Beghyn, K., 2006. TAFS 114, 637–656.

Mampaey, F., Xu, Z.A., 1997. TAFS 105, 95–103.

Mampaey, F., Xu, Z.A., 1999. TAFS 107, 529–536.

Mampaey, F., Habets, D., Plessers, J., Seutens, F., 2008. Int. Found. Res./Geissereiforschung 60 (1), 2–19.

Mampaey, F., 1999. TAFS 107, 425–432.

Mampaey, F., 2000. TAFS 108, 11–17.

Mandal, B.P., 2000. Indian Foundry J. 46 (4), 25–28.

Mansfield, T.L., 1984. In: Proc. Conf. "Light Metals". Met. Soc. AIME, pp. 1305–1327.

Manzari, M.T., Lewis, R.W., Gethin, D.T., 2000. Optimum design of chills in sand casting process. In: Proc. IMECE 2000 International Mech. Eng. Congr., Florida, USA. DETC98/DAC-1234 ASME, p. 8.

Marck, C.T., Keskar, A.R., 1968. TAFS 76, 29–43.

Marin, T., Utigard, T., 2010. Metall. Mater. Trans. B 41B (3), 535–542.

Mariotto, C.L., 1994. TAFS 102, 567–573.

Martin, L.C.B., Keller, C.T., Shivkumar, S., 1992. In: AFS 3rd Internat. Conf. on Molten Aluminum Processing. Orlando, Florida, pp. 79–91.

Mashl, S.J., 2008. ASM Handbook. In: Casting, vol. 15, pp. 408–416.

Maske, F., Piwowarsky, E., March 1929. Foundry Trade J. 28, 233–243.

Masuda, S., Toda, H., Aoyama, S., Orii, S., Ueda, S., Kobayashi, M., 2009. J. Jpn. Found. Eng. Soc. 81, 312–322.

Mathier, V., Grasso, P.-D., Rappaz, M., 2008. Metall. Mater. Trans. 39A, 1399–1409.

Mathier, V., Vernede, S., Jarry, P., Rappaz, M., 2009. Metall. Mater. Trans. A 40A (4), 943–957.

Matsubara, Y., Suga, S., Trojan, P.F., Flinn, R.A., 1972. TAFS 80, 37–44.

Matsuda, M., Ohmi, M., 1981. AFS Int. Cast. Met. J. 6 (4), 18–27.

Mazed, S., Campbell, J., 1992. Univ. Birmingham, UK, unpublished work.

Mbuya, T.O., Oduari, M.F., Rading, G.O., Wekesa, M.S., 2006. Int. J. Cast Met. Res. 19 (6), 357–360.

McCartney, D.G., 1989. Int. Mater. Rev. 34 (5), 247–260.

McClain, S.T., McClain, A.S., Berry, J.T., 2001. TAFS 109, 75–86.

McDavid, R.M., Dantzig, J., 1998. Metall. Mater. Trans. 29B, 679–690.

McDonald, R.J., Hunt, J.D., 1969. Trans. AIME 245, 1993–1997.

McDonald, R.I., Hunt, J.D., 1970. Trans. AIME 1, 1787–1788.

McGrath, C., Fischer, R.B., 1973. TAFS 81, 603–620.

McKim, P.E., Livingstone, K.E., 1977. TAFS 85, 491–498.

McParland, A.J., 1987. In: Sheffield Conf. "Solidification Processing" Institute of Metals, London, 1988, pp. 323–326.

Medvedev Ya, I., Kuzukov, V.K., 1966. Russ. Cast. Prod. 263–266.

Mehta, R., 1 April 2008. Mater. World.

Mejia, I., Altamirano, G., Bedolla-Jacuinde, A., Cabrera, J.M., November 2013. Metall. Mater. Trans. A 44A, 5165–5176.

Melendez, A.J., Carlson, K.D., Beckermann, C., 2010. Int. J. Cast Met. Res. 23 (5), 278–288.

Mendelson, S.J., 1962. Appl. Phys. 33 (7), 2182–2186.

Mertz, J.M., Heine, R.W., 1973. TAFS 81, 493–495.

Merz, R., Marincek, B., 1954. In: 21st International Foundry Congress. Paper 44, pp. 1–7.

Metcalf, G.J., 1945. J. Inst. Met. 1029, 487–500.

Meyers, C.W., 1986. TAFS 94, 511–518.

Mi, J., Harding, R.A., Campbell, J., 2002. Int. J. Cast Met. Res. 14 (6), 325–334.

Micks, F.W., Zabek, V.J., 1973. TAFS 81, 38–42.

Middleton, 1. M., Canwood, B., 1967. Br. Found. 60, 494–503.

Middleton, J.M., 1953. TAFS 61, 167–183.

Middleton, J.M., 1964. Br. Found. 57, 1–19.

Middleton, J.M., 1965. Br. Found. 58, 13–24.

Middleton, J.M., 1970. Br. Found. 64, 207–223.

Midea, A.C., 2001. TAFS 109, 41–50 and Foundryman March 2003 60–63.

Midson, S.P., September 2008. Die Cast. Eng. Semi solid casting of aluminum alloys - an update.

Miguelucci, E.W., 1985. TAFS 93, 913–916.

Mihaichuk, W., February 1986. Mod. Cast. 36–38 and (March) 33–35.

Mikkola, P.H., Heine, R.W., 1970. TAFS 78, 265–268.

Miles, G.W., 1956. Proc. Inst. Br. Found. 49, A201–A210.

Miller, G.F., 1967. Foundry 95, 104–107.

Minakawa, S., Samarasekera, I.V., Weinberg, F., 1985. Met. Trans. 16B, 823–829.

Minetola, P., Iuliano, L., Argentieri, G., 2012. Int. J. Cast Met. Res. 25 (1), 38–46.

Mintz, B., Yue, S., Jonas, J.J., 1991. Int. Mater. Rev. 36 (5), 187–217.

Mirak, A.R., Divandari, M., Boutorabi, S.M.A., Campbell, J., 2007. Int. J. Cast Met. Res. 20 (4), 215–220.

Mirak, A.R., Divandari, M., Boutorabi, S.M.A., August 2010. Mater. Sci. Technol.

Miresmaeili, S.M., Campbell, J., Shabestari, S.G., Boutorabi, S.M.A., 2005. Metall. Mater. Trans. A 36A, 2341–2349.

Miresmaeili, S.M., 2006. University of Birmingham, UK, unpublished work.

Miyagi, Y., Hino, M., Tsuda, O., 1985. "R&D" Kobe Steel Engineering Reports 1983 33 July (3). Also published in Kobelco Technical Bulletin, 1076 (1985).

Mizoguchi, T., Perepezko, J.H., Loper, C.R., 1997. TAFS 105, 89–94 and Mater. Sci. Eng. A 1997 A226–A228, 813–817.

Mizumoto, M., Sasaki, S., Ohgai, T., Kagawa, A., 2008. Int. J. Cast Met. Res. 21 (1–4), 49–55.

Mizuno, K., Nylund, A., Olefjord, I., 1996. Mater. Sci. Technol. 12, 306–314.

Mohanty, P.S., Gruzleski, J.E., 1995. Acta Metall. Mater. 43, 2001–2012.

Mohla, P.P., Beech, J., 1968. Br. Found. 61, 453–460.

Molaroni, A., Pozzesi, V., 1963. In: 30th Internat. Found. Congress, pp. 145–161.

Molina, J.M., Voytovych, R., Louis, E., Eustathopoulos, N., 2007. Int. J. Adhes. Adhes. 27, 394–401.

Mollard, F.R., Davidson, N., 1978. TAFS 78, 479–486.

Momchilov, E., 1993. Institute for Metal Science, Bulgarian Academy of Sciences, Sofia J. Mater. Sci. Technol. 1 (1), 5–12.

Mondloch, P.A., Baker, D.W., Euvrard, L., 1987. TAFS 95, 385–392.

Monroe, R.W., Blair, M., 1995. Am. Found. Soc. Trans. 103, 633–640.

Morgan, A.D., 1966. Br. Found. 59, 186–204.

Morgan, P.C., 1989. Met. Mater. 5 (9), 518–520.

Morthland, T.E., Byrne, P.E., Tortorelli, D.A., Dantzig, J.A., 1995. Metall. Mater. Trans. B 26B, 871–885.

Moumeni, E., Stefanescu, D.M., Tiedje, N.S., Larranaga, P., Hattel, J.H., November 2013. Metall. Mater. Trans. 44A, 5134–5146.

Mountford, N.D.G., Calvert, R., 1959–1960. J. Inst. Met. 88, 121–127.

Mountford, N.D.G., Sommerville, I.D., 1993. Steel Technology International. London, UK, pp. 155–169.

Mountford, N.D.G., Sommerville, I.D., From, L.E., Lee, S., Sun, C., 1992/1993. In: A Measuring Device for Quality Control in Liquid metals. Scaninject Conf., Lulea, Sweden.

Mountford, N.D.G., Sommerville, I.D., Simionescu, A., Bai, C., 1997. TAFS 105, 939–946.

Mroz, S.S., Goodrich, G.M., 2006. TAFS 114, 493–505.

Mufti, N.A., Webster, P.D., Dean, T.A., 1995. Mater. Sci. Technol. 11 (8), 803–809.

Muhmond, H.M., Frederiksson, H., 2013. Metall. Mater. Trans. 44B, 283–298.

Mukherjee, M., Garcia-Moreno, F., Banhart, J., 2010. Metall. Mater. Trans. B 41B (3), 500–504.

Mulazimoglu, M.H., Handiak, N., Gruzleski, J.E., 1989. TAFS 97, 225–232.

Mulazimoglu, M.H., Tenekedjiev, N., Closset, B.M., Gruzleski, J.E., 1993. Cast. Met. 6 (1), 16–28.

Muller, W., Feikus, F.J., 1996. TAFS 104, 1111–1117.

Muller, F.C.G., 1887. Zeit. ver dent. Ingenieure 23, 493.

Munoz-Moreno, R., Perez-Prado, M.T., Llorca, J., Ruiz-Navas, E.M., Boehlert, C.J., April 2013. Metall. Mater. Trans. A 44A, 1887–1896.

Mutharasan, R., Apelian, D., Romanowski, C., 1981. J. Met. 83 (12), 12–18.

Mutharasan, R., Apelian, D., Ali, S., 1985. Met. Trans. 16B, 725–742.

Muthukumarasamy, S., Seshan, S., 1992. TAFS 100, 873–879.

Myllymaki, R., 1987. In: Young, W.B. (Ed.), Residual Stress in Design, Process and Materials Selection. ASM Conf., USA, pp. 137–141.

Nadell, R., Eskin, D., Katgerman, L., 2007. Mater. Sci. Technol. 23 (11), 1327–1335.

Nai, S.M.L., Gupta, M., 2002. Mater. Sci. Technol. 18, 633–641.

Nakae, H., Shin, H., 1999. Int. J. Cast Met. Res. 11 (5), 345–349.

Nakae, H., Takai, K., Okauchi, K., Koizumi, H., 1991. IMONO 63, 692–703.

Nakae, H., Koizumi, H., Takai, K., Okauchi, K., 1992. IMONO Either 64, 34–39.

Nakagawa, Y., Momose, A., 1967. Tetsu-to-Hagane 53, 1477–1508.

Naro, R.L., Pelfrey, R.L., 1983. TAFS 91, 365–376.

Naro, R.L., Tanaglio, R.O., 1977. TAFS 85, 65–74.

Naro, R.L., Wallace, J.F., 1967. TAFS 75, 741–758.

Naro, R.L., Wallace, J.F., 1992. TAFS 100, 797–820.

1992. TAFS 100, 797–820.

Naro, R.L., http://www.asi-alloys.com/DocsPDF/Naro.pdf, the first 2 paragraphs of p. 6.

Naro, R.L., 1974. TAFS 82, 257–266.

Naro, R.L., 2004. TAFS 112, 527–545.

Nath, R.H., Grupke, C.C., Leonard, C., Johnson, M.K., 2011. In: Crepeau, P., et al. (Eds.), Shape Casting the John Berry Honorary Symposium. TMS, San Diego.

Navisi, S., 2010. Mod. Cast. 100 (9), 35–38.

Nayal, G.El, Beech, J., 1987. In: 3rd Internat. Conf. Solidification Processing, Sheffield, pp. 384–387 (Inst Metals Publication 421).

Nayal, G.El, 1986. Mater. Sci. Technol. 2, 803.

Nazar, A.M.M., Cupini, N.L., Prates, M., Daview, G.J., 1979. Metall. Trans. B 10B, 203–210.

Neiswaag, H., Deen, H.J.J., 1990. In: 57 World Foundry Congress, Osaka, Japan.

Nereo, G.E., Polch, R.F., Flemings, M.C., 1965. TAFS 73, 1–13.

Neumeier, L.A., Betts, B.A., Crosby, R.L., 1976. TAFS 84, 437–448.

Newell, M., D'Souza, N., Green, N.R., 2009. Int. J. Cast Met. Res. 22 (1–4), 66–69.

Nguyen, T., Carrig, J.F., 1986. TAFS 94, 519–528.

Nguyen, T., Loose, G.de, Carrig, J., Nguyen, V., Cowley, B., 2006. TAFS 114, 695–706.

Ni, P., Jonsson, L.T.I., Ersson, M., Johsson, P.G., December 2014. Metall. Mater. Trans. B 45B, 2414–2424.

Nicholas, K.E.L., Roberts, R.W., 1961. BCIRA J. 9 (4), 519.

Nicholas, K.E.L., 1972. Br. Found. 65, 441–451.

Nikolai, M.F., 1996. TAFS 104, 1017–1029.

Nishida, Y., Droste, W., Engler, S., 1986. Met. Trans. 17B, 833–844.

Niu, J.P., Yang, K.N., Sun, X.F., Jin, T., Guan, H.R., Hu, Z.Q., 2002. Mater. Sci. Technol. 18 (9), 1041–1044.

Niu, J.P., Sun, X.F., Jin, T., Yang, K.N., Guan, H.R., Hu, Z.Q., 2003. Mater. Sci. Technol. 19 (4), 435–439.

Niyama, E., Ishikawa, M., 1966. In: 4th Asian Foundry Congress, pp. 513–523.

Niyama, E., Uchida, T., Morikawa, M., Saito, S., 1982. In: Internat. Found. Congress 49, Chicago, paper 10.

Noesen, S.J., Williams, H.A., 1966. In: 4th National Die Casting Congress, Cleveland, Ohio, paper number 801.

Noguchi, T., Kamota, S., Sato, T., Sakai, M., 1993. TAFS 101, 231–239.

Noguchi, T., Kano, J., Noguchi, K., Horikawa, N., Nakamura, T., 2001. Int. J. Cast Met. Res. 13, 363–371.

Noguchi, T., Horikawa, N., Nagate, H., Nakamura, T., Sato, K., 2005. Int. J. Cast Met. Res. 18 (4), 214–220.

Nolli, P., Cramb, A.W., 2008. Metall. Mater. Trans. 39B, 56–65.

Nordland, W.A., 1967. Trans. AIME 239, 2002–2004.

Nordlien, J.H., Davenport, A.J., Scamans, G.M., 2000. In: ASST 2000 (2nd Internat. Symp. Al Surface Science Technology) UMIST Manchester, UK. Alcan + Dera, pp. 107–112.

Nordlien, J.H., 1999. Alum. Extrus. 4 (4), 39–41.

Noreo, G.E., Polich, R.F., Flemings, M.C., 1965. TAFS 73, 1–13 and 28–33.

Northcott, L., 1941. J. Iron Steel Inst. 143, 49–91.

Novikov, I.I., Grushko, O.E., 1995. Mater. Sci. Technol. 11, 926–932.

Novikov, I.I., Portnoi, V.K., 1996. Russ. Cast. Prod. 4, 163–166.

Novikov, I.I., Novik, F.S., Indenbaum, G.V., 1966. Izv. Akad. Nauk. Met. 5, 107–110 (English translation in Russ. Metall. Min., 1966 5, 55–59).

Novikov, I.I., April 1962. Russ. Cast. Prod. 167–172.

Nyahumwa, C., Green, N.R., Campbell, J., 1998. TAFS 106, 215–223.

Nyahumwa, C., Green, N.R., Campbell, J., 2000. Metall. Mat. Trans. A 31A, 1–10.

Nyamekye, K., An, Y.-K., Bain, R., Cunningham, M., Askeland, D., Ramsay, C., 1994a. TAFS 102, 127–131.

Nyamekye, K., Wei, S., Askeland, D., Voigt, R.C., Pischel, R.P., Rasmussen, W., Ramsay, C., 1994b. TAFS 102, 869–876.

Nyichomba, B.B., Campbell, J., 1998. Int. J. Cast Met. Res. 11 (3), 163–186.

O'Hara, P., February 1990. Engineering 41–42.

Ocampo, C.M., Talavera, M., Lopez, H., 1999. Metall. Mater. Trans. A 30A, 611–620.

Oelsen, W., 1936. Stahl Eisen 56, 182.

Ohnaka, 2003. Mould Coat Doubles the Back-Pressure in Moulds.

Ohno, A., 1987. Solidification: The Separation Theory and its Practical Applications. Springer-Verlag.

Ohsasa, K.-I., Takahash, T., Kobori, K.J., 1988. Jpn. Inst. Met. 52 (10), 1006–1011 and Ohsasa, K.-I., Ohmi, T., Takahashi, T., 1988. Bull. Fac. Eng. Hokkaido Univ. 143 (in Japanese).

Okada, Y., Fujii, N., Goto, A., Morimoto, S., Yusuda, Y., 1982. TAFS 90, 135–146.

Okamoto, S., August 2013. Personal Communication to JC 01. Ehime University, Matsuyama, Japan.

Oki, S., 1970. IMONO J. Jpn. Found. Soc. 42 (11), 9. English Translation in Bulletin of the Faculty of Engineering, Yokohama National University, March 1969, vol. 18, pp. 127–136.

Omura, N., Tada, S., TMS Annual Meeting, 2012.

Oper, C.R., 1981. TAFS 89, 405–408.

Osborn, D.A., 1979. Br. Found. 72, 157–161.

Ostrom, T.R., Trojan, P.K., Flinn, R.A., 1974. TAFS 82, 519–524.

Ostrom, T.R., Trojan, P.K., Flinn, R.A., 1975. TAFS 83, 485–492.

Ostrom, T.R., Frasier, D.J., Trojan, P.K., Flinn, R.A., 1976. TAFS 84, 665–674.

Ostrom, T.R., Trojan, P.K., Flinn, R.A., 1981. TAFS 89, 731–736.

Ostrom, T.R., Biel, J., Wager, W., Flinn, R.A., Trojan, P.K., 1982. TAFS 90, 701–709.

Outlaw, R.A., Peterson, D.T., Schmidt, F.A., 1981. Met. Trans. 12A, 1809–1816.

Owen, M., 1966. Br. Found. 59, 415–421.

Owusu, Y.A., Draper, A.B., 1978. TAFS 86, 589–598.

Pakes, M., Wall, A., 1982. Zinc Development Assoc., UK.

Pan, E.N., Hu, J.F., 1996. In: 4th Asian Foundry Congress, pp. 396–405.

Pan, E.N., Hu, J.F., 1997. TAFS 105, 413–418.

Pan, E.N., Hsieh, M.W., Jang, S.S., Loper, C.R., 1989. TAFS 97, 379–414.

Panchanathan, V., Seshadri, M.R., Ramachandran, A., 1965. Br. Found. 58, 380–384.

Pandee, P., Gourley, G.M., Beyakov, S.A., Ozaki, R., Yasuda, H., Limmanaevichit, C., September 2014. Metall. Mater. Trans. A 45A, 4549–4560.

Papworth, A., Fox, P., 1998. Mater. Lett. 35, 202–206.

Paray, F., Kulunk, B., Gruzleski, J.E., 2000. Int. J. Cast Met. Res. 13, 147–159.

Parkes, T.W., Loper, C.R., 1969. TAFS 77, 90–96.

Parkes, W.B., 1952. TAFS 60, 23–37.

Pattabhi, R., Lane, A.M., Piwonka, T.S., 1996. Part III, TAFS 104, 1259–1264.

Patterson, W., Koppe, W., 1962. Giesserel 4, 225–249.

Patterson, W., Engler, S., Kupfer, R., 1967a. Giesserei-Foschung 19 (3), 151–160.

Pattyn, R., 1967. In: 34th Internat. Foundry Cong., Paris, paper 26.

Paxton D.M., Dudder G.J., Reynolds J., Charron W., Cleaver T., TMS Annual Meeting Light Metals Division Session III, 2006.

Pechiney Aluminium, 26 April 1977. British Patent, 1574321.

Pekguleryuz, M.O., Lin, S., Ozbakir, E., Temur, D., Aliravci, C., 2010. Int. J. Cast Met. Res. 23 (5), 310–320.

Pelleg, J., Heine, R.W., 1966. TAFS 74, 541–544.

Pellerier, M., Carpentier, M., April 1988. Hommes et Fonderie 184, 7–14. In French. Abstract English translation BCIRA 1989 Abstr. 27.

Pellini, W.S., 1952. Foundry 80, 124–133, 194, 196, 199.

Pellini, W.S., 1953. TAFS 61, 61–80 and 302–308.

Pell-Walpole, W.T., 1946. J. Inst. Met. 72, 19–30.

Pennors, A., Samuel, A.M., Samuel, F.H., Doty, H.W., 1998. TAFS 106, 251–264.

Perbet, D., 1988. Hommes Fonderie (Mars) 15. French plus English Translation.

Perets, S., Arbel, A., Ariely, Venkert, A., Schneck, R.Z., 2004. Mater. Sci. Technol. 20 (12), 1519–1524.

Peters, T.M., Twarog, D.L., 1992. TAFS 100, 1005–1023.

Peterson, W.M., Blanke, J.E., 1980. TAFS 88, 503–506.

Petro, A., Flinn, R.A., 1978. TAFS 86, 357–364.

Petrzela, L., 31 October 1968. Foundry Trade J. 693–696.

Pettersson, H., 1951. TAFS 59, 35–55.

Phillion, A.B., Vernede, S., Rappaz, M., Cockcroft, S.L., Lee, P.D., 2009. Int. J. Cast Met. Res. 22 (1–4), 1–5.

Pillai, R.M., Mallya, V.D., Panchanathan, V., 1976. TAFS 84, 615–620.

Pitcher, P.D., Forsyth, P.J.E., November 1982. The Influence of Microstructure on the Fatigue Properties of an Al Casting Alloy. Royal Aircraft Establishment Technical Report 82107.

Piwonka, T.S., Flemings, M.C., 1966. Trans. Met. Soc. AIME 236, 1157–1165.

Plyatskii, V.M., 1965. "Extrusion Casting" Primary Sources. New York, USA.

Poirier, D.R., Yeum, K., 1987. In: Solidification Processing Conference, Sheffield. Institute of Metals, London.

Poirier, D.R., Yeum, K., 1988. In: Light Metals Conf., USA, pp. 469–476.

Poirier, D.R., Yeum, K., Maples, A.L., 1987. Met. Trans. 18, 1979–1987.

Poirier, D.R., Sung, P.K., Felicelli, S.D., 2001. TAFS 105, 139–155.

Poirier, D.R., 1987. Met. Trans. 18B, 245–255.

Pokorny, M.G., Monroe, C.A., Beckermann, C., 2009. In: Campbell, J., Crepeau, P.N., Tiryakioglu, M. (Eds.), Shape Casting: The 3rd International Symposium. TMS.

Polich, R.F., Flemings, M.C., 1965. TAFS 73, 28–33.

Pollard, W.A., 1964. TAFS 72, 587–599.

Pollard, W.A., 1965. TAFS 73, 371–379.

Polmear, I.J., 2006. Light Alloys, fourth ed. Butterworth Heinemann.

Polodurov, N.N., 1965. Russ. Cast. Prod. 5, 209–210.

Pope, J.A., 1965. Br. Found. 58, 207–224.

Popel, G.I., Esin, O.A., 1956. Zh. Fiz. Khim 30, 1193.

Porter, L.F., Rosenthal, P.C., 1952. Trans. Am. Found. Soc. 60, 125–136.

Portevin, A., Bastien, P., 1934. J. Inst. Met. 54, 45–58.

Portevin, A., Bastien, P., 1936. Inst. Br. Found. In: 33rd Annual Conf. Glasgow, pp. 88–116.

Pouly, P., Wuilloud, E., 1997. In: Light Metals (TMS Conference), pp. 829–835.

Pozdniakov, A.V., Zolotorevskiy, V.S., 2014. Int. J. Cast Met. Res. 27 (4), 193–198.

Prakash, M., Cleary, P., Grandfield, J., 2009. J. Mater. Process. Technol. 209 (7), 3396–3407.

Prates, M., Biloni, H., 1972. Met. Trans. 3A, 1501–1510.

Prible, J., Havlicek, F., 1963. In: 30th Internat. Foundry Congress, pp. 394–410.

Prodham, A., Carpenter, M., Campbell, J., 1999. CIATF Technical Forum.

Puhakka, R., Campbell, J., June 2009. Mod. Cast. 27–29.

Puhakka, R., 2009. Personal communication.

Puhakka, R., June 2009. Mod. Cast. 27–29.

Puhakka, R., November 2010. Foundry Trade J. 184 (3679), 277–279.

Puhakka, R., 2011. Shape Casting; the John Berry Honorary Symposium. TMS Annual Meeting, San Diego, CA.

Puhakka, R., 2011. www.castdifferently.com.

Pulkonik, K.J., Lee, W.E., Rosenberg, R.A., 1967. TAFS 75, 38–41.

Pumphrey, W.I., Lyons, J.V., 1948. J. Inst. Met. 74, 439–455.

Pumphrey, W.I., Moore, D.C., 1949. J. Inst. Met. 75, 727–736.

Pumphrey, W.I., 1955. Researches into the Welding of Aluminium and its Alloys. Research Report 27. Aluminium Development Association, UK.

Purdom, P., March 1992. Met. Mater. 169.

Purvis, A.L., Hanslits, C.R., Diehm, R.S., 1994. TAFS 102, 637–644.

Qian, M., Graham, D., Zheng, L., St John, D.H., Frost, M.T., 2003. Mater. Sci. Technol. 19 (2), 156–162.

Qingchung, L., Kuiying, C., Chi, L., Songyan, Z., 1991. TAFS 99, 245–253.

Rabinovich, A., March 1969. AFS Cast Met. Res. J. 19–24.

Ragab, Kh A., Bournane, M., Samuel, A.M., Al-Ahmari, A.M.A., Samuel, F.H., Doty, H.W., 2013. Mater. Sci. Technol. 29 (4), 412–425.

Ragone, D.V., Adams, C.M., Taylor, H.F., 1956. TAFS 64, 640–652 and 653–657.

Rahmani, Kh, Nategh, S., 2010. Metall. Mater. Trans. A 41A (1), 125–137.

Raiszadeh, R., Griffiths, W.D., 2008. Metall. Mater. Trans. 39B, 298–303.

Raiszadeh, R., Griffiths, W.D., 2011. Metall. Mater. Trans. 43B, 133–143.

Raiszadeh, R., Nateghian, M., Doostmohammadi, H., December. Metall. Mater. Trans. B 43B, 1540–1549.

Raiszadeh, R., Amirinejhad, S., Doostmohammadi, H., 2013. Int. J. Cast Met. Res. 26 (6), 330–338.

Ramesh, G., Prabhu, K.N., August 2013. Metall. Mater. Trans. B 44B, 797–799.

Ramrattan, S.N., Guichelaar, P.J., Palukunnu, A., Tieder, R., 1996. TAFS 104, 877–886.

Ramseyer, J.C., Gabathuler, J.P., Feurer, U., 1982. Aluminium 58 (10), E192–E194, 581–585.

Ransley, C.E., Neufeld, H., 1948. J. Inst. Met. 74, 599–620.

Ransing, R., Pao, W.K.S., Lin, C., Snood, M.P., Lewis, R.W., 2004. Int. J. Cast Met. Res. 18, 1–12.

Rao, Y.K., Lee, H.G., 1984. Met. Trans. 15B, 396–400.

Rao, G.V.K., Panchanathan, V., 1973. Cast. Met. Res. J. 19 (3), 135–138.

Rao, T.S.V., Roshan, H.Md, 1988. TAFS 96, 37–46.

Rao, G.V.K., Srinivasan, M.N., Seshadri, M.R., 1975. TAFS 83, 525–530.

Rapoport, D.B., 1964. Foundry Trade J. 116, 169.

Rappaz, M., Drezet, J.-M., Gremaud, M., 1999. Metall. Mater. Trans. A 30A (2), 449–455.

Rappaz, M., Drezet, J.-M., Mathier, V., Vernede, S., July 2006. Materials Science Forum, vols. 519–521. Trans Tech Publications, Switzerland p.1665–1674.

Rappaz, M., 1989. Int. Mater. Rev. 34 (3), 93–123.

Rashid, A.K.M.B., Campbell, J., 2004. Metall. Mater. Trans. 35A (7), 2063–2071.

Rashid, M.S., Hanna, M.D., 1993. North American Die casting Association (NADCA) Conf., Cleveland, paper T93-041. pp. 105–111 (US Patent 4990310, 1989).

Rassenfoss, J.A., 1977. TAFS 85, 583–596.

Rauch, A.H., Peck, J.P., Thomas, G.F., 1959. TAFS 67, 263–266.

Rault, L., Allibert, M., Prin, M., Dubus, A., 1996. Light Met. 345–355.

Reddy, D.C., Murty, S.S.N., Chakravorty, P.N., 1988. TAFS 96, 839–844.

Reding, J.N., 1968. TAFS 76, 92–98.

Rege, R.A., Szekeres, E.S., Foreng, W.D., 1970. Met. Trans. 1, 2652–2653.

Ren, X.C., Zhou, Q.J., Shan, G.B., Chu, W.Y., Li, J.X., Su, Y.J., Qiao, L.J., 2008. Metall. Mat. Trans. 39A, 87–97.

Ren, Y., Zhang, L., Yang, W., Duan, H., December 2014. Metall. Mater. Trans. B 45B, 2057–2071.

Revankar, V., Baker, P., Schultz, A.H., Brandt, H., 2000. Light Met. 51–55.

Rezvani, M., Yang, X., Campbell, J., 1999. TAFS 107, 181–188.

Richins, D.S., Wetmore, W.O., 1951. In: AFS Symposium on Principles of Gating, pp. 1–24.

Richmond, O., Tien, R.H., 1971. J. Mech. Phys. Solids 19, 273–284.

Richmond, O., Hector, L.G., Fridy, J.M., 1990. Trans. ASME J. Appl. Mech. 57, 529–536.

Rickards, P.J., 1975. Br. Found. 68, 53–60.

Rickards, P.J., 1982. Br. Found. 75, 213–223.

Ricken, H., Ostermeier, M., Hoffmann, H., Fent, A., 2009. Cast. Plant Technol. (1), 30–32.

Riposan, I., Chisamera, M., Stan, S., Toboc, P., Ecob, E., White, D., 2008. Mater. Sci. Technol. 24 (5), 579–584.

Rishel, L.L., Pollock, T.M., Cramb, A.W., February 1999. In: Proc. 1999 Internat. Symp. Liquid Metal Processing and Casting; Santa Fe, NM, pp. 287–299.

Rivas, R.A.A., Biloni, H., 1980. Zeit. Met. 71 (4), 264–268.

Rivera, G., Boeri, R., Sikora, J., 2002. Mater. Sci. Technol. 18, 691–697.

Roberts, T.E., Kovarik, D.P., Maier, R.D., 1979. TAFS 87, 279–298.

Roberts, R.J., 1996. TAFS 104, 523–526.

Roedter, H., September 1986. Foundry Trade J. Int. 6.

Roehrig, K., 1978. TAFS 86, 75–92.

Rogberg, B., 1980. Solidification technology in the foundry and cast house. Warwick Conf, pp. 365–371 (Metals Soc, 198).

Rogers, K.P., Heathcock, C.J., 1990. US Patent 5316070. Date: 31 May 1994.

Rohatgi, P., 1990. Adv. Mater. Process. 2, 39–44.

Romankiewicz, F., 1976. AFS Int. Cast Met. J. 1 (4), 13–17.

Romero, J.M., Smith, R.W., Sahoo, M., 1991. TAFS 99, 465–468.

Rometsch, P.A., Schaffer, G.B., Taylor, J.A., 2001. Int. J. Cast Met. Res. 14 (1), 59–69.

Rong De, L., Xiang, Y.J., 1991. TAFS 99, 707–712.

Rooy, E.L., Fischer, E.F., 1968. Am. Found. Soc. Trans. 76, 237–240.

Rosenberg, R.A., Flemings, M.C., Taylor, H.F., 1960. TAFS 68, 518–528.

Rossmann, M., February 1982. Giesserei 69 (4), 102–103.

Rostoker, W., Berger, M.J., July 1953. Foundry 81, 100–105 and 260–265.

Roth, M.C., Weatherly, G.C., Miller, W.A., 1980. Acta Met. 28, 841–853.

Rouse, J., 1987. 54th Internat. Foundry Congress, New Dehli, p. 16. Paper 30.

Roviglione, A.N., Hermida, J.D., 2004. Trans. Metall. Mater. B 35B, 313–330.

Ruddle, R.W., Cibula, A., 1957. Inst. Met. Monogr. Report. 22, 5–32.

Ruddle, R.W., Mincher, A.L., 1949–1950. J. Inst. Met. 76, 43–90.

Ruddle, R.W., 1956. The Running and Gating of Sand Castings. Monograph & Report Series No 19, Institute of Metals, London.

Ruddle, R.W., 1960. TAFS 68, 685–690.

Runyoro, J., Boutorabi, S.M.A., Campbell, J., 1992. TAFS 100, 225–234.

Runyoro, J., 1992. Design of the Gating System (Ph.D. thesis). University of Birmingham.

Ruxanda, R., Sanchez, L.B., Massone, J., Stefanescu, D.M., 2001. Trans. Am. Found. Soc. 109, 37–48 (Cast iron division).

Sabatino Di, M., Syvertsen, F., Arnberg, L., Nordmark, A., 2005. Int. J. Cast Met. Res. 18 (1), 59–62.

Sabatino Di, M., Arnberg, L., Brusethaug, S., Apelian, D., 2006. Int. J. Cast Met. Res. 19 (2), 94–97.

Sadayappan, M., Fasoyinu, F.A., Thomson, J., Sahoo, M., 1999. TAFS 107, 337–342.

Sadayappan, M., Sahoo, M., Liao, G., Yang, B.J., Li, D., Smith, R.W., 2001. TAFS 109, 341–352.

Saeger, C.M., Ash, E.J., 1930. TAFS 38, 107–145.

Safaraz, A.R., Creese, R.C., 1989. TAFS 97, 863–870.

Sahoo, M., Whiting, L.V., 1984. TAFS 92, 861–870.

Sahoo, M., Worth, M., 1990. TAFS 98, 25–33.

Sahoo, M., Whiting, L.V., White, D.W.G., 1985. TAFS 93, 475–480.

Saia, A., Edelman, R.E., 1968. TAFS 76, 189–195.

Saigal, A., Berry, J.T., 1984. TAFS 92, 703–708.

Sakakibara, Y., Suzuki, T., Hayashi, H., et al. 1988. Japan Patent 17578/88, Europe patent EP0306841 A2.

Sakamoto, M., Akiyama, S., Ogi, K., 1996. In: 4th Asian Found. Cong. Proc., Australia, pp. 467–476.

Samarasekera, I.V., Anderson, D.L., Brimacombe, J.K., 1982. Met. Trans. 13B, 91–104.

Sambasivan, S.V., Roshan, H.Md, 1977. TAFS 85, 265–270.

Samuel, A.M., Samuel, F.H., 1993. Met. Trans. A 24A, 1857–1868.

Samuel, F.H., Samuel, A.M., Doty, H.W., Valtierra, S., 2001. Metall. Mater. Trans. A 32A, 2061–2075.

Sandford, P., March 1988. Foundryman 110–118.

Sandford, P., 1993. TAFS 101, 817–824.

Santos, R.G., Garcia, A., 1998. Int. J. Cast Met. Res. 11, 187–195.

Sarazin, J.R., Hellawell, A., 1987. In: Beech, J., Jones, H. (Eds.), Solidification Processing, Sheffield University, UK, pp. 94–97.

Sare, I.R., 1989. Cast. Met. 1 (4), 182–190.

Saucedo, I.G., Beech, J., Davies, G.J., 1980. In: Conf. "Solidification technology in foundry and cast house". Warwick Univ., Metals Society Publication 1983, pp. 461–468.

Scarber, P., Bates, C.E., 2006. TAFS 114, 37–44.

Scarber, P., Bates, C.E., Griffin, 2006. TAFS 114, 435–445.

Schaffer, P.L., Dahle, A.K., 2009. Metall. Mater. Trans. A 40A (2), 481–485.

Schaffer, P.L., Dahle, A.K., Zindel, J.W., 2004. Light Met. 821–826.

Schaffer, G.B., 2004. Mater. Forum 28, 65–74.

Scheffer, K.D., 1975. TAFS 83, 585–592.

Schilling, H., Reithmann, M., 1992. Cast. Plant Technol. (3), 12–14.

Schilling, H., 1987. Patent PCT WO 87/07543.

Schmidt, D.G., Jacobson, A.E., 1970. TAFS 78, 332–337.

Schmied, H.-J., 1988. Giessereitechnik 34 (4), 133–135.

Schneider, W., 2006. Cast. Plant Technol. (1), 30–37.

Schrey, A., June 2007. Foundry Pract. (Foseco) 246, 01–07.

Schumacher, P., Greer, A.L., 1993. Key Eng. Mater. 81-83, 631.

Schumacher, P., Greer, A.L., 1994. Mater. Sci. Eng. A181/A182, 1335–1339.

Schurmann, E., 1965. Arch. Eisenh. 36, 619–631 (BISI Translation 4579).

Schwandt, C., Fray, D.J., December 2014. Metall. Mater. Trans. B 45B, 2145–2152.

Sciama, G., 1974. TAFS 82, 39–44.

Sciama, G., 1975. TAFS 83, 127–140.

Sciama, G., 1993. TAFS 101, 643–651.

Scott, W.D., Bates, C.E., 1975. TAFS 83, 519–524.

Scott, D., Smith, T.J., 1985. Personal communication.

Scott, A.F., et al., 1948. J. Chem. Phys. 16, 495–502.

Scott, W.D., Goodman, P.A., Monroe, R.W., 1978. TAFS 86, 599–610.

SCRATA, 1981. Hot Tearing – Causes and Cures, Tech. Bull. No. 1. Steel Castings Research and Trade Assoc, Sheffield, UK.

Seetharamu, S., Srinivasan, M.W., 1985. TAFS 93, 347–350.

SERF, June 2002. Foundry Trade J. 34.

Sexton, A.H., Primrose, J.S.G., 1911. The Principles of Ironfounding. The Technical Publishing Co, London.

SFSA (Steel Founders Society of America), May 2000. Foundry Trade J. 40–41.

Shabestari, S.G., Gruzleski, J.E., 1995. TAFS 103, 285–293.

Shafaei, A., Raiszadeh, R., December 2014. Metall. Mater. Trans. B 45B, 2486–2494.

Shamsuzzoha, M., Nastac, L., Berry, J.T., 2012. TAFS 116, paper 12-101.

Shen, P., Zhen, X.-H., Lin, Q.-L., Zhang, D., Jiang, Q.-P., 2009. Metall. Mater. Trans. 40A (2), 444–449.

Shendye, S.B., Gilles, D.J., 2009. TAFS 117, 305–311.

Sherby, O.D., May 1962. Met. Eng. Q. (ASM) 3–13.

Shi, Q., Ding, X., Chen, J., Zhang, X., Zheng, Y., Feng, Q., 2014. Metall. Mater. Trans. A 45A, 1665–1669.

Ship Department Publication 18, 1975. Design & Manufacture of Nickel-Aluminium-Bronze Sand Castings. Ministry of Defence (Procurement Executive), Foxhill, Bath, UK.

Shirey, D.R., Williams, D.C., 1968. TAFS 76, 661–674.

Shivkumar, S., Wang, L., Steenhoff, B., 1989. TAFS 97, 825–836.

Showman, R.E., Aufderheide, R.C., 2003. 111, paper 145, p. 12.

Showman, R.E., Aufderheide, R.C., Yeomans, N.P., 2006. TAFS 114, 391–399.

Showman, R.E., Aufderheide, R.C., Yeomans, N.P., August 2007. Foundry Trade J. 224–227.

Sicha, W.E., Boehm, R.C., 1948. TAFS 56, 398–409.

Sidhu, R.K., Richards, N.L., Chaturvedi, M.C., 2005. Mater. Sci. Technol. 21 (10), 1119–1131.

Sieurin, S.I., June 1974. Foundry, page 94 and following.

Sieurin, S.I., March 1975. Paper G-T75-T045. In: 8th Soc. Die Casting Engineers Internat. Congress, Detroit, MI (4 pages).

Sigworth, G.K., Engh, T.A., 1982. Met. Trans. 13B, 447–460.

Sigworth, G.K., Kuhn, T.A., 2007. Int. J. Metalcast. 1 (1), 31–40.

Sigworth, G.K., Wang, C., Huang, H., Berry, J.T., 1994. TAFS 102, 245–261.

Sigworth, G.K., Howell, J., Rios, O., Kaufman, M.J., 2006. Int. J. Cast Met. Res. 19 (2), 123–129.

Sigworth, G., Jorstad, J., Campbell, J., 2009. AFS Int. J. Metalcast. Corresp. 3 (1), 65–77.

Sigworth, G., 1984. Met. Trans. 15A, 227–282.

Sikorski, S., Groteke, D.E., October 2005. In: AFS Internat. Conf. High Integrity Light Metal Castings.

Simard, A., Proulx, J., Paquin, D., Samuel, F.H., Habibi, N., November 2001. In: Am. Found. Soc. Molten Al Processing Conf., Orlando, Florida.

Simensen, C.J., 1993. Zeit Met. 84 (10), 730–733.

Simensesn, C.J., 1981. Metall. Trans. 12B, 733–743.

Simmons, W., Trinkl, G., 1987. In: BCIRA Conf. British Cast Iron Research Assoc., UK.

Sin, S.L., Dube, D., Tremblay, R., 2006. Mater. Sci. Technol. 22 (12), 1456–1463.

Singh, S.N., Bardes, B.P., Flemings, M.C., 1970. Met. Trans. 1, 1383.

Sinha, N.P., Kondic, V., 1974. Br. Found. 67, 155–165.

Sinha, N.P., 1973. Ph.D. thesis. University of Birmingham, UK.

Sirrell, B., Campbell, J., 1997. TAFS 105, 645–654.

Skaland, T., 2001. TAFS 105, 77–88.

Skarbinski, M., 1971. Br. Found. 44, 126–140.

Skelly, H.M., Sunnucks, D.C., 1954. TAFS 62, 481–491.

Smith, D.D., Aubrey, L.S., Miller, W.C., 1988. In: Welch, B. (Ed.), "Light Metals" Conf., The Minerals, Metals and Materials Soc., pp. 893–915.

Smith, T.J., Lewis, R.W., Scott, D.M., 1990. Foundryman 83, 499–507.

Smith, C.S., 1948. Trans. AIME 175, 15–51.

Smith, C.S., 1949. Trans. AIME 185, 762–768.

Smith, C.S., 1952. Metal Interfaces. ASM, Cleveland, Ohio, pp. 65–113.

Smith, D.M., 1981. UK Patent application GB 2085920 A.

Smith, R.A., 1986. UK Patent GB2187984 A; priority date 21 February 1986.

Sokolowski, J.H., Kierkus, C.A., Brosnan, B., Evans, W.J., 2000. TAFS 108, 497–503.

Sokolowski, J.H., Kierkus, C.A., Brosnan, B., Evans, W.J., 2003. Mod. Cast. 93 (1), 39–42.

Sokolowski, J.H., 2006. University of Windsor, Canada, unpublished work.

Solberg, J.K., Onsoien, M.I., 2001. Mater. Sci. Technol. 17, 1238–1242.

Song, T., Cooman, B.C.de, April 2013. Metall. Mater. A 44A, 1686–1705.

Song, H., Hellawell, A., 1989. Light Met. 819–823.

Sontz, A., 1972. TAFS 80, 1–12.

Sosman, R.B., 1927. The properties of silica. In: Am. Chem. Soc. Monogr. Ser. The Chemical Catalogue Co., USA, pp. iv-45.

Southam, D.L., 1987. Foundry Manage. Technol. 7, 34–38.

Southin, R.T., Romeyn, A., 1980. In: Warwick Conf. "Solidification Technology in the Foundry and Cast House" Metals Soc. 1983, pp. 355–358.

Southin, R.T., 1967. The solidification of metals. In: Brighton Conf. UK. ISI Publication 110, pp. 305–308.

Spada, A.T., 20–24 February 2004. Mod. Cast. 94. June 2004, 48; July 2005, 18–22.

Speidel, M.O., 1982. In: Sixth European Non-ferrous Metals Industry Colloquium CAEF, pp. 65–78.

Spenser, D., Mehrabian, R., Flemings, M.C., 1972. Met. Trans. 3, 1925.

Spitaler, P., 1957. Giesserei 44, 757–766.

Spittle, J.A., Brown, S.G.R., 1989a. J. Mater. Sci. 23, 1777–1781.

Spittle, J.A., Brown, S.G.R., 1989b. J. Mater. Sci. 5, 362–368.

Spittle, J.A., Cushway, A.A., 1983. Met. Technol. 10, 6–13.

Spraragen, W., Claussen, G.E., 1937. J. Am. Weld. Soc. 16 (11), 2–62 (Supplement: Welding Research Committee).

Srimanosaowapak, S., O'Reilly, Keyna, 2005. In: Tiryakioglu, M., Crepeau, P.N. (Eds.), Shape Casting: The John Campbell Symposium. TMS, pp. 41–50.

Srinagesh, K., 1979. AFS Int. Cast Met. J. 4 (1), 50–63.

Srinivasan, A., Pillai, U.T.S., Pai, B.C., 2006. TAFS 114, 737–746.

Stahl, G.W., 1961. TAFS 69, 476–478.

Stahl, G.W., 1963. TAFS 71, 216–220.

Stahl, G.W., 1986. TAFS 94, 793–796.

Stahl, G.W., 1989. The gravity tilt pour process. In: Proc. AFS Internat Conf: Permanent Mold Castings, Miami. Paper 2.

Staley, J.T., 1981. Metals Handbook. In: Heat Treating, ninth ed., vol. 4. American Society for Metals, USA. pp. 675–718.

Staley, J.T., 1986. Aluminium technology. In: Inst. Metals UK Conference, vol. 86, pp. 396–407.

Stanford, N., Sabirov, I., Sha, G., La Fontaine, A., Ringer, S.P., Barnett, M.R., 2010. Metall. Mater. Trans A 41A (3), 734–743.

Starobin, A., Goettsch, D., Walker, M., 2011. AFS Int. J. Met. 5 (3), 57–64.

Steel Founders Society of America (Anon), May 2000. Foundry Trade J. 40–41.

Steel Founders' Society of America (Anon), 1970. Steel Castings Handbook, fourth ed.

Steen, H.A.H., Hellawell, A., 1975. Acta Met. 23, 529–535.

Stefanescu, D.M., Hummer, R., Nechtelberger, E., 1988. Metals Handbook. In: Casting, ninth ed., vol. 15. ASM, Ohio, USA. pp. 667–677.

Stefanescu, D.M., Giese, S.R., Piwonka, T.S., Lane, A.M., Barlow, J., Pattabhi, R., 1996. TAFS 104, 1233–1264.

Stefanescu, D.M., Wills, S., Massone, J., 2009. TAFS, pp. 20, paper 09-120.

Stefanescu, D.M., 1988. Metals Handbook. In: Casting, ninth ed., vol. 15. ASM, Ohio, USA. pp. 168–181.

Stefanescu, D.M., 2007. Trans. Metall. Mater. A 38A (7), 1433–1447.

Stefanescu, D.M., 2015. Thermal analysis – theory and applications in metalcasting. Int. J. Metalcast. 9 (1), 7–22.

von Steiger, R., 1913. Stahl Eisen 33, 1442.

Stein, H., Iske, F., Karcher, D., 1958. Giesserei Technisch Wissenschaftliche Beihefte 21, 115–1124.

StJohn, D.H., Qian, M., Easton, M.A., Cao, P., Hildebrand, Z., 2005. Metall. Mater. Trans. A 36A, 1669–1679.

StJohn, D.H., Easton, M.A., Cao, P., Qian, M., July 2007. Int. J. Cast Met. Res. 20 (3), 131–135 and In: Jones, H. (Ed.), SP07 Proc. 5th Decennial Internat. Conf. Solidification Processing. July 2007, Sheffield, UK, pp. 99–103.

StJohn, D.H., Easton, M.A., Qian, M., Taylor, J.A., July 2013. Metall. Mater. Trans. A 44A, 2935–2949.

Stolarczyk, J.E., 1960. Br. Found. 53, 531–548.

Stoltze, P., Norskov, I.K., Landman, U., 1988. Phys. Rev. Lett. 61 (4), 440–443.

Storaska, G.A., Howe, J.M., 2004. Mater. Sci. Eng. A 368, 183–190.

Stratton, P., June 2010. Mater. World 28–29.

Street, A.C., 1986. The Diecasting Book, second ed. Portcullis Press, Redhill, UK. (Chapter 8) on zinc alloys pp. 170–194.

Su, J.Y., Chow, C.T., Wallace, J.F., 1982. TAFS 90, 565–574.

Sugden, A.A.B., Bhadeshia, H.K.D.H., June 1988. Metall. Trans. A 19A. University of Cambridge, UK. pp.1597–1610.

Sullivan, E.J., Adams, C.M., Taylor, H.F., 1957. TAFS 65, 394–401.

Sumiyoshi, Y., Ito, N., Noda, T., 1968. J. Cryst. Growth 3 and 4, 327–339.

Sun, L., Campbell, J., 2003. TAFS 111 (paper 03-018).

Sun, G.X., Loper, C.R., 1983. TAFS 91, 841–854.

Surappa, M.K., Blank, E., Jaquet, J.C., 1986a. In: Sheppard, T. (Ed.), Conference "Aluminium technology". Inst. Metals, pp. 498–504.

Surappa, M.K., Blank, E., Jaquet, J.C., 1986b. Scr. Met. 20, 1281–1286.

Sutton, T., June 2002. Foundryman 95 (6), 223–231.

Suutala, N., 1983. Met. Trans. 14A, 191–197.

Suzuki, K., Nishikawa, K., Watakabe, S., 1996. Mater. Trans. Jpn. Inst. Met. 37 (12), 1793–1801.

Suzuki, S., October 1989. Mod. Cast. 38–40.

Svensson, I.L., Dioszegi, A., 2000. In: Sahm, P., Hansen, P., Conley (Eds.), Modelling of Casting, Welding and Advanced Solidification Processes IX, pp. 102–109.

Svensson, I., Villner, L., 1974. Br. Found. 67, 277–287.

Svoboda, J.M., Geiger, G.H., 1969. TAFS 77, 281–288.

Svoboda, J.M., Monroe, R.W., Bates, C.E., Griffin, J., 1987. TAFS 95, 187–202.

Svoboda, J.M., 1994. TAFS 102, 461–471.

Sweeney, V.D., 1964. TAFS 72, 911–913.

Swift, R.E., Jackson, J.H., Eastwood, L.W., 1949. TAFS 57, 76–88.

Swing, E., 1962. TAFS 70, 364–373.

de Sy, A., 1967. TAFS 75, 161–172.

Syvertsen, M., 2006. Metall. Mater. Trans. B 37B (6), 495–504.

Syvertsen, M., Engh, T.A., 2001. Light Met. 957–963.

Szklarska-Smialowska, Z., 1999. Corros. Sci. 41, 1743–1767.

Tadayon, M.R., Lewis, R.W., 1988. Cast. Met. 1, 24–28.

Tadayon, M.R., 28 September 1982. Finite Element Modelling of Heat Transfer and Solidification in the Squeeze Forming Process (Ph.D. thesis). University College, Swansea, UK.

Tafazzoli-Yadzi, M., Kondic, V., 1977. AFS Internat. Cast. Met. J. 2 (4), 41–47.

Taft, D.J., 1968. Br. Found. 61, 69–75.

Takeuchi, E., Brimacombe, J.K., 1984. Metall. Mater. Trans. B 15, 493–509.

Talbot, D.E.G., Granger, D.A.J., 1962. Inst. Met. 91, 319–320.

Talbot, D.E., 1975. J. Int. Metall. Rev. 20, 166.

Taleyarkhan, R.P., Kim, S.H., Gulec, K., 1998. TAFS 106, 619–624.

Tasaki R., Noda Y., Terashima K., Hashimoto K., Proc. 28th World Foundry Cong., February 2008, Chennai, India, pp. 121–126.

Taylor, K.C., Baier, A., 2003. Cast. Plant Technol. Int. 123 (2), 36–46.

Taylor Jr., C.M., Taylor, H.F., 1953. Fundamentals of riser behavior. AFS Trans. 686–693.

Taylor, L.S., 1960. Foundry Trade J. 109 (2287), 419–427.

Teague, J., Richards, V., 2010. AFS Int. J. Metalcast. 4 (2), 45–57.

Terashima K., Noda Y., et al. Proc. 28th World Foundry Cong., February 2008, Chennai, India, pp. 113–119.

Theuwissen, K., Lafont, M.-C., Laffont, L., Viguier, B., Lacaze, J., 2012. Microstructural characterization of graphite spheroids in ductile iron. Trans. Indian Inst. Met. 65 (6), 627–631.

Theile von, W., 1962. Alum. Ger. 38, 707–715, 780–786. (English Translation by Nercessian, H., 29 April 1963, 5229, Alcan Banbury, UK and Electicity Council Research Centre, UK.).

Thomas, B.G., 1995. J. Iron Steel Inst. Jpn. Int. 35 (6), 737–743.

Thomas, B.G., Parkman, J.T., 1997. In: Conf. "Solidification" Indianapolis Indiana. The Minerals, Metals and Materials Society, 1998, pp. 509–520.

Thomason, P.F., 1968. J. Inst. Met. 96, 360–365.

Thornton, D.R., 1956. J. Iron Steel Inst. 183 (3), 300–315.

Tian, C., Irons, G.A., Wilkinson, D.S., 1999. Metall. Mater Trans. B 30B (2), 241–252.

Tiberg, L., 1960. Jerkont. Ann. 144 (10), 771–793.

Tiekink, W., Boom, R., Overbosch, A., Kooter, R., Sridar, S., 2010. Ironmak. Steelmak. 37 (7), 488–495.

Tien, R.H., Richmond, O., 1982. Trans. ASME J. Appl. Mech. 49, 481–486.

Tigges, U., December 2010. Foundry Trade J. 310–311.

Timelli, G., Bonollo, F., 2007. Int. J. Cast Met. Res. 20 (6), 304–311.

Timmons, W.W., Spiegelberg, W.D., Wallace, J.F., 1969. TAFS 77, 57–61.

Tiryakioglu, M., Campbell, J., 2007. Int. J. Cast Met. Res. 20 (1), 25–29.

Tiryakioglu, M., Campbell, J., 2009. Mater. Sci. Technol. 25 (6), 784–789.

Tiryakioglu, M., Hudak, D., 2008. J. Mater. Sci. 43, 1914–1919.

Tiryakioglu, M., Askeland, D.R., Ramsay, C.W., 1993. TAFS 101, 685–691.

Tiryakioğlu, M., Tiryakioğlu, E., Askeland, D.A., 1997a. Int. J. Cast Met. Res. 9, 259–267.

Tiryakioğlu, M., Tiryakioğlu, E., Askeland, D.A., 1997b. TAFS 105, 907–915.

Tiryakioglu, M., Tiryakioglu, E., Campbell, J., 2002. Int. Cast Met. Res. J. 14 (6), 371–375.

Tiryakioglu, M., Hudak, D., 2008. Mater. Sci. Eng. A 498, 501–503.

Tiryakioglu, M., Hudak, D., Okten, G., 2009a. Mater. Sci. Eng. A 527, 397–399.

Tiryakioglu, M., Campbell, J., Alexopoulus, N.D., 2009b. Metall. Mater. Trans. A 40A, 1000–1007.

Tiryakioglu, M., Campbell, J., Alexopoulus, N.D., 2009c. Mater. Sci. Eng. A 506, 23–26.

Tiryakioglu, M., Campbell, J., Nyahumwa, C., 2011. Fracture surface facets and fatigue life potential of castings. In: Crepeau, P., et al. (Eds.), Shape Casting: The John Berry Honorary Symposium. TMS, San Diego, CA.

Tiryakioğlu E., 1964. A study of the dimensioning of feeders for sand castings (Ph.D. thesis). University of Birmingham, UK.

Tiryakioglu, M., 2001. Personal communication.

Tiryakioglu, M., 2008. Statistical distributions for the size of fatigue-initiating defects. A comparative study. Mater. Sci. Eng. A 497, 119–125.

Tiryakioglu, M., 2009a. The size of fatigue-initiating defects. Mater. Sci. Eng. 520, 114–120.

Tiryakioglu, M., Unpublished Work on the Derivation of Linear Equation for M.

Tiryakioglu, M., 2009b. Metall. Mater. Trans. 40A, 1000–1007.

Tiryakioglu, M., 2011. In: Shape Casting; "John Berry Honorary Symposium" IV. TMS Annual Congress, San Diego, CA, USA.

Tiwara, S.N., Gupta, A.K., Maihotra, S.L., 1985. Br. Found. 78 (1), 24–27.

Toda, H., Minami, K., Koyama, K., Ichitani, K., Kobayashi, M., Uesugi, K., et al., 2009. Acta Mater. 57, 4391.

Toda, H., Oogo, H., Horikawa, K., Uesugi, K., Takeuchi, A., Suzuki, Y., et al., February 2014. Metall. Mater. Trans. A 45A, 765–776.

Tokarev, A.I., 1966. Iz.v VUZ Chern. Met. 3, 193–200. BCIRA Translation T1190, January 1966.

van Tol, R., Katerman, L., van der Akker, H.E.A., 1997. Solidification processing. In: Beech, J., Jones, H. (Eds.), Conference Sheffield, UK (SP97). pp. 79–82.

Tomono, H., Ackermann, P., Kurz, W., Heinemann, W., 1980. Solidification technology in foundry and cast house. In: Warwick Univ. Conf. Metals Society Publication 1983, pp. 524–531.

Torabi Rad, M., Kotas, P., Bekermann, C., September 2013. Metall. Mater. Trans. 44A, 4266–4281.

Tordoff, E.G., Wolfgram, T., Talwar, V., Hysell, M., 1996. TAFS 104, 461–466.

Townsend, D.W., May 1984. Foundry Trade J. 24, 409–414.

Travena, D.H., 1987. Cavitation and Tension in Liquids. Adam Huger, Inst. Phys, Bristol.

Trbizan Katerina, 2001. Casting Simulation. World Foundry Organisation (paper 4) pp. 83–97.

Trikha, S.K., Bates, C.E., 1994. TAFS 102, 173–180.

Trojan, P.K., Guichelaar, P.J., Flinn, R.A., 1966. TAFS 74, 462–469.

Tsai, H.L., Chiang, K.C., Chen, T.S., 1988. In: Giamei, A.F., Abbaschian, O.J. (Eds.), Modeling of Casting and Welding Processes IV. The Minerals, Metals and Materials Society, USA, 1989.

Tschopp, M.A., Ramsay, C.W., Askeland, D.R., 2000. TAFS 108, 609–614.

Tsuruya, S., Ishikawa, Y., Sono, K., 1974. Trans. AFS 82, 27–34.

Tucker, S.P., Hochgraph, F.G., 1973. Metallography 6, 457–464.

Turchin, A.N., Eskin, D.G., Katgerman, L., 2007. Int. J. Cast Met. Res. 20 (6), 312–318.

Turchin, A.N., Eskin, D.G., Katgerman, L., 2007. Metall. Mater. Trans. A 38A (6), 1317–1329.

Turkdogan, E.T., 1986. Foundry processes, their chemistry and physics. In: Katz, S., Landefeld, C.F. (Eds.), International Symposium, Warren, Mich. USA. Plenum Press 1988, pp. 53–100.

Turner, A., Owen, F., 1964. Br. Found. 57, 55–61, 355–356.

Turner, G.L., 1965. Br. Found. 58, 504–505.

Twitty, M.D., 1960. BCIRA I., Report 575. In: Walton, C.F. (Ed.) (1971), Gray and Ductile Iron Casting Handbook. Gray and Ductile Iron Founders' Soc. Inc, Cleveland, Ohio, pp. 844–856.

Tyberg, B., Granehult, O., 1970. In: 37th Internat. Foundry Congress, p. 4.

Tyndall, J., 1872. The Forms of Water. D. Appleton and Co, New York.

Tynelius, K., Major, J., Apelian, D., 1993. Am. Found. Soc. Trans. 101, 401–413.

Unsworth, W., February 1988. Met. Mater. 83–86.

Uto, Y., Yamasaki, D., 1967. UK Patent Specification 1198700.

Valdes, J., King, P., Liu, X., 2010. Metall. Mater. Trans. A 41A (9), 2408–2416.

Valdez, M.E., Uranga, P., Cramb, A.W., 2007. Metall. Mater. Trans. B 38B, 257–266.

Van Ende, M.A., Guo, M., Proost, J., Blanpain, B., Wollants, P., 2010. Ironmak. Steelmak. 37 (7), 496–501.

Vandenbos, S.A., 1985. TAFS 93, 871–878.

Venturelli, G., Sant'Unione, G., February 1981. Alluminio 100–106.

Vernede, S., Jarry, P., Rappaz, M., 2006. Acta Mater. 54, 4023–4034.

Vernede, S., Dantzig, J.A., Rappaz, M., 2009. Acta Mater. 57, 1554–1569.

Versnyder, F.I., Shank, M.E., 1970. Mater. Sci. Technol. 6 (4), 213–247.

Vigh, L., Bennett, G.H.J., 1989. Cast Met. 2 (3), 144–150.

Villner, L., 1969. Br. Found. 62, 458–468. Also published in Giesserei 1970 57 (27), 837–844 and Cast Met. Res. J. 1970 6 (3) 137–142.

Vincent, R.S., Simmons, G.H., 1943. Proc. Phys. Soc. (London) 376–382.

Vogel, A., Doherty, R.D., Cantor, B., 1977. In: Univ. Sheffield Conf. "The Solidification and Casting of Metals". Metals Soc. 1979, pp. 518–525.

Voigt, R.C., Holmgren, S.D., 1990. TAFS 98, 213–225.

Vorren, O., Evensen, J.E., Pedersen, T.B., 1984. TAFS 92, 459–466.

Wakefield, G.R., Sharp, R.M., 1992. J Mater. Sci. Technol. 8, 1125–1129.

Wakefield, G.R., Sharp, R.M., 1996. J Mater. Sci. Technol. 12, 518–522.

Walker, J.L., 1961. Physical Chemistry of Process Metallurgy. II. Interscience, NY, p. 845.

Wall, A.J., Cocks, D.L., 1980. Br. Found. 73, 292–300.

Wallace, J.F., Hrusovsky, J.P., TAFS 1979. 87, 269–278.

Wallace, J.F., Kissling, R.J., December 1962. Foundry 36–39. January 1963, 64–68.

Wallace, J.F., 1988. TAFS 96, 261–270.

Walther, W.D., Adams, C.M., Taylor, H.F., 1954. TAFS 62, 219–230.

Wan, L., Nakashima, T., Kato, E., Nomura, H., 2002. Int. J. Cast Met. Res. 15, 187–192.

Wang, Q.G., Apelian, D., Lados, D.A., 2001a. Part I. J. Light Met. 1 (1), 73–84.

Wang, Q.G., Apelian, D., Lados, D.A., 2001b. Part II. J. Light Met. 1 (1), 85–97.

Wang, R.-Y., Lu, W.-H., Ma, Z.-Y., 2007. TAFS 111, 8, paper 124.

Wang, J., Lee, P.D., Hamilton, R.W., Li, M., Allison, J., 2009. Scr. Met. 60 (7), 516–519.

Wang, C., Gao, H., Dai, Y., Ruan, X., Wang, J., Sun, B., 2010. Metall. Mater. Trans. A 41A (7), 1616–1620.

Wang, L., Lett, R., Felicelli, S.D., Berry, J.T., 2011. TAFS, paper 11-039.

Wang, Q., Geng, H., Zhang, S., Jiang, H., Zuo, M., March 2014. Metall. Mater. Trans. A 45A, 1621–1630.

Wannasin, J., Schwam, D., Wallace, J.F., 2007. J. Mater. Process. Technol. 191 (1–3), 242–246.

Ward, C.W., Jacobs, I.C., 1962. TAFS 70, 332–337.

Wardle, G., Billington, J.C., October 1983. Met. Technol. 10, 393–400.

Warrick, R.J., 1966. TAFS 74, 722–733.

Warrington, D., McCartney, D.G., 1989. Cast Met. 2 (3), 134–143.

Watmough, T., 1980. TAFS 88, 481–488.

Waudby, P.E., George, G.H., June 1986. In: Twentieth EICF Conf. on Investment Casting Brussels, p. 19, paper 12.

Way, L.D., 2001. Mater. Sci. Technol. 17 (10), 1175–1190.

Weber, J.A., Rearwin, E.W., 1961. Foundry 2, 69–71.

Webster, P.D., 1964. Br. Found. 57, 520–523.

Webster, P.D., 1966. Br. Found. 59, 387–393.

Webster, P.D., 1967. Br. Found. 60, 314–319.

Webster, P.D., 1980. In: Fundamentals of Foundry Technology. Portcullis Press.

Weibull, W., 1951. J. Appl. Mech. 18, 293–297.

Weiner, J.H., Boley, B.A., 1963. J. Mech. Phys. Solids 11, 145–154.

Weins, M.J., Bottom, J.L.S.de, Flinn, R.A., 1964. TAFS 72, 832–839.

Weiss, D.J., Rose, D., 1993. TAFS 101, 1065–1066.

Wells, M.B., Oleka, J.T., 1988. TAFS 96, 913–918.

Welter, V.G., 1931. Z. Met. 23.

Wen, S.W., Jolly, M.R., Campbell, J., 1997a. In: Beech, J., Jones, H. (Eds.), Proc. 4th Decennial Internat. Conf. on Solidification Processing, Sheffield, pp. 66–69.

Wen, S.W., Jolly, M.R., Campbell, J., 7–10 July 1997b. In: Solidification Processing Conference SP97, Sheffield (Paper on "Promotion of directional solidification…").

West, T.D., 1882. American Foundry Practice, 11th ed. Wiley NY, Chapman & Hall London. 1910, p. 107.

Whittenberger, E.J., Rhines, F.N., 1952. J. Met. 4 (4), 409–420 and Trans AIME, 194, 409–420.

Wieser, P.F., Dutta, I., 1986. TAFS 94, 85–92.

Wieser, P.F., Wallace, J.F., 1969. TAFS 77, 22–26.

Wieser, P.F., 1983. TAFS 91, 647–656.

Wightman, G., Fray, D.J., 1983. Met. Trans. 14B, 625–631.

Wildermuth, J.W., Lutz, R.H., Loper, C.R., 1968. TAFS 76, 258–263.

Wile, L.E., Strausbaugh, K., Archibald, J.J., Smith, R.L., Piwonka, T.S., 1988. Metals Handbook. In: Casting, ninth ed., vol. 15. ASM International. pp. 240–241.

Williams, J.A., Singer, A.R.E., 1968. J. Inst. Met. 96, 5–12.

Williams, S.J., Bache, M.R., Wilshire, B., 2010. Mater. Sci. Technol. 26 (11), 1332–1337.

Williams, D.C., 1970. TAFS 78, 374–381, 466–467.

Wilshire, B., Scharning, P.J., 2008. Mater. Sci. Technol. 24, 1–9.

Winardi, L., Littleton, H.E., Bates, C.E., 2007. TAFS 04, 10, paper 062.

Winter, B.P., Ostrom, T.R., Sleder, T.A., Trojan, P.K., Pehlke, R.D., 1987. TAFS 95, 259–266.

Winzer, N., Cross, C.E., 2009. Metall. Mater. Trans. A 40A (2), 273–274.

Wittmoser, A., Hofmann, R., 1968. In: 35th Foundry Congress, Kyoto, Japan.

Wittmoser, A., Steinack, K., Hofmann, R., February 1972. Br. Found. 65, 73–84.

Wittmoser, A., 1975. Trans. AFS 83, 63–72.

Wlodawer, R., 1966. Directional Solidification of Steel Castings. English translation by Hewit, L.D., Riley, R.V. Pergamon Press.

Wojcik, W.M., Raybeck, R.M., Paliwoda, E.J., December 1967. J. Met. 19, 36–41.

Wolf, A., Steinhauser, T., 2004. Cast. Plant Technol. Int. 3, 6–11.

Woodbury, K.A., Chen, Y., Parker, J.K., Piwonka, T.S., 1998. TAFS 106, 705–711.

Woodbury, K.A., Piwonka, T.S., Ke, Q., 2000. In: Sahm, P.R., Hansen, P.N., Conley, J.G. (Eds.), Modeling of Casting, Welding and Advanced Solidification Processes IX, pp. 270–277.

Woodbury, K.A., Ke, Q., Piwonka, T.S., 2000. TAFS 108, 259–265.

Woolley, J.W., Woodbury, K.A., 2007. TAFS 115, 18 (Paper 075(01)).

Woolley, J.W., Woodbury, K.A., 2009. TAFS 117, 31–40.

Worman, R.A., Nieman, J.R., 1973. TAFS 81, 170–179.

Wray, P.J., 1976a. Acta Met. 76, 125–135.

Wray, P.J., 1976b. Met. Trans. 7B, 639–646.

Wright, T.C., Campbell, J., 1997. TAFS 105, 639–644 and Mod. Cast. June 1997 (see also Dimmick, T., March 2001. Mod. Cast. 91 (3), 31–33).

Wu, R., Sandstrom, R., 1995. Mater. Sci. Technol. 11 (6), 579–588.

Wu, C., Sahajwalla, Pehlke, R.D., 1997. TAFS 105, 739–744.

Wu, M., 1997. TAFS 105, 693–702.

Wu, M., Sahm, P.R., 1997. Trans. Am. Found. Soc. 105, 693–702.

Wuilloud, E., 1994. Light Met. 1079–1082.

Wurker, L., Zeuner, Th, 2004. Aluminium 80 (11), 1207–1213.

Wurker, L., Zeuner, Th, 2006. Cast. Plant Technol. (1), 38–42.

Xiao, L., Anzai, K., Niyama, E., Kimura, T., Kubo, H., 1998. Int. J. Cast Met. Res. 11 (2), 71–81.

Xiao, B., Wang, Q.G., Jadhav, P., Li, K., 2010. J. Mater. Process. Technol. 210, 2023–2028.

Xiong, S.-M., Liu, X.-L., 2007. Metall. Mater. Trans. A 38A, 248–434.

Xu, Z.A., Mampaey, F., 1994. TAFS 102, 181–190.

Xu, Z.A., Mampaey, F., 1997. TAFS 105, 853–860.

Xu, J., Liu, F., Xu, X., Chen, Y., 2012. Metall. Mater. Trans. A 43A, 1268–1276.

Xu, Z.A., August 2007. TMS U.S.A.

Xue, X., Thorpe, R., 1995. TAFS 103, 743–747.

Yamaguchi, K., Healy, G., 1974. Met. Trans. 5, 2591–2596.

Yamamoto, N., Kawagoishi, N., 2000. Trans. AFS 108, 113–118.

Yamamoto, S., Kawano, Y., Murakami, Y., Chang, B., Ozaki, R., 1975. TAFS 83, 217–226.

Yamamoto, Y., Iwahori, H., Yonekura, K., Nakmura, M., 1980. AFS Int. Cast Met. J. 5 (2), 60–65.

Yan, Y., Yang, G., Mao, Z., 1989. J. Aeronaut. Mater. (China) 9 (3), 29–36.

Yang, X., Campbell, J., 1998. Pouring basin. Int. J. Cast Met. Res. 10, 239–253.

Yang, X., Din, T., Campbell, J., 1998. Offset sprue. Int. J. Cast Met. Res. 11, 1–12.

Yang, X., Jolly, M.R., Campbell, J., 2000. Vortex flow runner. Alum. Trans. 2 (1), 67–80 and Sahm, P., Hansen, P., Conley, (Eds.), Modelling of Casting, Welding and Advanced Solidification Processing IX. 2000, pp. 420–427.

Yaokawa, J., Miura, D., Anzai, K., Yamada, Y., Yoshi, H., 2007. Jpn. Foundry Eng. Soc. Mater. Trans. 48 (5), 1034–1041.

Yarborough, W.A., Messier, R., 1990. Science 247, 688–696.

Yazzie, K.E., Williams, J.J., Kingsbury, D., Peralta, P., Jiang, H., Chawla, N., 2010. J. Met. 62 (7), 16–21.

Yonekura, K., et al., 1986. TAFS 94, 277–284.

Yoshimi, K., Nakamura, J., Kanekon, D., Yamamoto, S., Maruyama, K., Katsui, H., Goto, T., 2014. J. Met. 66 (9), 1930–1938.

Youdelis, W.V., Yang, C.S., 1982. Metal. Sci. 16, 275–281.

Young, P., April 2002. Quoted Anon. Foundry Trade J. 27–28.

Young, K.P., Kirkwood, D.H., 1975. Met. Trans. 6A, 197–205.

Yu, X.Q., Sun, Y.S., 2004. Mater. Sci. Technol. 20 (3), 339–342.

Yu, L.X., Sun, Y.R., Sun, W.R., Zhang, W.H., Liu, F., Xin, X., Qi, F., Jia, D., Sun, X.F., Guo, S.R., Hu, Z.Q., 2013. Mater. Sci. Technol. 29 (12), 1470–1477.

Yue, T.M., Chadwick, G.A., 1991. Personal communication relating to the effect of grain size on the 0.2PS of 7010 Al alloy, AZ91 Mg alloy and Al-4.5Cu alloy.

Yurko, J.A., Martinez, R.A., Flemings, M.C., June 2003. Metall. Sci. Technol. (Teksid Alum.) 21 (1).

Zadeh, A.H., Campbell, J., 2002. TAFS Paper 02-020, pp. 1–17.

Zang, Z., Bian, X., Liu, X., 2001. Int. J. Cast Met. Res. 14 (1), 31–35.

Zeitler, H., Scharfenberger, W., 1984. Aluminium (Germany) 60 (12), E803–E808.

Zemcik, L., 2015. Formation of oxide films in castings from nickel-base superalloys. AFS Int. J. Metalcast., in press.

Zhang, D.L., Zheng, L.H., St John, D.H., 1998. Mater. Sci. Technol. 14 (7), 619–625.

Zhang, C., Mucciardi, F., Gruzleski, J., Burke, P., Hart, M., 2003. Trans AFS paper 03-010, pp. 1–11.

Zhang, E., Wnag, G.J., Xu, J.W., Hu, Z.C., 2010. Mat. Sci. Technol. 26 (8), 956–961.

Zhao, L., Baoyin, Wang, N., Sahajwalla, V., Pehlke, R.D., 2000. Int. J. Cast Met. Res. 13, 167–174.

Zhong, H., Rometsch, P.A., Estrin, Y., 2013. Metall. Mater. Trans. A 44A, 3970–3983.

Zhou, Y.Z., Volek, A., 2007. Mater. Sci. Technol. 23 (3), 297–302.

Zhu, P., Sha, R., Li, Y., Effect of twin/tilt on the growth of graphite. In: Fredriksson, H., Hillert, M. (Eds.), The Physical Metallurgy of Cast Iron. Proc. Materials Research Soc., vol. 34. p. 3.

Zildjian Avedis Company, May 2002. Mod. Cast. 68.

Ziman, J., 2001. Non-Instrumental Roles of Science. Physics Department, University of Bristol.

Zuehlke, H.B., 1943. Trans. Am. Found. Assoc. 51, 773–797.

Zuithoff, A.1, 1964. Paper 29; and in Geisserei, (1965) 52 (9). In: 31st Internat. Foundry Congress, Amsterdam, pp. 820–827.

Zurecki, Z., Best, R.C., 1996. TAFS 104, 859–864.

Zurecki, Z., 1996. In: Saha, D. (Ed.), Gas Interactions in Nonferrous Metals Processing. TMS, pp. 79–93.

Index

Note: 'Page numbers followed by "f" indicate figures and "t" indicate tables'.

Printed and bound by CPI Group (UK) Ltd, Croydon, CR0 4YY

12/05/2025

01869339-0001